HANDBOOK OF
HUMAN FACTORS
IN WEB DESIGN
SECOND EDITION

Human Factors and Ergonomics

Series Editor

Gavriel Salvendy

Professor Emeritus
School of Industrial Engineering
Purdue University

Chair Professor & Head
Dept. of Industrial Engineering
Tsinghua Univ., P.R. China

HANDBOOK OF
HUMAN FACTORS
IN WEB DESIGN
SECOND EDITION

Edited by
Kim-Phuong L. Vu
Robert W. Proctor

CRC Press
Taylor & Francis Group
Boca Raton London New York

CRC Press is an imprint of the
Taylor & Francis Group, an **informa** business

CRC Press
Taylor & Francis Group
6000 Broken Sound Parkway NW, Suite 300
Boca Raton, FL 33487-2742

First issued in paperback 2017

© 2011 by Taylor and Francis Group, LLC
CRC Press is an imprint of Taylor & Francis Group, an Informa business

No claim to original U.S. Government works

ISBN 13: 978-1-4398-2594-5 (hbk)
ISBN 13: 978-1-138-07460-6 (pbk)

Visit the Taylor & Francis Web site at
http://www.taylorandfrancis.com

and the CRC Press Web site at
http://www.crcpress.com

Contents

SECTION I Background and Overview

SECTION II Human Factors and Ergonomics

SECTION III Interface Design and Presentation of Information

SECTION IV   Organization of Information for the Web

SECTION V   Information Retrieval and Sharing: Search Engines, Portals, and Intranets

SECTION VI   Accessibility and Universal Access

SECTION VII   Web Usability Engineering

SECTION VIII   Task Analysis and Performance Modeling

SECTION IX   Specific Web Applications

SECTION X   User Behavior and Cultural Influences

SECTION XI Emerging Technologies

SECTION XII Return on Investment for the Web

Foreword

When I recently got e-mail from Kim Vu and Robert Proctor that it was already 5 years since the publication of their *Handbook on Human Factors in Web Design*, I could hardly believe it. How reflective of the increased pace of information flow we have experienced as World Wide Web users during this period of technological history. The expanded content of the second edition likewise reflects how widely the Web has pervaded our working and personal lives.

The years 2005–2011 witnessed many changes in both Web technology and our collective usage behaviors. With the increase in throughput speed as users transitioned from dial-up networks to broadband cable and DSL came the ability to create much richer, more graphical, three-dimensional (3D), and animated user interfaces, including virtual realities. Web-based commerce, which had been a novelty as sellers searched for good business models, is now taken for granted, as we saw profitable Black Mondays (pre-Christmas shopping days) for major retailers in both 2008 and 2009. The Software as a service (Saas) model of sales and distribution has become standard for enterprise software, with thriving cloud computing vendors like salesforce.com and Workday. Nearly all major software vendors have Web-based user experiences provided by service-oriented architectures (SOA). It is easy to download upgraded features and functions to software applications from the Web.

In addition, new user services and user experience models have emerged. In my Foreword for the first edition, I discussed Wikis, blogs, mashups, and other Web 2.0 features as they were just entering the mainstream. In 2010, the latest technologies are centered around social networking and collaboration. With the advent of MySpace and LinkedIn (2003), YouTube (2005), and Twitter and Facebook (2006), interesting facts, images, videos, and animations can be shared easily by users around the globe. This has empowered end users to participate and influence world events, such as the election of the president of the United States in 2008 and the shooting of the Iranian student protester in 2009, which were viewed by millions around the world.

The rise of the Web also contributed to the decline of various legacy industries, such as the publishing of newspapers and magazines, many of which went bankrupt as readers switched to online news Web sites, and the decline in sales of music recordings on CDs in favor of iTunes downloads. Travel agencies, job-hunting services, and bookstores have felt the effects of Web-based services such as Expedia.com, Monster.com, and Amazon.com, respectively. Other industries experiencing dramatic impacts of the shift to Web-based consumption include movie making, which is now transitioning to 3D technology as a means of sustaining itself in theaters, and television, which is rapidly integrating Web-based participatory viewer experiences, such as online voting and in-depth news stories complementing broadcast news programming.

Reversing the usual trend in computing, the Web, a primarily software-based technology, has driven advances in computing hardware and network development. Most laptop computers sold today have wireless network capabilities. The proliferation of applications on the Web, including Google word processing, spreadsheets, and desktop Gadgets, have created a market for smaller, simpler, cheaper netbooks (laptop computers with limited capabilities). Paradoxically, user manuals to set up computers now have to be downloaded from the Web. Finally, the evolution of cell phones to smart phones, which have enhanced user interfaces and use 3G networks to reach the Web from virtually anywhere, are a direct result of users' strong desire to always stay connected.

This last point deserves further attention, because we are currently witnessing a revolution in end-user computing spawned by the popularity of smart phones. The promise of mobile computing, especially the iPhone and Blackberry, has been fulfilled through integration with Web technologies and content. For some users, it is almost possible in 2010 to replace a networked desktop, laptop, or netbook computer with a smart phone. New interaction models enabled by advanced touchscreen technologies have allowed end users to access many thousands of "apps" available for these phones, to navigate while walking, biking, and driving, to find a good restaurant nearby, and to perform many other practical, day-to-day tasks.

How do we account for the enthusiastic adoption of the World Wide Web? First, its evolution has really lived up to its name. It truly is worldwide, and its technologies have attempted to embrace the entire universe of users. Second, the free nature of many aspects of the Web has been a strong driver of this technology. This freedom has also influenced the software industry as a whole, as reflected in the Open Source movement. Finally, the Web has empowered users. Given business models that can literally embrace a significant portion of humanity existing on the planet as a target market, it has been worthwhile for technologists to overcome many of the impediments to ease of use that existed in pre-Web computing environments.

The table of contents of this second edition reflects these profound changes in many ways. There are eight new chapter topics in the book, covering areas that were just beginning to receive attention when the first edition was published. These include chapters on information design for collaboration (chapters 11 and 12), online portals (15), international accessibility standards (19), analyzing and modeling of user activity (31), mobile user interfaces for the Web (34), human factors of online games (35), and use of avatars for educational and organizational learning (36).

In addition, all chapters have been updated, and the chapters on search (14) and the value of human factors in Web design (37) have been rewritten by new authors. As in the first edition, the handbook still covers basic human factors relating to screen design, input devices, and information organization and processing, in addition to user-centered design methodologies, psychological and social issues such as security and privacy, and applications for academic, industrial, and medical settings.

The handbook is once again timely as we cross the threshold to the next generation of the Internet. Dubbed Web 3.0, it promises semantic webs and other extensions of Web technologies, in addition to the use of HTML 4, which will bring improved style sheets, printing, accessibility, support for bidirectional languages, forms, and tables. As radically fast as technology evolves, it is wise to keep in mind that human sensory, motor, perceptual, and cognitive capabilities and limitations have not changed very much since the emergence of *Homo sapiens* 200,000 years ago. Therefore, we need to pay attention to human factors as long as human beings are the primary users and beneficiaries of the Web and other advanced technologies.

Anna M. Wichansky, PhD, CPE
Applications User Experience
Oracle
Redwood Shores, California

Preface

Since the World Wide Web became widely available in the mid 1990s, Web-based applications have developed rapidly. The Web has come to be used by a range of people for many different purposes, including online banking, e-commerce, distance education, social networking, data sharing, collaborating on team projects, and healthcare-related activities. Effective user interactions are required for a Web site to accomplish its specific goals, which is necessary for an organization to receive a proper return on investment or other desired outcomes. Several handbooks exist that describe work in the fields of human factors and human-computer interaction, but they are not specific to Web design. Although several books on Web design are available, this handbook is the only comprehensive volume devoted to human factors in Web design. Given the user-intensive nature of the Web and the relatively unique usability issues associated with performing tasks on the Web, a comprehensive handbook of this type is a must for designers, researchers, students, and anyone else who has interests relating to Web design and usability.

The first edition was favorably received and has proved to be an indispensable reference for researchers and practitioners. However, much has changed since its publication. Probably no technology has progressed as rapidly over the past several years as computer technology, and developments involving use, application, and availability of the Web have been at the forefront of this progress. Because of the rapid growth of the Web since publication of the first edition in 2005 and the increased integration of the Web with emerging technologies, this second edition is needed to bring the coverage of human factors in Web design up to date. Twenty-nine chapters from the original edition have been revised for the second edition to reflect the current state of affairs in their topic areas. In addition, new chapters are included on topics that have emerged as important since the first edition. These include organization of information for concept sharing and Web collaboration, Web-based organization models, searching and evaluating information on the WWW, Web portals, human factors of online games, accessibility guidelines and the ISO standards, use of avatars, analyzing and modeling user activities for Web interactions, and mobile interface design for m-commerce.

The handbook contains chapters on a full scope of topics relevant to Human Factors in Web design. The chapters are written by leading researchers and/or practitioners in the field. The handbook is divided into 12 sections, beginning with background chapters on broad topics and narrowing down to specific human factors applications in Web design. Section I includes chapters that provide historical backgrounds and overviews of Human Factors and Ergonomics (HFE), Computers and the Internet, and Human-Computer Interaction (HCI). Chapter 1, by Roscoe, describes the development of the field of HFE. Although much of the work summarized in the chapter predates the Internet and the World Wide Web, the chapter highlights the fact that many issues and principles examined in this earlier work are still applicable to current-day systems, including the Web. In Chapter 2, Bernstein portrays the evolution of computers, placing emphasis on personal computers, first introduced in the early 1980s. He captures how, in less than 30 years, the Internet has revolutionized the tasks that we perform in our daily lives. Dix and Shabir survey the field of HCI in Chapter 3, emphasizing new developments in the field and how usability of new technology can be evaluated throughout the design lifecycle. One of the primary points of these three chapters is that much is known about designing products and systems for human use that is applicable to Web design.

Section II contains chapters devoted to specific subfields of HFE: Physical Ergonomics, Cognitive Ergonomics, and Team Cognition. In Chapter 4, Smith and Taveira discuss issues associated with users' sensory and perceptual-motor skills and provide recommendations for the design and use of Web systems and interfaces. Chapter 5 by Harvey et al. focuses on the characteristics of cognitive processes in performance of individual and team tasks and on methods for analyzing and modeling those processes involved in performance of specific tasks. Although the traditional emphasis in HFE is on individual users, in recent years, there has been increasing concern with team cognition and performance. Chapter 6 by Kiekel and Cooke describes perspectives on team cognition and methods for measuring it. They provide examples of different application areas that benefit from studying team cognition. Together, the three chapters in this section provide insight into how Web interface designs need to take into consideration physical, cognitive, and team aspects of the target user groups.

Section III is devoted to issues involved in content preparation for the Web. The three chapters in this section concentrate on how to organize and structure the information in a manner that is usable for humans. Chapter 7, by Altinsoy and Hempel, focuses on auditory and tactile displays, providing design guidelines and examples and considering multisensory interactions. In Chapter 8, Tullis et al. review how information can be organized and presented visually. Specifically, they discuss issues concerning how to effectively structure information on Web sites in general and on individual Web pages in particular. In Chapter 9, Stephanidis et al. describe a process for designing adaptive interfaces for the Web. They present two complementary case studies, namely, an adaptive Web browser and an adaptive information system, discussing for both cases the architectural frameworks and decision-making mechanisms used for implementing adaptations.

Chapter 9 leads into section IV, which deals specifically with organization of Information for the Web. In Chapter 10, Coffey et al. discuss the use of concept maps to represent and convey the structure of knowledge and research on ways in which concept maps can aid comprehension and reasoning, as well as reduce cognitive demands. They present examples of how concept maps can support knowledge sharing on the Web. Chapter 11 by Michaelis et al. provides an overview of the semantic Web, which through languages and ontologies can allow computers to understand and communicate with each other. The authors emphasize human factors issues associated with language development and design of tools to assist in use of the Semantic Web technology. In Chapter 12, Kojima describes how to implement an ontologically based structure to the design of information systems that can represent knowledge that is to be shared. Dignum and Vázquez-Salceda, in Chapter 13, discuss how Web services can be used to structure Web information for use within and between organizations. This chapter focuses on dynamic interaction and dynamic content. Taken together, these four chapters emphasize the importance of organizing information in a manner that is easily usable by both humans and computers.

Section V contains chapters devoted to the topics of information retrieval and sharing. In Chapter 14, Kammerer and Gerjets provide an overview of the cognitive processes involved in Web search and discuss how search engines can be structured to support Web users in searching for and evaluating information. Chapter 15 by Eisen describes how portals are used to integrate multiple sources of information. He presents best practices for the design of Web portal user interfaces. Chapter 16 by Jacko et al. illustrates how intranets can provide infrastructure for information retrieval and knowledge sharing by organizations. These chapters illustrate the importance of developing information architectures that allow users to quickly retrieve and share information relevant to their goals.

Section VI is concerned with designing for universal access and specific user populations. Chapter 17 by Stephanidis and Akoumianakis provides an introduction to the concept of universal access and discusses some of the representative efforts devoted to understanding universal access. Chapter 18 by Caldwell and Vanderheiden explores the relation between Web accessibility and usability. They discuss general principles of accessible design and go into detail concerning the Web Content Accessibility Guidelines 2.0 document. Chapter 19 by Pappas et al. extends coverage of accessibility to include international standards and design guidelines. Pappas et al. illustrate how accessibility tools can be used to test adherence to color and contrast guidelines.

The first two chapters of section VII are concerned with methods for designing and assessing Web usability, and the last chapter discusses how to incorporate usability into the design process. Chapter 20 by Volk et al. reviews quantitative and qualitative methods for understanding users, including focus groups, interviews, surveys, and contextual observation. Volk et al. emphasize the importance of using these methods early in a development process so that real user data can be collected and utilized to drive product direction. Chapter 21 by Vu et al. reviews methods available for evaluating Web usability, centering on user testing. They describe the advantages and disadvantages of each of several methods and advocate a multimethod approach for evaluating Web usability. Chapter 22 by Mayhew describes a structured, top-down approach to design that is part of the usability engineering life cycle. She provides a detailed case study illustrating how human factors considerations are taken into account at each phase of the Web site development cycle.

Section VIII focuses on task analysis, meaning analysis, and performance modeling. In Chapter 23, Strybel provides detailed descriptions of several approaches to task analysis and recommendations for their use in Web design. Chapter 24 by Flach et al. introduces an ecological perspective for analyzing meaning in complex environments. The authors relate this perspective to cognitive systems engineering, which has the goal of ensuring that the adaptive and creative abilities of humans can be fully exploited in their interactions with the Web as well as other complex environments. In Chapter 25, van Rijn et al. review different modeling techniques for describing how people perform various tasks and give example applications. Techniques are also described for modeling individual users in a way that allows adaptation of the computer interface to match the users' current knowledge and mode of processing. These chapters illustrate how simple and complex tasks can be decomposed into subcomponents that can be examined to understand, improve, predict, and model human performance.

Section IX contains four chapters devoted to specific Web applications in academic and industrial settings. Chapter 26 by Gualtieri and Miller provides an overview of the Web in distance education. The authors cover the state of the art for a variety of different types of e-learning and discuss the benefits and costs of implementing courses via the Web. Chapter 27 by Reips and Birnbaum illustrates how the Web can be used as a research tool and discusses issues relating to online research of various types. Chapter 28 by Najjar describes the steps involved in developing a user-centered e-commerce site. He discusses successful features of e-commerce sites, gives examples of Web sites that employ these features, and presents existing user interface guidelines. Chapter 29 by Sainfort et al. provides an overview of the e-health movement, discussing benefits of e-health and barriers to its proliferation. Given that healthcare has been at the forefront of many recent policy changes in the United States, there is no doubt that e-health will become an important aspect of our daily lives. Thus these chapters illustrate how teachers, researchers, and professionals can benefit from use of Web-based services, provided that those services are designed with usability in mind.

Section X contains four chapters devoted to user behavior and cultural influences. Chapter 30 by Volk and Kraft discusses factors that affect consumers' decision-making processes and behaviors with regard to e-commerce sites. The authors note that there are many factors that affect consumer online purchasing behaviors such as perceived risks and trust, and they provide design recommendations to promote a positive consumer experience. Chapter 31 by Zeng and Duan describes mathematical and computer modeling methods for user Web activity, including clicking activity, selective activity, and propagation activity. The authors discuss techniques for capturing user activity and provide different methods for modeling user activity. Chapter 32 by Schultz considers a largely overlooked issue—the relation between usability, Web security, and privacy. Considerations of Web security are necessary because security is a major factor affecting users' decisions regarding whether to use Web-based services. Schultz discusses human factors issues in several areas of Web security and suggests possible solutions. Chapter 33 by Rau et al. examines effects of cultural differences on Web usability and provides guidelines for cross-cultural Web designs. Rau et al. illustrate the importance of taking into account cultural considerations when designing Web sites. The factors described in these chapters are crucial ones that must be addressed satisfactorily by Web service providers.

Section XI contains three chapters devoted to emerging technological developments and applications for the Web. Chapter 34 by Xu and Fang provides an introduction to use of mobile devices and considers issues associated with interface design, usability of mobile Web applications, and designing for the mobile Internet. Chapter 35 by Steiner presents an overview of usability considerations in the design of online games. He provides detailed illustrations of how specific evaluation and testing techniques can be applied to games. Chapter 36 by Gamor gives an overview of avatars and how their attributes can impact educational and business communications. She discusses how avatars can be used to represent individuals and organizations in virtual worlds and offers practical recommendations. These chapters illustrate that Web-based technologies and applications continue to evolve. Each stage of development has its own associated usability problems, providing new challenges for human factors intervention.

The final section contains a chapter that focuses on analyzing the costs and benefits of incorporating human factors for the Web. Chapter 37 by Richeson et al. describes the value of incorporating human factors into Web design and illustrates how to calculate return on investment. The methods that the authors provide can be used to justify the expenditure on human factors for Web design projects.

In this edition, as in the first edition, we have tried to represent the varied backgrounds and interests of individuals involved in all aspects of human factors and Web design. Thus the handbook contributions are from a diverse group of researchers and practitioners. The contributors reside in nine different countries and have various backgrounds in academia, industry, and research institutes. It is our opinion that awareness of the wide range of views and concerns across the field is essential for usability specialists and Web designers, as well as for researchers investigating theoretical and applied problems concerning Web use and design. We hope that you find the second edition to be a valuable resource for your work in these areas.

We end by thanking the Editorial Board Members for their feedback regarding the handbook's organization and content. The editorial board members include: Helmut Degen, Siemens Corporate Research; Xiaowen Fang, DePaul University; Julie Jacko, University of Minnesota School of Public Health; Lawrence Najjar, TandemSeven, Inc.; Constantine Stephanidis, ICS-FORTH and University of Crete; Tom Strybel, California State University Long Beach; Huifang Wang, SAS Institute, and Anna Wichansky, Oracle Corporation.

Kim-Phuong L. Vu and Robert W. Proctor

Editors

Kim-Phuong L. Vu is associate professor of Psychology at California State University, Long Beach. She is associate director of the Center for Usability in Design and Accessibility and of the Center for Human Factors in Advanced Aeronautics Technologies at CSULB. Dr. Vu has over 75 publications in areas relating to human performance, human factors, and human-computer interaction. She is co-author of the book *Stimulus-Response Compatibility Principles: Data, Theory, and Application.* She is faculty advisor of the CSULB Student Chapter of the Human Factors and Ergonomics Society. Dr. Vu is the recipient of the 2009 Earl A. Alluisi Award for Early Career Contributions of Division 21 (Applied Experimental and Engineering Psychology) of the American Psychological Association.

Robert W. Proctor is distinguished professor of psychology at Purdue University, with a courtesy appointment in the School of Industrial Engineering. Dr. Proctor teaches courses in human factors in engineering, human information processing, attention, and perception and action. He is faculty advisor of the Purdue Student Chapter of the Human Factors and Ergonomics Society. Dr. Proctor's research focuses on basic and applied aspects of human performance. He has published over 150 articles on human performance and is author of numerous books and book chapters. His books include *Human Factors in Simple and Complex Systems* (co-authored with Trisha Van Zandt), *Skill Acquisition and Human Performance* (co-authored with Addie Dutta), *Stimulus-Response Compatibility: An Integrated Perspective* (co-edited with T. Gilmour Reeve), and *Attention: Theory and Practice* (with Addie Johnson). He is fellow of the American Psychological Association, the Association for Psychological Science, and the Human Factors and Ergonomics Society.

Contributors

Demosthenes Akoumianakis, PhD
Department of Applied Information
   Technologies & Multimedia
Technological Educational Institution
   of Crete
Heraklion, Crete, Greece

M. Ercan Altinsoy, PhD
Dresden University of Technology
Dresden, Germany

K. B. Bennett, PhD
Department of Psychology
Wright State University
Dayton, Ohio

Eugenie Bertus, PhD
Kenshoo Inc.
Leander, Texas

Ira H. Bernstein, PhD
Department of Clinical Sciences
The University of Texas Southwestern
   Medical Center
Dallas, Texas

Randolph G. Bias, PhD
University of Texas at Austin
Austin, Texas

Michael H. Birnbaum, PhD
Department of Psychology
Fullerton, California

Benjamin B. Caldwell
Trace R&D Center
Madison, Wisconsin

Yee-Yin Choong, PhD
National Institute of Standards and
   Technology
Gaithersburg, Maryland

John W. Coffey, EdD
Department of Computer Science
University of West Florida
Pensacola, Florida

Nancy J. Cooke, PhD
Arizona State University Polytechnic
Mesa, Arizona

Ashok Darisipudi, PhD
Illinois Tools Works Tech Center
Glenview, Illinois

Virginia Dignum, PhD
Delft University of Technology
Faculty Technology, Policy and
   Management
Delft, Netherlands

Alan J. Dix, PhD
Lancaster University
Lancaster, and Talis
Birmingham, United Kingdom

Jiangjiao Duan, PhD
Department of Computer Science
Xiamen University
Xiamen, People's Republic of China

Paul Eisen, PhD
TandemSeven Inc.
Toronto, Ontario

Xiaowen Fang, PhD
College of Computing and Digital
   Media
DePaul University
Chicago, Illinois

J. M. Flach, PhD
Department of Psychology
Wright State University
Dayton, Ohio

Keysha I. Gamor, PhD
George Mason University Center for
   Online Workforce Development
Fairfax, Virginia

Peter Gerjets, PhD
Knowledge Media Research Center
   (KMRC)
Tübingen, Germany

Jennifer Golbeck, PhD
College of Information Studies
University of Maryland
College Park, Maryland

Lisa Neal Gualtieri, PhD
Tufts University School of Medicine
Boston, Massachusetts

Craig M. Harvey, PhD
Department of Construction Management
   and Industrial Engineering
Louisiana State University
Baton Rouge, Louisiana

Thomas Hempel, PhD
Siemens Audiological Engineering
   Group
Erlangen, Germany

James Hendler, PhD
Tetherless World Constellation
Rensselaer Polytechnic Institute
Troy, New York

Richard Hodgkinson
FISTC Chandler's Ford
Hampshire, United Kingdom

Robert R. Hoffman, PhD
Institute for Human & Machine
   Cognition
Pensacola, Florida

Julie A. Jacko, PhD
The School of Public Health
The Institute for Health Informatics
University of Minnesota
Minneapolis, Minnesota

Addie Johnson, PhD
University of Groningen
Groningen, Netherlands

Yvonne Kammerer, PhD
Knowledge Media Research Center
   (KMRC)
Tübingen, Germany

Preston A. Kiekel, PhD
Cognitive Engineering Research Institute
Mesa, Arizona

Hiroyuki Kojima, PhD
Hiroshima Institute of Technology
Hiroshima, Japan

Richard J. Koubek, PhD
College of Engineering
Louisiana State University
Baton Rouge, Louisiana

Frederic B. Kraft, PhD
Department of Marketing
Seidman College of Business
Allendale, Michigan

V. Kathlene Leonard, PhD
Cloverture LLC
Smyrna, Georgia

Deborah J. Mayhew, PhD
Deborah J. Mayhew & Associates
West Tisbury, Massachusetts

Molly McClellan, BS, MS
Office of Occupational Health &
    Safety, and
The Institute for Health Informatics
University of Minnesota
Minneapolis, Minnesota

James R. Michaelis
Tetherless World Constellation
Rensselaer Polytechnic Institute
Troy, New York

Diane Miller, MEd
Aptima, Inc.
Woburn, Massachusetts

Kevin P. Moloney, BS
School of Industrial & Systems
    Engineering
Georgia Institute of Technology
Atlanta, Georgia

Lawrence J. Najjar, PhD
TandemSeven, Inc.
Plymouth, Massachusetts

Joseph D. Novak, PhD
Institute for Human & Machine
    Cognition
Pensacola, Florida

Frank Pappas, PhD
Alexandria, Virginia

Lisa Pappas, MS
SAS Institute
Cary, North Carolina

Alexandros Paramythis, PhD
Johannes Kepler University Linz
Institute for Information Processing and
    Microprocessor Technology (FIM)
Linz, Austria

Tom Plocher
Honeywell ACS Labs
Golden Valley, Minnesota

Robert W. Proctor, PhD
Department of Psychological Sciences
Purdue University
West Lafayette, Indiana

Pei-Luen Patrick Rau, PhD
Department of Industrial Engineering
Tsinghua University
Beijing, China

Ulf-Dietrich Reips
Departamento de Psicología
Universidad de Deusto, and
IKERBASQUE, Basque Foundation
    for Science
Bilbao, Spain

Andrea Richeson
TradeMark Media
Austin, Texas

Linda Roberts, PhD
SAS Institute
Cary, North Carolina

Stanley N. Roscoe, PhD
Late of University of Illinois at
    Urbana-Champaign and New
    Mexico State University

Ling Rothrock, PhD
The Harold & Inge Marcus
    Department of Industrial &
    Manufacturing Engineering
Pennsylvania State University
University Park, Pennsylvania

D. P. Saakes, PhD
Department of Industrial Design
Delft University of Technology
Delft, Netherlands

François Sainfort, PhD
Division of Health Policy and
    Management, School of Public Health
University of Minnesota
Minneapolis, Minnesota

Anthony Savidis, PhD
Institute of Computer Science, Foundation
    for Research and Technology – Hellas
Heraklion, Crete, Greece

E. Eugene Schultz, PhD
Emagined Security
San Carlos, California

Nadeem Shabir
Talis
Birmingham, United Kingdom

Marisa J. Siegel
Fidelity Investments
Boston, Massachusetts

Michael J. Smith, PhD
Department of Industrial Engineering
University of Wisconsin-Madison
Middleton, Wisconsin

P. J. Stappers, PhD
Department of Industrial Design
Delft University of Technology
Delft, Netherlands

Karl Steiner, PhD
THQ Inc.
Plano, Texas

Constantine Stephanidis, PhD
Institute of Computer Science, Foundation
    for Research and Technology – Hellas
Heraklion, Crete, Greece and
Department of Computer Science
University of Crete
Heraklion, Crete, Greece

Thomas Z. Strybel, PhD
Department of Psychology
California State University
Long Beach, California

Niels A. Taatgen, PhD
University of Groningen
Groningen, Netherlands

Jana Tate
School of Information
University of Texas at Austin
Austin, Texas

Alvaro Taveira, PhD
Department of Occupational and
    Environmental
Safety and Health
University of Wisconsin-Whitewater
Whitewater, Wisconsin

Fiona J. Tranquada
Fidelity Investments
Boston, Massachusetts

Thomas S. Tullis, PhD
Fidelity Investments
Boston, Massachusetts

Gregg C. Vanderheiden, PhD
Trace R&D Center
Madison, Wisconsin

Hedderik Van Rijn, PhD
University of Groningen
Groningen, Netherlands

Javier Vázquez-Salceda, PhD
Departament de Llenguatges I
    Sistemes Informàtics
Universitat Politècnica de Catalunya
Barcelona, Spain

Frederick A. Volk, PhD
Department of Psychology
Liberty University
Lynchburg, Virginia

Kim-Phuong L. Vu, PhD
Department of Psychology
California State University
Long Beach, California

Huifang Wang, MS
SAS Institute
Cary, North Carolina

Anna Wichansky, PhD
Oracle
Redwood Shores, California

Shuang Xu, PhD
Lexmark International
Lexington, Kentucky

Jianping Zeng, PhD
Fudan University
Shanghai, People's Republic of China

Wenli Zhu, PhD
Alibaba Cloud Computing, Alibaba Group
Hangzhou, China

Section I

Background and Overview

1 Historical Overview of Human Factors and Ergonomics

Stanley N. Roscoe

CONTENTS

1.1 CONTEXT

The terms *human factors* and *ergonomics* are closely associated with engineering psychology, the study of human performance in the operation of systems (Proctor and Vu 2010). Human factors psychologists and engineers are concerned with anything that affects the performance of system operators—whether hardware, software, or liveware. They are involved in the study and application of principles of ergonomic design to equipment and operating procedures and in the scientific selection and training of operators. The goal of ergonomics is to optimize machine design for human operation, and the goal of selection and training is to produce people who get the best performance possible within machine design limitations.

Because the Internet and the World Wide Web involve complex human–machine interactions, many of the lessons learned from human factors research in other areas, notably in aviation design, training, and operations, are applicable to issues in Web design. The goal of this chapter is to provide an overview of the pioneering contributions of the people who shaped the field of human factors in system design and to discuss issues and established principles that can be applied to Web design. The chapter is also intended to help the reader understand how this early history provided a serendipitous foundation for the goals of this handbook.

1.1.1 PRINCIPLES OF DESIGN

Human factors specialists are concerned first with the distribution of system functions among people and machines. System functions are identified through the analysis of system operations. Human factors analysts typically work backward from the goal or desired output of the system to determine the conditions that must be satisfied if the goal is to be achieved. Next, they predict—on the basis of relevant, validated theory or actual experimentation with simulated systems—whether the functions associated with each subgoal can be satisfied more reliably and economically with automation or human participation.

Usually, it turns out that the functions assigned to people are best performed with machine assistance in the form of sensing, processing, and displaying information and reducing the order of control.* Not only should automation unburden operators of routine calculation and intimate control, but also it should protect them against rash decisions and blunders. The disturbing notion that machines should monitor people, rather than the converse, is based on the common observation that people are poor watchkeepers and, in addition, tend to be forgetful. This once radical notion is now a cornerstone of modern system design.

1.1.2 Selection and Training

The selection and training of system operators enhances performance within the limits inherent in the design of the system. Traditional operator selection criteria have tended to emphasize general intelligence and various basic abilities believed to contribute to good psychomotor performance. Although individuals without reasonable intelligence and skill do not make effective operators, it has become evident that these abilities are not sufficient. To handle emergencies while maintaining routine operations calls for breadth and rapid selectivity of attention and flexibility in reordering priorities.

The more obstinate a system is to operate and the poorer the operator-selection criteria, the greater the burden on training. Modern training technology is dominated by computer-based teaching programs, part-task training devices, and full-mission simulators (Roscoe 1980). Human factors psychologists pioneered the measurement of the transfer of training in synthetic devices to pilot performance in airplanes starting in the late 1940s and demonstrated the effectiveness of these relatively crude machines (Williams and Flexman 1949a, 1949b). More importantly, some general principles were discovered that can guide the design of training programs for systems other than airplanes—principles that could reduce the trial and error in learning to use the Web, for example.

1.1.3 Application

Fortunately, improved human performance in system operations can come from all directions. Ergonomic design can make the greatest and most abrupt differences in performance, but improvements in selection and training can be made more readily by operational management. More immediate, though usually less dramatic, improvements in system effectiveness can be made through the redesign of the operational procedures used with existing systems. A brief history of how all this got started is best told by focusing on the trailblazing organizations that made it happen.

1.2 THE TRAIL BLAZERS

Soon after the turn of the twentieth century, psychologists started being concerned with the capabilities of aviators and the effects of their limitations on flight operations. Of course, there were no human factors specialists in those days, but general psychologists, along with physicians, were called on to help select the best candidates for pilot training. Soon psychologists would be studying the effects of oxygen deprivation, temperature, noise, and G-forces on human perception and performance in this strange new environment. Later, during World War II, psychologists would start recognizing the effects of airplane cockpit design features on the errors made by pilots and, later yet, the effects of circadian rhythms on the pilots themselves.

Among the earliest experimental studies of the human factors in equipment design were those made during World War II at the Applied Psychology Unit of Cambridge University, England, under the leadership of Sir Frederick Bartlett. In 1939 this group began work on problems in the design of aviation and armored-force equipment (Bartlett 1943; Craik 1940). Early contributions to human factors and ergonomics research at APL included studies of human vigilance and the effects of system design variables on manual control performance, including direction-of-motion relationships between controls and displays (Poulton 1974).

Also in 1939, in the United States, the National Research Council (NRC) Committee on Aviation Psychology was established. This committee stimulated a wide range of research in aviation psychology. With support from the NRC, Alexander C. Williams Jr. at the University of Maryland began flight research in 1939 on psychophysiological "tension" as a determinant of performance in flight training. These experiments, involving the first airborne polygraph, also appear to have been the first in which pilot performance was measured and correlated with physiological responses in flight.

In 1940 the U.S. Army launched a large aviation psychology program (Koonce 1984). With America's entry into the war in 1941, the original organization, the Applied Psychology Panel of the National Defense Research Committee (Bray 1948), was greatly expanded, and its work was extended into what was later to be known as the U.S. Army Air Forces (AAF) Aviation Psychology Program (Flanagan 1947). One of the projects started in 1942 was a study of Army antiaircraft

---

* For those not familiar with the term "order of control," zero order refers to direct position control, as in positioning a display cursor by moving a "mouse." First-order refers to controlling the rate or velocity of movement of an object, as in holding a throttle-pedal position to maintain a constant speed. Second-order control refers to the acceleration or deceleration of an object—changing its speed—as in advancing the throttle or applying brakes. Third order refers to the rate of change in acceleration, and so on. In general, the higher the order of control, the more difficult the task.

artillery at Tufts College, which led to the development of a gun-director tracking simulator (Parsons 1972). Early efforts to study manual control problems included the effects of friction and inertia in controls.

1.2.1 Human Engineering

While most of the psychologists in the British Royal Air Force and the U.S. Army and Navy were involved hands-on in aviator selection and training, others were occasionally called on to deal directly with the subtle problems aviators were having in operating their newly developed machines. During the war the term "pilot error" started appearing with increasing frequency in training and combat accident reports. It is a reasonably safe guess that the first time anyone intentionally or unknowingly applied a psychological principle to solve a design problem in airplanes occurred during the war, and it is possible that the frequent wheels-up-after-landing mishap in certain airplanes was the first such case.

It happened this way. In 1943, Lieutenant Alphonse Chapanis was called on to figure out why pilots and copilots of P-47s, B-17s, and B-25s frequently retracted the wheels instead of the flaps after landing. Chapanis immediately noticed that the side-by-side wheel and flap controls—in most cases identical toggle switches or nearly identical levers—could easily be confused. He also noted that the corresponding controls on the C-47 were not adjacent and their methods of actuation were quite different; hence C-47 copilots never pulled up the wheels after landing.

Chapanis realized that the so-called pilot errors were really cockpit design errors and that by coding the shapes and modes of operation of controls the problem could be solved. As an immediate wartime fix, a small, rubber-tired wheel was attached to the end of the wheel control and a small wedge-shaped end to the flap control on several types of airplanes, and the pilots and copilots of the modified planes stopped retracting their wheels after landing. When the war was over, these mnemonically shape-coded wheel and flap controls were standardized worldwide, as were the tactually discriminable heads of the power control levers found in conventional airplanes today.

1.2.2 Pigeons in a "Pelican"

None of the wartime "human engineers" had received formal training relating human factors to equipment design; indeed, the term "human factors" had not been coined yet. Those who became involved in the study of human factors came from various branches of psychology and engineering and simply invented the budding science on the job. B. F. Skinner stretched the concept a bit by applying his expertise in animal learning to the design of an air-to-sea guidance system for the "Pelican" bomb that employed three kamikaze pigeons who learned to recognize enemy ships and voted on which way to steer the vehicle they were riding (Skinner 1960). It worked fine (and still would), but there were moral objections.

1.2.3 Postwar Developments

In the summer of 1945, the AAF Aviation Psychology Program included about 200 officers, 750 enlisted men, and 500 civilians (Alluisi 1994; Flanagan 1947). In August of 1945, with the war about to end, the AAF Aero Medical Laboratory at Wright Field near Dayton, Ohio, established a Psychology Branch. Their wartime work was documented in 1947 in a series of 19 publications that came to be known as "the blue books." Volume 19, edited by Paul Fitts (1947) and titled *Psychological Research on Equipment Design,* was the first major publication on human factors engineering, or simply human engineering as it was referred to in those times.

Meanwhile the U.S. Air Force's Personnel and Training Research Center, commonly referred to as "Afpatrick," was growing into a huge research organization with laboratories at Mather, Sted, Williams, Tinker, Goodfellow, Lowry, Tyndall, Randolph, and Lackland Air Force Bases. Afpatrick focused on selection and training but also became involved in human engineering and simulator development. In 1958 this far-flung empire was dismantled by the Air Force. Most of the aviation psychologists returned to academia, while others found civilian research positions in other government laboratories.

In late 1945, human engineering in the Navy was centered at the Naval Research Laboratory (NRL) in Washington, DC, under Franklin V. Taylor. The stature of NRL was greatly enhanced by the originality of Henry Birmingham, an engineer, and the writing skills of Taylor, a psychologist. Their remarkable 1954 work, *A Human Engineering Approach to the Design of Man-Operated Continuous Control Systems,* had an unanticipated benefit: to understand it, psychologists had to learn about the electrical engineering concepts Birmingham had transfused into the psychology of manual control.

Another fortunate development in 1945 was the establishment of the Navy's Special Devices Center (SDC) at Port Washington on Sands Point, Long Island. SDC invented and developed many ingenious training devices on site and monitored a vigorous university program for the Office of Naval Research, including the original contract with the University of Illinois Aviation Psychology Laboratory. Task Order XVI, as it was known, was renewed for 20 consecutive years.

In 1946, the Human Engineering Division was formed at the Naval Electronics Laboratory (NEL) in San Diego under Arnold Small. Small, who had majored in music and psychoacoustics and played in the symphony, hired several musicians at NEL, including Wesley Woodson, who published his *Human Engineering Guide for Equipment Designers* in 1954. Major contributions were also made by John Stroud, known for his "psychological moment" concept, and Carroll White, who discovered the phenomenal effect of "visual time compression" on noisy radar and sonar displays.

1.2.4 Human Factors in Academia

On January 1, 1946, Alexander Williams, who had served both as a selection and training psychologist and as a naval aviator, opened his Aviation Psychology Laboratory at the University of Illinois (Roscoe 1994). The laboratory initially focused on the conceptual foundations for mission analysis and the experimental study of flight display and control design principles (Williams 1980). Soon a second major thrust was the pioneering measurement of transfer of pilot training from simulators to airplanes, including the first closed-loop visual system for contact landing simulators. And by 1951, experiments were underway on the world's first air traffic control simulator.

In May, 1946, Alphonse Chapanis (1999, p. 29–30) joined The Johns Hopkins University's Systems Research Field Laboratory in Rhode Island, and in February 1947, he moved to the Psychology Department in Baltimore. Initially, Chapanis concentrated on writing rather than building a large research program with many graduate students, as Williams was doing at Illinois. The result was the first textbook in the field, *Applied Experimental Psychology: Human Factors in Engineering Design*, a monumental work for its time and still a useful reference (Chapanis, Garner, and Morgan 1949). With the book's publication and enthusiastic reception, engineering psychology had come of age, and aviation was to be its primary field of application in the years ahead.

Strong support for university research came from the Department of Defense, notably from the Office of Naval Research and its Special Devices Center and from the Air Force's Wright Air Development Center and its Personnel and Training Research Center. The Civil Aeronautics Administration (CAA) provided funds for human factors research via the National Research Council's Committee on Aviation Psychology. The research sponsored by the CAA via the NRC committee was performed mostly by universities and resulted in a series of studies that became known as "the gray cover reports."

At Illinois, Alex Williams undertook the first experimental study of instrument displays designed for use with the new very high frequency omnidirectional radio and distance measuring equipment (VOR/DME). Gray cover report Number 92 (Roscoe et al. 1950) documented the first simulator evaluation of a map-type VOR/DME navigation display employing a CRT in the cockpit. Gray cover report Number 122 described the previously mentioned first air traffic control simulator (Johnson, Williams, and Roscoe 1951), which was moved to the CAA's facility at Indianapolis and integrated with the flow of actual traffic at the airport to probe the effective limits of controller workload.

Paul Fitts opened his Laboratory of Aviation Psychology at Ohio State in 1949. The laboratories of Williams at Illinois, Chapanis at Johns Hopkins, and Fitts at Ohio State produced the lion's share of the engineering psychologists during the late 1940s and early 1950s, while Neil Warren at the University of Southern California and John Lyman at UCLA were introducing advanced degree programs for many who would distinguish themselves in the aerospace field. Several prominent engineering psychologists were mentored by Ernest McCormick at Purdue in the late 1950s and early 1960s.

By the late 1950s, many companies engaged in the design and manufacture of user products were forming human factors groups or calling on human factors consultants. Various branches of the federal government, in addition to the Defense Department and the Federal Aviation Administration, were hiring human factors specialists to study and deal with problems involving people and machines. In 1957 the Human Factors Society of America was incorporated, later to become an international Human Factors Society and eventually the Human Factors and Ergonomics Society of nearly 5000 members.

1.2.5 Human Factors in Industry

Starting in 1953, several of the airplane and aviation electronics companies hired psychologists, but few of these had specialized in human factors and fewer yet in aviation. As the graduates of the universities with aviation human factors programs started to appear, they were snapped up by industry and by military laboratories as it became painfully apparent that not all psychologists were alike. In a few cases, groups bearing such identities as Cockpit Research, Human Factors, or Human Factors Engineering were established. In other cases the new hires were assigned to the "Interiors Group," which was traditionally responsible for cockpit layouts, seating, galleys, carpeting, and rest rooms. Managers in industry were gradually recognizing that human factors considerations were more than just common sense.

1.2.6 Troubleshooting System Problems

In the early 1950s, an unanticipated technological problem arose in the military community, one that obviously had critical human components. The new and complex electronics in both ground and airborne weapon systems were not being maintained in dependable operating condition. The weapon systems included radar and infrared guided missiles and airplanes with all-weather flight, navigation, target-detection, and weapon-delivery capabilities. These systems had grown so complex that more often than not they were inoperable and, even worse, unfixable by ordinary technicians. Few could get past the first step—troubleshooting the failures. It was becoming evident that something had to be done.

The first alert on the scale of the problem came from the Rand Corporation in 1953 in the form of "the Carhart report," which documented a host of "people" problems in the care of electronic equipment. The technicians needed better training, aiding by built-in test circuits, simulation facilities for practicing diagnoses, critical information for problem solving, and objective performance evaluation. To address these problems, the Office of Naval Research in 1952 contracted with the University of Southern California to establish an Electronics Personnel Research Group with the mission of focusing on the people aspects of maintaining the new systems coming online.

The reports published during the 1950s by this group, organized and directed by Glenn Bryan, had a major impact on the subsequent efforts of the military to cope with the problems of maintaining electronic systems of ever increasing complexity. The lessons learned from this early work were later set forth in Nick Bond's 1970 *Human Factors* article, "Some Persistent Myths about Electronic System Maintenance."

The problems encountered by maintenance personnel of the 1950s in troubleshooting faults in new weapon systems had much in common with the problems of debugging modern software programs. There is one notable difference, however. Today's population of computer users is far more technologically advanced than were the maintenance technicians of the 1950s. So much so, in fact, that some software companies rush to release new programs as soon as they are up and running and depend heavily on their users to detect the bugs and report them. Users can also post solutions on Web pages and blogs that others can reference and search for. In fact, many users initially "Google" for solutions rather than call technical support services.

1.2.7 Design and Consulting Services

In parallel with the above developments, several small companies were organized to provide design and consulting services to industry and the government. Early examples were Dunlap and Associates, Applied Psychology Corporation, Institute of Human Relations, and American Institutes for Research (Alluisi 1994, p. 16). Of these, the American Institutes for Research and Dunlap and Associates expanded or transitioned into fields other than engineering psychology. Still, Dunlap and Associates warrants extra attention here because of its predominant association with human factors over a long period and the importance of its contributions.

1.2.8 Course Setting Committees and Reports

During the 1950s, "blue ribbon" committees were frequently called on to study specific problem areas for both civilian and military agencies, and aviation psychologists and other human factors experts were often included in and sometimes headed such committees. Three of the most influential committee reports were:

- *Human Engineering for an Effective Air-Navigation and Traffic-Control System* (Fitts 1951a).
- *Human Factors in the Operation and Maintenance of All-Weather Interceptors* (Licklider et al. 1953).
- *The USAF Human Factor Engineering Mission as Related to the Qualitative Superiority of Future Weapon Systems* (Fitts et al. 1957).

The air-navigation and traffic-control study by the Fitts committee was of particular significance because, in addition to its sound content, it was a beautifully constructed piece that set the standard for such study reports. Today, original copies of that report are treasured collectors items. The study of all-weather interceptor operation and maintenance by

J. C. R. "Lick" Licklider et al. (1953), though not as widely known, marked the recognition by the military and the aviation industry that engineering psychologists in the academic community had expertise applicable to equipment problems involving human factors not available elsewhere at that time.

Not all of the reports of this genre were the products of large committees. Others written in academia, usually under military sponsorship, included:

- *Handbook of Human Engineering Data* (1949), generally referred to as "The Tufts Handbook," produced at Tufts College under a program directed by Leonard Mead for the Navy's Special Devices Center and heavily contributed to by Dunlap and Associates, followed by:
- *Vision in Military Aviation* by Joseph Wulfeck, Alexander Weisz, and Margaret Raben (1958) for the Wright Air Development Center. Both were widely used in the aerospace industry.
- *Some Considerations in Deciding About the Complexity of Flight Simulators,* by Alexander Williams and Marvin Adelson (1954) at the University of Illinois for the USAF Personnel and Training Research Center, followed by:
- *A Program of Human Engineering Research on the Design of Aircraft Instrument Displays and Controls,* by Alex Williams, Marvin Adelson, and Malcolm Ritchie (1956) at the University of Illinois for the USAF Wright Air Development Center.

Perhaps the three most influential tutorial articles in the field during the 1950s were:

- "Engineering psychology and equipment design," a chapter by Paul Fitts (1951b) in the *Handbook of Experimental Psychology* edited by S. S. Stevens, the major source of inspiration for graduate students for years to come.
- "The magical number seven, plus or minus two: some limits on our capacity to process information" in the *Psychological Review* by George A. Miller (1956), which encouraged quantification of cognitive activity and shifted the psychological application of information theory into high gear.
- *The Design and Conduct of Human Engineering Studies* by Alphonse Chapanis (1956), a concise, instructive handbook on the pitfalls of experimentation on human performance in equipment operation.

Taken as a whole, these key reports and articles—and the earlier research on which they were based—addressed not only pilot selection and training deficiencies and perceptual-motor problems encountered by aviators with poorly designed aircraft instrumentation but also flight operations, aircraft maintenance, and air traffic control. All of these problem areas have subsequently received serious experimental attention by human factors researchers both in the United States and abroad. There are now some established principles for the design, maintenance, and operation of complex systems that have application beyond the immediate settings of the individual experiments on which they are based—even to Web design.

1.3 HISTORICAL PERSPECTIVE

The early educators in the field had in common a recognition of the importance of a multidisciplinary approach to equipment and people problems, and their students were so trained. These early investigators and teachers could only be delighted by the extent to which all researchers and practitioners now have access to once unimagined information and technology to support creative designs based on ergonomics principles as applicable to Web design as to any complex system that involves human–machine interactions. These principles reach far beyond the specific topics and issues originally studied.

1.3.1 TOPICS

As we have seen, the topics addressed by the early human engineers were drawn from wartime needs and were mainly focused on aviation, although substantial work was done on battlefield gunnery and undersea warfare as well. Still the issues involved tended to cross modalities and missions and to be common to civilian as well as military activities. In all kinds of system operations, including human interactions with the Web, controls need to be compatible with population stereotypes, particularly in terms of direction-of-motion relationships, and displays need to be easy to identify and understand.

Not surprisingly, much of the early work was referred to as "knobs and dials" psychology. But human factors engineers are concerned with more than the design and arrangement of knobs and dials. Their approach is systematic, starting with the analysis of a system's goal or mission, followed by the division and assignment of functions among the people in the system and devices that support the performance of both manual and automatic functions: the sensors, transducers, computers, displays, controls, and actuators—the hardware and the software—all of which must do their jobs in some operating environment, whether hot or cold, wet or dry, friendly or hostile—including the Web.

1.3.2 ISSUES

Major issues that emerged in the early days of instrument flight included the following:

- Whether information is best presented pictorially or symbolically ("a picture is worth a thousand numbers" versus scale factor considerations).
- Whether related items of information should be presented on individual, dedicated instruments or

in an integrated display with a common coordinate system (some pilots actually argued for all flight variables to be presented individually on a bank of digital counters so no detail could be lost in the integration process).
- Whether information should be presented "inside-out" or "outside-in" (the worm's-eye view versus the bird's-eye view), with the consequent implications for control-display direction-of-motion relationships (should the world move or the airplane move?).
- Whether vehicle control should be arranged as a *compensatory* task in which a fixed display index is "flown" to a moving index of desired performance (referred to as "fly-to") or a *pursuit* task in which the moving part of a display representing the airplane is flown to a moving index of desired performance or an operator-selected fixed index (strangely referred to as "fly-from").

1.3.3 PRINCIPLES

1.3.3.1 Reduced Control Order

Out of the early experimentation emerged some design principles that have had largely unrecognized effects on the evolution of computers and the Web. The ubiquitous "mouse" and its cousin, the "rolling ball," with their one-to-one (zero-order) position control of a cursor or "marker," are direct descendents of radar hand controls and track balls. Although position control is a seemingly obvious control-display arrangement today, until 1953 radar "range-gate" cursors were velocity controlled by a knuckle-busting, five-position rocker switch—FAST IN, SLOW IN, STOP, SLOW OUT, FAST OUT—spring returned to STOP. The change reduced average "lock-on time" from seven seconds to two seconds.

1.3.3.2 Pictorial Integration

Control and display principles—in addition to reducing control order—that have found their way, but only part way, into computers and the Web are display integration and pictorial presentation. The integration of information to minimize the need to perform mental transformations and computations is the most obvious and dramatic, and the ease with which one can call up whatever is wanted at the moment depends on both integration and pictorial presentation. The use of easily recognizable icons is a form of mnemonic pictorial representation that descended logically from the meaningful shape-coding of aircraft control knobs and display symbology.

1.3.3.3 Prediction

A form of display integration not yet applied to computers or Web design involves the subsidiary principle of flight path prediction. With related information presented in a common coordinate system, the rate of movement of a display element can be shown in the same context as its position. Multiplying the rate of movement of the "arrowhead" or other "marker" symbol by a short time constant (average reaction time) and presenting the result on a small predictor dot that moves in advance of the marker by that amount would virtually eliminate overshooting and undershooting the desired spot on a document, a pull-down menu, a file list, or a tool bar.

1.3.3.4 Simplicity: Just the Facts, Ma'am

In Web-page design, simplicity rules. As the Internet evolved, it became apparent that if a Web page takes a long time to download, users become impatient, then frustrated, and are likely to surf on to another page. The obvious answer was to design pages to download faster, and to do that the recommended approach was to keep the use of graphics and multimedia effects to a minimum (Nielsen 2000). To illustrate the point about simplicity, a fancy PowerPoint presentation with animation may distract from what the speaker is saying, thus making it harder to convey the message (e.g., Savoy, Proctor, and Salvendy 2009).

The transfer of flight training in simulators to pilot performance in airplanes demonstrated the benefit of simplicity early on (Payne et al. 1954; see Roscoe 1980, 199–200). The earliest system to teach students to make visual approaches to landings consisted of a 1-CA-2 Link trainer with a closed-loop geometric outline of a landing runway rear-projected on a translucent screen in front of the trainer. This simple "visual system" reduced the error rate by 85%, with a 61% saving in the number of trials to learn to land the SNJ airplane. Today, not all complex and highly cosmetic visual systems do that well.

Subsequent research has isolated essential visual cues for landing an airplane, and they are remarkably skeletal. High-resolution detail improves the apparent literal fidelity and face validity of simulators, as well as their acceptance by flight instructors and training managers, but it does not improve their transfer of training to performance in airplanes. Essential cues are only those necessary and sufficient for the discrimination of position relative to the runway, flight attitude, projected flight path, and other traffic or obstructions. Additional detail falls in the category of expensive "bells and whistles." The same principle applies to software programs, Web page design, and use of the aforementioned Microsoft PowerPoint software.

The purpose of visual aids in teaching or convincing people is to facilitate communication, not to hold the listener's or reader's attention, not to entertain or impress, and certainly not to distract attention from the message. For best effects, some guidelines will help make images legible and understandable in the back of the room or on a Web site:

- Use black, upright letters (avoid italics) on a white background (not a pastel color that reduces contrast) and select an easily discriminated font that goes easy on the serifs and other squiggles).
- Use the entire screen for text and for simple, bold graphs or diagrams, not for session titles, company or institutional logos, or fancy borders.
- Restrict graphs to material that helps the listener or reader understand the message, with abscissas and

ordinates boldly labeled and experimental conditions or participant groups clearly identified on the graph rather than in a legend (apply the same idea to diagrams). Do not include anything that will not be discussed.
- Use saturated colors but only as needed to distinguish classes of things, not just to make the image "pretty."
- Avoid extraneous, distracting apparent motion, as in slide changes, and "cute" animations, such as sheep jumping over a low fence rail (the listeners or readers might start counting them).

1.3.3.5 Training Wheels

Another principle derived from transfer of training research is that "training wheels" added to flight simulators can induce correct responses early in the training sequence, following which the wheels are removed to avoid developing a dependency on them (Lintern 1978; Lintern, Roscoe, and Sivier 1990). In the case of simulators, the display augmentation takes the form of a flight path "predictor" symbol (a small airplane icon) that moves in immediate response to control inputs to show their imminent effects. This intentional departure from literal fidelity of simulation "steers" the trainee to the desired flight path and greatly facilitates learning.

Although the analogy is a bit of a reach, some features of word processing programs involve essentially the same principle, namely, the flagging of misspellings and spacing and usage errors as the words are typed in. The user is shown where he or she has probably erred (embarked on the wrong flight path), thereby inducing the immediate correction of the faulty response. When the correct response is made, the "training wheels" are removed. In the process, not only is the performance improved, but also a small increment of learning has presumably occurred, with more to follow as future errors are made.

1.3.3.6 Adaptive Interfaces

The removal of "training wheels" when a pilot starts making the correct steering responses is an example of the automatically adaptive training pioneered by aviation psychologists in the 1970s. Initially, the concept focused on increasing the difficulty of a task as the trainee learns and a bit later on reducing the error tolerances allowed (Kelley 1968; McGrath and Harris 1971). A logical extension of automatic adaptation, made possible by the advent of relatively small digital computers in the early 1970s (compared with the earlier analog devices), was the introduction of synthetic visual guidance to replace the flight instructor's almost useless "follow-me-through" routines (Lintern 1978).

A further extension of automatic adaptation is inching toward the goal of universal access to Web sites by tailoring the interface to the perceptual and cognitive capabilities of various user groups (Proctor and Vu 2004). To do this, the first step is to infer the capabilities and limitations of individual users based on their early search behavior. With a tentative person profile as a starting point, the interface is iteratively adapted to the user's characteristics as the individual continues the search process; no matter what question is asked or what response follows, what happens next has to be meaningful and understandable to the seeker.

1.3.3.7 Mission Analysis, Task Analysis, and Modeling

The earliest analyses of complex operations and the tasks involved in their performance are lost in antiquity. However, the formal analysis of aviation missions and tasks did not start appearing in published reports until the late 1940s, following the end of World War II. Certainly, one of the first of these was the "Preliminary Analysis of Information Required by Pilots for Instrument Flight" by Alexander Williams, submitted as an Interim Report to the Special Devices Center of the Office of Naval Research in 1947 and published posthumously in 1971 as "Discrimination and Manipulation in Goal-Directed Instrument Flight" (also see Williams 1980).

This trail-blazing analysis was followed in 1958 by "Aspects of pilot decision making," coauthored by Williams and Charles Hopkins, which surveyed various approaches to modeling human performance, and in 1960 by "Display and control requirements for manned space flight" by Hopkins, Donald Bauerschmidt, and Melvin Anderson. This, in turn, led in 1963 to a description by Hopkins of "Analytic techniques in the development of controls and displays for orbital flight." Incredible as it may seem, these were the only readily available early publications that directly addressed mission and task analyses and the modeling of human performance in system operations. There may have been others, but they were either classified or proprietary.

These early studies provided a systematic basis for the analysis of any human–machine operation, even one as complex as accessing specific information on the Web. Hopkins, in the early 1960s, analyzed the requirements, constraints, and functions of an orbital space mission followed almost exactly about a decade later by NASA's space shuttle. Although a shuttle mission and surfing the net would seem to have little in common, the analytical approach used by Williams and Hopkins is widely generalizable. With a little imagination, it can be extended to any operation involving a branching logic as found in modern operating systems and software programs.

1.3.3.8 Human Factors and the Web

The fact that the wonderful technological advancement during the second half of the twentieth century was largely an outgrowth of aerospace research and development may come as a surprise to many of today's Web designers. It is not generally recognized how many of the Web's most useful features were generically anticipated in flight displays and controls designed by engineering psychologists working in interdisciplinary teams with engineers, physicists, and computer scientists. Human factors experts with interdisciplinary training and research experience in aviation are still applying ergonomic principles to the design of Web sites and other Web-related innovations.

Issues specifically concerning the Internet and Web have been investigated by human factors specialists since the earliest days of the Internet. In 1998, a group of human factors specialists formalized the Internet Technical Group of HFES, which still exists today. Although initiated by a dozen or so individuals, interest in this technical group was sufficiently great that "Within a few weeks, so many people had indicated their interest in joining and participating that it was obvious that we had the necessary critical mass" (Forsythe 1998). The Web site for the group lists the following specific areas of interest:

- Web interface design and content preparation
- Web-based user assistance and Internet devices
- Methodologies for research, design, and testing
- Behavioral and sociological phenomena associated with the Web
- Privacy and security
- Human reliability in administration and maintenance of data networks
- Human factors in e-commerce
- Universal accessibility

These topics and others are all covered in detail in later chapters of this handbook.

ACKNOWLEDGMENT

With permission, this chapter draws heavily on *The Adolescence of Engineering Psychology*, the first issue in the Human Factors History Monograph Series, copyright 1997 by the Human Factors and Ergonomics Society. All rights reserved. Stanley Roscoe (1920–2007) prepared that article and the chapter for the first edition of the handbook. The editors have made minor updates and revisions to the chapter for the second edition.

REFERENCES

Alluisi, E. A. 1994. APA division 21: Roots and rooters. In *Division 21 Members Who Made Distinguished Contributions to Engineering Psychology*, ed. H. L. Taylor, 4–22. Washington, DC: Division 21 of the American Psychological Association.

Bartlett, F. C. 1943. Instrument controls and display—efficient human manipulation, Report 565. London: UK Medical Research Council, Flying Personnel Research Committee.

Birmingham, H. P., and F. V. Taylor. 1954. A human engineering approach to the design of man-operated continuous control systems. Report NRL 4333. Washington, DC: Naval Research Laboratory, Engineering Psychology Branch.

Bond, N. A., Jr. 1970. Some persistent myths about military electronics maintenance. *Human Factors* 12: 241–252.

Bray, C. W. 1948. *Psychology and Military Proficiency. A History of the Applied Psychology Panel of the National Defense Research Committee.* Princeton, NJ: Princeton University Press.

Carhart, R. R. 1953. A survey of the current status of the electronic reliability problem. Report RM-1131-PR. Santa Monica, CA: Rand Corporation.

Chapanis, A. 1956. *The Design and Conduct of Human Engineering Studies.* San Diego, CA: San Diego State College Foundation.

Chapanis, A. 1999. *The Chapanis Chronicles.* Santa Barbara, CA: Aegean.

Chapanis, A., W. R. Garner, and C. T. Morgan. 1949. *Applied Experimental Psychology.* New York: Wiley.

Craik, K. J. W. 1940. The fatigue apparatus (Cambridge cockpit). Report 119. London: British Air Ministry, Flying Personnel Research Committee.

Fitts, P. M. 1947. Psychological research on equipment design. Research Report 19. Washington, DC: U.S. Army Air Forces Aviation Psychology Program.

Fitts, P. M., ed. 1951a. *Human Engineering for an Effective Air-Navigation and Traffic-Control System.* Washington, DC: National Research Council Committee on Aviation Psychology.

Fitts, P. M. 1951b. Engineering psychology and equipment design. In *Handbook of Experimental Psychology*, ed. S. S. Stevens, 1287–1340. New York: Wiley.

Fitts, P. M., M. M. Flood, R. A. Garman, and A. C. Williams, Jr. 1957. *The USAF Human Factor Engineering Mission as Related to the Qualitative Superiority of Future Man–Machine Weapon Systems.* Washington, DC: U.S. Air Force Scientific Advisory Board, Working Group on Human Factor Engineering Social Science Panel.

Flanagan, J. C., ed. 1947. The Aviation Psychology Program in the Army Air Force. Research Report 1. Washington, DC: U.S. Army Air Forces Aviation Psychology Program.

Forsythe, C. 1998. The makings of a technical group. *Internetworking: ITG Newsletter, 1.1, June.* http://www.internettg.org/newsletter/june98/making.html.

Hopkins, C. O. 1963. Analytic techniques in the development of controls and displays for orbital flight. In *Human Factors in Technology*, eds. E. Bennett, J. Degan, and J. Spiegel, 556–571. New York: McGraw-Hill.

Hopkins, C. O., D. K. Bauerschmidt, and M. J. Anderson. 1960. Display and control requirements for manned space flight. WADD Technical Report 60-197. Dayton, OH: Wright-Patterson Air Force Base, Wright Air Development Division.

Johnson, B. E., A. C. Williams, Jr., and S. N. Roscoe. 1951. A simulator for studying human factors in air traffic control systems. Report 122. Washington, DC: National Research Council Committee on Aviation Psychology.

Kelley, C. R. 1968. What is adaptive training? *Human Factors* 11: 547–556.

Koonce, J. M. 1984. A brief history of aviation psychology. *Human Factors* 26: 499–508.

Licklider, J. C. R., G. C. Clementson, J. M. Doughty, W. H. Huggins, C. M. Seeger, C. C. Smith, A. C. Williams, Jr., and J. Wray. 1953. Human factors in the operation and maintenance of all-weather interceptor systems: conclusions and recommendations of Project Jay Ray, a study group on human factors in all-weather interception. HFORL Memorandum 41. Bolling Air Force Base, DC: Human Factors Operations Research Laboratories.

Lintern, G. 1978. Transfer of landing skill after training with supplementary visual cues. PhD Diss. Eng Psy78-3/AFOSR-78-2. Champaign: University of Illinois at Urbana-Champaign, Department of Psychology.

Lintern, G., S. N. Roscoe, and J. E. Sivier. 1990. Display principles, control dynamics, and environmental factors in pilot training and transfer. *Human Factors* 32: 299–317.

McGrath, J. J., and D. H. Harris. 1971. Adaptive training. *Aviation Research Monographs* 1(1): 1–130.

Miller, G. A. 1956. The magical number seven, plus or minus two: Some limits on our capacity for processing information. *Psychological Review* 63: 81–97.

Nielsen, J. 2000. *Designing Web Usability: The Practice of Simplicity.* Indianapolis: New Riders.

Parsons, H. M. 1972. *Man–Machine System Experiments.* Baltimore, MD: Johns Hopkins Press.

Payne, T. A., D. J. Dougherty, S. G. Hasler, J. R. Skeen, E. L. Brown, and A. C. Williams, Jr. 1954. Improving landing performance using a contact landing trainer. Technical Report SPECDEVCEN 71-16-11, Contract N6ori-71, Task Order XVI. Port Washington, NY: Office of Naval Research, Special Devices Center.

Poulton, E. C. 1974. *Tracking Skill and Manual Control.* New York: Academic Press.

Proctor, R. W., and K.-P. L. Vu. 2004. Human factors and ergonomics for the Internet. In *The Internet Encyclopedia*, ed. H. Bidgoli, vol. 2, 141–149. Hoboken, NJ: John Wiley.

Proctor, R. W., and K.-P. L. Vu. 2010. Cumulative knowledge and progress in human factors. *Annual Review of Psychology* 61: 623–651.

Roscoe, S. N. 1980. Transfer and cost effectiveness of ground-based flight trainers. In *Aviation Psychology*, ed. S. N. Roscoe, 194–203. Ames: Iowa State University Press.

Roscoe, S. N. 1994. Alexander Coxe Williams, Jr., 1914–1962. In *Division 21 Members Who Made Distinguished Contributions to Engineering Psychology*, ed. H. L. Taylor, 68–93. Washington, DC: Division 21 of the American Psychological Association.

Roscoe, S. N., J. F. Smith, B. E. Johnson, P. E. Dittman, and A. C. Williams, Jr. 1950. Comparative evaluation of pictorial and symbolic VOR navigation displays in a 1-CA-1 Link trainer. Report 92. Washington, DC: Civil Aeronautics Administration, Division of Research.

Savoy, A., R. W. Proctor, and G. Salvendy. 2009. Information retention from PowerPoint™ and traditional lectures. *Computers and Education* 52: 858–867.

Skinner, B. F. 1960. Pigeon in a pelican. *American Psychologist* 15: 28–37.

Tufts College and U.S. Naval Training Devices Center. 1949. *Handbook of Human Engineering Data.* Medford, MA: Author.

Williams, A. C., Jr. 1947. Preliminary analysis of information required by pilots for instrument flight. Interim Report 71-16-1, Contract N6ori-71, Task Order XVI. Port Washington, NY: Office of Naval Research, Special Devices Center.

Williams, A. C., Jr. 1971. Discrimination and manipulation in goal-directed instrument flight. *Aviation Research Monographs* 1(1): 1–54.

Williams, A. C., Jr. 1980. Discrimination and manipulation in flight. In *Aviation Psychology*, ed. S. N. Roscoe, 11–30. Ames: Iowa State University Press.

Williams, A. C., Jr., and M. Adelson. 1954. Some considerations in deciding about the complexity of flight simulators. Research Bulletin AFPTRC-TR-54-106. Lackland Air Force Base: Air Force Personnel and Training Research Center.

Williams, A. C., Jr., M. Adelson, and M. L. Ritchie. 1956. A program of human engineering research on the design of aircraft instrument displays and controls. WADC Technical Report 56-526, Dayton, OH: Wright Air Development Center, Wright Patterson Air Force Base.

Williams, A. C., Jr., and R. E. Flexman. 1949a. An evaluation of the Link SNJ operational trainer as an aid in contact flight training. Technical Report 71-16-5, Contract N6ori-71, Task Order XVI. Port Washington, NY: Office of Naval Research, Special Devices Center.

Williams, A. C., Jr., and R. E. Flexman. 1949b. Evaluation of the School Link as an aid in primary flight instruction. *University of Illinois Bulletin* 46(7); Aeronautics Bulletin 5.

Williams, A. C., Jr., and C. O. Hopkins. 1958. Aspects of pilot decision making. WADC Technical Report 58-522. Dayton, OH: Wright Air Development Center, Wright-Patterson Air Force Base.

Woodson, W. 1954. *Human Engineering Guide for Equipment Designers*. Berkeley: University of California Press.

Wulfeck, J. W., A. Weisz, and M. Raben. 1958. Vision in military aviation. Technical Report TR-WADC 58-399. Dayton, OH: Wright Air Development Center, Wright-Patterson Air Force Base.

2 A Brief History of Computers and the Internet

Ira H. Bernstein

CONTENTS

2.1 THE EXPANSION OF THE INTERNET

According to http://www.internetworldstats.com/stats.htm, there are roughly 1.7 billion Internet users in the world as of 2009 out of a total world population of 6.8 billion people. This means that one of every four people on the planet is now connected in some way. For example, many in the United States who cannot afford a computer go to their local library. This number of Internet users increased from 361 million a scant 10 years ago, a 4.6-fold increase. It is difficult to conceive of any technological development that has increased at such a rate for the world population as a whole. Even people who expressed minimal interest in computers now order merchandise, pay bills and do other banking chores, and send e-mail instead of letters or telephone calls on a routine basis. It is probable that even the smallest business has a World Wide Web (hereafter, simply Web) site. References to the development of the Web include Gillies and Cailliau (2000) and Orkin (2005).

From where did this phenomenon emerge? Recognizing that the functional beginning of the Internet depends upon defining what characteristic of it is most critical, the Internet officially began in 1983. It was an outgrowth of the somewhat narrowly accessible academic/military Advanced Research Projects Agency Network, ARPAnet (Moschovitis et al. 1999). ARPA was eventually prefixed with "Defense" to become DARPA, but I will refer to it as ARPA throughout this paper to keep it in historical context. The agency has a "total information awareness system" in the fight against terrorism, which has generated supporters and detractors, an issue that goes beyond this chapter.

Moschovitis et al.'s (1999) definition, as ours, is that the Internet is a collection of computers (*nodes*) that use TCP/IP (Transmission Control Protocol and Internet Protocol, to be defined more precisely below). The purpose of this chapter is to review the contributing ingredients of this technological phenomenon.

2.2 PARENT OF THE INTERNET

Because many, if not most, of those reading this book have at least some training in psychology, it may come as a very pleasant surprise to know that "one of us" played a vital role in the Internet's eventual development (though defining the origin of the parent of the Internet runs into the same criterion problem as defining its own beginning). His name was J. C. R. (Joseph Carl Roberts, nicknamed "Lick") Licklider (1915–1990),* and he was already well known for his research on psychoacoustics (see Roscoe, this volume).

Like many trained in the late 1950s and early 1960s, I read Licklider's famous chapters on hearing and on speech in Stevens' (1951) *Handbook of Experimental Psychology*. He was a professor at Harvard and then at MIT's Acoustics' Laboratory. In 1960, he wrote a paper entitled "man–computer symbiosis" in which he proposed that computers would go beyond computing to perform operations in advanced research, by no means the least of which I employed in using online sources for this chapter. Two years later, Licklider directed the Information Processing Techniques Office (IPTO) at ARPA where he pioneered the use of time-sharing computers and formed a group of computer users under the named Intergalactic Computer Network. The next year (1963) he wrote a memorandum in which he outlined the concept of an interacting network linking people together.

2.3 COMPONENTS OF THE INTERNET

In its most basic form, the Internet can be viewed simply as a way to transfer information (files) from one location to another, e.g., your specification of a pair of shoes from your home computer to a store that might be downtown or might not even exist as a physical entity beyond an order taking and filling mechanism. Though it is not implicit in this definition, one would normally require that the information transfer be made with sufficient speed so that it can allow decisions to be made, that is, in *real time*, though given our impatience, real time is often incorrectly interpreted as "instantaneously." Thus, people commonly use the Internet to decide whether or not to make a plane reservation given information about flight availability. Recognizing that its parts must function as an integrated whole, the Internet can be viewed in terms of three components: (1) the devices that store the information

---

* Dates of individual's will be provided where generally available. However, not all individual's dates are public.

and/or request it, i.e., nodes (that is, generally computers, as above noted), (2) how they are connected, and (3) the types of information that are transferred. After considering these in main sections, I will discuss a bit of the history of the role of human factors in the evolution of the Internet before concluding. Some of the recent historically oriented books on the Internet besides Moschovitis et al. (1999) include Comer (2007), Dern (1994), Hauben and Hauben (1997), and Nielsen (1995). There are a huge number of books that are general guides to using the Internet. These include Abbate (2000), Deitel (2000), Gralla (2006), Hahn (1996), Hall (2000), Honeycutt (2007), Connor-Sax and Krol (1999), Quercia (1997), and Sperling (1998). The Internet Society (2009) has several interesting chapters online, and Wikipedia.com is a useful starting point to all aspects of this chapter.

2.4 STORING AND OBTAINING INFORMATION

Thirty years ago, this section could have been limited to what are now called mainframe computers although the word "mainframe" would have been superfluous. Since then, we have seen the development of minicomputers, both in the Apple and personal computer (PC, Windows, or IBM based) traditions (there were also computers of intermediate size, known as "midicomputers" as well as a number of other devices, but they are less important to our history as they basically were absorbed into the growing minicomputer tradition). However, one cannot simply contrast mainframes and minicomputers because a variety of other devices such as telephones ("smartphones") are playing an increasingly important role in Internet communication. For a while, the personal digital assistant (PDA), which, by definition, does not have telephone capabilities, was extremely popular, but many of its functions have been taken over by smartphones. Ironically, some smartphones have more computing power than the earliest minicomputers because of their ability to utilize the Internet! One exception to the decline of the PDA is the hugely successful iPod Touch (as opposed to the iPhone) and some of its less popular competitors. While their main purpose is to play music in MP3 or related format, newer models, especially the Touch, have Internet connectivity and can run a wide variety of programs to do such things as look at the stock market or movie time for a local theatre. There is, of course, the iPhone, which combines the iPod's function with that of a smartphone, but some prefer the iPod because the iPhone is currently limited to a single wireless phone company. Other smartphones approximate, to some extent, the iPhone.

I will follow common practice in *not* simply thinking of a computer in the literal term of "something that computes." That is, it is common to think of it as a device that is capable of storing the instructions on which it operates internally in the form of a numeric code (program) so that one need not reenter instructions to repeat operations. This definition also allows a computer's program to be modified by its own operations. Consequently, one can recognize the historical importance of abaci, slide rules, desktop calculators, and,

in particular, the plugboard-based devices that were the mainstay of industry before mainframes became economically practicable even though these latter devices are not computers by this definition. However, before implementing this definition, I will consider perhaps the computer's single most important early ancestor, the Jacquard loom, and early devices that did not store programs.

In addition, talking about a PC as a "Windows" machine is an oversimplification that neglects the Linux operating system in its several dialects. However, because the vast majority of PCs use some form of Windows, this simplification is not unreasonable.

2.4.1 JACQUARD AND HIS LOOM

Joseph-Marie Jacquard (1752–1834) gave a fitting present at the birth of the nineteenth century with the development of his loom (see Dunne 1999, for a further discussion on this history). The era in which this was developed saw the industrialization of the weaving industry, but, until he developed his loom, machines could not generate the complex patterns that skilled humans could. His device overcame this limitation by use of what was basically a card reader (in principle, the same device that many of us used to enter programs into mainframes earlier in our careers). This device literally distinguished punched and nonpunched zones. Equally important is that the loom was an actual machine, not a concept; Dunne noted that a 10,000-card program knitted a black-and-white portrait of Jacquard on silk. An early Jacquard loom is housed in Munich's Deutches Museum.

Because the program defined by the punched cards controlled a weaving process external to itself, the loom does not meet the definition of a modern stored-program computer. One could clearly distinguish the program from its output. In contrast, both programs and the data that are input or output in a true computer are indistinguishable since both simply exist as strings of ones and zeros.

2.4.2 CHARLES BABBAGE AND ADA AUGUSTA BYRON

The next important figures in this history are more recorded as theoreticians than those providing a finished product, but the reasons were outside of their control. Babbage (1791–1871) extended (or perhaps developed, depending upon one's definition) the concept of the computer as a stored program device. He first started to develop the design of what he called the "difference engine" in 1833 to solve a class of mathematical problems but eventually shifted to his broader concept of the "analytic engine" in 1840. Whereas Jacquard developed his loom to solve a particular problem, Babbage was concerned with general mathematical calculation. Byron (1816–1852), Countess of Lovelace and daughter of the great poet, worked with Babbage and was responsible for many aspects of their work between 1840 and her death in 1852.

Babbage's device required a steam engine because electricity was not yet available and extended Jacquard's use of the punched card as a programming device. However, the British government withdrew its funding so the device was not built until 1991 when the British Scientific Museum showed that it would in fact solve complex polynomial equalities to a high degree of precision. This, of course, did little for Babbage personally who had died 120 years earlier. Indeed, for much of his life he was quite embittered. I am sure that those of us with a file drawer full of "approved but not funded" research proposals can commiserate with Babbage. Further information on Babbage is available in Buxton (1988) and Dubbey (2004).

2.4.3   George Boole and Binary Algebra

Boole (1815–1864) was a mathematician and logician who had little concern for what eventuated into calculators. In contrast, he was interested in the purely abstract algebra of binary events. As things turned out, all contemporary digital computers basically work along principles of what eventually came to be known as Boolean algebra. This algebra consists of events that might be labeled "P," "Q," "R," etc., each of which is true or false, i.e., 0 or 1. It is concerned with what happens to compound events, such as "P" OR "Q," "S" AND "T," NOT "R" AND "S," etc. One of the important developments required for the modern computer was the construction of physical devices or gates that could perform these logical operations.

Boole's realization of an abstract binary algebra is related in a significant way to the "Flip-Flop" or Eccles-Jordan switch in 1919. This binary device is central to computer memory devices. It has two states that can be denoted "0" and "1." It will stay in one of these states indefinitely, in which case one circuit will conduct and another will not, until it receives a designated signal, in which case the two circuits reverse roles. By connecting these in series, they can perform such important functions as counting. See Jacquette (2002) for a recent book on Boole.

2.4.4   Herman Hollerith

Hollerith (1860–1929) was the developer of the punched card used on IBM computers and a cofounder of that company. His work followed directly from Jacquard and was also directed toward a practical issue, the U.S. Census. He began with the U.S. Census Office in 1881 and developed his equipment to solve the various computational problems that arose. He had extensive discussions with John Shaw Billings, who was involved in the data analysis, to discuss mechanization. The outcome was the extension of the punched card and, equally important, a reader/sorter that could place cards into bins depending upon the columns that had been punched. He lost a major battle when a rival company took over his idea, but he eventually founded the Tabulating Machine Company that evolved into IBM, and his technology was still widely used through the 1960s. Material on Hollerith may be found in Bohme (1991) and Austrian (1982).

2.4.5   Alan Turing

Alan Turing (1912–1954) was a brilliant, albeit tragic figure whose short life was ended by his own hand. Clearly a prodigy,

he wrote a paper entitled "On Computable Numbers," which was published when he was only 24. This defined the mathematical foundations of the modern digital, stored program computer. After obtaining his doctorate, he worked for the British Government as a cryptographer. His work was invaluable to the first operational electronic computer, Colossus.

"On Computable Numbers" describes a theoretical device, which became known as a "Turing machine," in response to the eminent mathematician David Hilbert's assertion that all mathematical problems were solvable. A Turing machine is capable of reading a tape that contains binary encoded instructions in sequence. Data are likewise binary encoded, as is the output, which represents the solution to the problem. Humans do not intervene in the process of computation. Turing argued that a problem could only be solved if his machine could solve it, but he also showed that many problems lacked algorithms to solve them.

Perhaps he is most famous for the criterion to define artificial intelligence, "Turing's Test," which states that if a person asks the same question to a human and to a computer and if the answers cannot be distinguished, the machine is intelligent. Carpenter (1986), Millican and Clark (1999), Prager (2001), and Strathern (1999) provide material on Turing.

2.4.6   Early Computers

As has been noted, workable designs for what meets the general definition of a stored program computer go back to Babbage in the first half of the nineteenth century. However, more meaningful early computers, especially in their role as part of a network, also had to await such other inventions as the transatlantic cable, telegraphy and telephony, in general, and, of course, electricity. There are numerous informative books on the history of computers. The most recent include Campbell-Kelly and Aspray (2004), Ceruzzi (1998), Davis (2000), Rojas (2000, 2001), and Williams (1997). The *IEEE Annals of the History of Computing* is a journal devoted to topics of historical interest. There are also many Internet sites that present information such as White (2005).

The immediate pre-World War II era was vital to the development of usable computers. This time period illustrates how such developments often proceed in parallel. Specifically, Konrad Zuse (1910–1995) developed an interest in automated computing as a civil engineering student in 1934 and eventually built binary computing machines, the Z1, Z2, and Z3, from 1938 to 1941. However, Germany's annihilation limited the scope of Zuse's inventions and Germany's role in computer development.

John Atanasoff (1903–1995), a professor at what is now Iowa State University, and his colleague, Clifford Berry (1918–1963), developed what became known as the "ABC" (Atanasoff-Berry Computer). They devised the architecture in 1937 and completed a prototype in 1939, thus overlapping with Zuse. Vacuum tubes were available at that time, but they used relays rather than tubes because tubes of that era were relatively unreliable. However, they did use the recently developed capacitor as the basis of the computer's memory.

The ABC was error prone, and the project was abandoned because of the war, a somewhat paradoxical outcome considering its priority in the British defense and its later importance to the United States. Atanasoff won a court case in 1973 acknowledging him as the inventor of the electronic computer.

Shortly before, during (in England and Germany), and, especially, after the war, several computers appeared. Table 2.1 summarizes their various distinctions and limitations. Note that all of these computers were physically enormous in both size and electricity consumption, so their utility was quite often limited, but they wound up doing huge amounts of productive work. Dates and unique features are somewhat arbitrary. For example, both the EDVAC and the Manchester Mark I are credited with being the first internally programmed computers and some of the earlier computers were modified over their lives.

2.4.7 Grace Murray Hopper and COBOL

Grace Murray Hopper (1906–1992) was trained as a mathematician. She had a long naval career, reaching the rank of admiral, despite prejudice against her gender and having started her career at the unusually late age of 37. She played a pivotal role as a programmer of the Mark I, Mark II (its successor), and the Univac. She invented the compiler, a program that translates user-written programs into machine language. Note that compilers take an entire program, translate it *in toto*, and then act upon it, as opposed to interpreters that take an individual instruction, act on it, take the next instruction,

act on it, etc. Hopper was instrumental to the development of the Flow-matic language and, even more important, COBOL (*Common, business-oriented language*). The latter was one of the two mainstays of programming, along with FORTRAN (Formula translation), for decades.

While certainly not her most important contribution, she popularized the word "bug" in our language. She was working on the Mark II when it began to generate aberrant results. She discovered a dead moth in one of the computer's relays and noted "First actual case of a bug being found" in her notebook. While the term actually dated back at least as far as Thomas Edison, she did introduce it as a term applicable to a programming error and is probably responsible for the gerund, "debugging."

2.4.8 John Backus and Others (FORTRAN)

FORTRAN became the longest-lived program for scientific applications. Many began their computer work using it (I was one of them). FORTRAN dates to an IBM group headed by John Backus (1924–2007) in 1954. It took three years to complete what became known as FORTRAN I, which was basically specific to the IBM model 704 computer. The next year (1958) FORTRAN II emerged, which could be implemented on a variety of computers. Following a short-lived FORTRAN III, FORTRAN IV was developed in 1962 followed by ANSI FORTRAN of 1977 and, a decade later, by FORTRAN 90. It was as close to a common language for programmers of all interest as any has ever been, even though languages like "C" have probably achieved dominance for pure programming

TABLE 2.1
Early Computers

Name(s)	Completion Year(s)	Developer(s)	Characteristics
Z1, Z2, Z3	1938–1941	Konrad Zuse	Used relays obtained from old phones; could perform floating point operations
ABC	1939	John Atanasoff/ Clifford Berry	Credited by U.S. court as first computer; also used relays; abandoned due to U.S. war effort
Collosus	1943	Alan Turing	Used for deciphering; used vacuum tubes; 11 versions built
Mark I	1944	Howard Aiken (1900–1973)	Supported by IBM; used to create mathematical tables and to simulate missile trajectories
ENIAC (Electronic Numerical Integrator and Computer)	1945	John Eckert (1919–1995)/ John Mauchly (1907–1980)	Vacuum tube based; originally had to be externally programmed; 1000 time faster than the Mark I
Manchester Mark I	1949	Max Newman (1897–1984)/ F. C. Williams (1911–1977)	First true stored program computer
EDVAC (Electronic Discrete Variable Automatic Calculator)	1951	John von Neumann (1903–1957)/ A. W. Burks (1915–2008)/ H. Goldstine/John Eckert, and John Mauchly	Completely binary in internal operations; had floating point operations that greatly simplified complex calculations
ORDVAC (Ordnance Variable Automatic Computer)	1952	P. M. Kintner/G. H. Leichner/ C. R. Williams/J. P. Nash	A family that includes ILLIAC, ORACLE, AVIDAC, MANIAC, JOHNNIAC, MISTIC, and CYCLONE; parallel data transfer
UNIVAC (Universal Automatic Computer)	1951	John Eckert and John Mauchly; Remington Rand Corporation	Designed for general commercial sales; various models were sold for many years

and statistical packages like SPSS and SAS for statistical applications among social and behavioral scientists.

2.4.9 JOHN BARDEEN, WALTER BRATTAIN, WILLIAM SHOCKLEY, AND OTHERS (TRANSISTORS AND INTEGRATED CIRCUITS)

In 1947, John Bardeen (1908–1991), Walter Brattain (1902–1987), and William Shockley (1910–1989) of Bell labs started a project concerned with the use of semiconductors, such as silicon, which are materials whose conductivity can be electrically controlled with ease. The first working transistor, which used germanium as its working element and amplified its input signal, appeared in 1949. This *point-contact* transistor evolved into the more useful *junction* transistor. Transistors began to be used in computers in 1953 and, starting with the IBM 7000 series and competing computers in the late 1950s, initiated a new generation of computers. In some cases, e.g., the IBM 7000 series versus the 700 series it replaced, the major difference was the substitution of transistors for less stable vacuum tubes. Eventually, the ability to etch patterns onto silicon led to the integrated circuit, first built by Texas Instruments in 1959, in the present generation of computers. Robert Noyce (1927–1990) and Jack Kilby (1923–2005) of Texas Instruments were important in devising some of the relevant concepts. Shockley (1956), Bardeen (1956 and 1972), and Kilby (2000) eventually won Nobel Prizes, the only such awards thus far given to people connected with development of computers and the Internet.

This development had several obvious effects, increased speed and stability being probably the two most important. Quickly, computers shrank from their massive size. None could ever be thought of as "desktop" or even "personal" (unless you were a very rich person) until the 1970s, but their increased power made it possible for many to interact at a distance via a "dumb terminal." Moreover, there was a reduced need for the massive air conditioning that vacuum tubes required, although transistors and integrated circuits themselves require some cooling. As a note back to earlier days, many welcomed the trips they had to make to university computer centers in hot weather because they were generally the coldest place on campus! Finally, miniaturization became increasingly important as computer designers bumped into fundamental limits imposed by the finite, albeit enormously rapid, time for electrical conduction.

2.5 THE EVOLUTION OF THE PERSONAL COMPUTER

By common and perhaps overly restrictive usage, a personal computer (PC) is a descendant of the IBM Personal Computer, first released to the public in 1981, and it is common to distinguish those that are descendants of this platform from other minicomputers even though the latter are just as "personal" in that they are most frequently used by one or a small number of users. Two other important minicomputers

are the Apple II (1977), and the Macintosh or Mac (1984), both produced by Apple computers, which was founded by Steve Jobs (1955–) and Steve Wozniak (1950–). Preference between the PC and Mac platforms is still a topic that evokes great passions among adherents although PC users far outnumber Mac users. A major key to the PC's success was its affordability to a mass market (the Mac eventually became more economically competitive with the PC, but only after the PC had gained ascendance). Companies like Altair of Model Instruments Telemetry Systems (MITS), founded by Ed Roberts (1942–). MITS arguably offered the first truly personal computer for sale in 1974. Commodore, Radio Shack, and Xerox Data Systems (XDS) were also players in the minicomputer market, and their products were often excellent, but both offered less power for the same money or cost considerably more than a PC. The Apple Lisa (1983) was largely a forerunner to the Mac but, like the XDS Star, had a price of nearly $10,000 that made it unsuitable for mass purchase. One source for the history of the minicomputer is Wikipedia (2009).

It is easy to think a seamless connection between the minicomputer and the Internet because of the overlap in both time and usage (simply think of sending an e-mail or "surfing the Web" from a minicomputer, as I have done myself at several points in writing this chapter), but much of their evolution was in parallel. The early years of the Internet were largely devoted to communication among mainframes. However, the immense popularity of the minicomputer is what made the Internet such a popular success.

Minicomputers went through an evolution paralleling that of mainframes. The earliest ones typically had no disk or other mass storage. The first popular operating system was developed by Gary Kiddall (1942–1994) for Digital Research and was known as the Central Program for Microprocessors (CP/M). As one legend has it, a number of IBM executives went to see Kiddall to have him design the operating system of their first PC. However, Kiddall stalled them by flying around in his personal airplane. As a backup, they went to see Bill Gates at his fledgling Microsoft Company. You might have heard of Gates and Microsoft; Kiddall died generally unknown to the public, long after his CP/M had become obsolete.

All users are now familiar with the graphical user interface (GUI) or "point-and-click" approach that is an integral part of personal computing in general and the Web. Priority is generally given to the XDS Star, but it was also used shortly thereafter by the Lisa and therefore by Macs. Of course, it revolutionized PCs when Windows first became the shell for the command-line DOS. XDS withdrew from the computer market (in my opinion, regretfully, as their mainframes were superb), but Lisa evolved into the Mac, where it has remained a strong minority force compared to the PC tradition.

Of course, no discussion of minicomputers is complete without a discussion of Bill Gates (1955–), who some venerate as a hero for his work in insuring mass use of computers and others view in the same vein as vegetarians view a porterhouse. There are several biographies of this still-young person on the Internet such as http://ei.cs.vt.edu/~history/Gates.Mirick.html. He began with his colleague Paul Allen (1953–) in developing a BASIC interpreter, an easily learned language, for the Altair. BASIC stands for Beginner's All Purpose Symbolic Instruction Code and was developed at Dartmouth College in 1964 under the directory of John Kemeny (1926–1992) and Thomas Kurtz (1928–). Early computers were typically bundled with at least one version of BASIC and sometimes two. BASIC is a stepping-stone for richer languages like FORTRAN that exists in still popular languages like Visual BASIC, though it is no longer part of a typical computer bundle. Gates and Microsoft developed PC-DOS for IBM, but were permitted to market nearly the same product under the name of MS-DOS (Microsoft DOS) for other computers.

Apple and IBM made very different corporate decisions, and it is interesting to debate which one was worse. Apple used their patents to keep their computers proprietary for many years, so competing computers using their platform appeared only recently. In contrast, IBM never made any attempt to do so, so it became popular to contrast their computers with "clones" that attempted to fully duplicate the original and "compatibles" that were designed to accomplish the same end but by different means. The competition caused prices of PCs to plummet, as they are still doing, though most of the changes are in the form of increased power for the same money rather than less money for the same power. The PC still numerically dominates microcomputer use, which makes for greater availability of products (though many can be adapted to both platforms). Unfortunately for IBM, most of the sales went to upstart companies like Dell and Gateway, among others that fell by the wayside. In contrast, prices of Macs remained relatively high, so entry-level users, either those starting from scratch or making the transition from mainframes, often had little choice but to start buying PCs. Most would agree that early Macs had more power than their PC counterparts in such areas as graphics, but those individuals who were primarily text and/or computationally oriented did not feel this discrepancy. In addition, many who are good typists still relish the command-line approach because they do not have to remove their hands from the keyboard to use a mouse. However, nearly everyone appreciates the ease with which both computers can switch among applications, in particular, those that involve importing Internet material.

2.5.1 Supercomputers

A supercomputer is typically and somewhat loosely defined as one of extremely high power and containing several processing units. The Cray series, designed by Seymour Cray (1925–1996) was the prototype. These never played much role in the behavioral sciences, but they were extremely important in areas that demanded enormous storage and calculating capabilities. However, minicomputers have evolved to such power as to mitigate against much of the need for supercomputers, even for applications for which they were once needed.

2.5.2 Development of the Midicomputer

Midicomputers were basically smaller than mainframes but too large to fit on a desktop in their entirety. They had a modest period of popularity before minicomputers gained the power they now have. Two important uses in psychology were to control psychology experiments and to allow multiple users to share access to programs and data as an early form of Intranet. The DEC PDP series was perhaps the most familiar of these to control psychological experiments, and the Wang, developed by An Wang (1920–1990) was extremely popular in offices. Dedicated typing systems, popular in the 1980s, also fall in this general category for office use.

2.5.3 Recent Developments

In the past decade, the *tablet* computer, a laptop with a touch screen that can be operated with a pen instead of the now familiar mouse and keyboard has found a niche. It is designed for circumstances where a conventional computer would not be optimal such as making demonstration and other field work. Variants on this theme include the slate, convertible, and hybrid computer. The Apple IPad is the most recent example.

An even more significant development has been the *netbook*, which is ostensibly a very simple computer primarily designed simply for connection to the Internet rather than the more demanding graphics and other projects possible on a PC. They usually rely upon the Internet to download and install programs. As such, they typically do not have an internal CD/DVD reader/writer, though an external USB model is easily attached. These are extremely inexpensive, which has been vital to their recent success during the economic downturn that began around 2008. Equally important is that they are extremely light (typically weighing less than three pounds) and compact. Despite their power limitations, many are at least as powerful as the first laptops though this might not be noticed in trying to run today's computer programs like Microsoft Office, as the programs have gotten more complex. How Stuff Works (2009) is an excellent nontechnical introduction to many computer-related issues.

2.6 THE EVOLUTION OF COMPUTER OPERATING SYSTEMS

Operating systems, or as they were also commonly called in earlier days, "monitors," went through a period of rapid evolution. All stored program computers need a "bootstrap" to tell them to start reading programs when they are first turned on. Running a program, say in FORTRAN, required several steps. For example, a *compile* step would take the program and generate a type of machine code. Frequently, compilation was broken into several steps. Because programs often had to perform a complex series of operations like extracting square roots or other user-defined processes common to many programs, programmers would write *subroutines* to these ends. Some of these subroutines were (and are) part of a library that was a component of the original programming language; others might be locally written or written by third parties. The codes for the main program and the various subroutines were then merged in a second, *linkage* step. Finally, a *go* step would execute the linked program to generate the desired (hopefully) results (sometimes, the linkage editing and go steps could be combined into a single *execute* step). However, anyone who has written a program knows that numerous iterations were and are needed to get bugs out of the program.

Thanks to the ingenuity of programmers and their willingness to share information, there was rapid evolution in the creation of the necessary libraries of programs. It is convenient to identify the following evolutionary stages (Senning 2000). In each case, the high cost of computers kept a particular computer in service well after far better machines had been developed. For example, even though various forms of batch processing, which required a computer with a monitor, were introduced in the mid-1950s, older machines without a monitor were often kept on for at least a decade.

1. No monitor (roughly, pre-1950). Someone would physically enter what were card decks at each step. At the first step, one deck would contain the original program and a second deck would contain the programming language's compiler. Assuming there were no errors, this would provide punched cards of intermediate output. This intermediate output would then be placed with a second programming deck (the loader) at the second step to provide a final or executable program card deck. This executable deck would be entered along with the source data at a third step, though sometimes, there were additional intermediate steps. The programs were on card decks because the computer was bare of any stored programs. In "closed shops" one would hand the program and source data over to someone. The technicians, typically dressed in white to enhance their mystique, were the only ones allowed access to the sacred computer. In "open shops," one would place the program and data deck in the card reader and wait one's turn. Regardless, one went to a "gang printer" to print the card decks on paper. These printers were "off-line" in that they were not connected to the computer because of their relatively slowness.

2. Stacked batch processing (mid-1950s). As computers developed greater core memory, it became feasible to keep the operating system there. Auxiliary memory, in the form of disk or tape, was also introduced in this era. This auxiliary memory obviated the need for card decks to hold anything other than the user's source programs and data. Disk was much faster than tape but more limited because of its greater expense. Tapes were relatively cheap but had to be loaded on request. Many internecine battles were fought among departments as to which programs could be stored on disk for immediate (online) access versus which were consigned to tape and thus had to be mounted on special request. Disks could also hold intermediate results, such as those arising from compilation or from complex calculations. Users were often allowed to replace the main disk with one of their own disks. I recall my envy of a colleague in the College of Engineering who had gotten a grant to purchase his own disk, thus minimizing the number of cards he had to handle (and drop). The disk was the size of a large wedding cake and could store roughly 60,000 characters (roughly 4% of what a now-obsolete floppy disk can handle). A distinguishing feature of this stage of evolution was that only one program was fed in at a time, but similar jobs using this program could be run as a group.

3. Spooled batch systems (mid-1960s). Jobs were fed into a computer of relatively low computing power and stored on a tape. The tape was then loaded into the main computer. The term "Spool" is an acronym for *S*imultaneous *P*eripheral *O*peration *O*n*L*ine.

4. Multiprogramming (1960s). Jobs were submitted as a block, but several jobs could be run at the same time. The operating system would typically have several *initiators*, which were programs that shepherded user programs through the system. Half of these initiators might be dedicated to running programs that demanded few resources. These might be called A initiators, and there were initiators down to perhaps level "E" for progressively more demanding jobs. The user would estimate the system demands—amount of central (core) memory and time—and be assigned a priority by the operating system. Choosing core and time requirements became an art form unto itself. At first, it would seem to make the most sense to request as little in the way of computer resources as possible to stay in the A queue, since that had the most initiators and fastest turnover. However, toward the end of a semester, the A queue would contain literally hundreds of student jobs. One would quickly figure out that a request for more resources than one needed might place one in a queue with less competition; e.g., you might request 30 seconds for a job that only required 5 seconds. The various departments would lobby with the computer center over initiators. Departments like psychology that ran jobs that were relatively undemanding of the resources, save for the relatively few who did numerical simulation, wanted more "A" initiators dedicated to their jobs; departments that ran very long and complex computations wanted the converse.

5. Time sharing (1970s). Multiple users could gain broader access to the computer at the same time through "dumb terminals" at first and later through personal computers. In many ways, the dumb

terminal mode was even more dramatic a change from the past than the microcomputer because it freed the individual from the working hours of the computer center (early microcomputers were also quite limited). In turn, computer centers began to be accessible every day around the clock for users to enter and receive data and not just to run long programs. The connections were made at various *ports* or access points on the computer. They could enter programming statements without having to prepare them separately as punched cards. Sometimes, the prepared job was submitted in batch mode after preparation; in others cases, it could be run in the real time of data entry. Prior to the development of the personal computer, there were many companies offering time-sharing on their computers to small businesses and individuals. One important effect of time sharing and the physical separation of the user from the mainframe was to develop the idea of computing at a distance, especially via modems, discussed below.

Sometimes, the operating systems and other programs were part of the package one bought when purchasing a computer; often, they were developed independently. I am particularly indebted to a version of FORTRAN developed at the University of Waterloo in Canada, which many others and I found vastly superior to the native form used on our university mainframe. It was not very useful in generating programs that were to be used many times, such as one generating a payroll, but its error detection was superb for many scientists who had to write many different programs to suit the needs of diverse applications.

2.6.1 UNIX

A wide variety of operating systems for mainframes evolved, such as IBM's VM (virtual memory), which is still used. They also include X-Windows, VMS, and CMS, which users may look at fondly or not so fondly depending upon how proficient they became with what was often an arcane language. Perhaps the most important of these is UNIX, developed by Bell Laboratories, arguably America's most successful private research company, which lasted until it became a victim of the breakup of the Bell System into components. Development started in 1969 and the first version was completed in 1970. The intent was to make it portable across computers so that users need learn only one system. The disk operating system (DOS) so familiar to the pre-Windows PC world had a command set that was largely UNIX-derived. Besides the Macintosh's operating system, LINUX is perhaps the most widely used alternative to Windows, and it owes a great debt to UNIX. The "C" language, a very popular programming language, also owes much to UNIX. The UUCP (UNIX-to-UNIX-Copy Protocol) once played an important role in network data transmission of data. UUENCODE and UUDECODE were also familiar to early e-mail users who

were limited to sending text messages. The former would translate binary files such as graphics into text format and the latter would translate them back. Even e-mail is traceable to this project as it began as a vehicle to troubleshoot computer problems.

2.6.2 WINDOWS

According to Wikepedia.com, the Windows operating system was first introduced in 1985 and now controls an estimated 93% of Internet client systems. Like the Macintosh, its GUI performs such functions as connecting to the Internet. Windows has gone through numerous revisions with Windows 7 being the current version (although a sizeable section of the market still prefers its predecessor, Windows XP). Perhaps the major reason for Window's dominance is that it allows multiple programs to be used at the same time by creating what are known as *virtual machines*. Windows was originally an add-on to the DOS that previously dominated PCs but eventually became a stand-alone program.

2.7 MAJOR PERIPHERAL DEVICES IMPORTANT TO THE INTERNET

2.7.1 THE MODEM, ROUTER, AND WIRELESS CARD

A *mo*dulator-*dem*odulator (modem) is a device that can translate data between the digital code of a computer and analog code. At first, this was done over ordinary telephone lines, which have a maximum transmission of around 56 kilobytes/second that greatly limits its utility with the Web. Not surprisingly, the Bell Telephone Laboratories developed the original modem. This 1958 invention was accompanied by AT&T's development of digital transmission lines to facilitate data, as opposed to voice transmission. Increasingly, this translation is now being done by higher-speed ISDN (*I*ntegrated *S*ervices *D*igital *N*etwork) telephone lines or by the cable company. A great many people now use very high speed Internet connections from universities and businesses (known as T1 or T3 lines) and/or ISDN or cable access so that the original dial-up modems and associated technology is becoming obsolete, but high-speed modems such as cable modems are necessities.

The switch from dial-up to high-speed connections caused modems designed for use with a high speed telephone lines or cable provider have replaced the formerly ubiquitous dial-up (56k) modem. Whereas most dial-up modems were placed inside a computer, modems designed for high-speed connections are external to the computer because they need to match the provider. They may be connected to the computer by an *Ethernet cable* or as discussed below, wirelessly via a *router*.

A router is a device that takes the output of a modem and provides wireless connection over a limited area to a *wireless card* in the computer. These have become extremely popular since a router that costs roughly $100 or less can provide coverage over a wide area, say several rooms of a house or apartment, and a wireless card is usually a standard laptop item. As a result, a laptop can be used in a convenient location and moved around to suit the situation. Desktops typically do not have the wireless card because they are typically not moved from one location to another as they are usually connected directly to a modem (routers also allow for such direct connections), but they can be adapted inexpensively if needed.

2.7.2 Wireless Technology in General

The past decade has seen tremendous growth in other wireless technology. *Bluetooth* is a particular form that allows several devices to be used with a computer or other devices, most commonly a mouse, headphone, and microphone (the latter two are especially popular with phones). It is limited to fairly short distance connections.

Technology is now developing to allow other devices to be connected to a computer wirelessly to reduce the "rat's nest" of cables that are a part of most computer systems. This technology is clearly a priority in the computer industry.

2.7.3 Cameras and Camcorders

Digital cameras and camcorders (which were once more commonly analogue but are now universally digital), and Web-based camcorders (Webcams) are joining scanners and printers as common devices that connect to computers and provide information that can be uploaded to the Internet. Digital cameras have already caused a precipitous decline in film cameras and in the sale of film in general, to the point that sale of the film itself has virtually disappeared from ordinary stores save for some highly specialized applications. In fact, digital cameras are well on their way to totally surpassing film along technical dimensions like resolution to say nothing of the convenience of instant access of results. Moreover, small Webcams are also a ubiquitous feature of most computers, especially laptops.

The storage devices for cameras are an important consideration to upload to a computer for possible placement on an Internet site. Some store pictures in *compact flash* format, originated by SanDisk in 1994, and Sony developed its own *Memory Stick* whose use was largely limited to its own products. Increasingly, though, several smaller devices such as *multimedia* and, especially, *secure digital (SD)* format have largely replaced them.

Camcorders and Webcams typically need larger storage capabilities, although increased capacity of devices like SD has made them a possible choice. One alternative is tape storage on the camcorder itself that is usually connected to a computer via a *Firewire (IEEE 1394)* connection that is otherwise more popular on Apple computers than on PCs. Another alternative is a hard disk on the camera that is usually connected to the camera via a USB connection. Webcams usually either broadcast on the Internet in real time, e.g., http://www.vps.it/new_vps/index.php Webcasts from Florence Italy (because of the large amount of bandwidth required by continuous output, this, like many other such sites, effectively Webcasts a series of still pictures). Alternatively, the output from the Webcam may be stored in a compressed video format like MP4.

2.7.4 The USB Connection Itself

Even though it too is properly a device used generally on a computer rather than being specific to the Internet, the importance of the USB connection cannot be overestimated for two extremely important reasons. First, devices that connect via USB are normally "hot swappable," which means that they can be inserted and removed as needed without rebooting. Second, they have rendered a number of separate connections that were popular in the late twentieth century obsolete. These include parallel, serial, small computer system interface (SCSI, pronounced "skuzzy"), and personal computer manufacturer interface adaptor (PCMCIA).

2.7.5 Other Important Devices

Among the many once popular devices that are now essentially obsolete are the various format floppy disks. In their typical format, they could hold up to 1.44 megabytes of data. Storage of relatively small amounts of data (say, up to 16 megabytes, a value that is steadily increasing) is now typically accomplished by what is variously called a thumb drive, flash drive, or jump drive. Larger amounts of information can be stored and transferred by an iPod or similar device, and still larger amounts by an external hard drive or a portable hard drive. The difference between the latter two is that the external hard drive usually requires an external current source, whereas the portable hard drive does not. Both are extremely cheap (in the $100–$200 range), can be totally portable, and carry 1 terabyte or more of data. All of these typically use some form of USB connection.

A *compact disk (CD)/digital versatile disk (DVD) reader/ burner* is a virtual necessity to (1) archive data in a format that is more secure than on a hard disk, (2) load many programs (although this role is rapidly being replaced by the Internet), and (3) access data such as movies. Currently, the standard unit reads and burns CDs and DVDs but not the larger capacity Blu-ray disk, which requires an upgraded unit.

2.8 CONNECTING COMPUTERS: THE INTERNET PROPER

Very early in this article, I defined the Internet in terms of nodes that are connected by transfer control protocol (TCP) and Internet Protocol (IP). This will help you understand how it evolved and how it is different from the many other computer networks that were and are in use such as a company's internal network (intranet). Networks that are part of the Internet are generally provided and maintained regionally. The National Science Foundation funded a high-speed feeder, known as very high speed Backbone Network Services (vBNS), that can carry information to scientific, governmental, and education agencies. Large corporations also finance such backbones. There has been talk for several years about a second Internet for very important, noncommercial applications. Along with the development of multiuser operating systems, modems were the driving force behind computing at a distance. They freed users from a physical presence at the computer center, even though they were slower than terminals that were directly linked to a computer.

2.8.1 Transfer Control Protocol (TCP) and Internet Protocol (IP)

Internet connections may be along telephone lines and modems, satellites, or direct connections in the form of t-carrier lines: T1 lines, the more popular, carry information at roughly 1.5 million bits/second, and T3 lines carry it at roughly 45 million bits/second. The two protocols perform complimentary functions and are often treated as a single concept—TCP/IP. Assume that user A at node Hatfield.com wants to send a file to user B at McCoy.com. Both nodes have been made known to the Internet through a numeric address that consists of four parts that are each octally (0–255) coded. Thus the symbolic name Hatfield.com may correspond to the numeric address 124.212.45.93, etc., and the process of linking the two involves the uniform (formerly universal) resource locator (URL). An additional part of the address is the user, which is separated from the node by "@," so the complete symbolic address for the source may be Paul@Hatfield.com. Because nodes are simply computers with their own addressing scheme, additional information may be needed to identify the file. Files stored on mainframes or large capacity file servers typically separate levels by the forward slash "/"; files stored on minicomputers typically separate levels by the backslash, "\".

The TCP part breaks down the file into small units called *packets*, which are encoded and routed separately. A file-checking system is used to ensure that the transmission was accurate. If it was not, the packet is re-sent. The packets are placed into IP envelopes with the sender and destination address and other information. The packets are relayed through one or more routing devices until they reach their destination. The IP part consists of a process of decoding the address of each packet and selecting the route, which depends upon the nodes functioning at that moment. High-speed connections that form the Internet's backbone are critical to its efficient operation. At the destination, they are individually checked and reassembled, again using TCP.

2.8.2 Sputnik and ARPA

The July 1, 1957 to December 31, 1958 period had been designated the International Geophysical Year, with a planned launch of satellites to map the planet's surface. America had planned to build a satellite to be named the *Vanguard*. However, the former Soviet Republic's 1957 launching of two unmanned satellites, *Sputnik* and *Sputnik II*, set off a major reaction that was to affect the generation of science-oriented students of the late-1950s profoundly. One of the immediate effects was to cancel the Vanguard project in favor of a much more ambitious satellite, Explorer I, which was launched on January 31, 1958. As someone of that generation, I benefited by graduate support that was unmatched before and after as the United States got involved in the Space Race (engineers and physicists were the major, but not only, recipients). The United States was as united in its effort to reclaim the lead it had won in science following World War I as it was to become fractionated over the subsequent Vietnam War.

In 1958, President Dwight Eisenhower (1890–1969) created the ARPA as a Defense Department agency whose nominal mission was to reduce national insecurity over Russia's accomplishments in space. However, it also gave the president an opportunity to support his profound belief in the importance of science. Unfortunately for ARPA, the vastly more photogenic National Aeronautics and Space Administration (NASA) came along shortly thereafter to siphon off much of its budget. However, thanks to J. C. R. Licklider's above-noted leadership, ARPA focused on computers and the processing of information. Shortly thereafter (1965), Larry Roberts (1937–), who later also headed the agency, connected a computer in Boston with one in California to create what was, in effect, the first network. Although it is tempting to think that the resulting ARPAnet was a "Dr. Strangelove" type cold war scenario, it really emerged from the more mundane needs to transmit information simply. The idea that a network in which this information was distributed via packets would be more effective than one in which information traveled *in toto* from one point to another. In 1962, Paul Baran (1926–) of RAND had noted that the system would be more robust in case of nuclear attack, but a variety of additional reasons, primarily the simple desire to transmit information, dictated this important redundancy.

I have noted Licklider's development of several projects connected with time sharing. A 1967 first plan of ARPAnet included packet switching, a term coined by engineer Roger Scantlebury, which later evolved into TCP. Scantlebury introduced the ARPA personnel to Baran. Finally, ARPAnet was born in 1969 and employed a contract to Bolt, Beranek, and Newman (BBN), a small, but immensely respected company located in Cambridge, Massachusetts. The initial four sites were UCLA, the University of California at Santa Barbara, the University of Utah, and Stanford Research Institute (SRI). The network began operations on schedule connecting the four institutions and used four different model computers made by three different manufacturers. However, as noted in an interesting timeline developed by Zakon (2006), Charlie Kline of UCLA sent the first packets to SRI in 1969, but the system crashed as he entered the "G" in "LOGIN"!

A protocol was then developed to expand the network to a total of 23 sites in 1971. ARPAnet continued until it was decommissioned in 1990 after having been replaced by a faster National Science Foundation (NSF) network, NSFnet. The Internet was officially born under that name in 1982, but another way of defining its birth in 1977 is when TCP was used to transmit information across three different networks: (1) ALOHAnet, (2) the Atlantic Packet Satellite Experiment (SATnet), and (3) ARPAnet. ALOHAnet was founded in Hawaii by Stanford University Professor Norman Abramson (1932–) in 1970. It used radio connections. SATNet was a cooperative experiment of the United States and Europe incorporating many groups, including ARPA and BBN. It was founded in 1973 and used satellite technology. Finally, ARPAnet used telephone lines and modems. This transmission traveled a total of 94,000 miles. It started in San Francisco from a computer in a van, went across the Atlantic, eventually arrived in Norway, began its return through London, and eventuated *intact* at the University of Southern California in Los Angeles. Note that at this point only TCP was employed; IP was introduced the next year. As a result, if one's definition of the Internet requires both protocols, 1978 would mark a somewhat different birth.

2.8.3 OTHER EARLY NETWORKS

The general idea of networking was also something "in the air," partly because the ARPAnet concept was not a secret. In many ways, local area networks (LANs) became outgrowths of midicomputers like the Wang and time-sharing computers, and the idea of networking at a distance led to wide area networks (WANs). Some of these major networks and related concepts that have not previously been cited are:

1. SABRE was founded by American Airlines to make airline reservations in 1964. Other companies followed suit over the succeeding decades but could not participate in the Internet until 1992 when Representative Frederick Boucher (D-Virginia) amended the National Science Foundation Act of 1950. This initiated the present era of what is known as "e-commerce."
2. Ward Chapman and Randy Suess of Chicago invented Bulletin Board Systems (BBS) in 1978.
3. Roy Trubshaw (1959–) and Richard Bartle (1960–) developed the first multiuser dungeon (MUD), which is important to computer role-playing games, in 1979.
4. Tom Truscott and Jim Ellis (1956–2001) of Duke University and Steve Bellovin of the University of North Carolina created USEnet in 1979. This is a multidisciplinary network of various news and discussion groups. Although later somewhat upstaged by listservs as a vehicle for people of common interests to get together and exchange ideas, it remains a vital force. For example, rec.music.bluenote is a USEnet group dedicated to jazz that has 191,000 postings in its archives.
5. The City University of New York and Yale University started BITnet ("Because it's there network") in 1981. It served a similar purpose to USEnet. BITnet used a combination of modem and leased telephone lines with communication via terminals. By the late 1990s it had over 300 sites. However, it did not use TCP/IP but a simpler "store and forward" protocol in which a message was forwarded from one node to the next and stored until its successful arrival at the next node. Overloaded or crashed loads led to continued storage rather than the seeking of alternative paths.
6. Tom Jennings (1955–) introduced FIDOnet in 1983 as a bulletin board system dedicated to open and free speech. Its wide use in elementary and secondary schools led to greatly increased popularity of bulletin boards and, later, the Internet.
7. The Internet was "officially" born in 1983, by at least one relevant definition, and the White House went online in 1993 with the Clinton Administration. Shortly thereafter, users often got a rude introduction to the importance of understanding top level domain names (see next section) when www.whitehouse.com, basically a pornographic site, came on and was often addressed by people seeking www.whitehouse.gov (or, at least that is what they said when they were caught at the former). However, it presently deals with health-care issues.

Because one can now FTP, send e-mail, or make connections to another computer seamlessly, it should not be forgotten that routing was once manual. If John, who was at a site served by BITnet, wanted to connect to Marcia, who was at a DARPAnet site, he would have to manually route the message to a site that was connected to both, much like changing airlines at an airport when the two cities are not connected by direct flights.

2.8.4 NAMING

An important aspect of Internet usage is that one does not have to remember the numeric code for a site so that entering fictitious URL (uniform resource locator, i.e., Web address) "123.123.123.123" would be fully equivalent to entering "www.phonysite.com." The relevant conventions of this domain name system were established in 1984. Using a right-to-left (reverse) scan common to computer applications, name consists of a top level, which either identifies the type of organization (.com for commercial enterprise, .edu for educational institution, etc.) in the United States or country elsewhere, so "gb" at the end of a symbolic name would denote the United Kingdom (though not often used, "us" could denote a site in the United States).

In fact, contemporary browsers do not even ask that you prefix the URL with *http://* if you want to go to a Web site, as that is the default. Conversely, FTP (file transfer protocol) sites that simply copy files from one site to another usually have a lowest domain name of "FTP" so it too would be recognized by most browsers and the appropriate protocol used.

Collaboration between NSF and Network Solutions Incorporated (NSI) resulted in the 1993 creation of InterNIC to register domain names. Over the 5 year contract period the number of requests went from a few hundred requests, which could be easily handled, to thousands, which could not. A new body, the Internet Corporation for Assigned Names and Numbers (given the clever acronym of "ICANN") was formed. It formed a centralized authority, replacing additional sites, such as the Internet Assigned Numbers Authority (IANA), which had previously handled root-server (top level) management. Because of the explosion of requests for names, income from name requests has become a billion dollar a year business.

2.8.5 Recent Issues Involving the Internet

Much of the time, a computer does not use any resources beyond itself. For example, writing an essay in Microsoft® Word simply relies on programs stored inside an individual computer and would work perfectly well in the absence of an Internet connection (although it would not update without this connection, of course).

Computers may interact with other computers in two basic ways. In a *peer-to-peer* connection, the two computers have roughly parallel roles so each would perhaps have software to communicate with the other. Some chat programs allow such communication without going through the e-mail protocol. Alternatively, one computer may be thought of as the *server* and the other as the *client* so that the former has the bulk of control. It may have data that it shares with a variety of other computers, for example. This is especially important when a very complex program has to be run on a supercomputer and its results transmitted to a less powerful client. In some ways, this is reminiscent of the earlier days noted above when a terminal interacted on a time-sharing basis with a mainframe. This procedure is quite commonly used to install programs. A relatively small skeleton program is given to the user who then runs it, which involves downloading the bulk of the program from the Internet.

One important development that may involve even a casual user is the virtual private network (VPN) which is activated by a program like Cisco Systems VPN Client or Juniper. The basic idea is to link a client to a server through an open circuit (virtual connection) rather than a hard-wired single private network. This allows the client to log into a company's network from any point in the world and not merely at the network's home site but have his/her session kept private. I used this, for example, to submit grades for a course I was teaching while on vacation in Italy, maintaining security for myself and my students. I also make a VPN connection to the University of Texas Southwestern Medical Center even if I am on campus but want to use a wireless connection because of a medical school's need for privacy.

Cloud computing denotes a client-server network of computers that performs a task that may be too complex for a single computer to handle. It may provide an infrastructure, platform, and or software to complete the task. In some ways it resembles the combination of terminal and time-sharing computers of a generation ago, but it interconnects multiple entities, say Google, Amazon.com, and Microsoft. One possible reason for doing so is that software is kept on the server so it cannot be dissembled by a user interested in developing a competing product. While ordinary computer users may be quite satisfied with standard products like Microsoft Office, many small companies or individual users may find it difficult to afford highly specialized software that is not sold in sufficient volume to be profitable by the vendor.

Three distinct, but somewhat distinctive trends are (1) *grid computing*, where a supercomputer is in fact constructed of a series of networked computers, (2) *utility computing*, where services are administered and billed like any other utility, e.g., the telephone company, and (3) *autonomic computing*, which are computers capable of self-regulation.

2.8.6 Web Browsers

A major reason for the popularity of the Internet, in general, and the World Wide Web, in particular, is that *Web browsers* (programs to access and read or hear the resulting content) have become so easy to use. You don't even have to know the URL you want to go to. If, for example, you wanted to buy something from Amazon.com, you might guess (correctly) that its address is Amazon.com so you enter that in the area for the URL. You do have to know that spaces are not used so "American Airlines" can be reached from "AmericanAirlines.com" or, as you get more familiar with it, simply "AA.com" (if you choose the former, you are automatically *redirected*). Capitalization usually does not matter, so "amerICANairLines.com" also works. Of course, after you get there, you may find that the design of the site is far from optimal (some, for example, make it impossible to change your e-mail address). Companies increasingly rely upon contacts that are limited to computer interactions, which is probably just as well if you have gone through the often hellacious task of calling a company and listening to its excuse for being too cheap to hire enough phone personnel to answer the questions that their Web designers brought about by their ineptitude.

Web browsers date to Tim Berners-Lee (1955–) and his work on the development of the Web in 1989. Then, in 1993, Marc Andreessen (1971–) and Eric Bina (1964–) at the National Center for Supercomputing Applications at the University of Illinois introduced Mosaic, later known as Netscape Navigator, which was the dominant browser for several years. Microsoft licensed a version of Mosaic in 1995 and created Internet Explorer, which became the dominant tool, though it has several excellent competitors in Mozilla Firefox, Cello, and Opera. Google is a very recent entry with its Chrome.

2.8.7 Online References

One of the most useful functions of a Web browser is the ability of various online references to answer questions that may pop into your mind at odd times, e.g., who replaced Diana Rigg as the female lead on the 1960s British series, *The*

Avengers. One way to find this is to start *Wikipedia.com*, and enter "The Avengers." Because this is ambiguous, Wikipedia will list several options, one of which is the TV series. Click on it (isn't hypertext great?) to go to the relevant pages, and there is Linda Thorson's name and character (Tara King) in the contents of the section.

Starting with Wikipedia is useful because it illustrates a fairly traditional tool (encyclopedia) whose use is facilitated by hypertext. Basically, a "wiki" is a Web site that uses software allowing easy creation and editing of Web pages, which imply such features as interlinking. It is defined from the Hawaiian word for "fast." Ward Cunningham (1949–) developed the first Wiki software. Although there is some argument as to who should be credited as Wikipedia's cofounders, Jimmy Wales (1966–) and Larry Sanger (1968–) played undeniably important roles. Although one may argue that long sections of material are better read in a conventional book (or, perhaps, in a newer format like the Kindle electronic book (eBook), a computer screen is certainly a preferred vehicle for short material. Indeed, if you have trouble locating the keyword in a long section, you can always do a search for that word. Typically, keywords are linked to relevant entries.

Wikipedia is quite different from a conventional encyclopedia or even one designed for electronic format like Microsoft *Encarta*, which was based upon Funk & Wagnall's print encyclopedia. Conventional encyclopedia developers hire experts in the field. Wikipedia allows anyone to amend any entry. This has caused noticeable problems with highly emotional topics, but one can well argue that it has resulted in progressively clearer and easier to follow entries in areas that are less emotional. While it should not be the only source consulted, it is usually an excellent starting point. It is very much the intellectual marketplace at work. It is perhaps a major reason that Microsoft decided to discontinue its once-thriving Encarta in 2009. As of this writing, Wikipedia has at least one article in 271 languages. It also has nine associated reader-supported projects: (1) *Commons*, a media repository; (2) *Wikinews*; (3) *Wiktionary*, a dictionary and thesaurus; (4) *Wikiquote*, a collection of quotations; (5) *Wikibooks*, textbooks and manuals; (6) *Wikisource*, a library; (7) *Wikispecies*, a directory of species; (8) *Wikiversity*, learning materials and activities; and (9) *Meta-Wiki*, project coordination.

Hotsheet.com illustrates a very different Web tool. In effect, it is a Web "metareference" as it consists of a series of well-organized links to other sites. For example, you can use it as your Home page and go to a news source of your preference. It also has links to such desktop tools as a thesaurus and calculator, various almanacs and opinion sources, search engines, "yellow" and "white" pages to obtain phone numbers and e-mail and street addresses, and software sites among numerous others.

2.8.8 E-COMMERCE

The 1990s were a period when iffy "dot-com" businesses took advantage of the first flush of Internet business and, as is totally unsurprising in retrospect, a huge number failed in grand style (along with their incoherent commercials). Nonetheless, commercial ventures using the Internet (e-commerce) won the war even though they did not win that battle. Nearly all traditional stores now have a vehicle for online ordering and may make material that is not available in stores. You can also pay nearly all bills online, even setting up a recurring payment plan for large purchases. Airline companies are doing everything in their power to get people to purchase their tickets and check in online (at the expense of conventional travel agents whose knowledge may be invaluable, though at least some will purchase your tickets for you at a reasonable fee).

In contrast, virtual stores like Amazon.com have sprung up and, in some cases, thrived. As implied above, a "virtual store" is one that you cannot visit to look at your merchandise. Indeed, it might be a single person at a single computer. Typically, it has little or no inventory so it has remarkably low overhead; it turns to its suppliers to furnish items when requested by a buyer. This is especially efficient when it comes to items like books, CDs, and DVDs where there are enormous numbers of items for potential sale. Laws governing sales tax vary, but it is quite typically the case that you do not need to pay sales tax if the virtual store is in a different state than the buyer and the company does not have a store in your state of residence. Because a large company may have stores in nearly every state, its online store will typically have to charge sales tax, which is certainly not a trivial item.

2.9 SOCIAL NETWORKING

Until fairly recently, using computers was largely (though not necessarily exclusively) a solitary endeavor, although Internet sites for people to meet have been around for a long time (though often with clearly nefarious intent). Running a statistical analysis, buying a pair of shoes, viewing a movie clip, etc. are all things that can be done in privacy without any thought of social *interaction*. Perhaps the most important change this past decade has been the emergence of computers for social networking purposes, which takes them out of the hands of the social introverts that (perhaps) dominated earlier use.

2.9.1 VOICE OVER INTERNET PHONE (VoIP) AND VIDEO CHATS

Networking takes on many forms. Perhaps the simplest from the user's standpoint is the equivalent of the phone call known as *voice over internet phone* (VoIP). This is simply a phone call using the Internet rather than landlines. There are special Internet phones that can be used, but it is perhaps most common to use a computer's built-in or USB microphone and speakers. The fidelity is typically lower than on landlines (home phones) with a moderate probability of dropping, but this is typically no greater than that occurring with cell phones. It also contributes to the increasing demand on Internet bandwidth, but, right now, it is extremely cheap with long distance (even international) calls currently in the $.02/

minute range to a landline or cell phone and free to another computer. The software from companies like Skype is fairly simple to use with no more difficulty in setting up than with any other software and sign-on program. Conference calls are also easy to make.

The next step up in sophistication (but not necessarily difficulty) is the capability of programs like Skype to make videoconferences, i.e., perform a video chat. As a personal note, our granddaughter was adopted in China. Our daughter and son-in-law took her from the orphanage to the hotel, and we saw her about 30 minutes after they did using no hardware more sophisticated than our laptops, microphone, and Webcam (because our computer did not have one built in). We even took a passable photograph of her with the Skype software. The entire event was free! Needless to say, other companies, particularly Google, are competing for this obviously important market. Skype itself can be used on some smartphones and even the current generation of iPod touch devices using an easily obtained microphone/headphone combination.

2.9.2 Text-Based Chat

Chatting via text messages is another straightforward application. Some programs are designed for peer-to-peer conversations; others are designed for group discussion. The latter are known as IRC (Internet Relay Chat). Of course, this may be done using e-mail, or, for groups, some of the older mechanisms, but there are several freeware programs that can be used to this end. These include ICQ (I seek you), AOL Messenger, Yahoo Messenger, Instan-T, and (not surprisingly) Google Talk.

2.9.3 Blogging

The term "blog" is a contraction of "Web log" and has become one of the most universally noticeable computer-based terms. It has become so popular that the Web-based search engine, Technorati, tracked more than 112 million blogs as of 2007. A format known as *RSS* (alternatively defined as "real simple syndication" or "rich site summary") is a standard vehicle for feeding such information. Indeed, it is difficult to determine where a blog stops and full-featured online news service like the HuffingtonPost or National Review Online starts. Yahoo! and Google Groups allow people to find blog sites to their individual taste. Anyone with even passing familiarity with blogging is familiar with Twitter to make brief comments, Facebook, and MySpace to create what in effect are Home pages more simply than using a program like Dreamweaver to create a Web site. Virtually every newspaper has a Web site, which has raised fears that these will make it difficult to sell the printed page. Sites like Linkedin.com are popular for business-related information. Getting background information on people may be accomplished using Livespaces.com or, if the intent is simply to locate a business address, yellow-pages.com. Similarly, Whitepages.com serves a role similar to that of the white pages of the phone book.

2.9.4 Audio-Oriented Sites

Many sites offer audio files, usually in a compressed format like AAC, RealAudio, or MP3. Perhaps the best known is Itunes.com, which accomplishes numerous ends such as providing downloads of programs for iPods and iPhones. One of its important functions is to serve as a distribution site for *podcasts*, which are basically recorded blogs formatted for iPods that come out on a more or less regular basis, e.g., "The Amateur Traveler," "Travel with Rick Steves," and the monologue from "A Prairie Home Companion." Several of these are also politically oriented, e.g., "Countdown with Keith Olbermann." A particularly attractive feature is that the podcasts are usually free and are excellent to listen to while driving to work. Material that has become part of the public domain, which includes some newer music by bands seeking to advertise themselves, is available through suitable searches for the appropriate category.

2.9.5 Graphics-Oriented Sites

Several sites are designed to allow people to post pictures and short movie clips, e.g., Flickr (a service of Yahoo!), Picasa (a service of Google), snapfish.com, WebShots, and Photobucket.

2.9.6 The Dark Side of Computing

Although popular computer magazines now, as before, contain articles on the latest hardware and software along with associated "how to" articles, increasingly space is devoted to various malicious events. Aycock (2006) and Wikipedia (2009) are good sources of information for this topic.

There is, of course, a huge class of material that is annoying rather than malicious. For example, every store you have ever purchased anything from will put you on their e-mail list because, from their perspective, the advertising costs essentially nothing. From your perspective, you may have to delete 100 or so advertisements that come in daily at least as fast as your ability to enter them on your exclusion file (kill-file) unless they change URLs. You can, of course, request to be deleted, which may or may not be honored (and, unfortunately, may lead to more e-mail than you started with). This category is commonly called *adware*. Whenever you buy a new computer, you may have to get rid of the various trial editions of software you don't want (trialware). However, in both of these cases, an ostensibly legitimate business is making an ostensibly honest effort to get you to buy an ostensibly useful product (to at least someone) even if more of your time winds up being devoted to this source of material than the more straightforwardly dishonest things to be discussed. However, I will also include in this category all of the phony awards that you can only collect if you pay something upfront. This is, of course, fraud rather than honest commerce, but all you have to do is delete it and it is gone. One of the more socially acceptable terms for this general category is *junkware*.

Although many people use the term *virus* to denote any of the computer-based malware that has led to the nearly universal use of software to detect it, its stricter definition is a self-replicating program that is attached to a seemingly innocuous file (the *host*) such as an ostensibly graphic file or program and can enter computers and reproduce itself without knowledge or approval of the computer's owner. The theory of self-replicating programs dates at least as far back as Von Neumann (1966), who lectured on the topic in 1949. An essential feature is that it only affects the computer when the host file is *opened*, which is what separates it from computer worms. The virus may be transmitted via an e-mail attachment, a host file on a jump drive, CD, or DVD, or through the computer network. In other words, merely receiving a virus-infected file is insufficient to cause harm. However, clicking on an attachment that contains the virus will release it to the now-infected computer. What happens may or may not seriously affect the target, just as many biological viruses may not produce symptoms, but they typically do cause at least minimal harm by affecting files.

In contrast, a *computer worm* can send copies of itself to other computers without any user intervention. It too is self-replicating. At the very least, worms use some of the resources (bandwidth) of networks. Some of these are designed to be beneficial, e.g., one that is simply used to study how transmission occurs.

Trojan horses are programs that appear to be useful but may have adverse consequences by giving unauthorized computer access, e.g., to order merchandise on the victim's account. By definition, Trojan horses do not self-replicate, which sets them apart from viruses and worms. They require further intervention by a hacker.

The process of gaining this information is called *phishing*, and software designed to gain unauthorized information is called *spyware*. However, much phishing is attempted by sending official-looking documents impersonating one's bank or other important company asking for such information as one's social security number. There may be an accompanying threat—if the information is not given, one's online banking privileges may be withdrawn, for example (banks never ask for such sensitive information via e-mail).

A particularly evil situation exists when one program takes unwilling control over another, creating what is called a *zombie* or *botnet*. This can be done on an enormous scale given the ferocity with which malware can spread. At this point, the botnets can send a command to a given site and cause it to be shut down. This is known as a *denial-of-service (DoS) attack*. DoS attacks can also reflect the concerted efforts of individual users without the intervention of botnets. Several variants of DoS attacks exist (see Schultz, this volume).

2.10   INTERNET PROTOCOLS AND RELATED PROGRAMS

TCP/IP gets information from one node to another, which is certainly necessary to any network. However, a variety of other protocols were present at the birth of the Internet with still others added later.

2.10.1   Telnet

Much Internet communication is based upon a *client-server* model in which a server controls information flow to a client model in contrast to a *peer* model in which the two computers are equals, although a client in one context may be a server in another. According to Moschovitis et al. (1999), Telnet was a quickly formulated system for logging on to remote sites that was replaced by a superior program called Network Control Protocol (NCP) in 1971. However, Telnet also denotes a program that is still used to allow remote logins; both programs are therefore still in use. Users unfamiliar with its application may have used it, perhaps unknowingly, by logging into a library to access its electronic card catalog. In this specific case, Telnet might run a library "card file" program like PULSE, which is capable of processing, say, a command to find all books written by a particular author. The idea is that one computer, the server, has the necessary resources and controls the process, in this case initiating a copy command to the second computer or client. The name "daemon" was also coined early on to describe the control process.

One particularly important role for Telnet is terminal emulation, especially in older operating systems. As has been noted, "dumb terminals" were widely used to communicate with mainframes long before the development of the mini-computer. A minicomputer had to "fool" the mainframe into making it think that it really was a terminal, which gave rise to Telnet-based emulation programs.

In recent years, JAVA-based programming has greatly expanded client-server interactions though not without risk since JAVA, unlike Telnet, can modify the client computer.

2.10.2   File Transfer Protocol (FTP)

Quite simply, FTP is used to define what files are to be sent or received. Although most such transfers are probably now made using the hypertext transport protocol (HTTP) of the Web, FTP is the more efficient process. It basically was part of the original Internet's capabilities.

In the early days, commands to send or receive files were extremely arcane, but they had a modest amount of power. For example, a file one sent could be routed for printing instead of being stored on disk. When the Web became popular, its browsers incorporated FTP. In addition, programs like WS-FTP offered point-and-click simplicity.

2.10.3   E-Mail

Ray Tomlinson (1941–) of BBN wrote three early e-mail programs that were used to allow exchange of information along the ARPAnet (other note-sending systems had been in use) while using an earlier messaging program called SNDMSG and file transfer program called CYPNET. He popularized the "@" symbol when he developed the convention of defining an e-mail address in the form *User@Node*. He accomplished the first actual communication in 1971 but did not conceive of the system as being used for routine communications, let alone for applications like listservs, which trace back to USEnet and the University of Illinois' PLATO time-sharing system (their use was greatly stimulated by Don Bitzer, a pioneer in the educational use of computers). Samuel F. B. Morse's (1791–1872) first telegraph message was the legendary "What hath God wrought?" Alexander Graham Bell's telephone call to his assistant, while less imposing, was still also memorable: "Mr. Watson, come here; I want you." Unfortunately, Tomlinson does not remember the first e-mail, but he thinks it was something like "QWERTY"! In addition, this first test message was apparently to two machines that were adjacent to one another. It achieved instant popularity.

Larry Roberts (1937–), who also worked at BBN, contributed to the more modern format as seen in popular programs like Outlook and Eudora by developing procedures to forward, reply, and list mail in a program called RD. Unfortunately, this program built in the capability for spamming. Given the overall impact of e-mail, this can easily be forgiven. Mail commands were added to the Internet's FTP program in 1972. Improvements in RD included Barry Wessler's NRD, Marty Yonke's WRD and BANANARD, John Vittal's MSG, Steve Walker et al.'s MS and MH, Dave Crocker, John Vittal, Kenneth Pogran, and D. Austin Henderson's RFC 733 (a specification rather than a program). The process was inefficient, as a separate message had to be sent to each recipient until the Simple Mail Transfer Protocol (SMTP) protocol was added in the early 1980s. Vinton Cerf (1943–) of MCImail, who described important aspects of network communication in 1974 with Bob Kahn (1938–), introduced commercial e-mail in 1988. The following year, CompuServe followed suit. AOL, the largest provider of network services at that time, connected its own mail program to the Internet in 1993.

Whereas SMTP is used to send mail, a complementary program is needed to retrieve it. The first version of such a protocol was Post Office Protocol, Version 2 (POP2), introduced in the mid 1980s. It was shortly updated to POP3 (perhaps unfortunately, "POP," without a number, denotes Point of Presence, i.e., a telephone number for dial access to an Internet service provider). Internet Message Access Protocol is a similar but more powerful program that was developed at Stanford University in 1986. It is currently in its fourth version (IMAP4). Its advantage is that it allows you to search messages that are still on the mail server for keywords and thus decide which to download, i.e., it allows one to create "killfiles."

Most of the early e-mail transmission was limited to text messages, and some sites today still impose this limitation. The original designers of e-mail decided to use a text code (seven-bit U.S. ASCII, American Standard Code for Information Interchange). This code that can handle English alphanumerics but not accented characters (a bit of xenophobia?) nor various other special characters such as "‡," which can be handled by the eight-bit version. IBM computers used a somewhat different code (EBCDIC, Extended Binary Coded Decimal Interchange Code). In contrast, programs, graphics,

audio, and other files (including viruses and worms) use a binary code (for an interesting history of codes, see Searle 2002). However, users quickly learned how to employ to convert binary files to ASCII. One of the more popular was UNIX-to-UNIX encoding and decoding as reflected in the programs Uuencode and Uudecode, respectively. The sender would apply the former, and the receiver would apply the latter, the relevant programs being in the public domain.

Users today are generally familiar with the far simpler Multipurpose Internet Mail Extensions (MIME) which emerged in 1993 that allow binary messages either to be attached or to be part of the message itself. Of course, these binary messages take longer to transfer from the e-mail server. This, of course, is why most of the time spent waiting for e-mail to load over a modem connection involves junk e-mail; the advertising is usually in a binary form.

There are currently over 40 e-mail clients listed in Wikipedia.com, most of which offer free software. Some of the more popular are Microsoft® Outlook, Microsoft® Outlook Express, Eudora, and Gmail. The term "client" is used because they are designed to respond to an e-mail server. They are also called "e-mail agents" and "mail user agents (MUA)." Several offer free software, like Gmail, and the ability to attach files of various types is now nearly universal.

2.10.4 Mailing Lists and Listservs

Once SENDMSG made it possible to send multiple e-mail messages, it was possible to create mailing lists. The term "listserv" is commonly used to denote any e-mail-based dissemination to a group of interested parties. In stricter usage, it denotes a program that was conceived of at BITnet by Ira Fuchs and at EDUCOM (later EDUCAUSE) by Dan Oberst. Ricky Hernandez, also of EDUCOM, implemented the program to support communication within the BITNET academic research network. The first program was written for an IBM mainframe and it still maintains that look even though it was later adapted to a variety of formats, including microcomputers.

As noted in the LivingInternet (2002): "By the year 2000, Listserv ran on computers around the world managing more than 50 thousand lists, with more than 30 million subscribers, delivering more than 20 million messages a day over the Internet."

2.10.5 Hyperlinks, Hypertext Transfer Protocol (HTTP), Hypertext Markup Language (HTML), the Uniform Resource Locator (URL), the Web, Gopher

The hypertext and hyperlink concepts were introduced by Ted Nelson (1937–), a self-appointed "computer liberator." His group sought to bring computing power to the people and correctly foresaw the importance of nonlinear document navigation. Hypertext is material designed to be viewed in this manner. Hyperlinks are the connections by which one may go from one point to another. Hypertext transfer protocol is the procedure that allows one to jump from one section or document to another. Hypertext markup language is what provides the format of the text at Web sites. As previously noted, the uniform resource locator is what allows translation of a symbolic Web site's name to its octal code so a proper connection can be made. The language that Andries van Dam (1938–) of Brown University developed was responsible for much of its implementation as part of the 1967 Hypertext Editing System (HES). However, the single person most strongly identified with the development of the Web is Tim Berners-Lee. Berners-Lee developed the Web from this hypertext system, which debuted in 1991. In conjunction with programmers at CERN (Centre Européen pour la Recherche Nucléaire, a site also made famous for its high energy physics), he developed the relevant protocols noted above.

Despite its recency, the Web has made many people think that it is co-extensive with the Internet rather than simply one of its protocols. Of course, an important part of its wide acceptance comes from the development of browsers and search engines. Berners-Lee also wrote the first GUI-based browser, which was simply called the "World Wide Web." An important aspect of Berners-Lee's work was that he strived to make everything as open and as publicly accessible as possible, encouraging programs to be written for it. This contrasts sharply with the highly profit-oriented view of Bill Gates and Microsoft, which dated back to his development of the BASIC program for Altair. Berners-Lee's encouragement gave rise to Mosaic, which was developed in 1993 by Marc Andreessen and Jim Clark, and renamed Netscape in 1995. They clearly had the dominant browser until the development of Internet Explorer.

Paradoxically, Bill Gates and Microsoft had evinced relatively little initial interest in the Internet. Then, in 1995, Gates issued a memo called "The Coming Internet Tidal Wave," which effectively reoriented Microsoft and, in effect, declared war upon Netscape. At the time, Netscape was the dominant browser, but within 5 years it was forced to merge with AOL because of the success of Microsoft's Explorer browser (see Moschovitis et al. 1999, 192, for a short note on this period in Microsoft's history).

Although now largely forgotten, many people's systems could not run GUI-based browsers and HTTP was not employed as universally. Consequently, nongraphics browsers were widely used in the early days of the Web. These included Lynx and, in particular, Gopher, which was developed in 1991 at the University of Minnesota and named after the university's mascot. Nongraphics browsers would take you to a site and allow you to download files but, unlike modern browsers, would not allow you to look at the material at the site. Consequently, you would typically have to download several files to get the one you wanted.

2.10.6 Archie and His Friends

Being able to reach a Web site is of little value if one does not know where it is or that it contains the desired information.

The number of Web sites is staggering. For example, it grew by a factor of 10 every year from 1991 to 1994. Veronica (*very easy rodent-oriented network index to computerized archives*) was an accessory to Gopher that looked at key words and identified sites but did not provide information as to which file was relevant. Archie was a counterpart that provided information about FTP sites. A third device, the wide area information server (WAIS) was also Gopher-based but used an index of key words created at the site to go beyond the titles, which were often misleading.

2.10.7 Newer Search Engines

David Filo (1966–) and Jerry Yang (1968–) developed Yahoo! in 1994, originally as a public service to users of the Web. Their cataloging and content-based approach revolutionized search engines. Yahoo also grew to the point that it generated hundreds of user groups for interest groups like the "Travelzine," a listserv for those interested in travel. Many other search engines followed in the wake.

Perhaps the most widely used is Google (http://www .google.com), although there are many others of high quality. Google was founded by Larry Page (1973–) and Sergey Brin (1973–). The company is also responsible for a new verb: to "google" someone means to look for references to them on the Web! This poses a problem for Google because once a word or phrase becomes generic, it loses its protected value as designating a specific search engine. Of greater importance is the variety of innovations they have contributed such as Google Phone, which has the potential to radically increase the flexibility of the increasing multiplicity of phone numbers a person deals with. Vinton Cert, a major figure in the development of Internet technology, is currently a vice president at Google.

2.11   THE INTERNET AND HUMAN FACTORS

The mass use and acceptance of the Internet is a testimony to its increased usability. For example, no contemporary FTP program would dare ask its users to know its arcane symbolism. Go to a site like Schauble (2003), which contains a list of FTP commands, and conduct a brief quiz on a computer literate friend. Some terms, like "Bye" and "Quit" are innocuous enough, and they do, in fact, produce an exit. However, how many other programs use "End" or "Stop" instead? Recall what may not have been the wonderful days of yesteryear when Control-S would save a program in one language and scratch it in another (even more endearing is when this happened in the same type of application, such as two different word processors). Next, consider sending a file via FTP. "Send" would be a likely candidate but, unfortunately, no such luck—the correct answer is "Put"! Similarly, "LCD" is not a type of display but a change of directory on your own machine. When you had to learn this, you learned it; hopefully without too much retroactive and proactive interference from other arcane computer terms. True to cognitive dissonance theory, those who had learned the vocabulary would tut-tut innovations like the GUI that rendered memory for these terms unnecessary. Indeed, they would note how programming was going to the proverbial "hell in a hand basket" (in fact, I reveal my own cognitive dissonance below).

The relative standardization of common Windows menus, e.g., placing file-related commands (opening, closing, saving, etc.) at the leftmost position relates to computing in general rather than specifically to the Internet. However, let us give thanks for the many programs that follow this practice. Unfortunately, that does not keep things like preferences, in the generic sense (Options, Customize, etc.), from wandering around from menu to menu. In addition, although it makes sense to distinguish between closing a file in a program like a word processor and exiting the program altogether, other programs exit with a "close" command.

While on the topic of computing in general, it is important to contrast the command-line and GUI approaches from a human factors standpoint. Unfortunately, while GUI seems to have deservedly won (and the victory will be even more decisive when appropriate Darwinian mechanisms appear and provide a third or mouse hand), the issue is confounded by a number of other issues. Like many, I found Microsoft's initial efforts less than impressive (Windows 3.0 excelled at crashing at the least opportune times), as I took pride in my DOS fluency (not a marketable skill nowadays) and lacked the computer power to run many programs at once (to say nothing of the lack of present availability of such programs). I was thus limited in my ability to take advantage of what may be Windows' most powerful feature (yes, I know that Macs already had that ability).

The development of the Internet has an important general effect because almost by definition its products are intended for others. This raises the very fundamental problem that information presented for the benefit of the developer need not make sense to another user (this problem would, of course, exist without the Internet given the market for programs computer software). For example, back in the horrific days before statistical packages, I could write statistical programs that I could understand because only I had to know my mnemonics. This did not always lead to the best results when one of my research assistants had to use it.

The explosion of Web sites has obviously brought a worldwide market to anyone with a telephone connection and a computer. Most sites are workable, even if many can stand improvement. However, I am reminded by the wisdom, of all things, of an IBM television commercial of some years back showing a young Web designer who could create any visual effect known to humanity but could not provide a critical linkage between inventory and sales. To paraphrase an old musician's joke about the banjo (I also hope I will be forgiven for the gender-specific nature of the comment)—a gentleman is a Web designer who can use any possible effect but does not. Who among us has not visited a site with whirlies whirling everywhere and a color scheme of dark blue on slightly darker dark blue? That color scheme shows how elegant the contrast generated by (255, 255, 255) white against (0, 0, 0) black is. Likewise, how many sites have vital key commands like "login" that are buried and/or change position weekly as the design team earns their keep by constantly changing the interface? How about sites that simply do not work for the ultimate user?

Of course, much poor human factors work comes from programming laziness. Consider one of the most routine requests for information—your telephone number. Numbers can be written in a variety of ways, e.g., (555) 987–6543, 555.876.6543, etc. It does not take great programming prowess to provide a mask (most database programs have such an option). However, this is not critical as long as any of the nonnumeric information is stripped away, not exactly big league programming. Now, imagine that this has not been done. You are trying to submit the information. Obviously, a programmer who would not provide a mask nor strip the nonnumeric information would not tell you which offending field keeps you from having your information accepted—that too involves a bit of work (again, programs like FrontPage have this capacity). Perhaps, though, this is not laziness. I do not know if the Marquis de Sade left any descendants, but it seems as if at least some became Web designers.

While writing this chapter, I received an example of something that might fall under the heading of poor human factors. On the other hand, being an occasional visitor to http://www.darwinawards.com, it might fall into the industrial equivalent of same (for those unfamiliar with the site, Darwinawards posthumously honors "those who improve our gene pool by removing themselves from it"). I am referring to companies that go to the trouble of spamming you with offers and then provide the wrong URL. Frankly, the amount of spam I receive kept me from contacting them with the suggestion that they might attract more customers by providing the correct address, because I felt that they might be very clever spammers (or social psychologists) who simply wanted to see how many people would tell them they had made an error so that they could reel in my address to send me perpetual offers of (phrase deleted by author).

Finally, one of the major areas in which work has been put with considerable apparent success is improving the access of individuals with various handicaps. Indeed, a rather extensive set of features are built into operating systems such as Microsoft Windows. This is discussed in Chapter 18 of this volume.

2.12 THE INTERNET'S FUTURE

The last half of the twentieth century saw fear arising from the possibility of nuclear disaster, a fear that is still cited. However, few who were alive at both the beginning and end of this period would have seen how computers made a transition from an esoteric device to one now accessible by nearly everybody. Likewise, what was once a connection of four mainframe computers now includes over half a billion people with home Internet access as of late 2002 according to Nielsen-Netratings (Hupprich and Bumatay 2002), although there has been some recent abandonment of sites, perhaps because of the recent economy and loss of small businesses. A small department in an office or at a university now has far more computers than were once conceived to exist in the world. Paraphrasing what I noted above, even the cheapest commercial computer has more power than was needed to put a person in orbit. One comparison shops and buys over the Internet. One communicates with friends and make new ones anywhere in the world over the Internet, and one learns over the Internet.

What is going to happen in the future? Even the most limited of minds can foresee the commercial usages leading to various technical improvements, but perhaps even the most intelligent cannot foresee breakthroughs. After all, who would have known of the ramifications of connecting four computers a mere 30 years ago? Obviously, we have begun to take for granted the role of the Internet in our lives (unless there is a major crash at a critical time) along with other forms of communication. Perhaps it is safest to note that just as legal actions dominated technological innovations in the late 1990s, we will see control passing from those who specialize in technology to those who apply it and to others who are concerned with its content.

REFERENCES

Abbate, J. 2000. *Inventing the Internet*. Cambridge, MA: MIT Press.

Austrian, G. 1982. *Herman Hollerith, Forgotten Giant of Information Processing*. New York: Columbia University Press.

Aycock, J. D. 2006. *Computer Viruses and Malware*. New York: Springer Science & Business Media.

Bohme, F. G. 1991. *100 Years of Data Processing: The Punchcard Century*. Washington, DC: U.S. Department of Commerce, Bureau of the Census.

Buxton, H. W. 1988. *Memoir of the Life and Labours of the Late Charles Babbage Esq., F.R.S.* Cambridge, MA: MIT Press.

Campbell-Kelly, M., and W. Aspray. 2004. *Computer: A History of the Information Machine*. New York: Basic Books.

Carpenter, B. E. 1986. *A.M. Turing's ACE Report of 1946 and Other Papers*. Cambridge, MA: MIT Press.

Ceruzzi, P. E. 1998. *A History of Modern Computing*. Cambridge, MA: MIT Press.

Comer, D. 2007. *The Internet Book: Everything You Need to Know About Computer Networking and How the Internet Works*. Upper Saddle River, NJ: Prentice Hall.

Conner-Sax, K., and E. Krol. 1999. *The Whole Internet*. Cambridge, MA: O'Reilly

Davis, M. D. 2000. *The Universal Computer: The Road from Leibniz to Turing*. New York: Norton.

Deitel, H. M. 2000. *Internet and World Wide Web: How to Program*. Upper Saddle River, NJ: Prentice Hall.

Dern, D. P. 1994. *The Internet Guide for New Users*. New York: McGraw-Hill.

Dubbey, J. M. 2004. *The Mathematical Work of Charles Babbage*. Cambridge, MA: Cambridge University Press.

Dunne, P. E. 1999. History of computation—Babbage, Boole, Hollerith. http://www.csc.liv.ac.uk/~ped/teachadmin/histsci/htmlform/lect4.html (accessed Oct. 1, 2009).

Gillies, J., and R. Cailliau. 2000. *How the Web Was Born: The Story of the World Wide Web*. New York: Oxford University Press.

Gralla, P. 2006. *How the Internet Works*, 8th ed. Indianapolis, IN: Que.

Hahn, H. 1996. *The Internet Complete Reference*, 2nd ed. Berkeley, CA: Osborne McGraw-Hill.

Hall, H. 2000. *Internet Core Protocols: The Definitive Guide*. Cambridge, MA: O'Reilly.

Hauben, M., and R. Hauben. 1997. *Netizens: On the History and Impact of Usenet and the Internet*. Los Alamitos, CA: Wiley-IEEE Computer Society Press.

Honeycutt, J. 2007. *Special Edition: Using the Internet*, 4th ed. Indianapolis, IN: Que.

How Stuff Works. 2009. http://www.howstuffworks.com/ (accessed Oct. 1, 2009).

Hupprich, L., and M. Bumatay. 2002. *More Internet Browsers Convert to Purchasers in the UK than in 10 Other Major Markets*. New York: Nielsen Media Research.

Internet Society. 2009. Histories of the Internet. http://www.isoc.org/internet/history (accessed Oct. 8, 2009).

Jacquette, D. 2002. *On Boole*. Belmont, CA: Wadsworth/Thomson Learning.

LivingInternet. 2002. The living Internet. http://www.livinginternet.com/ (accessed Oct. 1, 2009).

Millican, P. J. R., and A. Clark, eds. 1999. *The Legacy of Alan Turing*. New York: Oxford University Press.

Moschovitis, C. J. P., H. Poole, T. Schuyler, and T. Senft. 1999. *History of the Internet: A Chronology, 1983 to the Present*. Santa Barbara, CA: ABC-CLIO.

Nielsen, J. 1995. *Multimedia and Hypertext: The Internet and Beyond*. Boston, MA: AP Professional.

Orkin, J. R. 2005. *The Information Revolution: The Not-for-Dummies Guide to the History, Technology, and Use of the World Wide Web*. Winter Harbor, ME: Ironbound Press.

Prager, J. 2001. *On Turing*. Belmont, CA: Wadsworth/Thomson Learning.

Quercia, V. 1997. *Internet in a Nutshell: A Desktop Quick Reference*. Sebastopol, CA: O'Reilly & Associates.

Rojas, R. 2000. *The First Computers: History and Architectures*. Cambridge, MA: MIT Press.

Rojas, R. 2001. *Encyclopedia of Computers and Computer History*. Chicago, IL: Fitzroy.

Roscoe, S. N. this volume. Historical overview of human factors and ergonomics. In *Handbook of Human Factors in Web Design*, 2nd ed., eds. K.-P. L. Vu and R. W. Proctor, 3–12. Boca Raton, FL: CRC Press.

Schauble, C. J. C. 2003. Basic FTP commands. http://www.cs.colostate.edu/helpdocs/ftp.html (accessed Oct. 1, 2009).

Schultz, E. E. this volume. Web security, privacy, and usability. In *Handbook of Human Factors in Web Design*, 2nd ed., eds. K.-P. L. Vu and R. W. Proctor, 663–676. Boca Raton, FL: CRC Press.

Senning, J. R. 2000. CS322: Operating systems history. http://www.math-cs.gordon.edu/courses/cs322/lectures/history.html (accessed Oct. 1, 2009).

Searle, S. J. 2002. A brief history of character codes in North America, Europe, and East Asia. http://tronweb.super-nova.co.jp/characcodehist.html (accessed Oct. 1, 2009).

Sperling, D. 1998. *Internet Guide*, 2nd ed. Upper Saddle River, NJ: Prentice Hall.

Stevens, S. S. 1951. *Handbook of Experimental Psychology*. New York: Wiley.

Strathern, P. 1999. *Turing and the Computer: The Big Idea*. New York: Anchor Books.

White, S. 2005. A brief history of computing. http://www.ox.compsoc.net/~swhite/history/ (accessed Oct. 1, 2009).

Wikipedia. 2009. Minicomputer. http://www.wikipedia.org/wiki/Minicomputer (accessed Oct. 1, 2009).

Williams, M. R. 1997. *A History of Computing Technology*. Los Alamitos, CA: IEEE Computer Society Press.

Zakon, R. H. 2006. Hobbes' Internet timeline v6.0. http://www.zakon.org/robert/internet/timeline/ (accessed Oct. 1, 2009).

APPENDIX: OTHER IMPORTANT NAMES

There are many important individuals who were not discussed in this chapter. They include the following:

- Vannevar Bush (1890–1974) was a visionary about information technology who described a "memex" automated library system.
- Steve Case (1958–) founded Quantum Computer services, which evolved into America Online (AOL). Although AOL has run into difficult times as its approach always shielded its many users from the actual Internet. Present computer users are becoming sufficiently sophisticated so as to want more direct interaction. Nonetheless, AOL deserved credit for being an important portal to the Internet for many who might have found actual interactions daunting before the development of the World Wide Web and the current generation of browsers made "surfing" easy.
- Herman Goldstine (1913–2004) pioneered the use of computers in military applications in World War II, which became a vital part of the war effort.
- Mitch Kapor (1950–) developed the earliest spreadsheets starting in 1978, of which Lotus 1-2-3 was perhaps the best known. He also founded the Electronic Frontier Foundation, which is concerned with protecting the civil liberties of Internet users.
- Marshall McLuhan (1911–1980) described the popular concept of a "global village" interconnected by an electronic nervous system and became part of our popular culture.
- Robert Metcalfe (1946–) outlined the Ethernet specification in 1973, which is a vital part of most current networks.
- On a less positive note, Kevin Mitnick (1963–) was a hacker par excellence who could probably break into any system. He was caught, arrested, and sentenced to a year in jail in 1989. He has resurfaced several times since then. Most recently, he used his experience hacking to set up an Internet security company. With a rather profound bit of irony, his own company became the victim of two hackers. To his credit, he viewed the indignity as "quite amusing." Indeed, he received "professional courtesy." His reputation evidently had gathered the respect of both hackers as they simply posted messages (one being a welcome from his most recent incarceration) rather than doing damage to his corporate files.
- Also on a less positive note: Robert T. Morris (1965–) unleashed a worm or self-replicating program that spread rapidly through the Internet in 1988, which was the first major virus attack affecting the Internet (others in the 1980s had affected LANs and other small networks; some were beneficial in that they performed needed complex tasks to do things like post announcements). He was eventually convicted of violating the Computer Fraud and Abuse Act. Although he did not go to jail, he paid a large fine and performed 400 hours of community service. In 1992, a virus named "Michelangelo" was predicted to cause massive harm, but it proved to be largely a dud. Its main effect was to create the booming anti-virus software market.
- Kristen Nygaard (1926–2002) and Ole-Johan Dahl (1931–2002) of the Norwegian Computing Center in Oslo, Norway, developed Simula in 1967. This was the first object-oriented programming language (OOP).
- Claude Shannon (1916–2001) wrote the highly influential book *The Mathematical Theory of Communications*. In conjunction with Warren Weaver (1894–1978), they formulated a theory of information processing that was and is widely used in psychology along with an algorithm to process categorical data known as *information theory*.
- Richard Stallman (1953–) began the free (open-source) software movement in 1983, writing a UNIX-based operating system called GNU (Gnu's Not UNIX).
- Robert W. Taylor (1932–) of ARPA integrated different computers, each with their own command set, into a network.
- Linus Torvalds (1969–), then 21 years old, began writing the Linux operating system at Helsinki University in 1991. Linux quickly evolved from a one-man project into a global project, perhaps stimulated by the many who despised Microsoft. Much of its popularity was with small Internet Service Providers (ISPs) and businesses that operated on miniscule budgets. In a more general sense, this was an important part of a tradition known as the Free Software Movement.
- Norbert Wiener (1894–1964) founded the science of Cybernetics, which deals with the role of technology in extending human capabilities.
- Spam would be high on the list of any survey asking about the things people like least about the Internet. Indeed, it would probably rank first if one excluded technological limitations (the Web is commonly dubbed the "World Wide Wait," especially, but not only, by people who use conventional modem access). As noted above, the development of SMTP allowed a single e-mail to be sent to an unlimited number of recipients, nearly all of whom, by definition, did not want it. The term "Spam" comes from a Monty Python skit about a diner that serves only dishes containing this product. Because people's criteria differ about what constitutes spam, it is difficult to locate the first spammer. However, the husband and wife legal team of Laurence Canter (1953–)

and Martha Siegel (1948–2000) hold a special place in this history. In 1994, they posted advertisements to nearly 6000 USEnet groups (two thirds of those that existed at the time). Their efforts resulted in a substantial number of negative editorials in sources like the *New York Times*, and they achieved double infamy by being perhaps the first major recipients of e-mail flames. For nearly a decade now, Congress and state legislatures have struggled with what to do with the spam nuisance to no apparent success. Related to this issue are such events like the 1996 submission of a "cancelbot" or automated program that cancelled 25,000 Usenet messages.

- Without naming them all or judging their merits, there has been a plethora of law suits and legal actions over the past decade that has arguably outstripped the number of innovations, if one excludes important but simple improvements in transmission speed and improvements in programs, e.g., the 1998 introduction of extensible markup language (XML). Some examples of these legal issues include many concerned with objections to the Internet's content, e.g., by the German Government in 1995 and the American Communications Decency Act of 1996, which was found unconstitutional the next year. Other recent legal actions include the class action suit against America Online by its dissatisfied customers in 1997; the Department of Justice's suit against Microsoft; a suit against GeoCities in 1998 for deceptive online privacy practices, to say nothing of countless suits of one software manufacturer against another.

3 Human–Computer Interaction

Alan J. Dix and Nadeem Shabir

CONTENTS

3.1   INTRODUCTION

On a Web site for a UK airline, there are two pull-down menus, one for UK departure airports and the other for non-UK destinations (see Figure 3.1). When you select a departure airport the destination menu changes so that only those with flights from the chosen departure airport are shown in the second menu. So if you live particularly close to a single airport, you can easily ask "where can I fly to from here?"

However, you may be willing to travel to an airport and what you really want to know is how to get to a particular destination such as "where can I fly from in order to get to Faro?" The Web site does not support this. You can select the second menu first, but the first list does not change, all the options are there, and, as soon as you select anything in the first list, your selected destination disappears and you are back to the full list.

Now, in retrospect it seems like common sense that it is reasonable to want to ask "how do I get to Faro," but the designer simply thought logically: "from" then "to." The execution was technically flawless. Many similar sites fail completely on some browsers because of version-specific scripts. This worked well but did the wrong thing. The site was well designed aesthetically and technically but failed to deliver an experience that matched what a reasonable user might expect. Even more surprising is that this problem was present at the time of the first edition of this book and is still there. Since then the Web site has been redesigned, and the appearance of the menus has changed, but the behavior is the same.

Human–computer interaction (HCI) is about understanding this sort of situation and about techniques and methods that help avoid these problems. The adjective most closely linked to HCI is "usability." However, it often has almost

Taylorist* overtones of efficiency and time and motion studies. This is not the only aspect that matters, and there are three "use" words that capture a more complete view of HCI design. The things we design must be

Useful: Users get what they need—functionality.
Usable: Users can do these things easily and effectively.
Used: Users actually do start and continue to use it.

Technical design has tended to be primarily focused on the first of these and HCI on the second. However, the third is also crucially important. No matter how useful or usable it is, if a system is not used, then it is useless.

For an artifact to be used it often needs to be attractive, to fit within organizational structures, and to motivate the user. For this reason, the term "user experience" is often used rather than "usability," especially in Web design, emphasizing the holistic nature of human experience. We will look at some of these issues in more detail for the Web later in this chapter.

The remainder of this chapter is split into three main parts. First, in Section 3.2, we consider the context of the Web, some of the features that make applications designed for the Web special. Then, in Section 3.3 we look at the nature of HCI itself as an academic and design discipline: its roots, development, links to other disciplines, and we look at a typical HCI design process and the way different techniques and methods contribute to it. Many of the human design issues of Web design can be seen as special cases of more general usability issues and can be tackled by the general HCI design process. However, as we discuss in Section 3.2, there are special features of the Web, and so in Section 3.4 we discuss a few more particular HCI issues for the Web. Of course, this whole book is about human factors and the Web, and some issues are covered in detail in other chapters; hence the latter part of the chapter tries to complement these. This chapter concludes with a brief view of the directions in which

FIGURE 3.1   Part of an airline Web site.

---

* Frederick Taylor wrote *The Principles of Scientific Management* in 1911, a seminal work that introduced a philosophy of management focused on efficient production. Taylorism has come to represent a utilitarian approach to the workforce including practices such as time and motion studies (Taylor 1911; Thompson 2003).

HCI is developing within the context of the Web and related networked and mobile technologies.

3.2 THE CONTEXT OF THE WEB

Since the first edition of this book, the Web has changed substantially, not least with the introduction of the combination of technological and social changes of Web 2.0. Some of the usability issues of the Web are similar whether one uses Web 2.0 or older Web technologies, and so this section deals both with more generic issues of the Web (Sections 3.2.1 and 3.2.3) and also some of the specific new issues for Web 2.0 (Section 3.2.2).

3.2.1 THE OPEN ENVIRONMENT

When a traditional application is delivered, it is installed in a particular organizational context if it is a bespoke system, or, if it is shrink wrapped, it comes complete in a box and is marketed to a known group of people.

In contrast, the Web is an open environment both in terms of the target audience and the application environment.

3.2.1.1 Who Are the Users?

The most widely preached and important UI design principle is to understand who your users are and what they want to do. With the Web there are so many users with so many different purposes. Typically, they all hit the same Home page. Think of a university department's Web site. There will be potential students: post-18, mature students, part-time, full-time. There may be commercial users looking for consultancy. There may be job applicants checking the department's research and teaching portfolio. The list continues. In fact, it is not quite as bad as it seems—often, it is possible to identify the most significant user group and/or design the site to funnel different types of user to different areas, but it is certainly a challenge! Some sites cope by having a parallel structure, one more functional, and one more personal based on "information for prospective students," "information for industry," and so forth. Volk, Pappas, and Wang (this volume) examine this issue in depth.

3.2.1.2 Who Is This User?

The transactional nature of HTTP means that it is hard to know where a particular user has been before or what they have done before on your site. One of the clearest examples of this is when content changes. Typically, change is at the leaves of the site, but people enter at the root. Repeated visits give the same content—it is not surprising that few sites are revisited! A traditional information system can accommodate this by highlighting areas that have changed since a user has last seen them, but this only works when the system knows who the user is. This is partly a technological issue—there are many means of authentication and identification (e.g., cookies). But a combination of technological limitations and (understandable) user worries about privacy means that few traditional sites, except explicit portals (with "my- . . ." pages) and e-commerce sites, adapt themselves to visitors' past behavior.

This picture is entirely different for many social networking sites such as Facebook, where the site is as much oriented around the user who is viewing content as those whom the content is about. As a designer of Facebook apps, one has to develop a different mindset, which is focused on the viewer. These sites have also tended to reduce the privacy barrier for many users, making sign-ins more acceptable, at least for the social networking demographic. You can leverage existing user bases by using some form of single sign-on service for your own site such as OpenID or Facebook Connect (OpenID Foundation 2009; Facebook 2010).

As more sites gather data about users, issues of ownership are beginning to arise. The DataPortability Project (DataPortability 2010) is seeking to allow users to take their digital identity with them between applications, for example, using the same friend lists. This has both a political dimension: who "owns" this information anyway, and also a usability one: avoiding reentering the same information again and again. However, it conflicts with the closed business models of some sites; this attitude is slowly changing.

3.2.1.3 Where in the World?

Because the Web is global, it is possible to create an e-commerce site in Orkney and have customers in Ottawa, again, one of the joys of the Web! However, having customers or users across the world means we have to take into account different languages, different customs, and different laws. This is discussed in detail by Rau, Plocher, and Choong (this volume).

There are two broad approaches to this issue: globalization and localization (sometimes called internationalization). Globalization attempts to make a site that, with the exception of language translation, gives a single message to everyone, whereas localization seeks to make variants that apply to particular national, cultural, or linguistic groups. Both have problems. For globalization, even something as simple as left-right versus right-left ordering depends on cultural backgrounds, not to mention deeper issues such as acceptability of different kinds of images, color preferences, etc. For localization, the production of variants means that it is possible for users to look at those variants and hence see exactly how you think of their culture! This may be a positive thing, but you run the risk of unwittingly trivializing or stereotyping cultures different from your own.

When the Web is accessed through a GPS-enabled mobile phone or laptop, it then becomes possible to track the user's location and use this to modify content, for example, to show local information or a position on a map. Similar information might be obtained by the device using the locations of WiFi base stations or phone cell masts; this is not normally available to a Web page in raw form, but, for example, the iPhone provides a JavaScript API (application programming interface) to provide location independent of the underlying tracking technology. For fixed devices, the IP address of the device can be used to give a rough location at the level of country or city/region, although this is largely used for localization.

3.2.1.4 Where Is the Start?

The programmer usually has the ultimate say on where a user enters the program and, barring crashes, where they leave. With a Web site we have no such control. Many Web designers naively assume that people will start at the Home page and drill down from there. In reality, people will bookmark pages in the middle of a site or even worse enter a site for the first time from a link and find themselves at an internal page. Just imagine if someone were able to freeze your program halfway through executing, distribute it globally to friends and acquaintances, who then started off where it was frozen. Even the easiest interface would creak under that strain!

Remember, too, that bookmarks and links into a site will remain even if the site changes its structure. Ideally, think of a URL structure that will pass the test of time, or as Berners-Lee (1998) put it, "cool URIs don't change."

3.2.1.5 Where Does It End?

When a user exits your program, your responsibility ends. On the Web, users are just as likely to leave your site via a link to a third-party site. Your clean, easy to understand navigation model breaks down when someone leaves your site, but, of course, for them it is a single experience. To some extent, this is similar to any multiwindow interface. This is why Apple's guidelines have been so important in establishing a consistent interface on the Macintosh (Apple Computer 1996), with similar, but somewhat less successful initiatives on other platforms. However, it would neither be appropriate, nor welcomed by the Web community, to suggest a single Web look and feel. In short, the difference between traditional interface design and Web design is that the latter seems totally out of control.

3.2.1.6 What Is It Anyway?

Sometimes one is designing a whole site including all of its content. However, the range of things that are developed on the Web or using Web technology is far broader than simply Web pages. Increasingly, sites make use of both content data

and active elements from different sources in mashups, for example, using a Google map. Figure 3.2 shows an example page with a puzzle included from a different site using JavaScript. Furthermore, there are now many forms of Web-delivered applications that need to fit into another Web page, or as a micro-app within some sort of framework such as a Facebook app or a Google Widget. Alternatively, raw data or services are delivered using Web APIs, and Web-like technology is used for desktop applications such as the MacOSX Dashboard.

As a designer of a generic component to be mashed into another person's site, you need to take into account the fact that it could be used in many different contexts. On the one hand, this may often include the need to allow developers to customize it and maybe style using cascading style sheet (CSS). On the other hand, you may be the one designing a page that will contain other elements, and so need to consider how they will fit together into a single user experience or, alternatively, just accept that you are creating a Web montage.

For the designer of Facebook apps and similar single-site micro-apps, life is perhaps a little easier, more like designing for a single desktop platform, a matter of conforming to guidelines and working out what can be achieved within the platform's constraints. Conversely, if you are attempting to design a platform-like system such as Facebook itself, the challenge is perhaps hardest of all, as you have little control over the applications that can be embedded, and yet may be seeking to maintain some level of "brand" experience.

3.2.2 Web 2.0 and Beyond

The phenomenon termed "Web 2.0" has been an enormous change in the way the Web is used and viewed (McCormack 2002; O'Reilly 2005). A single term, Web 2.0 describes a wide range of trends (including end-user content and social Web) and technologies (largely AJAX and DOM manipulation). But, Web 2.0 was a bottom up movement driven by pragmatics and only later being recognized and theorized. Semantic Web technologies have been on the agenda of Berners-Lee and others but only recently have found their

FIGURE 3.2 Web page including components.

way into mainstream products. Some look forward to the merging of semantic Web and Web 2.0 technologies to form Web 3.0 or even Web 4.0.

3.2.2.1 End-User Content

One of the core aspects of Web 2.0 is the focus on end-user content: from blogs, to YouTube videos, and Wikipedia. For users, it may not be as clear how to assess the reliability of end-user produced content (see, for instance, the critique by Andrew Keen 2007); one can judge the differing biases of a U.S. or Russian newspaper, but it is less clear when one is reading a blog. However, the issue for much end-user content is not reliability or accuracy but enjoyment. As a Web designer or developer, this adds challenges: can you harness the power of end-user content on a site or maybe simply make it easy for users to link to you through existing means such as using "Digg it" buttons? Some applications have made strong use of "human computation," for example, reCAPTCHA asks users to type in text as a means to authenticate that they are human but at the same time uses their responses to fill in parts of documents that are difficult for optical character recognition (OCR; von Ahn et al. 2008).

3.2.2.2 The Personal and Social Webs

For many the Web has become the hub of their social lives, whether through instant messaging, e-mail, Twitter, or Facebook. Sites such as YouTube are predominantly about personal lives, not work. Traditional usability is still important, for example, when uploading a file to Flickr, or connecting to a friend on Facebook, one wants the interaction to be effortless. However, if you attempt to evaluate these sites using conventional usability measures, they appear to fail hopelessly, except that with closer analysis often the "failures" turn out to be the very features that make them successful (Silva and Dix 2007; Thompson and Kemp 2009). For example, on the Facebook wall, you only see half of a "conversation," the comments left by other people; the person you are looking at will have left her comments on her friends' walls. However, this often leads to enigmatic statements that are more interesting than the full conversation, and in order to see the whole conversation one needs to visit friends' pages and thus explore the social network. While social networking and other Web 2.0 applications are being studied heavily (e.g., Nardi 2010), we do not yet have the same level of heuristics or guidelines as for conventional applications. Understanding social networks needs understanding of issues such as self-presentation (Sas et al. 2009), as well as those of cognitive and perceptual psychology.

3.2.2.3 Blurring Boundaries

Web 2.0 has blurred the boundaries of Web-based and desktop applications. Software, such as spreadsheets, that once only existed as desktop applications, are now available as software services; even full photo/image editing is possible with purely Web-based applications (Pixlr 2010). Furthermore, the desktop is increasingly populated by applications that are Internet oriented, either delivered via the Internet (e.g., in MacOS Dashboard widgets, Adobe Air apps) and/or predominantly accessing Internet services (e.g., weather, newsfeeds, and IM clients). Often these applications, while running on the desktop, are constructed using Web-style technologies such as HTML and JavaScript.

Equally, desktop applications increasingly assume that computers are Internet connected for regular updates and online help and documentation. However, this can lead to usability problems for those times when the Internet is not available, while traveling, or due to faults (even having Internet documentation on fault finding when you cannot connect to the Internet).

Of course, periods of disconnection are inevitable for any mobile computer and even during periods when networks are faulty. If you are designing online applications, such as Google docs, you also want them to be usable when disconnected. Various technologies including Java Web Start, Google Gears, and the offline mode of HTML5 offer ways for Web-based applications to store reasonable amounts of data on a user's machine and thus be able to continue to operate while offline.

However, when Web-based applications operate offline, or when desktop applications store data in the "cloud" (e.g., Apple's MobileMe), there is need for synchronization. There are well-established algorithms for this, for example, Google Wave used variants of operation transformation (Sun and Ellis 1998). However, it is still common to see faults; for example, if an iPhone is set to synchronize using both MobileMe and directly with a desktop computer, it ends up with two copies of every contact and calendar event!

3.2.2.4 Open APIs, Open Data, and Web Semantics

Many popular Web services make functionality available to other Web and desktop applications through open APIs; for example, if you are writing a statistical Web site, you can access spreadsheets entered in Google docs. As a developer of new services, this means you can build on existing services, offering new functionality to an established user base, and often avoid re-implementing complex parts of the user interface. In the above example, you can focus on creating a good interface for statistical functions and avoid creating a means for users to enter data.

Of course as a developer, you, too, can make some or all of your functionality available using APIs encouraging third-parties to augment your own services. Indeed, one way to construct a Web application is to adopt a Seeheim-like structure with the application semantics in Web-based services and a front-end accessing these services. It is then a simple matter to make some or all of those services available to others. For the user, not only do they make existing services more valuable, but API-based services are likely to spawn multiple interfaces that may serve different users' preferences or situation; for example, there are many desktop interfaces for Twitter.

While most APIs adopt some form of standard protocol such as REST or SOAP APIs (Fielding 2000; W3C 2007), the

FIGURE 3.3 **(See color insert.)** (Left) Project Cenote and (right) Vodafone 360.

types of data provided vary from provider to provider. The semantic Web allows providers to publish data in ways that can be interlinked through semantic markup (Berners-Lee, Hender, and Lassila 2001). This leads to a paradigm where the data become more central and the data from multiple sources may all be gathered in a single application; for example, Project Cenote (Figure 3.3, left) brings together information from multiple RDF sources relating to a book (Talis 2010). This form of data aggregation can exist without semantic Web technologies, for example Vodafone 360 (Figure 3.3, right) gathers data about friends on a mobile phone from multiple sources including Facebook and Twitter (Vodafone 2010). However, semantic linking makes the job easier.

3.2.3 Commercial Context

3.2.3.1 Multiple Decision Points

We have already discussed the importance of getting an application actually used. When producing a standalone application, this is largely about getting it purchased; not just cynically because once the user has parted with money we do not care, but because once users have chosen this product instead of another they will use it for its particular task unless it is really, really bad. This is also true in a corporate setting where the decision may have been made by someone else. Even if there are no monetary costs, simply taking the effort to download and install software is a major decision and predisposes the user to ongoing use.

In contrast, many Web products, e-mail services, portals, etc., are services. There are no large up-front costs and few barriers to change. If you think another search engine may give you better results, you can swap with very little effort. Open data and data portability make this even easier, as the former makes it possible to have many interfaces to the same underlying data, and the latter lets you take your data with you between applications. With such products, every use is a potential decision point. Instead of convincing a potential customer once that your product is good, it is an ongoing process.

This continual reselecting of services means that the usability and user experience offered by Web products is even more important for continued use than it is for traditional software.

3.2.3.2 Web-Time Development

The Web seems to encourage a do-it-yesterday mentality. This may be because it is perceived as malleable and easy to change; or because of the ambiguity of the media, somewhere between print and broadcast; or because of the association with computers and hence rapid change; or perhaps a legacy from the headlong commercial stampede of the dot.com years.

Whatever the reasons, Web design is typically faced with development cycles that are far shorter than would be expected of a typical computer product. Furthermore, updates, enhancements, and bug fixes are not saved up for the next "release" but expected to be delivered and integrated into live systems within weeks, days, or sometimes hours.

In Web 2.0 this has been characterized as the "perpetual beta" (O'Reilly 2005) and is seen as a strength. Software is continuously being updated, features added, bugs fixed, etc.; while users of Microsoft Word know when a new major version is released, users of Google docs will be unaware of what version of the software they are using; it simply evolves. Also, because software is no longer installed by each individual customer, maintenance is easier to perform because your customers are typically all on the same version of your software.

However, this deeply challenges traditional software engineering practice, which still pays lip service to the staged waterfall model (despite its straw man status) where requirements are well established before code design and implementation begin (Sommerville 2001). However, agile methodologies fit well with this because the agile ethos is to prioritize tasks into small incremental development iterations, which require minimal planning (Shore 2007). This allows agile teams to be more dynamic and flexible in order to respond to changing priorities. For this reason, agile methods have been adopted widely by Web 2.0 services such as Flickr.

User interface design is similarly challenged, and there is not time for detailed user needs analysis, or observational or other end-user studies. Instead, one is often forced into a combination of rapid heuristics and a "try it and see" mentality. The delivered system effectively becomes the usability-cycle prototype.

Not only does this mean that end users become test subjects, but the distributed nature of the Web makes it hard to observe them in actual use. However, the fact that the Web is intrinsically networked and that much of this goes through a single server can also have advantages. It is possible to use logs of Web behavior to search for potential usability problems. For example, in an e-commerce site, we may analyze the logs and find that many visitors leave the site at a particular page. If this is the post-sale page we would be happy, but if it is before they make a purchase, then we may want to analyze that page in detail. This may involve bringing in some test subjects and trying them out on a task that involves the problematic page; it may be to use detailed heuristics in that page; or it may be simply to eyeball it.

The fact that HTML is (semi-)standardized and relatively simple (compared to the interface code of a GUI) means that there are also several Web tools to analyze pages and point out potential problems before deployment, for example, WebAIM's online Web accessibility checker WAVE (WebAIM 2010).

Although more agile, incremental, or evolutionary development methods are better fitted to the Web time cycle, they can cause problems with maintaining more global usability objectives such as consistency and overall navigation structure. Many minor upgrades and fixes, each fine in itself, can fast degrade the broad quality of a site.

There is also a danger of confusing users when new features are introduced. When you install a new version of desktop software, you know that there will be changes in the interaction, but with perpetual beta you may wake up one morning and find your favorite site behaves completely differently. Once when Facebook introduced a new site style, there was a popular groundswell among their users that forced them to maintain the 'old' Facebook in parallel for some time. As a designer one should consider how to manage change so that users are aware when something new has been added, or when a feature has been changed or removed completely. For example, Google uses overlays in their user interface to inform users when new features have been added.

3.2.3.3 Branding and Central Control

There is a counter effect to the Web time development pressure that affects the more information rich parts of many corporate Web sites, including academic ones. Because the Web is a "publication" it is treated quite reasonably, with the same care as other corporate publications. This may mean routing all Web updates and design through a central office or individual, maybe in the IT department or maybe connected with PR or marketing, who is responsible for maintaining quality and corporate image. After all, the Web site is increasingly the public face of the company. However, this typically introduces distance and delays, reducing the sense of individual "ownership" of information and often turning Web sites into historical documents about the company's past. As the site becomes out of date and irrelevant, current and potential customers ignore it.

Of course, it is true that the Web site is this public face and needs the same care and quality as any other publication. However, if you visited the company's offices you would find a host of publications: glossy sales flyers, grey official corporate documents, roughly photocopied product information notes. The reader can instantly see that these are different kinds of documents and so do not expect the same level of graphic imagery in a product specification as a sales leaflet. In the Web we find it hard to make such distinctions; everything looks similar: a Web page on a screen. So, organizations end up treating everything like the glossy sales leaflet . . . or even worse the grey corporate report.

It is hard to convince senior management and the Web gatekeepers that total control is not the only answer. This is not just a change of mind, but a change of organizational culture. However, we can make this easier if we design sites that do not have a single format but instead have clear graphical and interactional boundaries between different kinds of material—re-create digitally the glossy flyer, grey bound report, and stapled paper. This does not mean attempting to reproduce these graphically but, instead, creatively using fonts, color schemes, and graphical elements to express differences. Visitors can then appreciate the provenance of information: does this come from senior management, the sales team, or the technical staff? They can then make judgments more effectively, for example, trusting the price of a new water pump listed on the sales page but the pump capacity quoted in the technical specification!

3.2.3.4 The Real Product

Think about Web-based e-mail. Your personal mail is received by a multinational corporation, siphoned into their internal data stores, and dribbled out to you when you visit their site. Would you do that with your physical mail? However, this is not how we perceive it. Users have sufficient trust in the organizations concerned that they regard the Web mailbox as "mine"—a small section of a distant disk is forever home.

The factors that build this trust are complex and intertwined but certainly include the interface style, the brand and reputation of the provider, the wording used on the site, the way the service is advertised to you, and newspaper and magazine articles about the site. A few years ago the Chairman of Ratners, a large UK jewelery chain, said, in an off-the-cuff remark, that their products were cheap because they were "total crap." The store's sales plummeted as public perception changed. Imagine what would happen if a senior executive of Microsoft described hotmail in the terms at the beginning of the previous paragraph!

It is clear that the way we talk about a product influences how well it sells, but it goes deeper than that. The artifact we have designed only becomes a product once it takes on a set of values and purposes within the user's mind—and these are shaped intimately not just by the design but also by the way we market the product and every word we write or say about it (Figure 3.4).

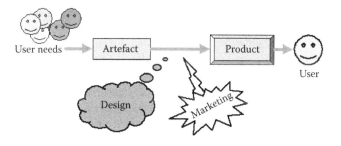

FIGURE 3.4 Artefact + marketing = product. (From Dix, A. 2001. Artefact + marketing = product. *Interfaces* 48: 20–21. http://www.hiraeth.com/alan/ebulletin/product-and-market/. With permission.)

As we address the needs of a networked society, we must go beyond the creation of useful, usable artifacts, and instead design products that will be used. To do this, we cannot rely solely on cosy relationships between users and designers but open up the design remit to consider every stage of product deployment from the first advert the user sees until the consumed product hits the bin, is deleted from the hard disk, or the URL is cleared from the favorites list.

3.2.3.5 Web Business Models and the Gift Economy

One of the reasons for the collapse of the dot.com industry in 2000 was the lack of clear business models for Web-based companies; investors had poured in billions but were seeing no revenue. There have been various solutions to this, and most depend on usability and user experience, given the multiple decision points noted previously, if people do not use a service it cannot make money.

For e-commerce sites such as Amazon, or those promoting a brand, there is a clear business reason for the site. For most others, the dominant means is through advertising. For brand sites the designer has the "purest" creative job—an enjoyable user experience that reinforces the brand image. For e-commerce the aim is a little more complex, to keep people on the site, but eventually lead them to purchase. For advertising sites, there is always some tension; they want visitors to stay on the site and return to it, but they also want them to click through to adverts. If advertising placement is too subtle, users may simply ignore it; if too "in your face," then users will not return.

However, as well as these more obvious business models, there has been a growth of newer ways to make money. Sites such as eBay and Amazon through their affiliates and marketplace have made use of the crowd, creating ways to profit through mass adoption. As a developer of such sites, you can consider ways to give value to others and yet also profit yourself. Apple's iTunes is also based on mass adoption, launched at a time when peer-peer distribution of copyright material appeared to threaten the survival of the music industry. Instead, iTunes simply made it easy and cheap to purchase, i.e., a seamless user experience. This micropayment platform also made possible the iPhone app store; however, the crucial things were that it offered small developers a means to distribute and profit, albeit often in small ways, from their work. With software the price of a chocolate bar, users graze applications, downloading and discarding equally quickly.

Perhaps the most radical change has been the way in which many sites offer data, APIs, and services for free, apparently putting immediate profitability low on their agenda. Theories of the gift economy (e.g., Hyde 1983; Mauss 1925) emphasize that there is usually a level of reciprocity: things are given free in the hope that something will come back. Sometimes this is monetary, as in "Freemium" services that offer a basic service for free with paid premium services. Other parts of this gift economy are more about reputation or a sense of contribution to the community, as in contributions to Wikipedia or the Open Source movement (Raymond 1999). However, the general lesson seems to be that while you do have to consider how a Web service or site will be economically sustainable, the most successful sites have been those that focus *first* on offering service to users and to other developers.

3.3 ABOUT HCI

3.3.1 WHAT IS HCI?

Human–computer interaction, not surprisingly, is all about the way in which people interact with computer systems. People here may mean individuals interacting with computers, groups of people, or whole organizations and social groups. They may be directly interacting with the computer, using a mouse or keyboard, or indirectly affected like the customer talking while the travel agent enters flight codes into a booking system. The computer, too, may be a simple screen keyboard and mouse, a mobile phone or personal digital assistant (PDA), or systems embedded in the environment such as car electronics.

And interaction? Well, in any system there are issues that arise because of the emergent properties of interaction that you would hardly guess from the individual parts. No one predicted the rise in text messaging through mobile phones—an interaction not just between each person and their phone, but one involving whole social groups, the underlying telecommunications infrastructure, pricing models, and more.

HCI involves physical aspects of this interaction (are the keys spaced right?), perceptual aspects (is the text color easy to see against the background?), cognitive aspects (will these menu names be understood?), and social aspects (will people trust each other on this auction site?).

HCI is a field of academic study (whether scientific is a matter of debate; Long and Dowell 1989; Carroll 2010; Dix 2010) that tries to understand these various aspects of interaction. Often, this study also gives insight into the constituents: shedding light on human cognition or forcing the development of new computational architectures. HCI is also a design discipline, using the lessons learned about interaction in order to create systems that better serve their users.

One reason HCI is exciting is because the boundary between the theoretical and the vocational is narrow. Today's theory is tomorrow's practice, and often today's practice also drives the theoretical agenda. Because of this it is also a discipline that is constantly struggling between short-term results and longer-term knowledge. The rate of technological change means that old design solutions cannot simply be transferred to the new. However, it also means that there is not time to relearn all the old lessons and innovative solutions can only be created based on fundamental knowledge.

3.3.1.1 Design

Design is about achieving some *purpose* within *constraints*. Within HCI often the purpose is not clear. In broad terms it may be "what users want to do," but we need a more detailed brief in order to design effectively. A substantial part of HCI effort goes into simply finding out what is wanted . . . and it is clearly not obvious as many expensive Web sites get it wrong.

In fact, the example that starts this chapter is all about not thinking of the user's likely purpose in coming to the site.

The other aspect is "within constraints." We are not magicians creating exactly what is wanted with no limits. There are different types of constraints including financial limits, completion deadlines, and compatibility with existing systems. As well as these more external constraints, there are also constraints due to the raw materials with which we are working. This leads to the "golden rule of design" (Dix et al. 2004): *understand your materials.* For an artist this would mean understanding the different ways in which water colors or oils can be used to achieve different types of painting. For a furniture designer it includes understanding the different structural properties of steel and wood when making a chair. In HCI the raw materials are the computers and the people.

It may seem rather dehumanizing to think of people as raw materials. However, one of the problems in design is that people are often given less regard than physical materials. If you design a chair such that it puts too much strain on the metal and it snaps, you say it is metal fatigue, and if predictable, you would regard it as a design failure. However, if an air crash is found to be due to a pilot under stress doing the wrong thing, we call it human error. The physical materials are frequently treated better in design than the humans!

It is also important to realize what it is we are designing. It is not just a Web site or an intranet. If we introduce an e-commerce system, we are changing the ways in which existing customers interact with the business. We are not just designing an artifact or product, we are designing *interventions*—changes from the status quo to something new. Not only may the changes be more extensive than the computer system itself, but they may also not even require any changes to the computer system. If the current intranet is not working, the solution may be an information leaflet or a training course.

Going back to the definition of design itself as achieving goals or purposes within constraints, this presupposes that not everything will be possible. There are decisions to be made. We may not be able to achieve all of the goals we have set ourselves; we may need to prioritize them, accepting that some desired functionality may not be present, that what we produce may be less beautiful, less useful, and less usable. We may be forced to reevaluate the constraints. Is the original time frame or budget sensible? Can this be achieved on the chosen platform?

Design is intimately concerned with *trade-off.* Utopian goals invariably meet pragmatic constraints. (In case this sounds too cynical, do remember that most utopian ideals when fully achieved become dystopian!)

3.3.1.2 At the Heart—The User

At the center of HCI is the user. In fact, it is often said the many techniques and methods used in HCI succeed only insofar as they focus the designer on the user. Good designers get to understand their users by watching them, talking to them, and looking at the things they produce.

Medical information system – expected users

Group 1: consultant	Group 2: trainee nurse
computer proficiency: low	computer proficiency: medium
medical expertise: high	medical expertise: low
education: university degree	education: school leaver
age: 35+	age: 18–25

FIGURE 3.5 User profiles.

Try producing a set of Web pages and then watch a user trying to navigate in them. It seems so clear to you what the links mean and how the various form fields ought to be completed. Why do these people not understand? Many years ago, the first author produced his first computer application for someone else to use. It was a simple command line system, and he watched the very first user. The first prompt asked for a name—that was easy. The second prompt asked for a title. The author was expecting the user to type a few words on a single line but watched with horror as she used the cursor keys to produce a wonderfully formatted multiline centered title that he knew the program could not understand. If you think this would not happen now, have you never filled out a text field on a Web form and used lines and spaces to lay it out like you would an e-mail message only to find that once you submit the form all the spaces and line breaks disappear leaving one long paragraph.

One technique that is used to help build this user focus is to produce profiles of expected users. Figure 3.5 shows a simple example. Note how this would make one ask questions like: "when doctors are using this system will they understand the word 'browser'?"

Some designers prefer richer profiles that create more of a character who becomes a surrogate for a real user in design. This is sometimes called a persona (see Figure 3.6). A design team may decide on several personae early in the design process typical of different user groups: Arthur on reception, Elaine the orthopaedic surgeon. When a new design feature is proposed someone may say "but how would Arthur feel about that?" The more real the description, even including irrelevant facts, the more the designers can identify with the different characters.

Betty is 37 years old. She has been Warehouse Manager for five years and worked for Simpkins Brothers Engineering for twelve years. She did not go to university, but has studied in her evenings for a business diploma. She has two children aged 15 and 7 and does not like to work late. She did part of an introductory in-house computer course some years ago, but it was interrupted when she was promoted and could no longer afford to take the time. Her vision is perfect, but her right-hand movement is slightly restricted following an industrial accident 3 years ago. She is enthusiastic about her work and is happy to delegate responsibility and take suggestions from her staff. However, she does feel threatened by the introduction of yet another new computer system (the third in her time at SBE).

FIGURE 3.6 Persona: a rich description of Betty the Warehouse Manager. (From Dix, A., J. Finlay, G. Abowd, and R. Beale. 2004. *Human–Computer Interaction,* 3rd ed. Englewood Cliffs, NJ: Prentice Hall, http://www.hcibook.com/e3/. With permission.)

3.3.2 Roots of HCI

3.3.2.1 Many Disciplines and One Discipline

The origins of HCI as an academic and professional discipline were in the early 1980s, with the move of computers from the cloistered machine room to the desktop. Retrospectively, earlier work can be seen as having an HCI flavor but would have been classed before as part of systems analysis or just doing computers.

Early researchers were predominantly from three disciplines: ergonomists concerned with the physical aspects of using computers in a work environment, psychologists (especially cognitive science) seeing in computers both an area to apply the knowledge they had of human perception and cognition, and computer scientists wanting to know how to make systems that worked when given to people other than the original programmers. Other disciplines have also made strong contributions including linguistics, sociology, business and management science, and anthropology. Because computer systems affect real people in real situations, at work, at home, they impinge on many areas of study. Furthermore, to understand and design these interactions requires knowledge from many areas.

However, HCI is not just an amalgam of knowledge from different areas; the special nature of technical interaction means that the pure knowledge from the different disciplines has not been sufficient or that the questions that HCI asks are just different. This is not to say that fundamental knowledge from these different areas is not important. A good example is Fitts' law (Fitts and Posner 1967; MacKenzie 2003; Seow 2005). This says that the time taken to move a pointer (e.g., mouse) to hit a target is proportional to the log of the distance (D) relative to the size of the object (S):

$$T = A + B \log (D/S)$$

That is, if you are 5 cm away from a 1 cm target, the time is the same as to hit a 4 cm target from 20 cm away.

This has very direct practical applications. If you design small link icons on screen, they are harder to hit and take longer than larger ones (this you could probably guess); furthermore, you can predict pretty accurately how much longer the smaller ones take to hit. However, typically in applying such knowledge, we find ourselves needing to use it in ways that either would not be of interest to the psychologist (or other area) or may be regarded as unacceptable simplifications (just like the complex mathematics of fluid dynamics gets reduced to simple tables for practical plumbing).

Taking Fitts' law, we can see examples of this. The constants A and B in the above formula are not universal but depend on the particular device being used. They differ between mouse and trackball and between different types of mouse. They depend on whether your finger is held down dragging or the mouse is used with the fingers "relaxed" (dragging times are longer). There is also evidence that they differ depending on the muscle groups involved so that a very large movement may start to use different muscles (arm rather than wrist) and so the time may cease to be a simple straight line and have different parts.

If we look at the interaction at a slightly broader level, more issues become important. If we design larger target images in order to reduce the time to hit them and so speed up the interaction, we will not be able to have as many on screen. This will mean either having a scrolling Web page or having to click through several Web pages to find what we want. Typically, these other interactions take longer than the time saved through having larger targets! But it does not end there. The process of navigating a site is not just about clicking links; the user needs to visually scan the page to choose which link to follow. The organization of the links and the size of the fonts may make this more or less easy. Large pages with dozens of tiny links may take longer to scan, but also very sparse pages with small numbers of links may mean the user has to dig very deeply to find things and so get disoriented. HCI attempts to understand these more complex interactions and so has become a discipline or area of study in its own right.

3.3.2.2 The Rise of the GUI

Since publication of the first edition of this handbook, access to Web-based information has shifted from predominantly the computer to mobile phones, as well as interactive TV. However, the Web (and for that matter the Internet) is still synonymous for many people with access of Web pages through browsers in windowed systems such as Microsoft Windows or Apple MacOS. This interface itself arose from a process of experience, experimentation, and design.

The Web is perhaps unusual in that it is one of the few major technical breakthroughs in computing that did not stem from research at Xerox PARC labs (although Ethernet, through which many of us connect to the Internet, did).

The Xerox Star released in 1981 was the first commercial system to include a window and mouse interface. It, in turn, built on the multiwindow interfaces of programming environments developed at PARC, including InterLISP and Smalltalk and on PARC's long experience with the Alto personal workstation used largely internally. Before that computers were almost universally accessed through command line or text menu systems now rarely seen by ordinary users except in techno-hacker films such as "The Matrix."

Although the Star was innovative, it was also solidly founded on previous experience and background knowledge. Where the Star team found that they did not have sufficient theoretical or practical background to make certain design decisions, they would perform experiments. Innovation was not a wild stab in the dark but clear sighted movement forward.

Unfortunately, the Star was a flop. This was due not to its interface but to commercial factors: Xerox's own positioning in the market, the potential productivity gains were hard to quantify, and the price tag was high. The essence of the Star design was then licensed to Apple who used it to produce the Lisa office system, which was also a flop. Again, it was too novel and too expensive. Remember it does not matter how useful or usable a product is; it needs to be used! The break-through came with the Macintosh in 1984, which included a full graphical interface at an affordable price. Some years later, Microsoft produced their first windows add-on to DOS, and the rest is history.

As well as giving some idea of the roots of the modern graphical user interface (GUI), this story has an important lesson. The design of the fine features of the GUI has come out of a long period of experience, analysis, and empirical studies. This is partly why Microsoft Windows lagged for so many years behind the Mac OS environments: it did not have the same access to the fine details of timing and interaction that made the environment feel fluid and natural.

Another thing to note though is that progress has not always been positive. One story that was only uncovered after several years of digging was the origin of the direction of the little arrows on a scroll bar (Dix 1998). These are now so standard it is hard to imagine them any other way, but if you are new to them there are two alternatives. They could point in the direction the scroll handle will move or they could point in the direction the page will move—these are always opposite ways round. It is not obvious. In the first version of the Star they were placed the way they are now, but the team were uncertain so started a series of user tests. They found that the other way round was clearer and more easily understood, so the revised version of the design changed the direction. Unfortunately, when the design was licensed to Apple they were given the wrong version of this part of the documentation.

For most developers of stand-alone applications the niceties of these fine features are immaterial. They are given, for good or ill, by the underlying windowing system and toolkits. Issues like choice of labels in menus belong to the developer, but the behavior of the menu is fixed.

Browsers, of course, are such applications with the same limitations, and if you use simple form element tags in a Web page, these follow whatever conventions are standard for that machine. However, these are fairly limited, and Web pages now create their own pull-down menus and other such features using JavaScript or Flash. For the first time in 20 years, ordinary inter-face developers find themselves designing widgets. Later in this chapter we will look in detail at an example of this.

3.3.3 The Interaction Design Process

For some people, usability is something added on at the end: making pretty screens. However, people do not just interact with the screen; they interact with the system as a whole. A user focus is needed throughout the design process. Figure 3.7 shows a sim-plified view of the interaction design process, and we will use this to survey some of the methods and techniques used in HCI design. We will not try to give an exhaustive list of techniques but give some flavor of what is there (see Dix et al. 2004 or Sears and Jacko 2008 for a more complete view).

3.3.3.1 Requirements—What Is Wanted?

We have already talked about the importance of understand-ing the real purpose of users and also of getting to know

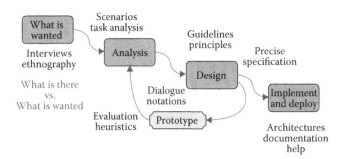

FIGURE 3.7 Interaction design process. (From Dix, A., J. Finlay, G. Abowd, and R. Beale. 2004. *Human–Computer Interaction*, 3rd ed. Englewood Cliffs, NJ: Prentice Hall, http://www.hcibook.com/e3/. With permission.)

exactly who your users are. There are many ways of finding out about what is wanted. The client will have some idea, but this will often be quite broad: "I want a Web-based hotel booking system and it needs to be really easy to use." The ultimate answer is in real users, talking to them, interview-ing them, and watching them. However, even here there is a problem. People are typically not able to articulate what they do (it was only with Eadweard Muybridge's time lapse pho-tography in the 1870s that people understood what happened when they walked or ran). Even less reliable are official doc-uments describing processes; numerous studies have shown that organizations work because of the many undocumented workarounds that people do.

Because of these difficulties, many in HCI would say that, in order to understand what is happening now, the only reli-able thing is to watch what people actually do, sometimes taking notes by hand or video recording. Ethnography, a technique from social anthropology, has become influential, especially when considering computer-supported cooperative work (CSCW; Crabtree 2003; Hughes et al. 1995; Suchman 1987; Volk, Pappas, and Wang, this volume). This involves detailed descriptions of situations, looking especially at the social interplay, including the way people use representations in the environment. Of course, in all such studies the very presence of the observer (or camera) changes the situation; there is no "pure" description of what is.

Knowing what people do now is essential, as it is easy to design a system that performs its functions perfectly but misses whole important areas (e.g., the story that starts this chapter). Even worse, if the system implements the docu-mented procedures, it may render impossible the workarounds that make systems really work. For example, a hotelier may know that a certain commercial traveler always stays on the first Monday of each month. This is not a formal reservation, but she might give him a ring before booking the last room for that day. An online system would be unlikely to cope with such nuances or tentative bookings and either have the room reserved or not.

3.3.3.2 Analysis

There is often some sort of analysis or formalization of requirements leading to some envisioning of new designs.

The traditional way to do this is using task analysis (Diaper and Stanton 2004; Strybel, this volume). The word task is very laden in HCI. It sometimes means the things that people want to do, although the word goal is perhaps better used for this, and sometimes the way in which people achieve these goals. The most common is some form of hierarchical task analysis, which involves a breakdown of tasks into subtasks (see Figure 3.8).

A task analysis like this can be used to describe the system as it is, or to describe how one imagines a new system would operate. One frequently used heuristic is to try to make the task structure using a new system match as closely as possible the existing way things are done. This makes it more likely that users will be able to do the right thing naturally. However, more radical changes may be needed.

Another less formal way to represent the way things are or to envisage ways of doing things is through scenarios (Carroll 2000). These are simply stories of someone (or people) using a system or doing an activity. These stories will include things connected with using the computer systems and things the users do outside of the computer system. While more formal representations like task analysis tend to make you think of abstract actions, the more concrete form of scenario makes you think about the surrounding context. Often, this makes you realize things that you miss when being more abstract.

There are forms of task analysis that include a more detailed view of the cognitive and perceptual demands of a task. This can include asking whether there is sufficient information for the user, whether it is clear what to do next, and whether the number of levels of tasks and subtasks is too deep to keep track of. For example, a common class of errors occurs when the goal corresponding to a higher-level task is satisfied part way through a sequence of subtasks. The classic case of this was the first generation of ATMs, which gave you your cash before the card. When the money was given the main goal of going to the ATM was satisfied, so people left, leaving their card behind. Now many ATMs return the card first and keep the money until the end of the transaction. This sort of detail can also be included in a scenario.

3.3.3.3 Detailed Design

In fact, there is no clear boundary between analyzing the system as it is and the detailed design of the new system. It is hard for users to envisage new things in the abstract and so almost always in investigating what is wanted one has to think about what is possible. However, there is a time when a more detailed and complete design is needed. This includes surface features like detailed screen design as well as deeper features such as the range of functionality and the navigation structure of the application.

In a stand-alone application the navigation structure is about the main screens and the order in which these appear depending on the user's actions. In a Web-based application the obvious equivalent is the structure of the site. The task structure can be used to help drive this design. Where there are a small number of very frequent and well-defined tasks, one may design parts of a site or application to directly reflect this. For example, the sort of step-by-step screens one often finds in the checkout part of an e-commerce site. Where the tasks are less well defined or very numerous one can use scenarios or a task hierarchy to check against a proposed site structure, effectively playing out the scenario and seeing how complicated it is. Often, a site may have a highly functional breakdown into separate parts, but you find that a frequent task involves moving back and forth between sections that are distant from each other. This may suggest restructuring the site or adding cross-links.

At the level of individual pages, the same holds true. Knowing how people understand a system can help guide choices about what to group on screen, or the type of language to use; and knowledge of the natural order of users' tasks can help guide the order of elements on a page. When a page includes applets or complicated scripts or where there are server-based applications then these dynamic parts of a Web-based application begin to resemble the screen-to-screen navigation structure of a stand-alone application. Various notations have been developed, or adopted from other areas of computing, to specify this structure. In HCI the ordering of user and system actions is called dialogue,

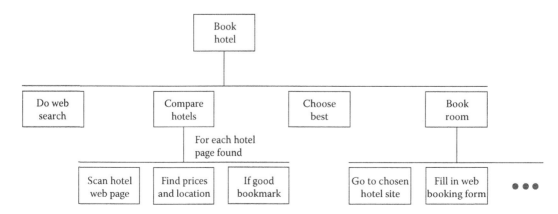

FIGURE 3.8 Hierarchical task analysis.

and so these are normally called dialogue notations. At a relatively informal level one can use simple network diagrams to show the order of screens or pages as in Figure 3.9. For more complex designs, formal state transition networks, state charts (as used in UML), formal grammars, and Petri Nets have all been used (Dix et al. 2004, chap. 16; Palanque and Paternó 1997).

In addition to the very specific knowledge about particular tasks of users, there is a wealth of more general knowledge in the form of design principles and guidelines. Some of these are general and there are specific Web-based guidelines such as the Yale collection (Lynch and Horton 2002). As an example, here are four simple rules for navigation design (Dix et al. 2004):

- Know where you are; e.g., use breadcrumb trails at the top of a Web page as in Figure 3.10.
- Know what you can do; e.g., make links clear by making graphic and text links obvious.
- Know where you are going or what will happen; e.g., make sure that the words used for links are very clear and ideally add a few words of explanation to avoid too many wrong paths.
- Know where you have been or what you have done; for normal Web pages the browser does this with its history, but in a dynamic application you need to make sure that the user has some confirmation that things have actually happened.

3.3.3.4 Iteration and Prototyping

Because interaction design involves people and people are complex, the only thing that you can be certain of in your initial design is that it will not be right! This is for two reasons:

1. It is hard to predict the consequences of design choices. Like the users who typed a multiline title that had never been considered, you often find that real users do not do what you expect. Understanding your users well, understanding their tasks, learning a bit about cognition and perception (e.g., what color combinations are most readable) can all make sure that the design is likely to work, but there will always be surprises.
2. It is hard to know what you want until you see it. As soon as users see the system, they will begin to realize what they really want. As we have already noted, it is very hard to envisage a new system, especially if

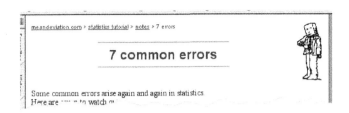

FIGURE 3.10 Breadcrumbs at the top of a page.

it is radically different from the past. It may be that only once a prototype of a system is available do you discover some missing functionality. Alternatively, you may find that users make unexpected use of the functionality they are given that you then need to support more effectively in future versions. The dramatic explosion in mobile phone text messaging among teenagers was just such an unexpected use.

Because of this all interaction design projects necessarily involve some element of iteration: produce prototypes, evaluate those prototypes, redesign, then do it all again. Note that this evaluation is *formative*; that is, evaluation designed to help suggest improvement or change. In addition, sometimes one wants *summative* evaluation—evaluation to give a measure or threshold of acceptability—but this is less useful except perhaps for contractual purposes. Evaluation for the Web is discussed in detail in the work of Vu, Zhu, and Proctor (this volume).

The prototypes used during iteration may range from paper prototypes and storyboards (pictures of the system to show users and step through by hand), through screen-based mock-ups (hand-produced Web pages, PowerPoint slides), to early versions of fully developed systems. Usability testing may be able to use systems that are incomplete or not yet fully engineered at the back-end (e.g., omitting proper database locking), as long as the parts that are missing are not encountered during the tasks being tested.

Ideally, one would like to test all designs with lots of different users in order to discover different patterns of use and different problems. However, this can take a long time, be very expensive, and with certain classes of user be very hard to arrange. There are a number of techniques developed to make the best use of small numbers of users during testing or even evaluate a system with no real users at all. An example of the former are think-aloud methods (Monk et al. 1993), where you ask someone to use the system and tell you what they are doing while they do it. This reflection makes the use unnatural but gives you more insight into what the user is thinking. Of course, the very act of talking about something changes the way you do it, so that has to be taken into account when interpreting the data. An example of evaluation with no users at all is heuristic evaluation (Nielsen 1994), which uses a small number of expert evaluators who look at the system using a number of heuristics. Early studies found that the first three or four evaluators would discover nearly all the usability problems; however, when heuristic usability is used in practice, there is frequently only one evaluator—and that

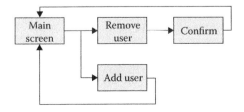

FIGURE 3.9 Network of screens/pages.

is the designer. This is better than nothing, certainly, but a little dangerous.

Because iteration is inevitable, it can lead to a try-it-and-see attitude. However, this is very bad practice. First, it is bad because it means that you do far more iterations than necessary. Many problems are easily foreseen, and so more careful thought as to why something is not working as expected (using knowledge of human cognition, the user's tasks, etc.) can mean that the resulting solutions are more likely to work. Second, and perhaps even more important, frequent small changes can lead to "local maxima" designs, which no small change will improve but are still not very good. If you start off with a bad design, small changes just make it not quite so bad. It is the deep and detailed analysis that leads to good initial designs.

3.3.3.5 Implementation and Deployment

Finally, one needs to actually build and deploy the system. To some extent the building part is "just writing software." However, producing user interface code is very different from other types of code. There has been extensive work, over many years, to develop architectures and frameworks to make this easier. The most influential has been the Seeheim model (Pfaff and ten Hagen 1985). This divides the interface software of a system into three main logical components corresponding to the lexical–syntactic–semantic levels of linguistics (Figure 3.11).

Presentation (lexical): How things actually look on screen, where they are placed, and what words appear on menus.

Dialogue (syntactic): The order in which the user can do things, what actions are available at different times, and what screens are displayed.

Functionality (semantic): the link to the underlying application or functionality of the system (originally called the application interface model).

Note that the extra box at the bottom of Figure 3.11 is to represent the fact that certain kinds of rapid semantic feedback (e.g., highlighting of applications and folders when you drag a file icon on a screen) need a direct connection between the presentation and semantic components to give acceptable response. This is under the control of the dialogue (hence the dotted line) but when activated can run independently.

Although Seeheim was originally designed before the Web and even before GUI interfaces, we can see the same

elements in a Web-based application. On the Web the actual layout is done by the Web browser and the presentation component is mainly concerned with producing the right HTML. In an XML-based architecture this would correspond with the XSLT or similar template mechanism to generate actual Web pages from abstract XML descriptions of content (Clark 1999).

The underlying application and functionality is often principally a database where the meaning associated with the data is distributed over lots of server scripts. This is the aspect that is often called business logic and in Java enterprise servers is packaged into Enterprise Beans, or similar objects in other architectures (Flanagan et al. 1999).

The dialogue component is perhaps most problematic on the Web. The code that corresponds to it is again distributed over many scripts, but perhaps more problematic is where the dialogue state is stored. We will return to this in the next part of this chapter.

Finally, the "switch" part is interesting on the Web because there is no rapid feedback except within the Web browser with JavaScript or similar code on the Web page itself. Even AJAX interactions require relatively slow HTTP transactions with the back end. That is, for rapid semantic feedback enough of the semantics has to be put in the page itself. In fact, this is an important point for all networked systems—wherever rapid feedback is needed the code and data needed to manage it must be close to the user. In particular, anything that requires hand-eye coordination, such as dragging, needs response times of the order of 100–200 ms and so cannot afford to have any network components except for very fast local networks.

In fact, this time of 100–200 ms occurs because the human body itself is networked and distributed. The cycle of seeing something, the visual stimulus being processed, and the signals going down your arm to your hand takes around 200 ms or so. So your brain and nervous system can cope with delays of this order.

As well as the actual coded system, the deployment of a system involves documentation, help systems (both electronic and perhaps in the form of a human support team), training, and so forth. Again, the form of this in a Web-based system is a little different in that it is easier to include more text and explanation in a Web page, so that the boundaries between application, documentation, and help system can become blurred. Furthermore, many stand-alone applications now have their documentation and help delivered as Web pages and may give user support through e-mail, Web FAQs, and Web-based user forums.

3.4 HCI AND THE WEB

3.4.1 IS THE WEB DIFFERENT?

In some ways designing for the Web is just like designing any computer application. Certainly, all the steps and techniques in the last section apply. However, there are also major differences, if not in kind, at least in the magnitude of certain

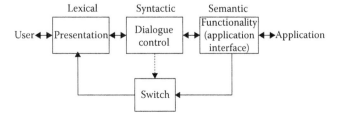

FIGURE 3.11 Seeheim model.

factors. Of course, many such issues will arise in other chapters of this book. In this section we will look at a few briefly and others in more detail.

3.4.2 Detailed Interface Issues

3.4.2.1 Platform Independence

The great joy of designing applications for the Web is that it works everywhere—write once and browse everywhere. However, this has a downside: you do not have fine control of layout, color, etc., as these depend on the computer system, version and make of browser, and window size. It is hard to find out enough about the system to decide what layout to use, and hence it is easy to develop the habit of designing for a particular platform, browser, and screen size and then hope for the best.

Anyone who has laid out a Web page is aware of the difficulty of getting it to look right. Because it is on-screen with graphics, one wants to arrange it in an almost desktop publishing or magazine style, but this has to be achieved using complex CSS or (however much deprecated for layout) tables, invisible images for spacing, and so forth. Then, when it looks as good as it can, you resize the window, or look at it on a different machine, but things are never quite as you expect. Each browser, each version of the same browser, and the same version of the same browser on different platforms—all behave differently. Even table layout tags, part of Web standards back to version 1.0 browsers, are not dealt with uniformly; for example, a width may be interpreted as meaning maximum on some or fixed size on others.

Some of these problems are because of bugs in browsers, different interpretation of standard tags by different vendors, or deliberate attempts to be nonstandard in order to preserve market dominance. Happily, many of these issues have gotten better over time as browsers become more standards compliant, at least for older features. However, even if we were in a bug free, fully standardized, open market, there would still be difficulties because of different window sizes, default font sizes, etc.

In fact, this is not a unique issue for the Web, as creating resizeable layouts and platform-independent user interface toolkits has been an important topic for many years. Those who have programmed in Java will have used layout managers to give approximate layouts ("put this above that") leaving it for the toolkit to resize the layout for the available screen space. This builds on similar techniques in X toolkits and their predecessors (Open Software Foundation 1995).

In all these technologies the resized layouts tend to be "good enough" but not "good." Graphic designers can be driven to despair by this and often resort to pages built entirely of graphics or to Macromedia Flash movies, both of which, without great care, reduce accessibility both for visually disabled users and for those with older browsers. A less restrictive alternative is to design a page using fixed-width layout; however, this means that users with wide screens see a small strip of page and lots of empty margin, while those with small screens may have to scroll horizontally. On the other hand, leaving the browser to resize text areas can lead to long lines spreading across the page that cannot be read.

When designing for platform independence, small differences do matter, and it is necessary to test your application on many browsers, operating systems, and versions of both. This is particularly a problem with scripted pages where the different browsers support different models of the Web document and have versions of JavaScript with different "features" and undeniable bugs. This may seem like an impractical counsel of perfection, as most companies do not have the physical or human resources to test on many different platforms and browsers. Happily, there are Web-based services to help; for example, BrowserCam (2009) allows you to see how a page looks on many different browsers, and John Resig's Test Swarm (Resig 2010) uses a Web 2.0 method, crowd-sourcing the testing of JavaScript test suites using plug-ins deployed on end-user machines.

These issues of size and platform features become even more apparent when designing Web pages that may be used on mobile phones. Although a certain amount of resizing can be accommodated using flexible layouts and CSS, in the end the user experience of using a 2 in. (5 cm) wide screen is very different from a desktop or laptop monitor. Phone-based browsers help deal with this problem, allowing you to view an entire page in miniature and then to zoom into particular columns. While useful for accessing any content, this solution is often far from ideal—even the best readers will require a combination of panning and/or zooming before you can read content, let alone click on a link.

Most large sites detect phones and produce content tailored to the smaller width. For example, on Facebook's mobile site, the personal Home page includes only the central column of the full-screen site's layout, with some of the sidebar content accessible by tabs and some not shown at all. The Facebook iPhone app is different, again making use of platform-specific features; although many of these are also available to the Web developer (e.g., location), it is at the cost of needing to design variants of a basic Web site for different specific devices.

3.4.2.2 Two Interfaces

Not only do different browsers affect the layout and behavior of pages, they are also part of the whole application that the user sees. That is, the user has two interfaces: the Web browser and the Web application itself.

Sometimes, this can be a help: we may rely on the Back button to avoid needing to add certain navigation paths; we can afford to have long pages because we know that the user will be able to scroll; we can launch new windows knowing the browser will manage this.

However, this can also be a problem. When you have a dynamic application the user may bookmark a Web page deep in an application, whereas a stand-alone application always starts "at the beginning." This may be worse if the user bookmarks or uses the Back button or history to visit a confirmation screen for an update transaction—depending on the particular browsers and methods used to code, the application this may lead to repeated updates. We will return to the importance of the Back button when we look at navigation.

In fact, in some circumstances, there may be three or even four interfaces simultaneously at work. When a user interacts with a Facebook app or Google gadget, there is the browser, the main page (e.g., Facebook, Google spreadsheet), and the micro-app within it. The outer platform may restrict the kind of things that the included micro-app can do, for example, Facebook sanitizes the HTML and JavaScript generated by apps, but as a developer, you still need to consider that you are operating within a larger framework and create an experience consonant with it.

The browser may also be augmented by plug-ins or extensions that change the behavior of pages. For example, some plug-ins scan for microformats and offer links to appropriate content, or create additional sections on popular sites. For the designer of the page, there is little one can do to design away potential usability problems that may be introduced by such plug-ins; however, you can use them as an opportunity, for example, creating plug-ins that augment the functionality of your site. Of course, as the designer of plug-ins, you need to take special care not to accidentally destroy the user experience of sites that are augmented.

3.4.2.3 UI Widgets

We have seen how the current GUI interfaces arose out of a long process of development. The key elements, menus, push buttons, etc., are usually implemented within the underlying windowing system. Although we may think that all menus are the same, in fact, they differ subtly between platforms. In each implementation is a wealth of experience built up over many years. But now as Web developers use scripting to produce roll-overs, menus, and other dynamic effects they find themselves having to design new widgets, or redesign old ones. Not surprisingly, this is difficult. Traditional application designers have not had to worry about the widgets, which were a given, so guidelines and usability information are sadly lacking or perhaps proprietary.

Let us look at one example, multilevel menus. The behavior is fairly simple. You click on a top menu label and a dropdown menu appears. Some items have an arrow indicating that there are further items, and as you move your mouse over these items, a submenu appears to the right. When you want to select something from the displayed submenu, you simply move to the item and click it. Figure 3.12 shows a simple multilevel menu. The user has pulled down the Insert menu and is about to select Face from the Symbol submenu.

In addition to the submenu, Figure 3.12 also shows the path the user's mouse took between the submenu appearing as it hovered over Symbol and finally settling over Face in the submenu. Notice that the mouse has "cut the corner." The user did not move across to the left and then down, but instead took a direct path to the desired menu item. Now if the menu system were written naively, the Symbol submenu would have disappeared as the mouse moved, to be replaced by the Picture and Object submenus in turn. Indeed, if you do the same movement slowly, this is exactly what happens.

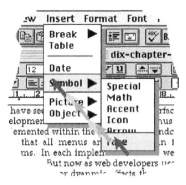

FIGURE 3.12 Multilevel menus—cutting corners.

Over years of design and redesign the toolkits delivered with major windowing systems have embodied careful choices of timing. When the mouse moves off an item (such as Symbol) with a submenu, the submenu does not close straightaway, but, instead, there is a timer and another submenu cannot appear until the time is complete. This is not an obvious design feature. Rounded window corners, three-dimensional (3D) icons, or animations are easy to see, but detailed timings have no immediate visual effect. However, it is this fine behavior that makes menus natural to use.

In contrast, when menus are produced on the Web, it is usually the case that the code says something like:

> on mouse_over (Symbol): show symbol_menu
> on mouse_over (Picture): show picture_menu

This would mean that if your mouse strayed even the slightest bit over the Picture item, the Symbol submenu would disappear. Menus like this are physically hard to navigate even if you understand what is going wrong; they just seem to have menus appearing and disappearing randomly.

Unfortunately, we do not know of any body of design advice for these standard elements such as menus, just because they are built into major windowing systems and so are invisible to the normal developer. At present the best advice is to just be aware of these fine timing issues and if in doubt choose an interaction mechanism (e.g., click based) that does not require fine timing.

As JavaScript user interface frameworks and libraries such as jQueryUI or script.aculo.us (jQueryUI 2009; Fuchs 2010) mature, some of these issues may fade for the ordinary Web UI designer and, as with desktop interfaces, become issues mainly for widget designers. These frameworks also make it easy for third parties to develop and publish new components, and this, together with the general mashup culture, is changing the granularity of user interface components. As well as menus and buttons, zoomable maps or image editors may be included in a Web page.

Finally, it should be noted that the term "widget" is often used in a slightly different way in Web-related interfaces than in the older user interface development literature; for example, Mac OSX and Yahoo! use the term to refer to downloadable desktop micro-apps. However, in a mashable world the distinction between a small application and a user interface component is fading.

3.4.2.4 Frames

Frames must have been the cause of more vitriol and argument than any other feature of the Web. Most well known is Nielsen's alert box on "Why frames suck" (Nielsen 1996). Frames were introduced for good reasons. They avoid refreshing parts of a page that do not change when you navigate to a new page. For example, the heading and navigation menu often stay the same; using frames allows these to be downloaded once and only the individual pages' changing content to update. This can significantly reduce the overall download time for a page, especially as headers often include graphics such as site logos. Another beneficial effect of this is that when you move to a new page, typically, only some of the frames refresh. So, even if the content of the page is slow to fully download, the user is not faced with a completely empty screen.

Frames also offer layout control that is similar to tables but differs in significant details. For example, with frames it is easier to specify exactly how wide a subframe is and have the browser adhere to it. Frames were designed initially for layout, whereas tables were designed for structured content; not surprisingly, some layout control is more precise.

In addition, because frames are defined relative to the screen rather than to the page it is possible to keep menus and headers continually visible while content subframes scroll. This is similar to the way "normal" computer applications divide up their windows. Unfortunately, frames have usability problems related directly to these strengths.

Working backward, let us start with scrolling subframes. Although they are similar to application windows, they do not correspond to the metaphor of Web page as a "page." This can in itself be confusing but especially so if (as is often the case) to improve aesthetics there is no clear boundary to the scrolling region. Similar problems are often found in Flash sites.

The ability to precisely specify a frame width or height works if the content is a graphic guaranteed to be of fixed size, but if there is any platform dependence, such as users using different fonts, or browsers using different frame margins, then the content can become ugly or unusable. If the frames are set not to allow scroll bars, then content can become inaccessible; if scroll bars are allowed in narrow or short columns, then one is left with a tiny column mostly consisting of scroll bar!

However, most problematic is the fact that what, for a user, is a single visible page consists of several pages from the system's point of view. The relatively simple view of a Web site as linked pages suddenly gives rise to a much more complicated model. Furthermore, the interaction model of the Web is oriented around the Web page and frames break these mechanisms. The page URL does not change as one navigates around the site, although happily the Back button does take this navigation into account. Then when you find the page you want, you bookmark it only to find that when you revisit the bookmarked page you end up back at the site's home page. Finally, it is hard to print the page or view the source. This is partly because it is not clear what the target of the request to print or view is: single frame or whole screen, and even when the target is understood, it is not clear what it means to print a framed page: the current screen display or what?

Despite these problems, there are good times to use frames. In particular, they can be used in Web applications to reduce flicker and where you explicitly do *not* want generated pages within a transaction stream to be bookmarked. For example, the authors were involved in the production of a community Web application, vfridge, which took the metaphor of leaving notes held on with magnets to a metal fridge door (Dix 2000a). The vfridge interface used frames extensively, both to keep control over the interface and also to allow parts of the interface to be delivered from separate servers (Figure 3.13). However, such applications do need careful coding as often several frames need to be updated and the knowledge of what frames are visible and need updating is distributed between the Web server and code on the page.

3.4.2.5 AJAX Update, iframes, and Hash URLs

Many of the reasons for using frames have been obviated by AJAX-based interfaces that update portions of the screen using DOM manipulation. Also, the mixing of content from multiple sites can be managed by using iframes or JSON-based Web services. However, although this may mean that frames become used less, many of the same problems arise afresh with new technology.

In particular, AJAX-based sites mean that, like frames, there is a single URL but the content may vary through interaction. For some applications this is acceptable. For example, in an online spreadsheet, you would expect a bookmark to give you the current state of a document, not the state when you created the bookmark (although having a facility to do this could be very useful). However, if you visit a news site, navigate to a story, and then bookmark it, you expect the link to take you back to the story, not the main site.

One solution for both AJAX and frames is to use the hash portion of the URL (e.g., "abc" in http://example.com/

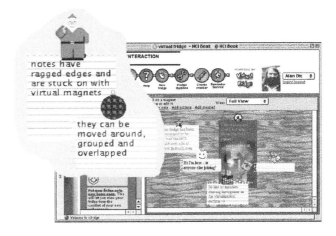

FIGURE 3.13 (See color insert.) The vfridge: a virtual fridge door on the Web.

apage#abc). While attempting to update the URL of the page leads to a reload, most browsers allow the hash portion of the URL to be updated. The hash is preserved when you bookmark a page or send the URL in an e-mail message. The Web server never gets to see the hash part of the URL; however, JavaScript on the page can read it and update the page accordingly. For example, on a tabbed page it can automatically select the correct tab.

AJAX-based live update is difficult to implement in a way that is completely accessible (see Caldwell and Vanderheiden, this volume). So, designers often fall back on the mantra: build the accessible version of the site first then augment it with AJAX using techniques like progressive disclosure. The problem with this is that building an accessible user interface is a very restrictive constraint. It is almost impossible to achieve the best aesthetic experience for the majority of users while catering for all accessibility guidelines, which is one reason why many Web sites completely ignore accessibility. The entertainment industry is a clear example: when was the last time you saw a fully accessible compliant Web site for a mainstream movie?

3.4.3 Navigation

3.4.3.1 Lost in Hyperspace

The power of the Web is linkage—instead of a predetermined linear or hierarchical path through information, each user is free to follow their own interests, clicking on links as they wish. The downside to this freedom is that after a period of browsing the user can feel utterly lost in this unruly virtual space.

There are two aspects to this lostness. One is about content. The Web encourages a style where each page is in some way contextless—you could find it through a search engine or via links from some other site or some other part of the same site. The designer cannot know what visitors will have seen before or where they have come from. However, it is impossible to communicate without a sense of shared context. Indeed, one of the things we do in conversation is to continually negotiate and renegotiate this common ground of shared understanding (Clark 1996; Monk 2003). Just as an interface designer always thinks "user," so also a writer always thinks "reader," trying to produce material suited to the expected reader's understanding and to produce a narrative that introduce material in a way that informs and motivates. In contrast, the Web encourages dislocated knowledge—rather like giving a child an encyclopedia on their first day at school and saying "learn." Interestingly, Nelson's (1981) classic exposition on the power of hypertext takes the hypertext paradigm back into print, giving us pages full of semirelated snippets, textual photomontage.

There is no simple solution to this problem of context. It is not possible or desirable to write without context, so the best option is usually to make the context of informational pages clear. You may have a reference to a particular page in a book, but one can skip back through the book to see the context of

the page, and read the cover to see whether it is an advanced theoretical treatment of a subject or a popular mass market publication. Similarly, breadcrumbs, headers, and menus can make it clear how the piece of information you are viewing fits within a wider context of knowledge.

The other type of lostness concerns spatial disorientation. Miller's (1956) classic paper showed that we have a relatively small working memory: 7 ± 2 "chunks" of information. Without external aids this is all we have to keep track of where we have been. In physical space the properties of physicality help us to navigate: for example, if we have gone down a path we can turn around and recognize the way back; however, in cyberspace, there is no easy turning back. Within sites a sitemap can help give users a global model of what is there and again breadcrumbs or other techniques let them know where they are. Between sites we have to rely on the browser's Back button and history.

3.4.3.2 Just Search—Does Lostness Matter?

While the Web is all about hypertext and linkages, for many users their first action is not to enter a URL or navigate to bookmark, but to use Web search. Instead of remembering the exact form of a Web site, users simply enter the main part of the site name "Facebook," or "YouTube" into a search box. Indeed, "Yahoo!" is one of the top searches on Google. Even within a site, users find pages through searches rather than navigation, and many sites include customized Google search boxes for this purpose.

There are two ways to look at this phenomenon. On the one hand, search-based navigation means that many users are less concerned about "where" they are, and are just focused on the content they see. Maybe lostness is no longer an issue for a generation used to serendipitous content? This certainly makes it even more essential that Web content can be understood either without its context or in some way that makes the context obvious. On the other hand, it may mean that one worries less about navigation and more about the findability (Morville 2005) of content, ensuring the site is easy to spider by search engines.

For some forms of Web sites, such as news or blogs, this may be an appropriate design strategy. However, this approach can also lead to laziness in site design. While people may come to your site via search, they can just as easily leave. If you want people to stay around in your site, then you need to make it attractive for them to stay; that is you want to make the site "sticky." So when someone lands on a page, make sure they can find other related information and things to see and do. The simple Web rules ("know where you are," "know what you can do," etc.) become more important still. However, "where you have come from" is now less of a challenge for the user and more for the developer—the user may have come from anywhere! If the user came from a popular search engine, then this and the query can be found in the referrer header of the HTTP request for the page, allowing you to customize the page depending on how the user got to it: for example, showing links to other pages on your site that satisfy the same search terms.

3.4.3.3 Broad versus Deep

Humans are poor at understanding complex information structures—we are cognitively built for the physical world—but, of such structures, hierarchies are best understood. However, even within a simple hierarchical structure, working memory limitations mean that deep structures are hard to understand. Where the size of the information space is very large, for example, Web portals such as the Open Directory (DMOZ 2010) or Yahoo! (2010), then there is no way to avoid deep structures. But, in fact, many much smaller sites adopt narrow deep structures by choice.

There are three pressures that have led to the frequent design of narrow–deep sites, where there are few options on each page, leading to long interactions. Two of these pressures are associated with the different schools of Web design and one with HCI itself. The first pressure is download time—less items shown results in a smaller page. The second is graphic design—a small set of headings to navigate looks so much nicer. The third is human processing capacity—based on a misapplication of Miller's (1956) famous 7 ± 2 result for short-term memory. Within HCI the Miller result is misapplied to many things, including the number of items on a page. So many designers mistakenly limit the number of choices or links on a page to 7 ± 2 leading to narrow–deep sites. However, the evidence is that, for the Web, broad–shallow structures are often better (Larson and Czerwinski 1998). This is because people can scan lists quite quickly by eye, especially if the lists have some structure (alphabetic, numeric, hierarchical), and so if the download time is at all slow it is better to get deeper into the site from a single page. (See also http://www.hcibook.com/e3/online/menu-breadth/.) Note that for CD-ROM and other interactive media the refresh time is faster and so a different time balance applies. Paradoxically, as noted previously, working memory is an issue for keeping track of where you have been—that is, the 7 ± 2 figure is more properly applied to menu depth. There is suggestive evidence that older users (for whom short-term memory is often impaired) find deep menus particularly difficult but are happy to scan long lists (Rouet et al. 2003).

3.4.3.4 Tags and Folksonomies

In many Web applications and even on the desktop, tags are being used alongside or instead of more hierarchical or structured classifications. For a single user a lot of the difference has to do with the way they are perceived. While creating a category seems like a major decision, simply adding a tag is not; categories feel techie and formal, tags feel friendly and informal. This difference in feel leads to different behavior. Although work on personal information management has often suggested that it would be useful to classify items under multiple headings, this facility is rarely used when provided. In contrast, it is common to add many tags to the same item. While items are seen as being "in" a category (and hence only in one), they are simply labeled by tags and hence many labels are possible. Note the difference is not so much to do with functionality, but the way it is perceived.

However, the power of tags is evident when used in social applications such as for tagging photos in Flickr, or hash tags in Twitter. While the informal nature of tags means that, in principle, *any* tag can be used, the fact that you want your photos or tweets to be found means that you are more likely to use tags that you have seen others use. No one decides on a standard vocabulary of tags, but over time conventions arise, a sort of language of tags or *folksonomy*.

Often, users add some level of structure to tags using conventions such as "sport/football," and some applications allow tags to have explicit "parents," effectively creating a hierarchical classification. Even if there is no explicit structure to tags, relationships are present as different tags may be used to label the same resource and this can be extracted via data mining techniques to suggest related tags (Dix, Levialdi, and Malizia 2006), rather like book recommendations in Amazon.

As a designer you have to consider whether you need more formal structuring with a hierarchical classification or the informality of tags. This may run counter to the normal standard in an area; for example, Google mail uses tags solely, even though this is problematic when used with offline IMAP-based mail clients. Where tags are used for public resources, can you enhance their social benefit?

3.4.3.5 Back and History

One way users can reassert their control over the Web is by the tools they use to browse. Web studies have shown that the Back button accounts for over 30% of the actions performed in a browser (Catledge and Pitkow 1995; Tauscher and Greenberg 1997), compared with 50% for link following. If you do the sums, this means that about two-thirds of the times a user visits a page, they leave by going back rather than following links forward.

So, why so much going back?

- Correcting mistakes: the user gets somewhere they do not want to be. Again, the curse of terse labels!
- Dead ends: the user gets where they want to go, but there is no way to go on.
- Exploratory browsing: the user is just taking a look.
- Depth first traversal: the user is trying to visit all of a site, so is expanding links one by one.

Given that the Back button is so common, one would like it to be easy to use, but, in fact, the semantics of Back are not entirely clear.

For one step, Back appears pretty easy—it takes you to the previous page. While this is fairly unambiguous for plain HTML, most of the Web is far more complicated with different types of interaction: frames, redirection, CGI scripts, applets, and JavaScript. Users may think they are following a normal Web link, but does the browser regard it as such? Of these interaction types, redirects are perhaps the most confusing (many browsers behave better in frames now). The user goes to a page, hits Back, and the same page reappears. What is really happening is that the browser has the extra redirect page in its history list: when the user presses Back, the browser goes back to the redirect, which then redirects them back to the page.

Multistep back is even less clear. Web use studies show few people using these or history mechanisms. One reason is that the Back menu depends on the visited pages having meaningful title tags. Some title pages are useful for distinguishing pages within a site, but poor at telling which site they refer to. Some sites have very similar titles on all pages. Some have no titles whatsoever.

Another reason is that the meaning of multistep back is very unclear even for hypertext browser designers. Although Web browsers are (reasonably) consistent in their model, a comparison of several different hypertext browsers showed that they all had different behavior when dealing with multistep back, especially when the path involved multiple hits to the same page (Dix and Mancini 1997). In particular, the Back button on several hypertext systems does not adequately support depth first traversal.

The semantics of full histories get even more confusing—do you record the backward paths? Do you record all the pages visited within a site? Do you record repeat visits to the same page? It is no wonder that users rarely use these features. However, when Tauscher and Greenberg (1997) analyzed revisitation patterns, they found that, although many pages are only visited once, a significant number are revisited. So, there is great potential for well-designed histories. See their paper for a short review of graphical history mechanisms. Browsers are still very inconsistent in their ways of listing and visualizing history (see Figures 3.14 and 3.15), although often users are more likely to simply search again for key terms, or use autocompletion in the URL entry area.

Studying your own site can suggest ways in which you can help users, perhaps keeping track of recently visited parts of the site so that users can make their way back easily or noticing which parts are the key entry points and hubs (they may not be the home page). Studies of revisitation suggest that there is a connection between the frequency and volume of updates and level of revisiting (Adar, Teevan, and Dumais

FIGURE 3.15   Firefox history list.

2009), so making sure that the hubs have up-to-date information is always good advice and is also likely to increase search engine ranking.

3.4.3.6   Understanding the Geometry of the Web

With something like 70 million Web sites (Netcraft 2009) and tens of billions of individual pages, how on earth does one get an overview of the material? No wonder so many people feel utterly overwhelmed by the Web—you know the information is there, but how to find it?

Although people may have a similar feeling about the Library of Congress, the Bibliothèque nationale de France, or the Bodleian, for some reason we feel less guilty about not having found all the relevant books on a subject than we do in missing a vital Web page. Electronic omniscience appears just within our grasp, but this dream is hubris. It could be that the challenge is the need not so much to access all available information, but to accept the incompleteness of information.

In fact, it is possible to capture the entire Web; for example, the Alexa project took snapshots of the entire Web for use in its navigation technology, and donated these to make the Wayback Machine at archive.org, the historical view of the Web (Lohr 1998; archive.org 2010). However, the problem is not to simply capture the Web but to understand the structure of it. You can see a beach at sunset, and somehow grasp it all, but it would be foolish to try to somehow understand each grain of sand, or even to find the smallest. Similarly, it is reasonable and valuable to view the overall structure of parts of the Web. Maps of the Web, both site maps and representations of larger bodies of pages, can help give us such an overview, but they are usually portrayed in 2D or 3D, and the Web just is not like that. We need to understand the geometry of cyberspace itself (Dix 2003a).

We are used to the geometry of 2D and 3D space—we have lived in it all our lives! However, it does not take much to confuse us. This has been part of the mystery and fascination of mazes throughout history (Fisher 1990). One of the biggest problems with mazes is that two points that appear close are, in fact, a long way apart. In cyberspace, not only does this happen, but also distant points can suddenly be joined—magic.

FIGURE 3.14   **(See color insert.)** Safari top sites.

The most obvious geometry of cyberspace is that of the links. This gives a directed graph structure. Actually, the directedness in itself is a problem. Just like driving round a one-way system! This is another reason why the Back button is so useful: it gives us the official permit to reverse up the one-way street after we have taken the wrong turn.

Lots of systems, including most site management tools, use this link geometry to create sitemaps. Different algorithms are used that attempt to place the pages in two or three dimensions so as to preserve some idea of link closeness. The difficulty (as with any graph layout algorithm) is twofold: (1) how to deal with remote links and (2) how to manage the fact that the number of pages distance N from a given page increases exponentially whereas the available space increases linearly (2D space) or quadratically (3D space). The first problem is fundamentally intractable, but, in practice, is solved by either simply repeating such nodes or marking some sort of "distant reference," effectively reducing a directed graph to a tree. The second problem is intractable in normal space, even for trees. The Hyperbolic Browser (Lamping, Rao, and Pirolli 1995) gets round this by mapping the Web structure into a non-Euclidean space (although beware: some papers describing this work confuse hyperbolic and projective geometries). Of course, they then have to map this into a 2D representation of hyperbolic space.

The second form of Web geometry is that defined by its content. This is the way search engines work. You look for all pages on or close to some hyperplane of a high-dimensional space (where the dimensions are occurrences of different words). Alexa operates on a similar principle, indicating the closest page to a given one using similar content as a distance metric (Lohr 1998), and there are several Web mappers, very similar to the link mappers, but using this form of semantic distance as the metric (Chen and Czerwinski 1998).

The third kind of geometry is that given indirectly by the people who view the pages. Two pages are close if the same people have viewed them. A whole battery of recommender systems have arisen that use this principle (Adomavicius and Tuzhilin 2005; Resnick and Varian 1997).

Of course, these are not independent measures. If pages share some common content, it is also likely that they will link to one another. If pages link to one another, it is likely that the people will follow these paths and hence visit the same pages. If search engines throw up the same two pages together for certain classes of query, it is likely they will have common visitors.

3.4.4 Architecture and Implementation

3.4.4.1 Deep Distribution

The Web is also by its nature a distributed system: a user's client machine may be in their home in Addis Ababa, but they may be accessing a Web server in Adelaide.

Networked applications are not unique to the Web. However, the majority of pre-Web networked applications used to be of two kinds. First, transaction-based systems with

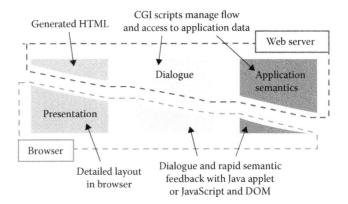

FIGURE 3.16 Seeheim for Web applications.

a minimal user interface, such as those used by travel agents, often with only a smart character-based terminal or PC emulation, where the real processing is all done on a large remote server. The other kind is client–server based systems, often operating over LANs or small corporate networks, where the majority of the work is done on the user's PC, which accesses a central database.

In contrast, Web applications are transaction based but are expected to have relatively sophisticated user interfaces often including client-side scripting. In terms of the Seeheim model this means that virtually all aspects of the user interface architecture get split between client and server (Figure 3.16). AJAX-based sites put more of the user interface into the browser and, at the extreme, can become purely client–server with the interface delivered entirely locally accessing back-end Web services.

This distribution also has important effects on timing. The Web transaction delay means that rapid feedback has to be generated locally. The Web designer has to be much more aware of timing issues, which affect both the choice of interface style and also the way the software is factored between client and server.

For example, imagine an application with a list of catalogue items. We may want to give the designer the ability to choose the order the items are displayed. In a stand-alone application we could allow the designer to select items to be moved and then simply press up and down icons to move the selected items up or down the list. For small lists this would be an intuitive way to reorder items. However, for a Web-based application this would involve a transaction for each arrow press and would be unacceptable. So, we can either use more client-side processing (using JavaScript and AJAX or Flash) or we could redesign the interaction—perhaps selecting the items to move with tick boxes and then pressing a Move Items button to give a second screen showing the remaining list items with a Move Here button between each, i.e., more of a cut and paste model.

If the application is closer to client–server with AJAX or Flash, then there are additional issues as the user interface might show that a change has occurred, but this might not yet be reflected in the server database. Alternatively, there may be delays when portions of the page are updated. The "A" in AJAX stands for asynchronous, that is, the script on the Web page can continue to work even while waiting for information from the server. As a designer, one has to design for the in between situations when the user has performed some action, but it has not yet been reflected in the server. Local storage available in Google Gears and HTML5's Web SQL Database and Web Storage (Google 2010; W3C 2010a, 2010b) makes this more likely as the user is able to interact with the page for a considerable amount of time offline, leading to the potential for conflicting updates, maybe from different devices or users, and then the need for synchronization.

3.4.4.2 UI Architecture for the Web?

We have already mentioned the way the deep distribution of the Web gives Web-based user interfaces different architectural challenges to conventional interfaces. We will look at this now in a little more detail focusing particularly on the issue of where dialogue state is held.

Although the Seeheim model is the conceptual root of most subsequent user interface architectures, it is probably the Model–View–Controller (MVC) architecture that has been most influential in actual code (Krasner and Pope 1998). It was developed for the graphical interface of early SmallTalk systems and is the framework underlying Java Swing. Whereas Seeheim is looking at the whole application, MVC focuses on individual components. In MVC there are three elements:

1. The *Model*, which stores the abstract state of the component, for example, the current value of target temperature.
2. The *View*, which knows how to represent the Model on a display, for example, one view might display the temperature as a number, another might show it as a picture of a thermometer.
3. The *Controller*, which knows how to interpret user actions such as keyboard or mouse clicks.

As MVC was developed in an object-oriented system and is often used in object-based languages, it is usually the case that each of these elements is a single object. One object holds the underlying state, one takes that and displays it, and one deals with mouse, keyboard, and other events.

The elements in MVC correspond roughly to parts of the Seeheim model:

Model   – semantics – application/functionality
View   – lexical   – presentation
Controller – syntax   – dialogue

However, there are some structural differences. Most important is that the Controller receives input directly from the interface toolkit and also that it influences the View only indirectly by updating the Model. If the user presses the plus key this goes to the Controller, which interprets this as increase temperature and so invokes the increment

Temperature method on the Model object. When the Model has updated its state, it notifies the View, which then updates the display.

User input → Controller → Model → View → Display

Note that the Controller needs to have some lexical knowledge to interpret the plus key. However, because it also needs to interpret, say, a mouse click on the thermometer, the pipeline process needs to break a little. In fact, the Controller "talks" to the View in order to determine the meaning of screen locations (Figure 3.17).

The Seeheim model too found itself in "tension" when faced with real applications and included the switch or fast path linking application to presentation. While the Seeheim model regarded this as the exception, in MVC this is the norm.

The structural differences between Seeheim and MVC are largely to do with the different environments they were developed for. When Seeheim was proposed, the underlying applications were largely quite complex and monolithic, for example, a finite element model of a new bridge. The actions would include "big" things such as calculate strain. Major aspects of the display would be recomputed infrequently. In contrast, MVC was developed in a highly interactive system where small user inputs had immediate effect on quite semantically shallow objects. MVC is optimized for maintaining a continuous representation of individually relatively simple objects.

Turning now to the Web, we see that the situation is different again, although perhaps closer to the old view. The equivalent of the MVC model or Seeheim application is typically the contents of a database, for example in an e-commerce system containing the product catalogue and customers' orders and account details. However, this database is "distant" from the actual interface at the browser and the relationship between the account details as displayed on the browser and as represented in the database is maintained on a per transaction basis, not continuously. If an item is dispatched while the user is looking at their order details, we do not normally expect the system to "tell" the browser straightaway, but instead only when the user refreshes the screen or commits an action will the change become visible. In fact, there are Web application frameworks that adopt an MVC breakdown, but these are fundamentally different. The View

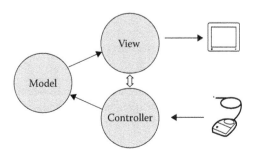

FIGURE 3.17   MVC model—model view controller.

in such a system embodies the knowledge of how to display the information in the Model as HTML but does not actively enforce the display.

3.4.4.3 Dialogue State on the Web

This distance becomes more problematic when we consider dialogue state. As well as the persistent state of application objects, there are many things that need to be remembered during an interaction, such as part-edited objects, current location in a list of items, etc. In the customer interface of an e-commerce system this would include the current shopping basket; in the stock-control interface to the same system, there will be temporary copies of stock records being updated.

In a conventional interface these would simply be stored as variables and objects in the running program. However, in the Web this state, information typically needs to be remembered explicitly. Because conventional programs hide this state, making it effortless, it is often very hard to get this right in a Web application.

A complicating factor is that there are many ways to store information in a Web application. One of the simplest is through hidden variables in Web forms or URL rewriting. The current state of the interaction is effectively held on the Web page on the user's machine! In fact, this is often a good place to store this information as it maintains the stateless nature of the application at the Web server end, so if the user goes away and does not complete the interaction, there is not dead state left at the server end. The downside is that the URL is also used to pass the values connected with the current user transaction, so information about the current event and current state are both encoded in a similar way. This is not a problem so long as the difference is well understood, but often it is clearly not! Small amounts of state are often stored in cookies too, especially for tiny applications, such as a color picker, which largely need to store semipersistent information such as favorite colors or last color model used.

Both cookies and URL encoding are often used to keep some sort of unique session identifier, which can then be used to index session state held in memory or in a database. Web frameworks often include some sort of session support, from simply keeping track of a session identifier to full support for storing data values and objects during one transaction, which are then available for the next transaction on the *same browser*.

Note, though, session support is almost always per browser, not per window/tab, nor per machine. In stand-alone applications it is fairly simple to keep track of data structures relating several windows to files, but on the Web windows may be closed, or the user clicks away to another Web site, yet the user can return to these "closed" pages via History or Back button. A stand-alone application would exercise control over this, but the Web application developer needs to keep track of these things or put mechanisms in place to prohibit them. Unfortunately, support from development frameworks at this level is virtually nonexistent.

3.4.4.4 Different Kinds of Web Applications

As already mentioned, Web applications can be of many kinds, and this offers new opportunities as well as new implementation challenges. Specific platforms often offer additional APIs; for example, Safari on the iPhone offers JavaScript location support and Facebook has an extended variant of HTML to allow easy access to the name of the current user, lists of friends, etc. However, this also requires learning new development paradigms, for example, a Facebook application is typically tailored to the person viewing the page and needs to interact with Facebook services such as creating news items. Rather than programming a complete Web application, the job of the developer is often more one of assembling components or fitting within an existing framework. Over time some of these may become more standardized, but currently each framework and each API needs to be learned afresh.

3.5 HCI IN FLUX

We have seen some of the ways in which HCI for the Web both share much in common with, but also differ from, traditional HCI for stand-alone systems. In fact, the Web is just one of a number of changes all of which are challenging those working in HCI. Sometimes, this is a challenge to think more clearly about what are the fundamentals of interaction and so can be taken over into new domains, as opposed to those things that are the ephemera of a particular technology. Sometimes, it is a challenge to consider new issues that were not apparent previously.

HCI grew up around the issues of personal computing. One man and his machine—and the gender is chosen carefully here because the vast majority of computer users until well into the 1990s were male. In the early days the machine may have been a terminal into a large central mainframe, but increasingly the pattern became, as we recognize it today, the personal computer on desktop, or laptop. E-mail was one of the earliest applications for the nascent Internet in the 1970s (originally, ARPANET connecting only a handful of U.S. universities and research sites), so from the beginning communication has been an important part of HCI, but it was in the late 1980s when networking became more widespread that groupware and the field of computer-supported cooperative work (CSCW) sprang up. That is, the focus changed from one man and his machine to many people (note the gender shift) collaborating each with their individual machines.

The Web has stretched this further as the natural place for computing has expanded from office to home, cafe, and airport departure lounge, and each user has at their finger tips information from anywhere the world and the potential to communicate with unseen friends. While this global network has been growing, there is also a shift at the fine scale. In talks the authors often ask, "How many computers in your house?"—sometimes the answer is 1, sometimes 2, 3, or 4. They then say, "Do you have a television, video recorder, or Hi-Fi?" The next question is, "how many computers do you carry with you?" They pull out PDAs and laptops. "OK, who has a mobile phone, digital camera, security car keys, or smart card?" Just as the Web has been connecting us in the large, we are being surrounded by computation in the small.

The changes in both the large and the small have included a movement from purely work-based computation to leisure, home use, entertainment, and fun. There are now workshops and books on user experience, domestic environments, humor, and even "funology" (Blythe et al. 2003). Adding to the areas listed in Section 3.3.2, HCI is now drawing on theories and practice from film criticism, philosophy, literary analysis, and performance arts (Wright, Finlay, and Light 2003). The multiple decision points on the Web (Section 3.2.3) mean that producing an engaging, enjoyable, possibly exciting user experience is more critical for Web-based applications than for conventional ones.

Increasingly, both the large and the small are being linked by wireless networks: WiFi, GSM, GPRS, Bluetooth; these technologies mean that every mobile phone is a Web access point and every smart card a sensor. This is changing the way we look at interfaces—there is no obvious link between an input device and the device it affects; there may be no explicit input at all, just sensed behavior and environment; as devices get smaller and cheaper they can be dedicated to places and purposes, and contrarily as devices are carried with us they become universal access points. Even the "computers" we carry with us day to day, in car keys or bank cards, are becoming connected. For example, some phones have RFID readers that can read the tags attached to clothes; or the camera built-in to the phone can be used to scan the barcodes on food or books. As everyday items become more intelligent, many talk about the "Internet of things" (ITU 2005). We are already used to interacting with broadband routers using Web interfaces, perhaps controlling our Hi-Fi using an iPhone, and Internet-connected refrigerators have been around for some years. In the future we may have Web interfaces to our car or can of coke. These issues are being investigated in the ubiquitous computing community and other areas in HCI, but there are certainly no definitive theoretical or practical models, methods, or theories for these new interactions (see Dix et al. 2004, chap. 18, for a more detailed discussion of some of these issues).

Small devices, both mobile phones and netbooks, are fast becoming the dominant means to access the Web. Indeed, for large parts of India, Africa, and China, the mobile phone is, for many, the first and only computer. HP Labs in India talk about the "Next Billion" customers who almost all will be phone-based Internet users (Hewlett Packard 2009). In the developed world the commoditization of smart phones such as the iPhone, and phone-based services such as Vodafone 360 is leading to a similar phenomenon where the Web is something in the hand not on the desk. For most persons working in Web design, the Web came after the desktop applications, and mobile Web came after the desktop Web. In the future the opposite will be the case.

HCI has long used cognitive science to help understand how users interact with particular devices and applications. However, with the Web, things are more complex: the whole of life, from early education to filling in a tax return, is increasingly being influenced or drawn into this global networked information structure. It is not just a technical artifact but part of the cultural backdrop of day-to-day life. Indeed, with social networking and user contributed content, for many the Web *is* culture and the Web *is* day-to-day life. This has far-reaching social implications and fundamental cognitive effects—we think differently and we are different people because of the Web.

If you look at the way we mathematically describe space, draw maps of countries or towns, and tell stories of the world, we can understand better how we understand and relate to physical space. This can be an important resource for designing electronic information spaces (Dix 2000b). However, as the information space becomes our first take on reality, it shapes our understanding of physical space. The authors were involved in the design of virtual Christmas crackers (Dix 2003b; vfridge limited 2003). Christmas crackers are a largely British phenomenon, paper tubes containing small gifts and a gunpowder strip that break apart with a bang. We get "fan mail" from people who are in far lands but are reminded of their childhood Christmases by using these virtual crackers. However, more strangely one woman first encountered the virtual crackers and then came to Britain; at Christmas she saw a real one and instantly knew what to expect.

Do you ever photocopy articles as a surrogate for reading them, or have a sense of accomplishment after an Internet search as you download, but do not read, PDF files? It is a truism of post-Internet society that it is not whether you know what you need to know but whether you know how to find out what you need to know. We use address books, diaries, and photo albums to aid our memory, but often because of this we grow lazy and use these instead of memory. As information becomes instantly globally available then this meta-cognitive knowledge, the about-information information, becomes increasingly important and it is not yet clear how this will change our cognitive engagement with the world (Dix, Howes, and Payne 2003; Mayer-Schönberger 2009). As interface designers, we need to be aware of this both because we design systems for this emerging cognitive demographic, but also because we are designing systems that shape it.

WEB LINKS AND FURTHER INFORMATION

Live links to many papers and sites mentioned in this chapter, together with any updates after publication, can be found at http://www.hcibook.com/alan/papers/web-hci-2011/.

ACKNOWLEDGMENTS

Work gathered for this chapter was originally supported by a number of sources including the UK EPSRC funded projects EQUATOR (http://www.equator.ac.uk) and DIRC (http://www.dirc.org.uk). Several illustrations are taken with permission from Human–Computer Interaction, 3rd ed. eds. A. Dix et al., Prentice Hall, 2004.

REFERENCES

Adar, E., J. Teevan, and S. Dumais. 2009. Resonance on the web: Web dynamics and revisitation patterns. In *Proceedings of the 27th International Conference on Human Factors in Computing Systems* (Boston, MA, April 04–09, 2009). CHI '09. New York: ACM Press, doi:10.1145/1518701.1518909.

Adomavicius, G., and A. Tuzhilin. 2005. Toward the next generation of recommender systems: a survey of the state-of-the-art and possible extension. *IEEE Transactions on Knowledge and Data Engineering* 17(6): 734–749. doi: 10.1109/TKDE.2005.99.

Apple Computer. 1996. Macintosh human interface guidelines. Apple Technical Library Series. Addison-Wesley Publishing Company, USA.

archive.org. 2010. Internet archive: wayback machine. http://web.archive.org/ (accessed Nov. 12, 2010).

Berners-Lee, T. 1998. Hypertext style: cool URIs don't change. W3C.org. http://www.w3.org/Provider/Style/URI (accessed Dec. 20, 2009).

Berners-Lee, T., J. Hendler, and O. Lassila. 2001. The semantic Web. *Scientific American Magazine*, May 17.

Blythe, M., K. Overbeeke, A. Monk, and P. Wright, eds. 2003. *Funology: From Usability to Enjoyment*. Dordrecht, Netherlands: Kluwer.

BrowserCam. 2009. BrowserCam—cross browser compatibility testing tools. http://www.browsercam.com/ (accessed March 6, 2003).

Caldwell, B. B., and G. C. Vanderheiden, this volume. Access to Web content by those with disabilities and others operating under constrained conditions. In *Handbook of Human Factors in Web Design*, 2nd ed., eds. K.-P. L. Vu and R. W. Proctor, 371–402. Boca Raton, FL: CRC Press.

Carroll, J., ed. 2000. *Making Use: Scenario-Based Design of Human–Computer Interactions*. Cambridge, MA: MIT Press.

Carroll, J. 2010. Conceptualizing a possible discipline of human–computer interaction. *Interacting with Computers* 22: 3–12, doi:10.1016/j.intcom.2009.11.008.

Catledge, L., and J. Pitkow. 1995. Characterizing browsing strategies in the World-Wide Web. In *Proceedings of the Third International World Wide Web Conference* (Darmstadt, Germany). http://www.igd.fhg.de/www/www95/papers/

Chen, C., and M. Czerwinski. 1998. *From Latent Semantics to Spatial Hypertext—An Integrated Approach, Hypertext '98*, 77–86. New York: ACM Press.

Clark, J., ed. 1999. XSL transformations (XSLT) version 1.0. W3C recommendation. http://www.w3.org/TR/xslt (accessed Nov. 16, 1999).

Clark, H. 1996. *Using Language*. Cambridge: Cambridge University Press.

Crabtree, A. 2003. *Designing Collaborative Systems: A Practical Guide to Ethnography*. New York: Springer.

DataPortability. 2010. The DataPortability Project. http://www.dataportability.org/ (accessed March 11, 2010).

Diaper, D., and N. Stanton, eds. 2004. *The Handbook of Task Analysis for Human–Computer Interaction*. Mahwah, NJ: Lawrence Erlbaum.

Dix, A. 1998. Hands across the screen—why scrollbars are on the right and other stories. *Interfaces* 37: 19–22. http://www.hcibook.com/alan/papers/scrollbar/.

Dix, A. 2000a. Designing a virtual fridge (poster). Computers and Fun 3, York, 13th December 2000. (Abstract in *Interfaces* 46: 10–11, Spring 2001.) http://www.vfridge.com/research/candf3/.

Dix, A. 2000b. Welsh Mathematician walks in cyberspace—the cartography of cyberspace (keynote). In *Proceedings of the Third International Conference on Collaborative Virtual Environments—CVE2000*, 3–7. New York: ACM Press. http://www.hcibook.com/alan/papers/CVE2000/.

Dix, A. 2001. Artefact + marketing = product. *Interfaces* 48: 20–21. http://www.hiraeth.com/alan/ebulletin/product-and-market/.

Dix, A. 2003a. In a strange land. http://www.hiraeth.com/alan/topics/cyberspace/.

Dix, A. 2003b. *Deconstructing Experience—Pulling Crackers Apart*. In *Funology: From Usability to Enjoyment*, eds. M. Blythe et al., 165–178. Dordrecht, Netherlands: Kluwer. http://www.hcibook.com/alan/papers/deconstruct2003/.

Dix, A. 2010. Human–computer interaction: a stable discipline, a nascent science, and the growth of the long tail. *Interacting with Computers* special issue, 22(1): 13–27, doi:10.1016/j.intcom.2009.11.007.

Dix, A., J. Finlay, G. Abowd, and R. Beale. 2004. *Human–Computer Interaction*, 3rd ed. Englewood Cliffs, NJ: Prentice Hall, http://www.hcibook.com/e3/.

Dix, A., A. Howes, and S. Payne. 2003. Post-web cognition: evolving knowledge strategies for global information environments. *International Journal of Web Engineering Technology* 1(1): 112–26. http://www.hcibook.com/alan/papers/post-web-cog-2003/.

Dix, A., S. Levialdi, and A. Malizia. 2006. Semantic Halo for Collaboration Tagging Systems. In *Proceedings of Workshops Held at the Fourth International Conference on Adaptive Hypermedia and Adaptive Web-Based Systems (AH2006)*, eds. S. Weibelzahl and A. Cristea, 514–521. Dublin: National College of Ireland.

Dix, A., and R. Mancini. 1997. Specifying history and backtracking mechanisms, In *Formal Methods in Human–Computer Interaction*, eds. P. Palanque and F. Paterno, 1–24. London: Springer-Verlag, http://www.hcibook.com/alan/papers/histchap97/.

DMOZ. 2003. Open Directory Project. http://www.dmoz.org.

Facebook. 2010. Facebook connect. http://developers.facebook.com/connect.php (accessed March 6, 2010).

Fielding, R. T. 2000. Architectural styles and the design of network-based software architectures. PhD diss. University of California, Irvine.

Fisher, A. 1990. *Labyrinth—Solving the Riddle of the Maze*. New York: Harmony Books.

Fitts, P., and M. Posner. 1967. *Human Performance*. Wokingham, UK: Wadsworth.

Flanagan, D., J. Farley, W. Crawford, and K. Magnusson. 1999. *Java Enterprise in a Nutshell*. O'Reilly.

Fuchs, T. 2010. Script.aculo.us—Web 2.0 javascript. http://script.aculo.us/ (accessed March 6, 2010).

Google. 2010. Gears: improving your Web browser. http://gears.google.com/ (accessed March 6, 2010).

Hewlett Packard. 2009. HPL India: innovations for the next billion customers. Hewlett Packard Development. http://www.hpl.hp.com/research/hpl_india_next_billion_customers/ (accessed Jan. 1, 2010).

Hughes, J., J. O'Brien, M. Rouncefield, I. Sommerville, and T. Rodden. 1995. Presenting ethnography in the requirements process. In *Proceedings of the IEEE Conference on Requirements Engineering, RE '95*, 27–34. New York: IEEE Press.

Hyde, L. 1983. *The Gift*. New York: Random House.

ITU. 2005. *ITU Internet Reports 2005: The Internet of Things*. Geneva, Switzerland: International Telecommunication Union.

jQueryUI. 2009. jQuery User Interface. http://jqueryui.com (accessed March 6, 2010).

Keen, A. 2007. *The Cult of the Amateur: How Today's Internet Is Killing Our Culture and Assaulting Our Economy.* Nicholas Brealey.

Krasner, G., and S. Pope. 1988. A cookbook for using the model-view-controller user interface paradigm in Smalltalk-80. *JOOP* 1(3).

Lamping, J., R. Rao, and P. Pirolli. 1995. A focus+context technique based on hyperbolic geometry for visualizing large hierarchies. In *Proceedings of the SIGCHI Conference on Human Factors in Computing Systems (CHI '95)*, eds. I. Katz, R. Mack, L. Marks, M. Rosson, and J. Nielsen, 401–408. New York, NY: ACM Press. doi:10.1145/223904.223956.

Larson, K., and M. Czerwinski. 1998. Web page design: implications of memory, structure and scent for information retrieval. In *Proceedings of CHI98, Human Factors in Computing Systems*, 25–32. New York, NY: ACM Press.

Lohr, C. 1998. Alexa Internet donates archive of the World Wide Web to Library of Congress. Alexa Internet Press Release, Oct. 13. http://www.alexa.com/company/inthenews/loc.html.

Long, J., and J. Dowell. 1989. Conceptions of the discipline of HCI: craft, applied science, and engineering. In *Proceedings of the Fifth Conference of the British Computer Society, Human–Computer Interaction Specialist Group on People and Computers V (Univ. of Nottingham)*, eds. A. Sutcliffe and L. Macaulay, 9–32. New York: Cambridge University Press.

Lynch, P., and S. Horton. 2002. *Web Style Guide: Basic Design Principles for Creating Web Sites*, 2nd ed. http://www.webstyleguide.com/.

MacKenzie, I. S. 2003. Motor behaviour models for human–computer interaction. In *HCI Models, Theories, and Frameworks: Toward an Multidisciplinary Science*, ed. J. Carroll. Morgan Kaufman.

Mauss, M. 1925. *The Gift: The Form and Reason for Exchange in Archaic Societies*. Originally entitled: Essai sur le don. Forme et raison de l'échange dans les sociétés archaïques.

McCormack, D. 2002. *Web 2.0: 2003-'08 AC (After Crash) The Resurgence of the Internet & E-Commerce.* Aspatore Books.

Mayer-Schönberger, V. 2009. *Delete: The Virtue of Forgetting in the Digital Age*. Princeton, NJ: Princeton University Press.

Miller, G. 1956. The magical number seven, plus or minus two: some limits on our capacity to process information. *Psychological Review* 63(2): 81–97.

Monk, A., P. Wright, J. Haber, and L. Davenport. 1993. *Improving Your Human Computer Interface: A Practical Approach*. Hemel Hempstead, UK: Prentice Hall International.

Monk, A. 2003. Common ground in electronically mediated communication: Clark's theory of language use. In HCI *Model, Theories and Frameworks: Towards a Multidisciplinary Science*, ed. J. Carroll, chap. 10, 263–289. Morgan Kaufmann.

Morville, P. 2005. *Ambient Findability: What We Find Changes Who We Become*. O'Reilly Media.

Nardi, B. 2010. *My Life as a Night Elf Priest: An Anthropological Account of World of Warcraft*. Cambridge, MA: MIT Press.

Nelson, T. 1981. *Literary Machines: The Report on, and of, Project Xanadu, Concerning Word Processing, Electronic Publishing, Hypertext, Thinkertoys, Tomorrow's Intellectual Revolution, and Certain other Topics Including Knowledge, Education and Freedom*. Sausalito, CA: Mindful Press.

Netcraft. 2009. November 2009 Web server survey. http://news.netcraft.com/archives/web_server_survey.html (accessed Dec. 19, 2009).

Nielsen, J. 1994. Heuristic evaluation. In *Usability Inspection Methods*. New York: John Wiley.

Nielsen, J. 1996. Why frames suck (most of the time). http://www.useit.com/alertbox/9612.html.

Open Software Foundation. 1995. *OSF/Motif Programmer's Guide, Revision 2*. Englewood Cliffs, NJ: Prentice Hall.

OpenID Foundation. 2009. What is OpenID? http://openid.net/get-an-openid/what-is-openid/ (accessed March 6, 2010).

O'Reilly, T. 2005. *What Is Web 2.0: Design Patterns and Business Models for the Next Generation of Software.* O'Reilly Media. http://oreilly.com/web2/archive/what-is-web-20.html.

Palanque, P., and F. Paternó, eds. 1997. *Formal Methods in Human–Computer Interaction*. London, Springer-Verlag.

Pfaff, P., and P. ten Hagen, eds. 1985. *Seeheim Workshop on User Interface Management Systems*. Berlin: Springer-Verlag.

Pixlr. 2009. Pixlr photo editing services. http://pixlr.com (accessed Jan. 5, 2010).

Rau, P.-L. P., T. Plocher, and Y.-Y. Choong, this volume. Cross-cultural Web design. In *Handbook of Human Factors in Web Design*, 2nd ed., eds. K.-P. L. Vu and R. W. Proctor, 677–698. Boca Raton, FL: CRC Press.

Raymond, E. 1999. *The Cathedral and the Bazaar*. O'Reilly.

Resig, J. 2010. Test swarm alpha open. http://ejohn.org/blog/test-swarm-alpha-open/ (accessed March 6, 2010).

Resnick, P., and H. Varian. 1997. Special Issue on Recommender Systems. *CACM* 40(3): 56–89.

Rouet, J.-F., C. Ros, G. Jégou, and S. Metta. 2003. Locating relevant categories in Web menus: effects of menu structure, aging and task complexity. *In Human-Centred Computing: Cognitive Social and Ergonomic Aspects*, vol. 3, eds. D. Harris et al. 547–551. Mahwah, NJ: Lawrence Erlbaum.

Sas, C., A. Dix, J. Hart, and S. Ronghui. 2009. Emotional experience on Facebook site. In *CHI '09: CHI '09 Extended Abstracts on Human Factors in Computing Systems* (Boston, MA, April 4–9, 2009), 4345–4350.

Sears, A., and J. Jacko, eds. 2008. *Human–Computer Interaction Handbook: Fundamentals, Evolving Technologies and Emerging Applications*, 2nd ed. New York: Taylor & Francis.

Seow, S. C. 2005. Information theoretic models of HCI: a comparison of the Hick-Hyman law and Fitts' law. *Human–Computer Interaction* 20: 315–352.

Shore, J. 2007. *The Art of Agile Development*. O'Reilly Media.

Silva, P., and A. Dix. 2007. Usability—not as we know it! In *Proceedings of BCS HCI 2007, People and Computers XXI*. BCS eWic.

Sommerville, I. 2001. *Software Engineering*, 6th ed. New York: Addison-Wesley. http://www.software-engin.com.

Strybel, T. Z, this volume. Task analysis methods and tools for developing Web applications. In *Handbook of Human Factors in Web Design*, 2nd ed., eds. K.-P. L. Vu and R. W. Proctor, 483–508. Boca Raton, FL: CRC Press.

Suchman, L. 1987. *Plans and Situated Actions: The Problem of Human–Machine Communication*. Cambridge, UK: Cambridge University Press.

Sun, C., and C. Ellis. 1998. Operational transformation in real-time group editors: Issues, algorithms, and achievements. *Proceedings of CSCW'98*, 59–68. New York: ACM Press.

Talis. 2010. Project Cenote. http://cenote.talis.com/.

Tauscher, L., and S. Greenberg. 1997. How people revisit Web pages: empirical findings and implications for the design of history systems. *International Journal of Human Computer Studies* 47(1): 399–406. http://www.cpsc.ucalgary.ca/grouplab/papers/1997/.

Taylor, F. 1911. *The Principles of Scientific Management*.

Thompson, J. 2003. What is Taylorism? http://instruct1.cit.cor nell.edu/courses/dea453_653/ideabook1/thompson_jones/ Taylorism.htm.

Thompson, A., and E. Kemp. 2009. Web 2.0: Extending the framework for heuristic evaluation. In *Proceedings of the 10th International Conference NZ Chapter of the Acm's Special Interest Group on Human–Computer Interaction CHINZ '09* (Auckland, New Zealand, July 6–7, 2009). New York: ACM Press. doi:/10.1145/1577782.1577788.

vfridge limited. 2003. Virtual Christmas crackers. http://www .vfridge.com/crackers/.

Vodafone. 2010. Vodafone 360. http://login.vodafone360.com/ (accessed Jan. 5, 2010).

Volk, F., F. Pappas, and H. Wang, this volume. User research: User-centered methods for the designing of web interfaces. In *Handbook of Human Factors in Web Design*, 2nd ed., eds. K.-P. L. Vu and R. W. Proctor, 417–438. Boca Raton, FL: CRC Press.

von Ahn, L., B. Maurer, C. McMillen, D. Abraham, and M. Blum. 2008. reCAPTCHA: Human-based character rec- ognition via Web security measures. *Science* 321(5895): 1465–1468.

Vu, K.-P. L., W. Zhu, and R. W. Proctor, this volume. Evaluating Web usability. In *Handbook of Human Factors in Web Design*, 2nd ed., eds. K.-P. L. Vu and R. W. Proctor, 439–460. Boca Raton, FL: CRC Press.

W3C. 2007. SOAP Version 1.2. dated 27th April 2007. http://dev .w3.org/html5/webdatabase/ (accessed March 6, 2010).

W3C. 2010a. Web SQL Database (Editors Draft), dated March 4, 2010. http://dev.w3.org/html5/webdatabase/.

W3C. 2010b. Web storage (Editors Draft), dated March 4, 2010. http://dev.w3.org/html5/webstorage/.

WebAIM. 2010. WAVE—Web Accessibility Evaluation Tool. http:// wave.webaim.org/ (accessed Jan. 5, 2010).

Wright, P., J. Finlay, and A. Light, eds. 2003. *HCI, The Arts and the Humanities*. York, UK: Hiraeth.http://www.hiraeth.com/conf/ HCI-arts-humanities-2003/.

Yahoo! 2010. Web site directory. http://www.yahoo.com (accessed Nov. 12, 2010).

Section II

Human Factors and Ergonomics

4 Physical Ergonomics and the Web

Michael J. Smith and Alvaro Taveira

CONTENTS

4.1 OVERVIEW

The purpose of this chapter is to present an overview of how the physical demands of interacting with Web devices must be considered by designers and users to reduce biomechanical and physiological strain on users. Problem areas will be identified using a work design framework and then possible solutions will be considered.

What does physical *ergonomics* have to do with the Web?

There are three basic areas where physical ergonomics can contribute to the effective design and use of the Web. These are (1) understanding the capabilities and limitations of people and using this knowledge to design the best possible Web technologies, (2) understanding how the design and use of the Web can lead to problems for people such as stress, musculoskeletal injury, and discomfort, and (3) understanding the environments of use of the Web and making accommodations to enhance Web access and use.

Human factors and ergonomics is concerned with fitting the environment, technology, and tasks to the capabilities, dimensions, and needs of people. The goal is to improve performance while enhancing comfort, health, and safety. Physical ergonomics deals with designing systems to minimize the physical loads (biomechanical and physiological) and forces on people to enhance comfort, to reduce pain, and to reduce sensory and musculoskeletal disturbances and disorders. There are many good resource books available to designers and technology users to help them identify and control problems due to physical ergonomics considerations. The National Research Council and the Institute of Medicine (2001) published *Musculoskeletal Disorders at the Workplace*. This book provides conceptual and pragmatic information for Web designers to understand how task demands can affect musculoskeletal discomfort and injury and provides guidance on how to prevent problems. Wickens et al. (2004) produced a text book, *An Introduction to Human Factors Engineering*, which describes basic human capabilities and performance considerations that can be used by Web designers to understand users' abilities, limitations, and needs when doing tasks. Lehto and Buck (2008) produced the text book *An Introduction to Human Factors and Ergonomics for Engineers*, which also addresses these critical considerations.

The two primary ideas in physical ergonomics are to define the factors that produce unwanted perceptual, motor, or cognitive strain, and then to design ways to eliminate or minimize the strain. Smith and Carayon (1995; Smith and Sainfort 1989; Carayon and Smith 2000) developed a model for understanding and controlling occupational stress and strain by focusing on five essential elements of a work system. These elements were the person(s), tasks, technologies, environmental features, and the organizational aspects of the work process (structural, administrative, supervision). The Smith and Carayon approach provides a systematic means for identifying salient features of a work system that produces loads and strains, and gives direction for its proper design. This Web handbook contains several chapters that

provide guidance on how to assess and design the various aspects of the "Web system" such as software interfaces and tasks. In addition, there are other handbooks that provide excellent information about technologies, interfaces, software, and ergonomic issues (Jacko and Sears 2003; Sears and Jacko 2008). Applying the technical information from these resources and using the approach provided by Smith and Carayon can lead to more effective and healthier Web system design and use.

Interaction with the Web is enabled through computers or related information technology (IT) devices. Personal IT devices now allow Web interaction while you walk, talk, sit, run, bike, swim, play tennis, or even when you are sleeping. Portability, universal access, enhanced usability, and expanded hardware and network capabilities have led to the use of computers in almost any conceivable activity or place. While this provides tremendous access to Web resources, it also introduces a host of ergonomic issues related to the design of interfaces and the environments in which interaction occurs. Several decades of research on human–computer interaction (HCI) have demonstrated that improper design of computer equipment, workstations, and environments of use can lead to user discomfort and even to serious health problems (sensory, musculoskeletal, and mental; e.g., see research by Grandjean 1979; Smith et al. 1981; NAS 1983; Smith 1984, 1987, 1997; Knave et al. 1985). Generally, poor design of human–technology interaction can lead to sensory disruption, musculoskeletal discomfort, pain, dysfunction, and psychological distress.

Most ergonomics experts agree that there are three primary considerations when using technology that can produce strain from biomechanical processes. The first and most important of these considerations is the frequency and duration of use of motor and sensory systems that may lead to fatigue, "overuse," and "wear and tear" of the perceptual-motor systems. The second consideration is the extent of loading and/or force that occurs on the perceptual-motor and musculoskeletal systems during the repetitive use. It is believed that the higher the demands on the perceptual-motor and musculoskeletal systems, then the greater the potential for fatigue, "overuse," and "wear and tear." The third consideration is the position or posture away from the neutral or "natural" positions of the perceptual-motor and musculoskeletal systems. For the musculoskeletal system it is believed that poor posture leads to the greater rubbing and abrading of tissues, constriction of blood flow, poor enervation, and/or the compression of tissues. For the perceptual-motor systems poor positioning may lead to intense attention and/or poor receptor acquisition of information that leads to fatigue and/or dysfunction. The interaction a person has with Web interfaces can create situations that produce user discomfort, pain, or fatigue when these three factors lead to strain. Chronic exposures of long durations may lead to injuries. Good design of technology, tasks, and rest breaks can reduce the effects of these factors on the extent of strain on the user.

The repeated and prolonged use of computer interfaces has been associated with perceptual discomfort, musculoskeletal discomfort, and to a more limited extent to musculoskeletal disorders (Grandjean 1979; Gerr, Marcus, and Monteilh 2004; Smith et al. 1981; Smith 1984, 1987, 1997; Stammerjohn, Smith, and Cohen 1981). These concerns are related to the general configuration of the workstation (work surface arrangement, chair design, placement of interfaces), to specific input devices (keyboard, mouse, touchpad), and to specific work regimens (levels of repetition, duration, speed, forces, and posture). The growing development and use of ubiquitous computing through personal digital assistants (PDAs), cell phones, and other "on-board" devices has led to Web access in a variety of new settings, which poses new challenges to fitting the tasks, environments, and technologies to peoples' capabilities and limitations.

We will explore the potential ergonomic problems when using the Web by looking at some of the components of the Smith and Carayon (1995; Smith and Sainfort, 1989) approach. We will start by looking at the person (user) component of this model.

4.2 PEOPLE CONSIDERATIONS

Wickens et al. (2004) and Lehto and Buck (2008) describe the process of how people function in the environments they live and work in. The basic concept is that people gather information from the environment using their senses; they perceive, remember, and process this information in the brain; and they respond to this information with their response mechanisms such as speaking, smiling, gesturing, or doing (walking, typing, and nodding). At each stage of this process the ability of people to perform well is influenced by the technology they use, their surroundings (environment), the directions they are given (organization), and the demands of the activities (tasks) they are doing. Physical ergonomics is interested in understanding the capacities and limitations of people so that the environment, technologies, and activities can be designed to enhance performance while reducing physical stress and strain. We will start our understanding of people's capabilities by looking at their sensory systems.

4.2.1 SENSORY ISSUES

The *Encyclopedia of Psychology* (Kazdin 2000) has several chapters that deal with the senses of vision, hearing, and touch (kinesthesis), and it is an excellent resource for basic information about how the senses work.

Vision is the primary sense used by sighted people to gather information from the environment and to engage in the activities. Vision has an advantage over other senses because the environment very often provides a continuous image of the information being sought and a trace of information previously encountered. Typically, what you see remains in place in the environment for a period of time (seconds, minutes, hours, days). This allows the person to refresh her/his understanding or memory of the information. The semipermanent characteristic of the traces tends to reduce the "load" placed on the "attention" function and the "memory storage"

mechanisms of information processing. In other words, it is an easier medium to use from an information processing perspective. Information processing is discussed in more detail in another chapter dealing with cognitive ergonomics (Harvey et al., this volume). The essential role of visual perception on human information processing and behavior has led IT designers to emphasize visual displays over other forms of stimuli.

The *Handbook of Visual Ergonomics* (Anshel 2005) and *Basic Vision: An Introduction to Visual Perception* (Snowden, Thompson, and Troscianko 2006) provide information on how the visual system works and the effects of vision overuse on human performance and health. Decades of research dealing with people reading from computer displays have shown that overusing the eyes leads to eye strain, diminished visual function, and poorer perceptual-motor performance (Grandjean and Vigliani 1980; Smith 1987, 1997).

Research has shown that the layout of Web pages, text characteristics, and color coding have influence on computer user performance and can affect older users differently than younger users (Grahame et al. 2004; Sheedy et al. 2005; Van Laar and Deshe 2007). In addition, workstation design characteristics such as the viewing distance and display height can affect visual discomfort and computer user's performance (Burgess-Limerick, Mons-Williams, and Coppard 2000; Jainta and Jaschinski 2002; Rempel et al. 2007).

A second major source of information gathering used by people is hearing. We interact socially through speech. Speech can provide "contextual" and "emotional" content to information through the modulation of the voice. Speech has a prominent information processing aspect because its comprehension derives from linguistic, social, emotional, and contextual elements of the message that are conveyed in it. Speech has a weakness when compared to vision because sounds do not leave traces in the environment that can be reviewed, and new sounds distract "attention" away from prior sounds.

Speech and auditory interfaces are gaining in popularity in many technology applications. Commarford et al. (2008) tested current design guidelines regarding speech-enabled interactive voice interfaces that call for shorter, more in-depth menus because of concerns over auditory memory load. Their research found that broad-structure menus provided better performance in an e-mail system and greater user satisfaction than shorter deep-structure menus. Thus, the current guidelines stressing deep menus may not be the best for some applications such as e-mail.

The environment in which speech interface use occurs is important. Many speech interfaces are used in environments where there are multiple talkers simultaneously speaking in separate and group conversations. Bolia, Nelson, and Morley (2001) found that as the number of simultaneous speakers increased in an interacting group, the spatial arrangement of the talkers and the target talker's speech hemi-field affected the extent of intelligibility for the listener. Rudmann, McCarley, and Kramer (2003) found that the greater the number of distracting simultaneous speakers, the poorer the message intelligibility of the primary message. However, providing a video display of the target speaker mitigated the effect of more simultaneous distracting speakers. Kilgore (2009) found that visual display identification of a primary speaker's location enhanced auditory accuracy and response time in multi-talker environments.

Roring, Hines, and Charness (2007) examined the effects of the speed of natural or synthetic speech on intelligibility and performance in young, middle-aged, and older users. They found an interaction of age, context of speech, voice type, and speech rate. The effects were due primarily to reduced hearing acuity with aging. The context of speech improved performance for all ages for the natural speech condition. Context also improved performance for synthetic speech, but not as much for older participants. Slowing speech impaired performance in all groups.

A third sensory modality that has some potential for use in IT applications is the tactile sense. It is a very "fast" sense for receiving stimulation, but it is difficult to provide complex, specific content, and context via this sense in an easy manner. Tactile sensation requires "translation" of the input into language, words, and concepts, which is a very slow process when compared to seeing or hearing.

Jones and Sarter (2008) have provided guidance for designing and using tactile displays. Hopp et al. (2005) demonstrated the effectiveness of employing tactile displays to shift user's attention from ongoing visual tasks to a different visual task. This type of application may be useful for users in complex multi-talking operations and in smart-phone technologies.

Many applications use a combination of vision, speech/audition, and tactile/spatial sensing. Generally, the research in this area indicates that the context of the application has a substantial influence on the effectiveness of using multimodal interfaces (Dowell and Shmuell 2008; Ferris and Sarter 2008).

In summary, vision is the primary means for engaging with the environment and tasks. Audition provides some important benefits when there is a need to deal with context and emotional content, as well as for providing additional feedback and redundancy. However, listening places greater load on information processing at the attention and memory stages than vision. Tactile sense is useful for alerting a user or when highlighting information but is limited in providing content and context. So what do we need to know about the Web users to be able to apply what we know about the senses?

People have a wide range of sensory capabilities, and therefore Web interfaces need to accommodate this range as much as possible and feasible. For example, visual function decreases with age, and by the time a person is 40 years old her/his ability to bring into focus images that are in the near visual field has diminished substantially for most people. This is a particular problem when the lighting conditions are poor because it severely degrades the user's ability to perceive visual information. When the objects and characters presented do not have adequate luminance, contrast, definition, size, and shape, vision is degraded especially as users

get older. This effect is exaggerated in very old users and may mean that designers have to provide redundant presentation modes (visual and auditory) for elderly users.

It has long been recognized that designers need to make technologies that can be "assistive" to the senses when these capabilities are reduced (TRACE 2009; ICS FORTH 2009). For example, having technology able to magnify the characters presented on a visual display or the loudness of an auditory message helps people with reduced visual or hearing capabilities. Likewise, users need to recognize their sensory limitations and select technologies that provide sensory assistance. In addition, users need to understand how poor environmental conditions can seriously limit their sensory capabilities, and either change or move to more suitable environments when necessary and/or possible.

Let us consider an example of the interactions between human limitations, technology characteristics, and environmental conditions as they affect a user's Web access and performance. Mr. Smith is traveling from Chicago to New York via an airplane. He keeps in touch with his office with a communication device that has capabilities for Internet connections. Mr. Smith can read/listen to e-mails, send messages, download files, search the Internet, and interact with colleagues or customers with his device. Mr. Smith is 50 years old and has presbyopia, which means he has a hard time reading the newspaper without glasses or some magnification of the images he looks at. Mr. Smith is sitting at O'Hare International Airport in Chicago connected to his office over the Web. The lighting is very bright, and there are many people in the area creating substantial background noise. It is very likely that Mr. Smith will have problems seeing a message on his screen because of the environmental glare and especially if he forgot to bring his reading glasses. If Mr. Smith decides to interact with his office using voice communications, he may have a hard time hearing and comprehending his messages because of the high level of environmental noise as a result of the many people near him talking on cell phones and the general noise of a busy airport.

Mr. Smith needs to move to a better visual environment with lower illumination, less glare, and less noise. He needs some personal space (privacy) so he can concentrate on his messages. He may also need a bigger screen with larger characters if he forgot his reading glasses. A fellow traveler informs Mr. Smith of the business kiosks just around the corner, and he goes there. The kiosks provide computer access to the Internet and control the environmental problems of bright illumination, glare, and noise. His sensory problems are solved by lower luminance lighting, a large screen with magnification capability, and no background noise. But he has to pay for this kiosk service, whereas he has already paid for his cell phone and his connection to the Internet. Because he has already paid for his technology and Internet connection, Mr. Smith may just struggle with his technology and the poor environment to save money rather than using the kiosk. This could lead to poor reception of his messages (on both ends), eye strain, intense attention load, and psychological distress.

While cell phones and other devices can provide "on-person" connectivity to the Internet and are widely used by the general population, the sensory-motor ease of use of these devices is low because of their small size, which produces a poor interaction with peoples' sensory-motor capabilities. Technology's small size makes it difficult to provide viewing and acoustic characteristics that can accommodate people with sensory limitations, and cell phones do not function very well in disruptive environments. Sure, their small size makes carrying these technologies easier and more convenient, and users like this convenience. Even so, designers and manufacturers need to recognize peoples' sensory-motor limitations and the adverse environmental conditions where these technologies are used and then make products that will assist the user when these limitations and adverse environmental conditions occur.

4.2.2 MOTOR SKILLS ISSUES

In a similar fashion to sensory capabilities the perceptual-motor skills, strength, and stamina of people diminish with age. The amount of skill reduction varies from person to person based on their constitution, health, natural ability, prior experience, current conditioning, and practice using devices. In general, peoples' perceptual-motor skills degrade much more slowly than their sensory capabilities. Most experienced users are able to maintain their computer operation dexterity well into old age, whereas by then their sensory capabilities can be substantially reduced. However, as we age, fine motor skills may become degraded, which affects our ability to make small, precise movements needed to control interfaces. This has implications for the design of technology, particularly interfaces for use by the elderly.

For example, touching a small icon on a screen becomes more difficult in old age. Carrying out tracking movements such as dragging an icon across a screen to a specific location becomes even more challenging to do as we get older. Prolonged, fast tracking (several seconds to minutes) quickly creates fatigue in the fingers and arms of older people. Discrete movements such as clicking or pushing a button hold up better into old age, but with reductions in speed and less accuracy. Actions that require sustained static contractions over time such as depressing a button for several seconds can become difficult if not impossible for older adults, especially as the force level and/or the time to hold increases. Designers need to understand the changes in peoples' skills and strength as they age and design technologies to assist those persons with changing capabilities.

4.2.3 FEEDBACK AND COMPLIANCE ISSUES

Let us look at another example with Mr. Smith. He interfaces with his cell phone using voice commands and touching icons on a screen. He "touches" the icons, but there is no acoustic or tactile feedback when he "presses" the buttons. This raises an important consideration because people perform better when they receive "reactive" feedback confirming the results of their action. In the absence of proper feedback, people have a tendency to push the screen button multiple times and harder than necessary to operate the button (Dennerlein and Yang 2001).

If Mr. Smith is using a pointing stick and a PDA, it is likely he will hold the stick with a "pinch grip" when touching the screen buttons. The pushing force is often increased when using the pointing stick to touch a button on a flat screen that does not give reactive feedback. Hard gripping of the pointing stick and harder than necessary "pushing" on the screen can lead to faster and greater levels of fatigue in the fingers, hand, and forearm when operating the pointing stick.

An interface that does not provide the traditional compliances of force (including feedback), space (movement, position), and time (instantaneous versus delayed) of actions with their effects on the control and display will cause user errors, reduce user performance in terms of speed and accuracy, increase fatigue, and lead the user to produce more force than necessary to operate the device. These noncompliances can lead to musculoskeletal discomfort, pain, and even injuries.

4.2.4 MUSCULOSKELETAL ISSUES

The musculoskeletal system operates best when there is movement of the muscles that leads to adequate blood flow. On the one hand, prolonged lack of muscle movement due to static postures can be detrimental for your musculoskeletal system, for example, if you are sitting and using a computer keyboard for more than one hour. On the other hand, too many frequent movements for a prolonged time can also be problematic for the musculoskeletal system in terms of fatigue, strain, and possibly injury. Basic risk factors that influence the probability of developing musculoskeletal disorders include repetitive motions for a long duration, excessive force, and poor posture of joints and appendages.

Repetitive motion can be exemplified by the daily use of computer keyboard and mouse interfaces for several hours each day at a high rate of typing and pointing. The likely result would be localized fatigue, discomfort, and pain. If this continues over years, there is the possibility for the development of musculoskeletal disorder. External loads on the muscles, tendons, ligaments, and joints are required to operate most controls. Higher external loads require greater internal force from the person, and excessive loads and forces can lead to fatigue and discomfort, and possibly injuries. Although force requirements to operate Web device controls are typically quite low, the forces actually applied by the users can vary substantially. Good interface design will reduce this variability and the level of force necessary to use interfaces.

The spine, joints, and appendages have "neutral" positions in which there is less internal force generated. As they are moved away from these "neutral" positions, the potential for greater force arises, especially when approaching their anatomic limits. For example, wrist postures when using a computer keyboard or a pointing device influence the internal force on the wrists and hands.

Prolonged static positions of the neck, back, arms, and legs require internal forces to maintain postures. This can lead to fatigue and diminished blood flow to muscles. For example, a fixed head and back position while surfing the Web could affect the level of muscle tension in the neck, shoulders, arms, and back.

4.2.5 Personal Susceptibility

There are some personal characteristics that can affect the susceptibility for musculoskeletal disorder. Gender is correlated with the incidence of some musculoskeletal disorders, and women have a higher incidence of carpal tunnel syndrome while men have a higher incidence of low back pain.

People who are older, obese, diabetic, and/or smoke also show increased propensity for neuromotor and musculoskeletal disorders. Older age has been associated with the slowing of nerve responses and poorer musculoskeletal conditioning. Chronic cigarette smoking has been associated with increased risk of carpal tunnel syndrome, arthritis, and low back pain. If you are a man who is obese, smokes cigarettes, and spends hours at a time surfing the Web, it is highly probable that you will have lower back pain.

In the 1980s, software designers developed software that warned users that they needed to stop using the computer and directed them to get up and move around after a prolonged period of interacting with the computer. The purpose was to relieve the physical, physiological, and psychological strain of intense and prolonged computer use. The software should also have warned the users to not smoke or consume calories during these breaks from computer use. Unfortunately, many computer workers used these breaks to smoke and/or eat or drink high calorie comfort food. This was a poor strategy for preventing low back pain.

A group of Web users who has received increased attention are children and adolescents. As computer use among this group has increased, the extent of musculoskeletal discomfort and disorders has become more prevalent. Jones and Orr (1998) examined 382 high school students in business education classes and found that 28, 40, and 41% reported hand discomfort, neck/back pain, and body pain, respectively, after prolonged computer use. Self-reported symptoms of carpal tunnel syndrome were found in 4% of the sample. Factors that increased reporting of symptoms were duration of computer use and type of environment (school, home, and work) of use. With the dramatic increase in typing in nonstandard keyboards (e.g., touch screens and miniaturized keyboards) associated with texting, and Web-based social networking among young users, an increase in musculoskeletal discomfort, pain, and disorders can be expected.

Some limited study findings suggest that musculoskeletal disorder morbidity may be attributed to a number of personal conditions including previous direct trauma to nerves, cartilage, muscles, and discs; wrist size and shape; and fractures that have not properly healed and have developed into arthritis or spurring. The findings suggest that technology and task

designers need to develop solutions that can accommodate personal characteristics to limit potential musculoskeletal problems.

4.3 ENVIRONMENTAL ISSUES AND PERCEPTUAL DISRUPTION

The focus of this section is on the role of the physical environment in human-computer interaction. Guidelines to reduce disruption caused by the environment are also provided. The increasingly widespread Web access through a number of portable or nondesktop (public access) fixed devices makes the fitting of the environment to the task and user's needs tremendously challenging. In this section, emphasis is given to conventional office environments because such environments are easier to control and still in high use. However, more and more Web interaction is taking place when users are in public environments, so some attention is also given to what users can do in these environments to improve their interaction with the Web.

4.3.1 Lighting and Glare

Lighting systems that meet the needs of users and the requirements of the job are vital for high performance and user comfort in visual tasks. Computer displays are prone to glare, which typically occurs in the form of light reflections on the screen or from unprotected light sources within the user's visual field shining directly on the screen. Excessive brightness contrasts within the user's visual field has been shown to lead to early onset of visual fatigue and discomfort (Smith 1984; Smith, Carayon, and Cohen 2008). Research has shown that the occurrence of glare depends primarily on the display positioning relative to the sources of light in the environment. Generally, excessive illumination leads to increased screen and environmental glare and poorer luminance contrast. Several studies have shown that screen and/or working surface glare are problematic for visual disturbances (see Smith, Carayon, and Cohen 2008).

The surface of the screen reflects light and images. The luminance of the reflections can decrease the contrast of characters on the screen and thus disturbs their legibility. Reflections can be so strong that they produce glare on the screen or in the general viewing environment. Screens may reflect environmental images in bright environments; for instance, the user's image may be reflected on the screen. The alignment of lighting source in relation to the computer workstation has been shown to influence reflections on the computer screen (Smith, Carayon, and Cohen 2008). Bright reflections on a device's screen are often a principal complaint of device users.

The readability of the screen is affected by the differences in luminance contrast in the work area. The level of illumination affects the extent of reflections from working surfaces and from the screen surface. Mismatches in these characteristics and the nature of the job tasks have been postulated to cause visual fatigue, strain, and discomfort. For instance, if the luminance on the working surfaces is much higher than the screen luminance, this can lead to visual fatigue (see Smith et al. Smith, Carayon, and Cohen 2008).

4.3.2 Reducing Glare and Improving Lighting

Basic recommendations for display positioning are to avoid having a user face sources of light such as an uncovered window, luminaries in direct viewing space, or light sources directly behind the user. Displays should be parallel to light sources. Ways to control brightness from light sources may be required to block direct glare (e.g., blinds, drapes, shades, screen covers, lighting louvers). Having proper luminance throughout the environment is important for good visual performance. Light fixtures that prevent direct view of the bulb and have large light emitting areas are preferred.

The illuminance in workplaces that primarily use computer screens should not be as high as in workplaces that use hardcopy on a regular basis. Lower levels of illumination will provide better computer screen image quality and reduced screen glare. Illuminance in the range of 500 lux measured on the horizontal working surface (not the computer screen) is normally preferable in screen intensive tasks. Higher illumination levels are necessary to read hardcopy (700 lux). Illumination from high-intensity luminance sources in the peripheral field of view (such as windows) should be controlled with blinds or shades. To reduce environmental glare, the luminance ratio within the user's near field of vision should be approximately 1:3 and approximately 1:10 within the far field of vision. For example, the working surfaces should not be more than 3 times brighter than the screen surface or the far window should not be more than 10 times brighter than the screen surface.

For luminance on the screen itself, the character-to-screen background luminance contrast ratio should be at least 7:1. That means the characters on the screen need to be at least 7 times brighter than the screen background. To give the best readability for each user, it is important to provide screens with adjustments for character contrast and brightness. These adjustments should have controls that are obvious and easily accessible from the normal working position.

For portable Web appliances the user generally has some discretion for moving the device around and positioning the display to avoid glare. However, these devices are most often used in environments where the user has little or no control over the illumination sources. This can make it very difficult to avoid glare. Glare control in these situations can rely on the display features and/or good positioning of the screen to avoid bright lighting. Possible technology solutions include luminance (brightness) adjustability of the display, the use of glare reducing films, and high-resolution characters with high contrast. The loss of brightness and contrast on the display are potential drawbacks of antiglare films.

4.3.3 Noise

Noise is any undesirable sound in the context of the activity being undertaken that interferes with communication. This is typically related to the physical aspects of the sound such as its loudness, pitch, duration, or suddenness. In some circumstances the information content of the sound, such as distracting conversations, make it undesirable. Research has demonstrated that attention, concentration, and intellectual activities can be disrupted by noise (Broadbent 1958). These findings indicate that noisy environments reduce efficiency, require more time to complete tasks, and cause an increased number of errors. Activities involving learning and sustained concentration are particularly vulnerable to loud, unexpected, high-pitched sounds that are common in public environments. Portable and fixed public access Web devices are exposed to all sorts of environments where little can be done to control such noise. In such environments when Web access under these conditions is brief and does not involve complex tasks, the amount of disruption and annoyance experienced by users may be acceptable. If the Web access requires sustained attention and concentration, then users will need to use ear pieces that control undesirable sounds or move to areas where the noise level is not disruptive.

Web-access devices located in offices or homes typically suffer less acoustic disruption because noise levels are lower and unexpectedly loud sounds are rare. However, in large, open plan offices, conversation often represents a significant source of distraction for users, and acoustic abatement through the use of partitions and sound-absorbing materials, as well as conversation masking strategies, are recommended.

4.3.4 Other Distractions

Privacy requirements include both visual and acoustical control of the environmental exposures. Visual control prevents physical intrusions, contributes to confidential/private conversations, and prevents the individual from feeling constantly watched. Acoustical control prevents distracting and unwanted noise (from machines or conversations) and permits speech privacy. While certain acoustical methods and materials such as free standing panels are used to control general office noise level, they can also be used for privacy. In public environments the user must find an area that provides privacy from visual and auditory intrusion to achieve privacy. Again, in public places this may be hard, but it is often possible to find a small secluded area free of intrusions.

4.4 TECHNOLOGY ISSUES AND USER COMFORT

4.4.1 Wearable Devices (On-Board, On-Body, and On-Person)

Wearable devices represent a fairly new attempt to merge seamlessly the user's physical and informational environments. As appropriately described by their name, these usually small digital devices are worn or carried by users and are available for constant computer and Web access (Gemperle et al. 1998). They allow users to interact with a device to the Web while performing other activities such as walking (Mann 1998). Wearable devices are useful for field activities like airline or other vehicle maintenance, rural/home health care, to on-the-job training, disaster assessment workers, when employees are traveling and even when recreating.

A wearable system should provide an interface that is unobtrusive and allows users to focus on the task at hand with no disruption from input or display devices. Traditional human–computer interfaces such as a keyboard, mouse, and screen may be inadequate for wearable systems because they require a physically constrained relationship between the user and the device. Some of the challenges of wearable computers include the nature of the display (visual, auditory, tactile, combination), how and where to locate the display, and the nature of input devices. User comfort, in particular, plays a critical role in the acceptance and overall performance of these devices (Knight and Barber 2005). Interfaces being used for wearable devices include head-mounted see-through and retinal projection displays, ear-mounted auditory displays, wrist- and neck-mounted devices, tactile displays (Gemperle, Ota, and Siewiorek 2001), and speech input devices. Wearable computers and their interfaces are critical for the feasibility of both virtual reality and augmented reality concepts.

The term virtual reality (VR), in its original intent, refers to a situation where the user becomes fully immersed in an artificial, three-dimensional world that is completely generated by a computer. Head-mounted visual displays are the most common interfaces utilized, but a variety of input and display devices relying on auditory (e.g., directional sound, voice recognition) and haptic stimuli (e.g., tactile and force feedback devices) are being used. The overall effectiveness of VR interactions is dependent on a number of factors related to the task, user, and human sensory and motor characteristics (Stanney, Mourand, and Kennedy 1998). Exposure to virtual environments has been associated to health and safety issues including musculoskeletal strain, dizziness, nausea, and physiological aftereffects (Stanney et al. 2003). The use of head-mounted displays especially has been shown to affect neck postures (Knight and Barber 2007) and has been linked to motion sickness or cybersickness (Mehri et al. 2007). Aftereffects of VR exposure are temporary maladaptive compensations resulting from changes in hand-eye coordination, vision, and posture, which are transferred to the real world (Champney et al. 2007). These aftereffects may include significant changes in balance and motor control, which can last for several hours and be dangerous as the individual resumes his or her real-world activities. Although virtual reality presents tremendous potential for an enriched Web interaction, significant technical (i.e., hardware and software), economic, and user compatibility (e.g., discrepancies between seen and felt body postures) obstacles remain.

Augmented reality (AR) refers to the enhancement of the real world by superimposing information onto it and creating a "mixed reality" environment (Azuma 1997, 2001). Developments in location-aware computing have made possible real time connections between electronic data and actual physical locations, thus enabling the enrichment of the real world with a layer of (virtual) information (Hollerer et al. 1999). AR systems commonly rely on head-worn, hand-held, or projection displays that overlay graphics and sound on a person's naturally occurring sight and hearing (Barfield and Caudell 2001). Visual input is usually provided by see-through devices relying either on optical (i.e., information projected on partially reflective eye glasses) or video (i.e., closed view system where real and virtual world views are merged) approaches (Milgram et al. 1994). Both optical and video see-through technologies present tradeoffs and situation- and task-specific aspects define the proper choice. On the one hand, Optical systems tend to be simpler, providing an instantaneous and mostly unimpeded view of the real world, thus ensuring synchronization between visual and proprioception information. Video systems, on the other hand, allow for a superior ability to merge real and virtual images simultaneously, but without a direct view of the real world (Rolland and Fuchs 2000). Flip-up devices may, under some circumstances, attenuate the issues associated with a lack of direct real world view in video see-through systems.

Although significant technical and human factors issues remain, including the possibility of information overload, research and commercial uses of AR are expanding apace. AR developments have been observed in multiple areas including health care (e.g., medical visualization), training and education, maintenance, aircraft manufacturing, driving assistance, entertainment, defense, construction, architecture, and others.

4.4.2 Ubiquitous Computing

Ubiquitous (omnipresent) computing allows the computer to adapt to everyday life to blend in and have an inconspicuous place in the background. The first attempts toward ubiquitous computing included devices such as tabs, pads, and boards (Weiser 1993) but in the future are expected to be significantly different from the common single screen-based interface we use today. Users will interact with a number of devices that are distributed and interconnected in the environment (Dey, Ljungstrand, and Schmidt 2001) in what is aptly called an "Internet of things" (Gershenfeld, Krikorian, and Cohen 2004). Computers and interfaces have a wide range of configurations including personal and mobile ones and others that will be part of the shared physical environment of a residence or a public space. These almost invisible devices will be aware and subordinate to users' needs. They will provide users with constant helpful indications about their surroundings.

Dey, Ljungstrand, and Schmidt (2001) assert that the essential goal of this approach is to make interfaces virtually disappear into the environment, being noticed only when needed. The realization of ubiquitous computing will require technologies that are capable of sensing environmental conditions, location of users, tasks being performed, physiological and emotional states, schedules, etc. Ideally, it will help free users from an enormous number of unnecessary chores.

Streitz (2008, p. 47) describes this "ambient intelligence" as a human centered approach where the computer becomes almost invisible but its functionality is ubiquitously available. In Streitz's words "the world around us is the interface to information." This notion implies that the environment will be infused with technology that allows the user to be continuously connected to the environment and beyond (the Web). It may even adapt itself to the specific needs of a particular user and accommodate her/his personality, desires, sensory and motor deficits, and enhance the environment for better Web communication with more effective Web use.

4.4.3 Web Interface Devices

This section addresses devices that allow user access and interaction with Web content. Currently, interaction between users and the Web occurs through a physical interface provided by a computer or other digital device. Human–computer interface technology is evolving rapidly and innovations are introduced into the marketplace at a pace that challenges evaluative research capacity. Someday "thought" interfaces may be widely available.

When designing or selecting physical interfaces for Web access one needs to focus on the users' anatomic, physiological, safety, and cognitive needs and capabilities. Interfaces must be flexible enough to satisfy the preferences of an enormous and increasingly diverse user population. Interface operation should be intuitive to users to keep training to a minimum. A well-conceived interface should allow quick interaction, be accurate, be convenient, be appropriate for environmental conditions, feel natural to users and not impose unacceptable physical, physiological, cognitive, or psychological loads. As emphasized by Hutchins, Hollan, and Norman (1985), interface design should minimize the gap between the users' intentions and the actions necessary to communicate them to the computer. Superior interfaces make the connection between users' intentions and the computer/ Web closer and more natural (Hincley 2003, 2008).

The "naturalness" of interfaces is often associated with the concept of direct manipulation (Shneiderman 1987) where familiar human behaviors such as pointing, grabbing, and dragging objects are used as significant analogies for human–computer interaction. Interfaces using natural behavior patterns and following popular stereotypes can reduce training needs and human error. Naturalness seems to be increased when the user can interact with the computer by touching and "feeling" virtual representations of the task at hand. As interface development proceeds, designers must be conscious of the inescapable asymmetry between user input and computer output. The former clearly characterized by his/her limited and narrow bandwidth and the latter by its large output capability (Hincley 2003, 2008). Alternatives have been examined to enrich user input through the use of

speech, voice emotion, gestures, gaze, facial expression, and direct brain-computer interface (Shackel 2000; Moore et al. 2001; Jaimes and Sebe 2007).

The basic function of an input device is to sense physical properties (e.g., behaviors, actions, thoughts) of the user and convert them into a predefined input to control the system process. These interaction devices allow users to perform a variety of tasks such as pointing, selecting, entering, or navigating through Web content. There is a growing number of input devices commercially available, some of them addressing the needs of distinct user populations (e.g., Taveira and Choi 2009). In this chapter a limited set of the most widely used input devices are addressed to illustrate ergonomic issues and principles. An extensive inventory of input devices can be found in Buxton (2009).

4.4.3.1 Keyboards

The keyboard is a primary interface still in wide use on many devices that interact with the Web. When operating keyboards users have a tendency to hold their hands and forearms in an anatomically awkward position (Simoneau, Marklin, and Monroe 1999; Keir and Wells 2002; Rempel et al. 2007). The linear layout leads to ulnar deviation and pronation of the forearm. It also produces wrist extension in response to the typical upward slope of keyboards and upper arm and shoulder abduction to compensate for forearm pronation (Rose 1991; Swanson et al. 1997). Today, many people are using cell phones and related handheld devices with very small keyboards to interact with the Web. They may pose many of the same postural problems and also new ones due to their small size and the design of the switches (Hsiao et al. 2009).

Although the evidence associating typing and musculoskeletal disorders (MSDs) of the upper limbs is somewhat mixed (Gerr, Marcus, and Monteilh 2004), the postural aspects of keyboarding have received sustained research attention and have motivated the development of multiple alternative designs. A few parameters to improve typing performance through keyboard redesign were suggested more than 30 years ago (Malt 1977):

- Balance the load between the hands with some allowance for right-hand dominance
- Balance the load between the fingers with allowance for individual finger capacity
- Reduce the finger travel distance with most used keys placed directly under digits
- Minimize awkward finger movement (e.g., avoiding use of same finger twice in succession)
- Increase the frequency of use of the fastest fingers
- Avoid long one-hand sequences

These recommendations are not relevant for the use of keyboards (keypads) on cell phones, PDAs, and other handheld devices because of the predominant use of only one hand and typically one finger of one hand to type on (point at) the keyboard.

4.4.3.2 Split Keyboards

Split keyboard design has been suggested to straighten the user's wrists when using a standard computer keyboard. This is usually accomplished in two ways: by increasing the distance between the right and left sides of the keyboard or by rotating each half of the keyboard so that each half is aligned with the forearm. Some alternative keyboards combine these two approaches. Split keyboards have been shown to promote a more neutral wrist posture (Marklin, Simoneau, and Monroe 1999; Nakaseko et al. 1985; Rempel et al. 2007; Smith et al. 1998; Tittiranonda et al. 1999) and to reduce muscle load in the wrist-forearm area (Gerard et al. 1991). Somerich (1994) showed that reducing the ulnar deviation of the wrist by means of using a split keyboard reduces carpal tunnel pressure. Yet, available research does not provide conclusive evidence that alternative keyboards reduce the risk of user discomfort or injury (Smith 1998; Swanson et al. 1997). Typing speed is generally slower on split keyboards, and the adaptation to the new motor skills required can be problematic (Smith et al. 1998).

4.4.4 Pointing Devices

Pointing devices allow the user to control cursor location and to select, activate, and drag items on display. Web interaction, in particular, involves frequent pointing and selecting tasks, commonly surpassing keyboard use. Important concerns relating to the usage of pointing devices are the prolonged static and constrained postures of the back and shoulders and frequent wrist motions and poor postures. These postures result from aspects pertaining to the device design and operational characteristics, as well as the workstation configuration, and the duration and pace of the tasks. Before reviewing some of the most popular pointing devices it is important to define two basic properties: (1) control-display (C-D) gain and (2) absolute versus relative positioning.

C-D gain is a ratio between the displacement or motion applied on a control, such as a mouse or a joystick, and the amount of movement shown in a displayed tracking symbol, such as a cursor on a screen. Usually, linear C-D relationships are used because it feels natural to the user, but nonlinear control-display gain has been considered as a potential way to improve performance. The optimum C-D gain is dependent on a number of factors including the type of control and the size of the display. For any specific computer system these are best determined through research and testing.

Absolute versus relative positioning can be explained most easily through examples. In absolute mode the cursor position on the display corresponds to the position of the pointing device. If the user touches a tablet in the upper right-hand corner, the cursor will move to the upper right-hand corner of the display. If the user then touches the bottom left side of the tablet, the cursor jumps to the bottom left side of the display.

Relative mode is typically observed in devices like the mouse; i.e., the cursor moves relative to its past position on the display rather than the pointing device's position on the tablet. This means the mouse can be lifted and repositioned on the tablet, but the cursor does not move on the display until specific movement input is received. The nature of the task usually determines the best mode of positioning (Greenstein 1997).

4.4.4.1 The Mouse

For desktop computers the mouse remains the primary input device for Web access (Po, Fisher, and Booth 2004) and the most commonly used nonkeyboard device (Atkinson et al. 2004; Sandfeld and Jensen 2005). A mouse requires little space to be operated, allows for good eye-hand coordination, and offers good cursor control and easy item selection. Mouse performance is high when compared to other pointing devices both in speed and accuracy.

The intensive mouse use has been associated with increased risk of upper extremity MSDs, including carpal tunnel syndrome (Keir, Bach, and Rempel 1999). An observational study conducted by Andre and English (1999) identified concerns regarding user posture during Web browsing. Those included constant finger clicking while scrolling through pages, keeping hands on the mouse when not in use (mouse freeze), and leaning away from the mouse while not using it, thereby placing stress on the wrist and elbow. Fogelman and Brogmus (1995) examined workers compensation claims between 1987 and 1993 and reported a greater prevalence of upper extremity symptoms (arm and wrist) among mouse users as compared to other workers.

Woods et al. (2002) in comprehensive multimethod study looking at pointing devices reported that 17% of 102 organizations participating in the study reported worker musculoskeletal complaints related to mouse use. More recently, Andersen et al. (2008) using an automated data collection on mouse and keyboard usage and weekly reports of neck and shoulder pain among 2146 technical assistants indicated that these activities were positively associated with acute neck and shoulder pain but were not associated with prolonged or chronic neck and shoulder pain.

4.4.4.2 Trackball

A trackball is appropriately described by Hinckley (2002) as an upside down mechanical mouse. It features usually buttons located to the side of the ball that allow the selection of items. Coordination between the ball and button activation can be an issue and unwanted thumb activation is a concern. Trackballs offer good eye-hand coordination, allowing users to focus their attention on the display. They require a minimal, fixed space, are compatible with mobile applications, can be operated in sloped surfaces, and can be easily integrated in the keyboard. Trackballs perform well in pointing and selecting tasks but are poor choices for drawing tasks. Trackballs are one of the most common alternatives to the mouse.

On the one hand, assessments conducted by a panel of experts on three different commercial trackballs were very critical of the comfort offered by these devices (Woods et al. 2002). On the other hand, trackballs may present advantages over the mouse for people with some form of motor impairment, including low strength, poor coordination, wrist pain, or limited ranges of motion (Wobbrock and Myers 2006).

4.4.4.3 Touch Pad

A touch pad is a flat panel that senses the position of a finger or stylus and is commonly found as an integrated pointing device on portable computers such as laptops, netbooks, and PDAs. These touch pads recognize clicking through tapping and double-tapping gestures, and accuracy can be an issue. In applications where touch pads are integrated with keyboards, inadvertent activation is a common problem.

Touch pads may feature visual or auditory feedback to provide users with a more direct relationship between control and display. They offer good display-control compatibility, can be used in degraded environments, can be positioned on most surfaces (e.g., sloped or vertical), and can be easily accessed. Touch pads are less comfortable than other input devices for sustained use and may lead to localized muscle fatigue under intense continual operation.

4.4.4.4 Touch Screens

Touch screens allow direct user input on a display. Input signals are generated as the user moves a finger or stylus over a transparent touch-sensitive display surface. The input may be produced through a number of technologies each with its own advantages and limitations.

On the positive side, touch screens offer a direct input-display relationship, good hand-eye coordination, and can be very space efficient. They are appropriate for situations where limited typing is required, for menu selection tasks, for tasks requiring constant display attention, and particularly for tasks where training is neither practical nor feasible such as public access information terminals, ATM machines, etc.

On the negative side, a poor visual environment can lead to screen reflections with possible relative glare and loss of contrast. Fingerprints on the screen can also reduce visibility. Accuracy can be an issue as separation between the touch surface and the targets can result in parallax errors. Visual feedback on current cursor location and on accuracy of operator's action helps reduce error rates (Weiman et al. 1985). In handheld applications the use of haptic or tactile feedback especially appears to reduce user error and cognitive demand while increasing input speed (Brewster, Chohan, and Brown 2007). Depending on their placement, touch screens may be uncomfortable for extended use and the user's hand may obstruct the view of the screen during activation. Among older users performing typing tasks in handheld devices (e.g., texting), touch screen-based keyboards performed worse than small physical keyboards with lower accuracy and speeds, with the latter being preferred by a wide margin (Wright et al. 2000). Touch screens also do not distinguish between actions intended to move the cursor over an item and drag the item itself, which may be bothersome. Touch screens are not recommended for drawing tasks.

4.4.4.5 Joystick

A joystick is basically a vertical lever mounted on a stationary base. Displacement or isotonic joysticks sense the angle of deflection of the joystick to determine cursor movement. Isometric joysticks typically do not move or move minimally. They sense the magnitude and direction of force applications to determine cursor movement. Joysticks require minimal space, especially isometric ones, and can be effectively integrated with keyboards in portable applications. The integration of joysticks into keyboards allows users to switch between typing and pointing tasks very quickly due to the reduction in the time required to acquire the pointing device. For purely pointing tasks, joystick performance is inferior to mice (Douglas and Mithal 1994) and requires significantly more practice for high performance. Joysticks are also very sensitive to physiological tremor (Mithal and Douglas 1996), and experience has shown that these can be hard to master on portable devices.

Provided that support is provided for the hand to rest, joysticks can be used comfortably for extended periods of time. However, intense and extended use of joysticks in computer games, especially with multifunction joysticks equipped with haptic displays (e.g., "rumble-pack") that vibrate to simulate game conditions, has prompted some concerns (Cleary, Mc Kendrick, and Sills 2002). Joysticks are best utilized for continuous tracking tasks and for pointing tasks where precision requirements are low.

The *trackpoint*, a small isometric joystick placed between the letter keys G, H, and B on the computer's keyboard, is commonly found in laptops and netbook computers. Despite its wide availability, research has shown that the trackpoint seems to be difficult to operate requiring a long time to master its use (Armbruster, Sutter, and Ziefle 2007).

4.4.4.6 Special Applications

4.4.4.6.1 *Voice Input*

Voice input may be helpful either as the sole input mode or jointly with other control means. Speech-based input may be appropriate when the user's hands or eyes are busy, when interacting with handheld computers with limited keyboards or screens, and for users with perceptual or motor impairments (Cohen and Oviatt 1995). With other input modes, voice recognition can reduce errors and allow for easier corrections and increase flexibility of handheld devices to different environments, tasks, and user needs and preferences (Cohen and Oviatt 1995). Although voice input may be an alternative for users affected by MSDs, its extensive use may lead to vocal fatigue.

4.4.4.6.2 *Eye-Tracking Devices*

Eye tracking is a technology in which a camera or imaging system visually tracks some feature of the eye and a computer then determines where the user is looking. Item selection is typically achieved by eye blinking. An eye-tracking device allows the user to look and point simultaneously. Eye-controlled devices offer the potential for users with limited manual dexterity to point and have potential application in virtual reality environments. They free up the hands to perform other tasks, virtually eliminating device acquisition time, and minimizing target acquisition time. Significant constraints to its wide application include cost, need to maintain steady head postures, frequency of calibrations, portability, and difficulty in operating. Other relevant problems include unintended item selection and poor accuracy, which limits applications involving small targets (Oyekoya and Stentiford 2004; Zhai, Morimoto, and Ihde 1999).

4.4.4.6.3 Head-Controlled Devices

Head-controlled devices have been considered a good choice for virtual reality applications (Brooks 1988) and for movement impaired computer users (Radwin, Vanderheiden, and Lin 1990). Head switches can also be used in conjunction with other devices to activate secondary functions. Unfortunately, neck muscles offer a low range of motion control that typically results in significantly higher target acquisition times when compared to a conventional mouse.

4.4.4.6.4 Mouth-Operated Devices

A few attempts to develop pointing devices controlled by the mouth have been made including commercial applications. Some of them use a joystick operated by the tongue or chin with clicking being performed by sipping or blowing. Typing tasks can be performed either by navigating and selecting keys through an on-screen keyboard or through Morse code. It is unlikely these will be primary input devices as they are much harder to interact with than other devices previously described.

4.4.5 Displays

4.4.5.1 Visual Displays

Visual displays in computers may rely on a number of different technologies such as liquid crystal display (LCD), light emitting diode (LED), plasma display panel (PDP), electroluminescent display (ELD), or other image projection technology. A brief summary of each technology can be found below. For a more complete review of these technologies the reader is directed to Luczak, Roetting, and Oehme (2003) and Schlick et al. (2008).

LCD technology allows for much thinner displays than CRT and uses much less energy than other technologies such as CRT, LED, or PDP. LCD technology offers good readability.

An LED is a semiconductor device that emits visible light when an electric current passes through it. LED technology offers good readability and is often used in heads-up displays and head-mounted displays.

In plasma technology, each pixel on the screen is illuminated by a small amount of charged gas, somewhat like a small neon light. PDPs are thinner than CRT displays and brighter than LCDs. A PDP is flat and therefore free of distortion on the edges of the screen. Unlike many LCD displays,

a plasma display offers a very wide viewing angle. On the negative side, PDPs have high-energy consumption, making this technology inappropriate for portable devices. Because of the large size of pixels it requires the user to be placed far from the display for proper viewing (Luczak, Roetting, and Oehme 2003; Schlick et al. 2008).

ELD technology is a thin and flat display used in portable devices. ELD works by sandwiching a thin film of phosphorescent substance between two plates. One plate is coated with vertical wires and the other with horizontal wires, forming a grid. Passing an electrical current through a horizontal and vertical wire causes the phosphorescent film at the intersection to glow, creating a pixel. ELDs require relatively low levels of power to operate, have long life, offer a wide viewing angle, and operate well in large temperature ranges. This latter characteristic makes this technology very appealing for mobile and portable applications.

Projection technologies represent a promising approach especially for large displays. Some of the issues related to this technology are the need to darken the room, the casting of shadows when front projection is used, and the reduction of image quality with increased angle of vision when back projection is used. Among the different types of projection technologies, laser-based ones have received increasing attention. An important application of lasers is the virtual retinal display (VRD). A VRD creates images by scanning low power laser light directly onto the retina. This method results in images that are bright, high contrast, and high resolution. VRD offers tremendous potential for people with low vision as well as for augmented reality applications (Viirre et al. 1998).

The critical questions about visual displays include the following: (1) are the characters and images large enough to be seen, (2) are the characters and images clear enough to be recognized, (3) is the display big enough to show enough of the message to provide context, and (4) do the displays characteristics deal with the surrounding environment?

4.4.5.2 Conversational Interfaces

Conversational interfaces allow people to talk to or listen to computers or other digital devices without the need of typing or using a pointing device. They have been successfully utilized in situations where users should not divert their visual attention from the task and where hands are busy performing other activities. Conversational interfaces can be beneficial to users who have low vision, who have motor deficits, or who are technologically naïve.

At its simplest level, speech-based interfaces will allow users to dictate specific instructions or will guide them through fixed paths asking predetermined questions, such as in touch-tone or voice response systems. Conversational interfaces can prove advantageous for tasks involving text composition, speech transcription, transaction completion, and remote collaboration (Karat, Vergo, and Nahamoo 2003; Lai, Karat, and Yankelovich 2008).

Technologies that provide the foundations for conversational interfaces include voice and gesture recognition and voice synthesis. Voice recognition has evolved quickly in part owing to advances in microphone devices and software. Microphones designed for voice recognition can be inexpensive, light, wireless, and have noise suppression capabilities allowing them to be used even in noisy environments such as airports. Effective voice recognition must be able to adapt to a variety of user characteristics such as different national accents, levels of expertise, age, health condition, and vocabulary (Karat, Vergo, and Nahamoo 2003; Lai, Karat, and Yankelovich 2008). Voice synthesis or text-to-voice systems enable computers to convert text input into a simulated human speech. Although most voice synthesis devices commercially available produce comprehensible speech, they still sound artificial and rigid. More advanced speech synthesis systems can closely simulate natural human conversation but still at high cost and complexity.

Dialogue between user and computer can be accomplished through three different approaches. In the "direct or system initiated" dialogue the user will be asked to make a selection from a set of given choices (e.g., "Say yes or no"). This is the most common dialogue style in current commercial applications. It emphasizes accuracy as it reduces the variety of words/sounds to be recognized by the system but with potential limitations on the interaction efficiency. In the "user-initiated" dialogue a knowledgeable user makes specific requests to the computer with minimal prompting by the system. This dialogue style is not intended to novice users and tends to have a lower accuracy as compared to the "system-initiated" dialogue. Finally, the "conversational or mixed initiative" combines the qualities of both system- and user-initiated dialogue styles allowing for a more natural and efficient interaction (Lai, Karat, and Yankelovich 2008). In the "mixed initiative" approach, open-ended and direct questions are alternatively employed by the system as the interaction evolves.

The recognition of body motions and stances, especially hand gestures and facial expressions, has the potential of making human–computer interaction much more natural, effective, and rich. Gesture recognition is deemed appealing for virtual and augmented environments and may in the future eliminate the need for physical input devices such as the keyboard and the mouse. Current research and development efforts have emphasized hand gesture recognition (Pavlovic, Sharma, and Huang 1997). Hand gesture recognition interfaces can be described as either touch screen or free form. Touch screens, which have been widely adopted in (smart) cell phones and PDAs, require direct finger contact with the device, thus limiting the types of gestures that can be employed. Free form gestural interfaces do not require direct user contact allowing for a much wider array of control gestures. Some free form applications rely on gloves or controllers for gesture input, but glove-based sensing has several drawbacks that reduce the ease and naturalness of the interactions, and it requires long calibration and setup procedures (Erol et al. 2007). Computer-vision–based free form gestural technologies have been preferred as they allow bare hand input and can provide a more natural, noncontact solution.

Vision-based gestural interfaces must take into consideration ergonomic principles when defining the inventory of control motions (i.e., gesture vocabulary). In addition to being intuitive and comfortable to perform (Stern, Wachs, and Edan 2006) the control gestures should avoid outstretched arm positions, limit motion repetitiveness, require low forces (internal and external), promote neutral postures, and avoid prolonged static postures (Nielsen et al. 2003). Saffer's (2008) *Designing Gestural Interfaces: Touchscreens and Interactive Devices* offers a review on the topic including current trends, emerging patterns of use, guidelines for the design and documentation of interactive gestures, and an overview of the technologies surrounding touch screens and interactive environments.

4.4.5.3 Haptic Interfaces

Haptics is the study of human touch and interaction with the external environment through the sense of touch. Haptic devices provide force feedback to muscles and skin as users interact with either a virtual or remote environment. These interfaces allow for a bidirectional flow of information, they can both sense and act on the environment. The integration of haptics has shown to improve human–computer interaction, with the potential of enhancing the performance of computer input devices such as the mouse (Kyung, Kwon, and Yang 2006) and the touch screen (Brewster, Chohan, and Brown 2007). Haptic feedback has been widely adopted in recent mobile applications including PDAs, cell phones, and game controllers. This limited form of feedback is provided through vibrotactile devices using eccentric motors.

Currently, most applications of haptics focus on hand tasks, such as manual exploration and manipulation of objects. This is justified because the human hand is a very versatile organ able to press, hold, and move objects and tools. It allows users to explore object properties such as surface shape, texture, and rigidity. A number of haptic devices designed to interact with other parts of the body and even the whole body applications are being used (Iwata 2003, 2008). Common examples of haptic interfaces available in the market are gloves and exoskeletons that track hand postures and joysticks that can reflect forces back to the user. These devices are commonly used in conjunction with visual displays.

Tactile displays excite nerve endings in the skin which indicate texture, pressure, and heat of the virtual or remote object. Vibrations, for instance, can be used to convey information about phenomena like surface texture, slip, impact, and puncture (Kontarinis and Rowe 1995; Rowe 2002). Small-scale shape or pressure distribution information can be conveyed by an array of closely spaced pins that can be individually raised and lowered against the fingertip to approximate the desired shape. Force displays interact with the skin and muscles and provide the user with a sensation of a force being applied, such as the reaction from a virtual object. These devices typically employ robotic manipulators that press against the user with the forces that correspond to the virtual environment. In the future, Web devices are expected to support multimodal interactions which will integrate visual, acoustic, and tactile input greatly improving user experience (Kwon 2007).

4.5 THE WORKSTATION

Workstation design is a major element in ergonomic strategies for improving user comfort and particularly for reducing musculoskeletal problems when using the Web. Task requirements can have a significant role in defining how a workstation will be laid out. The relative importance of the display, input devices, and hardcopy (e.g., source documents) depends primarily on the task, and this then influences the design considerations necessary to improve operator performance, comfort, and health.

Web tasks using the Internet require substantial time interfacing with the display and using input devices to select actions or respond to inputs. For these types of tasks the display devices and the input devices are emphasized when designing the workstation and environment.

4.5.1 Designing Fixed Computer Workstations

Thirty years of research have shown that poorly designed fixed computer workstations can produce a number of performance and health problems for users (see Smith, Carayon, and Cohen 2008). Over the past 30 years, much has been learned about design considerations for fixed computer workstations. Grandjean (1984) proposed the following features of workstation design:

1. The furniture should be as flexible as possible with adjustment ranges to accommodate the anthropometric diversity of the users.
2. Controls for workstation adjustment should be easy to use.
3. There should be sufficient knee space for seated operators.
4. The chair should have an elongated backrest with an adjustable inclination and a lumbar support.
5. The keyboard should be moveable on the desk surface. (This recommendation could be generalized to any input device.)

Following the general guidance below will be useful for designing and laying out fixed computer workstations in offices and home situations where users access the Web. A different approach will be proposed later for on-person and portable use of computers and IT/IS devices. See Smith, Carayon, and Cohen (2008) and ANSI/HFES-100 (2007) for specifics about workstation design.

The recommended size of the work surface is dependent upon the task(s) and the characteristics of the technology (dimensions, input devices, output devices). Workstations are composed of primary work surfaces, secondary surfaces, storage, and postural supports. The primary working surface (e.g., those supporting the keyboard, the mouse, the display) should allow the screen to be moved forward/backward and up/down

for comfortable viewing and allow input devices to be placed in several locations on the working surface for easy user access. There should be the possibility to adjust the height and orientation of the input devices to provide proper postures of the shoulders, arms, wrists, and hands. There should be adequate knee and leg room for the user to move around while working. It is important to provide unobstructed room under the working surface for the feet and legs so that operators can easily shift their posture.

The Human Factors and Ergonomics Society (HFES) developed an ANSI standard for computer workstations (ANSI/HFES-100 1998, 2007) that provides guidance for designers and users. This standard can be purchased through the HFES Web site (http://www.hfes.org). This standard provides specifications for workstation design. Knee space height and width and toe depth are the three key factors for the design of clearance space under the working surfaces. A good workstation design accounts for individual body sizes and often exceeds minimum clearances to allow for free postural movement of the user.

It is desirable for table heights to vary with the height of the user, particularly if the chair is not height adjustable. Height-adjustable working surfaces are effective for this. Adjustable multisurface tables encourage good posture by allowing the keyboard and screen to be independently adjusted to appropriate keying and viewing heights for each individual and each task. Tables that cannot be adjusted easily are a problem when workstations are used by multiple individuals of differing sizes, especially if the chair is not height adjustable.

4.5.2 What to Do When Fixed Workstations Are Not Available

Now let us move away from a structured office situation and look at a typical unstructured situation. Imagine our Mr. Smith again: this time he is sitting at the airport and his flight has been delayed for 2 hours. He has his laptop, and he decides to get some work done while he waits. As we discussed earlier, Mr. Smith could rent a kiosk at the airport that would provide him with a high-speed Internet connection, a telephone, a working surface (desk or table), a height-adjustable chair, and some privacy (noise control, personal space). Now imagine that Mr. Smith has been told to stay in the boarding area because it is possible that the departure may be sooner than 2 hours. Mr. Smith gets out his laptop, places it on his lap, and connects to the Internet. He is sitting in a waiting area chair with poor back support, and he has no table to place his laptop on. This situation is very common at airports. Clearly, Mr. Smith is not at an optimal workstation, and he will experience poor postures that could lead to musculoskeletal and visual discomfort.

Now imagine Mr. Smith is using his smart phone that provides access to the Internet. This device can be operated while he is standing in line at the airport to check in or sitting at the boarding gate. With the smart phone he can stand or sit and be pointing at miniature buttons (sometimes with a stylus because they are so small) and interacting with the interconnected world. Again, this scene is all too familiar in almost any venue (airport, restaurant, street, office). While the convenience and effectiveness of easy, lightweight portability are very high, the comfort and health factors are often very low because the person uses the laptop or smart phone in all manner of environments, workstations, and tasks that diminish the consistent application of good ergonomic principles.

While the convenience and effectiveness of easy, lightweight portability are very high, the comfort and health factors are often very low because the person uses the laptop or smart phone in all manner of environments, workstations, and tasks that diminish the consistent application of good ergonomic principles. The Human–Computer Interaction Committee of the International Ergonomics Association (IEA) produced a guideline for the use of laptop computers to improve ergonomic conditions (Saito et al. 2000). An important feature of the IEA laptop guideline (Saito et al. 2000) was to encourage conditions of use that mirror the best practices of ergonomic conditions for fixed computer workstations in an office environment. This is impossible at the airport for many people unless they pay for the use of a computer kiosk.

In situations where there is not a fixed workstation the device is typically positioned wherever is convenient. Very often, such positioning creates bad postures for the legs, back, shoulders, arms, wrists/hands, and/or neck. In addition, the smaller dimensions of the manual input devices (touch pad, buttons, keyboard, roller ball) make motions much more difficult and imprecise, and these often produce constrained postures. If the devices are used continuously for a prolonged period (such as one hour or more), muscle tension builds up and discomfort in joints, muscles, ligaments, tendons, and nerves can occur. To reduce the undesirable effects of the poor workstation, characteristics that lead to musculoskeletal and visual discomfort the following recommendations are given:

1. If you are using a laptop on your lap, find a work area where you can put the laptop on a table (rather than on your lap). Then arrange the work area as closely as possible with the recommendations presented for a standard office.
2. If you are using a handheld device such as a smart phone, you should position yourself so that your back is supported. It is preferable to use the device sitting down. Of course, if you are using the smart phone as you are walking, then this is not possible. If you are using a voice interface, then use an ear piece and a microphone so that you do not have to be constantly gripping the PDA in your hand.
3. Never work in poor postural conditions for more than *30 minutes* continuously. Take at least a *5 minute* break (preferably 10 minutes) away from the laptop/smart phone use, put the device down (away), get up, and stretch for 1 minute or more, and then walk for 2–3 minutes. If you are using a handheld smart phone in a standing position, then during your break

put it away, do 1 minute of stretching, and then sit down for 4 minutes. This may mean sitting on the floor, but preferably you will sit where you can support your back.
4. Buy equipment that provides the best possible input interfaces and output displays (screens, headphones, typing pads). Because these devices are small, the perceptual motor requirements for their use are much more difficult (sensory requirements, motion patterns, skill requirements, postural demands). Therefore, screens should provide easily readable characters (large, understandable), and input buttons should be easy to operate (large, properly spaced, easily accessible).
5. Only use these devices when you do not have access to fixed workstations that have better ergonomic characteristics. Do not use these devices continuously for more than 30 minutes.

4.5.3 POSTURAL SUPPORT

Postural support is essential for controlling loads on the spine and limbs. Studies have revealed that the sitting position, as compared to the standing position, reduces static muscular efforts in legs and hips, but increases the physical load on the intervertebral discs in the lumbar region of the spine (see Marras 2008).

Poorly designed chairs can contribute to computer user discomfort. Chair adjustability in terms of height, seat angle, backward tilt, and lumbar support helps to provide trunk, shoulder, neck, and leg postures that reduce strain on the muscles, tendons, ligaments, and discs. The "motion" of the chair helps encourage good movement patterns. A chair that provides swivel action encourages movement, while backward tilting increases the number of postures that can be assumed. The chair height should be adjustable so that the computer operator's feet can rest firmly on the floor with minimal pressure beneath the thighs. To enable short users to sit with their feet on the floor without compressing their thighs, it may be necessary to add a footrest. See Smith, Carayon, and Cohen (2008) and ANSI/HFES-100 (2007) for specifications on chair design.

The seat pan should be wide enough to permit operators to make shifts in posture from side to side. This not only helps to avoid static postures but also accommodates a large range of individual buttock sizes. The seat pan should not be overly U-shaped because this encourages static sitting postures. The front edge of the seat pan should be well-rounded downward to reduce pressure on the underside of the thighs that can affect blood flow to the legs and feet. The seat needs to be padded to the proper firmness that ensures an even distribution of pressure on the thighs and buttocks. A properly padded seat should compress about one-half to 1 inch when a person sits on it.

The tension and tilt angle of the chair's backrest should be adjustable. Inclination of the chair backrest is important for operators to be able to lean forward or back in a comfortable manner while maintaining a correct relationship between the seat pan angle and the backrest inclination. A backrest inclination of about 110 degrees is considered an appropriate posture by many experts. However, studies have shown that operators may incline backward as much as 125 degrees, which also is an appropriate posture. Backrests that tilt to allow an inclination of up to 125 degrees are therefore a good idea. The backrest tilt adjustments should be accessible and easy to use. Chairs with high backrests are preferred because they provide support to both lower back and the upper back (shoulder).

Another important chair feature is armrests. Armrests can provide support for resting the arms to prevent or reduce arm, shoulder, and neck fatigue. Adjustable armrests are an advantage because they provide greater flexibility for individual operator preference, as are removable arm rests. For specific tasks such as using a numeric keypad, a full armrest can be beneficial in supporting the arms.

4.6   GENERAL RECOMMENDATIONS

4.6.1   FOR DESIGNERS OF WEB SYSTEMS

Realize the wide range of sensory and perceptual-motor skills of the users of your Web system and provide means for universal access. Designers should provide the following:

1. Web systems that recognize a variety of input devices to provide options for users with different perceptual-motor capabilities and skills. Thus, users should be able to navigate the Web site using keyboards, pointing devices, tablets, etc. People with diminished motor capability can then use input devices most suited to their abilities.
2. Web systems that provide information through a variety of display devices for visual, auditory, and tactile output. People with diminished sensory capacity can then use those sensory modalities most suited to their abilities.
3. Web displays that have magnification capabilities. People in poor environments or with diminished or sensitive sensory capabilities can increase or decrease the gain as necessary to obtain a clear message.
4. Web navigation processes that minimize the frequency of input device usage. Reducing the frequency of actions required to navigate the Web site lowers the stress and strain on the sensory and musculoskeletal systems.
5. Web systems that minimize psychological strain as this will be beneficial for controlling biomechanical strain.

4.6.2   FOR DESIGNERS OF WEB INTERFACE TECHNOLOGY

Realize the wide range of sensory and perceptual-motor skills of the users of your Web system and provide means for universal access. Designers should provide the following:

1. Input devices that accommodate users with different perceptual-motor capabilities and skills. Understand that the users can range from highly skilled persons to novices and that each has different needs for exercising control over the Web system. Users may also have deficits in sensory-motor skills that need to be accommodated.
2. Input devices that promote skillful use by all users. Their actions should be intuitive, predictable, smooth, and require a minimum of force and nonneutral postures to operate.
3. A variety of displays for visual, auditory, and tactile output. People with diminished sensory capacity or who are in adverse sensory environments can then use those sensory modalities most suited to their abilities and environments.
4. Displays with magnification capabilities so that people with diminished or sensitive sensory capabilities or who are in adverse sensory environments can increase or decrease the gain as necessary to obtain a clear message.
5. Ways to deal with the miniaturization of the input devices and displays. Input devices and displays that are too small cannot be easily used by anybody, but are even more problematic for people with sensory and perceptual-motor deficiencies.
6. Input devices that provide proper feedback of action and/or actuation. This enhances performance and may also reduce the level of force applied by the user.

4.6.3   FOR WEB SYSTEM USERS

1. Take actions to have the best workstation possible in a given environment when you are using the Web.
   a. Fixed workstations with an adjustable height chair are superior to other situations as they provide postural support to reduce fatigue and enhance perceptual-motor skills.
   b. It is best to work at a table. The table and the chair should be set to appropriate heights that fit your physical dimensions. This means that the table and the chair need to be height adjustable.
   c. You should provide postural support for your back and preferably be in a seated position.
   d. When handheld or on-body interfaces are used you often lose support for your back and arms. In these situations, and a comfortable posture that provides support for your back and arms as best as possible.
   e. If you are walking and using a talking interface, it will be very hard to get good postural support. Take breaks every 30 minutes and sit down.
2. Do not interact with the interface devices for too long.
   a. Take a break at least every 30 minutes in which you allow the hands (voice) and eyes to rest for at least 5 minutes.

b. If your hands, legs, back, neck, or voice become tired after less than 30 minutes of use, then stop the interaction and rest for at least 5 minutes (or longer as needed) to become refreshed before continuing.

3. Highly repetitive motions for extended time periods without adequate resting will lead to muscle fatigue and a reduction in perceptual-motor skill. It may also lead to musculoskeletal discomfort, pain, and even dysfunction and injury. Take adequate rest breaks, and stop interaction when you have sensory or musculoskeletal fatigue or pain.

4. Rest your sensory system just like you rest your muscles.

a. For example, if you have been using a visual interface, when you take a rest break, do not pick up a newspaper or book to read. Rather than reading, let your eyes look off into the distance and enjoy the view.

b. If you have been using an auditory interface, it is best to rest in a quiet area to allow your ears (and brain) to rest.

5. Do not stay in static, fixed postures for very long.

a. It is a good idea to move from static positions at least every 30 minutes.

b. Stretching can be beneficial if done carefully and in moderation.

REFERENCES

Anshel, J. 2005. *Visual Ergonomics Handbook.* Boca Raton, FL: CRC Press.

Armbruster, C., C. Sutter, and M. Ziefle. 2007. Notebook input devices put to the age test: The usability of trackpoint and touchpad for middle-aged adults. *Ergonomics* 50: 426–445.

Andersen, J. H., M. Harhoff, S. Grimstrup, I.Vilstrup, C. F. Lassen, L. P. A. Brandt, et al. 2008. Computer mouse use predicts acute pain but not prolonged or chronic pain in the neck and shoulder. *Occupational and Environmental Medicine* 65: 126–131.

Andre, A. D., and J. D. English. 1999. Posture and web browsing: an observational study. In *Proceedings of the Human Factors and Ergonomics Society 43rd Annual Meeting*, 568–572. Santa Monica, CA: The Human Factors and Ergonomics Society.

ANSI/HFES-100. 1988. *American National Standard for Human Factors Engineering of Visual Display Terminal Workstations* (ANSI/HFS Standard 100–1988). Santa Monica, CA: The Human Factors and Ergonomics Society.

ANSI/HFES-100. 2007. *Human Factors Engineering of Computer Workstations.* Santa Monica, CA: The Human Factors and Ergonomics Society.

Atkinson, S., V. Woods, R. A. Haslam, and P. Buckle. 2004. Using non-keyboard input devices: Interviews with users in the workplace. *International Journal of Industrial Ergonomics* 33: 571–579.

Azuma, R. T. 1997. A survey of augmented reality. *Presence: Teleoperators and Virtual Environments* 6(4): 355–385.

Azuma, R. T. 2001. Augmented reality: Approaches and technical challenges. In *Fundamentals of Wearable Computers.*

Barfield, W., and Caudell, T. 2001. *Fundamentals of Wearable Computers and Augmented Reality.* Mahwah, NJ: Lawrence Erlbaum.

Bolia, R. S., W. T. Nelson, and R. M. Morley. 2001. Asymmetric performance in the cocktail party effect: Implications for the design of spatial audio displays. *Human Factors* 43(2): 208–216.

Brewster, S., F. Chohan, and L. Brown. 2007. Tactile feedback for mobile interactions. In *Proceedings of the SIGCHI Conference on Human Factors in Computing Systems*, 159–162. New York, NY: Association of Computing Macbbery (ACM).

Broadbent, D. E. 1958. Effect of noise on an intellectual task. *Journal of the Acoustical Society of America* 30: 84–95.

Brooks, F. P., Jr. 1988. Grasping reality through illusion: Interactive graphics serving science. *Proceedings of Computer–Human Interaction 1988: Association for Computing Machinery Conference on Human Factors in Computing Systems*, 1–11.

Burgess-Limerick, R., M. Mons-Williams, and V. Coppard. 2000. Visual display height. *Human Factors* 42(1): 140–150.

Buxton, W. 2009. A directory of sources for input technologies. http://www.billbuxton.com/InputSources.html (accessed Nov. 14, 2009).

Carayon, P., and M. J. Smith. 2000. Work organization and ergonomics. *Applied Ergonomics* 31: 649–662.

Champney, R. K., K. M. Stanney, P. A. K. Hash, and L. C. Malone. 2007. Recovery from virtual exposure: Expected time course of symptoms and potential readaptation strategies. *Human Factors* 49(3): 491–506.

Cleary, A. G., H. McKendrick, and J. A. Sills. 2002. Hand-arm vibration syndrome may be associated with prolonged use of vibrating computer games. *British Medical Journal* 324: 301.

Cohen P. R., and S. L. Oviatt. 1995. The role of voice input for human–machine communication. *Proceedings of the National Academy of Sciences USA* 92: 9921–9927.

Commarford, P. M., J. R. Lewis, J. A. Smither, and M. D. Gentzler. 2008. A comparison of broad versus deep auditory menu structures. *Human Factors* 50(1): 77–89.

Dennerlein, J. T., and M. C. Yang. 2001. Haptic force-feedback devices for the office computer: Performance and musculoskeletal loading issues. *Human Factors* 43(2): 278–286.

Dey, A. K., P. Ljungstrand, and A. Schmidt. 2001. Distributed and disappearing user interfaces in ubiquitous computing. http://www.cc.gatech.edu/fce/ctk/pubs/CHI2001-workshop.pdf (accessed Nov. 3, 2010).

Douglas, S. A., and A. K. Mithal. 1994. The effect of reducing homing time on the speed of a finger-controlled isometric pointing device. *Proceedings of the CHI 1994 Conference on Human Factors in Computer Systems*, 474–481.

Dowell, J., and Y. Shmuell. 2008. Blending speech output and visual text in the multimodal interface. *Human Factors* 50(5): 782–788.

Erol, A., G. Bebis, M. Nicolescu, R. D. Boyle, and X. Twombly. 2007. Vision-based hand pose estimation: A review. *Computer Vision and Image Understanding* 108(1–2): 52–73.

Ferris, T. K., and N. B. Sarter. 2008. Cross-modal links among vision, audition, and touch in complex environments. *Human Factors* 50(1): 17–26.

Fogelman, M., and G. Brogmus. 1995. Computer mouse use and cumulative disorders of the upper extremities. *Ergonomics* 38(12): 2465–2475

Gemperle, F., C. Kasabath, J. Stivoric, M. Bauer, and R. Martin. 1998. Design for wearability. In *The Second International Symposium on Wearable Computers*, 116–122. Los Alamitos, CA: IEEE Computer Society.

Gemperle, F., N. Ota, and D. Siewiorek. 2001. Design of a wearable tactile display. *Proceedings of the V IEEE International Symposium on Wearable Computer*, 5–12. New York: IEEE Computer Society.

Gerard, M. J., S. K. Jones, L. A. Smith, R. E. Thomas, and T. Wang. 1994. An ergonomic evaluation of the kinesis ergonomics computer keyboard. *Ergonomics* 37: 1616–1668.

Gerr, F., M. Marcus, and C. Monteilh. 2004. Epidemiology of musculoskeletal disorders among computer users: Lesson learned from the role of posture and keyboard use. *Journal of Electromyography and Kinesiology* 14(1): 25–31.

Gershenfeld, N., R. Krikorian, and D. Cohen. 2004. The Internet of things. *Scientific American* October, 291(4): 76–81.

Grahame. M., J. Laberge, and C. T. Scialfa. 2004. Age differences in search of web pages: The effects of link size, link number, and clutter. *Human Factors* 46(3): 385–398.

Grandjean, E. 1979. *Ergonomical and Medical Aspects of Cathode Ray Tube Displays*. Zurich: Federal Institute of Technology.

Grandjean, E. 1984. Postural problems at office machine work stations. In *Ergonomics and Health in Modern Offices*, ed. E. Grandjean, 445–455. London: Taylor & Francis.

Grandjean, E., and E. Vigliani. 1980. *Ergonomic Aspects of Visual Display Terminals*. London: Taylor & Francis.

Greenstein, J. S. 1997. Pointing devices. In *Handbook of Human Computer Interaction*, eds. M. Helander, T. K. Landauer, and P. Prabhu, 1317–1345. New York: Elsevier Science.

Harvey, C. M., R. J. Koubek, A. Darisipudi, and L. Rothrock, this volume. Cognitive ergonomics. In *Handbook of Human Factors in Web Design*, 2nd ed., eds. K.-P. L. Vu and R. W. Proctor, 85–106. Boca Raton, FL: CRC Press.

Hinckley, K. 2002. Input technologies and techniques. In *The Human-Computer Interaction Handbook: Fundamentals, Evolving Technologies and Emerging Applications (Human Factors and Ergonomics)*, eds. J. A. Jacko and A. Sears. Mahwah, NJ: Lawrence Erlbaum Associates.

Hinckley, K. 2003. Input technologies and techniques. In *The Human–Computer Interaction Handbook: Fundamentals, Evolving Technologies and Emerging Applications*, eds. J. A. Jacko and A. Sears, 151–168. Mahwah, NJ: Lawrence Erlbaum.

Hinckley, K. 2008. Input technologies and techniques. In *The Human–Computer Interaction Handbook,* 2nd ed., eds. A. Sears and J. A. Jacko, 161–176. Mahwah, NJ: Lawrence Erlbaum.

Hollerer, T., S. Feiner, T. Terauchi, D. Rashid, and D. Hallaway. 1999. Exploring MARS: Developing indoor and outdoor user interfaces to a mobile augmented reality system. *Computers & Graphics* 23(6): 779–785.

Hopp, P. J., C. A. P. Smith, B. A. Clegg, and E. D. Heggestad. 2005. Interruption management: The use of attention-directing tactile cues. *Human Factors* 47(1): 1–11.

Hsiao, H. C., F. G. Wu, R. Hsi, C. I. Ho, W. Z. Shi, and C. H. Chen. 2009. The evaluation of operating posture in typing the QWERTY keyboard on PDA. In *Ergonomics and Health Aspects of Work with Computers, Proceedings of the 13th International Conference on Human–Computer Interaction*, ed. B. T. Karsh, 241–249. Springer: Berlin.

Hutchins, E. L., J. D. Hollan, and D. A. Norman. 1985. Direct manipulation interfaces. *Human-Computer Interaction* 1(4): 311–338.

ICS FORTH. 2009. Institute for Computer Science. http://ics.forth.gr (accessed Nov. 2, 2010).

Iwata, H. 2003. Haptic interfaces. In *The Human–Computer Interaction Handbook: Fundamentals, Evolving Technologies and Emerging Applications*, eds. J. A. Jacko and A. Sears, 206–219. Mahwah, NJ: Lawrence Erlbaum.

Iwata, H. 2008. Haptic interfaces. 2008. In *The Human–Computer Interaction Handbook,* 2nd ed., eds. A. Sears and J. A. Jacko, 229–245. Mahwah, NJ: Lawrence Erlbaum.

Jacko, J., and A. Sears, eds. 2003. *The Human–Computer Interaction Handbook: Fundamentals, Evolving Technologies and Emerging Applications*, Mahwah, NJ: Lawrence Erlbaum.

Jainta, S., and W. Jaschinski. 2002. Fixation disparity: Binocular vergence accuracy for a visual display at different positions relative to the eyes. *Human Factors* 44(3): 443–450.

Jaimes, A., and N. Sebe. 2007. Multimodal human computer interaction: A survey. *Computer Vision and Image Understanding*, special issue, 108(1–2): 116–134.

Jones, C. S., and B. Orr. 1998. Computer-related musculoskeletal pain and discomfort among high school students. *American Journal of Health Studies* 14(1): 26–30.

Jones, L. A., and N. B. Sarter. 2008. Tactile displays: Guidance for their design and application. *Human Factors* 50(1): 90–111.

Karat, C. M., J. Vergo, and D. Nahamoo. 2002. Conversational interface technologies. In *The Human–Computer Interaction Handbook: Fundamentals, Evolving Technologies and Emerging Applications*, eds. J. A. Jacko and A. Sears, 286–304. Mahwah, NJ: Lawrence Erlbaum.

Kazdin, A. E. 2000. *Encyclopedia of Psychology*. Washington, DC: American Psychological Association.

Keir, P. J., and R. P. Wells. 2002. The effect of typing posture on wrist extensor muscle loading. *Human Factors* 44(3): 392–403.

Keir, P. J., J. M. Bach, and D. Rempel. 1999. Effects of computer mouse design and task on carpal tunnel pressure. *Ergonomics* 42: 1350–1360.

Kilgore, R. M. 2009. Simple displays of talker location improve voice identification performance in multitalker, spatialized audio environments. *Human Factors* 51(2): 224–239.

Knave, B. G., R. I. Wibom, M. Voss, L. D. Hedstrom, and O. V. Bergqvist. 1985. Work with video display terminals among office employees: I. Subjective symptoms and discomfort. *Scandinavian Journal of Work, Environment and Health* 11(6): 457–466.

Knight, J. F., and C. Barber. 2007. Effect of head-mounted displays on posture. *Human Factors* 49(5): 797–807.

Knight, J. F., and C. Barber. 2005. A tool to assess the comfort of wearable computers. *Human Factors* 47(1): 77–91.

Kontarinis, D. A., and R. D. Howe. 1995. Tactile display of vibratory information in teleoperation and virtual environment. *Presence* 4: 387–402.

Kwon, D. S. 2007. Will haptics be used in mobile devices? A historical review of haptics technology and its potential applications in multi-modal interfaces. *Proceedings of the Second International Workshop on Haptic and Audio Interaction Design*, 9–10. Springer.

Kyung, K. U., K. S. Kwon, and G. H. Yang. 2006. A novel interactive mouse system for holistic haptic display in a human–computer interface. *International Journal of Human–Computer Interaction* 20(3): 247–270.

Lai, J. L., C. M. Karat, and N. Yankelovich. 2008. Conversational speech interfaces and technologies. In *The Human–Computer Interaction Handbook,* 2nd ed., eds. A. Sears and J. A. Jacko, 381–391. Mahwah, NJ: Lawrence Erlbaum.

Lehto, M. R., and J. R. Buck. 2008. *An Introduction to Human Factors and Ergonomics for Engineers*. Mahwah, NJ: Lawrence Erlbaum.

Luczak, H., M. Roetting, and O. Oehme. 2003. Visual displays. In *The Human–Computer Interaction Handbook: Fundamentals, Evolving Technologies and Emerging Applications*, eds. J. A. Jacko and A. Sears, 187–205. Mahwah, NJ: Lawrence Erlbaum.

Malt, L. G. 1977. Keyboard design in the electric era. Conference Papers on Developments in Data Capture and Photocomposition, PIRA Eurotype Forum, September 14–15, London, 8pp.

Mann, S. 1998. Wearable computing as means for personal empowerment. Keynote Address for The First International Conference on Wearable Computing, ICWC-98, May 12–13, Fairfax, VA.

Marklin, R. W., G. G. Simoneau, and J. F. Monrow. 1999. Wrist and forearm posture from typing on split and vertically inclined computer keyboards. *Human Factors* 41(4): 5559–5569.

Marras, W. S. 2008. *The Working Back.* Hoboken, NJ: John Wiley.

Merhi, O., E. Faugloire, M. Flanagan, and T. A. Stoffregen. 2007. Motion sickness, console video games, and head-mounted displays. *Human Factors* 49(5): 920–934.

Milgram, P., H. Takemura, A. Utsumi, and F. Kishino. 1994. Augmented reality: A class of displays on the reality-virtuality continuum. *SPIE Telemanipulator and Telepresence Technologies* 2351: 282–292.

Mithal, A. K., and S. A. Douglas. 1996. Differences in movement microstructure of the mouse in the finger-controlled isometric joystick. *Proceeding of CHI* 1996: 300–307.

Moore, M., P. Kennedy, E. Mynatt, and J. Mankoff. 2001. Nudge and shove: Frequency thresholding for navigation in direct brain-computer interfaces. *Proceedings of Computer–Human Interaction* 2001: 361–362.

Nagaseko, M., E. Grandjean, W. Hunting, and R. Gierere. 1985. Studies in ergonomically designed alphanumeric keyboards. *Human Factors* 27: 175–187.

NAS. 1983. *Video Terminals, Work and Vision.* Washington, DC: National Academy Press.

National Research Council and Institute of Medicine. 2001. *Musculoskeletal Disorders and the Workplace.* Washington, DC: National Academy Press.

Nielsen, M., M. Storring, T. B. Moeslund, and E. Granum. 2003. A procedure for developing intuitive and ergonomic gesture interfaces for man–machine interaction. *Technical Report CVMT 03-01. CVMT,* Aalborg, Denmark: Aalborg University.

Oyekoya, O. K., and F. W. M. Stentiford. 2004. Eye tracking as a new interface for image retrieval. *BT Technology Journal* 22(3): 161–169.

Pavlovic, V. I., R. Sharma, and T. S. Huang. 1997. Visual interpretation of hand gestures for human–computer interaction: A review. *IEEE Transactions on Pattern Analysis and Machine Intelligence* 19(7): 677–695.

Po, B. A., B. D. Fisher, and K. S. Booth. 2004. Mouse and touchscreen selection in the upper and lower visual fields. *Proceedings of the CHI 2004,* 359–366.

Radwin, R. G., G. C. Vanderheiden, and M. L. Lin. 1990. A method for evaluating headcontrolled computer input devices using Fitts' law. *Human Factors* 32(4): 423–438.

Rempel, D., A. Barr, D. Brafman, and E. Young. 2007. The effect of six keyboard designs on wrist and forearm postures. *Applied Ergonomics* 38: 293–298.

Rempel, D., K. Willms, J. Anshel, W. Jaschinski, and J. Sheedy. 2007. The effects of visual display distance on eye accommodation, head posture, and vision and neck symptoms. *Human Factors* 49(5): 830–838.

Rolland, J. P., and H. Fuchs. 2000. Optical versus video see-through head-mounted displays in medical visualization. *Presence: Teleoperators and Virtual Environments* 9(3): 287–309.

Roring, R. W., F. G. Hines, and N. Charness. 2007. Age differences in identifying words in synthetic speech. *Human Factors* 49(1): 25–31.

Rudmann, D. S., J. S. McCarley, and A. F. Kramer. 2003. Bimodal displays improve speech comprehension in environments with multiple speakers. *Human Factors* 45(2): 329–336.

Rose, M. J. 1991. Keyboard operating posture and actuation force: Implications for muscle over-use. *Applied Ergonomics* 22: 198–203.

Rowe, R. D. 2002. Introduction to haptic display: Tactile display. http://haptic.mech.nwu.edu/TactileDisplay.html (accessed March 19, 2003).

Saffer, D. 2008. *Designing Gestural Interfaces: Touchscreens and Interactive Devices.* Sebastopol, CA: O'Reilly Media.

Saito, S., B. Piccoli, and M. J. Smith. 2000. Ergonomic guidelines for using notebook personal computers. *Industrial Health* 48(4): 421–434.

Sandfeld, J., and B. R. Jensen. 2005. Effect of computer mouse gain and visual demand on mouse clicking performance and muscle activation in a young and elderly group of experienced computer users. *Applied Ergonomics* 36: 547–555.

Schlick, C., M. Ziefle, M. Park, and H. Luczak. 2008. Visual displays. In *The Human–Computer Interaction Handbook,* 2nd ed., eds. A. Sears and J. A. Jacko, 201–245. Mahwah, NJ: Lawrence Erlbaum.

Sears, A. and J. A. Jacko, eds. 2008. *The Human–Computer Interaction Handbook,* 2nd ed. Mahwah, NJ: Lawrence Erlbaum.

Shackel, B. 2000. People and computers—some recent highlights. *Applied Ergonomics* 31(6): 595–608.

Sheedy, J. E., M. V. Subbaram, A. B. Zimmerman, and J. R. Hayes. 2005. Text legibility and the letter superiority effect. *Human Factors* 47(4): 797–815.

Shneiderman, B. 1987. Designing the User Interface: Strategies for Effective Human-Computer Interaction, Reading, MA.

Simoneau, G. G., R. W. Marklin, and J. F. Monroe. 1999. Wrist and forearm postures of users of conventional computer keyboards. *Human Factors* 41(3): 413–424.

Smith, M. J. 1984. Health issues in VDT work. In *Visual Display Terminals,* eds. J. Bennet et al., 193–228. Englewood Cliffs, NJ: Prentice Hall.

Smith, M. J. 1987. Mental and physical strain at VDT workstations. *Behaviour and Information Technology* 6(3): 243–255.

Smith, M. J. 1997. Psychosocial aspects of working with video display terminals (VDT's) and employee physical and mental health. *Ergonomics* 40(10): 1002–1015.

Smith, M. J., B. G. Cohen, L. W. Stammerjohn, and A. Happ. 1981. An investigation of health complaints and job stress in video display operations. *Human Factors* 23(4): 387–400.

Smith, M. J., and P. C. Sainfort. 1989. A balance theory of job design for stress reduction. *International Journal of Industrial Ergonomics* 4: 67–79.

Smith, M. J., and P. Carayon. 1995. New technology, automation and work organization: Stress problems and improved technology implementation strategies, *International Journal of Human Factors in Manufacturing* 5: 99–116.

Smith, M. J., B. Karsh, F. Conway, W. Cohen, C. James, J. Morgan, et al. 1998. Effects of a split keyboard design and wrist rests on performance, posture and comfort. *Human Factors* 40(2): 324–336.

Smith, M. J., P. Carayon, and W. J. Cohen. 2008. Design of computer workstations. In *The Human–Computer Interaction Handbook,* 2nd ed., eds. A. Sears and J. A. Jacko, 313–326. Mahwah, NJ: Lawrence Erlbaum.

Snowden, R., P. Thompson, and T. Troscianko. 2006. *Basic Vision: An Introduction to Visual Perception.* New York: Elsevier.

Somerich, C. M. 1994. Carpal tunnel pressure during typing: Effects of wrist posture and typing speed. *Proceedings of Human Factors and Ergonomics Society 38th Annual Meeting,* 611–615.

Stammerjohn, L. W., M. J. Smith, and B. G. F. Cohen. 1981. Evaluation of work station design factors in VDT operations. *Human Factors* 23(4): 401–412.

Stanney, K. M., K. S. Hale, I. Nahmens, and R. S. Kennedy. 2003. What to expect from immersive virtual environment exposure: Influences of gender, body mass index, and past experience. *Human Factors* 45(3): 504–520.

Stanney, K. M., R. R. Mourant, and R. S. Kennedy. 1998. Human factors issues in virtual environments: A review of the literature. *Presence: Teleoperations and Virtual Environments* 7(4): 327–351.

Stern, H., J. Wachs, and Y. Edan. 2006. Human factors for design of hand gesture human-machine interaction. *Proceedings of 2006 IEEE International Conference on Systems, Man, and Cybernetics*, 4052–4056. New York: IEEE Computer Society.

Streitz, N. 2008. Designing for people in ambient intelligence environments. In *Proceedings of the 2nd International Conference on Ambient Intelligence Developments*, eds. A. Mana and C. Rudolph, 47–54.

Swanson, N. G., T. L. Galinsky, L. L. Cole, C. S. Pan, and S. L. Sauter. 1997. The impact of keyboard design on comfort and productivity in a text-entry task. *Applied Ergonomics* 28(1): 9–16.

Taveira, A. D., and S. D. Choi. 2009. Review study of computer input devices and older users. *International Journal of Human Computer Interaction* 25(5): 455–474.

Tittiranonda, P., D. Rempel, T. Armstrong, and S. Burastero. 1999. Workplace use of adjustable keyboard: Adjustment preferences and effect on wrist posture. *American Industrial Hygiene Association Journal* 60: 340–348.

TRACE. 2009. TRACE Research Center. http://trace.wisc.edu (accessed Nov. 3, 2010).

Van Laar, D., and O. Deshe. 2007. Color coding of control room displays: The psychocartography of visual layering effects. *Human Factors* 49(3): 477–490.

Viirre, E., H. Pryor, S. Nagata, and T. A. Furness. 1998. The virtual retinal display: A new technology for virtual reality and augmented vision in medicine. In *Proceedings of Medicine Meets Virtual Reality*, 252–257.

Weiman, N., R. J. Beaton, S. T. Knox, and P. C. Glasser. 1985. Effects of key layout, visual feedback, and encoding algorithm on menu selection with LED-based touch panels. Tech. Rep. No. HFL 604–02. Beaverton, OR: Tektronix, Human Factors Research Laboratory.

Weiser, M. 1993. Ubiquitous computing. *IEEE Computer* 26(10): 71–72.

Wickens, C. D., J. D. Lee, Y. Liu, and S. E. G. Becker. 2004. *An Introduction to Human Factors Engineering,* 2nd ed. Upper Saddle River, NJ: Prentice Hall.

Wobbrock, J. O., and B. A. Myers. 2006. Trackball text entry for people with motor impairments. *Proceedings of the CHI 2006*, 479–488.

Woods, V., S. Hastings, P. Buckle, and R. Haslam. 2002. *Ergonomics of Using a Mouse or Other Non-keyboard Input Device.* Research Report 045. Surrey, UK: HSE Books.

Wright, P., C. Bartram, N. Rogers, H. Emslie, J. Evans, and B. Wilson. 2000. Text entry on handheld computers by older users. *Ergonomics* 43: 702–716.

Zhai, S., C. Morimoto, and S. Ihde. 1999. Manual and gaze input cascaded (MAGIC) pointing. In *Proceedings of Computer–Human Interaction 1999*, 246–253. Association for Computing Machinery Conference on Human Factors in Computing Systems.

5 Cognitive Ergonomics

Craig M. Harvey, Richard J. Koubek, Ashok Darisipudi, and Ling Rothrock

CONTENTS

5.1 INTRODUCTION

The next time you listen to an advertisement on the television or read an ad in the newspaper, take notice at the use of the Internet. You will find that in many cases, the only method of contact provided is the company's Web site. In fact, many companies are almost making it difficult to find their telephone information because they are requiring users to first seek out information "on our Web site." We are truly an information society, and the Web impacts how we pay bills, shop for merchandise, or even find the school lunch menus for our children.

In 1998, there were approximately 2,851,000 Web sites, and in 2002, there were approximately 9,040,000 (Online Computer Library Center 2003). As of January 2010, Netcraft reports there are over 206 million Web domains. As one can see, the growth of Web sites in the past twelve years has been exponential (Netcraft 2010). The number of users on the Internet varies depending on the report used; however, the U.S. government estimates that 54.6% of all households have Internet access (U.S. Department of Commerce 2004). The United States is estimated to have over 250 million users with some estimates putting the world Internet population at over 1.7 billion users (Internet World Stats 2009).

Additionally, the makeup of the user population is ever changing. Approximately 90% of children 5 to 17 use a computer, and many of these are using computers to access the Internet (U.S. Department of Commerce 2002). In addition, the senior population (65+) represented approximately 6.1 million users in 2002 and is expected to explode as the baby boomer generation ages (CyberAtlas 2003). Additionally, the mobile Internet is growing rapidly with the expansion of smart phones and other Web-browsing devices that let you take the Internet in your pocket. It is estimated that today that there are 89 million mobile Web users, which is approximately 30.6% of the mobile phone subscribers. This number is expected to grow to over 134 million users or 43.5% of mobile phone subscribers by 2013 (CircleID 2009). While the Internet growth is exploding, there are still segments of our population that cannot or do not take advantage of this technology. Lower income households and people with mental and physical disabilities are less likely to use the Internet than other Americans (U.S. Department of Commerce 2002, 2004).

It would be remiss if we also did not consider the use of the collaborative tools on the Web. Whether one is using social tools like Facebook or collaborative meeting tools like GoToMeeting®, the Web is being used to support collaborative work and interaction.

Given such a large population of users and variety of users, designing for the Web is anything but a trivial task. Alexander's (2003) site of Web bloopers and Johnson's (2000) book illustrate that Web design requires a science base. Designing Web sites requires an understanding of the users, their goals and objectives, their limitations, and how technology can augment them in their information quest.

Norman (1988) points out that users move through three stages when interacting with a product, whether it is the Web or some other product:

1. Goals: users form a goal of what they want to happen (e.g., find a Web site on fishing in Idaho).
2. Execution: users interact with the world in hopes of achieving their defined goal (e.g., use a search engine to find a Web site on fishing in Idaho).
3. Evaluation: users compare what happened to what they wanted to happen (e.g., did the user find a site on fishing in Idaho).

Norman illustrates that frequently users become lost in the gulfs of execution and evaluation. Users are not sure how to achieve their goal or the system does not correspond to their intentions. Likewise, the system may not provide a physical representation that is interpretable by the user or meets the expectations of the user. It is when the user falls into one of these gulfs that they are likely to become frustrated, angry, or give up using the product. The result of the user falling into these gulfs can ultimately affect a company's profitability. For example, Jacob Nielsen estimates that e-commerce sites lose half of their potential business because users cannot figure out how to use their site (Business Week 2002).

So how are these problems combated? Is there any way for one to understand users and how they interact with a Web site? The answer to both of those questions is a resounding yes. There are methods and models available through cognitive ergonomics that allow us to address Norman's three stages of user interaction.

Mitecs Abstracts (MITECS 2003) defines cognitive ergonomics as

> "the study of cognition in the workplace with a view to design technologies, organizations, and learning environments. Cognitive ergonomics analyzes work in terms of cognitive representations and processes and contributes to designing workplaces that elicit and support reliable, effective, and satisfactory cognitive processing. Cognitive ergonomics overlaps with related disciplines such as human factors, applied psychology, organizational studies, and human computer interaction."

Cognitive ergonomics attempts to develop models and methods for understanding the user such that designers can create technology humans can use effectively. While traditional ergonomics, as discussed by Smith and Taveira (this volume), focuses more on user physical abilities and limitations, cognitive ergonomics delves into human cognitive abilities and limitations and through that understanding attempts to influence the design process to improve user experiences with technology.

Figure 5.1 outlines a Human–Environment Interaction Model adapted from Koubek et al. (2003) that defines the elements that impact the user and his interaction with the Web. These include the following:

Understanding user tasks embedded within the environment
√ Task
A task has its unique goal description and attributes. The goal of a task delineates the purpose for the human's interaction with the environment.
√ Environment
The environment refers to social and technological work environment where the human interacts with the tool to accomplish her goal in a task.
Modeling the user(s) interaction with the Web and other users
√ Human (or user)
In order to complete a task, humans will utilize their knowledge of the task and how to use the tool while using cognitive resources such as perceptual, motor, and memory resources, etc.
Enhancing the designers and human abilities
√ Tools
Tools are a means of enhancing the human's ability to interact with the Web application and enhance designer's ability to create Web applications.

This model lays the framework for the chapter's organization. This framework is built on the idea that design is an

Understanding user tasks embedded within the environment

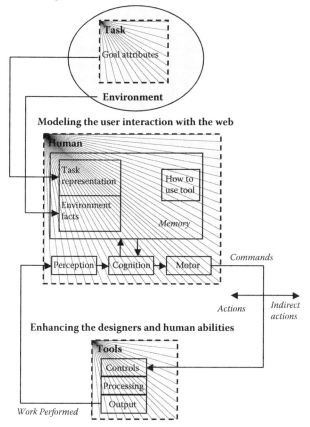

Modeling the user interaction with the web

Enhancing the designers and human abilities

FIGURE 5.1 Human–Environment Interaction Model and methods of cognitive engineering. (Adapted from Koubek, R. J., et al. *Ergonomics* 46, 1–3, 220–241, 2003.)

iterative process as portrayed in Figure 5.2, where, from bottom to top, the product becomes more concrete and usable after repeated testing. The tools and methods described in this chapter can be used from initial design concept through final design. Each brings to the table different perspectives

that allow cognitive ergonomists to better understand and model the user. Through this iterative process, designers move from a basic understanding of the task to a more complete understanding of the task. One may contend that it is when this process is not complete that users experience Norman's (1988) gulfs of evaluation and execution.

First, we discuss methods for understanding the user's task as embedded within his environment. Within this section, we review acquiring knowledge about users and their tasks and the methods to document those tasks. Next we describe methods that allow designers to model the users and their interactions with the Web including such techniques as goals, operators, methods, and selection rules (GOMS) and natural language GOMS (NGOMSL), along with computational models such as state, operator, and result (SOAR) and adaptive control of thought-rational (ACT-R). In addition, we characterize team interactions and the elements that impact distributed collaboration. Last, we discuss methods that can help designers enhance users' interactions with the Web. Although we use examples throughout the chapter to discuss the methods for individual tasks as well as collaborative team tasks, we encourage readers to seek out the original literature references for detailed explanations of each of the methods.

5.2 UNDERSTANDING USER TASKS EMBEDDED WITHIN THE ENVIRONMENT

Figure 5.3 identifies each of the methods we discuss in understanding the users in their environment. These methods help the designer understand what users are going to do (e.g., task) or want to do and where are they going to do it (e.g., environment). Although we have compartmentalized the methods for efficiency in covering the material, the use of these methods is typically iterative and focused at better understanding the user's needs.

5.2.1 KNOWLEDGE ACQUISITION

Norman (1986) stated that a "user's mental model" guides how the user constructs the interaction task with a system and how the computer system works. Owing to its diverse meanings in different contexts, the exact definition of "mental model" is difficult to address. In general, a mental model can be viewed as the users' understanding of the relationships between the input and the output. Users depend on

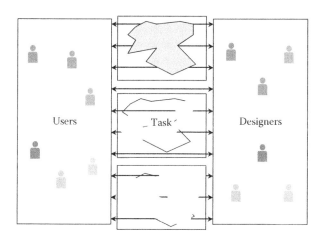

FIGURE 5.2 Iterative design process.

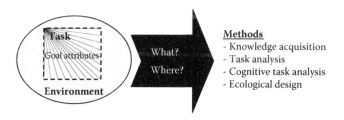

FIGURE 5.3 Methods for understanding the user.

their mental models to predict the output that would be produced for the possible inputs (Eberts 1994; van der Veer and Melguizo 2003; Payne 2008). The mental model is also called user model. While knowledge acquisition (KA) tools originally were developed to extract human expertise for the purpose of developing knowledge-based systems, these methods can be executed to extract this information that feeds further task analysis or cognitive task analysis that will be discussed later.

The user model serves as input to designers. Without consideration of the user model by designers, discrepancies between a Web site and a user's expectation will most likely result. To bridge the gap between a mental model and conceptual model, it is essential for designers to acquire knowledge from users about how their mental models respond to a task. Through the knowledge acquisition from users, explicit compatibility between a mental and conceptual model can be accomplished.

Knowledge acquisition techniques have traditionally been used to collect both declarative (facts) and procedural (operations) knowledge that is associated with how experts fulfill their goals in tasks (Lehto et al. 1997). As discussed earlier, the user population that interacts with the Web varies. They differ in knowledge and skills, and thus no one single user or user group represents a "true" expert. KA techniques, although originally derived to extract expertise from experts, also serve well in understanding the users and their task environments.

One of the first techniques introduced to acquire knowledge from experts is what Newell and Simon (1972) called the verbal protocol method. In this method, the experts are provided with a problem to solve and are asked to verbalize the knowledge they were using to complete a task (Ericsson and Simon 1980; Koubek, Salvendy, and Noland 1994; Preece et al. 2002). Throughout the years, many different methods have been used to extract information from users both expert and nonexpert. Methods include those listed in Table 5.1.

Each of these methods attempts to understand users and their interactions with the task and technology. It is from these techniques that designers extract information by which to define the goals and methods users execute to achieve their objective. While many authors will suggest that one method is better than another, in reality, each of the methods can be useful for different purposes in the design process. Most designers in the real world will use many of these techniques in order to narrow Norman's (1988) gulfs of execution and evaluation.

The main aim of knowledge acquisition is to construct a user knowledge base. But the quality of the knowledge base depends upon the skills of the knowledge engineer (KE), who plays a major part in the process of knowledge acquisition by obtaining the knowledge and then transferring this knowledge into a form that can be used by designers. While a knowledge engineer is trained to elicit knowledge from users, the techniques used ultimately rely on the user being

TABLE 5.1
Knowledge Acquisition Techniques

Technique	Description
Interviews (structured and unstructured)	Interviews allow designers to get a first account user perspective of the task.
Questionnaires	An excellent means to get information from many users.
Naturalistic observation	Allows the designer to see the user in her natural environment.
Storyboards	Used to walk a user through a concept by presenting the sequence of actions on separate boards where each board may represent a single or multiple actions.
Mock-ups	A physical representation of a preliminary design. This technique is very useful in allowing users to see several potential interface options prior to committing them to software.
Card-sorting task	Concepts are presented to users and they sort them into piles of similar concepts. This technique allows designers to understand how users classify concepts within a task.
Simulator experiments	A simulation of the interface that allows user interaction. The simulator can vary as to the level of realism depending on the specific design objectives being evaluated.

Note: In addition to the above, there are some other methods in the works by Randel, Pugh, and Reed (1996), Crandall et al. (1994), and Vicente (1999). (From Randel, J. M., H. L. Pugh, and S. K. Reed. 1996. Differences in expert and novice situation awareness in naturalistic decision-making. *International Journal of Human–Computer Studies* 45: 579–97; Crandall, B., G. Klein, L. G. Militello, and S. P. Wolf. 1994. Tools for applied cognitive task analysis (Contract Summary Report on N66001-94-C-7008). Fairborn, OH: Klein Associates; Vicente, K. J. 1999. *Cognitive Work Analysis*. Mahwah, NJ: Lawrence Erlbaum.)

able to verbalize their expertise. Eberts (1994) outlined several problems that can occur in the knowledge acquisition process:

√ Interpretation: KEs must be sure not to insert their biases into the data.
√ Completeness: KEs may leave out important steps in the problem-solving task.
√ Verbalization assumption: It is the assumption of most of the techniques that the user can verbalize the procedures and data used in accomplishing a task. In reality, some may not be amenable to verbalization.
√ Problem with users: User populations vary. Experts for example may automate many of their tasks and thus make interpretation difficult. Novices, however, may take unnecessary steps because of their lack of experience or to confirm their actions.

Owing to the large and cumbersome process, KA is not without its problems (McGraw and Harbison-Briggs 1989). Some include the following:

• A tiresome, time-consuming, and very expensive process
• Difficulty of finding representative user populations
• Problems in transferring the data

While problems exist, the designer must acquire the user's task knowledge in order to represent it through other methods that will be discussed later. Lehto et al. (1997) provide a detailed review of knowledge acquisition for further reading.

5.2.2 Task Analysis

Unfortunately, designers sometimes disregard the aspects of tasks from the perspective of users. For example, the hypertext in a Web site provides useful benefits to the user. The user is able to jump to multiple related articles with several clicks and convenient backtrackings. However, if designers build Web sites that are only biased toward system-oriented techniques with which they are familiar, the user might be lost in a Web site and frustrated. The user could encounter hyper-chaos in hypertext Web sites.

Kirwan and Ainsworth (1992) defined task analysis as the study of identifying the requirements of an operator (or a user) to accomplish goals in a system in terms of actions or cognitive processes. Task analysis is used to identify the details of specified tasks, like the required knowledge, skills, and personal characteristics for successful task performance and to use this information to analyze designs and systems (see also Strybel, this volume). We can define task analysis as a methodology to identify the mapping from task to cognitive human components and to define the scope of the knowledge acquired for designing any particular application or system (Wickens et al. 2004). Task analysis is a generic method that

will establish the conditions needed for a hierarchy of subtasks to achieve a system's goal.

The first step in any task analysis is to study the job or jobs to determine the task requirements. Understanding the task will take on different considerations depending on whether one is evaluating an individual task or a collaborative task. For an individual task, typically, the initial phase will consist of reviewing written material about the job including items such as training materials, job flowcharts, or procedure manuals. For collaborative tasks, one will still consider the elements of the individual; however, the group interaction must also be considered. Questions to be answered include: Is the work synchronous (e.g., conferencing) or asynchronous (e.g., e-mail)? Who are the participants? What is the current means of collaborating? Many issues are raised that must be considered including group makeup, technology, and team issues.

Once this familiarization phase is complete, many of the knowledge acquisition techniques discussed earlier are put into action. Typically, designers will interview users of the systems at many different organizational levels including the task workers, managers, and task support personnel. For example, if one were designing an e-commerce Web site, designers would talk to end use customers, business-to-business customers, and managers as an example of just some of the initial people interviewed. As more detailed information or types of information are needed, other techniques would be employed (refer to Table 5.1).

The second step is to identify through some representation the activities within the task and how they are interconnected. There are many ways one could go about this process. Fundamentally, however, a task analysis tries to link the interface (*Note*: Interface is used in a general sense here. It is anything with which a user interacts external to the user), elements (e.g., information displayed, colors of displays), and user's behavior (e.g., push the red button to stop the machine). Three fundamental approaches have been identified (Kirwan and Ainsworth 1992; Vicente 1999). These include (1) input/output constraints, (2) sequential flow, and (3) timeline. Input/output constraints identifies the inputs that are required to perform a task (e.g., information), the outputs that are achieved after the task is complete, and the constraints that must be taken into account in selecting the actions that are required (Vicente 2000). For example, let us assume a user wants to use one of the many Web-based atlases that will provide a user a trip route. Inputs into the process would include the starting point and the destination. Outputs that are possible include a route from the starting point to the destination along with other information including historical stops along the route, hotel locations, etc. Constraints would include that only interstate roads are possible for travel (as opposed to back roads or country driving), time allowed to complete the trip, etc. While the inputs, outputs, and constraints do not dictate the design, they start to define the functionality that such a Web site might include.

The sequential flow task analysis identifies the order to a sequence of actions that a user takes to accomplish a specific goal (Kirwan and Ainsworth 1992; Vicente 1999). A typical means of representing the sequential flow is through a flowchart or stepwise procedure description. For a simple task with a single goal, a single flowchart may be enough to describe all of the possible actions. However, for a complex task, there will most likely be many flowcharts where each chart represents a different goal. These multiple charts especially in computer interfaces are likely to be connected to a single decision point early in the flowchart where a user branches out to different flowcharts depending on the goal. For example, a user that enters a company's Web site may be faced with many different paths they can take depending on their specific goal/subgoal. If the user went to the Web site to seek out information on the company, they are likely to venture down one path. If they are there to purchase a product, they will venture most likely down another path. Thus one can see that there can be many varied task flowcharts based on the objective of the user. Hence the knowledge acquisition phase discussed earlier becomes critical to understanding the uses of your Web site.

A sequence flow analysis would describe each decision point and then the path that results based on the decision(s) made. In addition, a sequence flow analysis would describe alternative paths that may be executed to meet the same goal.

The last level of task analysis, timeline, identifies the temporally ordered sequence of actions along with the estimated durations. This is the most detailed form of a task analysis, and it is used heavily in manufacturing operations. Industrial engineers have used time-motion studies to describe work tasks and the timing of those work tasks to assist in the design of manufacturing lines (Niebel and Freivalds 1999). In addition, we discuss methods such as GOMS (Card, Moran, and Newell 1983) and NGOMSL (Kieras 1988) later that have been used in the human–computer interaction environment to model computer based tasks as well.

As mentioned, task analysis has been used for several different purposes including worker-oriented task analysis that deals with general human behaviors required in given jobs, job-oriented task analysis that deals with the technologies involved in a job, and cognitive task analysis that deals with the cognitive components associated with task performance. With the evolution of the tasks from more procedural to those that require higher cognitive activity on the part of users, we turn our attention to understanding how to clarify the cognitive components of a task through cognitive task analysis.

5.2.3 Cognitive Task Analysis

Cognitive task analysis (CTA) is "the extension of traditional task analysis techniques to yield information about the knowledge, thought processes, and goal structures that underlie observable task performance" (Chipman, Schraagen, and Shalin 2000, 3). The expansion of computer-based work domains has caused the generic properties of humans' tasks to be shifted from an emphasis on biomechanical aspects to cognitive activities such as multicriteria decision making or problem solving (Hollnagel and Woods 1999). There have

been increases in cognitive demands on humans with radical advances in technologies (Howell and Cooke 1989). Instead of procedural and predictable tasks, humans have become more responsible for tasks that are associated with inference, diagnosis, judgment, and decision making, while procedural and predictable tasks have been controlled by computerized tools (Millitello and Hutton 1998). For example, increased cognitive requirements may result because of:

1. Real-time decisions: a lack of clarity on how the decisions were temporally organized and related to external events requires operators to make real-time decisions.
2. Uncertainty: unpredictability and uncertainty of external events faced in task environment or even after having clear goals, it is unclear exactly what decisions have to be taken until the situation unfolds. Thus the operator must adapt both to the unfolding situation and to the results of actions taken.
3. Multitasking: the pace of events and uncertain processes require the decision maker to be prepared to interrupt any cognitive activity to address a more critical decision at any time. This will typically result in weak concurrent multitasking, in which the decision maker may have several decision processes underway at a time.
4. Indirect dialogue (e.g., computer-based, verbal interactions): the majority of information available to the user comes not from direct sensation of the task environment, but rather through information displayed at computer-based workstations and verbal messages from teammates. Similarly, decisions are implemented not through direct action, but as interactions with the computer workstation or verbal messages to other persons (Zachary, Ryder, and Hicinbothom 2000).

Hence, CTA moves beyond the observable human behaviors and attempts to understand the cognitive activities of the user that are many times invisible to the observer (e.g., the logic used to select one path of an activity over another. CTA identifies the information related to the cognitive, knowledge structures, and human thought processes that are involved in a task under study. CTA, while similar to general task analysis, focuses on how humans receive and process information in performing a task and how the task can be enhanced to improve human performance. The aim here is to investigate the cognitive aspects of tasks that may emphasize constructs like situational awareness, information processing, decision making, and problem solving.

CTA covers a wide range of approaches addressing cognitive as well as knowledge structures and internal events (Schraagen, Chipman, and Shalin 2000). In recent years, cognitive task analysis has gained more recognition with the transition to modern high-technology jobs that have more cognitive requirements. As mentioned, most of the time, these cognitive requirements or needs of the work will not be directly visible. Cognitive task analyses are conducted for many purposes including design of computer systems to help human work, development of different training programs, and tests to check and enhance the performance of humans.

The steps to completing a CTA are not much different than that of a task analysis. However, CTAs are more concentrated on what is internal to the user in addition to the external behaviors. In addition, many of the techniques identified in the knowledge acquisition section will additionally be used in conducting a CTA. However, some additional techniques specific to modeling knowledge of users have grown out of the expansion of CTA. A CTA generally consists of three phases: (1) task identification and definition, (2) identifying abstract user(s) knowledge, and (3) representing user knowledge (Chipman, Schraagen, and Shalin 2000; Klein 2000).

The task identification and definition phase identifies the tasks of the specified job that are important for detailed cognitive analysis. The second phase, identifying abstract user(s) knowledge, isolates the type of knowledge representation based upon the knowledge obtained and data gathered from the preliminary phase. Once the type of knowledge used within the task has been identified, the last step requires the use of knowledge acquisition techniques again to get at the underlying knowledge to complete the task so that the user's knowledge can be represented in a meaningful manner. A later chapter on task analysis provides a detailed description of CTA.

5.2.4 Ecological Design

Ecological design, which was adapted from the biological sciences, has been described as "the art and science of designing an appropriate fit between the human environment and the natural world" (Van der Ryn and Cowan 1995, p. 18). With its roots in biology, ecological design is concerned with design in relation to nature. Architect Sim Van der Ryn and coauthor Stuart Cowan describe several principles of ecological design, as described below, that can be carried through to human interface design:

1. *Solutions grow from place.* Ecological designs must address the needs and conditions of particular locations. Therefore, the designer must have significant knowledge of the "place" their design will be applied, and all designs must be "location specific."
2. *Ecological accounting informs design.* The designer must understand the environmental impacts of certain designs and consider the impact when determining the most ecologically sound choice.
3. *Everyone is a designer.* Each person has special knowledge that is valuable in the design process. Every voice should be considered.

As discussed in the introduction, the process of design and evaluation of usability requires designers to address four crucial components: (1) the environment, (2) the human,

(3) the tool, and (4) the task (Koubek et al. 2003). One of those elements, environment, is frequently not given much thought when designing products. Ecological design frames design problems with respect to their environment. An ecological approach to cognition believes that the situation has meaning in the design of the system. While systems are designed to try to meet the task demands of users, in complex tasks it is unlikely that a system can be designed for every possible activity. Therefore, ecological design tries to present the user a system that can meet with the complex rich environment in which the user operates (Flach 2000; Rasmussen and Pejtersen 1995).

Ecological design deals with the complexity involved in work demands by considering both cognitive constraints that originate with the human cognitive system and environmental constraints that originate based on the context in which people are situated like a collaborative work environment (Vicente 1999; Vicente and Rasmussen 1992). Design and analysis are done in accordance with the environmental impact on the work life cycle. Ecological design focuses on the user/worker mental model along with the mental model of the work environment. In other words, user mental models should also encompass the external work reality. However, the need for perfect integration of cognitive and environmental constraints depends on the domain of interest and the real need of design. Sometimes, there may also be need for social and cognitive factors in the human computer interaction (Eberts et al. 1990) and especially in team collaboration perspectives (Hammond, Koubek, and Harvey 2001).

The Skills, Rules, and Knowledge (SRK) taxonomy (Rasmussen 1990) states the knowledge that a user/operator possess about a system will make up his or her internal mental model. Rasmussen's taxonomy provides a good framework for understanding the user in her environment. It allows designers to consider the user's knowledge of the system to meet the uncertainty in any crunch or unexpected situation by improving decision-making efficiency as well as system management. In addition, it is useful for understanding the system itself and how it can be controlled within the environment.

Systems vary depending on their level of complexity. In some systems, tasks must follow certain physical processes that obey the laws of nature. Thus, the operator only has a limited number of actions (many times only one) that can be taken. For example, a light switch is either on or off. The laws of nature limit the flow of electricity (i.e., electricity will only flow across a closed circuit—on position). As a result, designers have created a simple interface to support users' interaction with the light switch (e.g., up is generally on and down is generally off in the United States).

In complex system environments the actions taken by users are typically very situation dependent. Thus, there are at best many different ways in which a situation may be dealt with and potentially even an infinite number. Therefore, trying to determine every potential situation and proceduralizing the steps that should be taken would be impossible. For example, one of the authors used to develop systems for a financial institution. This institution had many different forms that entered the company for customer transactions (e.g., loan request, notice of bankruptcy). In implementing a document imaging system, work queues were developed to handle each type of form received. When a form was received, it was scanned and routed to a specific work queue based on the form type where a clerk followed a very strict procedure governed by a combination of company policy, government regulation, and guarantor rules. However, one particular piece of mail could not be handled this simply. This piece of mail was a letter. Customers could write letters to the company for many different reasons (e.g., request a form, request a payment extension, request refinancing, inform the company of a death). While the piece of mail was still sent to a letters work queue, it became impossible to design a procedural interface that could handle all the situations that occurred. As a result, a more ecological design was followed. Instead of designing a system that is tightly coupled to a driven procedure, the interface was designed with consideration of the types of information needed to handle the varied tasks along with coupling the right type of user with the task.

In the simple light control example described earlier, a user with limited experience (e.g., a child) can operate the interface once they become familiar with functionality provided by the light switch. While understanding the laws of nature may add to their user ability, it is not vital to the task operation. Thus, even small children learn very quickly how to turn on and off the lights in their home provided they can reach them. Likewise, in the form work queues that followed company procedures, the user has to understand the system (e.g., form and interface) and at times deal with activities that are not the norm; however, most cases can have an interface designed to meet a majority of user activities. In the more complex letter work queue, we must not only consider the interface but also the user. Users in this environment must have a broader understanding of the work activities and be able to handle uncertainty. As well, their interface must be able to support the numerous types of activities the user may take in solving the customer inquiry.

5.3 MODELING USERS' INTERACTIONS WITH THE WEB

Increasingly, the World Wide Web is a medium to provide easy access to a variety of services and information online. However, according to Georgia Institute of Technology's GVU's 10th WWW Users Surveys (1998), only 20% of respondents answered that they could find what they are looking for when they were intentionally searching for products or service information. In this section, modeling techniques are discussed that allow designers to model who will accomplish the task and how they will accomplish the task (see also van Rijn, Johnson, and Taatgen, this volume). Figure 5.4 shows the different modeling techniques that will be reviewed in answering the *who* and *how* questions.

Understanding human cognitive functions would be helpful to design and model more interactive Web-based systems. Many of these models (Figure 5.4) allow designers to quantify (e.g., time to complete a task) a user's interaction with an interface. By quantifying this interaction, designers can make a more informed decision when choosing between alternative designs. Table 5.2 provides an overview of the many cognitive models that have been used extensively throughout the literature. We will review several of these in the discussion that follows. The models fit into two classifications: (1) user performance models and (2) computational cognitive models.

5.3.1 USER PERFORMANCE MODELS

5.3.1.1 Model Human Processor

The Model Human Processor (Card, Moran, and Newell 1983, 1986) affords a simplified concept of cognitive psychology theories and empirical data. It provides approximate predictions of human behavior through the timing characteristics of human information processing. This modeling technique has implicit assumptions that human information processing can be mainly characterized by discrete stages. The Model Human Processor is comprised of three subsystems: (1) perceptual system, (2) cognitive system, and (3) motor system. A set of complicated tasks can be broken down into individual

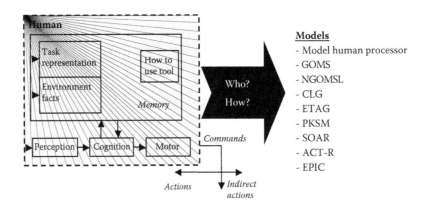

FIGURE 5.4 Modeling the user.

TABLE 5.2
Example Cognitive Models

Modeling Type	Uses	Examples	References
External tasks	Specification language for how to translate a task into commands in a given environment.	External Internal Task Mapping (ETIT)	(Moran 1983)
User knowledge	Represent and analyze knowledge required to translate goals to actions in a given environment.	Action Language, Task-Action Grammar (TAG)	(Reisner 1983) (Payne and Green 1986)
User performance	Describe, analyze, and predict user behavior and performance. Similar to User Knowledge models except that they provide quantitative performance measures.	Model Human Processor Goals Operators Methods and Selection rules (GOMS) Natural GOMS Language (NGOMSL) Cognitive Complexity Theory (CCT)	(Card, Moran, and Newell 1983, 1986) (Card, Moran, and Newell 1983) (Kieras and Meyer 1988) (Bovair, Kieras, and Polson 1990)
Task knowledge	Provide a specification for the full representation of the system interface and task at all levels of abstraction.	Command Language Grammar (CLG), Extended Task-Action Grammar (ETAG)	(Moran 1981) (Tauber 1990)
Computational cognitive models	Provide a means of simulating user cognitive activities through a real-time process.	Adaptive Control of Thought (ACT-R) State, Operator, and Result (SOAR) Executive Process-Interactive Control (EPIC)	(Anderson 1976, 1983) (Laird et al. 1987) (Kieras and Meyer 1995)

Source: Adapted from Koubek, R. J., et al. *Ergonomics* 46, 1–3, 220–241, 2003.

elements with relevant timing characteristics that would allow alternative interface designs to be compared based on the relative differences in task timings (Eberts 1994; see also Strybel, this volume).

5.3.1.2 GOMS

GOMS (Goals, Operators, Methods, and Selection rules) is a well-known task analysis technique that models procedural knowledge (Card, Moran, and Newell 1983). Procedural knowledge can be viewed as acquired cognitive skills for a sequence of interactive actions of the user. Kieras (1997) mentioned that a GOMS model is a representation of "how to do it" knowledge that is required by a system in order to accomplish a set of intended tasks. The GOMS model assumes that cognitive skills are a serial sequence of perceptual, cognitive, and motor activities (Lohse 1997). The four components of the GOMS model are:

√ Goals: target user's intentions (e.g., to search or retrieve information, to buy a digital camera online, or to make a payment of electronic bills through online transaction, etc.)

√ Operators: actions to complete tasks (e.g., to move the mouse to a menu, or to make several clicks, etc.)

√ Methods: an array of actions by operators to accomplish a goal (e.g., perform several menu selections to find a user's favorite digital camera, etc.)

√ Selection rules are described as choosing an appropriate method among competing methods. A

selection rule is represented in terms of if-then rules. Thus, it determines which method is to be applied to achieve a goal.

The benefits of GOMS models include the emphasis on user performance prediction and formalized grammar for describing user tasks (Eberts 1994). However, GOMS models have difficulty in dealing with errors, limited applicability to tasks associated with little or no problem solving, and its reliance on quantitative aspects of representing knowledge at the expense of qualitative aspects (De Haan et al. 1993; see also Strybel, this volume).

5.3.1.3 Natural GOMS Language (NGOMSL)

The Natural GOMS Language (NGOMSL) was first developed by Kieras (1988) in order to provide more specific task analysis than the GOMS model. The NGOMSL is a structured natural language to represent user's methods and selection rules. Thus, it affords explicit representation of the user's methods while the methods of users are assumed to be sequential and hierarchical forms (Kieras 1997). Like GOMS model, one important feature of NGOMSL models is that procedural knowledge ("how to do it" knowledge) is described in an executable form (Kieras 1997). The description of procedural knowledge to accomplish intended goals in a complex system can be useful fundamentals for learning and training documentation.

NGOMSL has two major features including learning time and execution time prediction. Kieras (1997) mentions that the learning time is to be determined by the total number and length of all methods and the execution time is predicted by the methods, steps, and operators that are necessary to accomplish a task. For learning time, the length of all methods indicates the amount of procedural knowledge that is necessary to be acquired so as to know how to use the system for all of the possible tasks to be considered (Kieras 1997; see also Strybel, this volume).

5.3.1.4 Command Language Grammar (CLG)

Moran (1981) introduced the Command Language Grammar (CLG) to represent the designer's conceptual model in the system interface implementation. The main purpose of the CLG is to allow the designer to fully build a conceptual model of a system that is acceptable to the user needs.

Moran (1981) addressed the features of CLG from three points of view such as linguistic, psychological, and design. From a linguistic view, the CLG describes the structural aspects of the system's user interface. This structural aspect indicates the intercommunication between the user and the system. From a psychological view, the CLG is to provide a model of a user's knowledge on a system. Thus, the user's mental model can be represented by the CLG, even though it is necessary to be validated. Finally, the CLG can make contributions to understand design specifications in a system. From this design view, the CLG affords a top-down design process. The top-down design process is to specify the conceptual model of a system to be implemented, and then it is possible to communicate with the conceptual model through the command language.

5.3.1.5 Extended Task-Action Grammar (ETAG)

Tauber (1990) proposed the Extended Task-Action Grammar (ETAG). The ETAG represents perfect user knowledge associated with the user interface. Even though it does not enumerate the user knowledge in the mental model, the ETAG exactly specifies the knowledge about how the interactive system works from the perspective of the user. It describes what has to be acknowledged by the user in order to successfully perform a task. De Haan and Van der Van (1992) state that an ETAG representation is a conceptual model that incorporates information that a user wants and needs of a computer systems. Thus, ETAG representations can assist in providing the fundamentals for intelligent help systems (De Haan and Van der Veer 1992).

5.3.1.6 Procedural Knowledge Structure Model (PKSM)

Benysh and Koubek (1993) proposed the Procedural Knowledge Structure Model (PKSM) that combines the characteristics of cognitive modeling techniques and a knowledge organization framework. The PKSM is a structural model of procedural knowledge. It incorporates structural aspects of the human cognitive representation and assimilates procedural aspects of the cognitive models (Benysh and Koubek 1993). Unlike CLG and ETAG, it delineates the procedural knowledge from task execution that is found in other applied cognitive models (Koubek et al. 2003).

A task in the PKSM is rendered as a three-dimensional pyramid. Task goals are decomposed into smaller goals, task elements, or decision nodes at the next lower level, much like GOMS or NGOMSL (Benysh and Koubek 1993). Each level has a flowchart representation of the task steps. The decision nodes are to control the flow through the chart. The most noticeable feature is that the PKSM is capable of defining parameters indicating psychological principles with skill and performance. Therefore, it is possible to differentiate the task performance of experts and novices. Moreover, the PKSM can assess the usability of the task (Koubek et al. 2003).

Aforementioned, it has been stated that the significant contribution of the PKSM is to afford the structural aspects on knowledge while simultaneously incorporating important procedural aspects of cognitive psychology and knowledge based systems (Benysh and Koubek 1993). The PKSM made deliverable contributions to model procedural aspects of knowledge to perform an array of tasks.

5.3.2 COMPUTATIONAL COGNITIVE MODELS

Cognitive computational models were initially developed to attempt to explain how all the components of the mind worked. Several models have been developed to meet this objective. We briefly discuss several of the major models; however, further reading is needed for a comprehensive understanding of these models.

5.3.2.1 SOAR

State, Operator, and Result (SOAR) is a cognitive architecture that delineates problems by finding a path from an initial state to a goal state (Laird, Rosenbloom, and Newell 1987; Newell 1990). SOAR utilizes heuristic search by which the decisions are to be made in problem spaces. A subgoal is generated when the decisions can be made within a set of problem spaces. Thus, the subgoal would be carried out in another problem space. SOAR's cognitive architecture has some features of working memory and long-term memory. SOAR does not have perception and motor components. However, it is assumed that a human through the perception cycle from Model Human Processor (Card, Moran, and Newell 1983) perceives physical visual stimuli with the cycle value of 100 ms (τ_p).

5.3.2.2 ACT-R

ACT-R was developed by Anderson (1976, 1983; Anderson and Lebiere 1998). ACT stands for Adaptive Control of Thought. R was added with the marrying of ACT with Anderson's (1990) rational basis. ACT-R has two types of permanent knowledge such as declarative and procedural knowledge. The declarative knowledge takes the form of chunks comprised of a number of slots with associated values. Procedural knowledge is represented as production rules.

5.3.2.3 EPIC

Kieras and Meyer (1995) developed Executive Process-Interactive Control (EPIC) architecture. EPIC incorporates various theoretical and empirical findings associated with human performance from the 1980s (Lin et al. 2001).

The production rules of the EPIC cognitive architecture were derived from more simplified NGOMSL (Lin et al. 2001). In EPIC, visual, auditory, and tactile processors are the elements of working memory with unlimited capacity. EPIC has four perceptual processors: visual sensory, visual perceptual, auditory perceptual, and tactile perceptual processors. Besides, EPIC has three motor processors: manual motor processor, ocular motor processor, and vocal motor processor. These motor processors receive motion related commands from the cognitive processor of EPIC. A strong feature of EPIC is that cognitive architectures can be easily built by following NGOMSL methodology (Lin et al. 2001).

5.4 CHARACTERIZING TEAM INTERACTIONS

One area in which the Internet has had a large impact is team collaboration. Because such collaboration involves interactions among multiple humans and technology, cognitive factors must be considered in the design of any collaborative Web technology. Many different forms for groups and teams interact via the Internet. Some are informal (e.g., Facebook), whereas others are more formal (e.g., business teams with specific purposes and objectives). Although there are various definitions for the term "team" (see Kiekel and Cooke, this volume), Arrow, McGrath, and Berdahl (2000) provide a good definition: A team is a complex, adaptive, dynamic, coordinated, and bounded set of patterned relations among team members, tasks, and tools. Arrow et al.'s definition of team is a comprehensive one based on a synthesis of the vast literature on teams and small groups. They included the complex, adaptive, and dynamic nature of teams along with coordination and relationships among team members to define teams. In considering team tasks, one must consider three elements: (1) relationships among team members, (2) tasks, and (3) tools. Therefore, a team is not merely a group of people who work together on a common objective and share the work responsibilities.

There exist many theories and models in the team literature, such as the input-process-output model (Hackman 1987; McGrath 1984), the Team Evolution and Maturation (TEAM) model (Morgan et al. 1986), the Forming-Storming-Norming-Performing model (Tuckman 1965), and the Team Adaptation Model (Entin and Serfaty 1999). Each of these models contributes to understanding team cognition. But most of these team theories and models are based on the classic systems theory of input-process-output in some manner (Ilgen 1999). Both McGrath (1984) and Hackman (1987) described the traditional small group's research of classic systems theory in terms of inputs, processes, and outputs. However, the other input-process-output approaches tended to focus more on the development of psychological process theories (e.g., Steiner's [1972] model of group process and Cooper's [1975] book on theories of group processes). Teams' tasks, contexts, and composition (on the input side) often were of interest only as boundary conditions, thereby restricting behaviors and contexts over which process theories generalized. Also, the theories mainly relied on subjective team performance measures and behavioral scales.

5.4.1 TEAM TASKS

Some early researchers such as Hackman (1969), Steiner (1972), and McGrath (1984) followed a typology of tasks approach. Because "teams" and "tasks" are often considered together, researchers were interested in the typology of task approach. Because task performance was central to these early researchers, a part of their research dealt with the effects of different types of tasks on performance of the teams (e.g., Kent and McGrath 1969; McGrath 1984; Steiner 1972).

Because teams engage in many different collective activities, a number of task typologies and descriptions have been presented in the team-related literature in an effort to better define and understand the critical role of the tasks and the associated team processes. Some task typologies include Intuitive Classification Method (Roby and Lanzatta 1958); Task description and classification method (Hackman 1969); Categorization scheme method (Steiner 1972); Laughlin's method (Laughlin 1980); and Task circumplex (McGrath 1984; see also Kiekel and Cooke, this volume).

Building on the task classification literature, Rothrock et al. (2005) discuss a team task complexity space model along with a framework for evaluating team performance. Darisipudi (2006) explored a revised version of Rothrock et al.'s model and found that the interaction of three dimensions of task complexity (scope, coordination, and uncertainty) contributed to team performance for interactive teams. Thus, the nature of the task and its components are important elements in evaluating team performance.

5.4.2 TEAM TOOLS

The loss of the $125 million NASA Mars Climate Orbiter in 1999 illustrates how technology can impact teams. In this case, one or both distributed teams' lack of communication over decisions made concerning unit measurements (i.e., English versus SI units) led to the ultimate destruction of the orbiter (CNN 1999). Understanding how team members communicate and interact with each other, and what makes a team successful, is important given that technologies available today allow organizations to take advantage of team collaborations. Plus team collaborations have become recognized as a competitive necessity (Hacker and Kleiner 1996).

Distributed team members, linked through technological interfaces, may vary in location, discipline, company loyalties, and culture (Forsythe and Ashby 1996; Hacker and Kleiner 1996; Hartman and Ashrafi 1996). Greiner and Metes (1995, 8) explained "Virtual teaming is an optimal way to work in the current environment of time compression, distributed resources, increasing dependence on knowledge-based input, premium on flexibility and adaptability and, availability of electronic information and communication through networks."

Current colocated work group research does not necessarily provide a sound theoretical framework for the understanding of virtual teams. This is due, in part, to the well-established finding that group interaction and communication differs when the participants are distributed (e.g., Carey and Kacmar 1997; Hiltz, Johnson, and Turoff 1987; Weeks and Chapanis 1976). Current research also does not take into account a number of sociotechnical issues introduced by the virtual team structure (e.g., Hacker and Kleiner 1996; Grosse 2002; Powell, Piccoli, and Ives 2004; Anderson et al. 2007).

Social demands, as well as task demands, are inherently increased by the nature of the virtual environment. Greiner and Metes (1995, 211) explained, "Many people are reluctant to use the electronic infrastructure to do their work, which may be highly communicative, knowledge intensive and ego-involved. They communicate best face-to-face . . . are resistant to wide-spread electronic information sharing and don't like the planning and formality that are required to make virtual processes work."

As the distributed setting serves to complicate the interactions of group members, the consequences of changed group dynamics cannot be underestimated. The types of interactions utilized by teams, including criticisms, opinions, clarifications, and summaries, are key to optimizing decisions. Changes in the normal social subsystem resulting in a narrower or less open exchange of ideas or information will obviously inhibit and hinder the success of design teams and require further study (Burleson et al. 1984; Olson et al. 1992).

The technical essentials consist of the actual technology, as well as the procedures and methods used by an organization in employing technology to complete tasks (Hacker and Kleiner 1996). In the case of virtual teams, the social subsystem may enjoy increased optimization by advances in the technical subsystem. Technical innovations, which support effective interactions between team members, may serve to reduce the complexity of the situation, as perceived by group members. Although technology can enable more teams to manipulate and discuss shared representations simultaneously, technological systems alone cannot provide an answer to the complexity problem presented by virtual teams. A review by Carey and Kacmar (1997) found that the introduction of technical systems, in most studies, has brought about operational and behavioral changes but not the desired increases in quality or productivity.

New technologies appear to further complicate the distributed environment in most applications. Carey and Kacmar (1997) stated that using such technologies within an already information-rich environment may be a poor choice. Their findings suggested rethinking the design of current collaborative technologies to ensure greater effectiveness.

Hammond, Koubek, and Harvey (2001) found that the literature through the 1980s and 1990s shows that most research relied on two ideas to explain media differences in communication: (1) efficiency of information transfer and (2) alteration of the communication process. Hammond, Koubek, and Harvey stated that when teams become virtual, their decision-making processes and potentially their end products are altered, as distributed participants compensate for reduced channels of communication and the altered "social presence" of the medium. Hammond et al. (2005) found that collaborative design teams had increased cognitive workload, perceived perception of declined performance, and even used coping mechanisms (e.g., limiting heuristics) to deal with virtual teamwork. In addition, virtual teams as compared to face-to-face teams had less frequent interactions, spent less time interacting, and took a greater time to reach consensus.

Kanawattanachai and Yoo (2007) found that MBA students performing a complex Web-based business simulation task spent much time initially during their interaction focusing on task-oriented communications. The frequency of interaction in the initial phase of the project was a significant determinant of team performance. These task communications allowed teams to eventually develop where expertise was located among team members as well as cognitive-based trust. Once teams developed a transactive memory system (Powell et al. 2004), task communication became less important. Both Hammond et al. and Kanawattanachai and Yoo indicate that frequent interactions are vital to improved team performance when that collaboration is mediated through technology.

5.4.3 THE TEAMS

Characterizing the team is another element in understanding the team process. People have tried to model social worker teams (Bronstein 2003), creative research teams (Guimerà et al. 2005), medical critical care teams (Reader et al. 2009), and all types of teams in between. Fleishman and Zaccarro (1992) prepared an extensive taxonomy of team functions that helps explain the many elements that define team interaction along with the need to explore the activities of a team in order to match the team needs to the Web-based technology. The seven functions include the following:

1. Orientation functions: the processes used by team members in information exchange needed for task accomplishment.
2. Resource distribution functions: processes used to assign members and their resources to particular task responsibilities.
3. Timing functions: organization of team activities and resources to complete the tasks within time frame and temporal boundaries.
4. Response coordination functions: coordination and integration and synchronized member activities.
5. Motivational functions: definition of team objectives/goals and motivational processes for members to achieve the proposed objectives.
6. Systems monitoring functions: error detection in the team as a whole and individual members.
7. Procedure maintenance: maintenance of synchronized and individual actions in compliance with established performance standards.

Team performance is likely to be impacted by several variables. So the question is: What team variables ultimately impact team performance? This chapter will not attempt to answer that question but will provide examples of work being done that may ultimately allow accurate modeling of team performance variables. We provide two examples that illustrate modeling of team performance.

Many military operations require team interactions to accomplish a task (Grootjen et al. 2007; Richardson et al., forthcoming). The issue is determining what attributes would affect team performance. One way to fit the task to the workers is to determine the work requirement and to measure the physical/mental ability of the worker, then match the team members to the proper jobs. This is a practical but time-consuming process. An alternative way is to develop a set of physical/mental test batteries, test every worker, and record the measurements in their personnel files. As a team task becomes available, workers can be assigned based on a match between their abilities and the abilities required for the job. Richardson et al. (forthcoming) applied Fleishman and Quaintance's (1984) taxonomy of 52 human abilities that can be required in various combinations and levels for the successful completion of jobs and tasks. The results showed a significant linear relation between human abilities composite test scores and loading time for teams completing the loading of the U.S. Navy's Close-in Weapon System.

Another variable that impacts team performance is team mental models. In much the same way that an individual's mental model may impact that individual's performance, team mental models can affect team performance. Cannon-Bowers, Salas, and Converse (1993) proposed four team mental models: (1) equipment—shared understanding of the technology and equipment to complete the task; (2) task—captures team members' perceptions and understanding of team procedures, strategies, task contingencies, and environmental conditions; (3) team interaction—reflects team members' understanding of team members' responsibilities, norms, and interaction patterns; and (4) team—summarizes team members' understanding of each others' knowledge, skills, attitudes, strengths, and weaknesses. Building on this categorization, Lim and Klein (2006) proposed that a team's taskwork mental model and teamwork mental model would impact team performance. Using a structural assessment technique, Pathfinder (Schvaneveldt 1990), to represent team mental models, Lim and Klein generated a network model that represented team's taskwork and teamwork mental models. Both mental models were significantly positively related to team performance. This result did not agree with previous work by Mathieu et al. (2000) that showed no relationship. Lim and Klein attributed these differences to the fact that they used military teams; thus, the finding may be a result of the training such teams receive. The question remains as how one might intervene in such teams to ensure such congruent mental models. Further work is needed to identify variables that impact team performance. That being said, much of the work on team modeling is exciting and provides hope that the mysteries of team interaction will be explained.

5.5 ENHANCING DESIGNERS' AND USERS' ABILITIES

Regardless of whether one is a manager of a business, researcher in academia, or a leader in public service, the number of information and knowledge sources with which to make informed decisions has been increasing dramatically. In fact, the amount of information available for a human decision maker is so large that one cannot possibly keep up with all of the different sources. While the Internet is helping with this proliferation, it is not providing the answer to the efficient utilization of information (Rouse 2002).

Because of the information explosion, people who interact with the Web or Web-based systems are expected to deal with increasing amounts of information concurrently. Users of Web-based systems could be in charge of processing large quantities of information while monitoring displays and deciding on the most appropriate action in each situation that is presented to them (Tsang and Wilson 1997). As in any system, if the demands placed on the user exceed his capabilities, it will be difficult to reach the goal of safe and efficient performance (Eggemeier et al. 1991). The main cause of a decrease in performance in a multiple-task environment, according to resource theory, is the lack of a sufficient amount of resources available to perform the required amount of tasks at one time (Eggemeier et al. 1991).

Applied cognitive models as discussed perform a number of functions. First, the task knowledge models, such as ETAG, provide formalisms to complete task specification from the semantic to the elemental motion level. The GOMS-type models demonstrate superior modeling of control structures and empirical, quantitative performance predictions of usability. Finally, the knowledge structure models provide a simpler, more intuitive, structural representation that can be subjected to quantitative and qualitative structural analysis. While each type of model captures a single element of the user interaction process, none completely represents the entire user's interaction.

Thus, in this section, we review modern tools that may change the way we design user interfaces and interact with them (Figure 5.5). First, we discuss a tool that can be used by designers on their desktop to model the user environment, and tools that may enhance a user's experience with an interface. Thus, we discuss two separate but related topics: enhancing the designer's ability and enhancing the user's ability. First, we describe a predictive user model, the User-Environment Modeling Language (UEML) usability framework and

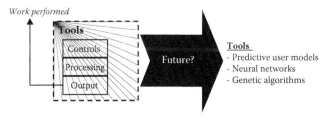

FIGURE 5.5 Human–computer interaction tools.

subsequent tool, Procedure Usability Analysis (PUA), which could provide designers two elements: (1) modeling dynamic interaction and (2) user knowledge (Koubek et al. 2003). Second, we discuss the use of algorithms that can be used to augment user cognition that may result in better human performance. Both of these areas are vital perspectives within cognitive ergonomics for improving human and technology interaction.

5.5.1 User-Environment Modeling Language

To better model human–computer interaction, the UEML framework was designed to evaluate the dynamic interaction between the human and machine interface. Because the machine can be a computer, vending machine, cellular phone, manufacturing equipment, car stereo, etc., the framework is adaptable to nearly any domain. There exists only one limitation; the task domain is limited to discrete control tasks. Thus, continuous control tasks that require constant monitoring of states or supervisory control tasks are not considered in the present case. Instead, a typical discrete control situation would involve performing part of the task directed toward a goal, assessing the resulting state, and deciding the next portion of task to be attempted. Furthermore, in terms of Rasmussen's (1985) taxonomy, the domain would encompass a wide assortment of skill- and rule-based tasks, but exclude knowledge-based tasks. As a result, the framework should have the capability to model tasks, including required actions, cognitive steps, decisions, perceptual inputs, and motor outputs.

The UEML framework represents the knowledge needed to work with the system being designed, including domain knowledge, knowledge of activities required to achieve goals, knowledge of how the tool's interface works, and how the system works in terms of internal actions invisible to the user.

For the unified UEML framework to model the user's knowledge structure and interactions, a hybrid model that contains Knowledge Structure (KS) and Cognitive Modeling methodologies was needed. Because the focus was on modeling tasks and procedures, the Procedural Knowledge Structure Model (PKSM) was selected as the initial model to represent the structural aspects of the task knowledge as well as procedures inherent in the interface. Since the core of PKSM is the structure of procedural knowledge, not the actual procedural elements, the syntax is fairly generic and can be applied to a wide variety of task domains. Consequently, the PKSM structural model can be used as an alternative representation of other models of cognition and behavior.

Although PKSM can be used as an alternative representation to either theoretic or applied cognitive models, PKSM suffers from formal definitions for how to specify task goals and levels. Applied cognitive models are able to better express the activities that occur, perceptual stimuli, cognitive steps, or motor responses, as well as provide a few quantitative measures of usability. Therefore, to operationalize the UEML framework NGOMSL is used. The use of NGOMSL allows for the procedural elements to be modeled by representing goals, subgoals, decisions, steps, sequences, input, output, etc. Therefore, the resulting model permits both NGOMSL and PKSM-types of analysis.

To implement the UEML framework, the Procedure Usability Analysis (PUA) tool was created. This Windows-based user-driven interface, shown in Figure 5.6, provides all the features common to graphical software, including the ability to insert, delete, modify, drag and drop, and copy and paste objects. In addition, it contains a search tool and the ability to view an object's properties.

PUA currently only implements declarative interaction; however, future versions will implement procedural-declarative interaction as specified in the UEML framework. Currently, it is left to the designer to ascertain the availability of these items. Further, upon "entering" a element within the task (node), its respective slow, medium, and fast times can be added to the node and the running totals for the task. PUA allows for the development of "composite" nodes that make the model creation easier by reducing the amount of elemental "internal cognitive process" nodes required in the model. This reduction simplifies the modeling process. Let's look at a case study where the PUA modeling tool was used.

U.S. Postal Service Postage and Mailing Center Case Example. Every day thousands of people interact with the post office in their local community. Many of the tasks are very simple, such as a request for a book of stamps or submission of an address change. Others are more complicated or require interaction with human clerks such as retrieving held mail or determining the quickest method to get mail to a select destination. As such, technology could potentially be developed to augment postal employees for many of the simple tasks and potentially minimize user wait time. Such a mechanism would need to be able to interact with many different types of users that vary on almost every characteristic imaginable (e.g., gender, age, cognitive ability, physical ability, language). As such, it would be helpful if a designer could evaluate these potential impacts of his interface prior to developing the technology. In implementing the UEML model, the PUA tool was

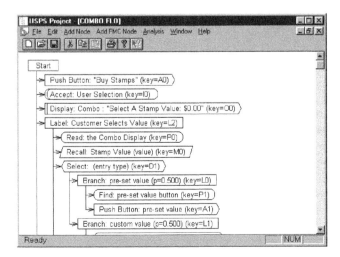

FIGURE 5.6 The PUA interface main screen.

designed with this in mind. Testing the UEML model and PUA tools were conducted using a recently designed vending system, the Postage and Mailing Center (PMC). To evaluate the interface designed through the PUA tool, both laboratory and field research was conducted. What is discussed is an overview of the findings and potential for such tools. See Koubek et al. (2003) for a complete discussion.

UEML measures collected found that correlations between predicted and actual performance were more accurate than those generated by the two techniques (NGOMSL and PKSM) combined. In addition, detailed UEML models predicted usability better than the technique of using composite modeling nodes. Finally, a field experiment revealed that the UEML technique transferred fairly well to a real-world setting.

Two potential limitations of the experimental instruments were found. The first limitation is that the laboratory participants' responses to workload assessment (e.g., NASA-Task Load Index) decreased over time. This indicated a practice effect that is not likely to occur in the real world. However, field data results found that the UEML model was still relatively accurate in predicting user workload.

The PUA tool, derived from the UEML, could be used in many real-world applications. The use of such a tool, incorporated into the design process, could result in practical benefits, such as decreasing development time and costs, as well as increasing product quality. Additionally, the hybrid model developed in this research provides the designer with immediate feedback.

In addition to direct time savings, the usage of this modeling technique results in three potential areas of cost savings. First, decreasing development time will also reduce the total costs associated with the project. Second, the feedback provided by the tool has the potential of reducing the need for an expert consultant and the associated costs. Finally, the tool provides an indirect savings in that more usable products result in higher potential utilization and customer satisfaction.

A cognitive modeling expert could most likely produce the resultant Process Usability Assessment (PUA) models in about the same amount of time it takes to produce NGOMSL or PKSM models, but not both. Therefore, PUA allows a modeler to build one model that produces a full range of measures. Furthermore, the analysis process is certainly more efficient using the UEML PUA when one considers design revisions, corrections, and system improvements that require further product remodeling.

Further research can proceed down two avenues, UEML expansion as well as evaluation within other domains. In general, expansion of the UEML would be directed toward making the representation of the machine and environment more complete and it would also involve making the analysis of the internal cognitive processes more robust. Exploration of the utility of the technique in other domains could include using the tool in the design process, comparison between alternative designs, and exploring the differences between the user's knowledge and the tool's representation. A natural extension of the use of this tool would be to Web-based applications.

5.5.2 MODELING THE USER(S) INTERACTION: NEURAL NETWORKS AND GENETIC ALGORITHMS

In addition to the models presented in previous sections, we now introduce soft computing methods that have been utilized in human–computer interaction and can be applied in modeling user interaction with the Web. In particular, we present artificial neural networks and genetic algorithms as key soft computing tools to model human abilities to pattern recognize, cluster and categorize, and apply rules. These tools stand in contrast to previously presented models that focus more on cognitive mechanisms (e.g., short-term memory store) and the process of cognition (e.g., execution times for task execution).

Artificial neural networks (ANNs) are a class of biologically inspired computational algorithms that assume:

- Information processing occurs at many simple neuron-like elements.
- Signals are passed between neurons over weighted connection links.
- Each neuron computes an activation function to determine an output signal.

Although most ANN systems were originally designed to model biological systems (Rumelhart and McClelland 1988), their application has spread to a multitude of disciplines to include computer science, engineering, medicine, and business (Fausett 1994; Tsoukalas and Uhrig 1997).

The two main distinguishing characteristics of a neural network include the network architecture and the method of setting the weights. The architecture of a neural network refers to the arrangement of the neurons into layers and the different patterns of connections between the layers. There are two modes of ANN learning—supervised and unsupervised—and corresponding methods for updating connection weights. In supervised training, the network is taught to associate input neuronal values and associated target values via weight adjustments. This type of training technique is particularly useful for pattern association and classification tasks. Unsupervised training networks also receive a series of input vectors but are not given target values. Rather, these ANNs modify the connection weights so that similar neurons are grouped together. Unsupervised training is particularly useful in clustering tasks (Fausett 1994). We focus on the use of supervised techniques such as the multilayer backpropagation network (Rumelhart, Hinton, and McClelland 1988) for extracting and recognizing patterns and the use of unsupervised techniques such as the adaptive resonance theory (Grossberg 1995) to create clusters of categorical data (Maren 1991).

5.5.2.1 Artificial Neural Networks

One area of ANN application is display design. According to Eberts (1991), the biggest bottleneck in designing an intelligent interface is acquiring knowledge from the users of the system. Because neural networks are capable of mapping stimuli to higher-level concepts, they prove to be useful in interface design; lower-level inputs such as keystrokes must be mapped to higher-level concepts such as a strategy. Both Eberts (1991) and Ye and Salvendy (1991) used the backpropagation learning method for building associations between keystroke-level commands and higher-order concepts to guide display design. Finlay and Beale (1990) also introduced an ANN system called ADAM to dynamic user interface activities for improving interface design.

A second area of application is modeling and aiding dynamic decision making. Sawaragi and Ozawa (2000) used recurrent neural networks (Elman 1990) to model human behavior in a naval navigation task. Gibson and his colleagues used two multilayer backpropagation networks to confirm experimental data gathered from human subjects in a sugar production task (Gibson, Fichman, and Plaut 1997). Rothrock (1992) used a backpropagation network to model decision making in a time and resource-constrained task.

The third area of ANN application is in clustering multiuser interactions. While the first two application areas have indirect ramifications toward Web interface design, this area is directly relevant to Web design. Berthold and his colleagues (Berthold et al. 1997) developed a supervised-training autoassociative (Rumelhart and McClelland 1988) network to extract typical examples from over 3000 postings to 30 Internet newsgroups. Through analysis of the connection weights, Berthold's group was able to create typical messages and find features common to the messages. Park (2000) presented a Fuzzy Adaptive Resonance Theory (Fuzzy ART) network using unsupervised training to categorize consumer purchases based on e-commerce sales data. Without extensive interviews with buyers, Park demonstrated how companies are able to extract and cluster buying preferences.

Although ANNs have been used often to aid interface design, few researchers have used genetic algorithms (GAs) in human computer interaction. We next provide some examples of GA use in manual control and dynamic decision making that may ultimately lead the way to future implementations of GAs in Web-based applications.

5.5.2.2 Genetic Algorithms in Human–Computer Interaction

With the proliferation of information, human decision makers need tools to deal with all of the information that is being presented to them. In order to be able to present a tool that will aid a decision maker, we must first know how humans use information and knowledge to make decisions. If it is known what types of information are important to people and what presentation methods are the best at conveying the desired message, it will be easier to design Web-based systems that enable people to make the most out of the information that is presented to them (Rouse 2002).

Determining how operators cope in a dynamic environment filled with uncertainty, complexity, and time pressure has been studied since at least the 1970s (Sheridan and Johannsen 1976). The first attempts to understand such domains were primarily prescriptive in nature; these attempts failed to provide insight into human behavior because of their failure to consider actions taken by the operator other than simply the selection of a decision alternative and the fact that they do not represent the effect of experience on human performance. However, studies have shown that experienced performers working in environments that are dynamic and uncertain almost always use shortcuts or heuristics that have been generated using knowledge from previous experiences in a quick and intuitive manner (Rothrock and Kirlik 2003).

Two methods are presented to generate potential heuristic strategies used by experienced operators in uncertain and dynamic environments based on previously collected behavioral data. Training, aiding, and other types of Web-based systems where humans interact in a dynamic manner would benefit from an increased understanding of performance in such environments. Once operator strategies and heuristics have been inferred, it would then become possible to determine where the user has misunderstandings about the task, and feedback can be targeted to these areas (Rothrock and Kirlik 2003).

In uncertain and dynamic domains, it is often difficult to explicitly explain the knowledge that the operator possesses in the form of rules (Greene and Smith 1993) because, often, this knowledge tends to be intuitive and based on experience (Filipic, Urbancic, and Krizman 1999). Genetic algorithms have proven to be robust concept learners, and can be used to modify a population of rule sets that represent possible judgment strategies. Each resulting classifier is in the form of an "if-then" rule that describes when and how the operator acted (Liepins et al. 1991). The set of rules that are generated to describe how the user interacts with the system can help provide an improved understanding of human decision making in dynamic and uncertain environments, which could lead to the improved design of training and display systems. Identifying the states of the environment where operators always make mistakes in judgment can help inform future operators how to avoid these errors. Web-based training programs can then be developed with the intent of reducing operator errors that have been experienced in the past. Let us explore a case where GAs has been applied. Discussion of a case study provides insight into how future Internet applications may apply GAs.

5.6 CASE EXAMPLE DISCUSSION

To conclude the chapter, we describe a scenario where cognitive ergonomics could be used to facilitate the development of Web-based applications to support an engineering design effort. The example we consider is the design and development of the NASA Orion space capsule. Note that this is a simulated case example and not based on any Web projects ongoing at NASA.

NASA is conducting the systems engineering and design work for the Orion Crew Vehicle that will eventually take a U.S. crew back to the moon and possibly Mars. Human factors professionals have been asked to consider the Web-based technology that will support this effort. To begin, researchers have to consider the human–environment interaction model presented in Figure 5.1.

The first step is to consider the tasks in which the team(s) will be engaging along with the task attributes. There are at minimum seven geographically separated teams interacting in the Orion design effort. At each location, there are numerous subgroups that contribute to the design. Thus, the tasks required to support this effort are not trivial. Any Web-based system would need to support both individual and team work. It would also have to support synchronous and asynchronous design, testing, and evaluation. This chapter hopefully allows the reader to understand such task complexities. Cognitive ergonomics tools (e.g., Cognitive Task Analysis, knowledge-acquisition tools) would help support defining the tasks needed to develop the appropriate Web-based tools.

Once the tasks have been identified, the human factors professional would need to work with the users to identify the means of achieving the tasks. What are tools (e.g., video conferencing, shared design databases) that would support the team in achieving their tasks? How do the tools help facilitate shared cognition among team members? How do we prevent teams from making errors like that with NASA's Mars climate observer (CNN 1999)? Is there a need for mobile applications to support the design members? How do we coordinate the work of different companies/contractors and NASA centers that share design data so that a common understanding of the design problems and solutions is accomplished?

The last phase would be to enhance the user's abilities through tools that are robust and intelligent. Can tools be developed that inform subsystem designers of changes to design assumptions and model changes? Can intelligent agents be designed that assist the human designer to recognizing problems? Munch and Rothrock (2003) began work using genetic algorithms as classifiers to design and develop an adaptive interface (Rothrock et al. 2002). In Munch and Rothrock's study, the display would allocate functions based on current goals and abilities of the user by monitoring the user status, the system task demands, and the existing system requirements. Sensing when the operator was overloaded, the system could automate some of the tasks so that the operator can focus his attention on other tasks. For such systems to know how to act when they take over control of some of the task requirements, rule sets that represent user behavior must be developed that will then be used to inform the system. If the system behaves as if the operator would, the operator will feel comfortable with the system performing some of the tasks and will also be able to take over again without losing a step once the task demands have been reduced to a more manageable level. Tools such as GBPC (Rothrock and Kirlik 2003) could be applied to infer, from operator actions, rule sets. These rules can then be verified and used to determine strengths and weaknesses of different operator strategies, as well as inform an adaptive interface that can alleviate workload when a user is overloaded. Research such as this could change the way individuals and teams interact in the future. Cognitive ergonomists will be needed to address such challenges in the future.

5.7 CONCLUSION

Cognitive engineering tools help designers and users identify the needs of the tools required to accomplish a task. Whether the application is a Web-based storefront or Web-based tools to design the next lunar capsule, cognitive ergonomics provides a variety of methods and tools to understand and model the user better. As discussed in the introduction, the design process is an iterative process. The use of these tools is not limited to a single phase of design. Instead, these methods should be viewed as a means to learn more information about the design throughout the design process. Depending on the organization and the product, one may find some methods more advantageous than others. Likewise, at times a method may be implemented in a limited fashion. For example, a complete GOMS analysis may be too tedious for the complete product interaction process. However, it may be fruitful for very essential elements of the design.

This chapter has also discussed elements of team interaction and how those may impact Web design along with new research areas investigated for improving the human–computer interaction process. Discussion included a review of the UEML model that may someday put a complete design tool on the designer's desktop to the application of algorithms (e.g., genetic algorithm) that may make interacting with a Web site or international design team easier in the future. Research on new methods as well as the application of existing cognitive ergonomic methods must continue if designers are to design future Web sites that are useful and friendly for the "typical" user.

REFERENCES

Alexander, D. 2003. Web bloopers, http://deyalexander.com/blooper/ (accessed Aug. 18, 2003).

Anderson, A. H., R. McEwan, J. Bal, and J. Carletta. 2007. Virtual team meetings: An analysis of communication and context. *Computers in Human Behavior* 23(5): 2558–2580, doi:10.1016/j.chb.2007.01.001.

Anderson, J. R. 1976. *Language, Memory, and Thought*. Mahwah, NJ: Lawrence Erlbaum.

Anderson, J. R. 1983. *The Architecture of Cognition*. Cambridge, MA: Harvard University Press.

Anderson, J. R. 1990. *The Adaptive Character of Thought*. Mahwah, NJ: Lawrence Erlbaum.

Anderson, J. R., and C. Lebiere. 1998. *The Atomic Components of Thoughts*. Mahwah, NJ: Lawrence Erlbaum.

Arrow, H., J. E. McGrath, and J. L. Berdahl. 2000. *Small Group as Complex Systems: Formation, Coordination, Development, and Adaptation*. Thousand Oaks, CA: Sage.

Benysh, D. V., and R. J. Koubek. 1993. The implementation of knowledge structures in cognitive simulation environments. In *Proceedings of the Fifth International Conference on Human–Computer Interaction (HCI International '93), Orlando, Florida*, 309–314. New York: Elsevier.

Berthold, M. R., F. Sudweeks, S. Newton, and R. D. Coyne. 1997. Clustering on the Net: Applying an autoassociative neural network to computer-mediated discussions. *Journal of Computer Mediated Communications* 2(4).

Bovair, S., D. E. Kieras, and P. G. Polson. 1990. The acquisition and performance of text-editing skill: A cognitive complexity analysis. *Human Computer Interaction* 5: 1–48.

Bronstein, L. R. 2003. A model for interdisciplinary collaboration. *Social Work: The Journal of Contemporary Social Work* 48: 297–306.

Burleson, B. R., B. J. Levine, and W. Samter. 1984. Decision-making procedure and decision quality. *Human Communication Research* 10(4): 557–574.

Business Week. 2002. Usability is next to profitability, *Business Week Online*. http://www.businessweek.com/print/technology/content/dec2002/tc2002124_2181.htm?tc (accessed Aug. 18, 2003).

Cannon-Bowers, J. A., E. Salas, and S. Converse. 1993. Shared mental models in expert team decision making. In *Individual and Group Decision Making*, ed. J. Castellan, 221–246. Hillsdale, NJ: Lawrence Erlbaum.

Card, S. K., T. P. Moran, and A. Newell. 1983. *The Psychology of Human–Computer Interaction*. Hillsdale, NJ: Lawrence Erlbaum.

Card, S. K., T. P. Moran, and A. Newell. 1986. The model human processor. In *Handbook of Perception and Human Performance*, vol. 2, eds. K. R. Boff, L. Kaufman, and J. P. Thomas, 1–35. Hoboken, NJ: Wiley-Interscience.

Carey, J. M. and Kacmar, C. J. 1997. The impact of communication mode and task complexity on small groups performance and member satisfaction. *Computers in Human Behavior* 13(1), 23–49.

Chipman, S. F., J. M. Schraagen, and V. L. Shalin. 2000. Introduction to cognitive task analysis. In *Cognitive Task Analysis*, eds. J. M. Schraagen, S. F. Chipman, and V. L. Shalin, 3–24. Mahwah, NJ: Lawrence Erlbaum.

CircleID. 2009. Mobile Internet Users to Reach 134 Million by 2013. CircleID: Internet Infrastructure. Retrieved from http://www.circleid.com/posts/mobile_internet_users_to_reach_134_million_by_2013/.

CNN. 1999. NASA's metric confusion caused Mars orbiter loss. Retrieved January 15, 2010: http://www.cnn.com/TECH/space/9909/30/mars.metric/.

Cooper, C. L. 1975. *Theories of Group Processes*. London: Wiley.

Crandall, B., G. Klein, L. G. Militello, and S. P. Wolf. 1994. Tools for applied cognitive task analysis (Contract Summary Report on N66001-94-C-7008). Fairborn, OH: Klein Associates.

CyberAtlas. 2003. Population explosion. Retrieved June 23, 2003: http://cyberatlas.internet.com/big_picture/geographics/article/0,1323,5911_151151,00.html.

Darisipudi, A. 2006. Towards a generalized team task complexity model, unpubl. diss., Louisiana State University, Baton Rouge.

De Haan, G., and G. C. Van der Veer. 1992. ETAG as the basis for intelligent help systems. In *Proceedings Human–Computer Interaction: Tasks and Organization,* eds. G. C. Van der Veer et al. (Balatonfured, Hungary, Sept. 6–11), 271–284.

De Haan, G., G. C. Van der Veer, and J. C. Van Vliet. 1993. Formal modeling techniques in human–computer interaction. In *Cognitive Ergonomics–Contributions from Experimental Psychology*, eds. G. C. Van der Veer, S. Bagnara, and G. A. M. Kempen, 27–68. Elsevier: Amsterdam.

Eberts, R. E. 1991. *Knowledge Acquisition Using Neural Networks for Intelligent Interface Design*. Paper presented at the International Conference on Systems Man and Cybernetics, Oct. 13–16, 1991, Charlottesville, VA.

Eberts, R. E. 1994. *User Interface Design*. Englewood Cliffs, NJ: Prentice Hall.

Eberts, R. E., A. Majchrzak, P. Payne, and G. Salvendy. 1990. Integrating social and cognitive factors in design of Human–computer interactive communication. *International Journal of Human–Computer Interaction* 2(1): 1–27.

Eggemeier, F. T., G. F. Wilson, A. F. Kramer, and D. L. Damos. 1991. Workload assessment in multi-task environments. In *Multiple-Task Performance*, ed. D. L. Damos, 207–216. London: Taylor & Francis.

Elman, J. L. 1990. Finding Structure in Time. *Cognitive Science* 14: 179–211.

Entin, E. E., and D. Serfaty. 1999. Adaptive team coordination. *Human Factors* 41(2): 312–325.

Ericsson, K. A., and H. A. Simon. 1980. Verbal reports as data. *Psychological Review* 87: 215–251.

Fausett, L. 1994. *Fundamentals of Neural Networks: Architectures, Algorithms, and Applications*. Englewood Cliffs, NJ: Prentice-Hall.

Filipic, B., T. Urbancic, and V. Krizman. 1999. A combined machine learning and genetic algorithm approach to controller design. *Engineering Applications of Artificial Intelligence* 12: 401–409.

Finlay, J., and R. Beale. 1990. Neural networks in human–computer interaction: A view of user modeling. Paper presented at the IEE Colloquium on Neural Nets in Human–Computer Interaction, London, UK, Dec. 14, 1990.

Flach, J. M. 2000. Discovering situated meaning: an ecological approach to task analysis. In *Cognitive Task Analysis*, eds. J. M. Schraagen, S. F. Chipman, and V. J. Shalin, 87–100. Mahwah, NJ: Lawrence Erlbaum.

Fleishman, E. A., and M. K. Quaintance. 1984. *Taxonomies of Human Performance: The Description of Human Tasks*. Orlando, FL: Academic Press.

Fleishmann, E. A., and S. J. Zaccaro. 1992. Toward a taxonomic classification of team performance functions: Initial considerations, subsequent evaluations and current formulations. In *Teams: Their Training and Performance*, eds. R. W. Swezey and E. Salas, 31–56. Norwood, NJ: Ablex.

Forsythe, C., and M. R. Ashby. 1996. Human factors in agile manufacturing, *Ergonomics in Design* 4(1): 15–21.

Gibson, F. P., M. Fichman, and D. C. Plaut. 1997. Learning in dynamic decision tasks: Computational model and empirical evidence. *Organizational Behavior and Human Decision Processes* 71(1): 1–35.

Greene, D. P., and S. F. Smith. 1993. Competition-based induction of decision models from examples. *Machine Learning* 13: 229–257.

Greiner, R., and G. Metes. 1995. *Going Virtual*. Englewood Cliffs, NJ: Prentice Hall.

Grootjen, M., M. A. Neerincx, J. C. M. van Weert, and K. P. Troung. 2007. Measuring cognitive task load on a naval ship: Implications of a real world environment. In *Foundations of Augmented Cognition*, eds. D. D. Schmorrow and L. M. Reeves, 147–156. Berlin: Springer-Verlag.

Grossberg, S. 1995. Neural dynamics of motion perception, recognition learning, and spatial attention. In *Mind as Motion: Explorations in the Dynamics of Cognition*, eds. R. F. Port and T. VanGelder, 449–489. Cambridge, MA: MIT Press.

Grosse, C. U. 2002. Managing communication within virtual intercultural teams. *Business Communication Quarterly* 65(4): 22–38.

Guimerà, R., B. Uzzi, J. Spiro, and L. A. N. Amaral. 2005. Team assembly mechanisms determine collaboration network structure and team performance. *Science* 308(5722): 697–702.

GVU's WWW User Surveys. 1998. Purchasing on the Internet. http://www.gvu.gatech.edu/user_surveys/survey-1998-10/ (accessed Aug. 25, 2003).

Hacker, M., and B. Kleiner. 1996. Identifying critical factors impacting virtual work group performance. *IEEE International Engineering Management Conference 1996*, 201–205. Piscataway, NJ: IEEE.

Hackman, J. R. 1969. Toward understanding the role of tasks in behavioral research. *Acta Psychologica* 31: 97–128.

Hackman, J. R. 1987. The design of work teams. In *Handbook of Organizational Behavior*, 315–342. Englewood Cliffs, NJ: Prentice Hall.

Hammond, J., C. M. Harvey, R. J. Koubek, W. D. Compton, and A. Darisipudi. 2005. Distributed collaborative design teams: Media effects on design processes. *International Journal of Human–Computer Interaction* 18(2): 145–165.

Hammond, J., R. J. Koubek, and C. M. Harvey. 2001. Distributed collaboration for engineering design: A review and reappraisal. *Human Factors and Ergonomics in Manufacturing* 11(1): 35–52.

Hartman, F., and R. Ashrafi. 1996. Virtual organizations—an opportunity for learning. *IEEE International Engineering Management Conference 1996*, 196–200. Piscataway, NJ: IEEE.

Hiltz, S. R., K. Johnson, and M. Turoff. 1987. Experiments in group decision making: Communication process and outcome in face-to-face versus computerized conferences. *Human Communication Research* 13(2): 225–252.

Hollnagel, E., and D. D. Woods. 1999. Cognitive systems engineering: New wine in new bottles. *International Journal of Human–Computer Studies* 51: 339–356.

Howell, W. C., and N. J. Cooke. 1989. Training the human information processor: a look at cognitive models. In *Training and Development in Work Organizations: Frontiers of Industrial and Organizational Psychology*, ed. I. L. Goldstein, 121–182. San Francisco, CA: Jossey-Bass.

Ilgen, D. 1999. Teams embedded in organizations: Some implications. *American Psychologist* 54(2): 129–139.

Internet World Stats. 2010. Internet world stats: Usage and population stats. http://www.internetworldstats.com/stats14.htm (accessed Jan. 7, 2010).

Internet World Stats. 2009. http://www.internetworldstats.com/ (accessed Dec. 12, 2009).

Johnson, J. 2000. *GUI Bloopers: Don'ts and Do's for Software Developers and Web Designers*. San Francisco, CA: Morgan Kaufmann.

Kanawattanachai, P., and Y. Yoo. 2007. The impact of knowledge coordination on virtual team performance over time. *MIS Quarterly* 31(4): 783–808.

Kiekel, P. A., and N. J. Cooke. this volume. Human factors aspects of team cognition. In *Handbook of Human Factors in Web Design*, 2nd ed., eds. K.-P. L. Vu and R. W. Proctor, 107–124. Boca Raton, FL: CRC Press.

Kent, R. N., and J. E. McGrath. 1969. Task and group characteristics as factors influencing group performance. *Journal of Experimental Social Psychology* 5: 429–440.

Kieras, D. E. 1988. Towards a practical GOMS model methodology for user interface design. In *Handbook of Human–Computer Interaction*, ed. M. Helander, 67–85. Amsterdam: Elsevier Science.

Kieras, D. E. 1997. A guide to GOMS model usability evaluation using NGOMSL. In *Handbook of Human–Computer Interaction*, 2nd ed., eds. M. Helander, T. Landauer, and P. Prabhu, 733–766. Amsterdam: Elsevier Science.

Kieras, D. E., and D. E. Meyer. 1995. An overview of the EPIC architecture for cognition and performance with application to human–computer interaction. Report 5, TR-95 ONR-EPIC-5, http://citeseer.nj.nec.com/rd/31449758%2C251100%2C1 %2C0.25%2CDownload/ftp%3AqSqqSqftp.eecs.umich .eduqSqpeopleqSqkierasqSqEPICqSqTR-EPIC-5.pdf (accessed Aug. 25, 2003).

Kieras, D. E., and D. E. Meyer. 1998. The EPIC architecture: principles of operation. Unpublished manuscript, ftp://www.eecs .umich.edu/people/kieras/EPIC/ (accessed Aug. 25, 2003).

Kirwan, B., and L. K. Ainsworth, eds. 1992. *A Guide to Task Analysis*. London: Taylor & Francis.

Klein, G. 2000. Cognitive task analysis of teams. In *Cognitive Task Analysis*, eds. J. M. C. Schraagen, S. F. Chipman, and V. J. Shalin, 417–430. Mahwah, NJ: Lawrence Erlbaum.

Klein, G. A., R. Calderwood, and D. Macgregor. 1989. Critical decision making for eliciting knowledge, *IEEE Transactions on Systems, Man, and Cybernetics* 19(3): 462–472.

Koubek, R. J., D. Benysh, M. Buck, C. M. Harvey, and M. Reynolds. 2003. The development of a theoretical framework and design tool for process usability assessment. *Ergonomics* 46(1–3): 220–241.

Koubek, R. J., G. Salvendy, and S. Noland. 1994. The use of protocol analysis for determining ability requirements for personnel selection on a computer-based task. *Ergonomics* 37: 1787–1800.

Laird, J. E., A. Newell, and P. S. Rosenbloom. 1987. Soar: An architecture for general intelligence. *Artificial Intelligence* 33: 1–64.

Laughlin, P. R. 1980. Social combination processes of cooperative, problem-solving groups as verbal intellective tasks. In *Progress in Social Psychology*, vol. 1, ed. M. Fishbein, 127–155. Mahwah, NJ: Lawrence Erlbaum.

Lehto, M., J. Boose, J. Sharit, and G. Salvendy. 1997. Knowledge acquisition. In *Handbook of Industrial Engineering*, 2nd ed., ed. G. Salvendy, 1495–1545. New York: John Wiley.

Liepins, G. E., M. R. Hilliard, M. Palmer, and G. Rangarajan. 1991. Credit assignment and discovery in classifier systems. *International Journal of Intelligent Systems* 6: 55–69.

Lim, B.-C., and K. J. Klein. 2006. Team mental models and team performance: A field study of the effects of team mental model similarity and accuracy. *Journal of Organizational Behavior* 27(4): 403–418.

Lin, H., R. J. Koubek, M. Haas, C. Phillips, and N. Brannon. 2001. Using cognitive models for adaptive control and display. *Automedica* 19: 211–239.

Lohse, G. L. 1997. Models of graphical perception. In *Handbook of Human–Computer Interaction*, 2nd ed., eds. M. G. Helander, T. K. Landauer, and P. V. Prabhu, 107–135. Elsevier: Amsterdam.

Maren, A. J. 1991. Neural networks for enhanced human–computer interactions. *IEEE Control Systems Magazine* 11(5): 34–36.

Mathieu, J. E., T. S. Heffner, G. F. Goodwin, E. Salas, and J. A. Cannon-Bowers. 2000. The influence of shared mental models on team process and performance. *Journal of Applied Psychology* 85: 273–283.

McGrath, J. E. 1984. *Groups: Interaction and Performance*. Englewood Cliffs, NJ: Prentice Hall.

McGraw, K. L., and K. Harbison-Briggs. 1989. *Knowledge Acquisition, Principles and Guidelines*, 1–27. Englewood Cliffs, NJ: Prentice Hall.

Militello, L. G., and R. J. G. Hutton. 1998. Applied Cognitive Task Analysis (ACTA): A practitioner's toolkit for understanding cognitive task demands. *Ergonomics* 41(11): 1618–1641.

MITECS. 2003. Cognitive ergonomics. In *The MIT Encyclopedia of the Cognitive Sciences*. http://cognet.mit.edu/MITECS/Entry/ gentner (accessed Aug. 18, 2003).

Moran, T. P. 1981. The command language grammar: A representation for the user-interface of interactive systems. *International Journal of Man–Machine Studies* 15: 3–50.

Moran, T. P. 1983. Getting into a system: External-internal task mapping analysis. In *Human Factors in Computing Systems: CHI'83 Proceedings*, ed. A. Janda, 46–49. New York, NY: ACM Press.

Morgan, B. B., Jr., A. S. Glickman, E. A. Woodward, A. Blaiwes, and E. Salas. 1986. Measurement of team behaviors in a Navy environment. NTSC Report 86-014. Orlando, FL: Naval Training System Center.

Munch, J., and L. Rothrock. 2003. Modeling human performance in supervisory control: Informing adaptive aid design. Paper presented at the Annual Conference of the Institute of Industrial Engineers, Portland, OR, May 18–23, 2003.

Netcraft 2010. January 2010 Web server survey. http://news.netcraft.com/archives/web_server_survey.html (accessed Jan. 7, 2010).

Newell, A. 1990. *Unified Theories of Cognition.* Cambridge, MA: Harvard University Press.

Newell, A. F., and H. Simon. 1972. *Human Problem Solving.* Englewood Cliffs, NJ: Prentice Hall.

Niebel, B. W., and A. Freivalds. 1999. *Methods, Standards, and Work Design.* Boston, MA: WCB McGraw-Hill.

Norman, D. A. 1986. Cognitive engineering, In *User Centered System Design: New Perspectives on Human–Computer Interaction*, eds. D. A. Norman and S. W. Draper, 31–61. Mahwah, NJ: Lawrence Erlbaum.

Norman, D. A. 1988. *The Design of Everyday Things.* New York: Basic Books Inc. Publishers.

Olson, G. M., J. S. Olson, M. R. Carter, and M. Storrøsten. 1992. Small groups design meetings: An analysis of collaboration. *Human–Computer Interaction* 7: 347–374.

Online Computer Library Center 2003. Web characterization project. http://wcp.oclc.org (accessed Aug. 18, 2003).

Park, S. 2000. Neural networks and customer grouping in e-commerce: A framework using fuzzy ART. Paper presented at the Academia/Industry Working Conference on Research Challenges, Buffalo, NY, April 27–29, 2000.

Payne, S. J. 2008. Mental models in human–computer interaction. In *The Human Computer Interaction Handbook: Fundamentals, Evolving Technologies, and Emerging Applications*, 2nd ed., eds. A. Sears and J. A. Jacko, 63–76. Mahwah, NJ: Lawrence Erlbaum.

Payne, S. J., and T. R. G. Green. 1986. Task-action grammars: A model of the mental representation of task languages. *Human–Computer Interaction* 2: 93–133.

Powell, A., G. Piccoli, and B. Ives. 2004. Virtual team: A review of current literature and directions for future research. *The DATA BASE for Advances in Information Systems* 35(1): 6–36.

Preece, J., Y. Rogers, and H. Sharp. 2002. *Interaction Design: Beyond Human–Computer Interaction.* New York: John Wiley.

Randel, J. M., H. L. Pugh, and S. K. Reed. 1996. Differences in expert and novice situation awareness in naturalistic decision-making. *International Journal of Human–Computer Studies* 45: 579–597.

Rasmussen, J. 1985. The role of hierarchical knowledge representation in decision-making and system management. *IEEE Transactions on Systems, Man, and Cybernetics* 15(2): 234–243.

Rasmussen, J. 1990. Mental models and the control of action in complex environments. In *Mental Models and Human–Computer Interaction*, vol. 1, eds. D. Ackermann and M. J. Tauber, 41–69. New York: Elsevier Science.

Rasmussen, J., and A. M. Pejtersen. 1995. Virtual ecology of work. In *Global Perspectives on the Ecology of Human–Machine Systems*, vol. 1, eds. J. M. Flach et al., 121–156. Mahwah, NJ: Lawrence Erlbaum.

Reader, T. W., R. Flin, K. Mearns, and B. H. Cuthbertson. 2009. Developing a team performance framework for the intensive care unit. *Critical Care Medicine* 37(5): 1787–1793.

Reisner, P. 1983. Analytic tools for human factors in software. In *Proceedings Enduser Systems and their Human Factors*, eds. A. Blaser and M. Zoeppritz, 94–121. Berlin: Springer Verlag.

Richardson, K. W., F. A. Aghazadeh, and C. M. Harvey, forthcoming. The efficiency of using a taxonomic approach to predicting performance of Navy tasks. *Theoretical Issues in Ergonomics Science.*

Roby, T. B., and J. T. Lanzatta. 1958. Considerations in the analysis of group tasks. *Psychological Bulletin* 55(4): 88–101.

Rothrock, L. 1992. Modeling human perceptual decision-making using an artificial neural network. Paper presented at the 1992 IEEE/INNS International Joint Conference on Neural Networks, Baltimore, MD, June 7–11, 1992.

Rothrock, L., and A. Kirlik. 2003. Inferring rule-based strategies in dynamic judgment tasks: Toward a noncompensatory formulation of the lens model. *IEEE Transactions on Systems, Man, and Cybernetics Part A* 33(1): 58–72.

Rothrock, L., R. Koubek, F. Fuchs, M. Haas, and G. Salvendy. 2002. Review and reappraisal of adaptive interfaces: Toward biologically-inspired paradigms. *Theoretical Issues in Ergonomics Science* 3(1): 47–84.

Rothrock, L., C. M. Harvey, and J. Burns. 2005. A theoretical framework and quantitative architecture to assess team task complexity in dynamic environments. *Theoretical Issues in Ergonomics Science* 6(2): 157–172.

Rouse, W. B. 2002. Need to know—information, knowledge, and decision making. *IEEE Transactions on Systems, Man, and Cybernetics* 32(4): 282–292.

Rumelhart, D. E., and J. L. McClelland, eds. 1988. *Parallel Distributed Processing: Explorations in the Microstructure of Cognition*, vol. 1, *Foundations.* Cambridge, MA: MIT Press.

Rumelhart, D. E., G. E. Hinton, and J. L. McClelland. 1988. A general framework for parallel distributed processing. In *Parallel Distributed Processing: Explorations in the Microstructure of Cognition*, vol. 1, *Foundations*, eds. D. E. Rumelhart and J. L. McClelland, 45–109. Cambridge, MA: MIT Press.

Sawaragi, T., and S. Ozawa. 2000. Semiotic modeling of human behaviors interacting with the artifacts using recurrent neural networks. Paper presented at the 26th Annual Conference of the IEEE Industrial Electronics Society (IECON 2000), Nagoya, Japan, October 22–28, 2000.

Schraagen, J. M., S. F. Chipman, and V. L. Shalin, eds. 2000. *Cognitive Task Analysis.* Mahwah, NJ: Lawrence Erlbaum.

Schvaneveldt, R. W. 1990. *Pathfinder Associative Networks: Studies in Knowledge Organization.* Norwood, NJ: Ablex.

Sheridan, T. B., and G. Johannsen, eds. 1976. *Monitoring Behavior and Supervisory Control.* New York: Plenum Press.

Smith, M. J., and A. Taveira. this volume. Physical ergonomics and the Web. In *Handbook of Human Factors in Web Design*, 2nd ed., eds. K.-P. L. Vu and R. W. Proctor, 65–84. Boca Raton, FL: CRC Press.

Steiner, I. D. (1972). *Group Process and Productivity.* New York: Academic Press.

Strybel, T. Z. this volume. Task analysis methods and tools for developing Web applications. In *Handbook of Human Factors in Web Design*, 2nd ed., eds. K.-P. L. Vu and R. W. Proctor, 483–508. Boca Raton, FL: CRC Press.

Tauber, M. J. 1990. *ETAG: Extended Task-Action Grammar—A Language for the Description of the User's Task Language*, *Proceedings of Interact '90*, 163–168. Amsterdam: Elsevier.

Tsang, P., and G. F. Wilson. 1997. Mental workload. In *Handbook of Human Factors and Ergonomics*, 2nd ed., ed. G. Salvendy, 417–449. New York: John Wiley.

Tsoukalas, L. H., and R. E. Uhrig. 1997. *Fuzzy and Neural Approaches in Engineering*. New York: John Wiley.

Tuckman, B. W. 1965. Developmental sequence in small groups. *Psychological Bulletin* 63: 384–399.

U.S. Department of Commerce. 2002. *A Nation Online: How Americans Are Expanding Their Use of the Internet*. Washington, DC: U.S. Department of Commerce, National Telecommunications and Information Administration.

U.S. Department of Commerce. 2004. *A Nation Online: Entering the Broadband Age*. Washington, DC: U.S. Department of Commerce, National Telecommunications and Information Administration.

Van der Ryn, S., and S. Cowan. 1995. *Ecological Design*. Washington, DC: Island Press.

Van der Veer, G. C., and M. C. P. Melguizo. 2003. Mental models. In *The Human Computer Interaction Handbook: Fundamentals, Evolving Technologies, and Emerging Applications*, eds. J. A. Jacko and A. Sears, 52–80. Mahwah, NJ: Lawrence Erlbaum.

van Rijn, H., A. Johnson, and N. Taatgen. this volume. Cognitive user modeling. In *Handbook of Human Factors in Web Design*, 2nd ed., eds. K.-P. L. Vu and R. W. Proctor, 527–542. Boca Raton, FL: CRC Press.

Vicente, K. J. 1999. *Cognitive Work Analysis*. Mahwah, NJ: Lawrence Erlbaum.

Vicente, K. J. 2000. Work domain analysis and task analysis: A difference that matters, In *Cognitive Task Analysis*, eds. J. M. Schraagen, S. F. Chipman, and V. L. Shalin, 101–118. Mahwah, NJ: Lawrence Erlbaum.

Vicente, K. J., and J. Rasmussen. 1992. Ecological interface design: Theoretical foundations. *IEEE Transactions on Systems, Man, and Cybernetics* 22(4): 589–606.

Weeks, G. D., and A. Chapanis. 1976. Cooperative versus conflictive problem solving in three telecommunication modes. *Perceptual and Motor Skills* 42: 879–917.

Wickens, C., J. Lee, Y. Liu, and S. Gordon-Becker, eds. 2004. *An Introduction to Human Factors Engineering*, 2nd ed. Englewood Cliffs, NJ: Prentice Hall.

Ye, N., and G. Salvendy. 1991. An adaptive interface design using neural networks. In *Human Aspects in Computing: Design and Use of Interactive Systems and Work with Terminals*, ed. H. J. Bullinger, 435–439. New York: North-Holland.

Zachary, W. W., J. M. Ryder, and J. H. Hicinbothom. 2000. Cognitive task analysis and modeling of decision making in complex environments. In *Decision Making under Stress: Implications for Training and Simulation*, eds. J. Canon-Bowers and E. Salas, 315–344. Washington, DC: American Psychological Association.

6 Human Factor Aspects of Team Cognition

Preston A. Kiekel and Nancy J. Cooke

CONTENTS

6.1 THE VALUE OF TEAM COGNITION

Teams think. That is, they assess the situation, plan, solve problems, design, and make decisions as an integrated unit. We refer to these collaborative thinking activities as *team cognition*. Why is team cognition important? A growing number of tasks take place in the context of complex sociotechnical systems. The cognitive requirements associated with emergency response, software development, transportation, factory and power plant operation, military operations, medicine, and a variety of other tasks exceed the limits of individual cognition. Teams are a natural solution to this problem, and so the emphasis on teams in these domains is increasing.

Because team tasks are widely varied, it follows that human factors applications involving team cognition are also widely varied. Of particular relevance to the topic of this book are the numerous software applications that involve collaborative activities. Team cognition is relevant to the design of computer-supported collaborative work (CSCW) tools, such as Group Decision Support Systems (GDSS), collaborative writing environments, and social networking to name a few. Many groupware applications (i.e., software designed for use by groups) are intended for Web-based collaboration (see van Tilburg and Briggs 2005).

The use of teams to resolve task complexity is a mixed blessing, however, as teams create their own brand of complexity.

In addition to assuring that each team member knows and performs her own task, it is now important to assure that the needed information is distributed appropriately among team members. The amount and type of information that needs to be distributed among team members depends on the task and the type of team.

6.2 SOME CHARACTERISTICS OF TEAMS

This leads us to our definition of "team." A team is almost always defined as a special case of a "small group" (Fisher and Ellis 1990, 12–22). Minimally, a team is defined as a special type of group, in which members work interdependently toward a common aim (e.g., Hare 1992; Beebe and Masterson 1997, 338; Kiekel et al. 2001). Additionally, teams are often defined as having "shared and valued common goals" (Dyer 1984; Salas, Cooke, and Rosen 2008).

In keeping with a large body of the human factors team literature, we will define team to include the additional characteristics of heterogeneous individual roles and limited life span (Salas et al. 1992, 4; e.g., Cannon-Bowers, Salas, and Converse 1993; Cooke, Kiekel, and Helm 2001).

6.2.1 HETEROGENEITY

The restriction of teams to mean *heterogeneous* interdependent groups is important for team cognition, because knowledge and/or cognitive processing may or may not be homogeneously distributed among members of a team. For *homogeneous* interdependent groups, it is assumed that task-related knowledge and/or cognitive processing is homogeneously distributed. That is, because everyone has the same role in a homogeneous group, the ideal is for every group member to know all aspects of the task. No individual emphases are required, and individual skill levels and knowledge levels are randomly dispersed among group members. These groups are essentially a collection of individuals, plus small group dynamics.

However, in cognitively complex tasks, specialization is where the real potential in teams resides. This is the motivation for much of the recent work on team cognition measurement (Cooke et al. 2000; Gorman, Cooke, and Winner 2006). Earlier efforts to measure team cognition have revolved around some sort of averaging of individual knowledge (Langan-Fox, Code, and Langfield-Smith 2000), which is most appropriate when knowledge is homogeneously distributed. Knowledge accuracy is often scored on the basis of a single referent, thereby assuming that shared mental models are identical or nearly identical among team members. Team cognition measurement has been weaker at addressing the needs of heterogeneous groups, although work has been done on measuring the extent to which team members are able to catalog their knowledge of who knows what and how to interact with one another (e.g., transactive memory: Wegner 1986; Hollingshead 1998; teamwork knowledge: Cannon-Bowers et al. 1995).

Team members need an understanding of the distribution of specialization with regard to expertise or cognition.

For example, in a heterogeneous group of company officers, everyone may need to talk to the treasurer to see if their plans are within a realistic budget. This added layer of role knowledge is critical for dividing the cognitive labor in complex tasks. However, it also makes heterogeneous groups more vulnerable to performance failures, because there is less redundancy in the system. In a completely heterogeneous group, each task or cognitive activity is handled by only one person. If that task is critical, then a failure of that one person to perform it is also critical. However, role heterogeneity also ensures that no single team member has to know everything. This is a trade-off of heterogeneity. In most instances, teams will not be completely heterogeneous with respect to role but will have some degree of specialization, along with some degree of overlap.

Finally, the presence of heterogeneous knowledge distribution raises questions such as how teams should be trained. Is it better if every team member be fully trained on their own role as well as the roles of other team members (i.e., full cross-training; Blickensderfer et al. 1993; Volpe et al. 1996; Cannon-Bowers et al. 1998; Cooke et al. 2003)? What if team members are only fully trained on their own roles and given a general overview of other team members' roles? Alternatively, what if team members are only trained on their own roles, so that complexity and training time can be minimized in complementary tasks? The answers to these and other questions are dependent on understanding team cognition in groups with different roles.

6.2.2 TEAM SIZE

Apart from role heterogeneity, another interesting aspect of Salas et al.'s (1992) definition is that "two or more" team members "interact dynamically." This would require that teams be small enough for team members to directly impact one another. This creates another interesting question. How much impact is required for a "team" to still be a "team?"

Social network research (e.g., Festinger, Schachter, and Back 1964; Fisher and Ellis 1990; Friedkin 1998; Steiner 1972) focuses on evaluating the impact of different interaction patterns among team members. Influence among team members is determined on a pairwise basis, such as by determining which team members are allowed to speak to which other team members. The global pattern of influence for the team is represented in a matrix or graphical network form. Topics of interest include evolution of gross patterns over time, effectiveness of various interaction patterns for particular task types, and so on.

Team size plays an import role in addressing these issues. For instance, conflict between dyads is more likely to result in a stalemate than is conflict among larger teams. Starting with triads, larger teams permit clique formation, majority decisions, disproportionate peer influence, and so on (Fisher and Ellis 1990). Amount of input by individual team members decreases with team size (Steiner 1972). This is both because communication time is more limited and because of diffusion of responsibility in larger teams (Shaw 1981).

Steiner (1972) outlines a taxonomy of types of teams and tasks, whereby individual contribution is combined in different ways to form a holistic outcome. Work of this nature is extended in the social decision schemes (SDS) literature (Davis 1973; Kerr et al. 1975; Gillett 1980a, 1980b; SDS for quantities, SDS-Q; Hinsz 1999). SDS research involves predicting how a team will combine their input to form a decision (e.g., by majority rule, single leader, etc.). Researchers create distributions of possible decisions under different decision schemes. Then they identify the team's decision scheme, by selecting the scheme whose distribution makes the observed team decision most probable.

It is important to ask which aspects of individual cognition carry over to teams and which aspects of team cognition carry over to individuals. The answer depends on which characteristics are of interest. An individual cannot encounter team conflict (though they can encounter indecision). A dyad cannot encounter disproportionate peer influence (though they can encounter disproportionate power roles). A triad cannot encounter subteam formation (though they can encounter a majority). So the number of team members required to "interact dynamically" depends on the dynamics of interest.

6.3 PERSPECTIVES ON TEAM COGNITION

The definition of team cognition starts with the definition of individual cognition. Let us define *cognition* as "the understanding, acquisition, and processing of knowledge, or, more loosely, thought processes" (Stuart-Hamilton 1995). Team cognition would have to be the team's ability to do the same. This raises the question of whether teams really have "cognition" or not because the team's mental faculties do not arise from a single, connected unit, such as a brain.

The what and where of the individual "mind" have long been a topic of debate. Perhaps an individual mind is not a single connected unit, regardless of whether or not a brain is. Our argument for teams having cognition is the same as our argument for an individual having cognition. One can only infer an individual's cognition from the observable actions that they display.

Similarly, teams take actions as unified wholes that reflect cognition at this level. That is, teams process, store, and retrieve information (Smith 1994; Wegner 1986). Teams behave in a coordinated manner, even if they do not intend to do so (e.g., Schmidt, Carello, and Turvey 1990; Sebanz, Knoblich, and Prinz 2003). These behaviors that occur at the team level lead us to question whether team cognition must be considered an aggregate of individual cognition, or if the thinking team can truly be treated as a distinct cognitive unit. The latter view of team cognition suggests that cognition exists external to a person's mind.

6.3.1 COLLECTIVE VERSUS HOLISTIC PERSPECTIVES ON TEAM COGNITION

More attention to context is needed when we start to look at team cognition. This is partly because team tasks tend to take place in complex environments, where outcomes, actions, and interactions take on numerous possibilities. This holds not only for team cognition but is generally true when researchers look at all complex systems and real-world applications. Several theories of cognition include both human and nonhuman aspects of the human–machine environment, such as computers, notepads, control panels, and so on (e.g., Hutchins 1995). Because team cognition involves more than a single information processor, we are forced to consider that the environment now includes other people, who are themselves information processors. For example, Salas et al. (2007) highlight the importance of coordination in forming shared mental models, e.g., by adaptive and supportive behaviors, closed loop communication, and mutual performance monitoring. If the team is to be thought of as a cognitive unit, then it is necessary to include a larger system in the account of cognition.

How external influences can be incorporated into cognitive theory is a question of debate. One major point of dispute is between symbolic information processing theories (e.g., Newell 1990; Smith 1994; Anderson 1995; Proctor and Vu 2006) and situated action/situated cognition theories (Clancey 1993, 1997; Nardi 1996; Hutchins 1991; Rogers and Ellis 1994; Suchman 1993; Vera and Simon 1993a, 1993b, 1993c) (see Green, Davies, and Gilmore 1996, for a human–computer interaction example). In the former, the primary focus is on information processing (IP), which is confined to the individual. In contrast, for situated action (SA) theorists, the focus is improvisational reaction to cues in a very rich environment (Norman 1993). The distinction lies mostly in the locus of information processing and degree of control given to individual goals versus the environment. According to SA theories, much of what symbolic theorists assign to the individual's information processing takes place outside of the confines of the individual. As a result of this distinction, information processing research tends to isolate general psychological principles and mechanisms from controlled laboratory research (e.g., Schneider and Shiffrin 1977), whereas SA research focuses on understanding specific contextual constraints of the natural environment.

There are other approaches in human factors that have a flavor similar to SA. Cognitive engineering (Norman 1986; Hutchins 1991, 1995) is a field of human factors that addresses cognition within complete environments as much as possible (see also Flach et al., this volume). Ecological psychology (Cooke, Gorman, and Rowe 2009; Gibson 1979; Rasmussen 2000a, 2000b; Torenvliet and Vicente 2000) (discussion of "affordances" in Gibson 1977; Norman 1988) suggests that perception and cognition are contextually determined, so that few specific principles will generalize across situations.

The implication of the SA approach for team cognition is that we need to consider the entire work domain as the unit, complete with all of the other people and machines. Work domain analysis (Hajdukiewicz et al. 1998; Vicente 1999, 2000) is a data collection method that supports this notion. In the case of teams, the design implications of a holistic treatment of this kind would mean providing awareness of the goals and constraints that each team member places on each other but no instructions to perform the team task. Dynamical systems theory (Guastello and Guastello 1998; Kelso 1999; Schmidt, Carello, and Turvey 1990; Vallacher and Nowak 1994; Watt and VanLear 1996) would argue that team behavior is an emergent property of the self-organizing system of individual behaviors. Gorman, Cooke, and Kiekel (2004) placed dynamical systems theory in a team cognition context, arguing that team communication acts as a mediational coupling mechanism to organize team members' behavior into behavior emergent at a team level.

In terms of team cognition, we would define conventional views of team cognition as "collective" (Cooke et al. 2000), in that they treat individual team members as the unit of analysis to be later aggregated into a team. Perspectives that extend the cognitive unit into a broader context would support a "holistic" view of team cognition. Such theories would view the team as a cognitive unit all its own. Cooke, Gorman, and Rowe (2009) present a view of team cognition that is derived from ecological psychology. Cooke, Gorman, and Winner (2007) further integrated various holistic approaches to team cognition into what they called the THEDA (Team Holistic Ecology and Dynamic Activity) perspective.

With regards to team cognition, a key defining distinction between these two perspectives has to do with the distribution of knowledge and expertise among team members. When team member expertise is highly diverse, the concept of "sharing" a mental model is best thought of in terms of sharing expectation, rather than sharing exact knowledge. Salas et al. (2007) found that situation assessment is predicted by forming common expectations during problem identification and conceptualization and by compatible interpretation and execution of plans. In evaluating the concept of the group mind, Klimoski and Mohammed (1994) note that the literature is usually vague and casual in defining the concept of shared. They suggest that a team mental model is an emergent property; i.e., it is more than the sum of individual mental models. Nevertheless, Rentsch and Klimoski (2001) found the collective, interpersonal agreement approach to measuring team cognition to be predictive of performance and to serve as a mediator variable. More recently, Miles and Kivlighan (2008) followed convergence over time of individual mental models into similarity and showed that interpersonal agreement improved group climate.

In general, collective approaches to team cognition are more appropriate when knowledge and/or information processing is distributed homogeneously among individuals. However, when cognitive specialization is part of the team's structure, holistic approaches are more appropriate.

Few studies have actually been designed conceptualizing teams at a holistic level (Curseu and Rus 2005). In a review article, Kozlowski and Ilgen (2006) juxtapose holistic and collective approaches to team cognition. They note that most work is done from an individual perspective, as we have a very individualist culture. However, they point out that there are practical constraints to expanding a classical information processing approach to team behavior, in that

the computational modeling rapidly becomes intractable. On the other hand, they add that macrolevel dynamical systems models disregard lower level units. They suggest that the two approaches be integrated to capture the best of both.

6.3.2 CONTEXT WITHIN A COLLECTIVE PERSPECTIVE

Both "camps" have proposed solutions to the problem of context (often similar solutions; e.g., Neisser 1982; Schneider and Shiffrin 1977). Several writers have argued against exclusively choosing either as a theoretical bent (Clancey 1997; Norman 1993; Greeno and Moore 1993; Kozlowski and Ilgen 2006; Rogers and Ellis 1994). For example, Derry, DuRussel, and O'Donnell (1998) attempted to bridge between situated cognition and information processing perspectives by asserting that each is appropriate to model team cognition during different phases of task performance. Proctor and Vu (2006) review the information processing perspective in the context of individuals and teams. They point out that distributed cognition (Hutchins 1991; Rogers and Ellis 1994; Nardi 1996) grows out of the information processing approach to team cognition and conclude that traditional cognitive models can be adapted to incorporate perception-action and context into team cognition.

Another attempt to introduce context into a conventional information processing framework is due to Gibson (2001). She reviews the team literature to develop a phase theory of team performance and argues that the individual behavior analogy does not map well onto team cognition, in that team efficacy and other processes are less consistent at the team level than at the individual level. However, her model is a meta-analysis that incorporates team-level cognitive processes into a conventional collective framework.

More recently, Cuevas et al. (2007) mention an information processing framework called the Team Effectiveness Model, which they juxtapose with Macrocognition, a more context-driven information processing model. Working with these two information processing frameworks, they integrate low-level cognitive processes with higher-level cognitive processes in order to augment team cognition with automation technology.

6.3.3 IMPLICATIONS OF PERSPECTIVES ON TEAM COGNITION

Design implications for the collective information processing perspective (i.e., team as a summation of individual cognitive units) would be centered on providing tools to facilitate planning and symbolic representation. Elements of the task might be represented at a gross level, reflecting the user's (in this case, the team's) need to support top-down processing. Task phases might be treated as distinct and supported as separate modules, permitting the team to make conscious shifts in activities. Modularization of individual team member actions might be supported by enforcing prescribed team member roles fairly strictly to support role distinctions as part of the team's plan of action. Traditionally, emphasis has been placed on assuring that all team members have the same information, so they can develop identical mental models of the task. Designs are generally geared toward building consensus.

One approach to collective cognition is to use a traditional social cognition approach to teams, i.e., applying interpersonal evaluation research to teams. Because social cognition's unit of analysis is the individual's perception of other persons, it is an easy fit for collective approaches to team cognition. For example, Leinonen and Järvelä (2006) used an experience-sampling questionnaire to facilitate a mutual understanding of team member knowledge. Hsu et al. (2007) also used social cognitive theory to study a collective form of efficacy with regards to computer skills. Their collective efficacy measure was based on team consensus, and it successfully predicted team performance. In this study, individual team members did not have distinct roles, so their homogeneous knowledge distribution would tend to make the collective approach appropriate.

Kim and Kim (2008) developed a coordination-based tool to facilitate shared mental models. They used mental model similarity as their measure. Hence, their focus was on explicit coordination, which is appropriate for early development of teams, in the same way that cognitive psychology research on expertise finds that early stages of expertise are marked by declarative knowledge.

Collective approaches to team cognition place more emphasis on explicit coordination and are therefore more appropriate for early stages of team development. Holistic approaches are more focused on implicit coordination and on following team evolution over time. Implicit coordination is characterized by team members anticipating and dynamically adjusting without explicit affirmation. For example, Rico et al. (2008) explored implicit coordination through team situation models. They defined team situation models as immediate/dynamic and emergent/holistic, compared to team mental models, which focus on static knowledge. They also differentiate team situation models from habitual routines by emphasizing the adaptability of team situation models. Their model is more focused on compatibility than exact agreement per se. However, they do hypothesize that more diverse knowledge sets make team situation models harder to form.

Working within a distributed cognition framework, Ligorio, Cesareni, and Schwartz (2008) used intersubjectivity as their unit of analysis. They defined intersubjectivity as complementary views and reciprocal perception of roles among team members.

Design strategies based on the ecological perspective have been posited as more appropriate for complex systems, and for environments in which rare, novel scenarios (such as disasters) are critical (Rasmussen 2000b; Vicente 2000; Flach et al., this volume). Team tasks tend to be of this nature because they tend to be too complex and/or dangerous for an individual to perform alone. Design implications (Rasmussen 2000a, 2000b; Torenvliet and Vicente 2000) are to reject a single "rational" model of good team behavior in favor of displaying the system state, and the natural constraints of the

workspace. System information flow is more important than individual actors. Rather than designing an interface to fit a preexisting mental model of the users, an ecological psychologist would design to constrain the team's mental model of the system. The lack of guidance to behavior is intended to facilitate adaptation and, in the case of teams, establishing idiosyncratic norms by social interaction.

A group decision support system (GDSS) is a particularly well-suited example to answer the question of design implications made by "holistic"-situated versus "collective"-symbolic theories. Problem-solving groups go through distinct phases, along the lines of orientation, conflict, resolution, and action (Shaw 1981; Tuckman 1965). So a collective view might say that these are different types of information that need to be conveyed and processed at different points in the task. The group would "plan" to finish one stage and move on (i.e., there is some degree of intentional choice to change subtasks). For design, this theoretical position implies relatively rigid GDSS to formally structure the group task (e.g., George and Jessup 1997).

A holistic approach would assume that the group just moves on and does not know it. They follow the cues provided to them by the system and each other, which leads them down a natural path toward the goal. They would use tools they value and/or need as the situation warrants. GDSS design from such a position would insist on allowing the group to take action on their own terms. Conversely, it would not permit any guidance to the team as to how they should progress through the task. Specific designs would be appropriate for specific groups-in-situations. The system would have to be made more flexibly or else designed specifically for a group at a task.

6.4 MEASURING TEAM COGNITION

How individuals measure team cognition is driven by their conceptualization of the construct as well as the perspective they take. In this section we distinguish between elicitation of team cognition and assessment and diagnosis activities based on information elicited.

Scaling individual cognitive measurement up to the team level has the potential for information overload, indicating a call for a higher unit of analysis and the need for newer methods to measure team cognition (Krueger and Banderet 2007; Cooke, Gorman, and Rowe 2009). Holistic approaches to measurement would rely less on elicitation of static knowledge, and more on dynamic observation of communication and behavior. For example, Gorman, Cooke, and Winner's (2007) CAST (coordinated awareness of situations by teams) measure of team situation awareness is not based on knowledge elicitation but on perception and evolution of holistic team behavior over time.

Cannon-Bowers et al. (1995) distinguish between taskwork and teamwork knowledge. Taskwork knowledge is knowledge about the individual and team task, and teamwork knowledge is knowledge about the roles, requirements, and responsibilities of team members. Others have distinguished between strategic, procedural, and declarative knowledge (Stout, Cannon-Bowers, and Salas 1996). When adequate measures exist to capture the constructs, we will be in a better position to test the validity of these theoretical distinctions.

6.4.1 Examples of Elicitation Methods

In an environment as complex as a team in a sociotechnical system, how can researchers know they are studying behaviors that really matter? Actually, measurement of team behavior is easier in some ways than measurement of individual behavior. For instance, abstraction is actually more stable in a unit of measurement with more components. That is, within the system as a whole, some actions are repeated by all components of the system. If one component tries to deviate from the norm, then other components will try to bring that piece back into agreement. It is the individuals within the group that are noisier to measure. There will be more small unpredictable actions, but the group will tend to "wash out" those effects.

Team cognition data can be collected and used in much the same way that individual cognition data are used in human factors. For example, walkthroughs (Nielsen 1993) can be adapted to teams. Teams would be walked through the task, and each team member expresses their expectations and needs at each step. Interviews can be conducted with team members, either individually, or in a group. Think-aloud protocols (Ericsson and Simon 1993) have their analogy in team dialogue, in that teams necessarily "think aloud" when they talk to each other during the team task.

There have been numerous calls for faster, more descriptive, and more contextual methods (Wickens 1998; Nielsen 1993). Some of these descriptive methods are already widely used, such as those discussed in the previous paragraph, data-rich ethnographic methods (Harper 2000; see Volk, Pappas, and Wang, this volume), or the purely descriptive task analysis methods (Jeffries 1997). Cooke (1994) catalogs a number of methods that have been used to elicit knowledge from individual experts. Three methods for eliciting team cognition are discussed in the next section, as examples. Mapping conceptual structure was chosen because it is a method that was developed to address individual cognition and has been altered to apply it to team cognition. Ethnography was included because of its popularity in CSCW design. Finally, communication research was included because communication can be thought of as the conscious "thought" of a team. Hence, these methods were selected for their relevance to team cognition.

6.4.1.1 Mapping Conceptual Structures

One method of eliciting individual knowledge is to focus on domain-related concepts and their relations. There are a variety of methods to elicit such conceptual structures (Cooke 1994). One that has been commonly used involves collecting from individuals' judgments of proximity for pairs of task-related concepts. Then a scaling algorithm is applied to reduce these ratings to a graphical representation of conceptual relatedness. This procedure highlights the rater's underlying conceptual structure and hence represents a view of the domain in question. Some common scaling algorithms include Pathfinder networks (Schvaneveldt 1990), multidimensional scaling (e.g., Anderson 1986), and cluster analysis (e.g., Everitt 1993).

Different approaches have been discussed for modifying this scaling procedure to assess team cognition (Cooke et al. 2004), such as the collective methods of averaging (or otherwise aggregating) individual pairwise ratings across team members. Carley (1997) used textual analysis to extract team cognition, based on aggregating individual cognitive maps.

One alternative, more holistic, method is to have the team members discuss their ratings and only make proximity judgments after a consensus is reached. With this method, one assumes that the consensus-building process is an important part of the team's cognition. It incorporates all the group biases and intrateam ranking that one would expect from such a decision making process.

6.4.1.2 Ethnography

Ethnography comes from the fields of anthropology and sociology (Harper 2000). The main idea is to make careful observations of people interacting in their natural environment in order to learn what meaning the observees assign to their actions. It is a program of study aimed at capturing the meaningful context in which actions are taken. In the case of a team task environment, one would trace the "life cycle" of information as it is passed among different team members. Artifacts that the team members use are of key importance, because they influence what the team members will do and what their actions mean. The method involves open-ended interviews of relevant personnel and enough engrossment in the context in question as to be taken seriously by the interviewees.

An ethnographic study of a collaborative writing team might entail recording what materials the expert on topic A uses to do their research (e.g., does she prefer Web sources to print, because they are more immediately updated?). Then the ethnographer would investigate the impact of writer A's decisions on writer B's input (e.g., does writer B write in a less formal style, because the references from section A are not as formal?). This process would go on throughout the life of the document to establish the team's writing process.

6.4.1.3 Communication Data

Another method of eliciting team cognition is to use team task dialogue, and other communication data, as a window to team cognition. Individuals express their thoughts to themselves during task performance by subvocal speech. In the case of team tasks, there is less need to amplify the "subvocalization," because team members naturally speak to one another during the task. This can be thought of as one form of team cognition. It can be directly observed and collected more easily than the awkward task of getting a person to (supposedly) say everything they are thinking. These dialogue data can be analyzed in a variety of ways, both qualitatively (e.g., "Are they arguing?" "Are they on task?" etc.) and quantitatively (e.g., "How long do speech turns tend to last?" "Who speaks the most?" etc.). A recent example of this approach is Ligorio, Cesareni, and Schwartz's (2008) assessment of team cognition by content analysis of transcripts.

6.4.2 ASSESSMENT AND DIAGNOSIS

The other side of measurement, apart from elicitation, is assessment and diagnosis. Elicitation should be seen as a precursor to assessment and diagnosis, as latter depends on the former. Assessment means measuring how well teams meet a criterion. Diagnosis means trying to identify a cause underlying a set of symptoms or actions, such as identifying a common explanation for the fact that specific actions fall short of their respective criteria. Diagnosis therefore involves looking for patterns of behaviors that can be summarily explained by a single cause (e.g., poor team situation awareness and poor leadership both being caused by an uninformed team member in a leadership position).

The particular approach to assessment and diagnosis one wishes to perform is tied to the types of measures one has taken. Ideally, the desired assessment and diagnosis strategy determines the measures, but there are often cases where the reverse is true. For example, if the communication data are recorded only as frequency and duration of speech acts, then one cannot assess aspects of communication that involve content.

There are many dimensions on which to classify measurement strategies. One of interest here is whether the measures are quantitative, qualitative, or some combination of the two. The decision should be based on the types of questions a researcher wishes to pose. Quantitative measures of team performance apply to relatively objective criteria, such as a final performance score (e.g., number of bugs fixed in a software team) or number of ideas generated by a decision-making team. More qualitative criteria will require (or be implied by) richer, more context-dependent data, such as interview or observational data.

So, for example, we may discover that an uninhabited air vehicle (UAV) team is missing most of their surveillance targets. Closer inspection will often consist of observation, interviews, reviewing transcripts, and so on. But the investigator does so with an idea in mind of what might go wrong on a team of this sort. If one is examining the transcripts, then they should have a set of qualitative criteria in mind that the navigator must tell the pilot where to go, the photographer must know where to take pictures, and so on. The investigator will also have some quantitative criteria in mind, such as the fact that better teams tend to speak less during high tension situations (e.g., Achille, Schulze, and Schmidt-Nielsen 1995).

There are also some challenges regarding assessment and diagnosis of team cognition that are related to some specific team task domains. Many team tasks that human factors specialists study are those in dynamic, fast tempo, high-risk environments (e.g., air traffic control, flight, mission control,

process control in nuclear power plants). Whereas in other domains, one may have the luxury of assessment and diagnosis apart from and subsequent to task performance; in the more dynamic and time-critical domains it is often necessary to be able to assess and diagnose team cognition in near real time. In other cases, real time is not timely enough. For instance, we would like to be able to predict loss of an air crew's situation awareness before it happens with potentially disastrous consequences. This challenge requires measures of team cognition that can be administered and automatically scored in real time as the task is performed.

The other aspect of teams that presents unique challenges to assessment is not specific to the task but inherent in the heterogeneous character of teams. When one assesses a team member's knowledge using a single, global referent or gold standard, one is assuming homogeneity. Typically, team member cognition needs to be assessed using role-specific referents.

6.5 USING TEAM COGNITION DATA IN HUMAN FACTORS

Three applications for the kind of information elicited from team cognition can be informing design, designing real-time assistance/intervention applications, or designing training routines. We discuss each of these in the following.

6.5.1 INFORMING DESIGN AND ERROR PREVENTION

A variety of design aids do exist for team tasks, but more translation is needed between research and application (Kozlowski and Ilgen 2006). One common design error is to provide all possible information, resulting in information overload. This mistake is easier to make in team tasks, because the diversity of team member roles means there is more potential information to display. From a naturalistic decision-making perspective, Nemeth et al. (2006) discussed how to design computer displays to facilitate team cognition in a hospital setting. They advocated concentrating on functional units called "cognitive artifacts," which can be shared within a distributed cognition framework.

Another common design error is the misuse or nonuse of automation. Technology has the potential to augment team behavior in such a way that it allows team members to interact more efficiently than they could if they were interacting without technology. Group decision support systems are a classic example of this.

If team usability methods are employed before the new interface is set in stone, then the data can indicate user needs, typical user errors, unclear elements of the current system, areas where the current system particularly excels, and so on. For example, if a team is trying to steer a vehicle toward a remote destination, then they should know their current distance to that destination. If their dialogue consistently shows that they refer to their distance in the wrong units, then designers may choose to change the vehicle's distance units or make the display more salient or otherwise match the environment to the users' expectations.

Also, by understanding team cognition and its ups and downs in a task setting, one can design technology to facilitate team cognition. For example, if a team appears to excel only when they exchange ideas in an egalitarian manner, then a voting software application employed at periodic intervals may facilitate this style of interaction.

Finally, the methods used to assess team cognition can also be used in the context of two or more design alternatives and would thus serve as an evaluation metric. In the case of the earlier example, different methods of representing distance units in the interface could be compared using team situation awareness or communication content as the criteria.

6.5.2 REAL-TIME INTERVENTION, ERROR CORRECTION, AND ADAPTATION

If the team cognition data can be analyzed quickly enough to diagnose problems in real time, then automatic system interventions can be designed to operate in real time. For example, suppose that same vehicle-operation team has hidden part of the distance display behind another window, so that its distance units are not salient. If the system can analyze their dialogue in time to determine this problem, then it can move the offending window, or pop up a cute paper clip with helpful pointers, or some other real-time attempt to correct the problem. With real-time interventions, as with other automatic system behaviors, it is important that the actual user not be superseded to an extent that they are unable to override (Parasuraman and Riley 1997). With teams, as opposed to individuals, designers have an advantage on monitoring and real-time intervention owing to the rich communication data available for analysis in real time. As is the case for the design example discussed above, team cognition data can also serve as a metric upon which we can evaluate the usefulness of the real-time intervention.

6.5.3 TRAINING

Cognitive data are particularly important in training because training is about learning. One can generate training content through a thorough understanding of the team cognition (including knowledge, skills, and abilities) involved in a task.

In designing training regimes, it is important to collect data on what aspects of training can be shown to be more effective than others. For example, if trainers wish to determine whether the benefits of full cross training justify the added time it requires (e.g., Cooke et al. 2003), then they would have to experimentally isolate those characteristics. Diagnostic data can be used to identify what is left to be learned or relearned. Teams can provide a special advantage here, in that their task dialogue can be used to assess what misconceptions they have about the task.

Comparison of learning curves can help identify teams who are learning more slowly or where a given team is expected

to asymptote. The data plotted may be outcome measures for task performance, in which case increased knowledge will be inferred from performance increases. If knowledge data can be collected at repeated intervals, then learning curves can be plotted of actual knowledge increase. Knowledge can be broken down into further components, such as the knowledge of technical task requirements for all team members (i.e., "taskwork"), versus knowledge of how team members are required to interact with one another in order to perform the task (i.e., "teamwork": Cannon-Bowers et al. 1995).

Research across several studies (Cooke, Kiekel, and Helm 2001; Cooke et al. 2004) has shown that taskwork knowledge is predictive of team performance, and teamwork knowledge does improve with experience, along a learning curve. There is also evidence that formation of teamwork knowledge is dependent upon first forming taskwork knowledge. It has also been found that fleeting, dynamically updated knowledge (i.e., team situation awareness) is predictive of team performance.

6.6 EXAMPLES OF TEAM APPLICATIONS

In this section we begin by discussing computer-mediated communication (CMC), a prominent Web-based domain application for team cognition. Then we discuss sports teams, which are a surprisingly underrepresented area of research on team cognition (Fiore and Salas 2006). Next we address emergency response teams, who are essential because of their criticality and increasingly prominent media attention. Finally, we conclude with command and control teams, because they are a classic area of research on team cognition. The examples we present are treated alternately from a conventional collective perspective or from a cognitive engineering or more holistic perspective.

6.6.1 APPLICATION TO COMPUTER-MEDIATED COMMUNICATION

CMC makes for a very general topic for the discussion of team cognition, particularly team communication. There is a large body of literature on CMC. This is because interconnected computers are so ubiquitous. Much of the literature is not on groups who share interdependence toward a common goal.

All of the points addressed in this chapter apply to CMC research. CMC may either involve heterogeneous groups (e.g., a team of experts collaborating on a book) or homogeneous groups (e.g., a committee of engineers deciding on designs). CMC can involve anything from dyads e-mailing one another to bulletin boards and mass Web-based communication. Similar diversity exists in the CMC literature on communication data elicitation, assessment, diagnosis, and how those data are applied. We will focus on one aspect of CMC, that of collective versus holistic interpretations of CMC research. Research and design methods for groupware rely more on anthropological methods than on traditional psychological methods (Green, Davis, and Gilmore 1996;

Harper 2000; Sanderson and Fisher 1994, 1997), which, as described earlier in the chapter, tend to be more holistic and to rely more on qualitative and/or observational techniques.

One important finding in CMC research is that, under certain conditions, anonymity effects can be achieved, leading to either reduced pressure to conform and an enhanced awareness of impersonal task details (e.g., Rogers and Horton 1992; Selfe 1992; Sproull and Kiesler 1986) or pressure to conform to norms that differ from those for face-to-face communication (e.g., Postmes, Spears, and Lea 1998). One theory to account for the former effect of anonymity is Media Richness Theory (MRT; Daft and Lengel 1986). Other, more contextual theories have been proposed to account for the latter effect of anonymity. We address this juxtaposition in the sequel.

In the context of CMC, MRT (Daft and Lengel 1986) has been linked to a symbolic framework (Fulk, Schmitz, and Ryu 1995). Medium "richness"/"leanness" is defined by its ability to convey strong cues of copresence. The theory states that, for tasks of a simple factual nature—for which lots of data need to be passed for very uncontroversial interpretations—"lean" media are most appropriate. For tasks requiring creation of new meaning (i.e., complex symbol manipulations), "rich" media are required. MRT has been related to symbolic theories because of the claim that users rationally formulate the appropriate plan (what medium to choose) and execute it. Further, all tasks can be defined by what symbols need to be conveyed to collaborators. The design implication of MRT is that one can maximize the task-team fit by incorporating media into the design that are appropriate for the task being performed. The team cognition data collected for MRT-based design would involve determining how equivocal the team members feel each communication task is.

MRT has been challenged repeatedly, in favor of more social and situational theories (Postmes, Spears, and Lea 1998; El-Shinnawy and Markus 1997). One interesting attempt to combine these two apparently disparate perspectives was due to Walther (1996). Walther made a compelling argument to treat MRT as a special case of the more social-situational theories. He argued that "rational" theories of media use, such as MRT, are adequately supported in *ad hoc* groups. But when groups get to know one another, they overcome technological boundaries, and their behavior is more driven by social factors. He cites couples who have met online as an example of overcoming these media boundaries. Hence, more contextual-social theories are needed for ongoing groups. Correspondingly, different design implications are in order, and media incorporation into the task environment will be dependent upon richness only for newly formed teams.

Harmon, Schneer, and Hoffman (1995) support this premise with a study of group decision support systems. They find that *ad hoc* groups exhibit the oft-cited anonymity effects (e.g., Anonymous 1998; Sproull and Kiesler 1986), but long-term groups are more influenced by norms of use than by media themselves. Postmes and Spears (1998) used a similar argument to explain the apparent tendency of computer-mediated groups to violate social norms. For example, "flaming" can be characterized as conformity to local norms of the immediate group rather than as deviance from global norms of society at large.

The design implication of this is that the incorporation of media into the task environment is dependent not only upon what the team task is like but also how familiar team members are with one another. So groupware that is intended to be used by strangers would be designed to fit media richness to task equivocality. Groupware for use by friends would be designed with less emphasis on media choice. The team cognition data collected for design based on a more social theoretical bent than MRT would involve determining such factors as how conformist team communication patterns are (e.g., by measuring position shift) team member familiarity (e.g., by measuring the amount of shared terminology in their speech patterns) and so on.

This broad example of a domain application for team cognition highlights the importance of how one approaches team cognition. On the basis of the accounts cited previously (e.g., Walther 1996), a collective-symbolic approach to team cognition is relevant for certain teams and situations, in this case, when teams do not know one another well. However, a holistic approach is more appropriate for other situations, in this case, when teams have interacted with one another for a longer time. We now turn to two more specific examples of team cognition applications. Among other things, these examples illustrate the issue of role heterogeneity and group size.

6.6.2 TEAM COGNITION IN SPORTS TEAMS

Often, when the term "team" comes up we think "sports teams." Some of the most salient examples of teams are sports teams. These are well-defined teams who operate under explicit rules and procedures. Interestingly, despite their salience, the team cognition literature has only been extended to sports teams recently (Fiore and Salas 2006). Most research on sports teams comes from sports science, which has frequently applied findings from social psychology to teams and has emphasized the physiology of teamwork over cognition. Though there is an undeniable physical component to the teamwork of sports teams, there is also a critical and often overlooked cognitive component to team performance in sports domains.

Sports teams, like military, business, and medical teams carry out cognitive activities as a unit such as planning, deciding, assessing the situation, solving problems, recognizing patterns, and coordinating. However, there are differences in the "teamness" of sports teams, with some teams having less independence and opportunity to coordinate (e.g., gymnastics teams) and others having extensive interdependence among team members and ample opportunity to coordinate implicitly or explicitly (e.g., basketball teams; Cannon-Bowers and Bowers 2006). Thus, one size is unlikely to fit all when it comes to applying team cognition theories and findings to sports teams. Indeed, this is a more general gap for the science of team cognition, in which it is unclear how theories and findings generalize from one type of team to another.

A successful application of team cognition to sports teams requires a mapping between types of sports teams and nonsports teams for which team cognition is better understood. For example, by understanding the nature of required interactions of specific types of sports teams, connections can be made to team tasks that have been studied with near-analogous interactions. Pedersen and Cooke (2006) drew this type of analogy between American football teams and military command and control teams. Examples of parallels are drawn in the context of heterogeneous team roles, teams of teams, team member interdependence, and the importance of communication, coordination, and team situation awareness.

With the caveat that one size may not fit all, there are a number of constructs from the team cognition literature that may be transitioned to some sports teams. Concepts such as shared mental models (Reimer, Park, and Hinsz 2006) and team member schema similarity (Rentsch and Davenport 2006) suggest that team members who share a conceptual background of the task or team make for more effective teams by virtue of their ability to anticipate and implicitly coordinate. However, there are little data to support this hypothesis in the sports arena, and some have challenged these ideas on the basis of their vague definitions and questionable assumptions (e.g., Ward and Eccles 2006).

Beyond adapting theoretical constructs from team cognition, sports science may benefit more broadly from the application of measures developed for assessing team performance, team process, and team cognition (Pedersen and Cooke 2006). These measures can be used to understand the dynamics of a team's interaction that are associated with particular outcomes (i.e., win or lose). This information could provide useful feedback to teams after competition about team roles, unique versus shared knowledge, and effective or ineffective team interactions. The topic of assessment is not new to team sports. Of particular importance are assessment measures that go beyond after-the-fact description of team performance to something more predictive that would aid in team composition (Gerrard 2001). Measures of team cognition that focus on team member knowledge and interaction as factors are predictive of team cognition and performance have potential to address this need.

Finally, findings from empirical studies of team cognition that provide guidance for team training or designing technology for team collaboration can similarly suggest training and design interventions for sports teams. Cannon-Bowers and Bowers (2006), for instance, describe training interventions that have succeeded in improving team effectiveness for other teams (e.g., cross training, training for adaptability, team self-correction) that may similarly benefit sports teams.

Although there are many potential connections between the work in team cognition and sports teams in terms of theories, findings, and measures, there has been little direct application of these ideas. It is likely that some types of sports teams will benefit from certain interventions not suited for other types of sports teams. The degree to which research on team cognition in one domain generalizes to a particular type of sports team or another nonsports team for that matter is not only a critical question for sports science but a significant gap in the team cognition literature.

6.6.3 Emergency Response and Team Cognition

A growing interest in emergency response systems since September 11, 2001, has resulted in application of theories, findings, and measures of team cognition to aspects of teamwork in this domain. There are many similarities between the military domains commonly simulated and studied as a context for team cognition and emergency response. The challenges, for instance, are nearly identical. Teams in both domains are often geographically distributed and require the collaboration of multiple organizations or agencies. Decisions are required at a fast tempo, though workload can be unevenly paced, and there is frequently considerable uncertainty. Less like traditional military teams, however, emergency response teams are by nature *ad hoc* with extensive requirements for adaptive and flexible response and emergent team structures.

The science of team cognition has been applied to emergency response in four main areas: (1) communication and coordination of emergency response teams, (2) the *ad hoc* nature of emergency response teams and need for adaptability, (3) training emergency response teams, and (4) designing to improve emergency response collaboration. Each of these applications is described in the remainder of this section.

As exemplified by the poor response to Hurricane Katrina, breakdowns in communication and inter- and intra-agency coordination can lead to failures in emergency response. DeJoode, Cooke, and Shope (2003) and Militello et al. (2005) conducted systematic observations of emergency response systems that have uncovered similar breakdowns in communication and coordination of emergency response teams. Some of these issues have been attributed to the structure and leadership of the emergency response organization. For example, Clancy et al. (2003) examined emergency response using a simulated three-person forest firefighting task. In the lab they were able to look at organizational and leadership differences in the teams, such as teams in which a leader issues commands and teams in which the leader states intent to be implemented by a less hierarchical organization. They found that the flatter organization associated with statements of intent from the leader led to a better distribution of workload and increased team effectiveness.

Models have also been developed to inform communication and coordination in emergency response organizations. For instance, Houghton et al. (2006) have shown how social network analysis of actual emergency communications can provide guidance on the organizational structure. Also, in an attempt to predict communication breakdowns, Nishida, Koisa, and Nakatani (2004) modeled communication within the context of emergency response. The model can highlight choke points and potential breakdowns in a communication network so that interventions can be taken to change organizational structure or to offer support for communication to thwart potential disasters.

The nature of emergencies makes well-defined teams and rigid plans impractical. Instead, emergencies require teams to be flexible in adapting to the needs of the emergency and the context in which it occurs. Teams and agencies may come together that have never worked together before, and events may occur that have never been imagined such as the events of 9/11. Butts, Petrescu-Prahova, and Cross (2007) analyzed the communications data (radio and police) from responders to the World Trade Center disaster. The results of this analysis emphasized the importance of emergent coordination, reinforcing the notion that teams must adapt to failures of the conventional system (e.g., severed lines of communication). Good teams will develop "workarounds" such as emergent hubs in which communication is filtered and routed to the appropriate individuals. Cognitive engineers (e.g., Ntuen et al. 2006) have proposed decision aids based on an analysis of emergency response as a cognitive system as a partial solution to the need for adaptive and rapid response to uncertain situations.

There has been significant application of team cognition research to the training of emergency response teams. A number of team-based synthetic task environments such as ISEE (Immersive Synthetic Environment for Exercises; McGrath, Hunt, and Bates 2005) and NeoCITIES (McNeese et al. 2005) have been developed in which training research can be conducted. Not only has the synthetic team environment been applied to emergency response, but Cooke and Winner (2008) have even suggested that principles and strategies from experimental design in psychology can be used to better structure and assess emergency response exercises. Although there is little in the literature on the application of measures of team cognition to emergency response, this area holds particular promise as assessment metrics with diagnostic feedback at the team level are not commonly used in emergency response exercises.

Not only can training lead to more effective emergency response teams, but technologies, especially communication and decision aiding technologies, can be designed to facilitate collaboration (Mendonca, Beroggi, and Wallace 2001). Sometimes the technology is nothing spectacular, but rather, simple, straightforward technologies are judiciously applied after an assessment of the system from a human-centered point of view. For example, Klinger and Klein (1999) were able to drastically improve team performance in a nuclear emergency response center by instituting fairly simple fixes such as white boards for increased situation awareness, role changes, and time out procedures for making sure everyone is on the same page.

In summary, since 2001 the research on team cognition has made significant contributions to the area of emergency response. There is much similarity between the military work that is the centerpiece for most team cognition research and emergency response. The opportunity for continued cross talk between these two areas is promising.

6.6.4 COMMAND AND CONTROL TEAMS

Cooke et al. (2001, 2003, 2007) have done several studies on team operation in a simulated ground control station of a UAV. This task involves heterogeneous teams of three members collaborating via six networked computers. The team flies the plane to different locations on a map and takes photographs.

Each role brings specific skills to the team and places particular constraints on the way the UAV can fly. The pilot controls airspeed, heading, and altitude, and monitors UAV systems. The photographer adjusts camera settings, takes photos, and monitors the camera equipment. The navigator oversees the mission and determines flight paths under various constraints. Most communication is done via microphones and headsets, although some involves computer messaging. Information and rules specific to each role is only available to the team member filling that role, though team members may communicate their role knowledge.

This command and control example is treated from a more holistic, cognitive engineering perspective. The research aim was focused on team cognition and addressed the complete environments as much as possible. The team members had heterogeneous role assignments, and there were three team members. The role of heterogeneity, team size, and task environment complexity allow for complex social dynamics.

Rich data were collected in great detail, including observational data. Several varieties of cognitive data were collected. Though these measures were related to normative definitions of ideal team knowledge, those definitions came in several diverse forms, addressing different aspects of knowledge. For example, individual taskwork knowledge was defined for each team member's global task knowledge, knowledge of their own task, and knowledge of other team members' tasks. In order to take more holistic measures, consensus metrics and communication data were also collected to capture team knowledge.

This is a complex task, and there are hence many criteria on which to assess and diagnose team performance. Foremost are the various performance measures. These include an overall performance score made up of a weighted average of number of targets photographed, total mission time, fuel used, and so on. That team performance score is useful to diagnose overall team behavior. To look at individual performance, it is necessary to create individual performance measures. This being a heterogeneous task, we cannot apply one measure to all team members and expect to aggregate. Therefore, other, more diagnostic performance measures for individuals include the three individual performance scores, each comprised of similar weighted averages but of individual behavior. Acceptability criteria are loosely defined for each of these variables, based on the asymptote of numerous teams as they learn the task. These performance measures represent acceptability criteria for other measures.

Because each team member has their own knowledge and their own knowledge dependencies, it is important to measure how well each team member knows their own role and how well they know each others' roles. We can call this knowledge of what to do during the task "taskwork knowledge." In team tasks, it is also important to know how each team member is expected to interact with the others. We will call this "teamwork knowledge." We discuss these in turn.

Teamwork knowledge was measured with questionnaires, with predefined correct answers. Individual team members were asked for a given scenario what information is passed and between which team members. The correct answers were separated out by individual role, as well as basic information that all team members should have. The tests were administered to each individual, and their accuracy score was calculated according to their own role. The scores now properly scaled, the proportion of accurate answers could then be aggregated. In addition, teams were asked to complete this questionnaire as a group, coming to consensus on the answers. This gives us an estimate of teamwork knowledge elicited at a more holistic level.

For taskwork knowledge, a criterion was first defined in the form of a network of pairwise links among domain relevant concepts. Then team networks could be collected and compared against this criterion. Like teamwork, taskwork knowledge was measured two ways. First, each individual rated their own pairwise proximities. Then teams were asked to give group ratings, in which they engage in group discussion, and reach consensus before rating any concepts. Again, this latter measure is more of a holistic measure of team cognition. The networks derived from Pathfinder analysis of the pairwise ratings (Schvaneveldt 1990) could be scored for accuracy by calculating the similarity of the teams' (or individuals') networks to an expert, referent network.

Because both the taskwork and teamwork measures yielded quantitative measures of fit to criterion, the accuracy could be used to predict team performance scores. This allows researchers to diagnose what may be wrong with a team that performs poorly. For example, a team where the taskwork scores are low would indicate that a training regime should concentrate on taskwork knowledge. Specific taskwork weaknesses can further be assessed by determining whether team members are weak in knowledge of their own role or on each others' roles. This diagnosis can go even further by examining specific links in the network that do not match up to the ideal referent. All of this information can be incorporated into a training regime or converted to interface design recommendations.

Another form of knowledge that was measured in the UAV task context was team situation awareness. This was measured using a query-based approach (Durso et al. 1998; for discussion of retrospective queries, e.g., Endsley 1990) in which individuals, and then the team as a whole, were asked during the course of a mission to answer projective questions regarding the current situation (e.g., how many targets will you get photos of by the end of the mission?). The responses were scored for accuracy as well as intra-team similarity.

Communication data were collected and analyzed extensively. Transcripts were taken to record actual utterances, and latent semantic analysis (Landauer, Foltz, and Laham 1998) was applied to analyze the content of the discourse. Speech acts were preserved in a raw form by specialized software to record quantity of verbal communication by each team member and to each team member.

These communication data were used in a wide array of measures, all aimed at predicting performance (Kiekel et al. 2001, 2002). For example, examining the transcripts revealed that a number of teams did not realize that they could photograph the target, as long as they were within a specific range of it. Kiekel et al. (2001) also used the communication log data to define speech events as discrete units then modeled the behavior of those units to determine complexity of team communication patterns. Teams that exhibited many diverse communication patterns were shown to be poorer performers. This was thought to indicate that the teams had not established a stable dialogue pattern and would perhaps imply more clear teamwork knowledge training.

As the above detail shows, this task involved a great deal of diverse measurement, both to assess multiple team cognition constructs and to assess individual and holistic team knowledge. Training or design recommendations based on these data would be very specific to the diagnostic findings. For example, the finding that better performing teams tended to have more stable communication patterns might imply a training regime aimed at stabilizing team discourse (of course, we must be cautious, in that we must avoid drawing causal implications from correlational data). Specificity of this sort was not needed for the first example, because the task was so much simpler.

A number of interesting findings were discovered in this series of studies (Cooke et al. 2007). The studies support the view that interaction is key to team cognition. In one of the studies, it was found that teams transferred expertise from one command and control task to another. This suggested that (1) team cognition emerges through the interactions of team members, (2) it is team member interactions that distinguish high-performing teams from average teams, and (3) these interactions transfer across different tasks.

One use of team cognition in this research context would be to design a real-time system monitor to assess the need to intervene. So, for example, suppose a communication monitor is running in the background, and it determines a point where the team communication pattern becomes erratic and unusually terse. This might indicate a red flag that something is wrong in the task. An appropriate intervention would then be called in to correct the problem.

This was a small-scale simulation task on an intranet. Similar real-world tasks occur in the context of network-centric warfare and distributed mission training, for which a critical issue is assessing team cognition and performance in distributed Web-based applications. The metrics of team cognition discussed in this example can be applied for that purpose and for real-time intervention.

6.7 CONCLUDING REMARKS

As noted at the beginning of this chapter, team tasks are extremely common and are being given increasingly greater focus within organizations. In particular, computer-mediated communication and decision making applications for teams are extremely varied and ubiquitous, ranging from e-mail to shared bulletin boards for classrooms to remote conferencing. As these applications are increasingly exploited on the Web, communication and coordination of teams will become even more widespread. This ubiquity is attributable to the self-encapsulated, cross-platform nature of Web-based collaboration tools.

With the growth of collaborative Web applications, an interesting ramification for team cognition will be the greater possibility of anonymity. Web-based applications make it much more possible for teams to form, interact, and perform tasks without ever having met. This creates a possibility to dramatically amplify issues such as interpersonal awareness, teamwork knowledge, task focus, telepresence, and so on.

More generally, as team tasks become an increasingly important part of life, it will become more important to consider the needs of teams. The interaction patterns among team members, including the cognitive processes that occur at the team level, add a second layer to behavior that is not present in individuals. However, human factors have long addressed systems, in which the human and the environment are treated as interacting factors. Much of the groundwork already exists for designing to meet the needs of teams.

Considerations of team cognition can be important in designing team tasks and environments, in much the same way that individual cognition is used in design for individuals. Team characteristics and abilities must be assessed, team task environments must be understood, and so on. The complexity is introduced when team cognition must account for the knowledge individuals have of their team members and the influence team members have on one another. Two major approaches to this are to either conceive of teams as a collection of individuals, in which each person's cognition is considered separately (collective team cognition) or as a single cognitive unit (holistic team cognition). The two approaches are not mutually exclusive, and some scenarios are better fitted to collective or holistic approaches, respectively.

To treat teams as holistic units, we transfer what is known from individual cognition and incorporate those features that individuals do not possess. For instance, team size and heterogeneity are issues that do not exist for individuals. When we treat teams holistically, say by using team communication data as our measure of cognition, we automatically incorporate the social dynamics intrinsic in team size, because the types of interaction we observe are inherently determined by this factor. Likewise, individual role expertise is incorporated in holistic measures such as consensus interviews, because team members with differential role expertise and/or influence will contribute differentially to the consensus formation. But issues unique to teams may also have their analogy in individual cognition. For instance, ambivalent deliberation during decision making appears analogous to team conflict.

As team cognition measurement becomes more adept at incorporating the added dimensions that teams bring, some of this advantage should transfer back to the measurement of individual cognition. For example, although individual cognition may have no such distinction as teamwork versus taskwork knowledge, methods developed to account for these constructs in teams may transfer back to individuals. It may, at least, help to enrich the complexity of our view of individual cognition. Hence, teams may raise new issues of complexity that exist in parallel for individual cognition but which might not have been addressed otherwise.

REFERENCES

Achille, L. B., K. G. Schulze, and A. Schmidt-Nielsen. 1995. An analysis of communication and the use of military terms in navy team training. *Military Psychology* 7(2): 96–107.

Anderson, A. M. 1986. Multidimensional scaling in product development. In *The Fascination of Statistics*, eds. R. J. Brook et al., 103–110. New York: Marcel Decker.

Anderson, J. R. 1995. *Cognitive Psychology and Its Implications,* 4th ed. New York: W. H. Freeman.

Anonymous. 1998. To reveal or not to reveal: A theoretical model of anonymous communication. *Communication Theory* 8(4): 381–407.

Beebe, S. A., and J. T. Masterson. 1997. *Communicating in Small Groups,* 5th ed. New York: Longman.

Blickensderfer, E. L., R. J. Stout, J. A. Cannon-Bowers, and E. Salas. 1993. Deriving theoretically-driven principles for cross-training teams. Paper presented at the 37th annual meeting of the Human Factors and Ergonomics Society, Seattle, WA, Oct 11–15.

Butts, C. T., M. Petrescu-Prahova, and B. R. Cross. 2007. Responder communication networks in the World Trade Center disaster: Implications for modeling of communication within emergency settings. *Journal of Mathematical Sociology* 31: 121–147.

Cannon-Bowers, J. A., and C. Bowers. 2006. Applying work teams results to sports teams: Opportunities and cautions. *International Journal of Sport and Exercise Psychology* 4: 447–462.

Cannon-Bowers, J. A., E. Salas, E. Blickensderfer, and C. A. Bowers. 1998. The impact of cross-training and workload on team functioning: A replication and extension of initial findings. *Human Factors* 40: 92–101.

Cannon-Bowers, J. A., E. Salas, and S. Converse. 1993. Shared mental models in expert team decision making. In *Current Issues in Individual and Group Decision Making*, eds. J. Castellan, Jr., 221–246. Hillsdale, NJ: Lawrence Erlbaum.

Cannon-Bowers, J. A., S. I. Tannenbaum, E. Salas, and C. E. Volpe. 1995. Defining team competencies and establishing team training requirements. In *Teams: Their Training and Performance*, eds. R. Guzzo and E. Salas, 101–124. Norwood, NJ: Ablex.

Carley, K. M. 1997. Extracting team mental models through textual analysis. *Journal of Organizational Behavior,* Special Issue, 18: 533–558.

Clancey, W. J. 1993. Situated action: A neuropsychological interpretation: response to Vera and Simon. *Cognitive Science* 17(1): 87–116.

Clancey, W. J. 1997. *Situated Cognition: On Human Knowledge and Computer Representations.* New York: Cambridge University Press.

Clancy, J., G. Elliot, T. Ley, J. McLennan, M. Omodei, E. Thorsteinsson, and A. Wearing. 2003. Command style and team performance in dynamic decision making tasks. In *Emerging Perspectives on Judgment and Decision Research*, eds. S. L. Schneider and J. Shanteau, 586–619. Cambridge, UK: Cambridge University Press.

Cooke, N. J. 1994. Varieties of knowledge elicitation techniques. *International Journal of Human-Computer Studies* 41: 801–849.

Cooke, N. J., J. C. Gorman, J. L. Duran, and A. R. Taylor. 2007. Team cognition in experienced command-and-control teams. *Journal of Experimental Psychology: Applied* 13(3): 146–157.

Cooke, N. J., J. C. Gorman, and L. J. Rowe. 2009. An ecological perspective on team cognition. In *Team Effectiveness in Complex Organizations: Cross-disciplinary Perspectives and Approaches*, SIOP Frontiers Series, eds. E. Salas, J. Goodwin, and C. S. Burke, 157–182. Mahwah, NJ: Lawrence Erlbaum.

Cooke, N. J., J. C. Gorman, and J. L. Winner. 2007. Team cognition. In *Handbook of Applied Cognition,* 2nd ed., eds. F. Durso et al., 239–268. New York: John Wiley.

Cooke, N. J., P. A. Kiekel, and E. Helm. 2001. Measuring team knowledge during skill acquisition of a complex task. *International Journal of Cognitive Ergonomics,* Special Section, 5: 297–315.

Cooke, N. J., P. A. Kiekel, E. Salas, R. Stout, C. Bowers, and J. Cannon-Bowers. 2003. Measuring team knowledge: A window to the cognitive underpinnings of team performance differences. *Group Dynamics* 7: 179–199.

Cooke, N. J., E. Salas, J. A. Cannon-Bowers, and R. Stout. 2000. Measuring team knowledge. *Human Factors* 42: 151–173.

Cooke, N. J., E. Salas, P. A. Kiekel, and B. Bell. 2004. Advances in measuring team cognition. In *Team Cognition: Process and Performance at the Inter- and Intra-individual Level*, eds. E. Salas and S. M. Fiore, 83–106. Washington, DC: American Psychological Association.

Cooke, N. J., and J. L. Winner. 2008. Human factors of Homeland Security. In *Reviews of Human Factors and Ergonomics*, vol. 3, 79–110. Santa Monica, CA: Human Factors and Ergonomics Society.

Cuevas, H. M., S. M. Fiore, B. S. Caldwell, and L. Strater. 2007. Augmenting team cognition in human–automation teams performing in complex operational environments. *Aviation, Space, and Environmental Medicine* 78(5), supplement, B63–B70.

Curseu, P. L., and D. Rus. 2005. The cognitive complexity of groups: A critical look at team cognition research. *Cognitie, Creier, Compartament (Cognition, Brain, and Behaviour)* 9(4): 681–710.

Daft, R. L., and R. H. Lengel. 1986. Organizational information requirements, media richness, and structural design. *Management Science* 32: 554–571.

Davis, J. H. 1973. Group decision and social interaction: A theory of social decision schemes. *Psychological Review,* 80(2): 97–125.

DeJoode, J., N. J. Cooke, and S. M. Shope. 2003. Naturalistic observations of an airport mass casualty exercise. In *Proceedings of the Human Factors and Ergonomics Society 47th Annual Meeting*, 663–667. Santa Monica, CA: Human Factors and Ergonomics Society.

Derry, S. J., L. A. DuRussel, and A. M. O'Donnell. 1998. Individual and distributed cognitions in interdisciplinary teamwork: A developing case study and emerging theory. *Educational Psychology Review* 10(1): 25–57.

Durso, F. T., C. A. Hackworth, T. R. Truitt, J. Crutchfield, D. Nikolic, and C. A. Manning. 1998. Situation awareness as a predictor of performance in en route air traffic controllers. *Air Traffic Control Quarterly* 6(1): 1–20.

Dyer, J. L. 1984. Team research and team training: a state of the art review. In *Human Factors Review*, eds. F. A. Muckler, 285–323. Santa Monica, CA: Human Factors Society.

Endsley, M. R. 1990. A methodology for the objective measure of situation awareness. In *Situational Awareness in Aerospace Operations* (AGARD-CP-478), 1/1–1/9. Neuilly-Sur-Seine, France: NATO–Advisory Group for Aerospace Research and Development.

El-Shinnawy, M., and M. L. Markus. 1997. The poverty of media richness theory: Explaining people's choice of electronic mail vs. voice mail. *International Journal of Human–Computer Studies* 46: 443–467.

Ericsson, K. A., and H. A. Simon. 1993. *Protocol Analysis: Verbal Reports as Data.* Cambridge, MA: MIT Press.

Everitt, B. S. 1993. *Cluster Analysis,* 3rd ed. New York: Halsted Press.

Festinger, L., S. Schachter, and K. Back. 1964. Patterns of group structure. In *Mathematics and Psychology*, ed. G. A. Miller. New York: John Wiley.

Fiore, S. M., and E. Salas. 2006. Team cognition and expert teams: Developing insights from cross-disciplinary analysis of exceptional teams. *International Journal of Sport and Exercise Psychology* 4: 369–375.

Fisher, A. B., and D. G. Ellis. 1990. *Small Group Decision Making*, 3rd ed. New York: McGraw-Hill.

Flach, J. M., K. B. Bennett, P. J. Stappers, and D. P. Saakes, this volume. An ecological perspective to meaning processing: The dynamics of abductive systems. In *Handbook of Human Factors in Web Design*, 2nd ed., eds. K.-P. L. Vu and R. W. Proctor, 509–526. Boca Raton, FL: CRC Press.

Friedkin, N. E. 1998. *A Structural Theory of Social Influence.* Cambridge, UK: Cambridge University Press.

Fulk, J., J. Schmitz, and D. Ryu. 1995. Cognitive elements in the social construction of communication technology. *Management Communication Quarterly* 8(3): 259–288.

George, J. F., and L. M. Jessup. 1997. Groups over time: What are we really studying? *International Journal of Human–Computer Studies* 47: 497–511.

Gerrard, B. 2001. A new approach to measuring player and team quality in professional team sports. *European Sport Management Quarterly* 1: 219–234.

Gibson, C. B. 2001. From knowledge accumulation to accommodation: Cycles of collective cognition in work groups. *Journal of Organizational Behavior* 22: 121–134.

Gibson, J. J. 1977. The theory of affordances. In *Perceiving, Acting, and Knowing*, eds. R. E. Shaw and J. Bransford, 67–82. Mahwah, NJ: Lawrence Erlbaum.

Gibson, J. J. 1979. *The Ecological Approach to Visual Perception.* Boston, MA: Houghton-Mifflin.

Gillett, R. 1980a. Probability expressions for simple social decision scheme models. *British Journal of Mathematical and Statistical Psychology* 33: 57–70.

Gillett, R. 1980b. Complex social decision scheme models. *British Journal of Mathematical and Statistical Psychology* 33: 71–83.

Gorman, J. C., N. J. Cooke, and P. A. Kiekel. 2004. Dynamical perspectives on team cognition. In *Proceedings of the 48th Annual Human Factors and Ergonomics Society Meeting*, 673–677. Santa Monica, CA: Human Factors and Ergonomics Society.

Gorman, J. C., N. J. Cooke, and J. L. Winner. 2006. Measuring team situation awareness in decentralized command and control environments. *Ergonomics* 49(12–13, 10–22): 1312–1325.

Green, T. R. G., S. P. Davies, and D. J. Gilmore. 1996. Delivering cognitive psychology to HCI: The problems of common language and of knowledge transfer. *Interacting with Computers* 8(1): 89–111.

Greeno, J. G., and J. L. Moore. 1993. Situativity and symbols: response to Vera and Simon. *Cognitive Science* 17(1): 49–59.

Guastello, S. J., and D. D. Guastello. 1998. Origins of coordination and team effectiveness: A perspective from game theory and nonlinear dynamics. *Journal of Applied Psychology* 83(3): 423–437.

Hajdukiewicz, J. R., D. J. Doyle, P. Milgram, K. J. Vicente, and C. M. Burns. 1998. A work domain analysis of patient monitoring in the operating room. In *Proceedings of the Human Factors and Ergonomics Society 42nd Annual Meeting*, 1038–1042. Santa Monica, CA: Human Factors and Ergonomics Society.

Hare, A. P. 1992. *Groups, Teams, and Social Interaction: Theories and Applications.* New York: Praeger.

Harmon, J., J. A. Schneer, and L. R. Hoffman. 1995. Electronic meetings and established decision groups: Audioconferencing effects on performance and structural stability. *Organizational Behavior and Human Decision Processes* 61(2): 138–147.

Harper, R. H. R. 2000. The organisation in ethnography—a discussion of ethnographic fieldwork programs in CSCW. *Computer Supported Cooperative Work* 9(2): 239–264.

Hinsz, V. B. 1999. Group decision making with responses of a quantitative nature: The theory of social decision schemes for quantities. *Organizational Behavior and Human Decision Processes* 80(1): 28–49.

Hollingshead, A. B. 1998. Retrieval processes in transactive memory systems. *Journal of Personality and Social Psychology* 74(3): 659–671.

Houghton, R. J., C. Baber, R. McMaster, N. A. Stanton, P. Salmon, R. Stewart, and G. Walker. 2006. Command and control in emergency services operations: A social network analysis. *Ergonomics* 49: 1204–1225.

Hsu, M., I. Y. Chen, C. Chiu, and T. L. Ju. 2007. Exploring the antecedents of team performance in collaborative learning of computer software. *Computers and Education* 48: 700–718.

Hutchins, E. 1991. The social organization of distributed cognition. In *Perspectives on Socially Shared Cognition*, eds. L. B. Resnick, J. M. Levine, and S. D. Teasley, 283–307. Washington, DC: American Psychological Association.

Hutchins, E. 1995. How a cockpit remembers its speed. *Cognitive Science* 19: 265–288.

Hutchins, E. 1996. *Cognition in the Wild.* Cambridge, MA: MIT Press.

Jeffries, R. 1997. The role of task analysis in the design of software. In *Handbook of Human–Computer Interaction,* 2nd ed., eds. H. Helander, T. K. Landauer, and P. Prabhu, 347–358. New York: Elsevier.

Kelso, J. A. S. 1999. *Dynamic Patterns: The Self-Organization of Brain and Behavior.* Cambridge: MIT Press.

Kerr, N. L., J. H. Davis, D. Meek, and A. K. Rissman. 1975. Group position as a function of member attitudes: Choice shift effects from the perspective of social decision scheme theory. *Journal of Personality and Social Psychology* 31(3): 574–593.

Kiekel, P. A., N. J. Cooke, P. W. Foltz, J. Gorman, and M. Martin. 2002. Some promising results of communication-based automatic measures of team cognition. In *Proceedings of the Human Factors and Ergonomics Society*, 298–302, Santa Monica, CA: Human Factors and Ergonomics Society.

Kiekel, P. A., N. J. Cooke, P. W. Foltz, and S. M. Shope. 2001. Automating measurement of team cognition through analysis of communication data. In *Usability Evaluation and Interface Design*, eds. M. J. Smith et al., 1382–1386. Mahwah, NJ: Lawrence Erlbaum.

Kim, H., and D. Kim. 2008. The effects of the coordination support on shared mental models and coordinated action. *British Journal of Educational Technology* 39(3): 522–537.

Klimoski, R., and S. Mohammed. 1994. Team mental model: Construct or metaphor? *Journal of Management* 20(2): 403–437.

Klinger, D. W. and G. Klein. 1999. Emergency response organizations: an accident waiting to happen. *Ergonomics in Design* 7: 20–25.

Kozlowski, S. W. J., and D. R. Ilgen. 2006. Enhancing the effectiveness of work groups and teams. *Psychological Science in the Public Interest* 7(3): 77–124.

Krueger, G. P., and L. E. Banderet. 2007. Implications for studying team cognition and team performance in network-centric warfare paradigms. *Aviation, Space, and Environmental Medicine* 78(5), supplement, B58–B62.

Landauer, T. K., P. W. Foltz, and D. Laham. 1998. An introduction to latent semantic analysis. *Discourse Processes*, 25(2&3): 259–284.

Langan-Fox, J., S. Code, and K. Langfield-Smith. 2000. Team mental models: Techniques, methods, and analytic approaches. *Human Factors* 42: 242–271.

Leinonen, P., and S. Järvelä. 2006. Facilitating interpersonal evaluation of knowledge in a context of distributed team collaboration. *British Journal of Educational Technology* 37(6): 897–916.

Ligorio, B. M., D. Cesareni, and N. Schwartz. 2008. Collaborative virtual environments as means to increase the level of intersubjectivity in a distributed cognition system. *Journal of Research on Technology in Education* 40(3): 339–357.

McGrath, D., A. Hunt, and M. Bates. 2005. A simple distributed simulation architecture for emergency responses exercises. In *Proceedings of the Ninth IEEE International Symposium on Distributed Simulation and Real-time Applications*, 221–228 (DS-RT 2005) (Montreal, Canada, Oct. 10–12).

McNeese, M. D., P. Bains, I. Brewer, C. Brown, E. S. Connors, T. Jefferson Jr., R. E. T. Jones, and L. Terrell. 2005. The NEOCITIES simulation: Understanding the design and experimental methodology used to develop a team emergency. In *Proceedings of the Human Factors and Ergonomics Society 49th Annual Meeting*, 591–594. Santa Monica, CA: Human Factors and Ergonomics Society.

Miles, J. R., and D. M. Kivlighan. 2008. Team cognition in group interventions: The relation between coleaders' shared mental models and group climate. *Group Dynamics: Theory, Research, and Practice* 12(3): 191–209.

Militello, L. G., L. Quill, E. S. Patterson, R. Wears, and J. A. Ritter. 2005. Large-scale coordination in emergency response. In the *Proceedings of the Human Factors and Ergonomics Society 49th Annual Meeting*, 534–538. Santa Monica, CA: Human Factors and Ergonomics Society.

Mendonca, D., G. E. G. Beroggi, and W. A. Wallace. 2001. Decision support for improvisation during emergency response operations. *International Journal of Emergency Management* 1: 30–38.

Nardi, B. A. 1996. Studying context: A comparison of activity theory, situated action models, and distributed cognition. In *Context and Consciousness: Activity Theory and Human–Computer Interaction*, ed. B. A. Nardi, 69–102. Cambridge, MA: MIT Press.

Neisser, U. 1982. Memory: what are the important questions? In *Memory Observed*, ed. U. Neisser, 3–18. New York: W. H. Freeman.

Nemeth, C., M. O'Connor, P. A. Klock, and R. Cook. 2006. Discovering healthcare cognition: The use of cognitive artifacts to reveal cognitive work. *Organization Studies* 27(7): 1011–1035.

Newell, A. 1990. *Unified Theories of Cognition.* Cambridge, MA: Harvard University Press.

Nielsen, J. 1993. *Usability Engineering.* New York: Academic Press.

Nishida, S., T. Koiso, and M. Nakatani. 2004. Evaluation of organizational structure in emergency situations from the viewpoint of communication. *International Journal of Human–Computer Interaction* 17: 25–42.

Norman, D. A. 1986. Cognitive engineering. In *User Centered System Design*, eds. D. A. Norman and S. Draper, 31–61. Mahwah, NJ: Lawrence Erlbaum.

Norman, D. A. 1988. *The Design of Everyday Things.* New York: Currency Doubleday.

Norman, D. A. 1993. Cognition in the head and in the world: An introduction to the special issue on situated action. *Cognitive Science* 17(1): 1–6.

Ntuen, C. A., O. Balogun, E. Boyle, and A. Turner. 2006. Supporting command and control training functions in the emergency management domain using cognitive systems engineering. *Ergonomics* 49: 1415–1436.

Parasuraman, R., and V. Riley. 1997. Humans and automation: Use, misuse, disuse, abuse. *Human Factors* 39(2): 230–253.

Pedersen, H. K., and N. J. Cooke. 2006. From battle plans to football plays: Extending military team cognition to football. *International Journal of Sport and Exercise Psychology* 4: 422–446.

Postmes, T., and R. Spears. 1998. Deindividuation and antinormative behavior: A meta-analysis. *Psychological Bulletin* 123(3): 238–259.

Postmes, T., R. Spears, and M. Lea. 1998. Breaching or building social boundaries? SIDE-effects of computer-mediated communication. *Communication Research* 25: 689–715.

Proctor, R. W., and K. L. Vu. 2006. The cognitive revolution at age 50: Has the promise of the human information-processing approach been fulfilled? *International Journal of Human–Computer Interaction* 21(3): 253–284.

Rasmussen, J. 2000a. Designing to support adaptation. In *Proceedings of the IEA 2000/HFES 2000 Congress*, 554–557, Santa Monica, CA: Human Factors and Ergonomics Society.

Rasmussen, J. 2000b. Trends in human factors evaluation of work support systems. *Proceedings of the IEA 2000/HFES 2000 Congress*, 561–564. Santa Monica, CA: Human Factors and Ergonomics Society.

Reimer, T., E. S. Park, and V. B. Hinsz. 2006. Shared and coordinated cognition in competitive and dynamic task environments: An information-processing perspective for team sports. *International Journal of Sport and Exercise Psychology* 4: 376–400.

Rentsch, J. R., and S. W. Davenport. 2006. Sporting a new view: Team member schema similarity in sports. *International Journal of Sport and Exercise Psychology* 4: 401–421.

Rentsch, J. R., and R. J. Klimoski. 2001. Why do 'great minds' think alike?: Antecedents of team member schema agreement. *Journal of Organizational Behavior* 22(2), Special Issue, 107–120.

Rico, R., M. Sanches-Manzanares, F. Gil, and C. Gibson. 2008. Team implicit coordination processes: A team knowledge-based approach. *Academy of Management Review* 33(1): 163–184.

Rogers, Y., and J. Ellis. 1994. Distributed cognition: an alternative framework for analyzing and explaining collaborative working. *Journal of Information Technology* 9: 119–128.

Rogers, P. S., and M. S. Horton. 1992. Exploring the value of face-to-face collaborative writing. In *New Visions of Collaborative Writing*, ed. J. Forman, 120–146. Portsmouth, NH: Boynton/Cook.

Salas, E., N. J. Cooke, and M. A. Rosen. 2008. On teams, teamwork, and team performance: Discoveries and developments. *Human Factors* 50(3): 540–547.

Salas, E., T. L. Dickinson, S. A. Converse, and S. I. Tannenbaum. 1992. Toward an understanding of team performance and training. In *Teams: Their Training and Performance*, eds. R. W. Swezey and E. Salas, 3–29. Norwood, NJ: Ablex.

Salas, E., M. A. Rosen, C. S. Burke, D. Nicholson, and W. R. Howse. 2007. Markers for enhancing team cognition in complex environments: The power of team performance diagnosis. *Aviation, Space, and Environmental Medicine* 78(5), supplement, B77–B85.

Sanderson, P. M., and C. Fisher. 1994. Exploratory sequential data analysis: Foundations. *Human–Computer Interaction* 9: 251–317.

Sanderson, P. M., and C. Fisher. 1997. Exploratory sequential data analysis: qualitative and quantitative handling of continuous observational data. In *Handbook of Human Factors and Ergonomics,* 2nd ed., ed. G. Salvendy, 1471–1513. New York: John Wiley.

Schneider, W., and R. M. Shiffrin. 1977. Controlled and automatic human information processing: I. Detection, search, and attention. *Psychological Review* 84(1): 1–66.

Schmidt, R. C., C. Carello, and M. T. Turvey. 1990. Phase transitions and critical fluctuations in the visual coordination of rhythmic movements between people. *Journal of Experimental Psychology: Human Perception and Performance* 16(2): 227–247.

Schvaneveldt, R. W. 1990. *Pathfinder Associative Networks: Studies in Knowledge Organization.* Norwood, NJ: Ablex.

Sebanz, N., G. Knoblich, and W. Prinz. 2003. Representing others' actions: Just like one's own? *Cognition* 88: 11–21.

Selfe, C. L. 1992. Computer-based conversations and the changing nature of collaboration. In *New Visions of Collaborative Writing*, ed. J. Forman, 147–169. Portsmouth, NH: Boynton/Cook.

Shaw, M. E. 1981. *Group Dynamics: The Psychology of Small Group Behavior,* 3rd ed. New York: McGraw-Hill.

Smith, J. B. 1994. *Collective Intelligence in Computer-based Collaboration.* Mahwah, NJ: Lawrence Erlbaum.

Sproull, L., and S. Kiesler. 1986. Reducing social context cues: Electronic mail in organizational communication. *Management Science* 32: 1492–1512.

Steiner, I. D. 1972. *Group Processes and Productivity.* New York: Academic Press.

Stout, R., J. A. Cannon-Bowers, and E. Salas. 1996. The role of shared mental models in developing team situation awareness: Implications for training. *Training Research Journal* 2: 85–116.

Stuart-Hamilton, I. 1995. *Dictionary of Cognitive Psychology.* Bristol, PA: J. Kingsley.

Suchman, L. 1993. Response to Vera and Simon's situated action: A symbolic interpretation. *Cognitive Science* 17(1): 71–76.

Torenvliet, G. L., and K. L. Vicente. 2000. Tool usage and ecological interface design. *Proceedings of the IEA 2000/HFES 2000 Congress*, 587–590. Santa Monica, CA: Human Factors and Ergonomics Society.

Tuckman, B. W. 1965. Developmental sequence in small groups. *Psychological Bulletin* 63(6): 384–399.

Vallacher, R. R., and A. Nowak, eds. 1994. *Dynamical Systems in Social Psychology.* San Diego, CA: Academic Press.

van Tilburg, M., and T. Briggs. 2005. Web-based collaboration. In *Handbook of Human Factors in Web Design*, eds. R. W. Proctor and K. L. Vu, 551–569. Mahwah, NJ: Lawrence Erlbaum.

Vera, A. H., and H. A. Simon. 1993a. Situated action: A symbolic interpretation. *Cognitive Science* 17(1): 7–48.

Vera, A. H., and H. A. Simon. 1993b. Situated action: reply to reviewers. *Cognitive Science* 17(1): 77–86.

Vera, A. H., and H. A. Simon. 1993c. Situated action: reply to William Clancey. *Cognitive Science* 17(1): 117–133.

Vicente, K. J. 1999. *Cognitive Work Analysis: Toward Safe, Productive, and Healthy Computer-based Work,* Mahwah, NJ: Lawrence Erlbaum.

Vicenter, K. J. 2000. Work domain analysis and task analysis: A difference that matters. In *Cognitive Task Analysis*, eds. J. M. Schraagen, S. F. Chipman, and V. L. Shalin, 101–118. Mahwah, NJ: Lawrence Erlbaum.

Volk, F., F. Pappas, and H. Wang, this volume. Understanding users: Some qualitative and quantitative methods. In *Handbook of Human Factors in Web Design*, 2nd ed., eds. K.-P. L. Vu and R. W. Proctor, 417–438. Boca Raton, FL: CRC Press.

Volpe, C. E., J. A. Cannon-Bowers, E. Salas, and P. E. Spector. 1996. The impact of cross-training on team functioning: An empirical investigation. *Human Factors* 38: 87–100.

Walther, J. B. 1996. Computer-mediated communication: Impersonal, interpersonal, and hyperpersonal interaction. *Communication Research* 23: 3–43.

Ward, P., and D. W. Eccles. 2006. A commentary on "Team cognition and expert teams: Emerging insights into performance for exceptional teams" 2006. *International Journal of Sport and Exercise Psychology* 4: 463–483.

Watt, J. H., and C. A. VanLear, eds. 1996. *Dynamic Patterns in Communication Processes*. Thousand Oaks, CA: Sage.

Wegner, D. M. 1986. Transactive memory: A contemporary analysis of the group mind. In *Theories of Group Behavior*, eds. B. Mullen and G. Goethals, 185–208. New York: Springer-Verlag.

Wickens, C. D. 1998. Commonsense statistics. *Ergonomics in Design* 6(4): 18–22.

Section III

Interface Design and Presentation of Information

define multisensory integration as "the synthesis of information from two or more sensory modalities so that information emerges which could not have been obtained from each of the sensory modalities separately." (p. 35)

Besides using multiple channels for obtaining pure information, this can lead to a more comfortable user experience by increasing the perceived overall quality of an event. For example, for reading a story in a book it could be sufficient if only vision would be addressed. But, in fact, we perceive the type of font, the texture of pages and cover as well as the "new" smell of our newly bought book. All these perceptual events additionally contribute to our overall quality assessment of the book.

From the viewpoint of design, purposely providing moderately redundant information in a user interface by a second or third modality can contribute much to higher task performance or the feeling of comfort. A simple example is the computer keyboard: although the fact of a letter typed can be seen on the screen immediately, the additional synchronous presentation of an auditory and tactile feedback is far more appreciated by users (among others, see, e.g., Pellegrini 2001). Especially when fast user reaction is intended, multimodal events are superior to monomodal ones (Ho, Reed, and Spence 2007; Spence and Ho 2008). So, when designing a multimedia system that is capable of addressing several modalities, it is important to provide the appropriate stimuli for the respective modalities at the right time for the purpose of perceptual integration.

Besides the considerations of comfort, the choice of modalities often is also determined by physical surroundings. For example, in adverse light conditions or for mobile applications, sound often is a suited feedback solution, whereas in noisy workplaces information is clearly preferred via the visual channel.

Now, a user interface designer is challenged by choosing the most appropriate way of presenting information by adequate media and modalities. Each of them must be carefully selected.* Since there are other chapters on the visual design of Web applications in this volume (see, e.g., Tullis, Tranquada, and Siegel, this volume), we focus on the properties and design for the auditory and tactile channels. Both are closely related because vibrations of a physical surface in our natural environment typically lead to both auditory and vibrotactile perception. Also, because the design of haptic devices for Web applications is just emerging, for developing new applications it is important to become acquainted with the basics of tactile perception. For both modalities, perceptual basics and design recommendations are given. In addition, aspects of interaction of modalities are considered,

including auditory-visual interaction because it plays a major role in improving existing multimedia applications.

7.1.1 Definitions

Although already noticed in the 2005 edition of this volume (Hempel and Altınsoy 2005), "multimedia" still is a buzz word for most people. It is used in a variety of contexts with only a loose representation of particular objects being meant. However, even in publications stemming from the field, different definitions of "media" or "multimedia" can be found. In others, the focus is just set on particular aspects of media, leading to the assumption that there would be another definition.

International Standards Organization (ISO 2002) provides a definition of *media* and *multimedia*:

- *Media* are different specific forms of presenting information to the human user (e.g., text, video, graphics, animation, and audio).

Analogously, *multimedia* is defined:

- *Multimedia* are combinations of static (e.g., text and picture) and/or dynamic media (e.g., video and music) that can be interactively controlled and simultaneously presented in an application.

Strictly speaking, a regular TV set could already be considered a multimedia device. However, the degree of interaction as demanded by the given definition is comparably low for the TV set in contrast to modern Web applications.

Interactive control of media requires that systems provide the possibility for interaction. ISO (1999) provides a definition of an *interactive system*:

- An *interactive system* is a combination of hardware and software components that receive input from, and communicate output to, a human user in order to support his or her performance of a task.

A system's support for a user performing a task brings us to the concept of usability of which the definitions are cited from ISO (1998). It is to be remarked that the ISO standard always refers to the term "product." However, in the Web context certain applications, sites, or services in fact are products, too. For this reason the author proposes the integration of the summarizing term "service" in the definitions as follows:

- *Usability* is the extent to which a product or service can be used by specified users to achieve specified goals with effectiveness, efficiency, and satisfaction in a specified context of use.

Here,

- *Effectiveness* means the accuracy and completeness with which users achieve a specified task.

---

* Which modality to choose depends on the intention of the message, the physical environment, and the properties of the channel. For example, with regard to spatial selectivity, our visual system uses the fovea for spatial acuity. Acute vision can only be obtained by a small area at a time. In contrast, the auditory system is able to receive information from all spatial directions simultaneously. Here, spatial selectivity can arbitrarily be focused on any perceived sound source whatsoever ("cocktail party effect"; Cherry 1953). Similarly, the olfactory system is able to perceive odors from all directions but typically from a shorter distance than sound. Tactile information and temperature are only able to be perceived on or close to the skin. Also, eyes can be shut but ears, nose, and skin cannot.

- *Efficiency* refers to the resources expended in relation to the accuracy and completeness with which users achieve goals.
- *Satisfaction* is the freedom from discomfort and positive attitudes to the use of the product or service.
- *Context of use* includes users, tasks, equipment (hardware, software, and materials), and the physical and social environments in which a product or service is used.

In contrast to *multimedia*, *multimodal* means something different. Because *mode* refers to our *sensory modalities* (vision, audition, touch, taste, etc.), *multimodal* perception means the perception of an event by more than one sensory modality. Accordingly, Brewster (1994, 8, based on Mayes 1992) provides a definition of multimodal interfaces:

- A *multimodal interface* is defined as one that presents information in different sensory modalities.

7.1.2 Consequences

As the previous definitions show, the concepts of multimedia and multimodality neither mean the same nor do they conflict each other. While multimodality focuses on the modalities used for the display of a desired event, multimedia focuses on the concept for presentation, independent of the use of specific modalities. For example, once it has been decided that a video clip would be the preferred *medium* for the presentation of certain information on a Web site, the modalities have to be considered that will optimize the design of the video clip. Thus, for example, temporal resolution of the visual presentation, or the technical bandwidth of the audio channel, as well as the threshold for audio-visual delays must be considered. However, considering all available combinations of modalities will lead to truly new media to be designed for future applications.

Vision surely is the most important modality regarding the most common output channel of today's stationary and mobile devices (PC, personal digital assistant [PDA], etc.). Nevertheless, owing to an increasingly widespread distribution of advanced sound reproduction equipment, auditory information is becoming increasingly important. Amazingly, although touch was very important for the input channel via keyboard, it has only been considered in recent years. Even more, the auditory-tactile feedback provided by a traditional, well-designed computer keyboard has been used for decades without being called multimodal. The challenges that occur using haptic devices as an output modality will be considered in Section 7.3. The requirements regarding the auditory channel are presented first.

7.2 DESIGN FOR THE AUDITORY CHANNEL

7.2.1 Motivation

Today, in many industrial branches, results from sound engineering already strongly influence the design of new products.

For example, in the automobile industry, because sound insulation between engine and passenger compartment has been improved over the years, lower interior sound levels could be obtained. As a consequence, sounds originating from interior devices became audible that have not been heard before. For example, the sounds of various small electric motors in today's vehicles' passenger compartment (used for adjustment of seats, windows, etc.) simply had been masked before by the noise of the engine. Nowadays, engineers do not try to hide such sounds any longer but rather tune them according to the image of the particular car brand. Although the car industry is clearly considered to be a pioneer in this field, at present, sound design in recent years became a selling point in other industries, too (e.g., train, aircraft, household appliances, switches, even food and beverages).

It is evident that there is no sound quality by itself. Rather, statements on the quality of sounds always must be considered in the context and system in which the sounds are used. Thus, when we speak of *sound quality* in this chapter we actually mean the *auditory quality of systems* (as introduced by Hempel and Blauert 1999 based on Blauert and Bodden 1994). So, the auditory quality of a system is the *suitability of the sounds used for the respective purpose*. In their article, as an example, they mention the nondesigned sound of a coffee machine, which should not be reduced to a zero sound level, because it is valuable to inform the user about the running status of the machine. Similarly, it should not be amplified to a maximum, because it would hinder communication between the persons who make coffee (also see Guski 1997).

Over the past 10 years, the well-designed use of sounds in user interfaces has spread widely throughout the software world, because it is an intuitive tool to provide users with additional cues that do not need to be displayed visually, and enhances the user experience emotionally. For a characterization of sounds in everyday software, see the articles of Wersény (2009) and Davison and Walker (2009).

The use of sounds in user interfaces may help to reduce the load of the user's visual system if the amount of information presented on the screen is very high. Especially, time-varying events that are important to get the attention of the user are suited for coding in the auditory domain. Omnidirectional presentation is another characteristic of sound that the sound designer has to keep in mind. While this is annoying if the information is unwanted, it is highly appreciated for desired information that otherwise easily would have been overlooked. This demands good discipline, experience, and careful use by the sound designer. Furthermore, some objects and actions can be presented much more natural when there is a perceptual correlation between different modalities, like vision and audition (on requirements regarding audio-visual presentation; see Section 7.4.2).

From the viewpoint of usability, a successful integration of auditory information in user interfaces leads to more intuitive understanding, improved productivity, and satisfaction of users. And from a marketing point of view, a well-suited sound design leads to a clearly perceived overall quality of the product or service and thus becomes a competitive advantage (see car industry, telecommunications, household appliances, and even design food).

However, if the physical environment allows the use of sound, it is recommended that the message to be displayed either is simple, is related to events in time, has omnidirectional capabilities (e.g., due to a mobile workplace), or requires immediate action by the user (Vilimek and Hempel 2005; Rouben and Terveen 2007; McGee-Lennon et al. 2007). Visual presentation should be used for messages of higher complexity.

There are several limitations on the use of sound. For absolute judgments, sound usually is not the preferred medium of presentation. In contrast, our auditory system is extremely sensitive to relative changes. This means that the absolute pitch or loudness of a sound presentation will not be remembered very well, but in contrast, small changes in pitch or loudness can be detected quite well. E.g., Peres, Kortum, and Stallmann [2007] make use of this for their study on an auditory progress bar. If sounds are of different perceived loudness but no intended information is connected with sequentially different sound levels, this leads to annoyance. Thus, if level is not intended as a means for coding, all sounds should be kept as equal in loudness as possible.

Another feature that one must be aware of when designing sound for user interfaces is the transience of information when sound is used: sound is a temporal medium—once information has been presented, it cannot be looked at again (unlike the visual domain).

7.2.2 Basics

7.2.2.1 Physics

Sound is mechanical vibrations transmitted by a physical medium (typically air) that contain frequencies that can be perceived by the human ear. The number of oscillations per second is measured in Hertz (Hz), commemorating physicist Heinrich Hertz (1857–1897). For adults, the range of audible frequencies typically is 16 Hz to 16 kHz. Sound propagates in waves of which the velocity is depending on the physical medium. In air the velocity of sound at 20°C (68°F) is 344 m/s (1128 ft/s). In water it is approximately 1500 m/s (4921 ft/s).

The minimum pressure p_0 necessary at the ear drums to perceive an auditory event (hearing threshold) is approximately 2×10^{-5}, whereas the threshold of pain requires pressures of ca. 10^2 Pa. The unit (Pa) refers to Blaise Pascal (1623–1662); $1 \text{ Pa} = 1 \text{ N/m}^2$.

For handling this large range the logarithmized pressure level related to p_0 is used as a measure: The sound pressure level L is defined as $L = 20 \log_{10} \dfrac{p}{p_0}$ dB. The unit dB indicates tenths of a Bel, referring to Alexander Graham Bell (1847–1922). Sound pressure levels of familiar environmental sounds are shown in Table 7.1.

TABLE 7.1

Approximate Sound Pressure Levels for Typical Environment Conditions

Sound Pressure Level (dB)	Environmental Condition
0	Threshold of hearing
20	Anechoic chamber
30	Bedroom in quiet neighborhood
40	Library
50	Quiet office room
60	Conversational speech
70	Car passing by
100	Symphony orchestra (fortissimo)
110	Rock band, techno club
120	Aircraft takeoff
130–140	Threshold of pain

7.2.2.2 Psychophysics

It is important to know about the physical foundations of sound in order to design for the dimensions and technical limitations of sounds and thus the respective playback equipment. In contrast to the physical domain, psychophysics, namely, psychoacoustics, covers the relation between the physical and perceptual auditory domain. As an illustration, when physical acoustics asks, "What sound signal has been emitted?" psychoacoustics ask, "What sound characteristics have been perceived?" Once the physical framework is known, it is important to know how to design for maximum audibility and efficiency. Psychoacoustics defines the perceptual limits within which auditory signs must be designed if they are to be effective.

First of all, it is important to know that the human auditory system is not equally sensitive to all frequencies. The drawn line in Figure 7.1 shows what is called the detection threshold for sinusoidal sounds of different frequency in an extremely quiet environment. This means that, e.g., a 1000-Hz sine wave can already be perceived at much lower levels than a 60-Hz hum. Thus, for the display of low frequencies much more energy in the amplification system at the user's site is needed than for higher frequencies in order to achieve the same perceptual loudness. In contrast, the threshold of pain remains comparably constant at levels of 120–130 dB for all frequencies.

As can be seen in Figure 7.1, the so-called hearing area provides information on the sensitivity of our auditory system and thus is important for the design of sound: Signals with high energy around 1000–4000 Hz will be detected much easier than signals with their main energy at very low frequencies. In contrast, reproduction of low frequencies contributes less toward an increase of, e.g., speech intelligibility than to the perceived quality of the sound (as can be seen by the larger area covered by music in contrast to speech).

The threshold in quiet and the threshold of pain form the perceptual auditory limits. Typical areas used by music and speech are also illustrated in Figure 7.1. Electronically designed sounds, of course, may leave the marked areas

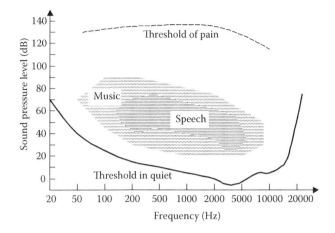

FIGURE 7.1   Hearing area. Typical areas for music and speech are displayed.

of music and speech, but except for special purposes, such extreme signals are not recommended because of annoyance (nevertheless, this may be required for the audibility of a warning signal like fire alarms). Also, always keep the characteristics of the users' reproduction equipment in mind (mobile phone or 5.1 setup?). If you design within the given limits of speech and music, you can be quite sure that the reproduction quality of the designed sounds will be basically acceptable if any specific reproduction equipment is not known in advance. If you design for virtual environments high-quality design and reproduction facilities are demanded, and thus much more considerations regarding the physical and psychoacoustic features must be taken in such case.

As you can see in Figure 7.1, the threshold in quiet is basically U-shaped. Taking into account this weighting of frequencies by the auditory system in a measure for perceived loudness, weighting curves are applied to the solely physical sound level measurements. This is the background for the widely used A-weighting curve dB(A): it weights the physically measured sound level according to the sensitivity of the human auditory system. Very low and very high frequencies thus are given less weight than well audible midrange frequencies (for psychoacoustic research, typically, the perceptual loudness N is calculated additionally to the A-weighted sound pressure level; see, e.g., Zwicker and Fastl 1999 for further reading).

The contour of the detection threshold changes over our lifetime. Elderly people will hardly hear frequencies higher than 10 kHz. But this deterioration process can already take place in earlier decades if the users formerly were frequently exposed to high sound pressure levels (e.g., factory noise, rock concerts, and military service). If you have no idea about the target audience to be addressed, try to design sounds in the indicated area of speech and music.

Another feature of the human auditory system that is useful to know is spectral masking. This means how the contour of the hearing threshold changes when other sound are present. Figure 7.2 shows the hearing threshold where a - kHz

tone is already present (at a level L_M of 70 or 90 dB). It can be seen that any "masking" tone leads to an increasing insensitivity of the auditory system toward frequencies higher than the masking tone. As can be seen, when the - kHz tone at a level of 90 dB is present for some reason, another tone at 2 kHz would need a level more than 50 dB above hearing threshold to be heard. For the 70-dB masking tone at least 30 dB is needed for the 2-kHz tone to be heard.

These results provide an idea of the problems present when the auditory display is to work in real acoustical environments where typically a lot of sounds are present. Thus it is good advice to use sounds containing a broad spectrum of frequencies (such as harmonics), minimizing the probability for not being perceived even if parts of the sound get spectrally masked.

The fact of spectral masking has been the basic idea for the development of perceptual coding algorithms like mp3. There, for short time frames the respective masking patterns are calculated. Information below the calculated audibility threshold will be omitted. This is the reason for the much lower file sizes of mp3 in contrast to a lossless file format (e.g., "wav"). However, at the users' site a decoding algorithm must be installed and the necessary calculation power must be available with respect to the overall performance of the system to be used. Nevertheless, regarding the quality of reproduction, for most applications in the Web context mp3 is completely sufficient for good-quality reproduction. For binaural displays and high-quality virtual environment applications, mp3 is not the file format of choice because for exact three-dimensional perception of sounds the auditory system needs further cues of the sound signal, some of which typically get lost in mp3-coded files.

7.2.2.2.1   Binaural Displays

It is an outstanding feature of our auditory system to be able to concentrate on a special sound in the presence of other "disturbing" sounds. This ability of localizing sound sources is enabled by simultaneous processing of the signals being present at both ears. Thus we are able to focus on one speaker in a group of concurrent speakers. This phenomenon is called the "cocktail party effect" and firstly was scientifically

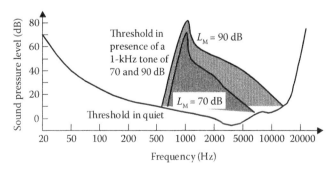

FIGURE 7.2   The threshold changes if a "masking" tone is present. The shaded area below the changed threshold thus is not audible and considered as redundant information to the auditory system and could be omitted for achieving lower bit rates for transmission (as in mp3 and other perceptual—"lossy"—codecs).

described by Cherry (1953).* However, most design considerations on auditory displays, sound design, and related fields in user interface design are implicitly done on a monaural basis. Nonetheless, using binaural cues, it is one of the exceptional advantages of the human auditory system to immediately provide the listener with spatial information on the physical environment the sound has appeared in (e.g., room size) by combining information obtained by two ears simultaneously. Acoustically, any environment is mainly characterized by the number, time, and spectral features of the reflections of the original sound wave. In combination with head movements the occurring interaural time differences and level differences present at the ear drums are processed and finally enable a most intuitive spatial display. Because of the high calculation power and requirements for reproduction equipment, binaural displays are mostly used in high-quality virtual environments.

7.2.2.2.2 Consequences

Strictly speaking, for the designer it should be more important what sound is perceived at the listeners' ears and not what waveform the loudspeakers emit. Thus, when easy localization must be obtained, small hard walled environments do not contribute to an easy localization because of the many reflections; instead they use sound-damped interior or larger rooms with low background noise. As it can be seen, also room acoustics and the background noise to be expected may not be negligible (particularly for Web applications in public spaces, factories, casinos, etc.).

As a design rule, if it is beneficial to provide information about the exact spatial position of a sound source, or if easy localization of the sound is required, the use of broadband signals that are spread widely across the audible frequency spectrum is recommended. Also, if your application is to be designed for no headphone usage, be aware that the room or environment in which the sound is played back is to be considered an important part of the transmission chain.

7.2.2.3 Semiotics

It goes without saying that the intentional use of specific sounds for presenting a certain type of information is a communication situation: thus the auditory percept is regarded as a sign by its listener. This prerequisite of regarding sounds as signs for something empowers an important field regarding the systemic description and design for auditory signs: *semiotics*. Semiotics (the science of signs) is the scientific discipline for analyzing and describing sign processes. Although semiotics by far is no new science, it has been established in the world of sound design not before the late 1990s.[†] A very short overview will be given here for the most basic classifications of signs and the respective benefit for the process of sound design.

TABLE 7.2
Model of the "Semiotic Matrix"[a]

	Impressive	Informative	Appellative	
Semantic classification		√		Iconic
				Symbolic
	Pragmatic classification			

[a] Enables the sound designer to become aware of the intended type of sign to use for a certain application (example here: an informative-iconic sign is to be designed, this could be realized by a wooden "click" when clicking on a button).

Charles S. Peirce (1839–1914), the founder of semiotics, proposed a threefold division of this relation:

- *Index*: a sign with a direct (often even physical) relation to the environment (e.g., footprints and thermometer).
- *Icon*: a sign that represents an object by similarity (e.g., pictograms).
- *Symbol*: a sign of which the meaning is widely fixed by interindividual convention (e.g., white dove and cross).

It is important to know that each sign basically has the capability to be each, either an index, an icon or a symbol, because such classification takes place at the user. But it is the art of any designer (be it by visual, tactile, or auditory means) to imply the intended meaning as clearly as possible.

Any of these sign-object relations, again, have the potential to be understood impressively, appellatively, or as a neutral source of information. This is illustrated by the semiotic matrix (Table 7.2).[‡] The semiotic matrix provides six elements in which any auditory sign can be categorized. (Because indexical relations are not so relevant here, iconic and symbolic relations are the ones mostly used in the context of sound design.) Because any sign has the potential to be represented by any element in the matrix, it is a tool for the sound designer to clearly define the intended category. Analogously, when carrying out usability tests with a designed interface, the auditory sign should be checked if the users share the aimed categorization.

Dürrer (2001) illustrates the variability of the sign-object relation as follows: an acoustic signal made up of Morse signs

---

* A comprehensive overview on the foundations of psychophysics in human sound localization is given by Blauert (1997).

† Mainly introduced into modern sound quality considerations and acoustic design by Jekosch and her group (Jekosch and Blauert 1996; Dürrer and Jekosch 1998, 2000; Jekosch 2001).

‡ Derived from the Organon-model of K. Bühler (1879–1963). For further reading on the development of the semiotic matrix, see Hempel (2001).

used in the context of an e-mail application could be used to indicate the arrival of a new e-mail at the inbox. Because typical e-mail users will not be accustomed to decoding Morse signals toward alphanumeric characters in real time, they may most likely recognize the presence of Morse signals knowing that they are typically used for electric signal transmission (due to pure convention reading of Morse signs would be a symbolic relation; e.g., " $\cdots$ – – – $\cdots$ " represents SOS, which is another symbol in itself by the way). Because receiving an e-mail is an electronic process and relates to the transmission of a message, too, the user establishes a relation that transforms the symbolic relation as intended by the use of Morse signs into an iconic one, indicating the arrival of new mail.

However, the use of auditory symbols could be advantageous in certain areas because a symbol can immediately trigger a large amount of similar associations in people. But auditory symbols must be avoided if the relation is likely to be hardly recognized by the listeners. This can be checked in usability tests.

Nonetheless, in auditory user interfaces icons rather than symbols are used. Auditory symbols are rare and sometimes even culturally dependent (e.g., a post horn as a symbol for the arrival of new e-mail). Exceptions may be symbols that are very well known among the target audience (e.g., police siren, ringing of a traditional land-line phone).

As it can be seen, designing auditory signs remains a creative process. The strength of the presented sign model is that it helps structuring the design of sounds in user interfaces. After all, semiotics provides no final answers, but it definitely helps to ask the right questions (Dürrer 2001).

7.2.3 SOUND DESIGN

7.2.3.1 Requirements

7.2.3.1.1 Physical Requirements

In Section 7.2.2.2 some basic psychophysical background was presented for the perceptual requirements on the design of sounds. Of course, owing to lack of experience, time, and costs, such extensive measurements and analyses can hardly be carried out for many applications. For this reason, even easier to measure design guidelines will be given in this section using typical sound pressure levels. The sound pressure levels given meet the requirements of the most often situations in practice.

In case the user does not typically use the application in a quiet room, it can be very useful to measure the stationary background noise level at a typical user's site with a regular sound level meter positioned at the typical position of the user's head. The obtained value may serve for reference purposes.

As a good guideline, present sounds approximately 10 dB higher than the background noise—that is, perceived double loudness. In situations where the sounds serve as a warning, make them 15 dB above the stationary background noise to ensure effective and consistent performance (see Table 7.3).

7.2.3.1.2 Semantic Requirements

It is the foremost goal of a sound designer to achieve a straightforward mapping of the intended meaning onto

TABLE 7.3

Recommendations for the Level of Auditory Signals

1	Signal levels around 5 dB above stationary background noise usually are detected. Atmospheric background sounds for some modern user interfaces often do not need to be louder.
2	Signal levels around 10 dB above stationary background noise are recommended for most applications.
3	Signal levels around 15 dB above stationary background noise usually are sufficient if clear detection must be guaranteed (e.g., warning signal).
4	Even warning signals should not be higher than 30 dB above stationary background noise for not annoying users unreasonably. If the sound pressure level needed comes close to 100 dB—according to the given requirements—think about other modalities for presenting information since hearing damage cannot be excluded.

sound. Concurrently, the sound must fit in the context of the overall user interface. A special challenge now is the design of sounds for abstract functionality (see the articles by Gaver 1997; Lashina 2001):

- Urgency mapping: In general, but particularly when it comes to warning sounds, be sure that an appropriate mapping between the perceived urgency of a sound and the needed urgency regarding the severity of a failure can be established by users. Incongruent urgency mapping is highly annoying (sometimes even dangerous) and makes many users switch off the sounds completely. Thus, usually no more than three urgency levels should be acoustically represented in a single application.

- Create for support: Sound designers tend to overoptimize their sounds as if they would be autonomously presented on stage or compact disk and in the end critically assessed by an expert audience. The opposite is the case, as sound in user interfaces always interacts with information obtained by the visual or any other channel, and the least part of users are experts in sound design. Also, sounds that are too autonomous would draw attention from the overall interface toward the auditory modality. This is not useful in the context of user interface design. And even a rather indefinite or ambiguous meaning of a sound influences the overall perception normally toward a much less ambiguous meaning. Finally, even a symbolic relationship between sign and object (see Section 7.2.2.3) may allow multiple settings depending on the context.*

---

* As Flückiger (2001) analyzes by the example of the audiovisual use of bells in various films, bells may represent a wedding, rural atmosphere, time of day (e.g., Sunday morning), religious service, and other information, depending on the context provided by visual means.

7.2.3.2 Design Guidelines

Typical issues that arise concerning the design of the auditory channel in user interfaces are general guidelines, auditory icons, earcons, and feedback sounds. In this section, authors from the respective fields are presented and recommended for further reading.

7.2.3.2.1 General: Auditory Signs

From a psychological point of view, Guski (1997) distinguishes three requirements:

1. *Suitability* or stimulus-response compatibility
2. *Pleasantness* of sounds (at least no unpleasantness)
3. *Identifiability* of sounds or sound sources

All three requirements must be met for design effectiveness, efficiency, and satisfaction.

7.2.3.2.1.1 Design process
Regarding a general design process for auditory displays, Dürrer (2001), p. 52 proposes the following five steps for the design of auditory signs:

1. *Analyzing the application*: An analysis regarding the task must be carried out that evaluates risks and potential disadvantages of the use of auditory signs. It must be taken into account if an auditory sign is suited at all for the transmission of the intended meaning.
2. *Defining priorities*: On the basis of the analysis, prioritization of appearing events must be decided. According to the defined priorities, levels of urgency must be established and be mapped to the events. In order to avoid ambiguities between signs, different levels of urgency could be represented with different acoustic parameters.
3. *Grouping of signs*: Depending on the sound emitting device (e.g., PC speaker, Hi-Fi system), timbre can be used to group auditory signs. Groups should be made up of events belonging to the same action/entity or logical task.
4. *Analyzing the acoustic environment*: It is required to analyze the acoustic environment. Especially, the frequency-dependent detection threshold (see 7.2.2.2) is important to know for the design of auditory signs. Thus, if possible, carry out a frequency analysis by an expert.
5. *Evaluation of signs*: The final set of auditory signs must be evaluated in usability tests (even sounds with strongly different acoustic parameters sometimes can be easily confused by listeners depending on the task); this depends on their previous knowledge and cognitive processes. Also, different cultural contexts must be considered (Kuwano et al. 2000). It must be kept in mind that sound designers, because of their daily experience and knowledge, are able to discern more auditory signs than end

users. Thus user tests are unavoidable. Usually, this last step leads to an iteration of prior steps depending on the test results.

7.2.3.2.1.2 Equalization of Loudness
Although the human auditory system is not capable of recalling absolute loudness precisely, it is very well designed to detect relative changes of loudness. In the context of Web applications this means that loudness differences between sequentially heard sounds can get the user's attention. Because in most applications there is no intended loudness coding (e.g., for important or less important sound events), make the loudness the same for all sounds. So, after finishing the design and implementation for all sounds of an application the sound designer should carry out the most typical users' tasks using the system, simultaneously relying on his experience in evaluating loudness differences and adjusting the individual sound levels accordingly. Once all sounds have the same loudness, there is no problem in amplifying all sounds equally, but, again, differences in loudness within the set of sounds have the high potential of becoming really annoying to the users. Also, be aware if you have sounds that are very different in their frequency spectra (e.g., a beep and a white noise), they may sound differently (and differently loud!) when played back on low budget equipment because of bad reproduction quality. Loudspeakers in combination with the listening room definitely are the most uncontrollable parts in the signal chain. Therefore, it is recommended to use studio-quality speakers for design as well as the most typical speakers of your target audience for fine-tuning in order to hear what the users will hear.

7.2.3.2.2 Feedback Sounds

With regard to telecommunication systems, feedback sounds basically were already present in early teletypes. For example, a bell tone informed users that a message was arriving. Until the 1980s, computers were mainly focusing on visual displays and usually just had a "beep" to acknowledge any action or to indicate a failure.

Although feedback sounds often are designed redundantly to a visual display they effectively can improve task performance or comfort of use. Or as Sweller et al. (1998) put it: sounds can be effectively used for providing redundant information, by giving the users complementary information that would otherwise overload a certain information channel. For the simple task of providing a supplemental auditory feedback for numerical data entry, the effect upon keying performance has been investigated by Pollard and Cooper (1979). They found that the presence of feedback sounds leads to a higher task performance. Hempel and Plücker (2000) could find similar results showing that the feedback conditions (e.g., multitone, single tone) were not as important as the presence as such. This means that the simple presence of feedback sounds can substantially improve the quality of user interfaces.

Proven guidelines for the design of feedback sounds are provided below (Bodden and Iglseder 2002):

- The feedback has to *meet the expectations and requirements* of the user. Although in the beginning a minor learning process might be acceptable for the user, this must not take too long. Otherwise, ineffective auditory signs have been chosen.
- The feedback has to be *meaningful, unmistakable,* and *intuitive.* If no clear relation between the sound and the denoted object respectively meaning can be established (see Section 7.2.2.3), the sounds quickly become annoying.
- The feedback sound has to *fit to the original product sound* and has to be perceived in it. It goes without saying that the feedback sound must not be masked by the regular sound of the product or the background noise. But it is even more difficult to integrate the specific kind of sound (using, e.g., timbre, iconicity) that fits to the overall product or service depending if it is to be marketed as an, e.g., exclusive, basic, or trendy one.
- Cost and realization aspects have to be considered. Think of the playback facilities at the users' site and how a clear design can overcome technical shortcomings.

7.2.3.2.3 Auditory Icons versus Earcons

Among those taking first steps designing for the auditory channel, often, there is a confusion about the terms "auditory icon" and "earcon." To clear it up, a short overview shall be given in this section. However, taking the semiotic matrix as a basis for the classification of auditory signs, auditory icons predominantly belong to the iconic level, whereas earcons typically go with the symbolic level.

7.2.3.2.3.1 Auditory Icons

Auditory icons were mainly introduced by Gaver (1989, 1997). The basic idea is to transfer existing sounds from another context into the context of a user interface representing certain objects or actions, thus creating an analogy between the known world and the world of the application to be used. Gaver describes them as "everyday sounds mapped to computer events by analogy with everyday sound-producing events," and "auditory icons are like sound effects for computers." There is a high communication potential for auditory icons. Because, as indicated above, a sound often implies conclusions about the sound source regarding physical material, room size, etc. This can effectively be coded in auditory icons for metaphors of any kind (saving a large file could be auralized by a long reverberation applied to the feedback sound, etc.). A disadvantage of auditory icons clearly is the lacking representation of abstract actions because they usually have no representation in everyday sounds familiar to the user. So, in general, auditory icons hardly limit the creativity of the sound designer. Nevertheless, a structured design and evaluation process is inevitable. For a methodology regarding the design of auditory icons, see Table 7.4.

Auditory icons often are used in film for auralizing objects or actions that are visually displayed but sound

TABLE 7.4

Basic Methodology for the Design of Auditory Icons

1	Choose short sounds that have a wide bandwidth and where length, intensity, and sound quality are roughly equal. (→ sounds should be clearly audible but not annoying.)
2	Evaluate the identifiability of the auditory cues using free-form answers. (→ sounds should clearly identify objects and actions.)
3	Evaluate the learnability of the auditory cues that are not readily identified. (→ sounds should be easy to learn.)
4	Test possible conceptual mapping for the auditory cues using a repeated measures design where the independent variable is the concept that the cue will represent.
5	Evaluate possible sets of auditory icons for potential problems with masking, discriminability, and conflict mappings.
6	Conduct usability experiments with interfaces using the auditory icons.

Source: Based on the proposal by Mynatt, E. D. 1994. Authors' comments in parentheses.

different than in the real world according to the intention of the producer. This could be, for example, an exaggeratedly audible heartbeat or sounds of starships that, in fact, would not be heard in space because of the missing physical medium for sound transmission (e.g., air, water). This means that sounds must be invented that cannot be authentic but are accepted as plausible, because a similarity between situations in the real world and the fictive world can easily be established by the viewer. Analogously, this concept can be transferred to user interfaces. Thus, the iconic relation is less action oriented than object oriented. Examples for applications are given in Table 7.5 based on guiding questions by Flückiger (2001) regarding the description of sound objects in film.

7.2.3.2.3.2 Earcons

Earcons have been introduced by Blattner, Sumikawa, and Greenberg (1989). In contrast to auditory icons (see previous section), earcons mostly are made of tone sequences that are arranged in a certain way to transmit a message. The sounds used are mainly of artificial/synthetic character or tonal sounds of musical instruments. Blattner et al. define earcons as "non-verbal audio messages that are used in the computer/user interface to provide information to the user about some computer object, operation or interaction." Because earcons often are designed like musical motifs (short melody-like tone sequences), they offer the advantage that such elements can be combined sequentially like words are combined to form a sentence. For example, the auditory element File can be combined with the element Download. But File could also be combined with Upload, resulting in an identical first auditory element and a different second one. Because such relations must be learned, earcons typically belong to the category of symbols. It is the power of symbols to convey much information by a single sign.

TABLE 7.5

Iconic Sound Objects: Parallels in Film and User Interfaces

Guiding Questions (Flückiger 2001)	Examples for Application in User Interfaces (Hempel and Altınsoy 2005)
What sounds?	A button, a menu, an object of educational interest (three-dimensional [3D] visualization of a car prototype)
What is moving?	A menu, a slider, a modeled object (a conveyor belt in a virtual 3D plant)
What material sounds?	If no object from the "real" world is concerned, invent a material for your abstract object. Objects of the same class could be given the same "materials" by certain sound characteristics (e.g., timbre), etc.
How does it sound?	Do verbal statements characterize the sound (e.g., "powerful," "soft," etc.)? Is the sound adequate to the action? Does it reflect the verbal attributes?
Where does it sound?	Like sound in film can indicate, e.g., dreams by means of reverberation, spectral filtering, selected background noise or others, different "scenes" or "rooms" can be created in your application (e.g., Edit, View, Preview, Download)

The most important features for earcons to be designed are rhythm, pitch and register, timbre, and dynamics. To improve the learnability of earcons, these features must be appropriately combined in order to achieve a high recognizability. An overview on valuable guidelines of the main design features of earcons is provided by Brewster, Wright, and Edwards (1994) and Brewster (2002); see Table 7.6.

7.2.3.2.3.3 Auditory Icons and Earcons: When to Use One or the Other? From the viewpoint of sign theory, auditory icons and earcons are groups of auditory signs characterized by a predominantly iconic or symbolic relation between the sign and its denoted object (see Table 7.7). Brewster (2002) considers them as poles of a continuum of which most auditory signs are found somewhere along the axis (see Figure 7.3), e.g., earcons could be designed by a sequence of auditory icons.

Because auditory icons basically are very intuitive to understand, they are recommended for applications that must be easy to use. These applications include those designed for beginners and less experienced users or users that use the application only once in a while. Generally, they tend to be preferred in situations where performance is most important (Wersény 2009). However, if there is a very comprehensive application being used by specialists, earcons can be a good solution owing to their structural ability. Again, it takes more time to get acquainted with earcons, but once they are intensively used, they can appear powerful. In any case, for design it is important to keep the overall user interface in mind as well as the aim of the application and the targeted user group.

7.2.3.2.4 Key Sounds

When it comes to *key sounds*, we leave the framework of "traditional" sound design for user interfaces. Introduced by Flückiger (2001), the concept is adopted from sound for films, but if applied thoughtfully, it can be powerful when the task of the designer is about creating emotional impact.

Key sounds are pure sound objects that become meaningful because of their frequent appearance, strategic placement—mostly in the exposition or key scenes of films—and an integration in the overall intention. They may not necessarily

TABLE 7.6

Guidelines of the Main Design Features of Earcons

Timbre	Most important grouping factor.
	Use musical timbres with multiple harmonics (as mentioned in Section 7.2.2.2, a broad frequency spectrum helps perception and can avoid masking by background noise).
Pitch and register	Absolute pitch should not be used as a cue on its own, preferably use in combination with other features.
	If register alone must be used then there should be large differences (two or three octaves) between earcons.
	Much smaller differences can be used if relative judgments are to be made.
	Maximum pitch should be no higher than 5 kHz and no lower than 125–150 Hz (also see Sections 7.2.2.1 and 7.2.2.2 on physical and psychoacoustic requirements for the design of auditory signs).
Rhythm, duration, and tempo	Make rhythms as different as possible.
	Putting different numbers of notes in each earcon is very effective.
Intensity	Should not be used as a cue on its own (cause of annoyance).

Source: Based on Brewster, S. A., P. C. Wright, and A. D. N. Edwards. 1994. A detailed investigation into the effectiveness of earcons. In *Proceedings of the International Conference on Auditory Displays ICAD'92*, 471–498. Santa Fe, NM: Santa Fe Institute. With permission; Brewster, S. A. 2002. Nonspeech auditory output. In *The Human Computer Interaction Handbook*, eds. J. Jacko and A. Sears, 220–39. Mahwah, NJ: Lawrence Erlbaum. With permission; author's comments in parentheses.

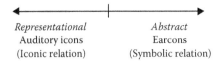

FIGURE 7.3 The presentation continuum of auditory icons and earcons from Brewster (2002), terminology added from Section 7.2.2.3 (in parentheses). (From Brewster, S. A. 2002. Non-speech auditory output. In *The Human Computer Interaction Handbook*, eds. J. Jacko and A. Sears, 220–239. Mahwah, NJ: Lawrence Erlbaum.)

exist outside of the respective film context but nevertheless characterize the feeling or atmosphere of the film. The viewer detects a certain structure that leads to the hypothesis that there is a deeper meaning linked to the sound. Thus, a relation gets built on a symbolic layer. Besides symbols that have been learned in everyday life, it is possible to generate symbols with their own specific meanings.

Transferred to user interfaces, this means that sound could be used more strategically, e.g., when starting up the application, being repeated at key actions like putting items in a virtual shopping cart, or others. This considerably increases the recognition of the application. This is particularly suited when the Web application or service is meant to be an experience like an exclusive shopping system where a kind of story is to be told. However, the application of key sounds is only suited for applications where generally any background score is accepted (e.g., image related sites). It is risky to use key sounds for a rather conservative user group.

The most natural and intuitive sound design tool is the human voice, of course. This aspect is already widely used in interactive speech dialog systems and brand design for marketing purposes. For a state-of-the-art overview, see Hempel (2008).

7.3 DESIGNING FOR THE TACTILE CHANNEL

The sense of touch is called the tactile sense. By touching a physical object, tactile perception provides us information about the size, shape, texture, mass, and temperature of the object. This information enables us not only to identify different objects but also to interact with these objects and our environment (e.g., open a door, drive a car, and play tennis).

Experimental psychological research on tactile sense began with Ernst Heinrich Weber (1795–1878). In his books, *De Tactu Weber 1834, 1936* and *Der Tastsinn und das Gemeingefühl* (Weber 1851) he reported upon some experiments that are related to fundamental aspects of our tactile sense, e.g., two-point threshold, weight discrimination, and temperature perception. Weber's theories bridged the gap between physiology and psychology. Research on the tactile sense grew out of work in the nineteenth century by Katz, who has concentrated mainly on the perception of roughness and argued that vibrations underlie the perception of texture (Katz 1969; Lederman, Loomis, and Williams 1982). Until today, different physiological and psychological aspects of

tactile sensation are studied by scientists. An overview of physical, physiological, and psychophysical aspects of the tactile sense will be given in the following two sections. The fundamental knowledge of the tactile sense is becoming more important with an increased interest to use haptic devices in multimedia applications.

Haptics comes from a Greek word "haptesthai" meaning "grasping or the science of touch" (Merriam-Webster Online Dictionary). In recent years, its meaning extended to the scientific study for applying tactile and force feedback sensations of humans into the computer-generated world. Haptic devices can be grouped into two categories: input and output. Until the late 1990s, in most Web applications, only the visual and auditory modalities were addressed. Tactile feedback was only an input modality using a keyboard, a mouse, or a joystick as an input device. But now it is being used to bring the sense of touch to Web applications. It increases the sense of presence in the Web and plays a role in getting more realistic and compelling Web applications for Web designers. Also, giving disabled people an additional input and output channel can greatly increase the amount of applications. This section of the chapter describes physical, physiological, and psychophysical aspects of the tactile sense, explains present haptic devices, and introduces design principles as well as some examples of haptic metaphors.

Also, physiology is described in a more detailed manner, because an understanding of physiological processes is fundamental (but until now not frequently documented for multimedia design) for developing suited applications for the tactile channel—a field cordially inviting today's designers.

7.3.1 BASICS

7.3.1.1 Physics and Physiology

There are two types of sensory receptors in the skin to be regarded at first: mechanoreceptors and thermoreceptors. Both types of cells located near the surface of the skin are responsible for our tactile sense. Mechanoreceptor cells are sensitive to vibration. Vibration is an oscillatory motion of a physical object or body that repeats itself over a given interval of time. Physical characteristics of vibration are described by amplitude (displacement), velocity, acceleration, and frequency. The other physical property that is sensed by mechanoreceptors is pressure, which is the ratio of force to the area on which it is applied.

TABLE 7.7

Overview on Features of Auditory Icons and Earcons

Auditory Icons	Earcons
Intuitive	Must be learned
Needs real-world equivalents	May represent abstract ideas
Sign ↔ object relation is iconic	Sign ↔ object relation is symbolic
Each icon represents a single object/ action	Earcons enable auditory structure by combination

TABLE 7.8
Mechanoreceptor Types

	Mechanoreceptor Cells			
	Rapidly Adapting		Slowly Adapting	
	Pacinian Corpuscle (PC)	Meissner Corpuscle (RA)	Merkel Disks (SA-I)	Ruffini Ending (SA-II)
Location	Deep subcutaneous tissue	Dermal papillae	Base of the epidermis	Dermis and deep subcutaneous tissue
Frequency range	50–1000 Hz	10–60 Hz	5–15 Hz	0.4 and 100 Hz
Sensitive to	Vibrations also when skin is compressed and the frictional displacement of the skin	Low-frequency vibrations, detection, and localization of small bumps and ridges	Compressing strain, does not have the capabilities of spatial summation	Directional stretching and local force

Mechanoreceptor cells can be grouped into two categories, *rapidly adapting* (RA) and *slowly adapting* (SA) (see Table 7.8), and they are responsible for important tactile features such as object surface parameters (roughness), shape, and orientation of an object.

The sensation of roughness is the principal dimension of texture perception. Some physiological studies have shown that RA mechanoreceptors are responsible for the sensation of roughness (Blake, Hsiao, and Johnson 1997; Connor et al. 1990; Connor and Johnson 1992). The RA response plays a role in roughness perception of surfaces such as raised dots of varying spacing and diameter. SA-I afferents are mainly responsible for information about form and texture, whereas RA afferents are mainly responsible for information about flutter, slip, and motion across the skin surface.

Temperature is one of the important tactile features. *Thermoreceptor* cells are sensitive to temperature. Temperature can be defined as the degree of hotness of an object that is proportional to the kinetic energy. By touching any object, there is a heat transfer between finger and object until they are in thermal equilibrium with each other. Thermoreceptors respond to cooling or warming but not to mechanical stimulation. Also, they are more sensitive to a change in temperature than to any constant temperature of the object that is touched. There are two different kinds of thermoreceptors, cold and warm, that are sensitive to specific ranges of thermal energy (Jones 1997). Warm thermoreceptors respond to temperatures of 29°–43°C (84°–109°F) and cold thermoreceptors respond to temperatures of 5°–40°C (41°–104°F) (Darian-Smith 1984).

7.3.1.2 Psychophysics

7.3.1.2.1 Vibration, Roughness, Shape, and Orientation

The most frequently employed method to measure tactile sensitivity is to find the smallest amplitude of vibration upon the skin that can be detected by an observer (Gescheider 1976) (Figure 7.4).

These thresholds depend on size of the stimulated skin area, the duration of the stimulus, and frequency of the vibration. Magnitude functions for apparent intensity of vibration were measured by Stevens (1959). Exponents of the power functions* relating subjective magnitude to vibration magnitude for 60 and 250 Hz on a finger are 0.95 and 0.6.

The principal characteristic of tactile texture is the roughness. When we move our hand across a surface, vibrations are produced within the skin (Katz 1969). Therefore, if a texture will be simulated in the Web application, roughness perception of humans should be taken into consideration. The psychophysical magnitude function of subjective roughness for different grits (grades of sandpaper) has been produced (Stevens and Harris 1962). The exponent of the function was −1.5.

The important physical parameters for roughness perception are fingertip force, velocity, and the physical roughness of the surface that the finger moves over. The results of the psychophysical experiments conducted to measure magnitude estimates of the perceived roughness of grooved metal plates show that the width of the groove and the land influenced

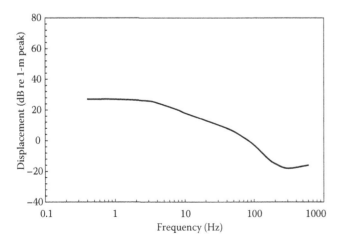

FIGURE 7.4 Psychophysical thresholds for the detection of vibro-tactile stimuli. (According to Bolanowski, S. J., Jr. *Journal of the Acoustical Society of America* 84, 1680–1694, 1988.)

---

* Stevens (1957) proposed that the sensation magnitude ψ grows as a power function of stimulus magnitude ϕ; $\psi = k\phi^n$, where n is the exponent of the power that characterizes the rate of growth and k is a constant. This relation is known as Steven's power law.

the perceived roughness. Increasing groove width resulted in increased roughness, whereas increasing land resulted in decreased roughness. Roughness is dependent on the force applied between the fingertip and the surface. Perceived roughness as a function of groove width increases faster with higher force than it does with low force (Lederman 1974; Taylor and Lederman 1975).

As stimuli, textured surfaces composed of dots of varying height and diameter were used to measure psychophysical roughness perception. Roughness sense increased with dot diameter. At small diameters, dot height has a large, almost linear effect on subjective roughness. At diameters of 2.0 and 2.5 mm, dot height has no significant effect on reported roughness (Blake, Hsiao, and Johnson 1997).

The other important tactile feature is the shape of an object. Tactile discrimination of straight edge was studied by Philips and Johnson (1981). Also, Wheat and Goodwin (2001) have conducted some experiments to quantify the human scaling and discriminating capacity of the curved edges of flat stimuli. The smallest difference in curvature that could be discriminated by subjects was about 20 m^{-1}. The threshold for the discrimination of the spatial interval between two bars was 0.3 mm.

For people who are visually impaired, the tactile sense is a very important channel in everyday life and increasingly in Web applications. It is often assumed that when input in one sensory modality is lost the remaining senses will be hypersensitive (Alary et al. 2009; Gougoux et al. 2005; Levänen and Hamdorf 2001). However, some studies assume that this advantage is not due to heightened sensitivity but rather to the development and refining of perceptual skills with practice (Alary et al. 2009; Sathian 2000). A comparative analysis of tactile sensitivity between blind and sighted people shows that blind persons present a higher tactile sensitivity (Barbacena et al. 2009).

Braille displays are the media of choice for blind people to access the information on the Internet (Braille displays are introduced in more detail below). Braille is a spatial code comprising raised dots arranged in "cells," each of which has three rows and two columns. At each of the six positions in the cell, a dot may be present or absent. Reading Braille seems to depend not on the outlines formed by the dots in a cell but on cues afforded by dot spacing (Sathian 2000). Compared to sighted subjects, blind Braille readers can identify some kinds of Braille-like dot patterns almost 15% more accurately (Van Boven et al. 2000). Another study has shown that blind subjects are not only initially better in Braille-like dot pattern discrimination tasks but also that they significantly outperform sighted ones in a haptic shape discrimination task (Alary et al. 2008).

The differences in the perception of virtual textures and objects by blind and sighted people were investigated by Colwell et al. (1998). The results of the psychophysical experiments showed that blind people are better at discriminating the roughness of textures than sighted people. Similar results can also be seen in the texture discrimination task of Alary et al. (2009). Blind persons not only have better haptic

capabilities, at the same time, they show enhanced speech recognition and auditory mapping capabilities. These results confirm that haptic and auditory modalities are suitable to convey the Web content to visually impaired persons.

7.3.1.2.2 Force and Pressure

Subjective scaling of apparent force was studied by Stevens and Mack (1959). They were able to show that the subjective force of handgrip grows as the 1.7 power of the physical force is exerted. The just-noticeable-difference (JND) for human force sensing was measured and found to be 7% (Jones 1989).

Burdea and Coiffet (1994) have found that for the average person the index finger can exert 7 N, middle finger 6 N, and ring fingers 4.5 N without experiencing discomfort or fatigue. Furthermore, the maximum exertable force from a finger is approximately 30–50 N (Salisbury and Srinavasan 1997).

Tan et al. (1994) conducted some psychophysical experiments to define human factors for the design of force-reflecting haptic interfaces. They measured the average pressure JNDs as percentages of reference pressure. The joint angle JNDs for the wrist, elbow, and shoulder were also measured. The JND values are 2.0°, 2.0°, and 0.8° for wrist, elbow, and shoulder. The other parameter that was measured by Tan et al. was the maximum controllable force that ranged from 16.5 to 102.3 N.* Von der Heyde and Häger-Ross (1998) conducted psychophysical experiments in a complex virtual environment. In these experiments, subjects sorted four cylinders according to their weight (between 25 and 250 g); the size of the cylinder could vary independently. The results confirmed classical weight-size illusions.

The exponent governing the growth of subjective magnitude for pressure sensation on the palm was measured using the magnitude estimation method by Stevens and Mack (1959) and found to be 1.1. The relationship between the physical magnitude and the subjective perception of applied pressure was also studied by Johansson et al. (1999). The pressure was judged to be higher at the thenar than at the finger and palm points. The mean slopes of the magnitude estimation functions were 0.66, 0.78, and 0.76 for the finger, palm, and thenar, respectively. Dennerlein and Yang (2001) have also measured the median discomfort pressure threshold and found 188, 200, and 100 kPa for the finger, palm, and thenar. The pain pressure thresholds were 496, 494, and 447 kPa.

7.3.2 HAPTIC DEVICES

In recent years a variety of customer products that have haptic input and output capabilities have been developed (for example, Apple iPhone, different touch-screen applications, Wiimote, etc.). Some of these devices bring new possibilities to interact with a PC or a PDA. Haptic devices could enable

---

* The values for the proximal interphalangeal joint are 16.5 N for females versus 41.9–50.9 N for males, for shoulder joints 87.2 N for females versus 101.6–102.3 N for males, for wrist 35.5 N for female versus 55.5–64.3 N for males, and for elbow (49.1 N for female versus 78.0–98.4 N for males).

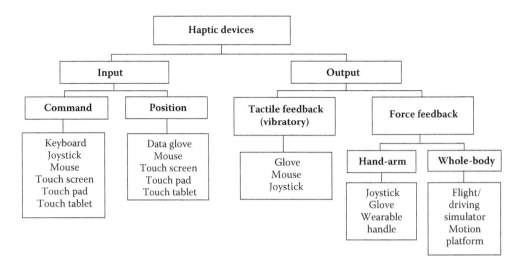

FIGURE 7.5 Categorization of types of haptic devices.

us to shake another user's hand over the Internet who may be up to 3000 miles or more away from us. It is also possible to play tennis with another person by using haptic devices with a real-time interactive tennis game simulation. At the impact time of the ball, we can get force feedback, which is generated by a glove (Molet et al. 1999).

Haptic devices may be categorized in different ways. A categorization scheme for haptic devices is shown in Figure 7.5.

7.3.2.1 Input Devices

If we want to introduce haptic input devices, again, we should begin with the most common device: a keyboard. A *keyboard* is the most familiar input device for a computer environment that enables us to enter data (e.g., text, numerical input) into the computer easily. Standard keyboards have 101 keys, which represent letters, numbers, symbols, and some functional commands. According to key arrangement, *qwerty*, *dvorak*, and *chiclet* type keyboards are available on the market. There are also some keyboards that enable users to surf more easily on the Web. Following the keyboard the most common input device of a computer is the *mouse*. As is well known, controlling the movement of a cursor on a computer display or positioning can be done using a mouse. According to button number and sensor types, there are different types of mice. Another input device that is typically used in gaming applications is the *joystick*. It contains a stick that enables the user to control either forward/back or left/right movement and buttons in different numbers that can be programmed by the user for different purposes. Joysticks can be desk based or handheld. A *data glove* is a type of glove that can be used to give spatial position or movement information to the computer. It contains some position sensors that have different functional principals, optical fiber, mechanical, and strain gage sensors. By using these sensors, hand gesture information can be received in a computer environment, and all information that is related to position and movement of hand and fingers can be used as input information.

In 2006, a new innovative input-controller Wii-remote was released by Nintendo. Most of the tracking technologies

are not suitable for mass customization because of their high prices. But Wii-remote has remarkable tracking capabilities as a low-cost product. It contains a three-axis accelerometer for motion sensing and a 1024×768 infrared camera with built-in hardware blob tracking of up to four points at 100Hz. These capabilities enable new interaction possibilities with PCs. Wii-remote was used in a variety of projects for finger/head tracking or for low-cost multipoint interactive whiteboards (Lee 2008).

The most commonly used touch-sensitive computer input device is the *touch screen*. It is based on a touch-sensitive sensor and a controller and enables the user to give any input to computer by touching the screen of computer. It is very usable for Internet and interactive applications. The use of touch screens, touch panels, and touch surfaces is growing rapidly because of their software flexibility and space and cost savings. They replace most functions of the mouse and keyboard and are used more and more in different technical devices like mobile phones, PDAs, navigation systems, customer electronics, etc. The iPhone is one of the most revolutionary interfaces since the mouse. It uses a touch screen that allows multitouch interaction. The technology behind of iPhone's touch screen plays an important role in its success.

There are four fundamental touch screen technologies. The *resistive systems* consist of several layers: the outer layer (durable hard coating), resistive, and conductive layers, which are very important for the functionality of the touch screen system. When the screen is touched, the conductive layer is pushed against the resistive layer, causing an electrical contact, and a touch is registered. Most PDAs and cell phones use resistive touch screens. They are cost-effective and durable, but they do not allow multitouch interaction. For the recognition of the position a significant amount of contact pressure is required. Therefore the smooth scrolling is not possible.

The *capacitive systems* have a layer that stores electrical charge. When the screen is touched, some of the charge is transferred to the user and the charge of the conductive layer decreases. Some circuits are located at the corners, and the relative differences in charge at the corners give information about the position. The advantages of the capacitive systems are the multiple points of touch recognition and easy scrolling interaction.

The *surface acoustic wave systems (SAW)* use ultrasonic waves. They consist of a transmitter transducer and a receiver transducer. The touch action disturbs the ultrasonic wave transfer and the position can be registered. This technology is very new, and there are few applications. The operation method of *infrared touch screens* is very similar to SAW technology. There is a matrix of infrared transmitter and receiver sensors on the touch screen. The disruption of the light array through touch will be recognized. The light conditions of the operating environment are very important for the good functionality of infrared systems.

Moreover, there is the *touch pad*. It is a rectangular touch-sensitive pad and allows us to control the movement of a cursor on a computer display like a mouse. A *touch tablet* is a device consisting of a flat horizontal surface divided into a matrix of smaller units. It is connected to a computer that registers when one of these units is pressed (Holmes and Jansson 1997).

7.3.2.2 Output Devices

There are two main haptic feedback information types that are used in computer haptic devices: *force feedback* and *vibratory feedback*. Force feedback devices apply physical forces and torques on the user's hand or finger (Burdea 1999). Different types of force feedback generation methods are used in haptic feedback devices (e.g., electromagnetic motors, hydraulics, and pneumatics). Vibratory feedback is generated with electric motors, loudspeaker coils, pins, or some piezoelectric materials.

Most of the cell phones, iPhone, Wiimote, and some PC mouse controls use low-cost pager motors (eccentric mass motors). A pager motor is a DC motor with an eccentric mass that is mounted to the motor shaft. The rotation of the eccentric mass causes an angular momentum. The motion of the eccentric mass motor is transferred to the housing and consequently the adjacent structure (Mortimer, Zets, and Cholewiak 2007). The quality of haptic feedback is limited (frequency response, amplitude, transducer latency, controlling, etc.). The motion is three dimensional and complex. It is possible to simulate vibratory sensations like texture, impulse, or pop. By using extra software, haptic sensations can be added to icons, menu items, or flash applications in standard office programs and Web pages.

The *glove* is used not only to provide force feedback but also vibratory feedback for the user. These two types of information enable the user more realistic interactions with objects. To generate force feedback by a glove, methods like pneumatic pistons are used (Cyber Glove). The *joystick* is a popular device of the computer environment, especially for entertainment applications. In the past 10 years, new joysticks have been developed that can apply force feedback and also vibration feedback generated by motors. Force feedback properties are changeable and are dependent on the type of joystick (desk based or handheld) and the method of force generation.

A *motion platform* is a platform that is driven by hydraulic, pneumatic, or electromagnetic power. Whole-body vibration and motion can be produced by these platforms. For different purposes, motion platforms are used, e.g., driving simulator and flight simulator.

Braille is a system of touch reading for the blind that employs embossed dots evenly arranged in quadrangular letter spaces or cells. *Braille displays* operate by raising and lowering different combinations of pins electronically to produce in Braille what appears on a portion of the computer screen.

7.3.3   Haptic Metaphor Design

In daily life, people mostly meet with unpleasant and unwanted vibrations from vibrating tools (drills, electric razor, vacuum cleaner, motor cycle handle, etc.), railway or road traffic induced building vibrations, travel sickness, etc. but also some pleasant and wanted vibrations, e.g., rocking chair, surfboard, shaking hand, musical whole-body vibrations in concert hall, or acupuncture. Griffin (1990) called such kind of vibrations "good vibrations."

Owing to the development of haptic devices and the application of new haptic devices in Web environments, haptic metaphor design is becoming more important. Haptic metaphor design has similar principles and thus big similarities with sound design. If the question is how can we design haptic metaphors for our applications, we should look at the experience of sound design (see Section 7.2.3). Three main aspects of sound design that are presented by Blauert and Bodden (1994) and Blauert and Jekosch (1997) are valid also for haptic metaphor design. The main aspects for haptic metaphor design are as follows:

1. Suitability of haptic metaphors: Multimedia users get information from different sensory channels related to their interaction with multimedia. The tactile channel is one of these sensory channels. Information that comes from the tactile channel like other channels inform the user about functional features of the interaction and also what the designer wants to transmit to the user. Haptic metaphor designers should be aware of the goal of the application when thinking about haptic metaphors.

2. Pleasantness of haptic metaphors: Although informative haptic metaphors are important they should not only be informative but also pleasant for users. Pleasantness with respect to a haptic metaphor is a complex task for the user interface designer. Here, psychoacoustics enables the designer to understand basic relationships between the haptic output signal and haptic perception. Psychophysical measurement methods (e.g., magnitude estimation, pair comparison, category scale, and semantic differential) as used for sound quality evaluations are summarized by Guski (1997). These methods can be also applied to haptic design evaluation. The

vibration and force intensities that produce discomfort and unpleasantness on the subject already have been investigated for different applications (see Griffin 1990).

3. Identifiability and familiarity of haptic metaphors: Another important aspect is the identifiability and familiarity of the metaphor to the user. We experience different haptic metaphors by interacting with the computer, and in the course of time, these metaphors could create a library (memory) for us. Tactile information that we get in our daily life also belongs to this library. When we experience a new haptic metaphor, we try to recognize it and understand its language by using the metaphor library. Typicality of haptic metaphors enables the user not only to understand the language of the haptic metaphor but also to communicate easier. Haptic devices are very new in their use, and therefore the size of our stored representations is very limited. Designers therefore still have a large amount of freedom to present new haptic metaphors.

7.3.3.1 Hapticons

Enriquez, MacLean, and Chita (2006) define a haptic phoneme as the smallest unit of a constructed haptic signal to which a meaning can be assigned. The maximum duration of a haptic phoneme is limited to 2 seconds It is possible to combine haptic phonemes to form haptic words, or *hapticons*, which can hold more elaborate meanings for their users. Hapticons are brief computer-generated signals. They are displayed to a user through force or tactile feedback to convey information such as event notification, identity, content, or state (MacLean and Enriquez 2003).

The requirements for the haptic phonemes are very similar to above mentioned criteria for haptic metaphor design. They should be differentiable, identifiable, and learnable (Enriquez, MacLean, and Chita 2006). Two approaches are proposed for building hapticons: concatenation (phonemes are combined serially) and superposition (phonemes are combined parallel). The results of Enriquez et al.'s study show that the associations between haptic phonemes and meanings can be learned after a 25-minute training period and remembered consistently for a relatively long period of time (45 minutes).

7.3.3.2 Tactons

Tactons are defined as structured, abstract vibrotactile messages for nonvisual information display (Brown 2007). According to Brown, a one-element tacton encodes a single piece of information in a short vibration burst or temporal pattern. Similar to the relationship between haptic phoneme and hapticon, one-element tactons can be combined to create compound tactons. Tactons can encode multiple types of information using tactile parameters, such as rhythm, frequency, duration, etc. Brown has given different examples for tactons:

- A vibration that increases in intensity could represent Create, while a vibration that decreases in intensity could represent Delete.
- A temporal pattern consisting of two short vibration bursts could represent a "file," while a temporal pattern consisting of three long vibration bursts could represent a "string."

Possible application areas of tactons are tactile alerts, aid navigation, and communication. The design recommendation for tactons (Brown 2007) is as follows:

- When designing two-dimensional tactons, encode one dimension in rhythm and a second dimension in roughness.
  - Use rhythm to encode the most important dimension.
  - Use roughness to encode the less important dimension.
- When designing two-dimensional tactons for phone motors, encode one dimension of information in rhythm and another in intensity.
- When designing three-dimensional tactons, encode the two most important dimensions in rhythm and spatial location. A less important dimension can be encoded in roughness, intensity, or a combination of both.

7.3.4 Examples for Web Applications

Regarding Web applications, the diversity of goals is big. Regarding the goals, haptic metaphors show considerable diversity, too. In some applications, haptic metaphors play a main role, and in other applications they support other modalities. In this part of the chapter, haptic metaphor examples for different Web applications are introduced. These examples may clarify some design principles of haptic metaphors.

- *Navigation*: Haptic metaphors are used for navigation purposes. Haptic pulse feedback (click) can be added to icons or buttons so that if users move the mouse over icons or buttons, they experience haptic feedback information.
- *E-learning*: In e-learning applications, designers often include some feedback signals to the user. If users are incorrect, they get feedback information. Mostly, auditory feedback is used for these types of information, but it is also possible to provide this information using haptic metaphors, e.g., very short impulse signals (like white noise) or resistance to movement. Designers also could implement congratulation feedback for a correct response by the user. Haptic metaphors could be very helpful to introduce some physical attributes from real life to adults and children, such as gravity force and pendulum oscillations.

- *E-trade*: To sell a product via the Web, haptic metaphors could provide some realistic information, e.g., texture, weight, or shape of the product.
- *Entertainment (games)*: In current driving games and flight simulators that are the first areas of frequent haptic interaction application, one can achieve very realistic effects. By fighting or shooting, force feedback in different intensities could excite or frighten the user. Haptics could serve as an "adrenaline pump" (Minsky 2000).
- *Musical instruments*: In our daily life we usually use our hands to play a musical instrument and also get tactile feedback from the instrument. It could be very exciting to play a musical instrument in a computer environment by experiencing different tactile information such as vibrations of the musical instrument, buttons of the musical instrument, etc., also allowing completely new interface concepts for new musical instruments.
- *Chat*: At present we can chat by using only visual or auditory modality. But, in near future it could be possible to utilize our tactile sense. Human–human touch could be an added value for chat applications.

7.4 MULTISENSORY INTERACTION

7.4.1 General: Selecting Media and Modalities

Often, for the user interface designer it is tempting to use a certain medium instead of another according to technical or conceptual benefits. Nevertheless, using a specific medium always remains a trade-off between several requirements to be met. General guidelines for media selection and combination for multimedia software interfaces can be found in ISO (2002). An overview based on the ISO standard is presented and commented in Table 7.9. However, a decision for selecting a certain medium directly affects the respective sensory modalities to be addressed. Hence, the table implies the use of modalities, too, although the text only refers to the related media (see Section 7.1.1). The given guidelines can be a tool for practical design or conceptual solutions.

When providing media, it is recommended not to rely on the user being equipped with the latest versions of hardware and software (relating to system performance and compatibility). Also, downloading a new plug-in for the purpose of watching a site is not always advantageous. Instead of simply demanding that users download a certain plug-in, give them an idea of what they will experience with the additional plug-in (preview of multimedia content).

7.4.2 Auditory–Visual Interaction

Although the focus of this chapter is on designing for the auditory and tactile channel, a very short overview on audiovisual interaction is included for the reason that the combination of the visual and auditory channel is becoming more of a standard for today's multimedia applications (videoconferencing presentations).

In psychophysics as well as cognitive psychology, current issues of interest are dealing with the interaction and integration of the auditory and visual channels. These issues

---

TABLE 7.9
Selection of General Guidelines for Media Selection and Combination

Supporting user tasks	Media should be selected and combined to support the user's task.
	Note: Some tasks benefit more from combination than others (sometimes even a single medium, e.g., text only, may be sufficient).
Supporting communication goals	Media should be selected to achieve the communication goal in the application (e.g., audio is well suited for warning signals).
Selecting media appropriate for users' characteristics	The characteristics of the user population should be considered when selecting the media (e.g., especially older users might need larger point size text, clear brightness contrasts, and audio enhancement).
Considering the context of use	Selection and combination of media should be appropriate in the context of use (e.g., speech output of bank account details should have been confirmed by the user first; otherwise, this could compromise the user's privacy).
Using redundancy for critical information	If important information is to be presented then the same subject matter should be presented in two or more media.
Avoiding conflicting perceptual channels	(For example, if speech output is used and a text presented on the screen that is not related to the voice, this would lead to a conflict by the user.)
Avoiding semantic conflicts	(For example, avoid information regarding a function while it is not accessible for some reason.)
Combining media for different viewpoints	Wherever appropriate to the task, different views on the same subject matter should be provided by media combination (e.g., musical notation and sound output).
Previewing media selections	Especially, when any restrictions are related to the media (e.g., estimated download time, specific hardware or software requirements that are not common in the respective use case), there should be a preview facility available, so users may decide on their own if the additional media would be a benefit regarding their task.

Source: From ISO, 2002. *Software Ergonomics for Multimedia User Interfaces—Part 3: Media Selection and Combination. ISO 14915-3.* Geneva, Switzerland: International Standards Organization. With permission; author's comments in parentheses.

TABLE 7.10

Mean Thresholds on Audio-Visual Synchronicity by Different Authors and Contexts

No.	Type of Stimuli	Audio Lead (ms)	Audio Lag (ms)	Study
1	Speech utterance	131	258	Dixon and Spitz (1980)
2	Hammer hitting a peg	75	188	Dixon and Spitz (1980)
3	TV scenes	40 (90 annoying)	120 (180 annoying)	Rihs (1995)
4	Repetitive events	—	172 (100 lowest)	Miner and Caudell (1998)
5	Speech	—	203	Miner and Caudell (1998)
6	Different	100	175	Hollier and Rimell (1998)

Source: As cited by Kohlrausch (1999).

include sensitivity to asynchrony between audio and video signals, cross-modal effects, and the perceived overall quality of audio-visual stimuli (e.g., Bulkin and Groh 2006; Suied, Bonneel, and Viaud-Delmon 2009; Vilimek 2007). As a basic outcome, it must be stated that a fair video quality may perceptually be enhanced by a good audio quality (which moreover is much cheaper to achieve). Additionally, there is the fact that humans tolerate an intersensory asynchrony that would not be tolerated for the temporal resolution of each of the separate modalities, depending on the used stimuli. For what regards audio-visual synchrony, according to Kohlrausch and van de Par (1999), a clear asymmetry in sensitivity exists for the two conditions of "audio lead" (with respect to the video) and "audio lag." Table 7.10 shows the results obtained by studies cited by Kohlrausch and van den Par.

As an outcome, there is the fact that the human auditory system is comparably sensitive for the audio lead condition compared to the audio lag condition. A reason for this may be our everyday experience of audio-visual stimuli and the fact that the velocity of light is much higher than the velocity of sound (light travels at approximately 300,000 km/s, which equals 186,411 miles/s, whereas the velocity of sound is 340 m/s = 1115 ft/s). However, for a professional audience, the video signal must not be later than 40 ms compared to the audio signal (a consumer audience may tolerate approximately 75 ms). Similarly, for a professional audience the audio signal must not be delayed more than 100 ms compared to the video signal (consumer may accept 175 ms). If discrete events are presented with a strong correlation between sound and vision (e.g., hits and abrupt change of movements), the audience is less tolerant than when speech is provided. In any case, do

not exceed 200 ms as delay between the presentation of the visual and auditory event. For an overview, see Table 7.11.

7.4.3 Auditory–Haptic Interaction

In our daily life, sound is usually produced by the vibration of a body. Therefore, there is a strong relationship between physical attributes of the sound and physical attributes of the vibration. This relationship plays a very important role in our integration mechanism of auditory and tactile information. If multimodal systems have been developed to take advantage of the multisensory nature of humans (Dix et al. 2004), then designers should take into consideration that auditory feedback and haptic feedback should be inherently linked to each other in multimedia applications. This objective requires a better understanding of the integration of the auditory and haptic modalities. We start with the physical coupling of auditory and tactile information and explain psychophysical aspects of auditory-haptic interaction.

7.4.3.1 Psychophysical Aspects

To gain a better understanding of the interaction of auditory and tactile information, it is necessary to be able to specify which criteria have to be met with respect to temporal factors, particularly synchrony. Temporal correlation is an important hint for the brain to integrate information that is generated by one event and obtained from different sensory channels and to differentiate information that is related with this event from other information that is not related with this event. In an event that generates information in different modalities, for example, by knocking on the door, we get tactile feedback on our hand and also hear a knocking sound. The time that it takes until the knocking sound arrives at the ear is related to the sound speed and the physical properties of the door. The time needed until the information arrives at the brain from our ear and hand is related to neuronal transmission properties. We learn this relationship in timing between multimodal information by our experience.

Technical constraints such as data transfer time, computer processing time, and delays that occur during feedback generation processes produce synchronization problems in multimedia applications (Vogels 2001). In multimodal-interaction research, there are several studies regarding the detection of

TABLE 7.11

Guidelines on Audio-Visual Synchrony

Audio Lead		Audio Lag	
Professional Audience	Consumer Audience	Professional Audience	Consumer Audience
≤40 ms	≤75 ms	≤100 ms	≤175 ms

synchronization thresholds that are important for multimodal interface designers. Multimodal synchronization threshold has been defined as the maximum tolerable temporal separation of the onset of two stimuli, one of which is presented to one sense and the other to another sense, such that the accompanying sensory objects are perceived as being synchronous (Altinsoy, Blauert, and Treier 2001). In order to measure this threshold, different psychophysical measurement methods can be used. Observers may be asked to report which of the two stimuli comes first (forced choice) or may be asked to make a three-alternative forced choice judgment whether the audio stimulus and the haptic stimulus were synchronous, the audio stimulus preceded the haptic stimulus, or the haptic stimulus preceded the audio stimulus. Observers may also be asked whether the audio and the haptic were synchronous or asynchronous. The obtained results vary, depending on the kind of stimuli and the psychometric methods employed. Perceptual threshold values for auditory-haptic asynchrony are 50 ms for audio lag and 25 ms for audio lead (haptic stimuli were presented at the tip of the index finger via a shaker; Altinsoy 2003) and perceptual threshold values for auditory-whole body vibration asynchrony are 39 ms for audio lag and 35 ms for audio lead (Altinsoy, Blauert, and Treier 2002). The results of the psychophysical experiments indicate that the synchronization between auditory and haptic modalities has to be at least within an accuracy of 25 ms. Thus the auditory-haptic delay is even more critical than the auditory-visual delay (see Section 7.4.2).

Content and physical properties such as level (intensity) or frequency characteristics of each single modality of information have influence on the overall perception of the multimodal event. If two modalities are combined, the resulting multimodal percept may be a weaker, stronger, or altogether different percept (McGee, Gray, and Brewster 2000). Information from two different modalities can also produce conflicting multimodal perception owing to their contents.

The effect of auditory cues on the haptic perception of stiffness was investigated by DiFranco, Beauregard, and Srinivasan (1997). Their investigation consists of a series of psychophysical experiments designed to examine the effect that various impact sounds have on the perceived stiffness of virtual objects felt by tapping with a force reflecting device. Auditory cues affect the ability of humans to discriminate stiffness. It was found that auditory cues are used for ranking surfaces when there is no difference in haptic stiffness between the surfaces. Current haptic devices have distinct limitations to produce high force feedback levels. The results of the DiFranco, Beauregard, and Srinivasan (1997) study indicate that properties of the sound cues can be used to expand the limitations of the current haptic devices.

The effect of the loudness of the drum sound on the strongness perception of the drum player by playing a virtual drum was investigated by Altinsoy (2003). The results confirm the DiFranco, Beauregard, and Srinivasan (1997) study. The magnitude of strongness increases with increasing loudness in spite of no change in force feedback, which is generated by a virtual drum and applied to the subject's hand.

Texture perception is a multisensory event. By scraping a surface with our finger, we get information from many of our sensory channels such as tactile, auditory, and visual simultaneously. People are able to judge the roughness of different surfaces by using the tactile feedback alone, using the sounds produced by touching the surfaces alone (Lederman 1979), or using visual information of the surfaces alone (Lederman and Abbott 1981). For multimodal integration we use all information available from our different sensory channels to judge (e.g., the roughness of surfaces). If tactile and auditory information are congruent, people tend to ignore sound and use only tactile information to determine the roughness of the surface (Lederman 1979). However, if they are incongruent, auditory information alters the tactile information (Altinsoy 2006; Jousmäki and Hari 1998).

7.4.3.2 Examples for Applications in the Web

Despite the growing use of haptic feedback as an input and output modality in computer environments, interactive sounds are an exciting topic for virtual reality and Web applications. An interactive sound can be defined as the sound that is generated in the real time according to the haptic interaction of the user with sound-producing objects in the Web environment or virtual reality environment. Sound-producing haptic interactions could be (Cook 2002):

- Striking, plucking, etc.
- Rubbing, scraping, stroking, bowing, etc.
- Blowing (voice, whistles, wind instruments, etc.)

The method to generate interactive sound is the physical modeling of the interactive event. Van den Doel and Pai (1998) proposed a general framework for the simulation of sounds produced by colliding physical objects in a virtual reality environment. General information about physical modeling of realistic interactive sounds and some examples are provided by Cook (2002). The physical synthesis of sound and haptic feedback at the same time was successfully applied to virtual reality and musical instrument applications (Altinsoy et al. 2009). Of course, Web designers use not only realistic sounds but also synthesized sounds for Web applications.

Unfortunately, most force and vibratory feedback information that is generated by current haptic devices are still audible, and therefore, users get also unwanted auditory information from the haptic device. By designing haptic metaphors and corresponding auditory metaphors, designers are challenged to combine the information generated by both devices.

Auditory–tactile interaction is a useful tool for visually handicapped people to cope with the Internet. Sighted people can visually process several Web contents simultaneously. They can build links between the texts and graphs or pictures and extract information. At the same time, they can handle navigation bars and skip unnecessary information on the Web page. But the Web contents should be differently structured for impaired people. WC3 (2008) covers a wide range of recommendations for making Web content more accessible. These guidelines are useful to make content accessible to a wider range of people with disabilities, including blindness and low vision, deafness and hearing loss, learning disabilities, cognitive limitations, limited movement, speech disabilities, photosensitivity, and combinations of these.

Graphs play an important role on Web pages; they can be used to summarize pages of textual information, compare and contrast between different data series, or show how data vary over time. Bitmapped graphs and particularly bar graphs are the most frequently found form of graph contained on Web pages (McAllister, Staiano, and Yu 2006). One strategy to make the graphs or the visual images on Web pages accessible for visually impaired users is to add an alternative text (ALT text) to the image. However, the length of the ALT texts should be limited and no longer than one sentence. It is quite difficult to summarize the multidimensional complex content of a graph with a sentence. Another problem is that the success of this summarization is strongly dependent on the skills of the Web page developer.

McAllister, Staiano, and Yu (2006) have developed an approach to make bitmapped graphs accessible to blind users on the Internet. The approach identifies the important regions of the graph and tags them with metadata. The metadata and bitmap graph are then exported to a Web page for sonification and exploration by the visually impaired user. Haptic data visualization is a growing research area, and two types of haptic interaction technique for charts are introduced by Panëels, Roberts, and Rogers (2009) to help the user get an overview of the data. The scatter plot technique models the plot by assigning a repulsive force to each point. The user explores the plot and greater force is felt for larger concentrations of data points. Different data sets can be felt successively. The line chart technique gives the user guidance tours, which means that the user is taken along a predefined path and stops at chosen points of interest.

HFVE Silooet software allows blind people to access features of visual images using low-cost equipment (Dewhurst 2009). The HFVE system aims to simulate the way that sighted people perceive visual features. The first step is the recognition of the specific entities (such as a person, person's face, a part of a diagram, etc.) of the visual image. Then, the borders of the specific entities are made noticeable via tracers and audiotactile effects. The apparently moving sounds, which are positioned in "sound space" according to location, and pitched according to height, are used to make the paths of the tracers perceivable. The paths are also presented via a moving force feedback device that moves/pulls the user's hand and arm. In both modalities the path describes the shape, size, and location (and possibly identity) of the specific entity. Users can choose which modality to use, or both modalities can be used simultaneously.

Another issue that is nowadays very popular in Web applications is navigation. The first Web-based software tool, Tactile Map Automated Production (TMAP), for rapid production of highly specific, tactile street maps of the United States was introduced by Miele, Landau, and Gilden (2006).

The talking TMAP system was enhanced by audio output in 2005. Aiming at providing a suitable navigation service for visually impaired pedestrians, Zeng and Weber (2009) have proposed a tactile map based on BrailleDis 9000 as a pre-journey system enabling users to follow virtual routes. BrailleDis 9000 is a pin-matrix device, which represent tactile graphics on a matrix of 60×120 refreshable pins. Users can locate their geographical position and touch maps on the tactile display to understand their position in relation to the context from tactile and speech output.

Most famous of the online virtual world applications is *Second Life*. A haptic-enabled version of the Second Life Client was proposed for visually impaired people by de Pascale, Mulatto, and Prattichizzo (2008). Two new input modes, "Blind Walk" and "Blind Vision," were implemented to navigate and explore the virtual environment. The haptic device (e.g., Phantom) used to control walking and flying actions in Blind Walk mode gives appropriate force feedback when collisions with obstacles occur. In Blind Vision mode, the haptic device is used to control virtual sonar, which feels objects as vibrations.

Haptic feedback in shared virtual environments can potentially make it easier for a visually impaired person to take part in and contribute to the process of group work (Moll and Sallnäs 2009). The haptic feedback can convey much more information than just the "feeling" of virtual objects. When two people (one visually impaired and one sighted) collaborate in a haptic interface, it is evident that the haptic guiding can be used by participants to communicate and as a way to navigate while at the same time exploring details of objects during joint problem solving. Haptic guidance is also useful in learning situations, such as training handwriting, for visually impaired pupils (Plimmer et al. 2008).

Touch-sensitive displays and touch surfaces are more and more replacing physical buttons. If a physical button is pressed, audio and tactile feedback confirms the successful operation. The loss of audiotactile feedback in touch-sensitive interfaces may create higher input error rates and user dissatisfaction. Therefore, the design and evaluation of suitable signals is necessary. Altinsoy and Merchel (2009) evaluated different haptic and auditory feedback signal forms and characteristics regarding their suitability to the touch screen applications. For the evaluation experiment, a dialing numbers task was used. The evaluation criteria were overall quality, suitability for confirmation feedback, and comfort. Execution time and the errors were also measured. The results of the study showed the advantage of tactile feedback in both quality and error rate for a dialing numbers task compared to no feedback. The results of the audiotactile experiments showed that if both modalities are combined, there are synergistic effects. The tactile signal can improve the audio only ratings, and almost all ratings get better. The potential benefits associated with the provision of multimodal feedback via a touch screen on older adults' performance in a demanding dual-task situation were examined by Lee, Poliakoff, and Spence (2009). The results showed that presentation of multimodal feedback (bi- or trimodal) with auditory signals via a touch screen device results in enhanced performance and subjective benefits for older adults.

The benefits of the audiotactile interaction are not limited to touchscreen applications. Audiotactile interaction can be a very promising tool for multimedia display of concert DVD reproductions, DVD film titles, and gaming applications. Sound and whole-body vibration perceptions are always coupled in live music experience. If concert recordings are played back with multimedia Hi-Fi systems at home, the vibratory information is missing in the majority of cases. The perceived overall quality of concert DVD reproduction can be improved by adding vertical whole body vibrations (Merchel and Altinsoy 2009). The vibration signals for the selected sequences could be generated by low pass filtering the audio sum signal. Similar quality enhancement was observed for multimedia display of action-oriented DVD films by adding whole-body vibrations and motions (Walker and Martens 2006). Custom motion programs successfully create in observers a sense of realism related to a virtual environment.

7.5   OUTLOOK

The use of audio-visual media is very common nowadays. However, because of low bit rate transmission, low-quality sound reproduction, or inadequate use of the medium, there is still a need for designing more adequate user interfaces using audio-visual media in the future.

For many simple Web applications (e.g., in administrative or business contexts), just the presence of auditory feedback could considerably improve task performance. For more sophisticated designs such as those used in the game and automobile industries, great progress has been made in recent years.

The use of haptic devices is emerging. In recent years, a lot of new consumer products (e.g., Wiimote, Apple iPhone, Phantom Omni, etc.) were developed. An increasing number of commercially viable products, particularly mobile devices, make use of haptic feedback as a supplement of auditory and visual feedback. Particularly, multitouch technology and user interfaces are promising areas for future development. A single device can transform itself into whatever interface is appropriate for the task at hand using multitouch technology. A multitouch interface can lead human–computer interaction beyond simple pointing and clicking, button pushing, and dragging and dropping that have dominated interactions (Elezovic 2008). The interest of companies like Microsoft (Windows 7 has multitouch support), Apple, and Mitsubishi in multitouch interfaces shows that they will be one of the key technologies of the future.

The development of haptic interfaces causes fundamental research on tactile perception and interaction techniques to become increasingly important. Researchers in human–computer interaction have recently begun to investigate the design of haptic icons and vibrotactile messaging. The benefits of the multimodal interaction and the design issues of auditory, visual, and tactile stimuli are very promising for engineers who design multimodal user interfaces.

It is clearly seen that adequate human–computer interaction is a key topic to provide better products, applications, and services (Wucherer 2001). Interface designers are challenged to provide the best solutions possible. At present, the tactile and auditory channel enables a large framework for creativity.

REFERENCES

Alary, F., M. Duquette, R. Goldstein, C. E. Chapman, P. Voss, V. LaBuissonniere Ariza, and F. Lepore. 2009. Tactile acuity in the blind: A closer look reveals superiority over the sighted in some but not all cutaneous tasks. *Neuropsychologia* 47(10): 2037–2043.

Alary, F., R. Goldstein, M. Duquette, C. E. Chapman, P. Voss, and F. Lepore. 2008. Tactile acuity in the blind: A psychophysical study using a two-dimensional angle discrimination task. *Experimental Brain Research* 187: 567–594.

Altinsoy, M. E., and S. Merchel. 2009. Audiotactile feedback design for touch screens. In *Haptic and Audio Interaction Design 2009* (LNCS 5763), eds. M. E. Altinsoy, U. Jekosch, and S. Brewster, 136–144. Berlin, Germany: Springer.

Altinsoy, E. 2006. *Auditory-Tactile Interaction in Virtual Environments*. Aachen, Germany: Shaker Verlag.

Altinsoy, M. E. 2003. Perceptual aspects of auditory-tactile asynchrony. *Proceedings of the Tenth International Congress on Sound and Vibration*. Stockholm, Sweden.

Altinsoy, M. E., J. Blauert, and C. Treier. 2001. Inter-modal effects of non-simultaneous stimulus presentation. *Proceedings of the 17th International Congress on Acoustics*. Rome, Italy.

Altinsoy, M. E., J. Blauert, and C. Treier. 2002. On the perception of the synchrony for auditory-tactile stimuli. In *Fortschritte der Akustik - DAGA'02*. Oldenburg, Germany: Deutsche Gesellschaft für Akustik. Originally published as Zur Wahrnehmung von Synchronität bei auditiv-taktil dargebotenen Stimuli (in German).

Altinsoy, M. E., S. Merchel, C. Erkut, and A. Jylhä. 2009. Physically-based synthesis modeling of xylophones for auditory-tactile virtual environments. In *Proceedings of Fourth International Workshop on Haptic and Audio Interaction Design, HAID 2009*. Dresden, Germany.

Barbacena, I. L., A. C. O. Lima, A. T. Barros, R. C. S. Freire, and J. R. Pereira. 2009. Comparative analysis of tactile sensitivity between blind, deaf and unimpaired people. *International Journal of Advanced Media and Communication* 3(1/2): 215–228.

Blake, D. T., S. S. Hsiao, and K. O. Johnson. 1997. Neural coding mechanism in tactile pattern recognition: The relative contributions of slowly and rapidly adapting mechanoreceptors to perceived roughness. *Journal of Neuroscience* 17: 7480–7489.

Blattner, M., D. Sumikawa, and R. Greenberg. 1989. Earcons and icons: Their structure and common design principles. *Human Computer Interaction* 4(1): 11–44.

Blauert, J. 1997. *Spatial Hearing. The Psychophysics of Human Sound Localization*. Cambridge, MA: MIT Press.

Blauert, J., and M. Bodden. 1994. Evaluation of sounds—why a problem? In *Soundengineering*, ed. Q.-H. Vo, 1–9. Renningen, Germany: Expert. Originally published as Gütebeurteilung von Geräuschen—Warum ein Problem? (in German).

Blauert, J., and U. Jekosch. 1997. Sound-quality evaluation—a multi-layered problem. *Acustica United with Acta Acustica* 83(5): 747–753.

Bodden, M., and H. Iglseder. 2002. Active sound design: Vacuum cleaner. *Revista de Acustica* 33, special issue.

Bolanowski, S. J., Jr., G. A. Gescheider, R. T. Verillo, and C. M. Checkosky. 1988. Four channels mediate the mechanical aspects of touch. *Journal of the Acoustical Society of America* 84: 1680–1694.

Brewster, S. A. 1994. Providing a structured method for integrating non-speech audio into human–computer interfaces. Doctoral diss., 471–498. University of York.

Brewster, S. A. 2002. Non-speech auditory output. In *The Human Computer Interaction Handbook*, eds. J. Jacko and A. Sears, 220–239. Mahwah, NJ: Lawrence Erlbaum.

Brewster, S. A., P. C. Wright, and A. D. N. Edwards. 1994. A detailed investigation into the effectiveness of earcons. In *Proceedings of the International Conference on Auditory Displays ICAD'92*. Santa Fe, NM, USA: Santa Fe Institute.

Brown, L. 2007. Tactons: Structured vibrotactile messages for non-visual information display. PhD thesis, Department of Computing Science, University of Glasgow, Glasgow, UK.

Bulkin, D. A., and J. M. Groh. 2006. Seeing sounds: visual and auditory interactions in the brain. *Current Opinions in Neurobiology* 16: 415–419.

Burdea, G. C. 1999. Haptic feedback for virtual reality. Paper presented at Virtual Reality and Prototyping Workshop, June 1999. Laval, France.

Burdea, G., and P. Coiffet. 1994. *Virtual Reality Technology*. New York: John Wiley.

Caldwell, D. G., and C. Gosney. 1993. Enhanced tactile feedback (tele-taction) using a multi-functional sensory system. *Paper presented at IEEE Robotics and Automation Conference*. Atlanta, GA, May 2–7.

Colwell, C., H. Petrie, D. Kornbrot, A. Hardwick, and S. Furner. 1998. Use of a haptic device by blind and sighted people: perception of virtual textures and objects. In *Improving the Quality of life for the European Citizen*, eds. I. Placencia Porrero and E. Ballabio, 243–247. Amsterdam: IOS Press.

Connor, C. E., S. S. Hsiao, J. R. Philips, and K. O. Johnson. 1990. Tactile roughness: neural codes that account for psychophysical magnitude estimates. *Journal of Neuroscience* 10: 3823–3836.

Connor, C. E., and K. O. Johnson. 1992. Neural coding of tactile texture: comparisons of spatial and temporal mechanisms for roughness perception. *Journal of Neuroscience* 12: 3414–3426.

Cook, P. R. 2002. *Real Sound Synthesis for Interactive Applications*. Boston, MA: Peters.

Darian-Smith, I., ed. 1984. Thermal sensibility. In *Handbook of Physiology: A critical, comprehensive presentation of physiological knowledge and concepts*, vol. 3, 879–913. Bethesda, MD: American Physiological Society.

Davison, B. K., and B. N. Walker. 2009. Measuring the use of sound in everyday software. In *15th International Conference on Auditory Display ICAD 2009*. Copenhagen, Denmark.

de Pascale, M., S. Mulatto, and D. Prattichizzo. 2008. Bringing haptics to second life for visually impaired people. Eurohaptics 2008 (LNCS 5024), 896–905. Berlin, Germany: Springer.

Dennerlein, J. T., and M. C. Yang. 2001. Haptic force-feedback devices for the office computer: Performance and musculoskeletal loading issues. *Human Factors* 43(2): 278–286.

Dewhurst, D. 2009. Accessing audiotactile images with HFVE Silooet. In *Haptic and Audio Interaction Design* 2009 (LNCS 5763), eds. M. E. Altinsoy, U. Jekosch, and S. Brewster, 61–70. Berlin, Germany: Springer.

DiFranco, D. E., G. L. Beauregard, and M. A. Srinivasan. 1997. The effect of auditory cues on the haptic perception of stiffness in virtual environments. In *Proceedings of the ASME Dynamic Systems and Control Division*, DSC vol. 61, ed. G. Rizzoni, 17–22. New York, NY: ASME.

Dionisio, J. 1997. Virtual hell: A trip into the flames. In *IEEE Computer Graphics and Applications*. New York: IEEE Society.

Dix, A., J. Finlay, G. Abowd, and R. Beale. 2004. *Human–Computer Interaction*, 3rd ed. New York: Prentice Hall.

Dixon, N. F., and L. Spitz. 1980. The detection of audiovisual desynchrony. *Perception* 9: 719–721.

Dürrer, B. 2001. Investigations into the design of auditory displays (in German). Dissertation, Berlin, Germany. dissertation.de - Verlag im Internet, D-Berlin.

Dürrer, B., and U. Jekosch. 1998. Meaning of sound: a contribution to product sound design. In *Designing for Silence— Prediction, Measurement and Evaluation of Noise and Vibration, Proceedings of Euronoise 98* (Munich), eds. H. Fastl and J. Scheuren, 535–540. Oldenburg, Germany.

Dürrer, B., and U. Jekosch. 2000. Structure of auditory signs: Semiotic theory applied to sounds. In *Proceedings of Internoise 2000* (The 29th International Congress on Noise Control Engineering, Nice, Côte d'Azure, France, Aug. 27–30 2000), ed. D. Cassersau, 2201. Paris: Societé Française d'Acoustique.

Elezovic, S. 2008. Multi touch user interfaces. Guided Research Final Report. Jacobs University Bremen, Germany.

Enriquez, M., K. E. MacLean, and C. Chita. 2006. Haptic phonemes: Basic building blocks of haptic communication, in *Proceedings of the 8th International Conference on Multimodal Interfaces, ICMI'06* (Banff, Alberta, Canada). New York, NY: ACM Press, 8 pp.

Flückiger, B. 2001. *Sound Design* (in German). Marburg, Germany: Schüren.

Gaver, W. W. 1989. The SonicFinder: An interface that uses auditory icons. *Human Computer Interaction* 4(1): 67–94.

Gaver, W. W. 1997. Auditory interfaces. In *Handbook of Human–Computer Interaction*, 2nd ed., eds. M. G. Helander, T. K. Landauer, and P. Prabhu. Amsterdam, Netherlands: Elsevier Science.

Gescheider, G. A. 1976. *Psychophysics Method and Theory*. Mahwah, NJ: Lawrence Erlbaum.

Gougoux, F., R. J. Zatorre, M. Lassonde, P. Voss, and F. Lepore. 2005. A functional neuroimaging study of sound localization: Visual cortex activity predicts performance in early-blind individuals. *Public Library of Science Biology* 3: 324–333.

Greenstein, J. S., and L. Y. Arnaut. 1987. Human factor aspects of manual computer input devices. In *Handbook of Human Factors*, ed. G. Salvendy, 1450–1489. New York: John Wiley.

Griffin, M. J. 1990. *Handbook of Human Vibration*. London: Academic Press.

Guski, R. 1997. Psychological methods for evaluating sound quality and assessing acoustic information. *Acustica United with Acta Acustica* 83: 765 ff.

Hempel, T. 2001. On the development of a model for the classification of auditory events. Paper presented at 4th European Conference on Noise Control (Euronoise), Patras, Greece, Jan. 14–17, 2001.

Hempel, T. 2003. Parallels in the concepts of sound design and usability engineering. In *Proceedings of the 1st ISCA Research Workshop on Auditory Quality of Systems*, eds. U. Jekosch and S. Möller, 145–148. Herne, Germany.

Hempel, T. 2008. *Usability of Speech Dialog Systems—Listening to the Target Audience*. Heidelberg, Germany: Springer.

Hempel, T., and E. Altınsoy. 2005. Multimodal user interfaces: designing media for the auditory and the tactile channel. In *Handbook of Human Factors in Web Design*, eds. R. W. Proctor and K.-P. L. Vu, 134–155. Mahwah, NJ: Lawrence Erlbaum.

Hempel, T., and J. Blauert. 1999. From "sound quality" to "auditory quality of systems." In *Impulse und Antworten*, eds. B. Feiten et al., 111–117. Berlin, Germany: Wissenschaft und Technik Verlag. Originally published as Von "Sound Quality" zur "Auditiven Systemqualität." Festschrift für Manfred Krause (in German).

Hempel, T., and R. Plücker. 2000. Evaluation of confirmation sounds for numerical data entry via keyboard. In *Fortschritte der Akustik—DAGA 2000*. Oldenburg, Germany: Deutsche Gesellschaft für Akustik. Originally published as Evaluation von Quittungsgeräuschen für die nummerische Dateneingabe per Tastatur (in German).

Ho, C., N. Reed, and C. Spence. 2007. Multisensory in-car warning signals for collision avoidance. *Journal of the Human Factors and Ergonomics Society* 49(6): 1107–1114.

Hollier, M. P., and A. N. Rimell. 1998. An experimental investigation into multi-modal synchronization sensitivity for perceptual model development. Paper presented at 105th Convention of the Audio Engineering Society, San Francisco, CA, Sept. 1998, Preprint 4790.

Holmes, E., and G. Jansson. 1997. A touch tablet enhanced with synthetic speech as a display for visually impaired people's reading of virtual maps. Paper presented at 12th Annual CSUN Conference: Technology and Persons with Disabilities, Los Angeles Airport, Marriott Hotel, Los Angeles, California, March 18–22, 1997.

ISO. 1998. Ergonomic requirements for office work with visual display terminals (VDTs)—Part II: Guidance on usability. ISO 9241-11. Geneva, Switzerland: International Standards Organization.

ISO. 1999. Human-centered design processes for interactive systems. ISO 13407. Geneva, Switzerland: International Standards Organization. Geneva, Switzerland: International Standards Organization.

ISO. 2002. Software ergonomics for multimedia user interfaces —Part 3: Media selection and combination. ISO 14915-3. Geneva, Switzerland: International Standards Organization.

Jekosch, U. 2001. Sound quality assessment in the context of product engineering. *Paper presented at 4th European Conference on Noise Control (Euronoise)*, Patras, Greece, Jan. 14–17.

Jekosch, U., and J. Blauert. 1996. A semiotic approach toward product sound quality. In *Noise Control—The Next 35 Years*, *Proceedings of Internoise 96, Liverpool*, 2283–2288. St. Albans, UK: Institute of Acoustics.

Johansson, L., A. Kjellberg, A. Kilbom, and G. M. Hägg. 1999. Perception of surface pressure applied to the hand. *Ergonomics* 42: 1274–1282.

Jones, L. 1989. The assessment of hand function: A critical review of techniques. *Journal of Hand Surgery* 14A: 221–228.

Jones, L. 1997. Dextrous hands: human, prosthetic, and robotic. *Presence* 6: 29–56.

Jones, L., and M. Berris. 2002. The psychophysics of temperature perception and thermal-interface design. *Paper presented at 10th Symposium on Haptic Interfaces for Virtual Environment and Teleoperator Systems*, Orlando, FL, 24–25 March 2002.

Jousmäki, V., and R. Hari. 1998. Parchment-skin Illusion: Sound-biased touch. *Current Biology* 8: 190.

Katz, D. 1969. *Der Aufbau der Tastwelt*. Darmstadt, Germany: Wissenschaftliche Buchgesellschaft. Originally published as *Zeitschrift für Psychologie*, 1925 (in German).

Kohlrausch, A., A. Messelaar, and E. Druyvensteyn. 1996. Experiments on the audio-video quality of TV scenes. In *EURASIP/ITG Workshop on Quality Assessment in Speech, Audio and Image Communication, Darmstadt, 11–13 März 1996*, eds. G. Hauske, U. Heute, and P. Vary, 105–106. Darmstadt, Germany: EURASIP.

Kohlrausch, A., and S. van de Par. 1999. Auditory-visual interaction: from fundamental research in cognitive psychology to (possible) applications. *Human Vision and Electronic Imaging IV*, eds. B. G. Rogowitz and T. N. Pappas. *Proc. SPIE* 3644: 34–44.

Kuwano, S., S. Namba, A. Schick, H. Hoege, H. Fastl, T. Filippou, M. Florentine, and H. Muesch. 2000. The timbre and annoyance of auditory warning signals in different countries. In *Proceedings of Internoise 2000*, The 29th International Congress on Noise Control Engineering (Nice, Côte d' Azure, France, Aug. 27–30), ed. D. Cassersau, 3201–206. Paris, France: Societé Française d'Acoustique.

Lashina, T. 2001. Auditory cues in a multimodal jukebox. In *Usability and Usefulness for Knowledge Economies, Proceedings of the Australian Conference on Computer–Human Interaction (OZCHI)* (Fremantle, Western Australia, Nov. 20–23), 210– 216.

Lederman, S. J. 1974. Tactile roughness of grooved surfaces: The touching process and effects of macro- and micro-surface structure. *Perception & Psychophysics* 16: 385–395.

Lederman, S. J. 1979. Auditory texture perception. *Perception* 8: 93–103.

Lederman, S. J., and S. G. Abbott. 1981. Texture perception: Studies of intersensory organization using a discrepancy paradigm and visual versus tactual psychophysics. *Journal of Experimental Psychology: Human Perception & Performance* 7(4): 902–915.

Lederman, S. J., J. M. Loomis, and D. A. Williams. 1982. The role of vibration in the tactual perception of roughness. *Perception & Psychophysics* 32: 109–116.

Lee, J. C. 2008. Talk: interaction techniques using the Wii remote. Paper presented at Stanford EE Computer Systems Colloquium, Stanford, CA, Feb. 13.

Lee, J. H., E. Poliakoff, and C. Spence. 2009. The effect of multimodal feedback presented via a touch screen on the performance of older adults. In *Haptic and Audio Interaction Design 2009 (LNCS 5763)*, eds. M. E. Altinsoy, U. Jekosch, and S. Brewster, 119–127. Berlin, Germany: Springer.

Levänen, S., and D. Hamdorf. 2001. Feeling vibrations: Enhanced tactile sensitivity in congenitally deaf humans. *Neuroscience Letters* 301: 75–77.

MacLean, K., and M. Enriquez. 2003. Perceptual design of haptic icons. In *Proceedings of Eurohaptics, Dublin, Ireland*.

Mayes, T. 1992. The 'M' word: multimedia interfaces and their role in interactive learning systems. In *Multimedia Interface Design in Education*, eds. A. D. N. Edwards and S. Holland, 1–22. Berlin, Germany: Springer.

McAllister, G., J. Staiano, and W. Yu. 2006. Creating accessible bitmapped graphs for the internet. In *Haptic and Audio Interaction Design 2006 (LNCS 4129)*, eds. D. McGookin and S. Brewster, 92–101. Berlin, Germany: Springer.

McGee, M. R., P. Gray, and S. Brewster. 2000. The effective combination of haptic and auditory textural information. In *Haptic Human–Computer-Interaction*, eds. S. Brewster and R. Murray-Smith, 118–127. Glasgow, UK.

McGee-Lennon, M. R., M. Wolters, and T. McBryan. 2007. Audio reminders in the home environment. In *13th International Conference on Auditory Display, ICAD 2007*, 437–444. Montréal, CA: Schulich School of Music, McGill University.

Merchel, S., and M. E. Altinsoy. 2009. Vibratory and acoustical factors in multimodal reproduction of concert DVDs. In *Haptic and Audio Interaction Design 2009* (LNCS 5763), eds. M. E. Altinsoy, U. Jekosch, U. and S. Brewster, 119–127. Berlin, Germany: Springer.

Miele, J. A., S. Landau, and D. Gilden. 2006. Talking TMAP: Automated generation of audio-tactile maps using Smith-Kettlewell's TMAP software. *British Journal of Visual Impairment* 24(2): 93–100.

Miner, N., and T. Caudell. 1998. Computational requirements and synchronization issues for virtual acoustic displays. *Presence* 7: 396–409.

Minsky, M. 2000. Haptics and entertainment. In *Human and Machine Haptics*, eds. R. D. Howe et al. Cambridge, MA: MIT Press.

Molet, T., A. Aubel, T. Capin, et al. 1999. Anyone for tennis, *Presence* 8: 140–156.

Moll, J., and E. Sallnäs. 2009. Communicative functions of haptic feedback. In *Haptic and Audio Interaction Design 2009* (LNCS 5763), eds. M. E. Altinsoy, U. Jekosch, and S. Brewster, 31–40. Berlin, Germany: Springer.

Monkman G. J., and P. M. Taylor. 1993. Thermal tactile sensing. *IEEE Transactions on Robotics and Automation* 9: 313–318.

Mortimer, B., G. Zets, and R. Cholewiak. 2007. Vibrotactile transduction and transducers. *Journal of the Acoustical Society of America* 121(5): 2970–2977.

Mynatt, E. D. 1994. Designing with auditory icons: How well do we identify auditory cues? In *Conference Companion on Human Factors in Computing Systems (Boston, Massachusetts, United States, April 24–28, 1994). CHI '94.* ed. C. Plaisant, 269–270. New York, NY: ACM.

Ottensmeyer, M. P., and J. K. Salisbury. 1997. Hot and cold running VR: adding thermal stimuli to the haptic experience. In *Proceedings of the Second PHANToM User's Group Workshop (AI Lab Technical Report 1617)*, 34–37. Dedham, MA: Endicott House.

Panëels, S., J. C. Roberts, and P. J. Rodgers. 2009. Haptic interaction techniques for exploring chart data. In *Haptic and Audio Interaction Design 2009* (LNCS 5763), eds. M. E. Altinsoy, U. Jekosch, and S. Brewster, 31–40. Berlin, Germany: Springer.

Pellegrini, R. S. 2001. Quality assessment of auditory virtual environments. In *Proceedings of the 7th International Conference on Auditory Display* (July 29 to Aug. 1), eds. J. Hiipakka, N. Zacharov, and T. Takala. Espoo, Finland: Helsinki University of Technology.

Peres, S. C., P. Kortum, and K. Stallmann. 2007. Auditory progress bars—preference, performance, and aesthetics. In *13th International Conference on Auditory Display*, 391–395. Montréal, CA: Schulich School of Music, McGill University.

Philips, J. R., and K. O. Johnson. 1981. Tactile spatial resolution: II. neural representation of bars, edges and gratings in monkey primary afferents. *Journal of Neurophysiology* 46: 1204–1225.

Plimmer, B., A. Crossan, S. Brewster, and R. Blagojevic. 2008. Multimodal collaborative handwriting training for visually-impaired people. In *Proceedings of CHI '08*, 393–402. New York: ACM Press.

Pollard, D., and M. B. Cooper. 1979. The effects of feedback on keying performance. *Applied Ergonomics* 10: 194–200.

Rihs, S. 1995. The influence of audio on perceived picture quality and subjective audio-video delay tolerance. In *Proceedings of the MOSAIC Workshop Advanced Methods for the Evaluation of Television Picture Quality*, eds. R. Hember and H. de Ridder, 133–137. Eindhoven, Netherlands: Institute for Perception Research.

Rouben, A., and L. Terveen. 2007. Speech and non-speech audio: Navigational information and cognitive load. In *13th International Conference on Auditory Displays, ICAD 2007*, Montréal, Canada.

Salisbury, J. K., and M. A. Srinivasan. 1997. Phantom-based haptic interaction with virtual objects. *IEEE Computer Graphics and Applications* 17(5): 6–10.

Sathian, K. 2000. Practice makes perfect: Sharper tactile perception in the blind. *Neurology* 54(12): 2203–2204.

Sweller, J., J. J. G. Merrienboer, and F. G. W. C. Paas. 1998. Cognitive architecture and instructional design. *Educational Psychology Review* 10(3): 251–295.

Spence, C., and C. Ho. 2008. Multisensory warning signals for event perception and safe driving. *Theoretical Issues in Ergonomics Science* 9(6): 523–554.

Stevens, J. C., and J. D. Mack. 1959. Scales of apparent force. *Journal of Experimental Psychology* 58: 405–413.

Stevens, S. S. 1957. On the psychophysical law. *Psychological Review* 64: 153–181.

Stevens, S. S. 1959. Tactile vibration: Dynamics of sensory intensity. *Journal of Experimental Psychology* 57: 210–218.

Stevens, S. S., and J. Harris. 1962. The scaling of subjective roughness and smoothness. *Journal of Experimental Psychology* 64: 489–494.

Strauss, H., and J. Blauert. 1995. Virtual auditory environments. In *Proceedings of the 1st FIVE International Conference*, 123–131. London.

Suied, C., N. Bonneel, and I. Viaud-Delmon. 2009. Integration of auditory and visual information in the recognition of realistic objects. *Experimental Brain Research* 194(1): 91–102.

Tan, H. Z., M. A. Srinavasan, B. Eberman, and B. Cheng. 1994. Human factors for the design of force-reflecting haptic interfaces. In *Dynamic Systems and Control*, DSC-Vol.55-1, ed. C. J. Radcliffe, 353–359. New York: American Society of Mechanical Engineers.

Taylor, M. M., and S. J. Lederman. 1975. Tactile roughness of grooved surfaces: A model and the effect of friction. *Perception & Psychophysics* 17: 23–36.

Tullis, T. S., F. J. Tranquada, and M. J. Siegel, this volume. Presentation of information. In *Handbook of Human Factors in Web Design*, 2nd ed., eds. K.-P. L. Vu and R. W. Proctor, 153–190. Boca Raton, FL: CRC Press.

Van Boven, R. W., R. H. Hamilton, T. Kauffman, J. P. Keenan, and A. Pascual-Leone. 2000. Tactile spatial resolution in blind Braille readers. *Neurology* 54: 2230–2236.

van den Doel, K., and D. K. Pai. 1998. The sounds of physical shapes. *Presence* 7(4): 382–395.

van den Doel, K., P. G. Kry, and D. K. Pai. 2001. FoleyAutomatic: Physically-based sound effects for interactive simulation and animation. In *Proceedings of the 28th International Conference on Computer Graphics and Interactive Techniques* (Los Angeles, CA, Aug. 12–17), 537–544. New York, NY: ACM Press.

Vilimek, R. 2007. *Gestaltungsaspekte multimodaler Interaktion im Fahrzeug—Ein Beitrag aus ingenieurpsychologischer Perspektive*. Düsseldorf, Germany: VDI-Verlag.

Vilimek, R., and T. Hempel. 2005. Effects of speech and non-speech sounds on short-term memory and possible implications for in-vehicle use. In *11ᵗʰ International Conference on Auditory Displays, ICAD '05*, 344–350. Limerick, Ireland.

Vogels, I. M. L. C. 2001. Selective attention and the perception of visual-haptic asynchrony. In *Eurohaptics 2001*, eds. C. Baber, S. Wall, and A. M. Wing, 167–169. Birmingham, UK: University of Birmingham.

von der Heyde, M., and C. Häger-Ross. 1998. Psychophysical experiments in a complex virtual environment. In *Proceedings of the Third PHANToM Users Group Workshop* (MIT Artificial Intelligence Report 1643, MIT R.L.E. TR 624), eds. J. K. Salisbury and M. A. Srinivasan, 101–104. Cambridge, MA: MIT Press.

Walker, K., and W. L. Martens. 2006. Perception of audio-generated and custom motion programs in multimedia display of action-oriented DVD films. *Haptic Audio Interaction Design* 2006: 1–11.

WC3. 2008. Web Content Accessibility Guidelines (WCAG) 2.0. http://www.w3.org/TR/WCAG20/ (accessed 04/11/2010).

Weber, E. H. 1834. *De Tactu*. Leipzig.

Weber, E. H. 1851. *Der Tastsinn und das Gemeingefühl*. Braunschweig: Vieweg.

Weber, H. 1996. *On the Tactile Senses*, 2nd ed. translated by H. E. Ross and D. J. Murray. Hove, East Sussex: Psychology Press.

Wersény, G. 2009. Auditory representations of a graphical user interface for a better human-computer interaction. In *Auditory Display*, eds. S. Ystad, M. Aramaki, R. Kronland-Martinet, and K. Jensen, 80–102. Berlin: Springer.

Wheat, H. E., and A. W. Goodwin. 2001. Tactile discrimination of edge shape: limits on spatial resolution imposed by parameters of the peripheral neural population. *Journal of Neuroscience* 21: 7751–7763.

Wucherer, K. 2001. HMI, the window to the manufacturing and process industry. *Paper presented at 8th IFAC/IFIP/IFORS/IEA Symposium on Analysis, Design, and Evaluation of Human–Machine Systems.* Kassel, Germany, Sept. 18–20.

Zeng, L., and G. Weber. 2009. Interactive haptic map for the visually impaired. In *Proceedings of the 4th International Haptic and Auditory Interaction Design Workshop 2009*, eds. M. E. Altinsoy, U. Jekosch, and S. Brewster, 16–17. Dresden, Germany: Dresden University of Technology.

Zwicker, E., and H. Fastl. 1999. *Psychoacoustics—Facts and Models*. Berlin, Germany: Springer.

8 Presentation of Information

Thomas S. Tullis, Fiona J. Tranquada, and Marisa J. Siegel*

CONTENTS

---

* This chapter is partly based on the version in the previous edition by Thomas S. Tullis, Michael Catani, Ann Chadwick-Dias, and Carrie Cianchette.

8.1   INTRODUCTION

The Web has revolutionized how people access information. Instead of picking up a telephone directory to look up a phone number, for example, many people prefer to simply do a quick lookup on the Web. Most students now turn to the Web for their research instead of their dictionaries or encyclopedias. Given that the Web has become such an important source of information for so many people, the significance of presenting that information in a way that people can quickly and easily use it should be obvious.

In this chapter, we present some of the key human factors issues surrounding the presentation of information on the Web. We have divided the topic into the following sections:

- Page layout (e.g., how users scan Web pages, page length and scrolling, depth versus breadth, fixed versus fluid layout)
- Navigation (e.g., presenting navigation options)
- Links (e.g., text versus image links, link affordance, link treatment, link anchors or terms, visited links)
- The browser window (e.g., use of frames, pop-ups, modal layers)
- Text and fonts (e.g., line length, font type and size, image polarity, color contrast)
- Graphics and multimedia (e.g., images, video, Flash, accessibility issues)
- Tables and graphs (e.g., "zebra striping" of tables, types of data graphs)
- Color (e.g., color coding, color schemes and aesthetics, color vision deficiencies)
- Forms and form controls (e.g., text entry, selection, feedback)
- Types of pages (e.g., Home pages, help pages, search results, product pages)

For each topic, our primary focus is on empirical human factors studies that have been conducted to address the issue. In cases where, to our knowledge, no relevant empirical studies exist, we have tried to summarize some of the common practices or recommendations. We hope that this discussion will help stimulate applied research into some of these human factors issues related to the Web.

8.2 PAGE LAYOUT

The layout of information on a computer screen clearly has a significant impact on its usability. This has been shown many times in the pre-Web world of displays (e.g., Tullis 1997), and there is every reason to believe it is just as important on the Web. With regard to page layout for the Web, some of the issues include how users scan Web pages, how much information to put on individual pages, adopting a fixed versus fluid approach to page layout, the use of white space, and what to put where on a Web page.

8.2.1 How Users Scan Web Pages

The way that a user views a page is affected by a myriad of factors, both internal and external. Yet, there is evidence that users have expectations about the location of information, as well as general patterns in their viewing habits. By being aware of this evidence, Web site designers may be better able to arrange their pages in a way that is both pleasing and efficient for users.

8.2.1.1 Expectations about Location of Information

There are several studies that provide insight into users' design expectations. Shaikh and Lenz (2006) showed participants a Web page with a grid overlaid on it. Participants were asked to identify the regions of the grid in which they would expect a variety of site features to be located. Forty-four percent of participants thought the "back to Home link" would be located in the top left corner of the page, while 56% thought internal links belonged on the left side of the page. The site search was expected to be in the top right corner by 17% of participants. The "About us" link was expected to be in the footer by 31% of participants, and with the internal links on the left side of the page by 9%. Finally, 35% of participants expected ads to be located in the top center of the page, and 32% expected them to be located on the right side of the page. The authors of this study noted that, over time, these expectations may evolve; therefore, it is necessary to be cognizant of contemporary design standards when applying these findings.

In an experiment that looked at the layout of Web pages, Tullis (1998) studied five different "Greeked" versions of candidate designs for a Home page. "Greeking" is a common technique in advertising, where the potential "copy," or text, for a new ad being developed is represented by nonwords so that the viewer will focus on the overall design rather than getting caught up in the actual details of the text. (Ironically, Latin is commonly used for this purpose.) In presenting these Greeked pages to participants in a study, Tullis asked them to try to identify what elements on the page represented each of a variety of standard elements (e.g., page title, navigation elements, "what's new information," last updated date, etc.). Participants also gave several subjective ratings to each Greeked page. He found that the average percentage of page elements that participants were able to identify across pages ranged from a low of 43% to a high of 67%. There was one page layout on which at least some participants correctly identified all of the page elements (Figure 8.1). As is often the case in behavioral studies, the design that yielded the highest accuracy in identifying the page elements (Figure 8.1) was

FIGURE 8.1 "Greeked" Web page studied by Tullis (1998) that yielded the highest accuracy in identification of page elements. (From Tullis, T. S. 1998. A method for evaluating Web page design concepts. Paper presented at the CHI 1998 Conference Summary on Human Factors in Computing Systems, Los Angeles, CA, April 18–23, 1998. With permission.)

FIGURE 8.2 "Greeked" Web page studied by Tullis (1998) that yielded the highest subjective ratings. (From Tullis, T. S. 1998. A method for evaluating Web page design concepts. Paper presented at the CHI 1998 Conference Summary on Human Factors in Computing Systems, Los Angeles, CA, April 18–23, 1998. With permission.)

not the design that the participants gave the highest subjective ratings (Figure 8.2). However, these results suggest that a well-designed page layout, even without content, may allow users to intuit the location of page elements.

It is important for designers to consider the way that users naturally look at pages. There is some evidence that users may follow habitually preferred scan paths when viewing Web pages (Josephson and Holmes 2002). For example, Buscher, Cutrell, and Morris (2009) identified what they called an "orientation phase" at the beginning of viewing a page during which users scanned the top left area first, regardless of the task assigned. They also found that the most important features were expected to be in this area of the page, both when users were browsing and when they were completing tasks. This was evidenced by a high median fixation duration, high fixation count, and low time to first fixation in this area.

There is also evidence that some areas of a Web page are typically overlooked by users. During the first second of viewing a page, there is almost no fixation on the right third of the page (Buscher, Cutrell, and Morris 2009). Fixations decrease as users read farther down the page (Granka, Joachims, and Gay 2004; Nielsen 2006; Shrestha et al. 2007) and, overall, there is more fixation above the fold of a page than below (Shrestha and Owens 2009). Goldberg et al. (2002) found that, during search tasks on Web portals, header bars were not viewed. As a result, they recommended that navigation features be placed on the left side of pages, which is consistent with the finding of Shaikh and Lenz (2006) that users expect internal links to be located on the left.

One must also consider the location of information within a Web site. Ozok and Salvendy (2000) studied whether the consistency of the layout of Web pages in a site actually makes a difference in terms of the usability of the site. Using selected pages of the Purdue University Web site, they manipulated three types of consistency across the pages of the site: physical, communicational, and conceptual. They found that these types of consistency did have a significant impact on

participants' error rates but not speed or satisfaction. One could interpret these results as indicating that users are accustomed to dealing with inconsistency across the Web in general but expect consistency across pages within a Web site.

8.2.1.2 Eye-Tracking Evidence

Eye tracking is a valuable tool in understanding the way that users view Web pages. Eye trackers measure fixations, or gazes, lasting 300 ms (Rayner et al. 2003). This is the threshold for visual perception, when a user consciously comprehends what he or she is seeing.

A common fixation pattern found on Web pages is known as the F-pattern, shown in Figure 8.3. This pattern is named for its shape—one long vertical fixation on the left side of a page, a long horizontal fixation at the top, and then a shorter horizontal fixation below. It should be noted that this pattern, first identified by Nielsen (2006), can vary a bit and is a "rough, general shape" that you may need to "squint" to see. Studies indicate that this pattern is upheld on several different page layouts, including one thin column, two columns, a full page of text, text with a picture on the left, and text with a picture on the right (Shrestha and Owens 2009). However, this pattern did not appear on image based pages when users were browsing or searching (Shrestha et al. 2007).

The visual complexity and hierarchy of a page is an important factor affecting how users scan a page. Pan et al. (2004) found that a more structurally or visually complex page, containing more elements, leads to more variability in users' scan paths. This may be mitigated by creating a visual hierarchy. A visual hierarchy is formed through the perceptual elements of a Web page, such as text and image (Faraday 2000). Through the proper arrangement of these elements, a designer can provide a path for users to scan. For example, larger elements are considered higher in the hierarchy than smaller elements, and images and graphics tend to be processed before text (Faraday 2000).

Several studies have examined the impact of images on visual patterns. Beymer, Russell, and Orton (2007) found that when images were placed next to text, they influenced the duration and location of fixations during reading. Furthermore, they found that the type of image (i.e., whether

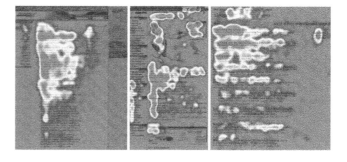

FIGURE 8.3   (See color insert.) Examples of heat maps showing the F-pattern; red areas received the most fixation. (From Nielsen, J. 2006. F-shaped pattern for reading Web content. Alertbox. http://www.useit.com/alertbox/reading_pattern.html (accessed November 6, 2010). With permission.)

it was an ad or related to the adjacent text) influenced the placement of fixations. Similarly, Tullis, Siegel, and Sun (2009) found that an image of a face led to significantly lower accuracy on tasks related to the area of the page in which the face was located. Additionally, the task completion time was longer when faces were there, and participants rated both the task ease and the ease of finding information significantly less positively. Interestingly, eye tracking revealed that the image of the face received almost no fixation. However, in that study, the image of the face was unrelated to the information adjacent to it. In another study, Tullis et al. (2009) found that placing images of authors next to their articles yielded similar results. Participants were significantly less accurate when completing tasks related to that section of the page, and tended to take longer. They were also less confident in their answers and, interestingly, trusted the accuracy of the information less. Similarly, Djamasbi et al. (2007) found that users fixated less and were less likely to click on a bricklet when the background was dark or the bricklet contained an image. Combined, these findings suggest that images, and possibly other embellishments, may actually lead to decreased fixation on areas, perhaps because they feel "ad-like."

The context in which users visit a Web page has a large impact on their viewing patterns as well. Shrestha and Owens (2008) found that page elements, such as ads, received higher levels of fixation when users were browsing than when they were searching. Another study (Shrestha et al., 2007) observed that fixations were uniformly distributed both above and below the fold during tasks but were focused above the fold during browsing.

Demographic differences may also play a role in users' viewing behavior. Pan et al. (2004) noted that the mean fixation duration of males is significantly longer than that of females. Another study found that females are comprehensive in their processing of information, looking at all areas of a page, while males tend to focus on only a few areas on a page (Meyers-Levy and Maheswaran 1991). Similarly, women are more thorough during page recognition, or browsing, tasks than males, fixating significantly longer on the page and every region on it (Buscher, Cutrell, and Morris 2009). The same study found the same pattern with age: users over 30 years of age had significantly longer fixation time on the page and every region on it (Buscher, Cutrell, and Morris 2009). This information is particularly useful for designing for specific demographic groups; for example, young males may fixate on fewer elements than older or female groups, which suggests that a page with fewer focal points may be more effective in communicating with this demographic.

8.2.1.3 Key Takeaways

- Some users tend not to scroll, so important information should be placed above the fold of a page. This will be discussed in more detail in the next section.
- Some embellishments, such as images or background colors, on areas of a page may actually decrease the amount of fixation that is given to those areas.

- The way that a site is viewed is affected both by the users' goals and demographics.

8.2.2 Page Length and Scrolling

As designers build out their pages, an important consideration is how much content to have on each individual page. For longer chunks of content, they must decide between breaking it out in shorter pages that require the user to click links to get to the next section of content, or one long page that requires the user to scroll. The advantages and disadvantages of each approach will be discussed in this section, but designers should always take into consideration the nature of the content and any specific user requirements for their site.

8.2.2.1 Scrolling versus Paging

As content has increased on the Web, users have become more willing and likely to scroll. Users generally expect some amount of vertical scrolling on most pages (Nielsen 2005). Even so, displaying all content on a single page is not always the best solution. As the eye-tracking research referenced earlier points out, the context of a user's task determines how they will view the page.

Several research studies have examined the differences in usability between presenting information in a long page that requires scrolling and presenting information in several pages that require clicking a link to proceed page by page. The research has yielded conflicting results. Dyson and Kipping (1998) found that participants read through paged documents faster than scrolled documents but showed no differences in reading comprehension. In their study, users spent about 13% of their online time scrolling within pages.

Baker (2003) examined reading time and comprehension using the three following conditions:

1. Paging: One passage displayed on four separate pages.
2. Full: One passage presented over two separate pages with no scrolling
3. Scrolling: One passage displayed on one page that required scrolling.

Baker found a significant difference in reading speed among the three groups. Contrary to Dyson and Kipping (1998), Baker found that reading time in the Paging condition was significantly slower than the Full or Scrolling conditions. Participants also showed no significant differences in their ability to answer comprehension questions correctly or in subjective responses to each of the reading conditions.

Both of these studies had indicated that comprehension was not affected by the content presentation. Sanchez and Wiley (2009) found something different when they compared the effects of scrolling on understanding the content being conveyed. They found that a scrolling format reduced the understanding of complex topics from Web pages, especially for readers who had lower working memory capacity. On the basis of these findings, they recommend that if comprehension is important, designers should present information in meaningfully paginated form.

Another element that can affect the length of the page, and ultimately how much scrolling is required, is the amount of white space on the page. Bernard, Chaparro, and Thomasson (2000) conducted a study in which participants completed tasks on Web sites with varying levels of white space (high, medium, and low). They found that, while there was no significant difference in performance, measured as completion time and accuracy, there was a significant user preference for the page with a medium level of white space. These findings suggest that designers should not reduce white space on a page in an effort to cut back the amount of scrolling required.

These studies suggest that if reducing reading time is important, designers should display content on just a couple of pages with no scrolling or one long scrollable page. If the topic is especially complex, research suggests that some amount of pagination can aid comprehension, especially for users who have lower working memory capacity.

8.2.2.2 The Fold and Avoiding a False Bottom

The phrase above or below "the fold" originates from newspaper publishing. Anything visible on the upper half of a folded newspaper was referred to as "above the fold." Newspapers put the most important information above the fold because this was the information that people could see without opening or unfolding the paper. Therefore, this was the information that sold the paper. On a Web page, "above the fold" refers to what is visible without scrolling.

Eye-tracking research has shown that the number of fixations decreases as users read farther down the page (Granka, Joachims, and Gay 2004; Nielsen 2006; Shrestha et al. 2007). This reinforces other research in the Web usability field on the importance of placing important information above the fold of a Web page so that it will be visible without vertical scrolling.

But unlike on a newspaper, the fold is not in one exact spot from user to user as there are different monitor and browser sizes. Clicktale (2007) identifies three peaks where the fold usually falls, corresponding to about 430, 600, and 860 pixels. They also found that users scroll to a relative position within the page regardless of the total page length, but that the most valuable Web page real estate was between 0 and 800 pixels, peaking at about 540 pixels.

Knowing this, designers can test their designs in a variety of browser sizes and make sure their designs avoid a false bottom. A false bottom refers to an illusion on a Web page that makes it seem like there is no additional information available below the fold. Spool (1998) identified horizontal rules and rows of text links as "scroll-stoppers," or visual elements that help create a false bottom. When users run into these elements, they do not scroll any further because they think they have reached the bottom of the page. In his later writings, he recommends that sites create a "cut-off" to provide a strong visual cue to users that there is additional content available beneath the fold (Spool 2006b).

Shrestha and Owens (2008) analyzed the fixation patterns for users of single- and double-column Web page layouts.

They found significantly more fixations on the right column of the double-column Web page than on the bottom half of a one-column article. On the basis of this, they suggest that using a double-column layout might also help reduce the problem of false bottoms.

8.2.2.3 Horizontal Scrolling

Horizontal scrolling is often caused when the text or graphics on a page assume a fixed width (in pixels) and cannot be displayed within the horizontal width of the user's browser. The browser window displays a horizontal scroll bar and the user must scroll back and forth to view the information. This makes it difficult for users to keep their place on a line and scan chunks of information. Horizontal scrolling, or the necessity to scroll horizontally to view all the information on a Web page, is strongly disliked by almost all Web users (Nielsen 2005) and should be avoided.

8.2.2.4 Key Takeaways

- For shorter reading time, display content on one long scrollable page or broken out across just two pages with no scrolling.
- For higher comprehension rates, use pagination to help group the content into meaningful chunks.
- Test designs and remove "scroll-stopper" design elements to reduce the likelihood of users encountering a false bottom.
- Avoid horizontal scrolling.

8.2.3 Depth versus Breadth

The trade-off between the amount of information to present on one screen or page in an information system versus the total number and depth of pages has been studied at least back to the 1980s (Kiger 1984; Miller 1981; Snowberry, Parkinson, and Sisson 1983; Tullis 1985). Most of those early studies compared various types of menu hierarchies, from broad hierarchies with many selections per screen and fewer levels, to deeper hierarchies with fewer selections per screen and more levels. They generally found that shallower, broader hierarchies are more effective and easier for users to navigate than deeper hierarchies. More recently, these issues have been revisited in the context of Web pages.

Zaphiris (2000) studied five different designs for a Web site containing information about Cyprus, ranging from two to six levels deep. He found that users could reach their desired target items significantly faster with most of the two-level designs. However, one of the two-level designs, which used a relatively unnatural breakdown of the information on the first page, took longer. This result points out the importance of the relationship between the site structure and the "natural," or perceived, structure of the information itself. Larson and Czerwinski (1998) similarly showed that users found target items faster in a two-level Web site than a three-level site. In calculating a "lostness" measure based on deviations from the optimal path to the target (Smith 1986), they found that the three-level site resulted in greater lostness. They also found that one of the two-level sites resulted in marginally greater lostness than the other, again reinforcing the importance of the relationship between the site structure and the structure of the information itself.

Tsunoda et al. (2001) studied four different Web hierarchies for accessing 81 product pages, ranging from only one level deep to four levels. Unlike many of the previous studies, they also manipulated the complexity of the user's task: simple tasks that did not require any comparisons and complex tasks that did. For the simple tasks, they found no differences in performance for the different hierarchies, although users preferred the four-level hierarchy. But for the complex tasks, users found products significantly faster with fewer levels (one- or two-level hierarchies), and they preferred the one-level hierarchy. Similarly, Miller and Remington (2002) studied two hierarchies (two levels and three levels) for organizing 481 department store items. Users were asked to find two types of target items: unambiguous (e.g., garage door remote) or ambiguous (e.g., bird bath). They found that unambiguous items were found faster in the three-level structure than in the two-level structure. However, ambiguous items were found faster in the two-level structure. The ambiguous items required more backtracking, thus increasing the penalty associated with more levels.

Bernard (2002) created six hierarchies varying in breadth, depth, and "shape" for finding merchandise, ranging from two levels to six levels. His results showed that users found items faster, and with fewer extra steps and "Back"s, when there were fewer levels: two levels was best and six levels was worst. But he also found that the shape of the hierarchy made a difference: hierarchies with fewer options in the middle levels did better than those with more options in those levels.

Galletta et al. (2006) studied the interaction of three different attributes of a Web site: page load delay, site breadth, and content familiarity. They found that users did worse with deeper sites, longer page load delays, and when they were unfamiliar with the site content. But they also found a significant interaction among all three of these variables. Basically, the negative effect of site depth was greater for the cases when the users were unfamiliar with the content and when the pages loaded slower. This outcome points out that other factors can influence the effects of depth and breadth.

Some researchers now focus on the "scent of information" that users encounter at each decision point along the way toward their objective (Spool, Perfetti, and Brittan 2004). Generally, the links at each point in a deeper site will tend to have poorer "scent" because they necessarily will be more general (or vague), but many other factors influence scent as well, including the exact wording of the links and the user's familiarity with the subject matter. Katz and Byrne (2003) manipulated both the scent of links (through their wording) and the depth of the site, primarily looking at how likely participants were to use site search. They found that participants were more likely to turn to site search when the site was deeper and when the links had low scent. There was also

a tendency for the breadth effect to be greater for the high-scent links.

8.2.3.1 Key Takeaways

- In complex or ambiguous situations, breadth still wins over depth, at least partly because it facilitates comparison.
- In very simple and clear situations, fewer choices per page win. Users are able to make choices quicker among fewer selections.
- Fewer choices in "middle" levels than in "top" or "bottom" levels of the hierarchy may be better.
- Other factors can mediate the effects of depth or breadth, especially the scent of the links.

8.2.4 Fixed versus Fluid Layout

One of the often-debated issues in page design for the Web is whether to use a fixed layout (which basically does not change with the size of the browser window) or a fluid layout (which adapts itself to the size of the browser window; also called variable-width or liquid layout). From a technical standpoint, this typically means using fixed pixel widths for tables and other page elements versus percentage widths for those elements. Figure 8.4 illustrates a fixed design of a page, while Figure 8.5 illustrates a fluid design. Both screenshots are using the same size browser window. Note the blank space on either side of the main content in the fixed-width design.

Bernard and Larsen (2001) studied three different approaches to the layout of multicolumn Web pages: fluid, fixed-centered, and fixed-left-justified. They also used two different window sizes: large (1006 pixels wide) and small (770 pixels wide). They found no significant differences in terms of the accuracy or speed with which users found the answers to questions. However, the fluid layout got significantly higher subjective ratings than either of the other two. Overall, 65% of the participants selected the fluid layout as their top choice. This is consistent with the recommendation

FIGURE 8.4 **(See color insert.)** Example of fixed-width design.

FIGURE 8.5 **(See color insert.)** Example of variable-width, or fluid, design.

of Nielsen and Tahir (2001, 23) to use a fluid layout because it adjusts to the user's screen resolution.

One problem with fixed-width designs is that they force the designer to choose a single resolution for which they optimize their design. But recent statistics (w3Counter.com, February 2010) show that Web users are running their systems in a wide variety of resolutions, as shown in Table 8.1. Many Web designers now adopt 1024 × 768 as that target resolution, so that only the 3.3% running in 800 × 600 will require horizontal scrolling, while the 56% running in a higher resolution are potentially presented with significant amounts of unused space.

Another major advantage of a fluid layout is that it automatically adjusts itself to the printed page when printing. A fixed-width design optimized for 800 × 600 resolution is too wide to print in portrait mode on most printers without cutting off content on the right.

8.3 NAVIGATION

Most Web designers would probably agree that the presentation of navigation options is of crucial importance to the

---

TABLE 8.1
Ten Most Popular Screen Resolutions as of February 2010

Resolution	Percent
1024 × 768	26.4%
1280 × 800	20.2%
1280 × 1024	10.9%
1440 × 900	8.6%
1680 × 1050	5.6%
1366 × 768	4.8%
800 × 600	3.3%
1152 × 864	2.3%
1920 × 1200	2.0%
1920 × 1080	1.7%

Source: From http://www.w3Counter.com. With permission.

---

usability of the site. However, very few of those designers would agree on the best way to present those navigation options. A myriad of techniques exist for presenting navigation options on a Web page, including static lists of links, expanding and contracting outlines, tab folders, pull-down menus, cascading menus, image maps, and many others. Given the importance of this topic, surprisingly few empirical human factors studies have been done to compare the effectiveness of the different techniques.

Zaphiris, Shneiderman, and Norman (1999) compared two different ways of presenting navigation options on Web pages: traditional lists of links on sequential pages versus an expanding outline style. In the expanding outline, after a link was selected, the subselections would appear indented under the link. In the traditional sequential approach, the subselections were presented on a new page, replacing the original list. They found that the expanding outline took longer to navigate and yielded more errors than the traditional approach, and the effect got worse with more levels.

Bernard and Hamblin (2003) studied three different approaches to presenting navigation options for a hypothetical online electronics store: index layout, in which all of the links were arrayed in a tabular form in the main part of the page; horizontal layout, in which menu headings and associated pull-down menus were arrayed across the top of the page; and vertical layout, in which the menu headings and associated fly-out menus were arrayed down the left side of the page. They found that users reached their targets significantly faster with the Index menus than with either of the other two menus. The index layout was also most often selected by the users as the most preferred of the three layouts.

Tullis and Cianchette (2003) studied four different approaches to presenting the navigation options for an online Web design guide. The four navigation approaches, illustrated in Figure 8.6, were as follows: table of contents (TOC), in which all of the menu items were listed down the left in a two-level scrolling list; vertical drop-downs, in which menu

FIGURE 8.6 Menu approaches studied by Tullis and Cianchette (2003): Table of contents (TOC), vertical drop-downs, horizontal drop-downs, and top and left. (From Tullis, T. S., and C. Cianchette. 2003. An Investigation of Web site navigation techniques. Usable Bits, 2nd Quarter 2003, http://hid.fidelity.com/q22003/navigation .htm, accessed Nov. 28, 2009. With permission.)

headers were arrayed across the top and associated vertical menus dropped down for each on mouse-over; Horizontal Drop-downs, in which menu headers were arrayed across the top and associated horizontal menus appeared under each on mouse click; and top and left, where tab folders were presented across the top for the main sections and when one of those was selected, the associated menu items were listed down the left.

They found that users were able to find the answers to questions in the design guide significantly faster with the TOC approach than with any of the others. The questions for which users were being asked to find the answers were relatively difficult, and users had to explore the site quite a bit to answer them. Consequently, the TOC approach, in which menu items for all of the topics in the guide appeared on the page, facilitated this kind of exploration. The authors point out, however, that this may only hold true for a relatively small site such as this one (27 pages total), in which it was practical to provide a full table of contents.

Through various sessions of user testing, Nielsen (2009) has found that a relatively new method of displaying navigation, "mega drop-down" navigation, may surpass the traditional method of dropdown navigation. Mega drop-downs have large panels with structured sections of navigation, so that users can see all options at once. Nielsen suggests that mega drop-downs are preferable to traditional drop-downs because they show all options without requiring scrolling, allow grouping, and have enough real estate to allow the use of graphics as needed.

Closely related to the issue of menus is the use of breadcrumbs. A breadcrumb is a textual representation of where a user is within a Web site, with links to content in sequential order of access. A carrot (">") typically indicates the hierarchy between the items in the breadcrumb. For example, the breadcrumb associated with a leather chair on the Office Max Web site might be: Home > Furniture > Chairs > Leather Chairs (Lida, Hull, and Pilcher 2003). Breadcrumbs are generally useful for orienting users to their location within a site and providing shortcuts through the hierarchy. In one study of breadcrumbs, Lida, Hull, and Pilcher had users complete search tasks on the Google Web site and e-commerce tasks on the Office Max Web site. Fifty-two percent of participants were identified as "breadcrumb users" and 48% as "nonbreadcrumb users," but there were no significant differences between these two groups in terms of efficiency, as measured by total clicks, Back button clicks, searches, and time. Another study (Maldonado and Resnick 2002) had participants complete tasks on an e-commerce Web site, with or without breadcrumbs. They found that the inclusion of breadcrumbs had a moderately positive effect, with a 12% decrease in the number of clicks, 14% decrease in task time, and 28% decrease in errors.

8.3.1 Key Takeaways

- Being able to see all options at once, without scrolling, facilitates Web site use.

- Stable methods of presenting navigation, such as a traditional site index and mega drop-downs, are preferred to dynamic methods, such as cascading menus, pull down menus, or collapsing/expanding lists, because they consistently present all options.
- Breadcrumbs may be useful for Web sites with clear and multileveled hierarchies.

8.4 LINKS

Although most people assume that the concept of hypertext is a relatively new one, it first emerged conceptually in the 1940s, when a U.S. science advisor named Vannevar Bush (1945) proposed a machine that could produce links between documents. In 1965, Ted Nelson coined the actual terms, hypertext and hypermedia, and proposed a worldwide hypertext system called "Xanadu" where individuals could contribute collective content (Moschovitis et al. 1999). Since then hypertext and links have become the primary navigation medium for the Web—it is how users navigate between pages. Many factors including appearance, placement, number, and type of links all influence how effective they are for users.

8.4.1 Text Links versus Image Links

A main distinction between links is whether they are textual or image-based. Research to date supports that text links are preferred by users. They download faster and provide the ability to distinguish between visited and unvisited links (Spool et al. 1998).

There are instances when using graphical or image-based links might be more effective than using textual links. For example, older adults often have difficulty clicking smaller targets, such as text links. Image-based links often provide a larger target area for users to click (Bohan and Scarlett 2003).

8.4.2 Link Placement

The placement of links on a Web page can directly affect whether users will see or click them. Research by Bailey, Koyani, and Nall (2000) demonstrated that for pages that require scrolling, users spend significantly more time scanning information at the top of the page and significantly less time on any information that fell "below the fold" or at the bottom of the page (requiring scrolling). Their research suggested that users spend about 80% of time spent scanning information that was on the top of the page (above the fold) and the remaining 20% of time scanning information on the rest of the page (below the fold). Therefore, it is critical to place the most important links higher on the Web page, above the fold.

8.4.3 Link Affordance and Visibility

Link affordance refers to the relative visibility and prominence of links on a Web page. Studies of link affordance typically have users look at static images of Web pages, often on paper, and ask them to identify each element they think is a link. The more evident that something is a link, the more quickly users will see and click it.

Research has demonstrated that when users are given a clear visual indicator that a Web page element (text, image, etc.) is a link, they find information faster (Lynch and Horton 2009). For textual links, the traditional visual treatment is blue, underlined text. For image-based links, Bailey (2000) provided the following guidelines based on his link-affordance studies:

- Do use meaningful words inside graphical links:
  - Target locations (Home, Back to Top, Next)
  - Common actions (Go, Login, Submit, Register)
- Do use graphical tabs that look like real-world tabs.
- Do use graphical buttons that look like real-world pushbuttons.
- Do use clear, descriptive labels inside tabs and pushbuttons.
- Do put clickable graphics close to descriptive, blue underlined text.
- Do use a frame (border) around certain graphical links.
- Do make all company logos clickable (to the Home page).
- Do not require users to do "minesweeping" to find links.
- Do not use stand-alone graphics that are not close to, or do not contain, text as links.

The primary goal is to design links that are clearly evident to users so that they do not have to move their mouse around the page (called minesweeping) to find where the links are located. Usability is improved by increasing link affordance.

8.4.4 Link Treatment

The past few years have seen a divergence from traditional blue underlined links and a move toward greater variety in the visual treatment of links, including not underlining them. In fact, an analysis of the Top 40 e-retail Web sites ranked by Foresee Results showed that only 32% primarily used underlined links while 68% primarily used links without underlining (Tullis and Siegel 2010). When the move away from underlining links began is not clear, but the use of non-underlined links has become almost synonymous with the "Web 2.0" style.

Most Web usability guidelines still recommend using color and underlining to indicate text links (Nielsen 2004b); others state that underlining may not be required in all cases but is highly recommended (Spool 2006a). To further investigate the issue, Tullis and Siegel (2010) studied underlined and nonunderlined blue links in the context of three different kinds of Web pages. On hover, underlined links became red and kept their underlining, while nonunderlined links became underlined and stayed blue. Participants in an online study were presented with two different kinds of tasks: navigation

tasks where the answer was found on a page linked to and tasks where the answer was contained within the text of a link itself. They found that for one particularly challenging navigation task, the participants were significantly more successful in completing the task when the links *were* underlined. For most other individual tasks there was no significant difference between the link treatments. But when the tasks involving answers within the links themselves were aggregated, they found that participants were significantly more efficient in finding the correct answers when the links *were not* underlined. These conflicting findings suggest that it may be helpful to underline links when their primary purpose is for navigation and not to underline them when their primary purpose is to convey data, with a secondary navigational role.

This finding that underlining links may have a detrimental effect in certain situations is supported by the work of Obendorf and Weinreich (2003), who found that underlined links, on pages where the answer being sought was sometimes in a link, yielded significantly fewer correct answers in comparison to the nonunderlined case. The readability of the underlined links was also rated as significantly worse.

8.4.5 Link Anchors or Terms

Hypertext links are anchored in text that the user clicks to navigate to the intended destination. It is important to make these anchors (or terms) as clear and concise as possible so that users understand, before clicking the link, where the link will take them. Research by Nielsen (2009) supports the concept that users should be able to predict what a link is going to do from the first 11 characters of the link, or about two words. If they are too long, they increase scanning time; if they are too short, they do not provide enough information to tell users where the link will take them.

Spool, Perfetti, and Brittan (2004) came to a slightly different conclusion from their analyses of clickstreams where users failed (did not find what they were looking for) versus clickstreams where users were successful. They found that the average success rate for all links was only 43%, but the links containing 7–12 words tended to be the most successful, with a likelihood of 50–60% that the clickstream will end successfully. Both shorter and longer links yielded lower success rates. Consistent with this finding that somewhat longer links are more effective are the results of Harper et al. (2004), who found that longer links were preferred by users in comparison to the case where the same text was present but with less of it as the actual link.

8.4.6 Link Titles

Links titles are small pop-up boxes that display a brief description of a link when you mouse over it. These titles provide additional information about the link destination and help users predict what will happen when they click the link. Nielsen (1998) provides a complete list of recommendations for creating effective link titles. Included in his recommendations are that link titles should include the name of the site

Method	Example	Accuracy
No Space	International Usability Guidelines in Design Accessibility for Special User Groups Human Factors	67%
Space	International Usability Guidelines in Design Accessibility for Special User Groups Human Factors	89%
Bullets	• International Usability • Guidelines in Design • Accessibility for Special User Groups • Human Factors	100%

FIGURE 8.7 Link-wrapping conditions studied by Spain (1999). (From Spain, K. 1999. What's the best way to wrap links? *Usability News* 1(1). With permission.)

to which the link will lead, that they should be less than 80 characters but should rarely go above 60 characters and that the link anchor and surrounding text should contain other descriptive information (not included in the link title) that helps users understand where the link will take them. Harper et al. (2004) found that users preferred links containing "preview" information in the title attribute, which was automatically generated from an analysis of the target page.

8.4.7 Wrapping Links

When presenting links that wrap on to a second line, it is important to carefully control how they wrap. The main usability problem is related to the fact that when you wrap links without clearly distinguishing between them, it is difficult for users to know which link terms belong together. Spain (1999) studied three different ways of presenting lists of links that wrap, as shown in Figure 8.7. Accuracy rates for the three conditions were 100% for bullets, 89% for spaces, and 67% for no spaces. All participants preferred either the bullets or spaces; no one preferred the no-space condition.

8.4.8 Visited Links

Distinguishing between visited links (that the user has already accessed) and unvisited links (that the user has not yet accessed) is widely believed to significantly improve usability (Nielsen 2004a). The main advantage appears to be that it allows a user who is searching for a piece of information in a site to readily identify those areas already checked. Further, the browser default colors (blue for active links, purple for visited links) appear to be the most recognizable to users.

8.5 THE BROWSER WINDOW AND BEYOND

Site designers can configure certain aspects of how their site interacts with and builds upon a browser window. They can decide when content should be opened within the same window, in just one part of a browser window, in a secondary

window, or in a layer that appears within the main browser window over its content. They can also decide the size of a browser window, whether or not it can be resized by the user, whether or not scrollbars or toolbars are included, and whether or not navigation controls are visible.

With all these options available, site designers need to consider how their users will be using the additional information provided outside of the main browser window.

8.5.1 Frames

Frames are an HTML construct for dividing the browser window into several areas that can scroll and otherwise act independently. Frames can display different Web pages; therefore, the content of one area can change without changing the entire window. This section covers the advantages and disadvantages of using frames.

8.5.1.1 Advantages of Frames

There seem to be at least two valid reasons for using frames in a site. One is that frames can provide separate areas for fixed content such as navigation. Such sites typically place a narrow frame along one side or along the top that contains a table of contents or other navigation area and a larger frame where the main content is displayed. This can mean that navigation is ever-present on a site even if the body area is scrolled. The second reason is that frames can be used as a mechanism for associating material from a specific author (such as comments) with other pages that are normally standalone (Bricklin 1998). Nielsen (1996) refers to this as using frames for metapages.

One common concern about frames is that they will make a site less usable. Spool et al. (1998) conducted an independent usability study of the Walt Disney Company's Web site. Partway through the study the site design changed from a version that did not use frames to one that did. On the framed version, a narrow pane on the right side featured the table of contents so it was always visible. They found that users performed significantly better with the framed version of the site than they did with the nonframed version. While they could not wholly attribute the improvement to the use of frames, they would claim with certainty that frames did not hurt the site (Spool et al. 1998).

Other research supports the use of frames for a table of contents or navigation. Bernard and Hull (2002) examined user performance using links within framed versus nonframed pages. They compared a vertical inline frame layout (i.e., frames dedicated to displaying the main navigational links within a site and which are subordinate to the main page) to a nonframed layout. Their study revealed that the framed version was preferred over the nonframed version. Interestingly, participants also suggested that the framed condition promoted comprehension.

Using frames for navigation can also make it faster for users to find information. Tullis and Cianchette (2003) studied four different navigation mechanisms (e.g., drop-down menus, a left-hand table of contents) for an online Web design guide. For each navigation mechanism, they studied both a framed version and a nonframed version. In the framed version, the navigation options were always available regardless of the scrolling state in the main window. In the nonframed version, the navigation options scrolled with the page and consequently could scroll out of view. Although the effect was not significant, users found their answers quicker with the framed version in three of the four navigation approaches studied.

A study by van Schaik and Ling (2001) investigated the impact of location of frames in a Web site. They studied the effect of frame layout and differential frame background contrast on visual search performance. They found that frame layout had an effect on both accuracy and speed of visual search. On the basis of their study, van Schaik and Ling recommend placing navigation frames at either the top or the left of the screen.

Many of these advantages also apply to using fixed positioning features within a site's cascading style sheets (CSS). Fixed positioning can imitate the behavior of frames such as maintaining a fixed header and footer on a page, while avoiding the bookmarking and Back button issues associated with frames (Johansson 2004).

8.5.1.2 Disadvantages of Frames

One disadvantage of frames is that they frequently "break" URLs, or Web addresses. URLs represent the basic addressing scheme by which Web pages are identified, giving each Web page a uniquely identifiable name. When a site uses frames, the URL appearing in the address bar may or may not refer to the page the user is viewing—it usually points to the page that set up the frames. Lynch and Horton (2009) warns that frames can confuse readers who try to bookmark a page because the URL will often refer to the page that calls the frames, not the page containing the information they thought they had bookmarked. Printing can also be problematic with frames, although this tends to vary between browsers. Most browsers require that you activate a frame by clicking in it (or tabbing to it) before you can print it (Johansson 2004). Just clicking print does not guarantee that the user will print what they are expecting to have printed.

Search engines can also break for the same reason. Some search engine spiders are unable to deal with frames appropriately. Some of them summarize Web pages that use frames with the following message, "Sorry! You need a frames-browser to view this site" (Sullivan 2000).

Another problem with frames is that users navigating the page using assistive technologies do not have the benefit of visual cues about the screen layout. Be sure to use the appropriate FRAMESET tags to explicate the relationship of frames (e.g., to indicate that one frame contains navigation links) and providing a NOFRAMES alternative (Lynch and Horton 2009).

8.5.2 Secondary Windows and Pop-Ups

Secondary windows can refer to additional windows that appear of their own accord (pop-ups) and information

requested by a user that opens in a secondary window. It is quite clear that users are annoyed by pop-up windows appearing when they have not been requested; the issue even made it to Nielsen's (2007) "Top 10" list of Web design mistakes for 2007. The number of "pop-up-stopping" software applications now available and even built in to Web browsers such as Firefox is also a testament to what users think of them. However, there do seem to be some appropriate uses of secondary windows for displaying information the user has requested.

Secondary windows are often used for presenting online help or additional, detailed, information. Ellison (2001, 2003) summarized the results of two studies of online help. The aim of the first study was to compare the ease of navigation of different Help interfaces. The second study again examined different methods of displaying help content and whether or not secondary windows could assist a user to navigate between topics. On the basis of these studies, Ellison suggests the use of secondary windows for linking to subprocedures or additional layers of detail, as long as the main window remains visible when the secondary window appears.

Ellison's (2001) study also revealed a problem with "Breaking the Back button." When users navigated from a topic in the main window to a procedure topic in the secondary window, they were then unable to use the Back button to return to the previous topic. Instead, the Back button returned them to the last procedure topic displayed within the secondary window, which caused frustration and likely undermined the students' confidence in the Help system (Ellison 2001). Storey et al. (2002) saw similar navigation difficulty in their comparison of two Web-based learning tools. One of the tools used secondary windows containing their own set of navigation buttons, and the students using this tool experienced many difficulties navigating between the secondary windows and the main window, such as simply getting back to the main window. A key takeaway for designers is to ensure that the secondary window is smaller than the main window so users recognize where they are in the site.

Ellison's second study examined the effectiveness of using secondary windows for displaying a subprocedure. Participants were asked to work through a procedure topic from beginning to end, and partway through the procedure, participants linked to a subprocedure. The subprocedure was contained either in a secondary window next to the main window or overwriting the content in the main window. They found that the users who viewed the subprocedure in the secondary window were better able to successfully resume the main procedure, and had lower task completion times, than the group who viewed the sub-procedure in the same window as the main procedure (Ellison 2003). Ellison's team surmised that because the main procedure remained visible on screen, users never lost track of where they were. They simply closed the secondary window containing the sub-procedure and returned to what they were doing in the main procedure.

While this research suggests that links opening secondary browser windows can be effective for displaying certain types of supplemental information, these links should not use JavaScript to force a pop-up window unless coded so a user choosing to open the link in a new window will still be successful. A recent study on Web usage showed a decrease in usage of the Back button but an increase in the number of pages that users open in either a new window or a separate tab (Obendorf et al. 2007). There are approaches to coding links using HTML and JavaScript that will work whether a user tries to open a link in a new window or just clicks it. Besides giving the user more control over their experience on the site, these approaches also make links more accessible to users of assistive technologies (Chassot 2004).

8.5.3 Layered Windows

Another approach to providing additional information is with a pop-up layer or overlay that opens over the main page, rather than as a separate window. This pop-up layer usually appears after the user has clicked on a link or hovered over a specific area of the screen. Layers that appear on hover usually close after the user has moved their mouse away from the area, whereas layers that appear on click require a user action (e.g., clicking a Close link) to close. An advantage to using layers is the elimination of windows management issues that arise with separate windows. But, for layers that appear on hover, there needs to be some type of delay between when the user's mouse hits the trigger area and when the pop-up appears, especially on a page where a lot of different areas are potential triggers. Netflix uses a 250-ms delay on its DVD browsing pages, which is long enough that it does not immediately trigger when the mouse is moved over an area, but not so long that the user never realizes the feature is there (Spool 2008a).

For layers that appear on click, a common treatment is to display them as modal. Modal layers require the user to take some kind of action in that part of the window before being allowed to interact with the main content "behind" it. These modals are often used for dialog boxes where the user provides some type of input without needing to access information on the page behind the window. One variation of a modal is a lightbox, which dims out the rest of the screen to help draw the user's attention to the dialog box or other modal content in the middle of the screen (Hudson and Viswanadha 2009; Nielsen 2008).

While there has not been a lot of empirical research focused on these techniques, several best practices have emerged based on usability testing and observation. Layered windows should not be used when the user might need to refer to information in the main content window, as that content may be hidden by the layer or, in the case of a lightbox, made difficult to read by the dimmed screen (Nielsen 2008). Variations that allow the layered window to be moved around can help address this, but using a true pop-up window would give the user more flexibility.

Also, similar to considerations with pop-up windows, the size of the layered window should be carefully chosen to accommodate the user, as the user will have no way to resize

it. If the modal overlay is so large that the main content page is obscured, users may click the Back button to try to get to the screen they were on before. Because these overlays are not new windows, the Back button will not work the way users expect.

8.5.4 Key Takeaways

- Frames and other methods of fixed content are most appropriate for providing a static location for navigation links.
- Secondary windows are appropriate for subprocedures or help content when the user is in the middle of a flow or process.
- A layered window such as a modal overlay or lightbox may be appropriate for focusing the user's attention on a specific subtask or piece of help content, but should be sized appropriately so the main content of the page is not completely obscured.
- If using secondary or pop-up windows, make sure the second window is a smaller size than the main window so users realize there are two windows.
- Be aware that users will not be able to correctly bookmark content that is in frames or layers.

8.6 TEXT AND FONTS

Because almost all Web pages include some type of text, the importance of understanding how to present that text effectively should be obvious. Consequently, this is one of the few Web design issues that has been studied rather extensively. The classic research in this area was done with printed materials, but many of the findings from those studies probably apply to the Web as well. The following sections summarize some of the key human factors evidence related to text presentation on Web pages, but first we will provide definitions of some terms unique to this topic:

- *Legibility* is generally considered to be a function of how readily individual letters can be recognized, although some researchers consider the focus to be at the word level. It is influenced by detailed characteristics of the individual letterforms. Legibility is commonly measured by very briefly displaying individual letters and measuring how often they are confused with each other (Mueller 2005).
- *Readability* is generally considered to be a function of how readily a body of text (e.g., sentence, paragraph, etc.) can be read and understood. It is influenced by legibility but also by higher-level characteristics such as line spacing, margins, justification, and others.
- *Serif/Sans Serif*: Serifs are the detailed finishing strokes at the ends of letters; sans serif fonts do not have these finishing strokes. Popular fonts with

serifs include Times Roman and Georgia. Popular sans serif fonts include Arial and Verdana.

- *Point size* is a way of describing the size of individual letters. It originally referred to the height of the metal block on which an individual letter was cast, so it is typically larger than the actual height of just the letter. There are 72 points to an inch.
- *Ascenders and descenders* refer to the portions of some lower-case letters that extend above or below the x-height. For example "y" has a descender while "t" has an ascender.
- *Leading* (pronounced "ledding"; rhymes with "heading") refers to the vertical distance between adjacent lines of text. If the bottoms of the descenders on one line almost touch the tops of the ascenders on the line below, there is no leading and the text is said to be set solid. The term originates from the use of strips of lead that printers would add between lines of text.
- *Tracking* refers to the amount of space between letters. It is sometimes adjusted to create even right margins for a block of text (justification).
- *Kerning* is closely related to tracking but is a more detailed adjustment of the spacing between individual pairs of letters that takes into consideration the actual shapes of the adjacent letters.

8.6.1 Letter Case

Studies of narrative text have generally found that mixed upper- and lowercase is read about 10–15% faster than all upper case, is generally preferred, and results in better comprehension (Moskel, Erno, and Shneiderman 1984; Poulton and Brown 1968; Tinker 1963; Vartabedian 1971; Wheildon 1995). One exception is the work of Arditi and Cho (2007), who found that all uppercase was read faster by users with vision impairment and by all users when the text was small. For search tasks or tasks involving individual letter or word recognition, all uppercase words are found about 13% quicker (Vartabedian 1971). Overall, the evidence supports the use of normal upper- and lowercase for most text on Web pages and the use of all upper case for headings or other short items that may need to attract attention.

8.6.2 Horizontal Spacing (Tracking) and Justification

The primary reason for adjusting the horizontal spacing of text is to create even right margins—a practice called full justification, which has been traditional in printed books since the beginning of movable type. In fact, the Gutenberg Bible printed in the 1450s used a two-column arrangement of fully justified text. This approach is thought to create a more "orderly" appearance to the page. With the monospaced fonts that were common on early computer displays (e.g., Courier), the addition of extra spaces between words to create an even right margin (full justification) generally slows reading (Campbell, Marchetti, and Mewhort 1981; Gregory and Poulton 1970; Trollip and Sales 1986). With the proportionally spaced fonts more commonly used today on Web pages (e.g., Arial, Times New Roman, Verdana), the effects of full justification are not quite as clear. Fabrizio, Kaplan, and Teal (1967) found no effect of justification on reading speed or comprehension, while Muncer et al. (1986) found that justification slowed reading performance. More recently, Baker (2005) found an interaction between justification and the width of the columns of text: justification slowed reading speed for narrow (30 characters) and wide (90 characters) columns but improved reading speed for medium-width columns (45 characters).

8.6.3 Vertical Spacing (Leading)

More generous vertical spacing between lines of text (e.g., space-and-a-half or double-spacing) generally results in slightly faster reading of narrative text (Kolers, Duchnicky, and Ferguson 1981; Kruk and Muter 1984; Williams and Scharff 2000). This effect seems to be greater for smaller font sizes (10 point) than larger (12 or 14 point) (Williams and Scharff 2000). One recommendation, primarily for users with vision problems, is that leading should be 25 to 30 percent of the font size (Arditi 2010).

8.6.4 Line Length

Several studies have investigated the effects of line length on reading speed and subjective reactions. Although the results are not totally conclusive, there is some evidence that very short line lengths (e.g., under 2.5") result in slower reading while longer line lengths (up to about 9.5") yield faster reading (Duchnicky and Kolers 1983; Dyson and Haselgrove 2001; Dyson and Kipping 1998; Ling and Schaik 2006; Youngman and Scharff 1998). No one has studied even longer lines, which are likely to cause problems. Alternatively, users seem to prefer lines about 3.5" to 5.5" long (Bernard, Fernandez, and Hull 2002; Youngman and Scharff 1998). And at least one study (Beymer, Russell, and Orton 2008) found that shorter lines were read slightly faster than longer lines and yielded greater comprehension of the material (as determined by a surprise multiple-choice test after reading the material).

8.6.5 Font Style

Joseph, Knott, and Grier (2002), in studying data displays with field labels nonbold and data values either bold or nonbold, found that the use of bold for the data values actually slowed down search performance. Hill and Scharff (1997) found a tendency for italic text to slow reading, while Boyarski et al. (1998) found that, at least for the Verdana font, users significantly preferred the normal over the italic version. Aten et al. (2002) found that participants were significantly less accurate at identifying words presented in italics compared to normally. Because underlining is easily mistaken for designating a hyperlink, obviously it should be avoided as a mechanism for highlighting. Overall, the evidence supports reserving underlining only for hyperlinks, and using other font styles such as bold and italics sparingly.

8.6.6 FONT TYPE AND SIZE

Several studies have investigated the effects of different on-screen fonts and sizes on reading performance and subjective reactions. The range of font sizes studied has generally been from about 6 point to 14 point. The smallest fonts (e.g., 6 to 8 point) appear to slow reading performance (Tullis, Boynton, and Hersh 1995). Studies of 10 and 12 point fonts have found either no difference in reading performance (Bernard and Mills, 2000) or a slight advantage for 12 point fonts (Bernard, Liao, and Mills 2001). One study with older adults found that they were able to read 14 point fonts faster than 12 point fonts (Bernard et al. 2001). Most of these studies also found that users generally prefer the larger fonts, at least for the range of sizes studied.

In looking at the effects of different fonts, Tullis, Boynton, and Hersh (1995) found that the sans serif fonts they studied (Arial and MS Sans Serif) yielded slightly better reading performance than the serifed font they studied (MS Serif) and the sans serif fonts were also preferred. In a series of studies, Bernard et al. (2001) found that three sans serif fonts (Arial, Verdana, and Comic Sans) were generally preferred over the other fonts they studied (Agency, Tahoma, Courier, Georgia, Goudy, Schoolbook, Times, Bradley, and Corsiva).

Tullis and Fleischman (2002) studied text presentation in Verdana or Times New Roman using three sizes: smallest, medium, and largest. For Times New Roman, the medium size was HTML size = 3. Since Verdana is a larger font, its medium size was HTML size = 2. In both cases, the smallest and largest conditions were derived from the medium condition using the browser's text size manipulations. They found that at the smallest size, users performed better with Times. At the medium size, there was no difference. At the largest size, users performed better with Verdana. They hypothesize that at the smallest size, the serifs of Times aid in distinguishing one letter from another, while at the largest size the looser kerning (spacing) of Verdana plays a more important role, allowing it to be read faster. Note that at all sizes users preferred Verdana over Times.

In an eye-tracking study, Beymer, Russell, and Orton (2008) tested three font sizes (10, 12, and 14 points) and both a sans serif (Helvetica) and serif (Georgia) font. They found that fixation durations were significantly longer for the 10-point font in comparison to the 14 point font. For the range of font sizes they studied, they found a "penalty" of roughly 10 ms per point as the font got smaller. This implies greater difficulty in processing the information in each fixation for the smaller font. They also found that participants given the 14 point font spent 34% more time in "return sweeps" than those given the 10 point font. A "return sweep" is when the users hits the end of one line of text and must move their fixation back to the beginning of the next line. This is not surprising because the larger font sizes also had longer text lines. Basically, they found that the return sweep associated with the shorter lines (10 point) could be done in one saccade (fast movement between fixations), while the return sweep for the longer lines (14 point) required two saccades. Finally, they found that the serif font, Georgia, was read about 8% faster than the sans serif font, Helvetica, although the difference was not statistically significant.

In an interesting study of the perceived "personality traits" of fonts, Shaikh, Chaparro, and Fox (2006) had participants rate 20 different fonts using 15 adjective pairs (e.g., flexible/rigid, exciting/dull, elegant/plain). Participants also rated appropriate uses of the fonts (e.g., e-mail, business documents, headlines on a Web site). Factor analysis of the adjective scores yielded five primary classifications of the fonts:

1. All purpose (e.g., Arial, Verdana, Calibri)
2. Traditional (e.g., Times New Roman, Georgia, Cambria)
3. Happy/creative (e.g., Monotype Corsiva, Comic Sans)
4. Assertive/bold (e.g., Impact, Agency FB)
5. Plain (e.g., Courier New, Consolas)

8.6.7 ANTI-ALIASING

Anti-aliasing, applied to fonts, is an attempt to make individual characters appear smoother on relatively low-resolution displays. As illustrated in Figure 8.8, this is done by introducing additional colors or shades of gray, particularly in curved parts of a character, to "fool" the eye into perceiving it as being smoother. Microsoft's ClearType® is an implementation of anti-aliasing at the subpixel level (i.e., the individual red, green, and blue components of a pixel) primarily designed for LCD screens.

Dillon et al. (2006) studied ClearType and regular versions of text presented either in a spreadsheet or an article. They found that visual search of the spreadsheet and reading of the article were both faster with ClearType. They found no differences in accuracy or visual fatigue. However, they found wide individual differences in performance, perhaps indicating that ClearType's apparent benefits may not hold for all users. Aten, Gugerty, and Tyrrell (2002) also found a superiority of ClearType when they tested the accuracy of

FIGURE 8.8 (See color insert.) Illustration of anti-aliasing applied to the letter "e." On the left is the original version, greatly enlarged. On the right is the same letter with anti-aliasing applied.

classifying briefly displayed words or nonwords. Likewise, Slattery and Rayner (2009), in an eye-tracking study, found that ClearType led to faster reading, fewer fixations, and shorter fixation durations.

Some studies have also found subjective preferences for ClearType fonts. For example, Sheedy et al. (2008) found a significant subjective preference for ClearType, as did Tyrrell et al. (2001).

Microsoft developed six new fonts specifically to make use of their ClearType technology: Corbel, Candara, Calibri, Constantia, Cambria, and Consolas. In a study comparing two of these fonts (Cambria and Constantia) to the traditional Times New Roman font (also with ClearType turned on), Chaparro, Shaikh, and Chaparro (2006) found that the legibility of briefly presented individual letters was highest for Cambria, followed by Constantia, and then Times New Roman. They also found that the digits 0, 1, and 2 in Constantia resulted in confusion with the letters o, l, and z. Times New Roman also had high levels of confusion for certain symbols and digits (e.g., ! and 1, 2 and z, 0 and o, $ and s).

8.6.8 Image Polarity

Most printed text is generally black on a light background. The earliest computer screens used light text on a dark background, but as GUI systems became more commonplace, this switched over to dark text on a light background, perhaps to emulate the printed page. Several studies have investigated the effects of image polarity on reading performance. (Note that there is some confusion in the literature about the terms positive and negative polarity and which combination of text and background each refers to. Consequently, we will avoid the use of those terms.) Several studies have found that dark characters on a light background are read faster and/or more accurately than light characters on a dark background (Bauer and Cavonius 1980; Gould et al. 1987; Parker and Scharff 1998; Scharff and Ahumada 2008; Snyder et al. 1990). There is also evidence that users prefer dark text on a light background (Bauer and Cavonius 1980; Radl 1980). However, studies by Cushman (1986) and Kühne et al. (1986) failed to find a statistically significant difference in performance between the two polarities. In addition, Parker and Scharff (1998) found that the effect of polarity was most pronounced for high-contrast displays and for older adults. Mills and Weldon (1987) suggest that the different findings from these studies could be a function of the CRT refresh rates used. Apparent flicker is potentially a greater problem with dark text on a light background. The studies that found an advantage for dark text on a light background generally used high refresh rates (e.g., 100 Hz), while those that did not used lower refresh rates (e.g., 60 Hz). Finally, Buchner, Mayr, and Brandt (2009) argued that the advantage commonly found for dark text on a light background is due to the higher overall level of display luminance associated with this polarity. When they matched overall levels of display luminance, they found no effect of polarity.

8.6.9 Color Contrast and Backgrounds

The conventional wisdom has been that the legibility of any particular combination of text and background color is largely a function of the level of contrast between the two, with higher contrast resulting in greater legibility (Fowler and Stanwick 1995; Tullis 1997; White 1990). While this has been supported in a general way by the empirical evidence (Hall and Hanna 2004), there is also evidence that there are significant interactions with other factors (Hill and Scharff 1997, 1999; Parker and Scharff 1998; Scharff and Ahumada 2003). For example, Hill and Scharff (1999) found that black text was read faster on a higher contrasting yellow or gray background than on a lower contrasting blue background, but the effect was greater when using a more highly textured background with less color saturation. And contrary to the conventional wisdom, Hill and Scharff (1997) found that black text on a medium gray or dark gray background was read faster than black text on a white background. Tullis, Boynton, and Hersh (1995) found no difference in reading speed for black text on a white background versus a light gray background. In studying the subjective ratings of readability that users gave to 20 different combinations of text and background colors, Scharff and Hill (1996) found that the highest rated combination was black text on a white background, followed closely by blue text on a white background and black text on a gray background. The combinations with the lowest ratings were fuchsia text on a blue background and red text on a green background. As shown in Figure 8.9, the ratings could be reasonably well predicted by looking at the simple contrast between the text and background (i.e., the difference in the gray values of the two colors), but the fit is certainly not perfect. In essence, the simple contrast between text and background is probably a reasonably good predictor of the legibility of that combination, but other factors enter in as

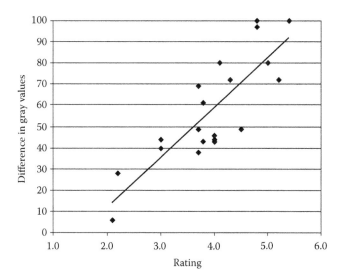

FIGURE 8.9 Correlation between data from Scharff and Hill (1996) giving ratings of legibility of various text and background colors, and the simple difference in gray values of the two colors: $r = .84$.

well. For a more detailed discussion of color contrast, see Section 8.9.

One of the techniques some Web designers like to use (because it is a feature supported by the major browsers) is to present a background image on their Web pages. Although this has not been widely studied, common sense would dictate avoiding any backgrounds whose images would interfere with the legibility of text being displayed on them. In studying the effects of four levels of textured backgrounds (plain, small, medium, and large) on the legibility of text, Hill and Scharff (1999) found that plain backgrounds yielded faster search times.

8.6.10 FONT IMPLEMENTATION ISSUES

The implementation of fonts is one of the more complicated parts of HTML. Although the technical details are beyond the scope of this chapter, two issues are so directly related to the user experience that they deserve mention: the use of scalable fonts and the use of style sheets. There are a variety of techniques that can be used to define font sizes in HTML. Unfortunately, some of them (e.g., the use of fixed pixel sizes) will usually defeat the ability of the user to adjust the font size dynamically when viewing the page (e.g., via the "View/ Text Size" menu in Internet Explorer). This is an extremely important feature to users with limited vision and should be supported whenever possible. The use of style sheets is really more of an advantage to developers because it is much simpler to change font characteristics in one place (the style sheet) than in all the places throughout the code that <FONT> tags might be embedded. Style sheets provide an indirect advantage to users because their use promotes greater consistency in the treatment of text.

8.7 GRAPHICS AND MULTIMEDIA

Few Web sites consist only of text. Graphics and multimedia elements are used to enhance the user experience but work best when certain guidelines are followed.

8.7.1 GRAPHICS

Graphics are almost as common as text on Web pages. The majority of interactive elements on a page, such as navigation bars and links, could be graphics. Graphics may also represent the smallest elements on a page such as bullets or lines. Additionally, they can be photographs, charts, or graphs of data. To the user, graphical items such as buttons or data graphs may not be thought of as being a graphic. These items are only graphics because they are represented by a graphic file (such as a JPEG or GIF) on the Web page. This notion that graphics may not always be perceived as being a "graphic" by the user may be the reason there is a lack of literature that specifically investigates graphics on Web pages. As seen in Section 8.9, a great deal of research exists that may speak to the content of a graphic, but little that focuses on graphics themselves. For instance, contrast and color blindness are

color topics, yet they may also apply to the visual content of a graphics file.

Although there is a lack of empirical data on the use of graphics on Web pages, a number of reputable publications exist that present guidelines for the use of graphics (Koyani, Bailey, and Nall 2004; Lynch and Horton 2009). A subset of these guidelines will be presented in this section.

8.7.1.1 Graphics and Bandwidth

As of May 2009, the Pew Research Center's Internet & American Life Project identified 94.65% of active Internet users in the United States as having broadband Internet access (Horrigan 2009). However, downloading graphics along with other items on a Web page can still take longer than many users want to wait. This will continue to be an issue especially on mobile Web devices (Lynch and Horton 2009). For this reason, it is still recommended to do two things: (1) produce fast-loading graphics and (2) use graphics wisely (Koyani, Bailey, and Nall 2004).

8.7.1.2 Produce Fast-Loading Graphics

Once the content of a graphic is developed, there are a number of things that can be done to the graphic to optimize the file's size and download times. Choosing the correct file format for the graphic can not only optimize size but may also help produce a better appearing graphic. The two predominant graphic file formats on the Web are the JPEG and GIF formats, although PNG is also sometimes used. The GIF format is usually best for icons and logos or other "drawn" images. The JPEG format is best for photographs, because that is what it was designed for. JPEG files allow for a varying degree of compression that can make the file size smaller but also degrade the image. Because most times a "perfect" image is not required for presentation on the Web, this compression control can be helpful in limiting the file size.

Reusing images already on a Web page can also help reduce page download times (Koyani, Bailey, and Nall 2004). If an image is already displayed, it exists in the browser cache and therefore does not have to be downloaded again.

When using the GIF or JPEG format, user satisfaction may be improved by using the interlaced GIF or progressive JPEG format as opposed to the noninterlaced/standard format (Lynch and Horton 2009). An interlaced GIF or progressive JPEG graphic renders progressively, allowing the user to identify the graphic content before it completely downloads (Figure 8.10). A noninterlaced/standard graphic renders from the top down, hiding the complete graphic content until its download is complete.

The PNG format was designed specifically for use on Web pages, but because it uses lossless compression, the resulting file is much larger than those with lossy JPEG compression (Lynch and Horton 2009). PNG might become more accepted in the future as its images look good and have a similar or smaller file size than GIFs (Lynch and Horton 2009).

FIGURE 8.10 Noninterlaced GIF/progressive JPEG. The image on the left shows what the picture looks like while progressively displaying.

8.7.1.3 Use Graphics Wisely

Graphics should only be used when they enhance the content of the Web page or improve the understanding of the information being presented (Koyani, Bailey, and Nall 2004). Though graphics take time to download, users tend to be more patient with useful graphics that they expect to be displayed. For example, if the user clicks on a link to show a map or an enlarged photo from a news article, a longer download time is acceptable as the user has requested the graphic and can anticipate a longer download time. However, displaying a large graphic on the Home page of a site may elicit a negative reaction from users because they had no part in requesting or expecting the large, slow-loading graphic. This could also discourage users from staying on the Home page, and hence possibly drive them away from the site.

8.7.1.4 Making Graphics Accessible

Though many visually impaired users may never see the graphics on a Web page, the tools that they use to access the page must consider all of the elements on the page. Furthermore, the content in some graphics may be important to users who cannot or have chosen not to view the graphics. For all of these reasons, it is important to supply ALT tags (alternate text tags) to every IMG (image) tag on the Web page (Lynch and Horton 2009). ALT tags serve a number of purposes:

- For visually impaired users with assistive technologies such as an audio screen reader, the ALT tag is what is read when the tool comes to a graphic. This allows the user to know something about the graphic.
- Some users choose to navigate the Web without graphics being displayed in the browser. When this occurs, the ALT text of the graphic is displayed on the Web page in the location where the graphic would have appeared.
- When graphics are used as navigation items such as buttons, the ALT tag should replicate the text displayed on the front of the button. This allows users

that cannot or do not view the images to navigate the site.
- The ALT tags for graphics with important content should be detailed enough (though short) to convey the meaning of the graphic to the users who cannot or do not view them.
- Unimportant graphics such as bullets and hard rules should have an ALT tag set to "" (as opposed to no ALT tag at all) so that users that cannot or do not view the graphics can avoid wasting time addressing them. If some of these items are important, short ALT tags can be assigned (e.g., "*" for bullets and "-" for hard rules).

Care should be taken when developing content for graphics to be sure that the colors or subject do not negatively affect the usability of the site. Additionally, care taken with the graphics files themselves and how they are implemented can greatly improve Web page usability.

8.7.2 MULTIMEDIA ELEMENTS

Most of the research done on multimedia has focused on its use in help systems or training. However, several findings can be extrapolated to this type of content on the Web, and there is a set of best practices that can guide how to make this content accessible to all users. The other chapter sections on Graphics and Color will also apply to the treatment of multimedia.

8.7.2.1 Using Videos, Animations, and Flash

Animation and videos can be very effective at communicating procedural information, over pure textual content (Shneiderman et al. 2009; Weiss, Knowlton, and Morrison 2002). But in many situations, the use of multimedia provokes an immediate reaction of either engagement or avoidance. For example, many users will describe immediately closing ads that appear as layers over the content of a page they are hoping to see. Or, when seeing a Flash movie loading before seeing a site's Home page, users will click Skip Intro to get more quickly to the desired content. Because of how easily these elements can capture the attention of users, it's important to have clear and useful reasons for using multimedia to avoid unnecessarily distracting users (Koyani, Bailey, and Nall 2004).

Shrestha (2006) investigated the effectiveness of banner ads, pop-up ads, and floating ads in terms of ad recall and recognition. He found that participants in the pop-up and floating ad condition had higher recall in the banner ad condition. The floating ad was recognized most (where recognition was measured by being able to recognize the advertisement seen during the study). Animation had no effect on the recall of the ads but did significantly bother participants more than the static ads. The prevalence of ads on Web sites make it likely that a user would immediately "swat" or close any content appearing in the same way as an ad.

Mosconi, Porta, and Ravarelli (2008) looked at different ways of embedding multimedia into news pages. They found that showing both a picture preview of the multimedia and providing a quick textual description worked best at drawing users' attention over just having the picture preview. This type of information can also help a user decide whether or not to play the multimedia and provide content that can be used for accessibility purposes.

Multimedia content should be user controlled rather than something that plays automatically when the user comes to a page. A possible exception is if the user has already clicked a link indicating that they want to view the content. Either way, a user should be able to pause, stop, replay, and ignore (Koyani, Bailey, and Nall 2004). Ideally, some type of closed captioning capability or transcript is also provided if the target audience may not be in an environment where they can listen to content (or if headphones are not available).

A good principle when working with multimedia like Flash is to use standard components whenever possible and to not break the standard expectations that a user brings to a Web site. For example, nonstandard scrollbars and other UI elements are not caused by Flash, but rather by how the designer has implemented their designs (Nielsen and Loranger 2006). A quick usability test can provide insight as to whether a designer's nonstandard UI inventions work better than the standard HTML elements.

8.7.2.2 Accessibility Considerations for Multimedia

When building a page with multimedia elements, it is important to consider accessibility issues from the very beginning. The key consideration is that designers need to provide an alternative that includes textual representations of the multimedia content, along with some comparable way to navigate within that content (IBM Corporation 2009). For example, if the Web page has a video on it, the designer could provide an audio description that complements the existing audio track with information that may be visible in the video. The W3C provides a comprehensive set of guidelines for audio and video considerations that should be reviewed early in the design process (W3C 2008).

For other types of multimedia elements, the fundamental principles of accessibility still apply, such as providing equivalent access for all users and ensuring that functionality can be accessed by users of assistive technology. For example, any nonstandard UI elements such as Flash scrollbars or buttons, or any other type of functionality that is mouse driven should be keyboard accessible. Bergel et al. (2009) also recommend that the page's information hierarchy be reflected in the code so assistive technologies can recognize the relationship between content elements and to support scalable fonts and relative sizing.

8.7.2.3 Page Load Time

Graphics and multimedia increase page size, which can result in slower download times than text-only pages. This section explores users' expectations for how pages will load and recommends techniques to make pages seem faster.

8.7.2.4 Perception and Tolerance

Some research has focused on how long users are willing to wait until they see a response from their computer system. Response time is the number of seconds it takes from the moment a user initiates an action until the computer begins to present results (Shneiderman et al. 2009). The expectations that a user has for how long she expects to wait for a page to load and how well a site matches her expectations affect her perception of quality and security in the site.

Many studies have focused on manipulating page download times and measuring user reactions. Ramsay, Barbesi, and Preece (1998) studied load times ranging from 2 sec to 2 minutes. They found that pages associated with delays longer than 41 sec were rated as less interesting and more difficult to scan. They also found that slower-loading pages resulted in lower ratings of quality for the associated products and an increased perception that the security of their online purchase was likely to be compromised.

Bouch, Kuchinsky, and Bhatti (2000) presented users with Web pages having load times that ranged from 2 to 73 sec and asked users to rate the "quality of service" being provided by each of these Web sites. They found a dramatic drop in the percentage of good ratings between 8 and 10 sec, accompanied by a corresponding jump in the percentage of poor ratings. In a second study, where users were asked to press an "Increase Quality" button when they felt that a site was not being sufficiently responsive, the average point at which the users pressed the button was 8.6 sec. In a third study, users were more tolerant of delays when the pages loaded incrementally. In addition, users' tolerance for delays decreased as they spent time interacting with a site, and their tolerance varied by task.

In fact, the longer a user waits for a response, the more physiological stress they experience. Trimmel, Meixner-Pendleton, and Haring (2003) compared page load times of 2, 10, or 22 sec while measuring skin conductance and heart rate, which are indicators of stress level. As response time increased on the Web page, there were significant increases in both heart rate and skin conductance. Selvidge (2003) examined how long participants would wait for Web pages to load before abandoning a site. She found that older adults waited longer than younger adults before leaving a site and were also less likely to leave a site even if its performance was slow. Participants with a high-speed connection were less tolerant of delays than those who used dial-up, but Internet experience level had no impact on delay tolerance.

Taken together, these studies indicate several factors can influence a user's acceptance of a particular page load time (such as their task, whether the page loads incrementally, their age, and how long they have been interacting with the site). The key takeaway is that users get more frustrated with a site as load time increases, and this frustration might cause them to leave the site entirely. To maintain the sense of continuity between the user and the Web page, a common recommendation is to have some kind of response provided to the user within 2 to 4 sec (Seow 2008). The longer the user waits, the more likely they are to give up and leave. Ideally there will be some kind of actionable response provided within about 10 sec, whether it's for a completed action or a visible change in status. Many users will abandon unresponsive Web sites after about 8–10 sec (Seow 2008). Beyond that point, there seems to be a significant increase in user frustration, perception of poor site and/or product quality, and simply giving up on the site.

8.7.2.5 Progress Indication and Buffering

Designers need to consider perceived time in addition to the actual time it takes to load a page. Seow (2008) distinguishes between actual duration, or objective time, and perceived duration, which reflects subjective or psychological time. The perceived duration is affected by things like how frequently a user has done something in the past and how similar it is to experiences using other sites or applications, as well as to how the Web page indicates the passage of time (e.g., with progress indicators or other UI elements). User frustration results when there is a mismatch between expectations and reality. While a designer may be unable to control how quickly a site loads, they have many tools at their disposal for making the site seem faster.

A Web page can indicate how long information will take to load or refresh through progress indicators. These indicators might be something like the rotating hourglass or a "loading" indicator that shows a visual update for how much longer the process will take. Providing a running tally of time remaining is helpful. In a study comparing types of progress indicators, users shown graphical dynamic progress indicators reported higher satisfaction and shorter perceived elapsed times than when shown indicators that have static text or number of seconds left (Meyer et al. 1996).

Another way to help with perceived time is to provide buffering for streaming video. Rather than requiring the user to wait until the entire video is downloaded, the page can store a certain amount of buffer and begin playback while the rest of the video is still being downloaded (Seow 2008). For pages that consist of many types of elements, the designer can explore ways of having the page load incrementally rather than requiring the user to wait until everything loads at once. This incremental display helps the user feel like progress is being made and lets them start exploring the page faster. In some cases, the order in which sections of the page load can help tell a story connecting the different elements together.

8.7.3 Key Takeaways

- When adding graphics or multimedia to a page, plan for accessibility from the beginning.
- Provide feedback to the user within 2–4 sec, ideally with an indication of how much longer content will take to load.
- Incremental loading and buffering can help reduce the perceived time it takes for multimedia to appear.

8.8 TABLES AND GRAPHS

Tables, graphs, and charts are among the most commonly used tools to display numeric information. They can appear in print or online and there are many ways each can be presented. Designers are often faced with the challenge of deciding how best to present information. For some information, it may seem equally plausible to employ a table or a graph to present the information. As reported by Coll, Coll, and Thakur (1994), a plethora of research exists extolling the superiority of tables over graphs (Ghani 1981; Grace 1966; Lucas 1981; Nawrocki 1972) as well as research showing graphs to be superior to tables (Benbasat and Schroeder 1977; Carter 1947; Feliciano, Powers, and Bryant 1963; Tullis 1981). Some researchers have even found tables and graphs to have no differences with regard to reader performance (Nawrocki 1972; Vicino and Ringel 1966).

Coll, Coll, and Thakur (1994) performed a study that highlighted the types of tasks where tables or graphs were the superior presentation tool. They found that when users were asked to retrieve relational information, graphs were superior to tables in performance. The opposite was true when users were asked to retrieve a specific value. Here, better performance was seen with information presented in tables. When users performed mixed tasks of searching for specific values and comparing relational information, they found that tables were superior to graphs in both performance measures (retrieval time and accuracy).

Similarly, Few (2004) makes several recommendations about the appropriate use of basic tables and graphs. Tables are advantageous when users are likely to look up values or compare related values. Tables are also useful because they enable the presentation of numbers at a high level of precision that are not possible with graphs, and they are able to display data in different units of measurement, such as dollars and units. On the other hand, graphs are visual by nature, allowing users to see the shape of the data presented. This can be particularly powerful, allowing users to identify patterns in information. Graphs are most useful when the shape of the information is telling, or when multiple values need to be compared.

8.8.1 TABLES

When presenting tables on a Web page, there are a variety of techniques that may be used. Tullis and Fleischman (2004) conducted a study to learn how to best present tabular data on the Web. The study focused on the effects of table design treatments such as borders, font size, cell background colors, and spacing. Over 1400 subjects performed specific value retrieval tasks using 16 different table designs. The tables had different combinations of the following attributes: horizontal lines separating rows (H), vertical lines separating columns (V), alternating row background colors (B), large text fonts (L), small text fonts (S), tight spacing within tables cells (T), and loose spacing in cells (L). Figure 8.11 shows the results from this study.

Inspection of Figure 8.11 reveals a clear winner: BLL (B: alternating row background colors, L: large font, L: loose

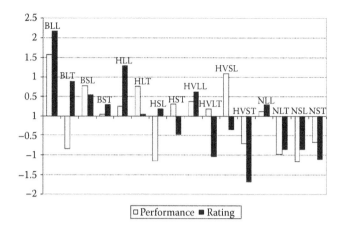

FIGURE 8.11 Data from Tullis and Fleischman (2004): Z-score transformations of performance and subjective data for all 16 table designs.

spacing). This table design was superior to all other designs in both performance and subjective ratings. Among the poorer-performing table designs, HVST (H, horizontal lines separating rows; V, vertical lines separating rows; S, small font; T, tight spacing) stands out as possibly being the poorest.

Tullis and Fleischman additionally analyzed the results based on the tables' individual design attributes. With regard to borders, they found that tables with alternating row colors, or "zebra striping," consistently performed better and had higher subjective ratings when compared to tables with horizontal lines separating rows, horizontal and vertical lines, or

no lines at all. With regard to fonts, tables with larger fonts performed significantly better than those with smaller fonts. Additionally, tables with looser spacing within their cells performed significantly better than those with tight spacing.

Enders (2008) examined the use of zebra striping on tables. Zebra striping is the practice of alternating the color of rows in a table, typically white and another color. Enders had people complete several tasks on tables that were plain, lined, or striped. Participants were significantly more accurate when tables were striped on three of the eight tasks and tended to be more accurate on an additional fourth task. In a separate study, Enders (2008) also had users rate six table designs, shown in Figure 8.12. Of these designs, users rated single striped tables as the most useful. These findings suggest that, at the very least, zebra striping does not hurt performance and, in some cases, may actually help improve it.

Few (2004), however, recommends that white space be used to delineate rows and columns whenever possible, and subtle fill color if necessary, but never use grids. Further, he provides several guidelines for general table design. For example, groups of data should be separated by white space, and column headers should be repeated at the beginning of each new group. Related columns, such as those that are derived from other columns or that contain data that should be compared, should be placed close together. Rows and columns containing summary data should be made visually distinct from other data in the table. These sorts of general guidelines are applicable to tables displayed over any medium, not just online.

A – Plain

Number	Name	Diameter	Dimensions	Dist from sun	Date discovered	Discoverer
1	Ceres	952	9.75×909	2.766	Janurary 1, 1801	Piazzi G.
2	Pallas	532	570×525×500	2.773	March 28, 1802	Olbers, H. W.
4	Vesta	530	578×560×458	2.361	March 29, 1807	Olbers, H. W.
10	Hygiea	407	500×385×350	3.137	April 12, 1849	de Gasparis, A.
704	Interamnia	326	350.4×303.7	3.067	October 10, 1910	Cerulli, V.
52	Europa	302	360×315×240	3.101	February 4, 1858	Goldschmict, H.
511	Davida	289	357×294×231	3.17	May 30, 1903	Dugan, R. S.

B – Double striped

Number	Name	Diameter	Dimensions	Dist from sun	Date discovered	Discoverer
1	Ceres	952	9.75×909	2.766	Janurary 1, 1801	Piazzi G.
2	Pallas	532	570×525×500	2.773	March 28, 1802	Olbers, H. W.
4	Vesta	530	578×560×458	2.361	March 29, 1807	Olbers, H. W.
10	Hygiea	407	500×385×350	3.137	April 12, 1849	de Gasparis, A.
704	Interamnia	326	350.4×303.7	3.067	October 10, 1910	Cerulli, V.
52	Europa	302	360×315×240	3.101	February 4, 1858	Goldschmict, H.
511	Davida	289	357×294×231	3.17	May 30, 1903	Dugan, R. S.

C – Lined

Number	Name	Diameter	Dimensions	Dist from sun	Date discovered	Discoverer
1	Ceres	952	9.75×909	2.766	January 1, 1801	Piazzi G.
2	Pallas	532	570×525×500	2.773	March 28, 1802	Olbers, H. W.
4	Vesta	530	578×560×458	2.361	March 29, 1807	Olbers, H. W.
10	Hygiea	407	500×385×350	3.137	April 12, 1849	de Gasparis, A.
704	Interamnia	326	350.4×303.7	3.067	October 10, 1910	Cerulli, V.
52	Europa	302	360×315×240	3.101	February 4, 1858	Goldschmict, H.
511	Davida	289	357×294×231	3.17	May 30, 1903	Dugan, R. S.

D – Triple striped

Number	Name	Diameter	Dimensions	Dist from sun	Date discovered	Discoverer
1	Ceres	952	9.75×909	2.766	Janurary 1, 1801	Piazzi G.
2	Pallas	532	570×525×500	2.773	March 28, 1802	Olbers, H. W.
4	Vesta	530	578×560×458	2.361	March 29, 1807	Olbers, H. W.
10	Hygiea	407	500×385×350	3.137	April 12, 1849	de Gasparis, A.
704	Interamnia	326	350.4×303.7	3.067	October 10, 1910	Cerulli, V.
52	Europa	302	360×315×240	3.101	February 4, 1858	Goldschmict, H.
511	Davida	289	357×294×231	3.17	May 30, 1903	Dugan, R. S.

E – Single striped

Number	Name	Diameter	Dimensions	Dist from sun	Date discovered	Discoverer
1	Ceres	952	9.75×909	2.766	Janurary 1, 1801	Piazzi G.
2	Pallas	532	570×525×500	2.773	March 28, 1802	Olbers, H. W.
4	Vesta	530	578×560×458	2.361	March 29, 1807	Olbers, H. W.
10	Hygiea	407	500×385×350	3.137	April 12, 1849	de Gasparis, A.
704	Interamnia	326	350.4×303.7	3.067	October 10, 1910	Cerulli, V.
52	Europa	302	360×315×240	3.101	February 4, 1858	Goldschmict, H.
511	Davida	289	357×294×231	3.17	May 30, 1903	Dugan, R. S.

F – Two colour striped

Number	Name	Diameter	Dimensions	Dist from sun	Date discovered	Discoverer
1	Ceres	952	9.75×909	2.766	Janurary 1, 1801	Piazzi G.
2	Pallas	532	570×525×500	2.773	March 28, 1802	Olbers, H. W.
4	Vesta	530	578×560×458	2.361	March 29, 1807	Olbers, H. W.
10	Hygiea	407	500×385×350	3.137	April 12, 1849	de Gasparis, A.
704	Interamnia	326	350.4×303.7	3.067	October 10, 1910	Cerulli, V.
52	Europa	302	360×315×240	3.101	February 4, 1858	Goldschmict, H.
511	Davida	289	357×294×231	3.17	May 30, 1903	Dugan, R. S.

FIGURE 8.12 (See color insert.) Table designs examined by Enders (2008). Single striped was rated the most useful. (From Enders, J. 2008. Zebra striping: more data for the case. *A List Apart*. http://www.alistapart.com/articles/zebrastripingmoredataforthecase (accessed November 6, 2010). With permission.)

8.8.2 Graphs

Unlike tables, graphs are typically not used as a primary display method on Web pages. Though there are exceptions, graphs on Web pages have the same properties as graphs displayed in any other medium, including those found in print. Their sole purpose is to convey a representation of data to the user. A number of studies and publications present guidelines for developing graphs (Carter 1947; Coll, Coll, and Thakur 1994; Few 2004; Harris 1999; Levy et al. 1996; Rabb 1989; Tufte 1983; Tullis 1981; Tversky and Schiano 1996). No known research exists, however, that investigates usability issues for graphs specifically presented on the Web. This may merely support the notion that presenting graphs on the Web is not that different from presenting them elsewhere. For accessibility purposes, an alternate tabular version of the data in a graph should be available.

8.8.3 Key Takeaways

- Although there is conflicting evidence about which is better in general, it is recommended that tables be used when values are likely to be compared or needs precision and graphs be used when the shape of data is of importance.
- White space, alternating row colors, and large font may improve the usability of tables.

8.9 COLOR

The effective use of color can significantly enhance most Web pages. Much of the early research on color in information systems focused on its use for coding purposes (Christ 1975; Kopala 1981; Sidorsky 1982), which showed that color can make it easier to find specific pieces of information. Care must be taken when applying color coding, as careless application could have detrimental effects on usability (Christ 1975; Christ and Teichner 1973; McTyre and Frommer 1985). This section will define what color is and review how it can affect the design of a Web page.

8.9.1 What Is Color?

Each color a human sees represents a different combination of perceived light from the color spectrum. From a Web design standpoint, each color can be constructed through a combination of differing levels of red, green, and blue light. To render a color in a Web browser, each of these levels is represented by a value from 0 to 255. The combination of the three levels is called a color's RGB (Red, Green, Blue) value. Table 8.2 shows some example RGB values.

On the Web, color definitions are often presented as a six-digit hexadecimal representation of a color's RGB value (Black = 000000, Red = FF0000, etc.). Although there are nearly an infinite number of different colors in nature, there are a little over 16 million colors in the hexadecimal system.

TABLE 8.2
Example RGB Values

Color	R Value	G Value	B Value
Black	0	0	0
White	255	255	255
Red	255	0	0
Violet	192	0	192

Though almost all of the 16 million colors can be rendered by the most popular Web browsers, it is possible that a user's computer settings could prevent them from viewing all of the colors. Some Web users may still have their computers set to lower color settings (e.g., 256 colors). However, recent data show that fewer than 1% of Web users are running their systems in 256-color mode (w3Schools.com 2010). In spite of this fact, some guidelines still recommend use of the 216 colors in the "Web-safe color palette" (Weinman 2004) to ensure that the colors display properly in all browsers and on all platforms.

8.9.2 Visual Scanning

Users will identify target items that differ in color more accurately than those that differ in other attributes such as size and brightness (Carter 1982; Christ 1975). Color can effectively be used to draw attention to an area of a page, a row in a table, a word in a paragraph, or even a single letter in a word. For example, while searching research abstracts online, the keywords used by the users during their search could appear as red text in the abstracts while the rest of the abstract is in black text. The red color of the keywords may allow users to more easily pick them out while scanning the abstract. For the red text in this example to truly be an emergent feature among the black text, there not only needs to be satisfactory contrast between all of the text and the background of the page (McTyre and Frommer 1985) but also between the emergent text (red) and the normal text (black). Additionally, the perceived contrast between these elements may differ among users, especially those that may be color blind. For these reasons, it is usually recommended that color be redundant to an additional emergent feature. In our example, the keyword text should not only be red but also perhaps bolded.

In an interesting study of the effectiveness of Google AdSense ads, Fox et al. (2009) manipulated the location and color scheme of the ads on a blog page. They found that ads with a high color contrast relative to the background color of the page resulted in better recall of the ads and received more visual fixations, especially when the ad was at the top of the page.

8.9.3 Color Schemes and Aesthetics

Many of the guidelines surrounding the use of color on Web pages are based on the color wheel. The theory behind the color wheel, and color theory in general, dates back to Sir Isaac Newton. A century later, in 1810, Johann Wolfgang von Goethe was the first to study the psychological effects of color (Douma 2006). The color wheel as we know it today was created by Johannes Itten in 1920 with his publication of *The Art of Color: The Subjective Experience and Objective Rationale of Color* (Itten 1997). According to color theory, harmonious color schemes can be created by combining colors that have certain relationships on the color wheel, as illustrated in Figure 8.13:

- *Monochromatic color scheme:* Created by combining variations in lightness and saturation of a single color.
- *Analogous color scheme:* Created by combining colors that are adjacent to each other on the color wheel.
- *Complementary color scheme:* Created by combining two colors that are opposite each other on the color wheel.

- *Split complementary color scheme:* Created by combining a color with the two colors adjacent to its complementary color.
- *Triadic color scheme:* Created by combining three colors that are equally spaced around the color wheel.
- *Tetradic color scheme:* Created by combining four colors that form the points of a rectangle on the color wheel.

A number of tools are available online for experimenting with these color schemes. Some, such as the Color Scheme Designer (http://colorschemedesigner.com/) will even dynamically construct an example of a "Greeked" Web page using the color scheme you have selected.

In studying the effects of color and balance on Web page aesthetics and usability, Brady and Phillips (2003) took the Home page of an existing site (http://www.CreateForLess .com) that uses a triadic color scheme and created a new

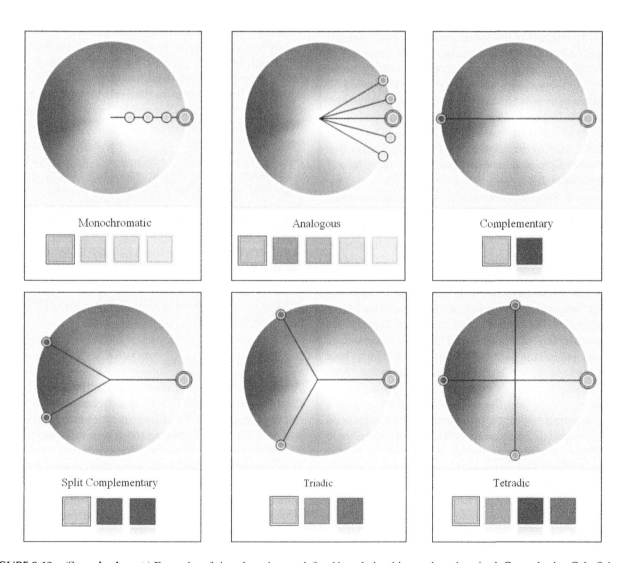

FIGURE 8.13 (See color insert.) Examples of six color schemes defined by relationships on the color wheel. Created using ColorSchemer Studio, Version 2.0 (http://www.ColorSchemer.com).

8.10.2 Form Controls

Radio buttons, text boxes, and dropdown menus are examples of HTML-based form controls that are used for inputting information. Eye-tracking studies show that form elements are very compelling and users often begin filling out the fields immediately rather than reading any introductory text (Jarrett and Gaffney 2009). Because of this salience, designers need to ensure that they are using the correct form controls for the situation and, as much as possible, that a user can complete the form without relying on introductory text. The two main challenges in form design are helping the user know what input a process needs in order to proceed and how exactly to enter data in the correct format (Pauwels et al. 2009). Designers should make sure that they are using the correct type of form control for the information that's being requested to reduce potential confusion and validation errors.

In some instances, a designer might wonder which form control is best for that situation. A text box has exclusive functionality (text entry), but other form controls seem interchangeable. For instance, you could use a list box to present two choices from which a user may select a single item, even though a pair of radio buttons is probably the preferred method for this type of presentation. To make the determination of which controls should be used, designers often turn to design guides such as the "Apple Human Interface Guidelines" (Apple Inc. 2009) and "The Windows User Experience Interaction Guidelines" (Microsoft Corporation 2009). These resources are important, as Tullis (1993) found that experienced software developers had difficulty choosing the best form controls for a particular task. Even though these publications are not specifically related to Web design, the user's basic interaction with form controls does not differ significantly among Windows, Macintosh, and Web-based applications.

Even after referring to published guidelines, some ambiguity may exist in determining which form controls should be used for certain applications. Very limited research has been conducted that compares form controls for certain tasks. The way the area around the form controls are formatted (labeling the controls, indicating required fields) can play a role in user's successful completion of the form. The research that is available in these two areas is reviewed in the following section.

8.10.2.1 Text Entry

The text box (text entry) control is unique among standard HTML form controls in its ability to allow a user to enter free-form text as opposed to making a selection. There are times when the functionality of a text box can be substituted by a selection control. An example of this would be the entry of a specific date. A user could type out a date (in a certain format) in a text box or the date could be selected with a series of drop-down lists or radio buttons. Gould et al. (1989) and Greene et al. (1992) both found that text entry was superior in speed and preference to selection controls for entry of dates.

8.10.2.2 Mutually Exclusive Selection

To allow a user to make a mutually exclusive selection on a Web page, a designer could use radio buttons, drop-down lists, or list boxes. All of these form controls allow a user to make a single selection from a set of choices. While the functionality of these elements is fairly interchangeable, they do not share the same level of visual prominence on a page. Penzo's (2006) eye-tracking study showed that drop-down lists were the most eye-catching form elements, and was the first fixation of all users on a simple form. When using a drop-down list as part of a form, designers should keep this salience in mind as it might cause users to skip over essential information and other form controls

Though all of these controls share a high degree of accuracy when manipulated by a user, research has shown that users made significantly faster selections when using radio buttons as compared to the other controls (Johnsgard et al. 1995; Tullis and Kodimer 1992). Additionally, Johnsgard et al. found that users completed tasks faster with radio buttons that were arranged in a meaningful, logical order when compared to a random order.

8.10.2.3 Nonmutually Exclusive Selection

Nonmutually exclusive selections (multiple selections) can be made with HTML form controls by using checkboxes, multiple-select list boxes, or a text box. Among these controls, Johnsgard et al. (1995) found that users performed tasks significantly faster with checkboxes compared to the other nonmutually exclusive selection controls. Additionally, checkboxes received the highest preference scores of the controls tested. Like radio buttons, participants performed better when the checkboxes were arranged in a meaningful and logical order instead of a random order. It is also noteworthy that unlike the other controls investigated with nonmutually exclusive selections, the text box had an exceptionally low accuracy rate (71%) compared to the checkbox (92%) or multiple-select list box (93%).

The above findings support the preferred use of radio buttons for mutually exclusive selections tasks and checkboxes for those tasks involving nonmutually exclusive selection. Johnsgard et al. (1995) suggest that radio buttons and checkboxes performed so well because all of the available options were visible to the users. This is supported by Bishu et al. (1991), who found that the best menu configurations were those that initially displayed all of the options.

8.10.2.4 Field Labels

All form controls are supported by some type of label that provides additional information or instruction to the user for what to do with that control. The most common placement on the Web today is left-aligned labels that are located to the left of input fields (Penzo 2006). Recent research on forms has focused on the best placement of labels for form elements to see if this left-aligned location is best or if another alternative helps users be more successful.

Penzo (2006) ran an eye-tracking study comparing task completion on forms where the only difference was the alignment of the form labels. He compared forms that had left-aligned labels placed to the left of the control (left-aligned), right-aligned labels located to the left of the control (right-aligned), and left-aligned labels located above the control (top-aligned). He found the longest saccade duration in the condition with left-aligned labels, as users had to spend more time looking between the label and the control because there was a variable amount of distance between the two. The right-aligned label condition had almost no saccadic activity between the labels and the input field (as shown in Figure 8.15) as users could quickly connect the meaning of the label to the field because of the ease of the lateral eye movement. The top-aligned condition also required fewer additional fixations and saccades, and users were able to quickly complete the form.

Das, McEwan, and Douglas (2008) conducted a similar eye-tracking study that replicated Penzo's findings. They found the longest average completion time in forms with left-aligned labels located next to the field and the shortest average completion times in forms with top- and right-aligned labels. They also found that users preferred right-aligned labels. Tan (2009) found that users filling out a set of existing Web forms preferred labels above the field as it was easier to scan down a column rather than having to scan from left to right.

These studies suggest that top-aligned and right-aligned labels are best, but left-aligned labels may be appropriate in some situations, such as when the labels can be located close to the fields (Jarrett and Gaffney 2009). This type of placement reduces the extra fixations required to move between the label and the field. Wroblewski (2008) recommends considering left-aligned labels when requesting unfamiliar data so a user can scan the left column of labels up and down without being interrupted by input fields. Although this increases the time users may need to spend on the form, the benefit for users in this situation may be worth it.

Penzo (2006) also looked at the effect of bolding the field label, which produced an almost sixty-percent increase in saccade time to move from the label to the input field. The input fields in his study had a dark border around them, so he theorized that there was an interaction between the bold text and the border and recommends removing heavy borders from input fields if a bold label is being used. He also cautions that

the forms used in his study had any visual design removed from them, which might have modified the impact of the bold labels. Both Tan (2009) and Wroblewski (2008) recommend using bold fonts to help emphasize the labels from the foreground of the layout, so the labels are not competing with the visual weight of the input fields.

8.10.2.5 Required versus Nonrequired Fields

Providing an indication of what form fields are required keeps users from encountering errors when they try to proceed. Approaches for presenting this information range from a message at the top of the page (e.g., when all fields on a page are required) and using an asterisk or "Required" label next to the fields.

Tullis and Pons (1997) compared different options for indicating required fields. They found that the fastest and most preferred method for indicating required fields was grouping the required fields separately but noted that logistically this may not always be possible. There were no significant differences between the other methods. Pauwels et al. (2009) tested two versions of a form, one using asterisks to indicate required fields and one using colored field backgrounds. Participants had significantly fewer errors when using the forms with colored fields indicating what was required and were also significantly faster and more satisfied. But the researchers note that screen readers are unable to distinguish the colored backgrounds, so it is best to use the color as an additional indicator of required fields.

8.10.2.6 Providing Feedback on Forms

It can be very frustrating when a user fills out a long form and clicks submit, only to be returned to the same page to correct a series of errors. Ideally, required field indicators, tips, and formatting guidelines have been provided in-context of the form (e.g., formatting requirements for dates) to limit the number of input errors, but even so, designers should ensure that errors can be handled gracefully.

JavaScript and Ajax can be used to provide immediate feedback to users as they are filling out forms. Ajax allows Web designers to provide feedback based on a user's actions without requiring a full page refresh (Shneiderman et al. 2009). Because they do not have to wait for a screen refresh, users are compelled to interact even more with the page. For example, a user can rate a movie on the Netflix site using the on-page rating system, but still maintain their context within the site.

Because the on-screen changes indicated by Ajax can be subtle, it can be necessary to provide some kind of additional indication that something has changed. Spool (2008b) mentions Kayak as an example that provides an interstitial message on some pages during the refresh to ensure that users notice something has changed. Spool notes that it is important to ensure that the experience degrades gracefully if the user's browser does not support JavaScript so users can still perform the same action.

Tan (2009) tested four existing registration forms, including two that provided instant feedback as users completed fields. The instant feedback was a tick at the end of the field

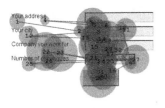
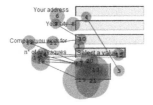

FIGURE 8.15 Eye-tracking results from Penzo (2006) showing the difference between scan paths for left-aligned (left) field labels and right-aligned (right) field labels. (From Penzo, M. 2006. Label placement in forms. *UXmatters*. http://www.uxmatters.com/mt/archives/2006/07/label-placement-in-forms.php (accessed November 7, 2010). With permission.)

that changed to a checkmark when the field was filled out. Participant response to this treatment was mixed, with some finding it helpful while others thought it was distracting. Tan also compared placements for animated tips and found that participants were more likely to notice and read the tips when displayed to the side of the relevant fields, and when the fields were aligned so it's easy to scan the page.

Wroblewski (2009) tested six variations of a typical Web registration form. The control version validated input after a user clicked the "create account button," and the five other versions providing different methods of inline validation. The inline validation provided was displayed next to the form input fields in the form of green checkmarks to indicate success and red triangles and text for error message. He found that users were more successful, less error-prone, and more satisfied on the forms with inline validation. His participants did not think the inline validation was necessary for fields such as name or address—questions that most people know the answers to. But for more difficult questions (such as selecting a username and password), the validation was reassuring and more likely to be noticed. The version of inline validation that had the lowest error rates and highest satisfaction ratings provided the validation after the user indicated that they were done with a question by moving on to a next one, instead of showing it before or while the user was answering the question. Also, success messages that were persistent performed better than those that faded away once the error was resolved.

8.10.3 Key Takeaways

- Keep forms in one column. Fields for names, dates, and time can be displayed on a single line as long as they are obviously grouped together.
- Whether formatting a form as one long page or several shorter pages, use principles of perceptual grouping to break the form into manageable chunks.
- Provide progress indicators when forms stretch across multiple pages, and provide early feedback that shows rapid progress.
- In general, field labels should be aligned either above the fields or right-aligned and to the left of the field. Consider left-aligned labels for information that is less familiar to users and if time spent on the form is less important.
- Use spacing to visually group each label and form control, and ensure that they're viewed as separate from the previous and following groups of labels and controls.
- Indicate required fields in some way, but only use color as an additional indicator rather than the primary one.
- Use inline validation for more difficult questions, such as selecting a username and password.
- Ensure that any treatments relying on JavaScript or Ajax degrade gracefully.

8.11 SPECIFIC TYPES OF PAGES

The principles described in this chapter can be applied to almost any type of Web page. This section details some additional considerations and common elements found on specific types of pages.

8.11.1 Home Pages

The most visited page of any Web site is the Home page. If a user knows the URL for a Web site, the Home page is likely to be their first interaction with the Web site. Users may arrive deeper on the site by following links from search engines or other sites, but the Home page is frequently the page that users will visit next (Nielsen 2002a). Recently, search engines such as Google have been linking to Web site Home pages directly from the search field, making it even more likely that users will begin at the Home page. Regardless, the Home page is typically accessible from most other pages in a Web site; users often expect a link to the Home page to be located in the top left corner of Web pages (Shaikh and Lenz 2006), and often it is directly linked to the logo of the Web site.

The Home page plays a vital role in communicating the purpose of a site and what a user will be able to accomplish. The goal of a Home page is to communicate what the company is, the value the site offers over the competition and the physical world, and the products or services offered (Nielsen and Tahir 2001). Further, the Home page should establish the tone and personality of the site while allowing users to start key tasks immediately and sending each user on the proper path (Redish 2007). The Home page also provides a basis for users' first impression of the Web site, which can be difficult to change and has a strong effect on users' perceived experience with and enjoyment of the rest of the site (Djamasbi, Siegel, and Tullis 2009; Lindgaard et al. 2006; Nisbett and Ross 1980).

No matter the domain, most Home pages consist of similar elements that make them distinguishable in layout from the other pages on the site. This content usually includes a masthead with a tagline, distinct and weighted category links listed in order of priority, and all major content categories (Koyani, Bailey, and Nall 2004). Home pages can be content rich, but they should not be wordy, as the information users are looking for is almost never on the Home page (Redish 2007). One study found that users' rating of the visual appeal of Home pages was negatively correlated to the total number of characters on the page and positively correlated to the size of the largest image on the page (Tullis and Tullis 2007). Another important layout consideration for Home pages is that some users are less likely to scroll, so important content should be located above the fold of the page (Djamasbi et al. 2008, 2009; Shrestha and Owens 2008).

To help users navigate the Web site, the Home page needs to provide a clear overview of the site contents. To do this, Spool (2006a) suggests a link-rich design that uses trigger words and is clustered together in a way that matches a user's

FIGURE 8.16   (See color insert.) A carousel of movies from Netflix.com—users can scroll left and right to view different movies. (From Netflix.com. With permission.)

mental model. An example of a link-rich version of a newspaper Home page might include the top level categories along with the headlines of the articles under each category.

In order to provide an overview of site content, some Web sites have started using carousels. A carousel is a fixed width space that shows a few images or links to content, with arrows on either side that a user can click to scroll through additional content, as shown in Figure 8.16. The user can also click on any of the items in the carousel, which will bring them to a page devoted to that topic. Little research has been done on the effectiveness of this type of interaction, but carousels seem to be most effective when there is not a need to compare the items being displayed and when there are alternative ways to get to the content linked to from within the carousel.

8.11.2   User Assistance, Help, and FAQs

There are various types of user assistance offered on Web sites and Web-based applications that are intended to help users find information or complete tasks. This information is rarely the first destination for users, as they typically only use help with the hope that they will be sent directly to the information that's needed to help them solve their specific problem (Shneiderman et al. 2009).

Relatively few research studies have been conducted to learn if and how these features are used. Even less research has been done to learn how best to design these features. Until recently, most Web-based user assistance has been essentially hard-copy documentation simply adapted for Web delivery. Previous research has demonstrated that this type of online help is infrequently used and not always helpful (Parush and Parush 2001). Technical communicators are now attempting to re-define and adapt how user assistance is designed for Web-based delivery. The emphasis is now on designing intelligent visually-based systems that give relevant, context-sensitive information to users, instead of text-based information that has been adapted from paper-based documentation.

8.11.2.1   Web-Based User Assistance

There are numerous variations of user assistance offered to users on the Web including field-level help, page-level help, general site help (not context-sensitive), and help simply built into the page. Additionally, the display of these user-assistance features varies: sometimes opening in a secondary

window, sometimes within the same window (a frame to the left or right), or simply displaying without a frame but in a certain region of the screen.

A study of the WinWriters online registration process for their annual conference tested five different types of user assistance systems (Ellison 2003):

- Condition 1: No user assistance
- Condition 2: Field-level context-sensitive Help (same window, frame to right)
- Condition 3: Page-level context-sensitive Help (secondary window)
- Condition 4: Field-level context-sensitive Help (small secondary window)
- Condition 5: Help built into the registration form (explanatory text on page)

They found an overall reluctance of most users to use the help provided, independent of condition, and that users tended to ignore information that is displayed in a right frame (condition 2). The most successful type of help was that which was built into the page (basically explanatory text provided within the form), with upfront information presented before the registration process began.

8.11.2.2   Embedded Assistance

Nielsen (2008) suggests that the current trend is to downplay user assistance as a separate feature while focusing on on-screen instructions, tooltips that display information automatically on mouse over, and click-tips that display information when the user clicks on a specific area of the screen. These approaches are thought of as embedded user assistance. This embedded user assistance provides help at the precise location needed, it keeps users "on task," and users see it as part of the application rather than as "help" (Ellison 2007).

One advantage of this type of assistance is that the information is clearly context sensitive. A common way to get field-level help on a form is to let the user click on a question mark icon or "What's this?" link located right next to the field in question. The simplest way to take context into account is to monitor the cursor location and provide helpful information about the object under the cursor, either after the mouse has rested over the object for a couple of seconds or after the user has pressed a help key (Shneiderman et al. 2009). Other ways to provide context-sensitive information is to position static descriptive text next to or under a field or control or to provide a dedicated user assistance pane that proactively displays useful information (Ellison, 2007).

8.11.2.3   Visual-Based Approaches

As the field of documentation has transitioned from hard-copy documents to Web-based documentation, there has been some effort to understand and design more visually based (instead of text-based) systems. Complex Web interfaces may require additional visual help such as demonstrations or tutorials to help a user successfully complete their task (Shneiderman et al. 2009).

Baecker (2002) described five visually-based solutions or sets of tools to provide help to Web users including screen linking (remote control of a user's screen), visual streaming (like video conferencing), animated icons (like bubble help and tool tips), screen capture (tools that allow easy capture for demo purposes), and structured video tools (multimedia that shows users how to do something). While the effectiveness of these approaches has not been studied, it is hypothesized that they will provide more relevant, timely help to users.

A more novel technique recently implemented by Microsoft and other vendors is the "Show Me" feature located within the help. This feature actively opens windows and controls in order to take the users where they need to go to accomplish a specific task. This type of solution is an example of how to more proactively provide user assistance to Web users that give users the information they need, when they need it.

8.11.2.4 Online Bulletin Boards

Online bulletin boards provide a way for a site to let users talk with and interact with other users to get assistance. Questions can be submitted directly to a help desk or staff member or they can be posted to a bulletin board (Shneiderman et al. 2009). Ideally, the responses are archived and made searchable for future users of the site. One of the best practices that has emerged for this type of information is to show the most recently answered questions first, while also providing a way to search the other answers by keyword or topic. Some bulletin boards also pull out Frequently Asked Questions as "sticky" posts or "best bets" that are always listed at the top of the question threads.

8.11.3 FAQs

Frequently asked questions (FAQs) are an outgrowth of Usenet groups where participants would frequently ask the same questions of the group administrator. In order to efficiently provide answers to these questions (without having to re-post repeatedly), FAQs were created.

FAQs are now a common feature on most Web sites and can be helpful to users if they actually reflect the most frequently asked questions users ask instead of questions the site designers wished users would ask (Nielsen 2002b). FAQs should be concise, simple questions with answers, designed in a way that allows users to quickly scan for target words. While the types of user-assistance offered may become more visually based, it is unlikely that anything will replace the importance or necessity of a well-designed set of the most frequently asked questions along with answers.

8.11.4 Gallery Pages

Gallery pages provide a list of links to content. These pages separate users who will find the content they are seeking from those who will not (Spool 2005). In the e-commerce world, these pages usually show multiple products within a category at once. For example, a gallery page for digital cameras would provide a listing of all the digital cameras offered by that site, possibly spread out across multiple pages, with different options on the page for sorting and comparison. Clicking in to a specific camera would bring the user to a page dedicated to that product. The goal of the gallery page is to get the user as quickly as possible to the product page that best meets their needs while reducing the likelihood that the user will need to click back to the gallery page and in to several other product pages before finding what they need (a behavior known as pogo-sticking). Spool (2005) shows that the more pogo-sticking a user does on a site, the less likely she is to make a purchase.

The gallery page needs to provide a listing of available products or content along with key features and details that can help users compare and contrast those items. If possible, pictures of each product can be useful to help with this decision making process. Hong, Thong, and Tam (2004) showed that product listing pages (their term for gallery pages) that includes images with text had shorter information search times, better recall of brand name and product images, and more positive attitudes toward the screen design and using the Web site than text-only versions of the page.

Redish (2007) recommends formatting textual information on gallery pages in fragments or bullets, but also to ensure that the information provided will be helpful for users in deciding how to move ahead. She describes these pages as "pathway" pages, because similar to Home pages, a gallery page is rarely a user's final destination. Spool (2005) also recommends that products be listed in order of importance. In addition, many sites provide more complex comparison tools such as additional ways of sorting the product listings and filtering capabilities.

8.11.5 Search Results

Search results pages can be thought of as a specialized type of gallery page (Spool 2008a). The goal of the search results page is to get the user to their desired destination in as few clicks as possible. A well designed search results page lets the user identify the best result, and continue on to the content they need. Ideally the user will not need to pogo-stick from an individual content page back to the search results page because the content did not match their needs. Spool (2008a) found that 66% of purchases on e-commerce sites happened without any pogo-sticking at all from the search results page and that the more pogo-sticking occurred, the less likely the users were to make a purchase.

Most sites use similar formats for presenting search results. These pages typically show a field repeating the initial search term, document titles, descriptive text for each result, the URL, and sometimes additional metadata or links. Eye-tracking research shows that most users view search results in a roughly linear order, with the most gaze activity directed at the first few items (Cutrell and Guan 2007).

Dumais, Cutrell, and Chen (2001) showed that search results pages that integrated category names and page titles, and reorganized the results by category, were significantly faster and preferred more than just a list of search results. This organization allowed users to quickly focus on areas of interest without needing to examine individual page titles. They also found that showing inline summaries improved performance more than showing summaries on-hover, as the additional scrolling time required on the page was offset by the cognitive effort required to decide which items to hover on for additional information.

Cutrell and Guan (2007) found that for informational tasks (where users had to find specific information that could be found in one or more places), having contextual snippets listed with each search result improved performance. The snippets also improved the accuracy of clicks in to search results. For navigational tasks (where users had to find a specific Web page), the contextual snippets both increased the time on task and reduced the accuracy. Eye-tracking data from their study showed that the attention paid to the snippets decreased the amount of time spent on the URL, which may explain the discrepancy across task types.

Another common question is how many search results to display at once. Bernard, Baker, and Fernandez (2002) conducted a study investigating the most advantageous way of presenting search results. They examined three sets of results pages, one which presented 10 results at a time, one that presented 50 results at a time and one that presented 100 results at a time. They found that the pages with 50 results had the fastest search time and were most preferred by participants. They concluded that participants preferred and performed best on results page layouts formatted for reduced paging and scrolling, and they recommend having moderate amounts of links (around 50) per page.

However, it is important to have as many of the search results visible above the fold of the page as possible. Hynes (2002) described changes made to the display of search results on the Staples.com site. In the previous design, whenever a user executed a search for a product, the design of the results page was such that very little of the results list fell above the fold. When they redesigned the results page to show more of the list above the fold, they found that user drop-off (leaving the site at that point) decreased by 10%.

8.11.6 Dashboard and Portal Pages

Unlike informational Web sites, portals are organized by channels that provide self-service options, customization options, and other self-contained utilities (Owens and Shrestha 2008; see Eisen, this volume). Most portals have limited layout options, typically with two or three columns available for customized content. Eye-tracking research shows that for both types of column layouts, users scan across columns of channels from top-to-bottom (Owens and Shrestha 2008). Other research showed that, despite column layouts, users read across rows instead of reading down columns (Goldberg et al. 2002). Owens and Shrestha (2008) found that users were faster to find information at the top left of a two-column layout, and top center left of a three-column layout. On the basis of these findings, they recommend placing the most important

on the page in these locations as users may be able to find it more quickly.

Social networking sites and intranet sites are examples of portal sites. The spread of social networking sites and mobile utilities has introduced new types of interactions. Fox and Naidu (2009) describe social networking sites as providing the ability to see others social networks, and offering users some level of customization such as uploading pictures and adjusting settings. Sites like Twitter and Facebook provide a running list of updates from a user's social networking and an entry form for a user to enter their own update.

While many principles of Web site design apply to these new interactions, the context in which they're being used has not yet been researched in depth. Fox and Naidu (2009) performed a usability evaluation of three social networking sites (MySpace, Facebook, and Orkut). Participants struggled on these sites with inadequate feedback and error messages and improper link location. These issues, however, are not unique to dashboard and portal Web pages and are discussed in other sections of this chapter.

A common usability issue on social networking sites is a lack of visibility into key functionality and insufficient feedback on the outcome of attempted actions. Owens, Lenz, and Speagle (2009) noted that Twitter users struggled with the lack of salient differences between modes, as a great deal of functionality is hidden until certain areas (e.g., messages) are moused over. In their usability study comparing three different social networking sites, Fox and Naidu (2009) found that certain key tasks did not provide adequate feedback when users took action, resulting in users repeating the action as they were not sure if their first attempt was successful.

8.11.7 Content and Product Pages

The page that is linked to from a gallery page is usually a product or content page. These content pages are usually a user's most frequent final destination as it contains the information the user came to the site to find (Perfetti 2005). This content might consist of something like a product description, the full text of a news article, and possibly some type of image or multimedia elements. In an e-commerce space, these pages should also include related or similar items, the product price, and a clear way to purchase the item above the fold.

The Poynter Institute has done a series of eye-tracking studies focused on online news article pages (Outing and Ruel 2004). Their studies have shown that single column text is better than paper-like multicolumn layout, and that the upper left content is watched before any other part of the page. Also, shorter paragraphs seemed to encourage reading, as paragraphs of one or two sentences received more than twice as many overall eye fixations than those with longer paragraphs.

Nielsen (2006) reviewed eye-tracking heat maps that included a product page for an e-commerce site. His analysis showed a similar "F" viewing pattern to the other pages, but with additional fixation time on the right-hand side where the price and an "add to cart" button were located. Some of his later research (Nielsen and Pernice 2010) also showed the importance of images, preferably from different angles and in different sizes, that let the user verify that the product has the desired features.

8.12 CONCLUSION

In this chapter we have tried to summarize some of the human factors issues surrounding the presentation of information on the Web. We hope that this will help Web designers make more informed decisions about how to present information and also help point researchers in promising directions for future investigation.

REFERENCES

Albert, W., W. Gribbons, and J. Almadas. 2009. Pre-conscious assessment of trust: A case study of financial and health care Web sites. *Human Factors and Ergonomics Society Annual Meeting Proceedings* 53: 449–453, doi:10.1518/107118109X 12524441082027.

Apple Inc. 2009. Apple human interface guidelines. http://developer .apple.com/mac/library/documentation/UserExperience/ Conceptual/AppleHIGuidelines/XHIGIntro/ (accessed November 7, 2010)

Arditi, A. 2010. Making text legible: designing for people with partial sight. http://www.lighthouse.org/accessibility/design/ accessible-print-design/making-text-legible/ (accessed March 6, 2010).

Arditi, A., and J. Cho. 2007. Letter case and text legibility in normal and low vision. *Vision Research* 47(19): 2499–2505.

Aten, T. R., L. Gugerty, and R. A. Tyrrell. 2002. Legibility of words rendered using cleartype. *Human Factors and Ergonomics Society Annual Meeting Proceedings* 46: 1684–1687.

Baecker, R. 2002. Showing instead of telling. Paper presented at the *Proceedings of the 20th Annual International Conference on Computer Documentation* (Toronto, ON, Canada, October 20–23, 2002).

Bailey, R. W. 2000. Link affordance. http://webusability.com/ article_link_affordance_11_2000.htm (accessed March 14, 2010).

Bailey, R. W., S. Koyani, and J. Nall. 2000. Usability testing of several health information web sites. Technical Report. National Cancer Institute.

Baker, J. R. 2003. Impact of paging versus scrolling on reading online text passages. *Usability News* 5(1).

Baker, J. R. 2005. Is multiple-column online text better? It depends! *Usability News* 7(2). http://www.surl.org/usabilitynews/72/ columns.asp (accessed November 7, 2010).

Bauer, D., and C. R. Cavonius.1980. Improving the legibility of visual display units through contrast reversal. In *Ergonomic Aspects of Visual Display Terminals*, eds. E. Grandjean and E. Vigliani, 137–142. London, UK: Taylor & Francis.

Benbasat, I., and R. Schroeder. 1977. An experimental investigation of some MIS design variables. *Management Information Systems Quarterly* 1(1): 37–49.

Bergel, M., A. Chadwick-Dias, D. Degler, and K. Walser. 2009. Applied design principles for rich Web interactions: full day tutorial. Paper presented at the UPA.

Bernard, M. 2002. Examining the effects of hypertext shape on user performance. *Usability News* 4(2).

Bernard, M., R. Baker, and M. Fernandez. 2002. Paging vs. scrolling: Looking for the best way to present search results. *Usability News*, 4.1.

Bernard, M., B. Chaparro, and R. Thomasson. 2000. Finding information on the Web: Does the amount of whitespace really matter? *Usability News*, 2(1), http://www.surl.org/ usabilitynews/21/whitespace.asp.

Bernard, M., M. Fernandez, and S. Hull. 2002. The effects of line length on children and adults' online reading performance. *Usability Journal* 4(2).

Bernard, M., and C. Hamblin. 2003. Cascading versus indexed menu design. *Usability News* 5(1). http://www.surl.org/ usabilitynews/51/menu.asp (accessed November 6, 2010).

Bernard, M., and L. Larsen. 2001. What is the best layout for multiple-column web pages? *Usability News* 3(2). http://www.surl.org/ usabilitynews/32/layout.asp (accessed November 6, 2010).

Bernard, M., and M. Mills. 2000. So, what size and type of font should I use on my website? *Usability News* 2(2). http:// psychology.wichita.edu/surl/usabilitynews/22/font.asp (accessed November 7, 2010).

Bernard, M., C. Liao, and M. Mills. 2001. Determining the best online font for older adults. *Usability News* 3(1). http:// psychology.wichita.edu/surl/usabilitynews/3W/fontSR.asp (accessed November 7, 2010).

Bernard, M., and S. Hull. 2002. Where should you put the links? Comparing embedded and framed/non-framed links. *Usability News* 4.1.

Bernard, M., M. Mills, M. Peterson, and K. Storrer. 2001. A comparison of popular online fonts: which is best and when? *Usability News* 3(2).

Beymer, D., D. Russell, and P. Orton. 2008. An Eye Tracking Study of How Font Size and Type Influence Online Reading. *Proceedings of the HCI08 Conference on People and Computers XXII 2008*, 15–18.

Beymer, D., D. Russell, and P. Orton. 2007. An eye tracking study of how pictures influence online reading. Paper presented at the Interact.

Bishu, R. R., P. Zhan, G. A. Sheeley, and W. P. Adams. 1991. Depth/ direction issues in menu design. Paper presented at the Annual International Industrial Ergonomics and Safety Conference (Lake Tahoe, Nevada, June 10–14, 1991).

Bohan, M., and D. Scarlett. 2003. Can expanding targets make object selection easier. *Usability News* 5(1).

Bouch, A., A. Kuchinsky, and N. Bhatti. 2000. Quality is in the eye of the beholder: meeting users' requirements for Internet quality of service. Paper presented at the *Proceedings of the SIGCHI conference on Human Factors in Computing Systems* (The Hague, Netherlands, April 1–6, 2000).

Boyarski, D., C. Neuwirth, J. Forlizzi, and S. H. Regli. 1998. A study of fonts designed for screen display. Paper presented at the CHI'98 (Los Angeles, CA, April 18–23, 1998).

Brady, L., and C. Phillips. 2003. Aesthetics and usability: A look at color and balance. *Usability News* 5(1).

Bricklin, D. 1998. When (and how) to use frames. http://www .gooddocuments.com/techniques/whenframes_m.htm (accessed Aug. 30, 2003).

Buchner, A., S. Mayr, and M. Brandt. 2009. The advantage of positive text-background polarity is due to high display luminance. *Ergonomics* 52(7): 882–886.

Buscher, G., E. Cutrell, and M. R. Morris. 2009. *What do you see when you're surfing?: Using eye tracking to predict salient regions of web pages*. Paper presented at the *Proceedings of the 27th International Conference on Human Factors in Computing Systems* (Boston, MA, April 4–9, 2009).

Bush, V. 1945. As we may think. *The Atlantic Monthly,* July 1945.

Campbell, A. J., F. M. Marchetti, and D. J. K. Mewhort. 1981. Reading speed and test production: a note on right-justification techniques. *Ergonomics* 24: 127–136.

Carter, L. 1947. An experiment on the design of tables and graphs used for presenting numerical data. *Journal of Applied Psychology* 31: 640–650.

Carter, R. L. 1982. Visual search with color. *Journal of Experimental Psychology: Human Perception and Performance* 8: 127–136.

Chaparro, B. S., A. D. Shaikh, and A. Chaparro. 2006. Examining the legibility of two new cleartype fonts. *Usability News* 8.1. http://psychology.wichita.edu/surl/usabilitynews/81/legibility.asp (accessed November 7, 2010).

Chassot, C. 2004. Accessible pop-up links. *A List Apart.*

Christ, R. E. 1975. Review and analysis of color coding research for visual displays. *Human Factors* 17: 542–570.

Christ, R. E., and W. H. Teichner. 1973. *Color Research for Visual Displays.* Las Cruces: New Mexico State University.

Clicktale. 2007. ClickTale scrolling research report V2.0 - Part 1: Visibility and scroll reach & Part 2—Visitor attention and Web page exposure. http://blog.clicktale.com/2007/10/05/clicktale-scrolling-research-report-v20-part-1-visibility-and-scroll-reach/ (accessed Dec. 1, 2009).

Coll, R., J. Coll, and G. Thakur. 1994. Graphs and tables: A four-factor experiment. *Communications of the ACM* 37(4): 77–86.

Conrad, F., M. Couper, R. Tourangeau, and A. Peytchev. 2005. Impact of progress feedback on task completion: First impressions matter. Paper presented at the CHI '05 Extended Abstracts on Human Factors in Computing Systems (Portland, OR, April 2–7, 2005).

Cushman, W. H. 1986. Reading for microfiche, a VDT, and the printed page: Subjective fatigue and performance. *Human Factors* 28: 63–73.

Cutrell, E., and Z. Guan. 2007. What are you looking for?: An eye-tracking study of information usage in web search. Paper presented at the *Proceedings of the SIGCHI Conference on Human Factors in Computing Systems* (San Jose, CA, April 28–May 3, 2007).

Das, S., T. McEwan, and D. Douglas. 2008. Using eye-tracking to evaluate label alignment in online forms. *Proceedings of the Fifth Nordic Conference on Human–Computer Interaction 2008,* 451–454.

Dillon, A., L. Kleinman, G. O. Choi, and R. Bias. 2006. Visual search and reading tasks using ClearType and regular displays: Two experiments. Paper presented at the Proceedings of the SIGCHI Conference on Human Factors in Computing Systems.

Djamasbi, S., M. Siegel, and T. Tullis. Forthcoming. Generation Y, Web design, and eye tracking. *International Journal of Human–Computer Studies* doi:10.1016/j.ijhcs.2009.12.006.

Djamasbi, S., M. Siegel, T. S. Tullis, and R. Dai. 2009. Tracking users' viewing pattern. Paper presented at the Eighth Pre-ICIS SIG HCI Workshop (Phoenix, Arizona, December 14, 2009).

Djamasbi, S., T. S. Tullis, J. Hsu, E. Mazuera, K. Osberg, and J. Bosch. 2007. Gender preferences in Web design: Usability testing through eye tracking. Paper presented at the AMCIS 2007, Keystone, Colorado, August 2007.

Djamasbi, S., T. S. Tullis, M. Siegel, F. Ng, D. Capozzo, and R. Groezinger. 2008. Generation Y & Web design: Usability testing through eye tracking. Paper presented at the Fourteenth Americas Conference on Information System (Toronto, ON, Canada, August 14–17, 2008).

Douma, M. 2006. Color vision and art. from http://www.webexhibits.org/colorart/bh.html (accessed March 14, 2010).

Duchnicky, J. L., and P. A. Kolers. 1983. Readability of text scrolled on visual display terminals as a function of window size. *Human Factors* 25: 683–692.

Dumais, S., E. Cutrell, and H. Chen. 2001. Optimizing search by showing results in context. Paper presented at the Proceedings of the SIGCHI Conference on Human Factors in Computing Systems (Seattle, WA, USA, March 31–April 5, 2001).

Dyson, M. C., and M. Haselgrove. 2001. The influence of reading speed and line length on the effectiveness of reading from a screen. *International Journal of Human–Computer Studies* 54: 585–612.

Dyson, M. C., and G. J. Kipping. 1998. The effects of linelength and method of movement on patterns of reading from screen. *Visible Language* 32: 150–181.

Eisen, P., this volume. Design of portals. In *Handbook of Human Factors in Web Design,* 2nd ed., eds. K.-P. L. Vu and R. W. Proctor, 303–328. Boca Raton, FL: CRC Press.

Ellison, M. 2001. Secondary Windows in online help: What do users really make of them? *Usability Interface: Newsletter of the STC Usability SIG* 7(3).

Ellison, M. 2003. A usability test of Web-based user assistance. http://www.winwriters.com/usability_test_analysis.htm (accessed Sept. 1, 2003).

Ellison, M. 2007. Embedded user assistance: The future for software help? *Interactions* 14(1): 30–31.

Enders, J. 2008. Zebra striping: more data for the case. *A List Apart.* http://www.alistapart.com/articles/zebrastripingmoredataforthecase/ (accessed November 7, 2010).

Fabrizio, R., I. Kaplan, and G. Teal. 1967. Readability of a function of the straightness of right-hand margins. *Journal of Typographic Research* 1(1).

Faraday, P. 2000. Visually critiquing Web pages. Paper presented at the 6th Conference on Human Factors and the Web. http://facweb.cs.depaul.edu/cmiller/faraday/Faraday.htm (Austin, Texas, USA, June 19, 2000).

Feliciano, G., R. Powers, and E. Bryant. 1963. The presentation of statistical information. *AV Communications Review* 11(13): 32–39.

Few, S. 2004. *Show Me the Numbers: Designing Tables and Graphs to Enlighten.* Oakland, CA: Analytics Press.

Fowler, S. L., and V. R. Stanwick. 1995. *The GUI Style Guide.* Cambridge, MA: AP Professional.

Fox, D., and S. Naidu. 2009. Usability evaluation of three social networking sites. *Usability News* 11.1.

Fox, D., A. Smith, B. S. Chaparro, and A. D. Shaikh. 2009. Optimizing presentation of AdSense ads within blogs. *Human Factors and Ergonomics Society Annual Meeting Proceeding* 53: 1267–1271, doi:10.1518/107118109X12524443346879.

Galletta, D. F., R. M. Henry, S. McCoy, and P. Polak. 2006. When the wait isn't so bad: The interacting effects of Website delay, familiarity, and breadth. *Information Systems Research* 17(1): 20–37.

Ghani, J. 1981. *The Effects of Information Representation and Modification of Decision Performance.* Philadelphia: Universtiy of Pennsylvania.

Goldberg, J. H., M. J. Stimson, M. Lewenstein, N. Scott, and A. M. Wichansky. 2002. Eyc tracking in web search tasks: design implications. Paper presented at the Proceedings of the 2002 Symposium on Eye Tracking Research & Applications (New Orleans, LA, USA, March 25–27, 2002).

Gould, J. D., S. J. Boies, M. Meluson, M. Rasammy, and A. Vosburgh. 1989. Entry and selection methods for specifying dates. *Human Factors* 31: 199–214.

Gould, J. D., L. Alfaro, R. Finn, B. Haupt, and A. Minuto. 1987. Reading from CRT displays can be as fast as reading from paper. *Human Factors* 29: 497–517.

Grace, G. 1966. Application of empirical methods to computer-based system design. *Journal of Applied Psychology* 50(6): 442–450.

Granka, L. A., T. Joachims, and G. Gay. 2004. Eye-tracking analysis of user behavior in WWW search. Paper presented at the *Proceedings of the 27th Annual International ACM SIGIR Conference on Research and Development in Information Retrieval* (Sheffield, United Kingdom, July 25–29, 2004).

Greene, S., J. D. Gould, M. R. Boies, M. Rasammy, and A. Meluson. 1992. Entry and selection-based methods of human–computer interaction. *Human Factors* 34: 97–113.

Gregory, M., and E. C. Poulton. 1970. Even versus uneven right-hand margins and the rate of comprehension in reading. *Ergonomics* 13: 427–434.

Hall, R. H., and P. Hanna. 2004. The impact of web page text-background colour combinations on readability, retention, aesthetics and behavioural intention. *Behaviour & Information Technology* 23(3): 183–195.

Harper, S., Y. Yesilada, C. Goble, and R. Stevens. 2004. How much is too much in a hypertext link?: Investigating context and preview—a formative evaluation. Paper presented at the *Proceedings of the Fifteenth ACM Conference on Hypertext and Hypermedia*.

Harris, R. L. 1999. *Information Graphics: A Comprehensive Illustrated Reference*: Oxford, UK: Oxford University Press.

Hess, R. 2000. Can color-blind users see your site? Microsoft MSDN Library. http://msdn.microsoft.com/en-us/library/bb 263953(VS.85).aspx (accessed November 6, 2010).

Hill, A., and L. Scharff. 1997. Readability of Websites with various foreground/background color combinations, font types and word styles. http://www.laurenscharff.com/research/AHNCUR.html (accessed November 6, 2010).

Hill, A., and L. Scharff. 1999. Readability of computer displays as a function of colour, saturation, and texture backgrounds. In *Engineering Psychology and Cognitive Ergonomics*, vol. 4, ed. D. Harris, 123–130. Germany: Springer.

Hong, W., J. Y. L. Thong, and K. Y. Tam. 2004. Designing product listing pages on e-commerce websites: An examination of presentation mode and information format. *International Journal of Human–Computer Studies* 61(4): 481–503.

Horrigan, J. 2009. Home broadband adoption 2009. http://www .pewinternet.org/Reports/2009/10-Home-Broadband-Adoption-2009.aspx (accessed November 6, 2010).

Hudson, J. M., and K. Viswanadha. 2009. Can "wow" be a design goal? *Interactions* 16(1): 58–61.

Hynes, C. 2002. Keeping users stuck to your site. White paper, May 15. Available from Human Factors International. http://www .humanfactors.com/training/webcastsrequest.asp?staples=yes (accessed Nov. 29, 2009).

IBM Corporation. 2009. IBM Web accessibility checklist. http://www-03.ibm.com/able/guidelines/web/accessweb.html (accessed November 6, 2010).

Itten, J. 1997. *The Art of Color: The Subjective Experience and Objective Rationale of Color*, rev. ed. New York: John Wiley.

Jarrett, C., and G. Gaffney. 2009. *Forms that Work: Designing Web Forms for Usability*. Burlington, MA: Morgan Kaufmann.

Johansson, R. 2004. Who framed the web: Frames and usability. 456 Berea St. http://www.456bereastreet.com/archive/200411/who_framed_the_web_frames_and_usability/ (accessed Dec. 10, 2009).

Johnsgard, T. J., S. R. Page, R. D. Wilson, and R. J. Zeno. 1995. A comparison of graphical user interface widgets for various tasks. Paper presented at the Human Factors Society 39th Annual Meeting (San Diego, CA, October 9–13, 1995).

Joseph, K. M., B. A. Knott, and R. A. Grier. 2002. The effects of bold text on visual search of form fields. *Human Factors and Ergonomics Society Annual Meeting Proceedings* 46: 583–587.

Josephson, S., and M. E. Holmes. 2002. *Visual attention to repeated internet images: Testing the scanpath theory on the world wide web*. Paper presented at the *Proceedings of the 2002 Symposium on Eye Tracking Research & Applications* (New Orleans, LA, March 25–27, 2002).

Katz, M. A., and M. D. Byrne. 2003. Effects of scent and breadth on use of site-specific search on e-commerce Web sites. *ACM Transactions of Computer–Human Interactions* 10(3): 198–220.

Kiger, J. I. 1984. The depth/breadth trade-off in the design of menu-driven user interfaces. *International Journal of Man–Machine Studies* 20: 201–213.

Kolers, P. A., R. L. Duchnicky, and D. C. Ferguson. 1981. Eye movements measurement of reaability of CRT displays. *Human Factors* 23: 517–527.

Kopala, C. J. 1981. The use of color coded symbols in a highly dense situation display. Paper presented at the Human Factors Society 23rd Annual Meeting (Rochester, NY, October 1981).

Koyani, S. J., R. W. Bailey, and J. R. Nall. 2004. *Research-Based Web Design & Usability Guidelines*. Washington, DC: U.S. Department of Health and Human Services.

Kruk, R. S., and P. Muter. 1984. Reading of continuous text on video screens. *Human Factors* 26: 339–345.

Kühne, A., H. Krueger, W. Graf, and L. Mers. (1986). Positive versus negative image polarity. Paper presented at the International Science Conference: Work with Display Units, Stockholm, Sweden, May 12–16.

Larson, K., and M. Czerwinski. 1998. Web page design: Implications of memory, structure and scent for information retrieval. Paper presented at the *Proceedings of the SIGCHI Conference on Human Factors in Computing Systems* (April 18–23, 1998).

Levy, E., J. Zacks, B. Tversky, and D. Schiano. 1996. Gratuitous graphics? Putting preferences in perspective. Paper presented at the SIGCHI Conference on Human Factors in Computing Systems (Vancouver, British Columbia, Canada, April 13–18, 1996).

Lida, B., S. Hull, and K. Pilcher. 2003. Breadcrumb navigation: an exploratory study of usage. *Usability News* 5(1).

Lindgaard, G., G. Fernandes, C. Dudek, and J. Brown. 2006. Attention web designers: You have 50 milliseconds to make a first good impression! *Behaviour & Information Technology* 25: 115–126.

Ling, J., and P. V. Schaik. 2006. The influence of font type and line length on visual search and information retrieval in web pages. *International Journal of Human–Computer Studies* 64(5): 395–404.

Lucas, H. 1981. An experimental investigation of the use of computer-based graphics in decision making. *Management Science* 27(7): 757–768.

Lynch, P. J., and S. Horton. 2009. *Web Style Guide*, 3rd ed. *Basic Design Principles for Creating Web Sites*. New Haven, CT: Yale University Press.

Maldonado, C. A., and M. L. Resnick. 2002. Do common user interface design patterns improve navigation. *Human Factors and Ergonomics Society Annual Meeting Proceedings* 46: 1315–1319.

McTyre, J. H., and W. D. Frommer. 1985. Effects of character/background color combinations on CRT character legibility. Paper presented at the Human Factors Society 29th Annual Meeting (Baltimore, MD, September 1985).

Meyer, J., D. Shinar, Y. Bitan, and D. Leiser. 1996. Duration estimates and users' preferences in human–computer interaction. *Ergonomics* 39(1): 46–60.

Meyers-Levy, J., and D. Maheswaran. 1991. Exploring differences in males' and females' processing strategy. *Journal of Consumer Research* 18: 63–70.

Microsoft Corporation. 2009. Windows user experience interaction guidelines. http://msdn.microsoft.com/en-us/library/aa511258.aspx (accessed Nov. 29, 2009).

Miller, C. S., and R. W. Remington. 2002. Effects of structure and label ambiguity on information navigation. Paper presented at the CHI '02 Extended Abstracts on Human Factors in Computing Systems (Minneapolis, MN, April 20–25, 2002).

Miller, D. P. 1981. The depth/breadth tradeoff in hierarchical computer menus. Paper presented at the Human Factors Society Annual Meeting (Rochester, NY, October 1981).

Mills, C. B., and L. J. Weldon. 1987. Reading text from computer screens. *ACM Computing Surveys* 19(4): 329–358.

Moschovitis, C. J. P., H. Poole, T. Schuyler, and T. M. Senft. 1999. *History of the Internet: A Chronology, 1843 to the Present.* ABC-CLIO, Oxford UK.

Mosconi, M., M. Porta, and A. Ravarelli. 2008. On-line newspapers and multimedia content: An eye tracking study. Paper presented at the *Proceedings of the 26th Annual ACM International Conference on Design of Communication* (Lisboa, Portugal, September 22–24, 2008).

Moskel, S., J. Erno, and B. Shneiderman. 1984. *Proofreading and Comprehension of Text on Screens and Paper.* College Park: University of Maryland.

Mueller, S. T. 2005. Letter similarity data set archive. http://obereed.net/lettersim/index.html (accessed March 6, 2010).

Muncer, S. J., B. S. Gorman, S. Gorman, and D. Bibel. 1986. Right is wrong: An examination of the effect of right justification on reading. *British Journal of Educational Technology* 17(1): 5–10.

Murayama, N., S. Saito, and M. Okumura. 2004. Are web pages characterized by color? Paper presented at the *Proceedings of the 13th International World Wide Web Conference on Alternate Track Papers & Posters* (New York, New York, May 17–22, 2004).

Nawrocki, I. 1972. *Alphanumeric versus Graphic Displays in a Problem-solving Task.* Arlington, VA: U.S. Army Behavior and Systems Research Laboratory.

Nielsen, J. 1996. Why frames suck (most of the time). Alertbox. http://www.useit.com/alertbox/9612.html (accessed Aug. 30, 2003).

Nielsen, J. 1998. Using link titles to help users predict where they are going. Alertbox, Jan. 1998 http://www.useit.com/alertbox/980111.html (accessed March 14, 2010).

Nielsen, J. 2002a. Top ten guidelines for Homepage usability. Alertbox. http://www.useit.com/alertbox/20020512.html (accessed Nov. 26, 2009).

Nielsen, J. 2002b. Top ten Web-design mistakes of 2002. Alertbox. http://www.useit.com/alertbox/20021223.html (accessed Nov. 26, 2009).

Nielsen, J. 2004a. Change the color of visited links. Alertbox, May 3. http://www.useit.com/alertbox/20040503.html (accessed March 14, 2010).

Nielsen, J. 2004b. Guidelines for visualizing links. http://www.useit.com/alertbox/20040510.html (accessed March 8, 2010).

Nielsen, J. 2005. Scrolling and scrollbars. Alertbox. http://www.useit.com/alertbox/20050711.html (accessed Nov. 29, 2009).

Nielsen, J. 2006. F-shaped pattern for reading Web content. Alertbox. http://www.useit.com/alertbox/reading_pattern.html (accessed November 6, 2010).

Nielsen, J. 2007. Top ten Web-design mistakes. Alertbox. http://www.useit.com/alertbox/9605.html (accessed Nov. 26, 2009).

Nielsen, J. 2008. 10 best application UIs. Alertbox. http://www.useit.com/alertbox/application-design.html (accessed Nov. 26, 2009).

Nielsen, J. 2009. First 2 words: A signal for the scanning eye. Alertbox, April 6. http://www.useit.com/alertbox/nanocontent.html (accessed March 14, 2010).

Nielsen, J., and H. Loranger. 2006. *Prioritizing Web Usability.* Berkley, CA: New Riders.

Nielsen, J., and K. Pernice. 2010. *Eyetracking Web Usability.* Berkley, CA: New Riders.

Nielsen, J., and M. Tahir. 2001. *Homepage Usability: 50 Websites Deconstructed.* Berkeley, CA: New Riders.

Nisbett, R. E., and L. Ross. 1980. *Human Inferences: Strategies and Shortcomings of Social Judgment.* Englewood Cliffs, NJ: Prentice Hall.

Obendorf, H., and H. Weinreich. 2003. Comparing link marker visualization techniques: Changes in reading behavior. Paper presented at the *Proceedings of the 12th International Conference on World Wide Web* (Budapest, Hungary, May 20–24, 2003).

Obendorf, H., H. Weinreich, E. Herder, and M. Mayer. 2007. Web page revisitation revisited: Iimplications of a long-term click-stream study of browser usage. Paper presented at the *Proceedings of the SIGCHI Conference on Human Factors in Computing Systems* (San Jose, CA, April 28–May 3, 2007).

Outing, S., and L. Ruel. 2004. The best of eyetrack III: What we saw when we looked through their eyes. http://www.poynterextra.org/eyetrack2004/main.htm (accessed November 6, 2010).

Owens, J., K. Lenz, and S. Speagle. 2009. Trick or tweet: How usable is Twitter for first-time users? *Usability News* 11(2).

Owens, J. W., and S. Shrestha. 2008. How do users browse a portal Website? An examination of user eye movements. *Usability News* 10(2). http://surl.org/usabilitynews/102/portal_column.asp (accessed November 7, 2010).

Ozok, A. A., and G. Salvendy. 2000. Measuring consistency of Web page design and its effects on performance and satisfaction. *Ergonomics* 43(4): 443–460.

Pan, B., H. A. Hembrooke, G. K. Gay, L. A. Granka, M. K. Feusner, and J. K. Newman. 2004. The determinants of web page viewing behavior: An eye-tracking study. Paper presented at the *Proceedings of the 2004 Symposium on Eye Tracking Research & Applications* (San Antonio, TX, March 22–24, 2004).

Parker, B., and L. Scharff. 1998. Influences of contrast sensitivity on text readability in the context of a graphical user interface. http://www.laurenscharff.com/research/agecontrast.html (March 14, 2010).

Parush, A., and D. K. Parush. 2001. Online help: too much of a good thing? *Usability Interface: Newsletter of the STC Usability SIG* 7(3). http://www.stcsig.org/usability/newsletter/0101-too much.html.

Pauwels, S. L., C. Hübscher, S. Leuthold, J. A. Bargas-Avila, and K. Opwis. 2009. Error prevention in online forms: Use color instead of asterisks to mark required-fields. *Interacting with Computers* 21(4): 257–262.

Penzo, M. 2006. Label placement in forms. *UXmatters.* http://www.uxmatters.com/mt/archives/2006/07/label-placement-in-forms.php (accessed November 6, 2010).

Perfetti, C. 2005. 5-second tests: Measuring your site's content pages. User Interface Engineering Articles. http://www.uie.com/articles/five_second_test/ (accessed November 6, 2010).

Poulton, E. C., and C. H. Brown. 1968. Rate of comprehension of an existing teleprinter output and of possible alternatives. *Journal of Applied Psychology* 52: 16–21.

Rabb, M. 1989. *The Presentation Design Book*. Chapel Hill, NC: Ventana Press.

Radl, G. W. 1980. Experimental investigations for optimal presentation mode and colours of symbols on the CRT screen. In *Ergonomic Aspects of Visual Display Terminals*, eds. E. Grandjean and E. Vigliani. London: Taylor & Francis, pp. 127–135.

Ramsay, J., A. Barbesi, and J. Preece. 1998. A psychological investigation of long retrieval times on the World Wide Web. *Interacting with Computers* 10: 77–86.

Rayner, K., S. P. Liversedge, S. J. White, and D. Vergilino-Perez. 2003. Reading disappearing text: Cognitive control of eye movements. *Psychological Science* 14(4): 385–388.

Redish, J. 2007. *Letting Go of the Words: Writing Web Content that Works (The Morgan Kaufmann Series in Interactive Technologies)*: Morgan Kaufmann Publishers Inc.

Rigden, C. 1999. 'The eye of the beholder'—Designing for colour-blind users. *British Telecommunications Engineering* 17: 2–6.

Sanchez, C. A., and J. Wiley. 2009. To scroll or not to scroll: Scrolling, working memory capacity, and comprehending complex texts. *Human Factors: The Journal of the Human Factors and Ergonomics Society* 51(5): 730–738.

Scharff, L., and A. Ahumada. 2008. Contrast polarity in letter identification. *Journal of Vision,* 8(6): 627–627.

Scharff, L. V., and A. J. Ahumada. 2003. *Contrast measures for prediction text readability*. Paper presented at the Human Vision and ElectronicImaging VII, Proceedings of SPIE. 5007.

Scharff, L. V., and A. Hill. 1996. Color test results. Retrieved March 14, 2010, from http://www.laurenscharff.com/research/surreslts.html.

Selvidge, P. 2003. Examining tolerance for online delays. *Usability News*. Retrieved 11/29/2009, from http://www.surl.org/usabilitynews/51/delaytime.asp.

Seow, S. C. (2008). *Designing and Engineering Time: The Psychology of Time Perception in Software*: Addison-Wesley Professional.

Shaikh, A. D., B. S. Chaparro, and D. Fox. (2006). *Personality of ClearType Fonts*. Paper presented at the *Proceedings of the Human Factors and Ergonomics Society 50th Annual Meeting*.

Shaikh, A. D., and K. Lenz. 2006. Where's the search? Re-examining user expectations of Web objects. *Usability News* 8(1).

Sheedy, J., Y.-C. Tai, M. Subbaram, S. Gowrisankaran, and J. Hayes. 2008. ClearType sub-pixel text rendering: Preference, legibility and reading performance. *Displays* 29(2): 138–151, doi:10.1016/j.displa.2007.09.016.

Shneiderman, B., C. Plaisant, M. Cohen, and S. Jacobs. 2009. *Designing the User Interface: Strategies for Effective Human–Computer Interaction*. Reading, MA: Addison-Wesley.

Shrestha, S. 2006. Does the intrusiveness of an online advertisement influence user recall and recognition? *Usability News* 8(1).

Shrestha, S., K. M. Lenz, J. W. Owens, and B. C. Chaparro. 2007. Eye gaze patterns while searching vs. browsing a Website. *Usability News* 9(1).

Shrestha, S., and J. Owens. 2009. Eye movement analysis of text-based Web page layouts. *Usability News* 11(1).

Shrestha, S., and J. W. Owens. 2008. Eye movement patterns on single and dual-column Web pages. *Usability News* 10(1).

Sidorsky, R. C. 1982. Color coding in tactical displays: Help or hindrance. *Army Research Institute Research Report*.

Slattery, T. J., and K. Rayner. 2009. The influence of text legibility on eye movements during reading. *Applied Cognitive Psychology* 24(8): 1129–1148.

Smith, P. A. 1986. Towards a practical measure of hypertext usability. *Interacting with Computers* 8(4): 365–381.

Snowberry, K., S. Parkinson, and N. Sisson. 1983. Computer display menus. *Ergonomics* 26: 699–712.

Snyder, H. L., J. J. Decker, C. J. C. Lloys, and C. Dye. 1990. Effect of image polarity on VDT task performance. Paper presented at the Human Factors Society 34th Annual Meeting (Orlando, Florida, October 8–12, 1990).

Spain, K. 1999. What's the best way to wrap links? *Usability News* 1(1).

Sperry, R. A., and J. D. Fernandez. 2008. Usability testing using physiological analysis. *Journal of Computing in College* 23(6): 157–163.

Spool, J., C. Perfetti, and D. Brittan. 2004. *Designing for the Scent of Information*. North Andover, MA: User Interface Engineering.

Spool, J. M. 1998. As the page scrolls. User Interface Engineering Articles. http://www.uie.com/articles/page_scrolling/ (accessed Nov. 29, 2009).

Spool, J. M. 2005. Galleries: the hardest working page on your site. User Interface Engineering Articles. http://www.uie.com/articles/galleries_reprint/ (accessed Nov. 26, 2009).

Spool, J. M. 2006a. Lifestyles of the link-rich Home pages. *User Interface Engineering Articles*. http://www.uie.com/articles/linkrich_home_pages/ (accessed November 7, 2010).

Spool, J. M. 2006b. Utilizing the cut-off look to encourage users to scroll. http://www.uie.com/brainsparks/2006/08/02/utilizing-the-cut-off-look-to-encourage-users-to-scroll/ (accessed Nov. 29, 2009).

Spool, J. M. 2008a. Producing great search results: Harder than it looks, part 2. User Interface Engineering Articles. http://www.uie .com/articles/search_results_part2/ (accessed Nov. 26, 2009).

Spool, J. M. 2008b. To refresh, or not to refresh. http://www.uie .com/articles/refresh-or-not/ (accessed November 7, 2010).

Spool, J. M., T. Scanlon, W. Schroeder, C. Synder, and T. DeAngelo. 1998. *Web Site Usability: A Designer's Guide*. San Francisco, CA: Morgan Kaufmann.

Storey, M.-A., B. Phillips, M. Maczewski, and M. Wang. 2002. Evaluating the usability of Web-based learning tools. *Educational Technology & Society* 5(3).

Sullivan, D. 2000. Search engines and frames. http://www.searchenginewatch.com/webmasters/article.php/2167901 (accessed Aug. 30, 2003).

Tan, C. C. 2009. Web form design guidelines: An eyetracking study. http://www.cxpartners.co.uk/thoughts/web_forms_design_guidelines_an_eyetracking_study.htm (accessed November 7, 2010).

Tinker, M. A. 1963. *Legibility of Print*. Iowa City: Iowa State University Press.

Tractinsky, N., A. S. Katz, and D. Ikar. 2000. What is beautiful is usable. *Interacting with Computers* 13(2): 127–145, doi:10.1016/S0953-5438(00)00031-X.

Trimmel, M., M. Meixner-Pendleton, and S. Haring. 2003. Stress response caused by system response time when searching for information on the Internet. *Human Factors* 45(4): 615–621.

Trollip, S. R., and G. Sales. 1986. Readability of computer-generated fill-justified text. *Human Factors* 28: 159–163.

Tsunoda, T., T. Yamaoka, K. Yamashita, T. Matsunobe, Y. Hashiya, Y. Nishiyama, et al. 2001. Measurement of task performance times and ease of use: Comparison of various menu structures and depth on the Web. *Human Factors and Ergonomics Society Annual Meeting Proceedings* 45: 1225–1229.

Tufte, E. 1983. *The Visual Display of Quantitative Information.* Chesire, CT: Graphics Press.

Tullis, T., M. Siegel, and E. Sun. 2009. Are people drawn to faces on webpages? Paper presented at the *Proceedings of the 27th International Conference on Human Factors in Computing Systems, Extended Abstracts* (Boston, Massachusetts, April 4–9, 2009).

Tullis, T. S. 1981. A computer-based tool for evaluating alphanumeric displays. *Human Factors* 23(5): 541–550.

Tullis, T. S. 1985. Designing a menu-based interface to an operating system. Paper presented at the *Proceedings of the SIGCHI Conference on Human Factors in Computing Systems* (San Francisco, California, April 14–18, 1985).

Tullis, T. S. 1993. Is user interface design just common sense? Paper presented at the *Fifth International Conference on Human–Computer Interaction* (Orlando, Florida, August 8–13, 1993).

Tullis, T. S. 1997. Screen design. In *Handbook of Human–Computer Interation*, 2nd ed., eds. M. Helander, T. Landauer, and P. Probhu, 503–531. Amsterdam: North-Holland.

Tullis, T. S. 1998. A method for evaluating Web page design concepts. Paper presented at the *CHI 1998 Conference Summary on Human Factors in Computing Systems* (Los Angeles, California, April 18–23, 1998).

Tullis, T. S., J. L. Boynton, and H. Hersh. 1995. Readability of fonts in the Windows environment. Paper presented at the *Proceedings of ACM CHI'95 Conference on Human Factors in Computing Systems* (Denver, Colorado, May 7–11, 1995).

Tullis, T. S., and C. Cianchette. 2003. An Investigation of Web site navigation techniques. *Usable Bits*, 2nd Quarter 2003, http://hid .fidelity.com/q22003/navigation.htm (accessed Nov. 28, 2009).

Tullis, T. S. and S. Fleischman. 2002. The latest word on font for the Web. http://www.eastonmass.net/tullis/publications/fonts.htm (accessed November 7, 2010).

Tullis, T. S. and S. Fleischman. 2004. Tabular data: Finding the best format. *Intercom*, June 2004: 12–14.

Tullis, T. S., and M. L. Kodimer. 1992. A comparison of direct-manipulation, selection, and data-entry techniques for re-ordering fields in a table. Paper presented at the Posters and Short Talks of the 1992 SIGCHI Conference on Human Factors in Computing Systems (Monterey, California, May 3–7, 1992).

Tullis, T. S., and A. Pons. 1997. Designating required vs. optional input fields. Paper presented at the CHI '97 Extended Abstracts on Human Factors in Computing Systems: Looking to the Future (Atlanta, Georgia, March 22–27, 1997).

Tullis, T. S., and M. Siegel. 2010. Does underlining links help or hurt? Paper presented at the CHI 2010 Conference on Human Factors in Computing Systems (Atlanta, GA, April 10–15).

Tullis, T. S., and C. Tullis. 2007. Statistical analyses of E-commerce Websites: Can a site be usable and beautiful? Paper presented at the 12th Annual Human–Computer Interaction International Conference (Beijing, China, July 22–27, 2007).

Tyrrell, R., T. Pasquale, T. Aten, and E. Francis. 2001. Empirical evaluation of user responses to reading text rendered using ClearType technologies. *Society for Information Display, Digest of Technical Papers,* 1205–1207.

van Schaik, P., and J. Ling. 2001. The effects of frame layout and differential background contrast on visual search performance in Web pages. *Interacting with Computers* 13: 513–525.

Vartabedian, A. G. 1971. The effects of letter size, case, and generation method on CRT display search time. *Human Factors* 13: 363–368.

Vicino, F., and S. Ringel. 1966. *Decision-making with Updated Graphic vs. Alpha-Numeric Information.* Washington, DC: US Army Personnel Research Office.

W3C. 2008. Web Content Accessibility Guidelines (WCAG) Overview. http://www.w3.org/WAI/intro/wcag (accessed Nov. 29, 2009).

w3Schools.com. 2010. Browser display statistics. http://www.w3 schools.com/browsers/browsers_display.asp (accessed March 13, 2010).

Weinman, L. 2004. The history of the browser-safe Web palette. http://www.lynda.com/resources/webpalette.aspx (accessed March 13, 2010).

Weiss, R. E., D. S. Knowlton, and G. R. Morrison. 2002. Principles for using animation in computer-based instruction: Theoretical heuristics for effective design. *Computers in Human Behavior* 18: 465–477.

Wheildon, C. 1995. *Type and Layout: How Typography and Design Can Get Your Message Across—or Get in the Way.* Berkeley, California: Strathmore Press.

White, J. V. 1990. *Color for the Electronic Age.* New York: Watson-Guptill.

Williams, S., and L. Scharff. 2000. The effects of font size and line spacing on readability of computer displays. http://www.laurenscharff.com/research/SWExp.html (accessed November 7, 2010).

Wroblewski, L. 2008. *Web Form Design: Filling in the Blanks.* Brooklyn, New York: Rosenfeld Media.

Wroblewski, L. 2009. Inline validation in Web forms. A List Apart. http://www.alistapart.com/articles/inline-validation-in-web-forms/ (accessed November 7, 2010).

Youngman, M., and L. Scharff. 1998. Text width and margin width influences on readability of GUIs. http://www.laurenscharff .com/research/textmargin.html (accessed March 14, 2010).

Zaphiris, P., B. Shneiderman, and K. Norman. 1999. Expandable indexes versus sequential menus for searching hierarchies on the World Wide Web. ftp://ftp.cs.umd.edu/pub/hcil/Reports-Abstracts-Bibliography/99-15html/99-15.html (accessed November 7, 2010).

Zaphiris, P. G. 2000. Depth vs breath in the arrangement of Web links. *Human Factors and Ergonomics Society Annual Meeting Proceedings* 44: 453–456.

9 Developing Adaptive Interfaces for the Web

Constantine Stephanidis, Alexandros Paramythis,* and Anthony Savidis

CONTENTS

9.1 INTRODUCTION

The advent of the World Wide Web as a global communication infrastructure enables concurrent access to heterogeneous and distributed information sources through a wide variety of media (hypertext, graphics, animation, audio, video, and so on) and access devices. This evolution has brought about fundamental changes in the way computer-mediated human activities are conceived, designed, developed, and experienced, giving rise to the progressive emergence of the Information Society.

This dynamic evolution is characterized by several dimensions of diversity that become evident when considering the broad range of user characteristics, as well as the changing nature of human activities, the variety of contexts of use, the increasing availability and diversification of information, knowledge sources and services, the proliferation of diverse technological platforms, etc. In this context, the "typical" computer user can no longer be identified: information artifacts are used by diverse user groups, including people with different cultural, educational, training, and employment background, novice and experienced computer users, children and older adults, and people with different types of disabilities. Existing computer-mediated human activities undergo fundamental changes, and a wide variety of new ones appear, such as access to online information, e-communication, digital libraries, e-business, online health services, e-learning, online communities, online public and administrative services, e-democracy, telework and telepresence, online entertainment, etc. Similarly, the context of use is changing. The "traditional" use of computers (i.e., scientific use by the specialist, business use for productivity enhancement) is increasingly being complemented by residential and nomadic use, thus penetrating a wider range of human activities in a broader variety of environments, such as the school, the home, the market place, and other civil and social contexts. Finally, technological proliferation contributes with an increased range of systems or devices to facilitate access to the community-wide pool of information resources. These devices include computers, standard telephones, cellular telephones with built-in displays, hand-held devices, television sets, information kiosks, information appliances, and various other "network-attachable" devices. In the years ahead, as the Information Society further develops and evolves, a

* Alexandros Paramythis is currently affiliated with the Institute for Information Processing and Microprocessor Technology of the Johannes Kepler University, Austria. The work reported in this chapter was conducted while he was affiliated with the Institute of Computer Science of FORTH.

new technological paradigm shift is anticipated to take place. The anticipated results will be the establishment of an environment, commonly named Ambient Intelligence, characterized by invisible (i.e., embedded) computational power in everyday appliances and other surrounding physical objects, and populated by intelligent mobile and wearable devices (Stephanidis 2001a).

The Information Society has the potential to improve the quality of life of citizens, to improve the efficiency of social and economic organization, and to reinforce cohesion. However, as with all major technological changes, it can also have disadvantages, e.g., introducing new barriers, human isolation, and alienation (the so-called "digital divide"). To overcome such a risk, the diverse requirements of all potential users need to be taken seriously into consideration, and an appropriate "connection" to computer applications and services needs to be guaranteed. In this context, the notion of universal access (Stephanidis 2001b) has become critically important for ensuring social acceptability of the emerging Information Society. Universal access implies the accessibility and usability of Information Society Technologies (IST) by anyone, anywhere, anytime. Its aim is to enable equitable access and active participation of potentially all citizens in existing and emerging computer-mediated human activities.

Therefore it is important to develop *universally accessible* and *usable* applications and services, embodying the capability to interact with the user in all contexts, independently of location, user's primary task, target machine, run-time environment, or the current physical conditions of the external environment. To this end, proactive and generic approaches are required, which account for all dimensions of diversity (i.e., abilities, skills, requirements and preferences of users, characteristics of technological platforms, relevant aspects of the context of use), and take into account the needs of the broadest possible end-user population from the early design phases of new products and services throughout all phases of the development life cycle. In the past two decades, the concept of intelligent adaptation has been investigated under the perspective of proactively supporting universal access by providing built-in accessibility and high interaction quality in applications and services in the emerging Information Society (Stephanidis 2001c). Adaptation characterizes software products that automatically configure their parameters according to the given attributes of individual users (e.g., mental/motor/ sensory characteristics, requirements, and preferences) and to the particular context of use (e.g., hardware and software platform and environment of use). In the context of the Web, universal access concerns both the interactive behavior and the content of applications and services, and requires a global approach to adaptation.

This chapter briefly discusses adaptation in the context of research efforts toward universal access (Section 9.2), and presents two case studies illustrating the contribution of adaptation-based techniques toward universal access to the Web. The first case study (Section 9.3) presents the development of the AVANTI universally accessible Web browser,

while the second case study concerns the development of the PALIO Adaptive Hypermedia Framework for building universally accessible Web services. While AVANTI constituted an adaptable and adaptive content viewing application that can view any type of content, adapted or not, PALIO supported the creation of adaptable and adaptive content that can be viewed with any kind of browser. Thus, the two complemented each other. Both case studies outline the design requirements that inform the adaptation space, the types of adaptations supported in each system, the architectural frameworks and decision-making mechanisms used for implementing adaptations at run-time, as well as concrete examples of adaptation scenarios. Finally, we discuss the experience gained in the course of these developments, as well as further research issues.

9.2 THE WORLD WIDE WEB: UNIVERSAL ACCESS AND ADAPTATION

Early approaches to the provision of accessibility to computer-based applications and services were mainly based on adaptations to existing systems or on "dedicated" developments targeted to specific user categories (Stephanidis and Emiliani 1999). Along the same lines, attempts to provide accessibility in the Web environment were usually based on approaches that can roughly be distinguished into three categories: alternative access systems, information content and structure, and user interface (Treviranus and Serflek 1996).

Support for alternative access systems involves the integration of special I/O devices (e.g., alternative keyboards, voice recognition systems, screen magnifiers, screen readers, and Braille displays), as well as adaptations of interaction techniques (or provision of alternative ones) in the operating system or the graphical environment (e.g., Microsoft 2003a, 2003b; Sun Microsystems). Adaptations at the level of the information content mainly concern the provision of guidelines for Web authors toward the development of more accessible HTML documents (W3C-WAI 1999). At the level of the user interface to Web browsers, adaptations mainly concern either the employment of the accessibility options provided by the operating system in conjunction with alternative input/ output devices or the development of special-purpose browsers for specific categories of disabled people. Specialized browsers in the literature aimed at supporting users with specific types of disabilities (most often visual impairments) but without adaptation capabilities included the pwWebSpeak browser (De Witt and Hakkinen 1998), and IBM's Home Page Reader (Asakawa and Lewis 1998). Considerable efforts have also been reported in the development of custom interaction techniques intended to be used in conjunction with mainstream browsers. Spalteholz, Li, and Livingston (2008), for instance, describe a specialized text input technique for users only capable of operating single switches.

However, the rapid evolution of technology restricts considerably the scope of such "reactive" approaches, because applications or services developed or adapted for specific categories of disabled users may incur technical complexity and high costs and address relatively small portions of the market, thus becoming impractical (Stephanidis and Emiliani 1999). These limitations become critically important when viewed in the context of the emerging information society. In such a context, accessibility can no longer be considered as a mere adaptation-based translation of interface manifestations to alternative modalities and media but as a quality requirement demanding a more generic solution. In light of the above, it became evident that the challenge of universal access needs to be addressed through more proactive and generic approaches, which account for all dimensions and sources of variation (Stephanidis et al. 1998b).

Under a different perspective, adaptable and adaptive software systems have been considered in a wide range of research efforts in the recent past. The relevant literature offers numerous examples illustrating tools for constructing adaptive interaction (e.g., Brusilovsky et al. 1998; Horvitz et al. 1998; Kobsa and Pohl 1995; Sukayariva and Foley 1993) and case studies in which adaptive interface technology has improved, or has the potential to improve, the usability of an interactive system (e.g., Benyon 1993, 1997; Dieterich et al. 1993). In particular, Adaptive Hypermedia Systems (AH systems, or AHS for short) are an area that has drawn considerable attention since the advent of the Web (which can be practically considered as a "universal," widely deployed hypermedia system). Major categories of adaptive hypermedia systems include *educational hypermedia, online information systems, online help systems, information retrieval systems,* and *institutional hypermedia.* Numerous adaptive systems are available today in various application domains with a great variety of capabilities (see, e.g., Ardissono and Goy 1999; Balabanovic and Shoham 1997; Brusilovsky et al. 1998; Henze 2001; Kobsa 2001; Oppermann and Specht 1998).

Until the past decade, however, adaptive techniques had limited impact on the issue of universal access (Stephanidis 2001c). For example, regarding accessibility in hypermedia browsing applications, DAHNI (Petrie et al. 1997) was a hypermedia system with a non-visual interface. This system supported a large variety of input and output devices but offered quite limited interaction options and exhibited no adaptivity capabilities at all. WebAdapter (Hermsdorf, Gappa, and Pieper 1998), developed at approximately the same time as AVANTI, employed adaptability to reactively support different categories of end users, focusing on users with disabilities. A more recent effort reported by Tan, Yu, and McAllister (2006) utilizes adaptation to provide access to graphics embedded in Web pages for blind users. The approach presented is particularly noteworthy in that it is based on an external, componentized architecture, which works in tandem with normal mainstream browsers (making use of Microsoft Active Accessibility*). However, the

solution is highly specialized (it is specifically tailored to the presentation of graphics for blind users), and appears to be limited to adaptability techniques only.

Moving to the domain of usability-oriented user interface adaptation in browsing applications, Henricksen and Indulska (2001) present an adaptive browser. With the collaboration of a customized Web server, the browser can adapt itself to characteristics of the network communication (e.g., throughput), and the availability of input/output devices.

A common characteristic of the aforementioned efforts is that they each addressed a specialized problem, rather than offering generic solutions that can concurrently cater for accessibility and usability. In contrast, the unified user interface development methodology† (Savidis, Akoumianakis, and Stephanidis 2001; Savidis and Stephanidis 2001a, 2001b; Stephanidis 2001d), and the related user interface development platform (Akoumianakis and Stephanidis 2001; Savidis and Stephanidis 2001c) have been proposed as a generic technological solution for supporting universal access of computer-based applications through automatic adaptation of user interfaces. The unified user interface framework provides an engineering methodology supporting the development of interfaces that exhibit both user awareness (i.e., the interface is capable of user-adapted behavior by automatically selecting interaction patterns appropriate to the particular user), and usage context awareness (i.e., the interface is capable of usage-context adapted behavior by automatically selecting interaction patterns appropriate to the particular physical and technological environment). A unified user interface is comprised of a single (unified) interface specification that (1) encompasses user- and context-specific information, (2) contains alternative dialogue artifacts in an implemented form, each appropriate for different user- and context-specific parameters, and (3) applies adaptation decisions, activates dialogue artifacts and is capable of interaction monitoring.

In the context of this chapter, adaptation characterizes software products that automatically modify (adapt) their interactive behavior according to the individual attributes of users (e.g., mental/motor/sensory characteristics, preferences) and to the particular context of use (e.g., hardware and software platform, environment of use), as well as the content of applications and services. Adaptation implies the capability, on the part of the system, of capturing and representing knowledge concerning alternative instantiations suitable for different users, contexts, purposes, etc., as well as for reasoning about those alternatives to arrive at adaptation decisions. Furthermore, adaptation implies the capability of assembling, coherently presenting, and managing at run-time, the appropriate alternatives for the current user, purpose, and context of use. Adaptation is a multifaceted process, which can be analyzed along three main axes, namely, the *source* of adaptation knowledge, the *level* of interaction at which it is applied, and the *type* of information on which it is based (Stephanidis 2001d).

---

* For information on the Microsoft Active Accessibility Application Programming Interfaces please see http://msdn.microsoft.com/en-us/library/dd373592(VS.85).aspx

---

† The unified user interface development framework has been developed in the context of the ACCESS project (see Acknowledgments section).

As far as the source of adaptation knowledge is concerned, one can identify two complementary classes: knowledge available at start-up, i.e., prior to the initiation of interaction (e.g., user profile, platform profile, and usage context), and knowledge derived at run-time (e.g., through interaction monitoring, or inspection of the computing environment). Adaptation behavior based on the former type of knowledge is termed *adaptability* and reflects the capability on the part of the interface to automatically tailor itself to the *initial* interaction requirements, as these are shaped by the information available to the interface. Adaptation behavior based on the latter type of knowledge is termed *adaptivity* and refers to the capability on the part of the interface to dynamically derive further knowledge about the user, the usage context, etc., and use that knowledge in order to further modify itself to better suit the revised interaction requirements.

The second axis of analysis of adaptation concerns the level of interaction at which adaptations are applied. In particular, it is possible to design and apply adaptations at all three levels of interaction:

1. At the semantic level of interaction (e.g., by employing different metaphors to convey the functionality and facilities of the underlying system)
2. At the syntactic level of interaction (e.g., by de/ activating alternative dialogue patterns, such as "object-function" versus "function-object" interaction sequencing)
3. At the lexical level of interaction (e.g., grouping and spatial arrangement of interactive elements, modification of presentation attributes, alternative input/ output devices).

The third main axis of analysis concerns the type of information being considered when deciding upon adaptations. Exemplary categories of information that can be employed include design constraints depending on user characteristics (e.g., abilities, skills, requirements, preferences, expertise, and cultural background), platform characteristics (e.g., terminal capabilities and input/output devices), and task requirements (e.g., urgency, criticality, error proneness, and sequencing), etc. Furthermore, information that can only be acquired during interaction can also play an important role in the decision process (e.g., identifying the user's inability to successfully complete a task, inferring the user's goal/intention, detecting modifications to the run-time environment).

In the next sections, this approach is illustrated in practice by showing how adaptation-oriented development techniques have been applied to facilitate universal access to the Web on both the Web browser side and the Web server side. Web browser adaptation is exemplified by the AVANTI universally accessible Web browser, while Web server side adaptation is exemplified through the PALIO hypermedia development framework for the development of adaptive Web services.

9.3 THE AVANTI WEB BROWSER

The AVANTI system* was, to the authors' knowledge, the first to employ adaptive techniques in order to ensure accessibility and high-quality of interaction for *all* potential users. AVANTI advocated a new approach to the development of Web-based information systems (Bini and Emiliani 1997; Bini, Ravaglia, and Rella 1997). In particular, it put forward a conceptual framework for the construction of systems that support adaptability and adaptivity at both the content (Fink et al. 1998) and the user interface levels.

The user interface component of the AVANTI system is functionally equivalent to a Web browser. In the AVANTI browser (Stephanidis et al. 1998a, 2001) user interface adaptability and adaptivity are applied to tailor the browser to the end-user abilities, requirements, and preferences both during the initiation of a new session and throughout interaction with the system. The distinctive characteristic of the AVANTI browser is its capability to dynamically tailor itself to the abilities, skills, requirements, and preferences of the end users to the different contexts of use, as well as to the changing characteristics of users as they interact with the system.

9.3.1 THE DESIGN OF THE AVANTI WEB BROWSER

The primary target for adaptability in the AVANTI Web browser was to ensure that each of the system's potential users is presented with an instance of the user interface that has been tailored to offer the highest possible degree of accessibility according to the system's knowledge about the user. Adaptivity was also employed to further tailor the system to the inferred needs or preferences of the user, so as to achieve the desired levels of interaction quality.

The AVANTI browser provides an accessible and usable interface to a range of user categories, irrespective of physical abilities or technology expertise. Moreover, it supports various differing situations of use. The end-user groups targeted in AVANTI, in terms of physical abilities, include (1) "able-bodied" people, assumed to have full use of all their sensory and motor communication "channels," (2) blind people, and (3) motor-impaired people, with different forms of impairments in their upper limps, causing different degrees of difficulty in employing traditional computer input devices, such as a keyboard and/or a mouse. In particular, in the case of motor-impaired people, two coarse levels of impairment were taken into account: "light" motor impairments (i.e., users have limited use of their upper limbs but can operate traditional input devices or equivalents with adequate support) and "severe" motor impairments (i.e., users cannot operate traditional input devices at all).

Furthermore, because the AVANTI system was intended to be used both by professionals (e.g., travel agents) and by general public (e.g., citizens, tourists), the users' experience

* The AVANTI information system has been developed in the framework of the AVANTI project (see Acknowledgments section).

in the use of, and interaction with, technology was another major parameter that was taken into account in the design of the user interface. Thus, in addition to the conventional requirement of supporting novice and experienced users of the system, two new requirements were put forward: (1) supporting users with any level of computer expertise and (2) supporting users with or without previous experience in the use of Web-based software.

In terms of usage context, the system was intended to be used both by individuals in their personal settings (e.g., home and office) and by the population at large through public information terminals (e.g., information kiosks at a railway station and an airport). Furthermore, in the case of private use, the front end of AVANTI was intended to be appropriate for general Web browsing, allowing users to make use of the accessibility facilities beyond the context of a particular information system.

Users were also continuously supported as their communication and interaction requirements changed over time, due to personal or environmental reasons (e.g., stress, tiredness, or system configuration). This entailed the capability, on the part of the system, to detect dynamic changes in the characteristics of the user and the context of use (either of temporary or of permanent nature) and cater for these changes by appropriately modifying itself.

From the above, it follows that the design space for the user interface of the AVANTI Web browser was rather large and complex, covering a range of diverse user requirements, different contexts of use, and dynamically changing interaction situations. These requirements dictated the development of a new experimental front end, which would not be based on existing Web browser technology, nor designed following traditional "typical-user" oriented techniques. In fact, the accessibility requirements posed by the user categories addressed in AVANTI could not be met either by existing customizability features supported by commercial Web browsers or through the use of third-party assistive products. Instead, to fully support the interaction requirements of each individual user category, it was necessary to have full control over both the task structure and the interaction dialogue, including the capability to arbitrarily modify them (Stephanidis et al. 2001).

The unified user interface development approach was adopted to address the above requirements, as it provides appropriate methodologies and tools to facilitate the design and implementation of user interfaces that cater for the requirements of multiple, diverse end-user categories and usage contexts (Stephanidis 2001d).

Following the unified user interface design method, the design of the user interface follows three main stages (Savidis, Akoumianakis, and Stephanidis 2001): (1) enumeration of different design alternatives to cater for the particular requirements of the users and the context of use, (2) encapsulation of the design alternatives into appropriate abstractions and integration into a polymorphic task hierarchy, and (3) development and documentation of the design rationale that will drive the run-time selection between the available alternatives.

Having defined the design space as well as the primary user tasks to be supported in the user interface of the AVANTI browser, different dialogue patterns (referred to as *instantiation styles*, or simply *styles*) were defined/designed for each task to cater for the identified interaction requirements. The second stage in the unified user interface design process concerns the definition of a task hierarchy for each task. This includes the hierarchical decomposition of each task in subtasks and styles (dialogue patterns), as well as the definition of task operators (e.g., BEFORE, OR, XOR, *, and +) that enable the expression of dialogue control flow formulas for task accomplishment. The task hierarchy encapsulates the "nodes of polymorphism," i.e., the nodes at which an instantiation style has to be selected in order to proceed farther down the task hierarchy. Polymorphism may concern alternative task subhierarchies, alternative abstract instantiations for a particular task, or alternative mappings of a task to physical interactive artifacts. Finally, the third stage consists in elaborating documentation forms recording the consolidated design rationale of each alternative design artifact. The aim is to capture, for each subtask, the design logic for deciding at run-time possible alternative styles by directly associating user-, usage-context-parameters, and design goals with the constructed artifacts (i.e., styles).

During the design of the AVANTI user interface, the definition/design of alternative instantiation styles was performed in parallel with the definition of the task hierarchy for each particular task. This entailed the adoption of an iterative design approach in the polymorphic task decomposition phase: the existence of alternative styles drove the decomposition process in subtask hierarchies, and the task decomposition itself imposed new requirements for the definition of alternative styles for each defined subtask. In practice, in the AVANTI user interface, the specific nodes in the task hierarchies were identified in which polymorphic decomposition was required (owing to the user and/or usage parameters), and alternative instantiation styles to accommodate these requirements were selected or defined. Figure 9.1 presents the result of the polymorphic task decomposition process for the "open location" task.

The polymorphic decomposition process is driven by the user and usage-context characteristics that impose certain requirements in the definition/selection of alternative styles and style components, and provide information and constraints on the physical appearance and interactive behavior of the employed interaction objects.

An initial requirements analysis phase led to the definition of two sets of user-oriented characteristics, which formed the basis of the decision parameters that drove the polymorphic decomposition process. The first set contained "static" user characteristics unlikely to change in the course of a single interaction session. These characteristics were assumed to be known prior to the initiation of interaction (retrieved from the user profile), and comprised: (1) physical abilities, (2) native language, (3) familiarity with computing, the Web in general, and the AVANTI system itself, (4) the overall interaction target (speed, comprehension, accuracy, error tolerance), and

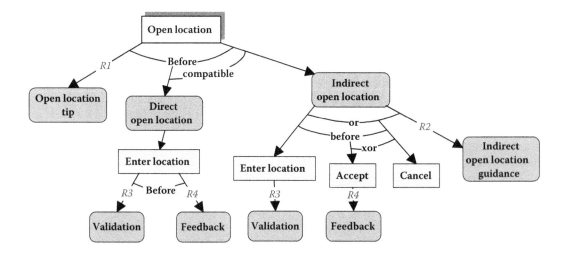

R1: (user *is* unable_to_initiate_task)

R2: (user *is* unable_to_complete_task) *or*
(user *is* disoriented)*or* (user *is* idle)

R3: (target *includes* (accuracy *or* ease_of_use)) *or*
(user *is* novice_in_computers) *or*
(user *is_not* good_at_typing) *or*
(user *has* high_error_rate)

R4: (target *includes* ease_of_use) *or*
(user *is* novice_in_networking) *or*
(user *is_not* frequent_in_AVANTI) *or*
user *is* novice_in_Web *or*
(user *is* disoriented)

FIGURE 9.1   Example of polymorphic task decomposition.

(5) user preferences regarding aspects of the application and the interaction. The second set of characteristics contained "dynamic user states and interaction situations," derived at run-time through interaction monitoring. The following "states" and "situations" were included in this set: (1) user familiarity with specific tasks (i.e., evidence of the user's capability to successfully initiate and complete certain tasks), (2) ability to navigate, (3) error rate, (4) disorientation (i.e., inability to cope with the current state of the system), and (5) user idle time.

A set of user and usage-context characteristics was associated with each polymorphic alternative during the decomposition process, providing the mechanism for deciding upon the need for and selecting different styles or style components. This set of characteristics constitutes the adaptation rationale, which is depicted explicitly on the hierarchical task structure (see Figure 9.1) in the form of design parameters associated with specific "values," which were then used as "transition attributes," qualifying each branch leading away from a polymorphic node. These attributes were later used to derive the run-time adaptation decision logic of the final interface.

9.3.2 The AVANTI Browser Architecture

The development of the AVANTI browser's architecture (see Figure 9.2) was based on the architectural framework of unified user interfaces (Savidis and Stephanidis 2001a). This framework proposes a specific way of structuring the implementation of interactive applications by means of independent

intercommunicating components, with well-defined roles and behavior. A unified user interface is comprised of (1) the dialogue patterns component, (2) the decision-making component, (3) the user information server, and (4) the context parameters server.

In the AVANTI browser architecture (see Figure 9.2), the adaptable and adaptive interface components, the interaction monitoring component, and the page presentation and

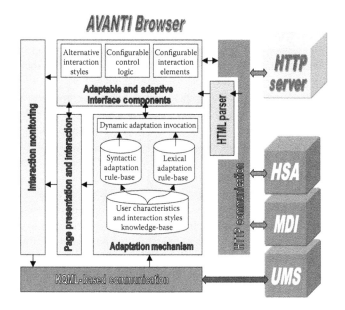

FIGURE 9.2   The AVANTI Web browser architecture.

interaction component were the functional equivalent of the dialogue patterns component in the unified user interface architecture. They encapsulated the implementation of the various alternative dialogue patterns (interaction/instantiation styles) identified during the design process and were responsible for their activation/deactivation, applying the adaptation decisions made by the respective module. Moreover, each style implementation had integrated functionality for monitoring user interaction and reporting the interaction sequence back to the user modeling component of the AVANTI system. The adaptation mechanism directly corresponded to the decision making component in the unified user interface architecture. It encompassed the logic for deciding upon and triggering adaptations, on the basis of information stored in its knowledge space (this information is, in turn, retrieved from user profiles and inferences based on interaction monitoring). Adaptations could be triggered both during the initiation of interaction or at run-time.

The role of the user information server in the unified user interface architecture was played by the user modeling server (UMS) of the AVANTI system. The UMS is implemented as a server, usually remotely located on the network, as it offers central, multiuser modeling functionality. The communication between the user interface and the UMS was bilateral: the interface sent messages regarding user actions at the physical and task levels (e.g., "the user pressed reload," or "the user successfully completed the loading of a new document," respectively). The UMS employed a set of stereotypes that store categorized knowledge about the users, their interactions and environment, and a set of rules to draw inferences on the current state of the user, based on the monitoring information.

9.3.3 Adaptation Logic and Decision Making

In the AVANTI browser, adaptation was based on a number of static and dynamic user characteristics. Static user characteristics are those that are unlikely to change in the course of a single interaction session (although they can change over longer periods of time). These characteristics are assumed to be known prior to the initiation of interaction (i.e., retrieved from the user profile). The second set of characteristics is termed dynamic to denote that the evidence they hold is usually derived at run-time through interaction monitoring.

Of particular interest in the context of realizing the polymorphic task hierarchy into a unified user interface is the transformation of the design parameters that guided the introduction of different instantiation styles into a form appropriate for the construction of a decision-making component. The approach taken in the AVANTI Web browser was affected by two complementary aspects of the system: the adaptation logic itself, as this had been captured in the polymorphic task hierarchy, and the capabilities of the UMS module. The UMS generated and communicated information in tuples of the form: <user related hypothesis, probability that the hypothesis holds>. This form is particularly suited to the development of "if . . . then" rules, which combine

```
task "open location" {
        if <user is unable to initiate task>
                        then activate_style "adaptive
prompting"
                if (<interaction target> includes <accuracy>
or                              <interaction target> includes
<ease of use> or               <user is computer
novice> or
                <user has high error rate> )
                        then activate_style "validation"
        if (<user is Web novice> or
                (not <user uses AVANTI frequently>) or
                <interaction target> includes <ease of
use> or
                <user is disoriented>)
                        then activate_style "extensive
feedback"
                if (<user is unable to complete task> or
                        <user is disoriented> or <user is idle>)
                        then activate_style "guidance"
}
```

FIGURE 9.3 Adaptation rules for the "open location" task.

the design parameter-oriented rationale captured in the polymorphic task hierarchy, and the inference capabilities of the UMS (i.e., each dynamically inferred parameter corresponds to a separate hypothesis). Along the above lines, the adaptation rationale was transformed into corresponding adaptation rules that were valuated at run-time. The revaluation of any particular rule was triggered by the addition or modification of inferred user- or context-related parameters in the interface's knowledge space (either from user profiles, or from the UMS). To illustrate the relationship between the design logic in the polymorphic task hierarchy and the rules, Figure 9.3 presents some of the actual rules derived from the polymorphic decomposition of the "open location" task presented in Figure 9.1.

9.3.4 Adaptation Scenarios from the AVANTI Browser

To illustrate some of the categories of adaptation available in the AVANTI browser, some instances of the browser's user interface are briefly reviewed below.

Figure 9.4 contains two instances of the interface that demonstrate adaptations based on the characteristics of the user and the usage context. Specifically, Figure 9.4A presents a simplified instance intended for use by a user unfamiliar with Web browsing. Note the "minimalist" user interface with which the user is presented, as well as the fact that links

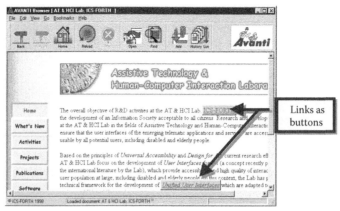

(a) Conventional, simplified instance of the interface

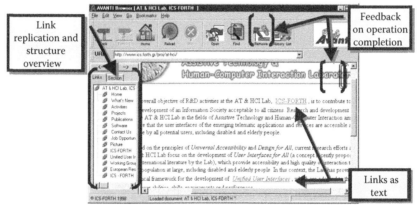

(b) Adapted instance for an experienced user

FIGURE 9.4 (**See color insert.**) Adapting to the user and the context of use.

are presented as buttons, arguably increasing their affordance (at least in terms of functionality) for users familiar with windowing applications in general. In the second instance, Figure 9.4b, the interface has been adapted for an experienced user. Note the additional functionality that is available to the user (e.g., a pane where the user can access an overview of the document itself or of the links contained therein, and an edit field for entering the URLs of local or remote HTML documents).

Figure 9.5 contains some sample instances demonstrating disability-oriented adaptations in the browser's interface. The instance in Figure 9.5a presents the interface when a special interaction technique for motor-impaired users is activated, namely hierarchical interface scanning (either manually or automatically activated). Scanning is a mechanism allowing to "isolate" each interactive object in the interface and to interact with it through binary switches. Note the scanning highlighter over an image link in the HTML document and the additional toolbar that was automatically added in the user interface. The latter is a "window manipulation" toolbar, containing three sets of controls enabling the user to perform typical actions on the browser's window (e.g., resizing and moving). Figure 9.5b illustrates the three sets of controls in the toolbar, as well as the "rotation" sequence between the sets (the three sets occupy the same space on the toolbar, to

better utilize screen real estate, and to speed up interaction; the user can switch between them by selecting the first of the controls). Figure 9.5c presents an instance of the same interface with an on-screen, "virtual" keyboard activated for text input. Interaction with the on-screen keyboard is also scanning-based.

The single interface instance in Figure 9.6 illustrates a case of adaptive prompting (Stephanidis et al. 1998b). Specifically, this particular instance is displayed in those cases in which there exists high probability that the user is unable to initiate the "open location" task (this would be the case if there was adequate evidence that the user is attempting to load an external document with unsupported means, e.g., using "drag and drop"). In this case, adaptive prompting is achieved through the activation of a "tip" dialog, i.e., a dialog notifying the user about the existence of the "open location" functionality and offering some preliminary indications of the steps involved in completing the task.

9.3.5 SUBSEQUENT EVOLUTION OF WEB BROWSING TECHNOLOGIES

To date, AVANTI remains the only case in the literature of both *adaptability and adaptivity* employed at the level of the user interface of a desktop application to improve both

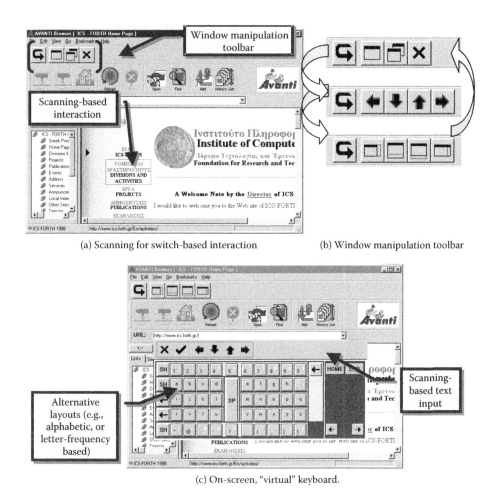

(a) Scanning for switch-based interaction (b) Window manipulation toolbar

(c) On-screen, "virtual" keyboard.

FIGURE 9.5 (**See color insert.**) Instances for motor-impaired users.

accessibility and usability for its end users. It represented a major development effort not only in terms of the adaptation capabilities described in the preceding sections but also in implementing HTTP communication and HTML rendering, designing, and developing the different styles for accessibility and usability, implementing the augmented navigation facilities, etc. This section considers how the AVANTI browser relates to more recent technologies related to Web browsers.

At the time the AVANTI project was started, there were only three Web browsers available for the Microsoft Windows operating system (out of the total four that were in active development at the time): NCSA Mosaic, Netscape Navigator (NN), and Microsoft Internet Explorer (MSIE). Of these, NCSA Mosaic was the only one for which source code was obtainable (the X Window System/Unix version publicly provided source code; source code for the other versions was

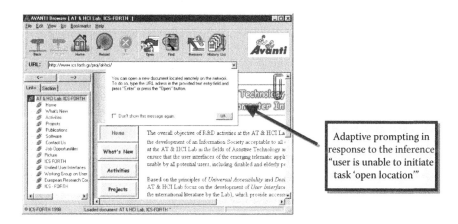

FIGURE 9.6 (**See color insert.**) Awareness prompting.

available after agreements were signed), although development had all but ceased in favor of Netscape Navigator, and the obtainable code was lagging behind the Web standards of the time (notably missing features required by the HTML 3.2 specification, such as nested tables). As a result, the NCSA Mosaic code base did not offer itself as a promising choice for basing a new Web browser on. The MSIE Web browser control, which enables the programmatic embedding of MSIE's HTTP modules and rendering engine in other applications, was not available yet (it was introduced with version 4 of MSIE, released in 1997).

At the same time, although there was some support for plug-ins in NN, this was rudimentary and was intended for the development of add-on components capable of handling MIME types other than HTML (e.g., PDF documents); there was practically no support for effecting modifications to the browser's user interface or in the rendering of HTML pages. A direct implication of this situation was that it was not possible to interact with the Document Object Model (DOM) of an HTML page unless doing so from within the rendering engine code of a browser.

Another technology that was missing at the time development in the AVANTI project begun was the Microsoft Active Accessibility (MSAA) Application Programming Interface (API). The MSAA API, which was introduced as an add-on to the Microsoft Windows 95 operating system in 1997, is designed to help Assistive Technology (AT) products (e.g., screen readers, on-screen keyboards, etc.) interact with standard and custom user interface elements of an application (or the operating system), as well as to access, identify, and manipulate an application's interface elements. The absence of this technology at the time of development meant that, even if an embeddable rendering engine were available, one would still not be able to inspect or modify the contents of the rendering pane in the ways that were necessary for the project's goals.

As stated already, things have changed dramatically in the intervening years. Starting from the last technology discussed, the MSAA API has enabled the development of sophisticated AT products that are specialized in assisting people with specific types of disabilities (e.g., the JAWS screen reader), which are often specifically tailored to facilitate the use of popular browsers. In fact, a lot of accessibility solutions (such as a workable on-screen keyboard) are now embedded into the Microsoft Windows operating system itself, in many cases obviating the need for external solutions. This is evidenced by the declining number of custom accessible browsers available as stand-alone applications. Older products, such as the pwWebSpeak browser (De Witt and Hakkinen 1998), and IBM's Home Page Reader (Asakawa and Lewis 1998) are not available any more, whereas other products, such as the WebbIE (King, Evans, and Blenkhorn 2004; see http://www.webbie.org.uk/webbie.htm for a 2008 updated version of the article) appear to be very specialized in nature (e.g., WebbIE is highly tailored for accessing the UK BBC's Web site).

But, whereas specialized browsers seem to be eclipsing, the opposite is true for browser plug-ins intended for

enhancing accessibility. This has been made possible by new plug-in architectures and APIs, which allow for much greater control over the presentation of, and interaction with, Web content, and specifically with a rendered page's DOM. A typical example is the Firefox Accessibility Extension developed the University of Illinois Center for Information Technology and Web Accessibility, and there exist many more.

Further to the above, accessibility of Web content in general has been facilitated by the emergence of sets of Web-related accessibility guidelines, such as those issued by the World Wide Web Consortium's Web Accessibility Initiative, which cover Web content, rich Internet applications, user agents (including browsers), and authoring tools, as well as of software to enable the automatic assessment of levels of compliance with the said guidelines.

Moving beyond accessibility, one can observe that the powerful and now standardized technologies that are available in all modern browsers (such as JavaScript, DOM access, rendering based on cascading style sheets, etc.), coupled with the aforementioned extended capabilities of browser "add-on" architectures and APIs, have had a significant impact on how adaptivity and personalization of Web content is approached. A characteristic example of this trend is the Firefox plug-in described in (Eynard 2008), which uses semantic data and a user model created from explicit and implicit modeling of users' browsing activities to augment the contents of visited pages with links to related pages, etc.

Against this background, were the AVANTI browser to be developed today, it is a relatively safe assumption that it would be implemented as an extensive browser plug-in or at least as a modified version of an open-source Web browser such as Firefox. In fact, to enable the full range of adaptations that AVANTI was capable of, such as augmentation of interactive dialogs, one would have to go with the later option of custom development "on top" of an existing browser.

9.4 THE PALIO SYSTEM FOR ADAPTIVE INFORMATION SERVICES

Building on the results and findings of the AVANTI project, PALIO* set out to address the issue of access to community-wide services by anyone, from anywhere, by proposing a hypermedia development framework supporting the creation of Adaptive Hypermedia Systems. PALIO (Stephanidis et al. 2004) supported the provision of tourist services in an integrated, open structure, while it constituted an extension of previous efforts, as it accommodated a broader perspective on adaptation and covered a wider range of interactive encounters beyond desktop access, and advanced the current state of affairs by considering novel types of adaptation based on context and situation awareness. The PALIO framework was based on the concurrent adoption of the following concepts: (1) integration of different wireless and wired telecommunication technologies to offer services through both fixed terminals

---

* The PALIO framework has been developed in the context of the PALIO project (see Acknowledgments section).

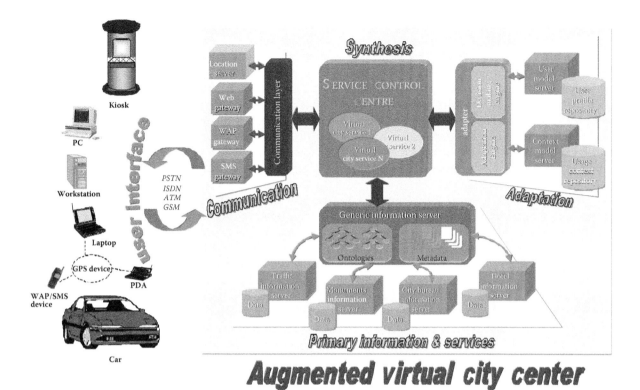

FIGURE 9.7 (**See color insert.**) Overall architecture of the PALIO system.

in public places and mobile personal terminals (e.g., mobile phones, PDAs, and laptops); (2) location awareness to allow the dynamic modification of information presented (according to user position); (3) adaptation of the contents to automatically provide different presentations depending on user requirements, needs, and preferences; (4) scalability of the information to different communication technologies and terminals; and (5) interoperability between different service providers in both the envisaged wireless network and the Web.

9.4.1 Overall PALIO Architecture

Figure 9.7 depicts the overall PALIO architecture. The *Augmented Virtual City Centre* (AVC) constitutes the core of the PALIO system. Users perceive the AVC as a system grouping together all information and services available in the city. It serves as an augmented, virtual facilitation point from which different types of information and services can be accessed. Context and location awareness, as well as the adaptation capabilities of the AVC, enable users to experience their interaction with services as *contextually grounded* dialogue. For example, the system always knows the user's location and can correctly infer what is near the user, without the user having to explicitly provide related information.

The main building blocks of the AVC can be broadly categorized as follows:

- The *Service Control Center* (SCC) is the central component of the PALIO system. It serves as the access point and the run-time platform for the system's

information services. The SCC is the framework upon which other services are built. It provides the generic building blocks required to compose services. Examples include the maintenance of the service state control, the creation of basic information retrieval mechanisms (through which service-specific modules can retrieve information from various distributed information sources/servers in PALIO), etc. Seen from a different perspective, the SCC acts as a central server that supports multiuser access to integrated, primary information and services; appropriately adapted to the user, the context of use, the access terminal, and the telecommunications infrastructure.

- The *User Communication Layer* (CL)* encapsulates the individual communication servers (Web gateway, WAP gateway, SMS gateway, etc.) and provides transparent communication independent of the server characteristics. This component unifies and abstracts the different communication protocols (e.g., WAP, and HTTP) and terminal platforms (e.g., mobile phone, PC, and Internet kiosk). Specifically, the CL transforms incoming communication from the user into a common format, so that the rest of the system does not need to handle the peculiarities of the underlying communication networks and

---

* The term "layer" is used in the PALIO project for historical reasons; the CL is not a layer in the sense of layered software architecture, but rather a component.

protocols. Symmetrically, the CL transforms information expressed in the aforementioned common format into a format appropriate for transmission and presentation on the user's terminal. In addition to the above, information regarding the capabilities and characteristics of the access terminal propagates across the PALIO system. This information is used to adapt the content and presentation of data transmitted to the user, so that it is appropriate for the user's terminal (e.g., in terms of media, modalities, and bandwidth consumption).

- The *Generic Information Server* (IS) integrates and manages existing information and services (which are distributed over the network). In this respect, the IS acts as a two-way facilitator. First, it combines appropriate content and data models (in the form of information ontologies and associated metadata), upon which it acts as a mediator for the retrieval of information and the utilization of existing services by the Service Control Center. Second, it communicates directly with the distributed servers that contain the respective data or realize the services. The existing information and services that are being used in PALIO are termed *primary,* in the sense that they already exist and constitute the building blocks for the PALIO services. The PALIO (virtual city) services, however, are synthesized on top of the primary ones and reside within the SCC.

- The *Adaptation Infrastructure* is responsible for content and interface adaptation in the PALIO system. Its major building blocks are the adapter, the user model server and the context model server. These are described in more detail in the next section.

From the user's point of view, a PALIO service is an application that provides information on an area of interest. From the developer's point of view, a PALIO service is a collection of dynamically generated and static template files, expressed in an XML-based device-independent language, which are used to generate information pages.

9.4.2 The PALIO Adaptation Infrastructure

In PALIO, user modeling is carried out through the dynamic personalisation server (DPS) developed by *human.* The DPS maintains four models: a *user* model, a *usage* model, a *system* model, and a *service* model. In general, user models consist of a part dedicated to users' interests and preferences, as well as a demographic part. In PALIO's version of the DPS, the principal part of a user model is devoted to representing users' interests and preferences. This part's structure is compliant with the information ontology, providing PALIO with a domain taxonomy. User models also incorporate information derived from group modeling by providing the following distinct probability estimates: *individual probability,* an assumption about a user's interests, derived solely from the user's interaction history (including information explicitly

provided by the user); *predicted probability,* a prediction about a user's interests based on a set of similar users, which is dynamically computed according to known and inferred user characteristics, preferences, etc.; and *normalized probability,* which compares an individual's interests with those of the whole user population.

The DPS *usage* model is a persistent storage space for all of DPS's usage-related data. It is comprised of interaction data communicated by the PALIO SCC (which monitors user activity within individual services) and information related to processing this data in user modeling components. These data are subsequently used to infer users' interests in specific items in PALIO's information ontology and/or domain taxonomy.

The system model encompasses information about the domain that is relevant for all user-modeling components of the DPS. The most important example of system model contents is the domain taxonomy. In contrast, the *service* model contains information that is relevant for single components only. Specifically, the service model contains information that is required for establishing communication between the DPS core and its user-modeling components.

The *Context Modeling Server* (CMS), as its name suggests, maintains information about the usage context. Usage context is defined, in PALIO, to include all information regarding an interactive episode that is not directly related to an individual user. PALIO follows the definition of context given by Dey and Abowd (2000) but diverges in that users engaged in direct interaction with the system are considered (and modeled) separately from other dimensions of context. Along these lines, a context model may contain information such as characteristics of the access terminal (including capabilities, supported mark-up language, etc.), characteristics of the network connection, current date and time, etc.

In addition, the CMS also maintains information about (1) the user's current location (which is communicated to the CMS by the location server, in the case of GSM-based localization, or the access device itself, in the case of GPS) and (2) information related to *push* services that users are subscribed to. It should be noted that in order to collect information about the current context of use, the CMS communicates, directly or indirectly, with several other components of the PALIO system. These other components are the *primary carriers* of the information. These first-level data collected by the CMS then undergo further analysis, with the intention of identifying and characterizing the current context of use. Like the DPS, the CMS responds to queries made by the adapter regarding the context and relays notifications to the adapter about important modifications (which may trigger specific adaptations) to the current context of use.

One of the innovative characteristics of the PALIO CMS is its ability to make context information available at different levels of abstraction. For example, the current time is available in full detail but also as a day period constant (e.g., morning); the target device can be described in general terms (e.g., for a simple WAP terminal: tiny screen device, graphics not supported, links supported, etc.). These abstraction capabilities also characterize aspects of the usage context that relate to the user's location. For instance, it is possible to identify the user's current location by its geographical longitude and latitude but also through the type of building the user may be in (e.g., a museum), the characteristics of the environment (e.g., noisy), and so on. The adaptation logic (that is based on these usage context abstractions) has the advantage that it is general enough to be applied in several related contexts. This makes it possible to define adaptations that address specific, semantically unambiguous characteristics of the usage context, in addition to addressing the context as a whole.

The third major component of the adaptation infrastructure is the adapter, which is the basic adaptation component of the system. It integrates information concerning the user, the context of use, the access environment and the interaction history and adapts the information content and presentation accordingly. Adaptations are performed on the basis of the following parameters: user interests (when available in the DPS or established from the ongoing interaction), user characteristics (when available in the DPS), user behavior during interaction (provided by the DPS, or derived from ongoing interaction), type of telecommunication technology and terminal (provided by the CMS), location of the user in the city (provided by the CMS), etc.

The adapter is comprised of two main modules, the decision making engine (DME) and the adaptation engine (AE). The DME is responsible for deciding upon the need for adaptations on the basis of (1) the information available about the user, the context of use, the access terminal, etc. and (2) a knowledge base that interrelates this information with adaptations (i.e., the adaptation logic). Combining the two, the DME makes decisions about the most appropriate adaptation for any particular setting and user/technology combination addressed by the project.

The AE instantiates the decisions communicated to it by the DME. The DMS and AE are kept as two distinct functional entities in order to decouple the adaptation decision logic from the adaptation implementation. This approach allows for a high level of flexibility, as new types of adaptations can be introduced into the system very easily. At the same time, the rationale for arriving at an adaptation decision and the functional steps required to carry it out can be expressed and modified separately.

9.4.3 Adaptation in PALIO

The term *adaptation determinant* refers to any piece of information that is explicitly represented in the system and that can serve as input for the adaptation logic. Seen from a different perspective, adaptation determinants are the facts known by the system that can be used to decide upon the need for, and the appropriate type of, adaptation at any point in time (Brusilovsky 1996). In general, three main categories of determinants can be identified: (1) information about the users themselves, (2) information about groups to which the users may belong, and (3) information about the context of use.

An important fact is that information about the group to which a user belongs can be used alongside (or, even better, in combination with) information about the individual user. This approach is actively pursued in PALIO, where, for instance, the interest of a particular person in a specific type of information can be inferred (with some degree of probability) from the system's knowledge about the related interests of other members in the user's group. However, this approach cannot be extended to all user-related information: user traits (affecting a person's personality, cognitive aptitudes, learning style, etc.), for instance, cannot be inferred from group modeling.

A second important point concerns the fact that knowledge about the user is not always sufficient to identify user needs. One should also consider the more general context in which the user interacts (including the user's location, the characteristics of the access terminal/device, etc.). In fact, with respect to context-related factors, users with varying characteristics can be expected to have similar needs. The PALIO framework takes full account of this fact by undertaking (through the CMS) the monitoring of observable context-related parameters. These parameters are available to the adaptation designer to be used as explicit adaptation determinants.

The primary decision engine implemented for PALIO is rule based. Although first-order logic rule engines can easily be plugged into the framework (the latter having been specifically designed to allow for that), it was decided that, to facilitate the wide adoption of the framework, a simpler and more accessible approach was in order. Along these lines, a new rule language was created, borrowing from control structures that are commonly supported in functional programming languages. The premise was that such structures were much more familiar to designers of adaptive systems, while, at the same time, they afforded lower degrees of complexity when it came to understanding the interrelations and dependencies between distinct pieces of adaptation logic.

An XML binding was developed for the aforementioned rule language, while a rule interpreter and a corresponding rule engine supported the run-time operation of the system. Adaptation rules expressed in such a rule language may be defined either in external files or embedded in the document to be adapted. Rules embedded in pages are constrained in that they can only be applied within the specific document in which they reside and therefore are not reusable. Rules external to documents are organized into rule sets. The framework currently supports three types of rules: *if-then-else* rules, *switch-case-default* rules, and *prologue-actions-epilogue* rules. If-then-else rules are the simplest type of conditional rules; they bind sets of adaptation actions with the truth value of a conditional expression. Following the example of functional languages, an if-then-else rule in PALIO is composed of a *condition* (containing an expression based on adaptation determinants), a *then* part (containing the adaptation actions to be taken whenever the condition is satisfied), and an optional *else* part (containing the actions to be taken when the condition fails). Switch-case-default rules can be used to relate multiple values (outcomes of run-time expression evaluation) to sets of adaptation actions. In this case, adaptation actions are executed if the value of a *variant* (expression) is equal to a *value*, or within the *range* of values, specified as a selection *case*. The switch-case-default rule construct supports the definition of multiple cases and requires the additional definition of an (optionally empty) set of *default* adaptation actions to be performed if none of the defined cases apply.

Finally, Prologue-actions-epilogue rules are intended mainly for the definition of unconditional rules. In other words, the specific rule construct is provided to support the definition of (sets of) actions to be performed at a particular stage in the sequence of adaptations. A very common use of the *prologue* and *epilogue* parts is the creation/assignment and retraction of variables that are used by the adaptation actions (e.g., in order to determine the point of application of an action).

9.4.4 Adaptation Scenarios in PALIO

To better illustrate the capabilities of the PALIO framework, this section presents an example from the PALIO information system, focusing on the following two fictional interaction scenarios.

First scenario: An English-speaking wheelchair-bound user is in Piazza del Campo (Siena, Italy) in the morning. She is interested in sightseeing and prefers visiting monuments and museums. She is accessing the system from her palmtop, which she rented from the Siena Tourist Bureau and which is fitted with a GPS unit. She accesses the **City Guide** service and asks for recommendations about what she could see or do next.

Second scenario: An Italian-speaking able-bodied user is also in Piazza del Campo (Siena, Italy) around noon. He has not used the system before, and therefore there is no information about what he might prefer. He is accessing the system from his mobile through WAP. He also accesses the **City Guide** service and asks for recommendations of what he could do next.

The result of the first user's interaction with the PALIO system is shown in Figure 9.8a. Relevant characteristics include (1) the presentation language is English, (2) the front-end is tailored for a small-screen terminal capable of color and graphics, (3) the system's recommendations are in accordance with the user's preferences as specified in the user profile, (4) recommended sites are in the immediate vicinity of the user, and (5) accessibility information is provided immediately (at the first level) because this information will impact on whether the user will decide to visit the place or not.

The result of the second user's interaction with the PALIO system is shown in Figure 9.8b. Relevant characteristics include (1) the presentation language is Italian, (2) the front-end is tailored for tiny-screen terminals without assumptions

 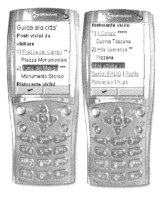

(a) GPS-enabled PDA   (b) WAP terminnal

FIGURE 9.8 Output on different devices from the PALIO Information System in Siena, Italy. Please note that the difference in language (English on the PDA screen, Italian on the mobile phone screens) is part of the adaptation as described in the two considered scenarios.

made about color and graphics, (3) the system's recommendations are derived from preferences for the user's group (asterisks next to each recommendation indicate other users' collective assessments of the venue recommended), (4) recommendations include a wider range of activities (e.g., sightseeing and eating), (5) one type of activity (eating) is relevant to temporal context (it's noon), (6) recommended sites are in the general vicinity of the user, and (7) accessibility information is not provided immediately (at the first level).

9.4.5 Subsequent Evolution of Adaptive Hypermedia Systems

The progress of the area of AHS in the years since the development of the PALIO framework has been explosive (Knutov, De Bra, and Pechenizkiy 2009; Brusilovsky, Kobsa, and Nejdl 2007). This has been due both to the increasing popularity of the Web, and the growing interest in supporting personalization in practically all application domains of computer-supported human activity. Perhaps the most well-known AHS framework in the literature that is still in active development, and also available in open source form, is AHA! (most recently described by De Bra, Smits, and Stash 2006). The AHA! framework has evolved significantly since its inception, and recent additions have introduced features that are on par with those of the PALIO framework (e.g., forward chaining of rules, a more extensive set of operand types and operations that can be used in rule expressions). Compared to AHA! and other systems in the literature, the PALIO framework retains, today, several of its advantages, the most important include the following:

- It has been specifically designed to be easy to combined with other Web-publishing frameworks, providing adaptation functionality "on top" of them, thus making the adaptation infrastructure more easily utilizable in new systems.

- It offers support for creating high-level declarative adaptation rules, separated entirely from the documents/fragments on which these are applied. To achieve the same level of functionality with AHA!, for example, one would have to use the latter in combination with the MOT authoring tool (Cristea and de Mooij 2003), which is capable of transforming high-level rules to a format compatible with AHA!.
- It features an explicit representation of basic adaptation actions that can be synthesized into most of the adaptation techniques in the literature (with the exception of some of the multimedia-oriented techniques, and ones involving Natural Language Generation).

However, there are additional areas of work in the PALIO framework that would have benefited from recent progress. One such area is the use of ontologies (see Michaelis, Golbeck, and Hendler, this volume). Although the framework did support the ontological representation and querying of domain data, the adopted approach was (1) custom and (2) not extended to the rest of the framework's modeling needs. Since the development of the PALIO framework, a significant amount of standardization activity has taken place in the area of semantically meaningful representations of data, spearheaded by the World Wide Web Consortium under the umbrella of the "Semantic Web" (Berners-Lee, Hendler, and Lassila 2001). Specifications such as the Resource Description Framework (RDF), the SPARQL Protocol and RDF Query Language (SPARQL), and the OWL Web Ontology Language (OWL) have provided the means for standardizing approaches to, and implementations of, representation and reasoning over semantic data. As one might expect, this has had a substantial effect in the area of AHS, with the aforementioned technologies being actively used for user modeling, domain modeling, model-based inferencing, etc. (see, e.g., Aroyo et al. 2007; Heckmann et al. 2005; Henze, Dolog, and Nejdl 2004). Furthermore, recent efforts have specifically addressed the challenge of interoperability of ontological data in adaptation models to allow for their sharing and reuse (Balík and Jelínek 2008). An in-depth overview of the use of semantic Web technologies in closed- and open-corpus AHS can be found in Dolog and Nejdl (2007) and several articles on recent developments in Angelides, Mylonas, and Wallace (2009). Against this background, if one were to revisit the PALIO framework today, the custom modeling solutions employed would likely be replaced with standards-based ones, with a view to interoperability and extended generality of the framework.

The PALIO framework also does not provide explicit support for newer paradigms of client-server communication on the Web, such as the use of Asynchronous JavaScript and XML (AJAX; see Dix and Shabir, this volume). This would be, in principle, an easier issue to address, because these paradigms necessarily still follow the request-response cycle for the communication between the client and the server. Consequently, the only modifications that would be potentially necessary in this direction would be (1) transformation of request data from the incoming formats to XML and (2) transformation of response data from XML to the required output formats. Note that these additions would only be necessary if the incoming or outgoing data need to be expressed in "nonstandard" formats, such as JavasScript Object Notation (JSON).

A distinctive trend in recent AHS research is adaptation in open corpus hypermedia (Brusilovsky and Henze 2007). Work in this direction aims to extend AHS with the possibility to operate on an open corpus of documents, which is not known at design time and in addition to this can be constantly changing and expanding. Although the PALIO framework can certainly be applied in open corpora, it suffers from the same shortcomings as other frameworks in its category: it provides neither mechanisms for concept and content alignment nor mechanisms for accessing and manipulating an expanding domain model.

A different case of adaptation "in the wild" (i.e., on arbitrary documents), where the PALIO framework would be more readily applicable is exemplified by the "Glue" system. The later comprises a plug-in for the Firefox browser, and a corresponding service, which, combined, augment Web pages that the user visits with links to information related to the page's content and of potential interest to the user. According to the system's documentation, part of these recommendations are of a social nature (based on, e.g., the browsing history or past ratings of the user's social network), and part are based on analysis of the page contents and their association with a number of categorical information sources, on the basis of the user's profile. In other words, the second type of recommendation consists in identifying "entities" and their type in a page (e.g., a musician, a restaurant, and a book), associating these entities with their semantic counterparts in the system's domain ontology, consulting the user model to determine whether the said entities are of interest to the user, and, if so, augmenting the entities with links to additional related information and pages. With additions of the type discussed earlier with respect to the introduction of support for AJAX, the PALIO framework could be easily tailored to play the role of the server-side component of the described system.

9.5 DISCUSSION AND CONCLUSIONS

R&D activities in the past years have provided valuable insights into the study of the contribution of automatic adaptation to universal access (Stephanidis 2001c; Stephanidis and Emiliani 1999). As accessibility and interaction quality have become global requirements in the Information Society, adaptation needs to be "designed into" the system rather than decided upon and implemented *a posteriori*. The experience gained in the development of AVANTI and PALIO has demonstrated the effectiveness of the use of adaptation-based methodologies, techniques, and tools toward the achievement of access to the Web by a wide range of user categories, irrespective of physical abilities or technology expertise, in a variety of contexts of use and through a variety of access devices, going far beyond previous approaches that rely on assistive or dedicated technologies.

The AVANTI browser constituted a large-scale, "real-world" interactive application capable of adapting itself to suit the requirements of able-bodied, blind, and motor-impaired users in a variety of contexts of use. Adaptability and adaptivity were used extensively to tailor and enhance the interface respectively, in order to effectively and efficiently meet the target of interface individualization for end users. The accessibility requirements posed by the target user groups addressed in the AVANTI development could not be met by employing then existing browsing and accessibility technologies. Although, as already discussed, technologies available today (e.g., browser "add-ons") would definitely influence the development of the browser, the level of adaptations supported in AVANTI cannot be supported solely through customization techniques (e.g., integrating guidance in system dialogues, dynamically modifying the interaction dialogue). On the other hand, meeting such requirements could not be supported by traditional user interface development methods, as they lack provisions for designing and implementing adaptation mechanisms. Such methods should be capable of capturing the adaptation behavior in the user interface design and encapsulate it accordingly in the implementation, and should be supported by appropriate tools (Stephanidis 2001d). As adaptation implies providing dialogue according to user- and context-related factors, suitable methods and tools should provide means for capturing user and context characteristics and their interrelationships with alternative interactive behaviors, as well as appropriate mechanisms for deciding adaptations on the basis of those parameters and relationships. Additionally, appropriate techniques for managing alternative implemented interactive behaviors and applying adaptation decisions at run-time are required. This is not the case in currently available interface development tools, which are mainly targeted to the provision of advanced support for implementing physical aspects of the interface via different techniques (e.g., visual construction, task-based, demonstration-based; Savidis and Stephanidis 2001b).

Given the ongoing trend toward ubiquitous access, mobile computing, and Internet appliances, the view that adaptable and adaptive behavior is a characteristic property of the interactive software may no longer suffice. PALIO constitutes a substantial extension over previous efforts toward universal access because it offers an innovative approach to the development of user- and context-aware Web-based information systems and supports novel types of adaptation and new interactive platforms beyond the desktop (Stephanidis et al. 2004). The architectural model of interaction of PALIO remains therefore widely applicable in service sectors other than tourism even today. One of the highly innovative characteristics of PALIO was the extensive support it offered for creating applications and services that are truly independent from the target device and computing platform. The framework supported a large repertoire of hypermedia adaptation techniques, thus enabling the easy creation of online systems. These systems could maintain high levels of consistency over different platforms, while, at the same time, allowing developers to make the best use of available resources. PALIO provided support for the strict separation of content and presentation, allowing each to be manipulated and adapted independently, thus making it possible to cater for the needs of nontraditional user groups (such as disabled users). Furthermore, users could enjoy the accessibility afforded by adaptations over the full range of devices supported by the system. Additionally, the PALIO framework was capable of selecting among alternative presentation modalities (when these were available), adopting the one that is most suitable for the user and device combination.

While AVANTI constituted an adaptable and adaptive content viewing/interaction application that supported any type of Web content, adapted or not, PALIO supported the creation of adaptable and adaptive content that could be viewed with any kind of browser. Thus, the two complement each other. Both AVANTI and PALIO make it possible to adopt a stepwise introduction of adaptation at different stages of development, thus enabling the progressive introduction of complex accessibility features and facilitating the incorporation of new user groups with distinct requirements in terms of accessibility.

ACKNOWLEDGMENTS

The ACCESS TP1001 (Development platform for unified ACCESS to enabling environments) project was partially funded by the TIDE Programme of the European Commission and lasted 36 months (January 1, 1994, to December 31, 1996). The partners of the ACCESS consortium are CNR-IROE (Italy), Prime contractor; ICS-FORTH (Greece); University of Hertforshire (United Kingdom); University of Athens (Greece); NAWH (Finland); VTT (Finland); Hereward College (United Kingdom); RNIB (United Kingdom); Seleco (Italy); MA Systems & Control (United Kingdom); and PIKOMED (Finland).

The AVANTI AC042 (Adaptable and Adaptive Interaction in Multimedia Telecommunications Applications) project was partially funded by the ACTS Programme of the European Commission and lasted 36 months (September 1, 1995, to August 31, 1998). The partners of the AVANTI consortium are ALCATEL Italia, Siette division (Italy), Prime contractor; IROE-CNR (Italy); ICS-FORTH (Greece); GMD (Germany), VTT (Finland); University of Siena (Italy), MA Systems and Control (UK); ECG (Italy); MATHEMA (Italy); University of Linz (Austria); EUROGICIEL (France); TELECOM (Italy); TECO (Italy); and ADR Study (Italy).

The IST-1999-20656 PALIO project (Personalised Access to Local Information and Services for Tourists) is partly funded by the Information Society Technologies Programme of the European Commission—DG Information Society. The partners in the PALIO consortium are ASSIOMA S.p.A. (Italy), Prime contractor; CNR-IROE (Italy); Comune di Firenze (Italy); FORTH-ICS (Greece); GMD (Germany);

Telecom Italia Mobile S.p.A. (Italy); University of Sienna (Italy); Comune di Siena (Italy); MA Systems & Control Ltd (UK); and FORTHnet (Greece).

REFERENCES

Akoumianakis, D., and C. Stephanidis. 2001. USE-IT: A tool for lexical design assistance. In *User Interfaces for All—Concepts, Methods, and Tools*, ed. C. Stephanidis, 469–487. Mahwah, NJ: Lawrence Erlbaum.

Angelides, M. C., P. Mylonas, and M. Wallace, eds. 2009. *Advances in Semantic Media Adaptation and Personalization*, vol. 2. Boca Raton, FL: Auerbach.

Ardissono, L., and A. Goy. 1999. Tailoring the interaction with users of electronic shops. In *Proceedings of 7th International Conference on User Modeling (UM99)*, ed. J. Kay, 35–44. Wien, Germany: Springer.

Aroyo, L., N. Stash, Y. Wang, P. Gorgels, and L. Rutledge. 2007. CHIP demonstrator: Semantics-driven recommendations and museum tour generation. In *Proceedings of the 6th International Semantic Web Conference, 2007*, eds. K. Aberer et al., 879–886. Berlin, Germany: Springer.

Asakawa, C., and C. Lewis. 1998. Home page reader: IBM's talking web browser. In *Proceedings of Closing the Gap Conference* (Minneapolis, Minnesota, October 22–24, 1998).

Balabanovic, M., and Y. Shoham. 1997. Fab: Content-based collaborative recommendation. *Communications of the ACM* 40(3): 66–72.

Balík, M., and I. Jelínek. 2006. Modelling of adaptive hypermedia systems. In *Proceedings of the International Conference on Computer Systems and Technologies (CompSysTech'06)* (University of Veliko Tarnovo, Bulgaria, June 15–16), http://ecet.ecs.ru.acad.bg/cst06/Docs/cp/sV/V.8.pdf.

Benyon, D. R. 1993. Adaptive systems: a solution to usability problems. *User Modeling and User-adapted Interaction* 3(1): 65–87.

Benyon, D. R. 1997. Intelligent Interface Technology to Improve Human–Computer Interaction (Tutorial). Paper presented at HCI International 1997 Conference, August 24–29, San Francisco, CA.

Berners-Lee, T., J. Hendler, and O. Lassila. 2001. The semantic Web. *Scientific American*, May 2001.

Bini, A., and P.-L. Emiliani. 1997. Information about mobility issues: The ACTS AVANTI project. In *Proceedings of 4th European Conference for the Advancement of Assistive Technology (AAATE '97)* (Porto Carras, Greece), 85–88. Amsterdam: IOS Press.

Bini, A., R. Ravaglia, and L. Rella. 1997. Adapted interactions for multimedia based telecommunications applications. Paper presented at Conference Neties 1997, Ancona, Italy.

Brusilovsky, P. 1996. Methods and techniques of adaptive hypermedia. *User Modeling and User-Adapted Interaction* 6(2–3): 87–129.

Brusilovsky, P., and N. Henze. 2007. Open corpus adaptive educational hypermedia. In *The Adaptive Web: Methods and Strategies of Web Personalization* (LNCS 4321), eds. P. Brusilovsky, A. Kobsa, and W. Nejdl, 671–696. Berlin, Germany: Springer.

Brusilovsky, P., A. Kobsa, and W. Nejdl, eds. 2007. *The Adaptive Web: Methods and Strategies of Web Personalization*. Berlin: Springer-Verlag.

Brusilovsky, P., A. Kobsa, and J. Vassileva, eds. 1998. *Adaptive Hypertext and Hypermedia*. Dordrecht: Kluwer Academic.

Cristea, A. I., and A. de Mooij. 2003. Adaptive course authoring: My online teacher. ICT'03 (International Conference on Telecommunications), Papeete, French Polynesia, 1762–1769. New York: IEEE Press.

De Bra, P., D. Smits, and N. Stash. 2006. The design of AHA! In *Proceedings of the 17th Conference on Hypertext and Hypermedia (Hypertext'06)*, 133–134. New York: ACM Press.

De Witt, J. C., and M. T. Hakkinen. 1998. Surfing the Web with pwWebSpeak. In *Proceedings of the Technology and Persons with Disabilities Conference*. http://www.csun.edu/cod/conf/1998/proceedings/159.htm (accessed November 2010).

Dey A. K., and G. D. Abowd. 2000. Towards a better understanding of context and context awareness. In the *Proceedings of the Workshop on the What, Who, Where, When and How of Context-Awareness, affiliated with the CHI 2000 Conference on Human Factors in Computer Systems*. ftp://ftp.cc.gatech.edu/pub/gvu/tr/2000/00-18e.pdf.

Dieterich, H., U. Malinowski, T. Kühme, and M. Schneider-Hufschnidt. 1993. State of the art in adaptive user interfaces. In *Adaptive User Interfaces: Principles and Practice*, eds. M. Schneider-Hufschnidt, T. Kühme, and U. Malinowski, 13–48. Amsterdam: North-Holland.

Dix, A., and N. Shabir, this volume. Human–computer interaction. In *Handbook of Human Factors in Web Design*, 2nd ed., eds. K.-P. L. Vu and R. W. Proctor, 35–62. Boca Raton, FL: CRC Press.

Dolog, P., and W. Nejdl. 2007. Semantic Web technologies for the adaptive Web. In *The Adaptive Web: Methods and Strategies of Web Personalization*, eds. P. Brusilovsky, A. Kobsa, and W. Nejdl, 697–719. Berlin, Germany: Springer-Verlag.

Eynard, D. 2008. Using semantics and user participation to customize personalization. Technical Report HPL-2008-197. Hewlet Packard Laboratories. http://www.hpl.hp.com/techreports/2008/HPL-2008-197.pdf.

Fink, J., A. Kobsa, and A. Nill. 1998. Adaptable and adaptive information provision for all users, including disabled and elderly people. *The New Review of Hypermedia and Multimedia* 4: 163–188.

Heckmann, D., T. Schwartz, B. Brandherm, M. Schmitz, and M. Von Wilamowitz-Moellendorff. 2005. GUMO—the general user model ontology. In *Proceedings of the 10th International Conference on User Modeling (UM'2005)*, Lecture Notes in Artificial Intelligence, vol. 3538/2005, 428–432. Berlin, Germany: Springer-Verlag.

Henricksen, K., and J. Indulska. 2001. Adapting the Web interface: an adaptive Web browser. In *Proceedings of the Second Australasian User Interface Conference, 2001 (AUIC 2001)* (Gold Coast, Qld., Australia, Jan. 29 to Feb. 1), 21–28.

Henze, N. 2001. Open adaptive hypermedia: an approach to adaptive information presentation on the Web. In *Universal Access in HCI: Towards an Information Society for All*, vol. 3, *Proceedings of the 9th International Conference on Human–Computer Interaction (HCI International 2001)* (New Orleans, Louisiana, Aug. 5–10), ed. C. Stephanidis, 818–821. Mahwah, NJ: Lawrence Erlbaum.

Henze, N., P. Dolog, and W. Nejdl. 2004. Towards personalized e-learning in a semantic Web. *Educational Technology and Society Journal*, Special Issue, 7(4): 82–97.

Hermsdorf, D., H. Gappa, and M. Pieper. 1998. WebAdapter: a prototype of a WWW-browser with new special needs adaptations. In *Proceedings of the 4th ERCIM Workshop on "User Interfaces for All"* (Stockholm, Sweden, Oct. 19–21), eds. C. Stephanidis and A. Waern, 15 pages, http://ui4all.ics.forth.gr/UI4ALL-98/hermsdorf.pdf.

Horvitz, E., J. Breese, D. Heckerman, D. Hovel, and K. Rommelse. 1998. The Lumiere Project: Bayesian user modeling for inferring the goals and needs of software users. In *Proceedings of the Fourteenth Conference on Uncertainty in Artificial Intelligence* (Madison, WI), 256–265. San Francisco, CA: Morgan Kaufmann.

King, A., G. Evans, and P. Blenkhorn. 2004. WebbIE: A web browser for visually impaired people. In Proceedings of the 2nd Cambridge Workshop on Universal Access and Assistive Technology (CWUAAT), 2004.

Knutov, E., P. De Bra, and M. Pechenizkiy. 2009. AH—12 years later: a comprehensive survey of adaptive hypermedia methods and techniques. *New Review of Hypermedia and Multimedia* 15(1): 5–38.

Kobsa, A. 2001. Generic user modeling systems. *User Modeling and User Adapted Interaction* 11(1–2): 49–63.

Kobsa, A., and W. Pohl. 1995. The user modelling shell system BGP-MS. *User Modeling and User-adapted Interaction* 4(2): 59–106.

Michaelis, J. R., J. Golbeck, and J. Hendler, this volume. Organization and structure of information using semantic Web technologies. In *Handbook of Human Factors in Web Design*, 2nd ed., eds. K.-P. L. Vu and R. W. Proctor, 231–248. Boca Raton, FL: CRC Press.

Microsoft. 2003a. Internet Explorer 6 Accessibility Information. http://www.microsoft.com/enable/products/IE6/default.aspx (accessed November 2010).

Microsoft. 2003b. Windows 2000 Professional Accessibility Information. http://www.microsoft.com/enable/products/windows2000/default.aspx (accessed November 2010).

Oppermann, R., and M. Specht. 1998. Adaptive support for a mobile museum guide. In *the Proceedings of the IMC '98—Workshop Interactive Application of Mobile Computing* (Rostock, Germany). http://fit.fraunhofer.de/~oppi/publications/NoWo Rost.NomadicComp.pdf (accessed November 2010).

Petrie, H., S. Morley, P. McNally, A. M. O'Neill, and D. Majoe. 1997. Initial design and evaluation of an interface to hypermedia systems for blind users. In *Proceedings of Hypertext '97* (Southampton, UK, April 6–11), 48–56. New York: ACM Press.

Savidis, A., D. Akoumianakis, and C. Stephanidis. 2001. The unified user interface design method. In *User Interfaces for All—Concepts, Methods, and Tools*, ed. C. Stephanidis, 417–440. Mahwah, NJ: Lawrence Erlbaum.

Savidis, A., and C. Stephanidis. 2001a. The unified user interface software architecture. In *User Interfaces for All—Concepts, Methods, and Tools*, ed. C. Stephanidis, 389–415. Mahwah, NJ: Lawrence Erlbaum.

Savidis, A., and C. Stephanidis. 2001b. Development requirements for implementing unified user interfaces. In *User Interfaces for All—Concepts, Methods, and Tools*, ed. C. Stephanidis, 441–68. Mahwah, NJ: Lawrence Erlbaum.

Savidis, A., and C. Stephanidis. 2001c. The I-GET UIMS for unified user interface implementation. In *User Interfaces for All—Concepts, Methods, and Tools*, ed. C. Stephanidis, 489–523. Mahwah, NJ: Lawrence Erlbaum.

Spalteholz, L., K. F. Li, and N. Livingston. 2008. Efficient navigation on the World Wide Web for the physically disabled. In *Proceedings of the 3rd International Conference on Web Information Systems and Technologies* (Barcelona, Spain, March 3–6), eds. J. Fillipe and J. Cordeiro, 321–326. Berlin, Germany: Springer.

Stephanidis, C. 2001a. Human–computer interaction in the age of the disappearing computer. In *"Advances in Human–Computer Interaction I," Proceedings of the Panhellenic Conference with International Participation on Human–Computer Interaction (PC-HCI 2001)* (Patras, Greece, Dec. 7–9), eds. N. Avouris and N. Fakotakis, 15–22. Patras, Greece: Typorama.

Stephanidis, C. ed. 2001b. User interfaces for all: new perspectives into human–computer interaction. In *User Interfaces for All—Concepts, Methods, and Tools*, 3–17. Mahwah, NJ: Lawrence Erlbaum.

Stephanidis, C. 2001c. Adaptive techniques for universal access. *User Modelling and User Adapted Interaction International Journal*, 10th Anniversary Issue, 11(1/2): 159–179.

Stephanidis, C. ed. 2001d. The concept of unified user interfaces. In *User Interfaces for All—Concepts, Methods, and Tools*, 371–388. Mahwah, NJ: Lawrence Erlbaum.

Stephanidis, C., and P. L. Emiliani. 1999. Connecting to the information society: A European perspective. *Technology and Disability Journal* 10(1): 21–44.

Stephanidis, C., A. Paramythis, M. Sfyrakis, and A. Savidis. 2001. A case study in unified user interface development: The AVANTI Web browser. In *User Interfaces for All—Concepts, Methods, and Tools* ed. C. Stephanidis, 525–568. Mahwah, NJ: Lawrence Erlbaum.

Stephanidis, C., A. Paramythis, M. Sfyrakis, A. Stergiou, N. Maou, A. Leventis, G. Paparoulis, and C. Karagiannidis. 1998a. Adaptable and adaptive user interfaces for disabled users in AVANTI project. In *Intelligence in Services and Networks: Technology for Ubiquitous Telecommunications Services—Proceedings of the 5th International Conference on Intelligence in Services and Networks (IS&N '98)* (Antwerp, Belgium, May 25–28), Lecture Notes in Computer Science, 1430, eds. S. Trigila et al., 153–66. Berlin, Germany: Springer.

Stephanidis, C., A. Paramythis, V. Zarikas, and A. Savidis. 2004. The PALIO Framework for Adaptive Information Services. In *Multiple User Interfaces: Cross-Platform Applications and Context-Aware Interfaces*, eds. A. Seffah and H. Javahery, 69–92. New York: John Wiley.

Stephanidis, C., G. Salvendy, D. Akoumianakis, N. Bevan, J. Brewer, P. L. Emiliani, A. Galetsas, et al. 1998b. Toward an information society for all: An international R&D agenda. *International Journal of Human–Computer Interaction* 10(2): 107–134.

Sukaviriya, P., and J. Foley. 1993. Supporting adaptive interfaces in a knowledge-based user interface environment. In *Proceedings of the International Workshop on Intelligent User Interfaces* (Orlando, FL), eds. W. D. Gray, W. E. Hefley, and D. Murray, 107–114. New York. ACM Press.

Sun Microsystems. YEAR. Sun Microsystems Accessibility Program. http://www.sun.com/access/general/overview.html.

Tan, C. C., W. Yu, and G. McAllister. 2006. Developing an ENABLED adaptive architecture to enhance internet accessibility for visually impaired people. *International Journal on Disability and Human Development* 5(2): 97–104.

Treviranus, J., and C. Serflek. 1996. Alternative Access to the World Wide Web. University of Toronto. http://www.utoronto.ca/atrc/rd/library/papers/WWW.html.

W3C–WAI. 1999. Web Content Accessibility Guidelines 1.0. http://www.w3.org/TR/WCAG10/.

Section IV

Organization of Information for the Web

10 Applications of Concept Maps to Web Design and Web Work

John W. Coffey, Robert R. Hoffman, and Joseph D. Novak

CONTENTS

10.1 INTRODUCTION

In 1990, Tim Berners-Lee described his vision of what would come to be called the World Wide Web (WWW, hereinafter referred to as Web). He based his proposal on the notion that a "web" or "mesh" of hyperlinked documents (papers, commentaries, etc.) might be more supportive of information preservation, information sharing, and collaboration activities than more restrictive fixed hierarchies or "trees." Furthermore, he argued that information would be more useful if the relations among documents could be given meaningful labels. He illustrated his ideas using a diagram (Berners-Lee 1990). In his diagram he used circles called "nodes" to depict individual documents and concepts (images, etc.), and he used labeled lines called "links" to express relations. Node-link node triples in his diagram expressed propositions. For example, a circle enclosing the words "This Document" was linked by a line to another circle containing the words "A Proposal," and appearing midway along the line was the relational term "describes." Hence this node-link-node triple expressed the proposition: *This Document describes A Proposal.* Berners-Lee used this meaningful diagram to express how documents, hypertext, and other media could be shared over a network of computers that would also support conferencing and other collaborative

activities. In referring to his diagram, he wrote (Berners-Lee 1990; reprinted in Berners-Lee 1999, pp. 213–216):

> When describing a complex system, many people resort to diagrams with circles and arrows. Circles and arrows leave one free to describe the interrelationships between things in a way that tables, for example, do not. The system we need [i.e., the Web] is like a diagram of circles and arrows . . .

The Web infrastructure for integrating and hyperlinking multimedia (Berners-Lee's "client-server" model) can support methods of knowledge sharing and collaboration that are limited perhaps only by the designer's imagination. Ironically, however, many "pages" on the Web, especially those that are intended for use in education and distance learning, appear very much like pages from the Gutenberg Bible: Somewhere near the top will be a graphic of some sort (e.g., a picture of a fish for course material in biology), and below it will simply be an outline and some scanned-in text, through which one proceeds just as if using a printed book. One might wonder if the current evolution of the Web is in some ways simply reinventing the wheel and doing so in a way that fails to capitalize on the capabilities and possibilities that the Web offers.

This chapter provides a discussion of how meaningful diagrams called "concept maps" can be used on the Web. First is a brief description of concept maps, including their origins, theoretical basis, and a brief comparison with other diagrammatical representations of knowledge. Next, is a capsule view of the process by which concept maps and knowledge models, Web-based hypermedia systems organized by concept maps, are made. Following this section is a discussion of highlights from the literature on concept maps in education, uses to foster collaborative knowledge creation and sharing, uses for knowledge elicitation and management, and uses as graphical browsers on the Web. The chapter closes with a more detailed consideration of a concept map-based system that provides Web-based access and organization of materials for a university-level Computer Science course.

10.2 CONCEPT MAPS

Concept maps are meaningful diagrams much like the one Berners-Lee created. That is, concepts are represented by one or a few words that are enclosed within geometric figures, with the relations among concepts expressed by one or a few words that label a directional, connecting line. The basic ideas of concept mapping were actually developed over two decades ago at Cornell University, where Joseph Novak and his colleagues sought to understand and follow changes in children's knowledge of science (Novak and Gowin 1984). Novak's work was inspired by the learning theory of David Ausubel (1962, 1968; Ausubel, Novak, and Hanesian 1978). The fundamental idea is that learning takes place by the assimilation of new concepts and relations into existing frameworks of propositions held by the learner. This is said to occur by processes of subsumption, differentiation, and reconciliation. Novak (2002, 542) reasoned that if this is how knowledge and learning work, why not represent it directly:

> Piaget popularized the clinical interview as a means to probe children's cognitive processes that they use to interpret events. We adapted his approach to serve a significantly different purpose, namely to identify the concept and propositional frameworks that people use to explain events. Working with almost 200 students in our 12-year longitudinal study, and interviewing these students several times during the first year of the study, we were soon overwhelmed with interview transcripts. Moreover, we found it difficult to observe specific changes that were occurring in the children's understanding of science concepts. We had to find a better way to represent the children's knowledge and their changing understanding of concepts. From these interviews we devised the technique of *concept mapping* to represent the interviewee's knowledge.

Over the past two decades, scores of studies conducted by educators and educational psychologists worldwide have demonstrated that concept maps are useful in learning (both individual and collaborative) in teaching, and in assessment.

Much of that research is summarized by Good, Novak, and Wandersee (1990), Mintzes, Wandersee, and Novak (2000), and Novak (1991, 1998, 2010). Additional representative studies are presented here after showing what concept maps look like and explaining how they are made. If it is true that concept maps are useful in representing and conveying meaning, then they should be useful to describe and illustrate concept maps, as in Figure 10.1.

Concept maps differ from other types of diagrams that utilize combinations of graphical and textual elements to represent or express meanings. For example, diagrams that Ackerman and Eden (2001) refer to as "cognitive maps" are large web-like diagrams with up to hundreds of "ideas" represented by the nodes. "Ideas" are typically expressed as sentences or even short paragraphs. Buzan and Buzan's (1996) "mind maps" have unlabeled links between nodes, so links tacitly represent "connections" among ideas. In Sowa's (1984) "conceptual graphs," the concept nodes are connected using only certain kinds of relational links, logical relations such as "is a kind of" and "has property." "Semantic networks," as described by Fisher (1990) are networks of nodes and labeled links arrayed in a centripetal morphology. The most basic concept is located in the center of the diagram and the subordinate concepts radiate out in all directions.

The presence of cross-links also makes concept maps unique relative to other forms of meaning diagrams. Cross-links express relations that cut across the clusters or regions within a concept map. Examples in Figure 10.1 are the proposition *CMaps as tools for learning and teaching facilitate assimilation and accommodation* and *Ausubel's Theory of meaningful learning postulates processes*. Why cross-links? In real-world domains of complexity, anything can relate to anything, and in some cases, everything does relate to nearly everything else. Cross-links capture the most significant of these relationships. Furthermore, creative insight can be defined as the result of a deliberate search for new relationships between concepts and/or propositions in one subdomain and others in another subdomain. This is facilitated by the fact that in a concept map one can see all of the important concepts in a single chunk of screen real estate. For more information on differences between concept maps and other types of meaningful diagrams, see Cañas et al. (2003).

10.2.1 Representing Evolving Knowledge

Concept maps are generally regarded as representations of domain knowledge of concepts and principles. But knowledge is, of course, never static, and concept maps should not be regarded as final or uniquely definitive or even "finished." In capturing the expert knowledge within an organization, for instance, practitioners can continually add to and modify the concept maps in the existing knowledge base. Older knowledge might become obsolete and have to be archived, just as new knowledge may emerge and entail modifications

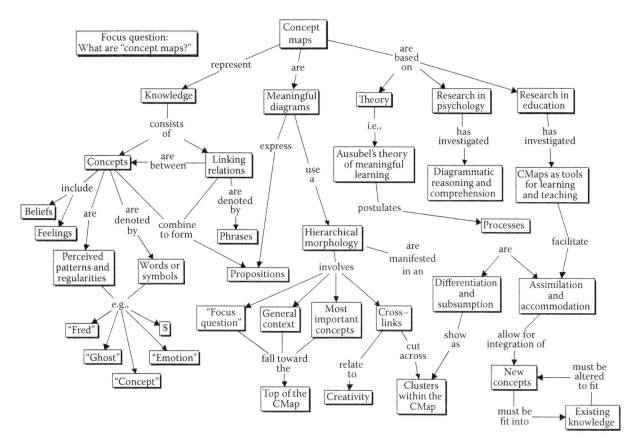

FIGURE 10.1   A concept map about concept maps.*

of existing representations. But there is another aspect to concept map dynamics. One of the things people know about is events, processes, and procedures. As topics of knowledge, these are fair game for concept mapping. The expressiveness of concept maps allows the inclusion of information about events, processes, and procedures. In this form of diagram a description of an event is embedded within a concept map. An example appears in Figure 10.2. The event description is subsumed under some higher-level nodes and links that provide the context, and lying below the event description are nodes and links that provide the "explanatory glue" that makes sense of the process.

In a review of the role of graphical displays in learning, Vekiri (2002) surveyed research spanning studies focused on maps, matrix representations, schematic diagrams (e.g., how lightning forms, how a machine works), decision trees,

semantic networks, etc. Vekiri sought generalizations about how diagrams aid comprehension and why they aid comprehension. Across applications (including knowledge elicitation, learning and assessment, communication, etc.), diagrams that work well rely on some notion of proximity: Information that is related is presented close in space. This spatial organization supports and encourages inference making. Diagrams that work well also reduce cognitive demands: They require fewer mental transformations (e.g., from images to text or vice versa), they can help mitigate or prevent memory overload by presenting information "at a glance," and they shift some of the burden of text processing over to the visual perception system. Finally, they can help "externalize" cognition in that they can help to guide reasoning. All these features can be said to characterize concept maps.

---

* The examples presented as figures should be viewed with the understanding that they are meant to be displayed as full-screen images. The size reduction necessary to make the figures has the inevitable consequence of reducing the legibility of the larger concept maps that are presented in this chapter. In addition, some of the concept maps that appear in figures use color, which reproduces in the book only as gray shades. Color versions of Figures 10.7 and 10.9 can be seen in the color inset and at http://www.ihmc.us/research/projects/StormLK/. Additional examples of concept maps can be seen at http://cmap. ihmc.us/.

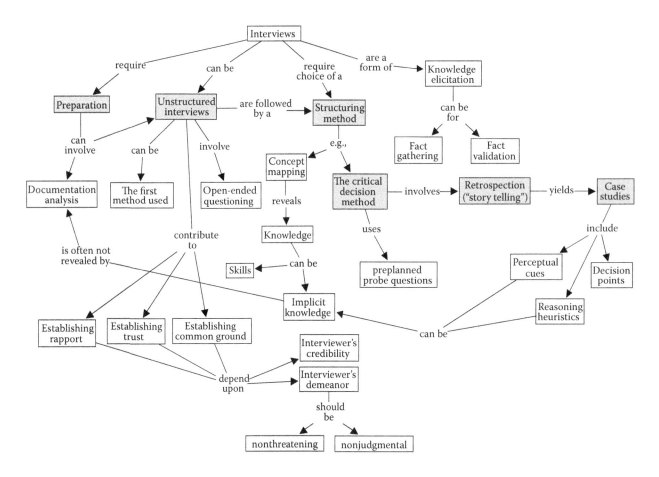

FIGURE 10.2 An example concept map that conveys process or event-full information, on the topic "Interviewing for knowledge elicitation." Process or event-structure information is highlighted in gray.

10.2.2 CREATING CONCEPT MAPS AND KNOWLEDGE MODELS

A number of commercially available software packages can support the creation of a variety of kinds of meaningful diagrams.* The concept maps presented in this chapter were all

made using CmapTools, a free downloadable software suite that was developed at the Florida Institute for Human and Machine Cognition. The CmapTools software was designed specifically to facilitate the creation and sharing of concept maps on the Web and to support the integration of concept maps with hypermedia. Aggregates of related, navigable concept maps and other electronic media are referred to as knowledge models. Concept mapping and knowledge modeling on the Web support distance learning, collaboration, and knowledge sharing (Cañas et al. 2003). Additional examples of concept maps and knowledge models can be seen at http://www.ihmc.us. Examples one can view there include both simple and complex concept maps and both well-formed and ill-formed ones. The process of creating concept maps using CmapTools is straightforward using the simple menu options, clicking to create concept nodes, click and drag to create links

---

* A number of commercially available software packages support the creation of meaning diagrams. Some differ considerably from concept maps, some are similar to concept maps but lack one or more of its key features (e.g., linking lines cannot be labeled, there is no principled morphology, there is no notion of propositional coherence, etc.). These systems go by such names as "Brain Maps," "Mind Maps," and "Mindjet." Intended applications range from creating Web pages having all sorts of graphical elements (e.g., pictures as the nodes) to the exploration of "personal issues" through the creation of diagrams that have a cartoon-like feel, to the management and planning of meetings, projects, and plans by creating tree-like structures with text notes appended to the branches. A number of academic research groups have built software to support the creation of meaning diagrams. Two in particular were created to support the creation of concept maps. One is TPL-KATS (Team Performance Lab–Knowledge Assessment Tool Suite) (Hoeft et al. 2002). This suite is designed to support research on concept mapping and card sorting as methods of knowledge elicitation. The system has an administrator mode that permits parameter-setting for the concept mapping task (e.g., whether or not the participants can add concepts, the maximum number of concepts, etc.), computerizes the logging of user actions and the scoring of

completed diagrams. When one conducts a concept mapping task, the system provides the concepts and the linking phrases. The system can also allow one to attach multimedia and comments. The system can produce output files that can be analyzed with standard statistics packages. A similar software suite is the "knowledge mapper" (Chung, Baker, and Cheak 2002), although this system allows the administrator to designate an existing concept map as the "expert map" to be used as a scoring criterion.

between concepts, and drag and drop to create hyperlinks to other documents. More information about CmapTools is available at http://www.ihmc.us.

A detailed discussion of how to make good concept maps appears in Crandall, Klein, and Hoffman (2006). Concept mapping is definitely a skill. It takes practice for most people to adapt to thinking in terms of propositions rather than in terms of sentences. There are also struggles with word choice in forming propositions. It is said that making a good concept map forces the "mapper" (the creator of the concept map) to reach for crystal clarity about the meanings to be expressed. Experience shows that there are a number of styles or approaches to the creation of concept maps. Some individuals prefer to just "dive down into the weeds," but others follow a rough sequence of steps that form the traditional, general procedure (Novak 1998):

Step 1. Domain and Focus

First, the mapper identifies a "focus question." Examples would be: What are the key characteristics of the turbine's hydroelectric shut-off valve? and What things are important to people when they look at a Web page about a forthcoming conference? Forming a "how" focus question typically results in concept maps that embed a process or event description. Examples would be: How do viruses reproduce? and What causes thunderstorms?

Step 2. Setting up the "Parking Lot"

Guided by the focus question, the mapper identifies 5 to 10 of the most important or broadest concepts related to the topic. The initial list may be generated through deliberative thought about what is most important or through a free-association. Thus one's train of thought about Mars might be, "It is a planet. It orbits the sun. Ah . . . it looks red," and so on. Figure 10.3 presents an example.

Step 3. Arrange the Concepts

Next, the concepts are moved around in the concept map space to form a rough ordering by placing

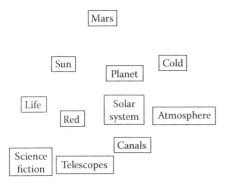

FIGURE 10.3 An example concept map in the "parking lot" stage of development.

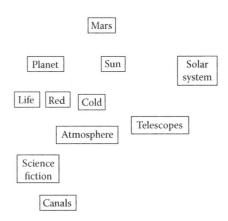

FIGURE 10.4 The Mars Concept Map as the parking lot is being arranged.

the most inclusive or most general concepts—those that seem to be most important or most closely related to the topic—toward the top of the concept map (see Figure 10.4). Sometimes this process leads to modification of the focus question and the writing of a new focus question. When concept maps are used to aid problem solving, developing a good focus question is imperative. It is common to revise the focus question several times as the mapping process proceeds and the individual or team gain a better understanding of the problem and its solution. Although this ordering process may be only approximate and will likely depart from a traditional hierarchical classification, the arranging process facilitates concept linking.

Step 4. Begin to Link the Concepts

This is often done by working downward from the top concept. The linking words should define the relationship between the two concepts so that the triplet reads as a proposition. It is necessary to be selective and as precise as possible in identifying linking words. In addition, one should avoid "sentences in the boxes" because this usually indicates that an entire subsection of a concept map could be constructed just from the statement in the node. There is no restriction on the sorts of relational terms that can be used as linking words or phrases.

Beginners sometimes comment that it is difficult to identify appropriate words to use as links between concepts. This is because it is simpler to identify the major concepts or ideas in a domain of discourse than it is to capture all the subtleties in how they interrelate. Once people begin to settle on good linking words and also identify good cross-links (see step 6, below), they can see that every concept could be related to many other concepts. This also produces some

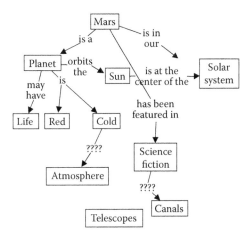

FIGURE 10.5 The Mars Concept Map as the concepts are being linked together.

frustration, and they must choose to identify the most prominent and most significant links. Figure 10.5 shows the Mars Concept Map at this stage of development.

Step 5. Refine the Concept Map

Refinement involves linking all the concepts and moving them around so as to minimize line overlaps and to make the concept map flow, as it is read from top to bottom. Refining the

concept map involves adding, deleting, and changing both the concepts and the link labels that express the various subsumption and differentiation relationships.

It is important to check that the same concept does not appear more than once on a concept map, because this produces ambiguities. High-quality concept maps usually are those that have undergone several waves of refinement, although with practice one can make a good concept map in a single pass through these steps.

Figure 10.6 shows the Mars Concept Map after it had undergone one wave of refinement. A number of heuristics are involved in making a high-quality concept map (see Novak and Gowin 1984). As a rule of thumb, if there are more than four concepts linked under a given concept (a "fan"), it is usually possible to identify some appropriate concepts of intermediate inclusiveness, thus creating another level of differentiation or subsumption.

The Mars Concept Map in Figure 10.6 shows that the mapper is working on fixing the fan that falls under "Planet," by trying to hang "Half as Big as Earth" under "Mars" rather than under "Planet." The mapper has also made a "string" that needs fixing: "Planet is Cold because of

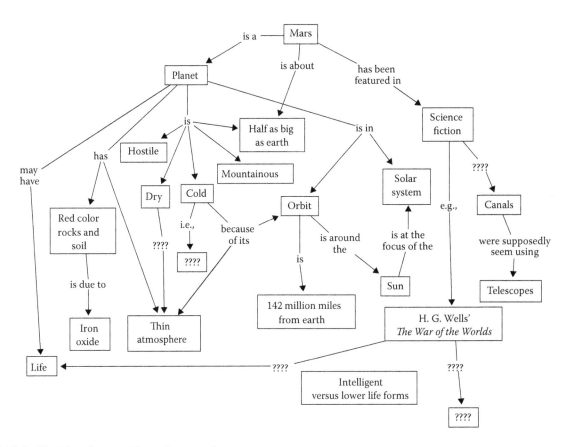

FIGURE 10.6 The Mars Concept Map undergoes refinement.

its Orbit" has the node-link-node triple "Cold because of its Orbit," which is not a well-formed proposition. The Figure 10.6 example also includes a node that is being used to formulate a nascent idea, and at this point it contains a phrase, "Intelligent versus Lower Life Forms." Eventually, that should be split up and appropriately integrated.

Alternatively, the mapper might decide at this point to carve out the material about life on Mars and make a separate concept map on that topic. Given the size of the typical computer display, 35 to 40 concepts seems to be a reasonable number for a single concept map (see Figure 10.1), although some concept maps are far larger. Overly complex concept maps overburden working memory as one seeks to make sense of the ideas.

Step 6. Look for New Relations

One should always keep an eye out to find "crosslinks" between concepts in different clusters or regions of the concept map. Deliberate search for cross-links can often help in the discovery of new, creative relationships in the knowledge domain. An example appears in Figure 10.6, where the mapper is considering a link between the concepts of life on Mars and H. G. Wells' depiction of life from Mars.

Step 7. Add Resources and Examples

Note in Figure 10.7 that there are small icons appended immediately beneath some of the concept nodes. These are created using the CmapTools software and hyperlink to electronic resources that elaborate the concepts in the concept maps. Adding supplementary resources to concept maps is the first step in knowledge model creation. Resources can be any form of digital media, including text pieces, detailed examples, images, charts, etc. They are stored in the concept map project folder, and through a simple "drag and drop" operation are appended to nodes as clickable icons. Resources can include URLs of relevant media on the Web that are accessed within the context of the concept map. Figure 10.7 presents a screenshot from a concept map about weather forecasting in the Gulf Coast. It is part of a knowledge model named STORM-LK (Hoffman, Coffey, and Ford 2000; Hoffman et al. 2000; http://www.ihmc.us/research/projects/StormLK). The opened resources include digital video, in this case the commanding officer of the weather forecasting facility explains the Gulf Coast winter regime.

Step 8. Stitch the Concept Maps together to Create a Knowledge Model.

Concept maps on a given theme can be hyperlinked together. The process of hyperlinking together concept maps on a given topic and organizing them with a high-level or overview "Map of Maps" is the final step in creating a concept map-based knowledge model. The concept map in Figure 10.7 is the "Top Map"—a global map that provides an overview of the topics covered in all of the other concept maps. Below some of the concept nodes in Figure 10.7 are small icons that look like little concept maps. Clicking on such an icon takes the user to the concept map that is indicated by the node. Thus clicking on the "Gulf of Mexico Effects" node in the Figure 10.7 concept map takes the navigator to the more specific map dedicated to Gulf of Mexico effects on the weather. Such a map might have concept nodes referring to such things as "fog" and "thunderstorms" and "hurricanes." Each of those might be the topic for its own concept map. One could navigate to and from the higher-level or overview concept map to each of the particular concept maps through hyperlinking. Concept map models of domain knowledge can be simple, consisting of a dozen or so concept maps, or they can be complex, consisting of hundreds of concept maps and other resources.

Concept maps can be made in many different forms for the same set of concepts. There is no one "correct" way to make a concept map, although concept mapping can be used to achieve group consensus (Cañas 1998; Gordon, Schmirer, and Gill 1993), and the concept maps made by domain experts tend to show high levels of agreement (see Graesser and Gordon 1991; Gordon 1992, 2000). Research suggests that one can expect a concept map that has been created

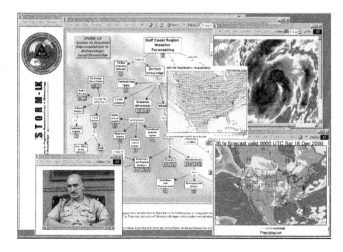

FIGURE 10.7 A screenshot of a concept map with some opened resources.

and refined by one expert to have about 10% of its propositions "altered" when the concept map is analyzed by some other expert (Hoffman, Coffey, and Ford 2000). This is not because experts disagree (although they can). Rather, it is wordsmithing—a reflection of their differing emphases, judgments of what is important, and the subtleties of word choice (e.g., "promotes" versus "causes"). As one's understanding of relationships between concepts changes, so will the concept maps. Conversely, the process of concept mapping almost always leads to new understandings and insights (e.g., "Gee, I never really thought about that in quite this way . . . ") on the part of the domain expert who is building the concept map.

10.3 APPLICATIONS OF CONCEPT MAPS

Concept mapping has been put to many uses, ranging from the preservation of traditional crafts and cultural knowledge, to uses in teaching and education, to applications in business management, and to many uses in government and the military. Daley et al. (2008) performed an analysis of the topics in the proceedings of the first two concept mapping conferences (CMC 2004 and CMC 2006). Their goal was to identify current areas of research in and applications of concept mapping. The major areas they identified were

- Concept mapping for teaching and learning,
- Concept mapping for knowledge and professional development,
- Research methods using concept maps,
- Assessment via concept maps including scoring of concept maps, and
- Software development for concept mapping.

For purposes of this chapter, a similar grouping of topics is used, which will be discussed in this order: educational research and applications, knowledge sharing and collaboration, problem solving and decision making, knowledge elicitation, and management, decision aiding, and concept maps as graphical browsers.

10.4 EDUCATIONAL RESEARCH AND APPLICATIONS

Concept mapping is being used as a teaching, learning, and assessment device in schools around the world. A great deal of literature speaks to the utility of concept mapping in educational settings (Cañas 1999; Cañas et al. 2000, 2001, 2003; Mintzes, Wandersee, and Novak 2000; Novak 1998, 2010). Research has clearly shown that concept mapping is far better than rote memorization. Memorization of isolated facts is encouraged by multiple-choice assessments and "teaching

for the tests." Concept mapping promotes meaningful learning and the integration of knowledge. Furthermore, concept maps are valuable in assessment since they support the pinpoint identification of gaps in students' conceptual understanding (e.g., Markham, Mintzes, and Jones 1994).

The research literature includes longitudinal studies, such as a 12-year study that used concept mapping to track student misconceptions in science and distinguish students who maintained a rote learning style from those who adopted a meaningful learning style (Novak and Gowin 1984; Novak and Musonda 1991). Examples of concept maps that show the evolution in children's knowledge and that illustrate concept maps made by meaningful versus rote learners can be seen in Novak (2002).

A study that is representative of the literature in this area is by Basconas and Novak (1985). Students in high school classes were given Raven's Progressive Matrices test of ability and were divided into ability groups based on these test scores. One group followed a traditional sequence of physics topics and textbook problems (the traditional group) while students in the experimental group made concept maps for the topics they studied. Evaluation of student performance was in terms of scores on physics problems that required some transfer of knowledge to novel settings (i.e., the problems were not merely superficial transformations of the problem exercises in the texts). The results showed that the experimental group outperformed the traditional group at the end of the first study unit, and the superiority of their performance was not only maintained but actually increased over the school year. While there was improvement in performance on unit tests over the school year for the traditional group, the improvement was only modest, and there was no acceleration of the improvement. The results suggested that students in the experimental group were not only learning physics better, they were building metacognitive skills that further fostered the continuing improvement in achievement. Also, student ability as measured had no significant relation to problem solving scores, indicating that the concept mapping was effective for all ability groups.

Basecones and Novak's findings are generally consistent across many studies that suggest that concept mapping is useful for both knowledge attainment and educational assessment, at all levels of education. Concept mapping is being used by curriculum designers in the US Navy's "Task Force Excel," which is in the process of reinventing many Navy jobs (Cañas et al. 2003). Concept mapping is also being used by university faculty for curriculum planning, collaborative course authoring, and the preparation of distance learning courses (Cristea and Okamoto 2001; Guimaraes, Chambel, and Bidarra 2000). Work on the utility of concept maps for assessment is presented by Markham and Mintzes (1994), Pearsall, Skipper, and Mintzes (1997), and Martin, Mintzes, and Clavijo (2000).

Across recent years, increasing numbers of studies of the application of concept mapping in science education have been reported. For instance, several studies of the application of concept maps in biology education (Fonseca and Extremina 2008; Mostrum 2008), mathematics education

(Afamasaga-Fuata'I 2008; Flores 2008; Grevholm 2008), and chemistry education (Francisco, Nicoll, and Trautmann 1998; Fechner and Sumfleth 2008; Pernaa and Aksela 2008; Schanze and Grub-Niehaus 2008) have been reported. Uses have included identifying content for courses, assessing student achievement, and identification and sharing of approaches to course material development.

Most of these studies utilize the capabilities and opportunities provided by the Web. This led Novak and Cañas to issue a call for "A New Model for Education" (Novak and Cañas 2004; Novak 2010), which will leverage improvements in CmapTools technology, bandwidth on the Web, and computer speeds. These technologies will increasingly make it possible to tailor instruction to the individual needs and aptitudes of learners or collaborating groups in school and corporate settings with no geographic bounds. The Internet and Web play a central role in the ideas put forth in this new model by facilitating collaborative learning and knowledge sharing at the individual school or facility, state, country, or worldwide level.

For example, Bowen and Meyer (2008) describe a large-scale effort in Washington State to foster new teacher improvement and retention, based upon ideas from the New Model of Education. The effort is sponsored by the New Teacher Alliance (NTA), a consortium of seven school districts and two Educational Service Districts in the state. The initiative has led to collaborative efforts by all involved parties to create a large-scale knowledge model pertaining to best practices to foster development and retention of new teachers. Since June 2008, the NTA knowledge model and partner's work products have been made available via the Web.

10.5 APPLICATIONS IN KNOWLEDGE SHARING AND COLLABORATION

Many projects have used concept mapping as a collaborative process to create and communicate knowledge. Tifi et al. (2008) described a number of collaboration models for concept mapping, which include

- Comparison of independently constructed concept maps,
- Alternating construction in the development of shared concept maps,
- Mixed alternation with peer review,
- Shared concepts with peer review, and
- Collaborative reading of texts with "Knowledge Soup" (described later in this section) sharing.

Tifi and Lombardi (2010) extend their earlier work in the area of "Web-mediated collaboration," describing several online collaborative concept mapping projects and initiatives to create EU-wide communities of learners building concept maps. The remainder of this section provides background on how the current state of the practice in collaborative concept mapping evolved and empirical evidence for the utility and efficacy of collaborative concept map development.

Kremer and Gaines (1994) described their groupware concept mapping tool in a report detailing its use by information technologists. Participants who had no prior experience in concept mapping used the tool to produce a concept map pertaining to the implications of introducing a networking system into an organization. The group created a 35-node concept map in 90 minutes. Kremer and Gaines reported that the concept mapping effort contributed in a positive way to the group's deliberations.

A number of attempts have been made to demonstrate experimentally the value of concept mapping for collaboration and knowledge sharing. Representative of this literature is a study by Esiobu and Soyibo (1995), which involved 800 secondary school students (said to be equivalent to tenth grade high school students in the United States). The subject matter was biology, ecology, and genetics. Students who were randomly assigned to the experimental condition made concept maps of each lesson, working either alone or in small groups, with the goal of integrating the lesson material. Students in the control condition did not engage in concept mapping. The concept maps were graded by the teachers to provide feedback to the students. Achievement following the instructional series was measured by three 40-item multiple-choice tests (biology, ecology, genetics) that had been reviewed by experts for their appropriateness for the classes. The results were that students in both concept mapping conditions greatly outscored those in the control condition.

Wallace and Mintzes (1990) conducted a study in which 111 elementary education majors participated in six concept mapping sessions of 75 minutes each. Results of the study indicated that the experimental group showed small gains on a traditional objective test after instruction, but they showed significant changes after instruction in the number of relationships, number of hierarchical levels, and in the branching and number of cross-links in their concept maps. There was also a significant increase in the number of critical concepts and propositions the students identified. A number of evaluation and map scoring methods exist; see Novak and Gowin (1984). The authors concluded that the students acquired knowledge, and not memorized facts, but their having acquired knowledge helped them remember facts.

At the Center for Mars Exploration at NASA Ames Research Center, it was decided to integrate information about Mars so that it might be more effectively distributed to the public via the Web, especially to school children. Over 100 concept maps were created by NASA scientists to organize the material into appropriate subtopics (e.g., Mars Lander missions, Mars in Science Fiction, Life on Mars, etc.). Using the resourcing capabilities of the CmapTools, all of the hundreds of resources (maps, photographs, URLs, etc.) were added into the Concept Map Project. These concept maps were used as the indexing structure for a CD-ROM released by NASA (Cañas 2003; Cañas et al. 2000; http://cmex.ihmc.us/CMEX/index.html).

Concept maps have been used for both collaborative learning and for meaning making (e.g., Coffey and Carnot 2003; Hall 1997; Salinas, Benito, and Garcia 2008). Collaborative concept mapping on the Web has been shown to be a highly effective activity for a community of learners. Afamasaga-Fuata'I and McPhan (2008) describe how a group of teachers banded together to form a community of learners to explore teaching practices before implementing them in the classroom. They collaboratively developed concept maps on new pedagogies and assignments. Their activities led to the introduction of concept mapping for students as a new part of their pedagogies.

The idea of online communities of learners utilizing concept maps was further explored by Conceicao, Baldor, and Desnoyers (2010). They reported results of a three year study that explored individual learning in a collaborative learning environment. Following asynchronous online discussions and other social learning activities, students individually created concept maps of their emerging understandings of the course content. Concept maps were constructed for each module in the course. The most interesting aspect of this study was that while concept maps are usually constructed in groups in collaborative learning environments, they were created individually to help students summarize and integrate collaborative learning experiences.

The idea of a "Knowledge Soup" as a useful adjunct to Web-based, collaborative concept mapping was reported by Cañas et al. (2001). This software is included in the CmapTools software suite. It allows students from distant schools to share claims (propositions) derived from their concept maps. Sharing takes place through the Knowledge Soup, which is a repository of propositions submitted by the students and stored on a server. Propositions in a soup that are found to be similar to those submitted by a student can be displayed on the student's screen. The student can use these propositions from other students to enhance his or her concept map. In addition, the student can question or criticize propositions submitted by other students, leading to a peer-review type of environment, where students themselves are responsible for the validity of the propositions in the soup.

Recent decades have witnessed a growth in Internet-based courses for secondary and tertiary learning. Although the quality of online courses varies widely, the fact is that enrollments in these courses have increased dramatically. For example, in 2002 some 1.6 million students were enrolled in online courses in post secondary degree granting programs and the number enrolled grew to some 3.5 million students by fall 2006 (Allen and Seaman 2007). In contrast, enrollment in regular programs grew from 16.6 million in 2002 to 17.6 million in 2006. In light of many of the results regarding concept maps for collaboration and knowledge sharing on the Web, it seems certain that Web-based electronic resources that are collaboratively developed, organized, shared, critiqued, and based upon or accessed by concept maps will play a larger role going forward.

10.6 COLLABORATIVE PROBLEM SOLVING AND DECISION MAKING

Concept maps have been used in a variety of planning, management, and other organizational processes, typically in group-oriented, collaborative settings (Ramirez, Florez, and Barros 2008). In the group context, concept mapping can scaffold collaborative problem solving and decision making (see Novak 1998, 2010) to support a process of achieving group consensus or to facilitate group creative problem solving (Gordon 2000; Gordon, Schmierer, and Gill 1993).

Buzan and Buzan (1996), and others who have proposed similar diagramming schemes, have suggested that the process of creating meaning diagrams can be employed as a "brainstorming" technique (by which they generally seem to mean creative problem solving and not necessarily the specific brainstorming method). Stanford (2003), refers to "strategy maps," which are diagrams used for planning and implementation purposes, and Eden (1988, 1992), used what he called "cognitive maps" to aid the process of strategic management. Dicheva and Aroyo (2000), and Gordon (2000), described the creation of diagram-based software tools that are intended to help people create representations of domain ontologies, present a visualization of a knowledge area, and provide support for decision-making and information sharing through Web navigation.

With regard to concept maps in particular, Kyro and Niskanen (2008) described their Learning Business Planning Project, which has a central goal of fostering creativity in business planning. They describe using concept maps both in the process of concept formation for planning and as an interface for Internet applications. In the latter capacity they state that concept maps aid the customers in understanding causal structures in the materials and the overall layout of content.

It is typical in uses of concept maps for decision making and management to have teams collaboratively build the concept maps. In an interesting study of a different sort, Fourie and van der Westhuizen (2008) designed an experiment in which he had three different constituents (the CEO of a company, a top-level executive team, and representatives of various categories of employees) build concept maps that pertained to the vision, mission, and strategic objectives of the organization. Five different firms with mappers in each of these categories participated. The goal of the study was to determine the degree of consistency of viewpoint regarding the goals of the company among these different constituencies. The concept maps presented rather stark evidence of disparity of viewpoint among the mappers regarding strategic objectives of the firms, with three of the five firms obtaining less than 50% alignment with regard to the firm's vision. The authors concluded that concept maps are useful to make explicit the vision, mission and strategic goals of a firm and to identify inconsistent viewpoints regarding these important planning elements.

Another example of generative planning and decision making that is facilitated by concept mapping is the effort at NASA to develop an interdisciplinary science of astrobiology (NASA 2003a), and establish a distributed institute. In a collaborative attempt to scope this new field, scientists including biologists, astrophysicists, and others each made concept maps expressing their own views and ideas about the field, and these were then collaboratively merged into a single concept map that captures the definitions, subject matter, and goals of astrobiology.

Illustrating the evolving use of concept mapping in problem solving are earlier studies by McNeese et al. (1990, 1993, 1995) and Citera, McNeese, and Brown (1995) on problem solving in avionics design. In one study, they conducted concept mapping interviews with U.S. Air Force pilots in order to support the creation of "The Pilot's Associate," an intelligent decision aid. A number of Air Force pilots created concept maps about the process of target acquisition. Pilots first identified decision points in the task of target acquisition from which a timeline and second concept map were constructed. Next, a storyboard and composite concept map was created by the researchers to express all the elements that were common to all of the individual concept maps and the elements that were unique to each of the individual concept maps.

In another study in this project (Gomes, Lind, and Snyder 1993), experienced USAF pilots, copilots, and flight engineers were engaged in concept mapping and storyboarding tasks in an attempt to redesign a helicopter cockpit, following-up on pilot complaints about workload. Concept mapping focused on describing the "typical" mission in terms of information requirements and sources (i.e., inside or outside of the cockpit). Storyboarding encouraged the pilots to reconfigure controls and displays and explain their reasons for their choices. The concept mapping revealed key points about the specific problems in the cockpit design (i.e., instrument illegibility and inaccessibility, the failure to group or centrally locate critical panels). Along with the mock-up design ideas, the researchers were able to recommend improvements (McNeese et al. 1995, 354).

A third study in this series is interesting in that it involved collaborative concept mapping about problems in the process of collaboration. Concept mapping was used to structure interviews with seven human factors specialists who worked in multidisciplinary teams in the USAF Aeronautical Systems Center. The results included a dozen rich concept maps (having 79–152 concepts) that described their collaborative design process. The concept maps revealed the importance of disagreements among team members concerning design elements, i.e., disagreements pointed to the design conflicts and trade-offs. The concept maps also made clear many specific reasons why the participants had a negative opinion about the traditional government-contractor approach to system acquisition, as opposed to the integrated team approach.

Gaines and his colleagues (Gaines and Shaw 1989, 1995a, 1995b, 1995c; Kremer and Gaines 1994, 1996) have reported an effort to use knowledge acquisition/representation tools to elicit and model the conceptual structures in a scientific community, with much of the work accomplished over the Web. They used concept mapping as a scaffold for conducting knowledge elicitation from expert scientists over the Web. The researchers studied the collaborative activities of scientists who were involved in the International Intelligent Manufacturing Systems Research Program (over 100 scientists in 31 industry and university organizations in 14 countries). The goal of the scientists in the IIMSRP pilot project was to develop a particular kind of mass production system and document their activities so that collaboration could be improved for the major 10-year project that would follow the pilot study.

In the study, concept mapping and other tools (interviews, ratings tasks, etc.) were used to support Web-based distance collaboration among scientists. The interactive refinement of a diagram created on the basis of an analysis of "The Technical Concept of IMS" document revealed that two major portions of the IIMSRP effort were not adequately described (i.e., product configuration management systems and configurable management systems). Hence descriptions of these "technical work packages" were subsequently added into the IMS research program. This documentation analysis also supported the creation of a concept map that described the IIMSRP mission statement. Gaines and Shaw argued that existing tools for the facilitation of scientific communication (e-mail, list servers, etc.) have been premised on the need to facilitate the processing of text. They argued that what is needed are tools including concept mapping to support knowledge elicitation and to facilitate knowledge sharing.

Some studies that have looked at team processes that occur as students jointly construct a meaning diagram over a computer network (Chung, O'Neil, and Herl 1999; Herl et al. 1999) have found no benefit for collaboration. In one group, students collaborated over a network to construct group concept map-like diagrams. In a second group, students worked individually to construct diagrams using information from Web searches. Students in the individual condition showed significant improvement in their diagrams (mapping scores) over the course of a year. Students in the collaboration condition did not show such change. However, the frequency of elaborative episodes in the discourse of concept mapping students was positively correlated with individual learning outcomes.

It seems likely that the nature of the interaction among participants has an influence on whether or not effects of collaboration are positive (Chinn, O'Donnell, and Jinks 2000; Van Boxtel, Van Der Linden, and Kanselaar 1997, 2000). Collaborative concept mapping promotes more debate and reasoning in the interaction among students (Baroody and Bartels 2000, 2001; Baroody and Coslick 1998). When done collaboratively, concept mapping promotes questioning, discussion, and debate (Stoyanova and Kommers 2002). Chiu, Huang, and Chang (2000) also looked at group interaction during collaborative Web-based concept mapping. Using a system for interaction analysis (Hertz-Lazarowitz 1990; Webb 1995), researchers found that "complex co-operation" correlated most highly with mapping performance. So, while the preponderance of evidence suggests that collaborative concept mapping on the Web typically has significant benefits, there is still more to learn about how to reliably produce the most positive outcome.

10.7 KNOWLEDGE ELICITATION, PRESERVATION, AND MANAGEMENT

Concept mapping and similar forms of meaningful diagramming have proven useful as a tool in expert knowledge elicitation (Ford, Cañas, and Jones 1991; Hameed, Sleeman, and Preece 2002; Hoffman et al. 1995; McNeese et al. 1995), as a knowledge representation scheme in cognitive science (Dorsey et al. 1999), and as a tool to support knowledge management (Cañas, Leake, and Wilson 1999). In concept map-based knowledge elicitation interviews, one interviewer acts as a facilitator while another uses a laptop to create a concept map "on the fly." Projected on a screen, the emerging concept map scaffolds the discussion.

McNeese et al. (1995) argued that concept mapping is an ideal knowledge elicitation interview procedure because:

- It requires little practice or training.
- It can be conducted in a reasonable amount of time.
- It allows for the identification and correction of inconsistency or incompleteness in the knowledge representation.
- The task encourages practitioners to organize and express previously tacit knowledge.
- It can effectively reveal both broad and deep knowledge.
- The concept map provides a focus for the discussion and allows the domain expert to know how the elicitor has interpreted their statements.
- The representation is one that both the domain practitioner and the system designer can understand, in contrast with other forms of knowledge elicitation.
- It results in a representation of knowledge (i.e., propositions) that can be readily instantiated in a form permitting logical operations.

In the knowledge elicitation context, concept mapping stands out because it permits the rapid creation of models of knowledge, whether it is the knowledge of a student/learner or the knowledge of an expert (Hoffman and Lintern 2006). Furthermore, concept map knowledge modeling scales up to hundreds of concept maps containing thousands of propositions and thousands of multimedia resources. Finally, concept mapping can be easily used in combination with other methods of knowledge elicitation, such as the Critical Decision Method (Hoffman, Crandall, and Shadbolt 1998) and storyboarding (McNeese et al. 1995), both of which can be used to generate rich case studies and design concepts and think-aloud problem solving (Ericsson and Simon 1993), which can help generate models of reasoning to complement concept map models of domain knowledge (see Hoffman and Lintern 2006).

Concept map knowledge models can be used in knowledge preservation and knowledge management. An example is work that was performed at NASA Glenn Research Center (Coffey and Hoffman 2003; Coffey, Moreman, and Dyer 1999). The motivation behind the work was the preservation of "lessons learned" by engineers who were about to retire. Another example is a project that demonstrated how concept maps might be used as the infrastructure for knowledge preservation on a national scale (Hoffman 2001; http://www.ihmc.us/research/projects/ThailandKnowledgeBase/).

In an effort to preserve traditional folk knowledge, a set of concept maps was made on the topic of Thai silk weaving. The goal was to capture and preserve the skills of elders in the craft villages of Thailand. Concept maps were made that described each of the many varieties of Thai silk and their methods of manufacture, and these were then resourced with photographs showing the silk patterns and illustrating various aspect of the weaving process. A next step would be to do the same for other traditional crafts, such as weaving, architecture, fishing, and the culinary arts.

Another example is a project funded by the Electric Power Research Institute and the Tennessee Valley Authority to capture and preserve the knowledge of experts in the utilities industry (Hoffman and Hanes 2003). Owing to downsizing in the 1980s, the electric utilities, like many sectors of the economy, face a situation where the senior experts are retiring and there is no cohort in the age range of 45–50 ready to assume the mantle of expertise. Hence, a project was begun to identify subdomains of expertise that were critical to the industry and that remained undocumented, "tacit" knowledge. Within organizations, the concept map knowledge base can be a living repository, undergoing regular refinement and expansion.

10.8   DECISION AIDING

A concept map knowledge model can serve as the interface for the explanation component of knowledge-based systems and performance support systems (Cañas et al. 1998; Dodson 1989). In this application, the model of the expert's knowledge becomes the interface for a performance support tool. This has been done for such domains as weather forecasting (Hoffman et al. 2000) and nuclear cardiology (Ford et al. 1996).

The use of concept maps to create navigable explanations of the operations of expert systems was demonstrated by Ford et al. (1992). In traditional, or first-generation expert systems, one could query the system about the reasoning chain that was used in making a decision. The result of the query would be a cryptic sequence of production rules, itself not terribly explanatory. Ford et al. used concept mapping to elicit the knowledge of an expert at first-pass nuclear functional imaging of ventricular function. After creating the inference engine they realized that the concept map representation of the expert's knowledge could be used to be the interface for the expert system, as a performance support tool. As one works a given case, one goes through the concept maps that lay out the various diagnostic features. Along the way one can access resources showing representative images that can be compared to those of the given case. Also included were digital videos in which the expert provided detailed discussion of various topics (e.g., subtle cues the journeyman might miss). This innovation represented a milestone in the evolution of expert systems.

A follow-up project involved capturing the knowledge of an expert at the maintenance and repair of a particular recording device used aboard US Navy vessels. Again, the concept map models of domain knowledge were used as the interface, and through a question-and-answer process it would guide others through a diagnostic procedure (Cañas et al. 1998; Coffey et al. 2003). This system was called El-Tech, standing for Electronics Technician. One idea this system embodied was that it can be used in both a learning mode and a performance support mode, that is, it is at once a training aid and a decision aid. Furthermore, it can be taken with sailors to support their performance at sea.

10.9 CONCEPT MAPS AS GRAPHICAL BROWSERS

In educational contexts, concept maps have been used as graphical browsers for navigating hypermedia (Hall, Balestra, and Davis 2000; Reynolds et al. 1991). The use of graphical browsers does not by itself necessarily lead to changes in knowledge structures (Nilsson and Mayer 2002). However, studies that have used meaning diagrams to support navigation of course material and that resulted in no evidence of a performance gain (e.g., Jonassen and Wang 1993; Stanton, Taylor, and Tweedie 1992) used impoverished diagrams, unfamiliar testing procedures, and did not control for

the amount of time students spent studying material in text versus diagram form.

Other studies have reported significant gain in student learning on the basis of concept maps and concept map-like diagrams used to navigate course material (e.g., Reader and Hammond 1994). Preliminary applications and demonstrations indicate that appropriate use of such browsers for navigation can help people find topics more easily (Carnot et al. 1999, 2000). Graphical browsers can provide easier, less frustrating access to information (Guimaraes, Chambel, and Bidarra 2000; Hall 1997; Jonassen and Wang 1993). Other studies have shown that the use of concept map-like diagrams not only facilitate navigation, but can result in a learning gain (McDonald and Stevenson 1998; Reynolds and Dansereau 1990; Reynolds et al. 1991).

A recent study by Tergan, Engelmann, and Hesse (2008) sheds further light on this issue. They compared a digital concept map to a digital concept list for efficacy in search and information retrieval. The concept map condition provided visual-spatial highlighting of category relationships and labeled links to highlight semantic relationships. The concept list condition provided visual-spatial highlighting only. When resources were to be identified simply by category, no differences between the groups were found. However, when

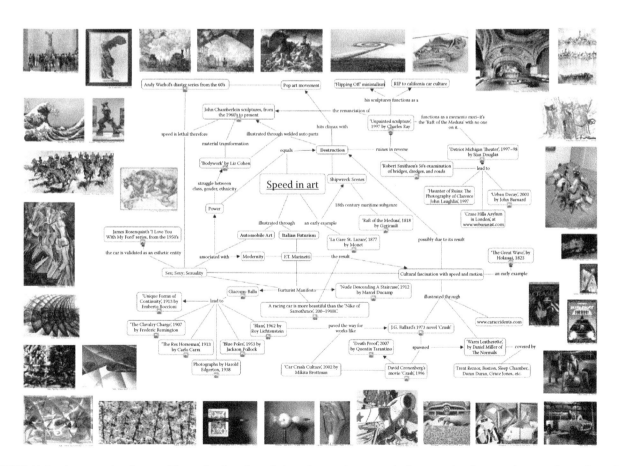

FIGURE 10.8 A composite diagram illustrating how knowledge of many forms can be integrated to form a coherent scheme for artistic productions, advertising campaigns, or other complex human activities involving skills from many fields. Inserts on periphery illustrate some of the inputs. Courtesy of C. Durocher.

functional (semantic) relationships among resources had to be understood as part of the search process, the concept map group performed significantly better than the concept list group. This result suggests important benefits in concept map interfaces to Web materials in resolving relevant versus irrelevant documents.

Another application of concept maps as browseable representations that is beginning to be developed is the creation of a "concept" artifact that can serve to guide complex operations such as theatrical productions. Figure 10.8 shows an example of a composite artifact for the concept of speed in art. Some resources tailored to fit this concept are on the periphery. Similar "concept" artifacts could be created to guide an advertising campaign or other complex ventures entailing coordination of many creative efforts.

The capability to hyperlink concept maps together into organized systems makes them useful as the infrastructure for sets of Web pages. The Web pages that are about concept mapping at the Institute for Human and Machine Cognition (IHMC) (http://www.ihmc.us) are all themselves concept maps. The capability to hyperlink multimedia onto the nodes of concept maps also lends utility to Web pages. The Sixth International Conference on Naturalistic Decision Making was hosted by the IHMC in Pensacola Florida in May 2003 (http://ihmc.us/research/projects/DecisionMakingConference/). The Web site for the conference was a single "page." That was all that was needed. It is shown here in Figure 10.9.

As with the concept map in Figure 10.7, one can see the resource icons attached to the nodes in Figure 10.9. Appended

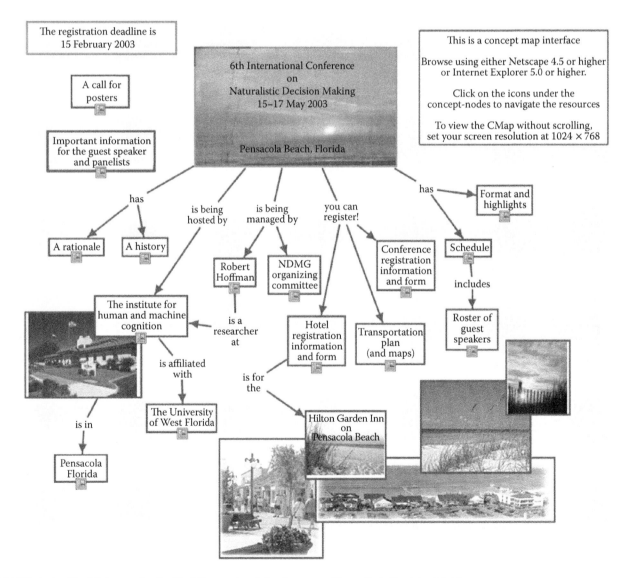

FIGURE 10.9 The Web page for the Sixth International Conference on Naturalistic Decision Making.

to the node "Conference Registration Information and Form," for example, was an informational document and a registration form. Each document could be viewed and downloaded in any of three formats: DOC, PDF, or HTML. Clicking on the icon under a node opened a small dropdown window that listed the attached documents, with the documents' formats included as a suffix to the document names. From this one concept map "page" the viewer could find directions, road maps, links to area hotels and restaurants, the conference schedule, biographies of the speakers, etc. Updates and special information for the guest speakers were also available.

Conference experiences can be painful when it comes to registration, travel, and other activities. In this light, it was telling that in the case of NDM, no one complained about the Web page or had problems in getting the information they needed. (A few participants ran into hardware-software compatibility problems when attempting to view the page. This occurred when individuals attempted to user older releases of a Web browser.) A number of Conference attendees spontaneously commented that they were struck by the page and wanted to learn more about concept maps. They found the page interesting and all reports were that people found the page effective and useful.

Overall, the results of work on the use of concept maps as graphical browsers suggest the need to distinguish between diagrams for navigation and diagrams for learning (see also Nilsson and Mayer 2002). Evidence suggests that if the goal is to foster easier access to a large corpus of content on the Web, concept maps can play a beneficial role. However, simply providing a diagram that gives easy access to information is not necessarily enough to promote knowledge of the structuring of the materials, or even knowledge of the content of the diagrams. In order to achieve these latter goals, it is necessary to engage the navigators actively as learners. As in all meaningful learning matters, it is critically important that the person using the diagram reflect on its content, how that content relates to currently held knowledge and understandings, and how accompanying resources provide rich detail to elaborate the concepts in the maps.

10.10 A CONCLUDING CASE STUDY

This chapter will conclude with a discussion of the design and use of a concept map-based knowledge model that illustrates a combination of features that have been discussed so far: educational resource organization, browsing, navigation, and distance learning. CmapTools was used to build a knowledge model for a Computer Science course entitled "Web-Enabled Applications" that is concerned with various Web development technologies including XML, XHTML, XSLT, JavaScript, AJAX, and PHP (Coffey 2008). The concept map-based, knowledge modeling approach was used to organize and present all resources associated with the course. No standard course delivery software (such as PowerPoint™) was used.

A key aspect of this course was the fact that the content pertained to either brand new, emerging technologies

or technologies that change so quickly that they are essentially new material at any given time. Consequently, no suitable textbook was available and the instructor of the course had to learn much of the subject matter in order to teach it. Mediating against these difficulties was the fact that a great many high-quality resources that were relevant to the course were available on the Web. Accordingly, some form of Web-based organization was necessary, and the one used proved to be intuitive and highly usable, as evidenced by student survey results reported later in this section. Additionally, the initial creation of concept maps served an extremely valuable role in assisting the instructor to develop well-structured, comprehensive knowledge of the subject to assess the topics to include and to provide a comprehensible representation of the structure of the concepts in the course to the students via the Web.

The knowledge model consisted of 10 concept maps, one for each of the major topics in the course and a course organizer CMap, a form of Map of Maps. The latter allowed access to the concept maps in the course and conveyed course sequencing. The concept maps contained a total of 308 unique concepts, and almost 400 accompanying resources, a significant majority of which were computer program code examples. The second most frequently employed resource type was links to other Web sites. The concept maps were developed to be "propositionally coherent," with each node-link-node triple comprising a self-contained proposition. The course materials were developed and/or linked together in approximately 5 weeks of work prior to the start of the course. An additional 40 external resources were created or linked into the knowledge model during the first semester in which the course was offered.

Figure 10.10 illustrates representative components in the knowledge model. It illustrates the modified map of maps (Coffey 2006)—the window entitled "Table of Contents" in the top left. In the scenario depicted in Figure 10.10, a concept map pertaining to XML (the window in the top right

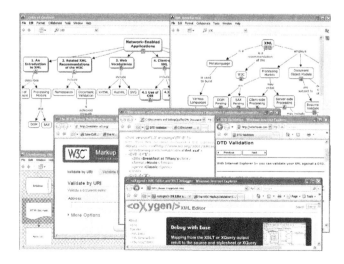

FIGURE 10.10 A graphic of the course organizer map of maps, a concept map, and several accompanying resources in a course entitled "Web-Enabled Applications."

TABLE 10.1

Selected Survey Response Rates[a]

Survey Item	Agree	Disagree
The knowledge model is confusing because there are so many documents in it.	17%	67%
The main value of the concept maps is as an organizing scheme for the other materials, rather than for the information they contain themselves.	33%	42%
I view the concept maps for the information they contain.	83%	17%
I used the concept maps to study for the exam.	58%	42%
I find it easy to find the information I want in a concept map-based knowledge model of the type used in this course.	75%	8%
I prefer the organization of materials in a book to the way that they are organized in the knowledge model.	25%	25%
I would prefer to have presentations from PowerPoint than from a concept map.	25%	67%

[a] The scale used was (strongly agree, agree, neutral, disagree, strongly disagree). Agree is the combined number of agree, strongly agree.

of Figure 10.10) was accessed from a node that is visible in the Map of Maps. From the XML concept map, information pertaining to that topic, including a validation Web site for XML, a sample XML document that might be validated, a link to a tutorial resource on validation from Document Type Definitions, and a link to a Web site where an XML editor is available, were accessed.

A survey was administered to students to identify attitudes regarding the use of this scheme to organize course resources. Table 10.1 contains the most significant results. Some of those results are aggregated when multiple questions pertained to the same issue. Percentages do not always sum to 100% because the five-point scale allowed neutral responses. Students were quite clear that they did not find the knowledge model of more than 400 resources difficult to navigate and that they could easily find materials they wanted to reference. An interesting result was that they had mixed opinions regarding whether the concept maps were more useful for the knowledge they contained or simply as an organizing scheme. The newness of the concept map representation to them certainly played a role in this result. A large majority thought the concept maps conveyed knowledge; however, a smaller majority used them to study for exams. This result was not surprising due to the newness of the representation and the fact that it was a highly detail-oriented course for which much of the time had to be devoted to the detailed particulars in the accompanying resources.

Despite the fact that the scheme was new to the students, they thought the knowledge model arrangement was an effective way to organize materials. However, when compared to more traditional course organization and presentation schemes, students did not express a clear preference either for the knowledge model or the traditional approach. They did show a clear preference for class presentations based upon concept maps when compared specifically to PowerPoint presentations. These results are encouraging, and it is thought that with more experience with concept maps and this type of course organization, student attitudes would be more favorable still.

10.11   CONCLUSIONS

All of the topics and application areas described here are rich with opportunities for further research. The comprehension of concept maps, their effects on learning, their influence on collaboration, their use on the Web—all of these should be subject to more, and more controlled, experimentation. Human factors considerations of user-friendliness, usefulness, and usability are also a part of the mix.

However, the pace of technological advance does not slow while the research is conducted. New uses and applications of concept maps are appearing all the time. Some feel that concept mapping may indeed represent an advance over Gutenberg's printed page. It is certainly true that people all around the world are using concept maps. As concept maps are used more and more in all levels of education, as concept mapping skill becomes assimilated as a part of the general culture, and as the concept mapping process is used more in all of the applications that have been discussed, concept mapping and its applications on the Web will become more common as well.

ACKNOWLEDGMENT

Robert Hoffman's contribution to this chapter was made possible by participation in the Advanced Decision Architectures Collaborative Technology Alliance, sponsored by the U.S. Army Research Laboratory under cooperative agreement DAAD19-01-2-0009.

REFERENCES

Ackerman, F., and C. Eden. 2001. Contrasting single user and networked group decision support systems for strategy making. *Group Decision and Negotiation* 10: 47–66.

Afamasaga-Fuata'I, K. 2008. Concept mapping and vee diagramming a primary mathematics sub-topic: Time. In *Proceedings of CMC 2008: The 3rd Concept Mapping Conference*, eds. A. J. Cañas et al., 181–188. Tallin, Estonia: OU Vali Press.

Afamasaga Fuata'I, K., and G. McPhan. 2008. Concept mapping and moving forward as a community of learners. In *Proceedings of CMC 2008: The 3rd Concept Mapping Conference*, eds. A. J. Cañas et al., 197–204. Tallin, Estonia: OU Vali Press.

Allen, E. I. and Seaman, J. 2007. Online Nation: Five years of growth in online learning. Needham, MA: Sloan Consortium.

Ausubel, D. P. 1962. A subsumption theory of meaningful verbal learning and retention. *Journal of General Psychology* 66: 213–224.

Ausubel, D. P. 1968. *Educational Psychology: A Cognitive View*. New York: Holt, Rinehart and Winston.

Ausubel, D. P., J. D. Novak, and H. Hanesian. 1978. *Educational Psychology: A Cognitive View*, 2nd ed. New York: Holt, Rinehart and Winston.

Baroody, A. J., and B. H. Bartels. 2000. Using concept maps to link mathematical ideas. *Mathematics Teaching in the Middle School* 5: 604–609.

Baroody, A. J., and B. H. Bartels. 2001. Assessing understanding in mathematics with concept mapping. *Mathematics in School* 30: 24–27.

Baroody, A. J., and R. T. Coslick. 1998. *Fostering Children's Mathematical Power*. Mahwah, NJ: Lawrence Erlbaum.

Bascones, J., and J. D. Novak. 1985. Alternative instructional systems and the development of problem solving skills in physics. *European Journal of Science Education* 7(3): 253–261.

Berners-Lee, T. 1990. Information management: A proposal. Technical report. CERN: Geneva, Switzerland. (Reprinted as Appendix. In W*eaving the Web: The Original Design and Ultimate Destiny of the World Wide Web*. New York: HarperCollins, 1999.)

Berners-Lee, T. 1999. W*eaving the Web: The Original Design and Ultimate Destiny of the World Wide Web*. New York: HarperCollins.

Bowen, B., and C. Meyer. 2008. Applying Novak's new model of education to facilitate organizational effectiveness, professional development, and capacity-building for the New Teacher Alliance. In *Proceedings of CMC 2008: The 3rd Concept Mapping Conference*, eds. A. J. Cañas et al., 108–113. Tallin, Estonia: OU Vali Press.

Buzan, T., and B. Buzan. 1996 *The Mind Map Book: How to Use Radiant Thinking to Maximize your Brain's Untapped Potential*. New York: Plume.

Cañas, A. 1998. Concept maps: new uses and the underlying technology. Report. NASA Ames Research Center, Mountain View, CA.

Cañas, A. 1999. Algunas Ideas sobre la Educación y las Herramientas Computacionales Necesarias para Apoyar su Implementación. Revista RED: Educación y Formación Profesional a Distancia, Ministry of Education, Spain.

Cañas, A. 2003. Concept maps for Mars. Report to the Human-Centered Computing Program, Intelligent Systems Project, NASA Ames Research Center. http://is.arc.nasa.gov/HCC/tasks/CptMaps.html.

Cañas, A. J., G. Briggs, D. Shamma, and J. Scargle. 2000. Mars 2000: multimedia-based knowledge representation of Mars. NASA Ames Research Center. http://www.coginst.uwf.edu/projects/conceptMaps/index.html.

Cañas, A., K. M. Ford, J. D. Novak, P. Hayes, T. Reichherzer, and S. Niranjan. 2001. Online concept maps: enhancing collaborative learning by using technology with concept maps. *The Science Teacher* 68: 49–51.

Cañas, A. J., J. W. Coffey, M. J. Carnot, P. Feltovich, R. Hoffman, J. Feltovich, and J. D. Novak. 2003. A Summary of literature pertaining to the use of concept mapping techniques and technologies for education and performance support. Report prepared for The Chief of Naval Education and Training by The Institute for Human and Machine Cognition, Pensacola, FL.

Cañas, A. J., J. Coffey, T. Reichherzer, N. Suri, and R. Carff. 1998. El-Tech: A performance support system with embedded training for electronics technicians, In *Proceedings of the Eleventh Florida Artificial Intelligence Research Symposium*, ed. D. J. Cook, 79–83. Menlo Park, CA: AAAI Press.

Cañas, A. J., G. Hill, R. Carff, and N. Suri. 2003. CmapTools: a knowledge modeling and sharing toolkit. Technical Report IHMC CmapTools 93-01, Institute for Human and Machine Cognition, Pensacola, FL.

Cañas, A. J., D. B. Leake, and D. C. Wilson. 1999. Managing, mapping and manipulating conceptual knowledge: Exploring the synergies of knowledge management and case-based reasoning. AAAI Workshop Technical Report WS-99-10. Menlo Park, CA: AAAI Press.

Carnot, M., B. Dunn, A. Cañas, G. Baker, and J. Bense. 1999. The effectiveness of computer interfaces in information search. Paper presented at the meeting of the Eastern Southeastern Psychological Association, Savannah, GA.

Carnot, M. J., B. Dunn, A. Cañas, J. Muldoon, and T. Brigham. 2000. Learning style, interface and question order effects on search performance. Paper presented at the meeting of the American Psychological Society, Miami Beach, FL.

Chinn, C. A., A. M. O'Donnell, and T. S. Jinks. 2000. The structure of discourse in collaborative learning. *Journal of Experimental Education* 69: 77–97.

Chiu, C.-H., C.-C. Huang, and W.-T. Chang. 2000. The evaluation and influence of interaction in network supported collaborative concept mapping. *Computers and Education* 34: 17–25.

Chung, G. K. W. K., H. F. O'Neil, Jr., and H. E. Herl. 1999. The use of computer-based collaborative knowledge mapping to measure team processes and team outcomes. *Computers in Human Behavior* 15: 463–493.

Chung, G., E. Baker, and A. Cheak. 2002. Knowledge Mapper Authoring System Prototype. Center for Research on Evaluation, Standards, and Student Testing, University of California, Los Angeles.

Citera, M., M. D. McNeese, and C. E. Brown. 1995. Fitting information systems to collaborating design teams. *Journal of the American Society for Information Science* 46: 551–559.

Coffey, J. W. 2006. In the heat of the moment . . . Strategies, tactics, and lessons learned regarding interactive knowledge modeling with concept maps. In *Proceedings of CMC 2006: The 2nd Concept Mapping Conference*, eds. A. J. Cañas and J. D. Novak, 263–271. San Jose, Costa Rica: Universidad de Costa Rica Press.

Coffey, J. W. 2008. A case study in organizing and presenting a course with concept maps and knowledge models. In *Proceedings of CMC 2008: The 3rd Concept Mapping Conference*, eds. A. J. Cañas et al., 1–8. Tallin, Estonia: OU Vali Press.

Coffey, J. W., A. J. Cañas, T. Reichherzer, G. Hill, N. Suri, R. Carff, T. Mitrovich, and D. Eberle. 2003. Knowledge modeling and the creation of El-Tech: Performance support and training system for electronic technicians. *Expert Systems with Applications* 25: 483–492.

Coffey, J. W., and M. J. Carnot. 2003. Graphical depictions for knowledge generation and sharing. In *Proceedings of IKS2003, IASTED International Conference on Information and Knowledge Sharing*, ed. W. Chu, 18–23. Anaheim, CA: ACTA Press.

Coffey, J. W., and R. R. Hoffman. 2003. The PreSERVe Method of knowledge modeling applied to the preservation of institutional memory. *Journal of Knowledge Management* 7: 38–52.

Coffey, J. W., D. Moreman, and J. Dyer. 1999. Knowledge preservation at NASA-Glenn Research Center. Technical Report. NASA Glenn Research Center, Cleveland, OH.

Conceicao, S. C. O., M. J. Baldor, and C. A. Desnoyers. 2010. Factors influencing individual construction of knowledge in an online community of learning and inquiry using concept maps. In *Handbook of Research on Collaborative Learning Using Concept Maps*, eds. P. T. Torres and R. C. V. Marriott, 100–119. Hershey, PA: IGI Global.

Crandall, B., G. Klein, and R. R. Hoffman. 2006. *Working Minds: A Practitioner's Guide to Cognitive Task Analysis.* Cambridge, MA: MIT Press.

Cristea, A., and T. Okamoto. 2001. Object-oriented collaborative course authoring environment supported by concept mapping in MyEnglishTeacher. *Educational Technology and Society* 4: 104–115.

Daley, B. J., S. Conceicao, L. Mina, B. A. Altman, M. Baldor, and J. Brown. 2008. Advancing concept map research: A review of 2004 and 2006 CMC research. In *Proceedings of CMC 2008: The 3rd Concept Mapping Conference*, eds. A. J. Cañas et al., 84–91. Tallin, Estonia: OU Vali Press.

Dicheva, D., and L. Aroyo. 2000. An approach to intelligent information handling in Web-based learning environments. Paper presented at the Proceedings of the 2000 International Conference on Artificial Intelligence (IC-AI 2000), Las Vegas, NV.

Dodson, D. 1989. Interaction with knowledge systems through connection diagrams: Please adjust your diagrams. In *Research and Development in Expert Systems V*, ed. B. R. Kelly, 35–46. New York: Cambridge University Press.

Dorsey, D. W., G. Campbell, L. Foster, and D. Miles. 1999. Assessing knowledge structures: Relations with experience and post-training performance. *Human Performance* 12: 31–57.

Eden, C. 1988. Cognitive mapping. *European Journal of Operational Research* 36: 1–13.

Eden, C. 1992. On the nature of cognitive maps. *Journal of Management Studies* 29: 261–265.

Ericsson, K. A., and H. A. Simon. 1993. *Protocol Analysis: Verbal Reports as Data*, 2nd. ed. Cambridge, MA: MIT Press.

Esiobu, G., and K. Soyibo. 1995. Effects of concept and vee mapping under three learning modes on students' cognitive achievement in ecology and genetics. *Journal of Research in Science Teaching* 32: 971–995.

Fechner, S., and E. Sumfleth. 2008. Collaborative concept mapping in context-oriented chemistry learning. In *Proceedings of CMC 2008: The 3rd Concept Mapping Conference*, eds. A. J. Cañas et al., 152–156. Tallin, Estonia: OU Vali Press.

Fisher, K. M. 1990. Semantic networking: the new kid on the block. *Journal of Research in Science Teaching* 27: 1001–1018.

Flores, R. P. 2008. Concept Mapping in mathematics: Tools for the development of cognitive and non-cognitive elements. In *Proceedings of the Third International Conference on Concept Mapping*, eds. A. J. Cañas, P. Reiska, M. Ahlberg, and J. D. Novak. Helsinki, Finland: University of Talinin (downloaded 1 November 2010 at http://cmc.ihmc.us/cmc2008papers/cmc2008-p213.pdf).

Fonseca, A. P., and C. I. Extremina. 2008. Concept maps, a tool for scientific research in microbiology: a case study. In *Proceedings of CMC 2008: The 3rd Concept Mapping Conference*, eds. A. J. Cañas et al., 245–251. Tallin, Estonia: OU Vali Press.

Ford, K. M., A. J. Cañas, J. W. Coffey, J. Andrews, E. J. Schad, and H. Stahl. 1992. Interpreting functional images with NUCES: Nuclear Cardiology Expert System. In *Proceedings of the Fifth Annual Florida Artificial Intelligence Research Symposium*, ed. M. B. Fishman, 85–90. Ft. Lauderdale, Florida: AI Research Society.

Ford, K. M., A. Cañas, J. Jones, H. Stahl, J. D. Novak, and J. Adams-Webber. 1991. ICONKAT: An integrated constructivist knowledge acquisition tool. *Knowledge Acquisition* 3: 215–236.

Ford, K. M., J. W. Coffey, A. Canas, E. J. Andrews, and C. W. Turne. 1996. Diagnosis and explanation by a nuclear cardiology expert system. *International Journal of Expert Systems* 9: 499–506.

Fourie, L., and T. van der Westhuizen. 2008. The value and use of concept maps in the alignment of strategy intent. In *Proceedings of CMC 2008: The 3rd Concept Mapping Conference*, eds. A. J. Cañas et al., 650–657. Tallin, Estonia: OU Vali Press.

Francisco, J. S., G. Nicoll, and M. Trautmann. 1998. Integrating multiple teaching methods into a general chemistry classroom. *Journal of Chemical Education* 75: 210–213.

Gaines, B. R., and M. Shaw. 1989. Comparing the conceptual systems of experts. In *Proceedings of the Eleventh International Joint Conference on Artificial Intelligence*, 633–638. San Mateo, CA: Morgan Kaufman.

Gaines, B. R., and M. Shaw. 1995a. Collaboration through concept maps. Paper presented at the Proceedings of CSCL95: The First International Conference on Computer Support for Collaborative Learning, Bloomington, IN.

Gaines, B. R., and M. Shaw. 1995b. Concept maps as hypermedia components. *International Journal of Human–Computer Studies* 43: 323–361.

Gaines, B. R., and M. Shaw. 1995c. WebMap: concept mapping on the Web. Paper presented at the WWW4: Fourth International World Wide Web Conference, Boston, MA.

Gomes, M. E., S. Lind, and D. E. Snyder. 1993. A human factors evaluation using tools for automated knowledge engineering. In *Proceedings of the IEEE National Aerospace and Electronics Conference,* vol. 2, 661–664. Dayton, OH: IEEE Aerospace and Electronic Systems Society.

Good, R., J. Novak, and J. Wandersee, eds. 1990. Special Issue: Perspectives on concept mapping. *Journal of Research in Science Teaching* 27.

Gordon, J. L. 2000. Creating knowledge maps by exploiting dependent relationships. *Knowledge-Based Systems* 13: 71–79.

Gordon, S. E. 1992. Implications of cognitive theory for knowledge acquisition. In *The Psychology of Expertise: Cognitive Research and Empirical AI*, ed. R. R. Hoffman, 99–120. New York: Springer-Verlag.

Gordon, S. E., K. A. Schmierer, and R. T. Gill. 1993. Conceptual graph analysis: knowledge acquisition for instructional systems design. *Human Factors* 35: 459–481.

Graesser, A. C., and S. E. Gordon. 1991. Question answering and the organization of world knowledge. In *Essays in honor of George Mandler*, eds. G. Craik, A. Ortony, and W. Kessen, 227–243. Mahwah, NJ: Lawrence Erlbaum.

Grevholm, B. 2008. Concept maps as a research tool in mathematics education. In *Proceedings of CMC 2008: The 3rd concept mapping Conference*, eds. A. J. Cañas et al., 290–297. Tallin, Estonia: OU Vali Press.

Guimaraes, N., Chambel, T., and Bidarra, J. 2000. From cognitive maps to hypervideo: Supporting flexible and rich learner-centered environments. Interactive Multimedia Electronic *Journal of Computer-Enhanced Learning*, 2, 3.

Hall, R. 1997. Guided surfing: development and assessment of a World Wide Web interface for an undergraduate psychology class. Paper presented at the North American Web Developers Conference (Fredericton, New Brunswick, October).

Hall, R., J. Balestra, and M. Davis. 2000. A navigational analysis of linear and non-linear hypermedia interfaces. Paper presented at the American Educational Research Association, New Orleans, LA.

Hameed, A., D. Sleeman, and A. Preece. 2002. Detecting mismatches among experts' ontologies acquired through knowledge elicitation. *Knowledge-Based Systems* 15: 265–273.

Herl, H. E., H. F. O'Neil, G. K. W. K. Chung, and J. Schachter. 1999. Reliability and validity of a computer-based knowledge mapping system to measure content understanding. *Computers in Human Behavior* 15: 315–333.

Hertz-Lazorowitz, R. 1990. An integrative model of the classroom: the enhancement of cooperation in learning. Paper presented at the American Educational Research Association Conference, Boston, MA.

Hoeft, R. M., F. Jentsch, M. E. Harper, A. W. Evans III, D. G. Berry, and C. A. Bowers. 2002. Structural knowledge assessment with the Team Performance Laboratory's Knowledge Analysis Test Suite (TPL-KATS). In *Proceedings of the Human Factors and Ergonomics Society 46th Annual Meeting*, 757–760. Santa Monica, CA: Human Factors and Ergonomics Society.

Hoffman, R. R. 2001. The Thailand National Knowledge Base Demonstration Project. Report. Institute for Human and Machine Cognition, Pensacola, FL.

Hoffman, R. R., J. W. Coffey, and K. M. Ford. 2000. A case study in the research paradigm of human-centered computing: local expertise in weather forecasting. Report on the Contract, Human-Centered System Prototype. Washington, DC: National Technology Alliance.

Hoffman, R. R., J. W. Coffey, K. M. Ford, and M. J. Carnot. 2000. STORM-LK: system to organize representations in meteorology. Institute for Human and Machine Cognition, Pensacola, FL. http://www.ihmc.us/research/projects/StormLK/.

Hoffman, R. R., B. Crandall, and N. Shadbolt. 1998. A case study in cognitive task analysis methodology: The Critical Decision Method for the elicitation of expert knowledge. *Human Factors* 40: 254–276.

Hoffman, R. R., and L. F. Hanes. 2003. The boiled frog problem. *IEEE Intelligent Systems*, July–August, 68–71.

Hoffman, R. R., P. J. Hayes, and K. M. Ford. 2001. Human-centered computing: thinking in and outside the box. *IEEE: Intelligent Systems*, September–October, 76–78.

Hoffman, R. R., and G. Lintern. 2006. Eliciting and representing the knowledge of experts. In *The Cambridge Handbook of Expertise and Expert Performance*, eds. K. A. Ericsson et al., 203–222. New York: Cambridge University Press.

Hoffman, R. R., N. Shadbolt, A. M. Burton, and G. A. Klein. 1995. Eliciting knowledge from experts: A methodological analysis. *Organizational Behavior and Human Decision Processes* 62: 129–158.

Jonassen, D., and S. Wang. 1993. Acquiring structural knowledge from semantically structured hypertext. *Journal of Computer-Based Instruction* 20: 1–8.

Kremer, R., and B. Gaines. 1994. Groupware concept mapping techniques. Paper presented at the SIGDOC '94: ACM 12th Annual International Conference on Systems Documentation, Association for Computing Machinery, New York.

Kremer, R., and B. Gaines. 1996. Embedded interactive concept maps in Web documents. Paper presented at the WebNet World Conference of the Web Society, San Francisco, CA.

Kyro, P., and V. A. Niskanen. 2008. A concept map modeling approach to business planning in a computer environment. In *Proceedings of CMC 2008: The 3rd concept mapping Conference*, eds. A. J. Cañas et al., 22–29. Tallin, Estonia: OU Vali Press.

Markham, K., and J. Mintzes. 1994. The concept map as a research and evaluation tool: Further evidence of validity. *Journal of Research in Science Teaching* 31: 91–101.

Markham, K., J. Mintzes, and G. Jones. 1994. The concept map as a research and evaluation tool: Further evidence of validity. *Journal of Research in Science Teaching* 31: 91–101.

Martin, B. L., J. J. Mintzes, and I. E. Clavijo. 2000. Restructuring knowledge in biology: Cognitive processes and metacognitive reflections. *International Journal of Science Education* 22: 303–323.

McDonald, S., and R. J. Stevenson. 1998. Navigation in hyperspace: an evaluation of the effects of navigational tools and subject matter expertise on browsing and information retrieval in hypertext. *Interacting with Computers* 10: 129–142.

McNeese, M. D., B. S. Zaff, K. J. Peio, D. E. Snyder, J. C. Duncan, and M. R. McFarren. 1990. An advanced knowledge and design acquisition methodology: Application for the Pilot's Associate. Report AAMRL-TR-90-060, Human Systems Division, Aerospace Medical Research Laboratory, Air Force Systems Command, Wright-Patterson Air Force Base, OH.

McNeese, M. D., B. S. Zaff, C. E. Brown, and M. Citera. 1993. Understanding the context of multidisciplinary design: Establishing ecological validity in the study of design problem solving. In *Proceedings of the 37th Annual Meeting of the Human Factors Society*, 1082–1086. Santa Monica, CA: Human Factors Society.

McNeese, M. D., B. S. Zaff, M. Citera, C. E. Brown, and R. Whitaker. 1995. AKADAM: Eliciting user knowledge to support participatory ergonomics. *International Journal of Industrial Ergonomics* 15: 345–363.

Mintzes, J. J., J. H. Wandersee, and J. D. Novak. 2000. *Assessing Science Understanding: A Human Constructivist View*. San Diego, CA: Academic Press.

Mostrum, A. M. 2008. A unique use of concept maps as the primary organizing structure in two upper-level undergraduate biology courses: Results from the first implementation. In *Proceedings of CMC 2008: The 3rd concept mapping Conference*, eds. A. J. Cañas et al., 76–83. Tallin, Estonia: OU Vali Press.

NASA. 2003a. Fundamental questions of astrobiology. http://cmex-www.arc.nasa.gov/CMEX/Origin%20of%20Life%20on%20E.html (accessed November 1, 2010).

NASA. 2003b. NASA Center for Mars Exploration. http://cmex-www.arc.nasa.gov/CMEX/Map%20of%20Maps.html.

Nilsson, R., and R. Mayer. 2002. The effects of graphic organizers giving cues to the structure of a hypertext document on users' navigation strategies and performance. *International Journal of Human–Computer Studies* 57: 1–26.

Novak, J. D. 1991. Clarify with concept maps: A tool for students and teachers alike. *The Science Teacher* 58: 45–49.

Novak, J. D. 1998. *Learning, Creating, and Using Knowledge: Concept Maps as Facilitative Tools in Schools and Corporations*. Mahwah, NJ: Lawrence Erlbaum.

Novak, J. D. 2002. Meaningful learning: the essential factor for conceptual change in limited or inappropriate propositional hierarchies leading to empowerment of learners. *International Journal of Science Education* 86: 548–571.

Novak, J. D. 2010 *Learning, Creating, and Using Knowledge: Concept Maps as Facilitative Tools in Schools and Corporations*, 2nd ed. New York: Taylor & Francis.

Novak, J. D., and A. J. Canas. 2004. Building on new constructivist ideas and the CMapTools to create new model for education. Closing Lecture. First International Conference on Concept Mapping: Theory, Methodology, Technology. University of Navarra, Pamplona, Spain.

Novak, J. D., and A. J. Canas. 2006. Theoretical origins of concept maps, how to construct them and uses in education. *Reflecting Education* 3(1).

Novak, J. D., and D. B. Gowin. 1984. *Learning How to Learn*. New York: Cambridge University Press.

Novak, J. D., and D. Musonda. 1991. A twelve-year longitudinal study of science concept learning. *American Educational Research Journal* 28(1): 117–153.

Pearsall, N. R., J. Skipper, and J. Mintzes. 1997. Knowledge restructuring in the life sciences: a longitudinal study of conceptual change in biology. *Science Education* 81: 193–215.

Pernaa, J., and M. Aksela. 2008. Concept maps as meaningful learning tools in a Web-based chemistry material. In *Proceedings of CMC 2008: The 3rd Concept Mapping Conference*, eds. A. J. Cañas et al., 282–289. Tallin, Estonia: OU Vali Press.

Ramirez, C., W. Florez, and R. Barros. 2008. Concept maps: A strategy for the development of technical reports in industrial engineering problems. In *Proceedings of CMC 2008: The 3rd Concept Mapping Conference*, eds. A. J. Cañas et al., 314–321. Tallin, Estonia: OU Vali Press.

Reader, W., and N. Hammond. 1994. Computer-based tools to support learning from hypertext: Concept mapping tools and beyond. *Computers and Education* 22: 99–106.

Reynolds, S., and D. Dansereau. 1990. The knowledge hypermap: An alternative to hypertext. *Computers in Education* 14: 409–416.

Reynolds, S., M. E. Patterson, L. Skaggs, and D. Dansereau. 1991. Knowledge hypermaps and cooperative learning. *Computers in Education* 16: 167–173.

Salinas, J., B. Benito, and M. Garcia. 2008. Collaborative construction of a concept map about flexible education. In *Proceedings of CMC 2008: The 3rd Concept Mapping Conference*, eds. A. J. Cañas et al., 165–172. Tallin, Estonia: OU Vali Press.

Schanze, S., and T. Grub-Niehaus. 2008. Supporting comprehension in chemistry education—the effect of computer-generated and progressive concept mapping. In *Proceedings of CMC 2008: The 3rd Concept Mapping Conference*, eds. A. J. Cañas et al., 595–602. Tallin, Estonia: OU Vali Press.

Sowa, J. F. 1984. *Conceptual Structures: Information Processing in Mind and Machine*. Reading, MA: Addison-Wesley.

Stanford, X. 2003. Stanford Solutions, Inc. http://www.knowmap.com/ssi (accessed November 1, 2010).

Stanton, N., R. Taylor, and L. Tweedie. 1992. Maps as navigational aids in hypertext environments. *Journal of Educational Multimedia and Hypermedia* 1: 431–444.

Stoyanova, N., and P. Kommers. 2002. Concept mapping as a medium of shared cognition in computer-supported collaborative problem solving. *Journal of Interactive Learning Research* 13: 111–133.

Tergan, S.-O., T. Engelmann, and F. W. Hesse. 2008. Digital concept maps as powerful interfaces for enhancing information search: An experimental study on the effects of semantic cueing. In *Proceedings of CMC 2008: The 3rd Concept Mapping Conference*, eds. A. J. Cañas et al., 351–358. Tallin, Estonia: OU Vali Press.

Tifi, A., and Lombardi, A. 2010. Distance collaboration with shared concept maps. In *Handbook of Research on Collaborative Learning using Concept Maps*, eds. P. T. Torres and R. C. V. Marriott, 120–150. Hershey, PA: IGI Global.

Tifi, A., S. S. Marche, A. Lombardi, and D. D. P. C. N. Ligure, 2008. Collaborative concept mapping models. In *Proceedings of CMC 2008: The 3rd Concept Mapping Conference*, eds. J. Cañas et al., 157–164. Tallin, Estonia: OU Vali Press.

Van Boxtel, C., J. Van Der Linden, and G. Kanselaar. 1997. Collaborative construction of conceptual understanding: Interaction processes and learning outcomes emerging from a concept mapping and a poster task. *Journal of Interactive Learning Research* 8: 341–361.

Van Boxtel, C., J. Van Der Linden, and G. Kanselaar. 2000. Collaborative learning tasks and the elaboration of conceptual knowledge. *Learning and Instruction* 10: 311–330.

Vekiri, I. 2002. What is the value of graphical displays in learning? *Educational Psychology Review* 14: 261–311.

Wallace, J., and J. Mintzes. 1990. The concept map as a research tool: Exploring conceptual change in biology. *Journal of Research in Science Teaching* 27: 1033–1052.

Webb, N. 1989. Peer interaction and learning in small groups. *International Journal of Educational Research* 13: 21–39.

11 Organization and Structure of Information Using Semantic Web Technologies

James R. Michaelis, Jennifer Golbeck, and James Hendler

CONTENTS

11.1 INTRODUCTION

Today's Web has millions of pages that are dynamically generated from content stored in databases. This not only makes managing a large site easier but also helps organize both Web-based user collaborations and content generation. These local databases, in one sense, are not full participants on the Web. Though they drive the presentation of Web sites, the databases themselves are not interconnected in the linked manner that traditional static Web sites are. Should a mashup be desired, one organization has no obvious online way of using or understanding another's data except through the human-readable presentation on a Web site. If these organizations want to share or merge information, any required database integration can be a significant undertaking. It is also a one-time solution. That is, if a third organization entered the picture, a new merging effort would have to be undertaken.

The trend toward dynamically driven Web sites, combined with the proliferation of nontext media files as a primary form of content, poses several problems to the current Web architecture. For example, search engines have difficulty indexing database-driven pages and cannot directly index raw data from the backend databases themselves. Media searches, such as video or mp3 searches, are notoriously poor, because there is no embedded text from which to extract key words that could be used to index the media. Additionally, it is difficult for a Web designer to use information from a database they do not directly control to drive their own site. Such databases are often not publicly accessible for queries nor are their underlying organization schema usually apparent (even when a direct interface is presented).

The HTML-driven Web was primarily received as a vehicle for conveying information in a *human* readable form—computers had no need to understand the content. However, as dynamic sources of information have become omnipresent on the Web, capable of being merged through content mashup, an effort has been made to make greater amounts of descriptive information *machine* readable. This technology, known as the Semantic Web, allows computers to better share metadata. That is, information about the data that they use in their applications, as well as to share the data themselves. This means data from other sites can be accessed and presented on your own Web site, and your own public data can be made easily accessible to anyone else. It follows that just as Web pages are currently hyperlinked, data can also be linked to form a second Web behind the scenes, allowing full across-the-Web integration of data.

The Semantic Web is a technology stack that increases the power of the World Wide Web by giving meaning to all of this data, as well as making it publicly accessible to anyone who is interested. Although some Web sites and designers will want to keep their backend data proprietary, many will find it in their interest to use semantic encodings and make their data more sharable. This can allow greater information use, transparency (especially for government data), and the linking of sources based on relations not explicit in human-readable content.

This chapter introduces the Semantic Web, explains how to organize content for use on the Semantic Web, and shows several examples of how it can be used. Throughout the discussion, we will describe how Semantic Web technologies affect the human factors in Web design and use.

11.2 WHAT IS THE SEMANTIC WEB?

The World Wide Web can be thought of as a collection of distributed, interlinked pages, encoded using languages such as HTML and XML (W3C 2003). Any person can create their own Web page, put it online, and point to other pages on the Web. Because the content of these pages is written in natural language, computers do not "know" about what is in the page, just how it looks. In addition, many dynamically generated Web pages obtain their content from databases not designed for integration with external data sources (often termed "data silos"). The Semantic Web makes it possible for machine-readable annotations to be created, linked to each other, and used for organizing and accessing Web content. Thus the Semantic Web offers new capabilities, made possible by the addition of documents that can encode the "knowledge" about a Web page, photo, or database in a publicly accessible, machine-readable form. Driving the Semantic Web are two mutually dependent efforts: (1) the organization of content using specialized vocabularies, called ontologies, which can be used by Web tools to provide new capabilities, and (2) the large-scale publication and linking of data to produce a web of data analogous to the current document web. In this section, we present a general introduction to both ontology-based vocabularies and the linking of Semantic Web data.

11.3 VOCABULARIES AND ONTOLOGIES

An ontology, as used in the Semantic Web context, is basically a collection of machine-readable terms used to describe a particular domain. Some ontologies are broad, covering a wide range of topics, while others are limited with very precise specifications about a given area.

The general elements that make up an ontology are as follows:

- Classes: general categories of things in the domain of interest
- Properties: attributes that instances of those classes may have
- The relationships that can exist between classes and between properties

Ontologies are usually expressed in logic-based languages, allowing people to use reasoners to analyze their relationships. XML (eXtensible Markup Language) exists to add metadata to applications, but there is no way to connect or infer information from these statements. For example, if we say "Sister" is a type of "Sibling" and then say "Jen is the sister of Tom," there is no way to automatically infer that Jen is also the sibling of Tom based on the specification inherent in the XML. By encoding these relationships in language with an underlying semantics, these and other more interesting inferences can be made. *Reasoners* are tools that use the semantics to make inferences about the terms defined in the ontologies. On the Web, these reasoners can be used in advanced applications, such as mashups. With these issues in mind, ontologies, and languages for developing them, constitute a key focus for Semantic Web development.

11.4 LINKED DATA

Many current Semantic Web projects focus on large quantities of data being publicly released and interlinked, yielding a web of "linked" data. The concept of linked data is analogous to that of Web pages connected via hyperlinks. On the Web, this linking between Web pages helped facilitate both content discovery and reuse by Web users, leading to a steadily increasing user base. As the number of participants on the Web increased, a network effect could be increasingly observed. Essentially, as the user base of the Web increased, its overall value to any given user, in turn, increased exponentially. For the network effect to apply on the Web, content integration through links between Web pages was necessary.

On the Semantic Web, the generation of linked data is now producing a network effect similar to the one observed on the early World Wide Web. To help maintain the viability of linked data, a set of best practices has emerged (Berners-Lee 2006), consisting of four main points.

First, every resource represented on the Semantic Web (e.g., classes, properties, and instances of classes) should be assigned Uniform Resource Identifier (URI). URIs serve as unique names for things on the Web, ranging from Semantic Web resources to HTML-based Web pages. A common URI format for identifying Semantic Web resources provides the Web address of a containing document, with a "#" and the name of the resource appended to the end. For example, a URI for New York City could read http://example.com/cities#NewYorkCity, where the resource "NewYorkCity" is defined in the document "http://example.com/cities."

Second, each URI should be dereferenceable through the HTTP protocol. Essentially, the act of dereferencing a URI entails looking up its corresponding information. On the World Wide Web, when a browser looks up a Web page corresponding to a given URL, it dereferences that URL to obtain information from the Web page. For linked data, the usage of a single protocol (HTTP) for managing dereferencing simplifies things for Web designers. Rather than face the possibility of managing multiple retrieval protocols, Web designers can instead rely on HTTP to dereference any linked data URI.

Third, each URI should correspond to useful information. For information to be considered useful, it should essentially be what a user expects to find upon URI dereferencing. Specifically, information should be in whatever format is expected and correspond to an expected topic. Typically, linked data will be encoded using Resource Description Framework (RDF) (W3C 2004b) based syntax (more information on RDF will be covered in Section 11.6). However, information in other formats (e.g., HTML, JPG) may be referenced as well. For instance, for the URI http://example.com/cities/page#NewYorkCity, a human-readable page in HTML could be expected. Likewise, RDF-based data could be expected for the URI http://example.com/cities/data#NewYorkCity. In both cases, the information should correspond to New York City and not another topic (e.g., Los Angeles).

Fourth, information on the Semantic Web should have embedded URI references to other linked data sources. An example of a linked data source that Web developers can link to is DBPedia (Auer et al. 2008), which derives Semantic Web based data from structured content on Wikipedia.* Here, the RDF-based information on New York City at the URI http://example.com/cities/data#NewYorkCity could include a URI reference to RDF-based information on DBPedia corresponding to the Empire State Building: http://dbpedia.org/data/Empire_State_Building.†

11.5 MOTIVATIONS

For Web designers, participation in the Semantic Web offers a number of new capabilities, enabled by the rich domain models of ontologies and the interlinking of ontology-backed data. However, at the same time, both of these components of the Semantic Web require a reasonable amount of work to manage. For ontologies, challenges often emerge in deciding how vocabularies should be structured. Likewise, the creation of linked data requires bringing databases into the formats of the Semantic Web and determining how to effectively link the resulting data to existing linked data sources.

To justify the effort required for these tasks, Web designers must understand the benefits that become available and doors that are opened. The following sections enumerate a listing of domains where Semantic Web technologies are actively being applied by Web developers. This list is by no means exhaustive; a more extensive collection of domains can be found at http://www.w3.org/2001/sw/sweo/public/UseCases/.

11.5.1 CONTENT MASHUP

A mashup is designed to integrate information from multiple Web-based sources, yielding new forms of functionality in the process. Often, mashups will leverage preexisting Web services as a way to access both their data and functionality. To enable this, many Web services provide application programming interfaces (or Web APIs) to facilitate application development. Currently, this Web service-oriented approach to constructing mashups faces a number of challenges, which include (1) a lack of interconnections between information in Web service backend databases (data silos) and (2) limited support for Web service information access outside of supported Web APIs.

The Semantic Web—and in particular, linked data—offers coping mechanisms that address the above issues in the mashup design process. First, the information connections provided in linked data can serve as a starting point for designing mashups. Second, as linked data are published

---

* Additionally, DBPedia-based information contains embedded links to alternate linked data sources. A listing of these linked data sources is provided at http://dbpedia.org/.

† In certain cases, URIs will lack "#," indicating that an entire document is devoted to the given resource.

directly on the Web, they can be accessed without the usage of specific Web APIs.

One example of a linked data-based mashup is the RDF Book Mashup (Bizer, Cygniak, and Gauß 2007), which is designed to make information about books (such as their authors and how much they cost at various online bookstores) available as linked data. The RDF Book Mashup draws upon established linked data sources (such as DBPedia), as well as linked data representations of information obtained from Web service APIs hosted by Amazon (http://aws.amazon.com/) and Google (http://code.google.com/apis/base/).

11.5.2 Semantic Search

Modern search engines are primarily driven by key words identified in user queries and lack significant support for semantic complexity. Semantic search engines are designed to interpret both key words and their context within the phrasing of a query. Even in very simple queries (e.g., one to two words long), a need for key word context is often apparent. Here, ontology-based context resolution can be readily utilized, as demonstrated through the search capabilities of Freebase (http://www.freebase/com/), an online knowledge base of structured data.

In the search bar, when a user types the phrase "Van Gogh," (with the intention of looking up Vincent van Gogh) a listing of matching phrases is returned. Because Freebase is unsure which Van Gogh the user is referring to, this disambiguation step is required to obtain relevant results. To aid the user to select correctly, each phrase is accompanied by a category and short description, both of which are derived from ontology-based definitions (Figure 11.1).

For the more complex query "What did Van Gogh paint?" basic natural language processing is additionally required for interpreting the given key word phrasing. To address this, the semantic search engine Powerset (http://www.powerset.com/) utilizes a combination of natural language processing and ontology-based information from Freebase. Through interpreting the phrasing of the query, Powerset can determine that the user has requested a list of paintings by "Van Gogh." Going back to the disambiguation capabilities of Freebase, it can be determined that paintings were only produced by Vincent van Gogh. From here, a listing of paintings

FIGURE 11.2 **(See color insert.)** Results of Powerset search "What did Van Gogh paint?"

can be returned from Freebase's ontology-backed knowledge base (Figure 11.2).

11.5.3 Provenance

Provenance is used to represent the history of artifacts—for instance, how they were created and used in a given setting. From a computational perspective, provenance has been applied in many data-intensive application domains (Moreau 2009). In situations where provenance is recorded in distributed locations, a Semantic Web–driven approach can be used for information integration (Pinheiro da Silva, McGuinness, and Fikes 2006). The ability to make inferences over provenance is also gained through use of Semantic Web technologies, observed in the context of scientific workflow systems (Zhao et al. 2008; Kim et al. 2008).

11.5.4 Ubiquitous Computing

Ubiquitous computing describes a movement toward extending computational power to devices embedded in the environment, to aid users in everyday tasks. Growth in the usage of "smart phones" (such as the Apple iPhone) has made them a focus area for ubiquitous computing research. Such devices will typically have access to personal information about a user, as well as information on the surrounding environment. Here, the use of Semantic Web technology has been explored in DBPedia Mobile (Becker and Bizer 2009), a location- aware application designed to run on the Apple iPhone. Using the GPS coordinates of a user, DBPedia Mobile can present information on nearby locations (such as text-based descriptions, photos, and online reviews)—derived from both DBPedia and other data sources it is interlinked with (Figure 11.3).

11.6 SEMANTIC WEB EXAMPLE

To demonstrate the Semantic Web in action, consider the task of making a Web page about a recent trip to Paris. The page would include some text describing the trip, when it happened, and undoubtedly a picture of the travelers in front of the Eiffel Tower. In addition, links could be provided to the hotel where the travelers stayed, to the City of Paris Home page, and some helpful travel books listed on Amazon. As the

FIGURE 11.1 Freebase disambiguation of "Van Gogh."

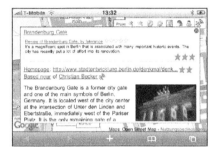

FIGURE 11.3 **(See color insert.)** DBPedia Mobile, displaying information from multiple linked data sources on the Brandenburg Gate, located in Berlin. (From Becker, C. and Bizer C. Exploring the Geospatial Semantic Web with DBPedia Mobile. *Journal of Web Semantic*, Vol 7(4) 278–286. Elsevier 2009. With permission.)

Web stands now, search engines would index the page by key words found in the text of the Web page and perhaps by the links included there. Short of that vague classification, there is no way for a computer to understand anything about the page. If the date of the trip, for example, were typed as "June 25-30," there would be no way for a computer to know that the trip was occurring on June 26, because it cannot understand dates in human-readable forms. For the nontextual elements, such as the photo, computers have no way of knowing who is in the picture, what happened in it, or where it occurred.

On the Semantic Web, all of this information, and more, can be made available for computers to understand. This is accomplished through encoding information in the Resource Description Framework (RDF) (W3C 2004b), as well as its extensions RDF Schema (W3C 2004a) and the Web Ontology Language (OWL). Using these languages, developers can encode the knowledge from the Web page and point to knowledge stored on other sites. To understand how this is done, it is necessary to have a general understanding of how Semantic Web markup works.

For Semantic Web data, developers define classes, much like classes in a programming language. These can be subclassed and instantiated. Properties allow users to define attributes of classes. In the example above, a "Photo" class would be useful. Properties of the Photo class may include the URL of the photo on the Web, the date it was taken, the location, references to the people and objects in the picture, as well as what event is taking place. To describe a particular photo, developers could create instances of the Photo class

and then define values for the Photo's properties. In a simple table format, the data may look like this:

> *Photo*
> > *Name:* ParisPhoto1
> > *URL:* http://www.example.com/photo1.jpg
> > *Date Taken:* June 26, 2001
> > *Location:* Parc Du Champ De Mars, Paris, France
> > *Person in Photo:* John Doe
> > *Person in Photo:* Joe Blog
> > *Object in Photo:* Eiffel Tower

As mentioned earlier, Semantic Web resources (collectively classes, properties, and instances of classes) are represented through URIs. Here, if an ontology describing the trip is at http://www.example.com/parisTrip, the URI of the photo would be http://www.example.com/parisTrip#ParisPhoto1.

Because each resource has a unique URI, it allows authors to make references to definitions elsewhere. In our trip document, the author can include definitions of the two travelers, John Doe and Joe Blog:

> *Person*
> > *Name:* JohnDoe
> > *First Name:* John
> > *Last Name:* Doe
> > *Age:* 23
> *Person*
> > *Name:* JoeBlog
> > *First Name:* Joe
> > *Last Name:* Blog
> > *Age:* 24

Then, in the properties of the photo, these definitions can be referenced. Instead of having just the string "John Doe," the computer will know that the people in the photo are the same ones defined in the document, with all of their properties.

> **Person in Photo:** http://www.example.com/parisTrip#JohnDoe
> **Person in Photo:** http://www.example.com/parisTrip#JoeBlog

Using linked data practices, it is also possible (and encouraged) to make references to Semantic Web data defined in other sources. In the simple table above, the property "Object in Photo" is listed as the simple string "Eiffel Tower." However, a URI reference to the RDF-based information on DBPedia for the Eiffel Tower could be used instead, yielding the following property value:

> **Object in Photo:** http://dbpedia.org/data/Eiffel_Tower

Again, the benefit of this linking is similar to why links are used in HTML documents. If a Web page mentions a book, a link to its listing on Amazon offers many benefits. Thorough information about the book may not belong on a page that just mentions it in passing or an author may not want to retype all of the text that is nicely presented elsewhere. A link passes users off to another site and in the process provides them with more data. References on the Semantic Web are even better at this. Though the travelers in this example may not know much about the Eiffel Tower, all sorts of interesting data may be defined elsewhere about its history, location, and construction. By making a reference to the RDF-based information from DBPedia, the computer understands that the Eiffel Tower in the photo has all of the properties described in that information. This means, among other things, that mashups and reasoners can automatically connect the properties defined in DBPedia to our document.

Once these data are encoded in a machine understandable form, they can be used in many ways. Of course, the definition of Semantic Web data is not done in simple tables as above. The next section gives a general overview of the capabilities of RDF, RDFS and OWL, which are used to formally express the semantic data. Once that is established, we discuss existing approaches for querying, creating, and using Semantic Web content.

11.7   ENCODING INFORMATION ON THE SEMANTIC WEB

The basic unit of information on the Semantic Web, independent of the language, is the triple. A triple is made up of a subject, predicate, and object or value. In the example from the previous section, one triple would have subject "JohnDoe," predicate "age," and value "23." Another would have the same subject, predicate "First Name," and value "John." On the Semantic Web, triples are encoded using URIs. Thus, for our Paris Trip ontology located at http://www.example.com/parisTrip, the two triples will be*

> **Subject**: http://www.example.com/parisTrip#JohnDoe
> **Predicate:** http://www.example.com/parisTrip#age
> **Value:** 23
> **Subject**: http://www.example.com/parisTrip#JohnDoe
> **Predicate:** http://www.example.com/parisTrip#firstName
> **Value:** John

In the two examples above, the predicates relate the subject to a string value. It is also possible to relate two resources through a predicate. For example:

> **Subject:** http://www.example.com/parisTrip#ParisPhoto1
> **Predicate:** http://www.example.com/parisTrip#objectInPhoto
> **Object:** http://dbpedia.org/data/Eiffel_Tower

---

* Actually, we slightly simplify the treatment of data types, as the details are not relevant to this chapter. Readers interested in the full details of the RDF encoding are directed to *Resource Description Framework (RDF)*. http://www.w3.org/RDF/.

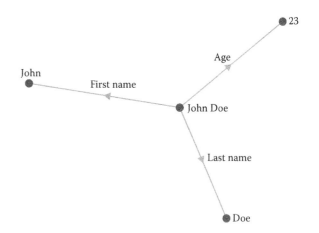

FIGURE 11.4   Three triples, rooted with JohnDoe as the subject of each, shown as the center node.

Each of these triples forms a small graph with two nodes, representing the subject and object, connected by an edge representing the predicate. In triple form, the information for John Doe is represented in Figure 11.4.

Taking all of the descriptors from the previous section and encoding them as triples will produce a much more complex graph (Figure 11.5). As documents are linked together, through joining of terms, these graphs grow to be large, complex, and interconnected webs. To create them, we need languages that support the creation of these relationships. For this purpose, the Semantic Web standards currently used are RDF, RDFS, and OWL.

11.7.1   RDF AND RDFS

The Resource Description Framework (RDF)—developed by the World Wide Web Consortium (W3C)—is the foundational language for the Semantic Web. Along with RDF Schema, it provides a basis for creating vocabularies and instance data. This section provides an introduction to RDF and RDFS syntax. The rest of this section will give a general overview of the syntax of RDF and OWL and their respective features but is not intended as a comprehensive guide. Links to thorough descriptions of both languages, including an RDF primer and an OWL Guide, each with numerous examples, are available on the W3C's Semantic Web Activity Web site at http://w3.org/2001/sw.

11.7.1.1   Document Skeleton

There are several flavors of RDF, but the version this chapter will focus on is RDF/XML, which is RDF based on XML syntax. The skeleton of an RDF document is as follows:

```
<rdf:RDF
xmlns:rdf="http://www.w3.org/1999/02/22-
rdf-syntax-ns#">
</rdf:RDF>
```

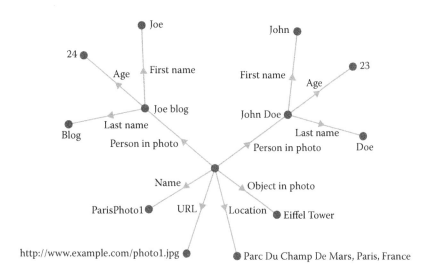

FIGURE 11.5 The graph of triples from the Paris example.

The tag structure is inherited from XML. The RDF tag begins and ends the document, indicating that the rest of the document will be encoded as RDF. Inside the rdf:RDF tag is an XML namespace declaration, represented as an `xmlns` attribute of the `rdf:RDF` start-tag. Namespaces are convenient abbreviations for full URIs. This declaration specifies that tags prefixed with rdf: are part of the namespace described in the document at http://www.w3.org/1999/02/22-rdf-syntax-ns#.

In the example detailed in previous sections, we made some assumptions. For creating triples about people, we assumed there was a class of things called People with defined properties, like "age" and "first name." Similarly, we assumed a class called "Photo" with its own properties. First, we will look at how to create RDF vocabularies, and then proceed to creating instances.

11.7.1.2 Defining Vocabularies

RDF provides a way to create instances and associate descriptive properties with each. RDF does not, however, provide syntax for defining classes, properties, and describing how they relate to one another. To do that, authors use RDF Schema (RDFS). RDF Schema uses RDF as a base to specify a set of predefined RDF resources and properties that allow users to define classes and restrict properties. The RDFS vocabulary is defined in a namespace identified by the URI reference http://www.w3.org/2000/01/rdf-schema# and commonly uses the prefix "rdfs:." As with the rdf namespace, the rdfs namespace is added to the rdf tag.

11.7.1.3 Describing Classes

Classes are the main way to describe types of things we are interested in. Classes are general categories that can later be instantiated. In the previous example, we want to create a class that can be used to describe photographs. The syntax to create a class is written:

```
<rdfs:Class rdf:ID="Photo"/>
```

The beginning of the tag "rdfs:Class" says that we are creating a Class of things. The second part, "rdf:ID" is used to assign a unique name to the resource in the document. Names always need to be enclosed in quotes, and class names are usually written with the first letter capitalized, though this is not required. Like all XML tags, the rdfs:Class tag must be closed, and this is accomplished with the "/" at the end.

Classes can also be subclassed. For example, if an ontology exists that defines a class called "Image," we could indicate that our Photo class is a subclass of that.

```
<rdfs:Class rdf:ID="Photo">
<rdfs:subClassOf rdf:resource=
"http://example.com/mediaOntology.
rdf#Image"/>
</rdfs:Class>
```

The rdfs:subClassOf tag indicates that the class we are defining will be a subclass of the resource indicated by the rdf:resource attribute. The value for rdf:resource should be the URI of another class. Subclasses are transitive. Thus, if X is a subclass of Y and Y is a subclass of Z, then X is also a subclass of Z. Classes may also be subclasses of multiple classes. This is accomplished by simply adding more rdfs:subClassOf statements.

11.7.1.4 Describing Properties

Properties are used to describe attributes. By default, properties are not attached to any particular class; that is, if a property is declared, it can be used with instances of any class. Using elements of RDFS, properties can be restricted in several ways.

All properties in RDF are described as instances of class rdf:Property. Just as classes are usually named with an initial capital letter, properties are usually named with an initial lower case letter. To declare the "object in photo" property that we used to describe instances of our Photo class, we use the following statement:

```
<rdf:Property rdf:ID="objectInPhoto"/>
```

This creates a Property called "objectInPhoto," which can be attached to any class. To limit the domain of the property, so it can only be used to describe instances of the Photo class, we can add a domain restriction:

```
<rdf:Property rdf:ID="objectInPhoto">
        <rdfs:domain rdf:resource="#Photo"/>
</rdf:Property>
```

Here, we use the rdfs:domain tag that limits which class the property can be used to describe. Here, the rdf:resource is used the same way as in the subclass restriction above, but we have used a local resource. Because the Photo class is declared in the same namespace (the same file, in this case) as the "objectInPhoto" property, we can abbreviate the resource reference to just the name.

Similar to the rdfs:subClassOf feature for classes, there is an rdfs:subPropertyOf feature for properties. In our photo example, the property "person in photo" is a subset of the "object in photo" property. To define this relation, the following syntax is used:

```
<rdf:Property rdf:ID="personInPhoto">
        <rdfs:subPropertyOf
        rdf:resource="#objectInPhoto"/>
</rdf:Property>
```

Subproperties inherit any restrictions of their parent properties. In this case, because the objectInPhoto property has a domain restriction to Photo, the personInPhoto has the same restriction. In turn, we can also add restrictions to these subproperties. In addition to the domain restriction that limits which classes the property can be used to described, we can add range restrictions for limiting what types of values the property can accept. For the person-InPhoto property, we should restrict the value to be an instance of the Person class. Ranges are restricted in the same way as domains:

```
<rdf:Property rdf:ID="personInPhoto">
        <rdfs:range
        rdf:resource="#Person"/>
</rdf:Property>
```

11.7.2 CREATING INSTANCES

Once this structure is set up, our instances can be defined. Consider the previous triple that we described as plain text:

Person
    Name: JoeBlog
    First Name: Joe
    Last Name: Blog
    Age: 24

Here, JoeBlog is the subject, and is an instance of the class Person. There are also properties for age, first name, and last name. Assuming we have defined the Person class and its corresponding properties, we can create the Joe Blog instance:

```
<Person rdf:ID="JoeBlog">
        <firstName>Joe</firstName>
        <lastName>Blog</lastName>
        <age>24</age>
</Person>
```

In the simplest case, the classes and properties we are using are declared in the same namespace as where our instances are being defined. If that is not the case, we use namespace prefixes, just as we used with rdf: and rdfs:. For example, if there is a property defined in an external file, we can add a prefix of our choosing to the rdf tag:

```
<rdf:RDF
xmlns:rdf="http://www.w3.org/1999/02/22-
rdf-syntax-ns#"
xmlns:rdfs="http://www.w3.org/2000/01/
rdf-schema#"
xmlns:edu="http://example.com/education.
rdf#">
```

Once this new namespace is introduced, it can be used to reference classes or properties in that file:

```
<Person rdf:ID="JoeBlog">
        <firstName>Joe</firstName>
        <lastName>Blog</lastName>
        <age>24</age>
        <edu:degreeEarned>PhD</
        edu:degreeEarned>
</Person>
```

11.7.3 OWL

The Web Ontology Language (OWL) is a vocabulary extension of RDFS that adds expressivity needed to define classes and their relationships more fully. Because OWL is built on RDF, any RDF graph forms a valid OWL ontology. However, OWL adds semantics and vocabulary to RDF and RDFS, giving it more power to express complex relationships.

OWL introduces many new features over what is available in RDF and RDFS. They include, among others, relations between classes (e.g., disjointness), cardinality of properties (e.g., "exactly one"), equality, characteristics of properties (e.g., symmetry), and enumerated classes. Because OWL is based in the knowledge engineering tradition, expressive power and computational tractability were major concerns in the drafting of the language. Features of OWL are well-documented online (McGuinness and van Harmelen 2004), and an overview is given here. Because OWL is based on RDF, the syntax is basically the same. OWL uses class and property definitions and restrictions from RDFS. It also adds the following syntactic elements:

Equality and Inequality:

- equivalentClass: This attribute is used to indicate equivalence between classes. In particular, it can be used to indicate that a locally defined class is the same is one defined in another namespace. Among other things, this allows properties with restrictions to a class to be used with the equivalent class.
- equivalentProperty: Just like equivalentClass, this indicates equivalence between properties.
- sameIndividualAs: This is the third equivalence relation, used to state that two instances are the same. Though instances in RDF and OWL must have unique names, there is no assumption that those names refer to unique entities or the same entities. This syntax allows authors to create instances with several names that refer to the same thing.
- differentFrom: This is used just like sameIndividual As but to indicate that two individuals are distinct.
- allDifferent: The allDifferent construct is used to indicate difference among a collection of individuals. Instead of requiring many long lists of pairwise "differentFrom" statements, allDifferent has the same effect in a compact form. AllDifferent is also unique in its use. While the other four attributes described under this heading are used in the definition of classes, properties, or instances, allDifferent is a special class for which the property owl:distinctMembers is defined, which links an instance of owl:AllDifferent to a list of individuals. The following example, taken from the OWL Reference (McGuinness and van Harmelen 2004), illustrates this syntax for distinguishing wine colors within an ontology of wines:*

```
<owl:AllDifferent>
    <owl:distinctMembers
    rdf:parseType="Collection">
    <vin:WineColor rdf:about="#Red"/>
    <vin:WineColor rdf:about="#White"/>
    <vin:WineColor rdf:about="#Rose"/>
    </owl:distinctMembers>
</owl:AllDifferent>
```

Property Characteristics:

- inverseOf: This indicates inverse properties. For example, "picturedInPhoto" for a Person would be the inverseOf the "personInPhoto" property for Photos.
- TransitiveProperty: Transitive properties state that if A relates to B with a transitive property, and B relates to C with the same transitive property, then A relates to C through that property.
- SymmetricProperty: Symmetric properties state that if A has a symmetric relationship with B, then B has that relationship with A. For example, a "knows" property could be considered transitive, because if A knows B, then B should also know A.
- FunctionalProperty: If a property is a FunctionalProperty, then it has no more than one value for each individual. "Age" could be considered a functional property, because no individual has more than one age.
- InverseFunctionalProperty: Inverse functional properties are formally properties such that their inverse property is a functional property. More clearly, inverse functional properties are unique identifiers, where an RDF object uniquely determines the corresponding subject.

Property Type Restrictions:

- allValuesFrom: This restriction is used in a class as a local restriction on the range of a property. While an rdfs:range restriction on a property globally restricts the values a property can take, allValuesFrom states that for instances of the restricting class, each value corresponding to the restricted property must be an instance of a specified class.
- someValuesFrom: Just like allValuesFrom, this is a local restriction on the range of a property, but it states that there is *at least one* value of the restricted property that has a value from the specified class.

Class Intersection:

- intersectionOf: Classes can be subclasses of multiple other classes. The intersectionOf statement is used to specify that a class lies directly in the intersection of two or more classes.

Restricted Cardinality:

- minCardinality: This limits the minimum number of values of a property that are attached to an instance. A minimum cardinality of 0 says that the property is optional, while a minimum cardinality of 1 states that there must be at least one value of that property attached to each instance.
- maxCardinality: Maximum cardinality restrictions limit the number of values for a property attached to an instance. A maximum cardinality of 0 means that there may be no values of a given property, while a maximum cardinality of 1 means that there is at most one. For example, an UnmarriedPerson should have a maximum cardinality of 0 on the hasSpouse property, while a MarriedPerson should have a maximum cardinality of 1.

Cardinality restrictions can be made for classes, and they specify how many values for a property can be attached to an instance of a particular class.

---

* http://www.w3.org/TR/2004/REC-owl-guide-20040210/wine.rdf

11.8 QUERYING SEMANTIC WEB DATA THROUGH SPARQL

For Web developers to leverage Semantic Web data in designing Web pages or applications, knowledge on how RDF-based data can be queried is necessary. The primary Semantic Web technology used for RDF-based data retrieval is a querying language known as SPARQL (W3C 2008). This section provides a brief overview of how typical SPARQL queries are structured—a more thorough reference on SPARQL can be found at http://www.w3.org/TR/rdf-sparql-query/.

Recalling our ontology for the Paris trip, consider the following SPARQL query for retrieving all objects in the photo:

```
PREFIX trip: <http://www.example.com/
parisTrip#>
SELECT ?object
FROM <http://www.example.com/parisTrip>
WHERE {     trip:ParisPhoto1
            trip:objectInPhoto
            ?object }
```

This query consists of four parts: a PREFIX statement, a SELECT clause, a FROM clause, and a WHERE clause. The PREXIX statement specifies a prefix for the namespace of the Paris Trip ontology. The SELECT clause indicates a set of variables to be returned in the final results. The FROM clause specifies that the triples in the Paris Trip ontology should be searched. Finally, the WHERE clause indicates the triple pattern to be used for generating results. Ultimately, the execution of this query will yield the following results:
?object: http://dbpedia.org/data/Eiffel_Tower

11.9 TOOLS FOR CREATING SEMANTIC WEB MARKUP

Although editing Semantic Web syntax by hand is an option, the use of editing tools is usually a more practical alternative. Editing tools, when properly used, can significantly reduce the time needed to complete tasks, as well as the possibility of syntax errors. The choice of what type of editing tool should be used will largely depend on the context of data creation. Currently, there are two main ways in which data on the Semantic Web is created: through (1) traditional "knowledge engineering" approaches and (2) "grassroots" collaborations online.

Typically, data creation through the knowledge engineering approach will be driven by a handful of experts in Semantic Web technology and closed off from the greater Web community. Such efforts are typical for projects with complex, semantically sophisticated terminologies. Here, OWL-based ontology editors are commonly used, which will be discussed in this section.

Grassroots collaborations, by contrast, are driven by the efforts of Web-based communities. Unlike knowledge engineers, these Web users will not necessarily have background in Semantic Web technology, or the time to focus on minute project details. Wiki-based services are commonly used in these settings, which will also be covered later in this section.

11.9.1 ONTOLOGY EDITORS

For knowledge engineering efforts, ontology editors are often used. An OWL-based Ontology Editor is a program designed to create and modify RDF, RDFS, and OWL-based data, as well as create instance data for ontology classes. Typically, ontology editors will be GUI-based applications, designed to present the information encoded in an ontology while hiding corresponding RDF-based syntax. Examples of currently existing ontology editors are SWOOP (http://code.google.com/p/swoop), Protégé (http://protege.stanford.edu/), and TopBraid Composer (http://www.topquadrant.com/products/TB_Composer.html). Web developers use these ontology editors to accomplish tasks that would be difficult to handle manually. An example of such a task is the generation of data statistics, such as the number of classes and property definitions, for a given ontology. Another example involves the visualization of relationships within an ontology, such as hierarchies of classes and properties (Figure 11.6).

11.9.2 SEMANTIC WIKIS

In contrast, for community-oriented efforts, semantic wikis can be used. Wikis are collections of interconnected documents, which Web users can both read and write to. As such, Web users can easily make contributions to documents on wikis, which would not be possible on the read-only Web. This has led to widespread adoption of wiki-based Web pages as a medium for community-based content generation.

To aid in creating and modifying documents, wikis provide editing syntaxes that help define how a document should be presented. Semantic wikis provide added editing syntax functionality for annotating documents with semantic markup, making them an application of interest for the creation of Semantic Web data.

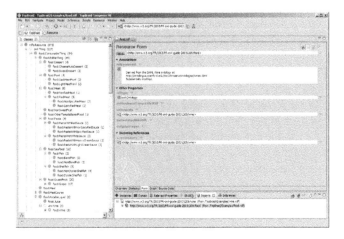

FIGURE 11.6   (See color insert.) TopBraid Composer visualization of the food ontology (http://www.w3.org/TR/2003/PR-owl-guide-20031209/food) class hierarchy.

One example of a semantic wiki platform is Semantic MediaWiki (Krötzsch, Vrandečić, and Völkel 2006), an extension of the MediaWiki platform used by Wikipedia. Given the established user base of Wikipedia, Semantic MediaWiki gains usability benefits by virtue of MediaWiki's familiarity (e.g., its editing syntax and interfaces). In addition, Semantic MediaWiki provides the capability to export documents containing semantic annotations as OWL ontologies, enhancing its relevance to the Semantic Web community.

On Semantic MediaWiki, semantic annotations are used as ways to interlink documents. For example, consider a document on a Semantic MediaWiki representing the city of Berlin. To express the fact that Berlin is in Germany, the Semantic MediaWiki annotation [[IsInCountry::Germany]] establishes a link to Germany via the relationship IsInCountry. When exported to RDF, a triple for representing this information would read something like:

Subject: http://www.example.com/Wiki/Berlin
Predicate: http://www.example.com/Wiki/IsInCountry
Object: http://www.example.com/Wiki/Germany

where the subject, predicate, and object represent URIs of content on the Semantic MediaWiki–driven Web site.

Inherently, the semantic annotations of Semantic MediaWiki offer less expressivity than ontology editors. However, given the comparative ease of use of Wikis for collaborative content generation, their usage makes more sense for community-based generation of Semantic Web data.

11.10 CASE STUDY: HTTP://DATA-GOV .TW.RPI.EDU

As the Semantic Web grows in a number of different ways—with an increasing technology base, more linked data becoming available, and more front-end development techniques explored—it becomes harder to generate specific recommendations for a single front-end methodology. Rather, current applications are tending to include a mix of Semantic Web and Web 2.0 approaches, combined with an increasing availability of new ways to view the mashups enabled by Semantic Web data. In this section we look at an example project that shows several different ways that Semantic technologies can be used to provide a new user experience.

The Web site http://Data-gov.tw.rpi.edu (Ding et al. 2010) aims to use Semantic Web technology to improve the usability and accessibility of information available on the government Web site http://Data.gov. To help achieve this goal, a Wiki-driven approach is used to aid in the publication and visualization of RDF-based data sets and linked data based on data provided by the United States (U.S.) government.

11.10.1 Background

In May 2009, the U.S. government launched http://Data.gov—a Web site for publishing data sets generated by its constituent agencies. The information on Data.gov is intended for public consumption, as a way of ensuring accessibility to information pertaining to the U.S. government. However, from the perspective of a Web designer seeking to reuse information from the Data.gov data sets, a number of human factors challenges present themselves. First, a multitude of data-encoding formats are used (such as CSV, TXT, XLS, and KML). Second, no common data set retrieval protocol is provided; while some data sets are available for direct download, others must be retrieved through form-based lookup. Third, large records of government data are often partitioned into multiple data sets, as a means of organizing data publication. However, only limited support (variable among data set collections) is provided on how to link these data sets to reform complete government records (or simply larger fragments). These three issues can easily complicate the task of working with multiple data sets. In addition, these issues can increase the amount of time required by a Web designer to learn how to work with a particular data set, reducing the odds that the data set will end up being used. As such, overcoming these limitations enhances the odds that published data sets can end up being used in applications developed by Web designers for the general public.

11.10.2 The Data-gov Wiki

The Data-gov.tw.rpi.edu (Data-gov) Wiki uses the Semantic MediaWiki system to address these limitations of Data.gov through the use of a set of best practices based on Semantic Web technologies (Figure 11.7). Data sets directly obtained from Data.gov are processed in three phases:

1. *Conversion Phase*: Data.gov data sets are converted from their original format to RDF; this phase ensures that a common data format is used across all data sets.
2. *Publication Phase*: RDF-based data sets, created in the conversion phase, are assigned URIs and made publicly available over the Web; this phase ensures that a common access protocol can be used across all data sets.

FIGURE 11.7 The Data-gov Wiki main page.

3. *Linking Phase*: Published data sets are linked to other data sources according to the linked data design principles discussed in Section 1.4. These links can interconnect the published data sets, as well as establish connections to outside data sources (e.g., DBPedia). The end result of this step is the synthesis of linked data from the original Data.gov data sets.

In the following sections, general overviews are given for each of these data processing phases. Following this, a discussion is given on how Web developers can use linked government data from the Data-gov Wiki to design Web applications.

11.10.3 Conversion Phase

As mentioned earlier, the purpose of the translation phase is to create versions of the Data.gov data sets for the Data-gov Wiki that share a common (RDF-based) format. During the creation of these RDF versions of the Data.gov data sets (known as converted data sets), steps must be taken to ensure their usability in Web development.

First, with each converted data set, Web developers should be able to find all of the information expressed in the original Data.gov version. The reason for this is that Data.gov remains the primary authority on its data sets. As such, Web developers will likely use documentation from Data.gov to determine what information *should be* contained within converted data sets.

Second, each converted data set should be structured around a set of ontologies for describing their classes and properties. Specifically, classes are based on a common data set definition ontology—defined on the Data-gov Wiki at http://Data-gov.tw.rpi.edu/2009/data-gov-twc.rdf. Likewise, properties, which are more variable across data sets, are defined in data set–specific ontologies.

Depending on the original data set format in question, different translation strategies end up being required for producing converted data sets. As the data sets on Data.gov are provided in many different formats, varying translation strategies are used, and a detailed discussion of their implementation is beyond the scope of this chapter. However, these details are made publicly available at http://Data-gov.tw.rpi.edu.

11.10.4 Publication Phase

Following their creation, converted data sets are made publicly accessible over the Web through the Data-gov Wiki. This is done in one of two ways, depending on the size of a data set:

1. For small data sets (under 1 MB in size), a single RDF document is used to store the entire data set and assigned a URI of the form: http://Data-gov .tw.rpi.edu/raw/X/data-X.rdf. Here, X denotes the data set number assigned to the original Data.gov data set.
2. For larger data sets, a collection of RDF documents (each under 1 MB in size) is used to store the data set in a partitioned form. Here, each RDF document is assigned a URI of the form http://Data-gov.tw.rpi .edu/raw/X/data-X-Y.rdf, where X denotes the Data .gov data set number and Y denotes a partition index number. To link these RDF documents together, a partition index file is defined (http://Data-gov.tw.rpi .edu/raw/X/link00001.rdf) with embedded links to each partition.

By partitioning large data sets across RDF files, RDF data on the Data-gov Wiki is guaranteed to be available in manageably sized documents. For Web designers, this ensures that large data sets on the Data-gov Wiki can be readily accessed in real time for consumption by applications.

11.10.5 Linking Phase

On the Data-gov Wiki, pages are written using semantic annotations and exportable as OWL ontologies—standard functionality through Semantic MediaWiki. As such, pages on the Data-gov Wiki can be used to generate additional RDF-based information on data sets following their creation during the conversion phase. These pages are designed to link back to the original RDF data sets, as well as to external data sources, resulting in the generation of linked data. Essentially, this means that anyone with write access to the Wiki page of a particular RDF data set may contribute

FIGURE 11.8 Wiki page corresponding to RDF version of Data.gov Dataset 10.

links to alternate data sources. It is important to keep in mind that the creation of links between data sources (e.g., from the Data-gov Wiki to DBPedia and other linked data sources) is a time-consuming process. Here, the "crowd sourcing" of link generation to a community of Web users offers a promising strategy for enabling the creation of linked data.

For each RDF data set on the Data-gov Wiki, a corresponding page is defined for managing links to alternate data sources. Each of these pages has two forms: a human-editable version and a corresponding RDF-based version. On each page, a link is provided back to a corresponding converted data set, which will take one of two forms:

1. A link to RDF document storing data set (for small data sets)
2. A link to an RDF-based partition index (for large data sets)

Additionally, these Wiki pages can contain references either to other data sets on the Data-gov Wiki or to external linked data sources (such as DBPedia). To illustrate how references to DBPedia can be made, consider the Wiki page http://Data-gov.tw.rpi.edu/Wiki/Dataset_10. This Wiki page corresponds to the RDF-based version of Data.gov Dataset 10, which provides data from the 2005 US Residential Energy Consumption Survey.

Among the listed properties on this page is one for defining geographic coverage (92/geographic coverage), with a corresponding value of http://Data-gov.tw.rpi.edu/Wiki/United_States. At this address is a Wiki page for describing the United States—much like the page used to describe Dataset 10. Here, a link is made to the DBPedia page for the United States through the statement [[owl:sameAs::http://dbpedia.org/resource/United_States]]* (Figure 11.8).

In the context of Web development, linked data versions of the data sets have significantly more utility than non-linked versions. The chief reason for this being that links provide embedded references to outside information, which a Web designer can leverage in the creation of new content.

11.10.6 Web Development through the Data-gov Wiki

For Web designers to leverage the data sets provided on the Data-gov Wiki, two conditions need to be met:

1. A mechanism must exist to enable the retrieval of information from the data sets.
2. In turn, this information must be formatted in a way that is usable by other Web development tools (such as Web APIs).

It should be noted that few Web APIs exist today that are designed to directly work with RDF-based data—formats

FIGURE 11.9 **(See color insert.)** Data-gov Wiki application: Worldwide earthquake viewer.

such as XML and JSON (http://www.json.org/) are more commonly used. To address both of the above points, the Data-gov Wiki hosts a SPARQL querying service for running queries against hosted RDF data sets (http://data-gov.tw .rpi.edu/ws/sparqlproxy.php), designed to return query results formatted in both JSON and XML.

While these formats are often used by Web APIs for reading in data over the Web, minor syntactic variations are often required across different APIs. Inherently, converting between different syntactic variations of JSON and XML will require time and effort on the part of Web developers. To address this issue, the Data-gov Wiki SPARQL endpoint is designed to return variations of JSON directly compatible with two popular data visualization Web APIs: the Google Visualization API (http://code.google.com/apis/visualization/) and Exhibit (http://www.simile-widgets.org/exhibit/). This provides Web developers a starting point in developing applications for leveraging the Data-gov Wiki RDF data sets.

One sample Web application based on the Google Visualization API is the Worldwide Earthquake Viewer,† designed to plot every detectable earthquake within the previous week based on intensity. This demo uses an RDF version of Data.gov Dataset 34, which provides a real-time, worldwide earthquake listing for the past 7 days.‡ This application starts by running the following query against the Data-gov Wiki SPARQL endpoint:

```
PREFIX dgp34: <http://Data-gov.tw.rpi.
edu/vocab/p/34/>
SELECT ?latitude ?longitude ?magnitude
?region
FROM <http://Data-gov.tw.rpi.edu/raw/34/
data-34.rdf>
WHERE {
        ?entry dgp34:magnitude ?magnitude.
```

---

* Configuration of Semantic MediaWiki is required to support the use of properties from specific external ontologies.

† http://Data-gov.tw.rpi.edu/wiki/Demo:_Worldwide_Earthquakes.

‡ On Data.gov, Dataset 34 is continually being updated. On the Data-gov wiki, a script is run once daily to generate a new RDF-based version of this data set.

FIGURE 11.10   **(See color insert.)** Data-gov Wiki application: Comparing U.S. (USAID) and U.K. (DFID) global foreign aid.

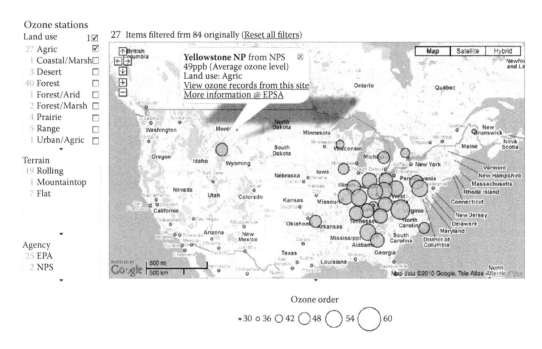

FIGURE 11.11   **(See color insert.)** Data-gov Wiki application: Castnet ozone map.

```
    ?entry dgp34:region ?region.
    ?entry dgp34:lat ?latitude.
    ?entry dgp34:lon ?longitude.
}
```

An English description of this query reads: for each data entry (earthquake record) in http://Data-gov.tw.rpi.edu/raw/34/data-34.rdf, fetch a corresponding latitude, longitude, magnitude, and region of occurrence. For this application, the call to the Data-gov Wiki SPARQL endpoint specifies that results be returned Google Viz formatted JSON, ensuring that the Google Visualization API can directly process them.

Passing this information directly to the Google Visualization API, earthquake data can be plotted on a world map (through use of latitude-longitude pairs). In addition, points corresponding to earthquakes can be assigned different sizes/

colors (based on magnitude value). Finally, each point can also be assigned a label corresponding to its region (based on the region value), accessible through a mouse click on the corresponding point (Figure 11.9).

The Worldwide Earthquake Viewer, while presenting both simple visualization and interaction mechanisms, demonstrates the data processing steps required to produce more sophisticated applications. Some examples, hosted on the Data-gov Wiki, include

- Comparing U.S. (USAID) and U.K. (DFID) Global Foreign Aid*: This presents a mashup of foreign aid data from the U.S. Agency for International

---

* http://Data-gov.tw.rpi.edu/wiki/Demo:_Comparing_US-USAID_and_ UK-DFID_Global_Foreign_Aid.

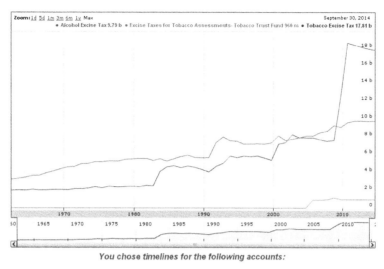

FIGURE 11.12   Data-gov Wiki application: Interactive government receipts timeline.

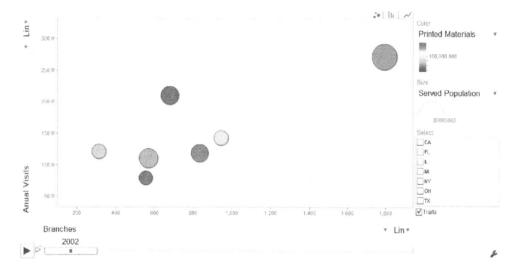

FIGURE 11.13   Data-gov Wiki application: Changes of public library statistics (2002–2008).

Development (USAID) and U.K. Department for International Development (DFID) for the 2007 U.S. Fiscal Year. Data corresponding to specific countries are retrieved through clicking on a provided world map (Figure 11.10).

• Castnet Ozone Map:* This displays Ozone readings taken within the United States on a map, using data derived from both Data.gov and alternate government sources. Faceted search is employed to allow users to display Ozone readings matching select criteria (e.g., only displaying readings from areas with agric land use) (Figure 11.11).

• Interactive Government Receipts Timeline:† This presents historical revenue figures for multiple government accounts (selected from a list by a user) on a common graph (Figure 11.12).

• Changes of Public Library Statistics (2002–2008):‡ This presents library statistics for public libraries gathered from several U.S. states and uses motion-based graphs to visualize statistical changes from 2002 through 2008 (Figure 11.13).

---

† http://Data-gov.tw.rpi.edu/wiki/Demo:_Interactive_Government_Receipts_ Timeline.

‡ http://Data-gov.tw.rpi.edu/wiki/Demo:_Changes_of_Public_Library_Statistics_ (2002_-_2008).

* http://Data-gov.tw.rpi.edu/wiki/Demo:_Castnet_Ozone_Map.

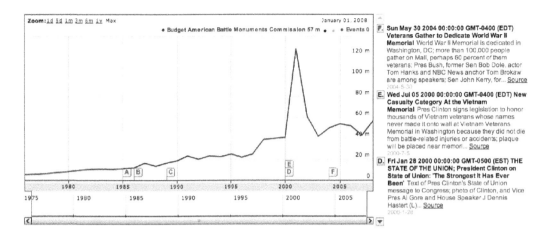

FIGURE 11.14 Data-gov Wiki application: Agency budget and *New York Times* news mashup.

• Agency Budget and *New York Times* News:* This application takes data on historical government agency budgets (derived from Data.gov) and data from the *New York Times* to produce budget graphs annotated with relevant news stories (Figure 11.14).

Ultimately, the interface elements of these Web applications are based on the functionality of Web APIs outside of the Data-gov Wiki. By providing representations of Data.gov data in easily processed formats (e.g., JSON), a wider selection of Web APIs (and hence, interface elements) can be called upon in the design of applications.

11.11 CONCLUSION

As the Web continues to incorporate more media, to become more dynamic in nature and to grow, it becomes harder and harder for humans to effectively find, sort, and manage all of the information that is available to them. Furthermore, it becomes increasingly difficult for Web designers to reuse information published on the Web for the design of new content (such as mashups). The Semantic Web offers promising solutions for the human factors challenges posed by these issues. Through the use of both ontologies and linked data, information on the Web can be put into a form that computers can "understand," process, and reason over.

This chapter presented a case study that illustrates how the Semantic Web can help overcome human factors challenges presented by conventional Web technology. To ensure that new human factors challenges are not created through using Semantic Web technologies, best practices for their usage are required (as shown in the case study). Ensuring the usability of Semantic Web technologies for Web designers will be necessary to ensure their continued adoption, as well as ensure the continued creation and reuse of data on the Semantic Web.

ACKNOWLEDGMENTS

The authors would like to thank both Amy and Ron Alford for their invaluable contributions as coauthors on an earlier version of this chapter appearing in the first edition of this book. Additionally, the authors would also like to thank Sarah Magidson for her work in developing the Worldwide Earthquake Viewer and Interactive Government Receipts Timeline applications mentioned in this chapter. Funding for this work was provided in part by gifts from Fujitsu Laboratories of America and Lockheed Martin Advanced Technology Laboratories.

REFERENCES

Auer, S., C. Bizer, G. Kobilarov, J. Lehmann, R. Cygniak, and Z. Ives. 2008. DBpedia: a nucleus for a Web of open data. Lecture Notes in Computer Science, 2007, Volume 4825, 722–735. Springer.

Becker, C., and C. Bizer. 2009. Exploring the geospatial Semantic Web with DBpedia mobile. *Journal of Web Semantics* 7(4): 278–286.

Berners-Lee, T. 2006. Linked data—design issues. http://www.w3.org/DesignIssues/LinkedData.html (accessed Nov. 9, 2009).

Bizer, C., R. Cygniak, and T. Gauß. 2007. The RDF book mashup: From Web APIs to a web of data. http://sunsite.informatik.rwth-aachen.de/Publications/CEUR-WS//Vol-248/paper4.pdf (accessed November 5, 2010).

Ding, L., D. DiFranzo, A. Graves Fuenzalida, J. R. Michaelis, X. Li, D. L. McGuinness, and J. Hendler. 2010. Data-gov Wiki: towards linked government data. Paper presented at AAAI Spring Symposium on Linked Data Meets Artificial Intelligence, Stanford, CA, March 22–24.

Kim, J., E. Deelman, Y. Gil, G. Mehta, and V. Ratnakar. 2008. Provenance trails in the Wings-Pegasus system. *Concurrency and Computation: Practice and Experience* 20(5): 587–597.

Krötzsch, M., D. Vrandečić, and M. Völkel. 2006. Semantic MediaWiki. Lecture Notes in Computer Science, 2006, Volume 4273, 935–942. Springer.

---

* http://Data-gov.tw.rpi.edu/wiki/Demo:_Agency_Budget_and_New_York_Times_News.

McGuinness, D. L., and F. van Harmelen. 2004. OWL Web ontology language overview. http://www.w3.org/TR/owl-features/ (accessed Nov. 9, 2009).

Moreau, L. 2009. The foundations for provenance on the Web. *Foundations and Trends in Web Science*. Submitted for publication.

Pinheiro da Silva, P., D. McGuinness, and R. Fikes. 2006. A proof markup language for Semantic Web services. *Information Systems* 31(4–5): 381–395.

W3C. 2003. Extensible Markup Language (XML). http://www.w3.org/XML/ (accessed Nov. 9, 2009).

W3C. 2004a. RDF Vocabulary Description Language 1.0: RDF Schema. http://www.w3.org/TR/rdf-schema/ (accessed Nov. 9, 2009).

W3C. 2004b. Resource Description Framework (RDF). http://www.w3.org/RDF/ (accessed Nov. 9, 2009).

W3C. 2008. SPARQL Query Language for RDF. http://www.w3.org/TR/rdf-sparql-query/ (accessed Nov. 9, 2009).

Zhao, J., C. A. Goble, R. Stevens, and D. Turi. 2008. Mining Taverna's semantic web of provenance. *Concurrency and Computation: Practice and Experience* 20(5): 463–472.

12 Organization of Information for Concept Sharing and Web Collaboration

Hiroyuki Kojima

CONTENTS

12.1 BACKGROUND AND AIMS

Technological innovations and a changing economic structure are forming a society in which knowledge dominates the core part of production capacity and requires workers to optimize their productivity (Nonaka and Takeuchi 1995). They are confronted with this requirement as it relates to work using knowledge. Therefore, in every business field, user support technologies are expected to aid and support the activities of workers. Support elements such as advisory, librarianship, reference, and training functions must sufficiently deal with the problems that workers face (Bramer and Winslow 1994). In the past, the functions of these elements were put together in the form of a manual. The electronic manual is an OJT (on the job training) oriented performance model that combines increased business performance and business instruction/training (Haga and Kojima 1993; Kojima 1992; Yamada and Kojima 1997). It should essentially satisfy the requirements of high understanding of a task procedure, supplying knowledge and know-how at the task execution site, and acting as a task instructor and a task manager.

Besides, workers must get timely and exact information for their task performance because this information may frequently change according to rapid progress in economic structures and network technology. So an information transfer method, in this case a dynamic electronic manual, must be able to cope with the above mentioned changes, instead of a typical business manual and execution manner that is systematized and arranged from a long historical standpoint.

Basic technology in which workers can change the situation, viewpoint, and view angle, as well as extract useful parts, change, and add comments to the information is expected. So it must contain an interface that properly presents information for business performance to each person and have an information transfer and acquisition mechanism that is oriented toward performance improvement.

Also, it is said that the first decade of this new millennium set out to provide society with a variety of wisdom that values knowledge, information acquisition, and self-improvement instead of emphasizing the growth rate of the economy, mass production, and mass spending. With the progress of information technology and a changing economic structure, knowledge has become one of the most important assets in an organization. On the one hand, for a business organization to remain competitive, there is a need to acquire knowledge (collective knowledge) according to the context of their business. This is particularly true in the production sector where the life cycle of a product is becoming shorter.

On the other hand, both information and knowledge in the business community are embodied because they depend on particular individuals or organizations that are identified as the senders or receivers in a collaborative form. A challenge is needed to transform this information into knowledge that can be reused and shared in some design sections or business communities.

This chapter describes two basic approaches, those of Information Organizing and Sharing Concept and Method for Knowledge Creation. Case studies of task support applications are also provided as illustrations. First, the information organizing and sharing method, which is composed of information watching and gathering, organizing, and viewpoint sharing, is proposed. This conception is adapted to a product design task as an implementation example (Kojima et al. 1999).

Second, in the product design field, designers frequently use images and drawings containing semantics and task models extensively. Therefore, it will be advantageous for performance of design tasks if the designer is able to describe the conceptual and semantic level of these pictures. This research has focused on the expression of task process and the structuring of related documents, as a result of task knowledge formalization, using information organizing and graphical user interfaces. This describes knowledge creation in the community through an information organizing system that is based on ontology to represent task models and the use of images as the key to information retrieval. XML (eXtensible Markup Language) is used as a means of building ontology (Kojima, Yamaguchi, and Woodworth 2004).

Third, in focusing attention on document retrieval, many document retrieval methods depend on matching a single key word, and the retrieval results consist of some documents that include merely the key word. These retrieval documents may be different sorts and are not the structuring or organizing information form based on the searcher's intention. Too often, both information and knowledge in the business community depend on particular individuals or organizations

that are identified as the senders or receivers in a collaborative form. Knowledge creation is dynamically generated in a context that includes a specified time, a particular location, a relationship with others, and a situation (Nonaka 2006). The challenge is to transform this information into knowledge that can be reused and shared in collaborative business communities. Research on this topic focuses on the expression of a task process model and the structuring of related documents, including both paper and electronic documents, as a result of task knowledge formalization. This research uses an information organizing method and radio frequency identification (RFID) tags. The present chapter describes knowledge creation in a community through an information organizing system that is based on ontology to represent task models and the use of RFID tags to link the physical location of retrieval information. XML is used as a tool to build the ontology (Kojima, Iwata, and Nishimura 2007).

12.2 INFORMATION ORGANIZING SYSTEM FOR PERFORMANCE SUPPORT OF PRODUCTS DESIGN TASK

It is effective for retrieval interaction to present an explicit description of a task procedure. In the product design field, design task knowledge (including design know-how and documents) is not always used to arrange or systematize, like manual contents because of the production cycle, etc. However, in joint product development with a collaborative design process formed in some project style, design task knowledge is needed to support the gathering and organizing of distributed design information along with a common product design concept, purpose, and work progress for related design workers. A designer needs to be able to quickly retrieve the design task information adapted from a large number design documents for design specifications, etc.

12.2.1 TASK PERFORMANCE SUPPORT USING AN INFORMATION ORGANIZING METHOD

Design information is usually stored in a database, and it is arranged according to each purpose of information creation. The retrieval of design information is divided into the following three cases as the result of a design task analysis:

1. Direct retrieval of a corresponding database with a known location of retrieval documents
2. Selective retrieval of some database with a rough idea of the location of retrieval documents
3. Overall retrieval of all related databases completely without an idea of the location of retrieval documents

Design workers are required in the three retrieval situations to rapidly retrieve some necessary document from many database files that are composed in different schemes. For this requirement, we propose the task support method using

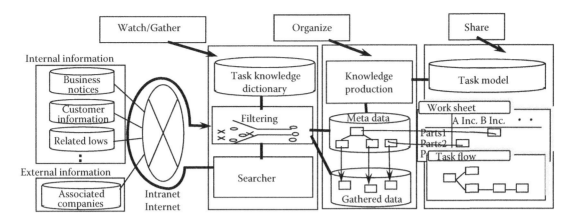

FIGURE 12.1 Basic concept of information organizing and sharing system.

metadata, which organizes relevant information adapted to the task.

Figure 12.1 shows the concept of a task performance support system that uses an information organizing and sharing method. In this system, the information searcher has selectively gathered material using relevant words defined by a task terminology dictionary. Metadata describes gathering data that are mutually connected using the procedures defined by the task model. Metadata refers to knowledge for the retrieval of design task information. Metaknowledge, which is composed of the above mentioned task model and related data gathering, is stored in a common database and results in viewpoint sharing of the internal and external company information.

12.2.2 INFORMATION ORGANIZATION AND RETRIEVAL USING A TASK MODEL

In task execution, which includes a creating process such as a design task, it is important to present workers with structured information along with a problem solution procedure for a task goal (i.e., task model). This is often needed to retrieve information, including documents for parts selection, design memorandums, mail logs, minutes of proceedings, and design drawings, as well as to relate this information. However, workers usually use normal retrieval tools only. Actually, workers must create a retrieval strategy that includes the selection of information sources and key words in order to obtain useful information for their

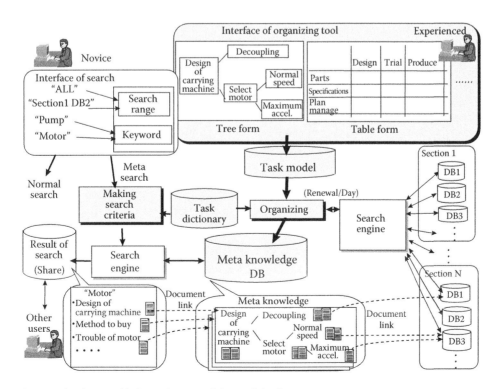

FIGURE 12.2 Implementation image of information organizing and sharing system.

work. Additionally, the scope of information sources can be so broad that workers cannot obtain retrieval results related to their work unless they repeat the retrieval process many times.

To solve the above mentioned problem, we use "task modeling" in organizing information for retrieval. The task modeling is based on the concept of "task ontology" (Mizoguchi 1994), and task models are defined and expressed in a form of tree structure or a form of table structure relating to some words of the task item. A task model is used for an input key of the retrieval process. A piece of metaknowledge is made by retrieving key words that are included in the task model and embedding links that direct the retrieval results into the task model. We developed a prototype system that enables design workers to share contents (such as know-how), create metaknowledge with a task model, and retrieve information related to task items. The information circulation and distribution image using this system is shown in Figure 12.2.

12.3 COOPERATIVE CREATION SUPPORT INTERFACE FOR COLLECTIVE KNOWLEDGE

This application research has focused on the expression of task process and the structuring of related documents, as a result of task knowledge formalization, using information organizing and graphical user interface. This chapter also describes knowledge creation in the community through an information organizing system that is based on ontology to represent task models and the use of images as the key to information retrieval. XML is used as a tool to build the ontology.

12.3.1 Knowledge Creation Process Support

12.3.1.1 Human–Computer Interaction for Knowledge Creation Process

Definitions and relations between data, information, and knowledge are discussed as follows (Tabata and Mitsumori 1999). On the one hand, both information and knowledge in the business task field are embodied because they depend on particular individuals or organizations that are identified as the senders and receivers, or as the owner. On the other hand, data are disembodied, because they are objective descriptions of facts. Informing is a proactive action to convey something to someone, and something received through informing is defined information. To know something means to understand it, and knowledge is defined as understanding about a subject that an individual or organization has. When someone understands data, information, or knowledge, it becomes his knowledge. Therefore, task knowledge may be tacit without any externalization for understanding. Knowledge creation process support is an important issue of human computer interaction (HCI) for collaborative works.

This research is focused on the expression of task process and the structuring of related documents, as a result of task knowledge formalization, using information organizing and graphical user interface. Information organization is assumed to be carried out timely and adaptively to a present problem, with a task model containing job processes and documents. The traditional manual succeeded in formalizing and transmitting workers knowledge and thus is "a good way to accomplish a task" as task knowledge creation.

12.3.1.2 SECI Model and Information Organizing

A distinction can be made between tacit knowledge and explicit knowledge (Polanyi 1966). Tacit knowledge is knowledge that is difficult to formalize and articulate. It often exists only in a person's mind or body. Explicit knowledge is knowledge that had been codified and formalized into procedures and rules. Transforming tacit knowledge to explicit knowledge is very important to an organization as it enables people to gain knowledge instead of just information regarding the task. Furthermore, it is not easy to share or learn tacit knowledge that is more instinctive rather than verbalizable as rules and procedures.

Personal tacit knowledge is shared and authorized in a group or organization, and the knowledge becomes larger and twice the original amount, through four transformation processes, according to the SECI (socialization, externalization, combination, and internalization) model (Nonaka and Takeuchi 1995). The SECI model represents four different modes of knowledge conversion in an organization. They are as follows, "tacit knowledge to tacit knowledge" called socialization, "tacit knowledge to explicit knowledge" called externalization, "explicit knowledge to explicit knowledge" called combination, and "explicit knowledge to tacit knowledge" called internalization. Though the SECI model is only conceptual, all progressive companies practice this model mechanism. Therefore, the computer system implementation of this model is very significant. This model is called the knowledge spiral because the interaction process between tacit and explicit knowledge will keep spiraling up as the new created knowledge continues to evolve. Thus, organization knowledge creation and sharing is a spiral process, starting at the individual level and moving up the organization. In addition to the people working in the organization, an information organizing tool should also drive this spiral. Codification of this knowledge puts the organization knowledge into a code that is accessible to all members of the organization.

The information organizing system architecture and activities are shown in Figure 12.3, corresponding with the SECI model. This research focused on the facility of information structuring support and visualization as knowledge spiral engine. For design workers to be able to share contents (such as know-how), create metaknowledge with a task model and retrieve information related to task items, we propose an information organizing system that will lead to the knowledge creation and sharing process.

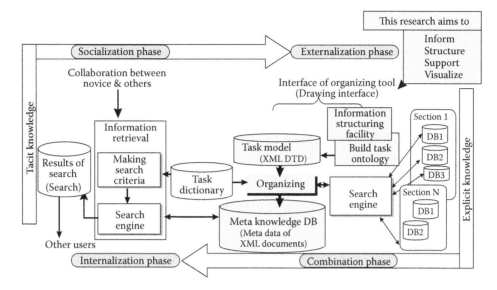

FIGURE 12.3 Information organizing systems architecture and knowledge creation process.

12.3.1.3 Task Ontology, Task Model, and XML for Knowledge Creation

Fikes (1996) has defined an ontology as a specification of concepts to be used for expressing knowledge (see also Michaelis, Golbeck, and Hendler, this volume). This would include entities, attributes, relations, and constraints. The most typical kind of ontology for the Web has a taxonomy and a set of inference rules. Task ontology consists of terminology, vocabulary, and ontology of a task. Task model consists of task procedure and task ontology. Our system task model will be based on the task ontology concept.

As ontology is a collaborative effort, workers with knowledge must be willing to share their knowledge in order for any knowledge management system to be useful. To build the ontology, workers with knowledge must decide the core structure of the ontology. Designers can formalize tacit knowledge into the form of a task model.

Task ontology is used as it provides a hierarchical representation of the task model, which a designer can understand easily. A designer has to be able to reflect her or his understanding of the conceptual and semantics of a domain through the ontology. Document Type Definition (DTD) of XML is used to represent the task model based on the task ontology. The concrete realization of the DTD is not important to the designer because he is only interested in the semantic level, irrespective of the document structure. It is important to provide a flexible interface where designers can describe the semantic level of a domain but strict enough to be able to implement the computational semantics.

The information gathered will be represented as a task ontology to be implemented in a task model. The analysis and decomposition of the task create knowledge for task performance. When this knowledge is changed through transformation processes consisting of socialization, externalization, combination, and internalization (Nonaka and Takeuchi 1995), the task ontology is changed and new knowledge is

created. This establishes an environment in which designers can externalize their own knowledge of their task and at the same time provide an environment in which new knowledge can be derived and shared.

A platform is needed to support the designers in transferring knowledge that they hold in their minds to a model where it can be shared across the organization. In the design field, task models are task procedures that have rules and procedures. Thus a designer can formalize her knowledge through the task model. The representation of design task model and the design knowledge description are discussed by citing an engineering design task example in the next section.

12.3.2 EXPERIMENTAL REPRESENTATION OF KNOWLEDGE IN ENGINEERING DESIGN COMMUNITY CASE

To build the task model based on task ontology, designers need to define the task, subtasks, constraints, and attributes. As an example, the strength design of pressure vessels such as a liquid oxygen (LOX) tank of a space shuttle is used. It is assumed that an expert designer understands the strength design procedure and analysis methods. This is exactly the designer's knowledge of a strength design task. The designer is also assumed to already know the structural parameter and loading condition. Figure 12.4 shows a list of the task procedures about the strength design of a pressure vessel.

The first step of creating the task ontology is to define the attributes and data type. This step creates the structure of the DTD documents and not the XML document itself. For example, the attribute *description* has the data type string. In DTD it is represented as

$$<!Element\ description(\#PCDATA)>$$

Once the attributes have been defined, the designer can annotate the class and its instances. Figure 12.5 shows the screen

1. Material data acquisition from material strength database
   - Material name: Titanium alloy
     * Chemical component, physical properties, mechanical properties, electrical properties and thermal properties

2. Overall strength design
   - Overall strength stress analysis
   - Stress analysis using FEM (getting stress values of all parts of tank)
   - Static failure evaluation

3. Local strength design
   - Local structure stress analysis
   - Strength evaluation using code rules
     * Stress categorization by ASME Code Sec. III
   - Local elastic-plastic stress analysis using FEM

4. Local fracture mechanics design
   - Fracture mechanics analysis

FIGURE 12.4   Example of strength design procedure.

image for creating a task. For example, in a space rocket, a *propellant tank* is a type of *tank*, and in this propellant tank, a LOX tank is used. Tank here would be the super class, while LOX tank is the subclass. Some of the LOX tank attributes are *material data acquisition* (*Mat_Data_Aqui*), *overall strength design* (*Ovr_Str_Dsg*), *local strength design* (*Local_Str_Dsg*), and *local fracture mechanics design* (*Local_ Frac_Mech_Dsg*), and they must occur only once. The DTD expression will be as follows using above notation:

<!Element LOX_tank_design (Mat_Data_Aqui, Ovr_Str_Dsg, Local_Str_Dsg, Local_ Frac_Mech_Dsg)>

The task model based on task ontology can be easily represented using an image of design drawing. It is assumed that a designer knows the design images very well and is able to divide the images semantically and conceptually. For example, the designer of the space shuttle engine will be able to tell which or where the propellant engine is and also to be

able to differentiate between a LOX tank and LH2 tank. In this system, a collection of images is used as a tool for organizing the task ontology. In our system, the designer will build a roadmap by using the image as a starting point and the relations between the tasks as roads, so that an information seeker will be guided to the information.

To define the task and its subtask, a designer can select an image and divide it into sections and associate the appropriate task to these sections. Figure 12.6 shows a screen image of using drawing image to represent the task model. The steps in associating information to the image of the task model are as follows:

1. Select or mark out a part in the drawing image that the information is going to be associated with using a mouse.
2. Select from the task dictionary information to be associated with the marked out part of the drawing image.
3. Input the value for the attributes.

For example, the designer can mark out the part of the propellant tank using a mouse and attached the information to it. If the propellant tank has more than one LOX tank, the designer could insert another *LOX_tank_design* as a subtask of *propellant_tank*. Therefore, images will have information represented in a hierarchical form.

Once the instances, attributes, and values have been added, DTD and XML documents will be generated and kept in a distributed database. A DTD and a XML document example describing the strength design of fuel tanks are shown in Figures 12.7 and 12.8, respectively. Although XML documents are generated, the distributed database can also contain files of other formats, such as .html, .doc, etc. This is because the attribute value might have links to one of these files. These files—XML, HTML, Microsoft Word documents—can be considered as a single large database where all these files are merged, which is different from the traditional Database Management System (DBMS).

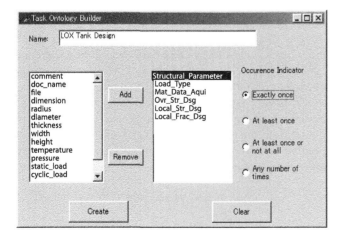

FIGURE 12.5   Screen image of the task model creation tool.

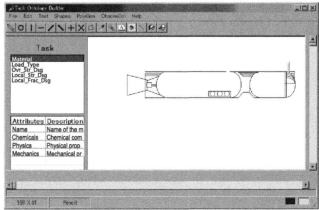

FIGURE 12.6   Screen image of drawing used to build task model with strength design procedure.

```
<!Entity % tank "tank | LOX_tank_design | LH2_tank | propellant_
tank">
<!Element LOX_tank_design (Mat_Data_Aqui,Ovr_Str_Dsg,
Local_Str_Dsg,Local_Frac_Mech_Dsg)>
<!Element description (#PCDATA)>
<!Element Mat_Data_Aqui (name, chemical*, Phy_prop, Mech_prop,
Elec_prop, Ther_prop)>
<!Element chemical (component,composition)>
<!Element Phy_prop (density, ......., ..........)>
...............................................
```

FIGURE 12.7   DTD document example of fuel tank design.

In a DTD document, entity is used to define the hierarchy relationship, for example, as follows:

<!Entity %tank "tank|LOX_tank_design| LH2_tank| propellant_tank">

The "%" defines a parameter entity. Parameter entities can be used to group together concepts that are related. In this example, a tank can be any normal tank, LOX tank, LH2 tank, or propellant tank. The structure entity is a concept of the structure of the rocket.

The system provides an intelligent retrieval tool, which is based on text, and a task model that will be further described in the next section. This is because with XML it is possible to know which data field types should be searched in order to match the query and return relevant results according to any structured design task process.

12.3.3   USING IMAGES AS A KEY TO RETRIEVE INFORMATION

Once the knowledge is formalized, it would not be of much use if other users have difficulty in locating the information they need. For an information seeker the most difficult part is the first step—where to start. With Internet search engines like Google, just enter the key word and hope that the information being searching for can be found in the many links returned by the search engine. As for the conventional relational database, the information seeker needs to know the structure of the database to be able to execute a SQL script. In this system, images are used as the key—the first step to start the search for information.

An assumption made is that designers or users are familiar with the design images and are able to differentiate the different parts of the image. In addition to the search engine, another way to retrieve information, for example, about the propellant tank, is by clicking on the propellant tank part of the image. As shown in Figure 12.9, information about the propellant tank will appear in a tree format. In Figure 12.9, an image of a space shuttle engine has the task model of designing the space shuttle associated with it. A user can retrieve information by traversing through the image, which is an alternative method to the text-based and concept retrieval tool.

When the user clicks on the LOX tank design, subclasses of *LOX_tank_design*, *Mat_data_aqui*, *Ovr_Str_Dsg*, *Local_Str_Dsg*, *Local_Frac_Mech_Dsg* will appear in a tree-like format according to the strength design method of a design knowledge. Clicking on the subelement of *Local_Str_Dsg* (local strength design), *Code_Str_Evl* (code rules strength evaluation) will open another window showing the stress category and values based on Pressure Vessel Code in diagram and table format.

A user can change the value or add another task to the LOX tank but cannot change the attributes of the task. Changing the attribute can only be done with tools used for task ontology building. For example, another designer can add the detail local nonlinear stress analysis data or fracture mechanics analysis data as a result of knowledge spiral.

```
<LOX_tank_design>
    <Mat_Data_Aqui>
        <name>Titamium Alloy/</name>
        <chemical>
            <component>Ti </component>
            <composition>98.9%</composition>
        </chemical>
        ...................................
        <chemical>
            <component> Ni </component>
            <composition>0.8%</composition>
        </chemical>
        ...................................
        <Mech_prop>
            <Elast_Mod>103Gpa</ Elast_Mod>
            <Shear_Mod>41Gpa</ Shear_Mod>
            <Fatigue_Str>...............................
        </Mech_prop>
        ...................................
    <Mat_Data_Aqui>
    <Ovr_Str_Dsg>
        ...................................
    <Local_Str_Dsg>
        <Code_Str_Evl>
        <description> stress_analysis.doc</description>
        ...........................
</LOX_tank_design>
```

FIGURE 12.8   XML document example of fuel tank design.

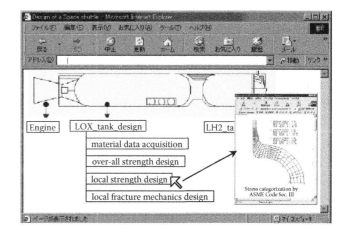

FIGURE 12.9   Using image as a key for information retrieval.

The advantages of this interface compared to a text-based page arranged alphabetically are as follows.

- Information is provided structurally as relations are shown between task items. This is especially useful to beginners who are not familiar with the task procedure, and it provides them a tool to understand the task model.
- Information retrieving is intuitive and faster because of self-evident effectiveness of visual interface like a tree structure expression of task procedure.
- Duplication of knowledge is prevented. The user only encounters new information and extensions to their knowledge. Usually, a knowledge worker may have to read all chapters of a document to acquire the knowledge that they need. Therefore, this tool represents knowledge in such a way that new knowledge can be integrated into the model by any knowledge worker, without any duplication.

12.4 DOCUMENT MANAGEMENT AND INFORMATION ORGANIZING

The research described in this section focuses on the expression of a task process model and the structuring of related documents, including both paper and electronic documents, as a result of task knowledge formalization. This research uses an information organizing method and RFID tags. This section describes knowledge creation in a community through an information organizing system that is based on ontology to represent task models and the use of RFID tags to link the physical location of retrieval information. Again, XML is used as a tool to build the ontology.

12.4.1 DOCUMENT MANAGEMENT ISSUES OF KNOWLEDGE SHARING

12.4.1.1 Need for Knowledge Sharing Facilities on Document Management

The essential problem of document management is the document retrieval function. In the case of document retrieval as aids for studying problems, a user stores tacit knowledge in his or her mind: for example, how to select key words, word orders, and combinations of words. This retrieval method is useful when shared with retrievers of the same problems in an organization and provides an effective assessment of the users' task performance.

Many information sharing systems and knowledge management systems are adopting retrieval systems with a search engine. However, all office workers in an organization may not have high-level retrieval skills. Information usage is dependent on their retrieval ability. The organization of information resulting from document retrieval should be shared as task knowledge in the business organization to improve the work performance. Therefore, information retrieval and a sharing interface are needed as facilities for adding semantic interpretation and task procedures to the retrieval results.

12.4.1.2 Problem-Solving for Knowledge Sharing of Document Management

From the standpoint of cooperative activities supported in both environments of inside office and mobile work, the issue is to be able to effectively provide information or data according to the task procedure and the task context of relevant documents. There are issues such as the details of the locations of needed paper documents and the most suitable document service for the task context, in accordance with the computer network environment or the physical environment of the workers.

A task model is defined as the framework for the task context, task procedure, and documents associated with the task. Therefore, the following system functions are the main issues concerned with task model data and the collective, organized content data for electronic documents and paper documents: the data acquisition, data description, data management, and data interpretation capability with the recognition function of the task process and task model.

There are two types of task practices in conjunction with cooperative work and knowledge sharing: (1) the task accomplishment following the work flow or task context and (2) the collection or arrangement of reference materials based on the task model, which contains the task purpose, mission, and strategy. To carry out type 2 following type 1, the task model is defined in accordance with the situation based on the factors of time, location, and progress state of the task. There are two kinds of processing data that contain the description data of the task model and document data source: paper documents with RFID tags and electronic documents. The issues are how the task model is tied to the document data source and how to arrange the data according to the task procedure and context. This applies to the intentional retrieval, the structural arrangement, and the information organizing method. This is the expert's knowledge of the task in the sense of the arrangement based on the know-how or method of work. The task model is expected to bring effective cooperative work support and knowledge sharing in the organization.

12.4.2 INFORMATION RESOURCE MANAGEMENT USING RFID AND INFORMATION ORGANIZING

The guidelines for solving the above mentioned issues are to study the usage of RFID tags, information organizing, and sharing on the premise that a cooperative workspace has paper documents and electronic documents. In this section, a task model is expressed as a typical table form that contains the task procedure and relative documents and uses an information retrieval interface.

XML has the ability to represent data. Because XML is a markup language, it allows the user to represent data in a tag-based tool scheme. A worker with knowledge has to be able to reflect his understanding of the concepts and semantics of a domain through an XML description that represents the task model. XML is used because it provides a hierarchical representation of the task model of the structuring of knowledge with organized documents and information for others' understanding. Another user can easily understand this representation. XML is also used for the registration of paper documents. The design specifications of the task model description and document retrieval using RFID tags for the management of paper documents are described as follows.

12.4.2.1 Description of Task Model and Document Retrieval

The information retrieval process attempts to sort out very complex, important issues as precisely as possible, by the appropriate selection of information resources and planning of the retrieval strategy. This process of sorting out issues is one good way of work and means the task model is a problem-solving method that forms task knowledge. Ontology is defined as the specification of concepts to be used for expressing knowledge (Fikes 1996). Our system task model is based on the task ontology concept. The most typical kind of ontology for the Web has taxonomy and a set of inference rules. Task ontology consists of the terminology, vocabulary and ontology of a task. The task model consists of the task procedure and task ontology.

In this chapter, a table form is typically considered as the expression of a task model of related document collection (Kojima et al. 1999). The table form is suitable for expressing a relation and presents an organizing structure to assist in understanding at a glance the relationship of items classified in the table. One example given in this chapter is mentioned about a survey problem of making progress of one's research. Figure 12.10 shows the task knowledge spiral cycle using a task model of the table form expression. Some experts could compose the task model with a table form using task items according to his task knowledge and register it in a database. The retrieval engine would collect documents using these items and organize them in the table. This information of organizing results with a table form of documents could be referenced, changed, and updated with the new ideas of other workers in the case of task progression or other expert participation.

For example, the task case in Figure 12.10 is a survey of literature documentation about the "method of installation and setting of computer server." The researcher selects "WWW server" and "mail server" for the server service software and "Windows Server 2003" and "Fedora Core Linux" for the related server OS. These selected key words are deployed in the table rows and columns. Document retrieval is carried out and produces an information organizing result with a combination of key words. The XML sample of this task model is also shown in Figure 12.10. The data can be represented as follows: <name>: the task name, <author>: the creator of the task model, <keyword>: a container of the selected key words, <row>: the arranged key words in a row of the table, <line>: the arranged key words in a column of the table, <task-model>: the container of the task model in the task model list <task-model-list>.

12.4.2.2 Electronic Management of Paper Documents Using RFID Tags

The trial sample RFID tag used in this experiment is a typical tag of the 13.56 MHz band with a relatively short read range. The memory on the microchip embedded in the tag usually can contain 48 bytes, and the memory capacity is 24 characters of kanji code or 48 alphanumerics of ASCII code. The RFID tag is expected ultimately to manage paper documents the same as electronic documents with a summary of minutes along the task context. Owing to the above mentioned memory restrictions, only short-length records can be stored in the RFID tag memory. Therefore, RFID tags carry a specified identification code number that represents the detail of the specified paper documents that include the content title name, location, author name, etc. These detail data with key identification codes are stored in a database designated for the retrieval of paper documents. A database update is executed corresponding to tag data in a read and write operation, in the case of a modification of a paper

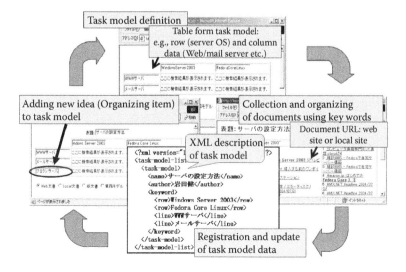

FIGURE 12.10 **(See color insert.)** Task model sharing using an example of a table form.

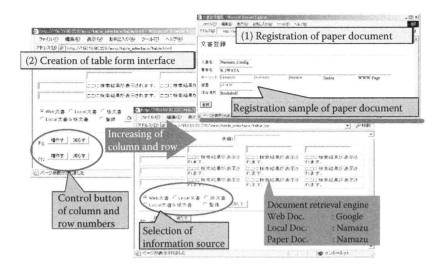

FIGURE 12.11 **(See color insert.)** (1) Registration of paper documents and (2) interface based on table form task model.

document. These detail data of the paper documents can be also described using XML to manage the paper documents together with the electronic documents, as follows:

```
<?xml version="1.0" encoding="Shift_JIS" ?>

<paper-list>
  <paper>
      <id>3</id>
      <title>Namazu_ConfigurationSetting_
            21-415_Bookshelf</title>
      <name>Namazu_ConfigSetting </name>
      <author>K_IWATA</author>
      <keyword>namzurc,mknmzrc</keyword>
      <room>21-415</room>
      <storage> Bookshelf </storage>
      <presence>Labo_WorkingDesk</presence>
  </paper>
</paper-list>.
```

12.4.3 PROTOTYPE SYSTEM FOR INFORMATION ORGANIZING AND DOCUMENT RETRIEVAL

The design specifications of the task model description and document retrieval using RFID tags for the management of paper documents were clarified in the above section, based on the problem-solution guidelines for the needs of task performance improvement and knowledge sharing of a task context. Seamless management between paper documents and electronic documents requires the following system facilities development.

12.4.3.1 Content Description of Paper Documents for the Registration of Paper Documents

A paper document registration function is needed to manage paper documents as well as electronic documents. Registration of paper documents handles the detail information of the document, according to abovementioned XML description form. A registration sample is shown in Figure 12.11. This example

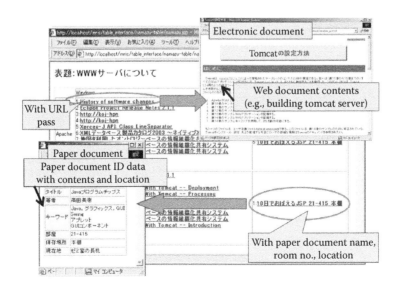

FIGURE 12.12 **(See color insert.)** Example of document retrieval result.

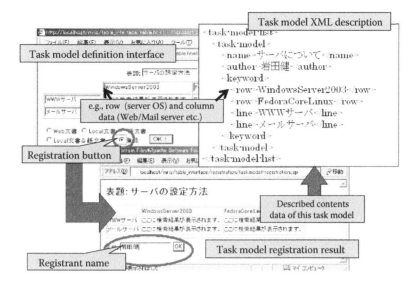

FIGURE 12.13 **(See color insert.)** Registration of task model.

shows the registration of the document's name, author, related key words, and storage location with an identification number, which is stored in the memory of the RFID tag attached to the paper document.

12.4.3.2 Retrieval Interface Based on the Table Form Task Model

In this chapter, the table form is typically adopted for representation of the task model. The table form task model is composed of row data and column data, which are viewed by the user with the task knowledge. The number of rows and columns is variable and depends on the task. Figure 12.11 shows a sample of the created table interface based on a task procedure. This table is also used as the information retrieval interface. A control button for the column and row numbers is used to add or subtract the number of items.

Information retrieval is carried out based on the combination of items in this table. Radio buttons are prepared for the selection of Web site documents, local site documents in an office, paper documents at a local site, or mixture of paper and electronic documents, in accordance with the classification of the information source. This retrieval processing is carried out in accordance with document characteristics, using an adequate search engine, such as Google for Web documents and Namazu (Baba 1998) for electronic documents, paper documents, and also the task model definition data, in a local site. The retrieval result sample for a local site and paper documents is shown in Figure 12.12.

12.4.3.3 Registration of Task Model and Referencing for Knowledge Sharing

It is necessary for task model sharing to store the task model definition data in a repository, as mentioned in the above section. A task model is described with XML. The registration interface screen sample is shown in Figure 12.13. Sharing of the task model created by some experts is effective for another person's task performance and is expected to promote

an upward knowledge spiral in an organization. Figure 12.14 shows a sample of task model retrieval and referencing flow.

12.4.3.4 Paper Document Management and Information Organizing and Sharing

RFID tag memory contains an ID number for tracking paper documents to manage them, and the detail data of paper documents is stored in a database with the ID number, document name, author, indexing key words of contents, and the physical location. RFID systems are arranged to provide automatic tag data acquisition through an RFID reading device, in the case that the location of the paper documents is displaced. In this experiment, the RFID reading/writing device is PTR631 (made by Toppan Printing Co., Ltd.). The RFID acquisition data are stored in a CSV (comma separated value) data format, with processing by a PC software tool. The RFID-enabled system provides an ID code of paper document acquisition through the CSV data format and a PC-processed database

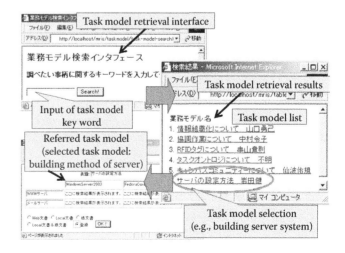

FIGURE 12.14 **(See color insert.)** Retrieval and referencing of task model.

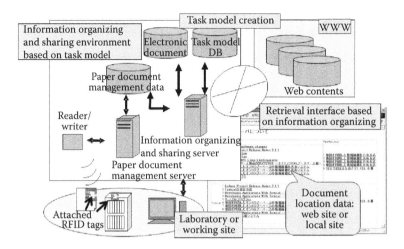

FIGURE 12.15 **(See color insert.)** Seamless document management systems overview.

update of paper document detail data. Figure 12.15 shows this seamless document management system overview.

An experiment of the system usability evaluation is carried out involving a sample office work. It is generally difficult to estimate the performance of intellectual tasks. It is assumed a simple task that is to search a reference paper in research materials and mean time to search is measured about two cases: case 1 is usual document retrieval in physical work space and case 2 is the system use retrieval based on the organized information of paper documents using RFID tags management. The experimental result shows that mean required time of case 1 is half again as much as case 2.

12.5 CONCLUSION

First, for distributed design task information, a method is proposed for design task performance support using the information organizing and sharing system. This system has the following functions and is expected to support the design task performance:

- Experienced designers are able to easily make meta-knowledge for task know-how transferring by using a task model definition interface tool.
- Beginners are able to obtain organized information in accordance with a task purpose with a few easy operations by retrieving the metaknowledge created and stored by an experienced user.

Second, on the basis of the proposed system architecture for information organizing and knowledge management, methods for cooperative activity supports are studied with the creation and sharing of information or knowledge and the system structure. Experimental description of knowledge, as generation of task model, is carried out for an application of XML, and the following results are obtained:

- The information organizing system consisting of four activities—information gathering, organizing and structuring, refinement, and retrieval provides

for knowledge creation and sharing. In this system, designers are able to formalize their knowledge of an image by associating tasks to the part of the image in the context that they understand. Task models based on task ontology are used to structure and organize the information.

- The system will generate a set of DTDs that is based on the task ontology that the designers have defined. These DTDs will provide the structure to the XML documents consists in a distributed database.

Third, from the standpoint of knowledge creation and sharing in a cooperative environment, an integrated method of information organizing is presented, in which a task procedure is structured and shared in table forms, and the document management system deals with paper documents with RFID tags as well as electronic documents. The following results were obtained.

- A method of organizing information as a means of knowledge representation uses a table form task model based on ontology, and the clarification of the correlation of collected information enables users to effectively understand the task knowledge to support users.
- The RFID tag system of paper documents allows the user to seamlessly collect every document, including paper documents in a computer network space, for information organizing.
- A retrieval prototype environment is implemented by using adequate search engines: Google for Web documents and Namazu for electronic documents, paper documents, and the task model in a local office site.

REFERENCES

Baba, H. 1998. *Development and Practical Use of Namazu System.* Japan: Soft Bank.
Bramer, W., and C. Winslow. 1994. *Future Work.* New York: Free Press.

Fikes, R. 1996. Ontologies: what are they, and where's the research? In *Proceedings of the 5th International Conference on Principles of Knowledge Representation and Reasoning*. Cambridge, MA, http://www-ksl.stanford.edu/KR96/FikesPositionStatement.html.

Haga, H., and H. Kojima. 1993. On the multimedia computer aided instruction system for novice programmers with exercise support facility. In *IFIP Transactions A-35, Computer Mediated Education of Information Technology Professionals and Advanced End-Users*, eds. B. Barta, J. Eccleston, and R. Hambusch, 155–62. New York: North-Holland.

Kojima, H. 1992. Self-learning electronic manual using multimedia-basic concept. In *Information Processing Society of Japan, 45th Conference Proceedings 5B-4*, 315–316. Tokushima, Japan.

Kojima, H., T. Yamada, Y. Mizuno, and T. Yuasa. 1999. Information organizing and sharing technique for task performance and training support. In *Educating Professionals for Network-Centric Organizations*, eds. P. Juliff, T. Kado, and B. Barta, 165–170. Boston, MA: Kluwer Academic.

Kojima, H., Y. Yamaguchi, and C. Woodworth. 2004. Cooperative creation support interface for collective knowledge. In *Proceedings of the 8th World Multi-Conference on Systemics, Cybernetics and Informatics (SCI2004)*. Vol. X, *Systemics and Information Systems, Technologies and Applications, International Institute of Informatics and Systemics (IIIS)*, 35–40. Orlando, FL.

Kojima, H., K. Iwata, and N. Nishimura. 2007. Document management and information organizing method using RFID tags. In *Human–Computer Interaction*, Part IV, *HCII 2007 Lecture Notes in Computer Science 4533*, ed. J. Jacko, 601–610. Berlin, Germany: Springer-Verlag.

Michaelis, J. R., J. Golbeck, and J. Hendler, this volume. Organization and structure of information using semantic Web technologies. In *Handbook of Human Factors in Web Design*, 2nd ed., eds. K.-P. L. Vu and R. W. Proctor, 231–248. Boca Raton, FL: CRC Press.

Mizoguchi, R. 1994. Current state of the art of knowledge sharing and reuse. *Journal of the Japanese Society for Artificial Intelligence* 9(1): 3–9.

Nonaka, I. 2006. The strategy of knowledge-based management. *Information Processing Society of Japan Magazine* 47(5): 547–552.

Nonaka, I., and H. Takeuchi. 1995. *The Knowledge Creating Company: How Japanese Companies Create the Dynamic Innovation*. New York: Oxford University Press.

Polanyi, M. 1966. *The Tacit Dimension*. London: Routledge.

Tabata, K., and S. Mitsumori. 1999. An email-based information system for online communities. *Information Knowledge Systems Management* 1: 249–265.

Yamada, T., and H. Kojima. 1997. Human interface design methodology for electronic manual system and its application. In *6th IEEE International Conference on Emerging Technologies and Factory Automation (ETFA'97)*, 165–170. Los Angeles, CA.

potential benefits but also serious challenges for how such systems and applications should be designed, managed, and deployed. In fact, until recently, existing technologies for the Web mostly ignore the organizational aspects of the application domain, providing designs of low abstraction level, based on (static) descriptions of tasks, or even, on the actual (remote) method invocations, losing track of the underlying aims and objectives that motivate the interaction among the different peers. That is, current Web technologies are not organization oriented but rather task or method centric. Even though some researchers treat workflows as "business logic," these are really static models that give no room for adaptation. Every single exception must be foreseen for the whole distributed system to operate without errors.

Moreover, existing approaches in some important areas (such as security, transactions, and federation) tend to cover only technology issues such as, for example, how to secure a protocol or connect federated directories, without considering the paradigm change that occurs when large numbers of services are deployed and managed over time. In particular, existing approaches do not offer satisfactory answers to the following questions:

- How to manage workflows in nontrivial open environments, where not all services are owned by the same organization? Because we cannot assume that all parties are either benevolent or that they will deliver results unless explicit obligations are defined and enforced, should workflows be agreed upon by all parties before they can be executed?
- How to align the configurations and settings needed by a service to operate with those of the operational environment?
- How is service execution affected by issues of trust, rights, obligations, and prohibitions?
- What if critical applications simply cease to function if services provisioned from third parties disappear or malfunction?
- How to deal with knowledge representation, when connecting or binding together two or more actual entities or services using different ontologies?

These issues point to the need for a *social layer* or *organizational layer* as part of the service interaction context. From an engineering perspective, new approaches are needed that take an holistic view of service environments and take into account not only the properties of individual applications but also the global objectives, structures, and dynamics of the system as a whole and of its context.

In this chapter, we introduce a layered model for the representation and analysis of service architectures that can be seen as a bridge between conceptual models designed for strategic organization concerns, and functional software descriptions of service architectures. This model thus enables the adaptation and monitoring of service-based systems at several levels from simple service substitution to task, plan, and goal reconsideration. In the remainder of this chapter,

we will describe how the aforementioned issues are tackled in the ALIVE* architecture, which extends traditional SOA architectures with a coordination and an organization layer.

13.2 BACKGROUND

The ability to seamlessly exchange information between companies, business units, customers, and partners is vital for the success of companies; yet most organizations employ a variety of disparate applications that store and exchange data in dissimilar ways and therefore cannot "talk" to one another productively. It is expected that soon most applications will require the integration of a large number of complex services. A fundamental principle of a service-oriented architecture (SOA) is the need for easy integration of sharable information, processes, and other resources through interactions among the shared components that are modeled as Web services. Composability of Web services is necessary to create (typically higher level) new services from a set of existing services. Choreography and orchestration are common ways of achieving such service composability. Work on choreography mostly focuses on specifying how the interacting Web services should behave globally. However, studies have shown that the relationships between global and local specifications of service interactions could be rather intricate.

In SOA working with abstractions is a necessity rather than a preference (Su et al. 2007). There are many reasons a service provider will not reveal the detailed information concerning the internals of a service. Given that services are proprietary, internal details cannot be assumed to be available, and therefore service design must rely on the abstract description of the needed services. Furthermore, as the SOA popularity grows, the number of available services also increases rapidly, requiring the automation of service design.

In order to deal with these issues, several authors have suggested a multilevel abstraction approach that separates the reasons for service interaction from the details of the interaction (Su and Rao 2004; Fluegge et al. 2006). At a higher level, specification focuses on how participating services should interact with each other while the lower level provides abstractions of services. Agent technology has begun to emerge as an integrated and holistic solution approach to distributed computing, communication, and data integration and is as such a likely candidate for Web services' composition architectures. In this section, we describe Web services compositional architectures in more detail and introduce the use of Organizational Theory constructs to support Web service composition.

13.2.1 Service-Oriented Architectures

Service-orientation presents an ideal vision of a world in which resources are cleanly partitioned and consistently

* The ALIVE research project (http://www.ist-alive.eu/) is funded by the European Commission within the 7th Framework Programme for RTD (contract FP7 215890).

represented. When applied to IT architectures service orientation represents a distinct approach for analysis, design, and development of solutions comprised of units of service-oriented processing logic. Service orientation introduces the following principles (Erl 2006):

- *Service reusability*: logic is divided into services with the intention of promoting reuse.
- *Service contract*: services adhere to a communications agreement, as defined collectively by one or more service description documents.
- *Service loose coupling*: services maintain a relationship that minimizes dependencies and only requires that they maintain an awareness of each other.
- *Service abstraction*: beyond what is described in the service contract, services hide logic from the outside world.
- *Service composability*: collections of services can be coordinated and assembled to form composite services.
- *Service autonomy*: services have control over the logic they encapsulate.
- *Service statelessness*: services minimize retaining information specific to an activity.
- *Service discoverability*: services are designed to be outwardly descriptive so that they can be found and accessed via available discovery mechanisms.

A *Service-oriented Architecture* (SOA) represents a collection of best practice principles and patterns in service-oriented design. The main drivers for SOA adoption are that it links services and promotes their reuse and decoupling. There are several initiatives that have developed the service-orientation approach. The OASIS Service-oriented Architecture Reference Model (SOA-RM) is a reference model for core service-oriented concepts developed to guide and foster the creation of more specific, service-oriented architectures. The W3C Web Services Architecture (W3C WSA) identifies the functional components of a Web service architecture and defines the relationships among those components to affect the desired properties of the overall architecture. The Open Gateway Service initiative, Grid, and JINI architectures are all service-oriented architectures, demonstrating how widely SOA is accepted as a paradigm.

One important limitation in most of current implementations of SOA comes from their initial focus on interoperability requirements and especially the aforementioned principle of services as stateless components offering very simple functionalities that composed may bring complex computation. Although this stateless approach eases interoperability, it makes it difficult (if not impossible) to have services that can dynamically detect and adapt their behavior to contextual changes or opportunities.

13.2.1.1 Web Services and Semantic Web Services

Web services represent one of the main current realizations of service-oriented architectures in practice. Web services offer an interoperable framework for stateless, message-based, and loosely coupled interactions between software components. These components can be spread across different companies and organizations, can be implemented on different platforms, and can reside in different computing infrastructures.

Web services are self-contained modular business applications that have open, Internet-oriented, standards-based interfaces. Web services are loosely coupled, communicating directly with other Web services via the Internet using standards-based technologies. This standards-based communication allows Web services to be accessed by customers, suppliers, and trading partners independent of hardware, operating system, or even programming environment. The result is intended to be an environment where businesses can expose their current and future business applications as Web services that can be easily discovered and consumed by external partners.

One of the key elements of the Web services technology is their so called composability. Web services specifications are being created in such a way that every specification is independent from the others; however, they can be combined (composed) to achieve more powerful and complex solutions. Reliability, security, transaction capabilities, and other features can be provided without adding unnecessary complexity to the specification. Moreover, the specifications are easily extended with new concepts, tools, and services by adding new layers and elements.

An emergent trend in Web services technology is Semantic Web Services (SWS) (McIlraith, Son, and Zeng 2001), services that are self-described and amenable to automated discovery, composition, and invocation. It is totally based and dependent on the evolution of the Semantic Web (Berners-Lee, Hendler, and Lassila 2001), which provides the infrastructure for the "semantic interoperability" of Web services.

The SWS approach is that Web services will be augmented with rich formal descriptions of their capabilities, such that they can be utilized by applications or other services without human assistance or highly constrained agreements on interfaces or protocols. Semantic Web Services have the potential to change the way knowledge and business services are consumed and provided on the Web.

According to (Cabral et al. 2004) SWS infrastructures can be characterized along three orthogonal dimensions or perspectives: usage activities, architecture, and service ontology. These dimensions relate to the requirements for semantic Web services at business, physical, and conceptual levels:

- From the *usage activities perspective*, Semantic Web Services are seen as objects within a business application execution scenario. The activities required for running an application using SWS include publishing, discovery, selection, composition, invocation, deployment, and ontology management.
- From the *architecture perspective*, Semantic Web Services are defined by a set of components which realize the activities above, with underlying security and trust mechanisms. The components

gathered from the discussion above include a register, a reasoner, a matchmaker, a decomposer, and an invoker.

- The *service ontology perspective* allows describing various aspects of the service and is an active research field. Semantic annotation of Web services description has been the issue of many initiatives, projects, and languages introduced, the most significant among them are the OWL-S Semantic Markup for Web Services [W3C 2004] and the Web Service Modeling Ontology (WSMO) (WSMO Working Group, n.d.] initiatives.

Apart from the limitations coming from the stateless service principle, another important limitation of both Web service and semantic Web service technologies is that they do not fully cover one of the identified requirements to support both the Web 2.0 and the Future Internet: *context awareness.* If services are to behave flexibly in dynamic, changing environments, they should be aware of their context in order to identify new opportunities, relevant changes, and adapt their internal behavior and/or the way they interact with others. In many cases correct, predictable behavior is arguably nearly impossible to guarantee without effective information about context.

13.2.1.2 Service Orchestration and Choreography

To bring context into a distributed service computation, current approaches are often based on the use of (static) business process models as a basic mechanism to support service composition. A business process specifies the potential execution order of operations from a collection of Web services, the data shared between these Web services, whose partners are involved and how they are involved in the business process, joint exception handling for collections of Web services, and other issues involving how multiple services and organizations participate. In other words, it defines the composition of Web services in order to provide higher functionalities.

There are competing initiatives for developing business process definition specifications that aim to define Web services composition. The terms "orchestration" and "choreography" have been widely used to describe business interaction protocols comprising collaborative Web services (Peltz 2003) that can provide more complex functionalities.

Orchestration defines the workflow between services from the "perspective of a single party," specifying the sequence and conditions in which one Web service invokes other Web services. Orchestration describes how services can interact at the message level, including the business logic and execution order of the interactions. These interactions may span applications and/or organizations and result in a long-lived, transactional process.

The standard approach to represent workflows in orchestration is the Business Process Execution Language (BPEL [OASIS, n.d.; IBM 2003]). Other specifications such as the Business Process Modeling Notation (BPMN [OMG Object

Management Group, n.d.]) or Business Process Management Language (BPML [BPMI, n.d.]) have been proposed for Business Process Management System (BPMS) technologies but have lost support in favor of BPEL. BPEL provides a language for the formal specification of business processes and business interaction protocols, extends the Web services interaction model, and enables it to support business transactions. BPEL is essentially a layer on top of the Web Services Description Language (WSDL), with WSDL defining the specific operations allowed and BPEL defining how the operations can be sequenced. A BPEL description can be interpreted and executed by an orchestration engine, which is controlled by one of the participating parties. The engine coordinates the various activities in the process and compensates the system when errors occur. But Singh (2003) claims that BPEL cannot accommodate the flexibility in dynamic business environments, as BPEL specifications only indicate the orderings of different tasks in a centralized and rigid way.

Choreography is more collaborative in nature than orchestration. It is described from the perspective of all parties (common view) and defines the complementary observable behavior between participants in a business process collaboration. A common view defines the shared state of the interactions between business entities and it can be used to determine specific deployment implementation for each individual entity. Choreography tracks the sequence of messages that may involve multiple parties and multiple sources where each party involved in the process describes the part they play in the interaction and no party owns the conversation.

The main approach to represent workflows in service choreography is the Web Service Choreography Description language (WS-CDL*). WS-CDL is an XML-based language that describes peer-to-peer collaborations of parties by defining, from a global viewpoint, their common and complementary observable behavior. WS-CDL's intended use is to model complex business protocols, such as order management, enabling interoperability between any types of application components, regardless of the supporting platform or programming model used. In that sense, WS-CDL complements orchestration languages that are concerned with executing the local view of the business process. WS-CDL specifies collaboration in terms of roles and work units. Work units consist of activities and ordering structures and especially important are interaction activities that result in exchange of information along a channel. A role enumerates the observable behavior a party exhibits in order to collaborate with other parties

In both orchestration and choreography, their instantiated process models (workflows) are the entities that keep the state of a given interaction and have the logic to decide what the next service call to be made is. In some sense it could be said that these workflows give some context of interac-

---

*Web Service Choreography Description Language (WS-CDL), http://www.w3.org/TR/ws-cdl-10/.

tion. However, the problem of current orchestration and choreography approaches to capture context is that

- They tend to be static, prone to failure (the failure of one service in the chain typically makes the workflow to fail), and very difficult to design and debug (the designer needs to foresee all possible execution paths and specify them in the workflow). This is thus clearly not the good approach to tackle the foreseen new generations of service technologies, which should be able to dynamically adapt and reconfigure in an ever-changing environment.
- They tend to model systems at a single level of granularity: services offered by individuals, corporations, multinationals, or departments within companies are modeled all with the same abstractions and with the same granularity. This is very limiting when modeling complex/rich multilevel interaction structures.
- They tend to have very little knowledge about the interaction context, lacking expressivity for many of the basic constructs arguably needed. For example, they lack explicit knowledge of holding norms and values in an organization and the boundaries of their application. This makes them very inflexible and hard to adapt to dynamic changes.

Therefore, there is a need for solutions that allow creating, maintaining, and managing evolving context-aware, longer-lived services. Our proposal is based on organizational models, which we will explain in the next sections.

13.2.2 Organization Theory

Recently, there is an increased interest in theories from human organizations as means to understand and model distributed complex domains. Organizations can be understood as complex entities where a multitude of entities interact, within a structured environment, aiming at some global purpose. In particular, organization-oriented approaches are well suited to the description of high-level abstraction for complex, distributed systems. In this chapter, we further investigate this approach to the design and specification of service architectures.

Models for organizational design also enable separation of organizational from individual concerns by providing an abstract representation of structures, environments, objectives, and participating entities that enables the analysis of their partial contributions to the performance of the organization. In this sense, the organization is a set of mechanisms of social order that regulate autonomous actors to achieve common goals (Dignum 2004). The performance of an organization in such framework is determined both by the structures of interactions between entities and by the individual characteristics of these entities. Organizational structures define the formal lines of communication, allocation of information processing tasks, distribution of decision-making authorities, and the provision of incentives.

That is, organizations describe roles, interactions, and rules in an environment. Social conventions are the ways people provide order and predictability in their environment, which have resulted in complex normative and organizational systems that regulate interaction. Roles* describe rights and obligations together with the capabilities and objectives of different parties. By knowing one's role, expectations can be defined and plans established. Organizational theory and sociology research have since long studied structure, dynamics, and behavior of human organizations and human interactions, based on the concept of role and relationships between roles. Research on Agent Organizations translates normative, social, and organization solutions coming from *human societies* into electronic distributed computational mechanisms for the design and analysis of distributed systems.

The definition of organizations as instruments of purpose implies that organizations have goals, or objectives, to be realized. Objectives of an organization are achieved through the action of the individuals in the organization, which means that an organization should employ the relevant actors, so that it can "enforce" the possibility of making its desires happen. Furthermore, one of the main reasons for creating organizations is efficiency, that is, to provide the means for coordination that enables the achievement of global goals in an efficient manner. This means that the actors in the organization need to coordinate their activities in order to efficiently achieve those objectives. Moreover, at any moment, unexpected change can also occur, which is not the result of the action of any of the actors in that world, leading to the need to evaluate the conditions of the environment and possibly decide to change the current strategy.

In its most simple expression, an organization consists of a set of *individuals* (which we will call *agents*) and a set of *objectives* to be achieved in a given *environment*. Organizational objectives are not necessarily shared by any of the individual participants and can only be achieved through combined action. According to contingency theory (Donaldson 2001), in order to achieve its goals it is thus necessary that an organization employs the relevant agents and structures their interactions and responsibilities to enable an efficient realization of the objectives. This leads to the need for at least the following three elements in an organization:

- An *organizational structure* consists of roles, their relationships, and predefined (abstract) interaction patterns. Organizational structure must reflect and implement the global objectives of the organization. Roles have objectives determined by the global aims of the organization and can be grouped into groups. Role objectives determine possible dependencies between different roles. Roles describe classes of agents, their activities, and possibly their norms and

---

* It is important to note than the concept of *roles* in organizational theories is far richer and expressive than the *roles* in Web services' choreography languages such as WS-CDL.

behavior rules. Roles are related to other roles by dependency relations. Desired interaction patterns between roles can be specified.

- An *agent* participates in the organization (system) by playing one or more roles. Role enactment is achieved either by allocation by the system developers that determine which available agent is the most adequate for a task or is decided by the agents themselves. In both cases, analysis techniques are needed to support enactment decision, by comparing and evaluating different role allocations. The set of agents that is active in an organization at a given moment is called the population. An agent population achieves the animation of organizational structures.

- The *environment* is the space in which organizations exist. This space is not completely controllable by the organization, and therefore results of activities cannot always be guaranteed. Two dimensions of environment are unpredictability and (task) complexity. The environment also includes the description of tasks and resources (such as size and frequency).

An organizational structure has essentially two objectives (Duncan 1979): First, it facilitates the flow of information within the organization in order to reduce the uncertainty of decision making. Second, the structure of the organization should integrate organizational behavior across the parts of the organization so that it is coordinated. The design of organizational structure determines the allocation of resources and people to specified tasks or purposes and the coordination of these resources to achieve organizational goals. Both in human enterprises as in multiagent systems the concept of structure is central to design and analysis of organizations, as discussed in this section.

13.2.2.1 Organizational Structures in Organization Theory

Organization Theory has for many decades investigated the issue of organizational structure. The key characteristic of an organizational structure is that it links the elements of the organization by providing the channels of communication through which information flows.

Notable is the work of Mintzberg (1993) on the classification of organizational structures according to which environmental variety is determined by both environmental

complexity and the pace of change. He identifies four types of organizational forms, which are associated with the four possible combinations of complexity and change. Each of the four forms of organization depends on fundamentally different coordination mechanisms. Table 13.1 summarizes Mintzberg's taxonomy of organizations.

Organizational design mostly adopts a multicontingency view, which says that an organization's design should be chosen based on the multidimensional characteristics of a particular environment or context. These characteristics include structural (e.g., goals, strategy, and structure) and human (e.g., work processes, people, coordination, and control) components. Contingency theory states that even if there is not one best way to organize, some designs perform better than others. In fact, Roberts states that the central problem of organization design is finding a strategy and an organization structure that are consistent with each other. This requires finding a dynamic balance in the face of strategic and organizational changes (Roberts 2004). Organization design is therefore about understanding the characteristics of the environment and the demands of this environment in terms of information and coordination. Given that there are no exact recipes to construct the optimal organization, means to evaluate certain design and determine its appropriateness given the organization's aims and constraints, is a main issue in Organization Theory.

An area of computer science where Organization Theory has been well exploited as a means to describe and structure software design is Multiagent Systems (MAS). Concepts from Organization Theory are well suitable to understand and structure MAS in ways that extend and complement traditional agent-centric approaches. Agent organizations can be understood as complex entities where a multitude of agents interact, within a structured environment aiming at some global purpose. In the next subsection, we will briefly introduce the application of Organization Theory to MAS.

13.2.2.2 Organization Structures in Multiagent Systems

Multiagent systems (MAS) are an important and innovative research area in artificial intelligence and computer science. They are viewed as a computational paradigm to specify, design, and implement software systems for solving complex tasks in terms of interacting computational entities. These computational entities, called agents, are usually assumed to be heterogeneous, autonomous, self-interested, and situated in an environment. They can communicate and coordinate

TABLE 13.1

Organizational Forms According to Mintzberg

Organizational Form	Machine Bureaucracy	Professional Organization	Entrepreneurial Startup	Adhocracy
Complexity	Low (simple)	High (complex)	Low (simple)	High (complex)
Change pace	Low (stable)	Low (stable)	High (dynamic)	High (dynamic)
Coordination mechanism	Standardized procedures and outputs	Standardized skills and norms	Direct supervision and control	Mutual adjustment of ad hoc teams

their activities to achieve their objectives. The environments in which individual agents are situated can be ordinary software environments (e.g., Internet), physical environments (e.g., robot), or normative mechanisms (e.g., virtual organization).

The multiagent system paradigm can be used to develop open and distributed applications such as multirobot systems, online auctions, virtual organizations, electronic institutions, market places, and virtual games. Some multiagent issues include how agents communicate and interact with each other, how they coordinate their activities, how their activities can be organized, and what characteristics of multiagent applications are. In Multiagent Systems, an organization has been defined as "what persists when components or individuals enter or leave an organization, i.e., the relationships that makes an aggregate of elements a whole." Establishing an organizational structure that specifies how agents in a system should work together helps the achievement of effective coordination in MAS (Barber and Martin 2001). A social structure may be explicitly implemented in the form of a social artifact existing independently of the implementations of the agents, may be realized as part of the implementations of the agents, or may exist only intangibly in the form of the policies or organizational rules followed by the agents during interaction.

Organization models have been advocated to deal with agent coordination and collaboration in open environments (Hübner, Sichman, and Boissier 2006). In this context the organization is the set of behavioral constraints adopted by, or enforced on, a group of agents to control individual autonomy and achieve global goals. Agent organizations rely on the notion of openness and heterogeneity of MAS and pose new demands on MAS models. These demands include the integration of organizational and individual perspectives and the dynamic adaptation of models to organizational and environmental changes. Organizational self-design plays a critical role in the development of larger and more complex MAS. As systems grow to include hundreds or thousands of agents, we must move from an agent-centric view of coordination and control to an organization-centric one (Dignum, Dignum, and Meyer 2004). Two main points characterize agent organizations: (1) agent organizations distinguish between the agents' goals and global goals, which may or may not be shared by the agents in the organization, and (2) agent organizations exist independently of agents and do not assume any specific internal characteristics of the agents.

The design of agent organizations starts from the social dimension of the system, and the system described in terms of organizational concepts such as *roles* (or function, or position), *groups* (or communities), *tasks* (or activities) and *interaction protocols* (or dialogue structure), thus on what relates the structure of an organization to the externally observable behavior of its agents. Organization knowledge and task-environment information is used to develop an explicit organizational structure that is then elaborated by the individual agents into appropriate behaviors. The society structure is determined by organizational design, which is independent of the agents themselves. Such structures implement the idea that agent interactions occur not just by accident but aim at achieving some desired global goals. That is, there are goals external to each individual agent that must be reached by the interaction of those agents. Desired behavior of a society is therefore external to the agents. Organizational hierarchies, teams, shared blackboards, global plans, and auction systems are all social structures. In closed domains, designed structures completely determine the communication primitives available and possibly describe the resources available in the environment. Furthermore, it may organize agents into groups or teams and specify joint action. A special case of designed systems are agent societies that include social concepts such as norms and ontologies. Besides the specification of action-oriented behavior as in designed MAS, agent societies enable the specification of normative behavior and as such provide agents with the possibility to reason about their own behavior, plans, and goals. These agent societies are norm-oriented structures, appropriate to model open domains, also due to the explicit specification of ontological aspects.

Studies on the range of relevant organizational characteristics for MAS (Carley and Gasser 1999; Horling and Lesser 2005) demonstrate the contingency of organization approaches: no single approach is necessarily better than all others in all situations. Design choices should be dictated by the needs imposed by the system's goals, the resources at hand, and the environment in which the participants will exist. In general, hierarchical, team-centric, coalition-based organizations and markets have proved to be most popular among multiagent researchers (Horling and Lesser 2005). *Hierarchies* are effective at addressing issues of scale, particularly if the domain can be easily decomposed along some dimension. *Teamwork* can be critical when working on large-grained tasks that require the coordinated capabilities of more than one agent. *Coalitions* allow agents to take advantage of economies of scale, without necessarily ceding authority to other agents. *Markets* take advantage of competition and risk to decide allocation problems in a fair, utility-centric manner.

13.3 ORGANIZATIONS FOR SERVICES

The increasing complexity of software applications requires modular, structured design approaches. As discussed in Section 13.2.1.2, conventional choreography descriptions are not able to capture the business goals of each participant and organizational dependencies motivating the interaction. These languages focus almost entirely on operational aspects such as data formats and control flow (Peltz 2003). Absence of this critical business knowledge makes it hard to reason if a particular customization satisfies the goals of participants. Mutual obligations of the participants are only specified in terms of constraints on the sequences of messages they can exchange.

This deficiency becomes critical when the choreography has to be customized to cater for emergent business needs. Approaches are needed that reflect the needs of the different partners and describe requirements and functionalities at different levels of abstraction in a way that makes design manageable and flexible to changes. Furthermore, requirements are usually left implicit in the design of applications. In dynamic environments, this results in brittle solutions that have difficulties coping with change. Even if internal regulations can be up to some extent statically modeled with business logics, external regulations can hardly be properly modeled in this way, as a change of context would need to redesign the static business model.

The above considerations indicate a need for modeling approaches that separate implementation and strategic concerns. Furthermore, experiences from human organizations benefit from autonomous decisions by their participants. Not only the formal organizational structure but also the informal circuit of communication and interaction determines the success of an organization. In human organizations a participant's contribution is evaluated based on the organizational requirements as well as the extra achievements. Someone who takes initiative and builds up a personal network is often more appreciated than someone who sticks to the official rules and does not do anything extra. The capability to act and make decisions in unexpected situations is usually perceived as a positive characteristic of human actors. This type of behavior is required from participants in service organizations. In this sense, services (or combinations of services) should be encapsulated by intelligent entities that are able to decide autonomously on a different course of action that will lead to the achievement of the high-level goals. This leads to the following requirements for service architecture approaches:

- Enable the explicit representation of internal and external requirements,
- Enable a modular approach to requirement specification, in which requirements are dealt with at different levels of abstraction, and
- Allow for autonomous interpretation of specification at run time, taking into account environment and actor possibilities.

Moreover, many applications benefit from the use of components that are individually autonomous (in their decisions on how to achieve a given goal) but corporately structured. High-level organizational rules and norms describe the desired global behavior of the organization such that choices are guided toward global goals but do not completely fix actor activities.

There is a growing recognition that a combination of organization structure and individual autonomy is often necessary. Selznick (1953, 251) states that "all formal organizations [follow] ordered structures and stated goals . . . and will develop an informal structure within the organization which will reflect the spontaneous efforts of individuals and subgroups to control . . . the [organizational] environment . . . Informal structures are indispensable to [. . .] the process of organizational control [. . .] and stability." Organization

systems must therefore be able to yield coordinated behavior from individually autonomous actions (Van Dyke Parunak and Brueckner 2001). This extends current approaches; that of models that limit the action of its agents to the strict realization of predetermined protocols are not agile enough and furthermore do not take full advantage of the potential of participating agents. Already in 1993, Wellman (1993, 20) noted that "combining individual rationality with laws of social interaction provides perhaps the most natural approach to generalizing the Knowledge Level analysis idea to distributed computations."

In the following section, we present the ALIVE framework that has been developed to implement the issues discussed above.

13.4 ALIVE MODEL

The ALIVE architecture combines Model Driven Development (MDD) (OMG, n.d.) with coordination and organizational mechanisms, providing support for live (that is, highly dynamic) and open systems of services. ALIVE's approach extends current trends in engineering by defining three levels in the design and management of distributed systems: the service level, the coordination level and the organization level, illustrated in Figure 13.1, and explained below.

- The *service level* supports the semantic description of services and the selection of the most appropriate service for a given task (based on the semantic information contained in the service description), effectively supporting higher-level, dynamic service composition. This comes in handy when highly dynamic services are present in the system, as the semantic description eases the process of finding equivalent services when a given services is not available, or when more suitable services have been registered recently. This layer is also responsible of low-level monitoring of service activity.
- The *coordination level* provides the means to specify, at a high level, the patterns of interaction between services, transforming the organizational representation (including information flows, constraints, tasks and agents) coming from *organizational level* into coordination plans. These plans are automatically synthesized to achieve organizational goals, taking into account possible human interventions. Means to enact these plans in a distributed fashion, making use of *service level*, are provided in this layer, as well as exception handling mechanisms to verify and analyze via logs the plans already executed.
- The *organizational level* provides context for the other levels, supporting an explicit representation of the organizational structure of the system. Mainly stakeholders and their interrelations are to be represented; this will allow the derivation of formal goals, requirements, and restrictions. Tools and

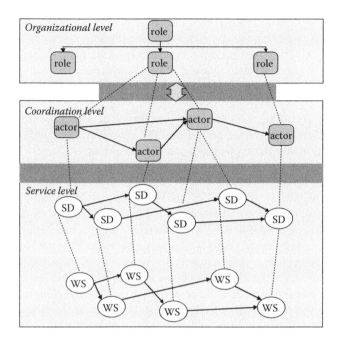

FIGURE 13.1 The overall structure of the ALIVE architecture and framework.

mechanisms for verification and analysis of provided organizational specifications are also provided. Changes in the organizational structure (reorganizational concepts or normative changes) are to be expressed too, and these changes can be proposed by either the *coordination* or the *service level*, so connections with both layers are present. Finally, this layer must support methods of norm-oriented organizational design, supporting scenarios where traditional approaches do not fit well.

This multilevel approach in the ALIVE framework makes it especially fit for scenarios where changes are likely to occur at either abstract or concrete levels, and where services are expected to be highly dynamic, with new services entering the system and existing services leaving it at run time. For example, when there is a significant change on a high level (e.g., a change in the organizational structure), the service orchestration at lower levels is automatically reorganized, effectively combining the existing services in new ways to reflect the changes at the higher levels. Another example is the (automatic) adaptation of higher levels when lower ones suffer significant changes (e.g., the continuous failure in some low-level services). Furthermore, the ALIVE framework allows lower levels to adapt in more efficient ways, keeping always clear track of the system's overall goals and objectives.

In the following sections, we introduce the different layers of the ALIVE's architecture using a simplified scenario taken from the ALIVE use case on crises management simulation.

13.4.1 RUNNING EXAMPLE

There are established procedures in the Netherlands to handle fire emergencies. These procedures address all steps in a fire emergency, all the way from when the fire is reported by a member of the public, to the aftermath of the fire being extinguished, when the area in which the fire took place is cleared. The diagram in Figure 13.2 depicts the various stakeholders in this situation, and their responsibilities.

When an emergency call is made to the *Emergency_Call_Centre*, it alerts the *Fire_Station* nearest the location of the fire, informing the *Fire_Station* about the location in which the potential danger has *been* reported. The *Fire_Station*, by means of its chief, is in charge of assembling an appropriate response, that is, a fire brigade; the *Fire_Station* chief depends on *Firefighter_Team* in order to locate, assess, and, ultimately, extinguish the fire. When the team arrives at the reported location, it has to decide (based on their expertise and previous experience) the severity of the problem. Only after this evaluation is made, an intervention decision is taken.

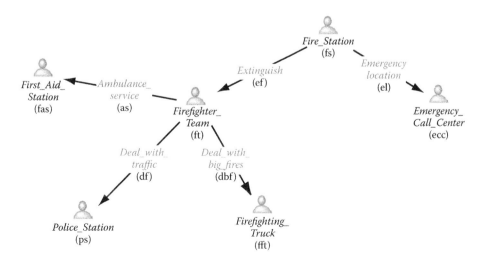

FIGURE 13.2 Stakeholders of the crises management scenario and their relationships.

According to current Dutch regulations, the evaluation should comply with standardized emergency levels, that is, the four GRIP emergency levels established by the Dutch Ministry of Internal Affairs. For the sake of simplicity, we assume that the *Firefighter_Team* sets up a strategic intervention (to achieve its ultimate objective which is to extinguish the fire) based on two evaluation criteria: damage evaluation and fire evaluation. From the former criterion, the *Firefighter_Team* checks how many wounded there are in order to come up with information about the necessity or not to ask for the ambulance service. Additionally, the *Firefighter_Team* checks if the damage involves building structures that may collapse, causing obstacles and risks for drivers on the roads; that is, they decide if police intervention is required to divert traffic to safer routes. On the basis of the fire evaluation criterion, the *Firefighter_Team* is able to decide whether or not to ask the *Fire_Station* for a *Firefighting_Truck* intervention.

13.4.2 ORGANIZATIONAL LEVEL

The *organizational level* aims to provide specifications of organizational patterns and rules in order to allow a structural adaptation of distributed systems over time. The main idea is that many of the strategies used today to organize the vastly complex interdependencies found in human social, economic behavior will be essential to structuring future service-based software systems. In order to apply existing techniques and methods in (human) societies, the approach taken in this framework is that a collection of services interacting with each other for some purpose and/or inhabiting a specific locality can be regarded as a *society*. Societies usually specify mechanisms of social order in terms of common norms and rules that members are expected to adhere to (Davidsson 2000).

In ALIVE, an *organization* is defined as a specific solution created by more or less autonomous *stakeholders* to achieve common *objectives* and *sub-objectives*. It provides the means to manage complex dynamics in societies, defining and controlling the social interactions among the stakeholders. An organization is viewed as a set of entities (the stakeholders) and their interactions, which are regulated by mechanisms of social order. The *organizational level* of ALIVE defines an *organizational model* that enables to represent organizations as a social system, and describes what the aims and the concerns of the organization are with respect to the social system. The Organizational Model is specified in terms of four structures:

- The *social structure* specifies objectives of the society, its roles and what kind of model governs coordination.
- The *interaction structure* describes interaction moments, as scene scripts, representing a society task that requires the coordinated action of several roles, and gives a partial ordering of scene scripts, which specify the intended interactions between roles.

- The *normative structure* expresses organizational norms and regulations related to roles.
- The *communicative structure* specifies the ontologies for description of domain concepts and communication illocutions.

By means of these four structures, the Organizational Model combines aspects from role theory with ontologies, normative description, and process specification.

13.4.2.1 Social Structure

The social structure of an organization describes the *objectives* of the society, its *roles*, and what kind of model governs coordination. The central concept in the social structure description is the *role*. In ALIVE, *roles* are abstractions, providing a means of generically addressing stereotypical (expected) behaviors (i.e., whoever takes up a role is expected to behave in a particular way) (Dignum 2004). Roles identify activities necessary to achieve organizational objectives and enable to abstract from the specific *actors* that will eventually perform them. That is, roles specify the expectations of the organization with respect to the actor's activity in the organization. Roles are described in terms of *objectives* (what an actor of the role is expected to achieve) and *norms* (how is an actor expected to behave). Furthermore, role descriptions also specify the *rights* associated with the role and the *role type*, that is, whether it is an institutional or external role. Actors of *institutional roles* are fixed and controlled by the society and are designed to enforce the social behavior of other agents in the society and to assure the global activity of the society. External roles can in principle be enacted by any actor, according to the access rules specified by the society (Dastani et al. 2003). In the emergency scenario, some of the roles that can be identified from the stakeholders are indicated in Figure 13.2, including *Fire_Station* (with the objective to coordinate fire extinction at all times), *Firefighter_Team* (with the objective to handle a given fire extinction mission), and *Police_Station* (with objectives such as securing the area or controlling the traffic), amongst others.

Roles can be organized into *groups*. In its most basic form, groups are just a way to collectively refer to a set of roles. This can be useful when, e.g., in an *interaction scene*, a participant is not an actor of a specific role but can be acting one of several roles. However, the most relevant feature of a group is that it can specify norms that must hold for enactors of roles in the group. For any society, the trivial group of roles is the group that contains *all roles* in the society.

Organization *objectives* form the background for the Organizational Model and are the basis for the definition of the *role objectives*, in the sense that the society objectives will be realized by the realization of role objectives. A *role objective* γ can be further described by specifying a set of *role sub-objectives* that must hold in order to achieve objective γ. Role sub-objectives give an indication of how an objective should be achieved, that is, describe the states that must be part of any *plan* (at *Coordination Level*) that an

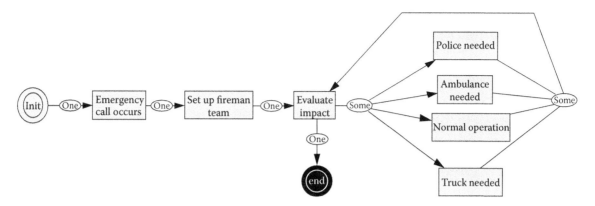

FIGURE 13.3 Interaction structure for the crises management scenario.

actor enacting the role will specify to achieve that objective.* Every *(sub-)objective* of the organization needs to be related in a *role (sub-)objective* (i.e., there should always be a role in the organization that wants to achieve (or is responsible for achieving) part of the organization's objective). The exact way organizational objectives are "split" into role objectives and what kind of ways are possible to realize those objectives depends on the requirements and characteristics of the domain.

For instance, in the crises management example, an overall objective of the organization (to handle all fire incidents—*handle_emergency_call*) becomes a role objective for the *Fire_Station_Role*. This objective is split into two subobjectives: to get the location for the fire emergency (*emergency_location*) and to extinguish the fire (*extinguish_fire*). Furthermore, the *extinguish_fire* objective is decomposed in two sub-objectives: *evaluate_severity* and *how_to_react*.

In ALIVE the notion of *role* is closely related to those of cooperation and coordination. Societies establish dependencies and power relations between roles, indicating relationships between roles. These relationships describe how actors can interact and contribute to the realization of the objectives of each other. The relations between roles are expressed in their *role dependencies*. The *role dependency* between two roles means that one role is dependent on another role for the realization of its objectives. How role objectives are actually passed between two roles depends on the *dependency type*:

- In *hierarchical relations* (or *hierarchies*), the *parent role* will delegate its sub-objectives to its *children roles*. In this case, the enactor of a child role cannot decide which objectives it will get but must accept whichever objectives are delegated to it by its parent role.
- In *market relations*, the *child role* can request the assignment of objectives from the *parent role*. That is, the enactors of a child role can choose to take the

objectives of its parent such that it best fits its own private goals.
- In *network relations*, both situations can happen. An objective can either be delegated by the parent role or requested by the child role, which defines an equivalence relation between related roles in a network.

In the emergency scenario, dependencies are based in network relations. Figure 13.2 depicts the five role dependencies between the above identified roles. For instance the *Fire_Station* role depends on the *Emergency_Call_Center* role to get the location of an emergency (*emergency_location* objective) and depends on the *Firefighter_Team* role to extinguish the fires (*estinguish_fire* objective). Similarly, the *Firefighter_team* role depends on the *First_Aid_Station* role for the *deal_with_wounded* objective.

13.4.2.2 Interaction Structure

The society objectives are realized by the interaction between actors. Roles represent different skills, entities, or interests relevant for those global objectives. Activities in a society are the composition of multiple, distinct, and possibly concurrent interactions, involving different actors, playing different roles. For each activity, interaction is articulated through *scenes*. Each scene serves as a blueprint for the actual interactions between actors. It wraps a subset of the interaction that has a meaning on itself. A scene is described by its players (*roles*), its intermediate states of interaction (*interaction patterns*), its desired results (*scene results*), and the norms regulating the interaction (*scene norms*). Scenes are partially ordered to form the interaction structure.

In the emergency scenario, some of the scenes that can be identified to reach the *handle_emergency_call* include the *Handle_Call* scene (with the part of the interaction concerning the reception of incoming calls), and the *Estinguish_Fire* scene (with the part of the interaction concerning the extinction of the fire itself) among others, as depicted in Figure 13.3.

The specification of expected interaction in scenes is based in the concept of landmark. A *landmark* is the description (by a declarative state expression) of an important (intermediate) state in the achievement of a scene. Landmarks can be connected through the specification of landmark patterns.

---

* It is important to note that the actor enacting a given (set of) role(s) might have other, individual, objectives; apart from the ones defined in the role(s). These actor objectives are not specified (or relevant) within the organizational model.

Roles: ft = Firefighter_Team, fas = First_Aid_Station,
   fs = Fire_Station, *f ft* = Firefighting_Truck
Results:
   r_1 = DONE ∀L ∈ AccidentLocation,
      evaluate_severity(ft, L);
   r_2 = DONE ∀L ∈ AccidentLocation,
      how_to_react(ft, L);
   r_3 = DONE ∀L ∈ AccidentLocation,
      inform(ft, fs, DONE('evaluate_severit`y', L));
   r_4 = DONE ∀L ∈ AccidentLocation,
      inform(ft, fs, DONE('how_to_react', L));
Interaction Patterns:
{ DONE(O, received_request(fs, ft, goal('extinguish_fire', L)))
      BEFORE DONE(ft, place_reached (L)),
   DONE(ft, place_reached(L)) BEFORE
      DONE(ft, fire_evaluated(L)),
   DONE(ft, place_reached(L)) BEFORE
      DONE(ft, wounded_evaluated(L)),
   DONE(ft, wounded_reached(L)) BEFORE
      DONE(ft, area_secured(L)),
   DONE(ft, area_secured(L)) BEFORE
      DONE(fas, wounded_rescued(L)),
   DONE(O, wounded_rescued(L) AND fire_evaluated(L)) BEFORE
      DONE(O, msg_sent(ft, fs, inform('evaluate_severity achieved', L));
   DONE(O, msg_sent(ft, fs, inform('evaluate_severity achieved', L)); BEFORE
      DONE(O, msg_sent(ft, fs, inform('how_to_react achieved', L)) }
Norms:
   . . .

FIGURE 13.4   Partial scene description for *Extinguish_Fire* scene.

Landmark patterns define a partial ordering by describing temporal relationships between landmarks. Landmark patterns give an indication of how a result can be achieved, that is, describe the states that must be part of any protocol that will eventually be used by actors to achieve the scene results. Landmarks and landmark patterns provide a flexible way to describe expected interaction.

A scene is related toward the achievement of a (set of) objective(s). Usually, all role objectives should be related to scene results. *Scene results* are declarative state expressions that describe the desired final states for the scene, that is, the states in which the scene ends and actors can leave it successfully.

An example of the *Extinguish_Fire* scene description can be seen in Figure 13.4. It includes the roles that are involved in the scene, the scene results, and the interaction patterns, which is a list of landmark patterns. Landmark patterns can also be represented graphically as depicted in Figure 13.5 to make them easier to interpret.

As in distributed organizations, where more complex activities can take place, scenes must be embedded in a broader context that allows representation of how the overall society objectives can be achieved. Therefore, scenes are organized into an *interaction structure* that specifies the coordination of the scenes. *Scene transitions* describe a partial ordering of the scenes, plus eventual synchronization

$λ_0$: DONE(received_request(fs,ft,goal('extinguish_fire',L)))

$λ_1$: DONE(ft,place_reached(L))

$λ_2$: DONE(ft,fire_evaluated(L))

$λ_3$: DONE(ft,wounded_evaluated(L))

$λ_4$: DONE(ft,area_secured(L))

$λ_5$: DONE(fas,wounded_rescued(L))

$λ_6$: DONE(O, msg_sent(ft,fs,inform('evaluate_severity achieved',L)))

$λ_7$: DONE(O,msg_sent(ft,fs,inform('how_to_react achieved',L)))

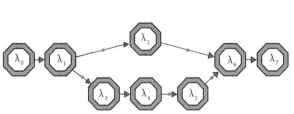

FIGURE 13.5   Landmark pattern for *Extinguish_Fire*.

constraints. Transitions between scenes specify which scenes can be reached from one given scene. Scene transitions describe the synchronization of interaction scenes. The definition of scene transitions is based on 1:1 relations between a source and a target scene specifying the partial ordering of the scenes. Scene transitions must also describe the conditions for the creation of a new instance of the scene. For each scene, the interaction structure also specifies an upper bound for the number of instances of that scene that are allowed simultaneously. Finally, each interaction structure definition must include the description of the *initial* and *final scenes*. As depicted in Figure 13.3, the graphical representation of interaction structures shows scenes as boxes labeled with the scene name, and transitions as directed arcs between two scenes. The initial and final scenes are represented by boxes with a black background.

13.4.2.3 Normative Structure

A part of the design process of an agent society must be the specification of the ethics of the society. One way to inspire trust in the parties is by incorporating the regulations (norms) in the society architecture that indicate the type of behavior expected from role enactors (Vázquez-Salceda 2003). The *normative structure* is the part of the Organizational Model that represents the collection of norms and rights related to the social and interaction structures. *Norms* in organizations capture the abstract normative statements and regulations that hold in the domain. Role *rights* (also *role rights* or *role capabilities*) indicate the capabilities that actors of the role receive when enacting the role. These are capabilities that an agent usually does not possess but which are inherent to the role.

In our framework, *norms* define the *obligations*, *permissions*, and *prohibitions* of the actors in the organization, related to the *roles* they play, or to a particular area of activity. ALIVE distinguishes between regulative and constitutive norms (Grossi 2007). *Regulative norms* regulate the behavior of agents enacting roles. They are expressed by means of deontic declarative expressions. *Constitutive norms* regulate the creation of institutional facts. Norms are classified into role norms, scene norms, and transition norms:

- *Role norms* specify the rules of behavior for actors performing that role, irrespective of the interaction scene.
- *Scene norms* describe the expected behavior of actors in a scene.
- *Transition norms* impose additional limitations to actors attempting to follow a transition between two scenes. These are typically used if a result in one scene limits the possibilities of accessible scenes afterward (e.g., buying something in the auction scene creates an obligation to pass the payment scene afterward).

An example of scene norms is shown in Figure 13.6. For example norms 1 and 2 show that Firefighter Team is obliged to call for the intervention of ambulance service and/or for

Roles: ft = Firefighter_Team, fas = First_Aid_Station,
       fs = Fire_Station, fft = Firefighting_Truck
Results:
   $r_1 = DONE \, \forall L \in Accident \, Location, \, evaluate\_severity(ft, L);$
   $r_2 = DONE \, \forall L \in Accident \, Location, \, how\_to\_read(ft, L);$
{...}
Interaction Patterns:
{...}
Norms:
(1) IF DONE(ft, severity_quantified(L, 'le_1'))
    THEN OBLIGED(fs, provide_help(fas, ft, L)))
(2) IF DONE(ft, severity_quantified(L, 'le_2'))
    THEN OBLIGED(fs, provide_help(fas, ft, L) AND provide_
    help(fft, ft, L))
(3) IF DONE(fs, authentication_failed(A, trusted_list(' fas',' ps',' f
    ft'))
    THEN FORBIDDEN(fs, provide_help(A, ft, L))
    . . .

FIGURE 13.6 Example of scene norms for the Extinguish_Fire scene.

firefighter truck service whenever it knows the level of emergency (i.e., le_1 and le_2).

13.4.2.4 Communicative Structure

Interaction in the Organizational Model is represented as communication between the interacting actors. The aim of the *communicative structure* component of the Organizational Model is to describe the communication primitives. It describes the set of *communicative acts* and the domain language specific to the society to be used in the communication between actors.

The domain language is defined by means of ontologies. An *ontology* formally captures a shared understanding of certain aspects of a domain. It provides a common vocabulary, including important *concepts*, *properties* and their definitions, and *constraints* regarding the intended meaning of the vocabulary, sometimes referred to as background assumptions. An *ontology language* is a formal language used to encode the ontology. Ontologies must be rich enough to cover the domain of application in order for actors in the society to be able to interact in several contexts, but, keep in mind the relevance of the concepts, because a voluminous ontology can lead to inefficiency because actors will spend too much time finding relevant concepts and meanings.

Communicative acts are the set of language primitives to enable dynamic, flexible communication between actors in an organization. Communicative acts are the basic building blocks of conversations and have a well-defined intentional semantics based in *Speech Act Theory* (Burkhardt 1990; Austin 1975; Searle 1975). One of the most important consequences of this theory is the fact that certain uses of language or messages can create *acts*, which means that saying (or exchanging a message) does something (Austin 1975).

As in speech acts, our communicative acts can be categorized in various types (Searle 1975):

- *Representatives* inform of something.
- *Directives* try to make another individual perform an action.

- *Commissives* commit the speaker to doing something in the future.
- *Expressives* express how the speaker feels about a situation.
- *Declarations* change the state of the world in an immediate way.

The way communicative acts are represented in ALIVE is defined by the *Agent Communication Language (ACL).* Following the ideas of the Speech Act Theory, Agent Communication Languages are computational language representations and dialogue handling mechanisms to allow for meaningful interactions between heterogeneous agents. Interactions are composed by *messages.* An Agent Communication Language makes a distinction between the intentional part of the message (the message container) and the propositional content (the message body). An ACL defines the syntax of the messages, in a way that allows the actor to communicate its intentions. Usually, this is done by adding a *performative* field in the message structure. The ACL defines a set of valid performatives and their semantics. These semantics can be used by the receiver actor to interpret, from the performative in the message, the intentions of the sender actor. Furthermore, the ACL defines also how the *content* of the message should be expressed.

The introduction of intentional semantics within the communication between actors allows for comparison between actual, and intended, actor behavior within a distributed scenario. Such a mechanism can help an actor to adapt its response according to the intentional meaning of a message received from another actor. Also, when an actor enacts a role, the objectives of a role can be related to communicative acts that may produce an expected effect in the society in a direction toward the intended objective.

For instance in the crises management scenario, an actor playing an *Emergency_Call_Center* role has as one of its objectives to keep the actor playing the *Fire_Station* role up to date to any new fire emergency. For such objective, the most useful communicative acts would be those that inform about the occurrence of a new emergency each time it happens. The diagram in Figure 13.7 summarizes the metamodel of Organizational Level. Note that some relations between concepts have (intentionally) not been shown for readability sake.

13.4.3 The ALIVE Toolset

The ALIVE architecture can be seen as a service-oriented middleware supporting the combination, reorganization, and adaptation of services at both design and run time. These activities follow organizational patterns and adopt coordination techniques. Furthermore, the Model-Driven Engineering approach taken offers significant assistance to system developers. It provides semiautomated transformations between models at the three levels described above, as well as the capacity for multiple target platforms and representation languages. The ALIVE approach, illustrated in Figure 13.8, extends current trends in engineering by providing support for design and management of distributed systems at three levels: the service level, the coordination level, and the organization level.

The framework provides graphical tools to support system administrators in the management of a distributed system.

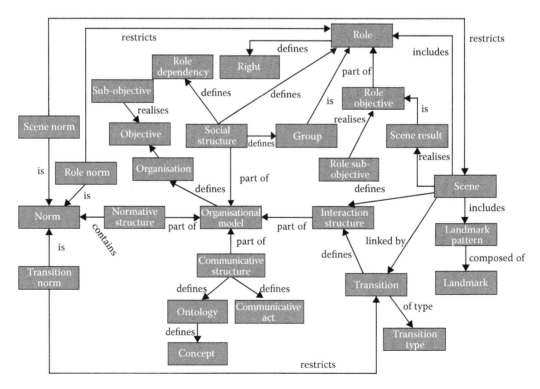

FIGURE 13.7 Organizational Level metamodel.

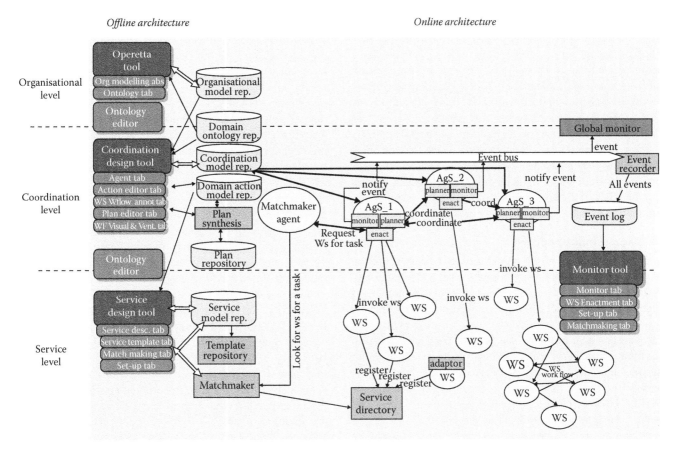

FIGURE 13.8   (See color insert.) The ALIVE toolset.

The *Monitor Tool* allows the administrator to inspect the status of a system's execution, getting information from the different components in the online architecture (especially from the monitoring components). In this way the administrator can keep track of the events generated at execution time and inspect how the system handles them. The *Service Set-up Tool* can be used by system administrators to check and modify the setup of the running environment. Finally, the *Service Matchmaking Tool* (also used off-line) allows administrators to search manually for services that match a given task description (by using the matchmaker component) when the system is not able to cope with a particular issue or the administrator wants to change manually an automatically selected service for another that is considered more suitable for reasons not modeled within the ALIVE framework.

13.4.3.1   Offline Support

The main objective of the offline subset of the ALIVE architecture is to support not only to the design phase of distributed systems based on the ALIVE methodology but also to support the continuous maintenance of deployed systems during their lifetime.

For the organization level, the *OperettA Too* (http://www.cs.uu.nl/research/projects/opera/) is used. OperettA is an open source modeling tool based on the OperA model described in Section 4.2. The functionality of this tool is extended by the *Dynamic Organizational Modeling* plug-in,

for those applications which need extra methods and components to handle dynamic organizational change. OperettA also includes a model checker used to verify some properties of organizational models. This component not only is used at design time but can also be used online during the execution of the system. The *Model Checker Tool* thus complements the OperettA Tool, and is linked directly to it through a button in the OperettA interface.

The *Coordination Design Tool* is used by designers and system administrators to create and manage the coordination model of a given application. The tool assists the user in the definition of actors and tasks, the generation of the agents that will perform the actual coordination tasks, and the inspection of predefined and generated plans. It also supports importing and inspecting the task descriptions defined in an application domain action model. A distinctive trait of the coordination models created by this tool (in comparison with other orchestration and choreography technologies; cf. Ghijsen, Jansweijer, and Wielinga 2008; Quillinan et al. 2009) is that the coordination models also abstract away from the low-level details of the services that may be invoked. The designer is able to design the whole coordination level of a distributed system by means of actors, tasks, plans, and plan coordination mechanisms. Agentified Services connect, at execution time, the abstract tasks with the actual services that are invoked. Apart from the dynamism this solution brings at execution time, this also allows end users to inspect and better comprehend coordination models. The *Service Design Tool* is used by designers and system administrators to generate or inspect service descriptions, edit service templates, and register them in the service directory. It also connects with the *Service Matchmaking Tool* (a human interface to the matchmaker component), allowing designers and system administrators to search for services matching a given task description or implementing a given service template and registering it in the service directory. Finally, the Service Setup Tool allows designers and system administrators to check and modify the setup of the running environment, including the URIs of different resources, facilitating components, predefined services, and service containers.

13.4.3.2 Online Support

The main objective of the online subset of ALIVE's architecture is to give support to the deployment and execution of distributed systems. This part also covers online monitoring mechanisms to detect failures or deviations from expected functionality and provides the mechanisms to correct, recover, or adapt to them. The *Service Directory* is one of the main components of the architecture. It is a repository for service interface descriptions, service process models, and service profiles. It thus supports several query mechanisms for the discovery of specific services based on their syntactic and/or semantic descriptions. The service directory can be used by any component or service, and its main clients are the matchmaker components. Depending on the requirements and size of the distributed system, one or many service directories may be deployed. Matchmakers are an important component for service recognition. They are responsible for the discovery of appropriate services that fulfill the requirements of a given task. Depending on the level of abstraction of the task to be fulfilled, matchmakers may query a service directory directly or by the use of service templates. The *Service Templates* are intermediary descriptions linking higher-level goals or tasks (as specified in the Coordination Model) with specific service interactions. A template includes a parameterized process model for a class of services in terms of pre- and post-conditions. If the described process is of some complexity, such description may include an abstract workflow fragment indicating required substeps in the process. Parameters in a template are specified as abstract types linked to the variables or parameters in the process model. The parameters are dynamically bound at execution time into concrete ontology instances and/or concrete service process models when the template is selected by a matchmaker. Service template definitions are stored in a Template Repository.

Agentified services can interact with normal Web services by means of standard SOAP and REST interfaces. Furthermore, agentified services communicate coordination-related issues to other agentified services using protocol-based conversations expressed in a coordination language (based on Generalized Partial Global Planning) implemented over SOAP. The exchanged plans are abstract workflows possibly with tasks referring to abstract services rather than to concrete ones. When a plan is agreed upon, an agentified service

will look (via the matchmaker component) for services that can fulfill the abstract tasks, binding them together.

External Services (i.e., existing, third-party services that have not been designed following the ALIVE framework) can be invoked at execution time according to their service description. Usually, external services are not consumed directly; instead, this is done via *Service Adaptors*. Service adaptors allow external services to be utilized for suitable organizational tasks. Typical examples of adaptation are type translation services to adapt a service interface to the entities and data types used by a given organization.

One or several *Monitoring Components* are able to aggregate and analyze events related to the execution of services, the fulfillment of coordination plans and the achievements of role and/or organizational objectives. During the online execution, events are generated by the components (namely, the agentified services), whenever deviations, exceptions, or failures are detected that cannot be handled by the agentified service itself or the existing coordination plan in place. In such situations the current organizational model is evaluated and then either (1) the objectives affected by the detected issue may be reevaluated (their priority may be lowered or they may even be dropped completely) or (2) more significant changes in the organization model may be required (for instance, changing the rights of a role). In case 1 the agent's coordination modules will create a new plan based on the updated organizational objectives. In case 2 the updated model is sent to the Agent Generator component to (re) generate the agentified services that populate the system. Depending on the set-up preferences of the administrator, the monitoring component may be a separate component used by several agentified services or may be a federation of several components inside the agentified services themselves.

Finally, there are graphical tools to support system administrators in the management of a distributed system. The *Monitor Tool* allows the administrator to inspect the status of a system's execution, getting information from the different components in the online architecture (especially from the monitoring components). In this way the administrator can keep track of the events generated at execution time and inspect how the system handles them. The *Service Set-up Tool* can be used by system administrators to check and modify the setup of the running environment. Finally, the *Service Matchmaking Tool* (also used off-line) allows administrators to search manually for services that match a given task description (by using the matchmaker component) when the system is not able to cope with a particular issue or the administrator wants to change manually an automatically selected service for another that is considered more suitable for reasons not modeled within the ALIVE framework.

13.4.4 ADAPTATION IN ALIVE

As explained in previous sections, we propose an approach that incorporates an organizational context (mainly organizational objectives, structure, and regulations) that can be used to dynamically select, compose, and invoke services, providing an organizational awareness to some components (such as agentified services at the coordination level or matchmakers at the service level) that can direct the system execution in order to maintain the focus on higher-level organizational objectives. One of the effects is that exceptions can be managed not only at the lower level (as in other SOA architectures) but at higher levels, looking for alternative ways to fulfill a task or even a full abstract workflow by agreeing a new plan of action.

Because of the connection among levels, a change in the service level can trigger changes on both coordination and organization levels; whereas a change in the organization level can also trigger changes in the other two. These changes can be automatically carried out (with the assistance of ALIVE tools). This provides the system with high dynamicity, making it capable to react to changes on the environment. Through the monitoring and exception system in the coordination level, workflow errors can be detected and treated properly. This provides the system with high stability, enabling it to fulfill its goals when a core component fails. The fact that service templates can abstract the service level and organization level from concrete Web services makes workflow specifications more stable through time, whereas service discovery provides continuous system adaptation to a highly dynamic environment. Finally, the high-level organization, connected to the coordination and, indirectly, to the service level, ensures systems compliance with organizational structure, as available services will be composed and used according to organization objectives, roles, norms, and restrictions.

13.5 CONCLUSIONS

Current visions about the Internet of the future require an evolution in the way distributed applications are designed, implemented, and deployed, moving from top-down approaches that generate partial, static (business) process descriptions to holistic approaches where both the participants and their surrounding environment are modeled. Such approaches will empower distributed applications with the ability to flexibly adapt their behavior to environmental changes, being able to identify opportunities and recover from unexpected failures or market switches.

In this chapter we have presented a holistic approach based on organizational theory. The ALIVE framework aims to support the design and development of distributed systems suitable for such highly dynamic environments, is based on model-driven engineering, and consists of three interconnected levels: service, coordination, and organization.

The crucial distinction of the ALIVE approach from existing ones is that it provides an organizational context (such as, for instance, objectives, structures, and regulations) that can be used to select, compose, and invoke services dynamically. ALIVE also provides a notion of organizational awareness to some components (such as the agentified services at the coordination level or the matchmaker component at the service

level) that can direct system execution in order to achieve higher-level organizational objectives.

One of the effects is that exceptions can be managed not only at the lower level (as in other service-oriented architectures) but at higher levels, looking for alternative ways to fulfill a task or even a full abstract workflow by agreeing upon a new plan of action. Furthermore, organizational and coordination models are defined at a level of abstraction that allows nonexpert end users to support better the design and the maintenance of the system. The first version of the ALIVE tool suite is now under development and will become available through the project's Sourceforge site (http://sourceforge.net/projects/ict-alive/).

REFERENCES

Austin, J. L. 1975. *How to Do Things with Words*. Cambridge, MA: Harvard University Press.

Berners-Lee, T., J. Hendler, and O. Lassila. 2001. The Semantic Web. *Scientific American* 284(5): 34–43.

BPMI, n.d. Business Process Modeling Language (BPML) Specification. http://www.bpmi.org (accessed Jan. 23, 2010).

Burkhardt, ed. 1990. *Speech Acts, Meanings and Intentions. Critical Approaches to the Philosophy of John R. Searle*. New York: de Gruyter.

Barber, K. S., and C. E. Martin. 2001. Dynamic reorganization of decision-making groups. In *Proceedings of the Fifth International Conference on Autonomous Agents* (Montreal Quebec, Canada), 513–520. AGENTS '01-New York, NY: ACM.

Cabral, L., J. Domingue, E. Motta, T. Payne, and F. Hakimpour. 2004. Approaches to Semantic Web services: An overview and comparisons. Paper presented at First European Semantic Web Symposium, Heraklion, Greece.

Carley, K. 2002. Computational organization science: A new frontier. *Proceedings of the National Academy of Sciences* 99(3): 7257–7262.

Castelfranchi, C. 2000. Engineering social order. In *Engineering Societies in the Agents World*, vol. 1972, eds. A. Omicini, R. Tolksdorf, and F. Zambonelli, 1–18. Heidelberg, Berlin: Springer.

Dastani, M., V. Dignum, and F. Dignum. 2003. Role Assignment in Open Agent Societies. Paper presented at of AAMAS'03, 2nd International Joint Conference in Autonomous Agents and Multi-Agent Systems, Melbourne, Australia.

Davidsson, P. 2000. *Emergent Societies of Information Agents* (LNAI 1860), Cooperative Information Agents IV, eds. M. Klusch and L. Kerschberg, 143–153. New York: Springer.

Davidsson, P. 2002. Agent based social simulation: A computer science view. *Journal of Artificial Societies and Social Simulation* 5(1).

Dignum, V. 2004. A model for organizational interaction: Based on agents, Founded in Logic. PhD thesis. SIKS Diss. Ser. 2004-1. Utrecht University.

Dignum, V., F. Dignum, and J. J. Meyer. 2004. An agent-mediated approach to the support of knowledge sharing in organizations. *Knowledge Engineering Review* 19(2): 147–174.

DMTF, n.d. Web Services Management. http://www.dmtf.org/standards/wsman/ (accessed Jan. 23, 2010).

Donaldson, L. 2001. *The Contingency Theory of Organizations*. London: Sage Publications.

Duncan, R. 1979. What is the right organizational structure: Decision tree analysis provides the answer. *Organizational Dynamics*, Winter 1979, 59–80.

Fluegge, M., I. J. Santos, N. P. Tizzo, and E. R. Madeira. 2006. Challenges and techniques on the road to dynamically compose Web services. In *Proceedings of the 6th international Conference on Web Engineering* (ICWE '06), vol. 263 40–47. New York: ACM Press.

Erl, T. (2006). Serviceorientation.org—about the principles, 2005–2006. http://www.soaprinciples.com (accessed Jan. 23, 2010).

Ghijsen, M., W. Jansweijer, and B. B. Wielinga. 2008. Towards a framework for agent coordination and reorganization, AgentCoRe. In *Coordination, Organizations, Institutions, and Norms in Agent Systems III*. vol. 4870, 1–14. New York: Springer.

Grossi, D. (2007). Designing invisible handcuffs—formal investigations in institutions and organisations for multi-agent systems. PhD thesis. SIKS Diss. Ser. 2007-16. Universiteit Utrecht.

Hübner, J., J. Sichman, and O. Boissier. 2006. S-MOISE+: A middleware for developing organized multi-agent systems. In *COIN I*, vol. 3913, ed. O. Boissier et al., 64–78. New York: Springer.

Horling, B., and V. Lesser. 2005. A survey of multi-agent organizational paradigms. *The Knowledge Engineering Review* 19: 281–316.

IBM. 2003. Business process execution language for web services version 1.1, July 2003. http://www.ibm.com/developerworks/library/ws-bpel/ (accessed Jan. 23, 2010).

Kendall, E. 2000. Role modeling for agent system analysis, design, and implementation. *IEEE Concurrency* 8(2): 34–41.

Mintzberg, H. 1993. *Structures in Fives: Designing Effective Organizations*. Englewood Cliffs, NJ: Prentice Hall.

OASIS, n.d. Web Services Business Process Execution Language (WSBPEL). http://www.oasis-open.org/committees/download.php/18714/wsbpel-specification-draft-May17.htm (accessed Jan. 23, 2010).

OASIS. 2002. Universal Description, Discovery and Integration (UDDI) version 3.0.2. http://www.uddi.org/pubs/uddi_v3.htm (accessed Jan. 23, 2010).

OASIS. 2006. The framework for eBusiness. April 2006. http://www.ebxml.org/ (accessed Jan. 23, 2010).

OGSI Alliance, n.d. OGSI specifications. http://www.osgi.org/Specifications/HomePage (accessed Jan. 23, 2010).

OMG, n.d. Model driven architecture. http://www.omg.org/mda/ (accessed Aug. 23, 2009).

OMG Object Management Group, n.d. Business Process Modeling Notation. http://www.bpmn.org/Documents/OMG Final Adopted BPMN 1-0 Spec 06-02-01.pdf (accessed Jan 23, 2010).

Peltz, C. 2003. Web services orchestration and choreography. *IEEE Computer* 36: 46–52.

Quillinan, T., F. Brazier, H. M. Aldewereld, F. P. M. Dignum, M. V. Dignum, L. Penserini, and N. Wijngaards. 2009. Developing agent-based organizational models for crisis management. *Proceedings of the Industry Track of the 8th International Joint Conference on Autonomous Agents and Multi-Agent Systems (AAMAS 2009)*. eds. Decker, Sichman, Sierra and Castelfranchi.

Roberts, J. 2004. *The Modern Firm. Organizational Design for Performance and Growth*. New York: Oxford University Press.

Searle, J. R. 1975. *Language, Mind and Knowledge*. Minnesota Studies in the Philosophy of Science, 344–369. Minneapolis: University of Minnesota Press.

Selznick, P. 1953. *TVA and the Grass Roots: A Study of Politics and Organization*. Berkeley: University of California Press.

Singh, M. 2003. Distributed enactment of multi-agent workflows: Temporal logic for web service composition. In *Proceedings of 2nd International Joint Conference on Autonomous Agents and Multi-agent Systems (AAMAS 2003)*, 907–914. New York: ACM Press.

Su, J., T. Bultan, X. Fu, and X. Zhao. 2007. Towards a theory of Web service choreographies. In *Proceedings of WS-FM'07, Lecture Notes in Computer Science,* vol. 4937, 1–16. New York: Springer.

Su, X., and J. Rao. 2004. A survey of automated Web service composition methods. In *Proceedings of First International Workshop on Semantic Web Services and Web Process Composition*.

Van Dyke Parunak, H., and S. Brueckner. 2001. Entropy and self-organization in multi-agent systems. *Agents'01* 124–130.

Vázquez-Salceda, J. 2003. The role of norms and electronic institutions in multi-agent systems applied to complex domains. The HARMONIA framework. PhD thesis, Universitat Politècnica de Catalunya.

W3C, n.d.(a). Simple Object Access Protocol (SOAP) 1.1. http://www.w3.org/TR/2000/NOTE-SOAP-20000508 (accessed Jan. 23, 2010).

W3C. n.d.(b). Web Service Choreography Description Language (WS-CDL). http://www.w3.org/TR/ws-cdl-10/ (accessed Jan. 23, 2010).

W3C. n.d.(c). Web Service Description Language (WSDL). http://www.w3.org/TR/wsdl (accessed Jan. 23, 2010).

W3C. 2004. OWL-S—Semantic Markup for Web Services. http://www.w3.org/Submission/OWL-S/ (accessed Jan. 23, 2010).

Wellman, M. 1993. A market-oriented programming environment and its application to distributed multi-commodity flow problems. *Journal of Artificial Intelligence Research* 1: 1–23.

WfMC, n.d. XML Process Definition Language (XPDL). http://www.wfmc.org/xpdl.html (accessed Jan. 23, 2010).

WSMO Working Group, n.d. Web Service Modeling Ontology. http://www.wsmo.org/ (accessed Jan. 23, 2010).

Section V

Information Retrieval and Sharing:
Search Engines, Portals, and Intranets

14 Searching and Evaluating Information on the WWW: Cognitive Processes and User Support

Yvonne Kammerer and Peter Gerjets

CONTENTS

14.1 INTRODUCTION

Since the advent of the World Wide Web (WWW, hereinafter referred to as the Web) in the early 1990s, the amount of information available online has grown exponentially. The technical development since then enables individuals worldwide to comfortably retrieve information on almost any topic and in almost any representation format imaginable (e.g., text, pictures, audio, video). Therefore, in recent years for many people the Web has become the major information source both in professional and personal life. To discover and access information on the Web, most people start by using a general search engine such as Google, Yahoo!, or MSN/Bing. According to the latest report from the PEW Internet & American Life Project (Fallows 2008), Web searching has become an extremely popular Internet activity, with almost half (49%) of American internet users employing search engines on a "typical day" (only exceeded by e-mail usage with 60%).

Besides searching for simple and uncontroversial facts or researching product purchases, the Web also serves as a rich information source for people conducting research on more complex academic or science-related topics (Horrigan 2006). In the context of personal or social concerns of individuals, such as medicine and health care, environmental, technical, or political issues, this type of Web research has achieved great popularity in recent years (Fox 2006; Fox and Jones 2009; Morahan-Martin 2004; Smith 2009; Stadtler and Bromme 2007). Searching for complex and science-related information on the Web increasingly supplements the interaction with experts, for instance, when a diagnosis or a treatment for a medical or technical problem is needed. Furthermore, Web search on science-related topics has also become a central part of formal research activities in educational contexts (i.e., in schools and universities).

It is important to note, however, that finding relevant and high-quality information on the Web is not an easy task because of the specific characteristics of the Web. First, the nonlinear hypertext structure and nearly unlimited amount of information require a high degree of user control during Web search. Web users steadily have to decide which paths to follow in order to access information (Dillon 2002). This can easily lead to information overload and disorientation ("lost in hyperspace"; Conklin 1987). Second, although a wealth of information can be accessed on the Web, there is no guarantee to its validity and reliability. As anyone can publish virtually any information on the Web, the Web is characterized by a high heterogeneity of information sources differing, for instance, with regard to Web authors' expertise and motives. Moreover, contrary to traditional information sources such as printed publications, documents on the Web other than electronic copies of traditional publications seldom have explicit editorial review policies or undergo quality controls. Thus, Web users themselves are responsible for "gatekeeping," that is, evaluating the relevance and quality of information found on the Web (Bråten and Strømsø 2006; Britt and Aglinskas 2002). Because quality-related source information is usually not displayed in a salient way, this personal gatekeeping is a very hard task.

To conclude, Web searching is a complicated cognitive skill, requiring adequate evaluation and regulation strategies. Thus, not surprisingly, empirical research has shown that searchers often face difficulties during Web search, impairing their search outcomes (Brand-Gruwel, Wopereis, and Vermetten 2005; Gerjets and Hellenthal-Schorr 2008). In recent years, researchers have focused on the development and evaluation of novel search interfaces and tools to support users during their Web search.

Accordingly, this chapter contains two main parts. The first part (Section 14.2) examines the cognitive processes involved in searching and evaluating information on the Web by means of search engines. The second part (Section 14.3) introduces different interface design approaches aimed at supporting users in searching and evaluating Web information, including examples of recently developed search systems. Finally, in Section 14.4 the chapter ends with some directions for future research. Typical methods to investigate users' Web search behavior – irrespective of whether support is given or not – include transaction log analyses, eye-tracking methodology, verbal protocols, and surveys and questionnaires. These methods are referred to at various points throughout this chapter.

14.2 COGNITIVE PROCESSES IN WEB SEARCH

14.2.1 Different Types of Web Search Goals

As the Web provides a huge variety of information, unsurprisingly the search goals that people pursue on the Web are many and varied. Search tasks range from looking up specific facts such as "What is the exact height of the Empire State Building?," to product researches such as finding out which is the camera with the best cost effectiveness, to knowledge acquisition about a general topic such as "What are Web Mashups and what are their benefits?," to science-

related researches about complex and potentially controversial issues such as "What are the causes of global warming?" or "What are the benefits and risks of different medical treatments for cancer?"

Therefore, a number of researchers have attempted to classify the various goals underlying Web searches. First of all, whereas the above mentioned examples all focus on fulfilling users' information needs or in other words on changing users' state of knowledge, Broder's (2002) "taxonomy of Web searches" reflects that in the Web context the "need behind the query" is not always to retrieve information. Broder distinguished three general types of Web searches: informational, navigational, and transactional searches. Whereas in informational searches the user's goal indeed is to obtain information, in navigational searches users aim at reaching a specific Web site they have in mind. The purpose of transactional searches is to reach a site where a Web-mediated activity (e.g., shopping, downloading files) will be performed. A similar search goal taxonomy was proposed by Rose and Levinson (2004), comprising informational queries, navigational queries, and resource queries (e.g., downloading files, getting entertained by using a resource).

On the basis of manual classifications of search engine query logs, Broder (2002) reported 48% of Web queries as informational and about 20 and 30% of queries as navigational and transactional, respectively. Rose and Levinson (2004) found a higher proportion of informational queries (about 60%), and somewhat smaller proportions of navigational and resource queries (about 14 and 25%). Jansen, Booth, and Spink's (2008) analyses revealed an even higher proportion of informational queries (more than 80%), whereas navigational and transactional queries each represented only about 10% of Web queries. Despite some variations across the different studies, the results clearly indicate that informational queries make up the biggest proportion of searches on the Web.

Informational searches can be further distinguished into specific fact-finding tasks and more complex or exploratory information gathering tasks (e.g., Kellar, Watters, and Shepherd 2007; Navarro-Prieto, Scaife, and Rogers 1999; Rose and Levinson 2004; Shneiderman 1997). Similarly, Marchionini (2006) grouped search activities into simple lookup and more complex learning and investigating activities (see Figure 14.1). Both types of search activities are goal driven and have to be distinguished from serendipitous browsing, in which Web users are traversing information in a wholly random and undirected way without having a concrete goal in mind.

Fact-finding tasks are characterized as specific, closed-ended tasks. They include simple lookup of concrete facts about a specific topic such as the length of the Nile River or the symptoms of a well-known disease. Such tasks are also known as well-structured problems, which usually possess one correct answer or a set of convergent answers (Jonassen 1997). In contrast, in complex or exploratory search tasks the user's goal is to learn something by collecting information from multiple sources, to form her own opinions, or to make

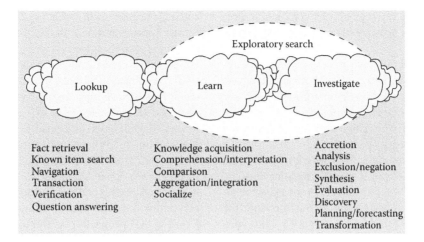

FIGURE 14.1 Search activities according to Marchionini (2006). (From Marchionini, G. *Communications of the ACM*, vol. 49: 41–46. ACM Press 2006. With permission.)

an informed decision about a controversial topic. Such search tasks are more unstructured and open ended and thus are also known as unstructured (Vakkari 1999) or ill-structured problems (Jonassen 1997). Ill-structured problems are the kind of decision-making problems encountered in everyday life, such as choosing one of several medical treatments (Pirolli 2007). They deal with controversial or contradictory issues characterized by divergent opinions and alternative viewpoints (Jonassen 1997; Simon 1973). In other words, ill-structured problems can possess multiple solutions or multiple solution paths or no solution at all. Therefore, complex or exploratory search tasks are usually more demanding and time-consuming than the aforementioned fact-finding tasks.

Aula and Russell (2008) propose a more detailed classification, in which they further differentiate between exploratory search and complex search by means of the complexity and clarity of the underlying search goal. The main characteristic of complex searches is the high complexity of the search goal or the topic of interest, which is often science related, such as, for instance, searching information about benefits and risks of different therapeutical treatments of a medical condition in order to decide about which therapy to undergo. Therefore, in such complex research tasks searchers need to consult multiple information sources and synthesize the information (Aula and Russell 2008). The main characteristic of exploratory searches is that the search goal is rather unclear or open ended. The underlying goal of exploratory search often is to learn more about a domain of interest, such as, finding out, for instance, what Web mashups are (see Kammerer et al. 2009a). White et al. (2007) propose that in exploratory search, as Web users may not have a well-defined idea of what information they are searching for, they usually submit a tentative query and then explore the documents they get returned, make decisions which paths to follow, select information, and try to get more relevant documents by refining their query. Therefore, in exploratory search, Web users generally combine querying and browsing strategies,

with learning taking place during this exploratory process (Marchionini 2006; White et al. 2007).

In summary, informational, navigational, and transactional searches can be distinguished as broad Web search categories, whereby in this chapter the focus lies on informational searches. A compilation from different taxonomies yields three different informational search tasks varying in their level of complexity and clarity: fact-finding tasks, complex research tasks (about science-related information problems), and exploratory tasks.

The remainder of this chapter illustrates these Web search tasks in terms of cognitive processes involved and discusses different approaches that support Web users in their Web searches.

14.2.2 Web-Based Information Seeking Models

The high-level process of information seeking has been described by numerous theoretical models (see Wilson 1999). Many of these models consider information seeking as a problem-solving process driven by an information problem (e.g., Brown 1991; Marchionini 1989, 1995; Kuhlthau 1993; Wilson 1999; Wopereis, Brand-Gruwel, and Vermetten 2008). An information problem arises when a discrepancy occurs between information needed to answer a question and information already known (Walraven, Brand-Gruwel, and Boishuizen 2009). Information problems, just as problems in general, can be described in view of four elements (see Newell and Simon 1972): (1) an initial state of the problem, which is an information need to acquire knowledge, (2) a goal state, which is the situation where this need is satisfied (information is found and processed), (3) a solution that enables the transition from the given state to the goal state, and (4) the problem solving process itself. Whereas the majority of the models address information seeking in general (e.g., Kuhlthau 1991, 1993; Sutcliffe and Ennis 1998; Wilson 1999), some researchers have tried to develop models specifically tailored for Web search. These models usually segment the complex process of Web search into several phases and propose an appropriate sequence of these phases.

In the following, three different process models of Web-based information seeking, which differ both in their level of granularity and their scope, will be outlined in detail. Subsequent to the description of the different process models, a computational model of Web search will be addressed; the Information Foraging Theory by Pirolli and Card (1999).

One of the earliest, more basic Web search models is the four-phase framework of Web text searches by Shneiderman, Byrd, and Croft (1997, 1998). The authors proposed a framework for general text searches on the Web comprising the following four phases: (1) formulation, (2) action, (3) review of results, and (4) refinement. After users have considered their information needs and have clarified their search goals, they are ready to perform the four phases of Web search by interacting with a search system. First, in the formulation phase, information seekers have to formulate a query according to their information need. Second, in the action phase, the search is started by clicking on a search button. Third, in the review phase, information seekers review the search results and decide which documents to select. Finally, in the refinement phase, information seekers consider how to improve the search results by refining the query. Equivalent steps are proposed by Broder's (2002) information retrieval model, which in addition explicitly addresses the interaction with a Web search engine, whereas Shneiderman et al.'s four-phase framework rather focuses on information seeking with electronic text databases.

Hölscher and Strube (2000) proposed a process model of Web-based information seeking, which is based on empirical research on experts' and novices' Web search behavior (see also Section 14.2.4). The model provides a detailed description of the subprocesses of information seeking via search engines: After having defined an information need (1), the information seeker has to decide whether to access a specific Web site directly or to use a search engine (2). In the latter case a specific search engine is selected and launched (3). Then, the information seeker generates and selects search terms (4) in order to formulate a query for the search system (5),

submits the query to the search system and receives the results (as the system's response to the query) (6). Subsequently, one or several search engine results pages (SERPs) are examined (7). In case of bad results the query has to be refined, otherwise a document is selected from the SERP (8) and the information seeker examines the document (9). The process can comprise several reiterations of (some of) the steps. Similar phases of Web-based information seeking are described by Marchionini and White (2008) based on Marchionini's (1995) framework of information seeking in electronic environments. In contrast to Shneiderman et al.'s and Broder's model, the models by Hölscher and Strube and by Marchionini and White also address the examination and use of the information contained in the documents (i.e., the Web pages).

With their IPS-I-model, Brand-Gruwel, Wopereis, and Walraven (2009) propose a comprehensive model that describes the process of information problem solving (IPS) on the Internet (I), or more precisely, the skills needed to solve an information problem via the Web. The model proposes five main steps of information problem solving: (1) define an information problem, (2) search information, (3) scan information, (4) process information, and (5) organize and present information; see Figure 14.2. Each of the five steps comprises several subskills. The process of information problem solving starts with the recognition of an information need, which has to be transformed into a concrete and comprehensive problem definition. While defining the problem, the information seeker is required to activate prior knowledge on the subject matter. The subprocess "search information" requires the information seeker to select an appropriate search strategy (e.g., using a search engine) and to specify the search terms, which then have to be entered into a search engine. The search results returned by the search engine have to be judged on both topical relevance and quality and those search results that seem useful have to be selected for further inspection. During the subprocess "scan information" the information seeker first scans the selected Web page in order to both get an idea of the provided content and to evaluate the information and the source. In case that information is judged to be useful, it might be stored by using bookmarks. Subsequently, during the subprocess "process information", which involves deep

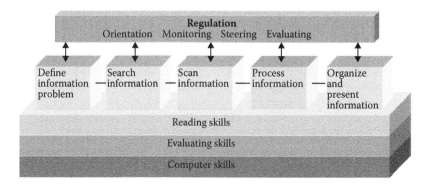

FIGURE 14.2 Information problem solving using the Internet model (IPS-I-model) by Brand-Gruwel, Wopereis, and Walraven (2009). (From Brand-Gruwel, S., I. Wopereis, and A. Walraven. *Computers and Education* vol. 53: 1207–1217. Elsevier 2009. With permission.)

processing, the information seeker reads the text, elaborates on it, structures the content, integrates the found information with prior knowledge, and judges the processed information with regard to quality and coherence. In the human–computer interaction (HCI) literature, this process is also referred to as "sensemaking," that is, the cognitive act of understanding information (Whittaker 2008). In summary, the first four skills are part of the analysis phase. Finally, the fifth subprocess "organize and present information" concerns the synthesis of information. The information has to be compared and integrated toward a solution of the information problem. The solution can be represented internally in the user's mind or externally for instance by means of an oral presentation or statement or a written product like an essay.

While executing all these skills in order to solve the information problem, according to the IPS-I model regulation activities have to be carried out (represented by arrows in Figure 14.2). Regulation activities such as orienting on the task, monitoring and steering performance, time management, and assessing the process as well as the final solution play key roles in the efficiency and effectiveness of the search process and its outcome. Furthermore, to appropriately perform the five problem solving skills as well as the regulation activities, Brand-Gruwel et al. propose that information seekers have to possess some basic prerequisite skills, that is, (1) hypertext reading skills, (2) computer skills, and (3) evaluating skills (see the three bottom layers in Figure 14.2).

With regard to evaluation skills, Gerjets, Kammerer, and Werner (forthcoming) distinguish between three different types of evaluation processes that should be carried out during Web search, namely, (1) the evaluation of search results, (2) the evaluation of Web pages, and (3) the evaluation of document collections (i.e., comparing and integrating multiple documents); see Section 14.2.3 for details.

Whereas the previous models proposed a set or sequence of subprocesses of information seeking, Pirolli and Card (1999) chose a different approach with the Information Foraging Theory. This theory is a predictive computational model about people's search and navigation behavior on the Web. The Information Foraging Model conceptualizes information seeking not as problem solving but as behavior analogous to food foraging. That is, the information seeker or information "forager" is seen as exploring, finding, and finally consuming information in different "patches" (information sources) following the "information scent" (see Section 14.2.3 for details) of the "prey" (information) similar to how animals forage for food. The model proposes that Web searchers perform a cost-benefit analysis during navigation, in which they weigh the costs and benefits of (1) evaluating and consuming already found information or (2) continuing the search for additional information. The model assumes that Web searchers aim at maximizing gains of valuable information in relation to effort (i.e., search time, cognitive effort), thus applying a "satisficing" strategy (see Simon 1955). This satisficing strategy implies, that at a specific point the searcher will decide that the found information is (good) enough and thus to stop searching and to "consume" the found information (Pirolli 2007).

To summarize, information seeking can be considered either as a problem-solving or as an information-foraging process. There are several theoretical models of Web-based information seeking that distinguish and analyze different subprocesses of Web search. In the following section, three subprocesses of Web search, namely, (1) formulating and refining search queries, (2) evaluating and selecting search results and Web sites, and (3) synthesizing Web information will be described in more detail. Both theoretical frameworks and findings from empirical research regarding the processes will be reported, also addressing difficulties users typically face in the different phases of Web search.

14.2.3 SUBPROCESSES OF WEB SEARCH

14.2.3.1 Search Query Formulation and Refinement

After having recognized an information need, this need has to be translated into a query posed to a search engine via a search entry field (e.g., Shneiderman, Byrd, and Croft 1997, 1998). However, the formulation of appropriate search queries is a challenging task for many Web users. In their reviews of search engine transaction logs research, Jansen and Pooch (2001) and Jansen and Spink (2006) found that Web searchers use short queries (approximately two terms per query) and only rarely include Boolean operators or other advanced query strategies (such as using double quotes or asterisks). Jansen and Spink (2006) state that apart from a slight trend of a decreasing percentage of one-word queries, users' strategies for search query formulation have hardly changed over the past decade. In a comprehensive transaction log study by Jansen, Spink, and Koshman (2007), however, the mean length of queries was found to be 2.8 terms, with only 18.5% as one-word queries. Specifically with regard to informational searches, results from Kellar, Watters, and Shepherd (2007) showed much higher numbers, with almost five query terms per query for fact-finding tasks and about three terms for exploratory search tasks. According to the authors, fact-finding queries are typically longer than queries used for exploratory searches, because when accomplishing fact-finding tasks users often submit very specific query strings, whole sentences, or specific questions. In contrast, queries submitted during exploratory search are usually vague and represent general topics.

In general, research shows that specifying search terms is difficult for searchers of all age groups (e.g., Bilal 2000; MaKinster, Beghetto, and Plucker 2002). Searchers often use too vague or too specific search queries, resulting either in millions of search results with a great number of false hits interspersed or in no search results at all (Spink et al. 2002). As a consequence, users then are required to spend much effort with refining their search terms, often with little success. According to a study by Jansen et al. (2007), searchers most often modified their query by changing query terms rather than adding or deleting terms. Furthermore, MaKinster, Beghetto, and Plucker (2002) could show that domain knowledge helps to specify better search terms (also see Section 14.2.4), and well-defined search queries lead to more successful search outcomes. Therefore, researchers have developed various techniques for spelling corrections, query completion, keyword suggestions, or query refinements (see Section 14.3.1) in order to support Web users in appropriately formulating and refining their queries.

14.2.3.2 Evaluation and Selection of Search Results and Web Sites: Relevance Judgments

To find relevant and high-quality information, evaluation processes are a crucial factor during Web search. The evaluation of Web information is often considered as a two-stage-process (e.g., Tang and Solomon 2001) taking place *before* and *after* selecting a search result from a search engine results page (SERP). Rieh (2002) and, subsequently, Crystal and Greenberg (2006) differentiate between *predictive judgments* regarding search results and *evaluative judgments* regarding Web sites.

Theoretical accounts for the evaluation of search results and Web pages usually refer to the concept of relevance judgments. However, the definition of relevance varies between different scientific disciplines concerned with Web search.

14.2.3.2.1   Relevance as Topicality

For researchers at the intersection of information retrieval (IR), HCI, and cognitive science, judging the relevance of Web information is mainly an issue of topical relevance or topicality, that is, the match between the topic of the information need and the topic of a search result or Web page (e.g., Blackmon, Kitajima, and Polson 2005; Brumby and Howes 2008; Miller and Remington 2004; Pirolli and Card 1999). Accordingly, ranking algorithms of search engines are built to return the most topical relevant pages with regard to the search query first (Gordon and Pathak 1999).

A theory that focuses on cognitive mechanisms of topicality evaluation for explaining and predicting Web search and navigation behavior is the Information Foraging Theory (Pirolli 2007; Pirolli and Card 1999; see also Section 14.2.2). The theory assumes that the selection of hyperlinks (e.g., from a SERP or Web page) is determined by the strength of the so called "information scent." Information scent is defined as the perceived semantic similarity between the *proximal cues* (i.e., keywords or trigger words) available in link labels or search results representing the content of a distal information source (i.e., the Web page associated with the link) and the current *search goal* of the user. The stronger the information scent, the higher is the likelihood that the distal source contains the desired information and thus the higher is the likelihood that the hyperlink is selected. On the basis of a theoretical integration of Information Foraging Theory and the cognitive architecture ACT-R (Anderson et al. 2004), Web search and navigation behavior have been modeled computationally in detail, purely based on topicality evaluation (e.g., Pirolli and Fu 2003). Similarly, other computational cognitive models of link selection are constrained to topical relevance evaluation as well (e.g., Blackmon, Kitajima, and Polson 2005; Brumby and Howes 2008; Miller and Remington 2004; Juvina and Van Oostendorp 2008).

Most of the work referred to so far has addressed relevance judgments in the context of rather simple fact-finding tasks or search tasks for which a selection of Web sources containing uncontroversial and consistent information of high quality was provided. Such tasks demand that users focus their attention mainly on the topical fit of available information. However, for many (complex) search tasks, for instance, in the domain of medical and health information, the Web additionally contains potentially contradictory and unreliable information. In this case, it might be assumed that Web searchers will also engage in evaluating additional quality-related aspects of Web information, at least when certain preconditions are given (e.g., the search task is sufficiently complex, the available information is highly variable with regard to its quality, the user has the cognitive prerequisites to evaluate the quality of Web information at his or her disposal).

14.2.3.2.2   Relevance as a Multidimensional Concept

In line with the important role of information quality for certain types of search tasks, the information science literature defines relevance as a multidimensional concept that cannot be reduced to topical fit but is based on a set of different evaluation criteria (Borlund 2003; Schamber 1994) also reflecting the quality of information. In the past decade, several studies that addressed the evaluation of Web search results and Web pages (e.g., Crystal and Greenberg 2006; Rieh 2002; Savolainen and Kari 2005; Tombros, Ruthven, and Jose 2005) yielded the following pattern of results: Although topical fit (e.g., topicality, topical interest, scope) is the evaluation criteria most often applied, searchers nevertheless use other evaluation criteria as well that refer to the quality of information (e.g., trustworthiness, credibility, author reliability, up-to-dateness). Moreover, according to Web credibility research (e.g., Fogg et al. 2001), Web page design (e.g., colors, clarity, usability, pictures), as well as publishing and source information (e.g., publication date, author's name or profession, information about the organization, contact information, commercial purpose) are important characteristics used to evaluate the quality of Web information. However, most of these studies are exploratory and there is no consensus on the importance of different relevance criteria and on the terms used for these criteria. An important caveat with regard to the abovementioned studies is that participants are usually instructed beforehand to mention or mark important features of search results or Web pages (e.g., Crystal and Greenberg 2006) or to mention which evaluation criteria they employ during Web search (e.g., Tombros, Ruthven, and Jose 2005). It can be expected that this type of explicit evaluation instruction will result in a distortion of users' spontaneous evaluation processes because users become much more aware of their selection and evaluation strategies due to the instruction used (see Ericsson and Simon 1993). In line with this assumption, a study by Gerjets, Kammerer, and Werner (2011) revealed that compared with a neutral instruction to think aloud during Web search, an explicit instruction to mention evaluation criteria applied during Web search resulted in more utterances of quality-related evaluation criteria. Thus, the rather high rate of quality-related evaluation processes in the above mentioned studies might be an overestimation. Moreover, some researchers argue that the concept of quality should be treated as independent from relevance (e.g., Mandl 2006): According to Mandl (2006), relevance describes the situational value of a Web page according to a user's current information need, whereas quality describes aspects of a Web page independent of a current information need. In line with this definition, in the remainder of this chapter, the term relevance will be used to refer to topical relevance or topicality only and the term quality will be used to refer to several quality-related criteria (e.g., credibility, trustworthiness, reliability, up-to-dateness).

Despite the methodological problems in studying the quality-related evaluation criteria proposed in the information science literature, unquestionably these criteria are crucial factors that should be considered by Web searchers in order to avoid the selection and use of incomplete, biased, or even false information (e.g., Taraborelli 2008). This is especially the case when dealing with complex and controversial issues, such as for instance the effectiveness of specific medical treatments. Contrary to this claim, studies investigating Web users' spontaneous evaluation processes during complex information gathering (i.e., studies without instructions to evaluate) revealed that searchers usually tend not to appropriately evaluate information during Web search by assessing qualitative aspects of search results and Web pages (Brand-Gruwel, Wopereis, and Vermetten 2005; Gerjets and Hellenthal-Schorr 2008).

With regard to the selection of search results, several Web search studies have shown by means of eye-tracking and/or log file analyses that users of popular search engines such as Google, Yahoo!, or MSN/Bing tend to inspect the search results in the order presented by the search engine (Cutrell and Guan 2007; Granka, Joachims, and Gay 2004; Joachims et al. 2005), with the second result page being only rarely visited (e.g., Lorigo et al. 2006). Moreover, searchers spend most attention to the search results on top of a SERP and also predominantly select these first few links (Cutrell and Guan 2007; Eysenbach and Köhler 2002; Granka, Joachims, and Gay 2004; Guan and Cutrell 2007; Joachims et al. 2005; Pan et al. 2007). Keane, O'Brien, and Smyth (2008) and Pan et al. (2007) used an interesting methodological paradigm to show that searchers conducting fact-finding tasks are heavily influenced by the ranking of the search engine. In this paradigm, the relevance order of 10 search results on a Google SERP was experimentally manipulated. When the top search results on a SERP were the least relevant ones, participants inspected more search results than when the top search results were the most relevant ones. As a consequence, they sometimes selected a highly relevant search result placed further down the list. However, participants generally still paid most attention to the search results on top of the SERP and selected these results most often, even when they were the least relevant results. Similarly, a study by Guan and Cutrell (2007) that varied the position of the target search result (i.e., the most relevant result) to accomplish a fact-finding task in

a MSN SERP showed that when the target result was placed lower on the SERP, participants still tended to click on one of the top search results, even though they inspected more search results than when the target search result was among the top positions on the SERP. These findings indicate that Web users searching for specific facts, although employing some degree of evaluations regarding the topical relevance of the search results presented on a SERP, often tend to neglect their own evaluations in favor of obeying the ranking determined by the search engine.

Analyses of verbal protocols from participants engaged in complex search tasks (Brand-Gruwel, Van Meeuven, and Van Gog 2008; Gerjets, Kammerer, and Werner, forthcoming; Walraven, Brand-Gruwel, and Boshuizen 2009) showed that the majority of participants' utterances referred to the topical relevance (i.e., connection to task) of search results and Web pages (as predicted by Information Foraging Theory). In contrast, evaluation criteria beyond topicality with regard to the quality of information (e.g., credibility, trustworthiness, reliability, accuracy, or up-to-dateness) were uttered only rarely. These findings are in line with results from other Web search studies that found that people often do not spontaneously verify the trustworthiness of information obtained via the Web or use inadequate criteria to evaluate information, for instance, when searching for medical information (e.g., Eysenbach and Köhler 2002; Metzger, Flanagin, and Zwarun 2003; Fox 2006, Stadtler and Bromme 2007; Wiley et al. 2009).

On the basis of these disillusioning results, researchers have started to develop search results interfaces that aim at fostering users' quality-related evaluations of Web information (see Section 14.3.2).

14.2.3.3 Integration of Web Information

In addition to the evaluation of search results and Web pages, in case that information is retrieved from various sources, a third evaluation process is required that aims at comparing and integrating information from different Web pages. This process is connected to the last step of the information seeking process that concerns the synthesis of information toward a solution of an information problem. This step is particularly important for complex or exploratory search tasks, where information is usually retrieved from multiple sources (i.e., Web pages), which might express diverse or even contradictory viewpoints. Thus, users are required to compare the information from different Web pages, to incorporate the found information into their prior knowledge base, and to integrate it toward a meaningful solution of the information problem. The skills required for this integration are referred to as multiple-documents literacy (Bråten and Strømsø 2010). For the synthesis of information, evaluations are required to resolve conflicts and incoherencies within the document collection retrieved. In case that diverse or even contradictory information is found in different sources, source evaluations (according to the quality-related criteria described in the previous section) are of primary importance.

The best-known framework for describing multiple-documents literacy is the "theory of documents representation"

14.2.4.4 "Net Generation"

One might assume that today's students, the so-called "net generation" or "digital natives" (Prensky 2001) are highly competent Web searchers, because they already grew up with the Internet. However, even though the net generation usually is highly tech-savvy and computer literate, empirical research revealed that many of today's school kids and adolescents are not able to solve information-based problems successfully on the Web (Brand-Gruwel, Wopereis, and Vermetten 2005; Breivik 2005; Gerjets and Hellenthal-Schorr 2008). Like other Web searchers, these users frequently employ superficial strategies for searching and processing information. They face problems with defining information problems, specifying search terms, evaluating search results, information, and information sources, and employing regulation activities in order to steer and monitor their search process (Brand-Gruwel and Gerjets 2008; Walraven, Brand-Gruwel, and Boshuizen 2008).

A specific characteristic of "digital natives" is their tendency to search quickly and take the first available information without wasting too much time and resources (Law 2008). They are quite uncritical of the information they find on the Web, not reflecting on the quality and credibility of information and sources (Metzger, Flanagin, and Zwarun 2003; Walraven, Brand-Gruwel, and Boshuizen 2009). Furthermore, they are often tempted to simply "cut and paste" information from a Web page (Law 2008).

To summarize, many correlational studies have shown that specific user prerequisites such as prior knowledge about the search topic, Web search experience, and sophisticated epistemological beliefs are crucial factors for successful Web searches. Thus, if one or more of these personal prerequisites is lacking, users are likely to face difficulties during Web search, which might impair their search outcomes. Furthermore, although today's kids and teenagers have grown up with the Internet and usually are highly tech-savvy, like older users they often show deficient Web search performance.

Therefore researchers have focused on the development and evaluation of novel search interfaces and tools to support users during their information seeking on the Web, which will be outlined in the remainder of this chapter.

14.3 INTERFACE DESIGN APPROACHES TO SUPPORT DIFFERENT SUBPROCESSES OF WEB SEARCH

To provide an overview of interface design approaches that aim to support users during different parts of the Web search process, we will use different search systems of current research and development as illustrations (also see Hearst 2009).

14.3.1 SUPPORTING QUERY FORMULATION AND REFINEMENT

Researchers have developed various techniques to support Web users in appropriately formulating and refining their queries. Most current search engines use highly effective algorithms for correcting spelling errors. Furthermore, they suggest query terms dynamically (on a character-by-character basis) while the user types the search terms into the entry field. Another form of query suggestion currently used by many popular search engines is to automatically suggest alternative, additional, or related terms after the user has sent his or her query to the search engine. By selecting one of the presented suggestions, which are usually presented at the top (e.g., Yahoo!), top and bottom (e.g., Google), or side (e.g., Bing) of the search results list, users can replace or augment their current query (e.g., White and Marchionini 2007). Yahoo!'s Search Assist interface, for instance, combines both types of query suggestions (i.e., dynamic and after-search suggestions) displayed in a single "sliding tray" (Anick and Kantamneni 2008) beneath the search entry field. In the left column of the tray (see Figure 14.3), the refinement tool offers dynamic term suggestions for a user's query string on

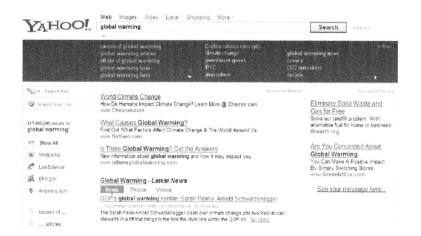

FIGURE 14.3   (See color insert.) Screenshot of Yahoo!'s Search Assist interface with the "sliding tray" including dynamic term suggestions (in the left column) and related concepts (in the middle and right columns) for the query "global warming."

a character-by-character basis. By clicking on one of these suggestions the query in the search box is replaced and the search is executed.

Additionally, once a search has been run, in the middle and right columns of the sliding tray (see Figure 14.3) the refinement tool offers related concepts derived from an analysis of the top-ranked search results. By clicking on such an after-search suggestion, this term is added to the current query (cf. the AND operator) and a new search result set is generated. However, the refinement terms are not merged into the search entry field but are moved to the left-hand side of the tray (see Figure 14.4). In a transaction log study measuring search performance on four distinct days over a period of 17 weeks right after the launch of Yahoo!'s Search Assist features, Anick and Kantamneni (2008) found that 30–37% of users clicked on the dynamic suggestions and about 6% on the after-search suggestion to refine their queries.

Such kind of query reformulation assistance is considered to be particularly useful in exploratory search, as in this case Web searchers usually submit rather tentative queries. To support exploratory search and to maximize searchers' rate of information gain, in recent years, so-called exploratory search systems have been developed. One important capability of these systems to support query reformulation is to allow for relevance feedback given by the user (see Baeza-Yates and Ribeiro-Neto 1999). After having typed in a tentative query, the system presents search results along with a list of additional term suggestions related to the original query. Relevance feedback is enabled by allowing searchers to mark some of the related terms and/or search results that seem particularly relevant to their information needs, thus communicating this information to the search system. The system then uses this relevance feedback to produce a modified query in order to retrieve only documents containing the relevant keywords or documents similar to the search results marked as relevant. Thus, relevance feedback aims to improve a searcher's query and to facilitate retrieval of information relevant to a searcher's personal information needs. However, these additional benefits come at the cost of greater interface complexity.

Because exploratory search systems present additional information and support rich interaction, they are likely to induce a higher cognitive load (Wilson and Schraefel 2008). Bruza, McArthur, and Dennis (2000) compared search effectiveness when using a standard Web search engine (Google or Yahoo!) or an experimental interface with relevance feedback features, a so-called Hyperindex browser. Results of the study showed that query reformulation using the Hyperindex browser improved the relevance of the documents users retrieved compared to the use of a conventional search engine. Yet, this advantage came at the cost of increased search time and increased cognitive load when perusing query refinements. Kammerer et al. (2009a) found similar results with regard to the use of a tag-based exploratory search system called MrTaggy. MrTaggy allows users to give three types of relevance feedback: (1) about tags related to the entered search terms presented to the left of the search results, (2) about tags related to a specific Web page presented right below the title of each of the search results, and (3) about the presented search results themselves. Both positive and negative relevance feedback can be given by clicking on the respective upward or downward thumbs (see Figure 14.5). In an experiment that compared MrTaggy to a baseline version of the system without the relevance feedback features, Kammerer et al. could show for exploratory search tasks (e.g., about the topic global warming), that users of the exploratory search system wrote summaries of higher quality compared to the baseline system. Furthermore, for ambiguous search topics (e.g., future architecture), users of the exploratory system generated more reasonable keywords than did the baseline users. At the same time, exploratory search system users experienced a higher cognitive load and spent more time on the search tasks than baseline users. Moreover, the results of the study indicated that the exploratory search system allows compensation for effects due to differences in prior knowledge, such that in some tasks differences between domain experts and domain novices found for users of the baseline system were absent for exploratory system users.

To conclude, query refinement tools such as spelling corrections and after-search query suggestions used by many current search engines seem to be an important part of the search interface. Furthermore, results of the studies testing exploratory search systems point to the promises of combining query suggestions and relevance feedback for supporting Web users during exploratory search.

14.3.2 SUPPORTING THE EVALUATION AND SELECTION OF SEARCH RESULTS

Web search engines usually present search results in a vertical rank-ordered list, with the presumably most relevant and most popular Web pages being the highest-ranked ones (Cho and Roy 2004). In order to summarize the corresponding Web pages, for each search result on the SERP, a title, an excerpt of the content of the Web page, and its URL are displayed. Thus users' evaluations in order to decide which

FIGURE 14.4 Screenshot of the search box and "sliding tray" of Yahoo!'s Search Assist interface after having added the related concept "climate change" to the query.

FIGURE 14.5 **(See color insert.)** Screenshot of the MrTaggy user interface with (left) related tags section, (right) search results list displaying tags related to the search results, and upward and downward thumbs to provide relevance feedback.

search results to select for further inspection have to be based on sparse information.

To support evaluations with regard to the topical relevance of search results, various interface features have been tested, such as increasing the length of the search result presentation, presenting additional terms or categories, and displaying thumbnail images of Web pages (previews). Cutrell and Guan (2007) explored effects of the length of the document excerpts of search results. They showed that adding information to the document excerpts, thus increasing their length, improved performance for informational tasks but hampered performance for navigational tasks. Furthermore, Woodruff et al. (2002) proposed to use an additional thumbnail representation of documents to improve the evaluation of a Web site's topical relevance prior to selection. Their results showed that a combined presentation of text and thumbnails leads to better task performance in simple as well as in complex tasks than either text-only or thumbnail-only representations of the search results. Other studies (Drori 2003; Dumais, Cutrell, and Chen 2001) have demonstrated that adding metadata to the search results such as subject matter categories (e.g., sports, shopping, travel) or keywords indicating contents of Web pages improves task efficiency in fact-finding tasks and increases users' sense of confidence that the search result they selected will yield the target information. Moreover, according to the study by Dumais, Cutrell, and Chen (2001) so-called category interfaces, with search results grouped according to subject matter categories, were more effective than conventional list interfaces without grouping, even when the lists were augmented with category names for each result. Similarly, Käki and Aula (2005) showed that Web users found relevant search results significantly faster with their category filtering search system Findex than with a Google-like search interface. Furthermore, the Findex system yielded more positive subjective ratings.

These approaches outlined so far address the support of topicality-related evaluations of search results. However, empirical research has shown that quality-related evaluations of search results (and Web pages) are employed only rarely

during Web search and might thus need additional support (see Section 14.2.3). By now, only a few studies have investigated how changes in the interface design of search engines or specific interface features might foster users to evaluate search results in terms of the expected quality and credibility of the corresponding Web pages and the information provided therein.

As outlined in Section 14.2.3, previous research by Keane, O'Brien, and Smyth (2008) and Pan et al. (2007) had shown that Web users are heavily influenced by the ranking order offered by the search engines. However, especially when dealing with complex information searches about controversial topics, the focus on selecting the highest-ranked search results may lead to a biased and incomplete selection of information. For instance, it is often the case that popular commercial Web sites, which might provide rather one-sided information, fit the search terms entered by the user exactly and thus are listed among the highest-ranked results.

Kammerer and Gerjets (2010) suggest that the strong impact of the search engine's ranking might be at least partly caused by the list format, in which search results are typically presented. It can be assumed that a list format provides a strong affordance for users to start reading at the top of the list and that it imposes a strict and nonambiguous order in which to read and select search results. As a result, searchers may be inhibited to engage in search result evaluations on their own or to trust their own evaluations. Thus, the question arises whether changes in the format of a search results interface may reduce the impact of the ranking order, increase the awareness of the selection process, and thereby stimulate users to engage in quality-related evaluations on SERPs.

Kammerer and Gerjets (2010) examined this question by comparing a standard list format to an alternative grid format when searching for information about two alternative therapies for Bechterew's disease. The grid format, which was recently used by some novel search engines (e.g., Viewzi or Gceel), presents search results in multiple rows and columns (see Figure 14.6). This organization implies that there is no strict and nonambiguous order in which to read and

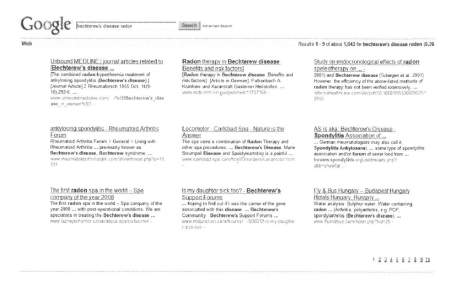

FIGURE 14.6 Screenshot of a grid interface with nine search results for the query "bechterew's disease radon" used in a study by Kammerer and Gerjets (2010).

select search results. It is unclear whether the ranking within a SERP is aligned horizontally (i.e., line by line, according to the regular western reading direction) or vertically (i.e., column by column, according to a list structure), or whether there exists a ranking at all. Therefore, in a grid interface the decision about the reading order is left open to the user and the ranking is less salient, which might encourage users to engage in their own evaluations in order to decide which search results to select. For the study, the methodological paradigm used by Keane, O'Brien, and Smyth (2008) and Pan et al. (2007) (see Section 14.2.3) was adapted by experimentally manipulating the trustworthiness order of the search results on a SERP. Nine search results, which were all of high topical relevance, were either presented in an *optimal* order, with the most trustworthy search results presented first and the least trustworthy ones presented last or in a reversed *suboptimal* order, with the least trustworthy search results presented first.* For both experimental order conditions, the list interface caused more homogeneous and linear viewing sequences on SERPs than did the grid interface. Furthermore, when using the list interface, most attention was given to the search results at the top of the list. In contrast, with a grid interface, nearly all search results on a SERP were attended to equivalently long. Consequently, when search results were presented in a *suboptimal* order (with the least trustworthy search results presented first), participants using the list interface attended longer to these least trustworthy search results and selected the most trustworthy search results significantly less often than participants using the grid interface. When search results were presented in an *optimal* order, in both interfaces the most trustworthy search results were predominantly selected. To conclude, at least in case that search results are not presented in an optimal trustworthiness order

(e.g., when the top ranked search results are linked to popular commercial Web sites or to contributions in blogs or forums), the presentation of Web search results by means of a grid interface seems to support users in their selection of trustworthy information sources.

Another approach to support users in their quality-related evaluation processes on SERPs is to provide additional quality-related information along with the search results. Search result descriptions used in standard search environments like Google are mostly confined to topical information, whereas quality-related source information is rather sparse and nonsalient. Thus, the interface design of standard search engines does not provide high affordances for users to spontaneously engage in quality-oriented evaluation processes when selecting search results during Web search. Providing additional quality-related source information on SERPs may stimulate searchers to engage in quality-related evaluations.

With regard to the online news service Google News, a study by Sundar, Knobloch-Westerwick, and Hastall (2007) showed that the availability of quality-related cues for news items (i.e., information on the source, the recency of a story, and the number of related articles; see Figure 14.7) indeed affected users' subjective evaluations of news leads, with the source information having the strongest effects on users'

Study suggests greater sea level rise from **warming**
The Associated Press - Malcolm Ritter - 16 hours ago
NEW YORK — **Global warming** in this century might raise sea levels more than expected in future centuries, says a study that looked at what happened at a time ...
Earth's Polar Ice Sheets Vulnerable to Even Moderate **Global Warming**. New ... Science Daily (press release)
Two degrees to disaster Sydney Morning Herald
Two-degree temperature rise could flood wide areas of planet, study says The Gazette (Montreal)
AFP - Scotsman
all 135 news articles » Email this story

FIGURE 14.7 Google news item for the query "global warming" including information on the source, the recency of the news, and the number of related articles.

---

* For the grid interface search results were arranged line by line, that is, from left to right in each of three rows.

credibility judgments regarding the news leads. Furthermore, when the source was judged as being of low credibility, the recency of the story and the number of related articles also had a strong impact on users' judgments regarding the credibility of the news lead.

Ivory, Yu, and Gronemyer (2004) enhanced the Google search engine interface with additional quality-related cues for each search result indicating the number of graphical ads, the number of words, and the estimated quality of the corresponding Web page. They used simple fact-finding tasks to demonstrate that adding this type of quality-related cues improved participants' ability to select appropriate search results. Rose, Orr, and Kantamneni (2007) conducted a series of experiments with rather complex search scenarios in which searchers had to answer questions about the quality of search results on the basis of search result descriptions. Search result descriptions were experimentally manipulated to make sure that they differed with regard to specific attributes (e.g., length, text choppiness, and cues indicating a source category). With regard to quality-related evaluations, results showed that providing cues about the source category of a Web page (e.g., a corporate homepage or a blog), influenced whether searchers trusted the information on the Web pages.

Similarly, Kammerer et al. (2009b) investigated whether increasing the salience of source information on a SERP fosters Web users' evaluation processes. They experimentally manipulated Google SERPs such that, in addition to the URL of each search result, source category cues were presented (see Figure 14.8) that indicated to which of five different source categories a search result belonged. The five source categories were "Science/Institutions," "Portals/Advisors," "Journalism/TV," "Readers' Comments," and "Shops/Companies" (translated from German). In the study university students used a standard Google search result list or an augmented list that contained additional source categories for each search result to solve a complex search task (about the pros and cons of different diet methods). The availability of source categories influenced students' evaluation and selection behavior, such that they spent less attention on commercial search results from the "Shops/Companies" category and were more likely to select search results from the "Portals/Advisors" category. Beyond that, the availability of source categories on SERPs stimulated students with sophisticated epistemological beliefs (i.e., critical Web searchers;

see Section 14.2.4) to engage in evaluation processes with regard to search results belonging to the categories "Portals/Advisors," "Journalism/TV," and "Readers' Comments," which were rather ambiguous in terms of their trustworthiness. In contrast, students with naïve epistemological beliefs did not focus on these ambiguous categories.

In summary, the results of these studies indicate that augmenting SERPs by presenting additional topical information or quality-related information, provide substantial affordances for searchers to engage in topicality-related or quality-related evaluation processes respectively, thus supporting them in their Web search. Hence, the development of automated Web page classification algorithms to organize Web information into a predefined set of topical categories (e.g., Dumais, Cutrell, and Chen 2001) or genre/functional categories (e.g., Chen and Choi 2008; Qi and Davison 2009; Santini, Power, and Evans 2006) seems to be a promising research direction. However, in contrast to topic classifications, genre or functional classifications based on dimensions such as type of source (e.g., forum, blog, university Web site), degree of author expertise, purpose of the Web page, and whether the information is based on facts or opinions, are highly subjective and thus difficult to generate appropriately (Chen and Choi 2008).

14.3.3 Supporting the Synthesis of Web Information

When search information is retrieved from more than one Web page, users are required to compare and integrate the different contents toward a solution of their current information problem. For the synthesis of information, trustworthiness evaluations and evaluations of the relations between different documents are required. Only few tools have been developed and evaluated that aim at helping users to relate different documents to each other, to corroborate and integrate information from different sources, or to evaluate information in the light of source characteristics. One example of a search system that displays the relationships between documents was the meta-search engine Kartoo (closed down in January 2010). Whereas conventional search engines such as Google do not signal any relationships between the search results presented, Kartoo displayed the search results by means of a graphical overview (i.e., a mind map). This kind of spatial arrangement indicates the relationships between the search results. When hovering

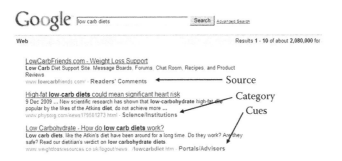

FIGURE 14.8 Screenshot of an augmented Google-like SERP showing 3 of 10 search results with additional source categories (Readers' Comments, Science/Institutions, and Portals/Advisors) for the query "low carb diets" used in a study by Kammerer et al. (2009b).

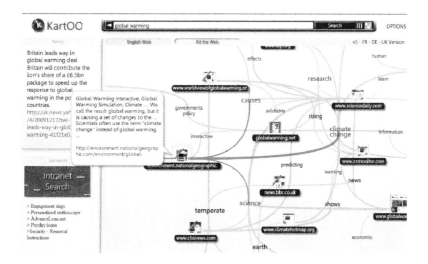

FIGURE 14.9   **(See color insert.)** Screenshot from the graphical-overview interface of the metasearch engine Kartoo for the query "global warming."

over an individual result a bunch of red lines appeared, which connected related links, thus indicating existing relations between Web pages (see Figure 14.9). Salmerón et al. (2010) investigated the effects of such a graphical overview search results interface on users' multiple documents comprehension in a complex search task about climate change. Results of the study showed that undergraduate students comprehended the information provided on the different Web pages better when search results were presented in a Kartoo-like format than when the same search results were presented in a standard list (i.e., Google-like) interface. The facilitative effect of the graphical-overview interface was reflected in intertextual inferential tasks, which required students to integrate key information from different Web pages.

Gwizdka (2009) compared Web users' search performance of simple fact-finding and complex search tasks in two types of search results interfaces, namely, a standard list interface and an overview interface. In the overview interface, a tag cloud with important terms associated with the returned search results was provided in addition to the search results presented in a vertical list (similar to the search engine Quintura, see Figure 14.10). Thus, in the tag cloud overview, those words (tags) that are most closely connected to the search query entered by the user were displayed. The tags served as links to refine the query. Results of the study revealed that the overview interface benefited searchers in several ways. Compared to the list interface, undergraduate students using the overview interface were faster in accomplishing simple

FIGURE 14.10   **(See color insert.)** Screenshot from the tag cloud overview interface of the search engine Quintura for the query "global warming."

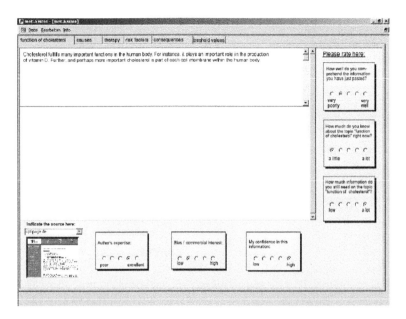

FIGURE 14.11 Screenshot of the experimental met.a.ware support tool with the note taking facility including (top) ontological categories, (bottom) evaluation prompts, and (right) monitoring prompts.

search tasks (but not complex search tasks) and were more efficient regarding the number of queries used. Furthermore, the overview interface facilitated query refinements and better supported the evaluation of search results.

Stadtler and Bromme (2007, 2008) used a different approach to signal relations between various documents on the Web. They did not change the design of the search results interface but developed an experimental support tool called met.a.ware, which provided users with a note taking facility including a set of predefined ontological categories (about the topic cholesterol, see tabs in Figure 14.11, top) in order to systematically store information found during Web search. Furthermore, met.a.ware provided prompts to evaluate the different Web pages users accessed during Web search regarding their trustworthiness (see Figure 14.11, bottom) as well as prompts to monitor their own understanding of the contents (see Figure 14.11, right).

In their study, in which university students had to conduct a Web search and to write an essay about a medical topic (cholesterol), Stadtler and Bromme (2007, 2008) showed that the note taking facility including categories improved the structure of students' essays compared to a control group that took notes without the facility. Furthermore, the prompts effectively supported users' metacognitive processes during Web research. Monitoring prompts enhanced students' factual knowledge and the comprehension of relations between documents. Evaluation prompts improved the knowledge about sources and produced more source-related justifications of credibility judgments compared to control groups not receiving the respective prompts. Thus, in terms of the "theory of documents representation" (Perfetti, Rouet, and Britt 1999; see Section 14.2.3) the met.a.ware tool fostered the formation of documents models when dealing with multiple

documents on the Web. As the categories used in this study were topic specific (about cholesterol), it is, however, unclear how met.a.ware could be applied for various topic domains and whether the effects would hold for different domains. To summarize, search results interfaces that indicate relations between different documents or provide an overview of important terms associated with the returned search results are promising measures to support users in the process of comparing and integrating information. Web browsers that provide a structured note taking facility and metacognitive prompts can also be expected to be beneficial for users.

14.4 CONCLUSIONS AND FUTURE RESEARCH DIRECTIONS

To conclude, as the Web is constantly growing, it can be assumed that information seeking on the Web will also be an important field of research in the future. Moreover, new application areas and interface features are constantly emerging that can be assumed to play an important role in future Web search. First, for instance, most major search engines currently offer the opportunity of a personalized Web search according to users' personal preferences and search habits (e.g., iGoogle). Second, with the increasing bandwidth of mobile devices and the development of larger high-resolution displays, Web search via mobile devices is becoming increasingly popular. This affords the development of appropriate search results interfaces dealing with the limited display space of mobile devices. Third, apart from general search engines, specialized search engines, for instance, on Web 2.0 contents such as Wikis or blogs, and on multimedia material such as videos (e.g., Blinkx or Truveo) are emerging. Finally, with the rise of the Web 2.0, social tagging and recommendation systems (e.g., Delicious or Digg), which use the "wisdom of the crowd" to rate the relevance of Web information, to answer questions, or to recommend products, become an increasingly popular way for users to find information.

With regard to all of these developments, research is needed to investigate how novel search systems and interface features are used during different types of Web search tasks in order to further expand our understanding on cognitive processes involved during searching and evaluating information on the Web and to further improve Web searching in the future.

REFERENCES

Anderson, J. R., D. Bothell, M. Byrne, D. Douglass, C. Lebiere, and Y. Qin. 2004. An integrated theory of mind. *Psychological Review* 111: 1036–1060.

Anick, P., and R. G. P. Kantamneni. 2008. A longitudinal study of real-time search assistance adoption. In *Proceedings of the 31st Annual International ACM SIGIR Conference on Research and development in information retrieval. SIGIR '08,* 701–702. New York: ACM Press.

Aula, A., and D. M. Russell. 2008, Complex and exploratory Web search. Paper presented at the NSF IS3 Information Seeking Support Systems Workshop, Chapel Hill, NC, June. http://www.ils.unc.edu/ISSS/papers/papers/aula.pdf (accessed November 3, 2010).

Baeza-Yates, R., and B. Ribeiro-Neto. 1999. *Modern Information Retrieval.* Reading, MA: Addison-Wesley.

Bhavnani, S. K. 2001. Important cognitive components of domain-specific search knowledge. In *NIST Special Publication 500-250: The Tenth Text Retrieval Conference,* eds. E. M. Vorhees and D. K. Harman, 571–578. Washington, DC: National Institute of Standards and Technology.

Bilal, D. 2000. Children's use of the Yahooligans! Web search engine. I. Cognitive, physical, and affective behaviors on fact-based search tasks. *Journal of the American Society for Information Science* 51: 646–665.

Blackmon, M. H., M. Kitajima, and P. G. Polson. 2005. Tool for accurately predicting Website navigation problems, non-problems, problem severity, and effectiveness of repairs. In *Proceedings of the SIGCHI Conference on Human Factors in Computing Systems. CHI '05,* 31–40. New York: ACM Press.

Borlund, P. 2003. The concept of relevance in IR. *Journal of the American Society for Information Science and Technology* 54: 913–925.

Brand-Gruwel, S., and P. Gerjets. 2008. Instructional support for enhancing students' information problem solving ability. *Computers in Human Behavior* 24: 615–622.

Brand-Gruwel, S., L. Van Meeuwen, and T. Van Gog. 2008. The use of evaluation criteria when searching the WWW: An eye-tracking study. In *Proceedings of the EARLI SIG Text and Graphics* (Tilburg, NL), 34–37.

Brand-Gruwel, S., I. Wopereis, and Y. Vermetten. 2005. Information problem solving by experts and novices: Analysis of a complex cognitive skill. *Computers in Human Behavior* 21: 487–508.

Brand-Gruwel, S., I. Wopereis, and A. Walraven. 2009. A descriptive model of information problem solving while using Internet. *Computers & Education* 53: 1207–1217.

Bråten, I., and H. I. Strømsø. 2006. Epistemological beliefs, interest, and gender as predictors of Internet-based learning activities. *Computers in Human Behavior* 22: 1027–1042.

Bråten, I., and H. I. Strømsø. 2010. When law students read multiple documents about global warming: Examining the role of topic-specific beliefs about the nature of knowledge and knowing. *Instructional Science* 38: 635–657.

Bråten, I., H. I. Strømsø, and M. A. Britt. 2009. Trust matters: Examining the role of source evaluation in students' construction of meaning within and across multiple texts. *Reading Research Quarterly* 44: 6–28.

Breivik, P. S. 2005. 21st century learning and information literacy. *Change* 37: 20–27.

Britt, M. A., and C. Aglinskas. 2002. Improving students' ability to identify and use source information. *Cognition and Instruction* 20: 485–522.

Broder, A. 2002. A taxonomy of Web search. *SIGIR Forum* 36: 3–10.

Brown, M. E. 1991. A general model of information-seeking behavior. In P*roceedings of the 54th American Society for Information Science (ASIS) Annual Meeting,* ed. J.-M. Griffiths, vol. 28, 9–14. Medford, NJ: Learned Information.

Brumby, D. P., and A. Howes. 2008. Strategies for guiding interactive search: An empirical investigation into the consequences of label relevance for assessment and selection. *Human–Computer Interaction* 23: 1–46.

Bruza, P., R. McArthur, and S. Dennis. 2000. Interactive Internet search: keyword, directory and query reformulation mechanisms compared. In *Proceedings of the 23rd Annual International ACM SIGIR Conference on Research and Development in Information Retrieval, SIGIR 2000,* 280–287. New York: ACM Press.

Chen, G., and B. Choi. 2008. Web page genre classification. In *Proceedings of the 2008 ACM Symposium on Applied Computing. SAC '08,* 2353–2357. New York: ACM Press.

Cho, J., and S. Roy. 2004. Impact of search engines on page popularity. In *Proceedings of the 13th International Conference on World Wide Web. WWW 2004,* 20–29. New York: ACM Press.

Conklin, J. 1987. Hypertext: A survey and introduction. *IEEE Computer* 20: 17–41.

Crystal, A., and J. Greenberg. 2006. Relevance criteria identified by health information users during Web searches. *Journal of the American Society for Information Science and Technology* 57: 1368–1382.

Cutrell, E., and Z. Guan. 2007. What are you looking for? An eye-tracking study of information usage in Web search. In *Proceedings of the SIGCHI Conference on Human Factors in Computing Systems. CHI '07,* 407–416. New York: ACM Press.

Dillon, A. 2002. Writing as design: Hypermedia and the shape of information space. In *Writing Hypertext and Learning: Conceptual and Empirical Approaches,* eds. R. Bromme and E. Stahl. New York: Pergamon.

Drori, O. 2003. Display of search results in Google-based Yahoo! vs. LCC&K interfaces: A comparison study. In *Proceedings of Informing Science 2003 Conference,* 309–320. http://proceedings.informingscience.org/IS2003Proceedings/index.html (accessed November 3, 2010).

Duggan, G. B., and S. J. Payne. 2008. Knowledge in the head and on the Web: Using topic expertise to aid search. In *Proceedings of the CHI 2008 Conference on Human Factors in Computing Systems,* 39–48. New York: ACM Press.

Dumais, S. T., E. Cutrell, and H. Chen. 2001. Optimizing search by showing results in context. In *Proceedings of the SIGCHI Conference on Human Factors in Computing Systems. CHI '01,* 277–284. New York: ACM Press.

Ericsson, K. A., and H. A. Simon. 1993. *Protocol Analysis: Verbal Reports as Data.* Cambridge, MA: MIT Press.

Eysenbach, G., and C. Köhler. 2002. How do consumers search for and appraise health information on the World Wide Web? Qualitative studies using focus groups, usability tests, and in-depth interviews. *British Medical Journal* 324: 573–577.

Fallows, D. 2008. Search engine use. Pew Internet & American Life Project. http://www.pewinternet.org/Reports/2008/Search-Engine-Use.aspx (accessed November 3, 2010).

Fogg, B. J., J. Marshall, O. Laraki, A. Osipovich, C. Varma, N. Fang, J. Paul, A. Rangnekar, J. Shon, P Swani, and M. Treinen. 2001. What makes a Web site credible? A report on a large quantitative study. In *Proceedings of the SIGCHI Conference on Human Factors in Computing Systems. CHI'01*, 62–68. New York: ACM Press.

Fox, S. 2006. Online health search 2006. Pew Internet & American Life Project. http://www.pewinternet.org/~/media//Files/Reports/2006/PIP_Online_Health_2006.pdf.pdf (accessed November 3, 2010).

Fox, S., and S. Jones. 2009. The social life of health information. Pew Internet & American Life Project. http://www.pewinternet.org/Reports/2009/8-The-Social-Life-of-Health-Information.aspx (accessed November 3, 2010).

Gerjets, P., and T. Hellenthal-Schorr. 2008. Competent information search in the World Wide Web: Development and evaluation of a Web training for pupils. *Computers in Human Behavior* 24: 693–715.

Gerjets, P., Y. Kammerer, and B. Werner. 2011. Measuring spontaneous and instructed evaluation processes during Web search: Integrating concurrent thinking-aloud protocols and eye-tracking data. *Learning & Instruction* 21: 220–231.

Gordon, M., and P. Pathak. 1999. Finding information on the World Wide Web: The retrieval effectiveness of search engines. *Information Processing and Management* 35: 141–180.

Granka, L., T. Joachims, and G. Gay. 2004. Eye-tracking analysis of user behavior in www search. In *Proceedings of the 28th Annual ACM SIGIR Conference on Research and Development in Information Retrieval*, 478–479. New York: ACM Press.

Guan, Z., and E. Cutrell. 2007. An eye tracking study of the effect of target rank on Web search. In *Proceedings of the SIGCHI Conference on Human Factors in Computing Systems. CHI '07*. 417–420. New York: ACM Press.

Gwizdka, J. 2009. What a difference a tag cloud makes: Effects of tasks and cognitive abilities on search results interface use. *Information Research* 14, paper 414. http://informationr.net/ir/14-4/paper414.html.

Hearst, M. 2009. *Search User Interfaces*. Cambridge: Cambridge University Press.

Hofer, B. K. 2004. Epistemological understanding as a metacognitive process: thinking aloud during online searching. *Educational Psychologist* 39: 43–55.

Hofer, B. K., and P. R. Pintrich. 1997. The development of epistemological theories: Beliefs about knowledge and knowing and their relation to learning. *Review of Educational Research* 67: 88–140.

Horrigan, J. 2006. The Internet as a resource for news and information about science. Pew Internet & American Life Project. http://www.pewinternet.org/Reports/2006/The-Internet-as-a-Resource-for-News-and-Information-about-Science.aspx (accessed November 3, 2010).

Hölscher, C., and G. Strube. 2000. Web search behavior of Internet experts and newbies. *Computer Network* 33: 337–346.

Ivory, M. Y., S. Yu, and K. Gronemyer. 2004. Search result exploration: A preliminary study of blind and sighted users' decision making and performance. In *CHI '04 Extended Abstracts on Human Factors in Computing Systems*. 1453–1456. New York: ACM Press.

Jansen, B. J., D. L. Booth, and A. Spink. 2008. Determining the informational, navigational, and transactional intent of Web queries. *Information Processing & Management* 44: 1251–1266.

Jansen, B. J., and U. Pooch. 2001. A review of Web searching studies and a framework for future research. *Journal of the American Society for Information Science and Technology* 52: 235–246.

Jansen, B. J., and A. Spink. 2006. How are we searching the World Wide Web? A comparison of nine large search engine transaction logs. *Information Processing and Management* 42: 248–263.

Jansen, B. J., A. Spink, C. Blakely, and S. Koshman. 2007. Web searcher interaction with the Dogpile Web search engine. *Journal of the American Society for Information Science and Technology* 58: 744–755.

Jansen, B. J., A. Spink, and S. Koshman. 2007. Web searcher interaction with the Dogpile.com metasearch engine. *Journal of the American Society for Information Science and Technology* 58: 744–755.

Joachims, T., L. Granka, B. Pan, H. Hembrooke, and G. Gay. 2005. *Accurately* interpreting clickthrough data as implicit feedback. In *Proceedings of the 28th Annual International ACM SIGIR Conference on Research and Development in information Retrieval. SIGIR '05*, 154–161. New York: ACM Press.

Jonassen, D. H. 1997. Instructional design models for well-structured and ill-structured problem-solving learning outcomes. *Educational Technology: Research and Development* 45: 65–94.

Juvina, I., and H. Van Oostendorp. 2008. Modeling semantic and structural knowledge in Web navigation. *Discourse Processes* 45: 346–364.

Kammerer, Y., and P. Gerjets. 2010. How the interface design influences users' spontaneous trustworthiness evaluations of Web search results: Comparing a list and a grid interface. In *Proceedings of the 2010 Symposium on Eye Tracking Research & Applications ETRA '10*, eds. C. Morimoto and H. Instance, 299–306. New York: ACM Press.

Kammerer, Y., R. Nairn, P. Pirolli, and E. H. Chi. 2009a. Signpost from the masses: Learning effects in an exploratory social tag search browser. In *Proceedings of the 27th Annual SIGCHI Conference on Human Factors in Computing Systems CHI '09*, eds. D. R. Olsen et al., 625–634. New York: ACM Press.

Kammerer, Y., E. Wollny, P. Gerjets, and K. Scheiter. 2009b. How authority-related epistemological beliefs and salience of source information influence the evaluation of Web search results—an eye tracking study. In *Proceedings of the 31st Annual Conference of the Cognitive Science Society*, eds. N. A. Taatgen and H. van Rijn, 2158–2163. Austin, TX: Cognitive Science Society.

Käki, M., and A. Aula. 2005 Findex: Improving search result use through automatic filtering categories. *Interacting with Computers* 17: 187–206.

Keane, M.T., M. O'Brien, and B. Smyth. 2008. Are people biased in their use of search engines? *Communications of the ACM* 51: 49–52.

Kellar, M., C. Watters, and M. Shepherd. 2007. A field study characterizing Web-based information-seeking tasks. *Journal of the American Society for Information Science and Technology* 58: 999–1018.

Kuhlthau, C. 1991. Inside the search process: Information seeking from the user's perspective. *Journal of the American Society for Information Science* 42: 361–371.

Kuhlthau, C. 1993. *Seeking Meaning. A Process Approach to Library and Information Services*. Norwood, NJ: Ablex.

Law, D. 2008. Convenience trumps quality: How digital natives use information. http://web.fumsi.com/go/article/use/2971 (accessed November 3, 2010).

Lazonder, A. W., H. J. A. Biemans, and I. G.-J. H. Wopereis. 2000. Differences between novice and experienced users in searching information on the World Wide Web. *Journal of the American Society for Information Science* 51: 576–581.

Lorigo, L., B. Pan, H. Hembrooke, T. Joachims, L. Granka, and G. Gay. 2006. The influence of task and gender on search and evaluation behavior using Google. *Information Processing and Management* 42: 1123–1131.

MaKinster, J. G., R. A. Beghetto, and J. A. Plucker. 2002. Why can't I find Newton's third law?: Case studies of students using of the Web as a science resource. *Journal of Science Education and Technology* 11: 155–172.

Mandl, T. 2006. Implementation and evaluation of a quality-based search engine. In *Proceedings of the Seventeenth Conference on Hypertext and Hypermedia. HYPERTEXT '06*, 73–84. New York: ACM Press.

Marchionini, G. 1989. Information-seeking strategies of novices using a full-text electronic encyclopedia. *Journal of the American Society for Information Science* 40: 54–66.

Marchionini, G. 1995. *Information Seeking in Electronic Environments.* New York: Cambridge University Press.

Marchionini, G. 2006. Exploratory search: From finding to understanding. *Communications of the ACM* 49: 41–46.

Marchionini, G., and R. W. White. 2008. Find what you need, understand what you find. *Journal of Human–Computer Interaction* 23: 205–237.

Mason, L., and N. Ariasi. 2010. Critical thinking about biology during web page reading: Tracking students' evaluation of sources and information through eye fixations. In *Use of External Representations in Reasoning and Problem Solving: Analysis and Improvement*, eds. L. Verschaffel, E. De Corte, T. de Jong, & J. Elen, pp. 55–73. London: Routledge.

Metzger, M. J., A. J. Flanagin, and L. Zwarun. 2003. College student Web use, perceptions of information credibility, and verification behavior. *Computers & Education* 41: 271–290.

Miller, C. S., and R. W. Remington. 2004. Modeling information navigation: Implications for information architecture. *Human–Computer Interaction* 19: 225–271.

Mizzaro, S. 1997. Relevance: The whole history. *Journal of the American Society for Information Science* 48: 810–832.

Morahan-Martin, J. M. 2004. How Internet users find, evaluate, and use online health information: A cross-cultural review. *CyberPsychology & Behavior* 7: 497–510.

Navarro-Prieto, R., M. Scaife, and Y. Rogers. 1999. Cognitive strategies in Web searching. In *Proceedings of the 5th Conference on Human Factors & the Web.* http://zing.ncsl.nist.gov/hfweb/proceedings/navarro-prieto/ (accessed November 3, 2010).

Newell, A., and H. A. Simon. 1972. *Human Problem Solving.* Englewood Cliffs, NJ: Prentice Hall.

Palmquist, R. A., and K. S. Kim. 2000. Cognitive style and online database search experience as predictors of Web search performance. *Journal of the American Society for Information Science* 51: 558–566.

Pan, B., H. Hembrooke, T. Joachims, L. Lorigo, G. Gay, and L. Granka. 2007. In Google we trust: users' decisions on rank, position, and relevance. *Journal of Computer-Mediated Communication* 12: 801–823.

Park Y., and J. B. Black. 2007. Identifying the impact of domain knowledge and cognitive style on Web-based information search behavior. *Journal of Educational Computing Research* 38: 15–37.

Perfetti, C. A., J.-F. Rouet, and M. A. Britt. 1999. Toward a theory of documents representation. In *The construction of Mental Representations during Reading*, eds. H. van Oostendorp and S. R. Goldman, 99–122. Mahwah, NJ: Lawrence Erlbaum.

Pirolli, P. 2007. *Information Foraging Theory. Adaptive Interaction with Information.* New York: Oxford University Press.

Pirolli, P., and S. K. Card. 1999. Information foraging. *Psychological Review* 106: 643–675.

Pirolli, P., and W.-T. F. Fu. 2003. SNIF-ACT: A model of information foraging on the World Wide Web. In *9th International Conference on User Modeling*, ed. P. Brusilovsky, A. Corbett, and F. de Rosis, 45–54. New York: Springer-Verlag.

Prensky, M. 2001. Digital natives, digital immigrants. *On the Horizon, 9.*

Qi, X., and B. D. Davison. 2009. Web page classification: features and algorithms. *ACM Computing Surveys, 41.*

Rieh, S. Y. 2002. Judgment of information quality and cognitive authority in the Web. *Journal of the American Society for Information Science and Technology* 53: 145–161.

Rose, D. E., and D. Levinson. 2004. Understanding user goals in Web search. In *Proceedings of the 13th International Conference on World Wide Web, WWW '04*, 13–19. New York: ACM Press.

Rose, D. E., D. Orr, and R. G. P. Kantamneni. 2007. Summary attributes and perceived search quality. In *Proceedings of the 2007 International Conference on the World Wide Web 2007*, 1201–1202. New York: ACM Press.

Salmerón, L., L. Gil, I. Bråten, and H. I. Strømsø. 2010. Comprehension effects of signalling relationships between documents in search engines. *Computers in Human Behavior* 26: 419–426.

Santini M., R. Power, and E. Evans. 2006. Implementing a characterization of genre for automatic genre identification of Web pages. In *Proceedings of the COLING/ACL 2006 Main Conference Poster Sessions*, 699–706. Morristown: Association for Computational Linguistics.

Savolainen, R., and J. Kari. 2005. User-defined relevance criteria in Web searching. *Journal of Documentation* 62: 685–707.

Schamber, L. 1994. Relevance and information behaviour. In *Annual Review of Information Science and Technology (ARIST)*, ed. M. E. Wiliams, 3–48. Medford, NJ: Learned Information.

Shneiderman, B. 1997. Designing information-abundant Web sites: issues and recommendations. *International Journal of Human–Computer Studies* 47: 5–29.

Shneiderman, B., D. Byrd, and W. B. Croft. 1997. Clarifying search: a user-interface framework for text searches. *DL Magazine.* http://www.dlib.org/dlib/january97/retrieval/01shneiderman .html (accessed November 3, 2010).

Shneiderman, B., D. Byrd, and W. B. Croft. 1998. Sorting out searching: A user-interface framework for text searches. *Communications of the ACM* 41: 95–98.

Simon, H. A. 1955. A behavioral model of rational choice. *Quarterly Journal of Economics* 69: 99–118.

Simon, H. A. 1973. Structure of ill structured problems. *Artificial Intelligence* 4: 181–201.

Smith, A. 2009. The Internet's role in campaign 2008. Pew Internet & American Life Project. http://www.pewinternet.org/Reports/2009/6--The-Internets-Role-in-Campaign-2008.aspx (accessed November 3, 2010).

Spink, A., B. J. Jansen, D. Wolfram, and T. Saracevic. 2002. From E-sex to E-commerce: Web search changes. *IEEE Computer* 35: 107–109.

Spyridakis J. H., and C. S. Isakson. 1991. Hypertext: A new tool and its effect on audience comprehension. *IEEE-IPCC, 91*, 37–44.

Stadtler, M., and R. Bromme. 2004. Laypersons searching the WWW for medical information: The role of metacognition. In *Proceedings of the 26th Annual Conference of the Cognitive Science Society*, eds. K. Forbus, D. Gentner, and T. Reiger, 1638. Mahwah, NJ: Lawrence Erlbaum.

Stadtler, M., and R. Bromme. 2007. Dealing with multiple documents on the WWW: The role of metacognition in the formation of documents models. *Computer-Supported Collaborative Learning* 2: 191–210.

Stadtler, M., and R. Bromme. 2008. Effects of the metacognitive tool met.a.ware on the Web search of laypersons. *Computers in Human Behavior* 24: 716–737.

Strømsø, H. I., and I. Bråten. 2010. The role of personal epistemology in the self-regulation of Internet-based learning. *Metacognition and Learning* 5: 91–111.

Sundar, S. S., S. Knobloch-Westerwick, and M. R. Hastall. 2007. News cues: Information scent and cognitive heuristics. *Journal of the American Society for Information Science and Technology* 58: 366–378.

Sutcliffe, A. G., and M. Ennis. 1998. Towards a cognitive theory of information retrieval. *Interacting with Computers* 10: 321–351.

Taraborelli, D. 2008. How the Web is changing the way we trust. In *Current Issues in Computing and Philosophy*, eds. A. Briggle, K. Waelbers, and P. A. E. Brey, 194–204. Amsterdam: IOS Press.

Tang, P., and P. Solomon. 2001. Use of relevance criteria across stages of document evaluation: On the complementarity of experimental and naturalistic studies. *Journal of the American Society for Information Science and Technology* 52: 676–685.

Tombros, A., I. Ruthven, and J. M. Jose. 2005. How users assess Web pages for information-seeking. *Journal of the American Society for Information Science and Technology* 56: 327–344.

Tsai, C.-C. 2004. Beyond cognitive and metacognitive tools: The use of the Internet as an "epistemological" tool for instruction. *British Journal of Educational Technology* 35: 525–536.

Vakkari, P. 1999. Task complexity, problem structure and information actions: Integrating studies on information seeking and retrieval. *Information Processing and Management* 35: 819–837.

Walraven, A., S. Brand-Gruwel, and H. P. A. Boshuizen. 2008. Information-problem solving: A review of problems students encounter and instructional solutions. *Computers in Human Behavior* 24: 623–648.

Walraven, A., S. Brand-Gruwel, and H. P. A. Boshuizen. 2009. How students evaluate sources and information when searching the World Wide Web for information. *Computers & Education* 25: 234–246.

White, R. W., S. M. Drucker, M. Marchionini, M. Hearst, and M. C. Schraefel. 2007. Exploratory search and HCI: Designing and evaluating interfaces to support exploratory search interaction. In *Extended Abstracts CHI '07*, 2877–2880. New York: ACM Press.

White, R. W., S. T. Dumais, and J. Teevan. 2009. Characterizing the influence of domain expertise on Web search behavior. In *Proceedings of the Second ACM International Conference on Web Search and Data Mining. WSDM '09*, eds. R. Baeza-Yates et al., 132–142. New York: ACM Press.

White, R. W., and G. Marchionini. 2007. Examining the effectiveness of real-time query expansion. *Information Processing and Management* 43: 685–704.

Whitmire, E. 2003. Epistemological beliefs and the information-seeking behavior of undergraduates. *Library & Information Science Research* 25: 127–142.

Whittaker, S. 2008. Making sense of sensemaking. In *HCI Remixed: Reflections on Works that Have Influenced the HCI Community*. Boston, MA: MIT Press.

Wiley, J., S. Goldman, A. Graesser, C. Sanchez, I. Ash, and J. Hemmerich. 2009. Source evaluation, comprehension, and learning in internet science inquiry tasks. *American Educational Research Journal* 46: 1060–1106.

Willoughby, T., A. Anderson, E. Wood, J. Mueller, and C. Ross. 2009. Fast searching for information on the Internet to use in a learning context: The impact of prior knowledge. *Computers & Education* 52: 640–648.

Wilson, T. D. 1999. Models in information behaviour research. *Journal of Documentation* 55: 249–270.

Wilson, M. L., and M. C. Schraefel. 2008. Improving exploratory search interfaces: adding value or information overload? In *Second Workshop on Human–Computer Interaction and Information Retrieval*, 81–84.

Wineburg, S. S. 1991. Historical problem solving: A study of the cognitive processes used in the evaluation of documentary and pictorial evidence. *Journal of Educational Psychology* 83: 79.

Woodruff, A., R. Rosenholtz, J. Morrison, A. Faulring, and P. Pirolli. 2002. A comparison of the use of text summaries, plain thumbnails, and enhanced thumbnails for web search tasks. *Journal of the American Society for Information Science and Technology* 53: 172–185.

Wopereis, I., S. Brand-Gruwel, and Y. Vermetten. 2008. The effect of embedded instruction on solving information problems. *Computers in Human Behavior* 24: 738–752.

15 Design of Portals

Paul Eisen

CONTENTS

15.1 INTRODUCTION

The term *portal* is often bandied about throughout the literature and conversation, with little or no consistency. When people talk about portals in the context of information technology, they are almost always referring to some form of Web-based application that aggregates content or information from multiple sources. But do not most Web sites do that already? What do people really mean when they refer to a portal, and what is the merit of focusing on this type of application in a human factors design handbook?

This chapter attempts to bring some clarity to the question of the definition and value of portals by focusing on a portal-based interaction model. The chapter then further provides a set of practical guidelines for the design of software using the portal interaction model. The goal is to instill in portal package developers, portal designers, and human factors practitioners a strong model of portal design, ultimately, to increase the likelihood that the portals they develop will achieve a high level of user acceptance and optimize the return on investment.

FIGURE 15.1 (**See color insert.**) Portal page showing five levels of user interaction.

tions in the suite. Figure 15.1 illustrates five levels of user actions on a sample portal (the Rogers™ | Yahoo!® personal Web portal).

15.1.1.3.2 Working Definition of Portal

As the focus of this chapter is on user-interface design guidelines (Section 15.3), the interaction model, which directly informs the user's conceptual model of the software, will serve as the basis for a working definition of a portal, as follows:

A portal is a software application that is page-based; portlet-based; and supports user interaction at multiple levels.

In addition to the framework of the interaction model itself, the guidelines address effective ways to capitalize on the primary attributes of portals described above.

15.1.1.3.3 Portal Ecosystem

In the role of consolidating systems and functions, described as trait 2 in Section 15.1.1.1, the environment encapsulating the full set of integrated components can be thought of as the *portal ecosystem*. A hub-and-spoke diagram, such as that shown in Figure 15.2, illustrates the conceptual relationship between the portal UI, as provided by the portal technology, and the connected systems. The ecosystem typically includes a subset of the following types of applications, varying by portal application:

- Search facility
- Workflow services (e.g., notifications and tracking of business process events)
- Business intelligence and reporting tools
- Communications and personal information management (PIM) services (e.g., calendar, e-mail, and instant messaging)
- E-learning applications
- Specialized applications (e.g., a people directory)

- Compliance processes
- Records management
- Messaging
- Collaboration tools
- Content and document management

15.1.1.3.4 Types of Portals

Portals are generally classified by audience and by domain focus, targeting specific user groups or communities, as follows:

- *Enterprise portals.* Enterprise portals are business-to-employee (B2E) applications, targeted to employees inside an organization or business entity. While portals designed for specialized areas, such as a human resources portal, may exist inside an enterprise, an enterprise portal typically targets all employees and thereby crosses, functional, geographical, business unit, and other boundaries within the organization. The population of employees may be defined broadly, to include retirees and pre-hires (i.e., those who have agreed to join the organization but have not yet started).

  Because portals are Web-based applications, these business-to-employee portals are known generically as intranets. While historically intranets have been conventional page-based Web sites, today portal technologies are very commonly used for intranets, with content being delivered via portlets. A separate chapter, Jacko et al. (this volume), contains an in-depth look at the benefits of, and approach to, intranets for internal communications.
- *Customer portals.* Customer portals are business-to-consumer (B2C) applications, targeting the clients,

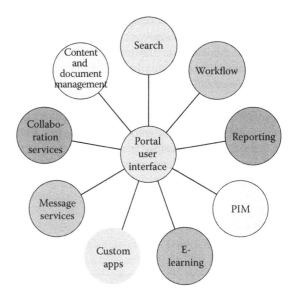

FIGURE 15.2 Conceptual portal ecosystem.

FIGURE 15.3 (**See color insert.**) *My msn* personal portal Home page.

customers, or patrons of organizations. Customer portals may be open and public but are more commonly extranets that are secured to specific audiences who have a relationship with the organization, helping the consumer to manage their side of the relationship. For example, an asset management firm may have a portal extranet that enables clients to view and download reports about their financial portfolio.

- *Partner, supplier, and vendor portals.* The terms *partner portal*, *supplier portal*, and *vendor portal* are commonly given to business-to-business (B2B) extranets. Similar to customer portals, these B2B portals enable an organization's business partners to manage communications, documents, and transactional data. For example, an organization's procurement department may provide a portal for their suppliers to submit invoices and track payments.
- *Personal portals.* Personal portals, sometimes referred to as *horizontal portals* (because they are not specific to one industry vertical), are public-facing portals and are typically universally accessible and offer broad content, such as news, weather, games, stocks, and sports information. Examples of these portals are *iGoogle*™ and *my MSN*® (Figure 15.3).

As one would expect, the optimal design approach to portals varies widely for each of these target user groups, and application domains. The diversity of needs of these differing populations and application domains can be so dramatic that the resulting portal solutions are vastly different. Still, there are recognizable and repeatable design patterns that emerge from well designed portals, all stemming from the

key characteristics of the portal interaction model described in Section 15.1.1.3.1.

15.1.2 Benefits and Limitations of Portals

With a working definition in hand, and application areas described, it is instructive to address the question of the benefits and limitations of portals versus traditional Web sites and other applications.

Valdes (2008, p3) offers a compelling moniker for portals: "The Swiss Army Knife of enterprise software." This term aptly encapsulates arguably the primary benefit of a portal: the bundling of a set of useful tools and utilities in a single environment, with a consistent look and feel, to support the varying needs of users. Like the cost and space efficiencies and the ease of operation gained by binding multiple knives and other physical tools inside a common casing, consolidating applications into a single environment that services multiple user groups can reduce the total cost of ownership (TCO) of the bundled systems, owing to reduced software license and maintenance fees, reduced hardware costs, reduced usage demands, and a host of other factors (see, for example, Compuware 2008 for an elaboration of these technology costs in a shared-services environment).

For enterprise and other employee-facing portals, the TCO can be reduced not only to technology and operational costs but also by improving the overall user experience. The user experience benefits contributing to an organization's bottom line stemming from well-designed portals are due to two main factors:

- Reduced learning demands of a common UI used to access common functions across tools and applications require lower training costs and facilitate faster times to proficiency (Caya and Nielsen 2008). This benefit is particularly noteworthy as people shift roles (e.g., job changes within an organization) but still use the same portal.
- Providing a consolidated environment to access all common functions, tools, and content through a sensible design scheme reduces the overall time to find information and perform work activities (Caya and Nielsen 2008).

For customer-facing portals, the above factors may not represent harvestable savings for an organization, but the improved user experience can directly contribute to user satisfaction and thereby elevate the good will of the sponsoring organization, a bottom-line asset. Gootzit (2009) notes that well-designed portals have led to measured productivity improvements inside enterprises, as well as enhanced revenue and profitability for customer portals.

The above proposes some financial and user-experience benefits of consolidating systems and content into a single, well-designed environment with a common user interface. It does not necessarily demand, however, that the common user interface is based on a portal interaction model. So, what is the benefit of that consolidated environment implementing a portal interaction model?

Five user-experience benefits of the portal interaction model are described:

1. The model uses a simple UI design approach that is easily understood (Lamantia 2006). Page-based design is pervasive in Web sites, while portlets conceptually mimic the *window* construct in a computer desktop environment, which should be familiar to users of popular personal computer operating systems such as Apple's Macintosh OS X and Microsoft Windows.

2. The portal model lends itself to breaking down multiple content items into consumable chunks that can be presented simultaneously (Lamantia 2006). Portlets provide the cells inside of which the chunks of information are presented.

3. The UI can be easily and extended by adding pages portlets inside the framework.

4. If properly implemented, the consolidation of systems into a single environment enables *single sign-on* and propagates user sign-on credentials among all participating applications. The net result is effective personalization of the portal experience wherever appropriate. Single sign-on is a feature whereby access is granted to all of the connected systems in the portal ecosystem once a user has provided their ID and password to gain access to the portal (or to a computer workstation that can provide secure access to the portal). With single sign-on, not only do the systems provide access, but they also inherently recognize the individual user, thus deepening the opportunities for personalizing the experience. Portal research continually identifies single sign-on as one of the most desired outcomes of system consolidation as expressed by end users (Caya and Nielsen 2008). Practically, however, Caya and Nielsen point out that this goal is still elusive, and portal developers are currently settling for a solution where there is a reduced demand for signing on but not reduced as far as just one sign-on.

5. The independence of portlets from pages facilitates content sharing and distribution across multiple sections and pages and enables users to easily select and arrange content to suit their needs.

Note that the above discussion focuses on the benefits of a well-designed portal itself and not on the benefits associated with any of the participating systems and applications.

On the down side, a drawback to the portal model is pointed out by Lamantia (2006): ". . . portlets are inherently flat, or two-dimensional . . ." and that success leads to content growth, which is inherently handled by portlet sprawl, with little or no inherent navigational solution to simplify a broadening page cluttered with portlets. Underlying Lamantia's

criticism is the fact that the portal model does not fully avoid the limitations of conventional page-based Web sites. Although the use of portlets introduces a great deal of flexibility to the UI, portlets are tied to pages, which are subject to the navigational limitations inherent in Web sites. To accommodate the growth within a portlet set, one alternative would be a richer UI, for example using an *application-model* approach.* However, an application model may not accommodate the diversity of features attributed above to portals.

15.2 PORTAL DESIGN PROCESS

As for all software solutions, a user-centered design (UCD) process for portals will reap benefits commensurate with the size of the end user population and the complexity of the UI. Much has been written in the literature about implementing UCD (e.g., Mayhew, this volume; Vredenburg, Isensee, and Righi 2002), and this content will not be repeated here. Rather, this section provides some portal-relevant insights that either supplement the ideal UCD process or at least highlight elements of the process that merit emphasis because they are particularly impactful and often ineffectively addressed in portal implementations.

15.2.1 IMPORTANCE OF DESIGN STANDARDS

Because portal solutions reach across many systems and areas of an organization, one of the critical success factors related to methodology is the documentation of design standards. This activity is very similar in approach and importance to the documentation of design standards for nonportal Web solutions. Documenting, monitoring, and enforcing design standards will help realize the benefits associated with a consolidated environment with a consistent look and feel. To this end, the following four design standards should be documented:

1. *Information architecture (structure and navigation).* The rationale for the assignment of features and content to the initial release of the portal needs to be defined and documented, so that as the portal grows the approach can be retained rather than diluted. Normally, this is done with a series of conceptual diagrams and textual descriptions.

   To illustrate, in an enterprise portal implementation for a retail bank, links to all of the internal tools are placed in a list on the "Tools & Applications" landing page. There is also a Tools & Applications menu showing only a subset of the tools, because during the design phase the full list of over 70 tools and applications was deemed excessive for the

---

* An application UI model is a windowed, page-less approach to design whereby menus, toolbars, and secondary windows provide access to functions. Most desktop software, such as Microsoft Word or Excel, use an application model. Using current Web technologies, Web-based software and Web sites are increasingly using an application model. These Web-based solutions are known as *rich internet applications*, or RIAs.

menu. A rule was established to surface only the cross-organizational tools in the menu, as in general these have the greatest use, and because those tools that are unique to a particular department are also accessible via a link on that department's landing page. Similarly, appropriate rules were established to govern the content in the other top-level menus. Over time, as new tools and applications have been added to the environment, and as content relevant to the other menus has evolved, the rules have been applied and adhered to. Consequently, the integrity of the information architecture remains intact and supportive of the natural growth of the portal.

2. *UI design (e.g., layout and controls).* To a large degree, UI guidelines address the "feel" aspect of the portal's "look and feel," as they govern the appropriate selection of UI widgets and describe how those widgets should respond to interaction. Documenting the appropriate usage and the behavior of UI widgets ensures consistency of the interaction design, helping to maximize the usability of all assets on the portal.

For example, a standard should be documented for "Print . . ." and all other actions assigned to the portlet menu and to an icon on the portlet title bar. This documented standard ensures that every portlet with a print function for its content will enable access to that print function in the same way; the same applies to all other common portlet actions. Mayhew (1999) describes a sound approach to documenting UI design standards. As portals are Web based, it is more time efficient to extend existing software UI design standards rather than starting from scratch. Popular and well-researched UI standards, such as those developed by Microsoft (Microsoft Corporation 2009) or Apple (Apple Inc. 2009), as appropriate to the platform being used for the portal, should provide a solid basis for the portal UI design and increase the likelihood of a usable portal solution. While some design standards have a basis in traditional graphic user interfaces (GUIs), over the past few years they have been adjusted to accommodate the convergence of browser-, page-, and link-based interaction with traditional GUI controls.

Designers should keep in mind that, although housed in Web browsers, the rich interaction and transactional capability offered in most portals more closely resemble those of a traditional GUI than conventional Web sites. End users interacting with portal content will be confused and frustrated if the UI does not offer the same cues necessary to communicate the depth of modes and states inherent in rich UI's as are available in traditional GUIs. Therefore, Web design standards that focus mainly on navigation and the display of content and that do not provide comprehensive coverage of interactive

controls should be either avoided or thoroughly supplemented.

One challenge when extending an industry UI guideline to a particular portal implementation is to determine what to keep from the industry guideline. The designer must carefully eliminate those sections in the standard that are superseded by the portal environment, such as the visual appearance of windows elements and positions of window controls, while adhering to those guidelines that are universal to virtually all software environments, such as the use and behavior of radio buttons and other GUI widgets.

3. *Visual style.* Among all four design standards, visual style guidelines—the "look" component of "look and feel"—are typically thought of first. These guidelines identify the various visual elements and describe the appropriate visual treatment under all designed themes. Figures 15.4 and 15.5 display small excerpts from a portal visual style guide. Notice the description of each visual element in all potential states. Portal UI elements normally may be presented in up to four states:

a. Normal state (referred to as "Off" in Figures 15.4 and 15.5).

b. Target state, when the cursor is directly over the element, indicating that it can be selected by

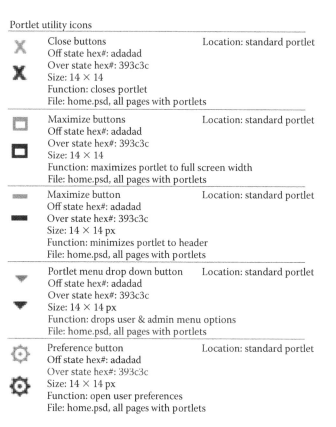

Portlet utility icons

Close buttons — Location: standard portlet
Off state hex#: adadad
Over state hex#: 393c3c
Size: 14 × 14
Function: closes portlet
File: home.psd, all pages with portlets

Maximize buttons — Location: standard portlet
Off state hex#: adadad
Over state hex#: 393c3c
Size: 14 × 14
Function: maximizes portlet to full screen width
File: home.psd, all pages with portlets

Maximize button — Location: standard portlet
Off state hex#: adadad
Over state hex#: 393c3c
Size: 14 × 14 px
Function: minimizes portlet to header
File: home.psd, all pages with portlets

Portlet menu drop down button — Location: standard portlet
Off state hex#: adadad
Over state hex#: 393c3c
Size: 14 × 14 px
Function: drops user & admin menu options
File: home.psd, all pages with portlets

Preference button — Location: standard portlet
Off state hex#: adadad
Over state hex#: 393c3c
Size: 14 × 14 px
Function: open user preferences
File: home.psd, all pages with portlets

FIGURE 15.4 Excerpt 1 from sample visual style guide for a portal.

Form buttons   **Location: preference panels**
Typeface: calibri bold   **and forms**
Size: 10 pt
Style: upper case
Small button fixed size: 80 × 22 px
Large button fixed size: 114 × 22 px
Button off: hex#: 3d3c3c
Background linear gradient hex#: cecdcd, ffffff
Outline: 1 pixel hex#: a4a9a9
Button over: hex#: 3d3c3c
Background linear gradient hex#: cce9f1, ffffff
Outline: 1 pixel hex#: a4a9a9
Button inactive: hex#: 3d3c3c
Background linear gradient hex#: e7e6e6, ffffff
Outline: 1 pixel hex#: d2d4d4

FIGURE 15.5 Excerpt 2 from sample visual style guide for a portal.

clicking (referred to as "Over" in Figures 15.4 and 15.5).

c. Selected state, indicating that either the function associated with the element is currently selected or the element is a visited link. The selected state often displays according to the default browser behavior and is not always explicitly described.

d. Inactive state, indicating that the visual element is not selectable (Figure 15.5).

In addition to these four common states, UI elements may have one or more special states to support any unique needs for content and functionality. The reader is, however, cautioned that additional states should be used judiciously; extending a set of four familiar, standard states to include any additional special states may be risking undue end-user confusion.

Also note that Figure 15.4 references a source file for the displayed icons. In practice, the visual elements will need to be copied ("cut") from the source files and placed into one or more HTML files, which themselves are reusable. A recommended enhancement to these guidelines is thus to provide a link to, or otherwise specify, the relevant HTML files, to facilitate reuse.

4. *Editorial style.* One final aspect of the design to standardize is the tone and voice of the writing and common terminology. Editorial style guidelines also address rules of capitalization, such as using sentence case for field prompts and labels.

Some attention to the often violated convention of the use of the ellipsis is worth noting here. Emanating initially from its use in traditional GUIs, appending an ellipsis to action labels continues to be a universal standard whenever the action requires further input from the user before the associated operation is performed (see, for example, UI guidelines from Microsoft [Microsoft Corporation 2009], Apple [Apple Inc. 2009], or Sun's Java guidelines [Sun Microsystems Inc. 2001]). To illustrate, when

a print action opens a window that requires input from the user before actually printing (the *print dialog*), the label on the associated push button or menu item would be "Print. . . ." This is contrasted with the label on the push button inside the print dialog, Print, whose selection triggers the print operation. The use of an ellipsis is not warranted or appropriate with navigation actions or elements, such as conventional links. Here, the direct result of the selection of the link action is the fulfillment of the requested operation, with no further input required.

15.2.2 PHASED ROLLOUT

Conventional wisdom for large software application projects recommends phasing in functionality incrementally over a series of product releases (Highsmith 2002), a fact that in part has led to the agile software movement. In general, the incremental-release approach offers many benefits over the so-called "big-bang" approach of implementing one very large release, including managing risk, learning from early releases and adjusting incrementally, and maintaining the morale of the project team through incremental successes. Because of the typical diversity and volume of content in portals, this approach takes on an even greater importance. Fortunately, as pointed out earlier, the extensibility of the portal architecture lends itself to incremental releases, without requiring rework.

To this end, a release roadmap is recommended to provide a high-level plan for the phasing in of content and features over time. An abstraction of a portal release roadmap is illustrated in Figure 15.6. In the diagram, each release is labeled as a stage number and a conceptual description. The user roles targeted in each release are identified as persona groups and must be elaborated elsewhere. Further, the set of content requirements and applications to be implemented at each release are specified as groupings in the high-level diagram; they, too, must reference details under separate cover. Once agreement is gained, the roadmap can then be expanded into detailed project plans for each release.

Of note in the roadmap diagram, the first releases of a portal should focus on the technology environment, UI framework, and the core services (e.g., security management, user role management, content management, and search, among others to be elaborated in Section 15.3.5). Normally, these releases will have only basic content that can be used to test the framework and services, released to a limited set of end users. Once the portal framework and core services are implemented, in practice, growing the content and functionality is relatively very fast and efficient.

15.2.3 PORTAL GOVERNANCE

Governance refers to the organizational structures and processes that are established to support work activities. Establishing ownership and implementing effective governance for software applications is always a critical

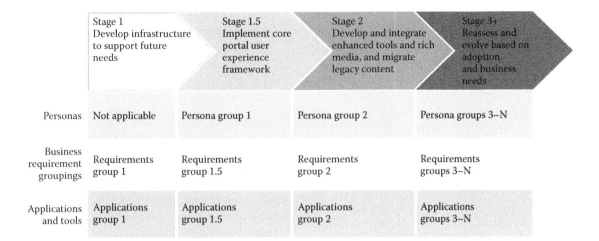

FIGURE 15.6 Conceptual portal release roadmap.

foundational step (Phifer 2008). Because of a typical portal's reach across multiple systems and content areas, establishing effective governance for portal projects is particularly challenging. Of course, with the multiplicity of portal configurations and applications, there is not one right governance model. Still, although each portal project will have a unique solution to governance shaped by organizational considerations as well as the scope of the content and functionality, some practical guidelines to avoid common pitfalls are described below.

- As discussed in Section 15.1.2 above, a portal consists of infrastructure components, including the portal framework and the environment in which it is housed and typically a host of participating applications and content areas. A common mistake is to consider the full portal initiative as purely infrastructure and assign ownership to an infrastructure group, such as information technology or a business operations department. Ownership and accountability for a portal needs to rest with an executive who oversees all business lines responsible for the substantive content and applications in the portal. In many cases, such as for an enterprise portal, this places ownership directly at the top of the organizational hierarchy. Anecdotal evidence indicates that when ownership is assigned to a subordinate or inappropriate organizational role, the effectiveness of the portal solution will suffer.
- With portals' broad reach of content and services, unlike many business applications, merely assigning accountability to a business owner or single business area is generally insufficient to direct the development efforts. One or more governance bodies with representation from each of the organization divisions with a stake in the portal should be formed. For enterprise portals, one effective model is to establish an operating committee consisting of working members (from each impacted business area) who have the time and insights to provide input to core project documents, and a steering committee of senior executives covering the same areas who have decision-making authority.
- Because of their broad nature—with the scope of content and features that they comprise—most portals are more susceptible to, and more impacted by, stagnancy than other business software. Whereas for some implemented software it is sufficient to maintain a skeleton staff to manage it, portals require active effort and attention to keep them relevant and useful. Establish governance at the start of the project that creates the initial release of the portal and keep the governance bodies and processes active over the full portal lifetime.

15.2.4 CHANGE MANAGEMENT

An oft-overlooked and underresourced area in technology development is managing the impact of change to the end users, their processes, and their environment. A well-designed and developed application, even one that if adopted will make life easier or more rewarding for the user, can and often does fail because of lack of user acceptance. The lack of acceptance may be due to cultural bias, lack of understanding, fear of change, lack of political support, and other social and political reasons. To address this phenomenon, the discipline of *change management* may be applied (Cameron and Green 2004). Change management attempts to mitigate the risks of lack of acceptance by applying a structured and proactive approach to managing transitions through communications, recruitment of change ambassadors, and other techniques.

A detailed discussion of the practice of change management is outside of the scope of this chapter. However, this challenge is worth mentioning in passing, as enterprise portal are particularly vulnerable to failure due to poor change management.

This vulnerability is due to the cross-organizational reach of enterprise portals. As Caya and Nielsen (2008, 8) note: ". . . just buying portal software doesn't guarantee a good portal; you must also manage internal company politics. Indeed, back-of-the-envelope math from successful portal launches suggests technology accounts for roughly one-third of the work, and internal processes account for the rest."

To increase the probability of adoption and success of an enterprise portal effort, the following suggestions are offered:

- Assign primary ownership of the enterprise portal to an executive whose jurisdiction covers all organizational areas participating. *Ownership* here refers to publicly stated accountability for success. If the portal is truly enterprise-wide, this implies the chief executive officer of the organization must take on ownership. One common mistake organizations will make is placing ownership with a support organization such as information technology or human resources. This often leads to difficulty getting buy-in from the client organizations whose employees will be impacted by the portal.
- Ensure the executive owner understands and buys into the value of the portal and is committed to communicating this value broadly, deeply, and frequently.
- Create a steering committee whose members manage all of the participating organizations. This steering committee should meet regularly to decide on critical matters related to the success of the portal, such as funding, schedule, scope, and design. Often these decisions will take the form of approving decisions made by the operating committee (described below).
- Create an operating committee whose members represent each member of the steering committee, on a one-to-one or many-to-one basis. These members should be able to represent the areas covered by their steering committee counterpart, and should have decision-making authority to recommend proposed approaches for approval by the steering committee.
- Communicate early and often with all impacted employees about the nature of the change and the associated value. Put additional emphasis on the content providers, as their participation is critical to keeping the portal thriving and valuable. Persuading the content providers can be relatively easy if the content-management processes are usable and effective, as described later in Section 15.3.5.
- Provide performance support within the portal (such as tips, tutorials, and online chat).
- On rollout of each major release of functionality, assign portal ambassadors in each physical workplace facility in prominent locations where they can offer support and encouragement for employees

adjusting to the change. Create incentive-based contests to encourage employees to explore new portal features so they can appreciate the value firsthand.

15.2.5 Measuring Success

As evidence of the value of a portal implementation, measures of success should be established during the portal planning stages and tracked at rollout and beyond. The measures of success will derive from the goals and objectives for the portal. For example, a portal whose primary goal is to consolidate and centralize technology infrastructure could emphasize costs of technology infrastructure to demonstrate the benefits and success of the effort. Alternatively, where the user experience is central to the objectives of a portal effort, user-experience-based metrics should be emphasized.

As is the case in other applications, the measurement of user experience can be challenging. From a qualitative perspective, there is no replacement for usability testing during the design and in the field after rollout to assess the impact of the portal on the end-user experience. Usability testing has the added advantage of identifying areas for improvement, and if carried out properly can also point to the relative impact of proposed improvements. The usability testing may also be set up to gather quantitative metrics, including objective measure of time on task and numbers of user actions; other methods that may reap valuable measures include questionnaires and surveys. A chapter in this book focuses on the cost justification of focusing on usability in Web applications (Richeson et al., this volume).

Some suggestions for appropriate objective metrics in portal design efforts include the following:

- *Portal usage* measures the number of users logging on to the portal in a designated time period. It can be helpful to track usage across user segments, locations, and other meaningful groupings, as these data can provide insights into variability in levels of user engagement across these groups.
- *Page and portlet hits* detect the amount of activity in a designated content area. Higher activity may indicate greater interest and engagement. However, it may also result from inefficient UI design. Therefore, qualitative user feedback is important to help interpret these data.
- *Portal, page, and portlet durations* track the amount of time spent by users actively interacting with individual portlets, pages, and the portal as a whole. Care must be taken to establish parameters distinguishing active interaction from idle time. When users are distracted by nonportal tasks, durations will be artificially inflated. Unfortunately, current technologies do not readily allow for the clean measurement of active interaction.
- *Search query analysis*, wherein search terms are analyzed for frequency, affinity, and meaning,

indicates content of high interest to users and the terminology used by users to find that content. Care must be taken in interpreting this measure as well, as content of high interest but also high prominence may demand fewer user searches. Still, search query analysis is very effective at pointing out the content users are compelled to search for, pointing to opportunities for design improvement.

- *Navigation tracking* reveals the paths that users take through content by such measures as the relative frequency of selecting various links from a particular location. The results of navigation tracking reveal insights about the effort required for users to carry out tasks on the portal and highlight opportunities for consolidating content (e.g., by adding portlets to pages or creating new pages with a combination of portlets on the heavily used paths).
- *Customization tracking* measures the frequency with which users take advantage of customization offered by the portal and the particular choices made by end users. Customization tracking reveals insights about the relative value of customizations, which are particularly interesting to compare with user requests for these customizations during requirements elicitation sessions.
- *Support tracking* measures the frequency and types of help queries in various support channels. Effective tracking of query types and user issues through root cause analysis can provide insights into the areas where redesign or training is needed, and the overall volumes provide clues into the adaption and effectiveness of the portal. These data are best analyzed together with portal usage statistics to interpret volumes in the context of overall usage rather than in isolation. For example, the relative amount of support queries per user hour is generally more meaningful to the design of the portal than the total number of support queries, although the latter is also important to gauge resourcing levels for the support area.
- *Content activity tracking* measures the frequency and types of content updates on the portal. Content updates may be the sole jurisdiction of the portal administrators or may be distributed among end users using collaborative tools. These content activity measures indicate the freshness of vibrancy of the portal, which, in turn, impacts the level of end-user engagement and experience.

15.3 PORTAL DESIGN GUIDELINES

As suggested in Section 15.2.1, as specialized software applications in Web environments, most portals will benefit tremendously from adherence to appropriate sections of a widely accepted UI design standard, such as Apple Inc. (2009) or Microsoft Corporation (2009). A UI standard will provide guidance for the proper use of all of the standard controls, such as push buttons and drop-down list boxes. Caldwell and Vanderheiden (this volume) provide a good overview of available UI design standards and their applicability to the Web. This chapter presents guidelines that are either unique to portals or deserve emphasis because of their importance in portal environments.

It should be noted that these guidelines, being general in nature, may not apply with equal validity to all portal applications. They are useful in providing direction for the design; however, they cannot replace the effectiveness of a comprehensive user-centered design process and the engagement of skilled user interface designers in achieving high-quality portal user experiences.

15.3.1 INFORMATION ARCHITECTURE IN PORTALS

An information architecture serves as the foundation to a UI design and therefore merits the starting point for a discussion of portal UI design guidelines. A coherent and comprehensive prescription for a portal information architecture is described in a series of articles by Lamantia (2006). The proposed information architecture is internally consistent, flexible to accommodate various portal applications, readily extensible in multiple dimensions, and compatible with the portal interaction model described in Section, 15.1.1.3. It therefore can serve as the basis for a sound UI framework for portals. It should be noted, however, that the architecture allows for very sophisticated configurations of content and functionality, which if fully fleshed out would be overengineered for many applications. Although Lamantia does not make this claim, to simplify, it is proposed that some of the components of the architecture not be mandatory.

Guidelines for portal information architecture—both portal structure and navigation—are presented in this section. The fundamentals of Lamantia's proposed architecture are described in each subsection. In the description, components are suggested as mandatory or as optional.

15.3.1.1 Portal Structure

To begin, Lamantia specifies a precondition to enable fine-grained control of role-based security and access. Authentication must take place at the user level. Further, each portlet, which represents the most atomic component in the framework, needs to be able to independently establish access to its content.

Lamantia's architecture itself consists of seven types of building blocks of content. Provided in order of granularity, from smallest to largest, the building blocks are as follows:

1. *Portlet* (mandatory).* A portlet is the most atomic level of the architecture and directly holds the content. Design guidance for portlets is presented in Section 15.3.3.

---

* Lamantia (2006) uses the term *tile* to refer to portlets or web parts. In this summary, the term portlet is used instead.

2. *Portlet group (optional).* This is a grouping of portlets, normally combined because of related content.
One or more portlets can combine to form a group.
One advantage of grouping portlets is to visually
emphasize their relationship. Grouping also may
enable for efficient interaction for users wishing to
invoke operations on all portlets in a group, such
as showing all of the portlets on a different page.
However, the introduction of group-level functionality can increase the UI complexity and is recommended only for portals where users are motivated
to achieve proficiency.

3. *View (optional).* A view is a grouping of related
portlets and portlet groups, pulled together to
enable one or more perspectives on a set of content.
In Lamantia's words, ". . . views allow Dashboard
or Portal users to see the most logical subsets of all
available [portlets] related to one aspect of an area
of interest" (Lamantia 2006). Although potentially
very powerful, this level of sophistication is not germane to frameworks of the popular portal packages
referred to earlier and is not discussed further in this
chapter.

4. *Page (mandatory).* A page, equivalent to conventional Web pages, houses one or more views, portlets groups, and portlets. Figure 15.7 provides a
conceptual illustration, in the form of a wireframe
diagram, of the relationship of a page to its subcomponents. Design guidance for portal pages is offered
in Section 15.3.2.

5. *Section (optional).* Corresponding to the same construct in conventional Web sites, sections logically
group one or more related pages. For example, in
an enterprise portal, all department pages may

be organized in a section labeled, *Departments.*
Sections are not strictly mandatory for portals and
may be skipped for portals of unusually small scope
without violating the portal interaction model. For
most practical portal implementations, however,
sections serve as an important organizational mechanism (Van Duyne, Landay, and Hong 2007).

6. *Portal (mandatory).* This corresponds to the *portal*
construct that is the subject of this chapter. Portals
consist of multiple sections and pages.

7. *Portal suite (optional).* To round out the architecture, Lamantia designates a portal suite as the context in which a portal may exist. Portal suites may
contain multiple portals.

These building blocks can be thought of as elements in a
hierarchy, wherein blocks of a lower number (or greater granularity) can be contained inside blocks of a higher number.
For example, one or more portlets can be contained inside a
portlet group or inside a view or even a page. In contrast, a
view cannot contain a page, a portlet group cannot contain
a view, and a portlet cannot contain any other blocks.

15.3.1.2 Tips for Structuring Content in Portals

The independence of portlets from pages enables a single
portlet to appear on multiple pages. This feature provides
flexibility for the section and page administrators to reuse
portlets in multiple locations and contexts. (For example, a
vendor portal may have a portlet with a calendar showing
important payment-processing dates. This single calendar
portlet may be of value on multiple pages of the portal, servicing vendors of different types.) Also, just like conventional
Web sites, HTML can easily support a structure where the
same page can be made to appear as a child in multiple sections. However, this flexibility comes with a risk: even when
adhering to the defined blocks and associated rules specified
above, the ability to surface multiple instances of portlets
and to link to pages from multiple sections can cause portal
structures to get unwieldy, leading to user disorientation. To
mitigate this risk, it is recommended that each page or portlet
have only one master, as described in the following:

- Every page should have one and only one absolute
  location in the portal hierarchy. This location is normally designated by the breadcrumb trail, reflecting
  that it can be visited by drilling down through the
  section and subsections designated in the navigation bars (Tidwell 2005; illustrated in Figure 15.8).
  The page can further be accessible via supplementary links from other pages (see crosswalk connector, below) anywhere in the portal where this
  is relevant and provides convenience. However, to
  support the user in learning to navigate the portal
  more efficiently, no other page should be treated as
  its parent.
- Every portlet should have one and only one primary
  source page, to which it is permanently affixed.

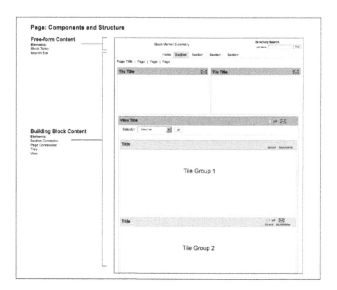

FIGURE 15.7 Wireframe of page and subcomponents in portal
information architecture. (From Lamantia, J. 2006. The challenge of
dashboards and portals. Boxes and Arrows. http://www.boxesandarrows
.com/view/the-challenge-of. With permission.)

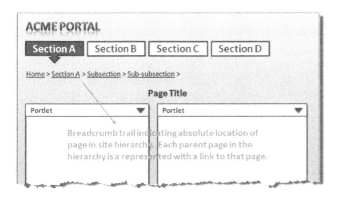

FIGURE 15.8 Portal page with breadcrumb trail showing location.

Additionally, the portlet may be shown on other, secondary pages if deemed appropriate by the page owners (e.g., administrators). These two rules ensure that page owners will be able to reliably find the portlet, while limiting restrictions for building their own pages. Methods for propagating portlets are discussed in Section 15.3.3.2 below.

As an example, consider the scenario illustrated in Figure 15.9. Imagine a financial services company has developed and published a financial calculator for use in a customer portal. Like all content, the calculator must be displayed inside a portlet. Also, imagine the portal designers designated a page for each tool in a section labeled Tools. On the Financial Calculator page, there might be two portlets: one portlet with the calculator and one with learning and support materials for the calculator. As the primary source page for these two portlets, this page can always be relied upon to have these two portlets. Now imagine that the same customer portal has a different page for featuring special content, Featured Stuff. To feature the financial calculator, the administrator of the Featured Stuff page might create a link to it, or for more impact might choose to *mirror* the Financial Calculator portlet directly on the featured content page, as shown in Figure 15.9. *Mirroring* is a feature provided by many portal packages. It enables multiple views of the same portlet. An end user browsing either the Featured Stuff page or the Financial Calculator page may decide to mirror the same portlet on their own customized page (My Stuff, of which the end user is technically the administrator). So now it is in at least three locations for that end user: one primary source page and two secondary pages. Over time, the portlet may be removed from the Featured Stuff page, and the user may tire of it on their My Stuff page and remove it, but it will always be affixed to, and therefore retrievable from, the Financial Calculator page.

Figure 15.9 emphasizes the nature of the mirrored portlets on the secondary pages by showing

a dotted line border and a close icon to remove the mirror. In reality, the administrator of the Financial Calculator page would also have the ability to remove the master portlet from this primary source page. But, once removed, all mirrors of it would disappear (naturally, an intervening warning message describing this consequence would be valuable). If a secondary page owner does not want this behavior, instead of mirroring the portlet on their page, they might choose to *clone*, or copy the portlet, and therefore have their page act as the primary source page for the cloned portlet. Some design guidelines for mirroring and cloning portlets are presented in Section 15.3.3.2.

15.3.1.3 Navigational Mechanisms

To enable clear and efficient movement through the portal, Lamantia (2006) offers a series of navigation elements, which he calls *connectors*. This navigation, much of which is illustrated in Figure 15.10, is presented in the following, along with guidance for appropriate use and implementation.

1. *Portal-suite navigation.* Portal-suite navigation, termed a *dashboard connector* by Lamantia, enables one-click access to any of the sibling portals in the suite from any place in the portal. For portals that are not part of a suite, this connector would simply not be present. Even for portals that are part of a portal suite, however, the portal-suite navigation is

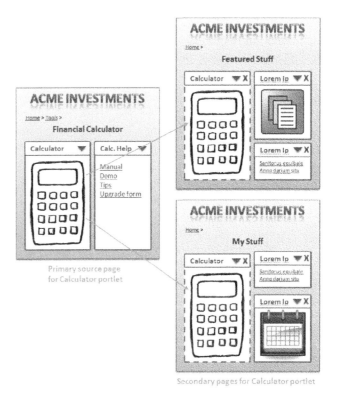

FIGURE 15.9 Portlet shown on primary source page and secondary pages.

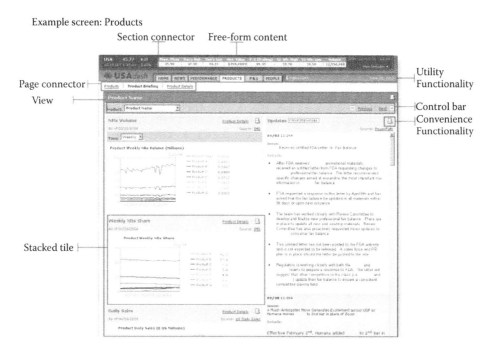

FIGURE 15.10 **(See color insert.)** Sample portal page with blocks and connectors. (From Lamantia, J. 2006. The challenge of dashboards and portals. Boxes and Arrows. http://www.boxesandarrows.com/view/the-challenge-of. With permission.)

an optional convenience, which may not merit the real estate it would have to occupy on every page of a portal, depending on the applicability of other portals in the suite to the portal's target end users.

For example, a suite of portals may include a customer portal for corporate clients, one for small business clients, and one for personal clients. Although technology efficiencies may motivate overlap of content and services, there are likely to be very few clients with a need to access more than one portal, thereby negating the value of presenting portal-suite navigation on every page in each portal.

2. *Global navigation.* Global, or *top-level*, navigation, which Lamantia calls *section connectors*, enables one-click access to the landing pages of all portal sections from anywhere in the portal (Figure 15.10). Although design guidance for section connectors does not differ from that for conventional global navigation, the results may differ insofar as role-based content presentation and customization may result in a unique view of section content.

3. *Secondary navigation.* *Secondary navigation*, which Lamantia calls *page connectors*, enables access to the sibling pages inside a section (Figure 15.10). UI design guidance for second-level navigation for portals is not distinct from conventional Web sites. Although, as they do for global navigation, portals present a unique angle because of the role-based presentation of content: When pages are not accessible to a user based on their role, links to those pages should be hidden. Most portal packages conform to

this guideline in their secondary navigation without any need for customization.

4. *Control bar.* A control bar operates at the content level, within a portal, portal group, or view (Figure 15.10). It contains controls specific to the contents to allow adjustment of the scope (e.g., filtering) and view (e.g., sorting and paging). Arguably, the control bar is not strictly a navigation construct, as the controls impact the view of the contents, not the location in the portal. Guidelines for the control bar in the context of portlet management are presented in Section 15.3.3.

Lamantia includes a unique connector type he calls a *geography selector* to enable users to specify the geography or locale to which content should pertain. This is essentially a specialized element of a control bar elevated to the status of a connector type due to its prevalence.

5. *Crosswalk connector.* A crosswalk connector is a link that traverses the boundaries of the building blocks, providing convenient access to content irrespective of the portal information structure. This is the traditional hyperlink that is part of the backbone of hypermedia. The crosswalk connector is a critically important mechanism to support portal usability and efficiency in portals, as it allows content publishers and end users to customize content and create collections of links based on their own unique needs and criteria, while respecting and not confounding the portal structure.

To illustrate, portal designers may choose to house all forms in a section called a forms library (that is, every form is placed as a child page of the forms library landing page). On the same portal, publishers of content in individual sections could readily provide links to any desired forms, and end users may even create a custom list of links to their favorite forms. The forms themselves belong only in one location—the forms library—while access to the forms is distributed via crosswalk connectors for convenience.

To this end, one very valuable portlet type worth implementing on most portals is a links portlet to enable page administrators and end users to create sets of customized links.

An important guideline for crosswalk connectors is to ensure the link navigates to the location on the page where the target content is positioned. For example, if a connector links to a portlet that is below the fold, the response to selection of that link should be to navigate to the target page and then automatically scroll down the page until the portlet is in view. This system behavior will avoid the frustrating and inefficient experience of the user having to search for the target content after selecting the link.

6. *Utility navigation.* Utility navigation provides connectors to portal-wide utilities, such as general search or specialized directories. Section 15.3.5 expands on the use of utilities in portals.

15.3.2 PAGE MANAGEMENT

15.3.2.1 Portal Pages versus Traditional Web Pages

Because of its unique interaction model, a portal lends itself to a somewhat different content design philosophy than traditional Web sites. In both cases, it is important to modularize the content for better user understanding. On the one hand, on traditional Web sites, this modularization of content tends to be realized by assigning the content modules to separate pages and then structuring those pages in a logical hierarchy. This spreading of content across pages leads to an experience for the user of clicking on links to get deeper content. In portals, on the other hand, portlets provide an effective mechanism for the presentation of modularized content within a page. Thus the portal interaction model lends itself to allocation of the full set content across fewer pages, with more content on each page. This design approach has led to the term "dashboard" being applied to portal pages, because the collection of related portlets into a single view or page functionally mirrors a traditional, high-density automobile dashboard.

15.3.2.2 Page Layout

Today, all popular portal packages support fixed grids, with cells to which portlets are assigned, to establish the layout of content on a page. Most portal packages, in fact, provide the ability to set up multiple page templates to support a variety of grid layouts, such as shown in Figure 15.11. Note that

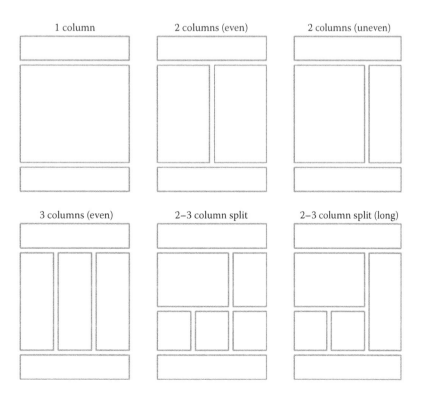

FIGURE 15.11 Sample grids for portal pages.

every grid includes a page-wide cell at the top and bottom. These cells are recommended to display page header and page footer content.

It is difficult to provide general guidelines for selecting grids, because the nature of the content and the user's goals and tasks are primary drivers of the layout. However, two general guidelines can be stated, addressing consistency and horizontal scrolling. With respect to consistency, when a grid is selected to present a specific layout of content for a page of a certain type, say a news story, all other pages of the same type (i.e., all other news story pages) should use the same grid and layout. This will help users form clear expectations about where to look for specific content. Also, select layouts to minimize horizontal scrolling inside portlets. Avoid a layout or implementations requiring users to scroll horizontally to see content when viewing the portal in a maximized browser on the lowest-target-resolution monitor. As with traditional Web sites, horizontal scrolling generally leads to a suboptimal user experience.

15.3.2.3 Page-Level Actions

The following guidelines are offered for enabling operations at the page level:

- Place all page-level controls in a group, for example, page-level controls may be rendered as a row of links, an icon set in a toolbar, or a menu.
- Place the page-level controls in a location and with a visual treatment that reinforces the scope of the controls.
- On each page, provide an action to edit the page, available and visible only to those users with authority to do so. For example, on a customizable page, such as the My Stuff page in Figure 15.9, each user should have an action on that page to allow the user to edit the page.
- This edit page operation may include features like changing the page layout, moving the portlets in the layout, selecting new content (portlets) to be added to the page, or removing portlets from the page. As an aid to the user, provide recommended grids and column widths for each portlet to minimize the need for horizontal scrolling.
- If the edit page action impacts the page for others besides the end user, provide a distinct label to make the scope of the edit page action explicit. For example, label the action, "Edit page for all users." This will remind the page administrator that their action will be impacting everyone who views the page and is not just a personal customization.
- Provide an action for users to print the page. Ideally, in this print operation, give the user the option to print the content in portlets that is scrolled out of view or just print the page as it appears.
- Provide an action for users to create a Portable Document Format (pdf) version of the page, with the same options as for printing.

- Provide an action to e-mail a link to the page. In the e-mail, append information about access restrictions, and mention that the page appearance and content may vary based on the access rights of the person who is viewing it. This will help avoid misunderstandings and missed expectations.
- Provide an action to bookmark the page, either using the browser bookmarks functionality or a portal bookmarks utility. A portal bookmarks utility has the advantage of portability appropriate for those applications where users may be expected to access the portal from multiple computer workstations.
- Page scrolling is a feature of all browsers rendering the portal content. The once valid concern of users missing content that was hidden below the fold no longer holds true (Tarquini 2009). Under normal circumstances, for knowledge workers it is not disadvantageous to have long pages that scroll; most users actually prefer this to paging through lists and other UI mechanisms used to minimize scrolling. The exception to this is a portal pages displaying content for time-sensitive operational tasks, where for example the user needs to be able to interpret content from a glance at the portal page. (Note that this type of task is rarely a particularly good application for portals, and is generally better served by an application-based interaction model.)

15.3.3 Portlet Management

All of the direct user interaction with portal content takes place at or inside the portlet. Thus, the UI design of portlet actions can have a dramatic impact on the portal usability.

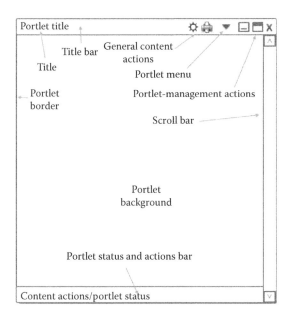

FIGURE 15.12 Portlet UI elements.

This section provides a list of considerations and guidelines in this regard.

15.3.3.1 Portlet Elements

User interface elements common to portlets are shown in Figure 15.12 and described in the following:

- *Portlet border and portlet background.* The portlet border and background are visual design mechanisms used to delineate the portlet content from that of other portlets. These elements emphasize the structure of the contents on the page, as well as reinforce the scope of portlet actions. The borders and backgrounds can be muted or even excluded if white space in the portal page grid is sufficient to provide visual separation.
- *Portlet title bar.* To visually ground the portlet, provide context for the contents, and house portlet controls, a portlet title bar can be used. The title bar is akin to the window title bar in a traditional desktop environment. As the title bar is the conventional location for the portlet controls, users will normally look there first, which is a compelling reason to display it. Another role of the title bar is the click target for a move action. Most portal platforms support portlet drag-and-drop as a supplement to page-edit capability. It is highly recommended to support the drag-and-drop action with animation to provide concrete feedback about the source portlet being moved and the target location.

    In some cases, the portlet does not share the page with any other portlets. Here, the page title will usually provide sufficient context. In this circumstance moving and other portlet-management actions (see below) are not required, and so it is not as important to display the title bar.
- *Portlet-management actions.* Portlet management actions enable portlet sizing and removal. The location and order of these controls should adhere to the convention on the Macintosh (left side of title bar) if the majority of users are accustomed to the Macintosh. Otherwise, the location and order should adhere to that of the Windows operating system (right side of title bar, as in Figure 15.12). In

both cases, show tooltips on mouseover of the icon to clearly communicate the associated action. The recommended actions are as follows:

- *Minimize.* Minimize typically removes the content area and leaves the title bar (more accurately, a *collapse* action). Minimizing a portlet allows other content below the portlet to move higher on the page, while leaving the title bar available to retrieve the content. When the portlet is in minimized state, the Minimize action should be replaced by a Restore action to enable the user to return the portlet to the normal state.
- *Maximize.* Most portal packages provide a Maximize action, which typically hides all of the portlets on the page and enlarges the current portlet to occupy the full page width and height. In maximized state, the Maximize action should be replaced by a Restore action to enable the user to return the portlet and the page to the normal state. The benefit of the Maximize action is that it enables the user to see a larger view of the portlet contents in a single click without navigating away from the page, which is particularly helpful for portlets with rich content such as reports.

    Unfortunately, the typical implementation of the maximize feature can easily disorient the user, because it shifts the view to a different mode while providing only subtle cues about the nature of the mode (e.g., the breadcrumb trail and page title may reinforce that the user is on the same page, and the portlet has a Restore icon where the Maximize icon was). This disorientation is particularly common when the user inadvertently or unknowingly clicks on Maximize. To avoid disorienting the user, the maximized portlet should be presented with a different skin—specifically, one that provides clear visual cues about the nature of the state and how to return the page to its normal state. Figure 15.13 presents one clever visual design solution that has been validated with end users in production environments. Here, the skin assigned to the maximized portlet feigns a

FIGURE 15.13 (**See color insert.**) Maximized portlet, feigning appearance of superimposed window.

window overlay on top of the page. Further, the button with the Close icon actually performs a restore function, because the user expectation for a Close operation for a superimposed window is to remove it and return to the page in normal state. This bit of visual trickery enables the designer to capitalize on the out-of-the-box portlet maximize function while ensuring an effective user experience.

- *Remove.* Where appropriate, provide a Remove action to allow the user to remove the portlet from their customized page. Before carrying out the remove operation, however, provide a dialog or other device to enable the user to confirm their action. A nice feature to include in the confirmation is information about how to retrieve the portlet. Use the remove action judiciously if the results will impact other users' view of the page. Here, at a minimum, emphasize in the confirmation dialog that the portlet will be removed for all users. Alternatively, instead of a remove action on the portlet title bar, enable the feature only through an edit page dialog (see Section 15.3.2.3).

- *General content actions.* General content actions are those actions that apply widely to portlets across the portal, for example, printing the portlet contents and modifying preferences. Provide quick access to frequently used general content actions as a convenience for users if real estate allows. Additionally, include equivalent choices in the portlet menu for these actions.

- *Portlet menu.* Although a portlet menu is not a required element, it is commonly used and effective at presenting relevant actions for the portlet content without taking up undue space. One common problem with portlet menus is that they are not easily noticed by novice users. Popular personal portals have addressed this problem in the past few years by using a text label, such as Menu or Options, as a salient cue to click on for opening the menu. Recommendations for the menu actions are as follows:
  - Follow industry standard guidelines for menus, such as using dividers to separate logical groupings of menu choices.
  - Duplicate general content actions showing on the portlet title bar and all actions on the status and action bar to provide a comprehensive set of actions for the portlet content in the menu. Duplicating the portlet-management actions is less important, since they are so pervasive on application and window title bars that most users expect to find them there. Still, include them for completeness if their addition does not excessively lengthen the menu.
  - Provide an action to print the portlet contents. At a minimum, provide the ability to print the

full contents of the portlet. Ideally, allow the user to choose whether to print the full contents, just what's in view, or a logical subset of the contents, filtered by meaningful variables.

- Provide an action to save the portlet contents in pdf format, with the same options as for printing.
- Provide an action to save (download) the portlet contents in a native format that can be further acted upon, for example, by word processing or spreadsheet software.
- Provide an action to show the portlet on another page, such as a customizable user page.
- Provide an action that enables users to adjust preferences for viewing the portlet content. For example, the choice can launch a preferences dialog wherein the user can specify the desired height of the portlet. Here, the best approach is to enable the user to specify the portlet height in natural terms, such as number of rows of content. However, as this is typically difficult to control due to variability of the user viewing environments, options such as short, medium, and tall, preconfigured for each portlet, are recommended. Also, unless the content is unusually voluminous, include an option to fit the height to the contents, which will allow the portlet to dynamically stretch as required without scrolling.
- Provide an action to subscribe to notifications of content changes. This function provides a convenient way for users to keep current with the portlet content, without having to explicitly visit the portlet to check for updates.
- Provide an action to get help on the portlet contents.
- Provide an action to get information on the portlet, such as the portlet administrator name and contact details and the time of last update. Alternatively, display this information in the portal content area.
- Provide an action for the user to send feedback on the portlet contents to the administrator.
- Provide an action to enable users to refresh the portlet contents, wherever appropriate. For example, in a portlet drawing content from an external database, the refresh interval may be controlled to manage system performance. Here, enabling the user to update the contents leaves the user in control.
- Avoid placing actions that allow users to affect how content is displayed in the portlet menu, if those actions are unique to the content. Instead, place those actions in controls inside the portlet, accompanying the content. For example, if a portlet contains a graph, and graphs are not pervasive in the portal, controls for adjusting the

graph parameters should be included inside the portlet with the graph.

- *Portlet status and action bar.* A portlet status and action bar may be used to supplement the display of portlet actions, particularly for novice users and when real estate is limited (e.g., on portals with narrow-column grids). This is helpful when there are several frequently used or important portlet actions—too many to display on the title bar. Use a consistent location for this bar, such as just below the title bar, or at the bottom of the portlet, as shown in Figure 15.12. Also use this bar to show relevant status, such as the time of last update of the content.
- *Portlet scroll bar.* Provide a scroll bar at the portlet level for all portlets housing an extremely large volume of content. Additionally, provide a scroll bar for any fixed-height user setting where content would otherwise be truncated.

15.3.3.2 Propagating Portlets—Cloning versus Mirroring

One important feature of portals is the ability to show portlets in multiple locations. This is normally controlled by portal and page administrators and is also a powerful feature for end users to configure their view of certain pages. Portals should support two forms of propagation: *mirroring* and *cloning.*

A *mirroring* operation creates a view of the original or master portlet, but not a separate instance, much like an alias or shortcut on a computer desktop provides a reflection of the master object. When the contents are updated on the master portlet, all mirrored views of that portlet immediately reflect the new content. Mirroring provides the ability to feature single-sourced content in multiple locations and in multiple contexts, while avoiding the effort and overhead needed to replicate the portlet and its properties. This benefit is especially noteworthy for portlets that are updated frequently, because those updates are automatically propagated to all mirrors.

For example, imagine that an administrator of an enterprise portal creates a portlet to display a list of events for the enterprise. In this portlet, a query to the events database that will result in the display of all scheduled events for the department associated with the page on which the portlet is placed. This portlet is then deployed on the enterprise portal, and, say, placed on the headquarters landing page. When deployed, it automatically displays all scheduled events for the headquarters. When an administrator sets up the Information Technology Department landing page, she may choose to mirror this events portlet. The result would be the display of all Information Technology Department events. Similarly, the portlet is mirrored on the Human Resources Department landing page, thereby showing all scheduled Human Resources Department events. At some point, the events portlet may get updated, perhaps to display the information with a different visual style and limit the list to the coming three months. This update is immediately reflected

in all mirrored portlets, whose visual style and lists are automatically updated accordingly. Here, the mirroring function demonstrates the powerful feature of portals to capitalize on single sourcing of content.

In contrast to mirroring by *cloning* a portlet, the user creates a separate instance of the portlet, assigning to the new instance the same initial properties as the original portlet. This allows the administrator of the cloned portlet to change the properties of that portlet however desired. The cloning has provided a convenient starting point for the new portlet but not restricted it to continually reflect the original portlet. Using the above example of the events portlet, should an administrator clone the events portlet and place the clone on the Information Technology page, she could then adjust the visual style or other properties of the portlet independently. If the original events portlet is updated in some fashion, that update is not propagated to the clone.

The following guidelines are offered for implementing mirroring and cloning of portlets in portal solutions:

- Implement mirroring to enable portal administrators to keep frequently updated portlets current wherever they appear.
- If the power of mirroring is desired for keeping a frequently updated portlet current wherever it appears, while at the same time individual administrators and users would benefit from some degree of individual control over the portlet, then create parameters in the portlet content that the end user can update. For example, provide a control inside the events portlet to allow each end user to specify the time horizon for viewing events (e.g., coming three months, coming year, or all future events). This enables the mirrored portlets to vary in their display of the content, while still capitalizing on updates to other features of the master portlet, such as the visual style.
- Provide the ability for administrators to clone a portlet as a convenient starting point for a new portlet instance.
- Provide clear labels and in-context tips to support portal administrators in making the appropriate choice of mirroring or cloning a portlet. Figure 15.14 displays a wireframe of a sample page-administration window, with two push button labels, "Add selected portlet" and "Create new portlet like selected . . ." for mirroring and cloning, respectively. Figure 15.14 also displays sample instructions to guide the administrator in making the appropriate choice.
- Provide a unique portlet skin for mirrored portlets to communicate that the portlet being viewed is a mirror of another, versus a master instance (e.g., the calculator portlets on the secondary pages in Figure 15.9 are mirrors and shown with a unique skin).
- Unless the portal end users are motivated to become highly proficient in customizing pages, avoid explicitly offering end users (that are not administrators)

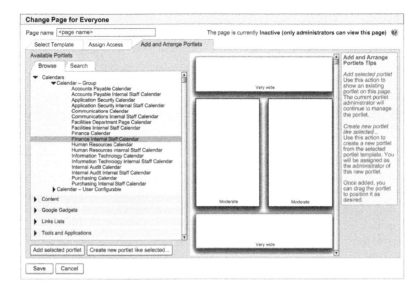

FIGURE 15.14   Wireframe diagram of sample page-administration window.

both choices of mirroring or cloning. Instead, use only mirroring as the mechanism for users to copy desired portlets and provide the user flexibility to customize the content in the mirrored portlet. For example, if a personal portal end user chooses to show a weather portlet on one of their customized pages, use mirroring to show the portlet when the user selects it. On the mirrored portlet, allow the user to specify the location of the weather forecast. When the master portlet is updated, this will update the mirrored portlet but still allow the end user to retain their location settings.

15.3.4   PERSONALIZATION, CUSTOMIZATION, AND SECURITY

The past decade has brought with it great expectations from end users for Web-based software to tailor the user experience—both the content and the presentation—to each individual user. With the proliferation of widely accessible personal portals and retail Web sites that tailor experiences, such as Amazon.com's book recommendations based on past searches, user demands for relevant content are growing. This is especially the case in enterprise portal applications, where employees are challenging their employers to deliver similarly tailored experiences (De Carvalho, Ferreira, and Choo 2005; Rudnick 2008). Rudnick (2008) reports that, in a 2008 survey of U.S. companies, over 80% of employees in 2008 demanded personalized experiences from their enterprise intranets.

15.3.4.1   Personalization versus Customization

Tailored experiences are normally delivered through two related mechanisms: *personalization* and *customization*. *Personalization* refers to the tailoring of the user experience based on what is known about the end user. In an enterprise

portal, using the known employee ID to display the employee's personal pay stub is an example of personalization. *Customization* refers to the tailoring of the user experience based explicit user input. Enabling the user to control whether the initial view of their paystub shows or masks their pay amount is an example of customization. From the user's perspective, personalization and customization are passive and active approaches, respectively, to achieving a tailored experience. In other words, personalization requires that the technology developers implement intelligent solutions that tailor user experiences, while customization implies that the end user will have to do the work to tailor their own experience.

In general, it is desirable to enable deeply tailored experiences in portals, as this will help to maximize the benefits and efficiency of interaction. Balancing the delivery of tailoring between personalization and customization depends on the portal application and the associated user tolerance for, or sensitivity to, personalization. For example, a desirable level of personalization in an enterprise portal may be considered personally invasive in a customer portal. To illustrate, consider an employee who participates in communities, each of which has a calendar stored in the portal. That employee may accept or even expect a global calendar grid to automatically present all of the employee's community calendars. However, that same person, when signed into a customer portal of a vendor organization, may be unwilling to accept a similar function automatically performed. Instead, they may prefer to explicitly identify those calendars they wish to show on a grid. As another example, a user may be more tolerant to an e-commerce portal making product recommendations after they've created an account and signed in, as opposed to while they are anonymously browsing.

Interestingly, with respect to intranet applications, Caya and Nielsen (2008) note that, although end users frequently ask for customization features, these features when implemented are vastly underutilized.

15.3.4.2 Role-Based Personalization and Security

Probably the most pervasive mechanism to support personalization, and almost certainly the most effective at this point in time for most elements of the portal user experience (Caya and Nielsen 2008), is *role-based personalization*. Role-based personalization tailors the user experience based on role or group memberships that are assigned to the user. When using role-based personalization, portal administrators will assign users to various groups. When the user authenticates (signs in to the portal), the portal will check the roles assigned to the user and tailor content and presentation elements accordingly. This same role-checking mechanism is typically used to control portal security by establishing authority for users to access the portal as a whole, a section or page of the portal, or even content inside individual portlets. The securing of restricted content is sometimes referred to as *exclusive personalization* (Caya and Nielsen 2008). Consequently, both security and personalization can be supported using the same role-based mechanism.

To illustrate, consider the following case study. A large global consumer goods corporation recently implemented an enterprise portal for the entire organization, supporting all of their business units and geographies around the world. In this portal, role-based authentication is used to control access to the portal and to various content areas of the portal, such as providing the ability to view one's own paycheck and no one else's paycheck. Further, roles are employed to ensure that all users, with all of their diverse needs, experience relevant and personalized content and presentation. For example, when an employee in the Asia-Pacific business unit signs into the portal, they are greeted with a personalized Home page whose visual skin is based on the Asia-Pacific business unit's visual brand identity. Further, their Home page automatically consolidates news from the corporate headquarters and from the Asia-Pacific business unit into a news portlet on their Home page. (If the employee wishes, she can customize her Home page news portlet by electing to view news from additional sources, such as other business units.) A correspondingly tailored experience is available to all employees in all business units of the corporation. This role-based approach enables the corporation to capitalize on the cost efficiencies of a single portal infrastructure supporting the full organization, while at the same time affording each employee a tailored user experience that leads to measurable productivity benefits.

When implementing role-based personalization and security, consider the following guidelines:

- Use role-based personalization to control the content inside a portlet, hiding content to which the user should not have access.
- When content is less relevant to a user, but not restricted, initially hide the less relevant content but provide the ability for the user to access the content. This form of customization, known as *inclusive personalization*, is desired because it leaves the user in control (Caya and Nielsen 2008).
- Define roles that segment your users into meaningful audiences based on the nature of the portal content. For example, in an enterprise portal, some employee benefits news may be relevant to retirees, while other content would likely be restricted to employees. In this circumstance, creating a separate role for retirees is warranted. A separate role for retirees would further enable sorting or filtering the benefits news uniquely for these two audiences.
- Develop a hierarchical, inheritance-based model for roles. This approach will allow the efficient implementation of role-based functionality. For example, in a customer portal for Company X, create a "user" role for all users, such as customers of Company X, and administrators and other employees in Company X. Also, create a role called "customer," a type of user, which is assigned to all customers and which could inherit access rights from the user role. A third role, called "purchaser," could be assigned as a type of customer, thereby inheriting all of the personalization features available to customer in addition to user. When a user assigned to the role of purchaser is on the portal, their view of specific content would be personalized to the role of purchaser if that is applicable. And through inheritance, this purchaser role would have all rights assigned both to customer and to users as applicable.

  If an inheritance-based role hierarchy is difficult to implement, an alternative approach is to apply roles cumulatively. This implies that an end user assigned to the roles user, customer, and purchaser would see content available to any one of those three roles.
- Enable users to view all content to which their roles allow access without having to reauthenticate. This approach, known as *role obfuscation* or *user obfuscation*, hides the complexity of multiple role assignments from the user. Role obfuscation should apply even to those users who are assigned to administrator and end user roles, as long as the guidelines for making administrative actions explicit (Sections 15.3.2 and 15.3.3) are followed.
- Implement role-based personalization gradually, starting with a few roles and then layering in additional roles to refine the personalization over time. This will allow you to validate the personalization approach without initially overcommitting to maintenance of a complex set of roles.
- When a portal block is not available to a user based on their assigned roles, hide all links to that portal block for that user. This will avoid the frustrating experience of selecting a link only to be informed that the linked content is not available to the user. This guideline applies to highly dynamic links, such as search results, in addition to relatively stable links that have been explicitly created by portal administrators.

- Wherever available, rely on existing information and systems to assign user roles. For example, if portal sign-on is controlled by a shared directory service for managing authentication, refer to this shared directory service to extract meaningful roles. This approach is known as *implicit* role assignment. For the roles where user information is not accessible, *explicit* role assignment must be done. Explicit assignment allows fine control to shape the roles optimally for the portal experience. The drawback, however, is that it usually places an additional operational burden on the portal administration to initialize and then maintain the role assignments for all users.

- When securing content, use the block hierarchy established in Section 15.3.1.1 and design the security with inheritance, assigning security at the highest level of the hierarchy as appropriate. In other words, if all content on a page should be restricted to a specific role rather than restricting each view, portlet group, or portlet individually, restrict the page and allow that restriction to cascade to the contained blocks. Similarly, if all pages in a section should have the same security assignment, apply that security assignment at the section level rather than for each individual page. Additionally, allow security assignment at a lower level of the hierarchy to override assignment at a higher level for that lower-level block and all of its contained blocks.

  This security-inheritance approach is illustrated conceptually in Figure 15.15, where the portal is secured as a whole to Role 1 (represented by blue text and borders), by setting the security at the top level and allowing inheritance by subordinate blocks (the inheritance function is represented by blue arrows). The figure further shows the override of Role 2, set at Page 3, illustrated with red coloring.

Notice that the overriding role, Role 2, is inherited by all of its subordinate blocks.

Note the importance of ensuring compatibility of role assignments when designing the security. In Figure 15.15, if a user assigned to Role 2 is not also assigned to Role 1 (either through an inheritance-based role hierarchy or separately), that user would not have access to any of the blocks assigned to Role 1 only, which would make navigation to Page 3 cumbersome at best. To avoid this problem in a role hierarchy, ensure that Role 2 is the more restrictive role, inheriting access from Role 1. If there is no role hierarchy, ensure that both Roles 1 and 2 are assigned to the appropriate users.

The benefit of this inheritance-based approach to security assignment is that it is very efficient to administer.

- In the portal administrator's user interface, display the role assignment of each block. If a role assignment is inherited, clearly indicate that it is inherited and show the topmost initiating block from which the role is inherited. This information will help the administrator manage security without having to traverse the hierarchy looking for the source of a security setting.

- In general, use single sign-on (SSO), as described in Section 15.1.4, to provide authenticated access to all areas of the portal. The exception to this guideline applies to the unusual condition where the portal might not be secured by a private workstation in a controlled environment. As mentioned above, a number of practical considerations limit the ability of portals to achieve SSO. For those applications not integrated into a SSO, be explicit in showing this. For example, show an icon with a tooltip by each link to the separately secured content and provide a tip to explain the meaning of the icon and the timing for integration if appropriate.

15.3.5 CONTENT MANAGEMENT

A frequently overlooked factor in the development of portals is the user experience of those responsible for publishing and maintaining portal content. A portal loaded with valuable content and whose end user experience is well designed may still be considered a failure if the content publishers must struggle to keep the content current. Unfortunately, in this author's experience, this circumstance is quite common; otherwise well-designed portals that do not provide an effective user experience for the content publishers quickly stagnate and become underutilized.

One way to address the risk of stagnating content is to increase the areas in the portal where end users can actively contribute to the content through *collaboration tools*, such as document-sharing, blogs, wikis, and threaded discussions. If these tools are well designed, they afford end users the ability

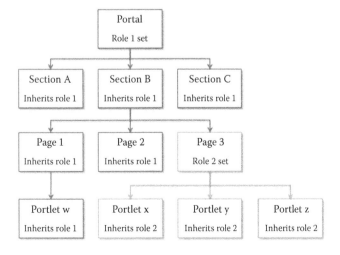

FIGURE 15.15 Inheritance of role assignments for securing portal blocks.

to keep the portal environment fresh and valuable for one another. Although in the past few years, portals have been more frequently offering tools for collaboration; a discussion of the design of collaboration tools is outside of the scope of this chapter. Applications, utilities, and any other specialized tools that stagnate can similarly have a material impact on the portal end-user experience but also owing to their broad nature, guidelines for these tools are also omitted from this chapter. Instead, this section contains some guidelines addressing the user experience for portal administrators in publishing *general content*—that is, text-based content that embeds images and other multimedia assets, such as streaming audio and video—to be consumed by portal end users.

To support the initial publication and maintenance of general content, *content management systems* (CMS's) have arrived in the marketplace. CMS's enable the storage and management of content for consumption by one or more systems (e.g., one or more portals). They typically allow the distribution of publishing privileges, assigning various rights to create, edit, approve, and release content to the portal. CMS's further support the creation of templates for content standardization.

For example, a news template may offer content creators the ability to craft a headline, byline, body content and to supply images, audio, or video, as well as keywords to support cataloguing and search of the news item. The news template will communicate consistently to all content creators the type of content expected in each news story. Portlet owners can then point to one or more news items. The visual style guide to which the portal adheres will specify the appearance of the news item in the portal. Conveniently, this allows a single news item to be presented differentially depending on the design rules of the calling portlets. A news overview portlet may show just the headline and primary image of each news story, while a news story portlet could show the full content of each news story.

With this explanation, it's easy to imagine the power and convenience of a CMS to support content management on a portal. Despite this, a survey of 48 organizations with enterprise intranets found that CMS's are growing in popularity but still are not universal and, further, that usability issues and lack of ease of use is common and thereby necessitates training for effective use of CMS's (Caya and Nielsen 2008). A CMS that requires training for effective use will either lower productivity or limit the effective pool of content publishers for the portal.

The following guidelines are offered for the effective integration of CMS's in portal environments:

- Provide the ability for the user to identify the primary content author (or owner) in each portlet, showing the name or title and contact information. This information should not need to be provided by the content owner or portal administrator; rather, it should be provided automatically based on the user profile of the content owner. This contact

information will enable the end users to provide feedback and input to the author and also publicly exposes accountability of content, improving the likelihood that it will be kept current.

- Capture the date and time of the most recent version of content and pass this information to be presented in the portlets pointing to the content.
- Avoid securing content in the content management system. Instead, secure access to the content by controlling access to the portlets where the content appears, using the guidelines offered in Section 15.3.4.1. This approach avoids the need to manage content security in both the portal and the CMS.
- To ensure the integrity of the above security approach is maintained in portal searches, show and provide links to only the portlets in which the content is presented, rather than to the content itself. This restriction, along with adherence to the earlier guideline to hide links to content that the user does not have rights to view, will ensure that only the content to which the user has access will be presented in the search results.
- Provide the ability for the content publishers to manage the content from the portal. At a minimum, this involves initiating all relevant actions based on the user rights and the status of the content, as well as surfacing the CMS features (such as the content templates and approval workflow) in or via the portal.

15.3.6 Integrating Applications

As described in detail in the discussion on portal definition (specifically, the functional description in Section 15.1.1.1), two primary attributes of a portal are consolidation of systems and functions and gateways to applications. There are a variety of ways in which systems and applications can be integrated in portals, with varying benefits and limitations. Four approaches to integration, ordered roughly from most to least effort required, are described below.

1. *Replace or build from scratch.* An application that is built and integrated directly inside the portal environment can capitalize on the existing services inside the portal, such as search, document and people directories, visual rendering, navigational infrastructure, and availability of common page and portlet functions, for example, printing or copying to user-customizable pages. The result is a user experience that has a rich set of features and many aspects of user experience consistent with the rest of the portal purely owing to sharing of a common infrastructure. Building an application inside the portal further provides the opportunity to shape the content and functionality inside the application using the documented portal standards to enhance the

consistency of the user experience. This approach to integration is appropriate for most new applications that do not require specialized front-end platforms and for existing applications in need of replacement or front-end redesign because of obsolescence or other inadequacies. The drawback to this approach is the relatively large amount of time and effort required to build and deploy the applications.

2. *Subscribe to external source.* Rather than present content generated from internal sources, portlets can subscribe to sources outside of the portal. This method allows content, data, and information that are controlled by an external application to be presented inside a portlet, adapting the look and feel of the portal. For example, a personal portal administrator may want to put a portlet on a page that allows users to select stocks on global stock exchanges and display their share prices. Rather than create an application to store and manage the content, the administrator may find an existing service in the public domain to provide this set of capabilities. The administrator can then create a portlet and develop some code to subscribe to the service, which then displays the relevant content in the portal user interface. An industry-standard communications protocol and syntax has been developed* to enable content that is housed in portlets on external portals, or otherwise available through external Web services, to be presented inside a portlet. This protocol, known as Web Services for Remote Portlets (WSRP), defines an interface for interacting with presentation-oriented Web services. The advantage of this approach is that the end user can experience a rich application embedded in the portal, with the appearance and behavior consistent with the rest of the portal. This application, with much less effort than would be required to create it from scratch, can capitalize on the portal infrastructure and associated common functions. The drawbacks to this approach are the limitation of available application services (particularly those that are WSRP compliant) and the subscription costs. Also, the portal administrator does not usually have control over the application, so undesirable changes to the remote application content or syntax may occur without notice.

3. *Embed.* An existing application can be managed internally and surfaced through a portlet, thereby integrating into the portal. This is mechanically similar to an external source subscription; however, an embedded application can be part of the portal ecosystem and thereby more readily capitalize on the portal infrastructure, such as the recognition of

roles for security and personalization. Embedding applications is appropriate when the application already exists and would be costly to rewrite as a native portal application. Embedded applications can be *standalone*, where a single application (such as a library collection search tool) is presented in a portlet. Alternatively, an embedded application can combine content from multiple sources toward a common purpose (termed a *mashup*). For example, combining a database of library branches with a separate database of reading groups allows the portlet to highlight locations where both exist. It is recommended that, like external applications, embedded applications use the portal visual styles or skins to create a more consistent experience for the user.

4. *Link to external application.* The quickest and easiest way to integrate an application is merely to create a link to it from inside a portlet. This enables rapid integration of multiple applications; however, it generally brings with it a fragmented user experience. Even if the linked application conforms to similar interaction and visual style guidelines, the fact that it is not embedded in the portal UI framework will negatively impact the user, because it cannot provide the benefits of the navigation, bookmarking, portlet-copying, and other portal functions. To avoid disorienting the user, linked applications should be launched in a new tab or a new window that is spatially staggered to allow the portal to still be partially in view behind it.

15.4 CONCLUSIONS

The breath and flexibility of portal technologies and applications provide a world of promise for improved end-user experiences of applications and content. However, as noted throughout these pages, creating ideal user experiences requires not only custom design effort but also significant code customization. Efforts to baseline and track the maturity of this software sector are underway (e.g., Hawking and Stein 2003; De Carvalho, Ferreira, and Choo 2005) but are few and far between. It is safe to say that, with only about a decade of development, portals are still in their infancy. Many of the specific recommendations in this chapter are not readily achievable using portal package technologies, without significant customizations. This in of itself is evidence of the lack of maturity of the sector.

One particularly important growth area for portals is the consolidation of role-based security, content management, document management, and administrative functions into the portal package technologies. As time passes, the quality gap between the end-user experience and the administrative-user experience continues to widen. The need to create integrated and effective user experiences for the administrative users and content producers of the portals continues to be underserved. Yet it is only through improvements to the

---

* Through the joint efforts of the Web Services for Interactive Applications (WSIA) and Web Services for Remote Portlets (WSRP) OASIS Technical Committees.

experiences for these user groups that the portal applications can achieve significant advances in effectiveness. To this end, there needs to be a greater emphasis on content activity tracking metrics in portal development projects, forcing an increased focus on the administrator and content-producer's user experience.

Another growth area for portals will be in the subspecialization of portal packages to address niches that currently are being served by general purpose portal technologies. For example, almost every enterprise portal application has a similar challenge of providing quick and flexible access to organizational forms and documents, employee contact information, company tools, and other common content types. Elegant design solutions to providing access to these content types exist and can be prepackaged, allowing more efficient and consistent user experiences.

With the rate of growth of portal implementations, and the fantastic advances occurring in Web-based software technologies today, there continues to be a great deal of untapped opportunity for portal applications. Many of these opportunities, such as the improvement of role-based security and personalization, represent challenges specifically for those companies that develop portal package technologies. These opportunities will be better served by shaping their growth around a set of well-defined, internally consistent, and end-user driven design guidelines such as those offered in this chapter.

ACKNOWLEDGMENTS

The author wishes to acknowledge the thorough review and comments from Lawrence Najjar.

REFERENCES

Apple Inc. 2009. Mac OS X Reference Library: Text. http://developer.apple.com/mac/library/documentation/UserExperience/Conceptual/AppleHIGuidelines/XHIGText/XHIGText.html#//apple_ref/doc/uid/TP30000365-TP6 (accessed December 2009).

Cameron, E., and M. Green. 2004. *Making Sense of Change Management: A Complete Guide to the Models, Tools & Techniques of Organizational Change*. Glasgow, UK: Bell and Bain.

Caya, P., and J. Nielsen. 2008. *Usability of Intranet Portals—A Report from the Trenches: Experience from Real-life Portal Projects*, 3rd ed. Freemont, CA: Nielsen-Norman Group.

Compuware. 2008. *Collaborative Portal Decisions That Make "Cents": Reducing Total Cost of Ownership by 50% or More with Saas Solutions*. Detroit, MI: Covisint.

De Carvalho, R. B., M. A. T. Ferreira, and C. W. Choo. 2005. Towards a Portal Maturity Model (PMM): Investigating social and technological effects of portals on knowledge management initiatives. In *Proceedings of the 2005 annual CAIS/ACSI Conference*, eds. L. Vaughan, 1–12. Toronto, CA: Canadian Association for Information Science (CAIS).

Gootzit, D. 2008. The portal doctor is IN: A portal health check and the cures you need. Report PCCC5_129. Stamford, CT: Gartner.

Gootzit, D. 2009. Key issues in enterprise portals. Research G00164910. Stamford, CT: Gartner.

Hawking, P. and A. Stein. 2003. B2E portal maturity: An employee self-service case study. Paper presented at Ausweb2003, Gold Coast, Australia: Norsearch Conference Services, July 2003.

Highsmith, J. 2002. *Agile Software Development Ecosystems*. Boston: MA: Pearson Education.

Jacko, J. A., M. McClellan, F. Sainfort, V. K. Leonard, and K. P. Moloney, this volume. Intranets and intra-organizational communication. In *Handbook of Human Factors in Web Design*, 2nd ed., eds. K.-P. L. Vu and R. W. Proctor, 329–356. Boca Raton, FL: CRC Press.

Lamantia, J. 2006. The challenge of dashboards and portals. Boxes and Arrows. http://www.boxesandarrows.com/view/the-challenge-of (accessed December 2009).

Mayhew, D. 1999. *The Usability Engineering Lifecycle: A Practitioner's Handbook for User Interface Design*. San Francisco, CA: Morgan Kaufmann.

Mayhew, D. J., this volume. The Web UX design process—a case study. In *Handbook of Human Factors in Web Design*, 2nd ed., eds. K.-P. L. Vu and R. W. Proctor, 461–480. Boca Raton, FL: CRC Press.

Microsoft Corporation. 2009. Microsoft developer network windows development center. Command buttons: UI text guidelines. http://msdn.microsoft.com/en-us/library/bb226792(VS.85).aspx. (accessed December 2009).

Phifer, G. 2008. Use best practices to ensure your portals' success. Report PCC4_106. Stamford, CT: Gartner.

Richeson, A., E. Bertus, R. G. Bias, and J. Tate, this volume. Determining the value of usability in Web design. In *Handbook of Human Factors in Web Design*, 2nd ed., eds. K.-P. L. Vu and R. W. Proctor, 753–764. Boca Raton, FL: CRC Press.

Rudnick, M. 2008. Realizing the promise of an HR portal on SharePoint. Paper presented at HRM Strategies, 2008. International Association for Human Resources Information Management, Orlando, FL, June 2–4, 2008.

Sharp, H., Y. Rogers, and J. Preece. 2007. *Interaction Design: Beyond Human–Computer Interaction*. New York: John Wiley.

Sun Microsystems Inc. 2001. Java look and feel design guidelines, 2nd ed. http://java.sun.com/products/jlf/ed2/book/ (accessed December 2009).

Tarquini, M. 2009. Blasting the myth of the fold. Boxes and Arrows. http://www.boxesandarrows.com/view/blasting-the-myth-of (accessed December 2009).

Tatnall, A. 2005. Portals, portals, everywhere. In *Web Portals: The New Gateways to Internet Information and Services*, ed. A. Tatnall, 1–14. Hershey, PA: Idea Publishing Group.

Tidwell, J. 2005. *Designing Interfaces*. Sebastopol, CA: O'Reilly.

Torres, R. J. 2002. *User Interface Design and Development*. Upper Saddle River, NJ: Prentice Hall.

Van Duyne, D. K., J. A. Landay, and J. I. Hong. 2007. *The Design of Sites: Patterns for Creating Winning Web Sites*, 2nd ed. Upper Saddle River, NJ: Prentice Hall.

Vredenburg, K., S. Isensee, and C. Righi. 2002. *User-centred Design: An Integrated Approach*. Upper Saddle River, NJ: Prentice Hall.

Valdes, R. 2008. Constructing and Reconstructing the External Facing Portal. Report PCC5_115. Stamford, CT: Gartner.

Wojtkowski, W., and M. Major. 2005. On portals, a parsimonious approach. In *Web Portals: The New Gateways to Internet Information and Services*, ed. A. Tatnall, 15–39. Hershey, PA: Idea Publishing Group.

16 Intranets and Intra-Organizational Communication

Julie A. Jacko, Molly McClellan, François Sainfort,
V. Kathlene Leonard, and Kevin P. Moloney

CONTENTS

16.1 INTRODUCTION

Intranets capitalize on World Wide Web technology within the boundaries of an organization. Among the various types of information technologies used within organizations, intranets are poised to make the largest impact, especially in terms of organizational learning. Since their infancy (circa 1995), intranet technologies have evolved from being viewed as internal information publication and posting tools to more sophisticated organizational knowledge management and learning platforms. The power of intranet technology is illustrated by its ability to restructure and fortify internal communications throughout an organization. Interconnectivity of

people and organizational entities strengthens the communications network to promote innovation, collaborative work, and organizational learning.

The human element is essential to the success of an intranet, as the system's users contribute the knowledge that is necessary for the dynamic expansion of the company's learning organization. In turn, information technology (IT) tools can facilitate or induce changes in an individual's work, as well as impact the productivity, effectiveness, quality, and job satisfaction within an organization. Not only should the intranet provide a collaborative environment, it should also actively connect all relevant content in order to be presented to the users when relevant (Danson 2009). From a human factors perspective, the integration of IT within an organizational structure must consider the attributes of its users, rules, goals, and structures within the context of the organization.

At the onset of 1996, popular business/technology press, such as *Business Week,* touted the advent of intranets (with statements such as "Here comes the Intranet" and "No questions, intranets are coming"). In the spring of 1996, *CNet* noted the emergence of a new penchant in the technical community for corporate intranet applications, rather than consumer-based applications. These "internal World Wide Webs" or "corporate Internets" were poised to take over the role of groupware applications and revolutionize the way organizations were doing business. Not only have these predictions come to fruition, but the success of intranets has even surpassed some of the estimates for success. Intranet success rates (whether users can complete their tasks with the user interface) have been demonstrated to be at least 80% and are typically about 33% higher than the Web, thus saving companies both time and money (Nielsen 2007).

This chapter focuses closely on the interplay between intranets and intra-organizational communication and how an organization can exploit the benefits of these information technologies. Specifically, we integrate the concepts of intra-organizational communication and intranets technologies to examine how these concepts can augment or impede organizational growth. We then discuss how principals of organizational structure and user-centered design relate to the planning, design, and support of intranets for intra-organizational communication.

16.2 KEY CONCEPTS IN ORGANIZATIONAL LEARNING

The quest for organizational learning has prompted significant reform within organizations. Akin to intranet, the topic of organizational learning has received much attention in the popular management and information technology media. Unlike intranets, however, organizational learning is accompanied by a substantial amount of theoretical and academic research. This section presents an overview and extraction of key concepts associated with organizational learning that influence the development of intra-organizational communication and the role of intranets within organizations.

The concept of organizational learning has been prominent in the organizational theory literature for at least four decades. It appeared several decades ago in the developmental stages of organizational behavioral theory (e.g., see Cyert and March 1963). Although most organizational theorists and practitioners currently agree that organizational learning is an essential factor in the evolution of organizations, there is still no universally accepted definition for this term. In addition, controversy still remains over the differences between "organizational learning," the "learning organization," "organizational knowledge," and "knowledge management." Miner and Mezias (1996) noted that, despite being quite sizeable, the literature on organizational learning

suffers from a lack of precise definition, measurement, and estimation. The majority of the historical literature is a philosophical debate about whether organizational knowledge is like or unlike scientific knowledge (Spender 2008). However, since 1996, empirical research has grown substantially and constitutes much of the organizational learning research published (Bapuji and Crossan 2004).

Definitions of organizational learning can be broadly categorized into one of two overall perspectives: behavioral or cognitive. The behavioral perspective states that an entity learns if, through its processing of information, experience, and knowledge, the range and nature of the entity's potential and realized behaviors are changed (e.g., see Cangelosi and Dill 1965; Levitt and March 1988). Therefore, the behavioral tradition views learning as the result of change in behavior, with learning taking place through the modification of an organization's programs, goals, decision rules, or routines. However, the cognitive tradition views organizations mainly as interpretive systems. Hence, learning is considered to be a process of revising assumptions through reflection and through continuous interpretation of the environment (e.g., see Argyris and Schon 1978; Daft and Weick 1984; Senge 1990).

March and Olsen (1975, 147–8), recognizing the importance of bounded rationality (that human rationality is restricted by limitations of their cognitive information-processing system) in adaptation and organizational learning, further proposed that organizational intelligence is built on two fundamental processes. The first is a process of "rational calculation, by which expectations about future consequences are used to choose among current alternatives. . . . The second process is learning from experience."

Argyris (1982) defined organizational learning as a process of detection and correction of errors and proposed that organizations learn through individuals acting as agents of learning. He also noted that an organization, itself, operates as an ecological system, which defines the environment in which individuals' learning activities take place and hence can influence the process and nature of learning. Easterby-Smith and Lyles (2003, 9) define organizational learning as "the study of the learning processes of and within organizations" and building on Argyris's ecological system, the implication is made that organizations can learn in ways that are "independent of the individuals within."

Levitt and March (1988) viewed the organizational learning process as a result of encoding inferences from history into routines that can guide behaviors. This definition is built on three observations. First, behavior in an organization is based on routines, whereby individuals in organizations behave and make decisions on the basis of a process of matching procedures to recognized situations rather than through using logic of consequentiality of intention. Second, organizational actions are history dependent, on the basis of the interpretation of the past rather than anticipations of the future. Finally, organizations are target oriented, meaning that behavior mainly depends on relations between observed outcomes and aspirations for these outcomes.

Weick (1991) argued that the defining properties of learning reside in different responses being generated from the same stimulus. He suggested that few organizations have such properties. In fact, he suggested that most organizations are not built to learn; rather, they are designed to produce the same response to different stimuli, a pattern that may conflict with the very essence of learning. He concluded that organizations do not learn in this traditional, intended way. Miner (1990) further noted that most organizational learning models assume that aspirations and targets precede and drive experimentation. He also demonstrated that organizations can change and adapt over time without top-down, goal-driven variation, and proposed an evolutionary model to organizational learning. Therefore, although organizational learning can be intended and designed, it can also be unintended and evolutionary.

In fact, experts sometimes make a distinction between organizational learning and learning organizations. Although some authors do not make this distinction or consider that organizational learning refers to the process of learning and learning organization refers to the structure in which learning takes place, others see important distinctions. An important characteristic of learning organizations is the adoption of "generative" or "double-loop" learning, "adaptive" instead of "single-loop" learning (Argyris 1982; Senge 1990). Double-loop learning focuses on solving problems in the present without examining the appropriateness of current learning behaviors. The double-loop feedback allows existing errors to be discovered and corrected and connects the errors to organizational values, thus changing values, strategies, and assumptions (Argyris and Schön 1996). It is believed that knowledge is built through double-loop cycles that are often initiated by reflections that deem past decisions as mistakes (Marabelli and Newell 2009). Therefore, learning organizations continuously learn and unlearn (Hedberg 1976), designing and redesigning themselves through experimentation and feedback.

In a comprehensive synthesis of organizational learning, Huber (1991) emphasized that learning is not necessarily conscious or intentional, does not always increase the learner's effectiveness, and does not always result in observable changes in behavior. In his review, rather than defining organizational learning, he adopted a more operational perspective and defined four important concepts of organizational learning that provide a useful framework for designing and improving organizational learning activities:

- *Knowledge acquisition*: The process by which knowledge is obtained.
- *Information distribution*: The process by which information is shared.
- *Information interpretation*: The process by which distributed information is given one or more interpretations.
- *Organizational memory*: The means by which knowledge is accumulated and stored for the future.

Huber (1991) asserted that the construct of organizational memory is central to the idea of organizational learning because the basic processes that contribute to the occurrence-recurrence (knowledge acquisition), breadth (information distribution), and depth richness (information interpretation) of organizational learning depend on organizational memory. Therefore, these four constructs are linked to one another as shown in Figure 16.1.

Stein and Zwass (1995) first proposed the use of information technology in order to accomplish the four processes related to organizational memory: acquisition, retention, maintenance, and search and retrieval. They also propose a design for an organizational memory information system (OMIS) that includes a "mnemonic functions layer" to support the above four processes. This proposed layer has the ability to encapsulate and symbolize the knowledge in OM, the ability to communicate knowledge, and the conservation of the contents of the OM. Despite the goodness of fit for IT and OM, there are legitimate reasons why this is a difficult endeavor. Nevoa and Wand (2005) listed the amount of contextual knowledge in OM, the different locations of OM, and the difficulties in combining memories, the difficulty of tracking and maintaining tacit knowledge, frequent changes to the contents of the OM, and finally the fact that because some knowledge is retained outside the organization or from unfamiliar sources and should be attached to OM to facilitate retrieval and use as reasons for why the application of IT to OM remains complex.

In a text on organizational learning, Argote (1999) focused on "factors explaining organizational learning curves and the persistence and transfer of productivity gains acquired through experience." She reported that organizational learning might explain significant performance variations that are evident at the firm level of analysis. She identified three broad categories of organizational factors that appear to influence the rate at which organizations learn and their subsequent productivity: (1) proficiency of individuals, (2) technology, and (3) an organization's routines, structures, and processes.

Consequently, the concepts of organizational memory, knowledge, and information are central to organizational learning. Organizational memory can be thought of as the repositories for knowledge acquired through experience, and

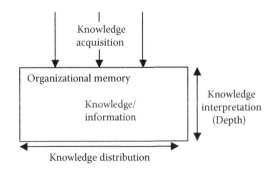

FIGURE 16.1 Four constructs of organizational learning. (Adapted from Huber, G. P., *Organizational Science* 2(1), 89–115, 1991.)

other means. As described by Cross and Baird (2000), such repositories consist of (1) the minds of individual employees, (2) the relationships between employees, (3) paper, electronic databases, and information, (4) work processes and technologies, and (5) products or services offered. Therefore, organizational memory takes many forms, from memory "embrained" in individuals to memory embedded in technologies, structures, and routines. Intuitively, the transmission of the many forms of organizational memory merits further attention.

According to Hills (1998), capturing knowledge and expertise so that a larger community can reuse the information is the foundation of the knowledge management infrastructure. To this end, intra-organizational communication and its supportive tools need to fit well with previously existing people, processes, and content to encourage organizational growth and expansion.

To characterize and understand intra-organizational communication, the classification of shareable knowledge is valuable. Lam (2000) proposed a typology of knowledge that integrates organizational and cognitive dimensions of knowledge. The organizational dimension describes knowledge as individual or collective. The cognitive dimension categorizes knowledge as explicit or tacit.

Explicit knowledge is highly codifiable, whereas the codification of tacit knowledge is more problematic. Explicit knowledge is more easily abstracted, aggregated, stored, and shared because it is structured and generated through logical deduction. Tacit knowledge, obtained through practical experiences, is more intuitive, implicit, and unarticulated. Tacit knowledge is difficult to aggregate because it is mostly personal and contextual. Knowledge management research primarily focuses on tacit knowledge. Organizations that desire to spread knowledge or inspire innovation desire to capture knowledge (Mooradian 2005). Unlike explicit knowledge, tacit knowledge must be processed and converted into a new form in order to be utilized.

Together, the organizational and cognitive dimensions define four categories of knowledge, as suggested first by Collins (1993). Table 16.1 itemizes the four categories of knowledge with attributes of organizational dimension, cognitive dimension, and the distinctive qualities of each.

Lam (2000) further suggested that each organization contains a mixture of knowledge types but that the relative importance of knowledge types can differ. Some organizations might be dominated by embrained knowledge, where other organizations might be dominated by encoded knowledge. Lam asserted that different types of knowledge correspond to different organizational forms and structures. As a result, the ability with which different organizations can harness and utilize knowledge, whether explicit or tacit, individual or collective, may vary greatly. As a result, the ways in which organizations learn also vary greatly.

The domains and forms of knowledge, explicit to an organization, may bear great weight on its organizational learning and growth. For example, although organizations dealing primarily with explicit, highly structured knowledge adopt formal structures of control and coordination, other organizations dealing with tacit knowledge may tend to be decentralized. Lam (2000) proposed that each knowledge type is associated with an ideal-typical organizational form with specific characteristics.

Following the work of Nonaka and Takeuchi (1995) documenting the Japanese type organization, Lam (2000) referred to the "J-form" as the organizational form associated with embedded knowledge. Here, the "innovative form" or I-form is referred to, which was initially developed by Sainfort (1987). Sainfort defined an I-form organization as one that has innovation as a core cultural value and capitalizes on organizational knowledge through continuous problem solving. This overview of knowledge types and their associated ideal-typical organizational structure is outlined in Table 16.2.

Information technologies, such as intranets, can give an organization the means to support four knowledge types listed in Table 16.2. Although specific applications will be discussed later on in the chapter it is important to first be aware of these categories of knowledge. The organizational characteristics associated with a knowledge type can have great bearing on the capacity of an organization to successfully incorporate different types of information technology in addition to the amount of transition and training time that will be necessary. Although information technologies can help an organization evolve to different structures, the existing knowledge types and characteristic are deep rooted and

TABLE 16.1
Four Categories of Knowledge

Knowledge Category	Organizational Dimension	Cognitive Dimension	Defining Characteristics
Embrained	Individual	Explicit	Formal, abstract, or theoretical knowledge, dependent on an individual's cognitive skills and abilities.
Encoded	Collective	Explicit	Information: Knowledge conveyed by signs and symbols.
Embodied	Individual	Tacit	Practical, action oriented, dependent on a person's practical experience.
Embedded	Collective	Tacit	Organizationally accepted routines and norms.

Source: Adapted from Collins, H. M. 1993. *Social Research* 60(1): 95–116.

TABLE 16.2

Types of Knowledge and the Associated Ideal-Typical Organizational Structure

Knowledge Type	Organizational Characteristics
Professional bureaucracy and embrained knowledge	• The knowledge structure is individualistic, functionally segmented, and hierarchical.
	• The sharing across functional boundaries is limited.
	• Formal demarcation of job boundaries inhibits the transfer of tacit knowledge.
	• Power and status inhibit interaction and the sharing of knowledge.
	• Learning focus tends to be narrow and constrained within the boundary of formal specialist knowledge.
Machine bureaucracy and encoded knowledge	• Key organizing principles are specialization, standardization, and control.
	• The knowledge agents reside in a formal managerial hierarchy.
	• High reliance on management information systems for knowledge aggregation.
	• The knowledge structure is fragmented, collective, functionally segmented, and hierarchical.
	• The organization seeks to minimize the role of tacit knowledge.
	• Learning occurs by correction through performance monitoring.
Operating adhocracy and embodied knowledge	• Highly organic form of organization with little standardization of knowledge or work process.
	• Knowledge structure is individualistic but collaborative, diverse, varied, and organic.
	• Tend to learn through interaction, trial and error, and experimentation.
	• Organization capable of divergent thinking and creative problem solving.
I-form organization and embedded knowledge	• Driven by knowledge and embedded in operating routines, team relationships, and shared culture.
	• Attempts to combine efficiency of a bureaucracy with flexibility and team dynamics of an adhocracy.
	• Key knowledge agent is neither autonomous individual expert nor the controlling managerial hierarchy but the semiautonomous cross-functional team.
	• Learning occurs through shared work experiences and joint problem solving.

Source: Proposed by Lam, A. 2000, *Organization Studies* 21: 487–513.

should be taken into consideration in the planning stages of organizational intranets.

Although the four constructs of organizational learning, presented in Table 16.2, are applicable to all types of knowledge, the actual processes of acquiring, storing, sharing, interpreting, and using the knowledge can vary greatly by knowledge type and organizational form. Figure 16.2 presents this separation of knowledge type and flow according to organization type. In terms of information technology, while the organizations who are classified in the upper two quadrants of Figure 16.2 seem most readily able to incorporate technology to store and communicate explicit knowledge, those organizations in the lower two quadrants who effectively implement IT can experience significant returns in terms of organizational learning. An organization's knowledge structure and form are compellingly correlated to the communication infrastructure. Therefore, tools and technologies used to augment communications should be purposefully chosen for an organization and integrated into an organization for effective coordination with existing structures.

The Technology Acceptance Model (TAM) is commonly used when considering the use of new technologies. The purpose of TAM is to track the impact of external factors on users' internal beliefs, attitudes, and intentions (Davis 1989). Perceived usefulness (PU) and perceived ease of use (EU) are the measures of IT usage. PU is a user's subjective assessment of an application's potential contribution to his or her job performance. EU is a user's subjective assessment of the amount of effort required to learn to use the application. The model assumes that both PU and EU affect users' attitudes toward new computer applications, which, in turn, would affect usage. PU has been demonstrated in research to be a good predictor of users' intentions to use computers. In a study applying TAM to electronic commerce adoption in businesses, it was found that four factors influence adoption. Those factors are organizational readiness, external pressure, perceived ease of use, and perceived usefulness (Grandon and Pearson 2004). Use of organizational theory measurements such as TAM allows researchers to quantify the usage of new technology in business applications.

16.2.1 Intra-Organizational Communication

Communication of an organization's knowledge is a mission critical component of their learning. In the information age, knowledge is an organization's foremost asset. The capital of

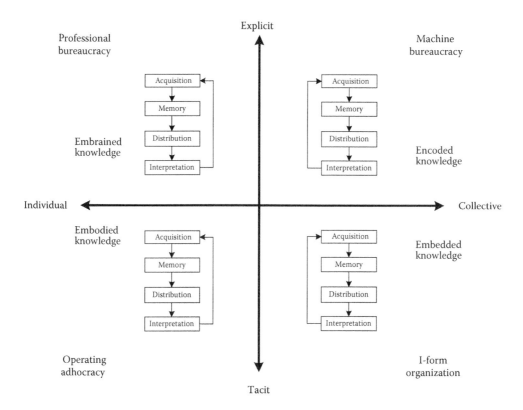

FIGURE 16.2 Knowledge, organizational learning, and organizational form. (Adapted from Lam, A. 2000. *Organization Studies* 21: 487–513; Huber, G. P. 1991. *Organizational Science* 2(1): 89–115.)

a company includes not only physical assets, but economic value is allocated to the employees' intellectual and social abilities (Carlson 2000; Drucker 1988). The sustainability and longevity of an organization have evolved to be highly contingent on the efficient and effective flow of information. In an interview with *Business Week Online* in 1998, Jack Welch stated, "An organization's ability to learn, and translate that learning into action rapidly, is the ultimate competitive business advantage." Literature from organizational theory from the past four decades has flooded the gates of popular business culture with terms such as organizational learning, knowledge management, and learning organization, among others. Organizational learning and knowledge management are critical tools for corporations of the twenty-first century and should not be overlooked.

The 1990s was a time during which organizations that strove to maintain a viable market presence restructured (sometimes drastically) the way they worked. Organizations transitioned away from a Taylor-reminiscent, or top-down model of workflow, to a more process-oriented, nonlinear model of workflow. The top-down flow of information from management to employees does not match well to the paradigms for organizational learning. The ability for an organization, at any level, to (1) remove physical and organizational imposed boundaries, (2) acquire and distribute information, and (3) produce new information, enables its successful evolution as a learning organization.

Internal communication, according to Hills (1997), can take several forms. Two intuitive groupings of communications emerge: formal broadcasts and intercolleague information exchange and collaboration. Table 16.3 summarizes each classification, information sources, common formats, and content. As organizations have evolved to emulate models of organizational learning, the flow of internal information communication with respect to the number of available sources and formats has changed. This freedom of information flow that characterizes organizational learning effectively enables knowledge distribution and acquisition improvements.

Traditionally, information flow within an organization was linear. The greater part of an organization's information was shared among a select group of individuals, and information between departments was typically kept separate. Information trickled down, pushed onto employees through the organizational hierarchy. The individuals who regulated information flow have often been referred to as *gatekeepers* (Gonzalez 1998; Koehler et al. 1998). The use of the Internet has brought with it a flood of new products and services as companies try to compete in a progressively more global and competitive arena. The Internet has also made it possible for smaller corporations to stake their ground on the international stage, thus affecting the role of the gatekeeper. In these smaller firms we are starting to see the gatekeeper role become more dispersed and less defined. In fact, the smaller the firm, the more people are likely to display gatekeeper

TABLE 16.3
Internal Forms of Organizational Communication

Communication Type	Communication Source(s)	Common Forms	Content
Formal broadcasts	• Executives • Corporate Communications department • Intra-organizational departments	• Announcements • Memos • Newsletters • Bulletins • Company-wide e-mail	• Policies • Values • Formal event announcements • Processes • Organization charts
Informal intercolleague information exchange	• Employees • Colleagues	• E-mail • Memos • Telephone calls • Newsgroups	• Gossip • Best practices • Problem solution • Practical job knowledge

communication behavior in their roles (Ettlie and Elsenbach 2007). Despite the changes in role, gatekeepers will continue to powerfully influence the information that is available in decision-making situations, thereby directly influencing organizational learning.

Most communication in traditional organizations was also point-to-point (Koehler et al. 1998); people communicated information person to person or over the phone. Wider messages were distributed through channels to regulate recipients (e.g., memos). The information technology explosion and

organizational learning revolution have clearly changed the communication paradigm. This has primarily been achieved through the expansion of possible channels for information transmission.

Koehler et al. (1998) summarized three traditional roles of individuals in intra-organizational communication: the gatekeeper, the filter, and the grapevine. Figure 16.3 presents a hypothetical organization chart. The chart depicts a traditional flow of information through an organization, where each symbol corresponds to a piece of organizational

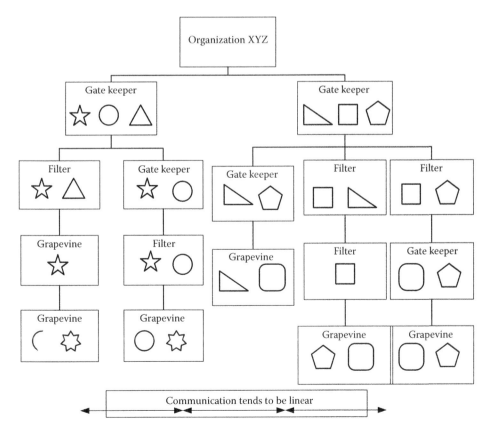

FIGURE 16.3   The linear flow of intra-organizational communication; each symbol represents a different piece of information. Information is often altered by members as it moves down the hierarchy.

knowledge that particular individual possesses. This figure illustrates the linear flow of the organizational knowledge, where gatekeepers control information flow, filters alter information, and grapevines informally share information. In particular, notice how each piece of information (or "symbol") tends to dwell in one classification/department of the organization and is distributed in a point-to-point manner. In this type of hierarchy, cross-organizational communication is rare, especially between hierarchies (Gonzalez 1998).

Information moves up and down within a hierarchical structure in a pattern sometimes called a *stovepipe* or silo. Information may be accurately transmitted but may also be inaccurately conveyed through the insertion or extraction or facts by different entities in the hierarchy. The absence of checks and balances to determine information redundancy, and/or accuracy, between classifications in an organization, makes it difficult to determine the utility of information to other departments.

Once (and if) the gatekeeper releases information, it often travels through filters (Koehler et al. 1998). Filters often put a spin on information as it passes through them. The sugarcoating of intentionally strong messages can cause those at the end of the communication chain to see a different picture of things than was intended. Grapevines, as shown in Figure 16.3, represent a more complex communication network that informally spread information with the underlying goal of knowledge verification. Entities in the organizations are sometimes left out of knowledge sharing (intentionally and unintentionally) by gatekeepers and filters. As a result, grapevines often transmit a distorted picture of the actual information to each other. Without intra-organizational communication channels, the information transmitted is often inaccurate, untimely, and counterproductive to organizational goals because grapevines are limited in their information and verification sources.

Barriers to intra-organizational communication can restrict an organization's potential for success in today's information age. Cost, globalization, time, information expirations, and information source platforms are all barriers to intra-organizational communication. Another barrier is an individual's willingness to contribute to the flow of information. In an age where information and knowledge are seen as capital investments, information ownership can give individuals a sense of power and job security, which they are reluctant to relinquish.

These challenges to open communication within an organization remind us that an intranet application that enables online publishing will not automatically return the desired results in terms of organizational growth. Information technology such as intranets is only a means to success. The groundwork of new models of intra-organizational knowledge sharing must first be put into operation. An infrastructure that encourages, supports, and documents knowledge sharing across the entire organization, and creates organizational memory, can better facilitate organizational learning (Harvey, Palmer, and Speier 1998). Organizations need to effectively reinvest in their human resources (Senge 1990),

TABLE 16.4
The Role of Technology in Organizational Communication

Technology Enhanced	Technology Facilitated
Improve coordination with internal business units.	Coordination of work and information across a variety of applications.
Promote the efficient exchange of information.	Promotes overall organization competitiveness with external forces.

not just information technology resources, because the employees are a core element to the learning organization.

Technology has the ability to both enhance existing organizational structures and facilitate new relationships and information flow (Riggins and Rhee 1998). Riggins and Rhee expanded this taxonomy to illustrate aspects of work affected by the facilitation and enhancement (e.g., integration and upgrade) of technology. Table 16.4 provides a summary of this expansion.

The human element in these technology facilitations and enhancements should not be disregarded. Carlson (2000) summarized four human-centered concepts for evolving organizations to recognize:

- The value of individual initiative and insight.
- The need for ethicality, responsibility, and commitment.
- The essential nature of teamwork, both formally and informally structured.
- The breakdown of hierarchical order through innovations such as matrix management. (p. 210)

The plausible impacts of an IT, such as an intranet, on the evolutionary process of organizational learning, merits judiciary implementation of the technology. An understanding of the hardware and software components associated with intranets can guide effective decisions and impact long-term effectiveness.

16.2.2 Defining Intranets

An intranet utilizes Internet technologies and applications from the World Wide Web within an organization. An intranet is a corporate network designed around Internet metaphors, protocols, and technology. Fundamentally, an intranet integrates several information systems that support various functional areas of an organization that would otherwise be functionally incompatible owing to differences in technological platforms. Although the components used to define intranets in this section may appear to be older in terms of technology, they should be regarded as core components, critical to the underlying functionality of intranets.

An intranet presents an organization with the functionality and benefits of the Web, with the added advantages of separation and security away from the rest of the Internet.

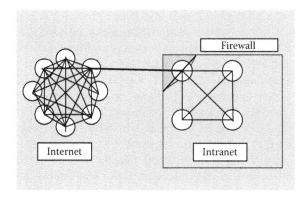

FIGURE 16.4 Intranet and Internet configuration.

Anyone outside of the organization is restricted to access of the intranet through the security of a firewall. People within the organization are given a channel through which to tap into the Internet. Figure 16.4 illustrates this association. Intranets mimic several aspects of the Internet, but with the security of a firewall and restricted access to organization-only members. An intranet is a distributed hypermedia system within an organization that retains the capability to enable people to access information, communicate with each other, collaborate in a geographically distributed manner, share knowledge, and learn from each other with innovative modes of communication.

Technically speaking, this distributed hypermedia system could be any private transmission control protocol that supports intranet applications, including Web protocols (HTTP) and others. Specifically, three Web standards (URL's, HTTP, and HTML) enable an intranet to simplify locating, displaying, sharing, and accessing information. As long as an operating system has a browser, users across different departments can access the same information and applications, independent of their individual respective platform.

At conception, intranets differed from other IT tools and shared applications because an intranet has the ability to integrate across different servers, improving the capacity of the information creator/information user relationship. Instances of intranet use in organizations may range from the simple to the complex, depending on an organization's resources. It challenged programs like Lotus Notes'™ key market in corporate information and message systems to support work flow. An intranet's Web-based cross-platform and cross–vendor compatibility were much-needed changes from existing products and systems (Dix et al. 1998).

Intranets enable normalized information accessibility among separate departments. Engineering, marketing/creative, and administrative departments can have comparable access to information, independent of the differences between operating system platforms (e.g., UNIX, Macintosh, or Windows). Intranets facilitate intra-organizational communication without requiring individuals to depart from the tools they prefer to use in their work, which could compromise efficiency, productivity, and worker satisfaction.

Several authors have identified the building blocks of an intranet (Gonzalez 1998; Hills 1997; Rosen 1997). Two of the most commonly cited components include networks and Web technologies. As some aspects of these components are beyond the technical scope of this text, we provide a brief overview of each. The following section summarizes each of these elements and explains key attributes with respect to intranet technologies.

16.2.1.1 Networks

Networks have been described as the most complex component of technology implementation (Hills 1997). Networks are the core of intranet and Internet communication, for they dictate how information systems are interconnected. There are several ways through which networks can be connected, and innovative products to access networks are continually being developed and released. The security issues related to intranets are compounded by increases in network connectivity, bridging multiple networks on a single device, shared knowledge bases, and Web 2.0 integration of legacy applications. The security issues are compounded (Landry et al. 2009) owing to the ever-changing technology and security risks. For this reason, we will not delve into the technological options for networks. Rather, we highlight attributes of networks that can have an impact on the implementation of intranets and intra-organizational communication.

Different network technologies make use of different protocols for information transfer within networks. One example of a type of network technology is Ethernet. Factors that impact the network efficacy within the context of an organization's intranet include speed of transmission, network capacity, and localization of networks. Greer (1998) identifies five types of networks in his work: Sneaker, Peer, Local Area Network (LAN), Metropolitan Area Network (MAN) and Wide Area Network (WAN). A sixth type of network is Wireless Local Area Network (WLAN). These networks span the technical spectrum from simple to complex. Table 16.5 clarifies the first five forms as interpreted by Greer (1998), along with a description of the sixth type. Crucial factors for networks with respect to intranets are ease of use, scalability, and interoperability.

System responsiveness is one technical attribute that can impact user interactions. In networked computing environments

TABLE 16.5

Types of Networks

Type of Network	Description
Sneaker	Walk information storage devices (e.g., disks, CDs, media cards) from computer to computer
Peer	Two or more computers share the same software and peripherals
LAN	Servers, computers, and peripherals organized for efficient information transmission
MAN	Campus or citywide network of LANs
WAN	Network of several LANs
WLAN	Links two or more devices using wireless distribution; provides an access point to the wider internet

such as the Internet or intranets, every user action may be followed by a delay as information is transmitted across the network. Research has demonstrated that the length, variability, and predictability of delays can affect the way users interact with such systems, and in the context of the Internet, has demonstrated that network delays can also significantly affect perceptions (e.g., Borella, Sears, and Jacko 1997; Jacko, Sears, and Borella 2000; Otto et al. 2003; Sears, Jacko, and Borella 1997a, 1997b; Sears, Jacko, and Dubach 2000). Consumers who experience increased delay times are likely to switch Web sites in a behavior similar to television viewers switching channels during long commercial breaks (Otto et al. 2003). Therefore, it is critical that the delays users experience be carefully considered as researchers investigate the issues that affect the usability of distributed multimedia documents accessed via an intranet (Sears and Jacko 2000), for these delays can impact organizational productivity and user satisfaction.

16.2.2.2 TCP/IP

Transmission Control Protocol/Internet Protocol (TCP/IP) is a principal protocol in the Internet because it transmits information across networks. The flow of information is managed from place to place in the network (Gonzalez 1998; Hills 1997; Rosen 1997). TCP/IP enables computers between departments to use the types of browsers and servers used to access the World Wide Web-based toolkits.

16.3 WORLD WIDE WEB TOOLKIT

The key element that distinguishes intranets from other organizational IT tools is the implementation of World Wide Web technologies and protocols that reside at its core. As previously mentioned, the intranet can benefit from the tools of the World Wide Web. Hyperlinks, search engines, site maps, and information repositories are all instruments successfully carried over from the Internet to intranets. Servers and browsers are recognized as two mission critical components for the start-up of an internal Web (Hills 1997).

16.3.1 Servers

Servers are the hubs of the intranet. The number of servers required for an intranet depends on the amount of information to be held in the system. Hills (1997) reported that companies have implemented between one to one hundred servers and, in unique cases, over 2000 servers to help support their intranet. Currently, the number of servers that some megacorporations have is estimated to be greater than 50,000 (Miller 2009). An important distinction is that servers use hypertext transfer protocol (HTTP). HTTP is the standard communication protocol between servers and browsers (Rosen 1997) and is another contributing factor to the consistency of Web applications in intra-organizational communication.

16.3.2 Browsers

Web browser platforms serve as the software to access and display information. This universal interface is a tool for access to all organizational information. In many organizations, specific browser platforms, brands, and versions utilized (e.g., Internet Explorer 8.0) are well documented, so that intranet pages can be designed to maximize effective usability throughout the entire organization. The objectives of information access and editing capabilities are independent from the information's source and format.

A browser requests information from the servers, and displays pages to the user in a graphical user interface (GUI). Browsers, along with hypertext transfer protocol (HTTP), ensure that users will see the same view of the information as it is translated from the server. Browser selection should be of utmost importance to an organization, as browsers are fundamental tools for intranet platform independence. Browsers impact scalability, usability, and usefulness of an organization's intranet.

Historically, only two browsers had permeated the marketplace, Netscape Navigator and Microsoft Internet Explorer. Netscape was discontinued early in 2008 because of low market share. As of December 2009, the most popular browser was Mozilla Firefox with a usage average of 46%, with Internet Explorer (versions 6.0–8.0) following closely at 37%, and browsers such as Chrome, Safari, and Opera rounding out the top five (Browser Statistics 2009). Owing to the increase in browser options, the browser decision still merits careful consideration. An incorrect match for an organization's needs could impact efficiency, employee productivity, and employee satisfaction, as well as quality. Guengerich et al. (1996) related that five dimensions to assess a browser's ability to meet organizational (corporate) requirements should to be (1) performance capacities, (2) multimedia support (e.g., sound, video, or plug-ins), (3) which computer languages are supported, (4) usability of the browser's functions, and (5) vendor support (Guengerich et al. 1996).

16.4 INTRANET APPLICATIONS

The extent of functionality gleaned from intranet use by an organization ranges from basic document management to a more sophisticated toolkit for the support and enrichment of knowledge within an organization. The extent of intranet technologies implemented in an organization and the extent of their use are interrelated with the growth of that organization, its learning activities, and knowledge sharing. Figure 16.5 illustrates this influential relationship. This section introduces and expounds on these uses, in terms of function exclusively. The evolution and interaction of an intranet amid layers of intra-organizational communication will be discussed subsequent to the discussion of organizational learning.

An intranet has a high capacity to force a company to focus on core information technology strategies and enforce strict technical and procedural standards at beginning stages. Traditionally, an intranet has been studied with respect to five fundamental work applications (Dascan 1997; Hill 1996; Greer 1998):

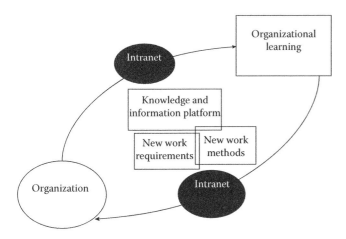

FIGURE 16.5 The extent to which intranets change the nature of organizations today with respect to organizational learning. (From Jacko, J. A. et al. 2002. *International Journal of Human Computer Interaction* 14(1): 93–130. With permission.)

1. *E-mail*: In terms of intra-organizational communication, comprises both person-to-person exchanges and person to group broadcasts.
2. *File sharing*: May be present as a person-to-person or a person-to-group interchange of files.
3. *Directories*: In a functional sense, directories are comprised of organizational information and are built to facilitate user access to the information.
4. *Searches*: The search functionality is one of the most important components of an organizational intranet. In a working environment, where time is money, a user's ability to find what they want, when they want it, the way they expect it can directly affect the bottom-line finances. Intranet search should reflect available knowledge and location of the information.
5. *Network management*: Network management deals with the maintenance and modification of intranet infrastructures for the modification and compilation of different organizational and work components.

A more categorical view of intranet applications lends itself to a three-dimensional classification: (1) information creation, (2) information dispersion, and (3) information manipulation (evolution). Furthermore, each of these activities may be executed in one of three ways:

1. Asynchronous or synchronous
   a. Are related work functions (especially information activities) performed in a serial or parallel manner?
2. Collaborative or independent
   a. Is the information/work activity assembled by an independent organization member or by group(s) of organization members in collaboration?

3. Information creator initiated or information user initiated
   a. Is the user of the information pushing or pulling information and knowledge through the intranet/organizational system?

An intranet-mediated function may be capable of a range of these attributes and functionalities, depending on the nature of the users, work, and existing organizational infrastructure. Figure 16.6 illustrates the three dimensions along which a function of intranet technology may contrast under different contexts of use. The boundaries within which functionality are defined represent the range of activities in support of that work. Dimensions associated more with organizational learning will be positioned in the area of the diagram with synchronous, collaborative, information-seeking characteristics.

The application of intranet tools to an organization can serve multiple functions or just one. The flexibility of the functionality will reflect the associated changes in the organization, whether it is due to differences between working projects or due to differences in the organizational structure as they develop into a learning organization. This list of applications in Table 16.6 has been assimilated from popular literature as well as research. Innovative strategic uses of intranet technologies continue to enter into the domain. For this reason we will refrain from detailing the specifics of each application but instead wish to lay emphasis upon the far-reaching impacts intranet technologies can create within an organization.

From a human factors perspective, each application is likely to have a unique set of users, contexts, rules, and goals for which it needs to be optimized. Consider, with each intranet-empowered activity listed, how it could potentially be classified within the organization, for this dictates the amount of planning, integration, and management. The chapter proceeds next to a discussion of the ways the application

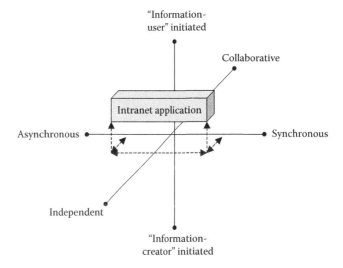

FIGURE 16.6 Classifying functionality of intranet tools.

TABLE 16.6

Examples of Intranet-Empowered Activity

• Company news distribution	• Departmental information	• Just-in-time training	• Stock prices
• Newsletters	• Project management	• Quality information	• Discussion groups
• Press releases	• Workflow management	• Product/pricing information	• Employee surveys
• Corporate policies	• Knowledge repositories	• Shipping and inventory data	• Multipoint authoring of documents
• Company goals	• Regulatory compliance status	• Materials ordering	• Distributed learning/ training
• Employee and community activities	• Employee locator	• Sales reports	• Corporate policies and procedures
• Employee recognition	• Skills directory	• Claims processing	• Health and safety regulations
• Organizational charts	• Job posting	• Benefits enrollment	• Sales automation

of intranet technologies evolve in an organization and the impacts upon intra-organizational communications.

16.5 ASSOCIATED INFORMATION TECHNOLOGIES

The capacity of intranets can be better understood through an examination of closely related (and often interrelated) information technologies. These information technologies include Internet, extranets, and portals. Both portals and extranet are important aspects of intra-organizational communication and intranets. A detailed account of these forms of IT is beyond the scope of this text. However, a macroscopic understanding of what each entails, as well as their relation to and association with intranets, each other (when applicable), and intra-organizational communications, lends additional insight into the role intranets play with these fundamental information technology structures.

16.5.1 DIFFERENTIATING INTRANETS FROM THE INTERNET

Although intranets and the Internet share the same general attributes, their context of use, and designated users deviate, meaning that implementation and utilization support of each is significantly different. As introduced in Section 16.2.1.1 (Networks), bandwidth, or system delays imposed by information quantity can affect users' perception of a system, as well as their productivity in the organization. Technological nuances aside, the type of content and fundamental applications differ between the Internet and intranets (Scheepers and Damsgaard 1997). It is especially important not to draw absolute parallels between acceptance and use of an intranet versus acceptance and use of intranet technologies (Scheepers and Damsgaard 1997). Stenmark (2004, 2) pointed out that although "the technology used in intranets is the same as in Internet, usage and content are different." The extent of utilization ultimately lies in the hands of the individuals within the organization, no matter the level technical sophistication an application may capacitate.

An intranet is an internal information system that employees might be encouraged (or even required) to use to accomplish their regular work functions. This is different from the Internet, where users may neither work within the scope of one organization, nor share common characteristics, and where users ultimately choose whether they want to visit specific Web sites (Lazar 2001). An assortment of specific organizational needs and requirements emerge with the integration of an intranet. An organization's intranet is unique because it narrows the scope of potential users. Also, the functional capacities of an intranet typically are far more complex and context specialized than those needed of the Internet. It is important to remember that although intranets are used for information storage and retrieval, they also play a major role in the flow of information and communication within the organization (Lehmuskallio 2006).

The specific needs and differences for an organization's intranet and Internet are compared and contrasted in Table 16.7. The fundamental differences between intranets and the Internet are observed on four dimensions: (1) the nature of the user, (2) the nature of work, (3) the nature of information, and (4) the nature of the technology. In consideration of user centered design principles, the level of difference between intranet and Internet applications signifies the necessity for distinct approaches to the design and implementation of each.

Despite the similarity in technologies used to build intranets and the Internet, the boundaries of use and goals for implementation create discord between the two. The fundamentals for motivating end users to use the intranet solutions provided to them differ innately from marketing an Internet site for external use. The quality of work and knowledge sharing produced by an intranet site relies heavily on employee awareness, comfort, and productivity with the tools.

Although marketing tools and practices are typically applied to Internet site launch, human resources and strategic management guide intranet launch and evolution within an organization. In the context of Internet sites, market research determines when, where, and how a launch should proceed in order to realize a high return on investment (Rosen 1997).

TABLE 16.7

Intranets Compared to the Internet

Dimension of Comparison	Intranets	Internet
User	Users have knowledge of organizational aspects such as structure, terminology, and circumstances.	Users are customers who possess less knowledge and have less investment in the organization (e.g., time and effort). Tasks involve primarily information retrieval.
Work	Tasks are related to daily work and involve complex applications.	Tasks involve primarily information retrieval.
Information	Information exists in multiple forms (e.g., progress reports, human resource information).	Information is primarily market-oriented.
Technology	Technology is more uniform and controlled to allow richer forms of media and to allow cross platform compatibility.	Technology capabilities arrange from low to high, measures taken to ensure performance on multiple computer platforms and software versions.

Source: Adapted from Nielsen (1997).

For intranet launches, the strategy is different. The launch will have a heavy impact on how the user community reacts to it (Chin 2005). Traditionally, some of the end users (early adopters) will be delighted at the intranet launch and feel that it is an integral part of their toolset. Others may find it to be an unnecessary nuisance (they already have the Internet) or as another annoying mandate from their IT department. The best strategies for the sale of intranet to its end users have been identified to include (1) executive support, (2) employee awareness and involvement with development, (3) training and education, and (4) continued management and support with evolving intranet use. Intranet designers must capitalize on their ability to characterize and understand users' needs and working environment.

16.5.2 Intranets and Extranets

It is rare to read about intranets without some mention of extranets. An extranet allows portions of an organization's intranet to be accessed by outside parties, vendors, and customers. Though there are several overlapping attributes, discrimination of intranet and extranet is accomplished through consideration of the location of the users with respect to the network firewall (Folan et al. 2006; Riggins and Rhee 1998). Intranets and extranets both capitalize on the independent protocol provided by Internet applications, which could cause issues with interorganizational compatibility. Figure 16.7 illustrates the placement of extranets relative to the location of the user with respect to the origination's firewall. The large break between intranet and Internet is crossed by intranet technologies accessed from Internet sites.

Typical information gained through an extranet would include order status, vendor lists, billing processes, and account information. It is a blend of the public Internet and closed intranet, incorporating fundamentals of each (Nielsen 1997, 1999). By using an extranet, a corporation creates a virtual organization. It is anticipated that with increased research

in the area, companies will eventually initiate a virtual supply chain (Scott and Mula 2009). Extranets can optimize communications to strengthen relationships with customers, suppliers, and partners for potential quality improvements and cost savings. In terms of the potential for costs savings, Riggins and Rhee (1998) assert that extranets can improve coordination with existing trading partners, as well as market to reach new customers.

Interactive communication of select internal information to privileged business partners may produce gains in productivity and for an organization's Just-in-Time processes (Koehler et al. 1998). Extranets are added value to secure business-to-business information sharing and transactions (Hope 2001). Riggins and Rhee (1998) assert that extranets associate an organization's Internet and intranet infrastructures. It is anticipated that this connection would extend the meaning of electronic commerce beyond point of sale applications.

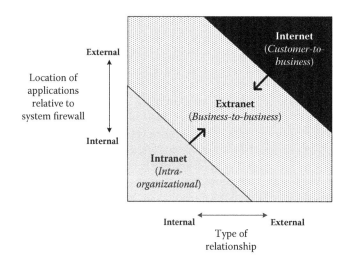

FIGURE 16.7 Position of extranet relative to intranet and Internet.

Extranets share some design requirements of intranets, with a few identifying attributes. This is indisputably attributed to extranets' placement as a customer relation tool. The goals and the underlying nature of the work to be accomplished via extranet do, in fact contrast with those of intranet. Hope (2001) lists in her work the top 10 benefits that companies expect intranets to yield:

1. Enhance competitiveness or create strategic advantage.
2. Enable easier access to information.
3. Provide new products or service to customers.
4. Increase the flexibility of information requests.
5. Improve customer relations.
6. Enhance the credibility and prestige of the organization.
7. Provide better products or services to customers.
8. Increase the volume of information output.
9. Align well with stated organizational goals.
10. Enable the organization to respond more quickly to change.

Items 2–6 and 10 are specific to customer relations and service quality. This list illustrates that the public relations aspect of the intranet is substantial. One area not directly addressed by Hope's (2001) work is the importance of extranets in the facilitation or relationships with vendors and other important business partners. Extranets provide an important bridge two the world outside the organization, in attempt to produce a more streamlined organization.

16.5.3 Intranets and Portals

The literal meaning for portal is a door or gateway. In the technical context, the term portal is nebulous. A few years younger than the concept of intranet (information on portals pre-1998 is sparse), the concept of portals began as a simple concept. Portals were principally viewed as Home pages that presented Web information applications and access to their use. Customization of a portal by an organization member enables the user to have, readily accessible, their most commonly used tools (see Eisen, this volume). A present-day example of a portal is MyYahoo!, an Internet portal that allows users to choose which resources to present on their start page. Like the concept of intranet, portals are associated with several terms, between which boundaries of distinct definitions are weak. Knowledge portals, enterprise portals, information portals, intranet portals, Internet portals, Web portals, and so on; these terms have been used synonymously while at other times used to define the different infrastructures of portals in limited research on portals, and especially in the popular technology periodicals and newsgroups. This is due to the high rate of development of portal creation applications by software companies and the slower rate of research in this area. The names associated with portals are more likely a product of marketing, and not research. At a less granular level of analysis, portals can be viewed as a

gateway to information resources. Two differing, yet overlapping research classifications of portals are presented within.

Tkach (1999) identified three basic typologies of portals: intranet Web portals, information portals, and knowledge portals. Table 16.8 differentiates between the three different types of portals. Another categorization of portals includes general portals, community portals, vertical industry portals, horizontal industry portals, enterprise information portals, e-marketplace portals, personal/mobile portals, information portals, and niche portals (Davison, Burgess, and Tatnall 2003). At the turn of the twenty-first century, portals were viewed as the best strategic IT move. No longer just a catalog of Web resources, portals are presently integrated with organization specific applications such as enterprise resource planning or ERP (discussed later in this chapter), along with other IT resources internal and external to an organization.

Collins (2001) divided portals into two classes: *corporate portals* and *enterprise portals*. A corporate portal builds an infrastructure for needs within an organization, and an enterprise portal spans the corporate portal as well as customers and vendors external to the originations. The differentiation between corporate portal and enterprise portal is analogous to intranet and extranet, respectively. The corporate portal assembles navigation services in one main location made available for employees to find information, launch applications, interact with corporate data, identify collaborators, share knowledge, and make decisions.

One definition of a corporate portal is:

> "A browser-based application that allows knowledge workers to gain access to, collaborate with, make decisions, and take action on a wide variety of business-related information regardless of the employee's virtual location or departmental affiliations, the location of the information, or the format in which the information is stored." (Collins 2001, 7)

Vendors create enterprise portals that allow end users to customize information access to their preferred internal *and* external resources. The intranet portal is a layer within the infrastructure of an organization (Collins 2003).

TABLE 16.8
Types of Portals

Portal	Description
Intranet Web portals	Provide links to all internal content providers and some external success providers.
Information portals	Present a uniformed, common look for access through system log in.
Knowledge portals	Make available all facilities of information catalog and collaborative facilities, management tools, and a knowledge repository.

Source: Adapted from Tkach, D., 1999, Knowledge portals. IBM. http://www.ibm.com/software/data/km/advances/kportals.html (accessed March 31, 2003).

Recently, the terms corporate and enterprise are used interchangeably when referring to portals. The term Enterprise Information Portals (EIP) was an offshoot of ERPs, which may explain the lack of distinction between corporate and enterprise portals. The term EIP refers to the access points to the corporate intranets that are used to manage the knowledge within an organization (Tatnall 2005). Business-to-employee (B2E) processes occur within the EIP, allowing employees to access and share data and information within the enterprise.

The functional role of an intranet portal may encompass six features:

1. Retrieve information from corporate IT systems and present the results according to the roles, specific tasks, and preferences of individual employees.
2. Present employees with information relevant to their daily tasks without making them search for it.
3. Gather information about each employee, facilitating communication between the people who need information and the people who can supply the information.
4. Allow employees to act on the information presented in the desktop without requiring them to switch to a different system or interface for the purpose of sharing the information and collaboration with other employees.
5. Present a desktop interface through a Web browser that requires minimal technical training.
6. Support multiple business processes for a single department, a single process across multiple departments, or multiple processes across multiple departments. (Collins 2001, 55)

Portals take the concept of intranet one stage further, by allowing individuals to integrate their personalized information and organizational information resources. Gains from portal use can be realized in terms of better decision-making abilities, improved organizational understanding of terminologies and information infrastructures, and more accessible intra-organization information and resource retrieval, direct links to reports, analyses, and queries, as well as personalized access to content for each employee/worker.

Selection of a portal vendor and implementation of the applications are complex decisions that can affect adoptability. One of the most common problems that businesses encounter when implementing an EIP is lack of buy-in from minority stakeholders. As with most applications, a lack of support from end users will lead to failures or to a low adoption rate. Tsui et al. (2007, 1026) reported the "common problems that hinder portal adoption include lack of an overall governance model, mis-alignment with business processes, poor or non-existent content management (process, tools, and governance), and technical problems associated with the development and configuration of portlets." Delay times can also be harmful to the implementation of a portal. If users experience unnecessary delays they may be inclined to use the internet directly. Many employees find enterprise portal

capabilities to be mediocre when compared to the Internet/Web portal with which they were familiar (Weiss, Capozzi, and Prusak 2004). Another issue with portal implementation involves allocation of funds. According to Murphy, Higgs, and Quirk (2002) sufficient funds are often lacking for portal content owing to miscalculations on data migration, content management, and feature upgrades. Many of these problems can be avoided by determining the user requirements for a new portal in advance of implementing one.

A need for a corporate portal is demonstrated when there are too many applications that the user has to start up and switch between. Portals are becoming a key facilitator of information technology, following in the footsteps of intranet technologies.

16.5.4 Instant Messaging and Intranets

Many companies are now seeking to utilize emerging technologies such as Instant Messaging (IM) as part of their corporate communication package. Originally introduced in current form in 1996, IM is now becoming a normal mode of business communication. IM is a type of computer mediated communication (CMC) that is essentially a hybrid of chat and e-mail. IM occurs in real time between two or more people through the use of typed text. URLs can easily be sent, along with files and images. Messages generally arrive as a pop-up window on the computer screen notifying the user of a communication request. Instead of being stored on servers, IM are stored on the user's computers and the server provides information for routing. Unlike chat and e-mail, IM cannot be accessed without the permission of all parties involved in the message. The possibility of some enhanced degree of privacy is appealing to many businesses.

IM has a number of applications in business. Most important, IM provides almost instantaneous singular communication and confirmation of messages. The only other technology that is close in comparison is the telephone, but IM is less intrusive. Unlike verbal conversations, IM allows for multitasking. Users can type an IM and return to their work while waiting for the message recipient to respond. This could increase productivity. Another advantage of IM over a phone call is the ability to tell when the person is available. Quite often employees' calls result in a transfer to voice mail. According to Rice and Shook (1990), 60% of business phone calls do not reach their intended recipient. IM users can tell whether or not someone is online and available, saving the time wasted on a phone call switched to voice-mail and the likely subsequent game of "telephone tag." IM applications also allow the individual to "be unavailable" without truly being otherwise occupied. A user can send an IM to someone and receive an immediate answer such as "out to lunch," "be right back (BRB)," or "on the phone." These answers can be preset in order to screen out individuals the recipient may not wish to interact with on a frequent basis. This is the CMC version of sending incoming calls to voice mail. In the pre-IM world this screening function was also handled by secretaries and receptionists.

Unlike telephonic communication, IM offers the capability of preserving a chat history for documentation. Also, IM can result in a dramatic reduction in long distance telephone charges. Many global companies could rapidly recover the cost of the intranet software and training via the reduced long distance bill. Although the privacy, convenience, and screening functionality of IM is attractive to corporations, the cost savings is arguably the best reason most businesses implement this software on their intranet.

Currently, most research regarding CMC has focused on e-mail because of its popularity and history in business. Very few studies have addressed other CMC applications, including IM. Studies that have focused on IM have been revealing. One study demonstrated that users of IM believed it conveyed more information and was more useful than conventional communication tools. They saw it as a rich medium able to conveniently exchange video and audio clips as well as photographs, drawings, charts, and graphs (Huang and Yen 2003). Other researchers have found that users replace other conventional communication tools with IM because of its polychronicity (users can talk to someone on the phone while sending a hyperlink or file via IM). Employees felt IM increased their level of privacy because cubicle/office mates can overhear conversations but not IMs (Cameron and Webster 2005). In spite of these advantages, management must insure that IM chatter in the workplace remains business related and not social, or potential productivity gains will not ensue. This may be more of a challenge to supervisors than limiting personal phone calls at work. However, it is possible to audit employee IMs, just as companies audit Internet usage. Some employees will view this as a breach of privacy. Before implementing IM services on their intranet companies should develop and define their IM policies.

16.6 INTRANETS AND INTRA-ORGANIZATIONAL COMMUNICATION

Immediate impacts of intranet use can be observed in the introduction of new communication mediums. The way people within the organization communicate can be altered through the implementation of intranets and Web-based technologies. Yet, despite these direct, more tangible returns, an intranet can translate into more indirect changes in business infrastructures. An intranet, while propagating information through an organization in the long-term triggers a metamorphosis of the way work is accomplished and organizational goals are met (Koehler et al. 1998).

Figure 16.8 provides a schematic view of several of the drivers for the use of intranets and similar information technologies within an organization. Additionally, this figure also illustrates a number of the positive outcomes that may result from the implementation of these technologies to the organizational structure.

Huber (1991) recognized that IT could play a critical role in supporting, storing, organizing, and accessing organizational

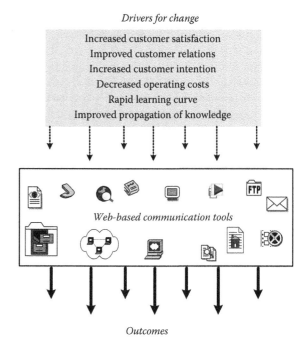

FIGURE 16.8 Interaction of Web-based communication tools in organizational change. (Based on Gonzalez, J. S. 1998. *The 21st Century Intranet.* Upper Saddle River, NJ: Prentice Hall.)

memory. Malhotra (1996) went on to suggest that IT could also help in knowledge acquisition (e.g., market research systems, competitive intelligence systems, scenario planning tools, and search tools), information and knowledge distribution (e.g., groupware tools, e-mail, bulletin boards, Web casting, and e-training). Although the use of IT for organizational memory is an excellent suggestion, designing such a system can be complex. According to Nevoa and Wand (2005), there are five reasons for the difficult design: (1) much of the knowledge in organizational memory is contextualized, (2) the multiple locations of knowledge, (3) knowledge is often tacit, (4) volatility in organizational knowledge results in frequent changes to organizational memory, and (5) some knowledge is retained from outside sources of which most of the organization is unaware.

Gonzalez (1998, 102) pointed out that Web-based technologies "can have a profound, long-lasting impact on organizations and can improve companies' chances to compete." The same message was iterated in books, news articles, and Web sites throughout the nineties—"Technology can help your organization achieve anything."

Gonzalez (1998) also asserted that the business model for IT and its positive changes can be characterized by a set of drivers and outcomes in the organization. The catalyst between the two is Web-based communication technologies.

Figure 16.8 is based onto Gonzalez's drivers and outcomes and illustrates these relationships.

Intranets, as with other IT solutions, initiate new ways of intra-organizational communication. Unlike many other information technologies, however, intranets can be molded from a generic solution into a support tool for a specified management strategy (Damsgaard and Scheepers 1999). Another term for intranet is *organizational Internet* (Scheepers and Damsgaard 1997), implying Web technologies in a closed, defined system. This gives structure to a more complex, dynamic dimension for the flow of information between creator and user (Telleen 1997).

Figure 16.9 depicts a departure from the traditional corporate view of communication introduced earlier, in Figure 16.3. This is a representation of the dispersion of information supported by an interconnected communication structure of a learning organization. Information is shared between departments at several hierarchical levels of the organization. The removal of organizational barriers stimulates better informed decision making and better development of organizational knowledge and learning. Communication of information and knowledge abandon the traditional linear structure for a network-based structure (Gonzalez 1998; Koehler et al. 1998). Information channels throughout the organization reflect both content and structure, as opposed to organizational power or hierarchy.

Roles of the individual within this model of organizational communication are less intrusive on the knowledge management process. Individuals often take on multiple roles, as they become both information seeking entities, as well as information providing entities, or gatekeepers. In fact, new, vital communication roles emerge: Webmaster and content providers (Koehler et al. 1998). The Webmaster may work alone or in conjunction with a staff, coordinating the structure of the intranet, hands-on. The role of Webmaster is broad and has grown to encompass such tasks as Perl programming, maintenance of site maps, maintenance of mirror sites, assistance in site promotion, writing glossary entries, providing user support, maintaining the search engine index, HTML validation, monitoring the error logs and reporting, verification of the links, checking usability in different browsers, editing content, maintenance of the quality and style of the site, and finding, creating, and installing tools to develop Web content (Stjean 2010). Content providers are anyone in the organization holding knowledge to share.

The drive for organizational learning and dispersion of information drives this model. The information flows within a more complex network of relationships, moving away from the linear flow of traditional organizations. Strategic management and information technology are often cited as the keys to unlock the potentials of informed organizations to achieve this level of communication.

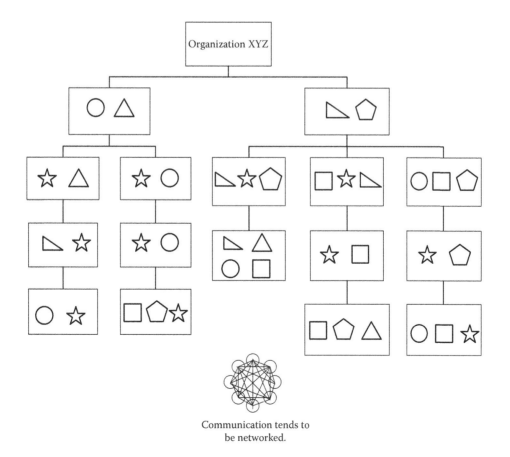

Communication tends to
be networked.

FIGURE 16.9 The nonlinear flow of intra-organizational communication: Each block in the organization chart represents one employee. Each symbol represents knowledge in the organization.

Intranet technologies better enable users to *pull* the information they want from the technology, instead of waiting for the information to be placed in front of them. Equally, intranets better enable users to push information, via the use of technology, instead of relying on the other, less efficient, less reliable models of communication. Furthermore, when access is only internal, a company has the potential for better control of information security than that of paper documents. Intranets hold great promise for organizational learning, for along with nonlinear flow of knowledge, intranets enable real-time exchanges of knowledge that users have the ability to shape in order to meet their working and information needs.

An intranet relates to learning organizations not only because it is a powerful communication medium but also because it has the potential to serve as a knowledge base. The intranet more easily captures and handles unstructured and implicit knowledge, a benefit over other IT systems. Intranets' usage and organizational learning are interrelated; as each expands within an organization, the development of one influences the other (Curry and Stancich 2000; Duane and Finnegan 2000; Harvey, Palmer, and Speier 1998; Jacko et al. 2002).

16.7 INTRANET AND ORGANIZATIONAL GROWTH

Several authors have discussed the waves of intranet implementation within an organization (Curry and Stancich 2000; Damsgaard and Scheepers 1999; Duane and Finnegan 2000;

TABLE 16.10
Four Phases of Intra-Organizational Learning

Phase 1

Learning: Stimulate learning within functional units

Intranet: Dissemination of existing documents, and user initiated sharing of tacit and embedded knowledge.

Phase 2

Learning: Expansion of learning between organizational hierarchies

Intranet: Use of collaborative technologies such as chat rooms, group e-mails to support knowledge sharing dialogues. Formulation of standards and techniques to support efficient information retrieval.

Phase 3

Learning: Extension of learning between division /strategic business units

Intranet: Focus on efficient, novel modes of information. Increased standardization of interfaces and components. Increased need for security measures with the amount of content shared on the intranet.

Phase 4

Learning: Encourage learning between connected organization

Intranet: Increased formality and standardization, including collaborative support. Increased management of resources.

Source: Based on Harvey, M. G., J. Palmer, and C. Speier. 1998. *European Management Journal*, 16(3): 341–54.

Harvey, Palmer, and Speier 1998; Jacko et al. 2002). The consensus is that intranet utilization and applications evolve from communication tools into communication infrastructures. An organization initiates its intranet adoption with a limited number of applications and can increase the technology's usefulness as they activate further intranet functionality (Damsgaard and Scheepers 1999). The different phases of intranet utility start with simple tools (posting static documents) and evolve into more complex solutions, or comprehensive toolkits. Organizations, in the latter stages of intranet implementation, realize the malleability of the technology (Damsgaard and Scheepers 1999), enabling the spectrum of functions. Stages of intranet development have been identified to be

- Publishing of static information
- Informal, asynchronous collaboration applications
- Transaction-oriented applications
- Formal collaboration applications

Several, more specific "stage models" of intranet growth have been generated in the literature, with significant overlap between. In their model, Duane and Finnegan (2000) have developed and validated their six-stage model of intranet growth. The authors traced the evolution of intranet adoption in a large corporate organization, Hewlett Packard in Ireland. Six stages of growth emerged: introduction, customized growth, collaborative interactions, process and systems integrations, external value chain integration, and institutional absorption. Table 16.9 further describes this growth process. In addition to the stages of growth, five

TABLE 16.9
Six Stages of Intranet Growth

Stage	Description
Introduction	Explore the use of an intranet and educate other departments of its potential benefits.
Customized growth	Acquire information, publish it in an organized and structured way, and provide rapid access to information via a search engine.
Process and systems integration	Integrate an intranet with computer-based systems and network applications to integrate work processes.
External value chain integration	Integrate the external supply and distribution benefits of the strategic alliance as well as share information.
Institutional absorption	Institutionalize an intranet in the organization and continue to explore ways of capturing sharing, storing, and managing tacit information.

Source: Based on Duane, A., and P. Finnegan. 2000. Managing intranet technology in an organizational context: toward a "stages of growth" model for balancing empowerment and control. *Proceeding of the 21st International Conference on Information Systems*, 242–258.

TABLE 16.11

Research Aimed at Extending the Role and Functionality of Intranets

Intranet Functionality	Reference
Unify dispersed computer based information systems in an organization into one rich system.	Baines (1996); Scheepers and Damsgaard (1997); Hazel and Jones (2000).
Document/information management and retrieval (e.g., forms, news, policies, phone directories, product specifications, pricing) for internal and external users.	Coleman (1997); Gonzalez (1998); Lai (2001); Nielsen (1999); Scheepers and Damsgaard (1997); Taylor (1997); Zhang and Chen (1997).
Web search and File Transfer Protocol (FTP): access to organizational memory: job descriptions, employee benefits, operating manuals, organizational policies for search.	Zhang and Chen (1997); Nambisan and Wang (1999).
Data mining and data access processes; search and manipulate data without leaving a home page.	Young (2000); Zhang and Chen (1997).
Real time transfer of information exchanges between individuals and groups, internal and external.	Lai (2001); Nambisan and Wang (1999).
Well-established technology to solve intra-organizational information sharing problems. Easy and straightforward implementation.	Dix et al. (1998); Scheepers and Damsgaard (1997).
E-mail and workgroup support; integrated functions, maintenance, grouping, sorting, calendaring/scheduling, eliminate geographic limitations, including external partners.	Coleman (1997); Greer (1998); Riggins and Rhee (1998); Zhang and Chen (1997).
Transparent interface to e-mail, file transfer, and discussion groups.	Scheepers and Damsgaard (1997).
Decision support and decentralized decision making.	Suresh (1998).
Interactive programs and learning labs where users manipulate systems dynamics models.	Zhang and Chen (1997); Bullinger, Müeller, and Kempf (2001).
All phases of the training and learning process for end users; at any time, any place, from any location.	Greer (1998); Mahapatra and Lai (1999).
Collaborative design, concurrent engineering, and workflow support.	Akoumianakis and Stephanidis (2001); Dix et al. (1998); Scheepers and Damsgaard (1997); Gill (2001); Martin (1999).
The added value of knowledge management activities and Web applications.	Rademacher (1999); Gonzalez (1998); Martin (1999).
Intranet as a component of strategic management and organizational learning.	Curry and Stancich (2000); Riggins and Rhee (1998), Koehler et al. (1998), Martin (1999).
Evolutionary stages of the adoption of intranet technology.	Damsgaard and Scheepers (1999); Duane and Finnegan (2000); Harvey, Palmer and Speier (1998); Lai (2001).
Strategic management of organizational knowledge through portals.	Collins (2001, 2003); Ji and Salvendy (2001); Nielsen (2000).

Source: Adapted from Jacko, J. A., et al. 2002. *International Journal of Human Computer Interaction* 14(1): 93–130.

interrelated roles during the intranet implementation process have been identified: the technology champion, organizational sponsor, intranet coordinator, intranet developer, and content provider (Scheepers 2003). Through this growth process, organizational infrastructures and intranet applications change concurrently with respect to abilities and requirements.

Similar to intranets, organizations and organizational learning evolve in phases and components, and the intranet can be utilized to support the different components. First, consider Huber's (1991) classification of organizational learning as introduced in Figure 16.1. Activities supported by an intranet can draw parallels between the classifications of knowledge acquisition, information distribution, and information interpretation, providing a repository of sorts for the organizational memory to reside.

Cangelosi and Dill (1965) asserted that organizations exhibit adaptive behavior over time. Hence, in their view, organizational learning must be viewed as a series of interactions between adaptation at the individual or subgroup level and adaptation at the organizational level. They suggested that adaptation occurs as the result of three kinds of stress: discomfort stress (related to the complexity and uncertainty of environment), performance stress (related to perceptions of past successes and failures, outcomes of past decisions, and aspirations levels and expectations), and disjunctive stress (divergence and conflict in individual behaviors). It assumes that learning is sporadic and stepwise rather than continuous and gradual.

Harvey, Palmer, and Speier (1998) explain that intranets support four phases of intra-organizational learning. Table 16.10 expounds on these phases, which include stimulation, expansion, extension, and encourage.

The significant interplay between intranets, organizational learning, and intra-organizational communication influences workflow, employees, and organization quality. Each should be viewed as a mission critical element of the organization. When observed collectively, these tools and ideologies impact the organization, in ways that may not always be positive.

Table 16.11 represents an extraction from the literature, representative of work conducted in involving intranet technologies, from 1996 to the present. The nature and scope of this work range from simple to complex and represent the wide range of functionality explored. This is shown in the left-hand column of the table where intranet functionality evolves from document production and management to knowledge management.

16.8 COSTS AND BENEFITS OF INTRANET UTILIZATION

The introduction of an intranet, with the necessary support, can spark transformation and generate benefits that have an impact at several levels within an organization. A systematic examination of benefits and costs associated with intranet utilization is useful to create awareness of their far-reaching impacts on intra-organizational communication.

Intuitively, a high return on investment is one of the main motivators for adopting intranet technologies. However, it should be emphasized that a focus on cost as justification for the use of a new technology has, many times, not been an indicator of success (Duffy and Salvendy 1999). Benefits expected from IT implementation, including productivity, managed change, and enhanced human abilities (Carlson 2000) could be influenced in positive ways with intranet technologies (Jacko and Duffy 1997). Benefits, however, are not without their costs. These costs are financial in nature, as well as in terms of resource expenditure, process redevelopment, and employee training and retraining, among others.

16.8.1 PRODUCTIVITY GAINS VERSUS INFORMATION OVERLOAD

Platform independence and cross-functional operations are two benefits of implementing intranets. This independence unifies "islands" of information that may have otherwise remained separated by technology, time, or physical space (Scheepers and Damsgaard 1997; Zhang and Chen 1997). The unification of these islands results in more informed actions on the part of the user.

Still, new freedoms that are granted to users by these technological applications (e.g., publishing, editing, developing, and exchanging information) could also have destructive consequences if left completely unmonitored. For an intranet to support successful knowledge sharing, the information needs to be timely, up to date, maintainable, and cost-effective (Curry and Stancich 2000). There is a point of diminishing returns, when the knowledge available for retrieval has to be filtered for relevance, appropriateness, and timeliness in order to fulfill the user's task. For example, it is imperative to closely monitor

information volume. If information builds up in uncontrollable quantities, valuable information may not be retrievable, user frustration and dissatisfaction with the system may increase, and use of the intranet may decline (Koehler et al. 1998).

Many individual group or department sites within an organization become grouped in different portals by an organizational IT group. IT groups may not have the facilities to maintain all of these individual sites to standard. The IT group usually receives the blame when these sites are not functional, are incorrect, or are outdated. Often times sites are abandoned when divisions become obsolete (a common occurrence in the past few years), leaving no one around who is accountable for removing them from the server. Additionally, it is difficult to upgrade and make changes to portals that experience high traffic. When a user interface is changed, the users often demand a "mirror" site that maintains the old style, for they are reluctant to accept the change.

Coordination of a system for accessible archives should be determined by the Webmaster and dictated by the uses of this information. An additional solution is the enforcement of standards for information and knowledge on the intranet. A standard specifies required, recommended, and optional design elements (Nielsen 2000) for navigational structure to page layout, icons, and logos. Nielsen (2000) identified several guidelines for standards.

- Every intranet page should have a search button. (Robust search functionality should support the search button.)
- Provide illustrated examples that fully comply with standards.
- Provide checklists for verifying standards are met.
- Provide consistent support to all questions about standards.
- Actively enforce and support the use of standards through "evangelism programs."
- Maintain changes to standards as needed.
- Comply with standard usability design standards, bringing attention to discrepancies, when these digressions are necessary.
- Support development tools and templates to facilitate compliance.
- Post standards to the intranet, with appropriate search function, and hypertext rules.
- Printed standards should have a sound index for quick reference.

A disclaimer should accompany these standards, for as an organization grows, the user interfaces of intranet applications become more difficult to standardize and maintain. The root of the impediments to intranet standardization parallels issues with Internet standardization: too much development, too fast. Developers and information technology specialists likely become decentralized and are dispersed throughout the organization. Additionally, the standardization of intranet pages may be given less priority than the standardization of externally facing Web pages (e.g., extranet and Internet).

However, in fact, a lack of continuity between intranet pages could negatively impact the bottom line of productivity.

It should also be noted, that too much control and standardization can undermine the intranet's ability to attract information content providers, as well as suppress the content provider's creativity. Ultimately, less information will be available, which will result in fewer interested users and eventually smother productivity gains. Alternately, allowing for continued "organic growth" of the information system, with a hands-off approach, could lead to a mass amount of information, interconnected in unproductive, confusing, or nonimplicit ways. This also leads to diminished productivity and a negative user perception of the tool. In their case study of a South African phone company's implementation of intranet technology, Scheepers and Rose (2001) pointed to the necessity of balancing control and individual ownership through strategic management.

16.8.2 Employee Empowerment versus Strategic Management

Intranet technologies allow for a varied "locus of control" for different activities, individuals, and contexts (Mahapatra and Lai 1999). Users are becoming empowered and beginning to track their own information needs (Harvey, Palmer, and Speier 1997). One compelling reason for the implementation of intranet technology is this motivation of employee initiative. This empowerment has potent effects, often creating a stronger sense of community among many individuals and groups within the organization (Greer 1998). Duane and Finnegan (2000) reported the effects of an intranet on employee empowerment, which illustrated the achievement of improved satisfaction and productivity by means of the implementation of an organizational intranet.

Although employees can share more knowledge with the intranet, the transition of many internal services becoming more "self-service" in nature, means less face time spent with people outside of a given department, and increased social isolation. Many groups, such as corporate travel, have been virtually replaced by online Web sites and forms. There are often critical exchanges that can be lost with the migration to Web sites and forms. Zuboff (1988, p. 141) painted a graphic picture of the psychological consequences of simplification, isolation, and computer mediation of clerical work. In a study of two offices migrating to computer-based clerical work, Zuboff found clerks complaining of physical and mental discomfort (eye strain, nervous exhaustion, physical strain, irritability, enervation, sedentariness, back pain, short tempers, intolerance, and a host of other concerns). Automation removed the need for "bodily presence in the service of interpersonal exchange and collaboration now required their bodily presence in the service of routine interaction with a machine." The resultant sentiments of the users could be prohibitive to bottom-line productivity.

In the sense that this technology will be used for human resources planning, and learning within the organization, it seems relevant that some lessons from manufacturing can be carried into this discussion. In manufacturing and product development, it is clear that one needs to consider some organizational aspects, human aspects, and social aspects to gain the expected benefits of a new technology (Duffy and Salvendy 2000). Furthermore, it must be understood how the organization's required structure supports the goal.

Strategic management is the most commonly cited factor of successful intranet adoption for intra-organizational communication (Damsgaard and Scheepers 1999; Duane and Finnegan 2000; Gonzalez 1998; Koepler et al. 1998). Damsgaard and Scheepers (1999) caution against the use of intranet technologies as a change agent, because it tends to replicate existing structures and may morph into a barrier for further change. Instead, they caution, the intranet should be institutionalize of support the adoption of a new management strategy. Intranet technology and management strategies could work against each other if not planned for effectively.

Active management is a mission critical component to optimizing intranet development efforts. While strategic planning is an important pre-intranet adoption and development tool (Wachter and Gupta 1997), intranet related management has been experienced to intensify in the latter stages of intranet evolution. The intranet, at this stage, is a critical component to workflow (Duane and Finnegan 2000). Strategic management must address issues of intranet growth, data ownership, data content, and intranet work-group coordination (Lai 2001). Scheepers and Rose (2001) point to the following guidelines for managing intranets:

- Balance control and individual ownership
  - Continually evaluate the balance of employee empowerment and standardization as the intranet and organization infrastructure have been shown to evolve.
- Cultivate intranet as a medium, not a system
  - Using collaborative and facilitative managerial style to garnish the individual user and task requirements with the system. Nielsen (2000) refers to this as *evangelism outreach,* because every department needs to be brought on board in the development of standards and intranet usage paradigms they are expected to follow, else risking the disregard of management mandates.
- Enforce tactics for a self-sustaining intranet
  - Convincing knowledge workers of the potential value of the intranet to their productivity and job satisfaction. This is often accomplished with the introduction of a "killer application" for users or subgroups of users that will draw in users (e.g., employee phone directories, Web enabled legacy systems).

16.8.3 Enterprise Resource Planning

Enterprise Resource Planning (ERP) is an industry term for the broad set of activities that helps a corporation manage the integral parts of its business. An alternate definition states that ERP is a "business management system that comprises integrated sets of comprehensive software, which can be used, when successfully implemented to manage and integrate all the business functions within an organization" (Shehab et al. 2004). This type of system provides visibility for key performance indicators (KPIs) and can also be used for product planning, parts purchasing, inventories, interacting with suppliers, providing customer service, and tracking orders. In addition, ERPs often include application modules for finance and human resources. Most commonly, an ERP system uses or is integrated with a relational database system and has modules available via the company intranet. Traditionally utilized in areas such as manufacturing, construction, aerospace, and defense, ERP systems are now seen across the board in banking/finance, health care/hospitals, hotel chains, education, insurance, retail, and telecommunications sectors. It is predicted application software, with ERP and office suite software being the largest proportion of purchases, will grow at a rate of 9.9% from 2010 through 2015 (Lohman 2009).

Globally, both private and public corporations are implementing ERP systems to replace legacy systems that are no longer compatible with the modern business environment. These implementations can be both cost and time prohibitive. Business processes must often times be changed with ERP implementation and require new procedures, supplementary employee training, and augmented managerial and technical support (Shang and Seddon 2002). Despite these drawbacks, recent literature has defined factors that influence success. Key factors for successful ERP implementation include top management support, business plan and vision, reengineering of business processes, retaining an effective project management overseer or champion, teamwork and composition, ERP system selection, user involvement, and finally end-user education and training (Al-Fawaz, Al-Salti, and Tillal 2008).

ERP platforms now exist in parallel with company intranets. They allow employees the functionality of a dashboard for decision support as well as provide a portal for business interactions. Despite the inflated cost of implementation, many organizations are finding ERP platforms to be a necessary and beneficial part of their intranet functionality. It is anticipated that as e-commerce continues to spread globally, even the smallest firms will require ERPs in order to compete.

16.8.4 Internet Collaborations—Enterprise 2.0

Web 2.0 is an all encompassing term for the next generation of Web applications. Instead of using the desktop to launch applications, Web 2.0 uses the Internet. Web 2.0 applications include social networking sites (SNS), blogs, Wikis, and mashups. Much like Web 2.0, Enterprise 2.0 (E2.0) refers to the next generation. The definition of E2.0 is "the use of emergent social software platforms within companies, or between companies and their partners or customers" (Andrew McAfee's Blog 2006). Businesses using E2.0 applications create new interfaces in a variety of ways. One study demonstrated that

51% use more flexible forms of cooperation with suppliers (Web services) and 75% use more flexible forms of internal communications such as wikis or blogs and other forms of external communication with customers, i.e., social networks (Bughin 2008). Companies that utilize E2.0 concepts are reshaping the way business is conducted, altering communication tools, cost structures, and business relationships.

There are some significant debates occurring over the benefits and future of Enterprise 2.0. One common argument regards the inherent nature of corporations. Historically, businesses were top down because someone at the top had information that those at the bottom did not have. Raises and promotions were given to the individuals with inside knowledge or insight. The Internet has virtually eliminated insider knowledge because anyone who can operate Google can search for the latest information. Cristobal Conde, the CEO of Fortune 500 company Sungard, promotes collaboration via Enterprise 2.0 technologies. He stated:

> "The answer is to allow employees to develop a name for themselves that is irrespective of their organizational ranking or where they sit in the org chart . . . recognition from their peers is, I think, an extremely strong motivating factor . . . By creating an atmosphere of collaboration, the people who are consistently right get a huge following, and their work product is talked about by people they've never met. It's fascinating" (Andrew McAfee's Blog 2010).

Sungard has adopted and implemented a Microblogging tool on their intranet called Yammer. Yammer was launched for businesses in September 2008, offering a private network. Yammer functions in a manner similar to Twitter—instead of posting status updates, users post business updates about new leads, novel ideas, or updates on current projects. Software applications like Yammer are starting to replace previous methods of communicating with both clients and employees. Examples of business activities that can be done on a Yammer/Twitter type of software include chat, e-mail, identifying trending topics, broadcasting breaking news, marketing and brand building, mining consumer sentiment, and engaging in customer service (Andrew McAfee's Blog 2009).

Companies can experience cost savings by using social networking applications to replace customer service employees. According to Natalie Petouhoff, an analyst at Forrester Research, online user groups conform to the 1-9-90 rule (Lohr 2009). This rule states that 1% of posters are superusers who supply the best answers or points of view, 9% are responders who do the majority of postings and ratings, and the other 90% are readers searching for answers. Petouhoff believes that the 90% will go to your Web site if you have the 1% (Lohr 2009). Companies can benefit if these superusers are not paid customer service employees. Rather, they could be retirees with expertise in the field or product users with expertise in the specific area of interest. Several suppliers including Help Stream, Jive Software, and Telligent offer companies social networking software with customer service applications.

There are as many critics as there are champions of E2.0. Critics' most common concern is that this type of technology only works for companies that are knowledge driven such as software or tech companies, law firms, architectural concerns, etc. Some people believe large enterprises have to work in structures and hierarchies and that E2.0 is anti-hierarchical and hence counterproductive (Howlett 2009). Poor financial indicators may also contribute to a negative perception of E2.0. It is unlikely that companies will invest in this new technology for their intranets in a time of financial uncertainty. This is especially true given the paucity of supportive evidential research. Enterprise 2.0 may not be a viable option for many entities until more case studies and research emerges.

16.9 LOOKING FORWARD: NEW APPLICATIONS

The constantly changing IT marketplace leaves some room for predictions. By the time this edition goes to press, the Google Wave application will still be in beta testing. Despite the fact that it has not yet been universally adopted, or even universally tested, this technology appears to be the next step for intra-office communication. Google Wave is an online tool for real-time communication and collaboration, allowing both a conversation and a document to be created simultaneously adding to it richly formatted text, photos, videos, maps, and possibly even more (Google Wave 2009). As far as technological predictions go, we anticipate that if the next best thing in organizational collaboration is not the Google Wave product, it will be something quite similar.

The premise of Google Wave is organized brainstorming. Individuals who currently function in a collaborative environment will instantly see the benefits from this type of software. Those who do not function collaboratively currently will have a new mechanism to encourage their participation in an open environment. The first user would begin at the Google Wave inbox in order to create a new wave. The user then selects the coworkers they want to be able to participate. The user writes their ideas or plans and asks for input from their colleagues and sends the wave. The selected coworkers can then be able to comment, add files or photos, and even work collaboratively in real time through this wave. The wave participants all have equal rights and can add or edit the document, add links, notes, or attachments. The waves reside in the first user's inbox and can allow the user to see contacts that are currently online by checking the presence indicators.

Invitations to test the beta version of Google Wave were sent out to 100,000 people in September 2009. It is unknown at this point when Google Wave will complete its beta testing and be released to the general public for consumption. In the meantime we predict that other similar software will begin to be developed and consumers will be left with many choices. The most difficult choice that will remain for organizations is to figure out where this technology works. Do you choose e-mail, chat, IM, or the "Wave" for your next idea? With the ever-increasing number of tools and options we do hope that users will base their choices more on content rather than format.

16.10 HIGH RETURNS ON INVESTMENT VERSUS HIGH COSTS OF INEFFECTIVE DESIGN

The return on investment (ROI) of an intranet has been calculated, as has the cost of poor intranet design to an organization. An intranet is a relatively inexpensive means of connecting multiple information system platforms. Use of an organization intranet can provide a significant cost savings with respect to administrative simplification (e.g., pre- versus post-intranet operating costs), that a 100% ROI should be realized within a period of weeks after successful intranet implementation (Carlson 2000; Lai 2001).

However, these are often the immediate gains electronic document conversion to reduce document duplication and distribution costs. Curry and Stancich (2000) advise against measuring intranet progress using only the ROI value. They advise that several companies calculate their ROI based on intranet-based publishing functions, as opposed to intranet-based applications.

Nielsen (1999) stated that it is common for intranet improvements to have 3:1 payout ratio. For example, an organization that invested $3 million in its intranet's usability saved an estimated $10 million per year for its 7000 users (Nielsen 1999). The 3:1 payout ratio for usability improvements and the 100% ROI show how potentially sensitive an organization's success can be to the stability and usability of its internal network.

Intranet designers have the keen ability to comprehensively scope user hardware, experience, skills sets, and workflow needs. Human resources can provide much of the tangible information, but additional information needs to be gathered in order to effectively support the organization's rules, norms, and structures and more comprehensive user requirements. Field studies or ethnographic studies are fairly easily facilitated within the confines of an organization (Nielsen 2000).

Essentially, the goal is to gather information on existing communication infrastructures and workflow, through observations of employees during actual work. Accordingly, after initial planning of intranet application development, the exploitation of low-cost usability testing methods should be employed throughout the development of intranet applications to ensure employee productivity and mitigate changes and misuse with the post-launch application. Given the severe headcount reductions in many companies in recent years, one may wonder who has the time or resources to accomplish these activities. Any investigation into existing communication infrastructures is better than none and will moderate the people hours and other resources required throughout the design and implementation process.

That said, tremendous care should be taken to ensure end user productivity and satisfaction with the intranet and its associated tools. It is of great importance to understand and track the skills, experience, and expertise of the people who will be creating, sharing, using, and disseminating information through the use of intranet. Research has demonstrated that in specific contexts of use, such as hospitals, different personnel have accepted and used Internet technologies much more readily than others (Jacko, Sears, and Sorensen 1998, 2001; Sorensen, Jacko, and Sears 1998a, 1998b). This is especially true for time critical work environments that may be highly sensitive to a possible decay in the value of information when issues of design, information presentation, and organization limit information accessibility.

Usability heuristics identified to influence intranet design include efficiency, memorableness, and error reduction (Nielsen 2000). These heuristics and several others have been proposed and explored as analytical tools for assessing the usability or effectiveness of an interface. These attributes directly impact a user's productivity and level frustration when working with the intranet.

16.11 CONCLUSION

Intranets, a form of IT, have demonstrated the capacity to promote gains in productivity, efficiency, and communication infrastructures within organizations. However, these applications of technology do not work independently. "Corporate efficiency goes up dramatically with clearer communication, and the intranet can be the infrastructure for the communication if—and only if—it is designed to make it easy for people to find information when they need it" (Nielsen 2000, 276–77).

According to Zuboff (1988), IT may induce innovative business initiatives and create an organization where knowledge should be used to make knowledge accessible to anyone with the capacity to understand it. IT is the product of human actions as well as material assets and the effective integration of the two can lead to success. Huber (1990) recognized that IT could increase the speed and quality of intra-organizational decisions in organizations supporting infrastructure of decentralized decision-making.

Intranet technology, unlike other forms of information technology, is more forgiving, flexible, and customizable to

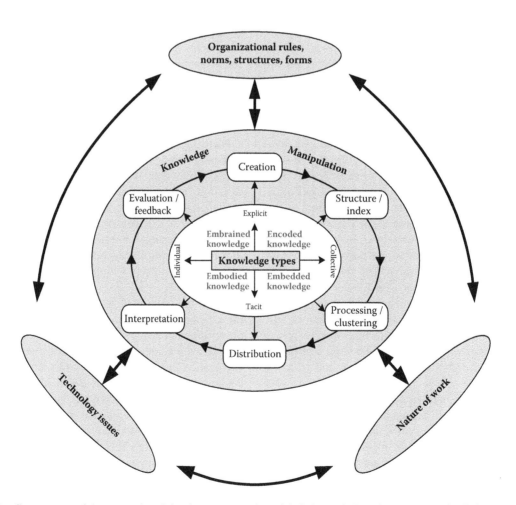

FIGURE 16.10 Components of the research and development agenda and their interrelationships. (From Jacko, J. A., et al. *International Journal of Human Computer Interaction* 14(1), 93–130, 2002. With permission.)

an organization's needs throughout the layers of an organization. Intranet technology and its utilization by organizations have matured over the past decade. The early adopters of intranet technologies are experiencing issues grounded in the later stages of intranet evolution, distanced from the problems of document publishing and altering. In their white paper on the topic of intranets and organizational learning, Jacko et al. (2002) pooled the expertise of researchers in the areas of intranets and organizational learning. The panel of experts was surveyed on their expectations for the future role of intranets, and ensuing research and development efforts to address their future use.

Experts indicated that the role of intranets in driving a new form of organization is an evolving tool upon which an organizational knowledge infrastructure can be built. Intranets were also identified to drive an organization to new forms of collaborative work, such as virtually distributed teams. Justification of this driver is the potential for the teams to possess optimal subject matter expertise and limited time spans for collaboration, which serve are motivators for fast, efficient knowledge exchange.

Four major categories for the research and development of intranets and organizational learning were ranked as follows: (1) organizational rules, norms, and structure, (2) changes in the nature of work, (3) knowledge manipulation, and (4) technology issues. Figure 16.10 presents a conceptual framework that resulted from this effort, depicting the interrelationships of the four categories of targets, with *knowledge* at the center of the figure to demonstrate its centrality at the core of this topic. Lam's (2000) proposed typology of knowledge, integrating the cognitive and organizational dimensions of knowledge, is adapted for this figure and shown at its center.

"Knowledge Manipulation" encapsulates knowledge in this diagram and is composed of knowledge creation, knowledge structuring, knowledge processing, knowledge distribution, knowledge interpretation, and evaluation and feedback. It is left to the researchers and developers to investigate the nature and degree of the interrelationships between the targets within each of the four categories of research and development. It is clear from this conceptual depiction and the experts' opinions that the next generation of intranets have the potential to serve as powerful tools for organizational learning and that the existing scientific literature serves as a solid foundation from which future research and development efforts can be launched.

Intranets provide new communication tools that organizations can use to support their evolution into learning organizations. Intranet use evolves over time to support the changing needs of intra-organizational communication. At the foundation of any organization are the users of information, whom possess different needs, abilities, and responsibilities at different times and places. As in any human-centered system, understanding user characteristics, and the role these characteristics play in the nature of the user's work flow, can inform the types of tasks and structures that should be supported with intranet applications (and the way in which these organizational aspects can be recorded).

The comprehensive, direct impacts of intranets on organizational learning are not yet known. The complexity of measuring organizational learning remains a challenge as well as the difficulty for researchers to gain access to private intranets for research. Intranets only address a small area within the overall knowledge space of organizational learning. Research is now focused on ERP implementation and success rates, Enterprise 2.0 technologies, and new collaborative applications on the horizon such as Google Wave. As organizations integrate broader levels of intranet functionality and more sophisticated technologies, intranet applications will become fully intertwined in the evolutionary process of organizational learning.

ACKNOWLEDGMENTS

The 2nd edition version of this chapter was made possible through a graduate fellowship awarded to the second author by the Institute for Health Informatics at the University of Minnesota.

REFERENCES

Akoumianakis, D., and C. Stephanidis. 2001. Computational environments for organisational learning. In *Proceeding of the 9th International Conference on Human–Computer Interaction*, eds. M. J. Smith and G. Salvendy, 301–305. Mahwah, NJ: Lawrence Erlbaum.

Al-Fawaz, K., Z. Al-Salti, and E. Tillal. 2008. Critical success factors in ERP implementation: A review. In *Proceedings of the European and Mediterranean Conference on Information Systems* (Dubai, May 25–26, 2008). http://bura.brunel.ac.uk/handle/2438/3336 (accessed March 4, 2010).

Andrew McAfee's Blog. 2006. Enterprise 2.0, version 2.0. May 27. http://andrewmcafee.org/2006/05/enterprise_20_version_20/ (accessed Jan. 28, 2010).

Andrew McAfee's Blog. 2010. Signs of Intelligent Life in the Corner Office. January 18. http://andrewmcafee.org/2010/01/signs-of-intelligent-life-in-the-corner-office/ (accessed Jan. 28, 2010).

Argote, L. 1999. *Organizational Learning: Creating, Retaining, and Transferring Knowledge*. Boston, MA: Kluwer Academic.

Argyris, C. 1982. *Reasoning, Learning, and Action: Individual and Organizational*. San Francisco: Jossey-Bass.

Argyris, C., and D. A. Schon. 1978. *Organizational Learning: A Theory of Action Perspective*. Reading, MA: Addison-Wesley.

Argyris, C., and D. A. Schön. 1996, *Organizational Learning II: Theory, Method, and Practice*. Reading, MA: Addison-Wesley.

Baines, A. 1996. Intranets. *Work Study* 5: 5–7.

Bapuji, H., and M. Crossan. 2004. From questions to answers: reviewing organizational learning research. *Management Learning* 35(4): 397–417.

Borella, M. S., A. Sears, and J. A. Jacko. 1997. The effect of Internet latency on user perception of information content. *Proceedings of IEEE Global Telecommunications Conference: GlobeCom 97* (Phoenix, AZ, Nov. 3–8), 1932–1936.

Browser Statistics. 2009. *W3Schools Online Web Tutorials*. Web (accessed Nov. 12, 2010). http://www.w3schools.com/browsers/browsers_stats.asp.

Bughin, J. 2008. The rise of enterprise 2.0. *Journal of Direct, Data and Digital Marketing Practice* 9(3). Bullinger, H. J., K.

Mueller, and F. Kempf. 2001. Using integrated platform solutions in work-oriented learning environments. In *Proceeding of the 9th International Conference on Human–Computer Interaction*, eds. M. J. Smith and G. Salvendy, 282–286. Mahwah, NJ: Lawrence Erlbaum.

Cameron, A. and J. Webster. (2005). Unintended consequences of emerging communication technologies: Instant messaging in the workplace. *Computers in Human Behavior* 21: 85–103.

Cangelosi, V. E., and W. R. Dill. 1965. Organizational learning: Observations toward a theory. *Administrative Science Quarterly* 10: 175–203.

Carlson, P. 2000. Information technology and emergence of a worker centered organization. *ACM Journal of Computer Documentation* 24(4): 204–212.

Chin, P. 2005. Delight or dismay: Intranet launch methods. *Intranet Journal*, May 26. http://www.intranetjournal.com/articles/200505/ij_05_26_05a.html (accessed March 4, 2010).

Coleman, D. 1997. Collaborating on the Internet and intranets information systems-collaboration systems and technology. *Proceedings of the Hawaii International Conference on System Sciences* 2: 350–358.

Collins, H. M. 1993. The structure of knowledge. *Social Research* 60(1): 95–116.

Collins, H. M. 2001. *Corporate Portals: Revolutionizing Information Access to Increase Productivity and Drive the Bottom Line.* New York: AMACOM.

Collins, H. M. 2003. *Enterprise Knowledge Portals: Next Generation Portal Solutions for Dynamic Information Access, Better Decision Making, and Maximum Results.* Chicago, IL: AMACOM.

Cross, R., and L. Baird. 2000. Technology is not enough: Improving performance by building organizational memory. *Sloan Management Review*, Spring.

Curry, A., and L. Stancich. 2000. The intranet—an intrinsic component of strategic information management? *International Journal of Information Management* 20: 249–268.

Cyert, R. M., and J. G. March. 1963. *A Behavioral Theory of the Firm.* Englewood Cliffs, NJ: Prentice Hall.

Daft, R. L., and K. E. Weick. 1984. Toward a model of organizations as interpretation systems. *Academy of Management Review* 9: 284–295.

Damsgaard, J., and R. Scheepers. 1999. A stage model of Intranet technology implementation and management, in *Proceedings of the 7th European Conference in Information Systems* (Copenhagen, Denmark, June 23–25), eds. J. Pries-Heje et al., 100–116. Copenhagen, Denmark: Copenhagen Business School.

Danson, N. 2009. The Intranet Model: A strategy for success, Part 1. *Intranet Journal*, Oct. 15. http://www.intranetjournal.com/articles/200910/ij_10_15_09b.html (accessed March 1, 2010).

Davis, F. 1989. Perceived usefulness, perceived ease of use and user acceptance of information technology, *MIS Quarterly* 13: 319–340.

Davison, A., S. Burgess, and A. Tatnall. 2003. *Internet Technologies and Business.* Melbourne, Australia: Data Publishing.

Dix, A., J. Finlay, G. Abowd, and R. Beale. 1998. *Human–Computer Interaction*, 2nd ed. England: Prentice Hall.

Drucker, P. F. 1988. The coming of the new organization. *Harvard Business Review*, 45–53.

Duane, A., and P. Finnegan. 2000. Managing intranet technology in an organizational context: Toward a "stages of growth" model for balancing empowerment and control. *Proceeding of the 21st International Conference on Information Systems*, 242–258. Atlanta, GA: Association for Information Systems.

Duffy, V. G., and G. Salvendy. 1999. Relating company performance to staff perceptions: The impact of concurrent engineering on time to market. *International Journal of Production Research* 37(4): 821–834.

Duffy, V. G., and G. Salvendy. 2000. Concurrent engineering and virtual reality for human resource planning. *Computers in Industry* 42: 109–125.

Easterby-Smith, M., and M. A. Lyles. 2003. Introduction: Watersheds of organizational learning and knowledge management. In *The Blackwell Handbook on Organizational Learning and Knowledge Management*, eds. M. Easterby-Smith and M. A. Lyles, 1–15. Malden, MA: Blackwell.

Eisen, P., this volume. Design of portals. In *Handbook of Human Factors in Web Design*, 2nd ed., eds. K.-P. L. Vu and R. W. Proctor, 303–328. Boca Raton, FL: CRC Press.

Ettlie, J. E., and J. M. Elsebanch. 2007. The changing role of R&D. *The Free Library*. Retrieved November 12, 2010 from http://www.thefreelibrary.com/ThechangingroleofR&Dgatekeepers:from primarily a first-line . . . -a0168511633.

Folan, P., P. Higgins, and J. Browne. 2006. A communications framework for extended enterprise performance measurement. *International Journal of Computer Integrated Manufacturing* 19(4): 301–314, doi:10.1080/09511920500340858.

Gill, Z. 2001. Webtank design—intranet support for collaborative problem-solving. In *Proceeding of the 9th International Conference on Human–Computer Interaction*, eds. M. J. Smith and G. Salvendy, 292–296. Mahwah, NJ: Lawrence Erlbaum.

Gonzalez, J. S. 1998. *The 21st Century Intranet.* Upper Saddle River, NJ: Prentice Hall.

Google Wave. 2009. Google. http://wave.google.com/help/wave/closed.html (accessed March 2010).

Grandon, E. E., and J. M. Pearson. 2004. Electronic commerce adoption: An empirical study of small and medium US businesses. *Information & Management* 42: 197–216.

Greer, T. 1998. *Understanding Intranets.* Redmond, WA: Microsoft Press.

Guengerich, S., D. Graham, M. Miller, and S. McDonald. 1996. *Building the Corporate Intranet.* New York: John Wiley.

Harvey, M. G., J. Palmer, and C. Speier. 1997. Intranets and organizational learning. In *Proceedings of the 1997 SIGCPR Conference, The Impact of the New Connectivity,* 100–116. New York: ACM Press.

Harvey, M. G., J. Palmer, and C. Speier. 1998. Implementing intra-organizational learning: A phase-model approach supported by intranet technology. *European Management Journal* 16(3): 341–354.

Hazel, H., and A. M. Jones. 2000. Show off the corporate library. *International Journal of Information Management* 20(2000): 121–130.

Hedberg, B. 1976. How organizations learn and unlearn. In *Handbook of Organizational Design*, eds. P. C. Nystrom and W. H. Starbuck, 3–27. New York: Oxford University Press.

Hills, M. 1996. *Intranet Business Strategies.* New York: John Wiley.

Hills, M. 1997. *Intranet as Groupware.* New York: John Wiley.

Hills, M. 1998. *Managing the Corporate Intranet.* New York: John Wiley.

Hope, B. 2001. Service quality in the virtual world; The case of extranets. In *Managing Internet and Intranet Technologies in Organizations: Challenges & Opportunities*, ed. S. Dasgupta, 21–34. Hershey: PA: Idea Group Publishing.

Howlett, D. 2009. Irregular Enterprise mobile edition. *ZDNet Blogs mobile edition*, Aug. 26. http://blogs.zdnet.com/Howlett/?p=1228 (accessed Jan. 28, 2010).

Huang, A. H., and D. C. Yen. 2003. Usefulness of instant messaging among young users: Social vs. work perspective. *Human Systems Management 22.*

Huber, G. P. 1991. Organizational learning: The contributing processes and the literatures. *Organizational Science* 2(1): 89–115.

Jacko, J. A., and V. Duffy. 1997. Interface requirements: An analysis of technologies designed for individual and group work. In *Proceedings of the 7th International Conference on Human–Computer Interaction* (San Francisco, CA, Aug. 24–29), 347–350.

Jacko, J. A., G. Salvendy, F. Sainfort, V. K. Emery, D. Akoumianakis, V. G. Duffy, et al., 2002. Intranets and organizational learning A research and development agenda. *International Journal of Human Computer Interaction* 14(1): 93–130.

Jacko, J. A., A. Sears, and M. S. Borella. 2000. The effect of network delay and media on user perceptions of web resources. *Behaviour and Information Technology* 19(6): 427–439.

Jacko, J. A., A. Sears, and S. J. Sorensen. 1998. The effect of domain on usage patterns and perceptions on the Internet: Focusing on healthcare professionals. In *Proceedings of the Human Factors and Ergonomics Society 42nd Annual Meeting* (Chicago, IL), 521–525.

Jacko, J. A., A. Sears, and S. J. Sorensen. 2001. A framework for usability: healthcare professionals and the Internet. *Ergonomics* 44: 989–1007.

Ji, Y. G., and G. Salvendy. 2001. Development and validation of intranet organizational memory information system for improving organizational learning. In *Proceeding of the 9th International Conference on Human–Computer Interaction*, eds. M. J. Smith and G. Salvendy, 297–300. Mahwah, NJ: Lawrence Erlbaum.

Koehler, J. W., T. Dupper, M. D. Scaff, F. Reitberger, and P. Paxon. 1998. *The Human Side of Intranets: Content, Style, & Politics*. Boca Raton, FL: St. Lucie Press.

Lai, V. S. 2001. Intraorganizational communication with intranets. *Communications of the ACM* 44(7): 95–100.

Lam, A. 2000. Tacit knowledge, organizational learning and societal institutions: An integrated framework. *Organization Studies* 21: 487–513.

Landry, B. J., M. S. Koger, S. Blanke, and C. Nielsen. 2009. Using the Private-Internet-Enterprise (PIE) model to examine IT risks and threats due to porous perimeters. *Information Security Journal: A Global Perspective* 18(4): 163–169.

Lazar, J. 2001. *User-Centered Web Development*. Sudbury, MA: Jones and Bartlett.

Lehmuskallio, S. 2006. Decision making and gatekeeping regarding intranet content in multinational companies. *In Proceedings of CCI Conference on Corporate Communication*, ed., C. M. Genest and M. B. Goodman, 211–224.

Levitt, B., and J. G. March. 1988. Organizational learning. *Annual Review of Sociology* 14: 319–340.

Lohman, T. 2009. Enterprise software spend to rebound in 2010: Gartner—IT spending, Gartner, enterprise software—Computerworld. *Computerworld*, Nov. 9.

Lohr, S. 2009. Unboxed—Verizon's Experiment in Volunteer Customer Service. *New York Times*, April 25.

Mahapatra, R. K., and V. S. Lai. 1999. Evaluation of intranet-based end-user training. *Proceedings of the 20th International Conference on Information Systems*, 524–527.

Malhotra, Y. 1996. Organizational learning and learning organizations: an overview. http://www.brint.com/papers/orglrng.htm (accessed March 31, 2003).

Marabelli, M., and S. Newell. 2009. Organizational learning and absorptive capacity in managing ERP implementation projects. *International Conference on Information Systems*. http://aisel.aisnet.org/cgi/viewcontent.cgi?article=1069&context=icis2009 (accessed March 2, 2010).

March, J. G. and J. P. Olsen. 1975. The uncertainly of the past: organizational learning under ambiguity. *European Journal of Political Research* 3: 147–171.

Martin, F. T. 1999. *Top Secret Intranet: How U.S. Intelligence Built Intellinet—The World's Largest, Most Secure Network*. Upper Saddle River, NJ: Prentice Hall.

Miller, R. 2009. Who has the most Web servers? Data Center Knowledge, May 14. http://www.datacenterknowledge.com/archives/2009/05/14/whos-got-the-most-web-servers/ (accessed Jan. 26, 2010).

Miner, A. S., and S. J. Mezias. 1996. Ugly duckling no more: Pasts and futures of organizational learning research, *Organization Science* 7: 88–99.

Miner, A. S. 1990. Structural evolution through idiosynercratic jobs: the potential for unplanned learning. *Organization Science* 1: 195–210.

Mooradian, N. 2005. Tacit knowledge: Philosophic roots and role in KM. *Journal of Knowledge Management* 9(6): 104–113.

Murphy, J., L. Higgs, and C. Quirk. 2002. The portal framework: The new battle for the enterprise desktop. AMR Research Report. Stamford, CT: Gartner, Inc.

Nambisan, S., and Y. Wang. 1999. Technical opinion: Roadblocks to Web technology adoption? *Communications of the ACM* 42(1): 98–101.

Nevoa, D., and Y. Wand. 2005. Organizational memory information systems: A transactive memory approach. *Clinical Decision Support* 39: 549–562.

Nielsen, J. 1997. Intranet vs Internet Design (Alertbox). Useit.com: Jakob Nielsen on Usability and Web Design. Sept. 15, 1997. Web. Nov. 12, 2010. http:/www.useit.com/alertbox970ab.html.

Nielsen, J. 1999. Intranet portals: The corporate information infrastructure. *Jakob Nielsen's Alert Box*, April 1999. http://www.useit.com (accessed March 31, 2003).

Nielsen, J. 2000. Intranet design. In *Designing Web Usability*, 263–94. New York: New Riders.

Nielsen, J. 2007. Intranet usability shows huge advances (Jakob Nielsen's Alertbox). *useit.com: Jakob Nielsen on Usability and Web Design*, Oct. 9. http://www.useit.com/alertbox/intranet-usability.html (accessed March 3, 2010).

Nonaka, I., and H. Takeuchi. 1995. *The Knowledge Creating Company*. New York: Oxford University Press.

Otto, J., M. Najdawi, and W. Wagner. 2003. An experimental study of Web switching behavior. *Human Systems Management* 22: 87–93.

Rademacher, R. A. 1999. Applying Bloom's taxonomy of cognition to knowledge management systems. In *Proceedings of the 1999 ACM Conference on Computer Personnel Research*, 276–278. New York, NY: ACM.

Rice, R. and D. Shook. 1990. Voice messaging, coordination and communication. In Intellectual teamwork: Social and technological bases of cooperative work, ed. J. Galegher, R. Kraut, and C. Egido, 327–350. New Jersey: Erlbaum.

Riggins, F., and H. S. Rhee. 1998. Toward a unified view of electronic commerce. *Communications of the ACM* 41: 88–95.

Rosen, A. 1997 *Looking into Intranets & the Internet: Advice for Managers*. Chicago, IL: AMACOM.

Sainfort, F. 1987. Innovation and organization: Toward an integration. Diss. Thesis, Ecole Centrale des Arts et Manufactures, Paris, France:

Scheepers, R. 2003. Key roles in intranet implementation: the conquest and the aftermath. *Journal of Information Technology* 18(2): 103–119.

Scheepers, R., and J. Damsgaard. 1997. Using Internet technology within the organization: A structurational analysis of intranets. In *Proceedings of the International ACM SIGGROUP Conference of Supporting Group Work: The Integration Challenge*, 9–18. Phoenix, AZ.

Scheepers, R., and J. Rose. 2001. Organizational intranets: cultivating information technology for the people, by the people. In *Managing Internet and Intranet Technologies in Organizations: Challenges & Opportunities*, ed. S. Dasgupta, 1–17. Hershey, PA: Idea Group Publishing.

Scott, A. H. S., and J. M. Mula. 2009. Contextual factors associated with information systems in a virtual supply chain. In *4th International Conference on Cooperation and Promotion Resources in Science and Technology (COINFO'09)* (Beijing, China, Nov. 21–23). Washington, DC: IEEE.

Sears, A., and J. A. Jacko. 2000. Understanding the relationship between network quality of service and the usability of distributed multimedia documents. *Human–Computer Interaction* 15: 43–68.

Sears, A., J. A. Jacko, and M. Borella. 1997a. The effect of Internet delay on the perceived quality of information. In *Proceedings of the 7th International Conference on Human–Computer Interaction* (San Francisco, CA, Aug. 24–29), 335–338. New York, NY: ACM.

Sears, A., J. A. Jacko, and M. Borella. 1997b. Internet delay effects: How users perceive quality, organization, and ease of use of information. In *Proceedings of the ACM Conference on Human Factors in Computing Systems* (Atlanta, GA, March 22–27), 2: 353–354.

Sears, A., J. A. Jacko, and E. M. Dubach. 2000. International aspects of WWW usability and the role of high-end graphical enhancements. *International Journal of Human–Computer Interaction* 12(2): 243–263.

Senge, P. M. 1990. *The Fifth Discipline: The Art and Practice of the Learning Organization*. New York: Doubleday.

Shang, S., and P. Seddon. 2002. Assessing and managing the benefits of enterprise systems: The business manager's perspective. *Information Systems Journal* 20(12): 271–299.

Shehab, E., M. Sharp, L. Supramaniam, and T. Spedding. 2004. Enterprise resource planning: An integrative review. *Business Process Management Journal* 10(4): 359–386.

Sorensen, S. J., J. A. Jacko, and A. Sears. 1998a. A characterization of clinical pharmacists' use of the Internet as an online drug information tool. In *1998 Spring Practice and Research Forum of the American College of Clinical Pharmacy* (Palm Springs, CA, April 5–8), 846.

Sorensen, S. J., J. A. Jacko, and A. Sears. 1998b. Hospital pharmacists' use, perceptions, and opinions of the Internet. *Pharmacotherapy* 18(2): 438.

Spender, J. 2008. Organizational learning and knowledge management: Whence and whither? *Management Learning* 39(2): 159–176.

Stein, E. W., and V. Zwass. 1995. Actualizing organizational memory with information systems. *Information Systems Research* 6(2): 85–117.

Stenmark, D. 2004. Intranets and organisational culture. *Proceedings of IRIS-27*, doi:10.1.1.125.6935.

Stjean, M. 2010. Role of a Webmaster in creating and maintaining a Web site. Ezine, Feb. 12.

Suresh, S., 1998. Decision support using the Intranet. *Decision Support Systems* 23: 19–28.

Tatnall, A. 2005. Portals, portals, everywhere. *IRMA International*.

Telleen, S. L. 1997. *Intranet Organization*. New York: John Wiley.

Tkach, D. 1999. Knowledge portals. IBM. http://www-.ibm.com/software/data/km/advances/kportals.html (accessed March 31, 2003).

Wachter, R., and J. N. D. Gupta. 1997. The establishment and management of corporate intranets, *International Journal of Information Management* 17(6): 393–404.

Weick, K. E. 1991. The nontraditional quality of organizational learning. *Organization Science* 2(1): 41–73.

Weiss, L. M., M. M. Capozzi, and L. Prusak. 2004. Learning from the Internet giants. *MIT Sloan Management Review*, 79–84.

Young, K. 2000. Entre nous. *The Banker*, London, February, 84–85.

Zhang, R., and J. Chen. 1997. Intranet architecture to support organizational learning. In *Proceedings—Annual Meeting of the Decision Sciences Institute* 2: 729–731.

Zuboff, S. 1988. *The Age of the Smart Machine*. New York: Basic Books.

Section VI

Accessibility and Universal Access

17 A Design Code of Practice for Universal Access: Methods and Techniques

Constantine Stephanidis and Demosthenes Akoumianakis*

CONTENTS

17.1 INTRODUCTION AND BACKGROUND

In the emerging Information Society, computer-mediated human activities are no longer bound to a particular execution context. Increasingly, they involve a multitude of personal, business, and residential encounters in a variety of domains, realized via networked terminals and appliances. Such a growth in information processing capability has created new opportunities for diverse user groups to access distributed digital resources and to appropriate the benefits of a broad type and range of information services. Among the wide implications of the prevailing highly distributed and communication-intensive computing era is that it challenges

traditional models of system development and creates a compelling need for more effective practices to cope with changing requirements and patterns of use. One important consequence of the new information-processing paradigm is on *accessibility*. Traditionally, the notion of computer accessibility had a narrow connotation, implying primarily the lack of physical or cognitive ability required to interact with computers. Nevertheless, the advent of the Internet, the proliferation of network-attachable interaction devices and the growth in the type, range, and scope of computer-mediated human activities raises accessibility issues for all potential users of applications and services (Olsen 1999) and demands a more encompassing account of the concept and a broader connotation.

This chapter is concerned with access to information and services by anyone, anywhere, and anytime. The normative perspective is that access is a contextual issue that is determined by at least three general clusters of parameters, broadly attributed to the target user, the access terminal (or

* Demosthenes Akoumianakis is currently affiliated with the Department of Applied Information Technologies & Multimedia, Faculty of Applied Technologies at the Technological Education Institution of Crete, Greece. The work reported in this chapter was conducted while he was affiliated with the Institute of Computer Science of FORTH.

interaction platform), and the context of use under which the task is executed. All three parameter clusters, with each one being a multifaceted construct, comprise a task's execution context. Thus, given a task, it stands to argue that the relationships linking the task to the three constructs increasingly become more complex. In contrast with previous information processing paradigms, where the vast majority of computer-mediated tasks were business oriented and executed by office workers using the personal computer in its various forms (i.e., initially alphanumeric terminals and later on graphical user interfaces), the Information Society signifies a growth not only in the range and scope of the tasks but also in the way in which they are carried out and experienced. In other words, humans are compelled to cope with a situation where an increasing range of activities become computer mediated and is no longer restricted to a particular device (e.g., the personal computer) or context of use (e.g., the office). Thus the notion of "one execution context used to host hosted many different tasks" denoted by a "one-to-many" or (1:N) relationship is progressively replaced by the conception of "many tasks being executed across many different execution contexts" denoted by a "many-to-many" or (M:N) relationship. In relational data modeling, it is common practice to attempt to normalize such M:N schemes to facilitate more structured and systematic analysis. It is also common that such normalization is achieved by introducing new entities and reducing the M:N relationships to 1:N. Using the above as a guiding principle, one can arrive at a revised scheme which does not include any M:N relationships, but one additional entity, which for the purposes of the present work, is referred to as *TaskExecutionContext*, as shown in Figure 17.1.

It stands, therefore, to argue that in the context of an Information Society, tasks may have several execution contexts distinguished either by the target users group, the designated context of use, or the access platform and/or terminal. In such a context, universal access implies an understanding of, and a conscious effort to, design for the *global* execution context of tasks. Alternatively, using a metaphor from the relational database vocabulary, universal access can be conceptualized as the population of the *extension* of the entity *TaskExecutionContext*. For each instance of *TaskExecutionContext* a particular style of interaction is designated, which need not be (and probably will not) be the same for all task execution contexts. Thus designing for universal access is partly related to identifying plausible (competing or alternative) styles, defining all together the global execution context of a specific task.

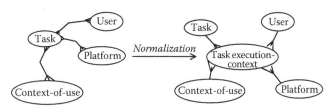

FIGURE 17.1 A contextual definition of TaskExecution-Context.

In light of the above, the methodological and engineering challenge posed by universal access amounts to defining the process and the techniques needed to design and develop interactive (software) artifacts that can exhibit alternative interactive manifestation (e.g., realization of different interaction styles), depending on situational parameters characterizing and determined by the task's execution context. This chapter attempts to provide an insight and contribution toward an improved understanding of the processes and methods suitable for universal access, by first reviewing related work (Section 17.2) and then proceeding to present, in Section 17.3, the results of a Thematic Network funded by the Information Society Technologies (IST) Programme of the European Commission and aiming to promote universal access in Health Telematics. Specifically, the methodology followed and the scenario-based techniques used to develop proposals for universal access to electronic medical records are described. Section 17.4 consolidates recent experience in a collection of universal access design benchmarks. Finally, Section 17.5 summarizes the contributions of this research and presents a number of open questions and directions for future work.

17.2 UNIVERSAL ACCESS: RELATED WORK AND CHALLENGES

Two prevalent approaches, namely *reactive* and *proactive*, have emerged in the recent past to address the issue of accessibility. They are distinctively characterized by the timing of appreciating and responding to the problem of accessibility (Stephanidis and Emiliani 1999), which, in turn, gives raise to alternative engineering practices and corresponding solutions. In what follows, a brief review will be presented of the baselines of each approach, emphasizing their relative merits and drawbacks in the context of accessibility.

17.2.1 THE REACTIVE APPROACH

The first approach assumes a *reactive* perspective according to which accessibility is an "a posteriori" concern, remedied by developing add-ons or adapting the original implementation of a product or service to accommodate new requirements as they arise. The reactive approach, which is the oldest and most explicitly related to disability access (Stephanidis and Emiliani 1999), has given rise to several methods for addressing accessibility, including techniques for the configuration of input/output at the level of user interface, the provision of alternative access systems, such as screen readers for blind users, scanning techniques for motor impaired, as well as an extensive body of knowledge in the form of human factors and ergonomic guidelines (ISO TS 16071; Nicolle and Abascal 2001; W3C-WAI guidelines).

Experience with these techniques reveals several shortcomings that amount to fatal consequences for generic accounts of accessibility. Some of them emerge from the fact that reactive methods effectively facilitate the reproduction instead of the redesign of the dialogues, which, in turn, requires extensive configuration of physical interaction parameters to be usable. Furthermore, there are inherent problems, which frequently cannot be overcome at the implementation level (e.g., cannot reproduce graphical images in a nonvisual form). Most importantly, however, reactive methods—being programming intensive and lacking suitable tools to expedite development—exhibit no upward compatibility, e.g., to new interaction technologies and terminals. These shortcomings have brought about a revision of the reactive approach toward more generic accounts of accessibility, which is the premise of the proactive approach.

17.2.2 THE PROACTIVE APPROACH

The second and more recent approach is *proactive*, treating accessibility from the early phases of concept creation and design and throughout the development life cycle (Stephanidis and Emiliani 1999; Stephanidis et al. 1998). According to the proactive approach, designers need to invest effort in anticipating new/changing requirements and accommodating them explicitly in the design of the product or service, in such a way as to allow for continuous design updates from the start. Additionally, developers require tools offering extended facilities for the management of interaction elements and object classes (Savidis and Stephanidis 2001a).

Several efforts in the past two decades aimed to promote proactive accounts of accessibility particularly in the context of human computer interaction (HCI). For example, in the mid-1990s the concept of *user interfaces for all* (Stephanidis 1995) was the first systematic effort to provide a methodological and an engineering base for the development of universally accessible user interfaces. *Unified user interface development* (Savidis and Stephanidis 2008) is the methodology proposed to facilitate this effort, while a collection of dedicated user interface software development tools (Savidis, Stephanidis, and Akoumianakis 1997, Savidis and Stephanidis 2001b) and design environments (Akoumianakis and Stephanidis 1997; Antona, Savidis, and Stephanidis 2006) comprises the engineering instruments for realizing user interfaces for all. With the progressive move toward an Information Society, the notion of *universal access* (Stephanidis et al. 1998, 1999), *information society for all* (Stephanidis et al. 1998) and *universal usability* (Shneiderman 2000) have became prominent research topics and acknowledged thematic areas of research and development activities within academic communities.*

Finally, under the cluster of proactive approaches, one should also acknowledge some early research and devel-

opment activities, either at national or international level, such as the FRIEND21 project (Ueda 2001) funded by the Japanese MITI, the AVANTI project (Stephanidis et al. 2001), which was funded by the European Commission, but also industrial initiatives, such as Active Accessibility by Microsoft, Java Accessibility by Sun Microsystems (Korn and Walker 2001), etc.

17.3 TOWARD A CODE OF PRACTICE FOR UNIVERSAL ACCESS DESIGN

More recent developments have emphasized the need to consolidate progress by means of establishing a common vocabulary and a code of design practice, which addresses the specific challenges posed by universal access. To this end, IS4ALL was a 3-year European Commission-funded project defined to serve this purpose. This work in IS4ALL was motivated by several reasons: First, universal access is a relatively new concept, not extensively studied and frequently confused with more traditional means toward accessible design. Second, it becomes increasingly obvious that prevailing conceptions (e.g., human-centered design), although useful, do not suffice to explicitly address universal access goals in the context of the Information Society. Third, universal access increasingly becomes a global quality attribute and a prominent factor of product/service differentiation in the public and private sectors. Therefore, a genuine and compelling need arises to consolidate existing experiences into a body of knowledge that can guide designers concerned with universal access through the various steps involved and provide concrete examples of good practice.

17.3.1 ELEMENTS AND SCOPE OF THE UNIVERSAL ACCESS CODE OF PRACTICE

In the context of IS4ALL, a *universal access code of practice* implies methods and techniques to facilitate two main targets, namely, an understanding of the global execution context of a task and the management of the design artifacts (or styles) suitable for different execution contexts. A useful starting point when considering universal access is the distinction between content (of tasks) and presentation. Tasks refer to computational operations (or functions) in a machine-oriented language. Examples include file management, data storage operations, database functions, electronic communication, etc. In computer-mediated human activities, tasks are represented by symbols in a user-oriented language, so that human can perceive and interact with the machine. However, the symbols used may vary depending on the choice of interaction platform. In conventional graphical environments these symbols take the form of primitive or composite interaction object classes, which are assembled to provide the artifacts through which humans perform a designated set of tasks. Typically, for any particular task there may be alternative interactive embodiments within the same interaction platform (e.g., a

---

* The term "community" is used in the present context to reflect the fact that research programmes on universal access, information society for all and universal usability are scaling-up to obtain international recognition, having their own and separate research agendas, technical and scientific forums (i.e., International Scientific Forum toward "An Information Society for all"—ISF-IS4ALL, http://ui4all.ics.forth.gr/isf_is4all/; ERCIM Working Group on "User Interfaces for all"—UI4ALL, http://ui4all.ics.forth.gr/), publication channels (e.g., http://www.hci-international.org/), and archival journals (i.e., http://www.springer.com/computer/user+interfaces/journal/10209).

selection in a Windows machine may be represented either through a menu, a checkbox, a listbox, etc.) but also across interaction platforms (e.g., a link in the case of Web documents). Moreover, different users may be familiar with, have a preference for, or be able to use certain types of artifacts.

Consequently, the way in which a task is (to be) performed is determined by at least three key categories of parameters, namely the current user, the interaction platform on which the task is to be executed, and the context of use (e.g., location, surroundings). In the light of the above, universal access implies the following:

1. Understanding how designated tasks are carried out by different users, across different interaction platforms and diverse contexts of use, and
2. Devising suitable artifacts for each relevant task execution context.

Attaining the above technical objectives entails design support at both the *macro-level* (e.g., a process-oriented protocol to explain to practitioners the steps and phases involved) and the *micro-level* (e.g., definition and examples of techniques to be used to attain specific targets, such as universal access requirements gathering, design, development, etc). The compound collection of validated macromethods and micromethods, along with the validation case studies, compile the IS4ALL Universal Access code of practice in Health Telematics. The primary target audience of such a code of practice is designers and developers of Health Telematics products and services, who are concerned with the incorporation of universal access principles into the development cycle of products and services. To this effect, the code of practice aims to offer targeted and validated support as to how to structure and manage the development process, what techniques can be used, as well as how to attain specific design and development targets.

17.3.1.1 Macrolevel Approach

The approach to compile a code of practice to facilitate universal access (to electronic medical records) in Health Telematics builds on an analytical perspective, which makes use of scenario-based design as the primary macro-level methodology (Akoumianakis and Stephanidis 2003). This brings scenarios into the forefront not only as a tool for documenting and reflecting upon existing practices but also as a mechanism for envisioning, studying, and understanding new execution contexts for the tasks designated in a reference scenario. To this end, a complete scenario is defined, according to Nardi (1992, p. 13) as

"... a set of users, a context and a set of tasks that users perform or want to perform ... it blends a carefully researched description of real on-going activities with an imaginative futuristic look at how technology could support those activities better."

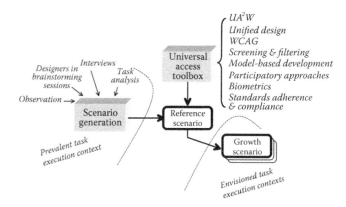

FIGURE 17.2 Overall approach for working with scenarios.

Figure 17.2 depicts a process for generating and articulating scenarios to understand and gain insight into universal access. As shown in Figure 17.2, there are a variety of techniques for generating scenarios. For example, one way is to gather designers in one or more brainstorming sessions, as reported by Tognazzini (1995). Alternatively, scenarios may be generated using material from observations (Beyer and Holzblatt 1998) or traditional techniques such as user profiling, task analysis, protocol analysis, and interviews (Suri and Marcsh 2000) or techniques such as SPES (Situated and Participative Enactment of Scenarios) sessions (Iacucci and Kuutti 2002).

One important parameter for selecting an appropriate method to generate the reference scenario is the availability of an existing system or prototype. If a system is already available, then scenario formulation entails a reflection on designated tasks as carried out by actual users. In this case, the scenario should effectively document existing behaviors of end users. If the system is new, thus not available, then the scenario should act as a source for generating new user behaviors and unfold envisioned tasks to be supported. In either case, the description compiled is subsequently peer reviewed, refined, and agreed upon by professionals or end-user communities. This peer review acts both as a consensus creation mechanism as well as a type of validity check to ensure that the scenarios are realistic, valid, and accepted by all parties. In the course of this iterative phase, any system mock-ups, prototypes, or other artifacts, which reveal aspects of the scenario's real execution context, are taken into account. This is in line with the prevailing conceptions of designers that scenarios and prototypes are frequently—and indeed should be—used together (Weidenhaupt et al. 1998). The rest of this section discusses the universal access toolbox (see Figure 17.1), i.e., an elaboration of methods used to facilitate universal access insight, as well as its validation in the context of accessing electronic medical records.

17.3.1.2 Micromethods

Various micromethods have been recruited to address specific design challenges posed by universal access. The focus on design, in this context, does not aim to convey the view that universal access is only a matter of design. On the contrary, the challenges posed to other phases in systems engineering (i.e., development phase, evaluation, etc.), are substantial and fully appreciated, but they are beyond the scope of the present work. The methods being addressed here come from different scientific disciplines such as Software Engineering, Human Computer Interaction, Industrial Engineering, Software Ergonomics, etc. Some of the methods have been in use for several years now but not specifically in service of universal access, while others are refined and extended versions of existing methods, fine-tuned so as to provide specific and targeted input toward universal access. Finally, some new methods have been developed to facilitate specific targets in the course of design activities. All methods are complete in the sense of Olson and Moran (1996, p. 269), who state that a complete design micro-method is "a systematic, repeatable way to design . . . " and as something that "includes (1) a statement of the problem that the method addresses, (2) a device (a tool, technique, or model), (3) a procedure for using the device, and (4) a result, or rather a statement of the nature of the result . . . "

It is practically impossible to review in any reasonable depth the micromethods studied and developed in the context of IS4ALL. A summary of some representative methods is provided in Table 17.1. The methods are represented by the rows in the table. The four columns depict an elaboration of each method in terms of the four basic criteria for design micromethods proposed by Olson and Moran (1996). For further details on the development of each method and its application and validation in the context of Health Telematics, the reader may follow the references provided in the table or consult the book (Stephanidis 2005).

An important characteristic of the IS4ALL methods is their complementary nature within a scenario-based inquiry. This is typically translated into portfolios of methods spanning all phases of development from requirements engineering to prototype/system evaluation. For example, at early stages (e.g., concept formation or requirements gathering), the UA²Ws method (Akoumianakis and Stephanidis 2005a) may be used to identify key universal access requirements, such as for example adaptability to different users, portability to different platforms, while, at a latter stage, specific design- and evaluation-oriented methods may be chosen to develop early prototypes of interaction styles and to assess the end result against the specified requirements. Another important feature of the methods is that they are generic; thus they can be reused across application areas. This makes the task of reapplying the methods across application domains easier, provided that there are examples (e.g., method validation) describing the details of how this is to be done.

17.3.2 METHOD VALIDATION

To validate the methods and to consolidate the experiences into a code of practice, IS4ALL compiled reference scenarios in which different designated methods were applied. All reference scenarios are related to accessing electronic patient records, but they reveal alternative usage contexts. Thus, there are scenarios addressing access to medical records by doctors in duty moving around in the ward of a hospital, paramedics attending an emergency situation with an ambulance, patients accessing parts of their medical record from a residential environment, etc. In all cases, the objective has been to make use of the methods described earlier to analyze, revise, and extend the initial formulations of the scenarios, so as to address designated universal access challenges. Thus, there have been studies addressing:

- The *accessibility* of selected parts of a system by using the W3C Content Accessibility Guidelines (Emiliani and Burzagli 2005).
- The *adaptability* of a system to user roles (e.g., view of medical record by pathologists, surgeons, paramedics), by applying model-based approaches to user interface development (Stary 2005).
- The *usability* of a system by making use of heuristics and feature inspections (Karampelas, Akoumianakis, and Stephanidis 2005).
- The *platform independence and/or portability* of a system to alternative platforms and devices, such as Web pages, iPAQ, WAP pages using style reengineering (Akoumianakis and Stephanidis 2005b).
- *Patients' access* to medical data using participatory analysis methods (Pieper and Stroetmann 2005).
- The use of biometrics for universal access (Fairhurst et al. 2003).

The code of universal access design practice is currently documented in a comprehensive handbook (Stephanidis 2005).

17.4 UNIVERSAL ACCESS BENCHMARKS

The accumulated experience indicates that designing for universal access requires a conscious effort and commitment to analytical and exploratory design to reveal and construct new virtualities. Universal access designers need to analyze tradeoffs between multiple conflicting design criteria in order to satisfy an increasing range of functional and nonfunctional requirements. In this context, there is a compelling need to study and define both process-oriented benchmarks but also features in the resulting product/service, which can be used to assess the dimension of "universality of access" and to inform designers and developers of the degree to which a particular design exhibits the required and/or recommended features.

17.4.1 PROCESS BENCHMARKS

Process benchmarks in the present context entail an almost exclusive focus on trying to rationalize the universal access design processes in terms of certain nonfunctional requirements. Specifically, it is claimed that in addition to conventional benchmarks of user-centered design (e.g., clear

TABLE 17.1
Micromethods

	Statement of the Problem	Device (a Tool, Technique, or Model)	Procedure for Using the Device	Statement of the Nature of the Result
Requirements engineering	**Universal Access Assessment Workshop** (Akoumianakis and Stephanidis 2005a) UA²W sessions are held to: • Facilitate an insight into the current tasks in a reference scenario • Identify new plausible task execution contexts • Formulate growth scenarios	• Scenario screening • Growth scenarios	In cases, where the resources are available, at least one UA²W should be devoted to scenario screening, followed by a separate UA²W targeted toward the compilation of suitable growth scenarios.	• Universal Access Assessment Form (UA2F) • Universal Access Quality Matrix (UAQM)
Design and development	**Unified user interface design** (Savidis, Antona, and Stephanidis 2005) Fuse all potentially distinct design alternatives into a single unified for without requiring multiple design phases.	Polymorphic task hierarchy (REF)	Polymorphic task decomposition (REF)	• Polymorphic task hierarchy • Styles of interaction • Style design rationale
	W3C Content Accessibility Guidelines (Emiliani and Burzagli 2005) Ensure accessibility of WWW content for people with disabilities	• Guidelines • Checkpoints • Priority levels	Three main steps: 1. Decide level of accessibility suitable for designated application 2. Select relevant guidelines 3. Verify check points	• Accessibility problems • Possible corrective actions • Conformance level A, AA, AAA
	Style reengineering (Akoumianakis and Stephanidis 2005b) Generate a new style of interaction (suitable for another user group, access terminal or context of use), from a given prototype	• Role-based interaction object modeling • Abstraction-based techniques	1. Identify semantic roles of interaction elements 2. Generate abstract object model 3. Instantiate alternative style from abstract object model	• Abstract object model • New styles generated from the abstract object model
	Model-based user interface development (Stary 2005) Develop design representations which allow the development of role-adapted user interfaces	Models and design representations of • Users • Tasks • Interaction • Domain	Two main steps: 1. Populate models into an executable specification using computer-aided design tools 2. Execute the specification to generate the user interfaces	• Design models and representations • User interface specification • Prototypes of role-adapted user interfaces
Evaluation	**Feature inspection** (Karampelas, Akoumianakis, and Stephanidis 2005) Use of prototypes to inspect compliance to certain criteria or features	Choice of designated: • Features • Standards criteria • Web accessibility guidelines • Heuristics	1. Planning and finding experts 2. Deciding on the specific instrument to be used 3. Conducting the evaluation	Usability/accessibility report
	Screening artifacts (Karampelas, Akoumianakis, and Stephanidis 2005) An inspection-based technique which seeks to identify potential barriers to use (according to filters) in a scenario and develop proposals to eliminate them	• Accessibility filters • Heuristics • Design criteria or principles	Four main phases: 1. Agreement on screening instruments (e.g., filters, heuristics) 2. Screening 3. Identification of break downs 4. New concept formation	The filtering grid, which consolidates • Set of universal access filters • Universal access problem descriptions • Rationale for new artifacts

TABLE 17.2

Intertwining of NFRs for Universal Access

Scenario Extract	Change (Examples)		Dimension of Adaptability	NFR
	State 1	State n		
...x number of messages per second...	10 messages/s	1000 messages/s	System throughput	Scalability
...manifested on the user's interactive terminal...	Desktop PC	WAP phone	Interaction platform	Platform independence
...to initiate an operation effectively and efficiently...	Ordinary user	Motor-impaired user	System's user communities	User tolerance
...from current location...	Office environment	Residential environment	System's context of use	Ubiquity

statement of requirements, multidisciplinary perspective, iterative process, tight design-evaluation feedback loop), universal access designers should make sure that more complex nonfunctional requirements (NFRs) or quality attributes are sufficiently addressed in a timely and systematic fashion in the course of the various design stages. Moreover, emphasis should be on accounting for such NFRs earlier rather than later in the design process. Unfortunately, there is not a formal definition or a complete list of NFRs or quality attributes. Nor is there a consolidated account of how they are intertwined. Thus, any effort to designate NFRs relevant to universal access is bound to be immature or incomplete.

In the past, there have been efforts from the software engineering community in the direction of identifying NFRs or quality attributes (Boehm 1978), introducing NFR taxonomies (Barbacci et al. 1995), developing representational notations for using NFRs (Mylopoulos, Chung, and Nixon 1992), proposing processes for working with them (Barbacci et al. 2002), and studying their relationship to software architecture (Bass, Clements, and Kazman 1998). However, the field is far from compiling a detailed, consolidated and codified body of knowledge on NFRs. Of course, this does not diminish the value of previous work but simply points out that NFRs and their incorporation into software systems remains a research issue.

Notwithstanding the difficulties, some NFRs can be identified that raise serious implications upon designing for universal access, as conceived in the present work. These are adaptability, scalability, user tolerance and individualization, and platform independence. It should be noted that there are other NFRs relevant to a software system, such as performance, reusability, security, etc., but these are beyond the scope of the present work.

17.4.1.1 Adaptability

According to ISO 9126 (1991), adaptability is defined as "...attributes of software that bear on the opportunity for its adaptation to different specified environments without applying other actions or means than those provided for this purpose for the software considered" (ISO 9126: 1991, A.2.6.1). An alternative formulation is that adaptability refers

to the capability of an interactive system to tolerate changes in its target execution context without external intervention. A system that supports adaptability is capable of undertaking context-sensitive processing to *recognize* the need for change, *determine* the plausible alternatives that satisfy the change and *effect* the optimal option in order to generate new behavior (Akoumianakis, Savidis, and Stephanidis 2000).

The criticality of adaptability is twofold. First, a commitment to support adaptability frequently imposes decisions which do not only affect the entire life cycle of an interactive system but are also hard to revise at later stages. For example, adaptability has implications on the architectural abstraction of the system, while commitment to an architecture that does not support adaptability may render universal access design impossible or very resource demanding and consuming. Second, adaptability assumes or correlates with other (first-order) NFRs, such as scalability, user tolerance, platform interoperability, and individualization, all of which are very important for universal access. To assess how the above are intertwined with adaptability, one may examine the type of changes in the system's execution context that need to be identified and accounted for through adaptation. To this effect, let us consider the general scenario summarized as follows:

"Periodic messages arrive at a system in a rate of x num-ber of messages per second. The system must respond to a message within a specified time interval. The system's response should be manifested on the user's interactive terminal to allow him/her to initiate an operation effectively and effi-ciently from the current location..."

It should be noticed that the above general and abstract scenario* makes no explicit or implicit claim on functional properties of the system, yet it pinpoints some of the (potential) changes—see text in bold, italics and underlined—that may take place and that need to be accounted through adaptation. Thus, as shown in Table 17.2, adaptability has several dimensions, which, in turn, point to additional NFRs. In what follows, a brief account of each NFR is provided, emphasizing

---

* Abstraction makes the scenario relevant to a broad range of interactive software systems, including those considered in the context of IS4ALL.

their intertwining with adaptability and pointing out in each case the scope of the desirable adaptable behavior a software system should exhibit in order to support the NFR.

17.4.1.2 Scalability

A scalable system is one that can undertake the required context-sensitive processing to cope with changes of scale in the system's target execution context. Changes of scale in the execution context of a system, necessitating adaptation, may result from increase in the number of inputs to the software, e.g., from 10 inputs per second to 1000 inputs per second (scale up), or variation in the modality of the input, e.g., from multimedia input steams to textual strings (scale down). Scalability is also relevant to the system's outputs, and in particular the way in which they are manifested to the user. Thus, for instance, a system may have to scale upward to present a piece of multimedia information (e.g., Web content) on a powerful desktop terminal or scale downward to present the same (Web) content in a unimodal terminal (e.g., a kiosk).* Scaling up or down as needed will allow the system to accommodate such changes in the course of the required adaptable behavior. It is therefore claimed that in the context of universal access, an adaptable system needs to be scalable to cope (certain types of) changes in its execution context.

17.4.1.3 User Tolerance and Individualization

Another source of change in the execution context of an interactive system may result from the target (or current users). Specifically, users differ with regards to levels of expertise, interests, requirements, preferences, roles assumed, etc. ISO 9241–Part 10 identifies individualization as an important principle (or software ergonomic criterion) for dialogue design, pointing out explicitly that dialogue systems should be designed to support suitability for individualization, by exhibiting the capability to adapt to the user's needs and skills for a given task. Accordingly, an adaptable system could foresee such variation and modify its behavior accordingly. For example, lack of the required (physical or cognitive) resources on behalf of the current user to interact with the software may force interaction to take place through an alternative channel or communication medium (e.g., nonvisual for blind users or using alternative interaction techniques such as switch-based interaction known to be appropriate for motor-impaired users). This is the case with novice users, elderly, and people with disabilities but also users in circumstances that cause situational loss of certain abilities. Thus, an adaptable system should be capable of identifying and responding to situations where the user is motor-impaired and does not possess certain abilities (e.g., ability for fine spatial control, finger isolation) sufficiently in order to operate an input device such as the mouse, or the user is vision impaired or his/her visual channel is preoccupied (e.g., driver), thus not being able to perceive easily information presented in a visual

modality. It can therefore be concluded that in the context of universal access an adaptable system should also exhibit user tolerance and individualization.

17.4.1.4 Platform Independence

Platform independence as a NFR of a software system in the context of universal access refers to the degree of tightness of the system, or units of the system, but mainly its interactive elements, to a particular class of target environments (or platforms). Platform independence can be realized either as a sort of portability, which is typically programmed, or as a result of compiling a more abstract specification. In the software engineering literature, portability constitutes a NFR indicating the easiness of porting a software system across a class of target environments. Though important, this type of system adaptability (regarding the target platforms) does not suffice for the purposes of universal access, because current implementations maintain a tight link with the runtime environments of the target platforms. In previous work (Savidis et al. 1997, 2001a, 2001b), it has been demonstrated that the later approach (e.g., compiling specifications to realize this sort of adaptability), though demanding, seems more appropriate and relevant for universal access. This is due to the facilitation of the automatic selection of designated platform-specific artifacts (from a repository of implemented artifacts and dialogue patterns registered to various target platforms), while avoiding direct "calls" to the platform. In this manner, the specification can be updated (with regards to target platforms) and recompiled to utilize new interaction facilities. However, despite its usefulness, the specifications approach has been shown to be demanding in terms of underlying tool support and corresponding engineering practices. Nevertheless, it remains the only comprehensive proposal that adequately addresses the development challenges posed by universal access. Thus it is claimed that, in the context of universal access, an adaptable system should also exhibit platform independence.

17.4.2 Product Benchmarks

Having pointed out some of the process-oriented criteria to be addressed by universal access engineering, this section considers some of the product-specific aspects to be exhibited by universally accessible artifacts, drawing upon recent experiences to point out a nonexhaustive list of attributes (or qualities) characterizing such artifacts, and emphasizing primarily end user-perceived qualities. The NFRs considered above are a useful basis, but one should bear in mind that not all of them are easily propagated into measurable end-user perceived qualities (e.g., presence or absence of some of them is not necessarily or easily perceivable by end users). For example, in contrast with the NFR of performance, which can be measured using state of the art user-based evaluation techniques, it is not obvious how a user can distinguish that a system is truly platform independent, as opposed to simply portable across a class of target platforms. Similar remarks hold for other NFRs (e.g., reusability), thus making their

---

* It is important to distinguish this type of scalability, which amounts to adaptability with regards to the way Web content is presented, from platform independence (discussed later on), which is more demanding.

subjective assessment by end users difficult, and, in several cases, inappropriate.

For some other NFRs, it is possible to assess their presence or absence in the target system by means of a subjective account of the system's capacity to cope with changes in its execution context. As already pointed out earlier, for the purposes of universal access such changes may result from variation in the users of the system, the context of use, and/or the platform or access terminal used to carry out a designated set of tasks. Thus, one can formulate a template for assessing subjectively the system's capability to respond to changes in each one of these three dimensions. An example of such a template is depicted in Table 17.3, providing a nonexhaustive list of statements, which could be part of an instrument aiming to elicit the user's subjective opinion on designated system qualities, given a specific test task or use case.

The result of such an assessment could be not only an indication of the presence or absence of certain architectural components in the target system (e.g., user model) but also an indication of the scope of adaptations that the system can recognize, reason about, and eventually put into effect.

17.4.3 UNIVERSAL ACCESS VERSUS PREVAILING PRACTICE

From the discussion thus far, it becomes evident that universal access requires a commitment toward design processes which explicitly facilitate the integration of (some critical) NFR into the software system. This is not a novel objective in itself. As already pointed out, the software engineering community but also HCI have been concerned for several years with the issue of accounting for NFRs. As a result, an evolving body of knowledge has become available, which enhances current understanding of NFRs, either as goals to be achieved in the course of development or as measurable qualities of a system. The present work adds to this knowledge an alternative perspective upon the treatment of NFRs, which is inspired by the technical challenges posed in the effort to advance universal access to interactive software. Nevertheless, such

TABLE 17.3

Subjective User Assessment of a System's Capability to Adapt to Execution Context Changes[a]

C1 = Adaptability, C2 = Scalability, C3 = User Tolerance and Individualization, C4 = Platform Independence		C1	C2	C3	C4
The system's model of the user	The system implements mechanisms to elicit information about the user.	√			
	The system keeps track of my personal preferences.			√	
	The system provides sufficient guidance and support for the user, as needed.			√	
	The system takes reasonable time to become aware of my personal preferences.	√		√	
	The system makes the right assumptions about myself.			√	
	The system's user model is sufficiently visible.	√			
	The system's user model can be inspected.	√			
	The system's user model is sufficiently transparent.	√			
	The system prohibits unauthorized access to the user model.	√			
	The system consults the user to confirm some of its assumptions.			√	
	I can inform the system about misuse of assumptions.			√	
	The system asks my approval before it updates the user model.	√			
	I can instruct the system to bypass the user model.			√	
Execution context	The system "observes" usage and accordingly modifies its behavior.	√		√	
	This system detects changes in the execution context and accordingly modifies its behavior.	√	√	√	√
	The system's design does not presuppose a particular context of use.	√	√	√	√
	I can use the system from different geographic locations.	√	√	√	√
	I can easily modify and individualize this system.			√	
	This system detects changes in the context of use and accordingly modifies its behavior.			√	
	The system allows me to undo adaptations.	√		√	
	The system allows me to disallow the future occurrence of adaptations.	√		√	
	The system provides informative explicit feedback on task progress.	√		√	
	The system provides informative implicit feedback on task progress.	√		√	
	The level of security is adequate.				√
Platform	I can access the system through different terminals.				√
	The system initiates appropriate interaction styles, which allow me to complete the task.		√		√
	This system is aware of the terminal I am using to access the content.		√		√
	The system seems aware of my personal preferences regarding the terminal in use.	√		√	√
	The system modifies its interactive behavior according to the terminal used by the user.	√	√	√	√

[a] It should be noted that Table 17.3 is presented only as an example rather than as a validated and standardized instrument.

a perspective upon NFRs raises broader implications on the engineering of universally accessible systems.

Table 17.4 provides a summary of such implications by contrasting premises of prevailing practices (mainly in HCI) against premises pertaining to universal access design. The table seeks to convey the shift (along various criteria) that is necessitated by a commitment to universal access. Some of these shifts are nontrivial but possible, given the current state of knowledge and the accumulated know-how, while others require additional research to become viable and feasible. Specifically, there are methodologies addressing NFRs both in terms of underlying processes needed to attain designated goals corresponding to NFRs, and in terms of measurable product attributes. Examples of the former are reported by Barbacci et al. (1995, 2002), Bass, Clements, and Kazman (1998), and Mylopoulos, Chung, and Nixon (1992), while the latter has been the driving objective in recent efforts to develop usability metrics and performance measures. Nevertheless, their relevance to universal access or their refinement to facilitate universal access targets is still pending. Also, there are issues raised in the table, which are still in a research phase and the only evidence is through experimental laboratory prototypes. This is the case for most of the transitions described under the development and implementation category.

With regards to HCI design, which has been the primary focus of the present work, it is important to acknowledge that universal access necessitates a blending between traditional and new methodologies, which can result in new design methods, tools, and engineering practice. The prominent target should be the commitment to an analytical perspective to unfold insights regarding the global execution context of (a set of) tasks. Then, the engineering challenge amounts to the management of design spaces (e.g., the broad range of styles and the accompanying rationale), which are bound to be appropriate or plausible for alternative execution contexts. Managing complex design spaces entails enumerating plausible alternative styles, justifying their role in the design space, relating them to the global execution context of the associated task, specifying conditions for style initiation and termination, style reengineering, etc. As discussed earlier, some of the above require new methods to address the corresponding challenge, while other challenges may be facilitated by refinement and extension of existing methods.

TABLE 17.4
Assessment of Universal Access versus Traditional Practice

Phase	Criterion	Traditional Practice	Universal Access Design
Requirements engineering	Scope	Capturing the data and the functions a system should support	In addition to the traditional scope, emphasis is on non-functional requirements (NFRs) such as adaptability, scalability, platform independence, usability, etc.
	Approach	A variety of techniques for data-, function- or object-oriented analysis, e.g., entity-relationship diagrams, data flow diagrams, object-oriented models	Goal-oriented driven by NFRs to complement traditional approaches
Design	Focus	Generating a single artifact that meets the designated requirements	Exploration of design alternatives to address the NFRs and analysis of trade-offs
	Unit for dialogue design	Physical interaction objects	Abstract interaction object classes
	Outcome	Single artifact depicting physical user tasks	Collection of alternative task structures (as opposed to a single interaction object hierarchy) represented by multiple dialogue styles Accompanying rationale for each task structure and style
	Representation	Interaction object hierarchy	Abstract object classes Rationalized design space
	Process	User-centered design	Enumerate–Retool–Rationalize cycles
Development and implementation	Implementation model	Programming as the basis for generation of the run-time environment of the system	Generation of run-time environment by compiling an abstract specification
	Premise of run-time code	Making direct calls to the platform	Linking to the platform
	Platform independence	Generalization across platforms belonging to a certain class (e.g., GUI window managers)	Platform abstraction mechanism which mediates between the specification and the target platforms
	Platform utilization	Multiplatform environments	Multiple toolkit environment

17.5 SUMMARY

This chapter provides an overview and summarizes recent efforts in the direction of consolidating current design practices on universal access to the effect of constructing a validated code of practice to guide designers in identifying universal access challenges, and subsequently planning and devising appropriate processes and methods to cope with them. The provided account of the topic has not been exhaustive but rather illustrative of the type of methods considered appropriate for universal access and the way in which specific challenges can be addressed.

The contributions of the present work are threefold. First, the scope of the universal access inquiry has been defined as an effort toward understanding and designing for the global execution context of tasks. Second, a collection of design-oriented methods has been described, covering the major phases of HCI design inquiries, such as requirements gathering, design, and evaluation, emphasizing their contribution to the aforementioned challenge. Finally, process- and product-oriented benchmarks have been described (e.g., nonfunctional requirements or quality attributes) in an effort to provide guidance as to the extent to which a design inquiry (in terms of process and outcomes) complies with the universal access code of practice. Devising processes, methods, and tools to meet the suggested benchmarks is neither an easy nor an exhausted task. The above constitute the basic elements of the IS4ALL code for universal access design practice. Validation of the code of practice has taken the form of practical application of the methods to allow universal access insights in the context of specific reference scenarios from the domain of Health Telematics and, in particular, access to electronic patient records.

ACKNOWLEDGMENTS

The work reported in this paper has been carried out in the framework of the European Commission funded Thematic Network (Working Group) "Information Society for All"—IS4ALL (IST-1999-14101) http://www.ui4all.gr/isf_is4all/. The IS4ALL Consortium comprises one coordinating partner, the Institute of Computer Science, Foundation for Research and Technology—Hellas (ICS-FORTH), and the following member organizations: Microsoft Healthcare Users Group Europe (MS-HUGe), European Health Telematics Association (EHTEL), Consiglio Nazionale delle Ricerche—Institute for Applied Physics "Nello Carrara" (CNR-IFAC), Fraunhofer-Gesellschaft zur Foerderung der angewandten Forschung e.V.—Institut für Angewandte Informationstechnik (FhG-FIT), Institut National de Recherche en Informatique et Automatique—Laboratoire lorrain de recherche en informatique et ses applications (INRIA) and Fraunhofer-Gesellschaft zur Foerderung der angewandten Forschung e.V.—Institut fur Arbeitswirtschaft und Organization (FhG-IAO). Several other cooperating organizations participate as subcontractors.

REFERENCES

Akoumianakis, D., and C. Stephanidis. 1997. Supporting user adapted interface design: the USE-IT system. *Interacting with Computers* 9(1): 73–104.

Akoumianakis, D., and C. Stephanidis. 2003. Scenario-based argumentation for universal access. In *Universal Access: Theoretical Perspectives, Practice, and Experience*, vol. 2615, *Lecture Notes in Computer Science* eds. N. Carbonell and C. Stephanidis, 474–485. Berlin: Springer.

Akoumianakis, D., and C. Stephanidis. 2005a. The Universal Access Assessment Workshop (UA2W) method. In *Universal Access in Health Telematics—A Design Code of Practice*, vol. 3041, *Lecture Notes on Computer Science Series*, ed. C. Stephanidis, 99–114. Berlin: Springer.

Akoumianakis, D., and C. Stephanidis. 2005b. Screening models and growth scenarios. In *Universal Access in Health Telematics—A Design Code of Practice*, vol. 3041, *Lecture Notes on Computer Science Series*, ed. C. Stephanidis, 156–174. Berlin: Springer.

Akoumianakis, D., A. Savidis, and C. Stephanidis. 2000. Encapsulating intelligent interaction behavior in unified user interface artifacts. *Interacting with Computers* 12: 383–408.

Antona, M., A. Savidis, and C. Stephanidis. 2006. A process–oriented interactive design environment for automatic user interface adaptation. *International Journal of Human Computer Interaction* 20(2): 79–116.

Barbacci, M., M. Klein, T. Longstaff, and C. Weinstock. 1995. Quality attributes. *Technical Report CMU/SEI-95-TR-021 ESC-TR-95-021*, available at http://www.sei.cmu.edu/pub/documents/95.reports/pdf/tr021.95.pdf.

Barbacci, M., R. Ellison, A. Lattanze, J. Stafford, C. Weinstock, and W. Wood. 2002. Quality Attribute Workshops, 2nd ed., *Technical Report CMU/SEI-2002-TR-019 ESC-TR-2002-019*, available at http://www.sei.cmu.edu/pub/documents/02.reports/pdf/02tr019.pdf.

Bass, L., P. Clements, and R. Kazman. 1998. *Software Architecture in Practice*. Reading, MA: Addison-Wesley.

Beyer, H., and K. Holzblatt. 1998. *Contextual Design: Defining Customer-Centered Systems*. San Francisco, CA: Morgan Kaufmann.

Boehm, B. 1978. *Characteristics of Software Quality*. New York: Elsevier.

Emiliani, P. L., and L. Burzagli. 2005. W3C-WAI content accessibility auditing. In *Universal Access in Health Telematics—A Design Code of Practice*, vol. 3041, *Lecture Notes on Computer Science Series*, ed. C. Stephanidis, 175–196. Berlin: Springer.

Fairhurst, M., R. Guest, F. Deravi, and J. George. 2003. Using biometrics as an enabling technology in balancing universality and selectivity for management of information access. In *Universal Access: Theoretical Perspectives, Practice, and Experience*, vol. 2615, *Lecture Notes in Computer Science Ser.*, eds. N. Carbonell and C. Stephanidis, 249–262. Berlin: Springer.

Iacucci, G., and K. Kuutti. 2002. Everyday life as a stage in creating and performing scenarios for wireless devices. *Personal and Ubiquitous Computing* 6: 299–306.

Karampelas, P., D. Akoumianakis, and C. Stephanidis. 2005. Usability inspection of the WardInHand prototype. In *Universal Access in Health Telematics—A Design Code of Practice*, vol. 3041, *Lecture Notes on Computer Science Series*, ed. C. Stephanidis, 197–208. Berlin: Springer.

Korn, P., and W. Walker. 2001. Accessibility in the Java™ platform. In *User Interfaces for All—Concepts, Methods, and Tools*, ed. C. Stephanidis, 319–338. Mahwah, NJ: Lawrence Erlbaum.

Mylopoulos, J., L. Chung, and B. Nixon. 1992. Representing and using non-functional requirements: A process-oriented approach. *Software Engineering* 18(6): 483–497.

Nardi, B. 1992. The use of scenarios in design. *SIGCHI Bulletin* 24(3): 13–14.

Nicolle, C., and J. Abascal, eds. 2001. *Inclusive Design Guidelines for HCI*. London: Taylor & Francis.

Olsen, D. 1999. Interacting in chaos. *Interactions*, September/October, 43–54.

Olson, J. S., and T. P. Moran. 1996. Mapping the method muddle: Guidance in using methods for user interface design. In *Human–Computer Interface Design: Success Stories, Emerging Methods, and Real-World Context*, eds. M. Rudisill et al., 101–121. San Francisco, CA: Morgan Kaufmann.

Pieper, M., and K. Stroetmann. 2005. Participatory insight to universal access: Methods and validation exercises. In *Universal Access in Health Telematics—A Design Code of Practice*, vol. 3041, *Lecture Notes on Computer Science Series*, ed. C. Stephanidis, 271–296. Berlin: Springer.

Savidis, A., M. Antona, and C. Stephanidis. 2005. Applying the unified user interface design method in health telematics. In *Universal Access in Health Telematics—A Design Code of Practice*, vol. 3041, *Lecture Notes on Computer Science Series*, ed. C. Stephanidis, 115–140. Berlin: Springer.

Savidis, A., C. Stephanidis, and D. Akoumianakis. 1997. Unifying toolkit programming layers: A multi-purpose toolkit integration module. In *Proceedings of the 4th Eurographics Workshop on Design, Specification and Verification of Interactive Systems (DSV-IS '97, Granada, Spain, 4–6 June)*, eds. M. D. Harrison and J. C. Torres, 177–192. Berlin: Springer-Verlag.

Savidis, A., and C. Stephanidis. 2001a. Development requirements for implementing unified user interfaces. In *User Interfaces for All—Concepts, Methods, and Tools*, ed. C. Stephanidis, 441–468. Mahwah, NJ: Lawrence Erlbaum.

Savidis, A., and C. Stephanidis. 2001b. The I-GET UIMS for unified user interface implementation. In *User Interfaces for All—Concepts, Methods, and Tools*, ed. C. Stephanidis, 489–523. Mahwah, NJ: Lawrence Erlbaum.

Savidis, A., and C. Stephanidis. 2008. Unified user interface development: New challenges and opportunities. In *The Human–Computer Interaction Handbook—Fundamentals, Evolving Technologies and Emerging Applications*, 2nd ed., eds. A. Sears and J. Jacko, 1083–1105. Boca Raton, FL: CRC Press.

Shneiderman, B. 2000. Universal usability: Pushing human computer interaction research to empower every citizen. *Communications of the ACM* 43: 84–91.

Stary, C. 2005. Role-adapted access to medical data: Experiences with model-based development. In *Universal Access in Health Telematics—A Design Code of Practice*, vol. 3041, *Lecture Notes on Computer Science Series*, ed. C. Stephanidis, 224–239. Berlin: Springer.

Stephanidis, C. 1995. Towards user interfaces for all: Some critical issues. In *Symbiosis of Human and Artifact—Future Computing and Design for Human–Computer Interaction*, eds. Y. Anzai, K. Ogawa, and H. Mori, 137–142. Amsterdam: Elsevier Science.

Stephanidis, C., ed. 2005. *Universal Access in Health Telematics—A Design Code of Practice*. Berlin: Springer.

Stephanidis, C., and P.-L. Emiliani. 1999. 'Connecting' to the information society: A European perspective. *Technology and Disability Journal* 10(1): 21–44.

Stephanidis, C., A. Paramythis, M. Sfyrakis, and A. Savidis. 2001. A case study in unified user interface development: The AVANTI web browser. In *User Interfaces for All—Concepts, Methods, and Tools* ed. C. Stephanidis, 525–568. Mahwah, NJ: Lawrence Erlbaum.

Stephanidis, C., G. Salvendy, D. Akoumianakis, A. Arnold, N. Bevan, D. Dardailler, P. L. Emiliani et al. 1999. Toward an information society for all: HCI challenges and R&D recommendations. *International Journal of Human–Computer Interaction* 11(1): 1–28.

Stephanidis C., G. Salvendy, D. Akoumianakis, N. Bevan, J. Brewer, P. L. Emiliani, A. Galetsas et al. 1998. Toward an information society for all: An international R&D agenda. *International Journal of Human–Computer Interaction* 10(2): 107–134.

Suri, J., and M. Marsh. 2002. Scenario building as an ergonomics method in consumer product design. *Applied Ergonomics* 31:151–157.

Tognazzini, B. 1995. *Tog on Software Design*. Reading, MA: Addison-Wesley.

Ueda, H. 2001. The FRIEND21 framework for human interface architectures. In *User Interfaces for All—Concepts, Methods, and Tools*, ed. C. Stephanidis, 245–270. Mahwah, NJ: Lawrence Erlbaum.

Weidenhaupt, K., K. Pohl, M. Jarke, and P. Haumer. 1998. Scenarios in system development: Current practice. *IEEE Software* 15(2): 34–45.

18 Access to Web Content by Those with Disabilities and Others Operating under Constrained Conditions

Benjamin B. Caldwell and Gregg C. Vanderheiden

CONTENTS

18.1 INTRODUCTION

Web accessibility has traditionally been defined as a separate topic from usability. In fact, they are two aspects of the same dimension. Products that are very inaccessible are also very unusable. It is also important to consider that Web content that is easily usable by one individual may turn out to be difficult to use by a second individual, and impossible to use by a third. In fact, the same content may be easy to use by a person in the morning sitting at their desk using a standard monitor and yet be unusable by the same individual later in the day when driving their car or trying to access Web content using a mobile device.

Even Web content that conforms to Web accessibility guidelines or standards is still unusable by individuals with some types, degrees, or combinations of disabilities. In looking at the topic of Web content accessibility, therefore, there are a number of underlying principles that are important to keep in mind. Among them are the following:

1. There is no content that is accessible without qualification. (That is, it is accessible to all people.)
   - Content can meet accessibility standards. It can also be more accessible or less accessible than other content; however, there is no content that is absolutely accessible (i.e., usable by everyone).
2. The only time one can talk concretely about something being "accessible" is if one is talking about a particular individual or a group of very similar individuals.
   - Even then, describing something as accessible can only be done if one can specify the environment in which users are operating and sometimes also the tasks they are trying to carry out. Another important factor is the time they have available to complete tasks. For example, you can say that a Web site is accessible to Joe, who uses screen reader x with browser y, but not that the Web page is accessible to all screen reader users. In fact, the site may not even be accessible for Joe when he is using another screen reader or carrying out another task.
   - All other statements of accessibility must be constrained to statements of conformance to one or another accessibility standard.
3. Usability and accessibility are just two parts of a continuum with "extremely easy to learn, understand, and use" at one end and "can't be used" at the other.
   - Nothing that is truly usable by someone is inaccessible to them. Nothing that is inaccessible to someone is usable by them.
4. Something that is only operable with time and effort should not be labeled accessible.
   - Accessibility implies that content is reasonably usable, not just theoretically operable. For example, a building should not be considered accessible if a person must cross the street, go through the parking garage to the loading dock, up the loading dock lift and then through the kitchen in order to get to the registration desk—even though the person can technically get to the registration desk in their wheelchair.

     Almost all of the measures that we associate with accessibility for people with disabilities have parallels in usability for individuals who do not have disabilities but who are operating under constrained conditions. For example, features that allow individuals who are blind to access Web content can also allow individuals who are driving their car to be able to have Web content presented to them audibly. Similarly, features that allow a Web site to be operated by individuals who are deaf also allow the site to be operated by individuals in a very noisy or a silent (e.g., library) environment. Figure 18.1 shows a listing of some of the parallelisms. This leads us to a fifth principle:
5. The needs of individuals with disabilities parallel the needs of individuals (without disabilities) who are operating under constrained conditions (e.g., in a very noisy or silent environment, when eyes are busy, when in a bouncing car, or when using a small-screen device).

   Another interesting parallel worth noting up front is the similarity between individuals operating under constrained conditions (disability, environment, etc.) and intelligent user agents. Increasingly, we are using intelligent user agents to wander, search, compile, abstract, and otherwise locate or pre-process Web content for us. These agents, however, do not currently "hear" very well nor can they interpret pictures or complex layouts very well. Although this will improve over time in some areas, in other areas (e.g., visual interpretation of pictures)

Operable without vision	- people who are *blind*	- people whose *eyes are busy* (e.g., using a product while driving a car or accessing information via a phone) - who are *in darkness.*
Operable with low vision	- people with *visual impairment*	- people using a *small display* - or in a *smoky environment.* - or who just *left their glasses in the other room*
Operable with no hearing	- people who are *deaf*	- people in *extremely loud environments* - or whose *ears are busy listening to someone or something else* - or are in *forced silence* (library or meeting)
Operable with limited hearing	- people who are *hard of hearing*	- people in *noisy environments*
Operable with limited manual dexterity	- people with a *physical disability*	- people in *vehicle on a rough road* - or who are in a *space suit or environmental suit* - *or* who are *tired*
Operable with limited cognition	- people with a *cognitive disability*	- people who are *distracted* - or *panicked* - or under the *influence of a drug* (prescription or otherwise)
Operable without reading	- people with a *cognitive disability*	- people who just *haven't learned to read or can't read 'that' language,* - people who are *foreign visitors,* - people who *left reading glasses behind*

FIGURE 18.1 Parallelisms between people with disabilities and without disabilities but operating under constrained conditions.

it is likely to be an issue for quite some time. This leads us to the final principle.

6. Many of the same measures that make Web content accessible to individuals with disabilities also make them more accessible to intelligent user agents and those operating under constrained conditions.

This train of thought leads to an interesting hypothesis.

Hypothesis: Content that is machine accessible (e.g., perceivable, operable, and understandable by intelligent information agents, see Klusch et al. 2005) could by definition be accessible to (most) people with disabilities (because such agents could represent the information in a form suitable to individuals with different disabilities; Vanderheiden 2002).

Thus, the topic of this chapter is not solely about making Web content accessible to individuals who have disabilities. Rather, it is about creating Web content that allows for flexible presentation and operation, allowing users to view and operate the content in different ways to match the constraints and abilities that they may be experiencing on either a temporary or a permanent basis as well as making them more accessible to intelligent user agents.

18.2   BASIS FOR THIS CHAPTER

The contents of this chapter are largely derived from work carried out by the World Wide Web Consortium's Web Content Accessibility Guidelines Working Group. This work had its origin in a Birds-of-a-Feather (BOF) session at the WWW2 Conference in Chicago in 1994. The results of that session were compiled in a document titled "Design of HTML (Mosaic) pages to increase their accessibility to users with disabilities: strategies for today and tomorrow" (Vanderheiden 1995). Over the years, a wide variety of groups and individuals developed different accessibility guidelines. In 1997, most of the leading guidelines were brought together into a single set of guidelines titled "Unified Web Site Accessibility Guidelines version 7.2" (Vanderheiden et al. 1997). These unified guidelines were later used (version 8) as the first working draft of guidelines of the W3C Web Content Accessibility Working Group.

In May 1999, the W3C released the Web Content Accessibility Guidelines (WCAG) 1.0 (Chisholm, Vanderheiden, and Jacobs 1999). These guidelines have been used as the basis for accessibility standards and regulations in the United States (Section 508 was based on these guidelines) as well as in over a dozen other countries.

The WCAG 1.0 guidelines focus primarily on HTML. To address the ever-expanding technologies in use on the Web, the W3C released a newer version of the guidelines (WCAG 2.0) in December 2008. The WCAG 2.0 guidelines (Caldwell et al. 2008a) describe the characteristics of accessible Web content and are not technology specific. WCAG 2.0 is based on general principles of accessibility. The principles are divided into a set of general guidelines and testable success criteria (requirements that define conformance) that are then supplemented by supporting documents that describe the success criteria in greater detail and with technology-specific techniques. The goal of WCAG 2.0 is to make it easier to understand how accessibility principles apply to different technologies and for authors to understand what they should be achieving in order to address the needs of individuals with disabilities as well as understand how to achieve it.

18.3 ESSENTIALS OF WEB CONTENT ACCESSIBILITY AND USABILITY

The Web Content Accessibility Guidelines 2.0 document is focused around three underlying principles for accessibility (Vanderheiden 2001). Basically, they state that in order for content to be usable or accessible to someone, the content must be *perceivable*, *operable*, and *understandable* by that individual in whatever circumstances and with whatever constraints they are experiencing.

Because there is a very wide range of abilities and constraints (including different types, degrees, and combinations of disabilities), it can lead to the impression that Web content would need to be delivered in an equally wide range of presentations or, alternatively, in a very constrained fashion. However, a relatively small number of strategies can, if carefully chosen, be used to provide content that is accessible to users experiencing widely varying constraints.

For example, ensuring that Web content is machine readable can make it more usable by a wider range of users, both with and without disabilities. It can make Web content accessible to individuals who cannot see the content (due to blindness or their eyes being busy, such as while driving a car) as well as to individuals who cannot see the content well (due to low vision or using a small mobile display) and would benefit from it being presented in a larger form. Machine-readable content can also be used by individuals with cognitive, language, or reading disabilities (because it could be presented orally) and even by individuals who are deaf-blind (because it could be presented in Braille). If content is machine readable, individuals with physical disabilities (or whose hands are busy) could instruct Web browsers to jump to locations or to operate buttons with verbal commands. Individuals who are having difficulty understanding the content will soon be able to have it re-rendered in simpler terms. Machine-readable content would also allow individuals who are deaf and use sign language as their primary means of communication to someday soon have the information presented in their native (sign) language.

The parallels between accessibility and usability again are evident here. Content that is machine readable can be represented verbally to individuals while they garden, walk, or do other things that occupy their vision. It can also be reformatted to fit on a small-screen mobile device, can be accessed by search and summarization engines, and so on. Refer to Figure 18.1 for additional examples.

18.3.1 OVERVIEW OF WCAG 2.0

The WCAG 2.0 Guidelines document is designed to be a stable, referenceable, technology-independent technical standard. A series of supporting documents provide additional layers of guidance for the many different audiences that WCAG 2.0 addresses. These include the following:

1. *How to Meet WCAG 2.0*: A customizable quick reference to WCAG 2.0 that lists all of the guidelines, success criteria (requirements), and techniques for authors to use as they are developing content. The How to Meet document allows authors to filter the requirements and techniques according to the technologies that they are using as well as the level of conformance they are interested in (Caldwell et al. 2008b).

2. *Understanding WCAG 2.0*: A guide to understanding and implementing WCAG 2.0. This resource provides a short "Understanding" document for each guideline and success criterion in WCAG 2.0. It includes additional explanation of the intent of the requirement, examples, related resources, key terms, and a listing of techniques (or combination of techniques) and common failures (Caldwell et al. 2008c).

3. *Techniques for WCAG 2.0*: A collection of techniques and common failures, each in a separate document that includes a description, examples, code, and tests. This collection includes both "sufficient techniques" (those that have been identified and documented by the WCAG Working Group as sufficient to meet specific success criterion in the guidelines) as well as "advisory techniques" (techniques that are best practices or can enhance accessibility but did not qualify as sufficient techniques) (Caldwell et al. 2008d).

Technology independence at the guidelines level provides a number of advantages for WCAG 2.0. Most importantly, it makes it possible for the WCAG Working Group to update the supporting documents and techniques to address the new and changing technologies that are in use on the Web. This allows the guidelines to be stable, yet accommodate changes in support for technology features and incorporate new accessibility techniques as they are identified.

18.3.2 ACCESSIBILITY AND IMPACT ON DESIGN

A concern that is commonly expressed regarding Web accessibility is the impact of accessibility regulations on freedom of expression and design. A common misconception about accessible sites, in fact, is that making sites accessible stifles creativity and innovation. These concerns are based on a number of common myths including

- You cannot use JavaScript, Flash, PDF, and other modern technologies on accessible sites.
- Accessible sites are boring and uninteresting.
- Accessible sites cater to the "lowest common denominator."
- Creating accessible content is complex and expensive.

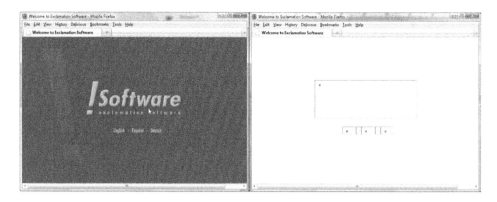

FIGURE 18.2 The left side shows a page that uses graphics to provide a pleasing visual effect. The right side shows what a screen reader or other nonvisual interface "sees" when it encounters such a page when no text alternatives are present. No useful information is provided, and the page is unusable.

The argument is that if making things more accessible means that (1) only certain types of presentation or technologies can be used, (2) only simple words and sentence structures are allowed, or (3) highly efficient visual presentations are not an option, then accessibility would be a barrier to freedom of expression and creativity on the Web. However, none of this reflects the reality of what it means to meet WCAG 2.0.

In discussing the various accessibility guidelines below, you will note that they fall into two basic categories. The first are requirements that are transparent to users who do not need them and that do not have an impact on the author's intended design or presentation. For example, when an image is included on a Web page, the `alt` attribute is added to the markup to describe the function of the graphic. "Alt text" (short for alternative text) is not visible to most users unless the user configures their browser so that images are not displayed (see Figures 18.2 and 18.3). Similarly, when a form is included on an HTML page, the `LABEL` element can be used to create an explicit association between a text field and its label. The use of this element does not change the presentation of the page in any way and provides assistive technologies with a clear understanding of how the form should be presented to users.

Another example is closed captions. As with television, people viewing video on the Web do not see the captions unless they turn them on by setting preferences within their media players or by using controls that are provided by the content directly.

In both of these examples (alt text and captions), the alternatives provide information when the individual is not able to perceive the primary presentation (in the first case, because they could not see the graphics and in the second case, because they could not hear the audio of the video presentation).

It is useful to note that, in both cases, the alternative text is also of benefit to search engines and people using them. In the first case, it allows searchers to locate content that was presented in a graphic but not in the text. In the second case, it allows searchers to not only locate media containing certain words or phrases but to actually search the audio visual presentation for particular text and then to jump to that location in the presentation.

FIGURE 18.3 Shows the same page as Figure 18.2, with text alternatives (alt text) added. As can be seen, the page on the left side is identical and the alt text is invisible. In this case, however, when graphics are turned off or when the page is viewed with a screen reader or other nonvisual interface, the user encounters the text alternatives (right), and the page is completely usable.

The second type of accessibility/usability provision includes those that do have an effect on the presentation of content. These requirements can impact the presentation by adding constraints or by requiring that the presentation be done in certain ways. For requirements in this category, addressing them as part of the design process (rather than after the fact) is essential to ensuring effective access (Guarino Reid and Snow-Weaver 2009). Examples of requirements in this category include visual and audio contrast, restrictions on the use of images of text, and the use of language that does not require advanced reading levels. In most cases, requirements in this category are only applicable to sites that strive to achieve the highest level of conformance. In other words, requirements for basic conformance with WCAG 2.0 are those that can be met with little or no change to the presentation of the content.

18.3.3 CONFORMANCE TO WCAG 2.0

Like WCAG 1.0, WCAG 2.0 includes three levels of conformance. Contributing factors for the inclusion of a success criterion at a given level include (1) whether content can be made accessible by assistive technologies, (2) whether it is possible to satisfy the success criterion on all Web sites and types of content, (3) whether meeting the criterion requires skills that content creators have or can reasonably obtain, (4) how it affects the intended design or functionality of the content, and (5) whether workarounds are available to users when the success criterion is not met. In some instances, accessibility topics in WCAG 2.0 are covered at more than one level.

There are a total of 61 success criteria listed in WCAG 2.0. To conform at a specific level, the success criteria listed for that level and lower levels must be met (e.g., to conform at level AA, authors must meet all of the level A and level AA success criteria). In addition, there are five overarching conformance requirements that must. An in-depth description of these requirements can be found in Understanding WCAG 2.0 (Caldwell et al. 2008c). Two of these conformance requirements merit special discussion: the use of conforming alternate versions, and the concept of "accessibility support."

18.3.3.1 Conforming Alternate Versions

To address some of the concerns related to design impact, authors can provide conforming alternate versions of content in cases where they find themselves unable to meet requirements with a particular technology or page design. Basically, a conforming alternate version is a second version of the nonconforming content that can meet the WCAG success criteria and that provides the same information and functionality and is as up to date as the nonconforming version. When conforming alternate versions are provided, authors must ensure that users can reach the conforming version from the non conforming version and that the nonconforming content does not prevent individuals with disabilities from accessing the alternate version. For example, the nonconforming content cannot include content that would be likely to induce a seizure or that traps the keyboard.

Making it possible for authors to conform to using this strategy has a number of advantages, particularly when it comes to the inclusion of newer technologies that may not have a robust set of accessibility features when they first begin to gain popularity. In addition, it is possible for authors to implement conforming alternate versions in ways that allow the same source content to be presented in different modes. Style switchers, for example, can easily be used in a way that allows an author to present content using a specific font size and color combination by default with an option for users to modify the presentation to increase font size or improve contrast.

18.3.3.2 Accessibility Support

One of the most important requirements in the WCAG 2.0 Guidelines is Conformance Requirement 4:

> Only accessibility-supported ways of using technologies are relied upon to satisfy the success criteria. Any information or functionality that is provided in a way that is not accessibility supported is also available in a way that is accessibility supported.

Simply put, any technique used to make a page accessible (and to satisfy a WCAG provision) must work with the users' assistive technologies and the accessibility features of their browsers. Because WCAG 2.0 is technology neutral, any Web technology can be used. However, all technologies (even HTML) can be used in both accessible and inaccessible ways. Authors need to have an understanding of which features of the technologies they are using will work with assistive technologies and which will not. If assistive technologies do not generally support a given way of using a Web technology, then authors cannot rely on that method to claim conformance to WCAG 2.0. Authors can consult the WCAG 2.0 Techniques documents and other resources related to best practices in accessible Web design for information related to accessibility support. These resources can be expanded over time to include new information about support as the technologies change. Additional information about accessibility support can be found in Understanding WCAG 2.0.

18.4 ACCESSIBILITY/USABILITY ESSENTIALS FOR CONSTRAINED USERS/SITUATIONS

As discussed above, there are three accessibility/usability principles. The content must be *perceivable*, *operable*, and *understandable*. At the end of this section, we also talk about a fourth principle, that content is *robust*. That is, that content remains accessible/usable with evolving technologies.

This section provides a high level overview of each of these principles, followed by a summary of the guidelines and a brief description of the success criteria (requirements) for each guideline. The goal is to provide a basic understanding of the types of barriers faced by those with disabilities as well as those who are operating under constrained conditions.

Note that this chapter is not meant to be a comprehensive or authoritative look at the requirements for WCAG 2.0. Rather, it provides a brief introduction to the requirements and describes the range of barriers to access faced by users who may be accessing content in ways that authors are not expecting. Readers interested in achieving any level of conformance to WCAG 2.0 should consult the official WCAG documentation.

18.5 PERCEIVABLE

The first essential to Web content accessibility/usability is that it must be possible to present content in ways that are perceivable to the user. This means that users must be able to recognize and interpret the information that is being presented by the content. As previously described, authors should consider the full range of senses and constraints that users may encounter when attempting to read and interact with the information that is being presented and ensure that the information they are presented is not invisible to their senses.

18.5.1 Ensuring that Nontext Content Is Accessible

Guideline 1.1 Text alternatives: Provide text alternatives for any nontext content so that it can be changed into other forms people need, such as large print, Braille, speech, symbols, or simpler language.

Electronic text has a unique property in that it has no particular presentation format. It can be presented visually, spoken aloud, presented as Braille, presented as Morse code tones or tactile taps, etc. It can also be presented in different sizes or at different volumes. Finally, it is machine readable, making it easy for intelligent agents to accurately perceive it. As a result, it is universally perceivable because it can be presented in whatever form best suits the user and their situation.

In contrast, nontext content such as pictures, color coding, audio recordings, and video is not universally perceivable. As a result, alternatives or supplements are needed in order for all users to be able to access and use these information formats in all situations.

The first (and only) requirement for Guideline 1.1 is

18.5.1.1 Success Criterion 1.1.1: Nontext Content

All nontext content that is presented to the user has a text alternative that serves the equivalent purpose, except for the situations listed below. (Level A)

- *Controls, input*: If nontext content is a control or accepts user input, then it has a name

that describes its purpose. (Refer to Success Criterion 4.1.2 for additional requirements for controls and content that accepts user input.)
- *Time-based media*: If nontext content is time-based media, then text alternatives at least provide descriptive identification of the nontext content. (Refer to Guideline 1.2 for additional requirements for media.)
- *Test*: If nontext content is a test or exercise that would be invalid if presented in text, then text alternatives at least provide descriptive identification of the nontext content.
- *Sensory*: If nontext content is primarily intended to create a specific sensory experience, then text alternatives at least provide descriptive identification of the nontext content.
- *CAPTCHA*: If the purpose nontext content is to confirm that content is being accessed by a person rather than a computer, then text alternatives that identify and describe the purpose of the nontext content are provided, and alternative forms of CAPTCHA using output modes for different types of sensory perception are provided to accommodate different disabilities.
- *Decoration, formatting, invisible*: If nontext content is pure decoration, is used only for visual formatting, or is not presented to users, then it is implemented in a way that it can be ignored by assistive technology.

This requirement includes a few exceptions because there are some types of information that cannot be expressed across sensory modalities. These fall into the category of "sensory experience." For example, a performance by a virtuoso or a great work of art cannot be expressed in words in a way that provides a true equivalent to the original. The feelings that an individual gets and some of the character of the piece might describe the content to some degree, but not the sensory experience itself, which is the purpose of the information. Also, a description would be one impression or interpretation rather than the desired experience.

Care must be taken in exercising the exceptions. Sometimes, a decorative flourish may, in fact, also function as a divider between sections of content. In this case, the flourish need not be described in detail, but the fact that there is a division between sections of content does need to be presented. In such a case, a simple description of the graphic can provide a cue to individuals who are listening to the page in auditory form.

In providing text alternatives, it is also important to focus on the *function* and not the appearance of the nontext content. For example, if a button that looks like a magnifying glass represents the "search" button on a page, then the alternative text for it should be "search" rather than "magnifying glass," which would be a description of the picture rather than its function. Similarly, if the purpose of a chart is to show declining revenues, then this purpose should be stated clearly in the description along with a focus on the trend followed by a more detailed description of the data rather than just describing what the data look like alone.

18.5.1.2 Design Impact for Guideline 1.1

For most technologies, the inclusion of text alternatives does not have a noticeable effect on the design of the content or on the user experience for individuals who interact with the content visually.

18.5.2 Ensuring that Alternatives Are Available for Time-Based Media

When time-based or synchronized media are included in content, such as with a movie or an animation, it becomes important not only to make the information available in an accessible form but also to ensure that the different streams of information can be presented together. For example, it would not be very helpful to someone who was deaf if they had a transcript of the dialogue for the movie in front of them while they were watching the movie because it would be very difficult to determine when specific audio events occurred. The second guideline under Perceivable is

---

Guideline 1.2 Time-based media: Provide alternatives for time-based media.

Similar to the previous guideline, Guideline 1.2 addresses the need to include alternatives for information presented in time-based media. This includes audio-only, video-only, and audio-video. It also includes media (audio and/or video) that incorporate interactive elements. For example, links could be embedded in an alternative for time-based media.

18.5.2.1 Success Criterion 1.2.1: Audio-Only and Video-Only (Prerecorded)

For prerecorded audio-only and prerecorded video-only media, the following are true, except when the audio or video is a media alternative for text and is clearly labeled as such (Level A):

- *Prerecorded audio only*: An alternative for time-based media is provided that presents equivalent information for prerecorded audio-only content.
- *Prerecorded video only*: Either an alternative for time-based media or an audio track is provided that presents equivalent information for prerecorded video-only content.

Typically, with information that is pure audio or pure video (e.g., a recording of a speech or a video with no audio track), there is no need to synchronize the alternative version. In these cases, a simple text alternative or a short transcript is often provided instead of captions. There are cases, however, where there is little or no benefit in describing the visual track of a multimedia presentation. For example, in a video of someone delivering a speech with the video showing only a fixed head and shoulders shot of the speaker, there may be no significant visual information and or little or no opportunity to inject a description of what is presented visually. In this case, a text transcript would be sufficient (note that the transcript in this example could include significant visual information, such as the individual pounding on the lectern if it was important).

A separate case occurs when an audio program is provided and the user is required to respond in some way to the audio presentation. In this case, synchronization between the text alternative and the audio presentation is important. Because the interaction is mixed with the audio presentation, this should be considered "time-based" media and a synchronized text presentation should be provided to make it possible for users to be able to interact at the appropriate times.

18.5.2.2 Success Criterion 1.2.2: Captions (Prerecorded)

Captions are provided for all prerecorded audio content in synchronized media, except when the media is a media alternative for text and is clearly labeled as such (Level A).

18.5.2.3 Success Criterion 1.2.4: Captions (Live)

Captions are provided for all live audio content in synchronized media (Level AA).

The combination of increased broadband speeds and improvements in compression technologies has made Web-based viewing of multimedia content increasingly common. With a variety of Web sites available where individuals can view a television episode they may have missed earlier in the week or queue up a movie from a Web-based movie rental service, it is critical for those who are deaf and hard of hearing to have access to captioned content. One challenge in this area is that caption information has often been left out in the process of converting broadcast media to formats that can be streamed over the Web. In WCAG 2.0, captions are required at Level A for prerecorded media content and at Level AA for live content.

18.5.2.4　Success Criterion 1.2.3: Audio Description or Media Alternative (Prerecorded)

An alternative for time-based media or audio description of the prerecorded video content is provided for synchronized media, except when the media are media alternatives for text and are clearly labeled as such (Level A).

18.5.2.5　Success Criterion 1.2.5: Audio Description (Prerecorded)

Audio description is provided for all prerecorded video content in synchronized media (Level AA).

18.5.2.6　Success Criterion 1.2.7: Extended Audio Description (Prerecorded)

Where pauses in foreground audio are insufficient to allow audio descriptions to convey the sense of the video, extended audio description is provided for all prerecorded video content in synchronized media (Level AAA).

There are three requirements related to audio description, one at each conformance level. Their purpose is to augment the audio information in a media presentation with descriptions of visual events. They are important for those who cannot see and would otherwise be able to understand the content (due to not seeing critical information that would otherwise only be presented visually). All three of the requirements are focused on prerecorded media. At Level A, authors can provide alternatives to time-based media instead of audio descriptions, but audio descriptions are required at Level AA. At Level AAA, the requirement is to provide extended audio descriptions (audio description that is added to an audiovisual presentation by pausing the video so that there is time to add additional description).

Because most audio description can only be added in gaps that occur during the dialogue of an audio program, there are some limitations with regard to how much of what is happening on screen can be effectively described. The goal is to include any important visual details that cannot be understood from the main soundtrack alone.

Alternatives for time-based media are similar to screenplays in that they include correctly sequenced text descriptions of the information presented in the audio and visual tracks of a media presentation. This type of alternative can be used to meet the audio description requirement at Level A and is important for individuals who are deaf-blind, where

including captions and audio descriptions will not provide the needed information. It is also important for individuals who cannot see or hear sufficiently to use either all visual or all auditory presentations that this type of alternative should be available.

18.5.2.7　Success Criterion 1.2.6: Sign Language (Prerecorded)

Sign language interpretation is provided for all prerecorded audio content in synchronized media (Level AAA).

For many deaf and hard of hearing individuals who are fluent in a sign language, written text such as the text found in captions is a second language. Sign language can include a richer and more nuanced alternative to synchronized media and is therefore included (at Level AAA).

18.5.2.8　Success Criterion 1.2.9: Audio-Only (Live)

An alternative for time-based media that presents equivalent information for live audio-only content is provided (Level AAA).

Text-based presentations should be available for those who need access to live (real-time) information and events that take place on the Web. Providing an alternative text transcript that includes all of the dialogue plus a description of the important visual events in a correctly sequenced format (similar to a screenplay or the script of a play) will provide the best access for these individuals and is required at Level AAA.

18.5.2.9　Design Impact for Guideline 1.2

Most of the requirements that fall under Guideline 1.2 are transparent to most users when authors either provide controls or choose media formats that allow users to turn captions and audio descriptions off. Audio description can be provided as an additional audio track and thus can also be turned on and off in most situations.

18.5.3　Ensuring that the Information, Structure, and Function of Web Content Can Be Presented in Different Ways without Losing Information or Structure

Because access to Web content requires that information be presentable in different ways, it is important that Web content design does not restrict any of the meaning, function, or structure of a page to a specific presentation format. For example, the information and structure of a document should not be encoded visually in a way that would be lost if the content were presented via speech. Similarly, information presented via color (e.g., "press the red button") must also be available in other nonvisual ways. The next guideline under Perceivable is

---

Guideline 1.3 Adaptable: Create content that can be presented in different ways (for example simpler layout) without losing information or structure.

The basic requirements under this guideline center primarily on the use of the structure and semantics of markup in a given technology. It also covers situations where content or the characteristics of the content's presentation can have an impact on meaning or understanding when users configure their browsers or assistive technologies to present the content to them in different ways.

18.5.3.1 Success Criterion 1.3.1: Information and Relationships

Information, structure, and relationships conveyed through presentation can be programmatically determined or are available in text (Level A).

In HTML, a number of elements and attributes can be used to convey structure. At the most basic level, this simply means using markup such as HTML headings (h1 through h6) or ordered and unordered lists in a way that is consistent with the specification. With Cascading Style Sheets (CSS), the presentation of structural markup can be modified in a number of ways and can, in fact, be used to make the presence of structure invisible to users. Similarly, Portable Document Format (PDF) includes structures (called tags) that allow authors to include information about the organization and structure of the document and its parts. Regardless of the technology in use, the key is that the structure can be easily detected by software. This allows browsers and assistive technologies of all types to make it possible for users to get an overview of the document, navigate by moving from header to header, or skip over long lists in much the same way that sighted users can visually scan through a document to orient themselves within a document.

Using proper machine-readable constructs for headers, lists, tables, and so on is important for achieving this objective. In addition, any functionality of the page (e.g., links or other controls) should also be easy to locate and identify without having to rely on the visual presentation of the page (e.g., a control and its label should be programmatically linked rather than just positioned near each other on a page if they are related).

18.5.3.2 Success Criterion 1.3.2: Meaningful Sequence

When the sequence in which content is presented affects its meaning, a correct reading sequence can be programmatically determined (Level A).

Some technologies make it possible for authors to present content in ways that vary from the underlying structure of the document. For example, CSS can be used to present information that is at the end of the source code in a way that appears visually at the top of the page. In some cases, this does not present an accessibility barrier to users. However, the ability for presentation technologies to rearrange content and present it in a different order can sometimes change the meaning of the content. For example, an author could use CSS positioning to arrange blocks of text into what looks like a table, but when read by a screen reader would be a linear set of blocks that does not make sense.

18.5.3.3 Success Criterion 1.3.3: Sensory Characteristics

Instructions provided for understanding and operating content do not rely solely on sensory characteristics of components such as shape, size, visual location, orientation, or sound (Level A).

The last topic covered by this guideline has to do with the way in which information is referenced within content. For example, if instructions on a Web page said something like, "Click on the red button in the lower left hand corner of the shaded box below to begin" it would create a great deal of confusion for a number of users including those who are color-blind, screen reader users and individuals with low vision who may be viewing the content in high-contrast mode or are using magnification. Rather than referencing the button in the example above by using characteristics of its presentation, something like, "Click on the red 'compare' button to begin" makes it possible for users who would otherwise be unable to find the button to search for the button that is labeled "compare."

18.5.3.4 Design Impact for Guideline 1.3

There are some Web content and presentation technologies that do not have any structural markup. For these technologies, structural information is conveyed only with formatting such that differences in font face or font size are the only cues that one section of content is different from another. Using this type of technology creates a number of problems and can restrict the options an authors has when it comes to finding techniques that can be used to effectively meet this guideline. Fortunately, technology developers are increasingly realizing the importance of making structure available.

Note that the requirements in this section do not mean that authors cannot use technologies and formats that do not have rich support for document semantics, only that they may need to provide alternative versions of content where important information would not be available to all users.

18.5.4 Ensuring that Foreground Content (Visual and Auditory) Is Differentiable from the Background

Part of being able to perceive information is being able to separate that information from background noise or background presentation. The guideline deals with ensuring that people with disabilities can distinguish the foreground information from the background. This barrier occurs both in visual and in auditory realms.

Guideline 1.4 Distinguishable: Make it easier for users to see and hear content including separating foreground from background.

18.5.4.1 Success Criterion 1.4.1: Use of Color

Color is not used as the only visual means of conveying information, indicating an action, prompting a response, or distinguishing a visual element (Level A).

Note: This success criterion addresses color perception specifically. Other forms of perception are covered in Guideline 1.3 including programmatic access to color and other visual presentation coding.

Color is a powerful presentation technique and its use is highly recommended as an accessibility/usability technique. However, because of the high percentage of individuals who have a color vision deficiency, it is important that information that is presented in color be presented such that it can also be perceived by those who do not have full color vision. For example, text is commonly presented over the top of an image. A user *without* a visual impairment may have no problem viewing the text because the user's mind can separate the lines from the background picture from the letters themselves. However, an individual who has very low vision and is viewing the letters and words with magnification (and sees them as blurry shapes) may be confused by parts of the image that intersect with the individual letters and words where the background image appears to be part of the letter shape when viewed under magnification. These users do not have the ability to view the text as a whole and therefore cannot mentally exclude the background image as well as those without visual impairment. Similarly, someone who has perfect color vision may

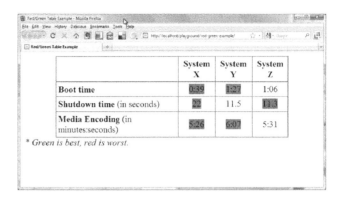

FIGURE 18.4   An example Web page illustrating a common barrier for individuals who have difficulty perceiving color.

be able to easily read red letters on a green background. However, for someone who is red/green color-blind, both colors can look essentially the same, perhaps something like looking at brown letters on a brown background. Depending on the darkness of the two browns, it may be impossible to read the letters. Figure 18.4 illustrates a common problem that occurs when authors use color alone to convey information.

18.5.4.2 Success Criterion 1.4.2: Audio Control

If any audio on a Web page plays automatically for more than 3 seconds, either a mechanism is available to pause or stop the audio, or a mechanism is available to control audio volume independently from the overall system volume level (Level A).

Note: Because any content that does not meet this success criterion can interfere with a user's ability to use the whole page, all content on the Web page (whether or not it is used to meet other success criteria) must meet this success criterion. See Conformance Requirement 5: Non-Interference.

Because audio that plays automatically on a page can compete with screen reading software in a way that can make it difficult for users to hear the screen reader over the background audio. Further, because the user cannot hear the screen reader very well over the background audio, they cannot use it to find a control on the page to turn off the audio. A requirement that audio that plays automatically for more than 3 seconds is therefore included. Techniques to meet this requirement include turning sounds off automatically after 3 seconds, providing controls near the beginning of a page that allow the user to turn off the sound (least desirable) and only including audio that plays on user request.

18.5.4.3　Differentiating Visual Content (Levels AA and AAA)

18.5.4.4　Success Criterion 1.4.3: Contrast (Minimum)

The visual presentation of text and images of text has a contrast ratio of at least 4.5:1, except for the following (Level AA):

- *Large text*: Large-scale text and images of large-scale text have a contrast ratio of at least 3:1.
- *Incidental*: Text or images of text that are part of an inactive user interface component, that are pure decoration, that are not visible to anyone, or that are part of a picture that contains significant other visual content have no contrast requirement.
- *Logotypes*: Text that is part of a logo or brand name has no contrast requirement.

18.5.4.5　Success Criterion 1.4.6: Contrast (Enhanced)

The visual presentation of text and images of text has a contrast ratio of at least 7:1, except for the following (Level AAA):

- *Large text*: Large-scale text and images of large-scale text have a contrast ratio of at least 3:1.
- *Incidental*: Text or images of text that are part of an inactive user interface component, that are pure decoration, that are not visible to anyone, or that are part of a picture that contains significant other visual content have no contrast requirement.
- *Logotypes*: Text that is part of a logo or brand name has no minimum contrast requirement.

Two of the success criteria for addressing Guideline 1.4 focus on providing sufficient contrast between the foreground content and the background content. If the background image is faint enough, then it does not interfere with the foreground content substantially. WCAG 2.0 introduced a measurable requirement for establishing a sufficient contrast level between foreground and background information called "contrast ratio." The contrast measure specified in WCAG 2.0 is a solution that works even if an individual is color-blind (Arditi, Vanderheiden, and Gordon 2005).

Another solution is to provide some mechanism for separating the foreground from the background so that the foreground information can be presented independently from the background information. This is also a viable technique but a less effective one because many individuals will not know how to carry out the separation or may not have the tools necessary to do so. Also, some individuals may believe

they see the foreground information clearly but simply read it incorrectly and then operate on the erroneous information. This can be true for individuals operating under functional constraints (e.g., using a phone) as well as for those who simply do not look carefully. The general approaches for meeting this requirement are as follows:

1. Provide a mechanism for viewing foreground information without the background, or
2. Ensure that the contrast between the foreground and background information exceeds a minimum standard.

The second approach is by far the more effective. Prior to WCAG 2.0, a standard for measuring contrast had not been available. A number of tools designed to help authors measure the contrast ratios of visual and audio content are available (Juicy Studio 2007; WAT-C and The Paciello Group 2008).

18.5.4.6　Success Criterion 1.4.4: Resize Text

Except for captions and images of text, text can be resized without assistive technology up to 200% without loss of content or functionality (Level AA).

The use of relative text sizes has been a best practice for usability and accessibility for some time. A new requirement in WCAG 2.0 at Level AA addresses the needs of not only those with vision impairment, but also individuals who are using small screen and mobile devices. Users who have mild vision impairments fall into a gray area where their needs are not often significant enough to justify the use of an assistive technology such as magnification, yet they would benefit substantially from presentations with larger font sizes. In many cases, the browser provides built-in mechanisms that allow users to zoom or increase the text size, but scenarios remain where an author can present information in a way that makes it difficult for users to increase the size of the text they are trying to read.

18.5.4.7　Success Criterion 1.4.5: Images of Text

If the technologies being used can achieve the visual presentation, text is used to convey information rather than images of text except for the following (Level AA):

- *Customizable*: The image of text can be visually customized to the user's requirements.
- *Essential*: A particular presentation of text is essential to the information being conveyed.

Note: Logotypes (text that is part of a logo or brand name) are considered essential.

18.5.4.8 Success Criterion 1.4.9: Images of Text (No Exception)

Images of text are only used for pure decoration or where a particular presentation of text is essential to the information being conveyed (Level AAA).

Note: Logotypes (text that is part of a logo or brand name) are considered essential.

Similar to text resizing, presenting text as an image presents barriers to those who employ magnification strategies to access Web content and to those who need to increase the size of text. While many modern Web browsers make it possible to scale images of text along with text (e.g., zoom the entire page), text that is part of a bitmapped image will often become blurry and distorted when resized in this manner. Where possible, authors should choose text (not images of text) that can be scaled in ways that do not decrease the quality of the text when content is resized.

18.5.4.9 Success Criterion 1.4.7: Low or No Background Audio

For prerecorded audio-only content that (1) contains primarily speech in the foreground, (2) is not an audio CAPTCHA or audio logo, and (3) is not vocalization intended to be primarily musical expression such as singing or rapping, at least one of the following is true (Level AAA):

- *No background*: The audio does not contain background sounds.
- *Turn off*: The background sounds can be turned off.
- *20 dB*: The background sounds are at least 20 dB lower than the foreground speech content, with the exception of occasional sounds that last for only 1 or 2 seconds. Note that per the definition of "decibel," background sound that meets this requirement will be approximately four times quieter than the foreground speech content.

The issue with regard to auditory information is very similar to that of visual is that the goal is to ensure that the foreground information is sufficiently differentiable from background sounds such that it is perceivable even by individuals who are hard of hearing or by individuals operating in noisy environments, etc. There are, however, several interesting distinctions.

First, visual information is often provided in components that overlay each other so that it is possible either to turn off the background or to copy the foreground text and present it separately. With auditory information, various layers are typically all mixed into a single sound track before the author has access to it. As a result, it is not possible to easily drop out the background sounds at will or to "lift" the foreground dialogue and present it separately.

Second, according to Guidelines 1.1 and 1.2, any auditory information will already have a text alternative available. Thus most individuals who have difficulty hearing would have captions or a transcript available that they could view.

Third, the auditory presentation is transitory. Although individuals can study visually presented text for a longer period of time if part of it is ambiguous, auditory presentation goes by quickly and often cannot be easily repeated. (Note that later requirements require that moving content be paused or stopped.)

Techniques to address this requirement include the following:

1. Providing a parallel text presentation of audio information (required by previous guidelines).
2. Ensuring that background sounds are at least 20 dB lower than the foreground audio content with the exception of occasional sounds.

Note that a 20-dB difference in sound level is roughly four times quieter (or louder). The exception for occasional sounds is provided because there are cases where brief sounds, such as an explosion, that cause audio content to be obliterated for all users. Such content can generally be reconstructed from context and surrounding audio. While WCAG 2.0 was in development, the Trace Center partnered with Eramp.com and Audacity, a popular open-source audio editing tool, to create an audio contrast analyzer (MacDonald 2008).

18.5.4.10 Success Criterion 1.4.8: Visual Presentation

For the visual presentation of blocks of text, a mechanism is available to achieve the following (Level AAA):

1. Foreground and background colors can be selected by the user.
2. Width is no more than 80 characters or glyphs (40 if CJK).
3. Text is not justified (aligned to both the left and the right margins).
4. Line spacing (leading) is at least space-and-a-half within paragraphs, and paragraph spacing is at least 1.5 times larger than the line spacing.
5. Text can be resized without assistive technology up to 200% in a way that does not require the user to scroll horizontally to read a line of text on a full-screen window.

For individuals with vision impairment or cognitive, language, and learning disabilities, there are many factors that can interfere with a user's ability to read the content. Examples include presentations that limit a user's ability to represent content using preferred color combinations, long line lengths that can make it difficult for users to read a line of text and find the beginning of the next line, text that is justified (aligned to both the left and right margins) sometimes causes people with reading disabilities to see "rivers of white" flowing down the page between the words (Hudson, Weakley, and Firminger 2005), text where the leading (line spacing) is so small that it becomes difficult to read the text, and text that requires users to scroll horizontally to read a single line of text when the user increases the font size. Strategies to address this Level AAA requirement advise authors to avoid presenting content in these ways.

18.5.4.11 Design Impact for Guideline 1.4

The requirements in this guideline have a greater impact on design at higher levels of conformance. While many of them may cause authors to adjust their design in various ways in order to meet the requirements, it is often possible to meet them without making significant changes. One area that would have an impact on design intent is the use of alternate visual presentation in parallel with a color presentation. For example,

- A line graph with red, blue, and green lines could use red solid lines, green dotted lines, and blue dashed lines.
- The use of differing patterns for pie charts and bar graphs or to simply attach the legend text directly to the pie slices.

Another area that impacts design is that of requirements related to images of text and contrast. Historically, the ability for authors to reliably deliver fonts of their choosing and to use typography effectively over the Web has been restricted unless authors resort to presenting text to users in image form. However, technologies and specifications are in development to address this problem, which should lessen the impact on design in the future.

18.6 OPERABLE

The second essential to Web accessibility/usability is the ability to operate Web content. Most Web content at least has links on it, but other content may have a variety of other controls and the number and complexity of controls within Web content is increasing as next-generation Web technologies make it possible for sites to operate in ways that often resemble software applications. For example, it is often necessary for users to be able to easily move around within content, sometimes filling out forms, manipulating information, etc.

Again, authors must think about the full range of constraints that a user might be under and the fact that they need to not only be able to "barely use" something but be able to use it efficiently and effectively enough to carry out the intended function of the content whether it is an information page, a form, an order entry system, or interactive media.

18.6.1 Ensuring that Content Can Be Operated from a Keyboard

Just as electronic text is the fundamental building block for access in the Perceivable section, keyboard access is the fundamental for making content "operable." Although many different mechanisms may be provided for operating Web content, the ability to navigate, manipulate, and operate the content using the keyboard or a keyboard interface (e.g., any software or device that emulates keystroke input) is the most universal or adaptable mechanism for control. Individuals who cannot see can use the content via the keyboard when they would be completely unable to use a mouse. Individuals who do not have sufficient physical control to use a mouse can use a keyboard or keyboard substitute. Individuals who are unable to use keyboards can use other interfaces (i.e., sip and puff, eye gaze, and thought-controlled interfaces) that can create keystrokes and keyboard functionality instead of directly using a keyboard to provide access to content.

Speech recognition interfaces all provide the ability to simulate keystrokes, allowing access in this fashion as well. Even mobile and pen-based devices have keyboard equivalent mechanisms for input (e.g., handwriting recognition and onscreen keyboards) and most all support the connection of external keyboards. Thus even these devices have the ability to be controlled by keyboards and devices that simulate keyboard input.

To address this need, the next guideline is

Guideline 2.1 Keyboard Accessible: Make all functionality available from a keyboard.

18.6.1.1 Keyboard Operation (Levels A and AAA)

18.6.1.2 Success Criterion 2.1.1: Keyboard

All functionality of the content is operable through a keyboard interface without requiring specific timings for individual keystrokes, except where the underlying function requires input that depends on the path of the user's movement and not just the endpoints (Level A).

Note 1: This exception relates to the underlying function, not the input technique. For example, if using handwriting to enter text, the input technique (handwriting) requires path-dependent input, but the underlying function (text input) does not.
Note 2: This does not forbid and should not discourage providing mouse input or other input methods in addition to keyboard operation.

18.6.1.3   Success Criterion 2.1.3: Keyboard (No Exception)

All functionality of the content is operable through a keyboard interface without requiring specific timings for individual keystrokes (Level AAA).

These two requirements are very straightforward. Simply put, authors must ensure that it is possible to operate their content from a keyboard. Note that this guideline does not preclude operation of content via use of a mouse, stylus, etc. In fact, content that is also operable in this fashion is often more accessible. However, all functionality must be operable by the keyboard or keyboard interface. At Level A, there is an exception for content that requires input that depends on the path of a user's movement and cannot reasonably be controlled from a keyboard. For example, a drawing program where the brush strokes change depending on the speed and duration of the mouse movement would be covered under the exception. At Level AAA, there are no exceptions.

18.6.1.4   Success Criterion 2.1.2: No Keyboard Trap

If keyboard focus can be moved to a component of the page using a keyboard interface, then focus can be moved away from that component using only a keyboard interface, and if it requires more than unmodified arrow or tab keys or other standard exit methods, the user is advised of the method for moving focus away (Level A).

Note: Because any content that does not meet this success criterion can interfere with a user's ability to use the whole page, all content on the Web page (whether it is used to meet other success criteria or not) must meet this success criterion. See Conformance Requirement 5: Non-Interference.

This is a new requirement for WCAG 2.0 designed to address situations where conforming content and nonconforming content are present on the same page. With certain technologies, it is possible for content to "trap" the keyboard focus in a specific control or section of the page in a way that makes it impossible for keyboard-only users to continue using a given page or to navigate to other pages. If content requires users to use nonstandard keystrokes or combinations to move focus to the next item, then authors are required to also provide instructions for users to "untrap" the keyboard focus.

18.6.1.5   Design Impact for Guideline 2.1

With the exception of content that depends on the path of a user's movement at Level AAA, the requirements in this guideline do not have an impact on the visual presentation or design of the content beyond providing instructions in cases where inaccessible content may trap keyboard focus.

18.6.2   TIME TO OPERATE

The second essential in operability is having enough time to access and use the page. There are a number of factors that can cause an individual to need more time to access the page. Individuals with low vision may need more time to find and read the content, including pop-up dialogues, transitory messages, etc. Individuals who are blind may take longer to detect and navigate to an item in order to have it read. Individuals with physical disabilities may physically take longer to enter information or respond to a cue. Individuals with language or reading problems, including individuals who are simply reading a page that is not written in their native language, may take longer. Also, anyone could be interrupted or distracted while reading and need additional time.

Guideline 2.2 Enough Time: Provide users enough time to read and use content.

18.6.2.1   Success Criterion 2.2.1: Timing Adjustable

For each time limit that is set by the content, at least one of the following is true (Level A):

- *Turn off*: The user is allowed to turn off the time limit before encountering it; or
- *Adjust*: The user is allowed to adjust the time limit before encountering it over a wide range that is at least ten times the length of the default setting; or
- *Extend*: The user is warned before time expires and given at least 20 seconds to extend the time limit with a simple action (for example, "press the space bar"), and the user is allowed to extend the time limit at least ten times; or
- *Real-time exception*: The time limit is a required part of a real-time event (for example, an auction), and no alternative to the time limit is possible; or
- *Essential exception*: The time limit is essential and extending it would invalidate the activity; or
- *20 hour exception*: The time limit is longer than 20 hours.

Note: This success criterion helps ensure that users can complete tasks without unexpected changes in content or context that are a result of a time limit. This success criterion should be considered in conjunction with Success Criterion 3.2.1, which puts limits on changes of content or context as a result of user action.

A number of options are available to authors for addressing this success criterion. Sometimes, making it possible to turn time limits off before users encounter them can address the problem. In other cases, the solution is to require that the response time be adjustable so that people can operate at the pace that works for them. On the basis of clinical experience, approximately 10 times the value one would ordinarily set as the default is an appropriate adjustment range for those who require additional time.

In some cases, time limits are used for security reasons and extending those limits by 10 times would not be a good idea. For example, the timeout that occurs if there is no activity at a public terminal such as an ATM and the person forgot to log out. In this case, a better strategy would be to ask the user if they need more time and then give them at least 20 seconds to respond. Again, clinical experience indicates that 20 seconds is enough time for most people to make a simple one-movement response (e.g., press enter or click "OK") to show that they are present and would like more time to complete their task.

There are some situations where time limits cannot be avoided. In some cases, the time limits are created by real-time events. For example, a time limit imposed by an auction where it is not possible to provide extended response periods for those who need them because the action happens when it happens. Other times, a time limit is imposed for competitive reasons or for test design. In these cases it may also not be possible to extend response time for some individuals without invalidating the competition or the standardization of the test. Thus there are a number of exceptions provided for this requirement.

These exceptions, again, should be used with care. Often, alternative test designs could be used that do not rely on a time factor for their validation. Consideration should also be given to designing competition (particularly in educational situations) that does not rely on timed responses.

18.6.2.2 Success Criterion 2.2.2: Pause, Stop, and Hide

For moving, blinking, scrolling, or auto-updating information, all of the following are true (Level A):

- *Moving, blinking, or scrolling*: For any moving, blinking, or scrolling information that

(1) starts automatically, (2) lasts more than 5 seconds, and (3) is presented in parallel with other content, there is a mechanism for the user to pause, stop, or hide it unless the movement, blinking, or scrolling is part of an activity where it is essential; and
- *Auto-updating*: For any auto-updating information that (1) starts automatically and (2) is presented in parallel with other content, there is a mechanism for the user to pause, stop, or hide it or to control the frequency of the update unless the auto-updating is part of an activity where it is essential.

Note 1: For requirements related to flickering or flashing content, refer to Guideline 2.3.

Note 2: Because any content that does not meet this success criterion can interfere with a user's ability to use the whole page, all content on the Web page (whether it is used to meet other success criteria or not) must meet this success criterion. See Conformance Requirement 5: Non-Interference.

Note 3: Content that is updated periodically by software or that is streamed to the user agent is not required to preserve or present information that is generated or received between the initiation of the pause and resuming presentation, as this may not be technically possible and in many situations could be misleading to do so.

Note 4: An animation that occurs as part of a preload phase or similar situation can be considered essential if interaction cannot occur during that phase for all users and if not indicating progress could confuse users or cause them to think that content was frozen or broken.

Another aspect of timed content deals with text or other elements that blink, move, scroll, are auto-updated, or are otherwise animated. For all of the reasons discussed previously, some individuals will find it difficult to read the text before it disappears from view. Others are physically unable to track text that moves or is constantly changing.

There are two possibilities for addressing this criterion. The first and most common strategy is to make it possible for the user to stop moving, blinking, and scrolling content. The second is to allow users to control the presentation or (for auto-updating content) to control the frequency of updates by specifying their presentation preferences.

It should, of course, be possible for users to pause, stop, or hide content by using the keyboard. It is also good practice to allow it to be done with a mouse or pointing device as well.

18.6.2.3 Success Criterion 2.2.4: Interruptions

Interruptions can be postponed or suppressed by the user, except interruptions involving an emergency (Level AAA).

Interruptions can present challenges to a variety of users including those with visual impairments and those who have cognitive limitations or attention deficit disorder, which can make it difficult to focus on content. Authors can address this requirement by providing mechanisms to postpone or turn off interruptions, by allowing users to request updates, or by using scripts to make alerts optional. An exception is emergencies, including messages related to danger to health, safety, and property including data loss.

18.6.2.4 Success Criterion 2.2.5: Re-Authenticating

When an authenticated session expires, the user can continue the activity without loss of data after re-authenticating (Level AAA).

Similar to previous requirements related to timing, this requirement makes it possible for users who have experienced a time out to return to their activity at a later date without loss of data. For security reasons, meeting this requirement may not always be possible for some sites, but it is nonetheless an important consideration for authors when designing content. The benefits to users for allowing re-authentication extend far beyond those with disabilities. For example, mobile users, who may be more likely than most to experience interruptions in their Internet connections, often benefit when a site is designed in a way that allows them to resume their activities or transactions without having to spend time getting back to where they were and without data loss.

18.6.2.5 Design Impact for Guideline 2.2

For this guideline, the requirement related to animated text and graphics has the most significant impact on the visual design of the content. The other requirements do not have as great an impact on visual design but may require authors to have a more advanced understanding of scripting or server-side technologies in order to address them effectively.

18.6.2.6 Seizure Prevention

The next essential is the ability to use Web content without experiencing a seizure. Many individuals are sensitive to content that flashes more than three times in any one second period. This requirement is usually easy for authors to meet because most Web content does not cause the type of flashing that individuals with photosensitive epilepsy find provocative. However, as we move to higher bandwidth Web

connections, which allow increased use of streaming high-quality video content such as television shows and movies, seizure prevention is something that is increasingly important for authors to keep in mind. In one instance, an animated television cartoon sent hundreds of children to the hospital with seizures in Japan (Clippingdale and Isono 1999). To make things worse, the evening news coverage of this event rebroadcast the provocative footage, sending another wave of children and adults to hospitals. (It is also believed that many more individuals experienced seizures and didn't end up in the hospital.)

Guideline 2.3 Seizures: Do not design content in a way that is known to cause seizures.

18.6.2.7 Success Criterion 2.3.1: Three Flashes or Below Threshold

Web pages do not contain anything that flashes more than three times in any one second period, or the flash is below the general flash and red flash thresholds (Level A).

Note: Because any content that does not meet this success criterion can interfere with a user's ability to use the whole page, all content on the Web page (whether it is used to meet other success criteria or not) must meet this success criterion. See Conformance Requirement 5: Non-Interference.

18.6.2.8 Success Criterion 2.3.2: Three Flashes

Web pages do not contain anything that flashes more than three times in any one second period (Level AAA).

Past Web accessibility guidelines related to seizure prevention centered on a prohibition of certain frequencies. WCAG 2.0, however, includes a specific measure that can be tested with automated tools. While WCAG 2.0 was in development, the University of Wisconsin-Madison's Trace Center released a tool called the Photosensitive Epilepsy Analysis Tool (PEAT) (Trace Center 2007) that is adapted for use with computers and Web content. This tool is based upon the Harding Flash and Pattern Analyzer Engine developed by Cambridge Research Associates in conjunction with Graham Harding, an international leader in photosensitive epilepsy research. The Harding FPA Flash and Pattern Analyzer (Cambridge Research Systems 1995) is used in England and other countries to analyze video content. The Trace Center worked with Cambridge Research Associates and Graham Harding to recalibrate and adjust the algorithms so that they

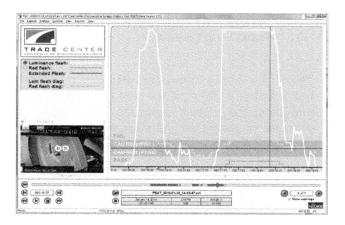

FIGURE 18.5 Screenshot of the Trace Photosensitive Epilepsy Analysis Tool (PEAT). The line chart illustrates results from a site that fails to meet the luminance flash threshold. Authors can use the tool to capture and analyze video content on their sites to determine which portions of their video content need repair.

could be applied to Web content. Figure 18.5 shows a screenshot of the PEAT software.

This tool takes into account not only the frequency but also the intensity and area of view. Thus, a small flashing area on the screen would not cause the page to fail even if it were flashing in frequencies that would have previously failed. The PEAT tool also separates types of flashing that are more provocative from those which are less so.

The existence of a tool for testing content is essential since complying with highly technical rules specific to the types of patterns, frequencies, and colors that have the potential to induce seizure would otherwise be extremely difficult for most Web authors.

18.6.2.9   Design Impact for Guideline 2.3

Clearly, changing the presentation to avoid doing something that is provocative for photosensitive epilepsy would require a change to the intended design. However, as noted above, it is often possible to make small changes to the sections of content that exceed the flash thresholds. For example, it may be possible to remove a frame (or small number of frames) and reduce the intensity of the transitions between dark and light without impacting the presentation in a way that most users would notice the change.

18.6.3   Facilitation of Orientation and Movement

The previous guidelines focus on ensuring that any structure in the document is available to all users. It is also useful to design documents in a way that ensures that there is adequate structure within them to facilitate orientation and navigation. In many cases, adding structure will have an impact on the design of content.

Guideline 2.4 Navigable: Provide ways to help users navigate, find content, and determine where they are.

18.6.3.1   Success Criterion 2.4.1: Bypass Blocks

A mechanism is available to bypass blocks of content that are repeated on multiple Web pages (Level A).

While sighted users can skip over blocks of repeated information and focus on the main content of a page, blocks of repeated text can mean that those individuals who access content in a linear fashion will find it difficult to quickly and efficiently reach the content that they are interested in. Similarly, individuals with physical disabilities who rely on the keyboard to navigate through a page may find themselves having to tab through a large number of links or invoke an in-page search on each page before they reach a link that is part of the main content of the page. Examples of repeated information include content such as navigation menus and layout regions that include advertising, repeated links, disclaimer text, etc.

Some of the techniques for addressing this criterion include

- Providing hierarchical structure to the document (e.g., sections, titles, subsections, etc.)
- Including a table of contents

There are also techniques that can help individuals who must navigate through documents by using the keyboard and do so more efficiently. These include

- Allowing users to skip over large blocks of repetitive material (e.g., navigation bars or other repetitive blocks of links)
- Providing controls to allow users to show and hide (or expand and collapse) sections of content

These strategies make it easier for individuals who are reading with a speech synthesizer as well as those who are navigating using the keyboard. When they encounter a block of links that they are already familiar with, they can hop over it and continue reading on the other side.

18.6.3.2   Success Criterion 2.4.2: Page Titled

Web pages have titles that describe topic or purpose (Level A).

Page titles are an important part of helping users orient themselves within page content. The use of descriptive titles gives users an idea of what content can be found on a page when following links or reviewing search engine results. The use of descriptive titles is also important for users who commonly have multiple browser windows or tabs open because it makes it possible to efficiently switch between them.

Web page titles should

- Identify the subject of the Web page
- Make sense when read out of context, for example by a screen reader or in a site map or list of search results
- Be short

Also, where possible, Web page titles should

- Identify the site or collection of pages to which the Web page belongs (e.g., Contact Us—Example Inc.)
- Be unique within the site or other resource to which the Web page belongs

18.6.3.3   Success Criterion 2.4.3: Focus Order

If a Web page can be navigated sequentially and the navigation sequences affect meaning or operation, focusable components receive focus in an order that preserves meaning and operability (Level A).

The focus order or navigation sequence of Web content may be defined in a variety of ways and will vary by technology. This success criterion is closely related to those for keyboard operation and meaningful sequence described previously in this chapter. For HTML, the focus order is generally based on the use of tab or arrow keys to move through content and the order in which items receive focus often mirrors the source order of the original content. However, technologies such as CSS can be used to modify the way in which focus is presented to users. Also, scripting can be used to change the focus order of content based on user actions in a way that is more similar to desktop applications than to what most people think of as Web pages. An example of a focus order problem would be specifying a tab order that separates a search box from the button used to initiate the search can be extremely confusing for keyboard users.

For keyboard users, the expectation is typically that focus order will follow the visual layout of the page. Users who use the tab key to move from one link or control to another generally expect that the order in which focus moves will parallel the presentation of content. When considering focus order, authors need to ensure that the order in which items receive focus will make sense to users who are looking at the page with their eyes as well as to those reading the page from the source code with a screen reader. The visual order and the source code order should match where possible.

It can also be helpful for authors to enhance the presentation of items that have focus so that users can more easily track changes in focus across content.

18.6.3.4   Success Criterion 2.4.4: Link Purpose (In Context)

The purpose of each link can be determined from the link text alone or from the link text together with its programmatically determined link context, except where the purpose of the link would be ambiguous to users in general (Level A).

18.6.3.5   Success Criterion 2.4.9: Link Purpose (Link Only)

A mechanism is available to allow the purpose of each link to be identified from link text alone, except where the purpose of the link would be ambiguous to users in general (Level AAA).

Links are a fundamental building block of Web content and are used in a multitude of ways throughout Web content. Both the content of a link and the way it is presented are key factors in whether or not users decide to follow that link. There are a number of situations where not having a clear picture about the purpose or destination of a link can be disadvantageous for a user. For example, it can be confusing if links are ambiguous or require a person to see the page as a whole in order to understand what the link will do when activated. Additionally, individuals with disabilities want to be confident about choices to follow links because they may require additional keystrokes to interact with each page or because they rely on complex (and sometimes time-consuming) strategies in order to understand each page.

Screen readers often include features that allow users to view all of the links on a given page as a list. For those who are already familiar with content on a page, using these features makes it possible to more quickly and accurately find the information they are looking for. However, when content includes link text that is not descriptive or that is identical to other link text found elsewhere on the page but has a different destination, this strategy is no longer effective.

Requirements related to link context are present at both Level AA and Level AAA in WCAG 2.0. At Level AA, the requirement is that the purpose of each link can be determined either from the link itself or from "programmatically determined link context" (such as text near a link that a screen reader user can easily listen to without leaving the link). At Level AAA, authors cannot rely on link context to meet the requirement, which means that it must be possible for users to determine the purpose of the link from link text alone. At both levels, an exception is provided for links that would be ambiguous to all users (i.e., readers without disabilities would not know what a link would do until they activated it).

18.6.3.6   Success Criterion 2.4.5: Multiple Ways

More than one way is available to locate a Web page within a set of Web pages except where the Web Page is the result of, or a step in, a process (Level AA).

Because the Web consists of pages connected in different ways to a variety of other pages, services, etc., it can be difficult for authors to predict the path(s) users may take to find a given page. Creating content in a way that makes it possible for it to be found via disparate links, tools and paths is an intentional part of the design of some content because it allows users to find content more quickly and according to their needs and preferred navigation strategies. For example, some users will prefer to start at the home page of a site and follow the navigation and links to find what they are looking for. Others will instead start with the search function. A number of techniques can be employed to address this requirement, including the inclusion of search features on a site, adding a table of contents or site map, and providing links to related pages.

18.6.3.7   Success Criterion 2.4.6: Headings and Labels

Headings and labels describe topic or purpose (Level AA).

For those who have trouble reading, who rely on structure to get an overview of content, or who have limited short-term memory, the use of descriptive headings and labels can help users learn more about what information is contained in a given page or section and how the content is organized. Providing clear and descriptive headings make it possible for all users to more quickly locate the information they are seeking, and descriptive labels help provide context and identification for components within content.

18.6.3.8   Success Criterion 2.4.7: Focus Visible

Any keyboard operable user interface has a mode of operation where the keyboard focus indicator is visible (Level AA).

In addition to focus order, the ability for users to clearly identify which component on a page has focus is an essential part of being able to effectively interact with content for some users. To address this requirement, authors need to ensure that at least one mode of operation is available where the keyboard focus can be visually located. Best practices here are for authors to provide highly visible focus indicators, which can reduce the likelihood that users will make mistakes by activating links or controls unintentionally (e.g., because they think a different control has focus). Enhancing the visibility of focus also helps those who are easily distracted, allowing

them to more quickly pick up where they left off, for example, when returning to filling out a multipart form after looking up information in another location.

18.6.3.9   Success Criterion 2.4.8: Location

Information about the user's location within a set of Web pages is available (Level AAA).

This AAA requirement focuses on the importance of providing information to users that helps them orient themselves within a site or set of pages. Because users may land on a page through a number of different paths (e.g., search, link from another site, by following a link in an e-mail, or tweet), it is often helpful to provide orientation clues that help users understand what it is they are interacting with. Strategies to address this success criterion include the use of site maps that are linked from each page on a site and breadcrumb navigation and navigation menus that indicate where within the navigation structure the page that the user is viewing fits.

18.6.3.10   Success Criterion 2.4.10: Section Headings

Section headings are used to organize the content (Level AAA).

Note 1: "Heading" is used in its general sense and includes titles and other ways to add a heading to different types of content.
Note 2: This success criterion covers sections within writing, not user interface components. User Interface components are covered under Success Criterion 4.1.2.

Although similar to requirements that headings be marked up using appropriate structure and that they appropriately describe the sections to which they relate, this requirement is about adding section headings and titles to content. When content is organized into sections, it becomes easier to interact with for those who rely on structure for navigation. It also makes it easier to understand for those with cognitive, language, and learning disabilities. By default, the addition of structure will introduce changes to the default presentation of the content, but authors can easily override default styles using technologies such as CSS.

18.6.3.11   Design Impact for Guideline 2.4

Similar to other guidelines, most of the requirements here can be met without significant changes to the intended design of the page. In many cases, the addition of structure and features that facilitate orientation make it much easier for everyone, including users without disabilities, to effectively interact with the content.

18.7   UNDERSTANDABLE

Probably the most controversial and difficult to codify area of accessibility is understandability. Although it is clearly evident that a page must be understandable in order to be usable, requiring that pages be made understandable by people with disabilities is a difficult requirement. This is especially true because cognitive disabilities are varied in both type and degree. While it is possible to make most Web content accessible to people without any vision and most audio-visual material accessible to someone without any hearing, it is not possible to make most Web content accessible to someone without any cognition. This area of disability is therefore qualitatively different.

However, there are many things that can be done to facilitate the accessibility of Web content to individuals with cognitive disabilities. This includes many techniques that do not affect the default presentation of the page. Furthermore, there are things that can be done to Web content to facilitate its understandability by individuals with visual, hearing, or language disabilities (not cognitive) as well.

A wide range of techniques have been suggested for making content simpler. Not all of them apply to all sites. Some of the items that should be considered when trying to make sites easier to understand include

- Familiarity of terms and language structure
- Length and complexity of sentences
- Restricting paragraphs to a single topic or subtopic
- Clarity of headings (and text links) when read out of context
- Accuracy and uniqueness of page titles
- Use of page design, graphics, color, fonts, animations, video, or audio to clarify complex text as necessary

18.7.1   Ability to Identify Language, the Meanings of Words, Abbreviations, and Acronyms

A fundamental step in facilitating understanding is having some mechanism for unambiguously decoding the meaning of ambiguous words, phrases, acronyms, abbreviations, and so on. Providing such a capability facilitates understanding of the material by any reader who is not familiar with all of the terms used within the content. Providing this information also facilitates translation of the material into other languages, more accurate reading and pronunciation of the material to individuals using voice synthesizers, and automatic translation of the materials into simpler language and forms.

Guideline 3.1 Readable: Make text content readable and understandable.

18.7.1.1   Success Criterion 3.1.1: Language of Page

The default human language of each Web page can be programmatically determined (Level A).

18.7.1.2   Success Criterion 3.1.2: Language of Parts

The human language of each passage or phrase in the content can be programmatically determined except for proper names, technical terms, words of indeterminate language, and words or phrases that have become part of the vernacular of the immediately surrounding text (Level AA).

These requirements are about providing information about the language(s) that are used within Web content. The first requirement is to include markup that cues assistive technologies and other user agents as to the primary language in use on a page. Because the use of foreign words and phrases on a page are examples of ambiguous text, the second requirement is about providing cues in markup as to the language represented by foreign words. Techniques to address this include

1. Specifying the default human language of the page
2. Identifying language changes for passages or phrases in the body of the content that are in a language other than the default language

Marking the language of the content is important not only to enable users to look up unknown words or phrases but also to facilitate the proper pronunciation of the words and phrases by voice synthesizers.

18.7.1.3   Success Criterion 3.1.3: Unusual Words

A mechanism is available for identifying specific definitions of words or phrases used in an unusual or restricted way, including idioms and jargon (Level AAA).

Providing a reference to the definition of words, phrases, acronyms, abbreviations, etc., is not well defined in many technologies. Such a mechanism would be of great value not only to individuals with disabilities but to all users. One potential mechanism to address this challenge would be to provide a set of cascading dictionaries that would be used with Web content. At the base would be some standard dictionary for the primary language (being used with the content). A tool could then compare all of the words in the content to this dictionary and highlight any words that were not found. It could also identify abbreviations or acronyms that may be ambiguous. On top of this dictionary could be a topic-specific dictionary dealing with the particular topic or topics covered by the particular site content. A final layer could be a dictionary that is maintained by the particular organization creating the Web content that includes the words or abbreviations that pertain to that organization or the words or abbreviations that they use commonly in their material. This dictionary could also be used to resolve ambiguity among abbreviations. For example, there may be multiple organizations with the same four initials; however, this dictionary would state that on this site, a specific set of four initials always has a particular meaning.

Again, a tool could be applied to the site that would use this cascading dictionary to identify any words that were not covered. These could then either be added specifically to that instance of the content or added to the general dictionary for the site. Tools using the cascading dictionary approach have not yet been developed. However, we propose that this approach could be an effective solution in this area.

This approach does not solve the problem of words like "read" that appear in the standard dictionary but have two different definitions and pronunciations. These instances may be left up to user agent tools to differentiate on the basis of an analysis of the grammar of the sentence containing the words and/or left up to the user to differentiate after calling up the definitions if there is ambiguity in the content (versus just ambiguity in the word by itself).

One instance where this type of ambiguity happens extensively is in Hebrew. In common usage, all of the vowels are left out of printed Hebrew. This makes the resulting words extremely ambiguous for users who are new to the language as well as for text to speech engines. However, reading the words in context, individuals who are reading Hebrew can determine the correct meaning of the words. A voice synthesizer would have great difficulty if it were trying to pronounce the words one at a time. Adding the vowels back in would solve the problem, but would require a significant change in the variation of content and essentially require that the content be presented in a form that is not common for the language.

For situations like this, specific user agent tools might be developed or deployed publicly that could carry out the disambiguation or that online tools be available that are linked to content sites and can automatically provide disambiguation.

18.7.1.4 Success Criterion 3.1.4: Abbreviations

A mechanism for identifying the expanded form or meaning of abbreviations is available (Level AAA).

Abbreviations and acronyms often present challenges for readers who are new or unfamiliar with a subject area. They also present difficulties for those who have cognitive, language, and learning disabilities. Providing a mechanism to display the expanded form of these abbreviations can greatly increase the user's ability to understand the content and helps users avoid situations where they have to frequently refer to secondary materials in order to understand the content. Strategies for addressing this include using features of markup languages that make it possible to associate supplementary text with abbreviations (e.g., the ABBR element and title attribute in HTML). Another approach would be to provide a link from the acronym or abbreviation to a glossary. An alternative approach would be to introduce the acronym in parenthetical form when the term is first used on the page. This would meet the requirement but may not be as effective because the expansions are not available in a machine-readable way.

18.7.1.5 Success Criterion 3.1.5: Reading Level

When text requires reading ability more advanced than the lower secondary education level after removal of proper names and titles, supplemental content, or a version that does not require reading ability more advanced than the lower secondary education level, is available (Level AAA).

For those who are unfamiliar with the language of the content they are using or who have a cognitive, language, or learning disability, the complexity of the content in use can be a significant barrier to understanding. In some cases (content that includes an in-depth discussion of string theory, for example), complexity will be difficult to avoid. However, for content that will be used by a wide range of users, writing content that does not require users to have advanced reading skills will increase understanding, decrease errors, and broaden the audience. The term "lower secondary education level" describes an education level roughly equivalent to 7–9 years of school. This provision can be met in a number of ways, including proving simplified text summaries, illustrations that help describe complex concepts, audio versions of the content, or simplified alternative versions of the content.

18.7.1.6 Success Criterion 3.1.6: Pronunciation

A mechanism is available for identifying specific pronunciation of words where meaning of the words, in context, is ambiguous without knowing the pronunciation (Level AAA).

In some languages, it is not possible for a voice synthesizer to correctly pronounce certain words where meaning in context is ambiguous without pronunciation information. Heteronyms, which are words that are spelled the same but have different pronunciations and meanings, such as the words lead (take charge) and lead (heavy metal). In other languages, characters and symbols can be pronounced in different ways. In Japanese, for example, there are characters like Han characters (Kanji) that have multiple pronunciations.

18.7.1.7 Design Impact for Guideline 3.1

All of the requirements discussed above could be implemented without changing the visual design or presentation of content. Rather than providing definitions of the words in visible text or as links to definitions, the location of definitions for the words could simply be referenced using metadata for the content. In fact, a machine-readable solution would be of more use to the users than the tradition of simply defining an abbreviation the first time it is used—although this is still encouraged. The machine-readable solution is superior in that any time the individual locates an acronym anywhere in the content, they would be able to have its definition electronically retrieved and presented.

18.7.2 Ensuring that Content Operates in Predictable Ways

Another consideration in making sites simpler directly is that you can only simplify the content so far before you begin to lose content as you invoke simpler and simpler language and presentation. As a result, to address the full range of cognitive, language, and learning disabilities, one would need to have a full range of different presentations in order to present the maximum amount that people at each level would be able to comprehend. Fortunately, we are approaching the point where user agents and/or server-based tools will be able to take many types of Web content and convert it to different language levels. Including provisions such as those described above and allowing for unambiguous machine translation will make such conversions possible.

In many cases, Web sites are more complex than they need to be. Authoring these sites in a more straightforward and easy-to-understand fashion would not decrease the content in them and would make them more usable by all users.

Using a consistent format and layout for the presentation of information facilitates accessibility for a number of different disabilities. People with low vision who look at a page one small area at a time using magnification (kind of like viewing a page through a soda straw or narrow tube) find it much easier to locate information, navigation bars, etc. if they are in a predictable location. Using consistent and predictable layout also facilitates navigation and use by people whose screen readers present information on the screen one word at a time. For individuals with cognitive disabilities, consistency makes it easier to understand where the information will be.

Although consistency is important, it is also important that the layout and format not be identical. If all houses looked exactly the same, most drivers would be lost most of the time and it would be a tremendous mental exercise to try to drive around and locate even often-visited homes and buildings. Similarly, if all pages look identical, it becomes difficult to tell when one has even left one page and moved to another. Also, one can often think that information should be on this page when in fact you found it last on a different page that looked just like it. The use of titles, images, and other visual and textual cues to help people maintain orientation is very important. Here again, the use of unique and easily understood page titles is very helpful. The guidelines for this area take the form similar to "make layout and behavior of content consistent and/or predictable."

Again, there are places where inconsistency or lack of predictability is essential to the function of the content (for example, mystery games, adventure games, tests, etc.). In these cases, the user might be notified in advance that they are entering a game or test in situations where it would not otherwise be obvious.

Guideline 3.2 Predictable: Make Web pages appear and operate in predictable ways.

18.7.2.1 Success Criterion 3.2.1: On Focus

When any component receives focus, it does not initiate a change of context (Level A).

18.7.2.2 Success Criterion 3.2.5: Change on Request

Changes of context are initiated only by user request or a mechanism is available to turn off such changes (Level AAA).

As defined in WCAG 2.0, changes of context are events that occur within Web content that are likely to disorient users who cannot view an entire page at the same time. Changes of user agent, focus, viewport (window or frame), and changes that alter the meaning of the overall content of a page are all examples of changes of context. When such changes occur without user awareness, it can be disorienting to introduce content the user is unaware has been added or changed. Authors can meet this requirement by avoiding situations where focus or changes to the settings of user interface components such as form controls trigger changes of context.

A common example of a change in context is when an individual is unexpectedly transported into a new window when they did not click on a link or do something else that usually takes them to a new page. For individuals using a screen reader or people with low vision, it may not be obvious to them that they have suddenly opened up a new window. If a new window does not have the standard navigation controls (back, forward, reload, etc.) at the top, it is even more disorienting because controls that are typically found at the top of their browsing windows are no longer present. This can make a person using a screen reader believe that their software is not functioning correctly because they can no longer find the controls at the top of the window. The individual with low vision may think that the controls are off the screen, and they are not able to determine how to get them back.

When a change of context exists, authors can either

1. Provide an easy to find setting to turn off such processes or features that cause changes in context.
2. Identify the change in context before it occurs so the user can determine whether they wish to proceed or so they can be prepared for the change.
3. Design content such that changes in content only occur on user request.

At Level A, success criteria is about not causing changes of context to occur when content receives focus or when changes to settings of user interface components occur. At Level AAA, designing content so that changes in context only occur on user request (e.g., when the user clicks a link, activates a submit button on a form) is required.

18.7.2.3 Success Criterion 3.2.3: Consistent Navigation

Navigational mechanisms that are repeated on multiple Web pages within a set of Web pages occur in the same relative order each time they are repeated, unless a change is initiated by the user (Level AA).

This Level AA requirement encourages the use of consistent layout and presentation for content that users will interact with in multiple places. Presenting repeated content in the same order each time a user interacts with it allows more effective keyboard interactions. For example, a frequent user of a shopping site may get in the habit of hitting the tab key three times on a navigation menu to get to items in "electronics." If the navigation menu does not have this same characteristic across pages, the user's ability to interact with the content efficiently is reduced.

18.7.2.4 Success Criterion 3.3.4: Consistent Identification

Components that have the same functionality within a set of Web pages are identified consistently (Level AA).

Much like consistent navigation, the labels for form components and other controls on a page are another important part of understandability. When components have the same functionality within a set of Web pages (i.e., the same results occur when they are used), they should be identified in the same way. This consistency applies to both text labels and links as well as to text alternatives that describe nontext interactive content that has functionality.

18.7.2.5 Design Impact for Guideline 3.2

Almost everything in this guideline would have an impact on design if it were done only to increase accessibility. However, the requirements in this section can also make the content more understandable and usable for all users.

18.7.3 Preventing Mistakes and Facilitating Error Recovery

Keys to any good interface design are provisions to help prevent users from making mistakes and to facilitate their efforts to recover. Facilitating these processes for people with disabilities involves the same strategies as for everyone else. However, particular emphasis has to be placed on ensuring that the warnings are obvious enough that they can be easily detected along with any indications that an error has occurred. Also, it can be difficult for some individuals to scan an entire page looking for text that is, for example, red. Including indicators at the top of the page that describe what the problem is and how to solve it are extremely useful.

Guideline 3.3 Input Assistance: Help users avoid and correct mistakes.

18.7.3.1 Success Criterion 3.3.1: Error Identification

If an input error is automatically detected, the item that is in error is identified and the error is described to the user in text (Level A).

Making it clear to users that an error has occurred is an essential component in understanding and interacting with content. For example, if a user does not realize that an error has occurred when submitting a form, she may abandon a page altogether, assuming that the page is not functional. Strategies for addressing requirements for error identification include the use of client-side validation and alerts as well as server-side validation and the use of text descriptions that clearly identify errors.

18.7.3.2 Success Criterion 3.3.2: Labels or Instructions

Labels or instructions are provided when content requires user input (Level A).

When content requires user action, it should be clear what the expectations for the user are, especially if the author is expecting that input will be provided in specific ways. In many cases, providing a descriptive label for a form field or control will address this requirement. Other strategies include the placement of instructions at the beginning of a page or (if necessary) providing supplementary information that describes the input in greater detail.

18.7.3.3   Success Criterion 3.3.3: Error Suggestion

If an input error is automatically detected and suggestions for correction are known, then the suggestions are provided to the user, unless it would jeopardize the security or purpose of the content (Level AA).

When input errors are automatically detected and suggestions for correction are known, authors should include features that make the suggestions available to users. The most common examples of this include functions that recognize errors in formatting and make suggestions that help users understand where their mistake was and providing suggested correction text. A common example of this can be found on the Google search results page, where users are prompted with "Did you mean . . ." questions when a search keyword is spelled incorrectly.

18.7.3.4   Success Criterion 3.3.4: Error Prevention (Legal, Financial, and Data)

For Web pages that cause legal commitments or financial transactions for the user to occur, that modify or delete user-controllable data in data storage systems, or that submit user test responses, at least one of the following is true (Level AA):

1. *Reversible*: Submissions are reversible.
2. *Checked*: Data entered by the user are checked for input errors and the user is provided an opportunity to correct them.
3. *Confirmed*: A mechanism is available for reviewing, confirming, and correcting information before.

18.7.3.5   Success Criterion 3.3.6: Error Prevention (All)

For Web pages that require the user to submit information, at least one of the following is true (Level AAA):

1. *Checked*: Data entered by the user is checked for input errors and the user is provided an opportunity to correct them finalizing the submission.
2. *Confirmed*: A mechanism is available for reviewing, confirming, and correcting information before.
3. *Reversible*: Submissions are reversible.

At Level AA, for content that causes data to be deleted or legal commitments or financial transactions to occur, the consequences to the user are much more significant than they are for other types of content. Where any of these situations occur, the requirement is that submissions are either (1) reversible, (2) checked for input errors in a way that allows users to correct submissions before the submission is made, or (3) confirmed by the user after review. At Level AAA, the error prevention techniques are required for all types of content that requires users to submit information.

18.7.3.6   Success Criterion 3.3.5: Help

Context-sensitive help is available (Level AAA).

The use of context-sensitive help can be a powerful way to help users understand content and use certain features effectively. Ajax and other scripting techniques make it possible for authors to provide help to users in a variety of ways. For example, a search field that provides a list of commonly submitted phrases as the user types, the addition of help links on pages, spell checking, and even interactive agents that help guide users through a process are all ways to provide context-sensitive help and meet this requirement.

18.7.3.7   Design Impact for Guideline 3.3

Many of the requirements in this guideline will not significantly impact the design and presentation of a page. However, some of the requirements, especially at Level AA and Level AAA present technical challenges that some authors may not be able to easily overcome without advanced knowledge of scripting or server-side technologies.

18.8   ROBUST

In addition to being perceivable, operable, and understandable, it is also important with the rapid evolution of technology to think about the accessibility (perceivability, operability, understandability) of a page over time. That is, are the technologies that authors rely on to provide accessibility going to continue to work with successive new versions of browsers or new versions of a technology? For the most part, user agents are becoming smarter and are providing the user with more capabilities. Thus, content has a tendency to become more accessible over time. However, as new browser and other user agent technologies are introduced or are updated, they often break the accessibility that was present with the previous version. The overall goal here is to use Web technologies that maximize the ability of content to work with current and future accessibility technologies and user agents.

18.8.1 Ensuring that Content Can Be Used with User Agents and Assistive Technologies

This is important because most assistive technologies and user agents make assumptions about how technologies are used. If technologies are used in unconventional ways, they will confuse or defeat strategies employed by assistive technologies to present content to people with disabilities. For example, in HTML if headers are created visually by simply creating large bold text rather than using a header element, they will not be recognized as headers by assistive technologies or user agents and cannot, therefore, be used as cues to the structure and organization of the page.

Other "creative" uses of technologies can cause similar problems, which leads to the following guideline:

Guideline 4.1 Compatible: Maximize compatibility with current and future user agents, including assistive technologies.

18.8.1.1 Success Criterion 4.1.1: Parsing

In content implemented using markup languages, elements have complete start and end tags, elements are nested according to their specifications, elements do not contain duplicate attributes, and any IDs are unique, except where the specifications allow these features (Level A).

Note: Start and end tags that are missing a critical character in their formation, such as a closing angle bracket or a mismatched attribute value quotation mark are not complete.

Perhaps one of the most significant challenges that browser and assistive technology manufacturers face is interpreting Web content that has been designed by a wide variety of authoring tools and by authors with an even wider range of skill levels. Different tools, authoring styles, and levels of conformance with Web standards mean that these tools need to be able to adapt to a variety of situations in order to effectively present content to users. In many cases, they need to include error correction and recovery algorithms in order to present content when it is not properly formed or cannot be parsed correctly. Historically, best practice advice from an accessibility perspective has been to recommend that authors validate their content by running conformance testing or quality

control tools against their content. While this is sound advice, such a requirement extends somewhat beyond requirements for accessibility because a number of validity errors that are commonly found on the Web today do not result in a problem for individuals with disabilities. Therefore, WCAG 2.0 specifies that it must be possible to parse content that is created using markup languages. This basically means that the markup contains complete start and end tags, that elements and attributes are nested according to specification, that attributes are not duplicated and that any IDs within content are not duplicated.

18.8.1.2 Success Criterion 4.1.2: Name, Role, and Value

For all user interface components (including, but not limited to, form elements, links, and components generated by scripts), the name and role can be programmatically determined; states, properties, and values that can be set by the user can be programmatically set; and notification of changes to these items is available to user agents, including assistive technologies (Level A).

Note: This success criterion is primarily for Web authors who develop or script their own user interface components. For example, standard HTML controls already meet this success criterion when used according to specification.

Another key issue arises around content that creates its own interface elements. If the content relies on the controls and interface elements presented by user agents that are presenting technologies used according to specification, then assistive technologies can be designed to work with these common user agents. However, when interface elements are created directly in the content, then the assistive technologies may not be able to provide access to them. There are guidelines on how to make accessible user interfaces. The User Agent Accessibility Guidelines 1.0 from the W3C (Jacobs, Gunderson, and Hansen 2002), for example provide specific guidelines to user agents. These could also be used by authors creating interfaces within content to ensure that their content interfaces are accessible.

18.8.1.3 Design Impact for Guideline 4.1

Requirements for this guideline do not significantly impact design or presentation. That is, they are focused more on the technologies used than on the specific way the information is presented. To a certain extent, however, the technologies you use may affect the options you have in presenting your information.

18.9 THE CHANGING FACE OF ACCESSIBILITY ON THE WEB

In comparing the approach to accessibility guidelines reflected in this chapter to the approach taken in the early

Web content accessibility guidelines (including WCAG 1.0), one can see that the same topics are covered except that they are addressed in a more general fashion. That is, rather than restricting the guidelines to specific technologies (such as HTML and CSS), the overall objectives are captured in the WCAG 2.0 Guidelines document and separate techniques documents are used to explain how to implement them within specific technologies. The goal was to describe the characteristics of accessible content in a way that was independent of the technologies in use on the Web today This is important both for addressing new technologies that are being introduced for Web content but also to address new strategies for accessibility that are emerging. Two strategies that are on the horizon include virtual assistive technologies and server-based assistive technologies.

18.9.1 Virtual Assistive Technologies and Services

There are a number of different approaches to providing access features for Web content. Most commonly, user agents (browser) and assistive technologies that are installed on a user's computer are used to directly interact with Web content. Other approaches that have been used include the use of proxy services that stand between a browser and the content on the Web. When an individual asks for content, the request goes through the proxy server to the Web site. The Web page is then sent back through the proxy server, which then passes it back to the user. As content passes through the proxy server, the proxy server can process the page in ways to make the page more accessible. A number of different researchers have explored this approach, with the most extensive work to date being done by IBM (Brown et al. 2001). With this approach, a page that would not otherwise be accessible to simple browsers and assistive technologies can be converted into a page that is more accessible to simple browsers and assistive technologies. For example, if the information is being viewed by somebody who is completely blind, the images could be removed and the alternate text substituted. Lists of links could have a header added, giving the number of links in the list. Text that is presented as an image could be automatically translated using OCR into electronic text, etc. It could also take a document such as an image-based PDF, which was not directly accessible, and convert into more accessible HTML formats. Current examples of this approach include WebAnywhere (Bigham and Ladner 2008) and System Access To Go (Serotek 2009).

This strategy could be used by users and suppliers of information. For example, a company with a massive archive of older documents may not have the resources to convert each of the documents into an accessible form. They may even question the need to do so if these archived documents are rarely accessed. However, if the archive is available to the public in general, they might use a special transcoding server to process any documents that are requested to ensure that the accessible form of the document was available. This transformation could be done completely automatically or it could include an on demand clean up by a human, if required.

As we develop more and more intelligent algorithms, some of the load that has been taken up by humans may be able to be transferred to such transcoding servers. This type of server might also be used to reduce the sophistication of the AT that the user must have to access content. That is, content that is accessible with very sophisticated assistive technology (but not simpler AT) could be passed through a server which would pre-process the document carrying out some of the operations normally done by the more sophisticated AT. The result could then be sent on to a user who had a simpler AT in a form which only required the simpler AT functionality. This can be an important advance for individuals who are not in employment situations or not otherwise able to afford the more sophisticated AT.

Another model is the use of Web services that could provide transformations of Web content on demand. Such services could be called automatically from content. For example, they may translate images of text into accessible text, shift colors to accommodate a specific type of colorblindness, or convert text into audio. Such services could be invoked automatically or at the user's discretion.

Another advance that has been proposed (Vanderheiden 2002) is virtual assistive technology. This is a special form of server-based assistance where the individual's browser would act as a very thin client. The actual browser and assistive technology functionality would take place on the server. For example, an individual who is blind might walk up to any computer in any location that had a simple browser on it. Logging into a Web site by rote (or asking someone to log them into the special Web site), they would connect to necessary assistive technologies on a remote server. The cloud-based assistive technology server would then send audio back to them that would be played to them by the standard computer and browser in front of them. They would then use the audio browser on the server to log into other Web sites and browse the Web. While browsing the Web, they would only be using the keyboard and speakers on the computer in front of them. The rest of the browser on the computer in front of them would simply be used as a mechanism to connect their speaker and keyboard to the talking browser on the remote server. The result would be that the individual who is blind would be able to browse the Web using keyboard and voice output from any standard browser without requiring that any assistive technology, screen reader, or otherwise be installed on the computer in front of them. Virtual assistive technologies could occur completely on the server or could be distributed between the server and the browser using JavaScript or another scripting technology within the browser.

This has particular implications for developing countries, as well as areas in developed countries that are not well served. An individual who is blind (or had other addressable disabilities) would not have to own any assistive technology or even own a computer. If their able-bodied peers had access to a computer in school or in the community, they would be able to access and use that same computer without requiring any modification or adaptation to it because they could invoke access technologies "from the cloud" (see Figure 18.6).

Multiple approaches

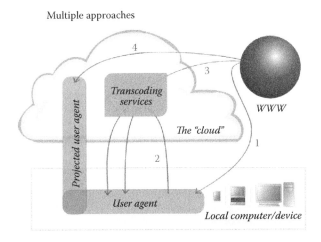

FIGURE 18.6 Multiple approaches for providing assistive technology features and functionality to users by using cloud-based services.

18.9.2 BEYOND CONTENT ACCESSIBILITY

Not all of the factors that contribute to the accessibility and usability of Web content can be addressed by guidelines alone. Despite the improvements that have been made to guidelines and content, the reality for many individuals with disabilities is that the tools they need in order to effectively interact with the Web are often not available, not affordable, and not keeping up with changes in technology in a way that meets their needs (National Task Force on Technology and Disability 2004; Vanderheiden et al. 2010; see Figure 18.7).

By describing the characteristics and behaviors of accessible Web content, WCAG 2.0 provides a useful set of requirements from which to work. Because it describes outcomes rather than specific tasks per technology, the effort required to meet the requirements can self-adjust over time based on the quality of the tools that are available to users. In other words, as browsers and assistive technologies improve, it will be easier for authors to meet the requirements. Therefore, the key to making WCAG 2.0 easier to meet is to make sure that users have access to the tools that they need to effectively interact with Web content.

A number of efforts have been launched to address these barriers:

- *Raising the Floor (RtF)* is an international coalition of individuals and organizations working to ensure that the Internet is accessible to people experiencing accessibility or literacy problems, even if they have very limited or no financial resources. The goal is to ensure that individuals who need special interfaces are as free and able as everyone else to use the new and emerging Internet tools and resources to live more independently and productively and to achieve their life's goals. More information can be found at http://raisingthefloor.net.
- *National Public Inclusive Infrastructure (NPII)* is a coalition of academic, mainstream, and assistive

Barriers to Robust and Affordable Access

Six barriers to effective and affordable access by people facing obstacles due to disabilities, aging, or low literacy are:

1. **Assistive technology costs:** Currently, because of the large development costs, the small market, and the need for individualization, many of the robust assistive technologies capable of handling modern Internet content are prohibitively expensive, even for many individuals who can afford computers and other technology devices. As a result, employers, educators, government agencies, businesses, and individuals face prohibitively high costs for accommodations that must be continually updated to provide access to mainstream technologies and services that are themselves continually changing.
2. **Limited selection of assistive technology and accessibility/usability features** that is unable cover the full range of needs, including people with vision, hearing, motor, language, literacy, and intellectual disabilities—and all combinations of them. Currently, assistive technology focuses on people with specific "disabilities" (e.g., a person who has vision disabilities but does not have intellectual disabilities) without addressing people with other impairments or multiple impairments. Solutions for those with deaf-blindness, for example, are few and expensive, and solutions for people with cognitive, language, and learning disabilities do not begin to cover the diversity of this group.
3. **Lack of awareness of assistive technology and accessibility/usability features:** Because of small markets and limited marketing capabilities of individual AT companies, awareness of the existence of access features by those who need them is limited. Even features that are built into operating systems and available to everyone without cost are mostly unknown to those who need them.
4. **Lack of tools and support** for the development of assistive technologies and extended-usability features, sufficient to ensure that developers can keep up with mainstream technology, continue to improve solutions, keep costs down, and still make a profit.
5. **Lack of cross-platform operability and portability of assistive technology and accessibility/usability** features. Computing is becoming ubiquitous. Computers are already beginning to be built into the environment around us (through cell phones, kiosks, information and transactions machines, and other devices), and this will continue and accelerate. We need ubiquitous accessibility/usability to match, or computing will evolve away from our accessibility solutions.
6. **Insufficient or weak free, public access features:** Because many individuals with disabilities, elderly people, and people with low literacy fall into the lowest income categories, robust commercial assistive technology is often not affordable for them. Many cannot even afford a computer or Internet connection and rely on public access to the Internet, often through libraries, public access points, government programs, or by using the computers of those around them. For these groups a (free) public, basic level of accessibility/usability is needed to use these (free) public Internet access points. Accessibility that is good enough to not be stymied by the new technologies being introduced continually on the Internet.

FIGURE 18.7 Six barriers to robust and affordable access. (From Vanderheiden 2009. With permission.)

technology companies, nongovernmental organizations, and individuals coming together to promote the creation of a National Public Inclusive Infrastructure (NPII). The purpose of the NPII is to build access features into the broadband infrastructure in order to (1) make it easy for individuals to learn about accessibility solutions and find out which solutions work best, (2) make it easy to invoke the access solutions they need and to use them on any computer, and (3) decrease the costs to develop, disseminate, and obtain access technologies and services so that they are affordable by all who need them. More information can be found at http://npii.org.

- *The ÆGIS project* seeks to determine whether third generation access techniques will provide a more accessible, more exploitable, and deeply embeddable approach in mainstream ICT (desktop, rich Internet, and mobile applications). More information can be found at http://www.aegis-project.eu/.

18.10 CONCLUSION

About a fifth of the population currently has some type of disability. In addition, two thirds of us who live beyond age 75 will acquire disabilities, the most common of which are vision, hearing, and physical (although Alzheimer's, decreasing memory, and the ability to deal with new situations also are common). Almost all of us will know someone or personally acquire disabilities at some point in our lives. It is therefore not only egalitarian but also in our own self-interest to ensure that the Web-based systems we are building will work well for all users both today and into the future. In the end, Web content that is more usable by individuals who have disabilities is also more usable by individuals who are interacting with Web content under constrained conditions. We must continue to work to overcome the barriers individuals with disabilities encounter as their reliance on the Web for information, social interaction, commerce, education, and employment continues to grow.

ACKNOWLEDGMENTS

Many people have contributed to the development of WCAG 2.0 over the years and without their hard work and dedication, this work would not have been possible. For a full listing of Working Group participants and contributors, see the WCAG 2.0 appendix. Special thanks to David MacDonald for his input and contributions in reviewing this chapter as it neared completion.

The contents of this chapter were developed with funding from the National Institute on Disability and Rehabilitation Research, U.S. Department of Education, grant number H133E080022. However, the contents do not necessarily represent the policy of the Department of Education, and endorsement by the federal government should not be assumed.

REFERENCES

Arditi, A., G. C. Vanderheiden, and A. R. Gordon. 2005. Proposal for an accessible color contrast standard for text on the World Wide Web. *Investigative Ophthalmology & Visual Science* 46, suppl., 2331.

Bigham, J., and R. Ladner. 2008. Webanywhere: A screen reader on the go. Computer software. University of Washington, Seattle. http://webanywhere.cs.washington.edu/.

Brown, F., S. Crayne, S. Detweiler, P. Fairweather, V. Hanson, J. Richards, R. Schwerdtfeger, and B. Tibbitts. 2001. Web accessibility for seniors. In *Universal Acccess in HCI*, vol. 3, *Towards an Information Society for All*, ed. C. Stephanidis, 663–666. Mahwah, NJ: Lawrence Erlbaum.

Caldwell, B., M. Cooper, L. Guarino Reid, and G. Vanderheiden, eds. 2008a. Web content accessibility guidelines 2.0. World Wide Web Consortium (W3C). http://www.w3.org/TR/WCAG20/ (accessed March 1, 2010).

Caldwell, B., M. Cooper, L. Guarino Reid, and G. Vanderheiden. 2008b. How to meet WCAG 2.0. W3C. http://www.w3.org/WAI/WCAG20/quickref/ (accessed March 1, 2010).

Caldwell, B., M. Cooper, L. Guarino Reid, and G. Vanderheiden. 2008c. Understanding WCAG 2.0: A guide to understanding and implementing web content accessibility guidelines 2.0. W3C. http://www.w3.org/TR/UNDERSTANDING-WCAG20/ (accessed March 1, 2010).

Caldwell, B., M. Cooper, L. Guarino Reid, and G. Vanderheiden. 2008d. Techniques for WCAG 2.0. W3C. http://www.w3.org/TR/WCAG20-TECHS/ (accessed March 1, 2010).

Cambridge Research Systems. 1995. Harding FPA Flash and Pattern Analyser. Computer software. Cambridge Research Systems Ltd., Kent, England. http://www.hardingfpa.co.uk.

Chisholm, W., G. Vanderheiden, and I. Jacobs. 1999. Web content accessibility guidelines 1.0. W3C. http://www.w3.org/TR/WCAG10/ (accessed March 1, 2010).

Clippingdale, S., and H. Isono. 1999. *Photosensitivity, Broadcast Guidelines and Video Monitoring*, II-22 to II-27. New York: IEEE Press.

Hudson, R., R. Weakley, and P. Firminger. 2005. Developing sites for users with cognitive disabilities and learning difficulties. Juicy Studio. http://juicystudio.com/article/cognitive-impairment.php (accessed March 1, 2010).

Guarino Reid, L., and A. Snow-Weaver. 2009. WCAG 2.0 for designers: beyond screen readers and captions. In *Universal Access in HCI*, Part III, vol. 516, 674–682. New York: Springer-Verlag.

Jacobs, I., J. Gunderson, and E. Hansen. 2002. User agent accessibility guidelines 1.0. W3C. http://www.w3.org/TR/UAAG10/ (accessed Feb. 9, 2004).

Juicy Studio. 2007. Luminosity colour contrast ratio analyser. http://juicystudio.com/services/luminositycontrastratio.php (accessed March 1, 2010).

Klusch, M., E. Andre, T. Rist, T. Ishida, and H. Nakanishi. 2005. Interactive information agents and interfaces. In *Handbook of Human Factors in Web Design*. Mahwah, NJ: Lawrence Erlbaum.

MacDonald, D. 2008. WCAG 2.0 audio contrast tool for help for success criteria 1.4.7. Ottawa, Ontario, Canada: Eramp Inc. http://www.eramp.com/WCAG_2_audio_contrast_tool_help.htm (accessed March 1, 2010).

National Task Force on Technology and Disability. 2004. Within our reach: Findings and recommendations of the national task force on technology and disability. http://www.ntftd.org/report.htm (accessed March 1, 2010).

Serotek. 2009. System Access To Go. Computer software. Minneapolis, MN. http://www.satogo.com/.

Trace Center. 2007. Photosensitive Epilepsy Analysis Software (PEAT). Computer software. Trace Center, Madison, WI. http://trace.wisc.edu/peat/.

Vanderheiden, G. 1995. Design of HTML (Mosaic) pages to increase their accessibility to users with disabilities strategies for today and tomorrow. Trace R&D Center, University of Wisconsin-Madison. http://trace.wisc.edu/archive/html_ guidelines/version1.html.

Vanderheiden, G. 2001. Fundamentals and priorities for design of information and telecommunication technologies. In *Universal Design Handbook*, eds. W. F. E. Preiser and E. Ostroff, 65.3–15. New York, NY: McGraw Hill.

Vanderheiden, G. 2002. Interaction for diverse users. In (Eds.) *Human–Computer Interaction Handbook*, eds. J. Jacko and A. Sears, 397–400. Mahwah, NJ: Lawrence Erlbaum.

Vanderheiden, G. 2009. National Public Inclusive Infrastructure: Building accessibility and extended usability directly into the Internet. http://npii.org/WhitePaper.html.

Vanderheiden, G., W. Chisholm, N. Ewers, and S. Dunphy. 1997. Unified Web site accessibility guidelines 7.2. Trace R&D Center, University of Wisconsin-Madison. http://trace.wisc .edu/archive/html_guidelines/version7.htm.

Vanderheiden, G., J. Fruchterman, L. Goldberg, D. Hatfield, E. Hill, K. P. Strauss, and J. Tobias. 2010. Before the Federal Communications Commission, supplemental comments in the matter of broadband accessibility for people with disabilities—Workshop II: Barriers, opportunities, and policy recommendations, GN Docket 09-47, 09-51, 09-137. http://fjallfoss .fcc.gov/ecfs/document/view?id=7020355648.

WAT-C and The Paciello Group. 2008. Contrast Analyser. Computer software. http://www.paciellogroup.com/resources/ contrast-analyser.html.

19 International Standards for Accessibility in Web Design and the Technical Challenges in Meeting Them

Lisa Pappas, Linda Roberts, and Richard Hodgkinson

CONTENTS

19.1 INTRODUCTION TO STANDARDS-CONFORMING ACCESSIBLE WEB DESIGN

As designers of Web entities develop graphical user interfaces and interaction workflows, they remain keenly aware of their end users, their goal-directed actions, and tasks they need to complete. Yet beneath the surface needs of users lie an equally critical set of requirements to meet the operational needs of functionally diverse people. An interface workflow must accommodate a range of input devices. For example, a mouse- or touch-screen driven interface might not be usable by a person with carpal tunnel syndrome or with arthritis in her hands. Rich multimedia without text transcripts or captioning has little value for someone with hearing loss.

Beyond meeting the functional needs of diverse users, however, successful Web designs must consider the context of the information provider, often the designers' employer or client. In an increasing number of regions, the capacity for Web interfaces to be used by people with disabilities is becoming a legal requirement or a prerequisite for procurement. Because the incidence of a physical disability increases with age, Web accessibility is increasingly important for sites with a significant portion of older users; examples include sites offering financial, health, or leisure services for retirees. On the basis of the target audience, the users' demographics, and the relevant policy drivers, Web designers must understand the legal and technical requirements for accessible Web development.

In this chapter, we introduce legislation and policies that affect Web design in major global markets. We identify the relevant standards and guidelines, address some of the challenges they present, and offer a strategy to conform to myriad guidelines. With that understanding of the framework, we next address the technical challenges of accessible design. Expanding user-centered design methodology, we address ways to include diverse functional needs. Finally, to demonstrate conformance to accessibility standards, an organization must validate the accessibility. We explore methods to measure conformance and discuss specific implications for usability testing with participants with disabilities.

19.2 POLICY AWARENESS: CHALLENGE OF MYRIAD GUIDELINES AND STANDARDS

Early accessibility guidelines were either region or technology specific. In 1998, the U.S. Section 508 Guidelines were the first attempt to tie accessibility compliance of software and technology products to public procurements; see Section 19.2.1. A year later, the World Wide Web Consortium (W3C) recommended guidelines for accessible Web content only (not desktop software), see Section 19.2.2 for details. Although both of these efforts were highly influential, global companies needed a unified set of standards for accessibility.

To support the need to address accessibility in international standards, in 2001 the International Organisation for Standardisation (ISO) and the International Electrotechnical Commission (IEC) published "ISO/IEC Guide 71—Guidelines for standards developers to address the needs of older persons and persons with disabilities" (hereinafter Guide 71). For European standardization activities, the European Committee for Standardisation (CEN) and the European Committee for Electrotechnical Standardisation (CENELEC) adopted this standard as CEN/CENELEC "Guide 6—Guidelines for standards developers to address the needs of older persons and persons with disabilities" (CEN/CENELEC 2002). A Japanese national standard version is also available (Japanese Standards Association 2004). Guide 71 is further supported by the publication of "ISO 22411:2008—Ergonomics data and guidelines for the application of ISO/IEC Guide 71 to products and services to address the needs of older persons and persons with disabilities" (ISO 2008c).

In recent years, international, regional (e.g., European), and national standards organizations have published a number of standards that address the accessibility of information and communication technology (ICT). On a national basis, Japan and Spain have been in the vanguard of producing ICT standards that address the accessibility of hardware, software, and Web content, including the provision of subtitles and sign language of multimedia. As European and international standards are published, Spain's practice has been to withdraw these standards, and many of the Japanese standards are being adopted into the international arena. While no conflicts exist between all of these standards, gaps and overlaps do occur. Recognizing the compliance challenges that such discrepancies have created for global companies, in 2004, the Joint Technical Committee 1 of ISO and the IEC established a Special Working Group on Accessibility (ISO/IEC JTC 1 SWG-A) to

- Determine and implement an approach for gathering user requirements.

- Gather and publish an inventory of all known accessibility standards efforts.
- Identify technological aspects where voluntary standards are not being addressed and suggest an appropriate body to consider the new work.
- Track public laws, policies, measures, and guidelines to ensure that the appropriate standards are available.
- Encourage the use of globally relevant voluntary standards through wide dissemination of the materials.
- Assist other organizations, if desired, in submitting their specifications to the formal standards process.

This activity resulted in the 2009 publication of "Accessibility considerations for people with disabilities," comprising:

- Part 1: User needs summary
- Part 2: Standards inventory
- Part 3: Guidance on user needs mapping

In parallel to this activity, the World Wide Web Consortium (W3C) and U.S. government produced complementary works. The W3C developed a suite of *de facto* Web accessibility standards (see Section 19.2.2: W3C Web Content Accessibility). The U.S. Congress, with Section 508 of the Rehabilitation Act, established a set of requirements for ICT that potential suppliers have to meet in order to be eligible for U.S. federal government contracts (see Section 19.2.1: U.S. Section 508).

Europe has closely followed the U.S. situation and, in 2006, issued "Mandate 376—European Accessibility Requirements for Public Procurement of Products and Services in the ICT Domain" to the three European standards organizations— CEN, CENELEC, and European Telecommunications Standardisation Institute (ETSI). Phase 1 of Mandate 376 initiated two reports on the current state of ICT accessibility procurement: implementation and conformity assessment. Phase 1 is now complete and the reports are available:

- ETSI TR 102 612—Human Factors (HF); European accessibility requirements for public procurement of products and services in the ICT domain (European Commission Mandate M 376, Phase 1) on the ETSI Web site (ETSI 2009).
- CEN/BT WG 185 Project Team Final Report for Approval—European accessibility requirements for public procurement of products and services in the ICT domain (European Commission Mandate M 376, Phase 1) on the CEN Conformity Web site (CEN 2008).

Phase 2 of Mandate 376 was expected to begin in late 2009 and to include the creation of a comprehensive European accessibility standard, a conformance scheme, and Web-based tools to assist procurers. The resulting work for Phase 2 is expected to be closely synchronized with the U.S. Section 508 refresh activity to create a set of "cross-

Atlantic" requirements and recommendations (U.S. Access Board 2008). The earlier European Commission report, "Measuring progress of eAccessibility in Europe" (MeAC), is also available for download (Cullen 2007).

Only published ICT accessibility standards with a relevance to Web interfaces are listed here. For information on all ICT accessibility standards that have been published and are currently under development, visit the Tiresias Web site (Royal National Institute for the Blind [RNIB] Digital Accessibility Team 2009). This site also contains information, tools, and design guidelines for enabling ICT accessibility.

19.2.1 U.S. SECTION 508

With the Rehabilitation Act of 1973, the U.S. Congress aimed to provide equal employment opportunities within the federal government to people with disabilities. Success of this law was hampered by the inaccessibility of information technology used in the government to those employees. In 1998, the U.S. Congress amended the Act with Section 508, adding provisions requiring Federal agencies to procure and provide electronic and information technology that is accessible to people with disabilities. The aim of Section 508 was to remove barriers to information technology for employees of the federal government and members of the public accessing federal information. Software and Web interfaces that agencies themselves develop must conform to Section 508 guidelines. When selling to the U.S. federal government, software vendors must show to what extent their software meets Section 508 guidelines. That conformance becomes a weighting factor in the purchasing agency's procurement decision.

From the suite of Section 508 standards, four documents that are germane to Web interface design include the following:

- 1194.21 Software applications and operating systems
- 1194.22 Web-based intranet and internet information and applications
- 1194.24 Video and multimedia products
- 1194.31 Functional performance criteria (U.S. Access Board 2000)

After a decade, the Web criteria have been outstripped by advances in technology and design. Prevalence of video and podcasting has heightened the need for accessible multimedia. Greater reliance on the Internet as a means of delivering government services and a growing population of computer users has driven demand for accessible Web interfaces. As of this writing, the Access Board has announced that it will release a Pre-Draft of revised Section 508 Guidelines in late 2009. On the basis of a report by the advisory Telecommunications and Electronic and Information Technology Advisory Committee (TEITAC), the revised guidelines will focus on functional performance and not on specific technologies (TEITAC 2008). Moreover, TEITAC endeavored to harmonize the emerging Section 508 revision with international standards,

such as the W3C Web Content Accessibility Guidelines, version 2.0 (see below).

19.2.2 W3C WEB CONTENT ACCESSIBILITY

The World Wide Web Consortium (W3C), through its Web Accessibility Initiative (WAI), developed a library of guidelines and techniques for developing and evaluating accessibility of Web-based information (W3C 2009f, 2009g). The guidelines that most directly apply to Web user interface designers are as follows:

- Web Content Accessibility Guidelines (WCAG) 2.0(W3C 2008d)
- Accessible Rich Internet Applications 1.0 (WAI-ARIA)(W3C 2009a)
- Authoring Tool Accessibility Guidelines (ATAG) (W3C 2009b)
- User Agent Accessibility Guidelines (UAAG) (W3C 2009d)

Note: As of this writing, WAI-ARIA is still a working draft. It is expected to reach W3C Technical Recommendation status in 2010.

WCAG addresses the information presented in a Web site, such as text, graphics, multimedia, forms, or other material. WAI-ARIA extends this, providing techniques to accomplish accessible interactions in dynamic Web interfaces and full applications that are served via a Web browser. Many governments, educational institutions, and private entities have adopted WCAG as a standard for Web development.

Currently, some dynamic functions are not available to some people with disabilities, especially those people who rely on screen readers or who cannot use a mouse. For example, consider a clock widget on a Web page. The widget might update "silently" each minute, only changing visual content but not moving programmatic focus to force a screen reader to acknowledge the change. Without some programmatic intervention, the screen reader would typically either ignore the update or refresh the page and begin reading from the top down, repeating even the content that had not changed (W3C 2009b).

WAI-ARIA defines device-independent interactions for Web-based interfaces and facilitates the communication of dynamic changes through the browser to assistive technologies (W3C 2009e). Recall the clock example above. Using WAI-ARIA techniques, the clock widget would be defined as a live region whose updates are either

- Off: Updates to the region will not be presented to the user.
- Polite: Assistive technology should announce the update at the next "graceful opportunity," such as at the end of speaking the current sentence or when the user pauses typing (W3C 2009a).
- Assertive: Notifies the user of a change immediately.

If the live region were a service monitor, then assertive may be appropriate. If a service or system has failed, the user needs to be alerted promptly. With WAI-ARIA live regions, designers can communicate semantic relevance to assistive technologies and enable nonsighted users to discover visual changes to the interface.

19.2.3 REGIONAL AND PER COUNTRY GUIDELINES

In addition to the U.S.-only Section 508 guidelines and the international WCAG standards, many countries and some regions have adopted their own measures to ensure the accessibility of public Web interfaces (W3C 2006b). In some places, compliance with WCAG or locale-specific guidelines is a qualifying criterion in procurements. In 2004, Italy adopted the Stanca Act to support access to information technologies for the disabled, obligating public procurers and their suppliers (CNIPA 2004). Technical Rules of Law 4/2004 followed, citing both W3C WCAG and Section 508 1194.22 as authoritative for Internet-based applications (CNIPA 2005). If the Web interface cannot be used by the purchasers' constituency, then they might opt not to buy the software. For a summary of countries and regions with these guidelines or legislation, see the W3C WAI "Policies relating to Web accessibility" (W3C 2006b). When selling in many locations, Web designers and software vendors face the challenge of multiple accessibility guidelines. While guidelines may complement one another, having to validate against and provide documentation for numerous sets of guidelines based on location can be tedious, providing little value to end users. For example, current Canadian law cites the 1999 WCAG 1.0; WCAG 2.0, which better addresses modern Web technology, has been final (a W3C Recommendation) since December 2008 (see Caldwell and Vanderheiden, this volume). A U.S.-based company might need to provide Canadian procurers with WCAG 1 checklists, European customers with WCAG 2 checklists, and U.S. government procurers with Section 508 1194.21 and 508 1194.22. One strategy for designers is to develop a composite list of functional requirements that addresses the range of guidelines in the target markets.

19.2.4 ISO/IEC 9241

This standard provides guidance on the human-centered design of software Web user interfaces with the aim of increasing usability. Web user interfaces address either all Internet users or closed user groups such as the members of an organization, customers and/or suppliers of a company, or other specific communities of users.

The recommendations given in this part of ISO 9241:2008 focus on the following aspects of the design of Web user interfaces: high-level design decisions and design strategy; content design; navigation and search; content presentation.

The user interfaces of different types of user agents such as Web browsers or Web authoring tools are not directly addressed in this part of ISO 9241:2008 (although some of its

guidance could apply to these systems as well). (See ATAG and UAAG for browser and tool guidance.)

Web user interfaces display on a personal computer system, mobile system, or some other type of network-connected device. While the recommendations given in this part of ISO 9241:2008 apply to a wide range of available front-end technologies, the design of mobile Web interfaces or smart devices could require additional guidance not within its scope; neither does it currently provide detailed guidance on technical implementation nor on issues of aesthetic or artistic design (ISO 2008). The W3C "Mobile Web best practices" may be helpful here (W3C 2008b).

19.2.5 ISO 9241-171:2008

Published in 2008, ISO 9241-171:2008 replaced ISO TR 16071:2003—"Ergonomics of human–system interaction— Guidance on accessibility for human–computer interfaces" (ISO 2008b). ISO 9241-171:2008 was developed in conjunction with ANSI/HFES 200.2—"Human Factors Engineering of Software User Interfaces—Part 2: Accessibility," and it contains almost identical content (Human Factors and Ergonomics Society [HFES] 2008). The ISO standard contains "requirements and recommendations" (with two conformance levels), while the ANSI/HFES standard contains three priority (conformance) levels.

ISO 9241-171 provides ergonomics guidance and specifications for the design of accessible software for use at work, in the home, in education, and in public places. It covers issues associated with designing accessible software for people with the widest range of physical, sensory, and cognitive abilities, including those who are temporarily disabled and the elderly. It addresses software considerations for accessibility that complement general design for usability as addressed by ISO 9241-110, ISO 9241-11 to ISO 9241-17, ISO 14915, and ISO 13407 (ISO 2009a).

ISO 9241-171:2008 applies to the accessibility of interactive systems and addresses a wide range of software (e.g., office, Web, learning support, and library systems). It promotes the increased usability of systems for a wider range of users. While it does not cover the behavior of, or requirements for, assistive technologies (including assistive software), the standard does address the use of assistive technologies as an integrated component of interactive systems. It is intended for use by those responsible for the specification, design, development, evaluation, and procurement of software platforms and software applications (ISO 2008b).

19.2.5.1 PAS 78:2006

Available from the British Standards Institution (BSI), PAS 78 helps site developers understand how people with disabilities use Web sites (British Standards Institution 2006). It provides guidance for defining an accessibility policy for the site, tips for accessibility testing, and advice for the contracting of Web design and accessibility auditing services. Rather than merely describing the final outcome, PAS 78 focuses on the iterative process of developing and maintaining accessibility conformant to W3C guidelines and specifications. It emphasizes the importance of involving people with disabilities throughout the development process. Aimed at those responsible for commissioning public-facing Web sites and Web-based services, PAS 78 is informative for any public or private organization wishing to observe good practice under the existing voluntary guidelines and the relevant legislation on this subject (British Standards Institution 2006).

PAS 78 was written because of a growing concern about the high number of prominent UK Web sites that did not meet even the most basic criteria for accessibility. When, in 2004, the Disability Rights Commission (DRC)—now The Equality and Human Rights Commission—conducted a formal investigation testing the accessibility of 1000 Web sites, their findings revealed that 81% of sites tested failed to provide sufficient usability for people with disabilities (Disability Rights Commission 2004). The investigation demonstrated that many Web managers lacked even a simple understanding of what Web accessibility meant for their organizations or how to achieve it and that detailed guidance on how to procure accessible services was desperately needed.

An attempt to provide such guidance can be found in a companion document to PAS 78, "Web accessibility: Making PAS 78 Work." Gathering lessons learned by working closely with a dedicated group of disabled Web users and through examining hundreds of Web sites together, the authors amassed a significant body of knowledge of how Web accessibility can be barred and how those barriers can be removed. "Web Accessibility: Making PAS 78 Work" helps Web developers understand how accessible services can be implemented and provides practical information on how to best obtain accessible deliverables from suppliers and development teams (Broome 2008). Note: At the time of writing, the BSI is revising PAS 78 as "BS 8878:2009 Web accessibility—Building accessible experiences for disabled people—Code of practice."

19.2.5.2   BS 7000-6:2005

An additional standard that can be very helpful for Web designers is BS 7000-6:2005—"Design management systems. Managing inclusive design. Guide." Part of the BS 7000 series on Design Management Systems, this standard provides guidance on managing inclusive design at both organization and project levels, though the inclusive approach, ultimately encompassing the whole of business and management. Inclusive design is comprehensive, integrating into the design process all aspects of a product used by consumers of diverse age and capability in a wide range of contexts. This standard applies to all levels of staff and management in all types of organizations operating in the manufacturing, process, service, and construction industries, as well as in public and not-for-profit sectors. It provides a strategic framework and associated processes by which business executives and design practitioners can understand and respond to the needs of diverse users without stigma or

limitations and is aimed at the following (British Standards Institution 2005):

- Top executives of all organizations offering products and services. It helps them to lead the introduction of an inclusive approach and evolve an appropriate corporate culture that nurtures inclusive success.
- Middle executives who set up and administer product and service development projects. It helps them formulate better focused and more enlightened briefs. It also assists in motivating project teams as well as the evaluation of solutions generated.
- Junior executives and specialists who are assigned to project teams that create and develop products and services. It helps them adopt more appropriate perspectives and approaches to inclusive design.
- Executives responsible for procuring outsourced product design services and supplies and for adhering to agreed specifications. It helps them to sustain the inclusive approach throughout the supply chain.

19.2.6   EU MANDATE 376: IMPLICATIONS FOR GLOBAL DESIGNERS

The synchronization of the refresh of U.S. Section 508 and EU Mandate 376 should lead to harmonized accessibility requirements for ICT products and services on either side of the Atlantic. Canada and Japan are understood to be closely involved with the Section 508 refresh activities, and Australia has already established a set of "standards" for accessible ICT. It remains to be seen how the non-Atlantic countries react to this situation and how globally pervasive the Section 508/Mandate 376 synchronization becomes.

19.2.7   THE PROMISE OF HARMONIZATION

For a decade, multinational firms have been challenged to design for and validate to multiple Web accessibility standards. Where accessibility compliance is tied to procurement (the software had to meet accessibility targets to be purchased), firms incurred added expense of testing and documenting against multiple standards while gaining little for end users with disabilities. Recognizing the negative impact to business, authors of standard and policy updates in recent years have attempted to harmonize the various standards. When the U.S. Access Board began the process of refreshing its Section 508 guidelines, industry and international representatives provided perspective from other guidelines. In particular, the Section 508 refresh (likely released in 2010) is expected to be complementary to the W3C WCAG version 2.0. Similarly, the authors of ISO 9241-171:2008 were cognizant of WCAG and potential conflicts. As a result, when the impending Section 508 refresh and Mandate 376 described above are fully deployed, Web content providers can design for, validate against, and demonstrate compliance against a unified set of functional criteria. Not only is this good for business but end users with disabilities benefit from a more consistent level of accessibility across sites and Web applications.

19.3 TECHNICAL CHALLENGES OF ACCESSIBLE WEB DESIGN

When one attempts to design for universal use, one faces the immediate challenges of understanding the needs of functionally diverse users and of trying to create a "one size fits all" user interaction. In this section, we present methods for meeting the interaction needs of users with disabilities and offer suggestions why a single interaction path is not only unfeasible but ill advised.

19.3.1 SPECIFIC DISABILITY CONSIDERATIONS FOR DESIGN

Successful Web interface design relies upon a good understanding of the intended users' goals in visiting the site (customer requirements), the circumstances under which they will visit (context, environment), and their interaction needs and expectations. The latter of these aspects most relates to accessible design. Within any group of potential users, a designer should understand the demographics, and the likelihood that a user will have a functional disability affecting his or her operation of the site. So, designers should become familiar with overall disability statistics and the incidence of specific disabilities among the target user base, where applicable. For example, when designing an e-commerce Web site for a firm that provides specialty keyboards or eye-tracking input devices, one could expect navigation not dependent on mouse input to be particularly important. When designing an e-government cite for a public entity, the design team must be aware of incidence of disability in the population as if affects public policy:

> Estimates of the number and percent of people with a disability would be used by federal, state, county and local governments to assess the impact of policies intended to reduce discrimination and improve participation in community activities. (Brault 2009)

If a firm delivers public-facing Web content where legislation compels accessibility, then disability demographics could guide allocation of design and development resources.

It is beyond the scope of this chapter to detail particular design considerations for specific types of disabilities. Rather, the authors refer you to the standard ISO 9241-171/ANSI HFES 200.2 as well as disability advocacy organizations for guidance. In addition, the non-profit Web Accessibility in Mind (WebAIM) provides a series of articles on disability types from the users' perspective, describing challenges in using Web assets and assistive technology that may aid such individuals. Disability categories addressed in the series include the following:

- Visual, covering blindness, low vision, and color-blindness.

- Auditory, covering hearing loss and its causes and deafness, with insightful information on deaf culture.
- Physical, resulting either from traumatic injury, such as spinal cord damage or limb loss) or from disease or congenital conditions, such as cerebral palsy, multiple sclerosis, etc.
- Cognitive, such as attention or memory challenges, or specific comprehension challenges, such as dyslexia.
- Seizure disorders, addressing design choices that can trigger photoepileptic seizures (WebAIM 2009a, 2009b).

In addition to these persistent disabilities, designers do well to consider the needs of people functionally impaired owing to situational factors. Consider a student engaged in online learning while in a noisy university lounge. This person might activate closed captions or view a concurrent transcript to perceive the audio portions. What about the commuter on a cramped train? With only a touchpad mouse, this person might resort to keyboard operation for ease of use. The point here is delivering flexible interfaces that accommodate users' changing needs or environments. With multiple modes of working and choice of layout and presentation, all users gain an interface that is more adaptable, and ultimately more usable, across devices and situations. This is discussed in greater detail in Section 19.3.4.

19.3.2 RESOURCES FOR AND DESIGN IMPLICATIONS OF DISABILITY STATISTICS

When making a business case for accessible development and to understand user demographics, statistics on occurrence of disabilities within populations can be informative. Assistive technology benefits more than just people who have severe impairments. In early 2003, Microsoft Corporation commissioned Forrester Research, Inc. to conduct a comprehensive study to measure the current and potential market of accessible technology in the United States. Phase 1 of the study (Microsoft and Forrester Research, Inc. 2003) focused on people, 18 to 64 years old, and the computer users among them. Phase 2 of the study (Microsoft Corporation and Forrester Research, Inc. 2004) focused on the accessibility of present technology. Researchers investigated native accessibility options—those built into products such as options to change font size and color—and assistive technologies such as screen readers and voice recognition products. The study concluded that

> 57 percent of working-age computer users in the U.S. between the ages of 18 and 64 (more than 74 million Americans) could benefit from accessible technology because of mild-to-severe vision, hearing, dexterity, speech, and cognitive difficulties and impairments that interfere with their ability to perform routine tasks—including their use of computers. Of the 74.2 million U.S. computer users who could benefit from accessible technology, 51.6 million have mild impairments and 22.6 million have more severe impairments. Another 56.2 million are unlikely to benefit from accessible technology because they experience either no difficulties or only minimal impairments. (Microsoft Corporation 2004)

For further evidence, numerous resources exist for obtaining disability statistical data. A compilation can be found at the Employment and Disability Institute Web site of Cornell University (Cornell University Employment and Disability Institute 2009). This site includes data collected by the Census Bureau through mail response, computer-assisted telephone interviewing (CATI), and computer-assisted personal interviewing (CAPI) (U.S. Census Bureau 2009a, 2009b). As part of the decennial (every 10-year) civilian population survey and the on-going frequent American Community Survey, the Census Bureau asks a series of questions about physical, mental, and emotional conditions. Indicative of the sensitive nature of disability terminology and issues introduced by self-selection (people identifying themselves, or not, as persons with disabilities), the U.S. Census Bureau, in 2008, significantly changed how such questions were asked in the survey, to better identify functional challenges among the population, independent of labels.

[The ACS Subcommittee on Disability Measurement, with guidance from the National Center for Health Statistics (NCHS)] recognized that as a concept, disability involves social factors that are both internal and external to the individual, often making its measurement in surveys difficult. (Brault 2009)

Researchers were looking to evaluate not only the incidence of disability but also access to employment and education for people with disabilities.

The subcommittee found that surveys could identify certain aspects of disability and estimate a population who would be likely to experience restrictions in participation due to physical, social and other environmental barriers. (Brown 2006)

The changes to the survey questions resulted in substantially fewer nonresponses by persons with disabilities.

Many other countries maintain statistical data on citizens with disabilities. For global information, the United Nations Statistics Division provides a statistical reference and guide to its data on human functioning and disability (United Nations Statistics Division 2009).

19.3.3 Diversity in Personas

Personas are created and used by a design team to help them as they design their products. Personas do not represent a *single* user; rather, they are typically an amalgamation of many users who would use the product. These personas are not real people, but they have real goals and can help content developers think about for whom they are developing products and Web sites so developers create products users need and not just what the developers think users need or developers want to provide.

For example, if you were creating an online banking Web site, the design team would probably develop many personas and may create one for a student account holder. This persona would not have the characteristics of a single student account holder but would contain characteristics from many student account holders that the design team has interviewed.

Personas are assigned very specific characteristics such as names (for example, Steve Student Account), jobs, experiences, goals, genders, and skills. For example, knowing that your persona has a specific computer skill (such as a high comfort level with online forms) is more useful to Web developers than just saying that your persona has used a computer or has used a database.

When developing a persona, one must remember to include some personas with disabilities, as people with disabilities comprise a significant portion of the population. In the United States alone, more than 33 million working-age Americans have (U.S. Department of Labor, n.d.). Globally, "around 10% of the total world's population, or roughly 650 million people, live with a disability." In *Just Ask: Integrating Accessibility throughout Design*, Shawn Henry offers solid advice on what information to include in your personas with disabilities. She states that the persona should "include a description of the limiting condition (disability or situational limitation) and the adaptive strategies for using the product, such as:

- Nature of limitation (for example, blind, unable to use mouse, operating in noisy environment)
- Special tools or assistive technology used (for example, uses a magnifying glass to read text smaller than 16 point, uses screen reader software, stops machinery to hear mobile phone)
- Experience and skills with the relevant tools or assistive technologies
- Frequency of use of relevant tools or assistive technologies." (Henry 2007)

To understand the impact of particular disabilities on someone's use of the Web, consider the persona that follows.

19.3.3.1 Example Persona

Betty Billpayer is a 53-year-old professional researcher who uses her bank's online banking system to pay her bills and to check her account's balances. Betty has a BS in French and an MLS in Library Science. She has excellent computer skills because, in her job as a researcher, she must mine the Internet, archives, and other sources to search for information for her customers. Betty is a busy woman, and she wants to spend most of her free time enjoying her three grandchildren and her book club and not figuring out her bank's online system. Betty has low visual acuity that she developed late in life. For the past three years, she has used screen magnification software to enlarge the text on her computer monitor.

19.3.3.2 Usability Testing with Participants with Disabilities

Web accessibility evaluation tools are available to help developers programmatically test their Web sites. A thorough list of tools can be found on the W3C WAI Complete List of Accessibility Evaluation Tools page (W3C 2006a). These tools help a developer to determine if his or her Web pages conform to accessibility guidelines. For example, tools exist to test the flicker rates of images so that a designer knows whether these images conform to stated guidelines. There are also tools that transform Web pages so that a developer can see how they would work for people with disabilities. For example, there are tools to render a Web page to look how it would look to a person with color blindness so a designer could determine whether the color and contrast in a Web page works and use tools to help a designer determine whether a Web page would make sense if it were read by screen reader software.

While these tools can help determine how a Web page conforms to guidelines and "looks like" to a person with disabilities, nothing is better than having persons with disabilities try out your product. As Slatin and Rush note, even if a content developer creates a Web site that conforms to all of WCAG's guidelines, that site might not be usable to a person with disabilities (Slatin 2003). Including people with disabilities in your testing can provide some great feedback to your efforts. The participants can tell you what works and what gets them stuck. Meeting their needs helps all users.

> In the end it all comes down to *people*, the human element can't be overlooked. People with disabilities who use the web can be your best source of accessibility testing . . . . Human review can help ensure clarity of language and ease of navigation. (Slatin 2003)

19.3.4 ADAPTIVE INTERACTION DESIGN

People with disabilities, like all computer users, prefer choice and flexibility in interactions. By designing to maximize flexibility, Web developers can increase not only the accessibility but also the overall usability of the software. For example, consider a mail application with multiple panes, perhaps a

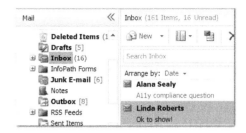

FIGURE 19.2 Web Inbox at 200% magnification.

navigation pane along the left side, a preview of e-mails at the top center, and an individual e-mail opened lower center. For someone with normal vision working on a full-size monitor, the layout may be fine; see Figure 19.1. Someone with low vision, who requires 200% screen magnification, the navigation pane might comprise too much horizontal real estate to render the e-mail preview useful, as shown in Figure 19.2.

Instead, the application needs to provide a flexible layout, enabling the user to choose what elements are visible and in what position. In this case, by selecting the double-chevron link, the user with low vision can hide the navigation pane. Then, by turning off the reading pane, the user regains much needed horizontal real estate for the magnified interface; see Figure 19.3.

Designers can leverage this flexibility when supporting ever-smaller devices. On a personal data assistant (PDA) or smart phone, a simplified, linear view of a single folder, such as the Inbox, at once, could meet most users' needs for mobile mail viewing.

19.3.5 COLOR AND CONTRAST

Visual deficiencies, aging, and color blindness are a few of the issues that can affect how people perceive certain color combinations. Color blindness is a condition where a person has faulty or missing color detectors (cone cells) and has "trouble distinguishing between combinations and/or pairs of colors" (Paciello 2000). About one in 10 U.S. males are colorblind; color blindness, however, rarely affects women. The most common type of color blindness creates a problem in distinguishing some shades of red and green. For persons with a visual deficiency, colors that are easy to distinguish to someone with full vision might be far less distinguishable to them.

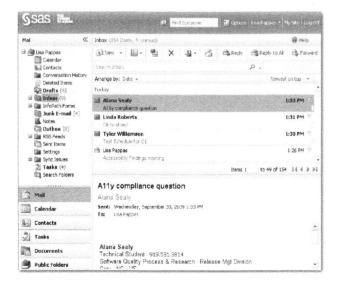

FIGURE 19.1 Microsoft Outlook Web Access InBox.

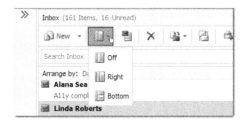

FIGURE 19.3 Web mail with navigation pane hidden and reading pane options.

FIGURE 19.4   (**See color insert.**) Screen capture showing insufficient text contrast.

WCAG 2.0 guideline 1.4.3 states that "the visual presentation of text and images of text has a contrast ratio of at least 4.5:1" (for text that is smaller than 18 points, if it is not boldfaced, and smaller than 14 points if it is boldfaced; W3C 2008). To conform to this guideline, you can select foreground or background text colors for your Web page that exaggerate or enhance the differences between them (for example, use dark colors against light colors that do not have contrasting hues) or by allowing the user agent to specify text colors because you did not specify the text color and background in your Web page.

To analyze a Web page for contrast, you can use a color contrast analyzer tool to help you determine good color combinations and to measure the difference between foreground and background elements. One example includes the Colour

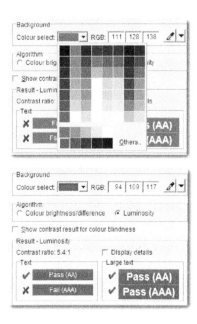

FIGURE 19.5   (**See color insert.**) Colour Contrast Analyser with color picker.

FIGURE 19.6   (**See color insert.**) RNIB screen capture in full color.

Contrast Analyser (The Paciello Group 2009). Consider the North Carolina State University Web site evaluated with the Colour Contrast Analyser, shown in Figure 19.4. Using the dropper tool, the authors sampled the white foreground text in the main content area and measured its contrast with the medium gray background. Note in the Results section that this color combination fails to meet W3C WCAG requirements for the AA level of conformance. Using the color picker drop-down, a designer can experiment with other combinations, as shown in Figure 19.5, to find a color choice that meets the desired level of contrast. Note that in the second image in Figure 19.5, the darker gray provides sufficient contrast with the white text to meet the WCAG 2—AA standard for text.

Visual simulators are another type of tool useful to content designers. For example, on the Web site Vischeck, designers can upload an image and Vischeck shows how that image appears to a colorblind user (Vischeck 2008). With roughly 1 in 20 people having some sort of color deficiency, misperceptions due to color choice can be significant. In particular, designers must guard against using color alone to indicate meaning. Consider the partial screen capture from the Web site of the Royal National Institute for the Blind (RNIB), a UK advocacy organization for people who are blind or partially sighted, shown in Figure 19.6.

After applying a deuteranope filter to the image with Vischeck, Figure 19.7 shows how the Web page might look to someone with red/green color blindness. Although the colors do appear significantly different, because the designers chose well, no information is lost to site visitors with red–green color blindness. Note: The Colour Contrast Analyser tool discussed previously also provides an option for some color

FIGURE 19.7   (**See color insert.**) RNIB with red–green simulation.

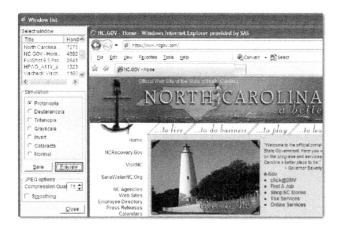

FIGURE 19.8 (**See color insert.**) Colour Contrast Analyser—color blind simulation.

blindness simulation and a choice to analyze a full window, as shown in Figure 19.8. Note that simulation options include the three primary types of color blindness; grayscale, for simulating low-light or noncolor conditions; color inversion, for a loose proxy of high contrast mode; and, especially useful for understanding older users, the cataracts simulation.

19.3.6 FLEXIBLE LAYOUT AND COMPONENTS

One key aspect of successful Web design is that of adaptability—it adapts to the users' needs. In terms of performance, this can mean a streamlined presentation for low-bandwidth or mobile users. In terms of accessibility, this can mean a liquid design, pages that reflow when the browser window is resized or resolution is altered. A multipart horizontal layout—if it is not flexible—may position key elements off-screen when rendered for low vision individuals using a screen magnifier. Designers can accommodate flexibility needs by enabling users to optionally display or obscure visual components as needed. Designers of flexible layouts gain the added benefit that their Web pages adapt more gracefully across different browsers or platforms than fixed size designs. To learn more about designing flexible layouts to achieve accessibility, see

- Flexible layouts with CSS positioning (Falby 2002)
- Flexible Web design: creating liquid and elastic layouts with CSS (Gillenwater 2008)

For any Web design aspect, there are accessibility considerations. In most geographies, standards or guidelines inform the accessibility requirements for a design. While designers may educate themselves about techniques to achieve accessibility, in all cases, to ensure success, they need to validate that accessibility has been achieved. While it is beyond the scope of this chapter to explore Web accessibility validation techniques, designers are encouraged to pursue avenues to obtain this guidance. Excellent sources include the following:

- Accessibility testing: case history of blind testers of enterprise software (Bayer and Pappas 2006)
- Conducting usability studies with users who are elderly or have disabilities (van der Geest 2006)
- Just ask: integrating accessibility throughout the design (Henry 2007)
- Evaluating Web sites for accessibility: overview (W3C 2008a)

19.4 CONCLUSION

While the array of Web accessibility standards may be daunting, designers should not be discouraged. Rather, the evolving harmonization of the standards is leading to a discrete set of functional criteria for design and evaluation of Web sites for accessibility. Moreover, by beginning the design process with accessibility in mind, designers garner substantial benefits beyond those of policy or standard compliance. Researchers with the Assistive Technology Research Institute concluded that accessible design benefits not only users with disabilities but also enhances overall usability (Anson 2004). Therefore, designers should not think of accessibility as accommodating a subset of users. By applying accessible design principles, they may achieve universal usability. The flexible layouts accommodate not only users with disabilities but also render interfaces that may be localized to other languages more readily. Successful Web design meets the needs of its user base, helping them to fulfill their purpose in visiting a site. Accessible, standards-based design helps to ensure that a Web site is usable for all visitors.

REFERENCES

Anson, D. E. 2004. *The Benefit of Accessible Design for Able-Bodied Users of the World Wide Web.* Dallas, PA: Assistive Technology Research Institute, Misericordia University.

Arditi, A. 2009. Effective color contrast. http://www.lighthouse.org/accessibility/effective-color-contrast/ (accepted Oct. 9, 2009).

Bayer, N. L., and L. Pappas. 2006. Accessibility testing: case history of blind testers of Enterprise Software. *Technical Communication* 32–38.

Brault, M. 2009. Review of changes to the measurement of disability in the 2008 American Community Survey. September 22. U.S. Census Bureau. http://www.census.gov/hhes/www/disability/2008ACS_disability.pdf (accessed Sept. 22, 2009).

British Standards Institution. 2005. *Design management systems. Managing inclusive design. Guide. (BS 7000-6:2005).* London, UK: BSI.

British Standards Institution. 2006. *Guide to Good Practice in Commissioning Accessible Websites (PAS 78:2006).* London: British Standards Institution.

British Standards Institution. 2009. BSI: standards, training, testing, assessment & certification. http://www.bsigroup.com/ (accessed Oct. 30, 2009).

Broome, G. 2008. *Web Accessibility. Making PAS 78 Work.* London: British Standards Institution.

Brown, S. C. 2006. *Developing Improved Disability Data.* Washington, DC: Interagency Committee on Disability Research Conference. http://www.icdr.us/disabilitydata/materials.htm (accessed Nov. 9, 2009).

Caldwell, B. B., and G. C. Vanderheiden, this volume. Access to Web content by those with disabilities and others operating under constrained conditions. In *Handbook of Human Factors in Web Design*, 2nd ed., eds. K.-P. L. Vu and R. W. Proctor, 371–402. Boca Raton, FL: CRC Press.

CEN. 2008. CEN/BT WG 185 Project Team Final draft report. October 1. http://www.econformance.eu/documents/BTWG185_N30_PT_Final_Report_for_Approval.pdf (accessed Oct. 30, 2009).

CEN/CENELEC. 2002. CEN/CENELEC Guide 6: Guidelines for standards developers to address the needs of older persons and persons with disabilities. January. ftp://ftp.cen.eu/BOSS/Reference_Documents/Guides/CEN_CLC/CEN_CLC_6.pdf (accessed Jan. 6, 2010).

CNIPA. 2004. Stanca Act (Law n. 4, January 9, 2004). http://www.pubbliaccesso.gov.it/normative/law_20040109_n4.htm (accessed Jan. 6, 2010).

CNIPA 2005. *The Technical Requirements and the Different Levels of Accessibility of Computer Tools.* July 8. http://www.pubbliaccesso.gov.it/normative/DM080705-A-en.htm (accessed July 6, 2010).

Cooper, A. 2004. *Inmates are Running the Asylum,* 2nd ed. Sams Publishing. http://www.amazon.com/Inmates-Are-Running-Asylum/dp/0672316498/ref=cm_cmu_pg_t#reader_0672316498

Cornell University Employment and Disability Institute. 2009. Disability statistics: online resource for U.S. disability statistics. http://www.ilr.cornell.edu/edi/DisabilityStatistics/ (accessed Oct. 30, 2009).

Cullen, K. E. 2007. Measuring progress of eAccessibility in Europe: assessment of the status of eAccessibility in Europe. http://ec.europa.eu/information_society/activities/einclusion/library/studies/meac_study/index_en.htm (accessed Oct. 30, 2009).

Disability Rights Commission. 2004. *Web Access and Inclusion for Disabled People.* London, UK: Disability Rights Commission.

Disablity World, n.d. Disability facts and statistics. http://www.disabled-world.com/disability/statistics/ (accessed Oct. 9, 2009).

ETSI. 2009. ETSI TR 102 612—Human Factors: European accessibility requirements for public procurement of products and services in the ICT domain. April 6. http://portal.etsi.org/STFs/STF_HomePages/STF333/STF333.asp (accessed Oct. 30, 2009).

European Telecommunications Standards Institute (ETSI). 2009. ETSI: world class standards. http://www.etsi.org/WebSite/homepage.aspx (accessed Oct. 30, 2009).

Falby, D. 2002. Flexible layouts with CSS positioning. November 15. http://www.alistapart.com/articles/flexiblelayouts/ (accessed Nov. 2, 2009).

Gillenwater, Z. M. 2008. *Flexible Web Design: Creating Liquid and Elastic Layouts with CSS.* Berkeley, CA: New Riders.

Henry, S. L. 2007. Just ask: Integrating accessibility throughout design. Lulu.com.

Human Factors and Ergonomics Society (HFES). 2008. ANSI/HFES 200.2—Human Factors Engineering of Software User Interfaces—Part 2: Accessibility. http://www.hfes.org/Publications/ProductDetail.aspx?Id=76 (accessed Oct. 30, 2009).

International Organization for Standardization (ISO). 2008a. ISO 9241-151: 2008—Ergonomics of human–system interaction—Part 151: Guidance on World Wide Web user interfaces. May 14. http://www.iso.org/iso/catalogue_detail.htm?csnumber=37031 (accessed Nov. 9, 2009).

International Organization for Standardization (ISO). 2008b. ISO 9241-171:2008—Ergonomics of human–system interaction—Part 171: Guidance on software accessibility. July 10. http://www.iso.org/iso/catalogue_detail.htm?csnumber=39080 (accessed Oct. 30, 2009).

International Organization for Standardization (ISO). 2008c. ISO/TR 22411:2008: Ergonomics data and guidelines for the application of ISO/IEC Guide 71 to products and services to address the needs of older persons and persons with disabilities. http://www.iso.org/iso/catalogue_detail?csnumber=40933 (accessed Jan. 6, 2010).

International Organization for Standardization (ISO). 2009a. ISO—International Organization for Standardization. http://www.iso.org/iso/home.htm (accessed Oct. 30, 2009).

International Organization for Standardization (ISO). 2009b. ISO/IEC TR 29138-1:2009. Information technology—Accessibility considerations for people with disabilities—Part 1: User needs summary. June 10. http://www.iso.org/iso/catalogue_detail.htm?csnumber=45161 (accessed Nov. 8, 2010).

International Organization for Standardization (ISO). 2009c. ISO/IEC TR 29138-2:2009. Information technology—Accessibility considerations for people with disabilities—Part 2: Standards inventory. June 10. http://www.iso.org/iso/catalogue_detail.htm?csnumber=51341 (accessed Nov. 8, 2010).

International Organization for Standardization (ISO). 2009d. ISO/IEC TR 29138-3:2009. Information technology—Accessibility considerations for people with disabilities—Part 3: Guidance on user needs mapping. June 10. http://www.iso.org/iso/iso_catalogue/catalogue_tc/catalogue_detail.htm?csnumber=51342 (accessed Nov. 8, 2010).

Japanese Standards Association (JSA). 2004. JIS X 8341-3:2004: Guidelines for older persons and persons with disabilities. June 20. http://www.webstore.jsa.or.jp/webstore/Top/indexEn.jsp?lang=en (accessed Jan. 6, 2010).

Microsoft Corporation. 2004. New research study shows 57 percent of adult computer users can benefit from accessible technology. February 2. http://www.microsoft.com/presspass/features/2004/feb04/02-02aging.mspx (accessed Jan. 28, 2010).

Microsoft Corporation and Forrester Research, Inc. 2003. The market for accessible technology—the wide range of abilities and its impact on computer use. http://www.microsoft.com/enable/research/phase1.aspx (accessed Jan. 28, 2010).

Microsoft Corporation and Forrester Research, Inc. 2004. Accessible technology in computing—examining awareness, use, and future potential. http://www.microsoft.com/enable/research/phase2.aspx (accessed Jan. 28, 2010).

Paciello, M. G. 2000. *Web Accessibility for People with Disabilities.* Berkeley, CA: CMP Books.

Royal National Institute for the Blind (RNIB) Digital Accessibility Team. 2009. Standards relating to the accessibility of ICT. September 10. http://www.tiresias.org/research/standards/index.htm (accessed Oct. 30, 2009).

Slatin, J. M. 2003. *Maximum Accessibility: Making Your Web Site More Usable for Everyone.* Reading, MA: Addison-Wesley.

Telecommunications and Electronic Information Technology Advisory Committee (TEITAC). 2008. Telecommunications and Electronic Information Technology Advisory Committee Report to the Access Board: Refreshed Accessibility Standards and Guidelines in Telecommunications and Electronic Information Technology. April. http://www.access-board.gov/sec508/refresh/report/ (accessed Sept. 29, 2009).

Thatcher, J. E. 2002. *Constructing Accessible Web Sites.* Birmingham, UK: Glasshaus.

The Paciello Group. 2009. Resource Center–Contrast Analyser 2.2. http://www.paciellogroup.com/resources/contrast-analyser.html (accessed Oct. 30, 2009).

U.S. Access Board. 2000. Electronic and Information Technology Accessibility Standards (Section 508). December 21. http://www.access-board.gov/sec508/standards.htm (accessed Sept. 30, 2009).

U.S. Access Board. 2008. Update of the 508 Standards and the Telecommunications Act Guidelines. April. http://www.access-board.gov/sec508/update-index.htm (accessed Sept. 29, 2009).

U.S. Census Bureau. 2009a. American Community Survey (ACS). October 26. http://www.census.gov/acs/www/ (accessed Oct. 30, 2009).

U.S. Census Bureau. 2009b. U.S. Census Bureau–Disability–Main. http://www.census.gov/hhes/www/disability/disability.html (accessed Oct. 2009).

U.S. Department of Labor, n.d. Frequently asked questions. http://www.dol.gov/odep/faqs/working.htm (accessed Oct. 9, 2009).

United Nations Statistics Division. 2009. Demographic and social statistics. http://unstats.un.org/unsd/demographic/sconcerns/disability/ (accessed Oct. 30, 2009).

van der Geest, T. 2006. Conducting usability studies with users who are elderly or have disabilities. *Technical Communication* 23–31.

Vischeck. 2008. Vischeck. July 4. http://www.vischeck.com/people/ (accesed Oct. 30, 2009).

W3C. 2006a. *Complete List of Web Accessibility Evaluation Tools.* March 17 Retrieved October 9, 2009, from http://www.w3.org/WAI/ER/tools/complete.

W3C. 2006b. Policies relating to Web accessibility. August 25. http://www.w3.org/WAI/Policy/ (accessed Sept. 29, 2009).

W3C. 2008a. Evaluating Web sites for accessibility: Overview. December 18. http://www.w3.org/WAI/eval/ (accessed Nov. 2, 2009).

W3C. 2008b. Mobile Web best practices. July 29. http://www.w3.org/TR/mobile-bp/ (accessed Jan. 4, 2010).

W3C. 2008c. Web content accessibility and mobile Web: making a Web site accessible both for people with disabilities and for mobile devices. October 14. http://www.w3.org/WAI/mobile/ (accessed Jan. 4, 2009).

W3C. 2008d. Web content accessibility guidelines. December 11. http://www.w3.org/TR/WCAG/ (accessed Jan. 4, 2010).

W3C. 2008e. Web content accessibility guidelines (WCAG) 2.0. December 11. http://www.w3.org/TR/WCAG20/ (accessed Oct. 30, 2009).

W3C. 2009a. Accessible rich Internet applications (WAI-ARIA) 1.0—W3C working draft. December 15. http://www.w3.org/TR/wai-aria/ (accessed Jan. 6, 2010).

W3C. 2009b. Authoring tool accessibility guidelines. December 15. http://www.w3.org/WAI/intro/atag.php (accessed Jan. 28, 2010).

W3C. 2009c. Shared Web experiences: Barriers common to mobile device users and people with disabilities. June 1. http://www.w3.org/WAI/mobile/experiences (accessed Jan. 4, 2010).

W3C. 2009d. User agent accessibility guidelines (UAAG). December 15. http://www.w3.org/WAI/intro/uaag.php (accessed Jan. 28, 2010).

W3C. 2009e. WAI-ARIA overview. February 24. http://w3.org/WAI/intro/aria.php (accessed Sept. 29, 2009).

W3C. 2009f. WAI guidelines and techniques. August 12. http://www.w3.org/WAI/guid-tech.html (accessed Oct. 30, 2009).

W3C. 2009g. Web Accessibility Initiative (WAI). http://www.w3.org/WAI/ (accessed Oct. 30, 2009).

WebAIM. 2009a. Articles: the user's perspective. http://www.webaim.org/articles/ (accessed Oct. 30, 2009).

WebAIM. 2009b. Creating accessible tables. http://www.webaim.org/techniques/tables/ (accessed Nov. 2, 2009).

Section VII

Web Usability Engineering

20 User Research: User-Centered Methods for Designing Web Interfaces

Fred Volk, Frank Pappas, and Huifang Wang

CONTENTS

TABLE 20.1

Research Proposal Recommended Sections

1. Review Process: Identifies primary contact(s) and contact information for both research team and survey sponsor. All assets, such as scripts, instruments, and reports should be reviewed and approved by the primary contacts throughout the course of the project.
2. Project Background: A brief, narrative summary of the project provides a bit of context to the research. This is helpful for anyone not intimately familiar with the project from its inception, describing in layman's terms why the research is being conducted as well as any specific objectives established by the client or researchers.
3. Assets for Review: Clients often have many assets on hand that help researchers design a successful project. This can include prior research reports (surveys, focus groups, etc.), competitive analyses, marketing, or other business materials that provide insights into markets, target audiences, or other areas. This may also include demo versions of products for the researchers to experience or other items unique to specific projects.
4. Tasks and Deliverables: A list of tasks and deliverables that will be completed and delivered throughout the project. This can include items delivered by the researchers to the client (such as the research design, sample design, interview script, and final survey report) as well as information requested by the researchers (lists of a client's subscribers for recruiting purposes or corporate logos to assist in branding the visual appearance of a survey).
5. Timeline: Tied closely to the list of tasks and deliverables, the timeline establishes target dates for implementation of major tasks, allocates time, and sets deadlines for client feedback and approval, delivery of specific reports, and completion of the project.
6. Anticipated Costs: The cost of implementing the project. This can include not only fees charged by the researchers, but pass-through costs for incentives (either cash or a cost for purchasing iPods, etc.), purchase of marketing lists from companies to facilitate recruiting, travel, facility rental (e.g., focus group facilities), per diems, and other anticipated costs. It's important to delineate what is included in the cost of the project versus those costs that may accrue throughout the course of the project.
7. Analytics/Research Focus: Defines the initial vision for how the researchers will approach the task of analyzing data. Clients may express a preference for particular qualitative or quantitative analysis or may request comparative analyses based on certain audience demographics or psychographics. While the ultimate array of analytics and the final focus may change, this identifies areas of concern to both client and researcher and provides a good starting point for analysis.
8. Sample Design: Establishes the characteristics of an effective and appropriate sample. This may include where potential respondents live based on a company's geographic operating area, length of time as a customer, purchasing history, accessibility (cell phone and/or e-mail address availability), or other factors. Also included should be how the sample will be drawn or recruited (e.g., random, stratified random, etc.), the ideal number of participants, and any other issues relating to the sample.
9. Additional Data Sources: Other data sources that may be tapped to facilitate sample design, analysis of data, or other tasks. This might include prior research data compiled by the client, the researchers, user logs maintained by the client, or publicly available data sets such as U.S. Census data.
10. Recruiting Strategy: Defines the approach for soliciting participation in the research. This may include the number of contacts to be sent throughout the project and what type (e.g., postal mail, e-mail, and telephone) will be used. It establishes an understanding of the logistics involved in delivering the invitations (e.g., e-mails sent by researchers, postal mailings prepared by researchers but mailed by client to take advantage of bulk mail rates), any special materials required (such as branded envelopes or letterhead provided by client).
11. Calling Scripts/Contact Letters: The actual draft text of calling scripts and contact letters that will be shared with potential participants. It is crucial to obtain client approval for these items before they are sent as they can damage a client's reputation (and a researcher's relationship with the client) if any aspects of the scripts or contacts diverge from established corporate communication standards, are sloppy, or otherwise reflect poorly on the client's organization.
12. Incentives: Identifies the recommended incentives to encourage participation, the total number required, any potential cost or logistical considerations, as well as how the incentives will be distributed to participants.
13. Instruments: A draft of the exact text of the instruments, including instructions, all questions, discussion guides, interviews, etc. If possible, any formatting that will be included (such as specific fonts, letterhead, or other characteristics) should be included.
14. Caveats: Any additional concerns, considerations, or other potential problems should be included in the document. Caveat any events or conditions that may negatively influence the success of the project and indicate mitigation strategies for any such threats.

---

users' ability to purchase books and the internal users' ability to provide the appropriate support.

20.2.2 Identifying the Users

Identifying the appropriate users for research is the most integral step in understanding the opportunity of your product or solution. Whereas primary users are typically the focus of user research, the degree that development and marketing teams can identify those users varies from effort to effort. There are times when the primary users are clearly defined and other times when a feature or service is conceptualized in such a way that there are no clearly articulated target users.

Each situation calls for a different approach to allocating resources and starting points for research.

20.2.2.1 Users Are Defined

The best situation for researchers is when the users are defined. This is often the case when a development effort is undertaken for an existing Web site, product, or service and the customer and user populations are well understood. In this scenario, researchers have an opportunity to focus on the users that are the least understood and are most important to the success of the product. The researchers should first focus on the primary end users and then identify the most meaningful user populations related to those end users' behavior.

Important user populations could include other primary end users (depending on the business opportunity of those populations), secondary end users, or internal users.

20.2.2.2 Many Types of Users

There are times when several distinct types of end users exist, and it is difficult to determine on which user population to focus your research resources. Similarly, when a user is clearly defined, a researcher must determine the user population or populations that present the best business opportunity for the product. In contrast to the clearly defined scenario, this situation places user research in the critical path to a product's success. Rather than having the luxury of choosing a specific user group for study, it becomes imperative to have a research plan that incorporates the study of a range of user groups. Without a clear vision regarding the way different types of users will engage the Web experience, it will be difficult to design a user interface that meets each of their needs effectively. Sometimes it is better to completely satisfy a few user groups than to satisfy every user group half-heartedly. Identifying those key groups early in the process will facilitate a higher return on research and development resources and maximize the business opportunity.

20.2.2.3 "Don't Know the Users"

While it seems counterintuitive that a company would produce a solution without adequate understanding of whom the users might be or how their current customers might incorporate it into their behavior, based on our experiences, this scenario occurs far more often than one might anticipate. This situation requires the user research to begin within the solution provider. The researchers must apply some of the same investigation methods to their own organization in an effort to identify the source of the requirements for building the product. This may require interviewing developers, product managers, development directors, and sales executives among others.

20.2.2.4 "Everyone Is Our User"

There is a tendency for each of us to think of users as similar to ourselves. Further, there also seems to be a tendency for us to think that if the product is good for us then, of course, everyone will want one. The claim that "everyone is our user," or that anyone who accesses the Web is our target audience, is equivalent to not knowing who your users are. At a minimum, Web-based products can be targeted in terms of household income and/or desired online experience (experiential versus utilitarian). Similarly, when the potential audience is unknown, the researcher must begin from within the organization to determine the source of the requirements.

20.2.3 GAIN ACCESS TO USERS

The decision of which users to sample for study is influenced by user access. In large corporations, customer-facing personnel, such as account managers, sales engineers, field engineers, field support managers, or product managers, can be a great starting point for gaining access to existing customers and users. Existing customers provide excellent opportunities to identify additional users or user groups of the product. Depending on how well the organization has adopted a user-centered approach, there may be account managers who will either directly assist you in contacting customers or actively prevent you from gaining access to customers. In marketing, the usual worries of nonmarketing individuals interacting with customers are that they may not have the skills to present the company positively, thus hurting sales. In our experiences with many customers with several large companies, usability professionals engaging customers enhance the relationship between customer and provider. These interactions, by their very nature, demonstrate the humility of the research sponsor by expending considerable resources to listen to user expertise and giving them a chance to influence the direction of the product in a nonsales situation. Mayhew (1999) and Wixon and Ramey (1996) report similar positive experiences as a result of user research.

If researchers do not have a pool of existing customer contacts, they may be able to gain access to user populations by: (1) working with professional recruiters, (2) publicizing a user volunteer page on the company's Web site, (3) contacting professional organizations for potential user groups, and (4) posting invitations in news groups, online forums, Web logs (blogs), and LISTSERVs. Usually, with a few rounds of user research, one can build a database of both internal contacts (e.g., marketing, support) and external customers' key users and stakeholders. Such a database can be an invaluable asset to initiate future research.

Recruiting users from the customer site is another challenge to overcome. Typically, we offer users something in return for their time and collaboration with our research. We have found the following to be great incentives for recruiting customers to participate in the research: (1) offering technical support to solve users' existing problems, (2) offering to do a workplace or process analysis to help identify areas of improvement on the customer site, and (3) offering monetary compensations to users (or noncustomer participants).

20.3 FOCUS OF RESEARCH

One important aspect of preparing for a study is to define the research goals and the scope of the study. A goal may begin as both broad and vague; however, it can become attainable in a research setting by assigning operational definitions. For example, Millen (2000) began with the broad research goal of understanding how Internet technology is revolutionizing the way people work and the ways in which they will conduct their business in the future. Then *technological pioneer* was operationally defined to identify the sites to be visited. As a further operational definition of "the way" people work and conduct their business, three elements were focused on (1) all forms of communications, (2) any artifacts that support communications, and (3) physical work environment.

Although what needs to be studied in a user research project is largely determined by the research goals, there are general categories of information that should be targeted. At a minimum, research can be structured into these categories: tasks, users, environments, satisfiers and dissatisfiers, and tools.

20.3.1 Tasks

The term *task* has been used in many different contexts, and its exact definition varies. In this chapter, we use this term loosely to refer to the set of activities that users need to perform in order to reach certain goals. A task description contains these key elements: goals, a series of steps or actions, and definitive starting and ending points. It is important to understand the relationship of each task with other tasks and the conditions under which each task becomes important. The task descriptions are arguably the information that provides the most direct and useful tool for a product team because developers can generate tangible use cases from them, testers can create test cases from them, and documentation writers can shape the organizations of their writing based on them. It is important to base the task descriptions on authentic and accurate user data.

20.3.2 Users

Descriptions and categorization of the users for an intended solution are important information to be obtained early in any product development cycle. It is important to understand users in terms of their Web experience, task experience, educational level, socioeconomic status, and demographics (Jones and Fox 2009). For product development, just having these data on the target users dramatically increases the chance of delivering a usable solution.

20.3.3 Environments

Environments are the physical or organizational settings in which users perform their jobs. Physical characteristics of the user environments play an important role in how the products need to function. There are a number of questions of interest: Is the physical space shared by more than one person? Is it usually loud there? How big is the space? Is there sufficient lighting? Does the user spend most of the time in the setting under study? Is it a stationary or a mobile environment? Organizational environments are often the powerful hidden force that determines what the users do. The answers to the following questions tell a lot about the organizational environment users find themselves in: Who are the colleagues? Who are the bosses? Who are the people to get help from? Who does the user spend the most time with at work?

20.3.4 Satisfiers and Dissatisfiers

User satisfaction refers to factors that make users feel satisfied with what they do. These factors are sometimes called *satisfiers*. For users who are looking for experiential usage (e.g.,

games) of an online service, delighters keep them exploring the service and excited about using it. For users who are looking to accomplish utilitarian goals (e.g., buy a computer), the delighters come in being able to finish the task in a timely manner. Identifying the satisfiers and dissatisfiers can help researchers understand what is important to the users, which may lead to solutions that motivate users to do their job.

20.3.5 Tools

Tools help users get their job done. Tools play an important role in defining what the users are doing and how they are doing it. This is of special interest to product development teams because they help identify opportunities for leveraging new technologies. Incorporating the function, form, and metaphor of an artifact into a new design enhances user performance and accelerate user acceptance. These tools can include physical objects (such as a specialized keyboard) as well as mental models or processes (such as understanding how the hiring process works in a company).

20.4 METHODS

This section describes four commonly used methods utilized for user research in the early stages of product development, namely focus groups, interviews, surveys, and contextual observations. Through the discussion of these four methods, we hope to shed light on how user research is conducted in the early stages of product development.

20.4.1 Focus Groups

The key to successful use of the focus group technique is understanding its role relative to other behavioral methods (Nielsen 1997; see Vu et al., this volume). Focus groups are good for eliciting comments that reflect users' attitudes and opinions. As an indirect method of user behavior, it requires that users recall and/or imagine Web site or product usage. Focus groups are not good for evaluating whether individual users can effectively interact with a system or interface. The advantage of the focus group dynamic is that it, with the guidance of a skilled moderator, encourages elaboration (depth) and variety (breadth) of discussions regarding user experiences, product concepts, and user needs (Kendall and Kendall 1993). Although best when coupled with other methods, focus groups may be used when a large range of potential users are difficult to identify or incent to participate in other more quantitative methods. When the incentive required to get 500 responses to a survey is too high, it may be more cost-effective to offer a greater incentive for 20 potential users to participate in focus groups (Childs 2004). In the early stages of user research, there are three ways to apply focus groups: user needs assessment, concept testing, and assessing the technical impact of a proposed solution. Each of these scenarios requires the researcher to identify the research goals, select the panelists to meet those goals, create a discussion guide based on those goals, conduct the group, analyze the data, and report results.

20.4.1.1 User Needs

One goal of a user needs focus group is to characterize the issues with a target user population. Each issue identified represents an opportunity for a solution provider to create a new product or Web-based application to meet the most pressing needs of the target population. The user needs discussion guide is built around tasks, environments, satisfiers, dissatisfiers, and tools. It is important for the research goals and the discussion to be grounded on the user's behavior (i.e., task). Organizing the discussion guide around the users' tasks provides users with an opportunity to visualize their own behavior and more easily articulate other aspects of the experience related to task performance. These groups can last as long as 4 hours, but because of the nature of the discussion (e.g., identification and elaboration of problems), it would be appropriate to plan a brief break at least every 90 minutes.

The *panelists* (focus group participants) should be selected from a single type of user for each group, but it is certainly appropriate to conduct multiple groups of both internal and external users. Additionally, if the secondary users are highly influential, it may be worthwhile to have a group that consists of secondary users. This homogeneity of user type within a single focus group is important for ensuring that a breadth of issues is captured for each user role and minimizing the natural conflict that can occur between the needs of the different user populations. These groups should have between six and eight panelists but can be accomplished successfully with as few as three (Kendall and Kendall 1993).

After discussing group norms, the moderator guides the group to describing the goals of the discussion and pace of the group. The moderator introduces a series of cues around the hypothesized goals of the users. For example, if a focus group goal is to assess high school teachers' needs relative to lesson plans, a moderator might cue the group with "About how much of your time do you spend preparing for class?" That is sometimes enough to lead to most of the issues around class preparation and lesson planning for those participants.

Once the data are collected, group observers should meet and share some of their first impressions. Keep in mind that the data from focus groups are typically used to generate stimulus materials for further research. The data for user needs focus groups should be organized in terms of tasks, environment, and overall themes. Capture the tasks that were proposed, the comments surrounding those tasks, and the relationship between the tasks. This is the point at which researchers should begin to understand some of the physical and environmental constraints that could potentially affect the development, delivery, and user acceptance of the proposed Web site. Summarize the data into generalized themes to ensure that the macroissues of the users' environments are captured. In our recent research with teachers, "lack of

resources" and "no time" emerged as themes. The "lack of resources" theme may indicate the inability of school districts to pay for new Web-based teacher solutions, and the "no time" theme may indicate that if a new Web-based solution can save teachers a substantial amount of time, teachers may be compelled to pay a fee out of pocket.

20.4.1.2 Concept Testing

The goal of a concept testing focus group is to assess the receptiveness of target users to a new product, Web site, or service. The new concept is illustrated in a manner that demonstrates its features and value. The discussion guide for concept focus groups is organized around the concept stimulus materials. These materials can include storyboards, vignettes, movies, or prototypes. Depending on the detail of the concept stimuli, a skilled moderator can cover one concept approximately every half hour.

Focus groups are typically accomplished with end users of the product or Web site. While it is unlikely that a statistically representative sample would be generated in focus group research, it is important to develop a research plan that ensures that a range of end users are represented in the set of panelists (Wan Hassan et al. 2008). For cases in which there is a range of target end users, it may be difficult to represent all of the potential user groups; this should be noted and considered when analyzing results. These focus groups should have no fewer than 6 participants and no more than 12.

For the concept testing focus group, the discussion is organized around the concepts. The moderator must ensure that the concept is understood by the participants and then obtain an initial vote with the panelists' rationale. During the discussion of the rationale, issues regarding potential pros and cons of the concept should be elaborated by the participants and then a second vote taken. Of particular interest to the researchers is a panelist's change in position on the second vote. Panelists are most accurate when the vote is regarding their own behavior and has the right level of specificity; for example, ask "If this site were available to you, would you make it your home page?" rather than "Do you think this site will be a successful home page?"

Concept test focus groups data are organized by concept and theme. The themes should be discussed within the concepts unless some underlying user need emerges during the discussion. It is important to note that focus group research does not stand alone with regard to understanding the potential usage and acceptance of a new Web site or product. Panelists can be unduly influenced and limited to their current experience and knowledge of emerging technologies. Presumably, the concepts proposed for focus group presentation have been conceived by the group sponsor with due diligence (e.g., review of the secondary and primary research, technical feasibility, and market opportunity). Recruiting *lead users*, or users on the leading edge of technology adoption, is a means of preventing limited user experience from suppressing the positive assessment of forward thinking concepts or Web sites (Von Hippel 1986).

20.4.1.3 Solution Assessment

The solution assessment focus group is the least common but is very important for understanding the cost of ownership and support of a new product or Web site. These groups are utilized when the effect of a new product or service may require a retooling of processes, procedures, or increased service capacity for the end users to interact with the new system in a satisfactory manner. This group should be conducted after concept testing and prior to development. It will help determine the development constraints and the cost side of the return on investment equation.

The discussion guide for these types of groups is organized around the processes and procedures of a provider that are accessed for supporting the current customer interaction models and services. For example, in their design of a customer support Web site for print troubleshooting, Pilsung et al. (2006) used three aspects (i.e., home, troubleshooting, and issue list) of the customer experience to structure focus groups with customer support group managers and help desk call agents.

The participants of these focus groups should include lead users from each of the internal organizations that might be affected by a particular customer contact. It is also useful if the participants have knowledge of how the new solution might be implemented from a technological perspective. The number of participants that attend this type of focus group varies depending on the complexity of the processes and the number of internal user groups that are involved; 12 panelists are the most a moderator should manage at a given time.

The discussion of this focus group is organized around processes and as a result, all participants need not be present at all times during the focus group session. Consider a situation where a customer contact point is a call center that triages incoming calls and places the customers in a queue according the customer request. When the discussion moves beyond the customer support processes that involve the call center, those representatives are no longer needed for the discussion. In these cases it is not unusual for the number of panelists in the focus group to expand and contract as the discussion progresses. Typically, at least two representatives from each of the groups that are involved in a given step of the process should be present. It is important for the moderator to keep the discussion flowing and not let the group become hindered by details. Moderators should identify the issue, articulate the issue back to the panelists, capture the issue, and proceed to another issue.

These groups generate a list of issues that need to be further explored, typically organized in terms of their severity. The actions that need to be taken in order to address the identified issues should be characterized as requirements to the development, implementation, and fulfillment teams.

As a methodology, focus groups can be used in a number of situations, but it is difficult to imagine a scenario where it should be the sole method for conducting user research and identifying early requirements. When used in coordination with other methods, they are highly useful and can add considerable value to the research process.

20.4.1.4 Brainwriting

Brainwriting groups are a less common but potentially useful way of eliciting more creative ideas from focus group participants. Though there are variations of brainwriting methodology, all are comprised of a period of silent writing in response to the facilitators probe statement or question followed by some form of cooperative elaboration (Heslin 2009). These two steps are completed prior to any oral evaluation of the ideas. This approach diminishes some of the potentially negative influences of the focus group dynamics such as cross-talking, premature evaluation of ideas, and blocking (i.e., discouraging others from speaking up).

Singh et al. (2008) employed brainwriting techniques in combination with traditional focus groups to assess U.S. Hispanics attitudes toward Web content and potential adaptation of content. They employed focus groups to generate original Web site content themes (e.g., importance of family) and then used brainwriting sessions to elaborate potential site adaptations.

20.4.2 INTERVIEWS

An interview is typically a face-to-face meeting between a user and a researcher. Often, the meeting topic is controlled by the researcher but variations are possible. To facilitate documentation, sometimes a voice or video recording device is used or another researcher, whose role is mainly to log the conversation, is present. Interviews can be used for various purposes such as

- To gather background information on users for the purposes of prescreening or profiling;
- To ask for further clarifications during observations (contextual inquiry) to set the topics of discussion in the context of the user's natural environments;
- To confirm previously generated task models and evaluate any existing product prototypes (Forlizzi and McCormack 2000); and
- To foster discussions and clarify any questions about the previous observations immediately after the observations.

Depending on the purposes of the interviews and the preexisting experiences of the researchers with the users, interviews can be grouped into two categories: structured interviews and semistructured interviews.

20.4.2.1 Structured Interview

With structured interviews, researchers prepare a list of specific questions beforehand. Structured interviews are generally used in the following situations, when

- There is a prescheduled fixed amount of time for the interview;
- Researchers are knowledgeable about the user group to know what specific questions to ask;

- Researchers need users to answer detailed questions and expect short answers; and
- Researchers are interested in gathering a consistent list of information from a larger group of users. Having similar data on similar questions makes the data analysis and identification of data patterns easier.

The questions are typically organized in certain categories such as the user's skill background or their job experiences with certain applications, similar to a structured survey. Although it is possible to use surveys to obtain similar data, there are several advantages of using interviews. First, the rate of response is higher, because the interviewer is present to ensure all questions are answered. Second, question phrasing can allow more room for interpretation because interviewers are able to clarify the questions if they are misinterpreted by the users. Last, interviews provide an opportunity for face-to-face meetings, which are important for establishing a stronger relationship with customers for further research opportunities.

20.4.2.2 Semistructured Interview

Unlike structured interviews, semistructured interviews are conducted with loosely defined talking points or broad questions. These can be employed any time during the development process but are often used iteratively with contextual observations (Kiewe 2008). The specific questions are generated during the interviews based on the user's answers. Therefore, semistructured interviews are more open-ended and the specific topics discussed can vary from one user to another. Semistructured interviews are typically used when (1) there is little preexisting knowledge about the user population, (2) asking a specific question can color the user's expectations, and (3) specific questions need to change from user to user on the basis of their job roles. Because of the loose organization, semistructured interviews tend to run longer when researchers uncover surprising answers that need to be drilled down to significant details.

20.4.2.3 User Walkthrough

One very common form of the semistructured interview is a user walkthrough. Taking a task-based perspective, interviews can be used to get the users to walk the researchers through their daily tasks, the detailed sequence of steps in completing the tasks, and the information and artifacts involved in each of the steps (Rouncefield et al. 1994).

A more commonly used form of this method is to ask users to walk through a product the researcher is interested in evaluating. The goal is to gain insights about how the product fits into the user's work practice and whether the product meets the user's expectations. This form of walkthrough is usually done at the end of interview sessions to avoid any bias that a specific product can introduce.

Similar to a usability test (see Vu et al., this volume), user walkthrough can be conducted around a few scenarios and tasks. Every step of the way, the researcher should pay attention to whether or not the user understands the product, how

to complete the task, sufficient feedback is given as a result of the user's actions, and the user knows where to go next or that the task is completed. In addition, when the product teams have some areas of concern about the prototype, these general areas can be used as talking points to obtain the user's opinions. Researchers must pay special attention to the type of questions that the users ask about the product and any relation they can make with their existing tools or work environments.

20.4.2.4 Conducting Interviews

20.4.2.4.1 Introduction

It is important to put the interviews in the appropriate context so that the user knows the general purpose of the interview. Ask the user for the amount of time he or she has for the interview, and be flexible if it differs from the time required for the interview. Make it clear to the users that the interview is voluntary. If there is any company confidential information involved, ask the user to sign the appropriate forms prior to the interview.

20.4.2.4.2 Break the Ice

It is important to make the user feel comfortable with talking to the researcher. After all, the data are only as good as what the user can and will tell the researcher. To help break the ice, it is often useful to ask the users to talk about a personal but nonthreatening topic. Such topics could include the personal work history or professional\personal path to their current role, their educational history, or where they grew up. During this initial discussion the interviewer should look to identify aspects of the user's answers that the interviewer can connect with in some way (Esterberg 2002). Because the interview is typically about what they do at work, users' answers might be a job title, their roles, or tasks. Communicate positively with users in an effort to put them at ease with the interview process and to open the lines of conversations with them.

20.4.2.4.3 Ask Neutral Questions

The way researchers ask questions can influence how the users answer the questions. Users have a tendency to agree with the researcher's point of view because they assume that researchers know more or have more invested in the specific topic. It takes more courage and energy for them to disagree with the researchers. The best way to avoid this practice is to phrase the interview questions and respond to the user's answers in a neutral manner; for example, phrase the question "What do you think . . . ?" instead of "Don't you think . . . ?" or "Isn't it easy to . . . ?" Also, the researcher needs to attend to his or her own body language when the user answers. Instead of showing either strong disagreement or agreement with the user's answer, the researcher should visibly pay close attention to what the user is saying.

20.4.2.4.4 Keep the Interviews Grounded

One concern about interview data is its self-report nature. It is well known that humans err and sometimes what they claim to do is not what they actually do. Users have a tendency to exaggerate, generalize, and skip over specific details, which may or may not be compatible with the generalization. It is important to keep interviews grounded to the specific factual incidents related to the topic being discussed. For example, Newman and Landay (2000) asked Web designers to walk through the entire process of designing a recently completed Web project. The conversations were grounded to the actual happenings in the recent past to promote fewer memory errors. Ask the user to present the specific document or tools related to what they are talking about. Seeing artifacts reminds users of the details they may have forgotten to mention. Also, focusing on the factual details allows researchers to reach their own conclusions about what was happening in addition to the user's own reflection.

When asking users about historical events, it is important to ground the users' memories in the timing of a known event. For low base-rate incidents of user behavior that grounding can be related to some other event that will assist the user in accurate recall. For example, if a user of a Web site recalls a particularly poor postpurchase interaction with the provider, an interviewer might ask the time of year. If the event happened in August, a contributor of the user's dissatisfaction may have been related to that time of year (i.e., returning to school).

20.4.2.4.5 Encourage Users to Answer Unasked Questions

Regardless of how well a researcher plans, there are always elements of surprise; many times, the nuances of research data come from these surprises more than any confirming data. It is important to allow for these unexpected elements to surface and be captured in interviews. One way of doing this is to ask questions such as "Is there anything else . . . " at the end of every section. Another way is to appropriately remind users during the interviews that the researcher is interested in seemingly unrelated details as well. Asking the users to expand on something they have touched on but did not explain in detail can also lead to unexpected topics researchers did not plan for but might find interesting.

20.4.2.4.6 Ending the Interview

Properly ending the interview can be a stepping stone into future research opportunities with the user. Debrief the user on the goals of the research, what will be done with the data, and what significance the interview will add to the research. Leave a card or contact information if the user has anything else to add to the interview conversations. Ask whether the user is interested in participating in future research efforts.

20.4.3 SURVEYS

Surveys are excellent tools for gaining invaluable insights into the wants and needs, attitudes, and experiences of almost any population of interest, from customers to subscribers, donors to employees. Traditionally accomplished by mail or

by telephone, the growing ubiquity of internet connectivity worldwide—coupled with the increasing availability of user-friendly online survey services and software—has dramatically decreased the cost, effort, time, and other resources required to field a successful survey.

The use of Web surveys for public opinion polling, customer satisfaction assessment, and graduate research has skyrocketed in recent years. But even with almost a decade of insight and experience with this mode, and supported by a growing body of professional and academic literature, Web survey methods are still very much in flux. While many of the practices useful to traditional survey methods can also be applied directly or indirectly to Web survey implementation, even a casual review of a handful of Web surveys reveals dramatic differences in visual design, interactivity, recruiting strategies, and other areas that influence response rates.

A survey—whether Web or mail or phone—can stand on its own as a valuable research tool. It can also be crafted, shaped, and enhanced through the learning and insights gained from other, preliminary research efforts (focus groups, interviews, etc.). Equally useful is that a survey can facilitate the creation of knowledge that allows researchers to later conduct more successful focus groups, interviews, contextual observations, or other research endeavors. But whether a particular project's needs demand a phone, Web, or postal mail mode, there are a number of considerations and steps that should be taken to improve a survey research project's design, implementation, response rate, data quality, and validity.

20.4.3.1 Crafting the Survey Experience

In many ways, fielding a successful survey also requires a bit of theatricality. Special considerations and attention to detail elevate what might otherwise seem like junk mail to a critical communication worthy of a respondent's scarce and valuable time and attention. An ideal survey experience requires a number of elements working in concert—from the questionnaire to the incentives to the content and timing of contact letters and more—to catch a potential respondent's eye, to make them want to open the recruitment letters, and to provide candid and complete responses to the questions (Dillman et al. 2008; Porter 2004). The goal is to make them feel, at least for a moment, that their opinions are invaluable and their effort in completing the survey is critical to achieving the important goals of the research.

20.4.3.2 The Questionnaire

Whether working with paper or online surveys, writing clear, concise, and appropriate questions will require a significant amount of time and effort. Poorly written questions can doom a project from the start, resulting in measurement error (where response data do not accurately assess the sample on the intended dependent measure), nonresponse error (where respondents skip items that are confusing or difficult), or—in the worst cases—result in much lower response rates when users abandon the survey midstream because of frustration, confusion, or boredom.

The types of questions asked by researchers, the specific order and wording used in each item, the response options (scales), and how the researchers intend to use the data (e.g., simple tabulation, regression models or other advanced statistical analysis) should all be considered carefully throughout the question design process.

20.4.3.3 Question Types

Three types of questions are most often utilized in self-administered questionnaires, including open-ended, closed-ended, and partially closed items (Dillman 1978, 2000; Dillman et al. 2008). The selection of question type for a particular item depends on the questions to be asked as well as any postsurvey constraints, such as coding, analysis, and reporting concerns.

Historically, open-ended questions have a higher nonresponse rate (Dillman et al. 2002). One recent Web survey reports an open-ended question at the beginning of the survey potentially contributes to higher abandonment rates, though respondents to open-ended items placed elsewhere in survey instruments often provide large amounts of information (Crawford, Couper, and Lamias 2001). The verbose and free-form open-ended responses provided by participants can be useful in gaining additional insight, context, and perspectives on their responses to closed-ended items (Converse and Presser 1986; Tourangeau, Reips, and Rasinski 2000). When reporting survey results, the use of respondent quotes from open-ended items can prove particularly powerful in framing findings, analysis, and recommendations (Dillman 2000; Fowler 2002; Rea and Parker 1997).

Closed-ended questions should be used whenever possible (Rea and Parker 1997), but researchers must consider both the content of the question and the content of the responses with great care (Dillman 2000). Such items are used when the researcher understands the range of potential meaningful answers and can communicate them in a response set that is easily understood by the user (Dillman 1978). Generally speaking, closed-ended questions are much easier for the user to answer. Ordered response categories are easier for users to manage than unordered (Dillman 2000). Closed-ended responses require little coding interpretation and the results are relatively simple to report.

20.4.3.4 Writing Items

As researchers work to draft survey questions, there are a number of best practices that should be kept in mind. Whether reviewing individual items, groups of questions, or the instrument as a whole, constant and repeated efforts should be made to refine survey items to their essential, most effective form. While these guidelines are arguably "common sense" concerns that all researchers should have in mind, they are certainly not universally intuitive and realistically, are easy to overlook under the pressures of deadlines, complicated survey designs, or other distractions (see Table 20.2). A more complete set of design principles can be found in Dillman et al. (2008) and Rea and Parker (1997).

20.4.3.5 Visual Design

While the questions included in a survey are arguably the single most important aspect of the instrument, it is crucial to appreciate the power of visual design elements to support or undermine the success of paper or Web surveys. Once the questions have been drafted, how they are layered into a final instrument (that is, either the paper questionnaire or an online form) will greatly influence the response rate and usefulness of the data collected (see Table 20.3).

20.4.3.6 Incentives

Both academic research and professional experience into avenues for increasing response rates have indicated that the offering of incentives—whether token or substantial—can decrease a survey's abandon rate (Lozar Manfreda and Vehovar 2002) and positively influence the number of potential respondents who complete a survey (Dillman et al. 2008; see Table 20.4).

TABLE 20.2

Question Writing Guidelines

1. Use the most basic, simple language to construct the question so that it can be easily understood by the target participants (Dillman et al. 2008; Rea and Parker 1997). For example, "use" in place of "utilize." Keep both language and sentence structure simple and well targeted to the survey's audience.

2. Write questions that are clear, concise, and lack ambiguity (Dillman et al. 2008; Rea and Parker 1997).

3. It is highly important to ask only one question at a time. If the item contains "and" or "or," there is a good chance that the researcher has inadvertently asked more than one question. Such compound items can frustrate users and cloud the meaning of their responses.

4. When possible, shorter questions are preferred. Longer sentences are more likely to contain complex phrases and sentence structure, often making such items difficult to comprehend while also increasing the respondent's cognitive workload (Dillman et al. 2008; Rea and Parker 1997).

5. Does the question or its response choices indicate any bias? Use extreme caution and multiple reviews by various researchers to ensure that neither items nor scales bias the users' potential response (Dillman et al. 2008; Rea and Parker 1997). For example, do not introduce the user to new facts, avoid mentioning one side of a semantic differential scale, or lead users through your choice of response categories (Dillman et al. 2008; Rea and Parker 1997).

6. Is the question appropriately specific? Response choices should not be so specific that the user cannot possibly determine the answer and yet specific enough to be useful for the study (Converse and Presser 1986; Dillman et al. 2008; Rea and Parker 1997).

7. Is there specific wording or the appearance of a "tone" in any of the items that may be explicitly or inadvertently objectionable to the target audience (Dillman et al. 2008; Rea and Parker 1997)? This is of particular concern when a survey is cross-cultural; pay particular attention to questions, sentence structure, and language choices that may be perfectly acceptable in one culture but offensive in another.

TABLE 20.3

Visual Design and Ordering Guidelines

1. Place instructions where they are likely to be read and understood (Dillman 1978; Kaye and Johnson 1999).
2. Make clear where each page starts, ends, and what to do upon completing a page.
3. Use contrast (e.g., smaller and larger type, bold fonts, or color) to make visual processing of the survey easier.
4. Keep question implementation similar to allow the user to focus on the content of the question instead of the implementation (Dillman and Bowker 2001).
5. White space should be used to provide visual separation between elements of the survey, such as between instructions and items.
6. Limit the number of fonts, colors, and contrasts used. Each extra element may hurt more than it helps (Dillman et al. 1998; Kaye and Johnson 1999).
7. Ask demographic questions first (Bosnjak 2001; Passmore et al. 2002; Reips 2002).
8. Make the first question interesting to the participant (Dillman et al. 2008; Dillman and Bowker 2001).
9. Do not require contact information such as phone number early in the survey (O'Neil and Penrod 2001; Solomon 2001).
10. Do not place open-ended questions at the beginning of the survey (Crawford, Couper, and Lamias 2001).

20.4.3.7 Recruiting Participants

Recruiting survey participants has never been an easy task. Traditional recruiting for mail surveys requires significant investments in time, labor, and money. Printing contact letters and questionnaires requires real dollars. Assembling and preparing (folding, stuffing, addressing) each round of invitations for the mail demands a significant amount of labor (or more money). Obtaining postage to distribute enough invitations to yield enough responses can be prohibitively expensive.

As Web surveys increase in popularity, even greater numbers of survey invitations will be sent by electronic mail (see Reips and Birnbaum, this volume). In some respects, recruiting by e-mail is a researcher's dream come true: there are no envelopes to stuff, postage is free, and batch mailing software (often) handles all the addressing and personalization of each invitation. But e-mail recruiting is hardly a panacea. Although there are certainly many efficiencies, e-mail in-

vitations compete for a potential respondent's attention in the inbox, possibly lost in a sea of spam messages, important business communications, and personal correspondence.

20.4.3.8 Testing the Survey

Conducting a limited test of a survey prior to wide distribution is an essential step in identifying and mitigating any lingering or unforeseen problems in a survey project. Depending on the type of survey to be implemented, the list of items to review will be slightly different.

20.4.3.9 Review by Colleagues, Experts, and Clients

Initial review of a near-final survey instrument and associated materials by trusted colleagues, subject-matter experts, and client representatives is an important, useful step that should be completed prior to pretesting the survey. These reviewers may prove invaluable in ensuring completeness

TABLE 20.4

Incentive and Recruiting Guidelines

1. For mail surveys, provide a small cash incentive (e.g., crisp dollar bill) (Dillman 1978; Dillman et al. 2008).
2. For Web-based surveys, incentives may include a random drawing, noncash incentives such as t-shirts, gift-memberships, or other braded material.
3. In situations where it is in the users' self-interest to respond to a survey incentives are less necessary.
4. Incentive choices should be based on (1) knowledge of and discussions with the survey sponsor, (2) knowledge of the target populations, what incentives (within the constraints established by the client) are most likely to motivate participation, (3) the mode and logistical realities of the survey project.
5. Invite participation by posting invitation to online sites and mailing groups that are targeted for your desired population (Kaye and Johnson 1999; Reips 2002).
6. State the target users in the invitation and in the introductory letter to the survey (Kaye and Johnson 1999).
7. Assign users a personal identification number (PIN) number in the invitation (Dillman and Bowker 2001; Heerwegh and Loosveldt 2002; Kaye and Johnson 1999).
8. Participant recruiting process should have multiple contacts including a preletter to cue respondents (Lozar Manfreda and Vehovar 2002), the e-mailing, a thank you reminder, a second e-mailing, and a final notice (Dillman et al. 2008).
9. Careful selection of the sponsor's name that appears on the invitation envelope, letterhead, or e-mail header can motivate or (dissuade) users from participating, based on the sponsor's perceived authority and relationship to the participant (Dillman et al. 2008; Porter and Whitcomb 2005).
10. Personalization of the invitations (signing letters or addressing envelopes by hand, using real stamps, or using actual names in salutations) underscores the uniqueness of the invitation and the perception that each participant's response matters (Dillman et al. 2008; Porter and Whitcomb 2005).
11. Timing of the survey matters. Know your users and when they are likely to respond to a solicitation. For example, college students are more likely to check their e-mail late in the day while business users are more likely to check their e-mail early in the morning. This timing can affect the position that an e-mail appears in the inbox when a user logs in.

and appropriateness of items and scales (Presser et al. 2004), may lend support or challenge assumptions about the wording of instructions or visual design decisions, or may provide insights that may prompt researchers to include questions not previously considered important or remove items deemed superfluous or biased.

20.4.3.10 Pretesting Surveys

In the most basic sense, pretesting a survey or a phone script with a few dozen participants provides an opportunity for the researcher to understand how the recruiting materials, survey experience design, and research instrument work together "in the real world." Respondents who may be less familiar with the survey's subject matter may explicitly vocalize problems or concerns with certain items, often spot overlooked typographical errors or poorly worded questions, may share that they are put off or confused by certain language within the contact letters or instrument, or may identify technological problems or challenges in completing the survey. Even beyond explicit comments from respondents, researchers may notice patterns in the data, such as survey or item nonresponse, or responses to certain pairs or groups of questions that seem unusual, which may necessitate additional revisions.

20.4.3.11 Capturing Data

No matter what mode a particular survey project will use, successful capture of survey responses is anything but point-and-shoot simple. Constant attention must be paid to the return of postal surveys, to the continued functionality of Web survey infrastructure, and to the incoming data itself to ensure the project is progressing satisfactorily and to identify and mitigate any potential problems.

20.4.3.11.1 Online Surveys

The number of "moving parts" that facilitate the fielding of a Web survey (e.g., Web server, survey server, and database)

should be checked and rechecked regularly throughout the project. While this can be done at random intervals to spot-check the survey, such tests should also be conducted immediately prior to sending each new round of invitations to potential respondents. Minimally, regular checks should include verifying that the Web server is operational, users can access the survey, any skip logic continues to work, and completed response data are being saved to the database.

20.4.3.11.2 Traditional Surveys

Surveys fielded using more traditional modes (e.g., postal mail) are usually much less complicated affairs to manage once they have been distributed to potential respondents. Even if researchers do not intend to enter data from traditional survey projects immediately upon receipt, some (if not all) envelopes returned should be opened and reviewed for any indication of potential problems with the recruiting letters, the incentives, the instrument, or any other aspects of the project. Consider this a much larger version of the pretest conducted on the survey package prior to fielding the project in its entirety; there may be feedback or other insights to be gained that can prompt changes to one or more components of the survey—thus dramatically improving the survey and the data collected—for subsequent recruiting rounds.

20.4.3.11.3 Privacy Issues

The advantage and disadvantage of survey research is that many hundreds or thousands of respondents share valuable information about sensitive or proprietary topics. Whether a survey focuses on sexual behaviors or drug use, customer satisfaction or product experiences, it is incumbent upon the researcher to treat respondents and their data with the utmost care and respect. This is important not only because it is the right thing to do but also because there may be significant legal or financial consequences for not properly handling such data (see Table 20.5).

TABLE 20.5
Privacy and Trust Guidelines

1. Be as transparent as possible with potential respondents about a survey's purpose, sponsor, and how data will be used.

2. Both implicit and explicit mentions in recruiting contact messages can (when appropriate) inform users that no individually identifiable responses will be shared (e.g., that no personally identifiable information will be collected, or perhaps that all results will be reported in the aggregate and that no identifying information whatsoever will be shared with third parties) (Best et al. 2001).

3. A phone number or e-mail address should be included should potential respondents have questions of the researchers before or after completing the survey.

4. Explain the following in the "motivational" introduction: (1) the credentials of the researcher or research group, (2) why the participant was selected, (3) the topic of study or purpose of the survey, (4) the benefits that the participant can expect from taking part in the study, and (5) clear instructions on how to participate (Dillman and Bowker 2001; Hewson, Laurent, and Vogel 1996; Passmore et al. 2002).

5. In many large organizations (e.g., school districts or major corporations), there may be privacy policies or other rules in place that limit or prohibit potential respondents from participating in surveys. These policies are often posted on organization Web sites and may offer insight or instructions for obtaining special permission for that audience to participate.

6. There are a variety of federal, state, and possibly local laws that may influence how a survey can be fielded. Two of the biggest, most well-known laws include the CAN-SPAM Act (http://www.ftc.gov/spam/) and the National Do Not Call Registry (https://www.donotcall.gov/), which respectively affect e-mail and telephone recruiting.

7. Obtain qualified legal advice concerning any potential restrictions, conditions, or other considerations that may influence the survey process and to reduce any potential liabilities.

20.4.4 CONTEXTUAL OBSERVATIONS

We use the term *contextual observation* to refer to the group of observation methods that require the researchers to systematically watch users in their normal or natural environments. The intention is to discover users as they normally appear in their day-to-day operations. There are two schools of thought in using this method: nonparticipant observation and participant observation.

20.4.4.1 Nonparticipant Observation

With nonparticipant observations, researchers try to be as nonintrusive as possible in order to minimize the effect of the researcher's presence on the observed events. This method is referred to as *naturalistic observation* (Shaughnessy, Zechmeister, and Zechmeister 2009) and *shadowing* in ethnographic literature (Bowers 1994). Researchers simply follow the users around and observe what they do, trying not to ask any questions as events occur. The advantage of this observation method is that it is less likely than participant observation for the researchers to have changed or manipulated the user's normal activities, thus better preserving the authenticity of the natural data. The disadvantage is that it may be difficult for researchers to understand the events, making it difficult to formulate these observations into tangible statements and recommendations later. Nonparticipant observations can be used for more observable events such as how many times the user opened a certain Web site on their computers daily. Nonparticipant observation can also be combined with follow-up interviews to allow researchers to ask any questions based on the recently observed events.

20.4.4.2 Participant Observation

With participant observation, researchers attempt to become a participant in the user's environment. The intention is for the researcher to be completely immersed in the user's environment and culture and to become its "natural" member over time. Through such thorough exposure to the user's environment, the researcher will develop a direct feeling and a deep understanding of the reasons behind events that may not be easily observed otherwise; for example, start to use the various Internet technologies used by the participants in the corporate sites (Millen 2000), hang out with the participants as much as possible, work together with the users through the different tasks in solving the problems, share the junk food together on the night shift in the network operations center, and attend their weekly meetings. The advantage of this type of observation is that the researcher is more likely to understand the hidden forces or unspoken rules behind the events being observed. The disadvantage is that it is more likely for the researchers to have changed the user's normal activities by being a direct contributor in the user's environment. Participation observations are more informative when the researcher has already had a well-established relationship with the users and for long-term research projects. A related but slightly different approach is for the participants themselves to become documenters of their environment.

In industry settings, researchers must almost always acknowledge themselves at the user's site. The very presence of external personnel might change the observed events regardless of the observation method used. Given enough time, the researchers will have some impact on the users and their environments. Sometimes, more than the method itself, it is the researcher's interpersonal skills and attitudes that make the users feel so comfortable that they behave as if the researchers were not around. Researchers must enjoy meeting new people, be open-minded, approach every user with a sense of humility, value what users have to say, and appreciate the expertise that users show.

More than any other methods discussed in this chapter, participant observation allows researchers to spend more time with the users in their natural environments. This is an excellent opportunity for the researcher to establish an ongoing partnership with the users and their organizations in general. We have experience with several companies where this type of customer-partner relationship has been used successfully as a reliable source of information for identifying user needs and piloting and codeveloping usable products.

20.4.4.3 Specific Observation Issues

20.4.4.3.1 Time Sampling

When it is impossible to spend weeks and months at the participants' site, it is important to carefully pick the times that researchers do spend on site so that they get to observe a spectrum of the events. For example, select both busy days and quiet days in a small office (Rouncefield et al. 1994), day and night shifts in a network operations center, or weekdays versus weekends and school days versus holidays in people's homes (Bell 2001).

20.4.4.3.2 Multiple Researchers

Instead of the one-on-one approach with one researcher and one participant, recent methods used for Web applications adopted a team approach, where two or more researchers team up to make the best of the time spent with the participants in person and on site. In European countries, Bell (2001) had teams of two go into homes, one being the main communicator who interacted with the participant, the other being the note taker who recorded and gathered artifacts. Wixon et al. (2002) had teams of two, one being a scribe or note taker, who wrote time-stamped notes on low-level tasks, and the other being a photographer who took photos and collected artifacts. Millen (2000) encouraged two or three researchers on site to conduct observations at different parts of the site. Contextual design (Beyer and Holtzblatt 1998) encourages research to be done by a design team, which consists of key stakeholders in a product development team such as engineering leads, marketing professionals, and documentation specialists. Members of the design team would then be divided into groups of two to spend time with a user on site. Again, one person is a note taker and the other the interviewer. Inclusion of multiple researchers/stakeholders that have different roles in the development process also facilitates the integration of usability findings into the design of the product (Sherman 2009). While a multiple researcher approach is best, a single researcher approach can yield meaningful results (Kiewe 2008).

20.4.4.3.3 Collect Artifacts

Artifacts can be an important part of what users are and do. Whenever possible, ask the users to show the researcher the artifacts. Make a point to collect them throughout the whole research period. They will become some of the most revealing and convincing tools you can use to convey the data back to the development team. The method of collection can be a physical copy of the object or a picture of what the object looks like in the user's environment.

20.4.4.3.4 Textual Observations

Internet technologies, such as online discussion groups, provide unique opportunities for researchers to observe the online communications between the members of the group (Millen 2000). Because much of the communication is already captured in text format, this method can alleviate any recording errors that may be present in the field notes. There are cases in which an online discussion group is the only form of communication among group members (Eichhorn 2001). Although Eichhorn tried to challenge "the assumption that ethnographic field work is necessarily dependent on physical displacement, and face-to-face encounters with our research participants" (p. 565), face-to-face meetings with the participants proved to be "an important turning point" (p. 571) in the very same study. However unique the opportunity is to be able to observe textual communications, it remains secondary and complementary to the face-to-face meetings and observations with the users.

20.4.4.3.5 Self-documentation by Participant

Asking the users to document what is important to them with regard to a specific topic can be another source of valuable data. Millen (2000) had informants create a personal history of their Internet use in trying to understand how technologies have affected them. Forlizzi and McCormack (2000) gave their participants, "logbooks and disposable cameras, asking them to write about their life goals they were currently trying to attain, to define what health, wellness, and fitness meant to them, and to catalog products that they relied on daily" (p. 276). Kumar and Whitney (2007) propose a model where the participants are equipped with disposable cameras and instructed to take as many pictures of the "people, objects, environments, messages, and services" (p. 51) of the most important aspects of the topic of study. Whenever necessary, the users could snap a quick photo of what was important to them. Sometimes users are asked to accomplish a set of tasks and document potential issues that could be solved with a new solution (Mattelmäki 2005; Obrist, Bernhaupt, and Tescheligi 2008). These self-documentaries can also be used as talking points during follow-up interviews with users.

20.4.4.3.6 Dealing with the Collected Data

Observation methods produce a vast amount of data, mostly qualitative, from various sources and media. Data include any information and artifacts gathered during the research stage (e.g., notes taken by the researchers, audio or video, photo stills, anything the participants created in response to or as a request of the researchers, text scraps of the online discussions, and chat rooms). Making sense of the data afterward can be a daunting task. In order to prevent a loss of data, it is important to conduct a daily review of the research notes prior to the next day of research. In addition, it is important to verify the data description with additional data that are obtained either from new users or more interviews from the same users (Wood 1996).

Although it is possible for one individual to analyze the data, it may be better to involve the entire research team. Typically, the goal is for the team to identify meaningful patterns and categories from the data. This can be done through a variety of activities such as constructing affinity diagrams (Hackos and Redish 1998), entering insights from the material into the database for clustering and querying (Kumar and Whitney 2007), and reviewing insights with users (Kramer, Noronha, and Vergo 2000). In addition, data analysis by the whole team is a way for the team to agree on the same vision of the customer data and what it means to the product design (Holtzblatt 2009).

20.5 DELIVERABLES

Although a lot of research results have strong implications for designing a product that will fit nicely into the user's world, they usually provide pieces of a puzzle rather than a neat and clear-cut solution. Research data do not get translated into a product design automatically, and it takes a team that is familiar with both the data and the technology to merge the two together in a way in which the technology facilitates what the user wants, needs, and desires to do. Deliverables from the user research are an important first step toward merging the user data with the technology in producing a usable product.

Ideally, the team that collects the user data involves representative leaders from the various aspects of the product development team, such as usability, marketing, engineering, testing documentation, and support. In reality, user data usually are collected by a small team of usability professionals. In this case, quality deliverables become essential tools to communicate the user data to the whole development team.

Among other factors, the form of deliverables or reports depends on the audience. For example, for designers who need to design the user interface, task scenarios and artifacts are useful because they need to design the specific user navigation and task flow in the product. For marketing that is in charge of positioning the products, user profiles and competitive product information are of special interest. For strategists who are in charge of setting future directions of products, the holistic account of the user, their environments, and existing artifacts that may be automated may be interesting.

TABLE 20.6

Deliverables for Different Audiences

Target Audiences	Deliverables
Usability—makes sure the product is useful, usable, and desirable. Specific roles include gathering user information, user interface design, and evaluation.	Complete report: users, tasks, environments, and artifacts.
Marketing—ensures the appropriate features in a product. Gather function requirements for future products, and set strategic directions for products in specific markets.	Presentation showing users, environments, artifacts (competitor's products), and any other user requirements.
Development—implements the product functions: from system architecture to user interface.	Presentation for engineering managers. Complete report for tasks, important sample artifacts, feedback on existing products.
Testing—ensures the quality of the products, makes sure that product does what it is designed to do, and does it reliably and at a reasonable speed.	Complete report for tasks, scenarios, feedback on existing products.
Documentation—writes the document or help information for the product.	Complete report for tasks, feedback on existing products. Especially interested in how help information is needed and used.
Support—answers technical questions from users when users encounter problems with the product.	Complete report for tasks and feedback on existing products. Especially interested in the problems users currently have and how they overcome the problems.
Training—hold classes and seminars to help users learn how to use the product.	Presentation and full report for tasks and real world scenarios. They are interested in making the training materials more relevant to the real world tasks and problems.
Customers/Users	Presentation for tasks, environments emphasizing on issues. They are more interested in recommendations for resolving identified issues.

Table 20.6 summarizes different stakeholders and what they might be interested in from user research.

Whereas it is obvious that deliverables need to be made to the various audiences of the product provider, deliverables back to the customers and users are often neglected. It is important to present or send the collected data back to the customers and users for several reasons: (1) to maintain an ongoing relationship with the customers, (2) to verify the data captured as being accurate to prevent recording errors or avoid outdated information, (3) to help the users and customers better understand each other, and (4) to provide customers with recommendations for issues or problems identified during the research.

20.5.1 A DELIVERABLE PACKAGE

To appeal to the wide range of audiences that may be interested in the user research data, we suggest that the researchers put together a deliverable package, which may be tailored to a specific audience if necessary. We include the following elements in the deliverable package: a complete report, a presentation deck, materials from the user, and materials designed to be persistent in the audience's environment.

20.5.1.1 Complete Report

This is a document that details the purpose of the research, the methods and process used to collect the user data, the detailed findings of the research, and discussions and conclusions that can be drawn from the collected data. This type of document is good for a complete and accurate description of the research, but usually receives less attention from the development team because of the effort needed to read a long document. With a bit of judicious editing, a well-written research design document (described above) can form a significant portion of a project's final report. This involves editing the content of the design document to reflect what actually transpired throughout the project (rather than what was initially envisioned). Once updated, this content can easily be repurposed as the "methods" section of the final report. Of course, simply writing about the project's methodology cannot tell the whole story, merely the mechanics of a project. Additional sections must be fleshed out and added to the report to make it a useful deliverable (see Table 20.7). The two sections that usually receive the most attention are the executive summary at the beginning and the findings and discussion section.

Even though this document does not get read cover to cover in fast-paced industrial settings, we argue that it is still important to have a full report for several reasons. First, all research has qualifiers and a specific method that clearly highlights strength and weaknesses of a research project. This will help consumers of the research understand when it is appropriate to apply the findings and where more research is needed. Clearly outline the results and discuss how they could be used so that other researchers can decide what items to include in their research projects. Second, it is important for the other three types of deliverables discussed below to be associated with this complete report to avoid overgeneralization of the data and to ensure that the useful details collected from the users are captured and available when needed.

20.5.1.2 A Presentation Deck

This is a more digestible version of the deliverable, normally used for presentations to answer questions like "In 30 minutes,

TABLE 20.7
Structure of the Final Report

Executive Summary	Clients—especially at the executive level—do not always have the time or the patience to read through 20 or more pages of long-winded statistical analysis and discussion about every last aspect of a research project. An executive summary at the beginning of the report—perhaps no more than 3 to a maximum of 5 pages—should encapsulate the most important takeaways from the project, including • A few paragraphs on the project's design (e.g., Why did we do this project, what were the goals? When and how did we field the project?) • The results of the project (e.g., How many people responded, what was the response rate? What were the most surprising, compelling, significant findings about the audience, its behaviors?, etc.) • The researchers' assessment (if appropriate) of the implications of the findings for the client's business, actionable recommendations for changes or enhancements to the client's operations, as well as "next steps" that the client should consider, such as additional research focused on particular topics.
Findings and Discussion	This section forms the heart of the final report, conveying detailed information, insight, and discussion about what was learned throughout the project. The demographic characteristics of the sample, distribution of responses to
About the Researchers	Brief biographies of the research team (or at least its leaders) can lend authority and weight to the findings, analysis, and recommendations.
Other sections (as needed)	Depending on the type of research conducted, other sections may be added to the final report should the research team feel it necessary (or if they are requested by the client). Such sections could include the full text of all open-ended responses to survey questions, observations made by the researchers during a focus group session or during individual interviews, or any other data or analysis that the researchers feel contribute to a better understanding of the research.

tell me what you found in your research?" This deliverable consists of a series of slide decks. In these slide decks, it is important to emphasize the limitations of the research and to caution against unwarranted generalizations beyond the data. If the findings are interesting or controversial, the slide deck is usually forwarded and integrated into other slide decks such as marketing's presentations. It is important to associate the slide deck with the complete report. For marketing and higher-level management, this presentation may be the only exposure they get to the user research; for development teams who need to design and implement the products based on the important findings from the research, this presentation can serve as an introduction.

20.5.1.3   Materials from the User

Researchers usually collect artifacts or samples from users that can serve as important reference materials for product teams. It is important to include such materials in the deliverable package so that development teams have real world samples as the target for their design, implementation, and testing efforts. For example, in research for online reporting tools, the researcher may collect example reports from the users. Often, users want to share these reports with the researchers because it is troublesome and time consuming to submit them. The format, content, and frequency of these reports may become very tangible and detailed requirements for a reporting product.

20.5.1.4   Persistent Materials

It is desirable to have hard copies of the major materials available to every member of the product team, such as the users and the major task scenarios. Communicating the research result is not a one-time activity. For large products, it is an

engineering necessity to break implementations into different parts, each being assigned to a team of development group. The physical presence of the research data that represent users, their goals, and tasks will help hold disparate development teams together. Forlizzi and McCormack (2000) established persistence of the user data by putting user profiles on large mobile boards in the development team offices. Cooper (1999) puts up the personas' pictures and information every chance they get in product meetings. Making sure that the design is actually based on the user data is a continuous effort that requires researcher's persistence and discipline. For a development team that is truly producing a product for a well-defined user audience, everyone in the team should be able to tell you who their users are and on what task scenario their work is based.

20.5.2   Deliverables that Answer Some Typical Questions

20.5.2.1   "Who Are the Users?"

It is a common goal of user research to find out who the current users or potential future users are. After the data are collected on this topic, there are various ways to deliver this information. It is usually the case that several types of users are targeted for a product. In addition to a description of each type of user, it is important to describe the relationships between these users in their workplace as well as whether they are primary or secondary users for a particular product. Hackos and Redish (1998) suggest a user list, which outlines the different types of users, including estimates of their percentage in the total user population, a brief description of each user, and a character matrix comparing several users

TABLE 20.8
User Matrix

Statistician Characteristics	User 1	User 2	User 3	User 4
Years of experience as a statistician	8	9	15	2
Computer platform	Windows	Windows	Unix	Windows
Microsoft Office Suite (Word, Excel, PowerPoint)	Frequent user	Frequent user	Use Word and Excel a few times. Never used PowerPoint.	Frequent user
Used previous version of the product	Yes, a little	Yes, tried a few times	Used it two years ago	Never
Familiar with SAS	Yes	Yes	Yes,	Base SAS
	SAS procedures	SAS data steps, Base SAS	SAS stat, QC, IML, Base SAS	
Other business intelligence and statistics tools	No	Used SPSS many years ago	No	No

across relevant characteristics. Table 20.8 is an example of a character matrix that we obtained for statisticians who participated in a study for an enterprise application.

Personas and user profiles are other ways of describing who the users are. Cooper (1999) advocates design by personas. A *persona* is a precise description of a user and what he wishes to accomplish; it is used as the user for whom the product is designed. Originating from abstractions of real users, the description is hypothetical in nature, striving to be precise rather than representative and specific rather than general (Cooper 1999). User profiles are narratives or visual descriptions of the users, including demographic, goals, lifestyle, attitudes information, job title, skills and technical knowledge, and their responsibilities relative to other users in the workplace.

20.5.2.2 "What Are the User Tasks?"

Most products aim at helping users perform certain tasks better, so the ability to answer this question is essential. The format in which the answers are presented varies a great deal in the literature. The simplest form is a list of tasks. Each item in the list is a short description of the task. For example, the following is a task list for a person whose primary job is to generate various reports in a corporate world. Figure 20.1 shows the decisions and steps needed for completing a task of "Create a report for the current quarter."

- Monitor data log.
- Extract data from main frame computers where raw data is collected.
- Pull together appropriate data for the current quarter by merging several data sources.
- Create current quarter report to publish in company intranet.
- Create quarterly year-to-date report for high-level executives.

It is important for tasks to be represented in the users' physical and organizational contexts and to be associated with important factors such as the possible users performing the tasks, the necessary objects or tools needed to accomplish the task, the time it takes to perform the task, and the

larger goals that the task is trying to accomplish. Above all this, it is important to make sure that all of the task information is based on the data actually collected at the user's site, which may include text, photos, videos, etc. It is challenging to discover the important threads in the data and still maintain a high degree of authenticity to the original data. Laakso, Laakso, and Page (2002) attempt to simplify the whole process of collecting and organizing data into tasks via a sequential overview of the whole session as well as task-based sequential views of specific events.

Often, the product team or the human-computer interaction professionals can take user tasks and create use cases around

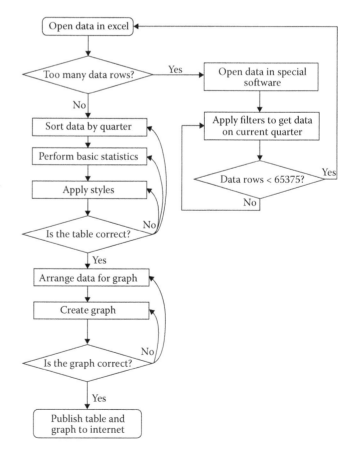

FIGURE 20.1 Task flow for creating current quarter report.

them. For this reason, deliverables representing tasks and scenarios speak well to developers and other technical audiences. The accuracy and coverage of these use cases play a large role in the way the products are designed, implemented, tested, and supported. Therefore tasks- and scenario-related deliverables may have greater impact on product development than any other form of deliverables from user research.

20.5.2.3 "What Tools or Aids Do the Users Have?"

In a sense, all the technologies, applications, and products are artifacts that play a part in people's career and life. In some cases, their whole career is defined by being able to produce proper artifact. Rouncefield et al. (1994) demonstrate that completing paperwork is the definition of working in a small office environment. Figure 20.2 shows the schematic of information replicated from the Enquiry Form.

Descriptions about artifacts include both usage and structure of the artifacts and how they are useful in accomplishing the user goals. In situations where there exist previous versions of a product or competitive products, the use of these products in users' work settings becomes essential to the design and improvement of the new version of the product. For example, in research for an enterprise application in the field of business intelligence, we consistently collected projects that users created with the previous versions of the product. By analyzing these projects, we found that most users concentrated on a small portion of the application function (data manipulation) while ignoring a large class of functions (statistical analysis) offered in the application. Through

follow-up interviews during site visits, we found that users needed to use the data manipulation functions first and the statistics function after that. Users spent so much time making sure the data were correct that they tended to just stick to the basic statistics they understood (e.g., mean and percentage) to simplify their lives. To encourage them to use more sophisticated statistical functions, we needed to (1) significantly improve the data manipulation function and (2) help the users understand what kind of statistics they could use in their professions. The first issue can be addressed by redesigning the user interface and the architecture to improve system performance. The second issue is more challenging and can be temporarily addressed through domain-specific training and will continue to serve as a source for generating new product ideas.

20.5.2.4 "What's the User's Environment Like?"

The environment as it is used in this chapter refers not only to the physical environment (e.g., the physical layout of an office) but also to the organizational environment (e.g., reporting chain or the personal network to get the job done) and cultural environment (e.g., cultural themes). Quite literally, the environment descriptions provide both an understanding and an explanation for the way things are in the user's world.

Perhaps the most obvious way of representing the user's physical environment is a well-chosen picture. The simple exercise of putting your product physically on the picture can give the product teams a sense of what the product must be able to accomplish. For example, if a product is to be used in

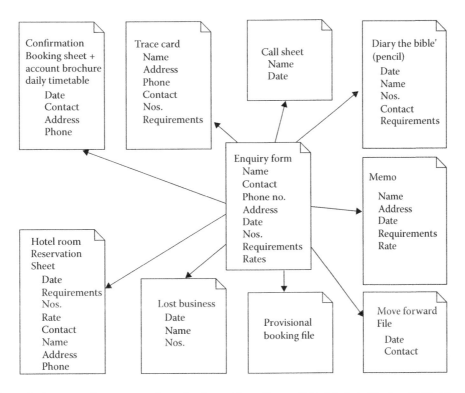

FIGURE 20.2 Schematic demonstrating context of enquiry form. (From Rouncefield, M., J. A. Hughes, T. Rodden, and S. Viller. 1994. Working with "constant interruption": CSCW and the small office. In *Proceedings of the ACM 1994 Conference on Computer Supported Cooperative Work*, 281. New York: ACM Press. With permission.)

the network operations center, the designer must consider the constant interruptions, and when something goes wrong, the place goes into a crisis mode. At such times, anything that is to be noticed (e.g., critical alarm) needs to cue to the user through multiple modes (e.g., color, shape, sound). Expecting the user to notice an icon changing colors on the computer in this environment is less than realistic.

Capturing the user's organizational and cultural environment is tricky. Beyer and Holtzblatt (1998) use a culture model as a tangible description of the intangible forces of influencers (people, organizations, and groups) and the things people think but do not say. Their flow model provides a bird's-eye view of the organization, demonstrating users and their responsibilities and the communication paths between users independent of time and things communicated.

20.6 CONCLUSIONS

Rather than driving the product decision by technology and opinion, user-centered design advocates setting the product direction based on user data. This calls for the application of research methods and processes to the problem of understanding users, their tasks, environments, satisfiers and dissatisfiers, delighters, and tools. We refer to these activities throughout the UCD process as user research. Although there are those who might suggest that the use of the term *research* puts the activity at risk (Sigel and Dray 2003), we think that it is imperative to accurately describe these methods as deliberate, timely, qualitative, and simultaneously scientific when applied with rigor. The goal of this early focus on user research is to provide marketing, product management, and development teams with the data that each needs to make decisions.

The positioning of a human-centered approach prior to or shortly after product conceptualization is fundamental to the efficient delivery of solutions that are more likely to meet needs of users. In other words, it is much more profitable to be the first to market acceptance than the first to market. Market acceptance is acquired by focusing on target users early in the process and prioritizing the potential features of a product. This ensures that the solution meets the minimum requirements relative to user acceptance and competitive solutions.

When using quasi-experimental and qualitative research methods, the application of multiple methods across instances of customer engagements and user populations becomes imperative. This multimethod multicustomer approach to understanding users increases the opportunity for user researchers to accurately assess users' needs. Whereas this approach is not necessarily articulated in this manner in other sources, it is recognized that applying multiple methods during the early stages of the UCD process can account for a greater set of variance with regard to the behavioral success of a proposed solution (Kirwan and Ainsworth 1992; Mayhew 1999). Furthermore, this approach has been advocated in the study of human behavior for nearly 50 years (Campbell and Fiske 1959).

Marketing teams can use early research on users to understand the competitive advantage of a new solution and begin characterizing the content of the marketing messages in a manner that are consistent with the desires and values of the target customers. Product management teams can use early user research to help determine which features are necessary to make the product successful in the marketplace. This feature prioritization can be used to advocate for additional resources to ensure that the necessary functionality to gain adoption and continued use of the solution in the marketplace is included in the product offering. User research data can be used by development teams to ensure that the performance of the new solution is consistent with how the users currently accomplish their goals. The successful transition of metaphors, mental models, language, and task flows from the real world to the online world is an important component of a given solution. The last and potentially most valuable consumer of user research is the corporate leadership team. The user research data can uncover that the proposed solution has little chance of succeeding in its current conceptualization and, at the same time, identify new opportunities.

Finally, it is necessary to (1) conduct user research to enable reasonable articulation of how the proposed solution will meet user needs, (2) get the data early in the process to maximize efficiency of time and resources needed for the development effort, (3) collect the data using multiple methods across multiple customer engagements to validate the findings, and (4) provide the stakeholders with meaningful and useful deliverables to ensure those findings are applied. It is important to note that these things need to occur within the context of an overall user-centered design process that employs user research methods from conceptualization to delivery. Furthermore, to maximize the effectiveness of a user research effort, it must be implemented within the context of a user research program. A programmatic approach ensures that user research is not tied to the success or failure of a single project. By applying these methods early in the process across projects, user-centered design and user research methods will gain credibility in as much as they contribute to the success of the organization.

REFERENCES

Bell, G. 2001. Looking across the Atlantic: Using ethnographic methods to make sense of Europe. *Intel Technology Journal Q3*: 1–10.

Bosnjak, M. 2001. Participation in non-restricted Web-surveys: A typology and explanatory model for item-non-response. In *Dimensions of Internet-Science*, eds. U. D. Reips and M. Bosnjak, 193–207. Lengerich: Pabst Science Publishers.

Best, S. J., B. Krueger, C. Hubbard, and A. Smith. 2001. An assessment of the generalizability of Internet surveys. *Social Science Computer Review* 19: 131–145.

Beyer, H., and K. Holtzblatt. 1998. *Contextual Design: Defining Customer-Centered Systems*. San Francisco, CA: Morgan Kaufmann Publishers.

Boivie, I., C. Åborg, J. Persson, and M. Löfberg. 2003. Why usability gets lost or usability in in-house software development. *Interacting with Computers* 15: 623–639.

Bowers, J. 1994. The work to make a network work: studying CSCW in action. In *Proceedings of the 1994 ACM Conference on Computer Supported Cooperative Work*, 287–298. Chapel Hill, NC. New York, NY: ACM Press.

Campbell, D. T., and D. W. Fiske. 1959. Convergent and discriminant validation by the multitrait-multimethod matrix. *Psychological Bulletin* 56: 81–105.

Childs, S. 2004. Developing health website quality assessment guidelines for the voluntary sector: Outcomes from the Judge Project. *Health Information and Libraries Journal* 21(2): 14–26.

Choe, P., C. Kim, M. R. Lehto, X. Lehto, and J. Allebach. 2006. Evaluating and improving a self-help technical support web site: Use of focus group interviews. *Journal of Human-Computer Interaction* 21(3): 333–354.

Cockton, G. 2006. Focus, fit, and fervor: Future factors beyond play with the interplay. *International Journal of Human-Computer Interaction* 21(2): 239–250.

Converse, J., and S. Presser. 1986. *Survey Questions: Handcrafting the Standardized Questionnaire*, No. 63. Thousand Oaks, CA: Sage Publications.

Cooper, A. 1999. *The Inmates Are Running the Asylum: Why High Tech Products Drive Us Crazy and How to Restore the Sanity.* Indianapolis, IN: Sams.

Crawford, S. D., M. P. Couper, and M. J. Lamias. 2001. Web surveys: perceptions of burden. *Social Science Computer Review* 19: 146–162.

Dillman, D. 1978. *Mail and Telephone Surveys: The Total Design Method.* New York: John Wiley.

Dillman, D. 2000. *Mail and Internet Surveys: The Tailored Design Method*, 2nd ed. New York: John Wiley.

Dillman, D., and D. Bowker. 2001. The Web questionnaire challenge to survey methodologists. In *Dimensions of Internet-science*, eds. U. D. Reips and M. Bosnjak, 159–178. Lengerich: Pabst Science Publishers.

Dillman, D., R. Tortora, J. Conradt, and D. Bowker. 1998. Influence of plain vs. fancy design on response rates for Web surveys. Paper presented at Annual Meetings of the American Statistical Association, Dallas, Texas.

Dillman, D., J. Eltinge, R. Groves, and R. Little. 2002. Survey non-response in design, data collection, and analysis. In *Survey Non-response*, eds. R. Groves et al., 3–26. New York: John Wiley.

Dillman, D. A., J. D. Smyth, and L. M. Christian. 2008. *Internet, mail, and mixed-mode surveys: The tailored design method.* Hoboken, NJ: John Wiley & Sons, Inc.

Eichhorn, K. 2001. Sites unseen: ethnographic research in a textual community. *International Journal of Qualitative Studies in Education* 14(4): 565–578.

Esterberg, K. G. 2002. *Qualitative methods in social research.* Boston, MA: McGraw-Hill.

Forlizzi, J., and M. McCormack. 2000. Case study: User research to inform the design and development of integrated wearable computers and Web-based services. In *Conference Proceedings on Designing Interactive Systems: Processes, Practices, Methods, and Techniques*, 275–279. New York: ACM Press.

Fowler, F. J., Jr. 2002. *Survey Research Methods*, 3rd ed. Thousand Oaks, CA: SAGE Publications.

Hackos, J. T., and J. C. Redish. 1998. *User and Task Analysis for Interface Design.* New York: John Wiley.

Heerwegh, D., and G. Loosveldt. 2002. Web surveys: The effect of controlling survey access using PIN numbers. *Social Science Computer Review* 20: 10–21.

Heslin, P. 2009. Better than brainstorming? Potential contextual boundary conditions to brainwriting for idea generation in organizations. *Journal of Occupational & Organizational Psychology* 82(1): 129–145.

Hewson, C. M., D. Laurent, and C. M. Vogel. 1996. Proper methodologies for psychological studies conducted via the Internet. *Behavior Research Methods, Instruments and Computer* 28: 186–191.

Holtzblatt, K. 2009. Contextual design. In *Human Computer Interaction; Development Process*, eds. A. Sears and J. A. Jacko, 55–70. Boca Raton, FL: CRC Press.

International Standards Organization. 1999. *ISO 13407: Human-centred Design Processes for Interactive Systems.* Geneva, Switzerland: ISO.

Jones, S., and S. Fox. 2009. Pew Internet project data memo. The Pew Internet and American Life Project. http://pewInternet.org/trends.asp (accessed March 10, 2010).

Kaye, B. K., and T. J. Johnson. 1999. Taming the cyber frontier: Techniques for improving online surveys. *Social Science Computer Review* 17: 323–337.

Kiewe, H. 2008. How may I help you? An ethnographic view of contact-center HCI. *Journal of Usability Studies* 3: 74–89.

Kendall, J., and K. Kendall. 1993. Metaphors and methodologies: living beyond the systems machine, *MIS Quarterly* 17: 149–171.

Kirwan, B., and L. K. Ainsworth (Eds.). 1992. *A Guide to Task Analysis.* London: Taylor & Francis.

Kramer, J., S. Noronha, and J. Vergo. 2000. A user-centered design approach to personalization. *Communications of the ACM* 43(8): 44–48.

Kumar, V., and P. Whitney. 2007. Daily life, not markets: customer-centered design. *Journal of Business Strategy* 28(4): 46–58.

Laakso, S. A., K. P. Laakso, and C. P. Page. 2002. DUO: A discount observation method. http://www.cs.helsinki.fi/u/salaakso/papers/DUO.pdf (accessed Jan. 13, 2010).

Lozar Manfreda, K., and V. Vehovar. 2002. Survey design features influencing response rates in Web surveys. Paper presented at the International Conference on Improving Surveys. Available at http://www.websm.org/index.php?fl=2&lact=1&bid=715&avtor=4&parent=12 (accessed April 23, 2010).

Mao, J., K. Vredenburg, P. Smith, and T. Carey. 2005. The state of user-centered design practice. *Communications of the ACM* 48(3): 105–109.

Mattelmäki, T. 2005. Applying probes—from inspirational notes to collaborative insights. *CoDesign* 1(2): 83–102. doi:10.1080/15719880500135821.

Mayhew, D. J. 1999. *The Usability Engineering Lifecycle.* San Francisco, CA: Morgan Kaufman Publishers.

Millen, D. 2000. Rapid ethnography: time deepening strategies for HCI field research. *Proceedings of the ACM 2000 Conference for Designing Interactive Systems: Processes, Practices, Methods, and Techniques*, 280–286. New York, NY: ACM Press.

Molich, R., and J. S. Dumas. 2008. Comparative usability evaluation (CUE-4). *Behaviour & Information Technology* 27(3): 263–281.

Nielsen, J. 1997. The use and misuse of focus groups. http://www.uscit.com/papers/focusgroups.html (accessed Jan. 15, 2010).

Newman, M., and J. Landay. 2000. Sitemaps, storyboards, and specifications: A sketch of Web site design practice as manifested through artifacts. In *Proceedings of Designing Interactive Systems: DIS '00*, 263–274. New York: ACM Press.

Obrist, M., R. Bernhaupt, and M. Tscheligi. 2008. Interactive TV for the home: An ethnographic study on users' requirements and experiences. *International Journal of Human-Computer Interaction* 24(2): 174–196.

O'Neil, K. M., and S. D. Penrod. 2001. Methodological variables in Web-based research that may affect results: Sample type, monetary incentives, and personal information. *Behavior Research Methods, Instruments & Computers* 33: 226–233.

Passmore, C., A. E. Dobbie, M. Parchman, and J. Tysigner. 2002. Guidelines for constructing a survey. *Family Medicine* 34(4): 281–286.

Pilsung, C., K. Chulwoo, M. Lehto, X. Lehto, and J. Allebach. 2006. Evaluating and improving a self-help technical support Web site: Use of focus group interviews. *International Journal of Human-Computer Interaction* 21(3): 333–354.

Porter, S. R. 2004. Raising response rates: What works? *New Directions for Institutional Research* 2004(121): 5–21. doi: 10.1002/ir.97.

Porter, S. R., and M. E. Whitcomb. 2005. E-mail subject lines and their effect on Web survey viewing and response. *Social Science Computer Review* 23(3): 380–387.

Presser, S., M. P. Couper, J. T. Lessler, E. Martin, J. Martin, J. M. Rothbeg, and E. Singer. 2004. Methods for testing and evaluating survey questions. *Public Opinion Quarterly* 68(1): 109–130.

Rea, L. M., and R. A. Parker. 1997 *Designing and Conducting Survey Research*, 2nd ed. San Francisco, CA: Jossey-Bass Publishing.

Reips, U. D. 2002. Internet-based psychological experimenting: five dos and five dont's. *Social Science Computer Review* 20: 241–249.

Reips, U.-D., and M. H. Birnbaum. this volume. Behavioral research and data collection via the internet. In *Handbook of Human Factors in Web Design*, 2nd ed., eds. K.-P. L. Vu and R. W. Proctor, 563–586. Boca Raton, FL: CRC Press.

Rouncefield, M., J. A. Hughes, T. Rodden, and S. Viller. 1994. Working with 'constant interruption': CSCW and the small office. In *Proceedings of the ACM 1994 Conference on Computer Supported Cooperative Work*, 275–86. New York: ACM Press.

Seffah, A., and E. Metzker. 2004. The obstacles and myths of usability and software engineering. *Communications of the ACM* 47(12): 71–76.

Shaugnessy, J. J., E. B. Zechmeister, and J. S. Zechmeister. 2009. *Research Methods in Psychology*, 8th ed. New York: McGraw-Hill.

Sherman, M. 2009. Getting started with contextual inquiries [Web log message]. http://dux.typepad.com/dux/2009/03/it-was-monday-january-5th-2009-business-had-just-resumed-after-a-two-week-break-and-product-design-was-reviewing-short-ter.html (accessed March 9, 2009).

Sigel, D., and S. Dray. 2003. Living on the edges, user centered design and the dynamics of specialization in organizations. *Interactions* 10(5): 18–27.

Singh, N., D. Baack, S. Kundu, and C. Hurtado. 2008. U.S. Hispanic consumer e-commerce preferences: Expectations and attitudes toward web content. *Journal of Electronic Commerce Research* 9(2): 162–175.

Solomon, D. J. 2001. Conducting Web-based surveys. *Practical Assessment, Research & Evaluation* 7(19): http://edresearch.org/pare/getvn.asp?v=7&n=19 (accessed Oct. 13, 2003).

Tourangeau, R., L. J. Reips, and K. Rasinski. 2000. *The Psychology of Survey Responses*. New York: Cambridge University Press.

Venturi, G., J. Troost, and T. Jokela. 2006. People, organizations, and processes: an inquiry into the adoption of user-centered design in industry. *International Journal of Human-Computer Interaction* 21(2): 219–238. doi:10.1207/s15327590ijhc2102_6.

Von Hippel, E. 1986. Lead users: a source of novel product concepts. *Management Science* 32: 791–805.

Vu, K.-P. L., W. Zhu, and R. W. Proctor. this volume. Evaluating web usability. In *Handbook of Human Factors in Web Design*, 2nd ed., eds. K.-P. L. Vu and R. W. Proctor, 439–460. Boca Raton, FL: CRC Press.

Wan Hassan, W., S. Hamid, A. Norman, and N. Yasin. 2008. Storyboards based on instructional design principles: A tool for creating e-learning content. *International Journal of Learning* 15(6): 163–178.

Weiss, M. 2008. Results-based interaction design. *Educause Quarterly* 31(4): 42–49.

Wixon, D. R., and J. Ramey. 1996. *Field Methods Casebook for Software Design*. New York: John Wiley.

Wixon, D. R., J. Ramey, K. Holtzblatt, H. Beyer, J. Hackos, S. Rosenbaum, C. Page, S. A. Laakso, and K. P. Laakso. 2002. Usability in practice: Field methods evolution and revolution. *Conference Proceedings on Human Factors and Computer Systems*, 880–884. New York: ACM Press.

Wood, L. 1996. The ethnographic interview in user-centered work/task analysis. In *Field Methods Casebook for Software Design*, eds. D. Wixon and J. Ramey, 35–56. New York: John Wiley.

21 Evaluating Web Usability

Kim-Phuong L. Vu, Wenli Zhu, and Robert W. Proctor

CONTENTS

Usability rules the Web.

—Jakob Nielsen
Designing Web Usability

21.1 INTRODUCTION

Usability is the most fundamental attribute that determines the ease with which users interact with a product. There have been numerous demonstrations of the value of incorporating usability into the design lifecycle for products in general and for Web applications in particular (see Bias and Mayhew 2005; Richeson et al., this volume). Bevan (2005) showed that the benefits of user-centered design include reduced development costs, increased sales on e-commerce Web sites, and increased employee productivity, efficiency, and retention. Because Web usability is an area within the general domain of computer software usability, many issues relevant to software usability also apply to Web usability (Nielsen 2000; Vu and Proctor 2006). Usability for Web design is typically defined to include the following factors (e.g., Brink, Gergle, and Wood 2002; Coutaz and Calvary 2008; Lazar 2006):

> Easy to use and learn: The Web site should be easy to use and contain features that are easy to learn.
> Efficient: The Web site should allow the user to complete a task as quickly as possible.
> Reliable: The Web site should be available and stable.
> Memorable: The functions and features of the Web site should be easy to remember. This will allow a person to use the site easily even after a period of not using it.
> Satisfying: The Web site should provide the user with a pleasant interaction so that the user's overall impression of the site is positive.
> Produce minimal errors: The Web site should have a low error rate. If an error occurs, an informative error message should be presented so that the user can make the appropriate corrections.
> Accessible: The Web site should be compatible with screen readers or other devices that allow persons with disabilities to use the site.

Portable: The Web site should also be adaptable so that it can be accessed from a variety of platforms, including mobile devices.

Evaluations of Web usability can be guided by and based on the aforementioned attributes. Tullis and Albert (2008) provide an overview of user performance metrics that can be used in evaluating Web usability. As emphasized by many books, articles, and online columns dedicated to Web usability (for examples, see Appendix A), the purpose, target audience, and core user tasks of a Web site must be considered and identified when evaluating Web usability.

21.1.1 PURPOSE OF A WEB SITE

Web usability evaluation is closely related to the goal or purpose of a Web site. When a site is designed, the designers and decision makers, also known as stakeholders (Kuniavsky 2008), typically have some purposes for the Web site in mind. These purposes may reflect the needs of an organization or specific tasks that the users need to accomplish. For example, from the point of view of a company's CEO, the purpose of an e-commerce Web site may be to promote sales of the company's products. From an end user's point of view, though, the goal of the site will be to help users find and purchase products. These two goals are not opposing but illustrate the different perspectives of the stakeholders. Table 21.1 contains examples of Web sites, with their stated purposes and target audiences for different types of Web sites.

Although a Web site that allows users to achieve their goals may not always be successful, Web sites that do not support user goals are rarely ever successful. Thus, Web sites should have clearly defined purposes. Some of the goals or statements of purpose may be general, but they provide guidance and benchmarks on how to evaluate the site. Ultimately, one can argue that usability is the means to an end, not an end in itself. The end is the goal or purpose of the Web site. Thus, to evaluate Web usability, the goals, purposes, and visions of a Web site must be explicitly stated. The goal statements

---

TABLE 21.1
Examples of Stated Purposes and Their Target Audiences for Different Web Sites

Web Site	Stated Purpose	Target Audience
Cancer.net	Cancer.net provides timely, oncologist-approved information to help patients and families make informed health-care decisions.	People living with cancer and those who care for and care about them. Source: http://www.cancer.net
Microsoft. com	Our mission and values are to help people and businesses throughout the world realize their full potential	Developers, IT decision makers, IT implementers, advanced business users, and general business users. Source: http://www.microsoft.com
CSULB.edu	California State University Long Beach is committed to providing highly valued undergraduate and graduate educational opportunities through superior teaching, research, creative activity, and service for the people of California and the world.	Faculty, staff, students, alumni, visitors, and community members Source: http://www.csulb.edu

can be transformed into specific and measurable values. For example, during the redesign of Sun.com, Nielsen (1998) summarized the goals of the redesign as increasing the download speed, providing a unified visual appearance, facilitating navigation, allowing users to access the site's search engine, updating the material, and ensuring high usability. Ensuring accessibility and portability have become goals of Web usability as well (e.g., Moreno, Martínez, and Ruiz-Mezcua 2009) to comply with government guidelines and meet users' increased reliance on mobile devices to access the Web (see Pappas et al., this volume; Xu and Fang, this volume). The articulation of the goals, purposes, and visions of a Web site also helps to identify the target audience and core user tasks, on which we elaborate in the next two sections.

21.1.2 Target Audience

The target audience of a Web site may include multiple user groups. Column 3 of Table 21.1 gives examples of target audiences, as identified by the different Web sites. Characteristics and descriptions of the target audience may include the following:

- Demographics: age, sex, education, occupation, geographical location, income, and language.
- Computer experience and skills: hardware, software, Web, and existing systems.
- User tasks and methods: roles, tasks, methods, strategies, and priorities.
- User constraints: hardware, software, connection speed, disabilities, and guidelines.
- User preferences: technology, style, habits, and concerns.

Characteristics of the target audience can be analyzed to identify representative users through the use of user profiles (Fleming 1998) or personas (Aldiin and Pruitt 2008; Cooper 1999; Thoma and Williams 2009). Cooper (1999) argued that the more targeted a design is for a specific type of user, the more likely it will succeed. One way to design the Web site is to look at user profiles, which are brief summaries of real users' characteristics. For example, the user profiles for a travel company would probably yield two major classes of users: business travelers and nonbusiness travelers. When looking at profiles of the two types of travelers, one would probably find key differences in their characteristics. For example, business travelers are likely to have higher income and travel more often than nonbusiness travelers. Their preferences may also be different. Business travelers may value efficient schedules more than price, but the reverse may be true for travelers on vacation. Thus, a good Web site can have features to accommodate the needs of their primary user groups such as allowing the user to search by flight schedule (for business travelers) or by price (for nonbusiness travelers) and by giving the user an option to save their preferences and billing information (for frequent users).

Personas are one way to describe user profiles. They are less detailed than user profiles and do not refer to real people.

Rather, they are concrete representations of a group or a category of users (Aldiin and Pruitt 2008; Cooper 1999). Personas are typically developed to mimic the characteristics of actual users and can be based on the designers' knowledge of who the users are likely to be. That is, personas can be considered as hypothetical archetypes of actual users. An example of a persona for a business traveler is as follows:

> John is a vice president of a chemical manufacturer. He lives in Boston and travels 3–4 times a month. He relies a lot on his iPhone for keeping in touch with his coworkers, scheduling meetings, and making travel arrangements. He mainly travels to industry conventions, client and supplier sites, and manufacturing plants. He likes XYZ Airlines very much and always travels with them. His travel priorities are timeliness, convenience, and cost. He flies business class most of the time. He is married, with two children. His daughter, Mary, is 9 and his son, Michael, is 6. His family uses his frequent flyer miles when they travel.

The purpose of using a persona is to help designers tailor the Web site to specific user groups that may otherwise be neglected. User profiles and personas can also be useful tools in evaluating Web usability because they help to focus the evaluation on representative users rather than leaving it open to all possible users.

21.1.3 Core User Tasks

Core user tasks are typical, high-frequency tasks performed by the target audience. For example, users may want to perform the following tasks on a company's Web site:

- Find product information.
- Purchase products.
- Track product delivery status.
- Read company news.
- Find out the location of a store.
- Get technical support.

Byrne et al. (1999) developed "a taskonomy of WWW use" (see summary in Table 21.2), which includes basic, general purpose tasks in which users engage while browsing the Web. The taskonomy does not include higher-level goals or tasks such as browsing, researching, searching, purchasing, communicating (i.e., e-mailing, chatting, or "twittering"), and accessing entertainment (such as listening to music and watching videos). Consequently, many of these tasks are still applicable even with the recent advances in Web design because higher-level goals can be decomposed into subgoals that correspond with the tasks in the "taskonomy." For example, the task of communication may include locating an e-mail address on a page and providing information. The tasks in the taskonomy can also be decomposed into subtasks, for example, by using the Goals, Operators, Methods, and Selection Rules (GOMS) analysis (Byrne 2008; Card, Moran, and Newell 1983). Therefore, when defining the core set of user tasks, it is helpful to keep in mind that there may

TABLE 21.2

Taskonomy of World Wide Web Tasks (Based on Byrne et al., 1999)

Category	Example Tasks
Use information	Download files
	Copy
	Print
Locate on page	Related concept
	Tagged information
	Image
Go to page	Hyperlink
	Back/forward buttons
	Bookmark page/History
Provide information	Search string
	Shipping address
	Credit card information
Configure browser	Add bookmark
	Change cache size
	Window management (scroll, resize, etc.)
React to environment	Respond to dialog
	Respond to display change
	Reload

be different levels at which the tasks can be defined. The taskonomy can be used as a reference point, but one must take into account the purpose and the target audience of the Web site. Different user groups may very well have different goals and tasks (Aldiin and Pruitt 2008).

21.2 METHODS TO EVALUATE WEB USABILITY

As mentioned in the Introduction, methods used to evaluate general software usability are applicable to Web usability. A survey of books, articles, and online columns dedicated to Web usability and evaluating user experiences (see Appendix A; see also Kuniavsky 2008; Lewis 2006; Vu and Proctor 2006) reveals several general classes of methods, listed below, although some of the specific methods may cut across categories:

1. Field methods/observation: ethnographic methods, diary studies, usability testing in the field, and field trials.
2. Card sorting, task analysis, and user modeling.
3. Prototyping: paper prototypes and interactive prototypes.
4. Interviews, focus groups, and questionnaires.
5. Usability Inspections: heuristic evaluation, cognitive walkthrough, and alternate viewing tools.
6. Usability testing: usability tests, verbal protocol analysis, and performance testing.
7. Web-based methods: automated sessions, Web logs, and opinion polls.

These methods are explained in the subsequent sections. As pointed out in other chapters in the handbook

(e.g., Mayhew, this volume; Volk, Pappas, and Wang, this volume) and emphasized by many Web usability resources (see Appendix A), Web usability requires a user-centered design process. Evaluations should be performed as early as possible and be carried out throughout the development cycle. Evaluations conducted only at the end, as a "stamp of approval" or a usability "bandage," are not sufficient. Many usability issues uncovered during evaluations can guide the design process. Because Web design should be an iterative process, methods to evaluate Web usability should also be used throughout the design process (see Mayhew, this volume; Volk et al., this volume).

Usability evaluations are well known in the computer software industry (e.g., Nielsen 1993). The thrust of the approach is to test a system with real users. Usability evaluations are a critical component of the software design process because it provides information about how people interact with the interface and the exact problems they encounter when doing so. The same holds true for Web design, perhaps even more so than for traditional graphical user interface (GUI) applications. There are two major reasons why it is important to design for usability on the Web. First, usability determines whether users will use a Web site (Vu and Proctor 2006). Whereas users' assessments of the usability of computer software occur after already having purchased the product, their assessments of the usability of a Web site occur prior to purchasing a product. Thus, the usability of a site can influence users' decisions regarding whether to proceed with using the site's services or to abandon it. Given that many competing Web sites offer similar products or services, users do not hesitate to go to a different Web site that they perceive to be more usable. Second, because a Web site may be used by millions of people each day, any potential usability problems can have serious consequences. Thus, evaluating the usability of Web sites is particularly important to e-businesses.

The next section describes some commonly used methods for testing and evaluating the usability of a Web site. Some of the methods are inspection based (see also Cockton, Woolrych, and Lavery 2008), whereas others are user based (see also Dumas and Fox 2008; Lewis 2006). We end the section with a Rapid Iterative Test and Evaluation (RITE) method that was proposed to help incorporate iterative usability testing into the design life cycle at lower costs (Medlock et al. 2005).

21.2.1 FIELD METHODS/OBSERVATION

To understand user requirements, it is often valuable to observe users in their natural environment. Moreover, formal usability evaluations can be conducted in the field, that is, at the users' sites, instead of in a usability laboratory (Yamada, Hong, and Sugita 1995; Zapf et al. 1992). The field methods can be used when (1) it may not be feasible or convenient for the participants to travel to the usability laboratory, for example, when it is too expensive (more hours billed) or burdensome (for participants with disabilities); (2) it is desirable to obtain data from a large number of participants in a "walk up and use" setting such as in a library; and (3) the goal of the test is to collect data from authentic users using existing systems.

The procedure and process of a usability test in the field is similar to that in the laboratory. However, the presence of an experimenter may not be feasible or necessary. More preparation time may be needed to set up the site, for example, to allow for videotaping or automated data collection. An advantage of conducting a usability test in the field is that the users may be more at ease and act more naturally. They may also be able to provide more in-context feedback as they are in an environment close to the one where the system will be used. With the availability of wireless access and Web-enabled mobile phones, it is even possible to record user browsing behavior (with user consent) throughout their day. Kamvar et al. (2009) found that mobile users were more concise with their Web searches than desktop users, with fewer queries and less diverse search topics. However, when a high-end iPhone was used, search behaviors more closely resembled those of desktop users. Although issues of privacy are a concern with this tracking method, it allows designers to capture naturalistic user activity.

21.2.1.1 Ethnographic Methods

The need to understand users' work contexts arose from the realization that many problems emerge because the design failed to consider the social context in which the system would be used. To accommodate the need to understand usability within the context of use, ethnographic methods have been proposed for the design and evaluation process (Hughes et al. 1995; see Volk et al. (this volume) for more detailed coverage). The idea of the ethnographic approach is for the usability expert to become immersed in the users' lifestyle. Ethnographic methods can provide a qualitative understanding of usability problems that arise with the actual uses to which the product is put. It has been argued that merely exposing usability experts and designers to the lifestyle of the users is sufficient to provide insights pertaining to usability that can be incorporated into the design (e.g., Wixon 1995). However, more elaborate ethnographic techniques have been developed (see Volk et al., this volume) to provide a more detailed understanding of how the users interact with the system, as well as a means for incorporating that understanding into the design.

As an example, Beyer and Holtzblatt (1998) proposed an approach called "Contextual Inquiry and Contextual Design," which uses a combination of observations and interviews to identify usability problems. The approach is based on a "master/apprentice model," which assumes that a design team can learn about its potential users by studying them in the same manner that an apprentice acquires a skill by studying a master. It emphasizes four principles in observing users:

- Context: Go to the users' site and see the work as it is performed.
- Partnership: Observer and user collaborate in understanding the user's work, the observer alternates between watching and asking probe questions.

- Interpretation: Interpret what the user's words and actions mean.
- Focus: Determine the goal of the observations.

The contextual inquiry and contextual design approach emphasizes the involvement of the entire team in the process. Data collection, analysis, and interpretation should all be done as a team. During the system design phase, work models drive the design. They are frequently revisited during the design, prototyping, and evaluation of the system. Holtzblatt (2008) identified several rapid contextual design processes that can be used in the design process. The emphasis of this approach is to get the designers to use customer data in guiding decisions and prioritizing tasks. It should be evident that the subjective interpretations of Web use provided by ethnographic approaches including contextual inquiry, although helpful to designers early in the design process, need to be supplemented later on with more systematic and controlled methods of data collection.

21.2.1.2 Field Trials

A combination of usability testing and ethnographic methods can be used to evaluate a system in the field when a system is available in an adequate functional state, such as a functional prototype, a Beta version, or a released version (e.g., Dray et al., 2002; Feldman, Pennington, and Ireland 2003). Zhu, Birtley, and Burgess-Whitman (2003) reported a study in which usability testing was conducted throughout the design and development of a note-taking application on a pen-based computer platform (the Tablet PC from Microsoft). Because note taking is context dependent (for example, note taking can happen in meetings, classes, brainstorming, etc.), the design and development team was concerned with the issue of whether the application would be useful in these different real world contexts. A field trial was conducted that used a combination of ethnographic methods and usability testing. Participants were recruited from the target population and provided with a Tablet PC and a pre-Beta version of the note-taking application. The participants were observed once a week for 8 weeks. Note taking was observed in a variety of situations and environments such as in classes and meetings, on a bus, and in a soccer field. Participants were also interviewed on each visit. To make sure that all the important features of the application were used by the participants, they were asked to complete specific tasks on some of the visits. First-time uses of some of the features were observed, and usability tests were conducted in a way similar to those conducted in a laboratory, which are described later.

The results from the field trials revealed that usability problems identified in pre-field-trial usability tests were also observed in the field. Although some initial usability problems identified in the laboratory did not persist after repeated usage, several lasted for an extended period of time and continued to cause difficulties even after the participants had learned how to use the tools. To summarize, field trials can be useful when introducing new products to the market. They may help to determine users' acceptance of the new product or Web site. Field trials can also help to distinguish usability problems that can be overcome through learning from those that cannot be. In addition, field trials allow researchers to discover problems that may arise from using a product in the daily environment (e.g., glare produced by light sources). As a result, field trials can be used in addition to systematic laboratory usability tests but should not replace them. Because of usability testing of tablet PCs, their usefulness has continued to improve the past 6 years and will probably continue to do so in the near future.

21.2.2 Card Sorting, Task Analysis, and User Modeling

Card sorting tasks can be used to help the Web designer organize information that the Web site will contain. With the card sorting procedure, each component of the Web site is written on an index card. The cards are handed to participants, who sort the cards into piles that represent a common purpose or theme (see Figure 21.1). This type of unguided card sorting (Goodman et al. 2007) allows designers to obtain information about how users understand the relations between the elements and functions of a Web site. Sometimes the users are asked to provide labels for each of their groupings. These labels can be used for higher-level menus or navigation links. Labels that users provide may be different from those given by Web designers because they can be context specific. An alternative to unguided card sorting is a guided process in which the researcher has the participant sort the cards into predefined categories (Goodman et al. 2007).

An example of the use of card sorting is a user test conducted for the department Web site at the university of one of the authors. Specifically, the listing of professors in the department was previously labeled as "Faculty" and that of supporting staff members as "Staff." During user testing, it was found that college freshman were looking for their professor's contact office hours in the "Staff" section, becoming frustrated when they were unable to locate their course instructor's name. A card sorting task showed that students grouped faculty and staff together in one pile and preferred

the label "People" rather than "Faculty" or "Staff." This example illustrates a benefit of unguided sorting, in that the participants created a superordinate category that designers likely would not have included. End users may have a different conceptualization of the task than that of the designer, and the card sorting technique can be used to examine their grouping and organization of information relating to functions of a Web site.

Once the targeted user groups are identified, they are usually interviewed or surveyed about the tasks that they need to perform with the Web site. Users can also indicate the features that the Web site should contain in order to support these tasks as well as what features that they would want for the Web site. This process is sometimes referred to as a user wants and needs analysis (Vu and Proctor 2006). After the list of tasks is compiled for each user group, the tasks can be prioritized and then broken down and analyzed.

Task analysis is a technique that allows the researcher to decompose the general tasks identified by the users into subtasks (see Strybel, this volume). Task analysis is a useful tool because users often list higher-level goals that are comprised of a series of smaller tasks. These subtasks can then be evaluated in terms number of steps needed to achieve the higher-level goal, time for completion of each step, and the types of errors made (Stanton and Baber 2005). Task analysis can help Web designers identify specific usability problems associated with a task, offer suggestions for alternative ways for completing the task, and be used to help guide future usability tests as well as to model user performance on a task.

Evaluating the usability of a new Web or interface design with end users can be time-consuming and costly. An alternative approach is to develop user performance models to predict their behavior (see van Rijn et al., this volume). User modeling can provide useful information about a site's usability in terms of performance with an interface and be used again to evaluate changes in the interface brought about through the iterative design process. User models also allow designers to trace the "user's" steps to help diagnose the cause of errors. That is, designers can examine the number of steps that would be needed for the user to achieve a goal and the different paths that can be taken to achieve the goal, using that information to optimize the site's organization. One drawback of user modeling, though, is that user impressions of the site's appeal and preferences for the site's features may not be revealed.

21.2.3 Prototyping

Prototyping is a tool that can be used in the design of a Web site (Lazar 2006). Alternative designs can be quickly mocked-up and tested in usability inspections or usability tests. Prototypes can be either high or low fidelity. High-fidelity prototypes try to mimic the look and feel, as well as the functionalities, of a real system. They are typically developed using a software tool such as Macromedia Director or Microsoft Visual Basic. The difference between a prototype and the real system is that the prototype does not have to

FIGURE 21.1 Picture of a user conducting a card-sorting task.

implement functional features; rather, it simulates the implementation of the features using "fake" data, perhaps based on scenarios. In a Web environment, it has been argued that it is not as cost-effective to develop high-fidelity prototypes using a desktop publishing layout tool because it would probably have taken the same amount of time to implement the Web site in HTML (Pearrow 2000). The ease with which a "skeleton" Web site can be quickly created using HTML makes the Web environment a "friendlier" environment for developing interactive prototypes. Also, the use of cascading style sheets (CSS) allows for changes to be made once but updated across all pages of a Web site.

Low-fidelity prototypes can be paper mock-ups, storyboards, or paper prototypes. These can be hand-drawn, without the details or the polished look of a high-fidelity design (see Figure 21.2a). Although, with word processors and programs such as PowerPoint, it is easy enough to quickly design a paper mock-up that can look polished (see Figure 21.2b). The goal is to convey the conceptual design and show the user scenarios that are important. A Web-based survey of practitioners of agile software development methods found that low-fidelity prototyping was the most widely used usability method, with 68% of the respondents indicating that they used that method in their work (Hussain, Slany, and Holzinger 2009).

21.2.3.1 Paper Prototypes

Paper prototypes can be hand-drawn or based on printouts of screen captures. Although it may be easy for designers to create prototypes in HTML, as mentioned above, sometimes experimenters may prefer to use paper prototypes. For example, if the network or server is down, the experimenter can continue the usability test. A hand-drawn prototype (see Figure 21.2a) is a quick and easy way to mock up several design ideas. It also allows one to focus on the "big picture" and not delve into the design and implementation details too early. Usability testing a paper prototype has several advantages: (1) it allows the designers to evaluate several designs early in the process, with a quick turnaround; (2) it allows the designers and users to focus on the paradigm and the general concepts, and thus may lead to more and higher-level feedback; (3) it is easier to make significant changes because the investment is low and it is easy to start over.

The buttons and links can be hand-drawn on paper of various sizes and colors such as index cards and sticky notes. For example, the experimenter can write all menus and lists on small sticky notes so that they can be changed on the fly to simulate the action of opening and closing a menu or a list. To give users a sense of working with a computer, an 8.5 × 11-inch image can be printed out and used as the desktop. If it is laminated, the experimenter can write on it using a dry-erase marker, to simulate the action of filling out text fields and other controls.

A usability test with a paper prototype differs in several ways from one using a computer. In a usability test with a paper prototype, someone has to "operate" the prototype, that is, to "act" as the computer. Users can interact with the paper prototype, using a finger or pen as a pointing device. They can also verbally express their actions such as to click, double click, type, etc. The person who acts as the computer has to move pieces of paper or write things down, accordingly, to simulate the response of the computer. If someone other than the experimenter can act as the computer, the experimenter can focus on observing the participants, taking notes of what they are saying and doing, and interacting with them. The experimenter will have to be careful in interacting with the participants, though. Sometimes it may be unavoidable to give them instructions or to remind them to think aloud; however, such interactions should be kept minimal just as in a regular usability test.

21.2.3.2 Interactive Prototypes

Interactive simulations can be done with off-the-shelf software such as Adobe Photoshop or through use of scripting languages (Beaudouin-Lafon and Mackay 2008). With this approach, different states of the Web site can be created in Photoshop and accessed in succession or a scripting language can be employed to write a small program to illustrate the basic functionality of the site. As mentioned, it is easier to build interactive prototypes using HTML in a Web

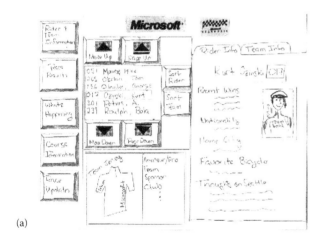

(a)

(b)

FIGURE 21.2 (a) A hand-drawn paper prototype (courtesy of Karl Melder) and (b) an electronic paper prototype (courtesy of Fred Garcia).

environment. Pearrow (2000) discussed two strategies in developing interactive prototypes: horizontal versus vertical. A horizontal prototype aims at enabling all the top-level functionalities, whereas a vertical prototype aims at enabling the functionalities of a particular path.

Aside from the purpose of testing a design that may be close to the real implementation, an interactive prototype may also be used to test particular aspects of a design. For example, Risden (1999) used an interactive prototype to evaluate Web hierarchies. Because the experimenter was only interested in finding out users' traversal paths when finding specific information on a Web site with several general categories, the prototype was implemented to display the categories in a simple hierarchical format. Users were thus led to focus on choosing between the categories and were not distracted by specific user interface problems that could be present if they were using an actual Web site or more elaborate prototype.

Another issue that should be considered when developing an interactive prototype is the system response time. With a real Web site, it may be slow to load a page, download a file, or execute a transaction. This delay is of particular concern when the Web site will be accessed using a wireless connection. If the goal of the usability evaluation is to test the overall experience of a Web site, realistic system response time should be built into the interactive prototype.

21.2.4 Interviews, Focus Groups, and Questionnaires

Several other subjective methods have been used to help in the design and evaluation of usability, including Web usability. These include interviews, focus groups, and questionnaires (see Volk et al., this volume). These methods are often used to understand user tasks and user needs, as well as their subjective opinions, attitudes, and preferences.

21.2.4.1 Interviews

Interviews are a well-known knowledge elicitation technique used in the design of knowledge-based systems (Newman 2005; Shadbolt and Burton 1995). In human-computer interaction, interviews are often used as a method of task analysis, or as a supplementary method to other techniques such as a usability test or a field study (e.g., Sebillotte 1995; Vu and Proctor 2006; Wood 1996). When gathering in-depth data, interviews are often conducted along with field observations (e.g., Beyer and Holtzblatt 1998; Isaacs 1997).

However, when field observations are not feasible, interviews may be the best method for gaining in-depth data of users and their tasks. When that is the case, it is sometimes helpful to ask participants to bring in "artifacts" from work places such as tools, documents, manuals, and so on, to the interviews. These artifacts help participants talk about the specific details of their work instead of summarizing at a high level. Another technique often used is to ask the participants to "walk through" examples or incidents that happened in the

past. This can provide more detailed information about what happened and the circumstances surrounding the incidents.

Interviews can also be used as a supplementary method. For example, in a usability evaluation, interviews can be conducted before and after a session to gather background information and probe for comments, opinions, and attitudes expressed during the session. When collecting quantitative data using questionnaires, interviews can help to formulate the questions and interpret the data. To conduct an interview, participants should be recruited from the target population. The focus and scope of the interview should be defined beforehand. Depending on the purpose, the interview can be structured, semistructured, or unstructured. A structured interview uses a predefined list of questions, whereas an unstructured interview allows participants to freely express their thoughts. Most of the time, interviews are semistructured, with predefined questions but allowing the interviewer opportunity to pursue issues raised by participants. Isaacs (1997) and Beyer and Holtzblatt (1998) describe good interviewing techniques (see also Volk et al., this volume), some of which are listed in Table 21.3. With the Web, interviews can also be conducted via a "chat" or dialogue box. The benefit with this type of interview is that there is a written record of the communication. However, drawbacks are that the communication can be short owing to typing requirements and many contextual cues are lost.

TABLE 21.3
Interviewing Tips

The interview is about them, not you!
- Don't tell them what you think.
- Don't answer questions until you know what they think.

Ask open, unbiased questions.
- Don't lead them.
- Don't present options for them to confirm.

Ask the question and let them answer.
- Keep the questions clean and simple.
- Ask one question at a time.
- Give them time to think.

Follow up.
- Make sure you fully understand.
- Paraphrase, but don't lead.
- If you don't find out what you want, try again.
- Be willing to seem a little naïve.

Adjust your questions to their previous answers.
- Don't blindly follow your planned script.
- Follow up on unexpected issues.

Ask questions in language they understand.
- Avoid jargon.

Be flexible.
- Let them cover the material in any order.

Listen to their complaints, but look for other problems.
- Find inefficiencies they "put up with" without noticing.

Pick on examples.

Source: Isaacs (1997).

Usually, a content analysis of the data collected from the interviews is conducted to identify themes and categories of information. Because categorization of responses may be open to interpretation, it is good practice to have at least two researchers perform independent content analysis for the same data and establish inter-rater reliability.

21.2.4.2 Focus Groups

Focus groups are commonly used in marketing research (Kuniavsky 2003; Morgan 1997). One benefit of focus groups is that problems and concerns can surface through interactions of the group members that would not have been identified otherwise. Focus groups can also be used to assess user needs and preferences. Examples of using focus groups include group task analysis (Dayton, McFarland, and Kramer 1998; Lafrehiere, Dayton, and Muller 1999) and user needs analysis (Delbecq and Van den Ven 1971; Graetz et al. 1997). Group task analysis is a knowledge elicitation technique in which experts describe their tasks and the steps to accomplish the tasks in a group format. User needs analysis uses a variety of group activities to elicit and generate user needs and product requirements.

The main advantage of a focus group is efficiency. Focus groups tend to generate a lot of data in a short amount of time. The main problems are (1) a lack of contextual information, that is, the context in which users perform a task may be unclear; (2) a lack of in-depth data about why users do certain things or have certain needs. As with interviews, techniques such as asking participants to bring in artifacts or walk through past incidents will help to make the conversations more concrete and specific.

Focus groups can also be used to gauge users' reactions, opinions, and feelings about a particular design (Greif 1992; O'Donnell, Scobie, and Baxter 1991; Sullivan 1991). Marketing research uses focus groups to gauge consumers' feelings toward product brands and advertisements. It provides complementary information that other methods such as a usability testing may not be able to provide. A danger is that users' usability perceptions about the design may change as their experience with a design changes (e.g., Vu and Proctor 2006).

To conduct a focus group, 6–10 participants are recommended (Morgan 1997). Again, the participants should be recruited from the target population. Recruitment may pose a particular challenge for focus groups because all participants must be available at the same time. After-hours appointments may be necessary for professionals who work regular hours. Overrecruitment is recommended as no-shows are quite common for these group sessions. The moderator should prepare an outline of the session to make sure all desired areas are covered. During the session, the moderator should keep the discussion on track and make sure that all participants contribute and no single participant dominates the discussion. Data analysis and interpretation are similar to that of interviews. One needs to look for themes, agreements, and disagreements among participants. As with interviews, content analysis of the data may be required.

21.2.4.3 Questionnaires

A questionnaire is a set of items (e.g., statements or questions) designed to elicit the opinions, beliefs, or preferences of users (Moroney and Cameron 2000). Questionnaires are often used to measure user satisfaction (Lewis 1995), attitudes toward computers (Harris 1999), usage (Teo and Lim 2000), and preferences (Proctor, Ali, and Vu 2008). The advantage of using a questionnaire is that quantitative data from a large sample consisting of several demographic groups or user populations can be obtained.

In evaluating Web usability, questionnaires can be used to develop personas or user profiles, evaluate usage of the Web site, and measure users' subjective opinions such as satisfaction and preferences. Questionnaires can also be used in combination with other usability evaluations to gather quantitative data. To develop a questionnaire, the goal must first be defined. What is the questionnaire trying to measure? Is it attitudes, preferences, and desires or is it beliefs, behavior, experience, and knowledge? Appropriate questions should be devised to accomplish the goal. The questions must be unambiguous and unbiased. Questionnaires need to be pilot tested to make sure the questions are clear, easy to understand, easy to answer, interpreted in the intended way, and finished in the time specified.

When evaluating Web usability, participants typically are the target audience of the Web site. The participants should be randomly selected, and the range of the sample size should be 50–1000 (Nielsen 1993). Questionnaires can be administrated by mail (including e-mail), face-to-face, in groups, or verbally. Mail-out questionnaires may not require participant recruitment. Mailing lists can be obtained from marketing research firms or other sources. Questionnaires can also be administrated on the Web. Participants can be notified by e-mail or solicited on the Web. Keep in mind that it may be difficult to verify or trust the identity of the participants solicited on the Web, but duplicate responses can be omitted by comparing the IP address of the respondent.

Commonly cited weaknesses of questionnaires include low response rate (typically 30% in the first pass), reliability and validity, and sample limitations. With Web-based questionnaires, portions of the survey can be collected when users quit in the middle of it, but in this case, care must be exerted in analysis and interpretation of the data. Concerns about anonymity and confidentiality should also be addressed with written explanation in the questionnaire in a cover letter or in an informed consent form. Return rates are usually higher when anonymity is promised, for example, 48 versus 22% (Moroney and Cameron 2000). Commonly used questions in a questionnaire include open-ended, binary/multiple/forced choice items, rating, ranking, checklists, and semantic differentials. Examples can be found in Table 21.4.

To summarize, interviews, focus groups, and questionnaires are useful methods to gather user characteristics and opinions. They are also commonly used in usability tests to provide complementary data and additional feedback (Vu and Proctor 2006). Although users' self-reports may be good

TABLE 21.4
Examples of Questions

Type	Examples
Open-ended	• What feature do you like the *most* about the Web site? Explain. • What feature do you like the *least* about the Web site? Explain.

Forced choice (options should be all inclusive and mutually exclusive)	• True or false questions
	• When you shop online, which of the following is *most* important to you?
	   ___ availability of products
	   ___ price
	   ___ shipping cost or shipping time
	   ___ level of security provided by the site
	• Age in years
	   • Under 20 years
	   • 20–24 years
	   • 25–29 years
	   • 30–34 years
	   • Over 34 years

Rating (5 or 7 choices allow for a neutral position; 4, 6, 8 choices force respondents to a positive or negative position)	• I found the "Help" page of the Web site to be:
	   ___ extremely useful
	   ___ of considerable use
	   ___ of use
	   ___ not very useful
	   ___ of no use
	• I found this Web site to be aesthetically pleasing:
	   ___ strongly agree
	   ___ agree
	   ___ undecided
	   ___ disagree
	   ___ strongly disagree

Ranking	Rank the following items in terms of importance to you. Assign a "1" to the most important, a "2" to the next most important, etc. Assign a different number to each of the items.
	   ___ quality of products
	   ___ efficiency of shipping
	   ___ price

Checklists	Check all the characteristics you consider important in selecting an e-commerce Web site:
	   ___ company name
	   ___ ratings from previous customers
	   ___ security of the Web site

Semantic differential	Place an "X" in each of the following rows to describe your feelings about the quality of the fakebakery.com's cake:
	   tasty___:___:___:___:___tasteless
	   fresh___:___:___:___:___stale
	   excellent___:___:___:___:___poor

Linear numerical scale	Extremely Important 1 2 3 4 5 6 Extremely Unimportant Downloading Speed ___ ___ ___ ___ ___ ___ Aesthetics ___ ___ ___ ___ ___ ___

Source: Adapted from Moroney and Cameron (2000).

indicators of their performance in some circumstances (e.g., Vu et al. 2000), there are well-known biases in self-report data (e.g., Isaacs 1997). For example, what users report they do may not be what they actually do when in the situation. Users may tend to summarize or not be able to articulate all of their thought processes. They may also tend to give socially acceptable answers. Users are also not good at predicting what they would do in a hypothetical scenario or estimating how often they do things. Further, studies have found that user preferences may not correlate with performance (e.g., Bailey 1993).

21.2.5 USABILITY INSPECTION METHODS

Usability inspection methods are employed by human factors professionals and software developers to inspect, evaluate, and assess the usability of a system, without testing actual users (see Cockton, Woolrych, and Lavery 2008). Inspection methods are typically considered to be less expensive and quicker than other methods because they do not require the costs associated with setting up and conducting usability tests with actual users. These methods are generally based on design guidelines, best practices, cognitive theories, and other theoretical frameworks. Among the inspection methods that have been proposed, the most well known and frequently used are heuristic evaluation (Nielsen 1994) and cognitive walkthrough (Lewis and Wharton 1997; Spencer 2000).

21.2.5.1 Heuristic Evaluation

Heuristic evaluation of a Web site is done by having a usability professional evaluate the site to determine whether the site's format, structure, and functions are consistent with established guidelines (e.g., Nielsen 1994; Nielsen and Molich 1990). Because some of these guidelines reflect basic properties that all products should incorporate for usability, many still apply today. Some of these guidelines are listed below:

- Use simple language.
- Minimize user memory load.
- Be consistent.
- Provide feedback.
- Provide clearly marked exits.

A detailed checklist for these and other heuristics is available (see Pierotti 1994).

In a heuristic evaluation, evaluators systematically evaluate a system using these heuristics. The system could be a design specification, a mock-up (see Figure 21.2), a prototype, or an implemented system. The evaluators are typically in-house personnel such as designers, developers, human factors professionals, and documentation writers, or outside consultants. Nielsen (1993) found that the following:

1. Different evaluators tend to find different problems. A single evaluator typically finds only about 35% of the usability problems. Aggregating the results from several evaluators tends to be more effective in identifying the majority of the problems. Aggregates of five evaluators can find about two-thirds of the usability problems, suggesting that three to five evaluators are needed to achieve a balance between finding the most usability problems and the most efficient use of resources (Nielsen and Molich 1990).
2. The more familiar the evaluators are with human factors and usability engineering, and the more experienced they are with evaluating the particular type of user interface (e.g., voice-based systems or Web-based systems), the more effective they are at identifying usability problems.

Different variants of the technique may involve the use of a different set of heuristics (see Cockton, Woolrych, and Lavery 2008, for an overview) such as design guidelines (Jeffries et al. 1991) and ergonomic criteria (Scapin 1990),

the involvement of end users in the process (Bias 1991), and the use of user scenarios and tasks (Carroll and Rosson 1992; Clarke 1991). Other authors (e.g., Pearrow 2000) have suggested additional heuristics specifically for the Web, including the following:

- Chunking
- Using the inverted pyramid style of writing
- Placing important information "above the fold"
- Avoiding gratuitous use of features
- Making the pages "scannable"
- Keeping download and response times low

In addition to using heuristics, usability inspectors can evaluate the Web site to determine whether the site contains common mistakes. For example, in 2005, Nielsen noted that the top 10 mistakes in Web design included the following (see http://www.useit.com/alertbox/designmistakes.html):

1. Legibility problems
2. Nonstandard links
3. Flash
4. Content that was not written for the Web
5. Bad search
6. Browser incompatibility
7. Cumbersome forms
8. No contact information or other company information
9. Frozen layouts with fixed page widths
10. Inadequate photo enlargements

In 2007, Nielsen revised the list to include the following problems (see http://www.useit.com/alertbox/9605.html):

1. Bad search
2. PDF files for online reading
3. Not changing the color of visited links
4. Nonscannable text
5. Fixed font size
6. Page titles with low search engine visibility
7. Anything that looks like an advertisement
8. Violating design conventions
9. Opening new browser windows
10. Not answering users' questions

It should be noted that since Web technology changes rapidly, new problems and mistakes will emerge, so user testing should supplement these heuristic evaluations. Yet, simply checking a Web site to see if it includes these common mistakes (and removing the mistakes) can improve the usability of the site.

21.2.5.2 Cognitive Walkthrough

Cognitive walkthrough is a method in which the usability evaluator goes through the Web site from the user's perspective and performs tasks that the user typically performs. The method focuses on ease of learning. It is designed to give designers the ability to identify problems associated with learnability before an implementation of the design, prior to having users test it (Polson et al. 1992). Smith-Jackson (2005) outlines six steps in the design of a cognitive walkthrough. These steps can be used in Web evaluation by having evaluators:

1. Develop a full understanding of the target users background and prior knowledge with the site
2. Identify tasks that users will actually perform when using the site
3. Create task-based scenarios for the walk through
4. Identify the correct sequence of events necessary to complete the task
5. Identify the cognitive processes that the users must employ to complete the task
6. Identify learning that is likely to occur while exploring the Web site

Evaluators may find it useful to ask the following questions (Wharton et al. 1992) during their implementation of the steps above:

- Will the user try to achieve the right effect? (Will the user form the right goal? How do they know to do this?)
- Will the user notice that the correct action is available? (What concepts must they have?)
- Will the user associate the correct action with the effect that the user is trying to achieve? (Can they find and use the correct control?)
- If the correct action is performed, will the user see that progress is being made toward the solution of the task? (Will they see feedback for progress?)

A cognitive walkthrough requires preparation of a set of tasks that are representative of those that users will perform using the Web site, along with the sequence of actions to accomplish the tasks and a description of the interface (a functional specification, a paper mock-up, a paper prototype, an interactive prototype or a working system). Consequently, the methodology uses a form-based approach to guide the evaluators through the process. It is intended to help evaluators who may not be familiar with the approach to implicitly use the walkthrough method by answering a series of questions.

Some disadvantages of the method include evaluators, especially software developers, who may not be familiar with the concepts or nuances of the approach; it may be difficult to choose the correct level of tasks; the walkthrough is a lengthy and slow process; it is tedious to answer the questions and fill out the forms; it is hard to analyze the voluminous output (Spencer 2000; Wharton et al. 1992) and select appropriate performance metrics (see Tullis and Albert 2008). Improvements have been suggested, and the method has evolved from its original form to be easier to apply to interface development (Polson et al. 1992; Spencer 2000). For example, Spencer proposed a streamlined version of the

cognitive walkthrough that could be more effectively implemented for larger software companies where the social environment (e.g., time pressure to produce a product) may not permit use of the traditional cognitive walkthrough method.

21.2.6 USABILITY TESTS

The methodology and procedure on how to conduct a usability test have been well documented (e.g., Brinck et al. 2002; Lewis 2006; Nielsen 2000; Appendix A). In essence, a usability laboratory test involves testing a product or system (in this case, a Web site) with targeted users performing intended tasks. The goal is to observe and analyze users' understandings and misunderstandings, successes and failures, correct actions and errors, and task completion times to infer potential usability problems of the product or system (see Tullis and Albert, 2008, for a comprehensive coverage of usability metrics).

The methodology of usability testing is based on psychological science but modified to be more applicable and feasible in practical situations. Nielsen (1989, 1993) coined the term "discount usability engineering" to emphasize the notion that "simpler methods stand a much better chance of actually being used in practical design situations" (Nielsen 1993, 17). A typical usability laboratory test uses a simplified version of verbal protocol analysis (verbal reports from the

(a)

(b)

FIGURE 21.3   (See color insert.) Examples of (a) standard usability laboratory test setup and (b) remote usability laboratory test viewed on projection screen.

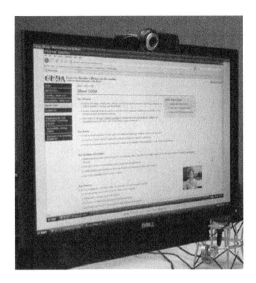

FIGURE 21.4   **(See color insert.)** Computer monitor with WebCam for user testing.

user as to what they are thinking and experiencing), a simple research design, and a small number of subjects, with a quick turnaround time. Usability testing is generally the most effective way of evaluating a Web site. Although a single usability test may improve the initial design, the maximum benefit comes from iterative testing.

Usability tests are usually conducted in an area that is specially set up for that purpose, such as a usability laboratory. Usability labs typically include a participant area, where the test participants are located, and an observation area, where the experimenters and observers can observe and record the test. These two areas are usually separated by a one-way mirror so that the participants cannot see and thus be distracted by the experimenters and observers (see Figure 21.3a). However, with Web recording and conferencing software, usability tests can be conducted remotely or even Webcasted to external locations (see Figure 21.3b). The user is also notified whether an intercom or conferencing software is used for communication during the test. In environments where two separate areas are not feasible, observers should not be allowed and experimenters should keep quiet and remain unobtrusive during the test. Cameras or a WebCam (see Figure 21.4) are usually set up to record the sessions so

that the experimenter can review the sessions at a later time. Nonobtrusive (e.g., infrared) eye trackers can also be used to capture user eye movement as another behavioral measure (see Figure 21.5).

Usability tests are the closest to controlled laboratory experiments, although they are not intended to be as rigorous as scientific experiments. For example, if the designer wants to evaluate whether a particular aspect or function of a Web site would increase its usability, the designer can measure users' performance with versions of the Web site that differ in terms of the function of interest. For a within-subject design, each user will use both versions of the Web site and their performance on each can be compared. With this method, half of the users should receive the test in one order and half in the other order to control for learning and fatigue effects. The advantage of a within-subjects design is that performance of the same individuals is compared for the two Web sites. With a between-subject design, one group of participants will use one version of the Web site and another group will use the other version. The advantage of a between-subject design is that there is no possibility of carryover effects. However, if the two groups differ in any systematic way (e.g., one group is more experienced than the other), the results cannot be interpreted unambiguously. Therefore, when possible, the experimenter should randomly assign users to the two groups or use other means to minimize any systematic differences between the groups that might influence the results (e.g., ensure that the two groups are matched in terms of experience using the Web).

21.2.6.1   Test Plan

A test plan helps the researcher or team conducting a usability test to think through the issues related to the test such as the following:

1. Test goals: Is the test a formative evaluation or summative evaluation? Is the goal of the test to assess the individual components of the Web site or its overall quality?
2. Problem statements: These are short summaries of the kinds of questions that should be answered by the test. For example, one question might be, "Can the users complete their online transactions without errors?"

FIGURE 21.5   Screenshot illustrating eye gaze movements in a Web usability test.

3. Methodology: This includes the users to be tested, the tasks they are to perform, and the procedures that are to be followed.
4. Data analysis: This includes issues regarding how the data will be analyzed and presented. Will the data collected allow you to answer the questions of interest?

21.2.6.2 Users to Be Tested

The users tested should be representative of the target audience. Participants should be selected on the basis of the test goals. For example, if the goal of the test is to evaluate the overall usability of a Web site, all user profiles of the target audience should be included. In contrast, if the goal is to evaluate whether users of an older system will be able to use the new Web-based system, users who have experience with the older system should be tested.

To determine the number of participants in a usability test, two factors should be considered: (1) the desired confidence level of the results and (2) the cost/benefit ratio (see Bias and Mayhew 2005). Nielsen (1993) used data based on a survey of 36 published studies to illustrate the relation between the confidence interval and the number of subjects for measures of expert user performance. He showed that when measuring expert user performance, about 13 participants are needed to reach a 90% confidence interval that the true value is in the range of +15% of the mean. If only 5 participants are used, the probability that the true value is in the range of +15% of the mean would be 70%.

With regard to the cost/benefit ratio, there is typically a diminishing rate of return on testing additional participants. Brinck et al. (2002) indicated that 80% of usability problems can be identified with only 4 or 5 users. However, they noted that it is better to test at least twice as many users. In most cases, 5–7 participants (per experimental condition) are tested in a usability test to balance the need of finding the most usability problems with the cost of running additional participants. Recruitment of participants can be accomplished by advertising on the Web site and through market research firms, referrals, or word of mouth. If resources allow, in-house databases can be set up to keep track of the participants and their history of participation. Users' participation should be voluntary. Most of the time, users are compensated for their time and effort, in the form of cash, gift certificates, or products from the company.

21.2.6.3 Tasks to Be Tested

Test tasks should be chosen based on the core user tasks that have been identified. They should be representative of the intended use of the Web site and answer the questions in the problem statements. Test tasks should be clear, realistic, and sensible to the participants. Test tasks should reflect user goals and should not be a simple mapping of the functionalities of the Web site. Test tasks should not contain instructions on how to perform the tasks. Normally, the test tasks should be written or typed on paper and handed to participants at the start of a test. Participants should work on the tasks one at a time and self-report the start and end of a task.

21.2.6.4 Procedures to Be Followed

A usability test typically starts with a pretest introduction, followed by the test tasks and a debriefing at the end. Frequently, questionnaires are used to collect background information and opinions from users. During the introduction, participants are given a tour of the usability laboratory so that they do not feel intimidated by the cameras and the one-way mirrors (if they are employed). Participants are also briefed on what the test session will be like and what to expect; they will also be advised that their participation is voluntary and confidential. Sometimes it may also be helpful to play back the video of a short video recording to the user that can be captured during the overview of the laboratory equipment that will be used, so s/he knows what information is being picked up by the Web camera or other measuring devices (e.g., eye tracker). Many organizations require participants to sign a nondisclosure agreement or an informed consent form. The experimenter's job, during the introduction, is to make the participants feel relaxed and comfortable.

During the test, the experimenter should remind the participants to think aloud (if the verbal protocol method is used) and to accommodate their special needs, such as a request to take a break. Other than that, the experimenter should not interact much with the participants. The experimenter should be careful in answering questions from the participants because the answers may bias the outcome of the tests. The experimenter should use neutral expressions such as "uh-huh," respond with neutral questions (e.g., if the user asks, "What is this icon supposed to be?" the experimenter could answer, "What do you think the icon represents?"), and tell participants that sometimes their questions may not be answered.

The experimenter should be empathetic when participants are frustrated. Participants should not leave the usability test thinking it was their fault if they had difficulties. The experimenter should reassure participants and provide help when appropriate. When the test is done, the experimenter can ask the participants for additional comments and suggestions. The experimenter can also discuss what happened in the test with the participants and answer any questions. The experimenter should let the participants know that their input is valued and their help is appreciated.

21.2.6.5 Apparatus

Usability tests can be conducted using paper prototypes, interactive prototypes, or operational systems. When testing an interactive prototype or operational system, the following equipment may be needed: a computer (with appropriate processing speed, memory, hard drive, and monitor size) with network connection (wired or wireless) or mobile device with wireless connection, the software or Web application, Web browser, and video camera or Web camera. The equipment should be set up to match that used by the target users. For example, if most users are likely to access the Web at home using a modem or a wireless connection, the computer in the laboratory should be hooked up to a modem or wireless connection instead of a cable connection. Special hardware and software may be needed for testing with special user groups, such as people with disabilities, who might need a table that can be adjusted for height or computer loaded with a screen reader.

21.2.6.6 Measurements and Data Analysis

With performance testing, quantitative data such as task completion time, success and failure rates, number of errors, number of clicks to complete tasks, and traversal paths should be measured and analyzed. Subjective methods such as verbal protocols and interviews, described later, can be used to supplement the performance data. These later methods may require transcription and content analysis, which can be time-consuming.

Data analyses include computation of statistics and categorization of usability problems based on observations and subject reports. Proper and adequate data analysis and data interpretation require skills and knowledge of the researchers. Usually, human factors professionals and usability specialists analyze and interpret the data and then work with the design team to come up with solutions to address the identified usability problems.

21.2.6.7 Comparison of Usability Testing and Usability Inspection

Several studies have compared the effectiveness of different usability inspections and empirical usability testing (Bailey, Allan, and Raiello 1992; Desurvire, Kondziela, and Atwood 1992; Jeffries et al. 1991; Karat, Campbell, and Fiegel 1992; Virzi, Sorce, and Herbert 1993). These studies have revealed the following:

1. Heuristic evaluations, when performed by a group of experts, are more effective than cognitive walkthroughs in identifying usability problems. The percentages of problems identified through heuristic evaluations are higher than those identified by cognitive walkthroughs.
2. Heuristic evaluations are better than cognitive walkthroughs at identifying serious problems, that is, problems that cause task failures.
3. Heuristic evaluations facilitate finding problems that go beyond the scope of the tasks, perhaps because the heuristics analyze more dimensions of the interface. However, heuristic evaluations also tend to find a large number of one-time and low-priority problems.
4. Cognitive walkthroughs are useful in forcing the team to think along the lines of user goals and explicitly state assumptions about user knowledge.
5. Usability testing is better than usability inspection methods at finding general, more severe, and recurring problems.

Usability inspection methods are typically less expensive than usability testing. However, changes based on usability inspections may not necessarily increase system performance. Bailey, Allan, and Raiello (1992) showed that

Rather than the 29 changes originally suggested by Molich and Nielsen [from a usability inspection], our results showed that only one change to each of the original screens was necessary to achieve the same performance and preference levels as those demonstrated by their 'ideal' system. The same task was repeated using a graphical user interface. A heuristic evaluation suggested up to 43 potential changes, whereas the usability test demonstrated that only two changes optimized performance. (p. 409)

It is generally agreed that although usability inspection methods cannot replace empirical usability testing, they can provide a good starting point and yield supplemental information. They are often cost-effective and generally have a quicker turnaround.

21.2.7 Web-Based Methods

In evaluating Web usability, the Web itself can be a valuable tool. Instrumented versions of a Web browser can collect data such as time on task, success rate, and user traversal paths from a relatively large sample of users. Web logs or blogs reveal when and how many users visit a Web site and where they come from (domains). As mentioned earlier, questionnaires can be directly posted on a Web site to solicit user feedback. There are many commercially available survey services (e.g., Survey Monkey) that allow quick implementation of Web-based surveys as well as providing the researcher with summary statistics and graphs. Couper (2008) provides an overview of the usability issues that need to be considered when conducting Web surveys.

21.2.7.1 Automated Sessions

Traditionally, instrumented versions of GUI software have been used to log the actual use of a system (Nielsen 1993), collect data automatically in a usability session (e.g., Nielsen and Lyngbæk 1990), and conduct remote usability tests (e.g., Hartson et al. 1996). Logged data can help to identify what commands and features are frequently used and what are rarely used. Frequently used features should be optimized for usability and rarely used features should be investigated. Nielsen (1993) gave examples of using logged data to identify erroneous commands (Bradford et al. 1990), study error messages (Chapanis 1991), and online help (Senay and Stabler 1987).

Instrumented versions can also be used to conduct remote usability tests. Hartson et al. (1996) and Castillo, Hartson, and Hix (1998) investigated the use of self-reported critical incidents (by clicking on a "report incident" button) along with a video clip showing screen activity immediately prior to clicking the button (captured by a scan converter) in identifying usability problems without conducting a laboratory test. Some programs (e.g., Microsoft Office) give users an option of sending a "bug" report to the organization when a program ends abruptly.

When evaluating Web usability, instrumented Web browsers can be used to conduct automated sessions of remote usability tests. Participants recruited for these tests are provided with the instrumented version of the browser. They can complete the sessions at any time of their choice, typically within a time period allowed. During a session, they will be prompted to complete tasks by visiting one or more Web sites. Performance measures such as time on task, successes and failures, links followed, and files opened are recorded. Users can also indicate when they have completed a task (either successfully or unsuccessfully) by clicking on a "finish" button and then be prompted to continue to the next task. During and after the session, users are usually asked to fill out questionnaires to verify their successes and give subjective opinions.

After the session, the data can be aggregated and analyzed, and statistical analysis can be conducted. The advantage of conducting a Web usability test with automated sessions is that a larger number of participants may be tested. An evaluator does not have to be present, and data collection and analysis can be done automatically. The disadvantage is the lack of verbal protocol. It may be hard to figure out why participants failed on a task or why they made errors.

Automated sessions may be best for summative usability evaluations. Jacques and Savastano (2001) reported a study in which automated sessions were compared with usability testing in the laboratory. They found that the performance data were very similar, demonstrating the viability of automated sessions for summative usability testing.

21.2.7.2 Web Logs

When a Web site is accessed, the Web server records information about the files sent to a browser, including the date, time, host (the Web address of the requesting browser), file name, referrer (the URL of the page that provided the link), the name and the version of the user's browser, and so on. Analysis of the log can reveal when users visit a Web site (frequency distribution), where the users are from (what domains), what files (i.e., Web pages) are most and least frequently requested, and how users navigate within a site using links provided (Drott 1998). These are unique data that are hard to obtain otherwise. For example, although Web users often indicate that privacy is an issue of central concern to them (see Vu et al. 2010), log files show that users seldom visit the privacy policy pages of Web sites (Jensen and Potts 2004). Many commercially available log analysis programs provide summary statistics, making this method appealing to some design teams.

Web log data may provide some clues as to possible usability problems. For example, log data may reveal that some pages are rarely visited or some links rarely used. However, log data alone cannot provide the explanations as to why some pages are rarely visited or some links rarely used. It could be because users do not know the page existed (it is in the wrong category) or the page is not properly linked to. Follow-up studies involving direct interaction with users such as a usability test should be conducted to investigate the issues identified.

21.2.7.3 Opinion Polls

Many Web sites have online interactive polls or questionnaires that solicit user input. The questionnaire design should be developed in the same manner as discussed earlier. These are typically informal, and it is hard to establish the unique identity of a participant. However, they are quick and easy to set up and allow designers to directly get a sense on users' opinions.

21.2.7.4 Alternative Viewing Tools

When evaluating Web usability, it is important to consider that a Web site may be viewed on a variety of devices and

(a)

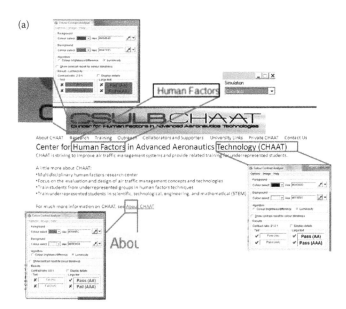

(b)

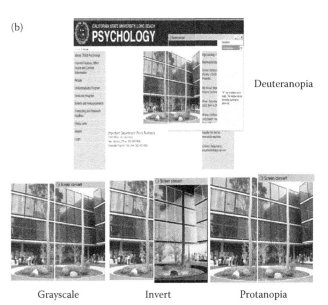

FIGURE 21.6   (See color insert.) (a) Use of the Web accessibility toolbar to test a Web site for accessibility concerns. Specific windows were used to test for contrast and to simulate cataracts. (b) Use of the Web accessibility toolbar to test for color deficiencies and color schemes. Courtesy of Ariana Kiken.

platforms such as different browsers, Web TV, PDAs, and mobile phones. To make sure that the Web site is usable when accessed from different devices, it is necessary to perform usability inspections on alternative viewing devices or platforms (Pearrow 2000). Alternative viewing tools are viewers, or software tools, that allow the designers to see a Web site as it would appear on a different device or platform. Viewing tools for accessibility can also provide information about color and contrast (see Figure 21.6). For example, specific tools allow the designer to see what the Web page will look like when displayed as "text only" or when viewed with earlier versions of a browser. The consideration of viewing Web sites on alternative devices may be particularly useful during the early phase of design when decisions are being made about the structure of the Web site, the categories of the content, and the navigation system. Xu and Fang (this volume) discuss specific requirements for mobile devices including the need to consider contextual factors associated with the use of mobile devices. Data about the target audience will help the designer to determine what other devices, platforms, and browsers are commonly used. Conducting usability inspections using alternate viewing tools will make sure that the Web site is usable on a variety of devices and platforms.

21.2.8 Iterative Testing and Evaluation

Almost every human factor and usability textbook or chapter points to the benefits of iterative testing and evaluation. Iterative usability testing should be a natural part of the design lifecycle from conception of the design to disposal of the system (Norman 2002). However, in practice, iterative testing is often prevented by time and monetary constraints. Medlock et al. (2005) noted that when usability problems are uncovered, they may not be fixed in the next iteration for several different reasons: The designer or other stakeholders may not think that the issues uncovered are really problems that need fixing; resources are better spent adding a new feature than fixing one that is currently functioning; the usability feedback arrives late in the design cycle; the design team is not sure that it can fix the problem based on the recommendation from the usability test.

To overcome these barriers, Medlock et al. (2005) endorsed a Rapid Iterative Test and Evaluation (RITE) method. The RITE method uses the usability methods described earlier in the chapter, with the main difference being that it emphasizes rapid changes to the design and verification of the effectiveness of implementing these changes. For example, changes can be made to the design as soon as the problem and solution are identified. In other words, these changes can be made after a single participant points out the problem rather than after testing multiple participants to verify that the problem is indeed a critical one. By making this change quickly, the "fix" can be tested with the next task performed by the same user if it can be completed quickly or tested in the next scheduled test session with a different user, reducing the iterative

design process time. For the changes to be implemented quickly the design team must work with the usability team and adequate time must be scheduled within or between sessions to review the usability test after each task or each participant. Furthermore, the usability team must schedule a sufficient number of participants so that the solutions to the usability problems can be verified, and no new problems emerge as a result of the fix.

The RITE method is only one technique that can be used to validate usability fixes in a timely manner. Although it should not replace traditional methods that are more thorough, it provides designers with a tool for rapid iterative testing and evaluation.

21.3 BENEFITS OF A MULTIMETHOD APPROACH

In Section 21.2, we described many methods for evaluating Web usability. Because the alternative methods have various strengths and weaknesses, a multimethod approach that uses several of the methods often provides the most complete understanding of user behavior (Eid and Diener 2006; Proctor and Capaldi 2006).

An example of a multimethod approach is provided by a study conducted by Proctor and Vu (2007) to examine issues associated with Web privacy. Because an Internet connection allows worldwide access to any computer connected to the Internet, security and privacy of information clearly are important matters. Because human interactions are crucial to most, if not all, security and privacy measures, usability is of utmost concern (see Schultz, this volume). Our research evaluated usability issues associated with two aspects of privacy: comprehension of privacy policies (which describe the information collected by a Web site, who will have access to this information, and what steps will be taken to ensure privacy of personal information), and configuration of privacy preferences by users for an online user agent (which checks a Web site's machine readable policy and signals to the user whether it is consistent or inconsistent with the user's privacy preferences). The research employed a variety of methods, including archival research, data mining, quantitative analysis, user surveys, and controlled experiments.

The archival component consisted of analyzing the content of existing privacy policies for many categories of Web sites to determine what information was collected by the sites and whether the sites displayed certification logos indicating adherence to good privacy practices. This archival analysis yielded the following information: (1) financial sites collected the most personal information; (2) within a category of Web sites, such as financial sites, there were large differences in the types and amount of information requested; (3) most sites had links to the site's privacy policy; (4) only half of the sites displayed privacy certification seals.

The next step was to use data mining and quantitative analysis methods to analyze the content of 100 privacy policies. The data mining technique identified the goals of the policies, which dealt with protection (user privacy rights protections) and vulnerability (practices that threaten consumer privacy), and standard readability metrics were also calculated for each policy. The policies varied greatly in length, with longer policies tending to state more goals in general. Consistent with Jensen and Potts (2004), the readability analysis indicated that the policies required a minimum of 1 year of college reading level.

Surveys were used to determine which privacy goals mattered most to college-age users. The users indicated that they were most concerned about companies sharing or selling their credit card and financial information to other organizations. In addition, the users did not want companies to use cookies to track their information, even if it was not personally identifiable. Participants indicated that they did not want organizations to share their contact information and welcomed the use of security logos, but they were not concerned about Web sites collecting profiling information in order to customize the presentation of information on the Web site for their use.

Privacy Bird® is a user agent designed to assist users in determining whether a Web site conforms to or violates their specified privacy preferences. Accordingly, it will only yield valid warnings if users configure it correctly to check for their specific concerns. We conducted user tests to evaluate Privacy Bird's interface and found that users were confused about what Privacy Bird could and could not check for. Finally, we evaluated the influence of Privacy Bird on user behaviors by examining whether its warnings influence users' purchasing behaviors. We found that users were more cautious when Privacy Bird warned them that the Web site violated their preferences. However, users interpreted sites with no violation warnings as being safe, when such warnings only indicated that the agent could not tell whether the privacy policy violated the user's preferences (e.g., the site had a privacy policy that was not machine readable).

As illustrated above, each method yielded unique findings regarding the topic of user privacy protection. The bad news from this research is that there is still a long way to go with regard to informing users accurately about the privacy policies of Web sites. The good news is that many of the findings converged to reveal specific privacy issues that need to be addressed. The main point is that multiple methods were needed to get a more complete picture of the situation than any one method could provide. Multiple methods can also be useful for other purposes. For example, Thoma and Williams (2009) described a multimethod process for development of personas in a large e-commerce business. The personas were developed from stakeholder interviews, user interviews and tests, data mining, customer surveys, and ethnographic research, allowing validation of the personas early in the project.

21.4 RELIABILITY, VALIDITY, AND ETHICAL CONSIDERATIONS

21.4.1 Reliability and Validity

As Salvendy and Carayon (1997, p. 1454) pointed out, "To ensure the appropriateness of conclusions drawn from an HFE intervention or some other process where HFE outcomes are important (e.g., experimental study in the laboratory), it is critical to ensure the 'goodness' of the outcome measures." Here, HFE refers to human factors and ergonomics. In particular, reliability and validity are two aspects of the "goodness" of the outcome measures.

Reliability is the consistency of measurements obtained when retested later with the same criterion measure or with a different test that is equivalent in nature. In essence, reliability addresses the notion of whether a measurement is consistent. However, obtaining the reliability of usability tests may be difficult owing to the use of a small set of users that may have large individual differences. The same probably holds true for many methods mentioned in this chapter. It is therefore important to keep reliability concerns in mind. Whenever possible, special care should be taken to use appropriate sampling methods, repeat tests or measures, and conduct statistical tests to measure reliability.

Validity refers to whether a measurement is capturing what it is intended to measure. A measure that is reliable is not necessarily valid. For example, a faulty cost calculator of an e-commerce Web site may add an extra $5 on every order. Thus, it gives a reliable measure of price (e.g., $5 + $5 = $15 every time), but not the valid price (e.g., $5 + $5 = $10, which is the correct price). It is important to examine both the reliability and validity of a measure to ensure accurate conclusions (Salvendy and Carayon 1997; Tullis and Albert 2008). The validity of usability tests can be compromised by many factors, including participation of nontargeted users in the usability tests and failing to provide the user with representative tasks or environmental constraints (e.g., time pressure and social influences).

A good way to address reliability and validity is perhaps the use of multiple methods. If results from multiple sources that are intended to measure the same outcome correlate well, the experimenter can have a higher level of confidence that the results are valid. For example, the study by Sellen and Harper (1997) on the use of paper in organizations could have conducted interviews, ethnographic studies, and questionnaires, in addition to the diary study. If the results from all different methods converged, then this would provide stronger evidence that the conclusions reached by the authors were valid.

21.4.2 Ethical Aspects of Working with Participants

Special care should be taken to handle the ethical aspects of testing with human subjects. This includes appropriate information disclosure about the nature and procedure of the study, voluntary nature of the participation, confidentiality

TABLE 21.5
Summary and Comparison of Methods to Evaluate Web Usability

Method	When to Use	Advantages	Disadvantages
Usability inspections	Early in design, by experts	Less expensive, quick turnaround can identify some usability problems	Not data from real users
Usability testing	Test with real users	Effective in finding usability problems	Costly and time-consuming
Field methods	Understand users and their tasks in the context of their work	Rich and in-depth data about users and their environment	Costly and time-consuming
Interviews, focus groups, questionnaires	Understand users and their tasks; supplementary methods to usability testing	Qualitative and quantitative data	Not good at identifying usability problems
Prototyping	Mock up designs	Allow iterative design and testing quickly	May not be the same as the final implementation, therefore may miss usability problems
Web-based methods	Automatic and remote data collection	Quantitative data of usage by real users	Cannot explain user behavior

and anonymity, and interacting with participants in a respectful, reassuring, and comforting way to ease their anxieties over being watched, observed, tested, and questioned. Many guidelines exist (e.g., American Psychological Association 2002). Whenever a method calls for participation of human subjects, the guidelines and recommendations should be followed.

21.5 SUMMARY

This chapter summarized the major methods that can be used to evaluate Web usability. Table 21.5 lists the different methods discussed in this chapter (see also Proctor et al. 2002), along with a brief description of when to use them and their major advantages and disadvantages. At different stages of design and evaluation, some methods may be more beneficial than others. It is up to the evaluators to choose when and how to use these methods, and in many cases, a multimethod approach may be needed.

Although this chapter is devoted to evaluating Web usability, we hope it is clear that the evaluations should be conducted as the Web product is being designed. When time and money constraints prevent thorough iterative testing, the RITE method (Medlock et al. 2005) can be used to guide the design process. One advantage of Web-based systems is that designers can implement changes even after the original product has been released. This flexibility provided by the Web may be taken to imply that usability considerations early in the design process are less of a concern because problems can be fixed as they arise, but this implication is false. Once users encounter problems with a Web site, they may not return to it since a competing Web site is just a click away. Evaluation of Web usability is more crucial than evaluation of other product usability owing to the fact that usability is a major determinant of whether a Web site will be successful (e.g., Nielsen 2000).

REFERENCES

Aldiin, T., and J. Pruitt. 2008. Putting personas to work: using data-driven personas to focus product planning, design, and development. In *The Human–Computer Interaction Handbook: Fundamentals, Evolving Technologies, and Emerging Applications*, 2nd ed., eds. A. Sears and J. A. Jacko, 991–1016. Boca Raton, FL: CRC Press.

American Psychological Association. 2002. *Ethical Principles of Psychologists and Code of Conduct*. Washington, DC: American Psychological Association. Available at http://www.apa.org/ethics/code2002.html.

Bailey, R. W. 1993. Performance vs. preference. In *Proceedings of the Human Factors and Ergonomics Society 37th Annual Meeting* (Seattle, WA, 11–15 Oct.), 282–286.

Bailey, R., R. Allan, and P. Raiello. 1992. Usability testing vs. heuristic evaluation: A head-to-head comparison. In *Proceedings of the Human Factors and Ergonomics Society 36th Annual Meeting* (Monterey, CA, 11–15 Oct.), 409–508.

Beaudouin-Lafon, M., and W. E. Mackay. 2008. Prototyping tools and techniques. In *The Human–Computer Interaction Handbook: Fundamentals, Evolving Technologies, and Emerging Applications*, 2nd ed., eds. A. Sears and J. A. Jacko, 1017–1039. Boca Raton, FL: CRC Press.

Beyer, H., and K. Holtzblatt. 1998. *Contextual Design: Defining Customer-centered Systems*. San Francisco, CA: Morgan Kaufmann.

Bevan, N. 2005. Cost-benefit framework and case studies. In *Cost-justifying Usability: An Update for the Internet Age*, eds. R. G. Bias and D. Mayhew, 575–599. London: Elsevier.

Bias, R. 1991. Walkthroughs: efficient collaborative testing. *IEEE Software* 8: 94–95.

Bias, R., and D. J. Mayhew (Eds.). 2005. *Cost-Justifying Usability: An Update for the Internet Age*. London: Elsevier.

Bradford, J. H., W. D. Murray, and T. T. Carey. 1990. What kind of errors do Unix users make? In *Proceedings of IFIP INTERACT'90 Third International Conference on Human–Computer Interaction* (Cambridge, U.K., 27–31 Aug.), 43–46.

Brinck, T., D. Gergle, and S. D. Wood. 2002. *Usability for the Web: Designing Web Sites that Work*. San Francisco: Morgan Kaufmann.

Byrne, M. D. 2008. Cognitive architecture. In *The Human–Computer Interaction Handbook: Fundamentals, Evolving Technologies, and Emerging Applications*, 2nd ed., eds. A. Sears and J. A. Jacko, 93–114. Boca Raton, FL: CRC Press.

Byrne, M. D., B. E. John, N. S. Wehrle, and D. C. Crow. 1999. The tangled Web we wove: A taskonomy of WWW use. In *Proceedings of ACM CHI 99 Conference on Human Factors in Computing Systems* (Pittsburgh, PA, 15–20 May), 544–551.

Card, S., T. Moran, and A. Newell. 1983. *The Psychology of Human–Computer Interaction*. Hillsdale, NJ: Lawrence Erlbaum.

Carroll, J. M., and M. B. Rosson. 1992. Getting around the task-artifact cycle: How to make claims and design by scenario. *ACM Transactions on Information Systems* 10: 181–212.

Castillo, J. C., H. R. Hartson, and D. Hix. 1998. Remote usability evaluation: can users report their own critical incidents? In *Proceedings of ACM CHI 98 Conference on Human Factors in Computing Systems* (Los Angeles, CA, 18–23 April), 253–254.

Chapanis, A. 1991. The business case for human factors in informatics. In *Human Factors for Informatics Usability*, eds. B. Shackel and S. Richardson, 21–37. New York: Cambridge University Press.

Clarke, L. 1991. The use of scenarios by user interface designers. In *People and Computers VI: Proceedings of the BCS HCI'91 Conference*, eds. D. Diaper and N. Hammond, 103–115. New York: Cambridge University Press.

Cockton, G., A. Woolrych, and D. Lavery. 2008. Inspection-based evaluations. In *The Human–Computer Interaction Handbook: Fundamentals, Evolving Technologies, and Emerging Applications*, 2nd ed., eds. A. Sears and J. A. Jacko, 1171–1190. Boca Raton, FL: CRC Press.

Cooper, A. 1999. *The Inmates Are Running the Asylum: Why High-Tech Products Drive Us Crazy and How to Restore the Sanity*. Indianapolis, IN: Sams.

Couper, M. 2008. *Designing Effective Web Surveys*. New York: Cambridge University Press.

Coutaz, J., and G. Calvary. 2008. HCI and software engineering: Designing for user interface plastisicity. In *The Human–Computer Interaction Handbook: Fundamentals, Evolving Technologies, and Emerging Applications*, 2nd ed., eds. A. Sears and J. A. Jacko, 1107–1128. Boca Raton, FL: CRC Press.

Dayton, T., A. McFarland, and J. Kramer. 1998. Bridging user needs to object oriented GUI prototype via task object design. In *User Interface Design: Bridging the Gap from User Requirements to Design*, eds. L. E. Wood, 15–56. Boca Raton, FL: CRC Press.

Delbecq, A. L., and A. H. Van den Ven. 1971. A group process model for identification and program planning. *Journal of Applied Behavioral Sciences* 7: 466–492.

Desurvire, H. W., J. M. Kondziela, and M. E. Atwood. 1992. What is gained and lost when using evaluation methods other than empirical testing. In *People and Computers VII: Proceedings of the BCS HCI'92 Conference*, eds. A. Monk, D. Diaper, and M. D. Harrison, 89–102. New York: Cambridge University Press.

Dray, S., D. Siegel, E. Feldman, and M. Potenza. 2002. Why do version 1.0 and not release it? conducting field trials of the tablet PC. *Interactions* 9: 11–16.

Drott, M. C. 1998. Using Web server logs to improve site design. In *Proceedings of ACM 16th International Conference on Systems Documentation* (Quebec, Quebec, Canada, 23–26 Sept.), 43–50.

Dumas, J. S., and J. E. Fox. 2008. User-base evaluations. In *The Human–Computer Interaction Handbook: Fundamentals, Evolving Technologies, and Emerging Applications*, 2nd ed., eds. A. Sears and J. A. Jacko, 1129–1150. Boca Raton, FL: CRC Press.

Eid, M., and E. Diener (Eds.). 2006. *Handbook of Multimethod Measurement in Psychology*. Washington, DC: American Psychological Association.

Feldman, E., E. Pennington, and J. Ireland. 2003. Tablet PC—Using field trials to define product design. In *Human–Computer Interaction: Theory and Practice II*, eds. C. Stephanidis and J. A. Jacko, 636–640. Mahwah, NJ: Lawrence Earlbaum.

Fleming, J. 1998. *Web Navigation: Designing the User Experience*. Sebastopol, CA: O'Reilly & Associates, Inc.

Goodman, J., S. Clarke, P. Langdon, and P. J. Clarkson. 2007. Designers' perceptions of methods of involving and understanding users. In *Universal Access in HCI, Part I, HCII 2007, LNCS 4554*, ed. C. Stephanidis, 127–136. Berlin: Springer-Verlag.

Graetz, K. A., N. Proulx, C. B. Barlow, and L. J. Pape. 1997. Facilitating idea generation in computer-supported teleconferences. In *Proceedings of GROUP'97: International Conference on Supporting Group Work* (Phoenix, AZ, 16–19 Nov.), 317–324.

Greif, I. 1992. Designing group-enabled applications: A spreadsheet example. In *Groupware'92* (San Jose, CA, 3–5 Aug.), ed. D. Coleman, 515–525. San Mateo, CA: Morgan Kaufmann Publishers.

Harris, R. W. 1999. Attitudes towards end-user computing: A structural equation model. *Behaviour and Information Technology* 18: 109–125.

Hartson, H. R., J. C. Castillo, J. Kelso, and W. C. Neale. 1996. Remote evaluation: The network as an extension of the usability laboratory. In *Proceedings of ACM CHI 96 Conference on Human Factors in Computing Systems* (Vancouver, BC, Canada, 13–18 Apr.), 228–235.

Holtzblatt, K. 2008. Contextual design. In *The Human–Computer Interaction Handbook: Fundamentals, Evolving Technologies, and Emerging Applications*, 2nd ed., eds. A. Sears and J. A. Jacko, 949–964. Boca Raton, FL: CRC Press.

Hughes, J., V. King, T. Rodden, and H. Andersen. 1995. The role of ethnography in interactive systems design. *Interactions* 2: 56–65.

Hussain, Z., W. Slany, and A. Holzinger. 2009. Current state of agile user-centered design: A survey. In *HCI and Usability for e-inclusion*, eds. A. Holzinger and K. Miesenberger. 416–427. Berlin: Springer-Verlag.

Isaacs, E. A. 1997. Interviewing customers: Discovering what they can't tell you. In *Proceedings of ACM CHI 97 Conference on Human Factors in Computing Systems* (Atlanta, GA, 22–27 Mar.), 180–181.

Jacques, R., and H. Savastano. 2001. Remote vs. local usability evaluation of Web sites. In *Proceedings of IHM-HCI 2001* (Lille, France, 10–14 Sept.), 91–92.

Jeffries, R., J. R. Miller, C. Wharton, and K. M. Uyeda. 1991. User interface evaluation in the real world: A comparison of four techniques. In *Proceedings of ACM CHI 91 Conference on Human Factors in Computing Systems* (New Orleans, LA, 28 Apr.–2 May), 119–124.

Jensen, C., and J. Potts. 2004. Privacy policies as decision-making tools: an evaluation of online privacy notices. In *Proceedings of the Special Interest Group on Human–Computer Interaction Conference on Human Factors in Computing Systems*, 6: 471–478.

Kamvar, M., M. Kellar, R. Patel, and Y. Xu. 2009. Computers and iPhones and mobile phones, Oh my! In *World Wide Web 2009*, 801–810. Madrid, Spain: ACM.

Karat, C.-M., R. Campbell, and T. Fiegel. 1992. Comparison of empirical testing and walkthrough methods in user interface evaluation. In *Proceedings of ACM CHI 92 Conference on Human Factors in Computing Systems* (Monterey, CA, 3–7 May), 397–404.

Kunivavsky, M. 2003. *Observing the User Experience*. San Francisco, CA: Morgan Kaufmann.

Kuniavsky, M. 2008. User experience and HCI. In *The Human–Computer Interaction Handbook: Fundamentals, Evolving Technologies, and Emerging Applications*, 2nd ed., eds. A. Sears and J. A. Jacko, 897–916. Boca Raton, FL: CRC Press.

Lafrehiere, D., T. Dayton, and M. Muller. 1999. Tutorial 27: Variations of a theme: card-based techniques for participator analysis and design. In *Proceedings of ACM CHI 99 Conference on Human Factors in Computing Systems* (Pittsburgh, PA, 15–20 May), 1–81.

Lazar, J. 2006. *Web Usability: A User-Centered Design Approach.* Boston: Addison-Wesley.

Lewis, C., and C. Wharton. 1997. Cognitive walkthroughs. In *Handbook of Human–Computer Interaction*, 2nd ed., eds. M. Helander, T. K. Landauer, and P. Prabhu, 717–732. Amsterdam: Elsevier Science.

Lewis, J. R. 1995. IBM computer usability satisfaction questionnaires: Psychometric evaluation and instructions for use. *International Journal of Human–Computer Interaction 7:* 57–78.

Lewis, J. R. 2006. Usability testing. In *Handbook of Human Factors and Ergonomics*, 3rd ed., ed. G. Salvendy, 1275–1316. New York: John Wiley.

Mayhew, D. J. this volume. The web UX design process—a case study. In *Handbook of Human Factors in Web Design*, 2nd ed., eds. K.-P. L. Vu and R. W. Proctor, 461–480. Boca Raton, FL: CRC Press.

Medlock, M., D. Wixon, M. McGee, and D. Welsh. 2005. The rapid iterative test and evaluation method: Better products in less time. In *Cost-justifying Usability: An Update for the Information Age*, eds. R. Bias and D. Mayhew, 489–517. San Francisco: Morgan Kaufmann.

Moreno, L., P. Martínez, and B. Ruiz-Mezcua. 2009. A bridge to Web accessibility from the usability heuristics. In *HCI and Usability for e-inclusion*, eds. A. Holzinger and K. Miesenberger, 290–300. Berlin: Springer-Verlag.

Morgan, D. L. 1997. *Focus Groups as Qualitative Research*, 2nd ed. Thousand Oaks, CA: Sage.

Moroney, W. F., and J. A. Cameron. 2000. Questionnaire design and use: a primer for practitioners. Workshop given at the 44th Annual Meeting of the Human Factors and Ergonomics Society (San Diego, CA, 29 July–4 Aug.).

Newman, L. 2005. Interview method. In, *Handbook of Human Factors and Ergonomics Methods*, eds. N. Stanton et al., 77-1 to 77-5. Boca Raton, FL: CRC Press.

Nielsen, J. 1989. Usability engineering at a discount. In *Proceedings of the Third International Conference on Human–Computer Interaction* (Boston, MA, 18–22 Sept.), 394–401.

Nielsen, J. 1993. *Usability Engineering.* San Diego, CA: Academic Press.

Nielsen, J. 1994. Heuristic evaluation. In *Usability Inspection Methods*, eds. J. Nielsen and R. L. Mack, 25–64. New York: John Wiley.

Nielsen, J. 1998. Sun's new Web design. http://www.sun.com/980113/sunonnet/index.html (accessed December 8, 2003).

Nielsen, J. 2000. *Designing Web Usability.* Indianapolis, IN: New Riders Publishing.

Nielsen, J., and U. Lyngbæk. 1990. Two field studies of hypermedia usability. In *Hypertext: State of the Art*, eds. R. McAleese and C. Green, 64–72. Norwood, NJ: Ablex Publishing.

Nielsen, J., and R. Molich. 1990. Heuristic evaluation of user interfaces. In *Proceedings of ACM CHI 90 Conference on Human Factors in Computing Systems* (Seattle, WA, 1–5 April), 249–256.

Norman, D. A. 2002. *The design of everyday things.* New York: Basic Books.

O'Donnell, P. J., G. Scobie, and I. Baxter. 1991. The use of focus groups as an evaluation technique in HCI evaluation. In *People and Computers VI: Proceedings of the BCS HCI'91 Conference*, eds. D. Diaper and N. Hammond, 211–224. New York: Cambridge University Press.

Pappas, L., L. Roberts, and R. Hodgkinson. this volume. International standards for accessibility in web design and the technical challenges in meeting them. In *Handbook of Human Factors in Web Design*, 2nd ed., eds. K.-P. L. Vu and R. W. Proctor, 403–413. Boca Raton, FL: CRC Press.

Pearrow M. 2000. *Web Site Usability Handbook.* Rockland, MA: Charles River Media.

Pierotti, D. 1994. Usability techniques: heuristic evaluation—a system checklist. Available at http://www.stcsig.org/usability/topics/articles/he-checklist.html.

Polson, P. G., C. Lewis, J. Rieman, and C. Wharton. 1992. Cognitive walkthroughs: A method for theory-based evaluation of user interfaces. *International Journal of Man–Machine Studies* 36: 741–773.

Proctor, R. W., M. A. Ali, and K.-P. L. Vu. 2008. Examining usability of Web privacy policies. *International Journal of Human–Computer Interaction* 24: 307–328.

Proctor, R. W., and E. J. Capaldi. 2006. *Why Science Matters: Understanding the Methods of Psychological Research.* Malden, MA: Blackwell.

Proctor, R. W., and K.-P. L. Vu. 2007. A multimethod approach to examining usability of Web privacy policies and user agents for specifying privacy preferences. *Behavior Research Methods* 39: 205–211.

Proctor, R. W., K.-P. L. Vu, L. Najjar, M. W. Vaughan, and G. Salvendy. 2003. Content preparation and management for E-commerce websites. *Communications of the ACM* 46(12ve): 289–299.

Proctor, R. W., K.-P. L. Vu, G. Salvendy, et al. 2002. Content preparation and management for web design: Eliciting, structuring, searching, and displaying information. *International Journal of Human–Computer Interaction* 14: 25–92.

Richeson, A., E. Bertus, R. G. Bias, and J. Tate. this volume. Determining the value of usability in web design. In *Handbook of Human Factors in Web Design*, 2nd ed., eds. K.-P. L. Vu and R. W. Proctor, 753–764. Boca Raton, FL: CRC Press.

Risden, K. 1999. Towards usable browse hierarchies for the Web. In *Proceedings of the Eighth International Conference on Human–Computer Interaction 1999* (Munich, Germany, 22–26 Aug.), 1098–1102.

Salvendy, G., and P. Carayon. 1997. Data-collection and evaluation of outcome measures. In *Handbook of Human Factors and Ergonomics*, 2nd ed., ed. G. Salvendy, 1451–1470. New York: John Wiley.

Scapin, D. L. 1990. Organizing human factors knowledge for the evaluation and design of interfaces. *International Journal of Human–Computer Interaction* 2: 203–229.

Schultz, E. E. this volume. Web security, privacy, and usability. In *Handbook of Human Factors in Web Design*, 2nd ed., eds. K.-P. L. Vu and R. W. Proctor, 663–676. Boca Raton, FL: CRC Press.

Sebillotte, S. 1995. Methodology guide to task analysis with the goal of extracting relevant characteristics for human–computer interfaces. *International Journal of Human–Computer Interaction* 7: 341–663.

Sellen, A., and R. Harper. 1997. Paper as an analytic resource for the design of new technologies. In *Proceedings of ACM CHI 97 Conference on Human Factors in Computing Systems* (Atlanta, GA, 22–27 Mar.), 319–326.

Senay, H., and E. P. Stabler. 1987. Online help system usage: An empirical investigation. In *Abridged Proceedings of the 2nd International Conference of Human–Computer Interaction* (Honolulu, HI, 10–14 Aug.), 244.

Shadbolt, N., and M. Burton. 1995. Knowledge elicitation: a systematic approach. In *Evaluation of Human Work*, 2nd ed., eds. J. R. Wilson and E. N. Corlett, 406–440. London: Taylor & Francis.

Smith-Jackson, T. 2005. Cognitive walk-through method (CWM). In *Handbook of Human Factors and Ergonomics Methods*, eds. N. Stanton et al., 82-1 to 82-7. Boca Raton, FL: CRC Press.

Spencer, R. 2000. The streamlined cognitive walkthrough method: Working around social constraints encountered in a software development company. In *Proceedings of ACM CHI 2000 Conference on Human Factors in Computing Systems* (The Hague, Netherlands, 1–6 Apr.), 353–359.

Stanton, N. A., and C. Baber. 2005. Task analysis for error identification. In *Handbook of Human Factors and Ergonomics Methods*, eds. N. Stanton et al., 37-1 to 37-8. Boca Raton, FL: CRC Press.

Strybel, T. Z. this volume. Task analysis methods and tools for developing web applications. In *Handbook of Human Factors in Web Design*, 2nd ed., eds. K.-P. L. Vu and R. W. Proctor, 483–508. Boca Raton, FL: CRC Press.

Sullivan, P. 1991. Multiple methods and the usability of interface prototypes: The complementarity of laboratory observation and focus groups. In *Proceedings of ACM Ninth International Conference on Systems Documentation* (Chicago, IL, 10–12 Oct.), 106–112.

Teo, T. S. H., and V. K. G. Lim. 2000. Gender differences in Internet usage and task preferences. *Behaviour and Information Technology* 19: 283–295.

Thoma, V., and B. Williams. 2009. Developing and validating personas in e-commerce: A heuristic approach. In *INTERACT 2009, Part II, LNCS 5727*, eds. T. Gross et al., 524–527. Berlin: Springer-Verlag.

Tullis, T., and B. Albert. 2008. *Measuring the User Experience: Collecting, Analyzing, and Presenting Usability Metrics*. Burlington, MA: Morgan Kaufmann.

van Rijn, H., A. Johnson, and N. Taatgen. this volume. Cognitive user modeling. In *Handbook of Human Factors in Web Design*, 2nd ed., eds. K.-P. L. Vu and R. W. Proctor, 527–542. Boca Raton, FL: CRC Press.

Virzi, R. A., J. F. Sorce, and L. B. Herbert. 1993. A comparison of three usability evaluation methods: Heuristic, think-aloud, and performance testing. In *Proceedings of the Human Factors and Ergonomics Society 37th Annual Meeting* (Seattle, WA, 11–15 Oct.), 309–313.

Volk, F., F. Pappas, and H. Wang. this volume. User research: User-centered methods for designing of web interfaces. In *Handbook of Human Factors in Web Design*, 2nd ed., eds. K.-P. L. Vu and R. W. Proctor, 417–438. Boca Raton, FL: CRC Press.

Vu, K.-P. L., V. Chambers, B. Creekmur, D. Cho, and R. W. Proctor. 2010. Influence of the Privacy Bird® user agent on user trust of different Web sites. *Computers in Industry* 61(4): 311–317.

Vu, K.-P. L., G. L. Hanley, T. Z. Strybel, and R. W. Proctor. 2000. Metacognitive processes in human–computer interaction: Self-assessments of knowledge as predictors of computer expertise. *International Journal of Human–Computer Interaction* 12: 43–71.

Vu, K.-P. L., and R. W. Proctor. 2006. Web site design and evaluation. In *Handbook of Human Factors and Ergonomics*, 3rd ed., ed. G. Salvendy, 1317–1343. New York: John Wiley.

Wharton, C., J. Bradford, R. Jeffries, and M. Franzke. 1992. Applying cognitive walkthroughs to more complex user interfaces: Experiences, issues, and recommendations. In *Proceedings of ACM CHI 92 Conference on Human Factors in Computing Systems* (Monterey, CA, 3–7 May), 381–388.

Wixon, D. 1995. Qualitative research methods in design and development. *Interactions* 2: 19–26.

Wogalter, M. S., and R. L. Frei. 1990. Social influence and preference of direct-manipulation and keyboard-command computer interfaces. In *Proceedings of the Human Factors Society 34th Annual Meeting* (Orlando, FL, 8–12 Oct.), 907–911.

Wood, L. E. 1996. The ethnographic interview in user-centered work. In *Field Methods Casebook for Software Design*, eds. D. Wixon and J. Ramey, 35–56. New York: John Wiley.

Xu, S., and X. Fang. this volume. Mobile interface design for M-commerce. In *Handbook of Human Factors in Web Design*, 2nd ed., eds. K.-P. L. Vu and R. W. Proctor, 701–724. Boca Raton, FL: CRC Press.

Yamada, S., J.-K. Hong, and S. Sugita. 1995. Development and evaluation of hypermedia for museum education: Validation of metrics research contributions. *ACM Transactions on Computer–Human Interaction* 2: 284–307.

Zapf, D., F. C. Brodbeck, M. Frese, H. Peters, and J. Prumper. 1992. Errors in working in office computers: A first validation of a taxonomy for observed errors in a field setting. *International Journal of Human–Computer Interaction* 4: 311–339.

Zhu, W., L. Birtley, and N. Burgess-Whitman. 2003. Combining usability with field research: designing an application for the tablet PC. In *Human–Computer Interaction: Theory and Practice II*, eds. C. Stephanidis and J. A. Jacko, 816–820. Mahwah, NJ: Lawrence Erlbaum.

APPENDIX A: WEB USABILITY RESOURCES

BOOKS

Badre, A. N. (2002). *Shaping Web Usability: Interaction Design in Context*. Boston, MA: Addison-Wesley.

Krug, S. (2006). *Don't Make Me Think! A Common Sense Approach to Web Usability*, 2nd ed. Berkeley, CA: New Riders.

Lawrence, D., and S. Tavakol (2007). *Balanced Website Design: Optimising Aesthetics, Usability and Purpose*. New York: Springer-Verlag.

Lazar, J. (2006). *Web Usability: A User-centered Design Approach*. Boston: Addison-Wesley.

Mander, R., and B. E. Smith (2002). *Web Usability for Dummies*. Indianapolis, IN: Hungry Minds.

Nielsen, J., and H. Loranger (2006). *Prioritizing Web Usability*. Berkeley, CA: New Riders.

Brink, T., D. Gergle, and S. D. Wood (2001). *Usability for the Web: Designing Web Sites that Work*. San Francisco, CA: Morgan Kaufmann.

Ratner, J. (Ed.) (2003). *Human Factors and Web Development*, 2nd ed. Mahwah, NJ: Lawrence Erlbaum.

ONLINE REFERENCES

http://usableweb.com/index.html: Keith Instone's guide to user interface, usability and human factors aspects of the Web.

http://www.useit.com: from Jakob Nielsen.

http://www.ibm.com/ibm/easy/: IBM Web site relating to user experience.

http://www.asktog.com: from Bruce Tognazzini, designer of Apple's user interface.

http://www.eshopability.com: Newsletter on current Web usability issues.

http://www.crocolyle.blogspot.com: Commentary on Web usability, Web design, information architecture, and related topics from a practicing usability professional.

http://www.csulb.edu/centers/cuda/support_ati/accessibility_database/: Online repository of Web accessibility materials from the Center for Usability in Design and Accessibility.

22 The Web UX Design Process—A Case Study

Deborah J. Mayhew

CONTENTS

22.1 INTRODUCTION

The Usability Engineering Lifecycle (Mayhew 1999) documents a structured and systematic approach to addressing usability within a software development process. It consists of a set of usability engineering tasks applied in a particular order at specified points in an overall development life cycle.

While it was written with a focus on traditional graphical user interface (GUI, a la Microsoft Windows and Apple Macintosh) desktop software applications, with adaptations it is equally applicable to the development of all kinds of Web sites.

Several types of tasks are included in the Usability Engineering Lifecycle, as follows:

- Structured *usability requirements analysis* tasks
- An explicit *usability goal setting* task, driven directly from requirements analysis data
- Tasks supporting a structured, *top-down* approach to *user interface design* that is driven directly from usability goals and other requirements data
- Objective *usability evaluation* tasks for iterating design toward usability goals

The chart in Figure 22.1 represents in summary, visual form, the Usability Engineering Lifecycle. The overall life cycle is cast in three phases: (1) requirements analysis, (2) design/testing/development, and (3) installation. Specific usability engineering tasks within each phase are presented in boxes, and arrows show the basic order in which tasks are carried out. Much of the sequencing of tasks is iterative, and the specific places where iterations would most typically occur are illustrated by arrows returning to earlier points in the life cycle.

Mayhew (1999) provides a detailed description of the full Usability Engineering Lifecycle. Other chapters in this volume specifically address many of the phases and tasks that are a part of the usability engineering life cycle in more detail (see Dix and Shabir, this volume), as well as general design principles and guidelines in the specific context of the Web (see, e.g., Tullis et al., this volume; Eisen, this volume).

Mayhew (2005) addressed adapting the usability engineering life cycle to Web development projects and focused primarily on the design/testing/development phase rather than the full life cycle, which also includes requirements analyses and installation. In that chapter, the focus was on *usability*, and not on other aspects of the *user experience (UX)*, including *graphic design* and—in the case of e-commerce and certain other types of Web sites—*persuasive design*.

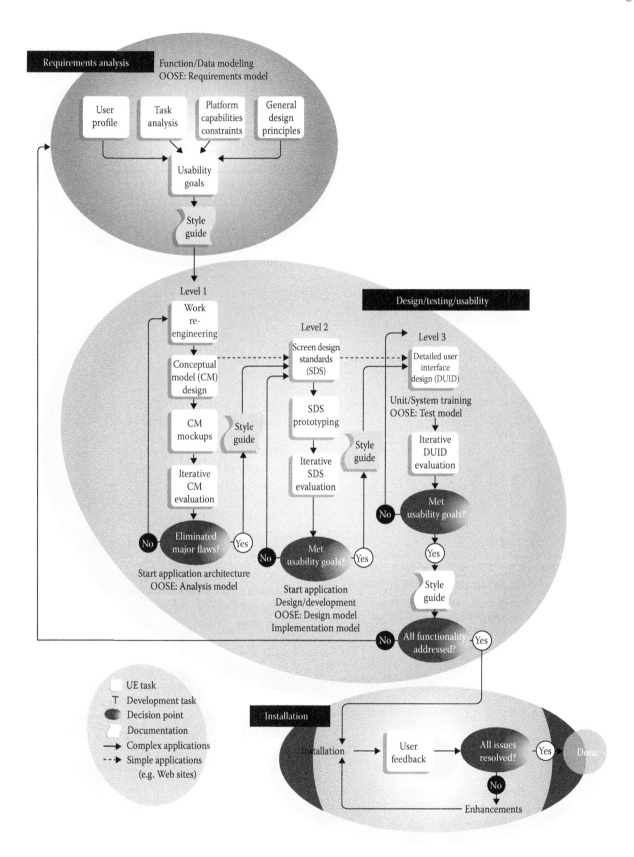

FIGURE 22.1 The Usability Engineering Lifecycle. (From Mayhew, D. J. 1999. *The Usability Engineering Lifecycle.* San Francisco, CA: Morgan Kaufmann Publishers. With permission.)

This chapter provides an update on how to adapt the design phase within the Usability Engineering Lifecycle for Web UX, as opposed to Web *usability*. UX is a relatively new term, which encompasses not just usability but also *graphic design* as well as a new specialty area known as *persuasive design*. Persuasive design applies a discipline known as *persuasion psychology*—a well-developed field with a literature of its own often applied in marketing (see http://www.persuasiondesign.info/ for a set of references for this topic)—to the design of Web sites whose goals include *conversions*, that is, influencing visitor behaviors such as sales, newsletter sign ups, lead generation, etc. Although design principles are touched upon, the focus of this chapter is on the design *process* rather than on the design *principles*.

This chapter presents a case study of Web site design and development to describe and illustrate a top-down design process for UX design within the overall Usability Engineering Lifecycle, which is now better referred to as the *UX engineering lifecycle* and will be throughout the rest of this chapter. It does not address the requirements analysis phase or the installation phases of the life cycle.

The main goals of this chapter are to communicate:

- A general *top-down process* for structuring UX design.
- How an *interdisciplinary UX team* should work together to accomplish the UX goals of usability, branding and aesthetics, and persuasion.

22.2 A UX DESIGN PROCESS FOR WEB DEVELOPMENT

22.2.1 OVERVIEW

Within the Usability Engineering Lifecycle (Mayhew 1999; see Figure 22.1) is a phase consisting of a structured, top-down, iterative approach to design. Design is driven by requirements data from the requirements analysis phase. The overall design phase is divided into three levels of design. Each level includes an iterative process of design, mock-up, and evaluation as follows (slight wording changes made to reflect Web site *UX* design in particular):

- Level 1
  - Information architecture
  - Conceptual model design
  - Conceptual model mock-ups
  - Iterative conceptual model evaluation
- Level 2
  - Page design standards
  - Page design standards prototyping
  - Iterative page design standards evaluation
- Level 3
  - Detailed UX design
  - Iterative detailed UX design evaluation

The UX engineering life cycle is flexible and can be expanded or contracted in a number of ways to accommodate the realities of specific development projects. Within the design/testing/development phase, these ways include combining some of the three design levels for the purposes of iterative evaluation. For example, when time and resources are limited, when Web site functionality is relatively simple, and/or when UX expertise is employed throughout the life cycle, testing for Levels 1 and 2, or Levels 2 and 3, or even Levels 1, 2 and 3, can be combined to just two or even one iterative testing process.

This does *not* mean, however, that the *design* tasks themselves should not be separately carried out as described below, just that *testing* can be collapsed into only two or even one round of iterations. The purpose of this chapter is to explain, justify, and describe a top-down *design* process.

The rest of this chapter is structured around the four distinct design tasks that arise across the three design levels:

- Information architecture
- Conceptual model design
- Page design standards
- Detailed UX design

A section is devoted to each of these design tasks. Each section is divided into two subsections:

- Design Issues
- Case Study

The purpose of a top-down approach to UX design is to address different sets of design issues separately and at different points in the process. The subsections on Design Issues within each design task section provide a brief overview of the design issues addressed in that design task. Recall, however, that the main focus of this chapter is on the design *process* rather than on the design principles. More detailed coverage of some of these design issues is offered in other chapters in this volume (e.g., see chapters by Tullis et al., this volume; Najjar, this volume; and Volk and Kraft, this volume).

The rationale behind a top-down approach to UX design is that it is more efficient and effective to address distinct sets of design issues independently of one another, and in a specific order, that is, from the highest level to the most detailed level. Because the design tasks address issues that are fairly independent of one another, focusing on one level of design at a time forces designers to address *all* UX design issues explicitly and consciously. It ensures efficiency in that lower level details are not constantly revisited and reworked as higher-level design issues are randomly addressed and reworked. Each level of design builds on the design decisions at higher levels, which might have already been validated through iterative evaluation.

The Case Study subsections in this chapter follow a particular Web site development project through each of the four design tasks in the three different design levels. The project reported in the case study involved the development of

FIGURE 22.2 Across-task information architecture for a shopping Web site: deep structure.

a Web site whose core purpose is to deliver a video-based e-learning curriculum on a fee basis. Secondary goals for the site included professional networking, resource sharing, and community building.

22.2.2 INFORMATION ARCHITECTURE

22.2.2.1 Design Issues

The top level in the top-down process for Web UX design is the *information architecture* for a Web site. The information architecture is a specification of the *navigational structure* of the Web site. It does *not* involve any visual design. It is often offered directly to site visitors in the form of a site map.

During the information architecture design task, UX designers must specify two broad types of navigational structure. First, distinct visitor tasks must be organized into a structure that determines how visitors will navigate *to each task*. Second, the workflow *within each task* must be designed.

Across-task navigational structures are often documented in a format we will refer to as *page flowcharts*. Page flowcharts look a lot like traditional organization charts or programming flowcharts, except that the boxes represent site pages and the lines between boxes represent direct navigational pathways between pages. Figures 22.2 through 22.5 all provide examples of page flowcharts.

Designers must design information architectures in a way that streamlines site visitor navigation across and within tasks and exploits the capabilities of automation (to enhance ease of *use*), while at the same time preserving best practices (in both task flow and design principles) and familiar structures that fit visitors' current mental models of their tasks (to enhance ease of *learning*).

For example, in an e-commerce application, key visitor tasks might include the following:

- Browse through categories of product offerings
- Search for products according to specific criteria
- Get products into a shopping cart
- Check out

Additional tasks might include the following:

- Contact the vendor
- Monitor one's account
- Request returns or exchanges

Designers must establish a navigational structure that determines how visitors will navigate to these different tasks. For example, will there be three top level categories called Shopping, Shopping Cart, and Customer Service such as specified in Figure 22.2?

Then, designers must define the navigational structure *within* each task, for example, the set of steps visitors must execute to conduct a search or complete a checkout. In the case of a search task, will there be a single page of potential search criteria at one level or will visitors answer one or more "screening" questions before being presented with a set of search criteria within a category? For example, will visitors simply get a search page on which they define product parameters, such as specified in Figure 22.4? Or will they first define whether they are interested in, say, Men's, Women's, or Kids clothes, then select a type of clothing within that broad category (e.g., Outerwear), then specify detailed criteria within that subcategory (e.g., fabric type, temperature rating, color, size, etc.) such as specified in Figure 22.5?

Information architecture can also be specified in a "wireframe" format such as that in Figure 22.6, showing a search sequence. Whereas the specification format in Figures 22.4 and 22.5 shows simple flow from page to page, wireframes are especially effective for expanding an information

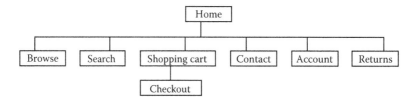

FIGURE 22.3 Across-task information architecture for a shopping Web site: broad structure.

FIGURE 22.4 Within-task information architecture for a search process: broad structure.

architecture specification to include within-task navigational detail as well as requirements for usability and persuasion which must be realized in the visual design. Figure 22.6 provides an illustration of a wireframe of a single page, but the description that follows specifies navigational flow from page to page as well.

Figure 22.6 shows the final step of a six-step process. Step 1 is navigating to the Search task (e.g., by clicking a Search link in a left-hand navigation frame). This brings up a page containing just a filter type A choice widget (e.g., clothing, housewares, sports equipment). Step 2 is selecting a filter choice from this widget. This brings up the filter type B widget next to the filter type A widget, as well as an execution widget ("GO"). In this design, the choices in the filter type B widget are unique to the choice made in the filter type A widget (e.g., Men's, Women's, Kid's). Step 3 is selecting a filter choice from the filter type B widget and executing it. This brings up a page displaying both the filter type A and filter type B widgets with the choices previously selected (but still modifiable), plus the search criteria display, which will be a fill-in form presenting some number of fields unique to the two filters already selected. Step 4 is filling out this criteria form and executing it. This brings up a set of search results on the search results display (filter types A and B are still visible on this page, and the search criteria fields are still filled in but obscured). Step 5 is selecting one of the search results and executing it, and this brings up the item detail display. Step 6 is executing some action on the content of the item detail display.

Note that while some visual design seems to be implied by this wireframe sketch (e.g., location of navigation bars, use of drop downs, tabs, Go buttons), it is really just meant to describe a task flow, and many different visual designs could be imposed upon it without changing the documented navigational structure. Also note that there is no specification

FIGURE 22.5 Within-task information architecture for a search process: deep structure.

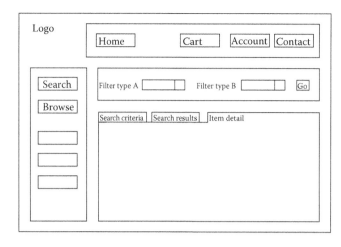

FIGURE 22.6 A one-page wireframe from a search task showing the last of a six-step process.

of individual page content here. That will also come later in the process; the focus at this design level is on navigational pathways.

In the information architecture design task, the lead should be a usability expert, but it is also important for a persuasion expert to be involved at this point if site goals include conversions. It is not yet necessary for a graphic designer to be involved.

A number of *usability* design issues come up in making decisions regarding navigational structure or information architecture. Perhaps the most obvious one is "depth versus breadth" (Nielsen 2000; Horton et al. 1996). That is, should the overall organization of tasks be broad (a shallow structure: many items at the same level, few levels, as in Figure 22.3) or deep (a deep structure: few items at the same level, many levels, as in Figure 22.2). Another issue is whether a structure is best cast as hierarchical (as in Figure 22.2), as linear (as in Figure 22.4), or as some other structure, such as network or matrix (Brinck et al. 2002).

Related to the question of overall structure *type* is the issue of insuring that the structure maps to natural and intuitive semantic categories (e.g., is Returns most appropriately organized under Shopping or under Customer Service?) and finding communicative and memorable labels for those semantic categories (e.g., does "Outerwear" or "Casual" communicate what you would find in that category adequately?). Finally, what is perhaps most important is that the navigational structure should map well to the actual or optimal task flow of visitors doing the tasks being automated in the site (Rosenfeld and Morville 1998).

Persuasion issues that come up in this design task for Web sites with conversion goals include *what* key pieces of information should be presented *when* and *where* to support the decision making process. For example, visitors arriving for the first time at the Home page of an unfamiliar vendor may need to see a brief and salient statement of the Web site *business* in order to determine if the site offerings may in fact meet their needs, and then some information establishing the credibility of the *vendor* before they are willing to go further.

FIGURE 22.7 Example of drop-down menus as a visual presentation of a hierarchical information architecture.

This is the purpose of what are known as a "value proposition" and a "tag line," the "About Us" pages, testimonials from past customers, and the like. Visitors may also want and need certain types of *product* information (e.g., cleaning instructions for clothing) on a product detail page before they will be willing to place a product in their cart, and they may need information regarding the vendor's privacy policy before they are willing to give out personal contact information during the check out process. Calls to action (e.g., Add to Cart, Submit Order) need to be strategically placed, as do things like returns policies, shipping costs, and other information visitors may need to reach a threshold of comfort for making a commitment.

Although it may seem difficult at first to separate *navigational/structural* issues from *visual design* issues, it is productive to learn to do so for at least three reasons. First, the two really are independent. You can have a valid and supportive information architecture and then fail to present it clearly through an effective visual design. Conversely, even a highly consistent and clear visual design cannot overcome the flaws of a poorly designed information architecture. Dealing with these issues separately forces designers to address them consciously.

To reiterate, *what* information visitors need to make decisions (e.g., privacy policy), and *where* and *when* it is placed in the task flow (e.g., privacy policy right next to the request for personal contact info, as opposed to on a totally separate page) is just as important as *how* that information is presented visually. The *how* is addressed in the design tasks that follow the information architecture task.

Second, different skill sets are relevant to information architecture design as opposed to visual design. In particular,

usability and persuasion skills are paramount to achieving optimal information architecture design, while graphic design skills are necessary to achieve effective appeal, atmosphere, tone, and branding, as well as to help realize and support many usability and persuasion goals.

Third, navigational structure (i.e., information architecture) is platform independent, while visual and behavioral design options will depend very much on the chosen platform. For example, a given information architecture may specify a hierarchical menu structure of categories and subcategories of products. Current Web platforms (i.e., browsers, browser versions, plug-ins) allow drop-down menus much like a GUI menu bar structure as an option for presenting second level navigational choices (see, e.g., Figure 22.7), while earlier browsers did not, requiring instead that different levels in a menu hierarchy be presented as sequences of pages with embedded links (see Figure 22.8 for an example).

These *visual* and *behavioral* techniques for presenting navigation are part of the *conceptual model design* task (see below), while the hierarchical structure of product categories and subcategories is the same regardless of presentation and behavior. There will always be *some* way to present a hierarchical structure on any platform. *How* it is actually presented—the visual and behavioral design—will be to some extent platform dependent.

Thus, it is useful to separate *structural* from *visual* and *behavioral* design issues to help focus design decision-making, structure the input of experts versed in all the different aspects of the UX, and be sure all critical design decisions

FIGURE 22.8 Example of separate page menus as a visual presentation of a hierarchical information architecture (clicking on "Shoes" from the top page takes the visitor to the bottom page, where subcategories of shoes are presented on the left).

are being consciously made on the basis of sound requirements data, consideration of relevant established design principles, and evaluation.

Evaluation techniques, such as heuristic evaluations by experts (assuming the designers themselves were not UX experts), user walkthroughs or formal usability testing can be productively applied at this early point in the design process to validate just the information architecture design, although it is more common for information architecture to be combined with conceptual model design before a first round of evaluation.

22.2.2.2 Case Study

The case study reported in this chapter to illustrate the top-down process for designing a Web site UX describes the design and development of a public Web site that can be found at http://www.ouxinstitute.com (the site is a work in progress). The business model for the Web site is e-commerce. The site's business is to market, sell, and deliver a curriculum of video-based online training courses in the discipline of Web site UX design to an audience with a wide variety of backgrounds and goals. The site also intends to provide a number of other products and services all aimed at supporting professional networking, resource sharing, and community building.

In the case study project, a small team of Web design and development professionals are designing and developing this public Web site. Their specialty areas include the following:

- Usability
- Persuasion
- Graphic design
- Web UX development

The project stakeholders started with a list of products and services they wanted to provide through the Web site; some would be offered on initial launch, others would be developed and added on over time. The long-term list included the following:

- A curriculum of video-based training courses on a broad set of topics in software and Web UX design aimed at an audience with a wide variety of backgrounds, level of experience, and goals
- A complete course catalog of the curriculum with comprehensive course descriptions and video previews
- Recommendations for "tracks" of courses depending on the characteristics and goals of the visitor/customer
- Registration and payment for the delivery of courses online
- Online delivery of video training courses
- Delivery of digital certificates upon course completion
- Ratings and reviews of individual courses by past enrollees available to prospective students

- Comprehensive information about the faculty teaching the courses, including self-introduction videos
- Scheduling of phone and Skype-based "office hours" provided to enrolled students
- Blogs and vlogs posted by faculty members
- Discussion groups for students enrolled in specific courses
- General forums for all site visitors
- A "Bookstore" portal that identifies relevant books, hardware/software tools, and souvenir items with the Web site logo and provides URLs for purchasing them
- A "Library" portal of free downloads of articles and papers and URLs to other online articles and papers of interest as well as to professional society Web sites
- A "Career Office" portal pointing visitors to relevant external job and resume posting sites
- Sign up for and delivery of a regular site newsletter
- Polls on various topics
- Profession-related humor and entertainment
- Web site guided tour/tutorial

The stakeholders made the following decisions regarding the "lowest common denominator" visitor platform on which to premise the UX design, based on traffic analytics from related Web sites:

- Operating systems
  - Microsoft Windows, versions XP, Vista, Windows 7
  - Apple Macintosh
- Browsers
  - Microsoft Internet Explorer, version 6.0 and up
  - Mozilla Firefox, version 2.0 and up
  - Google Chrome 4 and up
- Displays
  - Minimum 15" monitor
  - Minimum resolution: 1024 × 768

Starting with these functional and platform specifications, the information architecture for the Web site was then designed in two stages. First, an overall across-task architecture defining page flows was drafted by the usability team member in the format of a page flowchart, as shown in Figure 22.9.

At this stage in the project, the most likely scenario was that on initial launch the Web site would provide all information, but that course registration and delivery (and likely forums, discussion groups and faculty blogs and vlogs as well) would, in fact, be offered through third party Web sites. Thus, those pieces are not detailed in this draft information architecture. In Figure 22.9, green arrows specify the pathways from this site to the third party sites. In addition, while the page flowchart for the across-task navigational structure shows the basic organization that would be presented in primary navigation areas, it was anticipated that there would also be additional "cross-reference links" from the body of

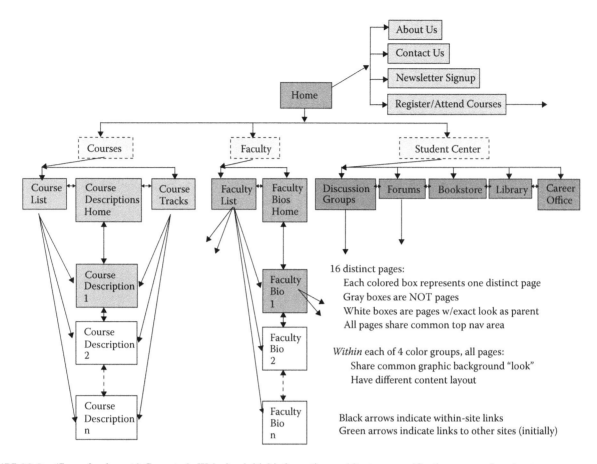

FIGURE 22.9 **(See color insert.)** Case study Web site: initial information architecture specification—page flowchart.

some pages directly to other pages to support optimal task flow. These would be identified later in this task.

Next, the usability team member teamed up with the persuasion team member to expand on the initial information architecture page flowcharts by beginning the construction of wireframes of individual pages. The focus was still on navigational structure, and the wireframes at this point documented initial ideas about the primary navigation for the site, as well as usability and persuasion requirements that would later be realized by the graphic designer.

Figure 22.10 shows an early information architecture wireframe for the Home page.

As illustrated in the page flowchart in Figure 22.9 and now again in Figure 22.10, there were two groups of top-level navigational links. These correspond to general types of information found on many Web sites (e.g., About Us, Contact Us)

and content categories more specific to this Web site's offerings (e.g., Course Catalog, Faculty).

The wireframe now goes beyond the page flowcharts in specificity by showing these groups of top-level links on the Home page. Top-level links were to be presented in distinct groups, shown in the wireframe as two horizontal lines of text, one right justified on the page and the other left justified.

The page flowchart in Figure 22.9 specified a hierarchical structure for some of the Web site content. The wireframe of an internal page shown in Figure 22.11 got a little more specific by indicating that second-level navigational links would be displayed associated with the top-level links.

Wireframes need to be accompanied by *text specifications* that describe intended *behaviors* as well as *requirements for the graphic design* that will impact usability and persuasion. One such description accompanying the wireframe in

FIGURE 22.10 **(See color insert.)** Case study Web site: information architecture wireframe of the Home page.

FIGURE 22.11 **(See color insert.)** Case study Web site: information architecture wireframe of an internal page.

Figure 22.11 indicated that the page represented by the first second-level link in the list of sublinks for a top-level link would be displayed by default whenever a visitor clicked on the top-level navigational link. That is, for example, when a visitor clicked on the Course Catalog link, they would not be taken to a Course Catalog page per se, but rather to the first of the three pages in the set of second-level links associated with the Course Catalog category: the Course List page.

The wireframe also specified that the page currently displayed would be indicated in the navigational links themselves by some special cue (known as a "You are here" cue) on the links representing the pathway to that page. In the wireframe, the requirement for this cue was indicated by gray or dimmed link text, but the actual visual look of this cue was to be created later by the graphic designer.

Similarly, the wireframe used blue underlined text to indicate links to other internal pages, and green underlined text to indicate links to external Web sites. Again, the intent was not that the two types of links should look specifically this way but that the graphic designer should be made aware of the two kinds of links and the need to design and consistently apply distinctive visual cues to make them distinguishable. In both these examples, the use of these types of cues and the need to apply them consistently are usability issues, but the design of the cues themselves is a graphic design issue. Thus, the wireframe shows examples of how the usability specialist captured important usability requirements for the look and feel that the graphic designer would need to realize later.

Another important aspect of the wireframe specification was that the entire top "banner" shown in Figures 22.10 and 22.11, which included the logo and navigational areas, was to be "fixed" (i.e., nonscrolling) at the top of every page of the site, while the page content below it would be scrollable. This is an important usability requirement that ensures that visitors maintain context (i.e., "Where am I?") and that they never have to scroll to find their navigational controls. It also reflects the persuasive goal of assuring visitors that they are still on the main site and the graphic design goal of effective branding.

Some important *persuasive* design requirements are specified in the wireframe specification as well. First, on the Home page (Figure 22.10), the wireframe indicated that there would be a logo for the business, as well as a prominent business name. Second, there would be a *value proposition*: "A Consortium of Experts Providing a Broad Curriculum of eLearning Courseware on UX Design." A value proposition provides a concise but clear statement of the business the Web site is in, in language that will appeal to the visitor's needs. First-time visitors who cannot quickly determine if a site is likely to meet their needs are less likely to stay and delve deeper into the site offerings, and more likely to "bounce" off the site and try another. Effectively written, positioned, and presented value propositions keep first time visitors engaged.

Also illustrated in the wireframes was a *tag line*: "Learn skills for your future from today's UX professionals." A tag line is similar to a value proposition in that its purpose is to capture the interest of visitors and keep them engaged. It is a little different though, in that its intent is to be a short, catchy, memorable phrase that can become part of the branding, like the logo. In addition, while the value proposition is focused on *describing* the site's offerings, the tag line is focused on the *benefits* of the site's offerings *to the visitor*.

In summary, at this stage in the project, the focus was completely on navigation and specifying ways to achieve usability and persuasion goals at this level, and any actual visual design, not to mention page content, were delayed until later steps in the design process.

On this project, because of the relative simplicity of the functionality, the expertise in the design team, and limited resources, any UX evaluation was deferred until a later point in the design process.

22.2.3 CONCEPTUAL MODEL DESIGN

22.2.3.1 Design Issues

In the conceptual model design task (Mayhew 1999; Nielsen 2000; Brinck et al. 2002; Johnson 2003; Sano 1996), the focus is still on navigation, but high-level design standards for visually presenting the information architecture are generated. Neither page content nor page design standards (i.e., visual presentation of page content) are addressed during this design task.

As an example of a conceptual model design, as illustrated in Figures 22.12 and 22.13, consider everything that is similar or the same when you use the three Microsoft commercial applications Word, Excel, and PowerPoint: the constant presence of the work product (a document, spreadsheet, or presentation) in the main application window, the fixed (i.e., nonscrolling) menu bar pull-down structure for accessing actions, pop-up dialog boxes for completing actions, the use of drag and drop for particular purposes, the use of dimming to indicate temporary inaccessibility of actions, etc.

FIGURE 22.12 Example of a conceptual model design: Microsoft Word.

All these visual presentation standards are part of the conceptual model—the visual presentation of the information architecture—of these applications. The consistency of these standards *within* one of these applications facilitates learning that one application, and the fact that these standards are the same or similar *across* applications makes each new application easier to learn.

The conceptual model for a Web site is similarly a set of visual design standards relating to the presentation of the information architecture. A good conceptual model design eliminates the need for the commonly seen "Site Map" page on Web sites. That is, the user interface itself reveals at all times the overall site structure and makes it clear where you currently are in it, how you got there, and where you can go from there. To the extent that it does this, it eliminates the need to go to a separate site map page to study and memorize the site structure (note that no "site map" is needed in Microsoft Windows applications, because the menu bar structure interface through which you navigate to actions is in effect its own map).

Visibility and clarity of the information architecture—an aspect of *usability*—is a large part of what we want to achieve in Web site conceptual model design. But as mentioned earlier, another key goal in a conceptual model design for a Web site is *persuasion*. Also, we want the *graphic design* to be aesthetically appealing as well as appropriate to the business, to create a particular atmosphere, and to provide strong branding.

A conceptual model design for a Web site would typically involve specifying standards that would cover the consistent use, location, and presentation of the following:

- Site title/logo
- Frames (e.g., for high-level links, context information, page content)
- Links to different levels in the information architecture (consistent within levels, distinctive across levels)
- "You are here" indicators on links (e.g., control attributes such as color, dimming, and/or breadcrumbs)

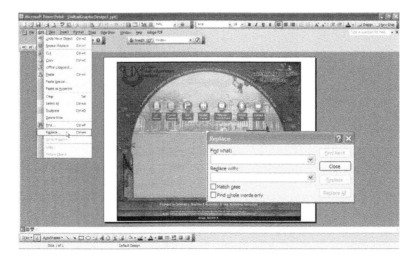

FIGURE 22.13 Example of a conceptual model design: Microsoft PowerPoint.

- "You are pointing to me now" indicators on links (i.e., visual indication that the cursor is currently positioned on the "hot spot" around the link)
- Use of pop-up dialog windows.

Consistency in the way all these things are used, located, and presented will greatly facilitate the visitors' process of perceiving, learning, and remembering their way around the site. This is particularly important on sites that will be used primarily by casual and discretionary visitors, as is the case with many e-commerce (and other types) of Web sites.

In summary, conceptual model design involves the development of visual design standards for presenting the information architecture. It does *not* address the visual design of page content, which is the focus of the next design task.

Again, for evaluation purposes, the conceptual model is often combined with the results of the information architecture, and sometimes with other design levels as well.

22.2.3.2 Case Study

Once the usability and persuasion team members had developed the page flowchart and started wire framing, it was time to bring the graphic designer on board. Her initial focus was on designing the branding and creating an overall atmosphere, as well as establishing visual design standards for presenting the primary navigational controls for the site.

The project stakeholders had a vision for the graphic design of the Home page, which they shared with the graphic designer. Figure 22.14 shows a mock-up sent to the graphic designer along with the page flowcharts and early wireframes to provide her with some direction regarding the branding and overall atmosphere they hoped to achieve on the site.

As can be seen in the mock-up, the stakeholders wanted to exploit the metaphor of a traditional bricks and mortar university or institute through which to present their offerings, as the intent of the site was, in fact, to provide a sort of "virtual" institute. The idea of the mock-up was to create the sense of entering a university or institute campus through a traditional-

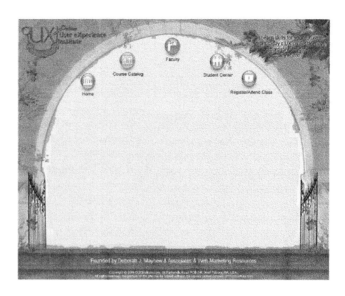

FIGURE 22.15 (See color insert.) Case study Web site: early iteration of Home page by graphic designer.

looking gated arch onto a campus green. Navigational links would be in a fixed (i.e., nonscrolling) area, the "steps" at the bottom containing the "footer" would also be fixed, and the middle area of the page would scroll to present all Home page content. The mock-up clearly communicates a vision, but it also makes clear that without the graphic design skills and the digital tools graphic designers are versed in, it's very hard to achieve what you want.

The graphic designer took the mock-up—as well as the page flowchart and early site wireframes—and developed a couple of high-level "looks" for the stakeholders to consider. One of these is presented in Figure 22.15. It makes it abundantly clear why you need a talented and skilled graphic designer on your UX team, and also illustrates the difference between a wireframe and a graphic design. This early design

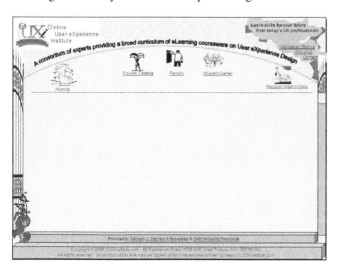

FIGURE 22.14 (See color insert.) Case study Web site: early stakeholder mock-up of the vision for graphic design.

FIGURE 22.16 (See color insert.) Case study Web site: mock-up providing feedback from stakeholders on early Home page graphic design.

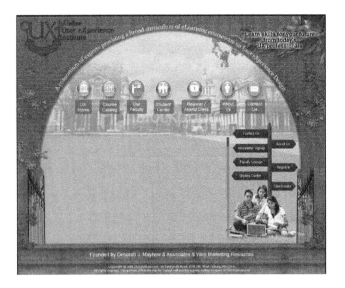

FIGURE 22.17 **(See color insert.)** Case study Web site: iteration of the Home page by the graphic designer.

from the graphic designer inspired some further visions in the stakeholders, who then sent back another mock-up premised on the design in Figure 22.15, shown in Figure 22.16.

Here it can be seen that the stakeholders evolved their vision of having a "campus green" scene inside the arch, although at this point they asked for it to be in some kind of watermark format, with the Home page content to be superimposed on it. In addition, this mock-up introduced the ideas of having navigational links presented as sign posts and placing people in the scene to promote a sense of community. Stakeholders also asked for some additional changes, such as changing the brick color to gray.

Figure 22.17 shows the graphic designer's next iteration. Note how much the design suggested in the early wireframe was now

evolving with the graphic designer's involvement, but *without* any change in the underlying information architecture. In Figure 22.17 it can be seen that the graphic designer implemented the stakeholders' requests and also showed them two alternatives for navigational controls within the same design: the requested sign posts and the graphic designer's earlier buttons. The scroll bar to the left of the sign post indicates the boundaries of the scrollable area to present the Home page content.

The stakeholders and team members continued to iterate on and evolve a Home page and internal page design, which embodied a set of design standards for presenting navigation that would then be carried out consistently across all pages in the site, that is, a conceptual model design. Figures 22.18 through 22.20 show a later (but still not final) version of this design.

The main conceptual model design standards embodied in this later design iteration illustrated in Figures 22.18 through 22.20 are as follows:

- The overall color palette for the Web site would be gray (brick), green (leaves), and teal.
- In general, a teal background would indicate that something is clickable (e.g., the Frisbee and Tour sign on the Home page and the second-level navigation bar on the internal pages).
- On the Home page (see Figure 22.18), content would be provided by rolling over the "speech bubbles," which would bring up call out windows (see Figure 22.19) similar to a technique that can be seen on http://www.netflix.com; callouts would stay up until the visitor either rolled over another speech bubble, clicked on a link within a callout, or clicked anywhere outside the callout.
- On the Home page, primary navigational links would be provided as signs on the sign post.

FIGURE 22.18 **(See color insert.)** Case study Web site: later Home page iteration by the graphic designer.

FIGURE 22.19 **(See color insert.)** Case study Web site: iteration of the Home page callout by the graphic designer.

FIGURE 22.20 **(See color insert.)** Case study Web site: later internal page by the graphic designer.

- Black arrows on images would indicate active links (e.g., on the Home page: Tour, the Frisbee, and News).
- The familiar recorder "play" icon (right pointing arrow) would indicate a link to a video, and all site videos other than the courseware itself would be displayed in pop-up windows that were smaller than the browser window the site was being viewed in, in order to maintain the context of being on the Web site.
- On the internal pages, a specific collage of images inside the brick "arch" would identify a page as belonging to one of the top-level navigational categories, such as Curriculum, Student Center, or About Us. For example, the collage shown in Figure 22.20 would be the same for the Course List, Course Descriptions, and Training Tracks pages.
- On internal pages, the logo, value proposition, and tag line would be reiterated from the Home page as shown in Figure 22.20.
- On internal pages, top-level navigational links would be presented as shown in Figure 22.18, bold white text in 14 point Arial font on a sign-shaped gray background. Business specific links (e.g., Curriculum, Faculty, Offices) would be on signs starting on the left side of the page and pointing to the right. Generic top-level links (e.g., About Us, Contact Us) would be on the right side of the page and pointing left. The top-level link representing the pathway to the currently displayed page would have a background of gray brick, be black, bold, Berling EF font in 14 point, and be inactive.
- On internal pages, second-level links would be on a teal background in bold white Arial 12 point text. The second-level link representing the currently displayed page would be in bold black Arial 12 point font on a white background with a "tab" look, as shown in Figure 22.20.
- On internal pages, everything from the second-level navigation bar on up would be fixed and would not scroll.
- On internal pages, the footer, which looks like stone steps, would scroll along with the rest of the page body.

Figure 22.21 shows an excerpt from an informal style guide put together by the team to document the conceptual

Conceptual model style guide			
Element	Style	Look	Behavior
Overall color palette	Medium gray (brick), dark green (leaves), dark teal		
Rollover callout image	White round speech bubble, no border		Mouse rollover invokes pop up callout
Rollover callout pop up	Gray brick heafer with white bold text title		Callout pop ups replace speech bubble; Pop up stays until visitor clicks anywhere OR rolls over another callout
Clickable objects	Teal background with white text and/or black arrow	Tours	Navigation to internal page
Video links	Familiar "play" icon from recorders (right pointint arrow)		Brings up secondary window with video player
Top level nav - Home page	Bold white text on signs on sign post		Navigation to internal page
Top level nav - Internal page	Dark gray with white bold text, sign look	Student Center Faculty Offices	Navigation to internal page
Second level nav	Bold white text on teal background	Course Descriptions Training Tracks	Navigation to internal page
You are Pointing to me Now cue for nav bar links	Underline	Curriculum	Text takes on underline as soon as mouse rolls into "hot spot"
You are Here cue on nav bar links	Black bold BerlingerEF text on gray brick background	Curriculum	Inactive
	Bold black Arial text on white "tab" background	Course List	Inactive
Internal page banner	Imagery in arch consistent within top nav categories (eg, Student Center, Curriculum), distinctive across categories		Fixed, non scrolling
Footer			Fixed, non scrolling on Home page, but scrolling on all internal pages

FIGURE 22.21 Case study Web site: style guide at the conceptual model level.

model design standards described above. The style guide was constantly updated as the site design evolved and was eventually implemented in cascading style sheets (CSS) in the Web site code.

Note again how very different from the initial wireframes this graphic design has become, without any change in the information architecture specified. For example, one of the main differences is how the content on the Home page is displayed. In the early wireframes, it was specified that there would be some kind of "watermark" scene in the background, but that content would be displayed in a fairly typical format (columns of text sections with headers and subheaders). With the graphic designer's input, however, the design evolved such that the Home page contained a fairly realistic scene, and content was displayed in rollover callouts indicated by "speech bubbles" next to some of the people in the scene. Similarly, in the early wireframes, a fairly traditional navigation bar on the Home page was suggested, while the graphic design (see Figure 22.18) evolved to present the main navigational controls on directional signs on a signpost such as you might see on a campus, and as other metaphorical controls such as a Frisbee, a computer, and a newspaper that were part of the scene.

On this project, evaluation was deferred until after the next design task, given tight time and resources, the moderate simplicity of the Web site, and the UX expertise available on the project team.

22.2.4 Page Design Standards

22.2.4.1 Design Issues

In this task, design standards for visually presenting and interacting with *page content* are generated. This new set of standards is designed in the context of both the information architecture and the conceptual model design standards that have already been generated and (in some cases) validated.

Page design standards for a Web site would typically include standards that would cover the consistent use and presentation of the following:

- Standard GUI widgets (e.g., check boxes, drop-down menus, radio buttons, text fields, pushbuttons, etc)
- White space
- Content headers and subheaders
- Fonts and point sizes
- Icons and images
- Forms design
- Color cues
- Context information (e.g., summaries of previous input to set context of later pages)
- Redundant/cross-reference links where appropriate
- Links versus other actions (e.g., "Submit")
- Links versus nonlinks (e.g., illustrations)
- Inter- versus intrapage links

It might include a set of templates illustrating design standards for different categories of pages (e.g., fill-in forms, information-only pages, product description pages, pop-up dialog boxes, etc.).

Page design standards may also include overall writing style standards (e.g., prose versus an easy to scan style such as bullet points) as well as more detailed writing style standards (e.g., column width, use of negative and passive voice, etc.).

Consistency in the way all these things are applied in the design will again, just as it does in the case of the conceptual model design standards, greatly facilitate the process of learning and remembering how to use the site. This is particularly important on sites that will be used primarily by casual and discretionary users, as is the case with many e-commerce and other types of sites.

It is important to make the disinction between the page design *standards* and the more general *principles* and *guidelines*. Figure 22.22 defines this distinction.

The literature offers many sources of general *principles* and *guidelines* for Web user interface design (Spool et al. 1999; Nielsen 2000; Nielsen et al. 2001; Horton et al. 1996; Brinck et al. 2002; Johnson 2003; Sano 1996; Travis 2003; National Cancer Institute 2003; Yale University 2003; Scott and Neil 2009; see chapters by Caldwell and Vanderheiden, this volume; Pappas et al., this volume; Vu et al., this volume; Najjar, this volume). However, designers must translate these principles and guidelines into specific *standards* for their site or sites to ensure usability, persuasion, and optimal graphic design within and across sites. Examples of standards are given in the Case Study section below.

Page design standards are just as important and useful in Web design as in traditional software design. Besides the usual advantages of standards in providing ease of learning and remembering, in a Web site they will also help visitors maintain a *sense of place within a site*, as each site's standards will probably be different from those on other sites.

Web design techniques (both good and bad) tend to be copied. Many de facto standards currently exist (e.g., blue underlined text for links) and more continue to evolve as browser capabilities improve (e.g., drop-down menus from navigation bar links). Perhaps someday we will even have a set of universal Web page content design standards supported by Web development tools, not unlike Microsoft Windows and Apple Macintosh standards for GUI platforms. This would contribute greatly to the usability of the Web, just as the latter standards have done for commercial desktop applications.

UX designers should develop a set of site-specific page design standards for all aspects of detailed page content design, based on any government and/or corporate standards that have been mandated by their organizations, the data generated during requirements analysis, general principles and guidelines for Web page design, and the unique conceptual model design arrived at during Level 1 of the UX engineering life cycle design process. Page design standards will ensure coherence and consistency, and the foundations of usability across the content on all site pages. They will also support persuasion, appropriate aesthetics, and branding.

	Principles	Guidelines	Standards
Definition	General high level goals	General rules of thumb	Specific rules customized translations of principles and guidelines
Example	Provide consistency and standards to communicate meaning	Color is the most salient visual cue. Use it consistently to communicate meaning	All text links will highlight by turning dark yellow when pointed to
Advantages	Drawn from basic research on human cognition/perception; help to focus design efforts	Drawn from applied research on human-computer interaction; help to focus design efforts	Directly applicable tailored to users and their tasks
Disadvantages	Very general not directly applicable; subject to interpretation	Very general, not directly applicable; subject to interpretation	So specific may need to be tailored for each application

FIGURE 22.22 Principles, guidelines, and standards.

Iterative evaluation is definitely in order at this stage, especially if it has not been conducted at earlier design levels. Again, if UX expertise has not been involved in the design process to date, heuristic evaluations can be a cost-effective technique to use. If UX expertise has been involved, then formal usability testing is advisable at this point.

22.2.4.2 Case Study

Just as in conceptual model design, the process for designing page standards started with the usability and persuasion team members collaborating on a wireframe, followed by the graphic designer generating visual design standards that realized and supported the usability and persuasion goals and in addition addressed the aesthetics, atmosphere, tone, and branding goals.

Figures 22.23 and 22.24 show early wireframes that illustrate usability and persuasion requirements for page design standards on the Home page. Figures 22.25 and 22.26 show wireframes that illustrated usability and persuasion requirements for page design standards on internal pages. These

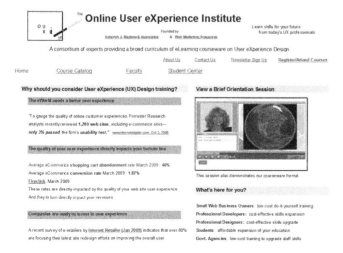

FIGURE 22.23 (See color insert.) Cast study Web site: page design standards wireframe—Home page above the fold.

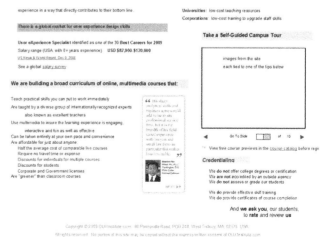

FIGURE 22.24 (See color insert.) Case study Web site: page design standards wireframe—Home page below the fold.

FIGURE 22.25 (See color insert.) Case study Web site: page design standards wireframe—internal page above the fold.

wireframes documented a number of requirements for page design standards that the graphic designer was asked to address in her graphic design for page content. The wireframe for the Course List page within the Course Catalog category (Figures 22.25 and 22.26) served as a template for all internal pages in that any design standards documented in that template were intended to carry over to all other internal pages wherever relevant. The content would of course be different on every page, and some other pages might introduce requirements for additional standards, but this template page was meant to establish most standards and thus drive the design of additional pages. Unaddressed opportunities for additional standards were to be resolved as they came up in the next design task: detailed UX design.

The most obvious requirements specified in Figures 22.23 through 22.26 involved content headers, subheaders, and bullet-pointed content. The usability principle driving these design requirements was simply that it should be easy for visitors to quickly scan page content, which Web visitors prefer to do rather than read dense text. Clear and salient headers and subheaders to quickly identify major topics, and short bulleted text to facilitate scanning major points, were specified to support this goal.

Headers were indicated in the wireframes as light blue background bars with bold black text. Subheaders were indicated as light yellow background bars—indented relative to the blue header bars—with bold black text. Most content was specified as either very short sentences or bullet points. The point was not to specify blue and yellow background bars and bold black text in particular but rather to specify the need to facilitate scanning by distinguishing between the two levels of headers, so the graphic designer could design and consistently apply cues for those purposes as a part of an overall look and feel and color palette.

Another design requirement embodied by these wireframes was to provide a cue to *distinguish between links* to other pages within this Web site, from links to external Web sites. The usability principle here was that visitors should be

able to predict the consequences of their interactions prior to taking action. In particular, visitors should know when a link would result in their leaving the site they are currently on. In the wireframes, internal links were identified as blue underlined text, while external links were identified as green underlined text. Again, the intent was not that these particular visual cues be used but simply to let the graphic designer know which were which so she could design and apply cues consistently.

Yet another usability requirement documented in these wireframes was the need to provide a special cue to *identify instructional text*, to make it easy to scan for instructions, and quickly identify text as instructions—as opposed to other kinds of content—when looking at it. In Figure 22.25, the word "Tip" in green superscript at the beginning of the instruction line is the specification for this. But again, it will be up to the graphic designer to design a particular cue for this purpose and apply it consistently.

The content wireframe for the Home and internal pages (Figures 22.23 through 22.26) also specified content that is largely *persuasive* in nature. Besides the logo and institute name, and the value proposition and tag line at the top of the navigational banner that were specified and designed in the conceptual model design step, the Home page wireframe content was almost entirely focused on the benefits of the site products and services to the visitor.

Credibility was addressed in the Home page wireframe design in a number of ways, including access to an "About Us" page, relevant statistics with references, a testimonial, and a video that would allow visitors to judge the quality of the course videos from an example video. The rest of the Home page content focused on the benefits of the curriculum to different categories of potential customers. Content that was not specifically persuasive in nature focused on being comprehensively descriptive of the Web site's offerings. Thus, the Home page content focused on three things: a clear description of site offerings, establishing site credibility, and highlighting benefits to the visitor.

FIGURE 22.26 (See color insert.) Case study Web site: page design standards wireframe—internal page below the fold.

Although not shown here, it was also specified on the Faculty Bio page wireframe that there would be extensive detail on faculty background and credentials, as well as video self-introductions, allowing potential customers to get a feeling for a faculty member's credibility and video presentation style. Course Description page wireframes also specified short video previews from actual courses. Finally, ratings and reviews of past customers of courses were specified on the Course Descriptions pages and summarized on the Course List page, as it is well known that reviews from past customers have significant influence on prospects. All these elements—only rarely seen to this extent on e-learning and other e-commerce Web sites—were aimed at establishing credibility, known to be a key aspect of persuasion.

Thus, the page content wireframes served the purposes of specifying both usability and persuasion requirements for visual standards for the graphic designer to build on.

The graphic designer then took these page content wireframes and began designing templates. The Home page went through many iterations, and one iteration close to final is shown in Figures 22.27 and 22.28. In these two figures it can be seen that the graphic design of the Home page evolved into a realistic scene of a campus, and instead of imposing the content on it, which would have been busy looking and distracted from the content, provided segments of the content in rollover callouts. The sample callout in Figure 22.28 illustrated some page design standards, some general and some specific to rollover callouts. These included the use of a gray-fading-to-brick pattern as a background to headers (in this case, the title of the callout), green text for subheaders and a leaf as a bullet point. These cues tied in nicely with the rest of the Home page design, which also included bricks and leaves and the colors gray and green. The standard of a white background and black text for the main content in the

FIGURE 22.28 (**See color insert.**) Case study Web site: late page design standards graphic design—Home page content callout.

callout served the usability goal of legibility. Here we can also see the cue established for external links: blue text (no underline) followed by a carat in superscript. As can be seen on the internal page template in Figure 22.29, internal links were presented as blue (no underline) text. While not seen here, the design also specified that the look of a link when the mouse was pointing to it (known as the "You are pointing to me now" cue) would be to simply add an underline to the link name text.

Further final page design standards can be seen illustrated in Figure 22.29, a late (though still not final) internal page template. Note that content headers had a background that was gray-fading-into-brick, and the subheaders were green-fading-into-leaves, matching those themes from the Home page and internal page banner as well as the Home page callouts. Tables had a particular look, which would be

FIGURE 22.27 (**See color insert.**) Case study Web site: late page design standards graphic design—Home page.

FIGURE 22.29 (**See color insert.**) Case study Web site: late page design standards graphic design—internal page template.

Page standards style guide			
Element	Style	Look	Behavior
Internal page body background	White		
Top level content header	Gray background fading into brick	Browse the Complete Course List	
	Bold black text, Arial 14		
Second level content header	Dark green background fading into leaves	Group: Introductory courses	
	Bold black text, Arial 12		
Text links to internal pages	Traditional blue text, no underline	Link	Links to internal page
Text links to external sites	Superscript right pointing caret at end of link text, also blue	Link^	External sites should open in separate tabs/windows rather than overlay OUXI.com pages
Tables	Light gray background with bold teal Arial 12 text for column caption row	Courses Course Title	Sortable by clicking on captions
	White background for other rows with normal black text	R 2 Learning Styles	
	Gray internal cell border lines		
	12 point Arial black normal text in cells		
Download links	Button, dark teal background, bold white text Arial 12		Takes on text underline when pointing to it, initiates download
Text representing instructions	Preceded by a superscript sized label "Tip" in bold dark teal, otherwise normal text	Instructions	
Callouts - header/title	Gray background fading into brick	Why should you consider User Design training?	
	14 point Arial bold black text		
Callouts - sub headers	A leaf as bullet point, text in 12 point dark green bold Arial font	TheWorld needs a better user	
Callouts - body text	12 point normal Arial font	To gauge the quality of a experiences, Forrester F reviewed 1,200 web adve	

FIGURE 22.30   **(See color insert.)** Case study Web site: page design standards style guide.

consistently followed on other site pages, including a very light gray header row with teal column header text.

Figure 22.30 shows some of these page design standards captured in the informal project style guide, eventually realized in CSS style sheets in the live site code. Remaining template pages were developed with wireframes and iterations of graphic design, and standards were added to the style guide as the opportunities arose and were addressed.

At this point in the project, the first and only planned round of prelaunch UX testing was to be conducted and changes made accordingly. These changes, and others as the site continues to evolve, can be seen in the live Web site at http://www.ouxinstitute.com.

22.2.5 Detailed UX Design

22.2.5.1 Design Issues

The standards documented during the conceptual model design and page design standards tasks, as well as the information architecture design, will dictate the detailed design of a large percentage of a site's functionality. The detailed UX design task is thus largely a matter of correctly and consistently applying standards already defined and validated to the actual detailed design of all pages and pop-up windows across the site.

However, there will always be unique details here and there across pages to which no particular established standard applies. These must still be designed and designed well. And, these design decisions should also be driven by requirements data and evaluated.

For example, sometimes there will be fill-in fields that are unique to a single page. Their captions must be clear. There may be no standard for these particular captions as they only

appear once. Or, there may be only a single page on which a certain status indicator is used. No standard will apply to it because it is unique to a single page, but again, it must be designed well and validated for clarity. These are the types of things that will not already have been designed and validated in previous tasks and must be addressed in this design task, with attention paid to all UX design principles.

In this task the UX designers carry out detailed design of the complete site UX based on the refined and validated information architecture, as well as conceptual model design and page design standards documented in the site style guide and in the Web site CSS style sheets. All aspects not specifically dictated by a design standard must be uniquely designed and should be driven by requirements data if there are any that are relevant, as well as by general UX design principles and guidelines (see Spool et al. 1999; Nielsen 2000; Nielsen et al. 2001; Ratner 2003; Horton et al. 1996; Brinck et al. 2002; Johnson 2003; Sano 1996; Travis 2003; National Cancer Institute 2003; Yale University 2003). The complete detailed design is then implemented, and back-end functionality is developed.

Iterative testing can continue at this point in order to test aspects of the UX design that were not addressed in any earlier rounds of evaluation.

22.2.5.2 Case Study

Figure 22.31 shows a page from the detailed UX design for the case study Web site. Note that this page design incorporated the information architecture as well as conceptual model design and page design standards applied to specific page content. The detailed UX design would include this level of specificity for every page and pop up in the Web site,

FIGURE 22.31 **(See color insert.)** Case study Web site: a page from the detailed UX design.

as well as text descriptions of any unique interaction with page elements not addressed in the Web site style guide.

On this project, because it met all the criteria given earlier—UX specialists led the design effort, the site content was moderately simple, and time and resources were tight—only one round of formal usability testing was planned, in which all design issues at all levels were to be addressed simultaneously. The prototype to be tested represented the information architecture and standards at both the conceptual model design and page design standards levels. It also covered some of the detailed UX design. No further testing was planned as the fully functional system was constructed. However, it was expected that after initial launch there would eventually be ongoing updates and that visitor feedback would be solicited to drive redesign in an ongoing process.

It is interesting to note that on this project the team was "virtual"—that is, a collection of independent consultants—spread across different geographic locations. Most of the collaborative design work was conducted over the phone and Internet. This can work perfectly well when the team is small and motivated, everyone has a cooperative spirit, and roles are clearly defined, as was the case on this project.

The actual project that this case study is based on, from inception to the time of this writing, occurred over about a year and a half. The level of effort for Web site design and development (not including course development) of the primary stakeholder, who was also lead designer and the usability team member, was a total of about 600 hours. The persuasion team member, also a stakeholder, spent about 75 hours, the graphic designer spent about 225 hours, and the developers spent about 15 hours.

22.3 SUMMARY

In summary, there are three key goals for Web site UX design:

1. Usability
2. Persuasion (in the case where the Web sites business goals include conversions)

3. Effective graphic design (e.g., branding, appropriate atmosphere, tone, and appeal)

The top-down process for UX design described in this chapter is comprised of the following design tasks in the order implied:

- Information architecture design (the design of the navigational structure for the site, both across task and within task; does not entail any visual design and does not address page content)
- Conceptual model design (the design of a set of visual design and behavioral, i.e., interaction standards for presenting the information architecture; does not entail the design of page content)
- Page design standards (the design of a set of visual design and behavioral, i.e., interaction standards for page content)
- Detailed UX design (the application of the information architecture, the conceptual model design standards, and the page design standards to the detailed design of the complete site UX)

The top-down process for Web UX design provides a structured approach to Web site UX design that:

- Allows designers to focus on manageable sets of design decisions and ensures that they address all UX issues during design.
- Streamlines the design process, minimizing rework.
- Defines a clear way for different roles (i.e., usability, persuasion, and graphic design professionals) to provide their input and negotiate design decisions with other disciplines.

REFERENCES

Brinck, T., D. Gergle, and S. D. Wood. 2002. *Usability for the Web.* San Francisco, CA: Morgan Kaufmann Publishers.

Caldwell, B. B., and G. C. Vanderheiden. this volume. Access to web content by those with disabilities and others operating under constrained conditions. In *Handbook of Human Factors in Web Design*, 2nd ed., eds. K.-P. L. Vu and R. W. Proctor, 371–402. Boca Raton, FL: CRC Press.

Dix, A. J., and N. Shabir. this volume. Human–computer interaction. In *Handbook of Human Factors in Web Design*, 2nd ed., eds. K.-P. L. Vu and R. W. Proctor, 35–62. Boca Raton, FL: CRC Press.

Eisen, P. this volume. Design of portals. In *Handbook of Human Factors in Web Design*, 2nd ed., eds. K.-P. L. Vu and R. W. Proctor, 303–328. Boca Raton, FL: CRC Press.

Horton, W., L. Taylor, A. Ignacio, and N. L. Hoft. 1996. *The Web Page Design Cookbook.* New York: John Wiley.

Johnson, J. 2003. *Web Bloopers.* San Francisco, CA: Morgan Kaufmann Publishers.

Mayhew, D. J. 1999. *The Usability Engineering Lifecycle.* San Francisco, CA: Morgan Kaufmann Publishers.

Mayhew, D. J. A Design Process for Web Usability, in Proctor, R. W. and Vu, K. L., 2005. Handbook of Human Factors in Web Design. Mahwah, NJ: Lawrence Erlbaum.

Najjar, L. J. this volume. Designing E-commerce user interfaces. In *Handbook of Human Factors in Web Design*, 2nd ed., eds. K.-P. L. Vu and R. W. Proctor, 587–598. Boca Raton, FL: CRC Press.

National Cancer Institute. 2003. Research based Web design & usability guidelines. http://www.usability.gov/guidelines/index.html.

Nielsen, J. 1993. *Usability Engineering.* Boston, MA: Academic Press.

Nielsen, J. 1996. Jakob Nielsen on Web site usability. *Eye for Design* 3(4): 1, 6–7.

Nielsen, J. 2000. *Designing Web Usability.* Indianapolis, IN: New Riders Publishing.

Nielsen, J., and R. L. Mack. 1994. *Usability Inspection Methods.* New York: John Wiley.

Nielsen, J., R. Molich, C. Snyder, and S. Farrell. 2001. *E-Commerce UX.* Freemont, CA: Nielsen Norman Group.

Pappas L., L. Roberts, and R. Hodgkinson. this volume. International standards for accessibility in web design and the technical challenges in meeting them. In *Handbook of Human Factors in Web Design*, 2nd ed., eds. K.-P. L. Vu and R. W. Proctor, 403–414. Boca Raton, FL: CRC Press.

Ratner, J., ed. 2003. *Human Factors and Web Development*, 2nd ed. Mahwah, NJ: Lawrence Erlbaum.

Rosenfeld, L., and L. Morville. 1998. *Information Architecture for the World Wide Web.* Sebastopol, CA: O'Reilly & Associates.

Sano, D. 1996. *User Designing Large-Scale Web Sites.* New York: Wiley Computer Publishing.

Snyder, C. 2003. *Paper Prototyping.* San Francisco, CA: Morgan Kaufmann Publishers.

Spool, J., T. Scanlon, W. Schroeder, C. Snyder, and T. DeAngelo. 1999. *Web Site Usability: A Designer's Guide.* San Francisco, CA: Morgan Kaufmann Publishers.

Tullis, T. S., F. J. Tranquada, and M. J. Siegel. this volume. Presentation of informtaion. In *Handbook of Human Factors in Web Design*, 2nd ed., eds. K.-P. L. Vu and R. W. Proctor, 153–190. Boca Raton, FL: CRC Press.

Travis, D. 2003. *E-Commerce Usability.* London: Taylor & Francis.

Volk, F. A., and F. B. Kraft. this volume. Human factors in online consumer behavior. In *Handbook of Human Factors in Web Design*, 2nd ed., eds. K.-P. L. Vu and R. W. Proctor, 625–644. Boca Raton, FL: CRC Press.

Vu, K.-P. L., W. Zhu, and R. W. Proctor. this volume. Evaluating web usability. In *Handbook of Human Factors in Web Design*, 2nd ed., eds. K.-P. L. Vu and R. W. Proctor, 439–460. Boca Raton, FL: CRC Press.

Yale University. 2003. Web style guide. http://www.webstyleguide.com.

Section VIII

Task Analysis and Performance Modeling

23 Task Analysis Methods and Tools for Developing Web Applications

Thomas Z. Strybel

CONTENTS

23.1 OVERVIEW

Broadly defined, task analysis refers to the activities associated with systematically describing and analyzing the tasks performed by humans to achieve one or more goals of a task. It is probably the most commonly used analytic method in the field of human factors: 92% of respondents in a recent survey of human factors professionals (Stone and Derby 2009) indicated that task analysis was important for their work. Task analysis may be the oldest technique in human factors, having been developed and applied for over 100 years.

The first methods of task analysis were developed early in the twentieth century, when most jobs consisted of repetitive, manual assembly tasks. Taylor (1911) and Gilbreth (1911) are usually credited for developing the first task analysis methods. With Gilbreth's method, repetitive tasks were decomposed into simple, observable motion elements called "therbligs." Approximately 17 therbligs were required to describe most manual tasks. Task completion times were computed as the sum of the times required to carry out each therblig. Reductions in task performance times were achieved by redesigning the layout of the workplace and resequencing therbligs (e.g., have both hands work together). Concurrent with Gilbreth's work, Munsterberg (1913) and Moss (1929) studied the psychology of work behavior and applied psychological principles to the design of jobs and task environments. These psychologists recognized the need for understanding the tasks carried out by an operator, but they relied on more informal observations of existing tasks. In fact, Munsterberg (1913) criticized Gilbreth's taxonomy as too narrow because it focused exclusively on observable physical motions and was unable to capture the critical mental aspects of performance.

During World War II, new task analysis techniques were developed and used to improve the design of human–machine interfaces (e.g., Fitts and Jones 1947a, 1947b). After the war, the application of task analyses and human factors was applied to many commercial and industrial concerns, although aerospace and other military industries continued to be the major application of human factors. The systems approach to design became the standard approach for aerospace and military systems, and new methods of task analysis were developed to fit with this approach. This approach is "system centered" because humans are viewed as a component of a person-machine system, and human factors was concerned with designing the operators' tasks in order to optimize system performance. The first formal task analysis method that was based on the systems approach was by Miller (1953): Tasks were initially described at a molar level and then decomposed into molecular task components. Miller created a behavior taxonomy for describing tasks at the lowest levels of analysis. According to Miller, task identification at the molar level may be arbitrary, but when the tasks were decomposed to the molecular level the task descriptions of different analysts should be similar based on the taxonomy. Although Miller's method primarily is of historical interest, the top-down approach to task analysis remains a major component of most subsequent methods developed for understanding tasks performed in a person-machine system.

With the rise of the software industry, the human factors' approach became more user-centered because users of both consumer and commercial software have more discretion over product use (e.g., Booth 1989). Individual consumers can decide not to purchase a product simply by not buying it. Employees can sometimes choose not to use a developed and installed software product (Mayhew 1999). The rise of the Internet and e-commerce has increased the amount of user discretion over product use. Nielsen (2000), for example, noted that with stand-alone software, the user purchases the product before working with it, but with Web-based applications, users can interact with a product before purchasing it and that makes it easier for consumers to reject an unusable or undesirable product. Another development affecting the need for task analysis was the revisions to the Americans with Disabilities Act of 1973. Section 508 of this law was passed in 2001. It created accessibility standards for electronic and information technology, requiring the federal government to develop, procure, maintain, and use electronic information technology that provides the same level of access to people with disabilities as to those without. States and local agencies have also adopted these standards, and task analysis is one tool for establishing that accessibility standards are being met (Henry 2006). These developments in software and Web-based development increased the importance of task analysis as a design tool, but at the same time, development cycles in the software industry have become very short, leaving less time for extensive task analyses.

In summary, task analysis has been an important tool for designing effective and usable products over the past 100 years. The methods of task analysis have changed in response to changes in the types of work done by humans, product design cycles, and expanded purposes for task analysis. Consequently, there are many methods of task analysis that can be found in the literature; yet most share some characteristics of the task analysis approach that remain unchanged, and these characteristics distinguish task analysis from other methods of understanding the user.

23.2 CHARACTERISTICS OF TASK ANALYSIS METHODS

Task analysis involves describing and analyzing "tasks," yet we have no agreed-upon definition of the term. For example, Meister (1985) defines a task as behaviors that usually occur together and are carried out to achieve a specific purpose. Shepherd (2001) defines a task as problems needing to be solved. In practice, tasks are defined in the context of the task analysis, but most would agree that tasks are goal directed, have a beginning and ending, and include the constraints and resources available to the user for achieving the goal. The idea that tasks are performed to achieve goals is critical to all task analysis methods. This assumption of goal-directed behavior implies that an understanding of the users' goals, tasks, knowledge requirements, and performance constraints are necessary to understand and improve task performance.

Task analysis also is more than a list or description of the tasks performed by the user, thus separating it from methods of job description commonly used in personnel selection. Although task description is a component of most task analysis methods, the description is accompanied by an analysis of the task as to the demands made on the user. A task analysis should produce conclusions, deductions, hypotheses, or recommendations for improving task performance derived from the original task descriptions (Meister 1985; Shepherd 2001). For most task analysis methods, the task description component is essentially a process of molecularization: tasks or task goals are systematically analyzed into simpler elements.

The differences between methods of task analysis lie in how the tasks are described and analyzed, particularly at the lowest levels of task description. Both Gilbreth's (1911) and Miller's (1953) methods of task analysis analyzed tasks by decomposing them at the lowest level using taxonomies of work behaviors. The potential benefits of using a taxonomy, or structured classification of work behaviors (Meister 1985), probably motivate the many attempts at creating them. A taxonomy provides a consistent language for task descriptions, enables comparisons of tasks across different work environments, and standardizes the training of task analysts. Although many taxonomies have been developed, none are sufficient to be universally applied. Current methods using taxonomies are focused on describing either specific types of information (e.g., expertise, cognitive demands) or specific task contexts (e.g., human–computer interaction). Consequently, the most flexible and widely applied methods of task analysis are not based on taxonomies of work behavior or knowledge representation schemes but create frameworks for analyzing any task information and in any task context.

23.2.1 Sources of Information for Task Analysis

The differences between methods of task analysis are found in how the tasks are described and analyzed, but, with all methods, information regarding the users' tasks must be obtained. The quality of this information is a major determinant of the effectiveness of the subsequent task analysis. Generally, one or more of the following information sources are utilized. These sources also contribute to other human factors' techniques such as user needs analysis or cognitive modeling described elsewhere in this volume. Because each of these sources has its shortcomings, it is generally recommended that multiple sources of task information be used.

- Task documents (e.g., training manuals, operating procedures, accident reports, user logs) are evaluated for information about how the task should be performed. Caution must be taken with training manuals and operating procedures, however, because these document how tasks should be done, not how they are actually performed.
- Direct observation or videotaping is used to get at observable task behaviors. These can provide useful information about the environment in which tasks are done and the sequence of observable tasks. Observation does not get at the cognitive aspects of the users' tasks, however, a characteristic of most modern jobs.
- Verbal protocols are used to learn about the user's thinking processes (e.g., expectations, strategies, rationale). With verbal protocols, users are asked to think aloud while doing a task. For some tasks, verbal protocols interfere with performance. In these cases, the user may be asked to comment on videotapes of previously completed tasks or on the performance of others doing the task. Nevertheless, for highly skilled or automatic tasks, requesting users to think aloud may change the way they perform the task to be consistent with their verbal construction of the task sequence.
- Interviews may be used to capture the user task knowledge. These may be structured knowledge elicitation techniques, or less structured, conversations with users about task knowledge.
- Surveys may be used to get at user attitudes, preferences, and critical incidents.

23.2.2 Applications of Task Analysis

Another reason for the many task analysis methods is that different methods are designed to elicit different types of information regarding task performance. Choosing an appropriate method of task analysis can be difficult, but the decisions regarding the choice of method, the level of detail of the task description, and the format of the information obtained should be based on the goals of the task analysis. Therefore, identifying the objectives of any analysis explicitly before making these decisions is essential. The common applications of task analysis information are described below.

23.2.2.1 Human–Computer Interface Design

Task analysis can be used to determine information and functional task requirements, optimal layouts of screen elements and functions, and workstation arrangements for optimizing task performance. The knowledge gained from an extensive task analysis can simplify subsequent design decisions regarding the user interface, because it provides a starting point for the definition of objects and methods in the new computer systems (Smith et al. 1982). Interface elements and functions can be grouped according to frequency of use or sequence of use for specific tasks. Displays can be arranged according to information needs for a given task. Task analysis can also be a point of departure for task metaphors in interface design, thus ensuring that interface objects and methods are developed from the standpoint of the users' tasks (Diaper and Stanton 2004; Moltgen and Kuhn 2000).

Probably the most widely cited successful application of task analysis in function identification, task metaphor, and interface design in the computer industry was in the design of the Xerox Star Operating System (Canfield et al. 1982; Smith et al. 1982). According to Smith et al. (1982), task analysis was conducted for 2 years before software development began to establish user characteristics, goals, and information requirements. Here, the added time spent on initial task analysis paid off handsomely, at least in terms of impact on future interface designs. Many standard features of today's computer interfaces (e.g., direct manipulation, integrated applications, desktop metaphor) were first introduced in the Xerox Star.

23.2.2.2 Function Identification and Allocation

Function identification is the process of listing all actions carried out by the system required to meet the objectives. In human factors, function allocation is a process of deciding which functions will be automated (done by machines) and which will be performed by humans. Function identification and function allocation are typically engaged in after system objectives, operational requirements, and constraints are identified and before specific design decisions are made. As with task analysis, there is a gulf between recommended procedures for function analysis and the actual procedures used. Meister (1985) notes that function allocation is usually based on a combination of formal methods, experience, heuristics, and other less formal techniques. Regardless of how function allocation is done, informed decisions regarding function allocation necessitate an understanding of the tasks or potential tasks that will be performed by the humans in the system. Consequently, task analysis should play a major role in function identification and allocation. Often task analyses of existing systems are conducted to contribute function-allocation decisions regarding new systems. Care must be taken, however, to avoid relying too heavily on low-level details of a task analysis done on an existing system because it can prohibit novel allocation strategies in the new system.

Function identification and allocation decisions occur throughout the design of software applications although the process is less formal. Nevertheless, understanding of the user's tasks is critical for determining what computer functions will be available to the user and when they become available (Kieras 1997). Goransson et al. (1987) describe several cases of computing systems that were difficult to use though the interface followed usability design guidelines. The difficulties were created because the functions available to the user were insufficient to allow him or her to complete

the task. Kieras (1997) advocates high-level task analyses as a means of identifying the functional requirements of a system in the early design stages.

23.2.2.3 Modeling

Models of how users perform tasks permit objective evaluations of existing interfaces and comparisons of design alternatives based on model-derived estimates of task performance. These models do not have to be complete to contribute to interface design. According to Kieras (2008; Kieras and Meyer 2000), user models are developed for different purposes, and not all contribute to interface design. Models developed for cognitive research purposes attempt to capture the overall cognitive architecture of the user. They are complex and more concerned with understanding human cognition. Consequently, they are rarely used to evaluate computer interfaces. Models developed for human factors purposes are approximate models aimed at improving the design of human–computer interfaces. These models offer unique contributions to the design of user interfaces. First, approximate human performance models help to organize the designer's thinking about how users engage in certain tasks and the system variables most relevant to human performance. Consequently, design trade-offs can be assessed in early design stages based on these models (e.g., Sheridan and Ferrell 1974). Second, a model of the task being performed on an existing or prototyped interface can be used to identify the reasons why the design supports or fails to support the users task performance. This is one advantage of models over empirical usability testing: usability tests identify problems with an interface but cannot determine the underlying reasons for user problems (Kieras 2008).

The most efficient use of model-based analysis in human interface design is during the initial design stages, before details regarding the layout and interface elements have been specified. The task-based models can be used to evaluate trade-offs prior to prototype development and subsequent user testing and may reduce the number of prototypes and user tests required. Kieras (2008) points out, however, that these performance models are only as good as the task analysis information used in developing the model. Therefore, model development is closely related to and depends on the use of task analysis methods, and sometimes separating these activities is difficult.

23.2.2.4 Accessibility Evaluation

Task analysis is one tool that is used to determine if a Web site meets accessibility standards. These standards were established when the Rehabilitation Act of 1973 was amended in 2001 to require that all electronic information and technologies be as accessible to those with disabilities as to those without (Henry 2006). Subpart C of Section 508 describes these rules in terms of functional performance requirements to ensure that the individual components work together to form a usable and accessible product. Task analysis is recommended as a first step in determining the critical tasks for establishing functional performance requirements.

Workflow analysis, describing the steps for completing a task, and workflow diagrams or flowcharts are some methods for representing tasks and establishing performance requirements. Scenarios are also suggested for putting context to the evaluation (see, e.g., http://accessibility.gtri.gatech.edu/aem/AEM1.html).

23.2.2.5 Testing and Evaluation

Testing and evaluation is a part of most approaches to design. Product design usually requires that performance requirements and performance tests be established during the early design stages and the performance tests are carried out during the final stages. Task analysis can contribute to the establishment of realistic human performance requirements and identify the critical tasks that should be evaluated in the later design stages. With the more iterative, object-oriented design approaches, establishing performance requirements and tests may occur over several design cycles. Nevertheless, the appropriate selection of tasks and establishment of realistic usability benchmark tasks and performance measures can greatly benefit from task analysis conducted early in the process. Without task analysis information, the establishment of performance requirements and the selection of tasks for performance evaluation can be based on incomplete task performance data, and it is possible that critical tasks are overlooked. For example, usability testing can be integrated into the object-oriented design process because usability tests can be done on a limited set of software components. As Mayhew (1999) points out, however, task analysis is necessary to obtain a valid user model and to establish usability goals and performance requirements. Without task analysis, the usefulness of the tests to the subsequent product design can be limited if the wrong tasks are tested.

23.2.2.6 Selection and Training

Task analysis is used to determine the necessary skills for performing new or redesigned jobs and establish selection requirements, job classifications, and pay scales. Taxonomies used for personnel decision making are usually empirically derived with factor analytic techniques and focus on skills needed for job performance. The most frequent purpose of task analysis, however, is to elicit task knowledge and procedures and use them for developing task training programs, and most task analysis methods were developed for this purpose. The outputs of the task analysis are easily translated into training modules with inputs, outputs, and performance specifications. For example, Instructional Systems Design, a systems-analytic approach to developing training systems mandated in military training systems and frequently used by civilian organizations, specifies that task analysis be performed in the early stages of training development. The outputs of task analysis for training purposes will include conditions under which the tasks are done, performance standards, and task inputs and outputs.

23.2.2.7 Measurement of Mental Workload

Mental or cognitive workload refers to the information processing resources required of an operator in achieving task goals. Knowledge of an operator's mental workload is required to ensure that it be within acceptable limits. A variety of workload measurement approaches have been developed, for example, subjective measures, performance measures, and physiological measures (e.g., Tsang and Wilson 1997) and task analysis is an important component of most of these. Task analysis forms the basis of some measures of mental workload or task analysis is used to determine the critical tasks requiring additional workload assessment.

23.2.3 Reliability, Validity, and Usability of Task Analysis

The wide variety of task analysis methods, besides reflecting differences in the various applications of task analysis, also reflects some dissatisfaction within the human factors' community over the existing methods. Formal methods of task analysis are criticized by both practitioners and behavioral scientists for diametrically opposing reasons. Practitioners often criticize these methods as not cost-effective. A complete task analysis (description and analysis using agreed-upon behavior taxonomy) would significantly and positively

TABLE 23.1
Some Task Plans in Hierarchical Task Analysis

Plan	Task Characteristics	Example
Fixed sequences	Operations carried out in a specific sequence	Saving a document
Contingent sequences	The subsequent action is contingent on either feedback from a previous action or the system itself	Completing online forms
Choice operations	Actions performed as a result of decision making operation	Diagnosing and fixing an operating system malfunction
Optional completion	Operations can be carried out in any order	Searching for information
Concurrent operations	Operations must be coordinated or carried out together	Clicking and dragging an icon with a mouse

Source: Shepard, A. 2001. *Hierarchical Task Analysis.* London: Taylor & Francis. With permission.

impact the design of a product only if the task analysis information was available when the design decisions are being made. A complete description using a formal method of task analysis can take considerable time, however, leading many practitioners to reject them as impractical when they are performed during the design process. This is less problematic in the design of large-scale systems given the greater time spent on development and the more linear structure of this design approach. Task analyses can be done in the early stages of design and be completed in time to impact later design stages. With the object-oriented design process, however, the iterative design phases may not provide sufficient lead time for a complete task analysis to impact design decisions fully. Consequently, task analyses for Web application design can be less formal, reflecting more the intuition of the analyst than the careful application of a specific method.

Behavioral scientists, on the other hand, express concerns about the reliability and validity of task analysis methods. Reliability refers to the consistency of the method, either between analysts or within the same analyst at different times. Assessing the reliability of a method is straightforward: compare the results of two or more analysts, for example, on the same task set. Validity, the extent to which the task analysis accurately captures or describes the task, is more difficult to measure. In the behavioral sciences, reliability and validity are considered essential characteristics of any objective measure of behavior, yet there are very few published reports of the reliability and validity of various task analysis methods. There are several possible reasons for the paucity of reliability and validity data. First, assessing reliability and validity increases the time required to complete the analysis, further contributing to perceptions of impracticality. Second, assessing the validity of a method requires a detailed level of analysis that may not be possible with new products early in design. Finally, reliability and validity are measures of the task description component only; they do not capture the usefulness of the analysis component to the success of the eventual product design. Meister (1985) refers to the usefulness of task analysis information as "utility"; recently this has been termed "usability." The utility or usability of a task analysis method measures how well the task analysis contributes to product design (Chipman, Schraagen, and Shalin 2000). Most discussions of task analysis usability focus on the ease of use of a method and how useful the output format is to product designers. Certainly, these are important components of task analysis usability, but usability is also affected by the quality of the information obtained from task analysis and ultimately the final design of the product. An easy-to-use and easy-to-understand task analysis method does not alone ensure usable products if the validity of the task description used to inform the design is questionable, because the task analysis was focused on either noncritical tasks or tasks not affected by the new design.

Of course, when task analysis is done informally without adhering to a specific method, issues of reliability and validity are moot, and the usability of the analysis is determined by the analysts and designers on a case-by-case basis.

23.2.4 METHODS OF TASK ANALYSIS

There are many task analysis methods, reflecting both differences in the purpose of a task analysis, and dissatisfaction with existing methods. Despite the many attempts at improving on existing task analysis methods, most of these newer methods have been used only in the laboratory or in limited numbers of investigations. Therefore, only three methods of task analysis will be discussed here. These were selected because they are well known and have a history of successful applications. They have been successfully applied to the design of Web applications and user interfaces, and they are relatively assumption free, meaning that they can be easily adapted for a variety of design purposes. These methods are hierarchical task analysis, critical incident technique (CIT), and Goals, Operators, Methods, and Selection Rules (GOMS) analysis. For each method, the assumptions and standard procedures are outlined. These are followed by examples of Web-based applications of the method and an evaluation of its usefulness. Finally, recent software tools developed for assisting the task analyst are reviewed, as these may affect the speed with which task analysis is done and the consistency of the output.

23.3 HIERARCHICAL TASK ANALYSIS

Hierarchical task analysis (HTA), developed by Annett and Duncan (1967) for designing effective training systems, is the most widely used method of task analysis. HTA is popular because it is extremely flexible and adaptable, and it is based on a system's approach to the description and analysis of tasks. It can be used to analyze both existing and new tasks; it allows quick, less formal analysis or detailed thorough procedures. Shepherd (2001) provides a detailed, readable account of the formal stages in HTA, and the following description is taken from his account. According to Shepherd, HTA is a framework for analyzing tasks. It is an exploratory process conducted to generate hypotheses for improving user performance. As hypotheses, additional validation is required before they can be recommended. HTA is compatible with the systems approach to design because both use a top-down strategy, starting with general goals and systematically detailing these goals into subgoals and components.

HTA also is widely used because it ensures breadth of coverage. If the analyst is constrained by time or budget, a breadth first approach will provide a high-level description of all tasks from which critical tasks needing detailed analysis can be identified and prioritized. For new systems, a breadth-first approach enables the analyst to distinguish between tasks that are unchanged from previous systems and those affected by the new design. Because HTA is not taxonomy based, HTA can serve as a framework for other methods of task analysis. For example, the subordinate goals and tasks most appropriate for cognitive analysis can be identified with a preliminary HTA. Moreover, HTA can be a valuable management tool for large-scale task analysis projects by establishing an overall framework for costing and scheduling activities.

23.3.1 HTA Procedures

In HTA, the analyst begins by identifying and evaluating task objectives and then determining whether each objective is being achieved. Those objectives that are not being achieved by the operators become the focus of subsequent analyses. If the user is not achieving task objectives, the analyst examines the task operations for possible solutions to the problems. These solutions become hypotheses that must be evaluated. If hypotheses cannot be generated at this stage, the analyst decomposes the task objective into subordinate objectives and task plans. This process is repeated for each subordinate objective until hypotheses are generated. This general framework, shown in Figure 23.1, is described in more detail below.

23.3.1.1 State the Goal

According to Shepherd, hierarchical task analysis begins with statements of the overall objectives of the system and descriptions of humans interacting with the system. It is important that all humans interacting with the device and their level of interaction be identified. For each user group, the analyst systematically evaluates task goals to determine whether they are being achieved. For goals that are not met, the analyst must determine whether operations can be examined or if additional goal refinement is necessary, as shown in Figure 23.1.

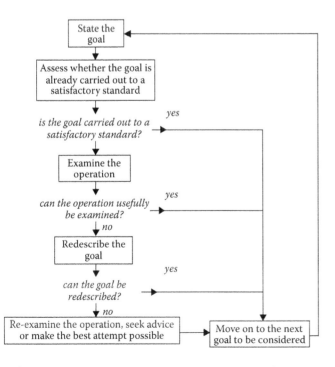

FIGURE 23.1 Basic decision cycle in HTA. (From Shepherd, A. 2001. *Hierarchical Task Analysis*. London: Taylor & Francis, Figure 2.1. With permission.)

23.3.1.2 Assess Whether a Goal Is Being Carried Out to a Satisfactory Standard

The decision to evaluate a goal is based on a consideration of practical constraints and an assessment of the consequences of not meeting the goal. The goal may be stated as a system concept or be specific to a human operator in the system. Constraints are factors that may determine how easily the goal is met but cannot be controlled by the designer or analyst. For example, the task may be done under severe environmental conditions such as high levels of background noise, task performance may be limited by safety regulations, or the technical background of users may be highly variable. Deciding to analyze a goal in more detail also depends on the importance of the individual goal to achieving system goals. If performance is acceptable, no further analysis is required and the next goal is considered. A common recommendation for determining whether to analyze further is the "P × C rule": evaluate the goal if the product of the probability of an incorrect response (P) and the cost of an incorrect response (C) is unacceptably high. Of course, obtaining accurate estimates of P and C is unlikely for most tasks, and the P × C rule is used heuristically after consulting with task experts.

23.3.1.3 Examine the Operation

If a task objective it is not being achieved satisfactorily, the analyst evaluates the operations used to achieve the task objective. An operation is a behavior that the operator performs to achieve a goal. The analyst can choose from a variety of methods for describing and evaluating the operations. The analyst may examine information processing or cognitive demands being made on the user, evaluate the operations against design guidelines, or use interviewing techniques to identify problems in carrying out the operations. If problems or limitations are identified, the analyst creates hypotheses for relieving them. As with any hypothesis they must be evaluated and subjected to additional cost-benefit analysis before specific recommendations can be made. If the analyst cannot generate reasonable hypotheses for improving performance, he or she may redescribe or refine the goal into operations and subgoals.

23.3.1.4 Redescribe the Goal

If hypotheses cannot be generated from a human-factors examination of the operations, the analyst may redescribe or refine the goal into task plans and subordinate objectives. A task plan is a description of the subordinate operations and the organization of these operations for meeting a goal. The development and format of a task plan should be determined by the nature of the task. Table 23.2 lists some plan formats designed to describe different tasks. Task plans may be developed with other task analysis methods. If a simple task plan cannot be determined at this point, the goal is either further refined into subordinate goals or subordinate goals are added to the task plan. That is, the analyst determines whether each subordinate goal is being achieved, and either develops hypotheses or decomposes the subordinate goal further. This

TABLE 23.2

Examples of CIT Applications and Questions

Purpose of CIT	Data Collection Method	Specific Question(s)
Identify cockpit display and control problems (Fitts and Jones 1947a, 1947b)	Interview and mail survey	"Describe in detail any error in the operation of a cockpit control (flight control, engine control, toggle switch, selector switch, trim tab, etc.) which was made by yourself or by another person whom you were watching at the time (332)."
Evaluate online and hard copy documentation for a computerized conferencing system (del Galdo et al. 1986)	Online (screen button with occasional reminders)	"1. Indicate the type of documentation involved in the incident . . . 2. Would you classify this incident as a success or failure? . . . 3. Briefly describe the details of the incident. Indicate the features of the computer interface which lead to your success or failure in completing the task. (p. 20)"
Evaluate whether medical information retrieved on Medline impacts medical decision making and patient care (Wilson, Starr-Schneidkraut, and Cooper 1989)	Phone interview	"I'm interested in recent MEDLINE searches that were especially helpful in your work or that were unsatisfactory. [For MD's, DDS's, RN's] I'm especially interested in searches that have had an impact on patient care. . . . I want to know what circumstance led you to do the search, what the context was, what specific question or issue you had in mind, what information you wanted and why. I also will want to know how you did the search, as nearly as you can recall, and what information you got. . . ."
Identify effective and ineffective assistive technology practices (http://natri.uky.edu/participation/incidents/cimenu.html) (Assistive Technology Research Institute 2003)	Print and interactive Web pages	"Describe a time that you provided, or observed someone else provide, assistive technology services in an EFFECTIVE way to a student with disabilities, or to someone who works with that student. The result of such action should have had a positive effect on the following general aim: When providing assistive technology devices or services, the general aim is to enable students with disabilities to improve their ability to function in the environment."
Develop a taxonomy for understanding people's activities on the World Wide Web (Morrison, Pirolli, and Card 2000)	Interactive Web page	"Please try to recall a recent instance in which you found important information on the World Wide Web, information that led to a significant action or decision. Please describe that incident in enough detail so that we can visualize the situation."
Assess the level of congruence between the critical incident perceptions of guests and contact employees in the gaming industry (Johnson 2002)	Interview	"Think of a time when playing slots at this or at another casino that you had a particularly satisfying/dissatisfying experience as a result of something an employee said or did. 1. When did the incident happen? 2. What specific circumstances led up to the situation? 3. What did the employee do? 4. What resulted that made you feel the experience was satisfying/dissatisfying?"

process is repeated until all goals have been analyzed to an acceptable level of detail or until the overall objectives of task analysis have been met.

23.3.1.5 Task Hierarchy

Shepherd states that this cycle of assessing goal achievement, examining operations, and refining goals into subordinate goals is a framework that directs task analysis procedures. However, he also notes that the essential characteristic of

HTA is the redescription of goals into subordinate goals. For quick, informal HTAs, decomposing tasks into subordinate tasks may be all that is done. Stanton (2006) provides additional recommendations that focus on the task hierarchy development and shows how this structure can be used for various analytic purposes. According to Stanton, the essential characteristics of HTA are the statements of task goals and the decomposition of the task goals into subgoals and operational plans. In essence, it is the task hierarchy that defines

HTA. Stanton provides guidelines for writing task goals: Task statements should be expressed as activity verbs and should include performance statements and conditions under which the task will be carried out. The number of immediate subgoals under any higher task goal should be kept to a small number if the hierarchy is to be maintained. Most goals should have between 3 and 10 subgoals. If the number of subgoals is outside this range, the description should be reanalyzed to see if any subgoals can be grouped under a higher task goal. In practice, the process of goal decomposition will require several iterations based on consultations with task experts followed by reviews of the hierarchy and further analyses. Once established, this task hierarchy becomes a useful first step for many other task analysis methods such as cognitive task analysis and analytic techniques such as function allocation and error analysis.

23.3.2 HTA Applied to Web Application Design

The flexibility of hierarchical task analysis can be seen in the many modifications made to the procedure. For example, Brinck, Gergle, and Wood (2002) developed a method of task analysis for Web applications that combines use cases with HTA. Use cases, developed in the early design stages, consist of descriptions of the users of a Web site and typical scenarios of usage. Performing HTA on tasks of critical users and critical use scenarios provides detailed descriptions of user goals and operations and hypotheses for improving performance. Note that use cases are consistent with Shepherd's recommendation that task analysis begins with identifying all user groups and for each group, the level of interaction with the system.

Van Dyk and Renaud (2004) applied the "Use Case + HTA" approach to the task "purchase product" for the design of e-commerce Web applications. This goal was refined into two subordinate goals, "Look, See, and Decide (LSD)," and "Checkout," as shown in Figure 23.2. The operations performed under LSD include searching for a product, searching for information about a product, comparing products and deciding to purchase a product. To achieve the second goal, "Checkout," the user pays for the product, arranges shipping, and confirms the purchase. From their task analysis, van Dyk

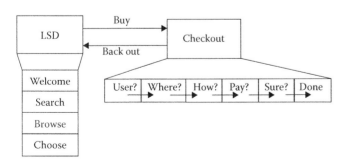

FIGURE 23.2 Task model of Web shopping. (From van Dyk, T., and K. Renaud. 2004. Task analysis for e-commerce and the Web. In *The Handbook of Task Analysis for Human–Computer Interaction*, eds. D. Diaper and N. Stanton, 247–262. Mahway, NJ: Lawrence Erlbaum. With permission.)

and Renaud (2004) determined that the LSD task is more user driven and less clearly defined. Several operations for achieving this goal were described. For example, the user could browse products, search and compare products having desired qualities, or simply find and select a product decided upon before entering the Web site. The subordinate goal checkout is more clearly defined and highly system driven because the user responds to system requests in a specified sequence. In Shepherd's terminology, LSD can be described as an optional completion task (see Table 23.1), under which the sequence of operations does not determine whether the goal is achieved. On the other hand, the operation plan for checkout is a contingent sequence in Table 23.1 because the user responds to information requests from the system. Van Dyk and Renaud also point out that hypotheses for improving performance on the LSD goal will be different from those for improving the checkout goal. During LSD, Web applications should be designed to keep the user interested and on the Web site. User interest is less important in the checkout phase, however. Prompt system feedback and clear instructions are more critical for achieving this goal.

23.3.3 HTA Software Tools

A thorough, broad HTA is time-consuming and may be tedious, a common criticism of task analysis in Web and human–computer interface design. Recently, software tools for facilitating HTA procedures have been developed that potentially can speed the process of goal description and refinement. In a recent review of HTA, Stanton (2006) lists several requirements for an effective HTA tool. First, it should support subgoal development and display the analysis in different commonly used formats. Second, the analyst should be able to edit and verify the analysis, in one representation, and have alternative representations updated. Third, the tool should allow the analyst to add more tasks or decompose existing tasks. Note that these requirements address the representation of the analysis made by the analyst but do not simplify or speed up the decomposition of tasks into subtasks. Nevertheless, these products represent significant advancements over commonly used general office tools for representing the task hierarchy, for example, Microsoft Excel or Microsoft Draw. Two HTA products are currently available: Task Architect (http://www.TaskArchitect.com) and HTA Tool (Human Factors Integration—Defence Technology Centre). These products share many features and should improve the usability of HTA in design. Both products simplify the process of entering task data and interactively developing a task hierarchy. HTA Tool and Task Architect have the following common features for facilitating the analysis.

Data from the task analysis can be viewed in several formats. Typically, task data would be input while in an outline mode ("Tabular Mode" in Task Architect and "List View" in HTA Tool; see Figures 23.3 and 23.4). Once tasks are entered, they can be easily viewed in either hierarchical or tabular formats. Changes made in any mode are automatically included in other representation modes. Both products

(a)
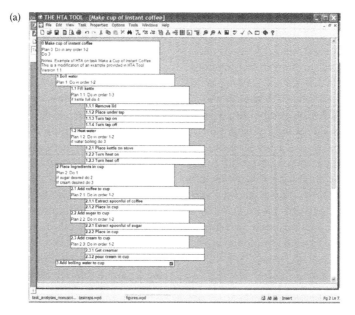

(b)
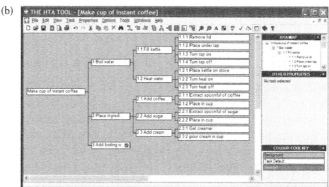

FIGURE 23.3   Hierarchical task analysis example from the HTA Tool: task description in (a) list view and (b) horizontal flowchart mode.

have automatic numbering features that simplify the organization of tasks and subtasks, and the number scheme can be tailored to a specific task analysis. Figures 23.3 and 23.4 show a task description of the task "make a cup of instant coffee" in HTA Tool and Task Architect as an outline (Figures 23.3a and Figures 23.4a) and a horizontal flowchart (Figures 23.3b and 23.4b). Additional features of these products are as follows:

- Properties can be assigned to each task in the analysis. These can be newly defined or developed from previous task analyses. Properties can be added to each task or to the analysis as a whole. In the tabular view of Task Architect, the selected properties are shown as columns filled in by the analyst. Properties may be text descriptions, time values, ratings (e.g., difficulty or workload), images, or hyperlinks.
- Operational task plans, either newly created or standard plans, can be added. Standard plans such as those listed in Table 23.1 are provided with each software tool.
- A tool is provided for checking whether the essential characteristics of HTA are met. Similar to grammar

checking, these tools check for task goals that are lacking operational plans and whether the number of subtasks is sufficient.

- Task data can be imported from standard software products such as Microsoft Excel and can be exported to common formats.
- Task analyses can be stored and maintained for future use. Both Task Architect and HTA Tool provide a method for adding all or some part of a previously completed task analysis or saving a part of a current task analysis for future use. This has the potential to improve the efficiency of HTA. For example, if an existing interface is currently being redesigned, those components of the product that are unchanged would not have to be reentered in a new analysis but could be simply linked to the new task analysis. In essence, this feature provides a method for standardizing and maintaining a historical database of previously completed analyses.

Each product has unique features. Task Architect provides several standardized report templates for task analysis. These provide a set of properties that are based on the purpose

(a)

(b)

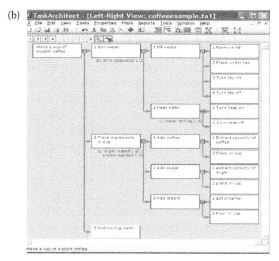

FIGURE 23.4 Hierarchical task analysis example from Task Architect: task description in (a) tabular view and (b) horizontal flowchart mode.

of the analysis. Among the special reports are Timeline, Link Analysis, and Systematic Human Error Reduction and Prediction Approach (SHERPA). Timelines show how tasks are related and done over time; link tables present how tasks carried out by different people interact during the analysis. In a recent review of Task Architect, Walls (2006, 29) noted that Task Architect is "well thought out and easy to use for recording HTA data, one of the most frustrating aspects of HTA."

HTA Tool contains property modules that allow a HTA analyses to be extended to other human factors analytic techniques. Some modules include Keystroke Level Modeling, P × C Stopping rules, SHERPA, Workload Assessment (TLX and SWAT), Interface Analysis, and Training Design. When the task analysis is being viewed in tabular format, the analyst selects a module and columns are added to the table based on the specific module. For example, if the TLX workload

module is selected, a column for each dimension of the TLX Workload Rating scale is added. One additional feature of HTA Tool is the ability to run in Novice or Expert Modes, which tailors the help function to the level of experience with HTA Tool. For novice task analysts, HTA Tool provides an Analysis Wizard that guides the user through the recognized process of developing a hierarchical task analysis by prompting the user for goals and subgoals. The user is asked for the overall goal followed by the first layer of subgoals. Each of these can be broken into subordinate goals until the analyst is satisfied that level of detail is appropriate. At each step the user can enter any additional notes concerning a task and select the appropriate operations plan for achieving a subordinate goal. The result is a complete analysis that an analyst can continue to modify if he or she wishes. The Analysis Wizard forces the analyst to concentrate on just one thing at a time and could be particularly useful for a novice analyst or new user of the application who might find starting their first analysis daunting.

In summary, tools such as Task Architect and HTA Tool should greatly improve the usability of HTA by providing a mechanism for quickly entering task data, interactively reviewing and revising the task description in several formats, and adding additional analyses based on human factors' analytic techniques. Moreover, by providing a standardized format for saving task analysis data, and allowing task modules to be inserted into an ongoing analysis, each tool can facilitate the establishment of a data base of task analyses for an organization.

23.3.4 CONCLUSIONS

HTA has been called "the nearest thing to a universal [task analysis] technique" (Ainsworth and Marshall 1998) because it serves as a framework for the process of task analysis rather than a taxonomy or model of task performance. In a survey of 30 task analysis applications in the defense industry, Ainsworth and Marshal (1998) found hierarchical task analysis was used most often and for a variety of purposes (e.g., system procurement, interface design, personnel requirements identification, and training). Stanton (2006) lists more than 200 reports of HTA in the armed services and nuclear power industries. Moreover, hierarchical task analysis underlies many other methods of task analysis (e.g., GOMS, see later sections of this chapter.) The cost of this flexibility is the steep learning curve for conducting HTA. The quality of a hierarchical task analysis is highly dependent on the training and expertise of the operator (Patrick, Gregov, and Halliday 2000; Stanton and Young 1998). In fact, Stanton (2006, 61) states that HTA is "something of a craft-skill to apply effectively. Whilst the basic approach can be trained in a few hours, it is generally acknowledged that sensitive use of the method will take some months of practice under expert guidance." Two developments may improve the training of task analysts. First, Ormerod (2000; Ormerod and Shepherd 2004) developed a subgoal template (SGT) approach to HTA. SGT developed a taxonomy of task elements required to achieve a subgoal for ensuring that the output of HTA be compatible with information requirements specifications and for standardizing the training and output format of HTA. Second, the aforementioned HTA software tools may provide a means of improving the training of HTA analysts. Both Task Architect and HTA Tool provide a "task analysis checker" that evaluates a task description according to a set of rules. If these rules were made available to educators and trainers, they could improve the training of HTA principles by providing immediate feedback based on the educator training objectives. Moreover, if new rules could be added to the checker, the educator could tailor the feedback to be consistent with concepts being taught.

23.4 CRITICAL INCIDENT TECHNIQUE

The critical incident technique (CIT) was developed during World War II to improve pilot selection, pilot training, and aircraft cockpit design. Subsequently, CIT was applied to many jobs in private industry: assembly line workers, supervisors, teachers, maintenance workers, bookkeepers, and psychologists (Flanagan 1954; Hobbes 1948). Today, CIT is most commonly used in the service industries and health professions, possibly because service incidents are easily defined and case studies are commonly used in the health professions (Urguhart et al. 2003). CIT is similar to hierarchical task analysis in that it is relatively assumption free and represents a framework for guiding task analysis. In most other respects, CIT is radically different from the top-down HTA approach. CIT is considered a bottom-up approach because the analysis proceeds inductively from descriptions of specific task incidents to general recommendations for improving task performance. Also, unlike hierarchical task analysis, CIT focuses almost exclusively on the task analysis component, not on task description. CIT has traditionally been used to analyze existing systems, but the process can be integrated in the iterative design process of new systems (del Galdo et al. 1986).

The basic unit of CIT is the "critical incident." Not surprisingly, the term critical incident (like the term task) is not precisely defined. An incident is similar to a task goal in that it should have a beginning and ending point, and an incident is critical if the intent and consequence are reasonably clear to the user (Flanagan 1954). Meister (1985) also notes that a critical incident must have "special significance" to the user. The term incident is used instead of goal because CIT is primarily a retrospective technique: the user or an observer reports on current or past incidents. When applied to human–computer interactions, incidents are defined as any interaction with an element of the interface that made completing a task particularly easy or difficult (e.g., del Galdo et al. 1987). The task analysis is essentially the administration of a few carefully designed questions to elicit specific critical incidents from either operators or nearby observers (e.g., supervisors) based on the goals of the task analysis. For example, Fitts and Jones (1947a, 1947b) used CIT to elicit pilot descriptions of critical incidents related to specific cockpit displays and controls. They developed three questions covering control problems,

display problems, and "pet peeves" with the cockpits, as shown in Table 23.2. The responses were classified according to the type of error and the control or display involved. Design recommendations developed from these responses established some basic cockpit-design principles that remain relevant in modern cockpits. CIT may at first seem similar to questionnaires or other surveys such as user satisfaction surveys but a CIT is different in one important aspect: the information obtained is specific to a problem experienced by the user while he or she was engaged in a task having well-defined goals. Similarly, CIT may resemble log reports obtained from users during beta testing. Here users report on problems experienced and attitudes toward the product. However, information obtained from CIT is usually more specific: the data consist of detailed information regarding a single incident that occurred while the task was performed and the interface elements contributing to or preventing the achievement of task goals (Castillo and Hartson 2000).

The unique approach taken by CIT offers several advantages over other methods of task analysis. It can be unobtrusive because operators can report incidents after they occur. Compared with other task analysis methods, greater numbers of users can be reached, especially if interviews are conducted online or by mail. Also, data can be obtained on rare events that would not otherwise be observed with other task analysis methods. Finally, CIT may be used to collect data over long time periods with less impact on cost. For example, Flanagan (1954) reported that the types of incidents reported by air traffic controllers varied with annual seasonal variations. On the other hand, CIT reports may be biased by the user's memory of the incident or differences between user groups with respect to task goals. Miller and Flanagan (1950), for example, administered CIT questionnaires to three groups of supervisors on an assembly line. One group was instructed to record incidents daily, one group weekly, and one group biweekly. The supervisors making daily reports generated twice as many incidents as those reporting weekly, and five times as many incidents as those reporting biweekly. When incidents are being reported by several user groups with different task goals, each group may report different incidents. Students, for example, reported different classes of incidents than professors, although both were evaluating the performance of the professor (Flanagan 1954).

23.4.1 CIT PROCEDURES

The specific procedures for CIT are described by Flanagan (1954) and Meister (1985). Generally, the procedure begins with a statement of task goals, and these drive the development of the reporting instrument. Data analysis is based on content analysis in which categories of incidents are developed empirically from a sample of responses.

23.4.1.1 Identify the Aims of the Task

A summary or brief statement of the purpose of the task agreed upon by most operators is necessary to determine the relevance of incidents to operator effectiveness. Specific aims can be determined from job or system descriptions, supervisor reports, training manuals, etc. Task operators should be included in this process because it is essential that users agree on the aims of the task. Specific criteria for successful performance or effectiveness of behaviors should be developed.

23.4.1.2 Develop Test Plan and Specifications

Precise instructions and reporting instruments must be developed for the critical incident observers because, although individuals doing the activity are probably the most qualified to report incidents, they are not trained observers. Therefore, instructions and instruments should be as specific as possible, and observers should be trained on reporting incidents. The instructions and training should include information on the place, persons, and conditions where the incidents will be recorded, the relevance and degree of criticality of behaviors to the general aims, the persons (e.g., operators, supervisors, or analysts) who will record the incidents, and the methods for recording data. Generally, only a few questions are developed. These are focused on specific critical incidents derived from the task aims. The questions should be pretested to ensure that operators agree upon the incidents being reported. Table 23.3 provides some examples of CIT questions for different task analysis purposes. Note that questions in each application address specific incidents in the performance of the task, whether the task aims were met, and the reasons why task aims were met or not met.

23.4.1.3 Collect Data

If possible, data should be collected while the events are still fresh in the mind of the respondent. However, when a more comprehensive survey of incidents is required, or when recording incidents as they occur is not practical, respondents may be asked to recall previous incidents. Memory for incidents can be improved by providing advanced notice that certain incidents will be recorded. Many standard instruments (face-to-face interviews, mail surveys, online forms, etc.) for recording data have been researched extensively, and recommendations for ensuring their effectiveness and objectivity apply here (e.g., see Volk, Pappas, and Wang, this volume). Obviously, more users can be reached at less cost with a CIT as a questionnaire. However, mail surveys may not elicit the specific information required for CIT. Respondents to a mail survey tend to respond in more general terms or say what they normally do instead of recalling specific incidents (Sweeney, Lapp, and Group 2000; Urquhart et al. 2003). Urquhart et al. (2003) also showed how an interviewing technique known as "explicitation" can facilitate users' recall of specific events. Explicitation helps respondents "relive" a critical incident by focusing on the sensory memories of the event.

23.4.1.4 Analyze Data, Interpret, and Report

Because of the empirical nature of this technique, data analysis proceeds from the specific incidents reported to general categories and recommendations. A frame of reference that creates categories most informative or relevant to the purpose of the analysis (e.g., interface design, training) should be selected. Categories are usually developed from a sample of incident reports. Reports in the sample are sorted into similar categories, and then each category is fully described. The level of specificity of the categories also should be determined from the purpose of the analysis. For example, in the analyses of aviation critical incidents by Fitts and Jones (1947a, 1947b), the purpose was to evaluate controls and displays. Critical incidents were subsequently classified in terms of errors made with a particular control or display (e.g., activating/reading the wrong control/display) as well as types of controls/display (e.g., multiple versus single pointer displays). Statements of requirements or recommendations from the CIT are developed and described along with any noticed limitations on the method and analytical techniques employed. Therefore, the analysis of task limitations is derived directly from the users, and not from a systematic description of the users' tasks.

TABLE 23.3
Summary of Different Versions of GOMS Models

Model	Description	Design Information Obtained
CMN-GOMS	Original formulation of GOMS Model	Operator sequences; execution times and error recovery for sequential tasks
KLM	List of keystrokes and mouse movements to perform a task	Execution times and error recovery for sequential tasks
NGOMSL	Procedure for identifying all GOMS components expressed in a programming language	Functionality consistency, operator sequences, execution times, procedure learning times, and error recovery for sequential tasks
CPM-GOMS	GOMS applied to tasks performed in parallel. Uses cognitive perceptual and motor operators	Operator sequences, execution times, and error recovery for sequential and parallel tasks

Source: John, B. E., and D. E. Kieras. 1996a. Using GOMS for user interface design and evaluation: Which technique? *ACM Transactions on Computer–Human Interaction* 3: 287–319; John, B. E., and D. E. Kieras. 1996b. The GOMS family of user interface analysis techniques: comparison and contrast. *ACM Transactions on Computer–Human Interaction* 3: 320–351. With permission.

23.4.2 Applications of CIT to Web Applications

Del Galdo et al. (1986) compared online versus hard copy documentation for an online conferencing system using CIT. Inexperienced users performed 19 tasks with the software and reported critical incidents after each task. Ninety-one incidents were reported. These were divided into categories of failure and success, and each of these was broken into the subcategories of online and hard copy documentation. Finally, for each category, the incidents were categorized according to the specific features that led to the incident. Thirty-one recommendations for improving both hard copy and online documentation were reported to the designers, and 71% of these recommendations were implemented. The newly designed product was reevaluated using the same CIT method on a new sample of users. Of the original 26 problems identified in the first evaluation, only three were reported in the follow-up evaluation. Moreover, the number of successful incidents increased significantly.

Sweeney, Lapp, and Group (2000) used CIT to understand the factors that lead to high or low service quality for Web-based services. Perceived service quality is a cognitive construct that is an important contributor to overall customer satisfaction. Satisfaction involves both cognitive and affective factors. A CIT questionnaire was mailed to 114 respondents requesting incidents that were perceived as high- or low-quality service. However, the mail surveys required a lengthy set of instructions and the respondents did not provide specific information to allow further analysis. Therefore, phone interviews were conducted and the interviewer asked follow-up questions regarding specific details of the reported incident. The incidents were initially screened according to whether they met the criteria for the analysis. A classification scheme was subsequently developed and incidents were classified according to the key factor that triggered the incident. These key factors were then sorted into three groups that represent the causes of high- and low-quality incidents: ease of use (e.g., instructions, navigation, screen layout), content (e.g., consistency, correctness, and currency of information), and process (perceived control and speed). The factors determining high and low service quality were shown to differ: Whereas speed, consistency, and control deficiencies led to perceptions of poor service quality, these factors, when adequate, did not produce perceptions of high service quality. High service quality was determined primarily by ease of use, especially structured design and layout. Thus, the criteria promoting perceptions of high service quality can be different from those creating perceptions of low service quality.

CIT also has been applied to the analysis of team cognition (e.g., Keikel et al. 2001; Keikel and Cooke, this volume). Interest in analyzing team cognition has emerged in recent years because of the realization that working teams are an important component of most organizations, and that many teams communicate via Web applications. A team can be defined as any small group of people working on a task; team cognition refers to the thoughts and knowledge of the team (see Keikel and Cooke, this volume). Information gathered from cognitive task analysis of teams would therefore inform decisions on the design of teams (e.g., structure, size), training methods, and support products (e.g., human–computer interfaces, communications). Keikel et al. obtained both physical measures of communication such as statistically distinct communication patterns and content measures such as the sequence of ideas expressed. No a priori assumptions are made regarding team cognition. Considering that communication data over the Web is almost always recorded (via e-mail or discussion groups), it may represent a promising method of evaluating team cognition for Web-based teams.

23.4.3 CIT Software Tools

Online CIT instruments can be easy to learn, relatively unobtrusive, and allow remote usability assessment (del Galdo et al. 1987). Although the incidents reported with online instruments has not been compared with those obtained by more traditional methods, Castillo, and his colleagues (Castillo and Hartson 2000; Castillo, Hartson, and Hix 1998) showed that users reported 75% of the incidents considered high in severity by expert usability evaluators, and the users' rankings of incident severity agreed with the usability evaluators on 83% of incidents reported. The flexibility of CIT also enables modifications to the method for improving the understanding of critical incidents. Castillo, for example, used a "contextualized" critical incident reporting tool that also captured the user's interface and actions once he or she initiated a critical incident report, as shown in Figure 23.5. Designers can evaluate the report along with the sequence of the user's actions and status of the user's interface to gain a better understanding of the problem. This tool was further extended to allow for the recording of "critical threads," because Castillo noted that many users did not initiate a critical incident report until 5 minutes after the incident occurred. This may reflect the user's need for time to understand the incident before making a report. The critical thread tool provides two buttons to the user: "Begin Incident Capture" and "Report Incident" (see Figure 23.5). When the user thinks an incident is occurring, he or she clicks "Begin Incident Capture" and the computer begins recording the user's actions and system status. When the user is ready to make the report, he or she clicks "Report Incident" and begins entering information. This tool should increase the likelihood that the user's actions at the time of the incident are recorded.

23.4.4 Conclusion

The CIT is unique among the methods of task analysis because it elicits problems directly from the users and it focuses on analysis rather than description. It can provide information regarding user difficulties with certain tasks but does not produce complete task descriptions that can be used for all decisions regarding product design. In one of the few empirical evaluations of CIT usability, del Galdo et al. (1986) showed that when design changes were made based on recommendations from a CIT analysis of software

FIGURE 23.5 Sample reporting form for CIT online. (From Castillo, J. C. 1997. The user-reported critical incident method for remote usability evaluation. Masters Thesis. Virginia Polytechnic Institute and State University, Blacksburg. With permission.)

documentation, 83–93% of the problems were not reported on a subsequent CIT analysis of the revised product. Moreover, a significant increase in the number of successful incidents was reported. However, the effectiveness of CIT is only as good as the quality of the content analysis for determining interface issues affecting usability. In a recent evaluation of the reliability of CIT content analysis (Koch et al. 2009) the reliability estimates were quite low, possibly because of the lack of adequate training materials on this method.

23.5 GOMS: A METHOD OF COGNITIVE TASK ANALYSIS

The methods of cognitive task analysis are potentially useful to the design of Web applications because Web-based tasks are primarily cognitive. The term cognitive task analysis describes a set of methods for obtaining information about the thought processes and goal structures that underlie observable task performance (Chipman et al. 2000). Early methods of cognitive task analysis were developed from theories of cognitive processing in the 1970s to contribute to instructional system design (Hall, Gott, and Pokorny 1995), but most cognitive task analysis methods are aimed at modeling user performance.

Many methods are being reported in the literature, and these vary in the stage of development (e.g., preliminary models to complete analytic techniques), usefulness to the design of Web applications, and ease of use. In fact, most cognitive task analysis methods have been applied only in artificial test environments or limited portions of real-world applications. The most widely used method of cognitive task analysis is GOMS, first developed by Card, Moran, and Newell (1983). GOMS is both a performance model and a cognitive task analysis method. A GOMS model represents user knowledge in a hierarchical stack structure consisting of Goals, the state of affairs to be achieved; Operators, elementary perceptual, motor, or cognitive acts whose execution is necessary to change any aspect of users mental state or to affect the task environment; Methods, sequences of operators that describe a procedure for accomplishing a goal; and Selection Rules, rules that determine which of the available methods is used to accomplish the goal. A GOMS task analysis is a process of creating a GOMS model by decomposing user task knowledge into GOMS components. The general strategy is similar to hierarchical task analysis: begin by identifying the top-level user goals, emphasizing breadth over depth. Then refine each goal into subgoals, methods, and operators. It differs from hierarchical task analysis in that a GOMS model has a specific format and taxonomy. GOMS is specifically designed to represent procedural knowledge of well-learned cognitive tasks. A GOMS model can be translated into executable programs for evaluating the consistency of the model and obtaining quantitative performance estimates.

A GOMS model can affect the design of Web applications. In early design stages a GOMS model of existing or proposed tasks can determine if the functional requirements are derived from the tasks performed by users: every task goal should have a specific method for achieving it. The GOMS analysis can also identify benchmark tasks and establish performance criteria for usability tests at later stages. A GOMS model also can be used to determine the consistency of procedures: similar goals should be accomplished with similar methods, and, when alternative methods for accomplishing a goal are designed, the GOMS analysis can determine whether the selection rules make sense. Quantitative assessments of performance based on the design can be made from assumptions about operator execution times. A GOMS model can be used to evaluate alternative design concepts such as time to learn to use the system, time to perform tasks, and time for recovering from errors.

GOMS analyses and models are most appropriate for user tasks having well-defined goals and requiring the application of learned cognitive skills (John and Kieras 1996a). They are not appropriate for open-ended, problem solving, or creative tasks such as composing a manuscript. However, even these problem-solving tasks without well-defined goals have task components consisting of highly learned cognitive skills that are suitable for GOMS. For example, composing a manuscript involves many subtasks such as text editing, a highly learned cognitive skill. GOMS also does not permit analysis of interface issues related to the layout of components, readability of text, etc. Instead, it is focused on the procedures that the user must learn and execute when doing cognitive tasks.

Four versions of GOMS have been developed, as shown in Table 23.3 (John and Kieras 1996a, 1996b). Each version of GOMS is suited for specific task types and for obtaining specific types of information. The keystroke-level model (KLM) and Card Moran Newell GOMS (CMN-GOMS) are the original models developed by Card et al. (1983). KLM represents task performance as a sequence of low-level operators and provides quantitative assessments of task performance times. CMN-GOMS is the GOMS model originally developed by Card et al. (1983) for describing text editing with a line editor and a teletype-output device. Natural GOMS Language (NGOMSL) is an extension of CMN-GOMS that describes tasks in English like statements, and Cognitive, Perceptual Motor, or Critical Path Method GOMS (CPM-GOMS) extends GOMS analysis to tasks done simultaneously (Johns and Kieras 1996a, 1996b). Specific procedures for KLM, NGOMSL, and CPM-GOMS are described below. CMN-GOMS is not described because it is similar to NGOMSL, and detailed procedures for NGOMSL have been published (e.g., Kieras 1997).

23.5.1 KLM Procedures

KLM was developed by Card et al. (1983) to predict task completion times for tasks involving interactive computer systems that does not require analyzing the underlying cognitive processes. By focusing on the lowest-level operators in a GOMS model, KLM is in many respects an extension of Gilbreth's motion analysis technique to the modern, human–computer interaction situation. Kieras (2001) provides a detailed description of KLM procedures. Very simply, KLM involves identifying the sequence of (mostly observable) keystroke operators for accomplishing a task. Keystroke operators are mostly standard user inputs such as a key press or mouse click. The execution times for each operator is determined, either empirically or from published tables (see Table 23.4, taken from Kieras, 2001). Task completion time is simply the sum of the times required for the execution of each operator. Kieras extended KLM by adding a mental operator to represent the cognitive activities of thinking or perceiving. When the execution times of some keystroke operators change between user classes (e.g., experienced versus novice users), performance times should be estimated separately for each user class.

To do a KLM analysis, a Web interface must be specified in sufficient detail to allow identification of the keystrokes needed to complete a set of tasks. According to Kieras, developing a KLM begins with the selection of representative task scenarios and a list of assumptions made about the user and task. For each task scenario the analyst determines either the best way to carry out the task with an existing design or the way that the analyst assumes the user will do the task with an incomplete design. These methods are decomposed into the keystroke-level operators for doing the task. Wait operators are inserted for situations where the user must wait for the system to respond and mental operators are inserted when the user must stop and think about the task. Wait operators may be especially important in the design of Web applications, given the variability in system response times. Once the operator sequence is properly developed, executions times for each operator are determined and the total execution time is simply the sum of the individual operator times.

Kieras (2001) observed that the most difficult part of this technique is deciding on the number and placement of mental operators, because the analyst must determine how the user thinks about the task. To assist in the placement of mental operators, Kieras offers the following suggestions. When comparing the execution times of alternative designs, consistency may be more important than accuracy. Therefore, developing explicit criteria for adding mental operators and applying them consistently is recommended. Also, the analyst

TABLE 23.4
Standard KLM Operators and Execution Times

Operator	Description	Execution Time
K: Keystroke	Press a key or button on the keyboard	0.12–1.2 sec (0.28 sec for most users)
T(n): Type sequence	Type a sequence of n characters (a series of K operators)	n × K
P: Point with mouse	Move the mouse to position the cursor over a specific target on the display	1.1 sec
B: Press or release mouse button	Press mouse button down or release mouse button that was previously pressed	0.1 sec
BB: Click mouse button	Rapid execution of two B operators; pressing and releasing mouse button	0.2 sec
H: Home hands	Moving hand between keyboard and mouse	0.4 sec
M: Mental act	Routing thinking/perceiving associated with highly-practiced, routine tasks (e.g., finding an item on the screen, deciding on a method, retrieving information from memory)	0.6–1.35 sec (1.2 sec for most users)
W: Wait	Wait for system to respond	Time must be determined for a specific system

Source: From D. Kieras, *Using the Keystroke-Level Model to Estimate Execution Times*, 2001. Retrieved from http://www.eecs.umich .edu/~kieras/goms.html (accessed December 30, 2003).

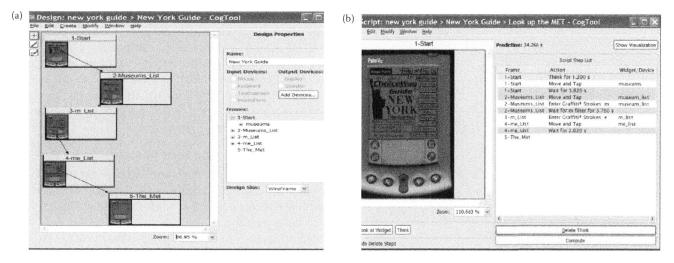

FIGURE 23.6 (**See color insert.**) Screenshots of CogTool for keystroke-level models of a task: (a) storyboard and (b) task script and computation of task performance time.

should focus on specifying the number of mental operators as opposed to their specific location in the sequence of operators, when estimating total task performance time. When different interface designs require different numbers of mental operators the analyst should consider removing the operators from the analysis by assuming that the operator performed them before accessing the Web page.

The simplicity of KLM has made it attractive for the evaluation of smaller, hand-held devices. For example, Myung (2004) developed a KLM for Korean text entry on mobile phones. Korean syllables are written either from left to right, top to bottom, or some combination, thus providing additional challenges for texting when a standard multipress keypad. Myung developed new estimates for keystrokes and mental operators, and showed that a KLM model predicted text entry times adequately. Moreover, the KLM model discriminated between two keypad layouts and predicted the reduction in task times with a new layout developed from the model. John and colleagues (e.g., Luo and John 2005; Teo and John 2006) developed a KLM for a handheld device and could predict task execution times with minimal error. Tonn-Eichstadt (2006) used a GOMS and KLM analysis to measure the efficiency of a Web-based task using a text-to-speech or Braille readers. Although the model was considered applicable and potentially useful in creating a usability efficiency measure, his analysis showed that new estimates of operator times are necessary for these systems. For example, estimates of the wait-time operator vary considerably because it depends on the rate of speech and brailed output.

23.5.2 SOFTWARE TOOLS FOR KLM

For large task analyses, identifying a series of keystrokes and computing task times can be time-consuming, especially if task completion times are used for comparing alternative design concepts. The simplicity of KLM, however, makes the development of task analysis and computation tools straightforward, and several tools are now available for this purpose.

The previously discussed HTA Tool, for example, can produce task performance times using either standard operator times or updated estimates of operator times input by the analyst. Another tool for KLM, "CogTool" (John 2009) combines an interface prototyping tool with a scripting tool to generate KLM task estimates from either interface descriptions or illustrations of an interface. CogTool automates the application of KLM to specific problems, providing an alternative to time-consuming and expensive user testing. User interfaces can be prototyped by creating a design story board consisting of individual screens ("frames") of the user interface (see Figure 23.6). These screens can be images of existing interfaces, or simple screen layouts developed in CogTool. For each frame, widgets are defined. Widgets represent simple user interactions such as mouse clicks and keyboard entries on the prototype interface. A storyboard of the frames is then created by identifying the action produced by user interactions with the widgets. Tasks are scripted by visually showing the series of user inputs in each frame and their effects ("transitions"). The script and computation of task performance time are automatically generated from an ACT-R cognitive model. The ACT-R model also inserts mental operators automatically based on the task demonstration and assumptions of the model. Although the model may at times be oversimplistic, standardizing the insertion of mental operators makes comparisons between alternative interface designs quicker and more straightforward.

23.5.3 NGOMSL PROCEDURES

Kieras (1988, 1997, 2004) developed this variation of Card et al.'s original GOMS model to provide an English like format for GOMS models, thus preserving the structure of GOMS, while making it less cryptic. NGOMSL is not intended to be a programming language, but it can be translated into a programming language and implemented as a computer program. NGOMSL can provide quantitative information on task performance times, cognitive complexity, and learning

FIGURE 23.7 Example NGOMSL model showing both qualitative and quantitative analysis. (From John, B. E., and D. E. Kieras. 1996b. The GOMS family of user interface analysis techniques: comparison and contrast. *ACM Transactions on Computer–Human Interaction* 3: 320–351. With permission.)

times based on assumptions regarding the time to execute each operator. GOMS and NGOMSL task analyses are top-down procedures, beginning with identification of overall task goals followed by specification of subgoals, methods, and operators.

The procedures for conducting NGOMSL analysis are summarized in Table 23.5. Normally, NGOMSL is conducted after obtaining information about the user's task. As with HTA, breadth is initially emphasized over depth: all top-level goals must be identified before they are detailed. Top-level goals are analyzed into methods for accomplishing each goal. At this stage, the steps making up a method

---

TABLE 23.5

Summary of Procedures for NGOMSL Analysis

I. Choose the top-level user's goals.

II. Do the following recursive procedure.

   A. Draft a method to accomplish each goal.

   B. After completing the draft, check and rewrite as needed for consistency and conformance to guidelines.

   C. If needed, go to a lower level of analysis.

     1. Change high-level operators to accomplish-goal operators.

     2. Provide methods for the corresponding goals.

III. Document and check the analysis.

IV. Check sensitivity to judgment calls and assumptions.

Source: Kieras, D. 1997. Task analysis and the design of functionality. In *Handbook of Computer Science and Engineering*, ed. A. C. Tucker, 1401–1423. Boca Raton, FL: CRC Press. With permission.

---

are higher-level operators. If further analysis is required, the high-level operators can be changed to subgoals and methods for achieving the subgoals can be determined. At the lowest level of analysis, keystroke-level operators are used. However, analyzing all high-level methods into this level of detail is not necessary. Additional information on each step in the processes summarized in Table 23.5 is provided in the following paragraphs.

23.5.3.1 Choose the User's Top-Level Goals

The top-level goals are high-level descriptions of the user's overall task that will be further analyzed in subsequent steps. With the assumed unit task, stack structure of GOMS models, the top-level goals will be met once all subgoals have been achieved. For example, if the overall goal in text editing is "edit a manuscript," the method for accomplishing this goal consists of two unit tasks, "find an edit" and "accomplish the edit," as shown in Table 23.6. Once all unit tasks are completed, the edit manuscript goal is met. If there are alternate methods for achieving a goal, selection rules are identified. These rules determine the conditions under which a specific method is applied. For example, there would be several methods for achieving the unit task "accomplish an edit" based on the type of editing (e.g., delete text, move text.) Kieras suggests that the structure of the edit manuscript task shown in Table 23.6 can be used as a template for beginning most NGOMSL analyses.

23.5.3.2 Draft a Method to Accomplish Each Goal

At this level, methods are simply high-level statements of what the user has to do. Methods described at this stage are answers to the questions "how would the user do this," omitting specific details in the procedure. For example, in the

---

TABLE 23.6

Top-Level User Goal and Control Structure for Accomplish Goals in the Edit Manuscript Task

Method for goal: edit the document.

   Step 1. Get next unit task information from marked-up document.

   Step 2. Decide: if no more unit tasks, then return with goal accomplished.

   Step 3. Accomplish goal: move to the unit task location.

   Step 4. Accomplish goal: perform the unit task.

   Step 5. Go to 1.

Selection rule set for goal: perform the unit task.

   If the task is deletion then accomplish goal: perform the deletion procedure.

   If the task is copying, then accomplish goal: perform copy procedure.

   . . . etc. . . .

   Return with goal accomplished.

Source: Kieras, D. 1997. Task analysis and the design of functionality. In *Handbook of Computer Science and Engineering*, ed. A. C. Tucker, 1401–1423. Boca Raton, FL: CRC Press. With permission.

leftmost column of Table 23.7, the steps for the goal "copy text" are listed. The first step is "select the text," without considering the details of selecting text. Low-level operators should be avoided at this stage because they may be hiding the underlying structure of the user's goals. Similarly, methods consisting of more than five steps may indicate that the detail level is hiding the underlying structure of the user's goals. If more than one method for accomplishing a goal is identified, selection rules for choosing a method are determined. To ensure that all high-level goals are described, new high-level operators may be defined and simplifying assumptions can be made explicit. These should be made explicit and recorded as part of a task description attached to the NGOMSL model.

23.5.3.3 Check for Consistency

Because of the breadth-first strategy used here, a complete high-level description of the users goals should be developed before further decomposition. At this point the overall model should be checked to ensure that assumptions are consistent, high-level operators correspond to a natural goal of the user, the level of detail of each method is consistent, and that terminology use is consistent across methods and goals. Inconsistencies should be dealt with by either redefining operators or modifying assumptions. By starting breadth first, the analyst can then decide, based on the goals of the GOMS analysis, which subgoals and methods need further refinement.

23.5.3.4 Detail Methods to the Next Level of Analysis if Necessary

For new interfaces, the high-level description of methods may be sufficient, especially if the details of the design have not been worked out. High-level procedures of existing products can be further refined by restating a high-level procedure as a goal, and developing a method for accomplishing the goal. For example, the leftmost column of Table 23.7 shows the method for achieving the user goal "copying text." This method can be further detailed by changing step one, "select text," into a goal, as in the middle column of Table 23.7. A method for accomplishing the select text goal is subsequently developed. Selection rules must be created because more than one select text method is available, "select word" or "select arbitrary text." Although each of these is stated as a goal, and methods for achieving them would be subsequently developed, one of which is shown in column 3. Note that the method for "select word" is at the keystroke level; no further detailing is necessary. The complete NGOMSL model is therefore a description of high-level user goals analyzed to some level of detail, based on the purposes of the task analysis.

23.5.3.5 Document the Analysis

A complete NGOMSL analysis also includes documentation of analyst-defined operators and judgment calls that were used to complete the analysis. Judgment calls are decisions and assumptions made by the analyst showing his or her thinking about the user's knowledge structure (e.g., how the task

---

TABLE 23.7

Example of High-Level Method for Copying Text and the Process of Analyzing It to Lower Keystroke Levels

High-Level	Pass 2 Level	Keystroke Level
Method for goal: copy text	Method for goal: copy text	
Step 1: *Select the text.*	Step 1: *Accomplish goal: Select the text.*	
Step 2: Issue Copy Command.	Step 2: Issue Copy Command.	
Step 3: Return with goal accomplished	Step 3: Return with goal accomplished	
	Method for goal: select text	
	Selection rule set for goal: select text.	
	If text is word then accomplish goal: *select word.*	
	If text is arbitrary then accomplish goal: select arbitrary text.	
	Return with goal accomplished.	
		Method for goal: select word
		Step 1: Locate middle of word.
		Step 2: Move cursor to middle of word.
		Step 3: Double-click mouse button.
		Step 4: Verify that correct text is selected.
		Step 5: Return with goal accomplished.

Source: Kieras, D. 1997. Task analysis and the design of functionality. In *Handbook of Computer Science and Engineering*, ed. A. C. Tucker, 1401–1423. Boca Raton, FL: CRC Press. With permission.

is decomposed into subgoals, how the users view the tasks). Some of these assumptions may be obtained at the outset by talking to users and designers or examining documentation, but sometimes the analyst is forced to make assumptions on limited data. By making this information explicit, it can be examined afterward by considering the impact on the recommendations and conclusions if the assumptions were changed. For example, if the analyst assumed that shortcut keys were not used by most users, would the recommendations change if the assumption were invalid?

An NGOMSL model of task performance can provide quantitative and qualitative assessments of an existing or proposed interface design. NGOMSL can be used to derive quantitative assessments of execution times and learning times. An NGOMSL model can be executed (either by hand or as a computer program) to determine how well the model predicts performance. NGOMSL is the only GOMS method that provides quantitative assessments of the learning time required for a system in order to compare design alternatives on ease of learning. From a qualitative standpoint, an NGOMSL model can be used to assess the procedural demands made on the user by providing a rigorous description of what the system requires the user to do. A sample NGOMSL model, shown in Figure 23.7, illustrates how both quantitative and qualitative assessments are made from the model. On the right-hand column, execution times can be used to estimate the total performance time. The analyst's comments alongside various operators show how qualitative assessments of the task procedures are developed. Qualitative assessments of interface designs based on NGOMSL or GOMS models provide information regarding the naturalness of the design, whether goals and subgoals make sense to the user, completeness of

the design, whether all goals have methods, cleanliness of the design, whether clear selection rules are available for choosing between alternative methods of achieving a goal, consistency whether similar goals are accomplished by similar methods, and efficiency, whether most important and frequent goals are accomplished by relatively short and quick methods (Kieras 2006).

23.5.4 NGOMSL Software Tools

There have been many software tools developed for GOMS analyses over the past 15 years. Most of these are aimed at developing quantitative models for predicting task performance times. Baumeister, John, and Byrne (2000) reviewed three of these tools, GOMS, CAT-HCI, and GLEAN3. They found considerable variation in the task performance times predicted in each model. For example, GLEAN3 is essentially a programming language for representing a GOMS model. Baumeister, John, and Byrne (2000, 5) concluded that none of these could be considered "the perfect GOMS tool." More important for task analysis work, these did not facilitate GOMS development or data entry, unlike the software tools for HTA. The authors suggest that to fit well with the design process, a GOMS tool should be integrated with a prototyping tool and other usability aids. It is noteworthy that Baumeister et al.'s review of GOMS tools, HTA was conducted first to identify the task procedures for the modeling tools. Possibly HTA development tools could serve as the starting point for NGOMSL models, and given the similarity in the (hierarchical) task structures, a GOMS module could be a feature of future versions of these tools. Another promising development, Williams (2005) has extended CAT-HCI

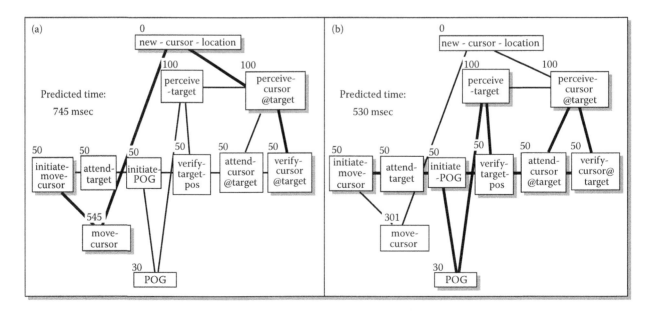

FIGURE 23.8 CPM model of moving a cursor over an icon with a mouse. (From Gray, W. D., and D. A. Boehm-Davis. 2003. Milliseconds matter: An introduction to microstrategies and their use in describing and predicting interactive behavior. *Journal of Experimental Psychology: Applied* 6: 322–335. With permission.)

to provide a query-based method for constructing GOMS models. CAT-HCI has a set of menus to describe classes of operations or activities based on interactions with human–computer interfaces. These menus guide the analyst to a level of detail such that operators and execution times can be automatically inserted into the model. Heuristics are applied to the lowest level methods of the analysis. For example, CAT-HCI will automatically insert cognitive operations for placing a subgoal into the GOMS stack structure. When 18 content experts with no NGOMSL or GOMS experience were trained on the software tool, and created cognitive models of their task, the models were 77% accurate on average (based on an expert-developed model of the same tasks) in specifying operators, and met or exceeded 80% agreement in task completion times. Therefore, CAT-HCI could be an important tool for improving the efficiency of GOMS modeling and task analysis.

23.5.5 CPM-GOMS PROCEDURES

KLM, CMN-GOMS, and NGOMSL assume a strict sequential execution of operators: the execution of any operator cannot begin until the previous operator is completed. CPM-GOMS was designed to analyze tasks having operators that can be operating in parallel. This method requires that user tasks be analyzed into cognitive, perceptual, and motor operators similar to keystroke-level operators. CPM also stands for critical path method, because task completion times are determined by the most time-consuming sequence of operations.

CPM-GOMS usually begins with a higher-level GOMS model such as NGOMSL. Operators are further decomposed as template-level goals that can be decomposed into templates. Templates, also known as microstrategies (e.g., Gray and Boehm-Davis 2003) are cognitive, perceptual, and motor acts for achieving a template-level goal. Usually, these are taken from models of cognition such as Card et al.'s (1983) Model Human Processor. CPM actions can be carried out concurrently if they use different resources (e.g., not requiring simultaneous eye movements to different locations) and if they are not constrained by the logic of the task (e.g., an icon cannot be clicked until after the cursor is positioned over it). In a CPM-GOMS analysis, each CPM action is placed on a scheduling chart with an estimate of execution time. Operations that use different resources are listed on separate rows and the dependencies between operators are indicated with connecting lines. The sequence of dependent operators having the longest total execution time (i.e., the critical path) is taken as the task completion time of the task template.

Although CPM-GOMS has not specifically been applied to the design of Web applications, the detail method of analysis makes models developed from stand-alone applications applicable to many Web-based tasks. For example, Figure 23.8 shows a portion of the CPM model of moving a cursor over a button with a mouse developed by Gray and Boehm-Davis (2003). This template action requires eleven CPM

operators, of which six are cognitive (e.g., attend target, initiate move cursor), two are perceptual (perceive target and perceive cursor at target) and two are motor (move cursor, POG, or "move eyes"). Each operator type is presented on a separate row and execution times are taken from the Model Human Processor of Card et al. (1983). Different move cursor times are used in the models of Figure 23.8, representing mouse movements toward distant (Figure 23.8a) and nearby buttons (Figure 23.8b). Note that differences in mouse movement time changed the critical path of the microstrategy: In Figure 23.8a, the total time is primarily determined by the mouse-movement time, because the cognitive and perceptual operators can be executed simultaneously with the movement. On the other hand, in Figure 23.8b, the shorter movement time shifts the critical path to the cognitive operators, especially the time to shift attention to the target.

CPM-GOMS has been criticized as too fine-grained to be useful in the design of Web applications and user interfaces overall. However, low-level processes can affect higher task methods (Gray and Boehm-Davis 2003). For example, decision-making strategies can be affected by the CPM operators required to gather information (Lohse and Johnson 1996); problem-solving strategies can be affected by the number of keystrokes required to make a move in a puzzle game. CPM-GOMS has been validated in a comparison of a newly designed telephone operator workstation with the existing product (Gray, John, and Atwood 1993), designed from usability guidelines. The proposed interface incorporated icons based on task metaphor, grouping interface elements based on user tasks, and a keyboard designed to minimize travel distance for common key sequences. The CPM-GOMS model predicted that the time to complete a task on the proposed interface (designed from usability guidelines) would be 3% slower than the time on the existing workstation. Subsequently, data collected on both designs showed a 4% increase in task completion times with the new interface. Moreover, the CPM-GOMS analysis showed that the new interface design either eliminated operations that were not on the critical path or added new operators to the critical path. In effect, although the new interface appeared more usable, the CPM-GOMS analysis showed that the changes did not affect the users' critical actions. Grey et al. (1993) suggested that the value of CPM-GOMS, besides predicting performance, lies in understanding why a counterintuitive result was obtained.

23.5.6 CPM-GOMS SOFTWARE TOOLS

CPM-GOMS models are often criticized because model development is tedious, time-consuming, and prone to errors (Vera et al. 2005). The most critical analysis is interleaving operations across different resources. It requires extensive knowledge of cognitive processing and the task operations. The interleaving requires attention to detail and involves many iterations when strategies fail and have to be undone. Moreover, the process is difficult to explain, making it dependent on the expertise of the analyst. As with other

```
(procedure
(index (slow-move-click ?target))
(step c1 (initiate-move-cursor ?target))
(step m1 (move-cursor ?target) (waitfor ?c1))
(step c2 (attend-target ?target))
(step c3 (initiate-eye-movement ?target) (waitfor ?c2))
(step m2 (eye-movement ?target) (waitfor ?c3))
(step p1 (perceive-target-complex ?target) (waitfor ?m2))
(step c4 (verify-target-position ?target) (waitfor ?c3 ?p1))
(step w1 (WORLD new-cursor-location ?target) (waitfor ?m1))
(step c5 (attend-cursor-at-target ?target) (waitfor ?c4))
(step p2 (perceive-cursor-at-target ?target) (waitfor ?p1 ?c5 ?w1))
(step c6 (verify-cursor-at-target ?target) (waitfor ?c5 ?p2))
(step c7 (initiate-click ?target) (waitfor ?c6 ?m1))
(step m3 (mouse-down ?target) (waitfor ?m1 ?c7))
(step m4 (mouse-up ?target) (waitfor ?m3))
(step t1 (terminate) (waitfor ?m4)))
```

FIGURE 23.9 PDL code for the CPM model. (From Vera et al., 2005. Figure 8A. With permission.)

assumption-free methods of task analysis, task analysis tools are needed to make such analyses practical and useful. Vera et al. (2005) have adapted software developed primarily at NASA known as APEX (Freed et al. 2003) for automating CPM-GOMS modeling. APEX is a toolkit for constructing software that behaves intelligently and responsively in demanding task environments. Among its components, a Resource Architecture (RA) can be used to represent the cognitive, perceptual, and motor operations necessary to carry out a task and an Action Selection Architecture (ASA) can be used for resource allocation. Together they provide the necessary tools for implementing the constraints and accomplish interleaving of operations.

A formal Procedure Description Language (PDL) is used to specify the sequence of tasks in CPM-GOMS. For example, Figure 23.9 shows the PDL code for moving the mouse to a distant target that is diagrammed in Figure 23.8a. When rules for establishing the sequence of operators, are entered, APEX determines the critical path and predicts task completion times. In a preliminary test of APEX-developed CPM-GOMS models on a simulated ATM interface, CPM-GOMS predictions were highly correlated with actual performance times. Moreover, with the use of standardized sequencing rules, learning to develop CPM-GOMS models is easier.

23.5.7 Conclusions

Kieras (2008) offers the following guidelines for choosing the appropriate GOMS model. In early design stages, where qualitative feedback is needed on alternative design decisions, a high-level, informal GOMS model should be sufficient. If the design specifies performance criteria for specific discrete tasks, KLM is recommended. If the design criteria include learnability, consistency execution speed, and a detailed interface is completed, a complete NGOMSL or CPM-GOMS model will be required. If design choices

depend on understanding low-level or subtle human interactions, CPM-GOMS may be the only choice.

GOMS has been criticized for being time-consuming, costly, and difficult to learn to use. However, estimates of GOMS development times usually include time spent validating the model, and validation is normally not done with other task analysis methods, and Kieras (1997) reports that students with little experience in task analysis can learn NGOMSL relatively quickly. Nevertheless, comparisons between task analysis methods solely based on learning time assume equivalence of training outcomes. The length of any training program should be weighed against the skills of the analyst at the completion of training. Recall that the quality of HTA information depends on the expertise of the analyst, meaning that learning continues long after formal training is completed. Automated tools such as APEX should also speed up model development. As more standard templates are being developed for interface interaction, newer interaction methods (e.g., hand-held devices, tablets, speech) and different classes of users (e.g., visually impaired), CPM-GOMS models should become more useful in predicting performance times without prototypes. By constraining CPM-GOMS and APEX to elementary task operations, and by creating templates of operators for standard tasks, the models can provide bracketed ranges of performance estimates of task times. Finally, the use of APEX highlights the idea that CPM-GOMS is not a model of cognitive processes but a model of performance for mostly cognitive tasks.

23.6 CONCLUSIONS AND RECOMMENDATIONS

Task analysis methods have been developed and applied in the workplace for roughly 100 years, yet there is little consensus regarding either the optimal methods or strategies for deciding among the methods. The diversity of methods is probably due to the diverging purposes for which task analyses are carried out and dissatisfaction over the practicality, reliability, validity, and usability of existing methods. The three methods described here, HTA, CIT and GOMS, are probably the most completely developed and widely applied methods of task analysis. All these are suitable for analyzing user tasks on Web applications, yet these methods are rarely used in their design. This probably stems from the shortened development times for Web-based software and the difficulty of fitting a task analysis into the iterative design approach of the software industry. However, recent tools that have been developed for each method promise to improve the efficiency of these methods and their contributions to Web application design: First, they should speed up the process of task analysis. The HTA Tools (Task Architect and HTA Tool) make data entry and editing easier and make it easy to view the analysis in different formats. Online CIT tools developed by Castillo and his colleagues can automate the data collection process itself. GOMS tools such as CogTool integrate a prototyping tool with task analysis so that a designer can explore design concepts and evaluate the impact on performance concurrently. Second, these software tools reduce the expertise required of task analysts by constraining the task inputs, utilizing dialogues and menus for leading the analyst to the correct level of detail of representation, and including task analysis checkers to ensure that the basic assumptions of the method are being met. Third, these tools will standardize the storing of task analysis data so that it can be reused during the design of major and minor software upgrades. The modularity of these products means that when upgrades are being developed, the analyst can identify the tasks that have changed with updated versions and focus his/her task analysis on this subsets of tasks. Tasks that have not been changed can be easily inserted to the new analysis so that the analyst can see the impact of changes to one function on the remaining design. If quantitative estimates of performance are required, these can be easily translated into new GOMS models and executed. Although most GOMS tools are focused on estimating performance at the perceptual-motor level, these activities take relatively fixed amounts of time, are heavily determined by the design of the interface, and frequently dominate the user's task (Kieras 2008). And, published estimates of task performance times for newer input and display types have been increasing (see Williams 2005) for a summary of perceptual, motor, and cognitive operators and published estimates of performance times, meaning that new analyses can take advantage of these estimates. Moreover, the focus on perceptual-motor operators will most likely drive GOMS analyses of design alternatives for improving accessibility of Web sites (e.g., Tonn-Eichstadt 2006).

Regardless of the tools used, the efficiency of task analysis in Web design will also be improved if it is applied early in the design of new systems. Task analyses done early will have the most impact on subsequent design stages and iterations. Essentially, early task analysis ensures that requirements and functions are identified that are based on the user's tasks. The information obtained from a task analysis can simplify subsequent decision making on design alternative, from a task-centered standpoint. Early task analysis also streamlines subsequent implementation and testing stages by ensuring that benchmark tasks and performance requirements reflect the most critical user tasks. Begin all task analyses with a flexible, breadth-first method such as HTA. This ensures that all system goals, user classes, and user goals are considered before any effort is expended describing tasks in detail. HTA's flexibility makes it easier to transfer hierarchical task data to other methods. Document carefully all decisions, assumptions, and task analysis outputs and to maintain and update the task analysis output for future design changes.

Task analysis is time-consuming, but often each analysis conducted independently without considering the results of previous analyses. In large organizations, product design is not a one-time process but continues after product release until either a new product is being designed, or upgrades to the existing product are undertaken. If previous task analyses are carefully and completely documented, these should be evaluated for the extent to which some or all of the analysis is still relevant. This would ensure that time and resources are spent only on tasks affected by the design changes that have not been previously analyzed. These records might contain previous models of task performance that need updating but not complete redevelopment. Note also that a carefully documented task analysis from an earlier design could be assessed, at least qualitatively, for validity, reliability, and usability.

REFERENCES

Ainsworth, L., and E. Marshall. 1998. Issues of quality and practicality in task analysis: Preliminary results from two surveys. *Ergonomics* 41: 1607–1617.

Assistive Technology Research Institute. 2003. Assistive technology training and technical assistance needs. http://natri.uky.edu/participation/incidents/ciovervw.html.

Annett, J., and D. Cunningham. 2000. Analyzing command team skills. In *Cognitive Task Analysis*, eds. J. M. Schraagen, S. F. Chipman, and V. L. Shalin, 401–416. Mahwah, NJ: Lawrence Erlbaum.

Annett, J., and K. D. Duncan. 1967. Task analysis and training design. *Occupational Psychology* 41: 211–221.

Baumeister, L. K., B. E. John, and M. D. Byrne. 2000. A comparison of tools for building GOMS models. *CHI* 2000: 1–6.

Booth, P. 1989. *An Introduction to Human–Computer Interaction*, Mahwah, NJ: Lawrence Erlbaum.

Brinck, T., D. Gergle, and S. D. Wood. 2002. *Usability for the Web: Designing Web Sites That Work*. San Francisco, CA: Morgan Kaufmann.

Canfield, D., C. Irby, R. Kimball, and B. Verplank. 1982. Designing the star user interface. *Byte* 243–280.

Card, S. K., T. P. Moran, and A. Newell. 1983. *The Psychology of Human–Computer Interaction*. Mahway, NJ: Lawrence Erlbaum.

Castillo, J. C. 1997. The user-reported critical incident method for remote usability evaluation. Masters Thesis. Virginia Polytechnic Institute and State University, Blacksburg.

Castillo, J., and H. R. Hartson. 2000. Critical incident data and their importance in remote usability evaluation. *Proceedings of the 40th Annual Meeting of the Human Factors and Ergonomics Society Conference* 44: 590–595.

Castillo, J. C., H. R. Hartson, and D. Hix. 1998. Remote usability evaluation: Can users report their own critical incidents? Summary of CHI'98. *Human Factors in Computing Systems* 253–254.

Chipman, S. F., J. M. Schraagen, and V. L. Shalin. 2000. Introduction to cognitive task analysis. In *Cognitive Task Analysis*, eds. J. M. Schraagen, S. F. Chipman, and V. L. Shalin, 3–23. Mahwah, NJ: Lawrence Erlbaum.

del Galdo, E. M., R. C. Williges, B. H. Williges, and D. R. Wixon. 1986. An evaluation of critical incidents for software documentation design. *Proceedings of the 30th Annual Meeting of the Human Factors and Ergonomics Society Conference* 19–23.

del Galdo, E. M., R. C. Williges, B. H. Williges, and D. R. Wixon. 1987. A critical incident tool for software documentation. In *Ergonomics and Human Factors: Recent Research*, eds. L. S. Mark, J. S. Warm, and R. L. Hutson, 253–258. New York: Springer-Verlag.

Diaper, D., and N. Stanton. 2004. *The Handbook of Task Analysis for Human–Computer Interaction*. Mahwah, NJ: Lawrence Erlbaum.

Flanagan, J. C. 1954. The critical incident technique. *Psychological Bulletin* 51: 327–358.

Fitts, P. M., and R. E. Jones. 1947a. Analysis of factors contributing to 460 "pilot-error" experiences in operating aircraft controls. Memorandum Report TSEAA-694-12, Aero Medical laboratory, Air Matereil Command, Wright-Patterson Air Force Base, Dayton, Ohio. Reprinted in H. W. Sinaiko, ed. 1960. *Selected Papers on Human Factors in the Design and Use of Control Systems*, 332–358. New York: Dover.

Fitts, P. M., and R. E. Jones. 1947b. Psychological aspects of instrument display. I: Analysis of 270 "pilot-error" experiences in reading and interpreting aircraft instruments. Memorandum Report TSEAA-694-12A, Aero Medical Laboratory, Air Matereil Command, Wright-Patterson Air Force Base, Dayton, Ohio. Reprinted in H. W. Sinaiko, ed. 1960. *Selected Papers on Human Factors in the Design and Use of Control Systems*, 359–396. New York: Dover.

Freed, M., M. Matessa, R. Remington, and A. Vera. 2003. How Apex automates CPM-GOMS. In *Proceedings of the Fifth International Conference on Cognitive Modeling*, Bamberg, Germany: Universitats-Verlag.

Gilbreth, F. B. 1911. *Motion Study*. Princeton, NJ: Van Nostrand.

Goransson, B., M. Lind, E. Pettersson, B. Sandblad, and P. Schwalbe. 1987. The interface is often not the problem. *Proceedings of Graphics Interface Conference* 133–136.

Gray, W. D., and D. A. Boehm-Davis. 2003. Milliseconds matter: An introduction to microstrategies and their use in describing and predicting interactive behavior. *Journal of Experimental Psychology: Applied* 6: 322–335.

Gray, W. D., B. E. John, and M. E. Atwood. 1993. Project Ernestine: A validation of GOMS for prediction and explanation of real-world task performance. *Human–Computer Interaction* 8: 287–319.

Hall, E. P., S. P. Gott, and R. A. Pokorny. 1995. Procedural guide to task analysis: The PARI methodology. Report Al/HR-TR-1995-0108. Brooks Air Force Base, TX: U.S. Air Force Materiel Command.

Henry, S. L. 2006. Understanding web accessibility. In *Web Accessibility: Web Standards and Regulatory Compliance*, eds. J. Thatcher et al., 1–49. Berkeley, CA: Apress.

Hobbes, N. 1948. The development of a code of ethical standards for psychology. *American Psychologist* 3: 80–84.

John, B. E., and D. E. Kieras. 1996a. Using GOMS for user interface design and evaluation: Which technique? *ACM Transactions on Computer–Human Interaction* 3: 287–319.

John, B. E., and D. E. Kieras. 1996b. The GOMS family of user interface analysis techniques: Comparison and contrast. *ACM Transactions on Computer–Human Interaction* 3: 320–351.

John, B. E. 2009. *CogTool User Guide, Version 1.1*. http://cogtool.org (accessed March 28, 2010).

Johnson, L. J. 1999. Critical Incidents in the Gaming Industry: Perceptions of Guests and Employees. Unpublished doctoral dissertation, University of Nevada, Las Vegas.

Johnson, L. 2002. Using the critical incident technique to assess gaming customer satisfaction. *UNLV Gaming Research & Review Journal* 6(2): 1–12.

Keikel, P. A., and N. J. Cooke. this volume. Human factor aspects of team cognition. In *Handbook of Human Factors in Web Design*, 2nd ed., eds. K.-P. L Vu and R. W. Proctor, 107–124. Boca Raton, FL: CRC Press.

Keikel, P. A., N. J. Cooke, P. W. Foltz, and S. M. Shope. 2001. Automating measurement of team cognition through analysis of communication data. In *Usability Evaluation and Interface Design*, vol. 1, eds. M. J. Smith et al., 1382–1386. Mahwah, NJ: Lawrence Erlbaum.

Kieras, D. 1988. Towards a practical GOMS model methodology for user interface design. In *Handbook of Human–Computer Interaction*, ed. M. Helander, 133–158. Amsterdam: North Holland.

Kieras, D. 1997. Task analysis and the design of functionality. In *Handbook of Computer Science and Engineering*, ed. A. C. Tucker, 1401–1423. Boca Raton, FL: CRC Press.

Kieras, D. 2001. Using the keystroke-level model to estimate execution times. http://www.eecs.umich.edu/~kieras/goms.html (accessed December 30, 2003).

Kieras, D. 2004. GOMS models for task analysis. In *The Handbook of Task Analysis for Human Computer Interaction*, eds. D. Diaper and N. Stanton, 83–116. Mahwah, NJ: Lawrence Erlbaum Associates.

Kieras, D. 2006. A guide to GOMS model usability evaluation using GOMSL and GLEAN4. ftp.eecs.umich.edu/people/keiras (accessed March 28, 2010).

Kieras, D. 2008. Model-based evaluation. In *The Human–Computer Interaction Handbook*, 2nd ed., eds. J. Jacko and A. Sears. Mahwah, NJ: Lawrence Erlbaum.

Kieras, D. E., and D. E. Meyer. 2000. The role of cognitive task analysis in the application of predictive models of human performance. In Cognitive Task Analysis, eds. J. M. Schraagen, S. F. Chipman, and V. L. Shalin, 237–260. Mahwah, NJ: Lawrence Erlbaum.

Kirwan, B., and L. K. Ainsworth. 1992. *The Task Analysis Guide*. London: Taylor & Francis.

Koch, A., A. Strobel, G. Kici, and K. Westhoff. 2009. Quality of the critical incident technique in practice: Interrater reliability and users' acceptance under real conditions. *Psychology Science Quarterly* 51: 3–15.

Lohse, G. L., and E. J. Johnson. 1996. A comparison of two process-tracing methods for choice tasks. *Organizational Behavior and Human Decision Processes* 68: 28–43.

Luo, L., and B. E. John. 2005. Predicting task execution time on handheld devices using the keystroke-level model. CHI *2005: Late Breaking Results: Posters*, 1601–1608.

Mayhew, D. J. 1999. The *Usability Engineering Lifecycle: A Practitioner's Handbook for User Interface Design*. San Francisco, CA: Morgan Kaufmann.

Meister, D. 1985. *Behavioral Analysis and Measurement Methods*. New York: John Wiley.

Miller, R. B. 1953. A method for man–machine analysis. Report 53–137. Wright Air Development Center, Wright-Patterson AFB, OH.

Miller, R. B., and J. C. Flanagan. 1950. The performance record: An objective merit-rating procedure for industry. *American Psychologist* 5: 331–332.

Moltgen, J., and W. Kuhn. 2000. Task analysis in transportation planning for user interface metaphor design. Paper presented at the 3rd AGILE Conference on Geographic Information Science, Helsinki, Finland, May 25–27, 2000.

Morrison, J. B., P. Pirolli, and S. Card. 2000. A taxonomic analysis of what world wide web activities significantly impact people's decisions and actions. Xerox Parc Tech Report UIR-R-2000-17. Xerox Parc.

Moss, F. A. 1929. *Applications of Psychology*. Cambridge, MA: Riverside Press.

Munsterberg, H. 1913. *Psychology and Industrial Efficiency*. Boston, MA: Houghton Mifflin.

Myung. R. 2004. Keystroke-level analysis of Korean text entry methods on mobile phones. *International Journal of Human–Computer Studies* 60: 545–563.

Nielsen, J. 2000. *Designing Web Usability: The Practice of Simplicity*. Indianapolis, IN: New Riders.

Ormerod, T. C. 2000. Using task analysis as a primary design method: The SGT approach. In *Cognitive Task Analysis*, eds. J. M. Schraagen, S. F. Chipman, and V. L. Shalin, 181–200. Mahwah, NJ: Lawrence Erlbaum.

Ormerod, T. C., and A. Shepherd. 2004. Using task analysis for information requirements specification. In *The Handbook of Task Analysis for Human–Computer Interaction*, eds. D. Diaper and N. Stanton, 347–366. Mahwah, NJ: Lawrence Erlbaum.

Patrick, J., A. Gregov, and P. Halliday. 2000. Analyzing and training task analysis. *Instructional Science* 28: 51–79.

Shepherd, A. 2001. *Hierarchical Task Analysis*. London: Taylor & Francis.

Sheridan, T. B., and W. R. Ferrell. 1974. *Man–Machine Systems: Information, Control and Decision Models of Human Performance*. Cambridge, MA: MIT Press.

Smith, D. C., C. Irby, R. Kimball, and B. Verplank. 1982. Designing the star user interface. *BYTE Magazine* 243–280.

Stanton, N. 2006. Hierarchical task analysis: developments, applications, and extensions. *Applied Ergonomics* 37: 55–79.

Stanton, N., and M. Young. 1998. Is utility in the eye of the beholder? A study of ergonomics methods. *Applied Ergonomics* 29: 41–54.

Stone, N. J., and P. Derby. 2009. *HFES 2009 Member Educational Needs Survey Results*. Santa Monica, CA: Human Factors and Ergonomics Society.

Sweeney, J. C., and W. Lapp. 2000. High quality and low quality internet service encounters. *ANZMAC 2000 Visionary Marketing for the 21st Century: Facing the Challenge* 1229–1233.

Sweeney, J. C., W. Lapp, and F. Group. 2000. High quality and low quality Internet service encounters. In *ANZMAC 2000 Visionary Marketing for the 21st Century: Facing the Challenge*, 1229–1233. Queensland, Australia: ANZMAC.

Taylor, F. W. 1911. *Principles of Scientific Management*. New York: Harper and Row.

Teo, L., and B. E. John. 2006. Comparisons of keystroke-level model predictions to observed data. *CHI 2006. Work-in-Progress* 1421–1426.

Tonn-Eichstadt, H. 2006. Measuring website usability for visually impaired people—a modified GOMS analysis. *ASSETS'06* 55–62.

Tsang, P. S., and G. F. Wilson. 1997. Mental workload. In Human Factors Handbook, 2nd ed., ed. G. Salvendy, 417–449. New York: John Wiley.

Urquhart, C., A. Light, R. Thomas, A. Barker, A. Yeoman, J. Cooper, C. Armstrong, R. Fenton, R. Lonsdale, and S. Spink. 2003. Critical incident technique and explicitation interviewing in studies of information behavior. *Library and Information Science Research* 25: 63–88.

van Dyk, T., and K. Renaud. 2004. Task analysis for e-commerce and the web. In *The Handbook of Task Analysis for Human–Computer Interaction*, eds. D. Diaper and N. Stanton, 247–262. Mahway, NJ: Lawrence Erlbaum.

Vera, H., B. E. John, R. Remington, M. Matessa, and M. A. Freed. 2005. Automating human-performance modeling at the millisecond level. *Human–Computer Interaction* 20: 225–265.

Volk, F., F. Pappas, and H. Wang. this volume. User research: User-centered methods for designing web interfaces. In *Handbook of Human Factors in Web Design*, 2nd ed., eds. K.-P. L. Vu and R. W. Proctor, 417–438. Boca Raton, FL: CRC Press.

Walls, M. 2006. Task architect. *Ergonomics in Design* 27–29.

Williams, K. E. 2005. Computer-aided GOMS: A description and evaluation of a tool that integrates existing research for modeling human–computer interaction. *International Journal of Human–Computer Interaction* 18: 39–58.

Wilson, S. R., N. S. Starr-Schneidkraut, and M. D. Cooper. 1989. Use of the critical incident technique to evaluate the impact of MEDLINE. http://www.nlm.nih.gov/od/ope/cit.html (accessed April 16, 2003).

24 An Ecological Perspective to Meaning Processing: The Dynamics of Abductive Systems

J. M. Flach, K. B. Bennett, P. J. Stappers, and D. P. Saakes

CONTENTS

> One fundamental challenge for the design of the interactive systems of the future is to invent and design environments as open systems in which humans can express themselves and engage in personally meaningful activities.—Fischer (2000, 283)

As reflected in the opening quote, a general premise of this chapter is that a target in the design of any technology, such as the Web (or other complex databases), is to enhance the meaningfulness of the human experience. Thus, a fundamental question for this chapter is the nature of "meaning" and the implications that this has for our understanding of cognitive processes (Flach 2009; Flach, Dekker, and Stappers 2008). In addressing this question, we first introduce the concept of "ecology" in order to explore the source of meaning and the relation of meaning to human experience. A key implication of the ecological perspective is that perception and action are coupled to create an abductive system for adaptive reasoning (Flach 2009). An abductive system is essentially a trial-and-error learning system, where validity is evaluated pragmatically (i.e., hypotheses that lead to successful actions are maintained). Second, we introduce Cognitive Systems Engineering (CSE) as a promising approach to abductive systems with an eye toward design. Third, we focus on the Web as a particular work domain to consider the special dimensions of this domain that might guide generalizations from previous research on human–machine systems. Fourth, we present two examples. The first example is the BookHouse interface (e.g., Pejtersen 1992). The second example is the Multidimensional Scaling Interface developed by Stappers. Last, we summarize an ecological approach to Web design and point to potentially interesting directions for research and design.

24.1 WHAT IS AN ECOLOGY?

> The notion of affordance implies a new theory of meaning
> and a new way of bridging the gap between mind and matter.
> To say that an affordance is meaningful is not to say that it
> is "mental." To say that it is "physical" is not to imply that
> it is meaningless. The dualism of mental vs. physical ceases
> to be compulsory. One does not have to believe in a separate
> realm of mind to speak of meaning, and one does not have to
> embrace materialism to recognize the necessity of physical
> stimuli for perception.—Gibson (1972/1982, 409)

It is impossible to consider any cognitive system (whether
human information processor, artificial intelligent agent, or
a multiagent human–machine system like the Web) without
facing the issue of meaning. What is meaning? Where does
it come from? Is it a property of the physical world? Or is it a
product of cognitive processing? On the one hand, it could be
argued that the world presents physical consequences that are
independent of the judgments of any cognitive agent (e.g., the
consequences of a high-speed collision). Thus, the meaning
of an event might be specified independent of any processing or judgment on the part of a cognitive agent. The collision will have the same consequences independent of what
an observer might think or believe. On the other hand, two
observers might interpret an event differently, depending on
individual differences, such as differing intentions (e.g., the
predator may be trying to create a collision, whereas the prey
may be trying to avoid collisions). In this case, the meaning
of the collision might be very different, depending on the
point of view (prey or predator).

Putting this in the context of information system design,
designers who associate meaning with the physical world are
likely to seek objective normative standards of meaning. For
example, are all the objects in the inventory correctly classified
in the Web database—where "correctly" is measured relative
to some objective normative criterion (e.g., Dewey decimal system or other convention for cataloging inventory)? The focus
of design efforts would be on the correspondence between
the representations in the Web and the physical objects (e.g.,
in our examples below, books or roller skates) that are being
represented. Typically, the degree of correspondence would
be measured against some model of the domain that specifies
both what belongs in the database and what differences make
a difference (i.e., the dimensions of the space). This model will
often reflect some degree of intentionality, but this will typically reflect the goals of professionals within a domain (e.g., a
librarian's need to catalog, store, and retrieve books) and may
not reflect the distinctions that matter to general users (e.g.,
library patrons looking for an interesting book).

Note that the whole idea of a normative system of classification implies a single approach based on some set of norms
(often defined by the experts) and often explicitly designed to
be "best" with respect to some standard. Thus, the diversity
among potential users is typically ignored. This reflects back
to the Scientific Management approach of Fredric W. Taylor
(1911, often referred to as Taylorism) that suggests a single
"best way" for design.

Designers who consider meaning to be the product of cognitive processing are likely to look to the human users for
subjective aspects of meaning. What is important (meaningful) to the human users? Designers who take this perspective
are likely to be interested in the mental models of their users
(including useful metaphors and semantic knowledge). They
will typically measure their designs in terms of their correspondence or match to the users' mental models. How do the
users think about the objects in the database? It is typically
assumed that the mental models will have some degree of
correspondence with the objective domain. However, designers who focus exclusively on what specific users think might
have difficulty differentiating between perspicacity and
superstition. Does the mental model reflect a well-tuned or a
naïve map to the actual work domain? Again, when the user
population is diverse, the designer must choose which of the
potentially numerous mental models to use as the guide for
design.

An ecological approach offers a middle path as illustrated
in Figure 24.1. The term "ecology" is used to reflect a relation
between human and environment, similar to the construct of
"Umwelt" used by von Uexküll (1957). Uexküll used Umwelt
to refer to the world in relation to a specific animal. For
example, the same object that is nourishment for one animal
may be toxic to another. It is important to note that this relation is not simply the sum of the two components (physical
world + animal), but it is the "fit" or "relation" between the
two. The ecology reflects emergent properties of the interaction of a human with an environment. As such, it has a
unique dimensionality that is different from those of either
the human or the environment. Figure 24.1 illustrates critical
dimensions that emerge from this relation between animal
and environment.

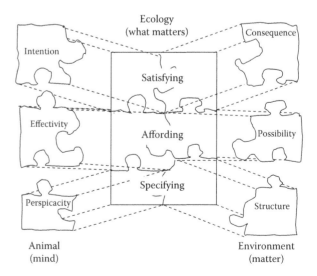

FIGURE 24.1 The ecology is a set of emergent properties resulting from the fit between an animal and an environment. Some of
these properties are represented by the verbs satisfying, affording,
and specifying.

The three dimensions of the ecology that are important for assessing the fit between human and environment are represented by the verbs satisfying, affording, and specifying. Satisfying represents the value constraints on the interaction. This dimension reflects the intentions or goals of the animal in relation to the consequences of the environment. In the context of Web design, satisfying reflects both the goals of the user searching the Web and the objective properties of the Web sites or the products represented by these sites. Do the hits from a search match the information needs of the person doing the search? Does the product selected through a Web site satisfy the needs of the user? From the perspective of design, this reflects a need for the designer to have insights both about what people want and about what can be offered.

Affording represents the action constraints on the interaction. It reflects the effectivities of the human in relation to the possibilities offered by the environment. This represents what the human can do. Specifying represents the information constraints on the interaction. It reflects the perspicacity (perceptual skill) of the human in relation to the structure (e.g., ambient optic array) in the environment. The confluence of satisfying, affording, and specifying creates the possibility for control and for competent action. That is, satisfying reflects the reference signal, specifying reflects feedback, and affording reflects the action to reduce discrepancies between the reference and feedback information.

In the context of database design, the possibilities represent the information and products available through the database and the "structure" would represent the measurable distinctions that might be used to organize or classify these possibilities. The effectivities and perspicacity of the user will largely be determined by the user interface. That is, the interface provides both the means for action and the feedback about the opportunities for and consequences of actions. The kinds of effectivities typically provided by the Web include pointing at, dragging, and clicking on graphical objects, selecting items from menus, and the Boolean logic used to initiate a search. The feedback includes graphical and text descriptions of data, products, and options. For the design to function as a control system the feedback must be comparable to a reference (reflecting the intentions of the users and consequences of action) and the comparison must result in a clear specification of the actions necessary to reduce any discrepancy (i.e., error) from the reference.

The choice of verbs to specify key dimensions of the ecology was intended to emphasize the fact that the ecology is not static. The ecology reflects a dynamic coupling of perception and action. Figure 24.2 was designed to emphasize this dynamic. The dynamic is depicted as two coupled loops (a figure 8). Movement from the top left to the lower right then looping back to the center represents the control problem where the center box (representation medium) functions as a comparator in which intentions and consequences are compared to produce an error signal that drives actions to reduce the error (satisfy the intention). This is essentially the cybernetic image of a cognitive system.

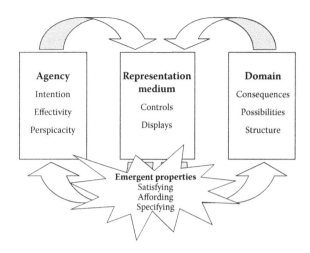

FIGURE 24.2 The dynamics of perception action coupling. The coupling of performance (control) with exploration (hypothesis testing) to form an adaptive control system.

The backward loop from the domain (e.g., consequences, possibilities) through the representation to an agent represents an abductive approach to the meaning processing problem. Abductive approaches focus on the generation of plausible hypotheses from experience rather than on conclusions deduced from existing accepted facts (Peirce 1878). The latter deals with evaluating necessary conclusions, and the former deals with generating possible premises. In an abductive approach, hypotheses are evaluated pragmatically—do they lead to satisfying actions (i.e., do they work successfully) rather than logically (as would be the case with inductive or deductive approaches). This loop reflects the observation process that allows the construction of expectations and hypotheses about the world. These beliefs can be tested, tuned, or otherwise improved by exploratory actions on the domain.

One of the key insights of an ecological approach is that the observation and control processes (i.e., perception–action) are dynamically linked in an abductive dynamic such that neither is logically prior to another. This is in contrast to classical information-processing models where the semiotic problem (perception) is often treated as logically prior to control (decisions and action). Many, though not all, approaches to cognition either ignore one aspect of this coupling (e.g., focus on behavior/activity without any theory of meaning, or focus on symbol processing [semiotics] without any theory of action) or parse the cognitive process in a way that the control and semiotic problems are treated as independent stages of processing. In contrast, an ecological approach is predicated on the assumption that the dynamic coupling is the heart of the problem and that understanding the emergent properties of this coupling should be the goal of cognitive science. This also reflects a vision of design where problem and solution, question and answer, evolve in mutual enhancement, not in succession (e.g., Schön 1983).

In fact, the case could be made that all of the dimensions specified in Figures 24.1 and 24.2 (those attributed to the human/agent, to the environment/domain, to the representation, and to the ecology) are emergent properties. This is consistent with William James' (1909) philosophy of radical empiricism, which states that meaning can only be measured relative to its impact on experience. This is in contrast to realist or idealist positions that posit extrinsic standards for meaning in either an objectively real world or some abstract, logical ideal. This radical empiricist view seems to be most consistent with our observations of Web use. For example, often a user will begin a Web search with only an ambiguously specified intention. However, this intention often is refined in the process of interacting with the Web to the point where the satisfactory end of the search may have little resemblance to the intention that initiated the interaction. Thus, the intention itself is an emergent property of the dynamic, not an extrinsic fact imposed on the dynamic. This is one of the key challenges of Web design—to help people to satisfy their intentions—even when these intentions are ill formed or impossible to articulate in advance. The users often will not be able to articulate a clear goal, even when they can reliably recognize a solution (when and if they get there). So, the key point here is that cognitive processes cannot be reduced to ordered sequences where intentions are refined into programs or plans that in turn direct actions. Cognitive systems are dynamic. They learn by doing, i.e., by trial and error (in Piagetian terms, by assimilation and accommodation; e.g., Ginsburg and Opper 1969). They must project their experiences into an uncertain future (assimilation), and they must revise their beliefs and expectations based on the experiences that result (accommodation). This active, iterative process for discovering meaning is also illustrated by Schrage (2000) with the concept of "serious play."

The dynamic coupling of perception and action illustrated in Figure 24.2 is our conceptual image of experience. The emergent properties of satisfying, affording, and specifying are dimensions of meaning. Thus meaning is neither an output (or product) of processing nor an input to processing. Rather, it is a property of the dynamic, similar to stability. When the representation provides rich structure that matches well with the consequences and opportunities in the domain and the effectivities and intentions of an agent, then the dynamic can be expected to stabilize in a fashion that is satisfying. This is what we mean by "skilled" interaction, and the sense is one of direct access to meaning—the experience is meaningful. When the representation does not match with these dimensions, then instability will be the likely result, or perhaps, stability in a fashion that is unsatisfying. The sense is one of confusion or meaninglessness.

24.2 COGNITIVE SYSTEMS ENGINEERING

The problem we face in modeling systems incorporating human actors is, however, that humans do not have stable input-output characteristics that can be studied in isolation. When the system is put to work, the human elements change their characteristics; they adapt to the functional characteristics of

the working system, and they modify system characteristics to serve their particular needs and preferences.—Rasmussen, Pejtersen, and Goodstein (1994, 6)

Cognitive Systems Engineering (CSE) is motivated by the simple question: What can designers do to ensure or at least increase the probability that the cognitive system will stabilize in a way that satisfies the functional goals motivating the design? This question is particularly difficult in today's environment where the work domain is changing at a rapid pace. These changes reflect the opportunities of evolving technologies such as the Internet and the competitive demands for innovation that result. When the pace of change is slow, stability can often be achieved by standardization around well-established procedures. But when the pace of change is fast, operators are often called on to complete the design by adapting to contingencies for which no well-established procedures exist. More recently, this goal has been reflected in terms such as "resilience engineering" (Hollnagel, Woods, and Leveson 2006), "joint cognitive systems" (Woods and Hollnagel 2006), "adaptive perspectives" (Kirlik 2006), and "macrocognition" (e.g., Flach 2008).

In a sense, the goal of CSE and resilience engineering is to design adaptive systems. In this context, the human factor takes on new and interesting dimensions. Early research on the human factor tended to focus on the human error problem. Designers tended to focus on ways to minimize human errors to ensure that the humans did not deviate from the well-established procedures. Today, however, focus is shifting to the adaptive and creative abilities that humans bring to the work place. The focus of these approaches tends to be on ways to leverage these abilities against the uncertainties of a complex and rapidly changing work environment. The focus is on providing the humans with the tools they need to invent new procedures as demanded by the changing work context. Figure 24.3 shows three fulcrum points that designers might use in leveraging the adaptive abilities of humans against the demands of complex work environments: work analysis, ecological interface design, and instruction design.

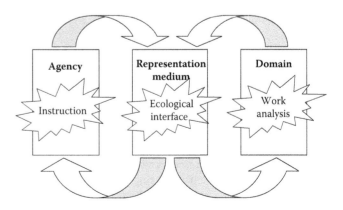

FIGURE 24.3 Illustrates promising opportunities for design interventions.

24.2.1 Work Analysis

Ironically, the first place to start in preparing to deal with change is to search for regularities (lawfulness, invariance, or constraints). The regularities become the landmarks that can guide exploration. Classically, primary sources of regularity were the standard procedures. Thus, classical work analysis focused on the activity of work, often in relation to one best way. However, in an environment where procedures must be adapted to changing work demands, the focus must shift from the procedures to the work demands themselves. In the terms of Simon's (1981) famous ant analogy, the focus shifts from the path of the ant to the landscape of the beach (see Vicente 1999; Flach, Mulder, and van Paassen 2004; Flach et al. 2008, 2010).

Flach et al. (2003) have made an analogy to the distinction between survey and route knowledge in the navigation literature. *Task analysis* describes the work domain in terms of routes. That is, sequences of actions or procedures. For example, "go to the menu and select the third option, then a dialogue box will appear, fill in the information in the dialogue box, and then click on the continue button, etc." In contrast to task analysis, *work analysis* attempts to provide a more complete map or survey of the field of possibilities or workspace (e.g., a tree diagram showing the complete menu structure making all options explicit or a flowchart that shows all possible ways to navigate through an interface).

Route knowledge has several attractive features relative to computers. First, it is easily encoded in a rule-based description. Thus, it fits with the computational image of information processing, and it is easily conveyed via text-based or verbal instructions. Second, it tends to reduce memory load (the number of rules) by focusing only on the optimal paths (e.g., minimum number of steps) to accomplish the intended functions; and at the same time it naturally inhibits deviations from the optimal paths (which may be especially important for safety critical systems). The major drawback of route knowledge is that it tends to be brittle. That is, it works only as long as the environment is static or stable. However, if the environment changes so that a prescribed route is blocked, if destinations/goals other than those specified in the formulation of the routes become desirable, or if unexpected events force an agent from the prescribed route, then route knowledge provides little support for adaptation.

Survey knowledge tends to require a higher degree of integration. That is, unlike route knowledge, which is nearly a direct analog to activity, survey knowledge requires the distillation of abstract relations (e.g., cardinal directions) from the activities. Survey knowledge is difficult to capture with rule-based descriptions or to communicate verbally. It typically requires a spatial form of representation (e.g., map, tree diagram, flow chart). The advantage of survey knowledge, however, is that it tends to be robust. That is, it provides a foundation for generating new solutions (e.g., a new path) when the environment changes (e.g., the normative route is blocked or a new goal emerges).

CSE was developed to address the demands for adaptation in complex work environments like the Web. Change is a universal in these environments. In such environments, rule-based or route descriptions will have limited utility. These environments demand survey representations; thus, work analysis represents an attempt to survey the work landscape. CSE is a belief that design decisions should reflect survey knowledge of the workspace.

There are two useful conceptual tools for surveying complex problem spaces: decomposition and abstraction. Decomposition attacks complexity by partitioning it into smaller chunks. The hope is that by addressing each chunk in turn we take small steps toward understanding the larger problem. The trick is to find a partitioning that preserves the meaning of the whole. That is, you hope that the steps will add up to a deeper understanding of the domain. There are many ways to break up a complex problem, but few ways that will preserve the integrity of the whole.

Abstraction attacks complexity by introducing conceptual structures. For example, the gist of a story could be preserved even though all the parts (words) were changed. In this sense, the gist is an abstraction. Another example is categorization. A category such as furniture reflects a relation among objects that cannot be reduced to a list of common parts. Abstractions generally function to highlight dimensions that are thought to be "essential" while blurring other dimensions that are considered less important. Abstraction is illustrated clearly in Hutchins (1995) discussion of different forms of maps. Maps are not simple iconic images of the layout of the Earth. Rather, they are carefully crafted abstractions designed to highlight some dimensions (e.g., line-of-sight directions) at the cost of blurring others (e.g., area).

Rasmussen (1986) emphasized that we need both strategies (abstraction and decomposition) and that the strategies are not independent. The dimensions highlighted and blurred by the abstraction process have implications for what decompositions will be appropriate. Figure 24.4 attempts to illustrate this interaction between abstraction and decomposition. Abstraction is illustrated as the vertical dimension of

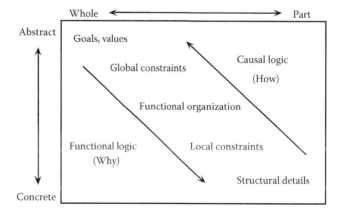

FIGURE 24.4 Interactions between abstraction and decomposition for attempts to understand means (how) and ends (why) within a complex work domain.

this space, and the level of decomposition is illustrated as the horizontal dimension.

Although five levels of abstraction are represented in Figure 24.4, consistent with Rasmussen's (1986) Abstraction Hierarchy, we will not attempt to define these levels or to make any claim about the absolute value of these particular levels. The claim that we do want to make is that no single level of abstraction provides a comprehensive survey of the meaning landscape. Abstractions that highlight the dimensions associated with satisfying (e.g., intentions and consequences) often will blur the dimensions associated with affording or specifying and vice versa. The diagonal regions of the matrix highlighted in Figure 24.4 reflect experiences from previous work analyses and from protocols of experts during problem solving in their domain (Rasmussen 1986; Rasmussen Pejtersen, and Goodstein 1994; Vicente 1999). These experiences suggest that questions associated with "why" (satisfying) are typically best seen with high degrees of abstraction and relatively gross levels of decomposition. This reflects a top-down view of work. Questions associated with "how" (affording and specifying) typically require more concrete descriptions and finer levels of decomposition. This reflects a bottom-up view of work. The bottom-up view helps to answer questions about how to carry out a particular procedure. The top-down view, however, becomes critical for deciding which of two procedures might be best in a given context.

Solving the adaptive control problem typically requires iterating between the bottom-up view and the top-down view. The bottom-up view helps to clarify the best way to carry out a particular strategy. The top-down view is useful for evaluating different strategies to assess which of several strategies might be most appropriate for a given context. Historically, the entry of CSE into many work domains has been a controversy over different possible strategies or over the appropriate criterion for switching from one strategy to another. This reflects the realization that there may be more than one best way and that the best way may vary with the changing context.

In summary, an important step in helping humans to respond in an adaptive way to meaningful properties of a work domain is to identify those properties. Fischer (2000, 284) refers to this as domain-oriented design. The goal is to "support not only human–computer interaction, but human problem-domain communication, making the computer invisible and bringing tasks to the forefront." A similar approach is reflected in the design of ontologies for guiding semantic search on the Web (e.g., Sheth, Ramakrishnan, and Thomas 2005; Thomas, Sheth and York 2007; see Michaelis, Golbeck, and Hendler, this volume). Work domain analysis is a process for surveying the work landscape to discover meaningful constraints. A comprehensive survey requires multiple perspectives in terms of the levels of abstraction and the degree of decomposition.

A final word of caution: understanding of a work domain is an asymptotic process. That is, it is never complete. Even for a simple system, such as a telephone, design does not end

with the specification of the product features. This is illustrated by a public phone in the Netherlands that featured a coin slot positioned at the top of the phone, just above the hook holding the horn of the phone. Unfortunately, in many places the phones were hung at eye-height (perhaps to make reading the numbers easy). However, the coin slot was now occluded from sight so that many people could not see where to put the money and thus could not operate the phone (as reflected on handwritten signs glued to the phones or mounted on the walls; Stappers and Hummels 2006). The point is that the people who hung the phones and the people who wrote the notes were participants in an evolving design.

Thus, for more complex work domains, there will always be new facets to discover. Work domain analysis is not just an informing phase that is completed and stops prior to a single design phase. It is an iterative component of an ongoing iterative design process that, as mentioned, Schrage (2000) called "serious play." Each new design is a step in an evolutionary process. It is a test of the current survey knowledge. It is a dialogue with a problem (Fallman 2003) that may continue through the life cycle of the product. With specific regard to the Web, this is reflected in the concept of Web 2.0 and the concept of semantic computing (Berners Lee, Hendler, and Lassila 2001; O'Reilly 2005; Thomas and Sheth 2006; Wegner 1997), where the focus is on interactions to create knowledge rather than simply access to data. It is important that there are mechanisms so that the survey knowledge can be refined based on the feedback from each test and to allow users to participate in the continuing design dialogue.

24.2.2 Ecological Interface Design

The goal of Ecological Interface Design (EID) is to explicitly represent the meaningful constraints that have been discovered through work domain analysis in the interface. An assumption of this approach is that agents who have an explicit survey representation (e.g., map) of the work landscape will be able to adapt their choice of routes to consistently satisfy the functional goals in the face of a changing world. Thus, the push for ecological interfaces comes from the increasingly fast pace of change in the work place and the demand for creative adaptations that results. The pull for ecological interfaces comes from the flexibility of graphical interfaces. As we noted in the discussion of survey knowledge above, it is difficult to communicate survey knowledge in the form of rules or verbal instructions. Typically, some form of spatial representation is required. Graphical interfaces provide the opportunity to build these representations.

The idea of EID is tightly coupled to the CSE approach to work analysis. The general notion is to make the nested hierarchy of constraints that has been identified through work analysis explicit to the operator. This has typically involved graphical representations using configural representations as illustrated in Figure 24.5. In some sense, the structure in the interface should mirror the structure of the work domain. Thus, the interface might have a geometry in which global properties reflect higher levels within the abstraction

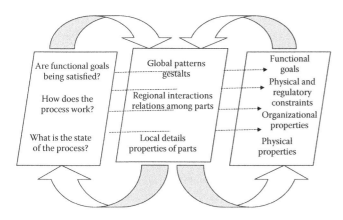

FIGURE 24.5 The goal of an ecological interface design is to make the hierarchical means-ends relations in the work domain explicit to the human operators who must navigate among these relations. Note that the items within boxes are assumed to operate in parallel, not in sequence as might be inferred from the arrows entering and leaving the boxes.

hierarchy (e.g., functional and global constraints). Nested within these global properties might be regional structures that reflect the organizational and physical relations within the work domain. Finally, the geometric details might map to specific states within the work domain. The purpose of this structure is to help the operator to "see" whether they are satisfying the functional goals; to see the logic of the process dynamic and to see the individual state variables in the context of the functional and process constraints.

The historical prototype for an EID is Vicente's (1999) DURESS interface. Vicente used the abstraction hierarchy framework to develop a detailed map of the means-ends relations associated with the thermodynamic DURESS process. He then developed a geometric structure to make these relations explicit in the interface representation. For example, he used a funnel metaphor to depict the flow of mass and energy through the system reservoir.

Vicente's interface represents an ideal that provides a valuable example of the ecological interface approach. However, it is an ideal that may be difficult to realize in work domains such as the Web, where there are many more degrees of freedom than in the DURESS example. The economic, political, and social constraints that play such an important role in the Internet are far less stable and less understood than the laws of mass and energy balance that constrain a thermodynamic process. Although computer graphics and virtual reality systems provide a rich palate for building representations, our understanding of many work domains remains greatly impoverished.

The attribute that distinguishes EID from other approaches to interface design is the role of the abstraction hierarchy for identifying the meaningful dimensions of the experience. However, the approach shares the goals of many other approaches (e.g., Fischer 2000; Hutchins 1995; Shneiderman 1983; Woods 1984); that is, to make the interactions with meaningful properties of the work domain as "direct" as possible. Where possible the interface should help the operator

to directly perceive the work domain; this involves not only seeing what is but seeing what is coming. Direct perception involves seeing the patterns of unfolding events that specify future outcomes (e.g., whether goals will be satisfied). This ability to anticipate the future is critical to stability in any control system where there are long time constants between actions and consequences. One way to think about the higher levels within the abstraction hierarchy is that they represent internal or hidden states of the process. Making these states explicit can be thought of as a form of prediction. For example, the shape of the funnels in the DURESS interface helps to specify the directions and rates of change of the mass and energy inventories. The interface should also help the operator to directly manipulate the work domain. This direct manipulation is important both for control and hypothesis testing.

Designers who are looking for an easy answer to the interface design problem will be sorely disappointed by the EID approach. Mapping out the abstraction hierarchy is a daunting task, and for domains that are evolving at the pace of the Internet, the meaning landscape is probably changing at a pace that is greater than our ability to keep up. Thus, the mapping problem will be a continuous battle. To the degree that the domain can be mapped, there is still the question of building a representation that captures meaningful aspects of the domain in a way that allows direct perception and direct manipulation. As usual, there are many more ways to do this wrong than right. There is no simple formula that guarantees success. The best that we can do is to provide some examples of promising interfaces in later sections of this chapter. Hopefully, these interfaces will be a source of inspiration. EID is not an answer but an invitation to engage in the difficult search for meaning.

This emphasis on the problem domain is in contrast to the typical emphasis on the user in other approaches to human–computer interaction (Norman and Draper 1986). Thus, some may wonder where the "awareness" side of the dynamic comes into the picture. In the ecological approach the emphasis shifts from matching a user model to shaping a user model. In this sense, the interface is designed with the explicit goal of shaping the user's understanding of the problem space. This leads naturally to the issue of instructional design.

24.2.3 INSTRUCTIONAL DESIGN

A misconception about Ecological Interface Design is that the resulting representation will be natural, where natural means that people will be able to immediately see the deep structure in the representation without instruction or practice, because it somehow taps into native perceptual abilities. Unfortunately, this will rarely be possible. The ecological interface design does not eliminate the complexity or requisite variety that is intrinsic to a work domain. Thus, it does not necessarily eliminate the need for extensive learning. However, it may have important implications for the design of training and instructional systems. By making the constraints explicit in the interface, ecological interfaces support users to work through perception/action rather than through inference in exploring situations outside the standard operating conditions. This creates the opportunity for learning by doing. In other words, it sets the conditions for an abductive dynamic.

To the extent that the interface allows direct manipulation and direct perception, it provides explicit feedback that should help agents to discover meaningful aspects of the domain through interaction. When meaning is clear, the interface should allow direct comparison between consequences and intentions in a way that specifies the appropriate action. However, even when meaning is not clear, the interface should allow interrogation through action, so that the actor can learn or discover the possibilities offered by the domain. That is, the actor should be able to test hypotheses out at the interface and get clear feedback to suggest new possibilities and alternative meanings.

Thus, learning is expected to be an integral part of any adaptive system. Learning becomes part of the job: practitioners are expected to be reflective (Schön 1983). A good interface should facilitate reflection. With an ecological interface the learning process might be more similar to learning to drive a car or to play a sport rather than learning facts or rules. In this sense, natural does not necessarily translate to easy. Rather, it says that the learning process might be more similar to learning the skill of kicking a soccer ball rather than memorizing the rules of soccer. Within CSE the associations acquired through interaction are typically illustrated as short cuts in a decision ladder. In this context, a goal of the design is to support the development of a rich network of skill-, rule-, and knowledge-based associations that complement the demands of the work domain (see Bennett and Flach [forthcoming] for more details about issues associated with awareness).

24.3   THE WEB AS A DOMAIN?

In principle, the Web is not a work domain, per se, but rather it is a medium or technology capable of supporting many different types of work. Just as "pen and paper" or "the computer" have been for a longer time. It is what Fischer (2000) has called a high functionality application (HFA). He noted that HFAs are complex systems in and of themselves because they serve the needs of large and diverse user populations. The HFA itself has little "meaning." The meaning arises when an HFA is used as a tool within a specific work domain. However, it may be valuable to speculate about the kind of work that might utilize this technology. The Web will most likely be a tool for domains of education, entertainment, and business.

Many of the early concepts of ecological interface, such as Vicente's (1999) DURESS, were designed for the process control industry. In these domains, physical laws (e.g., mass and energy balances) played an important role. However, the Web and the information systems upon which they depend are much less constrained by physical laws. Not that the physical laws are suspended, but they become transparent or uninteresting in relation to the functional goals for the work interactions. The structure that will give meaning to work over the Web is more likely to come from social conventions (which we will use broadly to refer to cultural, political, legal, and economic constraints).

In the context of work, the social, political, and economic sources of constraint are typically no less important than the physical laws that govern the flow of mass within a feed water control system. However, while physical laws tend to be invariant over the history of a process control system, social conventions tend to evolve along with the technologies they constrain (e.g., conventions of point and click operation, etiquette of using cell phones). Furthermore, it is likely that the physical constraints of most process control systems are both simpler (e.g., fewer degrees of freedom) and better understood than the social constraints that govern education, business, and entertainment. This poses a significant challenge for CSE and ecological interface design. The two examples in the following section were chosen specifically to illustrate systems dominated by social conventions; in both cases, people must search through complex databases to satisfy personal interests.

24.4   TWO CASE STUDIES

In the previous sections, we have tried to lay out a conceptual framework for an ecological approach to Web design. We hope that the conceptual abstractions will provide some insights into the approach. However, we also realize that many details have been glossed over and that there is a wide "gulf of execution" (to borrow from Norman 1986) between the conceptual abstractions and the details of a concrete design problem. In this section, we will try to bridge this gulf with two case studies that we think have particular relevance for the problem of Web design.

24.4.1   EXAMPLE 1: BOOKHOUSE FICTION RETRIEVAL SYSTEM

The BookHouse (Pejtersen 1979, 1984, 1988, 1992) was developed to help general patrons of a library including children find "interesting" books to read. In practice, the graphical interface that allows users to navigate through a virtual library is the most salient aspect of this design. However, from the point of view of CSE the more interesting question is associated with the development of a classification scheme for fiction. Traditional methods for identifying and locating books of fiction in a library are very simple and generally ineffective. Most libraries categorize these books by bibliographic data alone (e.g., author's name, title, ISBN number, etc.). Thus, the options for finding a book are rather limited unless the user knows of a specific book and at least some of its bibliographic data. If the user does not have the appropriate data or does not have a particular book in mind (often the case), he or she is left to browse the fiction shelves or to consult a librarian for professional help. The challenge for a CSE approach is to discover meaningful categorical distinctions

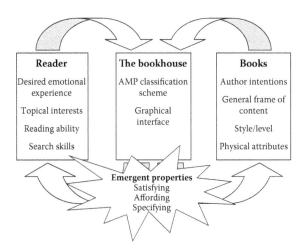

FIGURE 24.6 The challenge for the BookHouse was to develop a classifications system and interface to help a diverse group of users find books that they would enjoy.

for classifying fiction. That is, there is a need for a categorical structure that will address the goals of satisfying, affording, and specifying. This challenge is illustrated in Figure 24.6.

24.4.4.1 Domain

The key domain question was: What attributes of books made a difference with respect to the intentionality of users? To get insights into this problem, Pejtersen (1979) examined user queries during actual information retrieval negotiations with librarians. These observations led her to suggest five dimensions of books that reflected users needs: author's intention; literary school, style, and values; frame; subject matter; and accessibility. As a result ot this analysis, five general dimensions have been included in the BookHouse.

1. The Bibliographic data dimension includes the typical information used to classify books, including title, author, date, publisher, etc.
2. The Subject Matter dimension includes the "story line" of the book, essentially what the story is about. This includes the topic matter that is being addressed, the elements of the plot, the course of events and the social interactions and relationships that evolve.
3. The Frame dimension is essentially a description of the context within which the story occurs. The primary aspects of this dimension are the time (past, present, future), the place (country, geographic location), and the social milieu (mores, issues, reform movements, etc.) that form the setting for the story.
4. The Author's Intention dimension captures the reasons why (and to some degree how) the author wrote the book. Authors can write books to provide a variety of emotional experiences: among other reasons authors might strive to amuse, scare, or uplift their audience. Authors might also have ulterior motives: among other reasons authors might strive to inform, sway public opinion, or encite their readers to action.

Finally, the manner in which the authors write their books might adhere to particular literary schools, styles, or values.

5. The final dimension is Accessibility, which refers to a constellation of factors that can influence how well an author's story will be communicated. Some factors are straightforward factors determining whether or not a reader will be able to read the book, such as the reading level and physical characteristics of the book (e.g., size of the font type). Others will influence the degree to which readers will be intrigued by, or can relate to, the message in the story (e.g., literary form, the main characters, and their age).

This type of analysis, aimed at uncovering the sources of regularity in a particular domain, is the foundation of any enterprise that is to be referred to as ecological in nature. Thus, each of the books in the BookHouse were "indexed" according to how its content fit into the classification scheme; this constitutes the information about that book that will then be available during search. This constitutes the meaning landscape.

In addition to considering the meaning landscape, it is often useful to consider different strategies for navigating that landscape (Pejtersen 1979). How might people search for interesting books? One strategy is search by analogy. That is, the user specifies a book that was particularly satisfying and asks for a book that is similar. For example, "I want another book like *Gone with the Wind*." Note that "similar" can be problematic, in that everything is similar to everything else in many different ways (e.g., Dreyfus 1994). Within the BookHouse, the classification system provides a basis for judging similarity. Books that have many common dimensions (e.g., similar author intention, similar topic area, similar frame, etc.) will be judged more similar.

Another strategy is an analytic search. With this strategy the patrons specify their desires directly in terms of the classification dimensions: "I want a romance set in the South during the civil war." This strategy depends on the patron knowing the dimensions. An important design issue is who takes the responsibility for the "knowing." Will the patron be required to learn dimensions that are convenient for the database system? Or will the system be designed to match the natural dimensions used by patrons? For a system like the BookHouse, where user intentionality is critical to meaning, it makes sense to design the classification system to reflect distinctions that are meaningful to users.

Other strategies for search include empirical search and browsing. Empirical search refers to the situation where a librarian suggests books based on personal attributes of the patron: "Here is the type of book that teenage girls usually enjoy." Browsing could involve picking books out at random or based on cover illustrations (e.g., a woman in period costume in front of a southern mansion) and perusing the contents (e.g., reading the jacket).

Having a description of the meaning landscape and understanding some of the strategies for navigating that landscape provides the foundation for designing interfaces that will help patrons to achieve their functional goals.

24.4.4.2 Agency

One of the key elements in the design of any successful human–computer system is an explicit consideration of the capabilities, limitations, knoweldge, and other characteristics of those who will use it. Although the admonition to "Know Thy User" has become a platitude for design, it is, in fact, a very critical consideration. Field studies and unstructured interviews were conducted. It was found that the targeted user population had a profound lack of homogeneity, capabilities, and experiences. The results indicated that the user population ranged widely on a number of dimensions, including familiarity with computer systems, chronological age (and all of the associated differences in capabilities and limitations), familarity with fiction, and reasons for being there in the first place. Two broad subgroups with substantially different sets of capabilities and limitations were identified: children (ages 6 through 16) and adults (ages 16 and up). In particular, the general cognitive development, reading skills, and command of the language were quite different for these two subgroups. This translated into substantially different needs: children and adults tended to formulate their questions about books of fiction using different sets of linguistic terms and different types of semantic content (i.e., at different levels of the AMP classification scheme).

Consideration of these factors resulted in the development of two separate databases of fiction books for use in the BookHouse retrieval system. The children's database contained different types of accessibility information and different descriptions of the books and their content. These changes were aimed at matching the language and the concepts that children used in their natural interactions with librarians. A third database, esentially a combination of the previous two, was also developed to support "interdatabase" browsing (e.g., teenagers browsing the adult database or parents browsing the children database).

In summary, the targeted user population of the fiction retrieval system was extremely diverse with substantial differences in a number of dimensions. An additional consideration was that most users were likely to have only casual and infrequent interactions with the system. This stands in sharp contrast to the typical users of complex socio-technical systems (i.e., those systems controlled by the laws of nature, e.g., process control), who are highly homogeneous in capabilities and have extensive training. It was therefore essential that the system be both intuitive to learn and easy to use. A design strategy was needed that would capitalize upon general conceptual knowledge that was common to all library patrons. More specifically, the system would need to be designed using an interface metaphor that was familiar to this diverse group of patrons.

24.4.4.3 Ecological Interface

The traditional approach to searching a database is through linguistic commands issued to a computerized intermediary (i.e., keyboard-entered alpha-numeric searches). With the BookHouse system a user completes a search by navigating through a carefully crafted spatial metaphor: a virtual library. The user is initially presented with a view from the outside of this virtual library (see Figure 24.7a). A fundamental design feature of the system is that visual and verbal feedback is provided to alert the user of the potential for interaction, to specify the action required to initiate that interaction, and to outline the goal that the interaction will achieve. In this particular instance the user must enter the BookHouse to initiate a session. The user is informed of this requirement when he/she positions the pointer over the opening of the virtual library (the area framed by the doorway). A white graphical outline of the opening appears (i.e., the icon is highlighted), signifying the potential for action. At the same time a verbal description of the action required (i.e., "Press here") and the end result (i.e., "to get into the Book-House") is also provided in a text box located at the bottom of the screen. These conventions (point and click interaction style, visual and verbal feedback on potential interaction, and required acts of execution and goals) are used consistently throughout the interface.

(a)

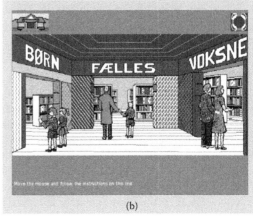

(b)

FIGURE 24.7 **(See color insert.)** The hierarchically nested spatial structure provided in the BookHouse interface. (a) The global metaphor: a virtual library. (b) One of several intermediate metaphors corresponding to locations in the virtual library: the hallway used to select a portion of the database.

Once inside the BookHouse, the user is presented with a view of three "virtual rooms" whose openings can be pointed at and clicked on to enter (see Figure 24.7b). Each room corresponds to one of the three databases (children, adult, and combined) that can be searched. The identity of each database is signified by verbal labels over the rooms, the composition of the library patrons at the entrance to the room, and the verbal instructions that appear at the bottom of the screen. Entering a room (i.e., pointing and clicking on the opening) is an action that signifies which of the three databases that the user wishes to search.

The user is then presented with a room that portrays various patrons engaged in four different activities shown in Figure 24.8a. Each of these people-activity icons corresponds to the four different types of strategies that can be executed to search a database; clicking on an icon indicates that the user wishes to initiate that type of search. The four search strategies are referred to as search by analogy, analytical search, browsing through descriptions of books, and

(a)

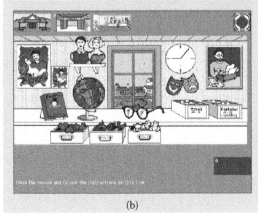

(b)

FIGURE 24.8 (See color insert.) Intermediate and local metaphors in the BookHouse interface. (a) The room used to specify a database search strategy: each person/activity combination represents a different strategy. (b) Another "scene" from the BookHouse. The local metaphors (i.e., the objects in the scene) represent tools for specifying dimensions in the AMP scheme form categorizing fiction. These tools can be used to "analytically" narrow the search to identify potentially interesting books.

browsing through a "picture thesaurus" that graphically represents constellations of books. The analytical search strategy allows the user to select specific terms from an assortment of characteristics that have been used to classify books in the databases. It will be used to demonstrate additional aspects of interaction with the BookHouse system.

As a result of clicking on the person-activity icon that corresponds to the analytical search (i.e., the person seated at the desk) the user adopts the perspective of that person and sees a number of icons arrayed on and around the desk as shown in Figure 24.8b. Each of the icons contain graphical representations that were designed to symbolically represent a particular dimension that was used to classify the books in the database. For example, the clock on the wall symbolizes the chronological date of a novel. The act of pointing and clicking on an icon is an indication that the user wishes to select an attribute of that dimension for use in searching the database. Upon activation of an icon the user is presented with a graphical depiction of an opened virtual book whose pages display a list of key words. The user can navigate to additional lists by either clicking on one of the "alphabet" icons (to jump to a distant list with terms organized according to the corresponding letter of the alpahbet) or by clicking on one of several arrows (e.g., to go to the next or the previous list).

The user specifies a key word for the search by pointing and clicking at an item on the virtual page. This action has several consequences. First, the key word appears in a text box at the bottom of the screen, indicating that it is now a term that will be used to execute the search. Second, additional icons appear at the top right of the screen, indicating that other actions may now be taken (in this case these are actions that can be used to modify the search). Third, the search is executed. Fourth, the results of the search are portrayed as an icon (i.e., a set of books) that appear in the lower right-hand corner, along with a number that indicates how many books met the search criterion.

The user has several options at this point. The results of the search can be examined by clicking on that icon. A verbal description of the first book appears as an entry on the pages of a virtual book. This description is organized according to the dimensions of classification that are used in the system; the description provides a detailed explanation of its contents (and in the process, just how this particular novel fits into the general classification scheme). Additional books meeting the search criteria can be examined by pointing and clicking on the arrows. The information about a book can be printed out (by clicking on the printer icon) or placed aside for future reference (by clicking on the icon with the hand holding a book).

Alternatively, a second primary option is to refine the criteria used in the search. Imagine that the user wanted to narrow down the results of the search by including a key word from another classification dimension. To do so, the user would point and click at the iconic representation of the analytical room located at the top and middle section of the interface. This returns the user to the analytical room;

he/she would then follow the steps outlined above to specify a key word in the second dimension. This key word is then added to the list at the bottom of the page and the system automatically executes a second search of the database using these two terms combined with a boolean "and."

One other fundamental aspect of the interface needs to be mentioned. A primary consideration in the design of the interface was to allow users to switch easily and seamlessly between databases and search strategies. There are a number of interface conventions that allow this to occur. A primary convention is the strip of navigational icons that appears across the top of the interface as the user navigates through the virtual library. For example, the user can easily choose an alternative database to search by clicking on the database icon in the navigation strip (second icon to the right) and then making the appropriate choice. Similarly, the user can easily choose an alternative search strategy by clicking on the strategy icon in the navigation strip (the third icon to the right) and selecting another person-activity icon. In some cases the possibility for alternative search strategies is even more direct. For example, when the results of a search are examined (i.e., when a book description is on the screen), the user can immediately switch to a search by analogy for the displayed book if they click on the icon with the appropriate symbol in the upper right.

Providing support for navigation is important for the design of Web applications, as evidenced by the numerous discussions of the topic in this handbook. In a particularly influential article, Woods (1984) described a concept that can improve the efficiency of navigation: visual momentum. Woods (1984, 231) defined visual momentum as "a measure of the user's ability to extract and integrate information across displays, in other words, as a measure of the distribution of attention." He also described several design techniques that can be used to increase visual momentum. These techniques help avoid several common problems including the "getting lost" phenomenon (i.e., users do not know where they are, where they have been, or where they can go to) and the need for cognitive reorientation following a visual transition (either between or within screens of information). The BookHouse interface will be used to illustrate some of these design techniques.

24.4.4.4 Spatial Cognition

This technique equates the actions required for system interaction with navigation through a spatial structure. The spatial structure in the BookHouse is a nested hierarchy of metaphors ranging from global (virtual library) and intermediate (rooms in the virtual library) metaphors to local (objects in the rooms) metaphors. The actions that are required to interact with the system (i.e., book searches) are related to commonplace, natural activities that are carried out constantly in everyday life (e.g., navigation through buildings). Using spatial structure in this fashion leverages powerful, perhaps innate, perception, action, and cognitive skills: interacting with an interface becomes much like interacting with a real ecology (e.g., wayfinding in an environment).

24.4.4.5 Overlap

The overlap technique places information of interest within a larger viewing context that explicitly illustrates meaningful physical or functional relationships. The navigation strip provides an excellent example of this technique. It is used to capture the user's navigational path through the virtual library: each time the agent leaves an area of the virtual library a small-scale, replica metaphor is added to the navigation strip (see Figures 24.7 and 24.8).

This is a particularly creative instantiation of the technique known as the "breadcrumbs" strategy (referring to the trail of breadcrumbs left by Hansel and Gretel). Pardue et al. (2009, 235) describe perhaps the most familiar use of this technique, one that is encountered in Web browsers:

> With the problem of becoming disoriented in large information spaces, orientation cues have been used to guide users, and found to be important for effective navigation of spatial maps, for example (Burigat and Chittaro 2008). One way to provide orientation cues in navigational hierarchies is with breadcrumbs (Bernstein 1988). A breadcrumb is a metaphor describing the practice of marking the path the user has taken. In web-based hypertext systems, the taken path has come to be represented by location breadcrumbs, defined as a text-based visual representation of a user's location within the navigational structure (Teng 2003). Location breadcrumbs are usually implemented as a list of links, each separated by a character such as an arrow or ">" to indicate the direction of the navigation. For example, You are here: Home > Grocery > Pasta & Grains > Pasta > Spaghetti.

Similarly, the replica metaphors in the navigation strip provide graphical breadcrumbs that specify the agent's navigational path through the virtual library. This provides one perspective of the spatial relations within the virtual ecology: a serially ordered, linearly arranged sequence of metaphors representing the physical interconnections between rooms. It provides powerful orientation cues, showing the user both where they are and where they have been.

24.4.4.6 Perceptual Landmarks

Siegel and White (1975, 23) identified the important role of landmark knowledge for navigation in real-world ecologies:

> Landmarks are unique configurations of perceptual events (patterns). They identify a specific geographical location. The intersection of Broadway and 42nd Street is as much a landmark as the Prudential Center in Boston. . . . These landmarks are the strategic foci to and from which the person moves or travels. . . . We are going to the park. We are coming from home.

The use of the spatial cognition design strategy provides the opportunity for perceptual landmarks in a virtual ecology and the associated benefits for navigation. The most prominent perceptual landmarks of the BookHouse interface are those associated with the areas of the virtual library that can be navigated to (i.e., alternative screens in the interface). Each of these physical locations (e.g., the entrance hallway) has a distinctive visual appearance (e.g., the three thresholds in the entrance hallway). The replica metaphors in the navigation strip (see Figures 24.7 and 24.8) preserve these distinctive visual features, thereby providing a set of perceptual landmarks (i.e., reminders of locations that can be navigated to) that increase visual momentum.

Note that these perceptual landmarks also serve as controls: the agent can navigate back to a previously visited room by pointing and clicking on a replica metaphor in the navigation strip. Thus, these perceptual landmarks provide an additional increase in visual momentum by facilitating transitions between screens: a visual preview of potential destinations in the interface is provided, one that orients the agent (visually, cognitively) to the ensuing transition.

The concept of visual momentum and the design techniques that can be used to increase it are directly relevant to the problem of navigation in the Web. Only three of the six techniques that Woods (1984) proposed were discussed in this brief synopsis. See Bennett and Flach (forthcoming) for a more comprehensive discussion of all techniques and their implementation in several different work domains.

The spatial metaphor of navigating through rooms and selecting icons was chosen to make the syntax of the interaction as intuitive as possible. However, it is important to keep in mind that the ecological aspect of this interface has more to do with the scheme for classifying books, than with the spatial metaphor. It is in the classification scheme that the question of meaning is directly addressed.

24.4.2 EXAMPLE 2: THE MULTIDIMENSIONAL SCALING INTERFACE

MDS-Interactive is an interface that uses multidimensional scaling techniques to help people to explore digital product catalogs. Originally, its design was developed to support designers querying a visual database of existing products in the early phases of a design problem (Pasman 2003). Designers often review precedent designs for form, aesthetics, style, and solution elements. But later it was realized that there are many situations in which people have to choose products in a domain with which they are not familiar. The solution for designers was found to hold well for other problems such as searching for a new home from the haphazard collection of houses on the market, finding a gift, selecting a holiday from among many options, or finding a TV show to watch.

MDS-Interactive was developed to counter some of the problems that designers (and consumers) have with regular database systems (Stappers and Pasman 1999). (1) The form of the interaction is mostly verbal, through discrete symbolic commands (e.g., search strings). (2) The user must be familiar with the jargon used in describing the content of the database and must be able to phrase the goal of the search in terms of this jargon. (3) The hierarchic form of the database often imposes an order in which different parts of the question must be addressed. (4) It is easy to lose track of your search steps and your decision process; for example, when using a Web search engine, many people find it difficult to retrace their steps or even to remember the decisions they took on

an item four steps back. To counter these problems, MDS-Interactive uses a mainly visual interaction style by letting the user interact with a small set of samples from the database, and presenting the samples in the form of an interactive map which presents similarity of samples by proximity of the sample's images.

In this section we illustrate MDS-Interactive by stepping through two small scenarios. Then we discuss the elements in the interface, and the technical underpinnings of the method (the left and right halves of Figure 24.9, respectively), indicating how and where it fits in the scheme of the theoretical sections above.

24.4.2.1 Scenario 1: "Finding a Gift for Joey"

Mark wants to buy a gift for his nephew Joey's 10th birthday. Although he knows the boy, he doesn't have a precisely described search question. He turns to an Internet shopping site using MDS-Interactive, and it shows him a small set of gift samples, including a teddy bear, a football, and a remote-controlled car. As Joey likes to play outside, he clicks near the football and receives some similar sample products: an inline roller skate, a skipping rope, and a fishing rod. Mark likes the skates most, drags the other samples off-screen and calls up some more roller skates by clicking near the skates. The MDS-Interactive screen now depicts a field of sample skates, which cluster into groups. Some groups appear to be visually similar (e.g., the off-road skates have only two or three wheels instead of the regular four and the speed skates have five). The visual groupings and the appearance of the skates convey some of these aspects directly or invite Mark to find an explanation of the grouping by inspecting the properties of each sample: as the mouse pointer rolls over a sample, the product info window shows a high-resolution picture of that sample, and lists a number of the key attributes about the sample. By comparing the elements in a group, Mark finds out that all the two-wheeled skates are used for off-road skating.

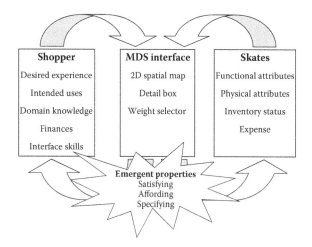

FIGURE 24.9 The challenge for the MDS interface was to help users with a vague notion about what they want (in terms of both what's available and implications of difference relative to functional goals) find a product that they will like.

Mark continues searching the database, removing uninteresting samples (by dragging them out of the circle), and calling up new samples in interesting parts of the sample map (by clicking at interesting positions, near desirable samples and away from undesirable ones). Mark narrows down the sample collection to find "a good gift for Joey." Along the way his originally vague goal gets more and more clarified, and Mark learns about the collection of skates in the database and their properties. Finally, he settles on a best gift sample. In the process he will have considered properties as diverse as application area, visual appearance, brand, quality, and price. He will probably have shifted search strategy a few times, at some times relying on the pictures of the samples, or the grouping of the samples, at other times comparing samples on their properties.

24.4.2.2 Scenario 2: "Find the Grind"

Joey has been skating for a year or so, knows a bit about skating, and wants to get a new pair of skates, but one that is better for cool stunt work. In MDS-Interactive's initial display, he quickly weeds out the skates he doesn't want, e.g., speed skates and off-road skates, and focuses on regular, compact four-wheeled skates. He is especially keen on questions of how the skates are used, so he sets the weight selector toward "application." As a result, the samples in the display arrange themselves on the basis of the properties in the database that have to do with application: properties of price and brand (quality) and the type of fixture and appearance (design) are ignored. The arrangement now reflects different clusters. The property list helps him to identify the groups as middle of the road, all-round, and grind skates.

He finds out (by asking the database or from sources outside) that "grind" is the jargon for doing stunts. After eliminating all nongrind skates, Joey resets the weight selector to bring back aspects of quality and design. Now he compares the collection of grind skates. By studying the remaining clusters, their property lists, he will find also that there are considerable price differences (which often concur with brands) and that there are differences in the ball bearings within this group. Prompted by this difference, he calls up information about ball bearings and finds out that these differ in hardness and price and that grinders care a lot about the quality differences in these.

24.4.2.3 Domain

As with the BookHouse, the graphical interface is the most salient surface feature of the MDS design. However, as with the BookHouse, the most important dimension, from the perspective of CSE, is the domain structure that can be accessed through this interface. What differences make a difference in terms of satisfying, affording, and specifying? At the top level of the search we have the product samples (which evolve from gift to outdoor products to skates in Scenario 1). This top level is found in the field of samples. Below that lies the level of aggregate skate properties quality, design, and application, which have direct implications for whether a particular skate will satisfy the goals of the users (in terms of use of

the skates and quality of the experience). Next there are the properties that identify the skate relative to other skates in the inventory (e.g., price and brand). Finally, there are the physical materials, such as the quality of ball bearings.

At the level of skates, much of the structure comes from physical inventory of skates and the intuitions of experts about what dimensions of this inventory make a difference in terms of how skates fit various uses or applications. The roller skate prototype contained 70 skates, described by 20 properties, organized in 3 groups, which are shown in the weight selector. The layout algorithms used all 20 properties, but only a few were explicitly presented in the detail window. The properties were selected, and the relative weights were set by a domain expert, so that the layout patterns were based on similarities that could be expected to be meaningful to the users of the interface.

At the level of an interesting gift for Joey, the structure comes from the user's experience with Joey. Building this into the interface structure would require a theory of what makes things interesting and perhaps a theory of "Joey." Without such a theory, it is still possible to let these constraints enter the process by enabling the searcher to explore freely (in other words, to let the searcher browse with minimal interference). By minimizing constraints associated with the interaction, we maximize the opportunity for the system to organize around constraints that the human brings to the system (e.g., such as a concept of what Joey might like).

24.4.2.4 Agency

As with the BookHouse, the target user population is a very heterogeneous group, both in terms of what they might be looking for and in terms of the skills they have for navigating through the Web. Again, a spatial metaphor is seen as a good solution because it taps into the general experiences of moving and manipulating objects that many people will bring to the interaction.

In each of the two scenarios we have a competent user conducting a search by negotiating his way through a set of samples. During the search, new samples and the way in which they are arranged implicitly prompt the user to make rough decisions in the visual display, to look deeper into the samples' properties, or to change the relative weights of the different properties. All these actions occurred out in the open, by direct manipulation of the samples, the space between the samples, the item property window, and the property weight selector.

As the user goes along, he or she not only learns about the skates in the database but also learns to relate the objective of the search to the jargon terms that are used in the database. Typically, the users learn about the jargon only when it affects their decisions. For example, if all the skates that the user is considering have the same quality ball bearings, there is no need for him or her to learn the ins and outs about ball bearings. This type of search occurs in many product choice decisions: when someone wants to buy a house, he often cannot phrase the question in terms as "it must have 4 rooms, one of 5' × 12', and a kitchen with two windows," even though these are relevant terms for deciding if a certain house fits. The searcher for a house is often looking for a best fit for a complex set of considerations, not for an aggregate of *optimal properties*, but for an *existing* sample that fits better than the others.

24.4.2.5 Ecological Interface

The interface shown in Figure 24.10 consists of three components:

1. A field of samples whose layout is continually adjusted to reflect their overall similarities. When new samples enter the field, old samples are removed, or the user tries to drag a sample to another location, the field adjusts and the samples settle to a new layout. The mathematical technique used to determine the optimal layout from the properties of the samples is called multidimensional scaling (MDS), a method that has been used in explorative statistics since the 1960s (see, e.g., Kruskal and Wish 1978; Borg and Groenen 1997). What is new in the interface is first its dynamic nature (new arrangements are reached by animations showing the motions of all samples), and foremost its interactive nature (Stappers, Pasman, and Groenen 2000). When the user clicks between the samples, a database query is performed to find the sample in the database that best matches the place of the click, i.e., the sample that is very similar to the samples that are close in the field and at the same time dissimilar to the samples that are on a larger distance. Through this, the field of samples becomes a field of meaning, where each position in the field expresses a query or a result.
2. A detail window listing a set of property values of the currently selected sample.
3. A weight selector that lets the user emphasize some properties and deemphasize others properties. In this way the display (and the queries) can be directed to a specific aspect.

This interface allows representation of meaning at many levels. At the top level of the search, there are the product samples (which evolve from gift to outdoor products to skates in scenario 1). This top level is found in the field of samples. Below that lies the level of aggregate skate properties quality, design, and application, which can be manipulated with the triangular weight selector. Finally, a selection of the properties that are expected to have a direct meaning for the user, such as price and brand, are shown in the detail inset. But most low-level properties, such as the materials of the ball bearings, are not normally on the display. Yet, as described in the second scenario, the hope is that attention will be drawn to these properties as they become meaningful in the course of the interaction.

24.4.2.6 Open-Endedness

MDS-Interactive is unlike most interfaces to databases in that its form does not incorporate the structure of the database. At most, it lists some labels of the properties that were attached to the objects, but the whole interaction with the user is on the level of similarity judgments on instances. The system itself does not take explicit questions of the form "which roller skate has the lowest price." An external observer looking at the screen (but not at the user) can only guess at the reasons, searches, and navigation steps that the user makes, unlike a search engine, which operates on explicit text queries, it merely mediates the collection of roller skates, in the form of an interactive map. In that sense the interaction is incomplete, adaptive, or metadesign (Fischer 2002).

In metadesign, designers create products in the knowledge that their final shape will be finished by the end user. This imposes different demands on the design process, anticipating unforeseen local circumstances and serendipitous uses. Keller et al. (2009) developed an interactive database in which designers could store, sort, and reuse their collections of visual imagery, and explored how this was used in a field study at three design offices. One of the main findings was that different users employed the enter, arrange, and query-based system in very different ways. In each case, however, they reported unanticipated (serendipitous) encounters with

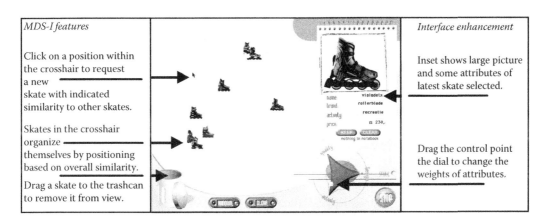

FIGURE 24.10 The interface for MDS-Interactive that allows search through a database of skates to discover one that satisfies your personal needs.

earlier visual material during their interactions with the system. In creative search tasks (exploratory search), the criteria for the search can radically change during the search itself. This fits in with Schön's notion of problem and solution co-evolving, as mentioned earlier in Section 24.1.

The MDS-Interactive interface has been user tested for a small range of application areas, product types, property types, ranging from whiskies (classified on subjective taste judgments), through consumer products such as roller skates (integrating technical properties and experiential judgments), to information services, such as a digital TV guide (classifying programs based on formal classifications and properties such as time and length of broadcast and type of audience appeal), and expert information systems (such as marketing evaluations of company portfolios). Findings from user testing indicate that the simplicity of the interface, and the fact that all of the interactions are visual and in the open, enable people to explore collections starting out with vague, taste-based, or expert-based questions. Moreover, people could quickly learn how to search by watching one other person use it, and people could use it in collaboration (whereas most search or browse systems provide very little support for collaborative use).

24.5   SUMMARY AND CONCLUSIONS

One can think of problem solving situations as composed of three basic elements; the world to be acted on, the agent who acts on the world, and the representation of the world utilized by the problem-solving agent. . . . There is an intimate relation between these three elements so that cognitive description is at once a psychological description and a domain description. . . . a complete description must specify the relationships between all three sets of factors.—Woods and Hollnagel (1987, 258)

Among those concerned about the "human factor" in complex systems, you will find almost unanimous agreement with this statement of Woods and Hollnagel. Thus there is not much controversy about the span of the problem. However, there are many disagreements about how meaning fits into this picture. What is the source of meaning? What is the nature of the underlying dynamic (e.g., the nature of causality)? What is the appropriate conceptual parsing for making sense of this dynamic? What are the implications for design? Our goal in this chapter was to provide an ecological context for framing these questions. For us some distinctive features of an ecological approach are as follows:

1. The focus on the ecology (or in James terms "experience") as the source for meaning. This approach acknowledges the contributions of both human and environment to meaning but gives priority to neither.
2. A circular, as opposed to a linear view of causality or rationality that reflects the dynamic of abduction. This circular dynamic, depicted in Figure 24.2,

reflects a coupling between perception and action and between control and exploration. Within the circular dynamic, neither perception or action, or control or exploration, takes precedence over the other. Each occurs by virtue of the other. When these processes are coupled in a symbiotic fashion, the result is normally an improving fit between organism and environment (i.e., an adaptive system).

3. In parsing this dynamic, the attributes of specifying, affording, and satisfying emerge as the objects of study. These are seen as critical dimensions of the "fit" of organism and environment that cognitive science should be exploring, both as a means to a richer cognitive science and as a bridge between basic questions about human performance and practical questions about designing to support meaning processing.
4. The foundation for design is first to ask the question, what is meaningful. What differences make a difference in terms of the functional goals of the work domain? The next question is to explore ways that these differences can be made more explicit to the humans in the system, through either the design of ecological interfaces, training, or both. We tried to provide a logic case for this in the discussion of CSE and tried to illustrate the design problem with the BookHouse and the MDS-Interactive interface.

In conclusion, we hope that this chapter will be appreciated by others, who are struggling to understand the nature of cognition and the implications for design, as an interesting approach—not an answer—but an interesting alternative perspective. This change in the approach to cognitive systems is mirrored by changing images of the Web itself reflected in such terms such as semantic computing (Berners Lee, Hendler, and Lassila 2001) and Web 2.0 (O'Reilly 2005). This alternative is described more fully in a forthcoming book (Bennett and Flach, forthcoming). The ecological framework is an attempt to scale-up theory to address the complexity of work environments, not an attempt to reduce the problems of work into a convenient theoretical paradigm. It is an attempt to make science and design phenomenon centered or problem centered. Above all, it is a search for meaning.

REFERENCES

Bennett, K. B., and J. M. Flach. Forthcoming. *Display and Interface Design: Subtle Science, Exact Art.* Boca Raton, FL: CRC Press.

Berners Lee, T., J. Hendler, and O. Lassila. 2001. The semantic web. *Scientific American* 284(5): 34–43.

Bernstein, M., 1988. The bookmark and the compass: Orientation tools for hypertext users. *SIGOIS Bulletin* 9(4): 34–45.

Borg, I., and P. J. F. Groenen. 1997. *Modern Multidimensional Scaling: Theory and Applications*, Berlin: Springer.

Burigat, S., and L. Chittaro. 2008. Navigation techniques for small-screen devices: An evaluation on maps and web pages. *International Journal of Human–Computer Studies* 66(2): 78–97.

Dreyfus, H. L. 1994. *What Computers Still Can't Do: A Critique of Artificial Reason.* Cambridge, MA: MIT Press.

Fallman, D. 2003. Design oriented human–computer interaction. In *CHI 2003* (Fort Lauderdale, FL, April 5–10), 225–232.

Fischer, G. 2000. Design, learning, collaboration and new media—A co-evolutionary HCI perspective. In *Proceedings of OZCHI 2000*, eds. C. Paris et al. (Sydney, Australia, December) 282–289.

Fischer, G. 2002. Beyond "couch potatoes": From consumers to designers and active contributors. *First Monday* 7(12). http://firstmonday.org/htbin/cgiwrap/bin/ojs/index.php/fm/article/view/1010/931.

Flach, J. M. 2008. Mind the gap: A skeptical view of macro-cognition. In *Naturalistic Decision Making and Macrocognition*, eds. J. M. Schraagan et al., 27–40. Aldershot, England: Ashgate.

Flach, J. M. 2009. The dynamics of experience: A search for what matters. In *Proceedings of the European Conference on Cognitive Ergonomics (ECCE 2009)* (Helsinki, Finland, 30 Sept. to 2 Oct.), 11–18.

Flach, J. M., S. Dekker, and P. J. Stappers. 2008. Playing twenty questions with nature (The surprise version): Reflections on the dynamics of experience. *Theoretical Issues in Ergonomics Science* 9: 125–145.

Flach, J. M., P. F. Jacques, D. L. Patrick, M. Amelink, M. M. van Paassen, and M. Mulder. 2003. A search for meaning: A case study of the approach-to-landing. In *The Handbook of Cognitive Task Design*, ed. E. Hollnagel, 171–191. Mahwah, NJ: Lawrence Erlbaum.

Flach, J., M. Mulder, and M. M. van Paassen. 2004. The concept of the "situation" in psychology. In A Cognitive Approach to Situation Awareness: Theory, Measurement, and Application, eds. S. Banbury and S. Tremblay, 42–60. Aldershot, England: Ashgate.

Flach, J. M., D. Schwartz, A. Bennett, K. Behymer, and W. Shebilski. 2010. Synthetic task environments: Measuring macrocognition. In *Macrocognition: Metrics and Scenarios: Design and Evaluation for Real World Teams*, eds. E. Patterson and J. Miller, 201–284. Aldershot, England: Ashgate.

Flach, J. M., D. Schwartz, A. Bennett, S. Russell, and T. Hughes. 2008. Integrated constraint evaluation: A framework for continuous work analysis. In *Applications of Cognitive Work Analysis*, eds. A. M. Bisantz and C. M. Burns, 273–297. Boca Raton, FL: CRC Press.

Gibson, J. J. 1982. The affordances of the environment. In *Reasons for Realism: Selected Essays of James J. Gibson*, eds. E. Reed and R. Jones, 408–410. Hillsdale, NJ: Lawrence Erlbaum.

Ginsburg, H., and S. Opper. 1969. *Piaget's Theory of Intellectual Development.* Englewood Cliffs, NJ: Prentice-Hall.

Hollnagel, E., D. D. Woods, and N. Leveson. 2006. *Resilience Engineering: Concepts and Percepts.* Burlington, VT: Ashgate.

Hutchins, E. 1995. *Cognition in the Wild.* Cambridge, MA: MIT Press.

James, W. 1909. *The Meaning of Truth: A Sequel to Pragmatism.* New York: Longmans, Green.

Keller, A. I., F. Sleeswijk Visser, R. van der Lugt, and P. J. Stappers. 2009. Collecting with cabinet: Or how designers organise visual material, researched through an experiential prototype. *Design Studies* 30: 69–86.

Kirlik, A., ed. 2006. *Adaptive Perspectives on Human–Technology Interaction.* Oxford, England: Oxford University Press.

Kruskal, J. B., and M. Wish. 1978. Multidimensional Scaling. Beverly Hills, CA: Sage.

Michaelis, J. R., J. Golbeck and J. Hendler. this volume. Organization and structure of information using semantic web technologies. In *Handbook of Human Factors in Web Design*, 2nd ed., eds. K.-P. L. Vu and R. W. Proctor, 231–248. Boca Raton, FL: CRC Press.

Norman, D. A. 1986. Cognitive engineering. In *User Centered System Design: New Perspectives on Human–Computer Interaction*, eds. D. A. Norman and S. W. Draper, 31–61. Hillsdale, NJ: Lawrence Erlbaum.

Norman, D. A., and S. W. Draper. 1986. *User Centered System Design.* Hillsdale, NJ: Lawrence Erlbaum.

O'Reilly, T. 2005. What is Web 2.9? Design patterns and business models for the next generation of software. http://oreilly.com/web2/archive/what-is-web-20.html (accessed Nov. 10, 2010).

Pardue, J. H., J. P. Landry, E. Kyper, and R. Lievano. 2009. Look-ahead and look-behind shortcuts in large item category hierarchies: The impact on search performance. *Interacting with Computers* 21: 235–242.

Pasman, G. 2003. Designing with precedents. Doctoral diss., Delft University of Technology. Delft, Netherlands.

Peirce, C. S. 1982. Deduction, induction, and hypothesis. In *The Writings of Charles S. Peirce: A Chronological Edition.* Compiled by the editors of the Peirce Edition Project, vol. 30. Bloomington: Indiana University Press.

Pejtersen, A. M. 1979. Investigation of search strategies in fiction based on an analysis of 134 user-librarian conversations. In *IRFIS 3 Conference Proceedings*, ed. T. Henriksen, 107–132. Oslo, Norway: Staten Biblioteks-och Informations Hoegskole.

Pejtersen, A. M. 1984. Design of a computer-aided user-system dialogue based on an analysis of users' search behavior. *Social Science Information Studies* 4: 167–183.

Pejtersen, A. M. 1988. Search strategies and database design for information retrieval from libraries. In *Tasks, Errors, and Mental Models: A Festschrift to Celebrate the 60th Birthday of Professor Jens Rassmussen*, eds. L. P. Goodstein, H. B. Andersen, and S. E. Olsen, 171–190. London: Taylor & Francis.

Pejtersen, A. M. 1992. The Bookhouse: An icon based database system for fiction retrieval in public libraries. In *The Marketing of Library and Information Services 2*, ed. B. Cronin, 572–591. London: ASLIB.

Rasmussen, J. 1986. *Information Processing and Human–Machine Interaction: An Approach to Cognitive Engineering.* New York: North Holland.

Rasmussen, J., A. M. Pejtersen, and L. P. Goodstein. 1994. *Cognitive Systems Engineering.* New York: Wiley.

Schrage, M. 2000. *Serious Play.* Boston, MA: Harvard Business School Press.

Siegel, A. W., and S. H. White. 1975. The development of spatial representations of large-scale environments. In *Advances in Child Development and Behavior*, vol. 10, ed. H. W. Reese, 10–55. New York: Academic Press.

Simon, H. 1981. *The Sciences of the Artificial*, 2nd ed. Cambridge, MA: MIT Press.

Sheth, A. P., C. Ramakrishnan, and C. Thomas. 2005. Semantics for the semantic web: The implicit, the formal and the powerful. *International Journal of Semantic Web Information Systems* 1(1): 1–18.

Shneiderman, B. 1983. Direct manipulation: A step beyond programming languages. *IEEE Computer* 16: 57–69.

Schön, D. A. 1983. *The Reflective Practitioner: How Professionals Think in Action.* New York: Basic Books.

Stappers, P. J., and C. C. M. Hummels. 2006. Form confusion in public spaces, or: How to lie with affordances. In *Design and Semantics of Form and Movement. Proceedings of DesForm 2006*, eds. L. Feijs, S. Kyffin, and B. Young, 104–108 (Eindhoven, Netherlands, Oct. 26–27).

Stappers, P. J., and G. Pasman. 1999. Exploring a database through interactive visualised similarity scaling. In *Human Factors in Computer Systems. CHI99 Extended Abstracts*, eds. M. W. Altom and M. G. Williams (Pittsburgh, PA, 15–20 May), 184–185.

Stappers, P. J., G. Pasman, and P. J. F. Groenen. 2000. Exploring databases for taste or inspiration with interactive multi-dimensional scaling. In *Proceedings of the Human Factors and Ergonomics Society* (San Diego, CA, July 29 to August 4), 3-575 to 3-578.

Taylor, F. W. 1911. *Principles of Scientific Management*. New York: Harper.

Teng, H. 2003. Location breadcrumbs for navigation: An exploratory study. Master's Thesis, Dalhousie University, Faculty of Computer Science, NS, Canada.

Thomas, C., and A. Sheth. 2006. On the expressiveness of the languages for the semantic web—making a case for a little more. In *Fuzzy Logic and the Semantic Web,* ed. E. Sanchez, 3–20. Amsterdam: Elsevier.

Thomas, C., A. Sheth, and W. York. 2006. Modular Ontology Design Using Canonical Building Blocks in the Biochemistry Domain. In *Proceeding of the 2006 conference on Formal Ontology in Information Systems: Proceedings of the Fourth International Conference (FOIS 2006)*, 115–127, Amsterdam, NL: IOS Press.

Vicente, K. J. 1999. *Cognitive Work Analysis*. Mahwah, NJ: Lawrence Erlbaum.

von Uexküll, J. 1957. A stroll through the worlds of animal and man. In *Instinctive Behavior*, ed. C. H. Schiller 5–80. New York: International Universities Press.

Wegner, P. 1997. Why interaction is more powerful than algorithm. *Communications of the ACM* 40(5): 80–91.

Woods, D. D. 1984. Visual momentum: a concept to improve the cognitive coupling of person and computer. *International Journal of Man–Machine Studies* 21: 229–244.

Woods, D. D., and E. Hollnagel. 1987. Mapping cognitive demands in complex problem-solving worlds. *International Journal of Man–Machine Studies* 26: 257–275.

Woods, D. D., and E. Hollnagel. 2006. *Joint Cognitive Systems: Patterns in Cognitive Systems Engineering*. Boca Raton, FL: CRC Press.

25 Cognitive User Modeling

Hedderik van Rijn, Addie Johnson, and Niels Taatgen

CONTENTS

The World Wide Web had an estimated 1.7 billion users in 2009 (http://www.internetworldstats.com/stats.htm) and was accessed by people of essentially all possible backgrounds. Each of these users had a goal in mind, whether it be trying to book a flight, search for information on a research topic, or just while away a few hours. Different users also have different knowledge, interests, abilities, learning styles, and preferences regarding information presentation. An increasingly important research area is how interfaces can be designed to recognize the goals and characteristics of the user and to adapt accordingly.

Companies, universities, and other organizations are becoming increasingly aware of the need to personalize Web pages for individual users or user groups. To offer personalized information, it is necessary to monitor a user's behavior and to make generalizations and predictions based on these observations. Information about the user that can be drawn on in this way is called a *user model* (see Fischer 2001, for a review and Pazzani and Billsus 2007, for a more technical introduction). Modeling the user may be as simple as fitting a *user profile* (e.g., single, young, female) or as complicated as discovering expert knowledge (e.g., how a chemist would classify a data set). The modeling system may acquire information explicitly by means of a user-completed questionnaire, or implicitly, by observing user actions and making inferences based on stored knowledge. The goal of user modeling may

be to predict user behavior, to gain knowledge of a particular user in order to tailor interactions to that user, or to create a database of users that can be accessed by others. The goal of user modeling may even be the creation of the model itself, when that model is used to create an autonomous agent to fill a role within a system. In this chapter, emphasis will be on cognitive user models (e.g., Carroll 1997; Ritter et al. 2000), that is, user models that take the cognitive properties (e.g., memory constraints) of the users into account.

We begin this chapter by describing the major modeling systems used to create predictive models of the user, that is, models that can be used to predict human behavior in a human–machine system. All of the models we discuss in this section are characterized by constraints that limit the computational power of the model. The purpose of the constraints is to enable the modeler to build models that operate within the same constraints as humans. The models produced within these systems can be used to test theories about how people learn and perform cognitive tasks. They can also be used for practical purposes such as testing the usability of different human–computer interfaces or inferring the knowledge structure of the user so that appropriate remedial help can be supplied.

The models discussed in the first section of the chapter have as their goal the description of how people, in general, perform a task. In the second section of the chapter we focus on the individual user. Our goal in this section is to describe techniques for gathering information about individual users and to describe how the computer interface can be adapted on the basis of that information. We end the chapter with a discussion of various applications ranging from tutoring programs to autonomous intelligent agents.

25.1 PREDICTIVE MODELS

The usual method of testing a new interface design is to perform a user evaluation study. Potential users are asked to carry out a number of tasks with the new interface and their performance is evaluated. Such user studies are typically time-consuming and costly. An alternative approach is to develop cognitive models of user performance and to use those models to predict behavior. Such models, or "synthetic users," have several advantages over human participants. First, once the model has been constructed, it can be used repeatedly to evaluate incremental changes in design. Second, a cognitive model may offer insight into the nature of the task. In a cognitive model, unlike in a human participant, each reasoning step can be traced, the contents of the model's memory can be inspected to see what the model has learned, and the errors made can be traced back to their source.

The generic user model that can be used to test any user interface is a holy grail of human factors. The current state of the art is that models are developed for specific tasks or aspects of tasks (e.g., menu search, icon identification, deployment of attention, or automatization) and then validated on a case-by-case basis using human data. Although no single generic model exists to test the complete range of applications, several attempts have been made to create models that can predict the outcomes of experiments rather than merely explaining outcomes after the experiments have been conducted (e.g., Salvucci and Macuga 2002; Taatgen, Van Rijn, and Anderson 2007).

The first step in developing a predictive user model is to perform a task analysis (see Strybel, this volume, for a more detailed overview of methods of task analysis). A task analysis gives a specification of the knowledge that is needed to perform the task and the sequence of operations required. Although a task analysis gives an indication of the complexity of the task, it does not generally take into account the details of human information processing. More accurate user models can be created by augmenting the task analysis with a specification of the constraints on human information processing that should be satisfied within the model (see Table 25.1 for a summary of how the major predictive models do this). For example, the Model Human Processor (MHP; Card, Moran, and Newell 1983), described below, provides a means of specifying the time to perform specific operations, the probability of errors, and the speed-up due to learning for the sequence of operations specified in the task analysis.

A more advanced method of incorporating human information processing constraints in task models is to embed the models in an architecture of cognition. A cognitive architecture is a simulation environment that can, given the necessary knowledge to do the task, mimic human behavior on that task. The knowledge specified in a task analysis within a cognitive architecture can be used to make predictions about various aspects of human performance, including reaction times, errors, choices made, and eye movements. A cognitive architecture is thus a simulation environment and also a theory of cognition. For example, cognitive architectures typically incorporate a theory about memory that specifies how knowledge is represented and how memory processes such as storage, retrieval, and forgetting function. This theory is embodied in the simulation environment and governs how the memory system will behave. Simulations of specific tasks are called models. Models are subject to the rules governing the cognitive architecture but also specify the knowledge needed for the task. Because all tasks require some knowledge, only models can be used to make specific predictions; the architecture provides the constraints within which the model is created (Taatgen and Anderson 2008). For example, although the architecture may incorporate a theory of memory search, a specific model of search applied to Web sites is needed to make predictions regarding human performance in Web navigation.

Although it is generally assumed that a model that produces the same behavior as people do is a valid model, several different models may produce the same behavior. In this case, a number of criteria can be applied to choose the "best" model. First, a model should have as few free parameters as possible. Many cognitive architectures have free parameters that can be given arbitrary values by the modeler. Because free parameters enable the modeler to manipulate the outcome of the model, increasing the number of free parameters diminishes the model's predictive power (see Roberts and Pashler, 2000, for a discussion and Shiffrin et al., 2008, for an overview of methods to evaluate models). Second, a model should not only describe behavior but should also predict it. Cognitive models are often made after the experimental data have been gathered and analyzed. A model with high validity should be able to predict performance. Finally, a model should learn or build upon its own task-specific knowledge. Building knowledge into a model increases its specificity but may decrease its validity. In an ideal situation, the model is provided with knowledge similar to the knowledge a novice user brings to a task and learns task-specific knowledge while performing the task.

Most of the current approaches to predictive modeling use task analysis to specify the knowledge that an expert would need to do the task. This violates the criterion that a model should acquire task-specific knowledge on its own. Moreover, basing a model on a task analysis of expert performance means that the model is of an expert user, whereas the typical user may not have mastered the task being modeled. Useful predictions and a complete understanding of the task require that models be built that start at the level of a novice and gradually proceed to become experts in the same way people do. In other words, many applications require models that not only perform as humans do but that also learn as humans do.

TABLE 25.1

Overview of Constraints that the MHP, SOAR, EPIC, and ACT-R Impose on Cognitive Processing

Process	Model	Constraint	Reference
Working memory			
	MHP	Working memory has a limited capacity of 5–9 chunks and decays in 900–3500 ms.	Card, Moran, and Newell (1983)
	SOAR	Limitations of working memory arise on functional grounds usually owing to lack of reasoning procedures to properly process information.	Young and Lewis (1999)
	ACT-R	Limitations of working memory arise from decay and interference in declarative memory. Individual differences are explained by differences in spreading activation.	Daily, Lovett, and Reder (2001)
Cognitive performance			
	MHP	The cognitive processor performs one recognize-act cycle every 25–170 ms, in which the contents of working memory initiate actions that are linked to them in long-term memory.	Card, Moran, and Newell (1983)
	SOAR	A decision cycle in SOAR takes 50 ms, multiple productions may fire but only when leading to a single choice.	Newell (1990)
	ACT-R	A production rule takes 50 ms to fire, no parallel firing is allowed. A rule is limited to inspecting the current contents or states of the peripheral buffers and memory-retrieval buffer and initiating motor actions (e.g., hand or eye movements) and memory-retrieval requests.	Anderson et al. (2004)
	EPIC	Production rules take 50 ms to fire, but parallel firing of rules is allowed.	Kieras and Meyer (1997)
Perceptual and motor systems			
	MHP	Perceptual processor takes 50–200 ms to process information, motor processor 30–100 ms. Duration of motor actions is determined by Fitts's law.	Card, Moran, and Newell (1983)
	EPIC	Perceptual and motor modules are based on timing from the MHP. Modules operate asynchronously alongside central cognition.	Kieras and Meyer (1997)
	ACT-R; SOAR	Both use modules adapted from EPIC.	Byrne and Anderson (2001); Chong (1999)
Learning			
	MHP	Speed up in performance is according to the power law of practice.	Card, Moran, and Newell (1983)
	SOAR	Learning is keyed to so-called impasses, where a subgoal is needed to resolve a choice problem in the main goal.	Newell (1990)
	ACT-R	Learning is based on rational analysis in which knowledge is added and maintained in memory on the basis of expected use and utility.	Anderson et al. (2004)

25.2 MODELS OF TASK PERFORMANCE

An established technique for describing the knowledge needed to perform a task at an expert level is GOMS (Card et al. 1983), in which a task is analyzed in terms of goals, operators, methods, and selection rules. In the GOMS methodology, goals form a hierarchy starting from the top goal that represents achieving the end result and proceeding to the so-called unit tasks, which are subgoals that cannot be further decomposed. For example, the top goal could be to fill out a Web form, while a goal at the unit task level might be to enter your street address in an input box. A general assumption is that unit tasks can be completed in the order of 10 seconds. To achieve a goal at the unit task level, methods (or interactive routines; Gray et al. 2006) are needed to specify what actions are to be carried out in terms of the operators that perform the actions. A method for moving to a certain input box in a Web form might specify applying the operator "press tab" until the desired box is reached. A different method might specify the operator "move the mouse to the required input box, and click." When a choice must be made

between alternative methods, selection rules are needed that specify when a certain method should be used (e.g., Use method 1 unless the cursor is more than three input boxes away, in which case use method 2).

A GOMS analysis enables a cognitive walkthrough of the task of interest. After specifying the top goal, the order in which subgoals are posed and attained, the operators used to attain them, and the selection rules applied to methods or operators, the number of steps and possible choices made to achieve the main goal can be described. This level of analysis allows a description of the order in which an expert user will execute actions. This will, in general, not be of much interest given that the GOMS analysis itself is based on the behavior of the expert. In order to make more interesting predictions it is necessary to augment the GOMS analysis with a psychological model. The first psychological model to be used in conjunction with GOMS was the MHP (Card et al. 1983). Figure 25.1 shows a simplified version of the MHP, comprising a memory system (working memory and long-term memory) and three processors (perceptual, cognitive, and motor). Each of the processors is assigned an approximate cycle time.

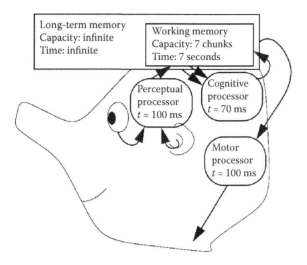

FIGURE 25.1 A simplified version of the MHP comprising a memory system (working memory and long-term memory) and three processors (perceptual, cognitive, and motor).

For example, the motor processor needs on the order of 100 ms to prepare a motor response and working memory has a limited capacity (about seven elements) and a limited retention time (about 7 seconds). The combination of the MHP and GOMS can be used to make simple predictions about quantitative aspects of human performance. The MHP can be used to annotate the analysis that is made with GOMS by supplying approximate execution times. It can also trace working-memory usage and signal potential capacity problems.

A limitation of MHP-GOMS analyses is that it must be assumed that behavior is serial in nature. Thus, the technique cannot explain any effects of multitasking at the methods level. For example, if the unit task is to enter your street address, a MHP-GOMS model cannot account for any parallelism between the memory-related methods required for retrieval of the to-be-entered information, and motor methods to move the mouse to the right input box. To incorporate effects of parallelism, CPM-GOMS (which stands for either Cognitive-Perceptual Motor GOMS or Critical Path Method GOMS) has been proposed (Gray, John, and Atwood 1992). CPM-GOMS enables the specification of multiple parallel paths and has a "relaxed" control structure that allows for interruptions or changes based on new information in the external world. After specifying all actions and order constraints that are relevant for a certain task, CPM-GOMS determines the sequence of actions that determines the shortest time possible to complete the task. This sequence is called the critical path.

Because of the parallelism and more relaxed control structure, CPM-GOMS can better account for embodied cognition phenomena such as the interleaving and interdependencies between cognitive, perceptual, and motor operations. In principle, the critical path, representing optimal behavior, can be determined by means of a CPM-GOMS analysis. However, for complex tasks, it is likely that trial-to-trial variability in the user's behavior influences this critical path. Patton and Gray (submitted manuscript) referred to this as "procedural

variability amidst strategic constancy." Their SANLab-CM system (Stochastic Activity Network Laboratory for Cognitive Modeling; Patton and Gray, submitted; Patton, Gray, and Schoelles 2009) allows for the modeling of complex activity networks as in CPM-GOMS and provides built-in tools for running Monte Carlo simulations that take into account the variability associated with the elements that make up the models.

25.3 PROCESSING ARCHITECTURES THAT INTERACT WITH THE OUTSIDE WORLD

GOMS-MHP approaches are limited in several important respects: They (1) predict task execution times only approximately (e.g., no differences are predicted in the latencies for difficult and easy memory retrievals), (2) cannot easily explain errors (but see Gray, 2000, for a notable exception), and (3) are vague regarding what can be done in one processing cycle. More precise and constrained predictions can be made with the more elaborate theoretical framework of the Executive Process-Interactive Control (EPIC) architecture (Kieras and Meyer 1997). Contrary to most GOMS-based systems, EPIC allows the implementation of process simulations of the user that can be run on a computer and that can be used to test the soundness of an analysis and to provide a concrete prediction of task performance.

The knowledge of EPIC models is represented using production rules. A production rule consists of a set of conditions that is tested against the current internal state and state of the modules, and a set of actions that is carried out once all conditions are satisfied. Production rules are a fairly universal way of representing knowledge: Although the exact details of the syntax differ, almost all cognitive architectures use production rules.

The main theoretical goal in an architecture such as EPIC is to constrain all possible simulations of behavior to only the behavior that in principle could be exhibited by users. As can been seen in Table 25.1, EPIC's main source of constraints is in the perceptual-motor aspects of the task. Central cognition is relatively unconstrained. The perceptual-motor modules in EPIC can accommodate only a single action at a time, and each of these actions takes a certain amount of time. Expert behavior on a task is exemplified by skillful interleaving of perceptual, cognitive, and motor actions. EPIC's modules incorporate mathematical models (e.g., Fitts's law) of the time it takes to complete operations that are based on empirical data.

EPIC has been applied in a number of contexts. For example, Hornof and Kieras (1997; described by Kieras and Meyer 1997) applied EPIC to menu search. The task modeled was to find a label in a pull-down menu as quickly as possible. Perhaps the simplest model of such a task is the serial search model in which the user first attends to the top item on the list and compares it to the label being searched for. If the item does not match the target, the next item on the list is checked; otherwise, the search is terminated. EPIC's predictions of search time using this method can be obtained by describing the strategy in EPIC production rules and performing a

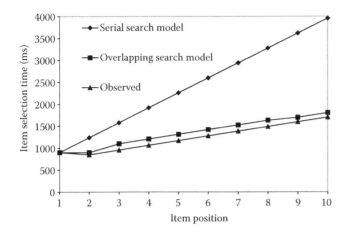

FIGURE 25.2 Actual search times (obtained with human subjects) and predictions of the serial search model and the overlapping search model.

simulation in a menu search test environment. As shown in Figure 25.2, such a model overestimates actual search time (obtained with human subjects), except when the target is in the first position to be searched.

In an alternative model, the overlapping search model, the parallelism of the cognitive system is exploited. Instead of waiting for the cognitive system to finish deciding whether or not the requested label is found, the eye moves on to the next item in the list while the first item is still being evaluated. A result of such a strategy is that the eye has to move back to a previous item in the list once it has been decided that the item has been found, but this is a small price to pay for the speed-up this parallelism produces (see Figure 25.2). Parallelism is allowed in EPIC as long as no two perceptual-motor modules are used at the same time. In practice, the most influential constraint is posed by the duration of actions. For example, within the serial search model, the parameter that influences the slope could be set to a value that produces output similar to human data. However, the value of this parameter would violate assumptions about the minimum time needed to make eye-movements and decisions as to whether the label is correct or not.

25.4 ARCHITECTURES THAT INCORPORATE LEARNING

As a cognitive architecture, EPIC states that the main sources of constraints on human performance are the perceptual and motor aspects of the task. However, the flexibility of the theory prevents the prediction of task performance before any data have been gathered. For example, although the theory allows a certain degree of parallelism for central cognition, it cannot predict a priori how much parallelism will be exploited in a given task. On the other hand, EPIC is overly constrained with respect to learning, as no learning mechanisms have been incorporated in the architecture. Because of this, a set of models that represents each of the different

levels of expertise of interest must be created to compare performance at different levels of expertise. One approach to developing a stronger theory capable of specifying the most plausible model of task performance is to incorporate learning mechanisms that make it possible for task models to emerge from the cognitive architecture rather than relying on knowledge supplied by the modeler.

An example of an architecture incorporating mechanisms of learning is SOAR (States, Operators, and Results; Newell 1990). The SOAR approach is to view all intelligent behavior as a form of problem solving. In SOAR, new knowledge is learned when impasses are encountered during problem solving. These impasses often take the form of a choice in which there are several possible actions available and no clear decision rule for selecting the appropriate one. Faced with such a choice problem, SOAR evaluates each of the possible actions, and selects the best one. The by-product of this evaluation process is a new rule that will, when SOAR is faced with a similar situation, enable the correct choice to be made without invoking evaluation processes. The focus of SOAR is on accounting for how humans interact with their external environment in terms of possible actions. Because of this emphasis, SOAR does not distinguish between factual, declarative knowledge (such as the fact "the Web site of the White House is http://www.whitehouse.gov") and procedural knowledge (such as routines for guiding initial eye movements on a Web page).

An example of a SOAR model in the domain of information localization is Ayn (Howes 1994). Ayn models the task of finding an item in a typical drop-down menu structure containing several labeled menus each of which contains several items. The model makes predictions regarding how many items will have to be searched before the correct item is found. Initially, the model searches all menus exhaustively. However, during a search it generates new rules that are subsequently used to decrease search times. Eventually, the model learns exactly where to look for each menu item. In one example, the model searches for a menu option that performs a certain action. The model starts out with no knowledge of where to find the information and thus is faced with an impasse: There are a number of potentially applicable menus, but the model cannot choose between them. In this situation, the model simply tries out the options successively, exploring all options until the correct one is found. Two new production rules result from success in reaching the goal item: a rule that specifies an incorrectly chosen submenu as the wrong menu for the current goal, and a rule that specifies the correct submenu.

SOAR's learning is purely symbolic and not subject to forgetting. Learning in SOAR can be characterized as an all-or-none process in which actions are learned immediately and remain available. It is obvious that not all human learning fits this characterization. Another limitation of the SOAR approach is that its learning is tied to impasses in which the model explicitly searches for new strategies. This seems at odds with the phenomena of implicit learning (Reber 1989) and statistical learning (Saffran, Aslin, and Newport 1996;

see also Perruchet and Pacton 2006), in which new knowledge may be acquired without the problem solver being conscious of it. Nevertheless, the constraints in SOAR's learning mechanism have empirical consequences for user modeling and improve the validity of models constructed within the SOAR framework (Howes and Young 1997). Furthermore, recent efforts have increased the cognitive plausibility of the SOAR architecture by including, for example, mechanisms to account for statistical learning (Nason and Laird 2005) and declarative memory phenomena (e.g., data chunking; Rosenbloom 2006). Laird (2008) gives a complete overview of the mechanisms recently added to SOAR.

25.5 HYBRID ARCHITECTURES

One way of overcoming the shortcoming of purely symbolic knowledge representations is to attach numeric quantities (e.g., "activation") to the knowledge elements in the architecture. Such an approach is sometimes referred to as *subsymbolic representation*. Architectures that use both symbolic and subsymbolic representations are referred to as hybrid architectures. Recent years have seen an increase in the number of hybrid architectures (see, e.g., Laird 2008; Sun 2006). Although these hybrid architectures provide new and interesting views on cognitive user modeling, the system that is most frequently used* in the context of cognitive user modeling is ACT-R.

ACT-R (Adaptive Control of Thought-Rational; Anderson 1993, 2007; Anderson et al. 2004; Anderson and Lebiere 1998) is an example of a hybrid architecture that supports learning and subsymbolic computations. The core of the ACT-R architecture is a production system similar to that of EPIC and SOAR. This production system is accompanied by a set of modules. A core module is the declarative memory system that is used to store facts. Facts in declarative memory have activation values that reflect how often a fact has been used in the past and its association with the current context. Activation determines how much time it takes to retrieve a fact from memory and whether it can be retrieved at all. The perceptual and motor modules that serve as an interface between ACT-R and the external world (Byrne and Anderson 2001) are similar to the modules used by EPIC. Multiple architectural refinements have been made over the past years to improve ACT-R's applicability to user modeling. For example, new mechanisms have been proposed to better account for short-term memory phenomena (Van Maanen and Van Rijn 2007, 2010; Van Maanen, Van Rijn, and Borst 2009) and declarative learning (Pavlik and Anderson 2005), the subsymbolic activations of production rules are now learned by reinforcement learning (Fu and Anderson 2006), and a new multitasking theory has been implemented that does not require a central executive (Salvucci and Taatgen 2008, 2010) but instead focuses on cognitive bottlenecks to explain multitasking behavior (e.g., Borst, Taatgen, and Van Rijn 2010). Apart from these changes to default modules, new modules have been proposed to account for temporal cognition (Taatgen, Van Rijn, and Anderson 2007; Van Rijn and Taatgen 2008) and eye movements (Salvucci 2001).

ACT-R modules communicate with the production system through buffers. For example, the visual buffer contains the currently attended visual element and the retrieval buffer contains the most recent retrieval from declarative memory. In addition to the buffers corresponding to the modules, a goal buffer is used to hold the current goal. Similarly, a problem state module is used to store the current state of the system, such as partial solutions of the current goal. Production rules have an associated utility value that reflects the success of the rule in the past. The utility is determined by reinforcement learning and expresses the reward resulting from past applications of a production rule. On each cycle of the production system, the rule with the highest utility is chosen from the rules that match the current contents of the buffers.

Byrne (2001) developed an ACT-R model of menu search on the basis of eye-movement data collected during menu search tasks. These data suggest that people do not use a single strategy to search menus, as predicted by previous models. Byrne modeled this by incorporating several competing strategies in the model, and by using ACT-R's utility learning mechanism to determine which strategy should be used. Byrne's model is a good example of a case where both learning and subsymbolic computation are needed to explain the full breath of behavior.

A more elaborate illustration of the type of modeling possible with ACT-R is based on a simplified air traffic control task (KA-ATC; Ackerman 1988). The model of the task is explained in detail by Taatgen (2002) and Taatgen and Lee (2003). In this task, participants land planes by choosing a plane that is waiting to be landed and designating the runway on which the plane should land. There are four runways, the use of which is restricted by rules that relate to the length of the runway, the current weather, and the type of plane that is to be landed. For example, a DC-10 can only be landed on a short runway if the runway is not icy and the wind is below 40 knots. Although participants receive an extended instruction period, they tend to forget some rules—especially the more complicated ones regarding weather, plane type, and runway length. The goal of the ACT-R model was to capture the learning in this task by predicting the improvement in performance of the participants at both a global level and at the level of individual keystrokes.

An example of a production rule from the air traffic control task follows:

IF	The goal is to land a plane and a plane has been selected that can be landed on the short runway (match of goal buffer)
AND	you are currently looking at the short runway and it is not occupied (match of visual buffer)
AND	the right hand is not used at this moment (match of manual buffer)

---

* See for an overview of ACT-R models related to user models http://act-r.psy.cmu.edu/publications/index.php?subtopic=49, and for related work on information search http://act-r.psy.cmu.edu/publications/index.php?subtopic=51.

THEN note that we are moving to the short runway (change to goal buffer)

AND push the arrow-down key (change to manual buffer)

AND move attention to the weather information (change to visual buffer)

This rule reflects the knowledge of an expert on the task at the stage in which a plane has been selected that has to be directed to the short runway. After checking whether the short runway is available, it issues the first motor command and also initiates an attentional shift to check the weather, information that might be needed for landing the next plane.

Although this example rule is very efficient, it is also highly task specific; rules like this have to be learned in the process of acquiring the skill. For novices, the model assumes that all the task-specific knowledge needed about air traffic control is present in declarative memory, having been put there by the instructions given to participants. This knowledge has a low activation because it is new and might have gaps in it in places where the participant did not properly memorize or understand the instructions. The production rules interpret these instructions and carry them out. Two examples of interpretive rules follow:

Get next instruction rule
> IF the goal is to do a certain task and you have just done a certain step (goal buffer)
>
> THEN request the instruction for the next step for this task (retrieval buffer)

Carry out a push key rule
> IF the goal is to do a certain task (goal buffer)
>
> AND the instruction is to push a certain key (retrieval buffer)
>
> AND the right hand is available (manual buffer)
>
> THEN note that the instruction is carried out (goal buffer)
>
> AND push the key (manual buffer)

A characteristic of interpreting instructions is that it results in behavior that is much slower than that of experts: Retrieving the instructions takes time and during this time not much else happens. Also, parts of the instructions might be forgotten or misinterpreted, leading to even more time loss. In such cases, the model reverts to even more general strategies, such as retrieving past experiences from memory:

Decide retrieve memory rule
> IF you have to make a certain decision in the current goal (goal buffer)
>
> THEN try to recall an experience that is similar to your current goal (retrieval buffer)

Decide on experience rule
> IF you have to make a certain decision in the current goal (goal buffer)

AND you have retrieved a similar experience that went well (retrieval buffer)

THEN make the same decision for the current goal (goal buffer)

This experience-based retrieval strategy retrieves the experience with the highest activation from declarative memory and is based on the assumption that experiences with a high activation are potentially the most relevant in the current situation.

The transition from novice to expert is modeled by ACT-R's mechanism for learning new rules, production compilation (Taatgen and Anderson 2002). This mechanism takes two existing rules that have been used in sequence and combines them into one rule, given that there are no buffer conflicts (for example, as would be the case when both rules specify using the right hand). An exception is requests to declarative memory: If the first rule requests a fact from declarative memory, and the second rule uses it, the retrieved fact is, instead, substituted into the new rule. This substitution procedure is the key to learning task-specific rules. For example, the two rules that retrieve an instruction and push a key, together with the instruction to press "enter" when the arrow points to the right plane during landing, would produce the following rule:

> IF the goal is to land a plane and your arrow points to the right plane (goal buffer)
>
> AND the right hand is available (manual buffer)
>
> THEN note that the instruction is carried out (goal buffer)
>
> AND push enter (manual buffer)

A rule that retrieves and uses old experiences can also be the source for production compilation. For example, in a situation in which the plane to be landed is a DC-10 and the runway is dry, and a previous example in which such a landing was successful on the short runway is retrieved, the following rule would be produced:

> IF you have to decide on a runway and the plane is a DC-10 and the runway is dry (goal buffer)
>
> THEN decide to take the short runway (goal buffer)

New rules have to "prove themselves" by competing with the parent rule, but once they are established they can be the source for even faster rules. Eventually, the model will acquire a rule set that performs like an expert. Comparisons with data from experiments by Ackerman (1988; see Taatgen and Lee 2003) show that the model accurately predicts the overall performance increase (in terms of number of planes landed) and the individual subtasks (e.g., how much time is taken to land a single plane; see Figure 25.3). The model is less accurate at the level of individual keystrokes because it cannot interleave actions as efficiently as do human subjects (but see Salvucci and Taatgen 2010, for a remedy for this).

An advantage of a model like that of the air traffic control task is that it can serve as a basis for a more general test bed for interface testing. Task-specific knowledge is entered into

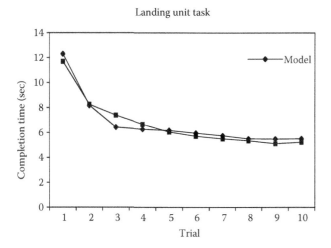

FIGURE 25.3 The actual (square markers) and predicted time (diamond markers) to perform the landing unit task for trials 1 to 10 in the air traffic controller task described in Taatgen and Lee (2003). Figure shows an updated and improved model fit of Figure 5a in Taatgen and Lee (2003).

the system as declarative knowledge, which is very close in form to the instructions provided to the learner. The model can consequently be used to study initial performance and the learning process. Accurate models of human performance also serve as a basis for more advanced forms of individualized user models, which we will discuss in the next section.

25.6 PERSONALIZED MODELS

Predictive models can provide helpful indications of how most people will approach and perform tasks. However, both the efficiency with which information can be accessed from the World Wide Web and the satisfaction of the user in doing so can be enhanced by adapting Web sites to individual users, taking into account their different preferences, knowledge, and goals (see Stephanidis et al., this volume, for more discussion of adaptive and intelligent interfaces). In this section, we describe techniques for determining user characteristics and the need for adaptation, the types of modifications that might be made to adapt Web sites to individual users, and the methods for carrying out the adaptations.

A distinction can be made between adaptable and adaptive systems. *Adaptable systems* (Scerbo 1996, 2007) allow the user to configure the system to suit individual needs. For example, the user may turn the task of checking spelling during typing over to the software or may turn "autofill" for passwords in Web forms on or off. Once a user allocates a task to the computer in such a system, the computer continues to perform the task until the user modifies the allocation. In *adaptive systems,* the system is designed such that it can modify its behavior itself. Such a system can detect the need to take over or relinquish control of certain tasks and can automatically reallocate them. An example would be a system that detects that many spelling errors are made and automatically invokes online spell checking.

Two things are crucial for an adaptive system to work: the existence of a means to adapt the task and the ability to detect the need for adaptation. It is, in principle, possible to adapt tasks to the observer's physical state. For example, cardiovascular measures or skin conductance responses could be used to detect that a user of a system is under stress and needs either additional support or needs to be relieved of some subtasks. However, most adaptive systems react on the basis of a user model that gives an indication of the user's current knowledge, interests, or activity. To construct these user models, information needs to be available that characterizes the current knowledge of the user.

25.6.1 DETERMINING USER CHARACTERISTICS

The starting point for any user-based adaptation is the user model (Fischer 2001). Many Web sites solicit information about the user directly by means of a questionnaire (e.g., Fink and Kobsa 2002). Questions are posed about demographic and personal characteristics, and this information is used to create a user profile. The profile of a given user can then be matched to user types stored in a database using a number of techniques. Generally, the determination of how or when to adapt to the user is based on grouping users on some set of features, assuming a degree of homogeneity within the group and then performing the adaptations that should benefit the average user.

Many user-modeling systems use the technique of predicting unknown characteristics of the user by comparing him or her to similar users. Many techniques have been developed for this "collaborative" or "clique-based" filtering, such as different types of clustering algorithms, correlation-based approaches, vector-based similarity techniques, or Bayesian networks or mixtures models (e.g., Huang and Bian 2009; Kleinberg and Sandler 2008). For example, in a correlation-based approach, known characteristics of the current user are used to match (as indicated by a correlation coefficient) the user to other users in the system. Predictions of unknown user characteristics are then computed based on a weighted aggregate of the stored users. One problem in using this technique is determining the number of matches to include in the computations of the predictions. Prediction accuracy tends to be a nonlinear function of increasing the number of matches (or "mentors"), with increments in accuracy decreasing or even reversing as the number of matches becomes large (e.g., Shardanand and Maes 1995). The optimal number of matches will depend on the strengths of the individual correlations. Although standard machine learning techniques can be applied to user modeling, Webb, Pazzani, and Billsus (2001) argue that the dynamic nature of user models needs to be taken into account for machine learning-based user modeling. One of the dynamic aspects discussed by Webb et al. is "concept drift" (see also Forman 2006; Koren 2009). Concept drift refers to the change in the meaning of certain labels over time (e.g., the meaning of "notebook" depending on whether one is shopping for stationary or a new computer) and the often sudden change in interest in certain labels (e.g., a drop in interest in "Lake Tahoe snow" after returning from a skiing trip).

Although user characteristics are most commonly determined on the basis of explicit measures of user interest such as answers to a questionnaire or commercial transactions, information or tasks may also be adapted on the basis of what the user appears to be trying to do. For example, the system may recognize that the user is experiencing difficulty performing a task and that help needs to be given. Research on adaptive systems in general (e.g., in aviation and process control) has focused on gathering physiological information about the user that can be used to infer the operator state. However, because these measures (e.g., electroencephalographic and cardiovascular measures) are not readily applied in Web settings, they will not be discussed here. Instead, we focus on behavioral measures of user performance such as actions taken by the operator and the manner in which they are made. Actions may range from mouse button clicks to eye movements, and the measures may range from simply registering that an action has occurred to measuring reaction time or more complex patterns of action. The most practical way to gather information about the user in most Web usage situations is by examining the actions performed by the user using a keyboard, touch screen, or mouse. Common measurements are links accessed and time spent viewing a page. On the basis of these measurements, the Web site attempts to infer the knowledge and interests of the user.

The knowledge the user possesses can be divided into what was known before visiting a Web site and what was learned from the Web site itself. For example, if the user has visited a particular page and spent considerable time on that page, it can be assumed afterward that he is familiar with the information presented there. Information about which sites have been visited can thus be used to modify the presentation of links on subsequent pages. Patterns of actions allow other inferences. For instance, rapid scanning through presentations of information can be assumed to reflect previous acquaintance with (or lack of interest in) the presented information.

A number of systems that use various sources of user information, including eye gaze direction, to determine user interests and information needs have been developed. These systems are "attentive" both in attending to what users do and in presenting information that users will want to attend to. The goal of such systems is to track user behavior, model user interests, and anticipate the actions and desires of the user. For example, the *Simple User Interest Tracker* (*Suitor;* Maglio et al. 2001) analyzes gaze direction and browsing behavior to determine the interests of a user (e.g., which headlines in a news window receive the most interest) and then provides information that should be of interest to the user in a timely manner. The new information is displayed in a peripheral display so that the user can decide whether or not to use the information without being distracted by sudden adaptations.

25.6.2 UPDATING THE USER MODEL ON THE BASIS OF INTERACTIONS

Many Web sites can be described as hypermedia environments (settings in which networks of multimedia nodes are connected by links that control information presentation and image retrieval). Such systems have been in use for a number of years, both on the Internet and off-line. Much of the early research on hypermedia environments took place in educational settings. Proponents of these systems emphasized that such environments enable learner control, thus increasing learner involvement and teaching learners to "learn how to learn" (i.e., to develop metacognitive skills; Large 1996). However, the randomness with which some learners move through hypermedia environments may limit the effectiveness of the learning (Federico 1999). User modeling can be an effective tool to increase the efficiency of hypermedia use. A simple example of this related to Web use is the tracking of choices in the search for information. Records of the navigational paths followed (so-called "audit trails") may be maintained as users search for information in hypermedia environments. For example, the browser or an independent application might keep track of all pages visited and of the sequences of mouse clicks or menu choices made. It has been argued that audit trails can provide insight into the cognitive and metacognitive processes of the learner that can be used to implement adaptive instructional strategies (Federico 1999; Milheim and Martin 1991).

Once an audit trail has been collected, the problem remains of what to do with it. One approach to using this information is to make comparisons across users to determine the search preferences of groups of users. One way of comparing paths is to compute a linearity function. This can be done by finding the ratio of (1) the number of visits to each node in the knowledge network from a parent node to (2) all node visits (Horney 1993). A hypermedia system that computes linearity functions online can then structure information to be compatible with the user's search preferences. Various classifications of users have been made on the basis of navigational paths. On the basis of frequency counts of the number of screens visited, buttons clicked, time spent on each screen, and the depth of search, it is possible to distinguish between users (1) who are driven by performance goals (*knowledge seekers*), (2) who like to find and explore features such as movies (*feature explorers*), (3) who are interested in determining the layout of the site (*cartographers*), and (4) who are not motivated to learn to use the site (*apathetic users*; Barab, Bowdish, and Lawless 1997; Lawless and Kulikowich 1996). Knowledge of the group to which the user belongs can be used to adapt the interface to the user's interests or to give users feedback to enhance their productivity. An example of a Web system that uses metacognitive support is met.a.ware (Stadtler and Bromme 2007). The met.a.ware system supports laypersons' Internet research for medical information by metacognitive prompting, for example, by asking evaluation questions on the quality of the source of medical information or how much information is still needed on a certain topic.

Patterns of information acquisition can be used in different ways to infer and adapt to user strategies. For example, a given user may wish to quickly get an overview of a site and then return to objects that were found especially interesting. Such a user should be provided with a bookmark or annotation tool that facilitates the return to visited objects. Deviations in search type might also signal a need for intervention. For example, a user who begins a session by searching for a target (as indicated by increasing depth in a hierarchy), and who then starts making erratic jumps outside of the hierarchy, could be assumed to be lost and in need of guidance. Much time can be wasted on fruitless searches or in trying to find one's way back to a previous point. A Web application that can detect that a user is lost and offer a way back (e.g., by presenting a list of recently visited sites or the front end of a search engine) might increase search efficiency. Given that many users underutilize history lists and other navigation functions in Web browsers, such a facility might be needed (Cockburn and Jones 1996, Cockburn et al. 2003).

25.6.3 ADAPTIVE ELEMENTS

The design of an adaptive system depends on the existence of aspects of the task that can be adapted or automated. Table 25.2 presents a summary of the kinds of tasks and information that might be adapted in a Web site. Choices regarding what to adapt will depend on the sort of information available on a Web site and the way in which it is to be used. As an illustration, consider a museum Web site. In addition to a gift shop and information about opening times and exhibition schedules, the Web site might contain pictures of, and detailed and specialized information about, the exhibited objects. Such a Web site could be used for different purposes. It might be used as a knowledge base, giving access to specific information for research purposes. It might also be used to provide a virtual tour, allowing users who are unable to physically visit the museum to view the objects. Users who have visited the museum and who want to look up background information about objects they found interesting might also visit it. Although the same database may contain all of the information needed to accommodate these different groups of users, it will have to be accessed and presented in different ways. Thus the success of the Web site will depend on the extent to which the wishes of the visitor can be detected and the information can be adapted.

The most obvious adaptation is to tailor the presentation of information. In our museum example, presenting all of the information pertaining to a certain object could be overwhelming and impractical. On the one hand, casual visitors are not likely to be interested in detailed technical information or references to scientific publications but probably would want to view a reproduction and might be interested in buying one. On the other hand, a researcher is likely to be more interested in detailed information and references and less interested in the souvenir shop. An adaptive interface might note the initial choices made by the user and adapt subsequent displays to highlight the most likely to be sought after information.

One might also adapt the format of the information presented to the user. Some users will prefer to view pictures of the objects, whereas others will only want to read background information. More generally, the type of question asked by the user may give the Web site information about how to best present the requested information. For example, "Who was Monet?" is obviously a bibliographic question and should be answered with bibliographic information. However, the query system may also be able to detect the generality of the question and could safely assume that the viewer is not familiar with the painting style of the artist and thus also present representative works (Smeulders et al. 2002; Van Maanen 2009).

Another aspect of adaptive automation is *task support*. This might involve the scheduling of tasks or automating the execution of tasks. This type of adaptivity is becoming more common in Web applications. For example, if a Web site detects that a visitor at the museum site repeatedly returns to a particular object, it might make a quick search of the gift shop and present any object-related products that are available.

The sequence and manner in which information is entered can also be adapted to the task or user. Fill-in forms are ubiquitous on the Web. By adapting the forms to the user, the filling in of these forms can become less tedious. A simple example is remembering details for specific users so that they need not be repeatedly entered.

The final element that might be adapted is the choices available to the user. As discussed in the applications section, below, choices may be informed or modified in a number of additional ways, such as by restricting the availability of links or by offering new links (e.g., informing a user who chooses to view a particular work of art which other works were chosen by people who enjoyed that one).

Another simple adaptation, "smart menus" (referred to as "personalized menus" in Microsoft products), has been incorporated into many programs. Smart menus reduce search times for finding a desired item and reduce movement time to select the item (Jameson 2002) by displaying only those menu commands that have been used most often and most recently. Because only the few options that the user is likely to consider are displayed, the need to search the full menu and move the cursor a long distance is eliminated. If the user wants to see the remaining commands, he must indicate that he wants them to appear by positioning the cursor on an "expand" cue to cause the full menu to appear so that the command can be searched for in the usual manner. Of course, the use of smart menus is to some degree limited by the inconsistency of the mapping of menu items to locations, the additional cognitive resources needed to realize that an option is available in the current menu, but currently hidden,

TABLE 25.2

Adaptable Aspects of Web Use

Information presentation	Selecting the right information
	Presenting information in the best format
Task support	Scheduling of tasks
	Automating tasks
Information input	Guiding information input
	Checking information input
Adapting choices	Presenting appropriate links
	Coding links

and the additional motor resources required to expand the reduced menu.

25.6.4 Empirical Evaluation of User Models

An important goal of user modeling is to aid users of a system. However, the literature is filled with proposals for user model-based innovations that have never been implemented. This might be due to difficulties involved in testing user models. For example, many user models require longer training periods than can easily be implemented in controlled experiments, or controlled experiments may not be suited for the free interaction styles of interest in the actual use of the proposed system. To aid researchers working in the field of user modeling, Chin (2001) presents rules of thumb for experimental design and useful tests for covariates and discusses common threats to experimental validity.

Another obstacle to implementing user models is that a particular user model-based adaptation can be useful in some, but not all, settings. An example of this is the smart menu adaptation discussed earlier. In an experiment on the reaction times associated with different menu adaptation methods (e.g., adaptable versus adaptive versus mixed-initiative menu interfaces), Al-Omar and Rigas (2009) showed that the total number of items in a menu determines which adaptation style performs best. Surprisingly, performance of the adaptable method was most efficient overall in menus with fewer menu items but least efficient in menus with more items.

A last word of caution with respect to the application of user models is that several studies (e.g., Findlater and McGrenere 2004; Jameson and Schwarzkopf 2002) have shown that personal interaction styles can to a large extent influence the perceived or objective benefit of adaptation. For example, Findlater and McGrenere reported that although most users preferred an adaptable interface to two alternatives (adaptive and mixed-initiative) interface, the users who preferred the adaptive interface expressed strong support.

25.7 APPLICATIONS

In this section we describe a range of applications that incorporate user models. The applications vary in their degree of adaptability and in the dynamism of the model.

25.7.1 Evaluating Web Pages

In order to be able to formally evaluate how the information on or structure of Web pages supports or hinders users in their search for information, we have to know how people seek information on the World Wide Web. A number of theories have been developed in recent years to explain human search behavior (e.g., CoLiDeS, Blackmon, Kitajima, and Polson 2005; WUFIS, Chi et al. 2001; MESA, Miller and Remington 2004; SNIF-ACT, Pirolli 2005). One promising theory is SNIF-ACT (Pirolli 2005; Fu and Pirolli 2007), which stands for Scent-based Navigation and Information Foraging in the ACT architecture. Central to the SNIF-ACT

theory (and WUFIS) is an account of how people use information scent cues, such as the text associated with links, to modify their behavior. A typical example of current day Web design that improves information scent is the presentation of snippets from Web pages shown by search engines such as Google and Bing. By returning a snippet of a Web page on which a hit to the query was found, searchers can better assess the information scent, that is, the potential usefulness of that page for their current goals.

According to information scent theory, people continuously assess the likelihood of finding the requested information by weighing proximal cues in relation to internal goals. Thus, given a certain starting page, the model attempts to select the best link to reach its goal. It does this by picking the link with the strongest information scent, that is, the link that receives the highest activation through spreading activation from the goal information. If the information scent is not strong enough, the model will abandon the current page and backtrack to earlier pages or give up all together.

Being based on the ACT-R cognitive architecture, SNIF-ACT explains Web browsing while adhering to the cognitive constraints identified by the ACT-R system, such as spreading activation and decay in declarative memory. Because of this embedding, SNIF-ACT gives principled accounts of phenomena such as concept drift. However, there are still aspects of human behavior that SNIF-ACT does not take into account. For example, SNIF-ACT does not take into account more detailed properties of human vision that might explain why some links are often overseen when quickly scanning a Web page. Another aspect that was missing in the 1.0 version of SNIF-ACT was the effect of sequential evaluation of links. According to SNIF-ACT 1.0, all links were equally likely to be selected, and the information scent of all links was evaluated in parallel before making a decision. SNIF-ACT 2.0 (Fu and Pirolli 2007) corrected this omission by implementing a Bayesian satisficing process. Instead of a parallel evaluation of all links on a Web page, SNIF-ACT 2.0 evaluates the links in order and decides after each evaluation whether additional links will be evaluated or a previously evaluated link will be selected. This improvement was partly driven by changes in the underlying ACT-R architecture (see, e.g., Fu and Anderson 2006). It is interesting to note that these changes to the architecture were, in turn, initiated in part because of the misfit observed in SNIF-ACT 1.0. This is a clear example of how cognitive architectures can inform user models and how user models can, in turn, inform theories of human cognition by trying to explain real-life behavior. Given that evaluations have shown that models such as SNIF-ACT predict searching behavior fairly well, it is reasonable to use them to evaluate the quality of Web sites: If the model is not able to find the desired information, users probably will not find it, either.

25.7.2 Adaptive Information Presentation

Another application of cognitive user models is information filtering and recommendation. This especially holds in science and healthcare. Although society expects that scientists and doctors be informed about all new developments in their fields, the number of papers that appear in many fields is higher than can be read by a single person. For example, even the relatively small scientific field of User Models saw more than 2000 scientific documents published in 2008 according to Google Scholar.* Although Google Scholar contains many documents more than once, this number is probably still an underestimate of the total number of papers a knowledge worker in the field of user modeling is potentially interested in. A number of information filtering systems have been proposed to ameliorate the problem of information overload. These systems promise personalized recommendations that can eliminate the need to search all information sources (e.g., Birukou, Blanzieri, and Giorgini 2006; Bogers and Van den Bosch 2008; Hess, Stein and Schlieder 2006). For example, the Personal Publication Assistant (PPA), proposed by Van Maanen et al. (2010), builds its initial user model by analyzing the scientific publications published by the user. On the basis of this analysis, a declarative memory structure is constructed similar to the memory system of ACT-R that contains all keywords associated with the user's publications. As these keywords are considered declarative memory elements, their importance is automatically scaled by their recency and frequency. Thus, if a scientist very recently published a paper on "adaptive menu structures," even if it is the user's only paper on that topic, the PPA will recommend papers that relate to the topic. Alternatively, if the scientist has published many papers on visual search over the years, the PPA will recommend new papers on visual search even though no recent papers on the topic have been published. This system has been tested with good results both in a controlled setting (Van Maanen et al. 2010) and in the context of a large conference with many parallel sessions (Taatgen and Van Rijn 2009; see http://cogsci.astrokraai.nl/).

25.7.3 Intelligent Tutors

One of the first user-model based applications was the intelligent tutor (e.g., Groen and Atkinson 1966; and for a review of early work, Sleeman and Brown, 1982). The basic tenet of intelligent tutors is that information about the user can be used to modify the presentation of information so that learning proceeds more efficiently. Intelligent tutors guide learning and adapt navigational paths to facilitate learner control of knowledge acquisition. A good human tutor could be considered intelligent in this sense if he or she adapted instructional material to the needs of the pupil. For example, a pupil asked to explain a concept might reveal deficiencies of knowledge that the tutor can remediate by means of new examples or repeated explanation. The better the tutor understands the pupil, the better the remediation. In the realm of human–computer interaction, the intelligent tutor needs to possess a user model. Having knowledge of the user is, of course, not enough. The tutor must also have at hand an arsenal of strategies for tailoring the information to the student.

Many tutors attempt to ascertain what the user knows, what they are trying to accomplish, and what specific help the user might need. These intelligent tutors seek to ascertain a student model and adapt the presentation of information to the student based on the current student model. *Student modeling* refers to the techniques and reasoning strategies incorporated in an instructional system to allow it to maintain a current understanding of the student and her activity on the system. The student model can then be used to adapt the learning system to the user by, for example, creating problem sets that reflect the interests and weaknesses (or strengths) of the student, or phasing out assistance in finding solutions to problems at the appropriate time (Federico 1999).

A distinction can be made between local modeling, which refers to system capabilities for carrying out online performance monitoring and error recognition (creating a "trace" of student performance), and global modeling, in which trace data serve as input to an inference engine that maintains a relatively permanent, evolving view of the student's domain knowledge (Hawkes and Derry 1996). In most intelligent tutoring systems, local modeling forms the foundation for both assessment of the student and of the tutoring strategy because both of these depend on inferences from the primitive elements detected by the local modeler. The key to effective systems is, then, to develop appropriate local modeling techniques.

Probably the most successful and influential approach to student modeling is the *model-tracing method* (e.g., Anderson, Boyle, and Reiser 1985; Heffernan, Koedinger, and Razzaq 2008). In this technique, knowledge is represented in terms of productions. A "generic" student model contains all the productions an expert problem solver would execute within the problem domain and may also contain a library of "buggy" productions that embody common student mistakes. The expert model and buggy productions form the knowledge base used by the tutor to diagnose student performance. Each step taken by the student is compared to the output of the productions that the generic model would fire under the same circumstances in order to distinguish between valid procedures and actions based on errors or misconceptions. Other student-modeling techniques include using imprecise pattern-matching algorithms that operate on separate semantic and structural knowledge representations (e.g., Hawkes and Derry 1996), model-tracing based on declarative knowledge (e.g., Pavlik and Anderson 2008; Van Rijn, Van Maanen, and Van Woudenberg 2009), and Bayesian models.

Bayesian models are user models based on the Bayesian methods widely used in classifier systems and speech and image recognition. Bayes's theorem states that the probability that hypothesis A is true given that evidence B is observed, is equal to the product of the probability of B being observed given that A is true and the probability of A being true, divided by the sum of the products of the probabilities that B is observed given that alternative hypotheses are true and the probabilities of these alternative hypotheses. An example of a

---

* Assessed by querying Google Scholar for documents published in 2008 that contained the phrase "user models" or "user modelling," Dec. 13, 2009.

system that incorporates a Bayesian model is Andes (Conati, Gertner, and Vanlehn 2002; Vanlehn et al. 2005; see http://www.Andestutor.org/), a tutoring system for helping students learn Newtonian physics. Andes models the process of mastering the laws of physics and learning how to apply these laws to various problems.

Students start to learn by applying a law (e.g., Newton's second law that mass times acceleration equals net force) in a specific problem context (e.g., the acceleration of a car given that its engine generates a certain amount of force). Andes observes that the student chooses a specific formula and computes the probability of that action being performed given that various knowledge has been mastered. Using Bayes's theorem, the probabilities in the model and the observed student actions, the probability that the user has mastered a certain piece of knowledge given his observed action can be calculated: The observed user actions are taken as the evidence and mastery of each piece of knowledge is taken as the various hypotheses.

In most cases, students will need to work examples from different contexts before they are able to apply a law in a consistently correct manner. Andes captures this with a mechanism that enables the distinction between context-specific and general rules. Within Bayesian modeling, mastery of the general rule is taken as the hypothesis, while mastery of the context-specific rule is used as the evidence. As students solve problems, they gain experience with general laws, increasing their knowledge. The probability that a law has been mastered can be used to deliver specific help. For example, if a student is unable to select a correct formula but has a high probability of mastering the context-specific rule, he might be given a hint (e.g., that a block resting on a table for which force must be calculated does not move and therefore has an acceleration of zero). However, if this probability is low, he might be given a more direct hint, such as which formula to apply.

25.7.4 ADAPTING TO PHYSICAL LIMITATIONS

Although the focus of this chapter is on acquiring knowledge and making inferences about cognitive strategies and adaptations based on them, other aspects of human–machine interaction may be also be improved by user modeling. For example, Trewin and Pain (1999; see also Keates and Trewin 2005) have demonstrated how personalized user modeling of keyboard operation can be used to dynamically adapt keyboard operation to people with various motor disabilities. Their model follows the keystrokes made by the user and detects difficulties caused by (1) holding keys depressed for too long, (2) accidentally hitting keys other than the intended one, (3) unintentionally pressing a key more than once, or (4) problems in using the "shift" or other modifier keys. For example, if the model observes that a typist uses the "caps lock" key to type just one capital letter, or presses and releases the "shift" key without modifying a key, or presses and releases the shift key more than once without typing any letter, it will invoke the "sticky keys" facility, which causes the shift key to stay active after having been pressed until

TABLE 25.3
Functions of a User-Modeling System

1. Timely identification of user interests and preferences based on observable behavior.
2. Assignment of users to groups so that subgroup stereotypes can be used to predict interests and preferences.
3. Use of domain knowledge to infer additional interests and preferences.
4. Storage and updating of explicitly provided information and implicitly acquired knowledge.
5. Guarding the consistency and privacy of user models.
6. Provision of user model information to authorized applications.

the next alphanumeric character is pressed. Trewin and Pain showed that the model performed well in differentiating between people who needed interventions to improve their typing and those who did not. Because it was able to make recommendations after just 20 key presses, the model may be of practical use in situations (such as libraries or internet cafes) where users use public terminals.

25.8 CONCLUSION

Essentially, every application of user modeling represents an effort to move beyond direct manipulation of computer software via commands entered using a keyboard, mouse, or touch screen to adaptive or even autonomous behavior on the part of the software (see Table 25.3 for an overview). In other words, they contribute to a situation in which the process of using a computer is cooperative, and both human and computer agents can initiate communication, monitor events, and perform tasks to meet a user's goals (Shneiderman and Maes 1997). Not all researchers are convinced that such a trend is desirable. For example, Shneiderman and Maes question whether the computer will ever be able to automatically ascertain the users' intentions or to take action based on vague statements of goals. They also suggest that users will be reluctant to relinquish the control over the system afforded by direct manipulation and warn that systems that adapt to the user or carry out actions on their behalf may be hard to understand. Despite such concerns, many people have argued that further increases in the usability of software require that the software itself infer some of the actions to be taken.

REFERENCES

Ackerman, P. L. 1988. Determinants of individual differences during skill acquisition: Cognitive abilities and information processing. *Journal of Experimental Psychology: General* 117: 288–318.

Al-Omar, K., and D. Rigas. 2009. Does the size of personalized menus affect user performance? *Journal of Computer Science* 5(12): 940–950.

Anderson, J. R. 1993. *Rules of the Mind*. Hillsdale, NJ: Lawrence Erlbaum.

Anderson, J. R. 2007. *How Can the Human Mind Occur in the Physical Universe?* New York: Oxford University Press.

Anderson, J. R., D. Bothell, M. D. Byrne, S. Douglass, C. Lebiere, and Y. Qin. 2004. An integrated theory of the mind. *Psychological Review* 111(4): 1036–1060.

Anderson, J. R., C. F. Boyle, and J. F. Reiser. 1985. Intelligent tutoring systems. *Science* 228: 456–468.

Anderson, J. R., and C. Lebiere. 1998. *The Atomic Components of Thought.* Mahwah, NJ: Lawrence Erlbaum.

Atkinson, R., and J. Paulson. 1972. An approach to the psychology of instruction. *Psychological Bulletin* 78(1): 49–61.

Barab, S., B. Bowdish, and K. Lawless. 1997. Hypermedia navigation: Profiles of hypermedia users. *Educational Technology Research and Development* 45: 23–41.

Birukou, A., E. Blanzieri, and P. Giorgini. 2006. A multi-agent system that facilitates scientific publications search. In *Proceedings of the Fifth International Joint Conference on Autonomous Agents and Multiagent Systems (AAMS'06)*, 265–272. New York: ACM Press.

Blackmon, M. H., M. Kitajima, and P. G. Polson. 2005. Tool for accurately predicting website navigation problems, non-problems, problem severity, and effectiveness of repairs. In *Proceedings of the 2005 SIGCHI Conference on Human Factors in Computing Systems*, 31–40. New York: ACM Press.

Bogers, T., and A. Van den Bosch. 2008. Recommending scientific articles using citeulike. In *RecSys '08: Proceedings of the 2008 ACM Conference on Recommender Systems*, 287–290. New York: ACM Press.

Borst, J. P., N. A. Taatgen, and H. Van Rijn. 2010. The problem state: A cognitive bottleneck in multitasking. *Journal of Experimental Psychology: Learning, Memory, and Cognition* 36(2): 363–382.

Byrne, M. D. 2001. ACT-R/PM and menu selection: applying a cognitive architecture to HCI. *International Journal of Human–Computer Studies* 55: 41–84.

Byrne, M. D., and J. R. Anderson. 2001. Serial modules in parallel: The psychological refractory period and perfect time-sharing. *Psychological Review* 108: 847–869.

Card, S. K., T. P. Moran, and A. Newell. 1983. *The Psychology of Human–Computer Interaction.* Hillsdale, NJ: Lawrence Erlbaum.

Carroll, J. 1997. Human–computer interaction: psychology as a science of design. *Annual Review of Psychology* 48(1): 61–83.

Chi, E., P. Pirolli, K. Chen, and J. Pitkow. 2001. Using information scent to model user information needs and actions and the web. In *Proceedings of the 2001 SIGCHI Conference on Human Factors in Computing Systems*, 490–497. New York: ACM Press.

Chin, D. 2001. Empirical evaluation of user models and user-adapted systems. *User Modeling and User-adapted Interaction* 11: 181–194.

Chong, R. S. 1999. Modeling dual-task performance improvement: casting executive process knowledge acquisition as strategy refinement. Dissertation. Ann Arbor: University of Michigan.

Cockburn, A., S. Greenberg, S. Jonesa, B. McKenzie, and M. Moyle. 2003. Improving web page revisitation: analysis, design, and evaluation. *Information Technology and Society* 3(1): 159–183.

Cockburn, A., and S. Jones. 1996. Which way now? Analyzing and easing inadequacies in WWW navigation. *International Journal of Human–Computer Studies* 45: 105–129.

Conati, C., A. Gertner, and K. Vanlehn. 2002. Using Bayesian networks to manage uncertainty in student modeling. *User Modeling and User-Adapted Interaction* 12: 371–417.

Daily, L. Z., M. V. Lovett, and L. M. Reder. 2001. Modeling individual differences in working memory performance: A source activation account. *Cognitive Science* 25: 315–353.

Federico, P.-A. 1999. Hypermedia environments and adaptive instruction. *Computers in Human Behavior* 15: 653–692.

Findlater, L., and J. McGrenere. 2004. A comparison of static, adaptive, and adaptable menus. In *Proceedings of the SIGCHI Conference on Human Factors in Computing Systems*, 89–96. New York: ACM Press.

Fink, J., and A. Kobsa. 2002. User modeling for personalized city tours. *Artificial Intelligence Review* 18: 33–74.

Fischer, G. 2001. User modeling in human–computer interaction. *User Modeling and User-Adapted Interaction, 11*, 65–86.

Forman, G. 2006. Tackling concept drift by temporal inductive transfer. In *SIGIR '06: Proceedings of the 29th Annual International ACM SIGIR Conference on Research and Development in Information Retrieval*, 252–259. New York: ACM Press.

Franklin, S., and A. Graesser. 1996. Is it an agent, or just a program?: A taxonomy for autonomous agents. In *Proceedings of the Third International Workshop on Agent Theories, Architectures, and Languages*, 21–35. Amsterdam: Springer-Verlag.

Fu, W. T., and J. R. Anderson. 2006. From recurrent choice to skill learning: A reinforcement-learning model. *Journal of Experimental Psychology: General* 135: 184–206.

Fu, W.-T., and P. Pirolli. 2007. SNIF-ACT: A cognitive model of user navigation on the World Wide Web. *Human–Computer Interaction* 22: 355–412.

Gray, W. D. 2000. The nature and processing of errors in interactive behavior. *Cognitive Science* 24: 205–248.

Gray, W. D., B. E. John, and M. E. Atwood. 1992. The precis of Project Ernestine or an overview of a validation of GOMS. In *CHI '92: Proceedings of the SIGCHI Conference on Human Factors in Computing Systems*, 307–312. New York: ACM Press.

Gray, W. D., C. Sims, W. Fu, and M. Schoelles. 2006. The soft constraints hypothesis: A rational analysis approach to resource allocation for interactive behavior. *Psychological Review* 113: 461–482.

Groen, G. J., and R. C. Atkinson. 1966. Models for optimizing the learning process. *Psychological Bulletin* 66(4): 309–320.

Hawkes, L. W., and S. J. Derry. 1996. Advances in local student modeling using informal fuzzy reasoning. *International Journal of Human–Computer Studies* 45: 697–722.

Heffernan, N. T., K. R. Koedinger, and L. Razzaq. 2008. Expanding the model-tracing architecture: A 3rd generation intelligent tutor for algebra symbolization. *International Journal of Artificial Intelligence in Education* 18: 153–178.

Hess, C., K. Stein, and C. Schlieder. 2006. Trust-enhanced visibility for personalized document recommendations. In *SAC '06: Proceedings of the 2006 ASCM Symposium on Applied Computing*, 1865–1869. New York: ACM Press.

Horney, M. 1993. A measure of hypertext linearity. *Journal of Educational Multimedia and Hypermedia* 2: 67–82.

Hornof, A. J., and D. E. Kieras. 1997. Cognitive modeling reveals menu search is both random and systematic. In *Proceedings of the CHI '97 Conference on Human Factors in Computing Systems*, 107–114. New York: ACM Press.

Howes, A. 1994. A model of the acquisition of menu knowledge by exploration. In *Proceedings of CHI '94: Human Factors in Computing Systems*, eds. B. Adelson, S. Dumais, and J. R. Olson, 445–451. New York: ACM Press.

Howes, A., and R. M. Young. 1997. The role of cognitive architecture in modeling the user: Soar's learning mechanism. *Human–Computer Interaction* 12: 311–343.

Huang, Y., and L. Bian. 2009. A Bayesian network and analytic hierarchy process based personalized recommendations for tourist attractions over the Internet. *Expert Systems with Applications* 36: 933–943.

Jameson, A. 2002. Adaptive interfaces and agents. In *The Human–Computer Interaction Handbook: Fundamentals, Evolving Technologies, and Emerging Applications*, eds. J. A. Jacko and A. Sears, 305–330. Mahwah, NJ: Lawrence Erlbaum.

Jameson, A., and E. Schwarzkopf. 2002. Pros and cons of controllability: An empirical study. In *Adaptive Hypermedia and Adaptive Web-Based Systems*, Vol. 2347, *Lecture Notes in Computer Science*, 193–202. Berlin: Springer.

Keates, S., and S. Trewin. 2005. Effect of age and Parkinson's disease on cursor positioning using a mouse. In *Proceedings of the ACM SIGACCESS Conference on Computers and Accessibility, ASSETS 2005*, eds. A. Sears and E. Pontelli, 68–75. Baltimore, MD: ACM Press.

Kieras, D. E., and D. E. Meyer. 1997. An overview of the EPIC architecture for cognition and performance with application to human–computer interaction. *Human–Computer Interaction* 12: 391–438.

Kleinberg, J., and M. Sandler. 2008. Using mixture models for collaborative filtering. *Journal of Computer and System Sciences* 74: 49–69.

Koren, Y. 2009. The BellKor Solution to the Netflix Grand Prize. http://www.netflixprize.com/assets/GrandPrize2009 BPC Bell Kor.pdf (accessed Dec. 13, 2009).

Laird, J. E. 2008. Extending the Soar cognitive architecture. In *Artificial General Intelligence 2008: Proceedings of the First AGI Conference*, vol. 171, *Frontiers in Artificial Intelligence and Applications*, eds. P. Wang, B. Goertzel, and S. Franklin, 224–235. Amsterdam: IOS Press.

Lawless, K., and J. Kulikowich. 1996. Understanding hypertext navigation through cluster analysis. *Journal of Educational Computing Research* 14: 385–399.

Large, A. 1996. Hypertext instructional programs and learner control: A research review. *Education for Information* 4: 95–106.

Maglio, P., C. Campbell, R. Barret, and T. Selker. 2001. An architecture for developing attentive information systems. *Knowledge-Based Systems* 14: 1–103.

Milheim, W., and B. Martin. 1991. Theoretical bases for the use of learner control: Three different perspectives. *Journal of Computer-Based Instruction* 18: 99–105.

Miller, C. S., and R. W. Remington. 2004. Modeling information navigation: Implications for information architecture. *Human–Computer Interaction* 19(3): 225–271.

Nason, S., and J. E. Laird. 2005. Soar-RL: integrating reinforcement learning with Soar. *Cognitive Systems Research* 6: 51–59.

Newell, A. 1990. *Unified Theories of Cognition*. Cambridge, MA: Harvard University Press.

Patton, E. W., and W. D. Gray. submitted. SANLab-CM: a tool for incorporating stochastic operations into activity network modeling.

Patton, E. W., W. D. Gray, and M. J. Schoelles. 2009. SANLab-CM—The Stochastic Activity Network Laboratory for Cognitive Modeling. In *53rd Annual Meeting of the Human Factors and Ergonomics Society*. San Antonio, TX: Human Factors and Ergonomics Society.

Pavlik, P. I., Jr., and J. R. Anderson. 2005. Practice and forgetting effects on vocabulary memory: An activation-based model of the spacing effect. *Cognitive Science* 29: 559–586.

Pavlik, P. I., Jr., and J. R. Anderson. 2008. Using a model to compute the optimal schedule of practice. *Journal of Experimental Psychology: Applied* 14: 101–117.

Pazzani, M. J., and D. Billsus. 2007. Content-based recommendation systems. In *Lecture Notes in Computer Science*, vol. 4321, *The Adaptive Web*, eds. P. Brusilovsky, A. Kobsa, and W. Nejdl, 325–341. Berlin: Springer-Verlag.

Perruchet, P., and S. Pacton. 2006. Implicit learning and statistical learning: One phenomenon, two approaches. *Trends in Cognitive Sciences* 10: 233–238.

Pirolli, P. 2005. Rational analyses of information foraging on the web. *Cognitive Science* 29: 343–373.

Ritter, F. E., G. D. Baxter, G. Jones, and R. M. Young. 2000. Supporting cognitive models as users. *ACM Transactions on Computer–Human Interaction* 7: 141–173.

Reber, A. S. 1989. Implicit learning and tacit knowledge. *Journal of Experimental Psychology: General* 118: 219–235.

Roberts, S., and H. Pashler. 2000. How persuasive is a good fit? A comment on theory testing. *Psychological Review* 107: 358–367.

Rosenbloom, P. S. 2006. A cognitive odyssey: from the power law of practice to a general learning mechanism and beyond. *Tutorials in Quantitative Methods for Psychology* 2(2): 38–42.

Saffran, J. R., R. N. Aslin, and E. L. Newport. 1996. Statistical learning by 8-month-old infants. *Science* 274(5294): 1926–1928.

Salvucci, D. 2001. An integrated model of eye movements and visual encoding. *Cognitive Systems Research* 1: 201–220.

Salvucci, D. D., and K. L. Macuga. 2002. Predicting the effects of cellular-phone dialing on driver performance. *Cognitive Systems Research* 3: 95–102.

Salvucci, D. D., and N. A. Taatgen. 2008. Threaded cognition: An integrated theory of concurrent multi-tasking. *Psychological Review* 115: 101–130.

Salvucci, D. D., and N. A. Taatgen. 2010. *The Multitasking Mind*. New York: Oxford University Press.

Scerbo, M. W. 1996. Theoretical perspectives on adaptive automation. In *Automation and Human Performance: Theory and Applications*, eds. R. Parasuraman and M. Mouloua, 37–63. Hillsdale, NJ: Lawrence Erlbaum.

Scerbo, M. W. 2007. Adaptive automation. In *Neuroergonomics: The Brain at Work*, eds. R. Parasuraman and M. Rizzo, 239–252. Oxford: Oxford University Press.

Shardanand, U., and P. Maes. 1995. Social information filtering: Algorithms for automating "word of mouth." In *Proceedings of CHI-95*, 210–217. New York: ACM Press.

Shiffrin, R. M., M. D. Lee, W. Kim, and E.-J. Wagenmakers. 2008. A survey of model evaluation approaches with a tutorial on hierarchical Bayesian methods. *Cognitive Science* 32: 1248–1284.

Shneiderman, B., and P. Maes. 1997. Direct manipulation vs. interface agents. *Interactions* 4: 42–61.

Sleeman, D., and J. S. Brown. 1982. *Intelligent Tutoring Systems*. San Diego, CA: Academic Press.

Smeulders, A. W. M., L. Hardman, G. Schreiber, and J. M. Geusebroek. 2002. An integrated multimedia approach to cultural heritage e-documents. *ACM Workshop on Multimedia Information Retrieval*. New York: ACM Press. Available at http://www.cs.vu.nl/~guus/papers/Smeulders02a.pdf.

Stadtler, M., and R. Bromme. 2008. Effects of the metacognitive computer-tool met.a.ware on the web search of laypersons. *Computers in Human Behavior* 24: 716–737.

Stephanidis, C., A. Paramythis, and A. Savidis. this volume. Developing adaptive interfaces for the web. In *Handbook of Human Factors in Web Design*, 2nd ed., eds. K.-P. L. Vu and R. W. Proctor, 191–208. Boca Raton, FL: CRC Press.

Strybel, T. Z. this volume. Task analysis methods and tools for developing web applications. In *Handbook of Human Factors in Web Design*, 2nd ed., eds. K.-P. L. Vu and R. W. Proctor, 483–508. Boca Raton, FL: CRC Press.

Sun, R. 2006. The CLARION cognitive architecture: Extending cognitive modeling to social simulation. In *Cognition and*

Multi-agent Interaction, ed. R. Sun, 79–102. New York: Cambridge University Press.

Taatgen, N. A. 2002. A model of individual differences in skill acquisition in the Kanfer-Ackerman Air Traffic Control Task. *Cognitive Systems Research* 3: 103–112.

Taatgen, N. A., and J. R. Anderson. 2002. Why do children learn to say "broke"? A model of the past tense without feedback. *Cognition* 86: 123–155.

Taatgen, N. A., and J. R. Anderson. 2008. Constraints in cognitive architectures. In *Handbook of Cognitive Modeling*, ed. R. Sun, 170–186. New York: Cambridge University Press.

Taatgen, N. A., and F. J. Lee. 2003. Production compilation: A simple mechanism to model complex skill acquisition. *Human Factors* 45: 61–76.

Taatgen, N. A., H. Van Rijn, and J. R. Anderson. 2007. An integrated theory of prospective time interval estimation: The role of cognition, attention and learning. *Psychological Review* 114(3): 577–598.

Taatgen, N. A., and H. Van Rijn, eds. 2009. *Proceedings of the 31th Annual Meeting of the Cognitive Science Society*. Austin, TX: Cognitive Science Society.

Trewin, S., and H. Pain. 1999. A model of keyboard configuration requirements. *Behaviour & Information Technology* 18: 27–35.

Vanlehn, K., C. Lynch, K. Schulze, J. A. Shapiro, R. Shelby, L. Taylor, et al. 2005. The Andes Physics Tutoring System: Lessons learned. *International Journal of Artificial Intelligence in Education* 15: 147–204.

Van Maanen, L. 2009. Context effects on memory retrieval: Theory and applications. Doctoral dissertation, University of Groningen, available at http://www.van-rijn.org/lab/thesis VanMaanen.pdf (accessed Dec. 13, 2009).

Van Maanen, L., and H. Van Rijn. 2007. An accumulator model of semantic interference. *Cognitive Systems Research* 8: 174–181.

Van Maanen, L., and H. Van Rijn. 2010. The locus of the Gratton effect in picture-word interference. *Topics in Cognitive Science* 2, 168–180.

Van Maanen, L., H. Van Rijn, and J. P. Borst. 2009. Stroop and picture-word interference are two sides of the same coin. *Psychonomic Bulletin & Review* 16: 987–999.

Van Maanen, L., H. Van Rijn, M. Van Grootel, S. Kemna, M. Klomp, and E. Scholtens. 2010. Personal publication assistant: Abstract recommendation by a cognitive model. *Cognitive Systems Research* (Special Issue on Brain Informatics) 11: 120–129.

Van Rijn, H., and N. A. Taatgen. 2008. Timing of multiple overlapping intervals: How many clocks do we have? *Acta Psychologica* 129: 365–375.

Van Rijn, H., L. Van Maanen, and M. Van Woudenberg. 2009. Passing the test: Improving learning gains by balancing spacing and testing effects. In *Proceedings of the 9th International Conference on Cognitive Modeling*, eds. A. Howes, D. Peebles, and R. Cooper, 108–113. Manchester, UK: University of Manchester.

Webb, G., M. Pazzani, and D. Billsus. 2001. Machine learning for user modeling. *User Modeling and User-Adapted Interaction* 11: 19–29.

Young, R. M., and R. L. Lewis. 1999. The Soar cognitive architecture and human working memory. In *Models of Working Memory: Mechanisms of Active Maintenance and Executive Control*, eds. A. Miyake and P. Shah, 224–256. New York: Cambridge University Press.

Section IX

Specific Web Applications

26 E-Learning 2.0

Lisa Neal Gualtieri and Diane Miller

CONTENTS

26.1 INTRODUCTION

Distance education existed long before the Internet, but technological advances have moved it from its humble roots in materials delivered by postal mail to delivery using sophisticated multimedia and online collaborative technologies. With the inclusion of Web 2.0 and social media technologies, e-learning is now often referred to as e-learning 2.0. E-learning growth has occurred in all sectors, including higher education, corporate training, and continuing and professional education, spurred by the availability of technology and, in some cases, by the perceived cost savings. More and more organizations are leveraging Internet technologies to provide education, training, and collaboration capabilities to geographically dispersed populations to enhance their educational experiences and enthusiasm for learning. However, the same issues of quality and effectiveness that exist in classroom settings occur in e-learning, in some cases compounded by the delivery mechanism. Design and development of Internet-delivered learning material is often more difficult and time-consuming, especially because materials may have to stand alone without the guidance and support of a teacher.

This chapter delves into many facets of the use of the Internet for e-learning, including definitions; reasons why e-learning has become so popular; and planning, designing, delivering, and evaluating an online course or program.

26.1.1 THE DEFINITION OF E-LEARNING

E-learning is most typically defined as education that takes place independent of location, in contrast to education delivered solely in the classroom, and that may be independent of time as well. In fact, the definition is quite fuzzy and is understood and interpreted differently by different groups and in different contexts. In particular, the definition varies with respect to the inclusion and role of teachers and peers. While traditional education places great emphasis on the role of the teacher as lecturer and mentor, in e-learning the teacher can be completely removed from the learning process. Traditional education also typically delineates duration for a learning situation, with a start and end date, while the appeal of e-learning for students is often the aspect of self-pacing.

ASTD, an education-focused professional organization, defines distance education in part by how it is delivered, as an "Educational situation in which the instructor and students are separated by time, location, or both. Education or training courses are delivered to remote locations via synchronous or asynchronous means of instruction, including written correspondence, text, graphics, audio- and videotape, compact disk-read only memory (CD-ROM), online learning, audio- and videoconferencing, interactive television, and facsimile (FAX). Distance education does not preclude the use of the traditional classroom. The definition of distance education is broader than and entails the definition of e-learning." (Kaplan-Leiserson, n.d.). E-learning, as seen in this definition, refers to education delivered over the Internet, while distance education can more broadly include videoconferencing and audioconferencing delivered over phone lines.

Commonly used terms delineate subsets of e-learning. Synchronous e-learning encompasses same-time interaction independent of location, while asynchronous e-learning includes any situation where learners are dispersed in time and location. E-learning 2.0 encompasses the use of Web 2.0 and social media, incorporating user-generated content and tools like Twitter, Facebook, and YouTube. Many approaches to e-learning blend synchronous, asynchronous, and classroom activities and interaction, all enhanced by social media.

One of the challenges in planning effective e-learning is selecting the appropriate mix of synchronous, asynchronous, classroom, and social media activities, as well as defining the role of teachers and peers. There is a considerable range in the sophistication, quality, cost, and development time of e-learning curricula. At the same time, outside of the confines of the classroom, there is the opportunity to creatively rethink and redefine formal and informal learning activities, materials, and courses.

26.1.2 E-Learning Compared to Traditional Education

Traditional classroom education serves a valuable role for children and young adults, where there are many benefits to face-to-face interaction and where maturity and self-discipline are still developing. For adults, the classroom is often necessary for supervised practice or for team-building activities that could not readily take place without face-to-face contact.

Aside from the lack of proximity, e-learning differs from education delivered in the classroom in many respects. Classroom education is traditionally developed and delivered by the same person, who may or may not have extensive

training in how to develop and deliver courses. In many universities, for example, a professor can be an excellent researcher but have little knowledge about how to plan and deliver a lecture. E-learning often separates the development and delivery processes. Development and delivery may be done by different people, or many people may be involved in different capacities, such as instructional designers, subject matter experts, and multimedia experts.

Many of the issues prevalent in classroom settings carry over to e-learning, occasionally amplified by delivery through technology. For instance, cheating has always been a problem for educators. Unless cameras or biometric devices are used, the opportunities to cheat are typically greater online. Other issues are similar; for example, a good teacher structures and teaches a small or large class differently, and the same approach applies to an online course. Arguably, the biggest opportunity in moving away from the confines of the classroom is the ability to redefine the notion of a "course": What type of interaction is most beneficial to students? What will help them learn most effectively? What are the opportunities for them to participate and be active learners? Educators grapple with how to rethink education in online settings. As John Maynard Keynes said, "The difficulty lies, not in the new ideas, but in escaping from the old ones, which ramify, for those brought up as most of us have been, into every corner of our minds."

26.1.3 The History of E-Learning

The rich history of e-learning has its roots in distance education, which started in the 1800s with a for-profit school developed by Sir Isaac Pitman for rural residents in Bath, England. Correspondence classes became an alternative for people needing education or training who were not able to attend or did not have access to a traditional program. There have been many notable instances of early distance education playing a major role in people's success; for example, in the 1920s, Edwin Shoemaker took a correspondence course in drafting and subsequently codeveloped the La-Z-Boy® recliner, which started an entire industry.

Following correspondence courses delivered by mail, radio became the next delivery vehicle, and it is still commonly used in developing regions where access to the phone or Internet is limited. Instructional television became common in the 1950s and 1960s and, like radio, is still in use. While instructional television never achieved the success anticipated in those decades, arguably the most successful offshoot have been the very popular Sesame Street–type shows and the cable in the classroom programs for children. And certainly, television is still a delivery vehicle in the sense that educational videotapes are still being produced. While radio and instructional television themselves have had limited success, their offshoots, podcasts and YouTube, play a major role in e-learning today.

As computers became more widespread, computer-based training (CBT) distributed on CD-ROMs became common. Web-based delivery followed in the 1990s. John Chambers,

Chief Executive Officer (CEO) of Cisco, predicted in 1999, that "Education over the Internet is so big it is going to make e-mail look like a rounding error." This growth, while not as rapid as predicted, has been fueled by economic factors including travel reductions, the increased need for professional development in the workplace, and perceived cost reductions.

26.1.4 Perspectives on E-Learning

For many years, popular media has created both falsely optimistic and gloomy scenarios about e-learning. Peter Drucker predicted that big university campuses will become relics. "Cyberuniversities will replace brick-and-mortar . . . It is as large a change as when we first got the printed book." (Lenzner and Johnson 1997). Other predictions have involved the demise of the classroom. However, e-learning does not remove or eliminate the need for the classroom, in the same way that e-commerce does not eliminate stores, e-mail does not eliminate phones, and information technology (IT) does not eliminate paper.

In epidemiology, after a critical mass is reached, a small change "tips" the system, leading to large effect (Gladwell 2000). In e-learning, this "tipping point" seems to have been reached in the past few years. It is now commonly accepted that e-learning is a viable alternative to classroom education and, despite spam advertising nonaccredited online programs, much of the déclassé impression left by e-learning's origins in correspondence schools no longer lingers. At the same time, there is considerable debate about whether online courses are effective and whether online degrees are valued. To address the former case, in Russell (n.d.), hundreds of studies since 1928 argue that teaching through remote instruction is as good as traditional methods. To address the latter, there are e-learning programs with excellent reputations, but, at the same time, there are many that offer a degree for a fee (Neal 2004).

There are a number of factors that affect the adoption rates of an idea, practice, or object (Rogers 1962): relative advantage, compatibility, complexity, trialability, and observability. Examining each of these factors with respect to e-learning, it is clear that it offers advantages and challenges. Relative advantage asks how much better the innovation is. E-learning at its best can offer education that surpasses most classroom education, and, even when not of the highest quality, can offer educational opportunities not otherwise possible. However, e-learning at its worst is as bad as the worst classroom experience and, unfortunately, the worst of e-learning includes unaccredited programs offering online degrees for a fee. Compatibility asks how consistent something is with what already exists. Much e-learning is quite consistent with the classroom, but it can often extend traditional norms, for instance, with simulations and discussions. Complexity asks how difficult something it to understand or use. With respect to e-learning, the primary difficulties are the additional skills and technical knowledge necessary to succeed in a virtual setting, which are beyond what is required in the classroom.

Trialability pertains to how well a capability can be tried on a limited basis, which is typically quite easy to support in distance learning because many programs offer sample or trial courses. Finally, observability asks how visible to others the results of use are. The outcomes of a well-designed online program should be as good as or better than the classroom equivalent.

26.1.5 The Popularity of E-Learning

Many factors have contributed to the popularity and growth of e-learning. Some are organizational, such as the desire to reduce cost or increase reach, and others are societal, including the all too pervasive enamorment with technology—the belief that just because it can be done, it should be done. Another is the desire to improve the quality of education, but there is little data to substantiate this to date (Russell, n.d.).

The desire for cost savings is one of the most common motivators for starting an e-learning program. Cost can be saved by reducing learner-related expenses, such as transportation to school, or by reducing institutional expenses, such as those for classrooms or buildings. However, it can be expensive to set up technology, develop appropriate courses, and compensate faculty, so the costs may just shift rather than decline substantially.

Time savings is another often-stated reason for e-learning's popularity. Certainly, for the learner, there can be a reduction in time to get to class. For asynchronous e-learning, there is greater flexibility to take a class at optimal times for the learner, based on preferences or schedule constraints. While this potentially removes "learner fatigue," when a student is in class based on the class schedule rather than personal preference, it can increase "time-bankruptcy," when what used to be leisure time is now spent on an online course. Time savings also crops up in the reduced time between identifying a need to learn something, and finding and taking a course. However, engaging students and promoting learning in an online environment can require a large investment of time and effort on the part of the instructor. Often, the online equivalent of a traditional course requires more "care and feeding," providing timely, substantive responses to all forms of student communication on an ongoing basis, in addition to grading and other typical teaching tasks (LeBaron and Miller 2004).

Increased accessibility is another factor contributing to the popularity of e-learning. Accessibility includes providing more learning opportunities to diverse students independent of location, provided the students have technology skills, access, and support. This includes students with disabilities for whom an online course is easier to take than one offered in a classroom. When an online course includes peer interaction and discussion, all students can benefit from the increased diversity that is possible. Online courses can also provide increased access to experts, for instance, when an acknowledged expert can easily offer a live guest lecture or a podcast to a course, and is more willing to do so if no travel is needed.

26.2 PLANNING AN E-LEARNING PROGRAM

Developing a small or large e-learning program requires planning in order to elucidate and meet goals. The planning steps include building a team, making sure the team is educated sufficiently about e-learning best practices, and engaging as a group in strategic planning, scenario building, and requirements analysis. Even if these steps are performed in a cursory fashion, they help with team building and with informing implementation so that it is clearer what is likely to lead to success and what the stumbling blocks might be. The most common mistake of e-learning initiatives is to start by evaluating, or even selecting and purchasing, technology; this should follow the planning phase.

A planning team should consist of people responsible for decision-making, budgeting, and managing and implementing the program, with representation from human resources, IT, etc., as appropriate for the organization. The facilitator may be an outsider, who offers a broader and unbiased perspective from being involved in overseeing other programs, or an insider, who is knowledgeable about the organization and its politics. Sometimes, a participatory approach, involving target students and faculty, is desirable, because both perspectives can be helpful. The team should be knowledgeable about e-learning and should develop a shared vocabulary.

A strategic plan describes how to bridge the gap from the current to the desired state and provides a well-structured, systematic view of an e-learning initiative (Driscoll 2001). The planning process is often driven by an external force, such as the expansion of an initiative or competitive pressures, but planning is beneficial in all situations. In the planning process, the team articulates and assesses the current state, looking at the organization's strengths, weaknesses, opportunities, and threats. It selects a planning horizon, which is typically about 2 years, to implement the plan. The team may also specify an additional, shorter horizon of perhaps 6 months, to pilot a few key initiatives in order to test the waters and get early feedback. The team defines the desired state it hopes to reach and describes this in mission and vision statements that relate to the institution's strategic direction. It determines the resources, constraints, and budget and develops a realistic schedule and project plan. If the e-learning program is internal, it is valuable to determine the readiness of the organization and the receptivity of employees and managers to e-learning.

Market research and a competitive analysis (UsabilityNet, n.d.). help in identifying "best of breed" programs and understanding what has been successful for other institutions—or what has failed and why. They help shape an e-learning program. A good example of this is the extensive assessment the Massachusetts Institute of Technology (MIT) did that led to OpenCourseWare (MIT, n.d.). In the analysis, they discovered that many for-profit online programs were failing to make a profit, which led to the idea of putting courses online for free (Shigeru Miyagawa, pers. comm., 2003). Not only have many of the programs they looked at since failed, but MIT ended up with more in grant money to fund their initiative than their most optimistic estimates of the revenue a for-profit program would generate (Shigeru Miyagawa, pers. comm., 2003). In addition, the publicity they received has been invaluable. By not being followers themselves, and making a courageous choice, they became a model for many other initiatives.

A needs assessment helps in identifying a target population and understanding what is likely to work for that target population. It can prevent failure based on erroneous assumptions. The very diversity among students that can enrich online learning also presents challenges with respect to designing courses that take into consideration the characteristics and needs of the student population. Age, background knowledge, technical knowledge, and comfort and experience using technology are just a few student characteristics that can affect an individual's online learning experience. While no two students are alike, in traditional classes variations among students are typically easier for an instructor to gauge. While differences in technology savvy may impact to some extent the range of tools a student might use to complete a task in a traditional course, students who are not comfortable with technology can feel intimidated and fall behind very quickly in the online world. These issues, compounded by cultural differences, potential language barriers, and potential wide variation with respect to geographic location (and the resulting time differences), can make defining and designing courses for a "target" population complex. Careful planning, however, can address many of these issues, while meeting additional challenges that may not be so easily addressed in a traditional setting (e.g., leveraging technology to accommodate disabilities).

Persona building is a facilitated, systematic process to define the experience of the various communities, learners, teachers, support staff, and other people, impacted by or impacting a e-learning program (Feldstein and Neal 2006; Neal, 2000). It does not focus only on what happens when taking or teaching a course but looks at everything that precedes, follows, and impacts what happens when taking or teaching a course and determines what is needed for members of these roles to be successful in their role. The process involves a number of steps: determining who will be involved in developing, delivering, supporting, administering, or taking courses and prioritizing these to order the next phase. Starting with the most important perspective, usually that of the student, the relevant demographics, characteristics, skills, needs, constraints, and motivators of members of that category are outlined. This may include computer and Internet literacy, work or study environment, and other time commitments. The group then determines what happens when a representative of that category, or persona, carries out their role, i.e., taking, teaching, or supporting a class. This process is repeated for the other personas. The result of the building personas and walking them through scenarios is a set of design decisions based on the realistic needs and requirements from the perspectives of the personas for the online program.

Media analysis follows persona building, providing recommendations for the delivery of capabilities that meet the needs of the target learner population, as well as the other constituents (IHETS, n.d.). These recommendations may include asynchronous technologies, synchronous technologies, and learning management, possibly integrated with existing educational and Internet capabilities. The media analysis task includes a detailed requirements analysis. Together with the strategic plan, this forms the basis for selecting the right capabilities to meet the vision and needs on an e-learning initiative. The process of media analysis requires careful selection of the optimal technology and training mode out of the myriad available. It is not enough to simply incorporate the most recent technological advances into a course—this does not guarantee an optimal learning environment. Media analysis can be performed in prioritized steps, where any program is constrained by cost, deadlines, market, technology/bandwidth issues, etc.

The first decision is whether face-to-face interaction is needed, and, if so, should it be for the entire course, to kick off and/or end the course, or on an intermittent or "as needed" basis. The reasons for having face-to-face interaction are for sensitive topics, developing relationships, or supervised practice.

The second decision involves identifying what the desired level of supervisory involvement should be and in what capacity. Possible roles include instructor, facilitator, moderator, subject matter expert, tutor, mentor, or coach. A person in one of these roles can provide or reinforce learning, support students, or facilitate interactions among students and can do so in an proactive manner or in a passive manner, for instance, by being available to answer questions. This is very different from the prominent role that a teacher plays in classroom education, and self-paced online courses may not provide anyone in a supervisory or teaching capacity.

The third decision identifies the appropriate amount and types of peer interaction. While the classroom is rarely a solitary place, some courses integrate more teamwork and collaboration than others. The range of interactions within an online course can vary from a solitary, self-paced course to a highly collaborative and interactive course, in which group projects and discussions play a major role. The advantages to projects and discussion are the support and context students provide for each other, as well as the enhanced learning that can result. The disadvantages, and one of the reasons that the self-paced style is sometimes preferred, is that the more interactive style requires technology to support collaboration, as well as ongoing human guidance and support.

The fourth decision, course delivery, leads directly from the previous one: determining whether this will be a synchronous, asynchronous, self-paced, or blended solution based on how much of and what types of interaction will be supported, as well as scheduling and time zone issues. It is also at this point that it makes sense to determine whether courses will have start and end dates, or be open-ended, and how content will be accessed—online, offline, or both. Some thought may go into what material students retain following completion of a course—whether synchronous sessions are archived, for example, and students retain access to these sessions. The final step is to determine the course style and components.

This includes how information is conveyed, for example, through lectures, discussion, and online materials; what the look and feel and interactivity these materials will have; the extent to which materials will be developed from scratch or repurposed from existing materials; the extent to which courses will be tailored or adapted to learning styles; and what students are expected to do, e.g., projects, simulations, and tests, and how these will be conducted. While many aspects of course design may be determined by the nature of the courses, assignments and testing may be determined in part by the type of program, because a degree-granting program requires evidence upon which to assign grades.

26.3 SELECTION, DEPLOYMENT, AND USE OF TECHNOLOGY TO SUPPORT E-LEARNING

Technology facilitates e-learning delivery. While initially many e-learning initiatives tried to replicate the classroom experience online, Laurillard (1993) offers the perspective that it is better to identify the unique qualities of a technology and exploit those so that one is doing what could not otherwise be done rather than seeking to replicate existing practices. However, technology is only the delivery mechanism and the selection process is not just to determine which features are needed and which technologies provide them but also to identify what will be most comfortable for students to learn and use, so that the focus is not on the technology itself. Bob Smith, Chief Operating Officer (COO) of Otis Elevator, said "When elevators are running really well, people do not notice them. Our objective is to go unnoticed." (Andersen 2002).

Some of the issues to consider in technology selection are the number of students and the need for scalability; any existing software, hardware, platform, or Internet access constraints; the cost to purchase, use, maintain, upgrade, and access a technology solution; security and reliability; and vendor preference, reputation, and the availability of training and support. A good decision is important because this can be a major investment. There are always new players coming into the market, as well as occasional mergers, buyouts, and failures, and the costs of switching technologies can be substantial. There are a number of current standards initiatives, including Shareable Content Object Reference Model (SCORM) and Aviation Industry Computer-Based Training Committee (AICC). These are important to know about when doing government work and for fostering interoperability between components.

The design and usability of e-learning technologies varies considerably, with some tools for similar purposes varying in the number of features and the usability of those features. Students should not be hindered in their learning of the subject matter by the need to spend large quantities of time learning to use, or troubleshooting the use of, a poorly designed technology. Technologies should be easy to use and provide adequate tutorials and help to enable students to become familiar with using the products to support their learning tasks. Additionally, students should be offered technical support and practice sessions, especially for complex or novel technologies. The technology utilized should also be appropriate for the tasks at hand; that is, it should be useful. In selecting technologies for e-learning, one should determine whether the capabilities provided can support the required level of student-system, student-student, student-instructor, and instructor-system interaction, while enabling instructors to deliver and students to carry out the required coursework. It is important to test tools to determine how features work and how the interface design impacts students' ability to learn. Because of the complexity of some tools, new models for teaching become necessary. For instance, when using many synchronous technologies, a teacher may need a support person to help with use of the technology and monitoring of student interactions.

26.3.1 Asynchronous Computer-Based Training and Web-Based Training

Computer-based training (CBT) and Web-based training (WBT) refer to a course that is distributed on CD-ROM or over the Web for students to take as a self-paced asynchronous course. They are often derogatively called "electronic page turners," because a standard layout is used where the mouse can be positioned and the learner can click through screens, often faster than he or she is reading. The advantages of WBT over CBT are that online materials can be updated and hence distributed more easily. The potential exists, because a student is online, for interaction with an instructor and other students. The disadvantages of WBT over CBT are that a student needs Internet access, either to be connected while taking the class or while uploading and downloading materials. This can be costly, especially for large multimedia files, and prohibitive if a student is traveling or access is expensive.

Many CBTs and WBTs are structured in a linear fashion, where a learner is expected to follow one path through the materials. Some offer more flexibility, so a learner can navigate on the basis of interests or needs. In the most sophisticated, each student follows a path tailored to his or her needs based on their responses to system prompts, testing, and progress. Some programs track what a student does or looks at, requiring, say, 80% of all screens to be looked at for the student to be considered "done." CBTs and WBTs have varying degrees of "interaction" or interactivity, which most often refers to the extent to which a learner is using the system passively versus actively (e.g., using a simulation, mousing over text to receive information, etc.). While development of CBTs and WBTs can be expensive, the costs can be recouped through broad distribution.

26.3.2 Technologies for Delivering Asynchronous Courses

There are many technologies that can be effectively employed to deliver asynchronous courses. These include many collaborative tools that can be leveraged for e-learning including

e-mail, discussion forums, blogging, and microblogging (Twitter). Although many learning management systems include the ability to upload and share documents, e-mail is often used for informal, "behind the scenes" exchange of resources (e.g., sharing draft versions of documents, Web links, etc.) in support of group work. E-mail, in its simplest use here, can provide a speedier version of the traditional correspondence course. More typically, it is used for teacher-student communication and student-student exchanges. Sometimes, it is even used by the system to provide automatically generated messages to either the teacher or the student (e.g., calendar reminders, request for service/help acknowledgments, etc.).

Discussion forums are often included in e-learning and provide a mechanism for managed discussion on specific course topics, as well as informal exchanges carried out asynchronously over time (e.g., days, weeks, months). More robust discussion forums might support the ability to attach documents, images, videos, and uniform resource locators (URLs), or send e-mail notifications when new posts are added. Threaded discussion forums are typically organized so that the exchange of messages and responses are grouped together and are easy to find and scan. Common ways to group/sort postings are by date, title, author, group, or by specific topics defined by the instructor or other participants. Often, threaded discussions are expandable and collapsible to allow participants to manage the number of posts shown on their screen at once and to facilitate browsing groups of posts.

Discussion forums, blogs, and microblogs can be extremely effective because they can promote a great deal of sharing of perspectives and insights among students; the challenge is to encourage the right amount and type of participation so that learning is enhanced through discussions. Many techniques can help, such as modeling appropriate discourse through an initial example dialog posted at the onset of a course, highlighting interesting student postings, assigning student moderators to summarize a discussion thread, and teacher participation and feedback (Neal and Ingram 2001). Often, an instructor or a designee will seed the discussion by posing a thought provoking question or statement to spark dialog. In academic settings, students are often motivated to participate in discussions when a portion of the course grade is awarded for participation (i.e., frequent, scholarly discussion posts) or when the discussion topics address real-world issues or concerns. LeBaron and Miller (2004, 2005) found that providing a detailed rubric for the scholarly asynchronous discussions required in an online course increased both the amount of participation and the quality of the comments made.

26.3.3 Technologies for Delivering Synchronous Courses

Similar to asynchronous courses, there are many collaborative technologies that can be effectively used to fully or partially support synchronous e-learning. Most of these have the advantage that they need not only be used for real-time interaction but can also be archived for subsequent review. These include audioconferencing, electronic whiteboards and screen sharing, instant messaging, text chat, virtual worlds, video communication, Webinars, and social media.

Audioconferencing, using the telephone or voice over Internet protocol (VoIP), allows a group to interact in real-time through sharing voice (audio) and other artifacts such as slides or text. In its simplest form, this can be accomplished using phone lines and previously downloaded meeting materials. Audio is a fairly simple and often inexpensive way of supporting lecture and discussion in a course. The biggest issue with effective audioconferencing is quality, because students are generally intolerant of poor quality audio. For telephony, speakerphones with mute capabilities aid participation and, for VoIP, headsets with microphones similarly make it easier for a student to participate. Sessions conducted with audio should not be more than 1–2 hours in length, because it is harder for a student than in the classroom. It is hard to replicate the classroom experience online, even with high-quality audio, and it is generally most effective to make material that would otherwise be provided in a lecture available as prework for class, and then place heavier emphasis on discussion during an online session. This also helps keep students accountable for their work outside of class. Audio sessions, just like the classroom, vary depending on class size, and what works for a small group is unlikely to work for a large group. Because people cannot see one another, simple protocols smooth interaction, such as prefacing a remark with one's name.

Audio works best when supplemented by other technologies to support information sharing and exchange. Whiteboard tools typically consist of an electronic version of a dry-erase board, which can be used by a group of people in a virtual classroom, either used alone or overlaid on a shared application. They are used for freehand writing and drawing in the former case and pointing to or highlighting information in the latter case. These range from very simplistic shared graphical editors to fairly sophisticated shared applications incorporating audio, slideshows, or applications through remote desktops. Some whiteboard tools may support graphing, polling, group Web browsing, and instructor moderation. Screen and application sharing allow a teacher to share an open application on his or her desktop with a class. Many Webcasting tools bundle application sharing, whiteboard, chat, a participant list, polling, and feedback indicators, such as hand-raising, with audio and video capabilities to provide a virtual classroom.

Probably the most frequently used form of synchronous interaction occurs via instant messaging (IM), text chat, and live-tweeting, technologies that have increased in usage as mobile devices become more prevalent. These tools provide the ability for synchronous conversations between people over the Internet by exchanging text messages back and forth at virtually the same time. IM typically involves pairs of individuals, while chat tools involve larger groups (sometimes whole classes). IM and chat provide a means for a teacher to hold online office hours. IM allows students to see when a teacher is available for questions and is quicker and easier than the phone or e-mail. It allows students to easily find when another is available, say, for collaboration on a project. It has the benefit that students can see who else is online, even if they do not communicate, which is helpful because online students may feel isolated. "One of the surprising properties of computing is that it is a social activity. Where I work, the most frequently run computer network program is the one called 'Where' or 'Finger' that finds other people who are logged onto the computer network" (Kiesler 1986). IM and chat tools can vary widely — some are simple, allowing the exchange of text messages with little else; others provide an ability for private messaging, ignoring specific participants, sharing files or URLs, or providing some structure for interaction (i.e., for students to ask questions and instructors to provide answers). Chats used within learning environments should be persistent (i.e., chat histories should remain available for review by students and instructors throughout the life of the course). Instructors should be able view chat logs (time-stamped) for student assessment.

Virtual worlds take IM and chat into a visual realm, where avatars, or representations of people, move in a two- or three-dimensional world and talk to each other. Virtual worlds have been used successfully for teaching languages and architectural design, where, in the latter case, students can construct buildings that others can tour. Virtual worlds are also useful for the informal social interaction that happens more readily on campus (Neal 1997). The popularity of virtual words increased dramatically with Second Life, which may now be on the decline.

Videoconferencing extends the capability of audioconferencing by including video. Such services enable instructors to either stream video from within the system, or else enable videoconferencing, between instructors and students, between students, or between multiple classrooms. While specialized technologies used to be needed to facilitate this, now videoconferencing is available through free, widely available tools such as Skype. High bandwidth connections are necessary to avoid debilitating delays or poor quality (i.e., choppy) audio and video, especially for education, where poor quality can distract. For multipoint videoconferencing, all participants must have access to video cameras—a requirement that is increasingly possible with cameras built in to many laptops and computers. Streaming video is becoming more common with the popularity of YouTube and is often replayed rather than viewed live. The all too common talking head generally does not provide useful information; often a high-quality photograph of the presenter and high-quality audio are more effective, because they require less bandwidth and the student can focus attention on the supplemental materials such as Microsoft PowerPoint slides. However, there are instances when video presentations can be quite riveting, when done well and when the visual information is of well-known people or of objects, so that useful information or authenticity or credibility is being provided (Draper 1998).

Webinars and Web conferencing technologies couple many of the above synchronous technologies into one package, using either the telephone or VoIP for audio. A few products even allow either to be used, providing maximum flexibility depending on location and phone charges. These technologies vary considerably in quality and cost, as well as feature set. As complexity increases, the number of capabilities becomes too much for one person to manage when teaching and an assistant or moderator becomes necessary, especially if a text chat is used, because it is very difficult to talk to a class and monitor a chat at the same time. In addition, extensive preparation and scripting is needed for a well-run session. For instance, if a polling tool is used, questions need to be entered in advance.

Social media couples many of the above capabilities into commonly used tools like Facebook and Twitter. While these are rarely used as the sole delivery technology, they are highly beneficial to supplement other delivery technologies. Furthermore, they have the advantage of being in common use and requiring no learning curve for many students beyond the shift from personal to educational use.

26.3.4 LEARNING MANAGEMENT AND LEARNING COMMUNITIES

Learning Management Systems (LMS) and Learning Content Management Systems (LCMS) provide the registration and administrative processes for e-learning. They can aid a student in determining learning needs through a skill gap analysis and personalized learning plan. They track student progress so that a teacher or manager can view what a student has accomplished. Typically, they provide multiple views and levels of security based on roles, including administrator, faculty, and student views. Many provide additional capabilities, such as authoring tools to create content. Authoring tools can be separate from an LMS, especially when used to develop multimedia.

Learning portals and learning communities are a means of bringing informal and formal learning together, using a virtual classroom for live sessions, and knowledge management techniques to facilitate informal knowledge locating and sharing. Communities of practice similarly support informal learning. Sometimes, this coupling of formal and informal learning is offered through a personalized learning portal. When students take a course, arguably much of the learning takes place following completion, when they are applying what they have learned within a realistic context that is often lacking in the physical or virtual classroom. There are many opportunities, especially when students are already used to communicating online, to facilitate sharing of the learning that follows a course in an online learning community.

As an example of a learning community, Cognitive Task Analysis (CTA) Resource was designed as an interactive community of practice dedicated to those interested in the field of CTA. CTA is a specialized field of practice with a relatively small number (a few thousand worldwide) of researchers and practitioners. Initially established and launched through funding from the U.S. Office of Naval Research, this not-for-profit Web site is intended to facilitate interaction among the CTA community worldwide. CTA Resource offers free membership and provides access to a variety of online resources including an extensive digital library of CTA literature and technical reports, a contact list for members, a directory of companies and academic institutions working in the field, and tools to enable members to communicate with each other. Interaction can occur at any time through the use of discussion threads or email between members.

Unlike a structured academic or training course, there is no explicit, shared motivation for members to visit the CTA Resource site. Members need encouragement to visit the site in groups and interact with each other. To this end, online seminars have been offered as a means to educate members while motivating large numbers to visit the site at the same time and interact (Neal, Miller, and Anastasi 2002). Survey responses from members indicate that there is perceived value in the online offerings: "The on-line seminar is an excellent method of sharing information in a participative manner without the expense and time of attending a physical conference."; "Actually, I like it better. It is a little more self-paced, and easier to get your questions asked." Members have also indicated the value of such events for those living and working overseas, who may not have many opportunities to network and collaborate with colleagues and experts in other countries. Another benefit of the virtual seminar format is the archival of seminars for those who could not attend live or wish to review the materials later.

26.4 HOW TO DESIGN, DEVELOP, AND STRUCTURE ONLINE COURSES

Most instructional development evolves by trial and error, where the instructor learns over the years what type of instruction seems to be most appropriate for his or her students without realizing that he or she is adjusting the educational instruction to accommodate such issues. Too often, in planning e-learning, the focus is on technology rather than on the needs and constraints of the learner and other stakeholders such as instructors, managers, and support personnel. In an effort to produce something different and highly "interactive," technology designers sometimes forget that the focus is learning. Instructional designers and technology experts must work together to minimize mismatches between instructional objectives and technology solutions. Research may suggest potential benefits in use of a particular approach or technology for instruction, but feedback from actual use may reveal little or no benefit in a particular context or situation. Further probing of stakeholders, e.g., instructors and students, might indicate that problems inherent in the design of the technology itself caused frustration and prevented instructors and learners from reaping the potential benefits of the technology.

In contrast to the development of classroom-delivered courses, e-learning is almost always developed by teams. While sometimes the teacher is the developer, a common model is for development and delivery to be done by separate people. Often, one or more subject matter experts are queried by instructional designers, who then work iteratively with the subject matter experts and media experts to develop a course. At the extreme, this process can be likened to movie production in terms of the number of roles involved. These roles include the following: project manager, who oversees planning, schedules, and purchasing; subject matter expert, who determines and validates content; instructional designer, who ensures that learning takes place; graphic designer, who uses authoring programs to implement instructional designer's storyboards; programmer, who enters course content, builds interactive exercises; quality assurance specialist, who tests and evaluates; interface designer, who determines look and feel to fit student needs; multimedia developer, who incorporates audio, video, and animation; writer/editor, who creates text and verifies grammar and style; Webmaster, who handles technical support and maintains hardware and network connections; librarian, who provides resources; instructor, who ensures the material can be taught effectively; and institutional stakeholder, who must be satisfied with course quality.

Martinez (2001) found that students did best, in terms of satisfaction, intentional learning performance, and learning efficacy, in environments that best suited their learning orientation; and time management was also a factor. Some of the key learning theories that are commonly applied to e-learning are as follows: implicit learning (intuition), for complex tasks; explicit learning, for salient noncomplex tasks; objectivistic, learn by being told; constructivistic, learn by doing, with the learning situated in an authentic or realistic context; problem or scenario based, in which a student is given a problem to solve and, in the process, has to delve into many aspects of the subject matter, with learning being self-directed; and activity theory, learning shaped by the physical. These theories can be applied individually or, ideally, are blended to develop an effective style of instruction and course design. For example, Ally (2008, p. 39) suggests that behaviorist, cognitivist, and constructivist theories contribute in different ways to the design of online learning: "Behaviorist strategies can be used to teach the facts (what); cognitivist strategies, the principles and processes (how); and constructivist strategies to teach the real-life and personal applications and contextual learning."

26.4.1 Course Templates, Development Plan, and Costs

Course templates can help instructors create the initial structure for an online course and can promote consistency in the design of course artifacts within and across online courses. Instructors can use templates to create announcements, course content, descriptions, objectives, policies, registration, syllabus, assignments, discussion forums, participant biographies, and postclass feedback (William Horton Consulting, n.d.). Templates typically include a "what you see is what you get" (WYSIWYG) content editor. Course content can be created from scratch using the editor or linked to or uploaded from existing files or a content repository. Instructors can typically copy and modify default templates or create new templates to meet the needs of a specific course.

The types of templates need to be carefully planned to maximize effectiveness of a course. Because online courses generate unique challenges, it is important to look at the social, not just educational, needs of students. For example, when dealing with different cultures, it is imperative that the learners share personal information about themselves so the other learners can assign an identity to the electronic communication (Klobas and Haddow 2000). LeBaron and Miller (2004) describe the use of expanded student biographies and an icebreaker exercise to promote student interaction and initiate a sense of community within a graduate level online academic course. Personal profiles or Home pages with photographs are an increasingly common way that students can get to know each other and their teacher and feel connected despite the distance.

A course development plan to analyze, design, build, and evaluate a course should be developed (Hall 2000). The typical steps are to analyze the learning problem, the organizational issues, and identify learner characteristics, knowledge, and skills; design what the instruction will look like (instructional goals, teaching/learning activities, and how the learning will be evaluated); develop materials to the design description and test; implement/ plan for how a class will be delivered, received, updated/revised, and maintained as well as any training and piloting; and evaluate throughout the design process and during implementation to make sure the class is effective for the learners. Iterative testing, even of prototypes or storyboards, should be performed to ensure that the course will be effective, and a course should be piloted with a representative audience in as close as possible to the actual setting. Evaluation is discussed in depth in a later section. Another part of the course development plan is to determine the "shelf life" of a course, i.e., how frequently it needs to be checked for accuracy (including working links) and how often it should be updated. Outdated links are a source of frustration for Internet users, and may also adversely impact the perceived credibility of an online course (http://en.wikipedia.org/wiki/Dead_link).

It is difficult to estimate the time and cost for producing an online course. ASTD uses very general ranges of 40 hours of development for 1 hour of classroom instruction and 200 hours of development for one finished hour of computer-based training. Some of the variables impacting time and cost are the length of the module because a longer module will cost more; complexity of the topic and resulting material to teach it; testing; graphics and multimedia, with audio, video, and animation being the most time-consuming and expensive to create; the simplicity or complexity of navigation because the more options included, the more complex and costly the development will be; and content, whether content already exists or needs to be developed or updated.

26.4.2 Instructional Design

E-learning typically relies heavily upon textual materials. Some data indicate that a continuum exists of lecture, reading, seeing, hearing, seeing and hearing, collaboration, doing, and teaching others, where lecture offers the lowest retention and teaching offers the highest. Gagne (1985) provides a nine-step model to help ensure effective learning. His "events of instruction" include the following:

1. **Capture learners' attention** with a thought-provoking question, interesting fact, or visual information.
2. **Inform learners of objectives:** initiates the internal process of expectancy and helps motivate them, objectives should form the basis for assessment and evaluation.
3. **Stimulate recall of prior learning:** associating new information with prior knowledge facilitates learning — ask about previous experiences or build upon an understanding of previously introduced concepts.
4. **Present the content:** content should be chunked and organized meaningfully, explained, and then demonstrated, and to appeal to different learning modalities, use a variety of media (text, graphics, audio narration, and video).
5. **Provide "learning guidance"** to help learners encode information for long-term storage with examples, non-examples, case studies, graphical representations, mnemonics, and analogies.
6. **Elicit performance:** learners practice new skill or behavior to confirm correct understanding, and repetition increases the likelihood of retention.
7. **Provide feedback** as learners practice and provide specific and immediate feedback of performance.
8. **Assess performance** with post-test or final assessment without receiving additional coaching, feedback, or hints; mastery of material is typically granted after achieving a 80–90% correct.
9. **Enhance retention and transfer to job** through repetition of learned concepts.

The above has led to the principles commonly used for instructional design; Keller (1987) developed the ARCS Model of Motivational Design, a systematic model describing four factors (Attention, Relevance, Confidence, and Satisfaction) intended to aid in designing instruction that is motivating to students. The ARCS Model categorizes the four factors as follows: Attention, arouse curiosity and sustain interest; Relevance, make tasks relevant to learner's needs and interests; Confidence, build on success and gain self-efficacy; and Satisfaction, provide extrinsic and intrinsic rewards.

New models more fully leverage the structural and communication potential of the Internet. Engagement theory (Kearsley and Shneiderman 1999) carries these notions further into online learning, stating that students must be meaningfully engaged in learning activities through interaction with others and worthwhile tasks, facilitated and enabled by technology. These learning activities occur in a group context, are project-based, and have an authentic focus. "The

difference between engagement and interactivity reflects the shift in thinking about computers in education as communication tools rather than some form of media delivery devices." However, this requires a group of students taking a course at the same time, under supervision, which not all e-learning includes. Even though the name "engagement theory" implies a more engaging design results from application of the theory, a student needs to be motivated for even the most engaging course to teach him or her.

In implementing the above theories, online course designers should keep in mind how the course will realistically be used by students. Many of the factors that need to be considered arise from persona building, such as low levels of comfort with technology and blocks of time to study that may be short or interrupted. Online courses can be designed to increase motivation by challenging students and providing tasks that are fun to accomplish. However, "fun and pleasure are elusive concepts" (Norman 2004, p. 106), and there is no consensus on how to design enjoyable experiences (Monk et al. 2002). Norman discusses the importance of emotion in design and that "you actually think differently when you are anxious than when you are happy . . . devices and software should be designed to influence the mood of the user; they will be more effective because they are more affective" (Gibbs 2004). Engagement is accepted as important in e-learning but is similarly elusive.

26.4.3 USABILITY PRINCIPLES

Usability principles increase the likelihood of designing a course that works from the learner's perspective. The five aspects of usability (Constantine and Lockwood 1999) are learnability, rememberability, efficiency in use, reliability in use, and user satisfaction. Because learners typically do not tolerate extensive training before taking an online course, it is important to keep these principles in mind during design to ensure that learners can focus on learning, not on using the learning application. A successful design facilitates learners reaching a "flow" state (Csikszentmihalyi 1990), in which learners are so focused on what they are doing that they are unaware of the passage of time. Inability to, for instance, navigate to the next screen easily disrupts flow. It is also important to remember that, when a course is delivered over the Internet, learners may have developed habits of skimming rather than reading material or quickly click to the next screen. Information in an online course needs to be designed to accommodate how people read online and make judicious use of highlighting, fonts, and formatting to increase readability. An appropriate balance is needed between boring and flashy in order to capture attention and enhance learning without detracting from their focus.

Progress indicators are a basic component of a well-designed interface and are especially important to help pace a student, set expectations, and offer feedback on progress. Some indicators to include are how long should something take to do or play, which should be exact for audio or video, and an estimate for a "lesson" or exam to set expectations and allow good allocation of time; and clear indications of where a student is in the course, what the student has accomplished so far, and what he or she should do next. To instill learner confidence, it is important to inform learners about learning objectives, course requirements, and assessment criteria and to provide feedback on their accomplishments (Keller 1987).

Another aspect of interface design that is critical for many online courses is appropriate internationalization and localization. Internationalization means designing the interface and writing for a global audience. Localization means making modifications to fit the language and culture of students. This includes being sensitive to the differences between, say, British English and American English, providing local references and examples, and understanding the differences between reactions to and acceptance of humor. While it could also include the use of different instructional styles to accommodate students, it rarely does.

Similarly, there are other student differences that impact design; the most common to accommodate are age, gender, background, or disabilities. In some countries or institutions, it is essential for legal reasons to make courses accessible to people with a range of disabilities. The key ones are visual, hearing, motor skills, and neurological disabilities. However, some of these, such as color blindness, are fairly common, and others may be important because of the environment a student is in, such as one that is noisy.

26.4.4 Selection of Media

The heart of a well-designed online course is appropriate selection and use of media. While the technologies to support the variety of media are described above, a designer needs to decide which media to use and then which technology to use to support it. Sometimes, in a constrained environment or within an established program, there is no choice about which technology to use, only the choice of which components to use and how to use them most effectively within the course.

Visual language refers to the tight integration of words and visual elements, including animation, cartoons, and diagrams (Horn 1998). Visual language is important in e-learning, because the online materials are crucial to student learning and there may be limited, if any, support available as a student uses these materials. A "standard" has developed in online courses that is often referred to as electronic page-turners, where a course consists of screens of material that a student clicks through. Each screen has some text and a graphic, with some optional interactive elements such as animations. An important principle for e-learning design is that graphics and interactive elements should be purposeful and aid in learning. At worst, they can be distracting, much like banner ads. But visual representations can be very powerful at conveying information (Tufte 1990, 1992), as can audio and video. The challenge to an online course designer is to determine how to most effectively convey the materials to students, given the nature of the topic and the demographics of the students, and to determine how to best use text, animations, graphics, audio, video, exercises, discussions, and projects to teach.

The Pennsylvania State University (n.d.) is based on learners' access to and familiarity with e-learning technology (Roblyer and Ekhaml 2000) and offers a rubric for assessing interactive qualities of distance learning courses that can be used to design an interactive course. Moss (2002) provides some guidelines and suggestions to design to increase learning, for instance, giving an example of how different exercises engage different (and additional) mental processes.

One of the most difficult aspects of course design is determining the type and design of testing. While there are some examples of e-learning where completing the course requires merely clicking on all or a high percentage of screens, most use some testing to gauge learning. Self-assessments give immediate feedback to a student, generally from multiple choice questions. While multiple choice questions offer the advantage of immediate feedback to the student, it is often difficult to develop questions that require deep thinking. Open-ended questions typically elicit more thoughtful answers and can more accurately gauge learning but require grading and individualized feedback. The best approach is a function of what is possible resource-wise and the value to the learner.

26.4.5 Before and After E-Learning

The above discussion pertains primarily to the design of courses. A last point on course design is the role of pre- and postclass activities. Before a class commences, students can become comfortable with any technology they will be using, can engage in pre-reading, either optionally or to help them acquire knowledge so that students have a level playing field in class, and can introduce themselves through profiles or a discussion forum. LeBaron and Miller (2004) describe use of an "icebreaker exercise" at the onset of their online course. The exercise served not only to help students introduce themselves and become more comfortable interacting with each other; it also provided an opportunity for the students to learn how to use the technology itself. An introduction, for first-time students, on how to be a successful online learner can be helpful, especially because many of the skills are different from what is needed to succeed in the classroom. Following an online course, there are opportunities for students to continue discussions. Often, the same tools used for class can be used to develop what is sometimes called "communities of learners" or "communities of practice." This may be especially valuable in environments where students are learning and using skills and can reflect upon and provide context to fellow learners.

26.5 TECHNIQUES AND BEST PRACTICES FOR DELIVERY OF E-LEARNING

Teaching online can be quite different from teaching in the classroom. While not all e-learning has a teacher—self-paced courses being an example—when they do, the teacher needs to be adept at using the delivery technology and at interacting with students effectively, without the physical presence and visual information that is so abundant in the classroom (Entin,

Sidman, and Neal 2008). It is not enough for an instructor to be expert in the content being taught—instructors must be adequately prepared to ensure their online courses will be a positive experience for students. This reaches beyond familiarity with the technology itself—an understanding of how teaching practices and behaviors need to be adjusted to succeed online is required. For example, adjustments might be needed regarding the number of assignments and due dates, given that text-based interaction and coordinating course work and feedback between peers and the instructor often takes longer than classroom-based situations.

Berge (1995) defines the roles of an online instructor to be pedagogical, social, managerial, and technical and points out that these may not all be carried out by the same person; in fact, they rarely are. Each type of medium (e.g., print, audio, video, Web based) requires specific instructional skills for effective use. For example, those teaching in video-based environments must learn on-camera behaviors and adapt to lack of learner feedback, while those using Web-based technology must adapt to the absence of nonverbal cues and learn to interpret online communication nuances (Schoenfeld-Tacher and Persichitte 2000). It is crucial for instructors to understand how to use technology, apply it to the discipline, generalize it to learn new applications, and guide students in applying the technology. Ideally, instructors should try to identify *opportunities to improve learning* via technology by developing a variety of teaching strategies and understanding how technology shapes new teaching roles and student roles (Coughlin and Lemke 1999). This understanding and ability to leverage technology in teaching is a learned skill—it does not come naturally to most people. Schoenfeld-Tacher and Persichitte (2000) describe areas in which instructors might require training for competence in online instruction including (1) time to prepare/teach, (2) communication methods (using tech), (3) build on experiences of other teachers, (4) strategies to increase student-student and student-teacher interaction, and (5) organizational details.

One cannot assume a simple transition from a face-to-face course to an online setting. It is not simply a matter of an instructor digitizing existing course materials and then "coasting" through the duration of the course while periodically checking student progress. Such an approach typically results in dry courses that are not engaging. Instead, existing course materials can form the basis for creating learning tools enhanced by technology including help, video and audio presentations, animations, and interactive models. These learning tools provide many benefits, not the least of which is a more engaging online learning experience. In addition, they provide an opportunity to reach more students. Other techniques that are effective in the classroom, such as storytelling, can be brought to online courses with great benefit to students but with some effort and planning on the instructor's part (Neal, n.d.).

Conveying information in multiple modalities can assist those with certain physical or learning disabilities as well as those who learn best in less traditional ways. Gardner's Multiple Intelligences (MI) theory, for example, states that every learner possesses certain dominant learning styles (or a single style) by which she or he best responds to learning situations (Gardner 1993). These styles include linguistic, musical, logical-mathematical, spatial, body-kinesthetic, intrapersonal (e.g., self-awareness), and interpersonal (e.g., social skills). Leveraging technology in the development of a diverse set of online course materials can help meet the needs associated with a variety of learning styles.

A successful instructor-led online course does not consist merely of well-designed standalone course materials. Care must be taken to embody a course with "personality." Anderson (2008) describes three aspects of "teacher presence," as it applies to online learning: designing and organizing the online learning context that includes a personalized tone, facilitating active discourse with each individual as well as the learning community as a whole, and provision of direct instruction. Enlivening the learning experience through frequent and meaningful student-instructor and student-student interaction facilitates a sense of course community, which is particularly important in online learning environments. For example, a perceived lack of feedback or contact with the instructor may impede a student's ability at self-evaluation, while a lack of interaction with peers can lead to feelings of alienation, isolation, and the lack of a sense of scholarly community (Galusha 1997). These outcomes, in turn, can lead to lower levels of learner motivation and engagement. It can be challenging to personalize online courses when technology can limit implicit communication (e.g., facial expressions, gesturing, etc.) and make it more difficult to convey and interpret emotions (Norman 2004).

There have been attempts to build ways that students can express emotion or that emotion can be detected, but none have even approached the success and timeliness of facial reactions and body language in the classroom. Yet instructors can compensate for those limitations through frequent dialog with students, timely responses to questions and concerns, seeding discussion forums to encourage a scholarly dialog among students, and providing meaningful assessments and feedback on students' work. In a nutshell—when teaching online it is important to maintain a steady presence by "being there" for the students. The ARCS Model (Keller 1987) identifies some key factors impacting learner motivation including the confidence level of the learner and sense of satisfaction derived from the learning experience. Learner satisfaction can be supported through extrinsic rewards such as positive reinforcement and motivational feedback.

Online grading tools help instructors track grading and provide feedback on student work. If an automated grading scheme is established, the grading tool can update the grade book whenever a new assignment, quiz, or test is completed. Grading tools should be flexible, allowing instructors to manually edit grades, assign partial credit for certain course activities or student answers, and add grades for assignments that require manual assessment. Online grade books allow the instructor to manage and view grades in a variety of ways including by assignment, by student, and for all students on all assignments. Instructors can also search the grade book based on specific criteria. If an assignment is added to the course, it is automatically added to the online grade book. The software can assist with analysis of grades by automatically calculating the minimum, maximum, and average assignment grade. Often, instructors can export versions of the online grade book to a spreadsheet or other data analysis application. Students can be informed of grades for individual assignments through an automated message or via more personalized feedback from the instructor.

Many learning management systems include Automated Testing and Scoring tools that allow instructors to create, administer, and score objective tests. These testing tools allow instructors to create test questions in a variety of styles including true/false, multiple choice, multiple answer, fill-in-the-blank, matching, short answer/essay, and calculated answer questions. Instructors can typically create feedback messages to be displayed when varying conditions occur. Many tools provide the ability for instructors to create personal course-specific test repositories, as well as tap into system-wide repositories, and then use that stored content to create tests for students. The system can randomize the questions in a test, and some tools can randomize on a per student basis to reduce the chance of cheating. Instructors can control access by assigning dates when tests can be accessed and/or setting time limits on a test.

Some instructors allocate a portion of a course grade to student participation. Tracking software allows the instructor to track the usage of course materials by students. Robust tracking software can report individual usage as compared to overall class usage of various materials and resources. However, often more important than the number of times or length of time that a student spends accessing course content is the quality of the time spent (i.e., the scholarly contributions a student makes to the overall course). Assessing the quality of each student's participation in a course is a challenge for online instructors—one for which tracking software offers little assistance.

Assessing the "quality" of online participation is very subjective, and a student's perception of his or her own participation may vary from that of his or her peers or the instructor. It is important for the instructor to describe the criteria upon which an assessment of participation is based, particularly if the assessment consists of more than a mere count of the number of posts each student makes. Example criteria defining high-quality student participation might include the level of reflection, thoughtfulness, depth of dialog, research quality, and consistency of participation in course discussion threads or within the context of a specific interactive exercise. It is often helpful for the instructor to model quality dialog by providing an initial example (e.g., some discussion posts or chat interaction) with a teaching assistant or student volunteer. The instructor's ongoing interaction within the course environment—to the class as a whole as well as with individual students—also sets the tone for the expected level of interaction within the course. When the instructor identifies students falling below the threshold of optimal course participation, it is important to express that concern to the student and reiterate expectations in a timely manner so that the student has enough time remaining in the course to remedy the situation. This is important not only for the benefit of the student in question but for the overall benefit of the class as the lack of participation on the part of one or more course participants can be discouraging and demotivating to the rest of the class.

With respect to all of the assessment methods described above, it is crucial for instructors of online courses to clearly describe the structure of the class and associated learning tasks and activities in the course syllabus. In addition, a detailed rubric describing major course assignments and breaking down how each factors in to the overall grade should be provided. Ambiguity can be a source of anxiety in traditional classroom settings. In distance learning situations, the perceived distance a student might feel between his or herself and the instructor can exacerbate feelings of anxiety stemming from ambiguous course objectives or instructor expectations. That said, merely summarizing each assignment and stating its weight in the overall course grade may not provide enough information for students to feel comfortable and prepared to tackle each assignment. Whenever possible, instructors should try to provide supporting information with respect to course assignments. Examples include more detailed descriptions of assignments, links to related resources, examples of exemplary work, a breakout of allocation of grade points to various aspects of the assignment, frequently asked questions (FAQ) about the assignment, etc. The trick is to think about each assignment, try to anticipate all of the questions and concerns students might have about it, and be proactive. Provide the answers using a combination of techniques such as those listed here. This approach can help students feel empowered to succeed while minimizing confusion and heading off a deluge of questions later on.

Effective online teaching that promotes engaged learning requires a large investment of time and effort on the part of the instructor. Facilitating active communication and community-based construction of knowledge in an online learning environment can take substantially more time and effort than comparable "traditional" efforts. For example, a face-to-face version of a graduate level academic course might consist of a weekly 3 hour course session supplemented with a handful of office hours on a weekly basis for students to use as-needed. An online version of the course would require much more of the instructor's time, albeit not necessarily on campus, to be successful. Providing timely, substantive feedback as well as providing an instructor "presence" for a group of online students on an ongoing basis, along with grading major course assignments, requires a significant instructional time commitment. Enlisting a teaching assistant—perhaps a graduate student—to assist in developing online course materials, moderating class discussions and chats, and managing ongoing course activities can yield a high level of responsiveness to students while allowing for collaboration on the design of course exercises and assessment materials. On the basis of their own positive experiences as an instructional team, as well as the perceived quality of instruction by their students, LeBaron and Miller (2004) recommend a collaborative instructional approach when feasible—particularly for novice online instructors.

26.6 EVALUATION OF E-LEARNING PROGRAMS

An evaluation plan for an e-learning program is essential to determine if the course meets the stated goals and, if not, to determine what to revise. Evaluation provides the data needed to determine the effectiveness of a program so that stakeholders can decide whether to accept, change, or eliminate any or all of its components. Good evaluation looks beyond the surface to inform decisions regarding content topics, organization of content, delivery methods, etc. It involves specifying a set of criteria to be evaluated, identifying appropriate measures to inform us about the criteria, and analyzing or otherwise examining the resulting data with respect to the criteria. Evaluation is not always done when developing e-learning or it is done in a cursory fashion to eliminate glaring problems. In many corporate environments, there is more of a focus on measuring return on investment (ROI) than in evaluating the effectiveness of a program. Certainly, quantitative and qualitative methods of collecting and processing data have their place in evaluation, and a preference for one over the other is largely based on how the data will be used.

Evaluation can be defined as the formal determination of quality, effectiveness, or value of a program, product, project, process, objective, or curriculum. Evaluation includes determining standards for judging quality, data collection, and applying the standards (i.e., criteria) to determine quality. Evaluation is the means for determining whether a program meets its goals; whether the instructional inputs match the intended/prescribed outputs. Evaluation can be a broad and continuous effort to inquire into the effects of utilizing content and process to meet clearly defined goals. All of these definitions have a common theme—evaluation can only be conducted after defining a set of criteria or standards based on underlying the educational goals and objectives.

Evaluation can take on a variety of forms and should ideally occur at various stages throughout the life cycle of a distance learning program. The first two definitions of evaluation in the preceding paragraph imply a summative evaluation—one that seeks to determine the level of quality or success after implementation (i.e., is the course as delivered working?). Formative evaluation, on the other hand, entails an approach that seeks to inform the design of a program as it evolves over time. Data are collected during design and development to fine-tune the program before it is implemented in an e-learning program. Evaluation, however, is situation specific and often takes place in a naturalistic setting.

Ornstein and Hunkins (1997) summarize five types of value questions regarding curriculum evaluation:

1. Intrinsic value pertains to goodness and appropriateness, as planned and as delivered; it asks whether a program incorporates the best thinking to date on the content, arrangement of the content, and its presentation. Intrinsic value is tied to philosophical and psychological views.
2. Instrumental value asks what the curriculum is good for (does it link back to the stated goals and objectives) and who the intended audience is (will students benefit from the program? Is it appropriate, given the context?).
3. Comparative value is concerned with determining whether the new or revised program is better than the original. Other factors include delivery, cost, necessary resources, and how it fits into the organization.
4. Idealization value is not concerned so much with evaluating expected outcomes but with how to continuously improve and fine-tune the program.
5. Decision value reminds us that the value of the decisions made regarding curriculum need to be assessed as the curriculum is being delivered.

It is important to maintain a historical perspective, capturing the key decision points and documenting the "whys," so those decisions can be revisited once data and experiences are available to ensure that the right path(s) was followed.

26.6.1 THE EVALUATION PROCESS

While the specifics involved in an evaluation (i.e., evaluation criteria, methods of measurement, analysis, etc.) vary depending on purpose of the evaluation and the philosophies and values of the stakeholders, evaluations must be systematic in order to provide useful results. That is, after all, the point—to inform the design or revision of a product. The following series of steps is common to many forms of evaluation:

1. Evaluation criteria: identify the phenomena to be evaluated and the design to use for the evaluation.
2. Data collection: identify information sources and the means of collecting necessary information.
3. Organize data: transform the data (e.g., via coding, organization, etc.) into interpretable form.
4. Data analysis: interpret the data.
5. Reporting: frame the results in meaningful fashion for the final audience by summarizing it, interpreting the findings, and providing recommendations.
6. Refining: use the results to provide iterative feedback regarding the evaluated product in order to continually fine-tune it.

The importance of appropriate reporting cannot be overemphasized. An evaluation can be based on a sound set of criteria and include meticulous data collection and thorough analysis, but if the results are not documented in a meaningful way for the target audience (i.e., stakeholders), it is all for naught. A thorough report should include not only a summary of the results and analysis, but an interpretation of the implications for the distance learning program. Whenever possible, a set of recommendations should also be provided for consideration by the stakeholders, keeping in mind that the type of and level of detail of the recommendations differs depending on the recipient's perspective and role.

26.6.2 USABILITY EVALUATION

Usability evaluation, as part of formative evaluation, looks at how well designed a course is from the learner's perspective (Feldstein and Neal 2006). The five aspects of usability discussed in Section 26.5 are learnability, rememberability, efficiency in use, reliability in use, and user satisfaction. It is important to test for these during evaluation to ensure that learners can be successful. For instance, if text is difficult to read or a learning application is difficult to navigate, learners may become frustrated and drop out. A consideration in technology evaluation and selection might be its appropriateness for both novice and more expert users (i.e., those with varying degrees of technical savvy). A well-designed, robust technology application should be usable and useful for learners with a diverse set of technical skills.

26.6.3 EVALUATION DURING AND FOLLOWING A COURSE

Feedback can be obtained during and following a course using both formal means, such as online surveys and data gathering, and informal techniques. The informal include using e-mail, phone calls, and office hours that are for student assistance as a way to seek feedback. Online survey design needs to balance the desire to receive feedback with student willingness to spend time filling in a form. Multiple choice questions are quicker to answer and compile but offer far less information than open-ended questions. Some online courses require a survey to be completed before the student receives course credit. While that ensures a high completion rate, students may be less willing to respond honestly and carefully under those circumstances. Capturing data through logs is usually easy and supported by many LMSs. At a very low level the data tend not be useful for improving a course or understanding how students are participating. Some data, however, can be quite helpful, such as learning which parts of a course students spend the most time on and when they access the course and for what duration.

Kirkpatrick (1994) provided the standard scale for measuring training effectiveness, which is used following an online course. Level 1 is how did the student react; level 2 is did learning take place, typically determined by a comparison of pre- and post-tests; level 3 is if behavior has changed as a result of the training, i.e., is the student able to perform his or her job better; and level 4 is if the training contributed materially to business results. It is easier to measure impact on productivity of a salesperson than a knowledge worker. Additional ways of measuring training effectiveness include tracking enrollments as an indicator that training is attracting students who perceive that it will meet a need; and tracking completion rate to indicate that courses were effectively

designed. In such cases, assumptions are often made that require need supporting data.

The topics, adapted from Evaluation for Distance Educators (n.d.), that are valuable to receive formal and informal feedback on to improve an online course or e-learning program include:

1. Use of technology: student familiarity, problems, positive aspects, attitude toward technology, usefulness of technology skills learned in class to other venues.
2. Class formats: effectiveness of lecture, discussion, question and answer; quality of questions or problems raised in class; encouragement and support given to students, pace of class.
3. Class atmosphere: conduciveness to student learning, social aspects.
4. Quantity and quality of interaction with other students, with instructor, and with any support staff.
5. Course content: relevancy, adequate body of knowledge, organization, pointers to supporting resources.
6. Assignments: usefulness, degree of difficulty and time required, timeliness of feedback.
7. Tests: frequency, relevancy, sufficient review, difficulty, feedback.
8. Support services: facilitator, technology, library services, instructor availability.
9. Student achievement: adequacy, appropriateness, timeliness, student involvement.
10. Student attitude: attendance, assignments submitted, class participation.
11. Instructor: contribution as discussion leader, effectiveness, organization, preparation, enthusiasm, openness to student views.

Some open-ended questions that provide in-depth feedback include the following:

1. What were the strongest aspects of this course and why?
2. What were the weakest aspects of this course and why?
3. What would you suggest to improve this course?
4. What were the most important things you learned in this course?
5. What would you have liked to have learned in this course that you did not learn?

ROI may not directly help improve a course or program but helps justify the funding and support for e-learning. Typically, ROI compares course development and delivery costs with student and teacher travel, classroom construction costs, and any additional expenses such as food. Some corporate ROI calculations include student salaries for the time they would have been in class but that may be a flawed measure because, with e-learning, a student still needs class time, it is just in smaller blocks and may be taken from the student's personal, rather than work, time. Measuring ROI necessitates assigning a cost structure to each objective and understanding effect on savings and impact on performance, and many are difficult to calculate, such as the value of an employee who is retained longer through better educational opportunities. Some of the more quantifiable measures include the decrease in production costs to update and distribute materials; the relative ease of tracking, monitoring, and measuring student performance; and the reduction in travel and lodging costs.

However, for all the evaluation techniques discussed, and without discounting the importance of the many measures, the most essential aspect of evaluation is to determine whether learning took place. This is commonly done through testing throughout a class and at the end of a class, by comparing the results of pre- and post-testing, and by evaluating student assignments. This is, of course, important for all education, not just e-learning, but the distance can make it more difficult to use more subjective measures such as seeing the spark of enthusiasm in a student or seeing a student have an "ah-ha" moment in class. Without those visual opportunities, students can still report their subjective satisfaction and understanding, and their online behavior can indicate their enthusiasm for and understanding of a course topic. Longitudinal testing can be used to measure retention.

26.7   SUPPORT FOR E-LEARNERS

"Teachers and students come to conventional higher education having already learned well-defined roles through years of common educational background and experience in the formal education system. During online learning this background and prior experience are less relevant to the context, which can invoke feelings of anomie" (Anderson et al. 2001). Students need help in advance to prepare to be successful distant students as well as support during a course. Students know what to do and how to behave in a physical classroom, for example, sit down, look attentive, and take notes. Online students, however, need to be taught what to do, how to act appropriately, and the importance of time management and distraction control.

Students who are more familiar with technology and distance learning are more likely to do better in a distributed instructional setting that requires computer use (e.g., see Gilbert and Werge 2000; Irani 2001). Students should feel comfortable enough with the technology that it does not become a barrier to successful learning. In addition, students need to acquire other necessary skills to support a positive learning experience including communication, collaboration, research, critical analysis, problem solving, and self-management. Students working online often work very independently; the lack of regularly scheduled classroom time may result in online students feeling less accountable for their work. It can be very easy for students to fall behind and very difficult for them to catch up later. Students must be highly motivated and able to work independently successfully.

Several of the barriers to distance learning described by Galusha (1997) pertain to the lack of student preparedness and support. With respect to online learning, first-timers have special needs. Study materials must reflect the fact that many students have little or no experience with online learning. Technical barriers must be made a nonissue. Selection of intuitive and flexible technology tools, coupled with technical training, can reduce technical barriers. However, similar to the challenges instructors face, students need to acquire more than a basic understanding of the use of the technology itself—they need to understand how to use it efficiently to support their learning tasks. For example, the Internet provides access to a wealth of resources with varying degrees of reputability. Students need to learn good research skills, including good Internet search skills. Because of the sometimes overwhelming amounts of data available, they also need guidance on where to look first (e.g., recommended data repositories) and to develop the ability to critically assess the information they find. A tutorial with an affiliated library service can help students hone these skills, and embedded links to reputable and relevant data resources can point students in the right direction for research projects.

Scaffolding can also help train students in the use of technology tools. For example, LeBaron and Miller (2004) describe the design and implementation of an icebreaker exercise for an online course. This loosely structured exercise was carried out using the teamwork capabilities provided by the Learning Management System used for the course. While this group exercise was intended to promote a sense of community among students at the onset of the course, it had the added benefit of "training" students in the use of the teamwork capabilities early in the course, as these same tools would be used later on in a more comprehensive mid-semester group exercise. Initial training such as this is advocated by Lin (1999), who found that practice time to acquire skills needed in an online course had positive impacts on self-efficacy, interest, and commitment.

Group work presents another challenge to online learners, who need to collaborate in novel ways to achieve group goals and work effectively. The use of technology tools changes the nature of interaction and group work. Students need to understand the dynamics of online interaction, including understanding "netiquette" (the etiquette of online behaviors). Students involved in group projects need to respect the accountability they have to their work partners as well as to the instructor. It can be more difficult to chase or prod an unresponsive "virtual" partner. Coordinating work on joint artifacts presents another challenge. Although in today's workplace such coordination is not uncommon, it cannot be assumed that all group members have similar technical skills or access to compatible versions of software tools (e.g., word processors, spreadsheets, data analysis software, presentation editors, etc.). Team members must learn how to coordinate and execute their group work.

Studying online requires a high level of motivation and an ability to work well independently. Students need to learn how to manage their study time, particularly when there are no face-to-face classes to reinforce a sense of accountability to the instructor and the class. In the online world, it can be very easy to let other responsibilities or distractions take precedence, and, once one has fallen behind, it can be very difficult to catch up. Course schedules integrated with academic planners can help, as can weekly course updates and reminders from the instructor. An instructor who presents an organized syllabus, provides timely feedback, and shows an overall commitment to the course can act as a role model demonstrating effective work practices.

The physical separation of students in programs offered at a distance may contribute to higher dropout rates (Rovai 2002). Separation has a tendency to reduce the sense of community among students, giving rise to feelings of disconnection (Kerka 1996), isolation, distraction, and lack of personal attention (Besser and Donahue 1996; Twigg 1997), which could affect student persistence in e-learning courses or programs. Course design, incorporating instructor and peer interaction, and providing adequate support for students, especially first-time online learners, can increase the sense of community, leading to greater student success as well as a more successful course or program.

26.8 CONCLUSIONS

E-learning exists in many forms, from self-paced asynchronous courses to highly interactive courses delivered using synchronous technology. The commonality, of course, is the lack of physical proximity, which is what prevents the familiar teaching styles and interaction that occurs in the classroom. Even social media, by its very nature of creating connections, can aid in bridging distance.

The range of existing e-learning also differs in the types and amount of support available to students. While the distance inherent to an online course does not per se necessitate that a student takes a course without a teacher or peer support, this is sometimes the case. Yet technology can facilitate rich discussion that takes place synchronously or asynchronously, and, ultimately, the selection of technology and media to support learning at a distance is dictated by technology access, availability and cost, student demographics and needs, and the objectives and constraints of the institution offering the courses. Even within a highly constrained situation, it is possible to offer high-quality online courses that meet students' needs and provide engaging learning experiences. The support of teachers and peers can enhance many learning experiences and can help students with the low points, when they are stuck or confused, as well as with the high points, when they are excited and enthusiastic about what they are learning. Students can benefit, not just from access to educational opportunities, but the increased technology skills and digital literacy that result from being online learners. It is important in planning, designing, or delivering e-learning for the focus to remain on learning: how can this topic be most effectively conveyed to the target learner population? This chapter has covered many aspects of technology and course design, delivery, and evaluation in support of this objective.

EXERCISE: EVALUATE ONLINE COURSE

Many free online courses are available. Try to take at least two or more for the sake of contrast. Answer the following questions for each course:

- What is the target audience and the topic?
- What works and does not work about the overall design; the interface, navigation, and usability; the content; and the assignments, activities, and tests? Be detailed in your response.
- Consider the delivery technology and media (audio, video, text, etc.): what works and does not work for the target audience and topic?
- How would you evaluate the effectiveness of the course?
- How would you redesign the course and what improvements would you expect as a result?
- What have you learned from contrasting the different approaches?
- What evaluation criteria do your answers to the above questions imply?

REFERENCES

Ally, M. 2008. Foundations of educational theory for online learning. In *Theory and Practice of Online Learning*, ed. T. Anderson, 15–44. Edmonton, Alberta, Canada: AU Press.

Anderson, T. 2008. Teaching in an online learning context. In *Theory and Practice of Online Learning*, ed. T. Anderson, 343–366. Edmonton, Alberta, Canada: AU Press.

Andersen, E. 2002. Infrastructure: The things we take for granted, *ACM Ubiquity* issue 18, http://ubiquity.acm.org/article.cfm?id=544737.

Anderson, T., L. Rourke, D. R. Garrison, and W. Archer. 2001. Assessing teacher presence in a computer conferencing context, *Journal of the Asynchronous Learning Network* 5(2), http://www.sloan-c.org/publications/jaln/v5n2/v5n2_anderson.asp.

Berge, Z. L. 1995. Facilitating computer conferencing: Recommendations from the field. *Educational Technology* 35(1): 22–30.

Besser, H., and S. Donahue. 1996. Perspectives on . . . distance independent education: Introduction and overview. *Journal of the American Society for Information Science* 47(11): 801–804.

Constantine, L., and L. Lockwood. 1999. *Software for Use: A Practical Guide to the Essential Models and Methods of Usage-Centered Design*. Reading, MA: Addison-Wesley.

Coughlin, E. C., and C. Lemke. 1999. *Professional Competency Continuum: Professional Skills for the Digital Age Classroom*. Santa Monica, CA: Milken Family Foundation.

Csikszentmihalyi, M. 1990. *Flow: The Psychology of Optimal Experience*. New York: Harper and Row.

Draper, S. 1998. When is video good for instruction? http://staff.psy.gla.ac.uk/~steve/vidInstr.html (accessed Feb. 17, 2004).

Driscoll, M. 2001. Strategic plans from scratch. *Learning Circuits* August.

Entin, E. B., J. Sidman, and L. Neal. 2008. Development of online distributed training: practical considerations and lessons learned. In *Computer-Supported Collaborative Learning*, eds. K. L. Orvis and A. L. R. Lassiter. Hershey, PA: Information Science Publishing.

Feldstein, M., and L. Neal. 2006. Designing usable, self-paced e-learning courses: a practical guide. *eLearn Magazine*, August.

Friedman, T. 1999. Next, it's E-ducation. *New York Times*, A29, November 17. p. A29.

Gagne, R. 1985. *The Conditions of Learning*, 4th ed. New York: Holt, Rinehart & Winston.

Galusha, J. M. 1997. Barriers to learning in distance education. *Interpersonal Computing and Technology: An Electronic Journal for the 21st Century* 5(3–4): 6–14.

Gardner, H. 1993. *Multiple Intelligences: The Theory in Practice.* New York: Basic Books.

Gibbs, W. W. 2004. Why machines should fear. *Scientific American* 290(1): 37.

Gilbert, S., and I. Werge. 2000. Education, technology, and change: Queries. IESD Report. http://www.tltgroup.org/gilbert/Quaker QueriesIWswg2-19-01.htm (accessed March 6, 2004).

Gladwell, M. 2000. *The Tipping Point: How Little Things Can Make a Big Difference.* New York: Little, Brown, and Company.

Hall, B. 2000. Resources for enterprise-wide e-learning initiatives. *E-Learning Magazine.*

Horn, R. E. 1998. *Visual Language: Global Communication for the 21st Century.* Bainbridge Island, WA: MacroVU, Inc.

IHETS, n.d. Guiding principles for faculty in distance learning, http://www.ihets.org/progserv/education/distance/guiding_principles (accessed March 10, 2004).

Irani, T. 2001. If we build it, will they come? The effects of experience and attitude on traditional-aged students' views of distance education. *International Journal of Educational Technology* 2(1).

Kaplan-Leiserson, E., n.d. E-learning glossary. http://www.learningcircuits.org/glossary.html (accessed Feb. 17, 2004).

Kearsley, G., and B. Shneiderman. 1999. Engagement theory: A framework for technology-based teaching and learning. http://home.sprynet.com/~gkearsley/engage.htm (accessed March 6, 2004).

Keller, J. M. 1987. Strategies for stimulating the motivation to learn. *Performance and Instruction* 26(8).

Kerka, S. 1996. Journal writing and adult learning. ERIC Document 399, 413.

Keynes, J. M. 2009. The General Theory of Employment, Interest and Money, p. 4, Classic Books America.

Kiesler, S. 1986. The Hidden Messages in Computer Networks. *Harvard Business Review* 64(1).

Kirkpatrick, D. L. (1994). *Evaluating Training Programs: The Four Levels.* San Francisco, CA: Berrett-Koehler.

Klobas, J., and G. Haddow. 2000. International computer-supported collaborative teamwork in business education: A case study and evaluation. *International Journal of Educational Technology* 2(1).

Latomaa, T., J. Pohjonen J. Pulkkinen, and M. Ruotsalainen, eds. *eReflections—Ten years of educational technology studies at the University of Oulu*, 109–125. http://herkules.oulu.fi/isbn9514276329/isbn9514276329.pdf (accessed March 10, 2010).

Laurillard, D. 1993. *Rethinking University Teaching: A Framework for the Effective Use of Educational Technology.* London: Routledge.

LeBaron, J., and D. Miller. 2004. The teacher as agent provocateur: Strategies to promote community in online course settings.

LeBaron, J., and D. Miller. 2005. The potential of jigsaw role-playing to promote the social construction of knowledge in an online graduate education course. *Teachers College Record* 107(8): 1652–1674.

Lenzner, R., and S. Johnson. 1997. Seeing things as they really are. *Forbes* March 10, 1997.

Lin, C.-J. 1999. The effects of self-efficacy and task values on students' commitment and achievement in web-based instruction for Taiwan higher education. *Dissertation Abstracts International Section A: Humanities & Social Sciences* 60(6-A): 1905.

Martinez, M. 2001. Mass customization: Designing for successful learning. *International Journal of Educational Technology* 2(2).

Massachusetts Institute of Technology, n.d. MIT OpenCourseWare. http://ocw.mit.edu/index.html (accessed Feb. 18, 2004).

MediaPro®, n.d. Readiness assessment. http://www.mediapro.com/html/resources/readiness.html (accessed Feb. 17, 2004).

Monk, A., M. Hassenzahl, M. Blythe, and D. Reed. 2002. Funology: Designing enjoyment, *SIGCHI Bulletin* September–October 2002.

Moss, C. 2002. Finding balance: The vices of our "versus." *First Monday* 7(1).

Neal, L., n.d. Storytelling at a distance, *eLearn Magazine.* http://www.elearnmag.org/subpage.cfm?section=research&article=1-1 (accessed March 6, 2004).

Neal, L. 1997. Virtual classrooms and communities. In *Proceedings of ACM GROUP '97 Conference* (Phoenix, AZ, 16–19 Nov.).

Neal, L. 2000. Scenario building to design a distance learning program. In *Proceedings of ED-MEDIA 2000—World Conference on Educational Multimedia, Hypermedia & Telecommunication* (Montréal, Quebec, Canada, June 28).

Neal, L. 2004. Degrees by mail: Look what you can buy for only $499!!! *eLearn* 2004(1): 2.

Neal, L., and D. Ingram. 2001. Asynchronous distance learning for corporate education. In *2001/2002 ASTD Distance Learning Yearbook*, eds. K. Mantyla and J. Woods. New York: McGraw-Hill.

Neal, L., D. Miller, and D. Anastasi. 2002. Launching a community of practice through an online seminar series. In *Proceedings of ED-MEDIA 2002.* Denver, CO, p. 72.

Norman, D. 2004. *Emotional Design.* New York. Basic Books.

Russell, T. L., n.d. http://www.nosignificantdifference.org (accessed Feb. 18, 2004).

Ornstein, A. C., and F. P. Hunkins. 1997. *Curriculum: Foundations, Principles, and Issues.* Boston, MA: Allyn & Bacon.

Pennsylvania State University, n.d. Media selection matrix draft. http://www.cde.psu.edu/de/id&D/media_selection_matrix.html (accessed Feb. 18, 2004).

Roblyer, M., and L. Ekhaml. 2000. How interactive are your distance courses? A rubric for assessing interaction in distance learning. *Online Journal of Distance Learning Administration* 3(2).

Rogers, E. M. 1962. *Diffusion of Innovation.* The Free Press. New York.

Rovai, A. 2002. Building sense of community at a distance. *International Review of Research, In Open and Distance Learning* 3(1).

Schoenfeld-Tacher, R., and K. Persichitte. 2000. Differential skills and competencies required of faculty teaching distance education courses. *International Journal of Educational Technology* 2(1).

Tufte, E. 1990. *Envisioning Information.* Cheshire, CT: Graphics Press.

Tufte, E. 1992. *The Visual Display of Quantitative Information.* Cheshire, CT: Graphics Press.

Twigg, C. A. 1997. The promise of instructional technology. *About Campus* 2(1): 2–3.

UsabilityNet. n.d. Competitor analysis. http://www.hostserver150.com/usabilit/tools/competitoranalysis.htm (accessed Feb. 17, 2004).

William Horton Consulting, n.d. Designing Web-based training. http://www.designingwbt.com (accessed March 6, 2004).

27 Behavioral Research and Data Collection via the Internet

Ulf-Dietrich Reips and Michael H. Birnbaum

CONTENTS

In the past 15 years it has become possible to collect data from participants who are tested via the Internet rather than in the laboratory. Although this mode of research has some inherent limitations owing to lack of control and observation of conditions, it also has a number of advantages over lab research. Many of the potential advantages have been well-described in a number of publications (Birnbaum 2000a, 2001a, 2004a, 2007; Krantz and Dalal 2000; Reips 1995, 2000, 2006, 2007; Reips and Bosnjak 2001; Schmidt 1997a, 1997b). Some of the chief advantages are that (1) one can test large numbers of participants very quickly, (2) one can recruit large heterogeneous samples and people with rare characteristics, and (3) the method is more cost-effective in time, space, and labor in comparison with lab research. This chapter provides an introduction to the major features of the new approach and illustrates the most important techniques in this area of research.

27.1 OVERVIEW OF INTERNET-BASED RESEARCH

The process of Web-based research, which is the most frequent type of Internet-based research, can be described as follows: Web pages containing surveys and experiments are placed in Web sites available to participants via the Internet. These Web pages are hosted (stored) on any server connected to the World Wide Web. People are recruited by special techniques to visit the site. People anywhere in the world access the study and submit their data, which are processed and stored in a file on a secure server. (The server that "hosts" or delivers the study to the participant and the server that receives, codes, and saves the data are often the same computer, but they can be different.)

The Internet scientist plans the study following guidelines while striving to avoid pitfalls (Birnbaum 2001a, 2004a, 2004b, 2007; Reips 2002b, 2002c, 2007; Reips and Bosnjak 2001; Schmidt 2007). The researcher creates Web pages and other files containing text, pictures, graphics, sounds, or other media for the study. He or she will upload these files to the host server (as needed) and configure the Web server to accept, code, and save the data. The researcher tests the system for delivering the experiment and for collecting, coding, and saving the data. The Web researcher must ensure that the process is working properly, recruit participants for the study, and finally retrieve and analyze the data. Although this process may sound difficult, once a researcher has mastered the prerequisite skills, it can be far more efficient than traditional lab methods (Birnbaum 2001a; Reips 1995, 1997, 2000; Reips and Krantz 2010).

27.2 PSYCHOLOGICAL RESEARCH ON THE WEB

To get an overall impression of the kinds of psychological studies that are currently in progress on the Web, visit studies linked at the following sites:

Web experiment list (Reips and Lengler 2005): http://wexlist.net (see Figure 27.1)

Psychological Research on the Net: http://psych.hanover.edu/research/exponnet.html

Web Experimental Psychology Lab (Reips 2001a, 2001b): http://wexlab.eu

Decision Research Center: http://psych.fullerton.edu/mbirnbaum/decisions/thanks.htm

The number of studies conducted via the Web appears to have grown exponentially since 1995 (Reips 2007), when psychologists began to take advantage of the new standard for HTML that allowed for convenient data collection (Musch and Reips 2000). Internet-based research has become a new topic and a new set of methods in psychology. The basics of authoring Web-based research studies will be described in the next sections.

FIGURE 27.1 **(See color insert.)** The Web experiment and Web survey lists at http://wexlist.net.

27.3 CONSTRUCTING STUDIES FOR THE INTERNET

There are many computer programs that allow one to create Web pages without knowing HTML. These programs include Adobe GoLive, Adobe Contribute, Adobe Dreamweaver, and Microsoft FrontPage (not recommended), among others. In addition, programs intended for other purposes, such as Open Office, Microsoft Word, PowerPoint, and Excel, or Apple Pages and Keynote allow one to save documents as Web pages. Although these programs can be useful on occasion, those doing Web research really need to understand and be able to compose basic HTML. While learning HTML, it is best to avoid these authoring programs. If you already know how to use these programs, you can study HTML by using them in *source code mode*, which displays the HTML rather than the "what you see is what you get" (WYSIWYG) or "layout" display.

There are many free, useful tutorials on the Web for learning about HTML and many good books on the subject. Birnbaum (2001a, chaps. 2–4) covers the most important tags (basic units in HTML) in three chapters that can be mastered in a week, with a separate chapter (Chapter 5) for the technique of Web forms, which is the technique that made Web research practical when this technique was supported by HTML 2, introduced in late 1994.

27.3.1 WEB FORMS

There are three aspects of Web forms that facilitate Internet-based research. First, forms support a number of devices by which the reader of a Web page can send data back to the server chosen by the author of a page. Forms support two-way communication of information, with the possibility for dynamic communication.

Second, Web forms allow a person without an e-mail account to send information from a computer, even if the computer is not configured to send e-mail. For example, a person at a local library, in an Internet café, or in a university lab could fill out a Web form on any Internet-connected computer, and click a button to send the data. This means that participants can remain anonymous.

Third, Web forms can deliver their data to a program on the server that codes and organizes the data and that saves them in a convenient form for analysis. In fact, server-side programs can even analyze data as they come in and update a report of cumulative results.

The Web form is the HTML between and including the tags, <FORM> and </FORM>, within a Web page. The response or "input" devices supported by forms allow the users (e.g., research participants) to type in text or numerical responses, click choices, choose from lists of selections, and send their data to the researcher. Table 27.1 shows a very simple Web form. You can type this text, save it with an extension of ".htm," and load it into a browser to examine how it performs. Table 27.1, along with other examples and links, are available from the following Web site, which is associated with this chapter: http://wexlab.eu/hcihandb/ (with a mirror at http://psych.fullerton.edu/mbirnbaum/handbook/).

In this example, there are four input devices: a "hidden" value, an input text box, a "submit" button, and a "reset" button. The "hidden" input records a value that may be used to identify the data; in this case, the value is "MyTest1." The "value" of the "submit" button or "reset" button is what is displayed on the buttons, but the "value" of a text box is whatever the viewer types in that field. When the "reset" button is clicked, the form is reset; i.e., any responses that were typed in or clicked are erased.

When the "submit" button is clicked, the *action* of the form is executed. In this example, the *action* sends e-mail with the two variables to the e-mail address specified. You should change this to your own e-mail address, load the form in the browser, fill in your age, and click the submit button. If your computer and browser are configured to send e-mail, you will receive an e-mail message with your responses in the message. The encryption type attribute can be erased (save the file and reload it in the browser), and you will see the effect that this attribute has on how the e-mail appears.

27.3.2 Server-Side Scripting to Save the Data

Although sending data by e-mail may be useful for testing Web forms or for small efforts, such as collecting RSVPs for a party, it is neither practical nor secure to collect large amounts of data via e-mail. Instead, we can let the Web server write the data to its log file for later analysis (Section 27.7), or we can use a computer program to code the data and save them in a file, in a form ready for analysis (Section 27.6). To do this, we use a Common Gateway Interface (CGI) script, e.g., written in Perl or PHP, that codes, organizes, and saves the data safely to a secure server (see Schmidt 1997a, 2000, 2007). The ACTION of the form is then changed to specify the URL address of this script.

For example, revise the FORM tag in Table 27.1 as follows:

```
<FORM METHOD='post' ACTION='http://
psych.fullerton.edu/cgi-win/polyform.
exe/generic'>
```

In this example, the ACTION specifies an address of a script that saves the data to a file named data.csv on the psych.fullerton.edu server. The script residing at this address is a generic one that accepts data from any form on the Web, and it arranges the data in order of the two leading digits in each input variable's NAME. It then redirects the participant to a file with a generic "thank you" message. This example is for instruction, demonstration, and testing purposes only. For real data collection, one should put the data file in a protected folder where it is only available to the researchers.

27.3.3 Downloading Data by FTP

To view the data, one can use the following link in a browser that supports File Transfer Protocol (FTP) or the more secure sFTP. This link specifies an FTP site with a username of "guest" and password of "guest99":

ftp://guest:guest99@psych.fullerton.edu

From this FTP site, you can download the file named data .csv. This file can be opened in a text editor or in Excel,

---

TABLE 27.1

Bare Bones Web Form (Example 2.1)

```
<html>
<head>
<title>
My First Form
</title>
</head>
<body>
<form method="post" action="mailto:mbirnbaum@fullerton.edu"
  enctype="text/plain">
<input type="hidden" name="00exp" value="MyTest1">
1. What is your age?
<input type="text" name="01age" size="4" maxlength="6">
<input type="submit" value="Send the Data">
<input type="reset" value="start over">
</form>
</body>
</html>
```

among other applications. At or near the end of the file will appear a line that contains the "hidden" value ("MyTest1") and the datum that you typed in for age.

27.3.4   OBTAINING AND USING A DEDICATED FTP PROGRAM

Although most browsers support FTP, it is more convenient to use a program dedicated to FTP that supports additional features. There are several FTP and sFTP (secure FTP) programs that are free to educational users, such as Fetch for the Mac and WS FTP LE for Windows PCs. These free programs can be obtained from CNET Download.com, which has the following URL: http://www.download.com

File transfer programs, FTP or sFTP, are not only useful for downloading data from a server, but they can also be used to upload files to a server, in the case of a server administrated by another person or institution. In a later section, we describe advantages of installing and running your own server. However, many academic users are dependent on use of a department or university server. Others have their Web sites hosted by commercial Internet service providers (ISP). In these cases, the academic researcher will upload his or her Web pages by means of FTP or a server-side program (e.g., a control panel) to the server and download data via sFTP, or WebDAV from the server. Like sFTP, WebDAV (Web-based Distributed Authoring and Versioning) allows for a secure transmission, so for transfers of the data one should always use one of these options.

Some academics find that they can get greater independence and control of their research by using a commercial ISP, rather than using a shared university server. At some campuses, the administrators or techs who control the server are very reluctant to allow faculty to collect data or to view the log files. Commercial sites can provide either free hosting (which may entail inclusion of commercial advertisements in the Web pages) or low cost service (typically about $5/month) without the advertisements. An example with educational materials of how to construct a Web site at a commercial ISP is provided in the following site:

http://ati-birnbaum.netfirms.com/

For this site, the URL of a generic script is as follows:

http://ati-birnbaum.netfirms.com/cgi-bin/generic.pl

This can be used in place of the ACTION in Section 27.3.2. The data will be added to the following site:

http://ati-birnbaum.netfirms.com/data/data.txt

Because the data in this case are freely available, this example is for instructional, demonstration, and testing purposes only. In real research, the file containing data should be protected, especially when data are not anonymous.

27.3.5   THE "HIDDEN" INPUT DEVICE

The display in the browser (Figure 27.2) shows the text in the body of the Web page, the text input box, *submit* button, and *reset* button. Note that the "hidden" input device does not display anything; however, one can view it by selecting Source (View Source, Frame Source or Page Source) from the View menu of the browser, so it would be a mistake to think that such a hidden value is truly hidden.

The term hidden unfortunately has the connotation that something "sneaky" is going on. The second author was once asked about the ethics of using hidden values in questionnaires, as if we were secretly reading the participant's subconscious mind without her knowledge or consent. In reality, nothing clandestine is going on. Hidden variables are routinely used to carry information such as the name of the experimental condition from one page to the next, to hold information from a JavaScript program that executes an experiment, or to collect background conditions such as date and time that the experiment was completed. In the current example, the hidden variable is used simply to identify which of many different questionnaires is associated with this line of data. This value can be used in Excel, for example, to segregate a mixed data file into subfiles for each separate research project (Birnbaum 2001a).

27.3.6   INPUT DEVICES

In addition to the text box, which allows a participant to type in a number or short answer, there are four other popular input devices, and one that makes use of images. The *textarea* input device is a rectangular box suitable for obtaining a longer response such as a paragraph or short essay. For multiple-choice answers, there are radio buttons, pull-down selection lists, and check boxes. Check boxes (usually displayed as squares on Web pages), however, should not be used in behavioral research. The problem with a check box is that it has only two states—it is either checked or unchecked. If a check box is unchecked, one does not know if the participant intended to leave it unchecked or perhaps just skipped over the item. For a "yes" or "no" answer, one must allow

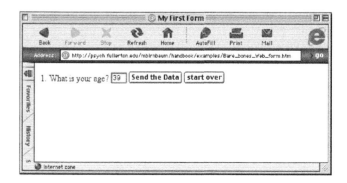

FIGURE 27.2   Appearance of basic Web form in the browser window.

at least three possibilities: "yes," "no," and "no answer." In some areas of survey research, one may need to distinguish as many as five distinct response options for a yes/no question, "yes," "no," "no response," "don't know," and "refuse to answer." Reips et al. (2010) studied the impact of the presence of such nonresponse options on active nonresponse like "I don't want to answer" and passive nonresponse (skipping). Combined active and passive nonresponse varied between 0 and 35% for the same item, when sensitive questions were asked.

We recommend avoiding the use of check boxes, except in situations where multiple responses to options in a list can be expected from a respondent, e.g., "check all music genres you were listening to today." Another use is when a person is required to check a box in order to continue with the study, as in "Check here to indicate that you have read the information and agree to participate in this study." Multiple-choice questions (including true-false questions) can be better handled by radio buttons than by check boxes.

With radio buttons (usually displayed as circles with round "dots" when clicked), one can construct a multiple-choice response device that allows one and only one answer from a potential list. The basic tags to create a yes/no question with three connected buttons are as follows:

```
<INPUT TYPE=radio NAME=02v2 VALUE=""
checked>Do you drive a car?
<BLOCKQUOTE>
<BR><INPUT TYPE=radio NAME=02v2
VALUE="0">No.
<BR><INPUT TYPE=radio NAME=02v2
VALUE="1">Yes.
</BLOCKQUOTE>
```

In this example, the first radio button will be already checked, before the participant responds. If the participant does not respond, the value sent to the data is empty (there is nothing between the quotes); SPSS, Excel, and certain other programs treat this null value as a missing value. To *connect* a set of buttons, as in this example, they must all have the same *name* (in this example, the name is 02v2). When one button is clicked, the dot jumps from the previously checked button (the nonresponse, or null value) to the one clicked. If the respondent clicks "No," the data value is "0" and if the respondent clicks "yes," the data value is "1."

We suggest that you follow this convention for yes/no responses: use larger numbers for positive responses to an item. In this case, the item measures driving. This convention helps prevent the experimenter from misinterpreting the signs of correlation coefficients between variables.

The selection list is another way to present a multiple choice to the participant, but this device is less familiar to both researchers and participants. Selection lists are typically arranged to initially display only one or two options, which when clicked, will expand to show other alternatives. The list remains hidden or partially revealed until the participant clicks on it, and the actual display may vary depending on

how far the participant has scrolled before clicking the item. In addition, there are really two responses that the participant makes in order to respond to an item. The participant must drag a certain distance and then release at a certain choice. Alternatively, depending on the Web browser and how the Web page was created, the participant may type a letter and jump to the first item beginning with that letter. Because of the complexities of the device, precautions with selection lists are recommended.

First, like any other multiple-choice device, it is important not to have a legitimate response preselected but rather to include an option that says something like, "choose from this list" and that returns a "missing" code unless the participant makes a choice (Birnbaum 2001a; Reips 2002b). If a legitimate answer has been preselected, as in the left side of Figure 27.3, the experimenter will be unable to distinguish real data from those that result when participants fail to respond to the item. The illustration on the right side of the figure shows a better way to handle this list. Reips (2002b) refers to this all too common error (of preselected, legitimate values) as "configuration error V."

Another problem can occur if the value used for missing data is the same as a code used for real data. For example, the second author found a survey on the Web in which the participants were asked to identify their nationalities. He noted that the same code value (99) was assigned to India as to the preselected "missing" value. Fortunately, the investigator was warned and fixed this problem before much data had been collected. Otherwise, the researcher might have concluded that there had been a surprisingly large number of participants from India.

Second, part of the psychology of the selection list is how the choice set and arrangement of options is displayed to the participant. It seems plausible that options that require a long scroll from the preset choice would be less likely to be selected. The experimenter communicates to the participant by the arrangement of the list (e.g., Smyth et al. 2006) and by the availability and placement of nonresponse options relative to the others (Reips et al. 2010). Birnbaum (2001a, Chap. 5) reported an experiment showing that mean responses can be significantly affected by the choice of options listed in a selection list. Birnbaum compared data obtained for the

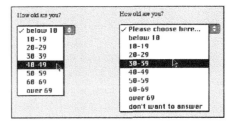

FIGURE 27.3 (Left) Improper drop-down selection list and (right) a better way. The preselected option should be neutral, and the respondent should be given an option to express a deliberate nonresponse. (From Reips, U.-D., *Soc. Sci. Comput. Rev.* 20(3), 241–249, 2002. With permission.)

judged value of the St. Petersburg gamble from three groups of participants who received different selection lists or a text input box for their responses. The St. Petersburg gamble is a gamble that pays $2 if a coin comes up heads on the first toss, $4 if the first toss is tails and the second is heads, $8 for tails-tails-heads, and so on, doubling the prize for each additional time that tails appears before heads, ad infinitum. One group judged the value of this gamble by means of a selection list with values spaced in equal intervals, and the other had a geometric series, with values spaced by equal ratios. A third group had a text box instead of a selection list and requested the participant to respond by typing a value. The mean judged value of this gamble was significantly larger with the geometric series than with equal spacing; furthermore, these values differed from the mean obtained with the text box method. The results therefore depend on the context provided by the response device.

In laboratory research on context effects (Parducci 1995), it has been shown that the response that one assigns to a stimulus depends on two contexts: the context of the stimuli (the frequency distribution and spacing of the stimulus levels presented) and the context provided by the instructions and response mode. Parducci (1995) summarized hundreds of studies that show that the stimulus that is judged as "average" depends on the endpoints of the stimuli, their spacing, and their relative frequencies. Hardin and Birnbaum (1990) showed that the response one uses to evaluate a situation depends on the distribution of potential responses incidentally shown as examples. It might seem therefore that one should try to "avoid" contextual effects by providing no other stimuli or responses; however, the head-in-the-sand approach yields even more bizarre findings.

To show what can happen when an experimenter tries to "avoid" contextual effects, Birnbaum (1999a) randomly assigned participants to two conditions in a Web study. In one condition, participants judged "how big" is the number 9, and in the other condition, they judged how big is the number 221. He found that 9 is significantly "bigger" than 221. Birnbaum (1999a) predicted that the result would occur based on the idea that each stimulus carries its own context, and even though the experiment specified no context, the participants supplied their own. So, one cannot avoid contextual effects by experimental design only.

In a study with (short) selection menus, Reips (2002a) found no significant difference between "drop-down" (preselected choice at top) versus "pop-up" (preselected choice at bottom) menus in choices made on the menu. Nevertheless, it would be dangerous to assume that this finding (of nonsignificance) guarantees immunity of this method from this potential bias, particularly for long selection lists.

Because the list of options, their spacing, order, and relative position of the initial value may all affect the results, we recommend that the selection list be used only for obtaining responses when the options are nominal, when there is a fixed list of possible answers, and the participant knows the answer by heart. For example, one might use such a device to ask people in what nations they were born. Although the list is a long one, one hopes that people know the right answer and will type the first letters correctly or patiently scroll to the right spot. Some investigators list their own country at the top of the list as well as at its place in the alphabetical sequence in order to facilitate this frequently chosen response in studies with participants from their own country.

There are many potential human factors that could be studied in selection lists, and much about their potential contextual effects is still unknown (see Birnbaum 2001b; Dillman 2001; Reips 2002a; Smyth et al. 2006, for early investigations, and Dillman, Smyth, and Christian 2008; Reips 2010, for more recent summaries).

One remaining category of input device to be discussed is that in which participants click on an image. This device allows the capture of the x and y coordinates of the spot on the image where the click occurred. One frequently used example is the visual analog scale. Visual analogue scales use a horizontal or vertical line segment. Participants are asked to click at a point that represents their judgment on a continuum defined by terms labeling the endpoints. For example, a participant may be asked to mark her degree of agreement with a political statement on a continuum ranging from "total disagreement" to "total agreement."

For the purpose of testing Internet-based visual analogue scales, Reips and Funke (2008) asked participants to repeatedly mark the points where 13 probabilities should be located on a 0% to 100% probability line segment (see also Funke and Reips 2007). They found that the average responses were approximately correct and linearly related to the nominal percentages, even for extreme scale lengths of 50 and 800 pixels. In previous research with this response mode using the subtractive model, it has been concluded that the mean responses are a sigmoidal function of subjective value (e.g., Rose and Birnbaum 1975; Stevenson 1993).

In early research using paper and pencil methods, the position on the line had to be measured individually for each response. This measurement can be automated by computer (e.g., Stevenson 1993). A description of the simple image method is given by Fraley (2004), and a method to construct such response scales using JavaScript is given by Baron and Siepmann (2000). They can even be constructed without the use of JavaScript, in pure HTML. See below an example by Reips (1999) from Internet-based research with children and adults, in which participants were asked to click on a "tunnel" (the visual analogue scale) at the point where "a snake's head would now be that had crawled into the tunnel." The scale was constructed by repeatedly linking the same image "pixel.gif" to the next page "next.html," each time with a different value "point" (here "37" and "38"). For a visual analogue scale with 100 pixels the code below would be repeated for values from 1 to 100 (full experiment available on the handbook Web site):

```
<A href="next.html?point=37"><IMG
src="pixel.gif" border="0"></A>
<A href="next.html?point=38"><IMG
src="pixel.gif" border="0"></A>
```

Reips and Funke (2008) developed VAS Generator (http://vasgenerator.net), a free tool to create visual analogue scales for Internet-based research. This tool automates construction of the response scales without requiring the user to do programming, and it allows the participant to mark and adjust the response before recording it.

Some findings on the "quality" of data obtained by browser-based methods are available (see e.g., Tuten, Urban, and Bosnjak 2002). Birnbaum (1999b) presented the same browser-based questionnaire on two occasions to 124 undergraduates in the laboratory. In the main body of the experiment, participants were to choose between pairs of gambles by clicking a radio button beside the gamble in each pair that they would prefer to play. In this part of the experiment, people agreed on average in 82% of their choices. With a multiple-choice response for gender, 2 of 124 participants switched gender from the first to second occasion; one switched from male to female and one made the opposite switch. In the choices between gambles, we assume that people were unsure of their choices or changed their minds, but in case of gender, it seems clear that people made errors. Every person agreed on his or her age, which was typed into a text box on each occasion. However, six people gave different answers for the number of years of education, which was also typed into a box.

Once it is realized that people can make errors with any response device, we see the need to design Web studies to allow for such inconsistent behavior. For example, suppose a researcher wants to study smoking behavior and suppose there are different questionnaires for people who are smokers, for those who never smoked, and for those who quit smoking and no longer smoke. If a participant makes an errant click on a branching question, that person might be sent to the wrong questionnaire, and most of the subsequent questions may be inappropriate for that person.

One approach to this problem of "human errors" is to build some redundancy and cross-examination into questionnaires and methods of linking people to different instruments. One device is the JavaScript prompt that provides cross-examination when a person clicks a link to identify his or her gender. If the person clicks "male," the prompt opens a new box with the question, "are you sure you are male?" requiring a yes/no answer before the person can continue. A similar check cross-examines those who click "female." A second technique is to provide paths that provide a "second chance" to link to the correct gender. Here, the person who clicks "male" then receives a page with a link to click "if you are female," which would send the person to the female questionnaire, which also has a second chance to revert.

Another approach has just been presented by Stieger and Reips (2010), who recorded via an application named "User Action Tracer" what Internet participants were actually doing on the study's Web page. They collected and recorded actions with their exact position (x and y coordinates): clicks (for all mouse buttons); double-clicks; clicks on checkboxes, radio buttons, list boxes, and submit buttons; choices in drop-down selection menus; inserted text in text boxes; keys pressed on the keyboard; and the position of the mouse pointer every half a second. This type of data is called "paradata" and it can be used to verify user entries.

27.4 CREATING SURVEYS AND EXPERIMENTS WITH SIMPLE DESIGNS

Typing HTML for a questionnaire can be tedious, and typing errors in HTML can introduce errors that would make it impossible to get meaningful data. Therefore, any Web page designed to collect data should be thoroughly tested before it is placed on the Internet for data collection. Birnbaum's (2000b) SurveyWiz and FactorWiz are programs that were written to help researchers avoid making mistakes in HTML coding. These programs are freely available on the Web, and they create basic sets of radio buttons and text boxes that will properly code and return the data. Still, when editing HTML files (for example, by importing them into Microsoft Word or by copying and pasting items without properly changing the names), there is potential for introducing errors that can ruin a study. More about checks to be conducted before collecting data on the Web will be presented in Section 27.8.

The leading digits on the name attribute are used by the default, generic CGI script that organizes and saves the data for SurveyWiz and FactorWiz. This particular script requires that leading digits on variable names be sequential and start at 00. However, the HTML need not be in that order. That means that one could cut and paste the HTML to rearrange items, and the data will still return to the data file in order of the leading digits (and not the order within the HTML file of the items). This device is also used by Birnbaum's (2000b) FactorWiz program. FactorWiz creates random orders for presentation of within-subjects factorial designs. Although the items can be put in as many random orders as desired, the data always return in the same, proper factorial order, ready for analysis by ANOVA programs.

27.4.1 USING SURVEYWIZ

Instructions for using SurveyWiz and FactorWiz are given by Birnbaum (2000b, 2001a) and within their files on the Web, which can be accessed from the following:

http://psych.fullerton.edu/mbirnbaum/programs/

SurveyWiz3 provides a good way to learn about making a survey for the Web. It automatically prepares the HTML for text answers and rows of radio buttons. One can add a preset list of demographic items by clicking a single button. The program is easy to use and is less likely to lead to errors than complex commercial programs. For example, suppose we want to calculate the correlation between the number of traffic accidents a person has had and the person's rated fear while driving.

In SurveyWiz, one simply enters the survey name and short name and then types the questions, one at a time. In

TABLE 27.2
Web Form Illustrating Input Devices

```
<HTML>
<HEAD>
<TITLE>Driving Survey</TITLE>
</HEAD><BODY BGCOLOR='papayawhip'>
<FONT FACE=Arial>
<H3> Instructions for Driving Survey</H3>
<BR>Please answer these questions honestly.
<BR>
<HR>
<FORM METHOD='post' ACTION='http://psych.fullerton.edu/cgi-win/
   polyform.exe/generic'>
<INPUT TYPE=hidden NAME=00exp VALUE=driving_srvy1 >
<INPUT TYPE=hidden NAME=01Date VALUE=pfDate>
<INPUT TYPE=hidden NAME=02Time VALUE=pfTime>
<INPUT TYPE=hidden NAME=03Adr VALUE=pfRemoteAddress>
<P>1. In how many accidents have you been involved when you were the
   driver?
<INPUT type=text NAME=04v1 SIZE=8 MAXLENGTH=20><BR>
<P><INPUT TYPE=radio NAME=05v2 VALUE='' checked>
2. I often feel fear when driving.<BR>
strongly disagree
<INPUT TYPE=radio NAME=05v2 VALUE=1>
<INPUT TYPE=radio NAME=05v2 VALUE=2>
<INPUT TYPE=radio NAME=05v2 VALUE=3>
<INPUT TYPE=radio NAME=05v2 VALUE=4>
<INPUT TYPE=radio NAME=05v2 VALUE=5>
strongly agree<BR>
<HR>
<INPUT TYPE=radio NAME=06sex value='0' CHECKED>
3. Are you Male or Female?<BR>
<BLOCKQUOTE>
<INPUT TYPE=radio NAME=06sex value='F'>Female<BR>
<INPUT TYPE=radio NAME=06sex value='M'>Male
</BLOCKQUOTE>
<P>4. What is your age?
<INPUT TYPE=TEXT NAME=07Age SIZE=2 maxlength=3>
   years.<BR>
<P>5. What is the highest level of education you have completed? <BR>
<SELECT NAME=08Ed >
 <OPTION VALUE='' SELECTED>Choose from this list
 <OPTION VALUE=11> Less than 12 years
 <OPTION VALUE=12> Graduated High School (12 years education)
 <OPTION VALUE=14> Some College (13–15 years education)
 <OPTION VALUE=16> Graduated from College (Bachelor's degree)
 <OPTION VALUE=18> Master's degree
 <OPTION VALUE=19> Advanced Grad School beyond Master's degree
 <OPTION VALUE=20> Doctoral Degree (Ph.D., M.D., J.D., etc.)
</SELECT><BR>
<P>6. Nationality (country of birth):
<INPUT TYPE=text NAME=09Cn SIZE=20 MAXLENGTH=30><BR>
<P>7. COMMENTS:<BR>
<TEXTAREA NAME=10CM ROWS=5 COLS=60 WRAP=virtual></
   TEXTAREA><BR>
<P>Please check your answers. When you are done, push the button
   below.
```

(continued)

TABLE 27.2 (Continued)
Web Form Illustrating Input Devices

```
<P><INPUT TYPE='submit' VALUE='finished'>
<H2>Thank You!</H2>
</FORM>
</FONT>
</BODY>
</HTML>
```

this case, the two questions are "In how many accidents have you been involved when you were the driver?" to be answered with a text box, and "I often feel fear when driving," to be answered with a category rating scale. Table 27.2 shows a Web form created by SurveyWiz3, which illustrates text input, rating scales, pull-down selection list, and a text area for open-ended comments.

27.4.2 FactorWiz and WEXTOR

Birnbaum's (2000b) FactorWiz program allows one to make within-subjects experimental factorial designs, with randomized order for the combinations. The program is even easier to use than SurveyWiz, once its concepts are understood. Information on how to use the program is contained in the works of Birnbaum (2000b, 2001a) and in the Web files containing the programs.

William Schmidt has written a freely available Perl script that works with SurveyWiz and FactorWiz. For instructions on how to install this generic perl script, see

> http://ati-birnbaum.netfirms.com/Install_Perl_script
> .htm

Göritz and Birnbaum (2005) have contributed a free PHP program for facilitating data collection with HTML forms that also works with SurveyWiz and FactorWiz. It parses the input from any HTML form and automatically creates a MySQL database. See http://goeritz.net/brmic/.

Both SurveyWiz and FactorWiz create studies that are contained within a single Web page. To construct between-subjects conditions, one might use these programs to create the materials in the various conditions and use HTML pages with links to assign participants to different conditions. However, when there are a large number of within- and between-subjects factors, and when the materials should be put in separate pages so that response times and experimental dropouts can be traced, it becomes difficult to keep all of the files organized. In these cases, it is helpful to use WEXTOR (Reips and Neuhaus 2002), a program available on the Web at http://wextor.org that organizes experimental designs and keeps the various pages properly linked (see Section 27.5). For creating surveys, also consider http://surveys.deusto.es/, which presents a new, free, open source survey tool.

27.5 CREATING WEB EXPERIMENTS WITH COMPLEX DESIGNS: WEXTOR

Reips and his group have built several tools available from the *iScience Server* at http://iscience.eu that help Web experimenters in all stages of the experiment process: learning about the method, design and visualization, recruitment, and analysis of data. All of Reips' tools are Web based and therefore platform independent and can be used from any computer that is connected to the Internet. If you prefer a multiple-page survey with dropout measure, individualized random ordering of questions, and response time measurement, then you may want to use WEXTOR, a program that is reviewed in this section.

WEXTOR by Reips and Neuhaus (2002) is an Internet-based system to create and visualize experimental designs and procedures for experiments on the Web and in the laboratory. WEXTOR dynamically creates the customized Web pages needed for the experimental procedure. It supports complete and incomplete factorial designs with between-subjects, within-subjects, and quasi-experimental factors, as well as mixed designs. It implements server-side and client-side response time measurement and contains a content wizard for creating interactive materials, as well as dependent measures (graphical scales, multiple-choice items, etc.), on the experiment pages.

Many human factors considerations are built into WEXTOR, and it automatically prevents several methodological pitfalls in Internet-based research. This program uses nonobvious file naming, automatic avoidance of page number confounding, JavaScript test redirect functionality to minimize dropout, and randomized distribution of participants to experimental conditions. It also provides for optional assignment to levels of quasi-experimental factors, optional client-side response time measurement, randomly generated continuous user IDs for enhanced multiple submission control, and it automatically implements the meta tags described in Section 27.9. It also implements the warm-up technique for drop-out control (Reips 2000, 2002a), and provides for interactive creation of dependent measures and materials (created via content wizard).

The English version of WEXTOR is available at the following URL: http://wextor.org/wextor/en/.

Academic researchers can sign up free and can then use WEXTOR to design and manage experiments from anywhere on the Internet using a login/password combination. Support is available from tutorials, a FAQ, via a feedback and bug report form, a hints file, and an associated mailing list (e-group), all linked at the site. Figure 27.4 shows WEXTOR's entry page.

The process of creating an experimental design and procedure for an experiment with WEXTOR involves 10 steps. The first steps are decisions that an experimenter would make whether using WEXTOR or any other device for generating the experiment, such as listing the factors, levels, of within- and between-subjects factors, deciding what quasi-experimental factors (if any) to use, and specifying how

FIGURE 27.4 (**See color insert.**) The entry page to WEXTOR at http://wextor.org.

assignment to conditions will function. WEXTOR produces an organized, pictorial representation of the experimental design and the Web pages and associated JavaScript and CSS files required to implement that design. One can then download the experimental materials in one compressed archive that contains all directories (folders), scripts, and Web pages.

After decompressing the archive, the resulting Web pages created in WEXTOR can then be viewed (even when not connected to the Internet), tested, and further edited in an HTML editor. Afterward the whole folder with all experimental materials can be uploaded to a Web server. This can be done by FTP as described above for the case of an experimenter who does not operate the server, or it can be done by simply placing the files in the proper folder on the server, as described in the next section.

Since the first edition of this book, WEXTOR (now in version 2.5) has received a complete, usability-checked redesign and many features were added. Experiments can now optionally be hosted on the WEXTOR server, so an experimenter has the choice of using WEXTOR's free hosting service or of setting up his or her own server (see later in this chapter). WEXTOR now generates a code plan for each experiment. Many techniques for Internet-based experimenting were built into WEXTOR to automatically avoid common errors found in Internet-based research. The seriousness check technique, the multiple site entry technique, and the high hurdle technique (Reips 2000, 2002b, 2007) are now implemented. Any object can be integrated with the experimental materials, including for example, videos or Flash content (see Krantz and Williams 2010 for more information on using media in Internet-based research). Flexible timing out of individual pages and soft form validation can now be applied to the Web pages. Skins are available to flexibly change the appearance of all Web pages in the entire experiment at once. Figure 27.5 shows options for the implementation of the high hurdle

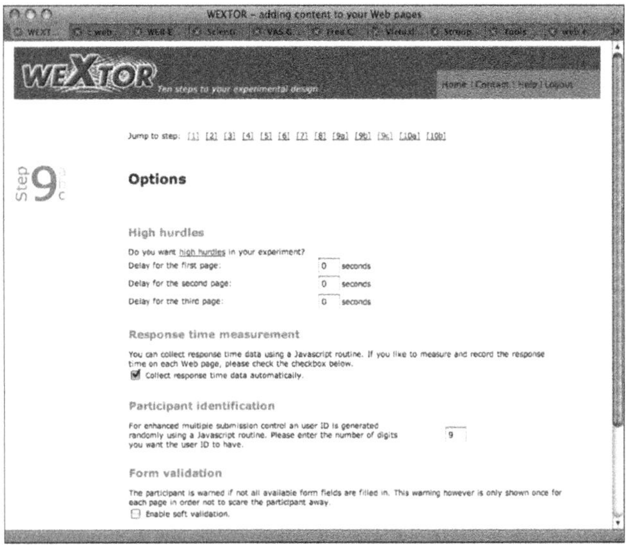

FIGURE 27.5 Options for the implementation of high hurdle technique, response time measurement, session ID, and form validation in step 9 in WEXTOR.

technique (Reips 2000, 2002c), response time measurement, session ID, and form validation in step 9 in WEXTOR.

For experiments hosted at the WEXTOR Web site, data preparation can optionally be done on the server, so that data can then be downloaded ready for analysis in spreadsheet programs like Excel or SPSS. The downloaded data file contains a column showing the path taken by each participant (e.g., to see use of the back button) and *two* measures of response time for each accessed page (server side and client side).

Some research projects require dynamic tailoring of an experiment to the participant's sequence of responses. Table 27.3 shows a list of various tools and techniques used to create the materials for running experiments via the Web. In Web research, there are often a number of different ways to accomplish the same tasks. Table 27.3 provides information to compare the tools available and help determine which application or programming language is best suited for a given task.

Certain projects require a programming language such as CGI programming, JavaScript, Java, Flash, or Authorware programs. Programming power is usually required by experiments that rely on computations based on the participant's responses, randomized events, precise control and measurement of timing, or precise control of psychophysical stimuli. The use of JavaScript to control experiments is described by Birnbaum and Wakcher (2002), Birnbaum (2001a), and Baron and Siepmann (2000). JavaScript programs can be sent as source code in the same Web page that runs the study. This allows investigators to openly share and communicate their methods. That way, it becomes possible to review, criticize, and build on previous work.

Java is a relatively advanced programming language, and like JavaScript, it is intended to work the same for any browser on any computer and system. However, practice has shown that different implementations of the Java engine in

different operating systems have not yet reached that goal. The use of Java to program cognitive psychology studies in which one can accurately control stimulus presentation timing and measure response times is described by Francis, Neath, and Suprenant (2000). Eichstaedt (2001) shows how to achieve very accurate response time measurement using Java.

Authorware is an expensive, but powerful application that allows one to accomplish many of the same tasks as one can do with Java, except it uses a graphical user interface in which the author can "program" processes and interactions with the participant by moving icons on a flow line. This approach has been used to good effect by McGraw, Tew, and Williams (2000; see also Williams, McGraw, and Tew 1999). However, participants are required to install a plug-in so this approach works best with participants in laboratories (where plug-ins can be preinstalled) or with online panels (where panelists will have installed them) rather than for experiments with open recruitment of participants at uncontrolled computers. Additional discussion of these approaches is given by Birnbaum (2000a, 2001a, 2004a, 2004b), Reips (2006, 2007), and Reips and Krantz (2010). Authorware was developed by Macromedia and acquired by Adobe; in 2010, Adobe announced that it will discontinue development of this program but will continue to support it.

JavaScript, Java, and Authorware experiments run client side. That means that the experiment runs on the participant's computer. This can be an advantage, in that it frees the server from making calculations and having a lot of traffic delays sending information back and forth. It can also be a disadvantage, if the participant does not have compatible script languages or plug-ins. At a time when Internet Explorer had a buggy version of JavaScript and JavaScript tended to produce error messages on people's computer screens, Schwarz and Reips (2001) found that JavaScript caused a higher rate of drop-out in Web studies compared with methods that did not require client-side programming. However, in recent years, JavaScript and Web browsers have become less error prone. Among the client-side programming options, JavaScript is probably the most widely used one.

There are certain tasks, such as random assignment to conditions, that can be done by HTML; by JavaScript, Java, and Authorware; or by server-side programs. By doing the computing on the server side, one guarantees that any user who can handle Web pages can complete the study (Schmidt 2000, 2007; Reips 2007). On the other hand, server side programs may introduce delays as the participant waits for a response from the server. When there are delays, some participants may think the program has frozen and may quit the study.

Server-side programs can analyze data as they come in and update a report of cumulative results. Perl and PHP are the two techniques most popular for such server-side programming.

The term "Ajax" (Garrett 2005) refers to a combination of techniques that allow continuous client-side updating with server-side information. AJAX stands for Asynchronous

TABLE 27.3

Purpose of Various Techniques and Tools to Create Materials for Web-Based Research

Technique	Purpose	Pros	Cons/Considerations
HTML (basics)	To create Web content	Basic to everything on the Web	Takes a day or two to learn the most important tags
HTML Forms	To transmit data to and from participant	Basic device used in Web research	Takes a day to learn the basics
WYSIWYG Web Editors	To create Web content, including HTML forms	Help people who do not know HTML to make Web pages. Easy to make good-looking pages without understanding. Easy to learn.	Can create headaches for researchers. Students often fall into traps from which they cannot escape or may even remain unaware of fatal errors that spoil their project.
SurveyWiz	Creates one-page HTML surveys	Easy to learn and use. Creates surveys with text boxes and scales of radio buttons.	Limited in what it does. No detailed response time measurement.
FactorWiz	Creates one-page, within-subjects factorial designs with up to six factors	Easy to learn. Creates random orders of factorial material.	Same limits as SurveyWiz. Must be used repeatedly to create between-subjects conditions.
Web Survey Assistant	Creates a variety of surveys and CGI scripts	More powerful than SurveyWiz.	More difficult to learn than SurveyWiz.
WEXTOR	Creates multipage, between-subjects, within-subjects, quasi-experimental, or mixed factorial designs; based on HTML and JavaScript.	Easy to learn. Many human factors considerations built-in, prevents methodological pitfalls in Internet-based research.	Cannot do certain things that can be done with Java or Authorware.
Scientific LogAnalyzer	For analysis of log files (turns raw server data into files in "one participant per line" format)	Handles any type of log file. Includes module for dropout analysis. Calculates response times. Highly flexible.	One must ensure that the server saves the key information in the log file.
CGI Scripting (e.g., Perl, PHP)	Control the server, save data, security (e.g., passwords, online exams). Server-side programming.	Automation of many necessary processes such as saving data. Works for any browser.	Not easy to learn. Constant version updating necessary, due to security risks. May function more slowly than client-side programs.
JavaScript	Powerful programming language. Include dynamic interaction with participant. Control and measure time.	Client side, hence requires JavaScript turned on in the participant's browser.	May not work properly in certain Web browsers.
Authorware	Construct experiments requiring timing, graphics, and dynamic interactions.	Products look good. Runs client side. Can do same tasks as Java but easier to learn. GUI.	Expensive. Plug-in download necessary on part of the participants.
Java	Powerful programming language, object-oriented.	Free. Usually preinstalled in browsers. Precompiled applets run on client side.	Very difficult to learn, may not work as expected with some Web browsers/operating systems.

JavaScript And XML. These techniques allow one to create dynamic, interactive Web applications that can retrieve data from the server asynchronously in the background without interfering with the display and functional behavior of the existing page (Wikipedia 2010).

Technically, Ajax uses (1) a combination of HTML and cascading style sheets (CSS) for marking up and styling information, (2) the standardized Document Object Model (DOM) accessed with JavaScript to dynamically display and interact with the information presented, and (3) a method for exchanging data asynchronously between browser and server, thereby avoiding page reloads. Thus, on a Web page created

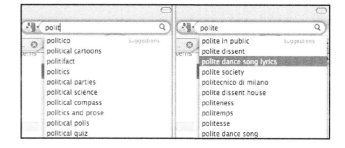

FIGURE 27.6 An example of Ajax scripting: Autocomplete. The list of suggestions changes with each letter that is typed, here from (left) "polit" to (right) "polite."

with Ajax technologies a visitor will see responses to his or her actions without having to wait for pages to reload.

An example can be seen in Figure 27.6, which shows an *autocomplete script* at work. The Web form provides a list of likely options that adapts to each new letter that is typed in the search field. Users not only can save time, because they don't need to fully type long options, but autocomplete also supports retrieval from memory. Autocomplete is a good human–computer interface implementation of a finding from memory research; namely, recognition is easier than recall. Of course, there is always the danger that autocomplete will cause the user to make errors by posing an unintended response.

Although some tasks can be done on either client or server side, there are certain tasks that can or should only be done by the server, such as saving data or handling issues of security (e.g., passwords, using an exam key to score an IQ test or an academic exam, etc.). Of course, to program the server, one needs to have access to the server.

27.6 RUNNING YOUR OWN SERVER

Running a Web server and making documents available on the Web has become increasingly easy over the years as manufacturers of operating systems have responded to demand for these services. Even if there is no preinstalled Web server on your system, installing one is neither complicated nor expensive. Thanks to contributions by an active "open source" community, there are free servers that are reliable and secure, along with free programming languages such as Perl and PHP that allow one considerable power for running and managing research from the server side (Schmidt 2000).

The free Apache Web server, available for a wide range of operating systems, can be downloaded from http://www .apache.org/. On this Web site there is also up-to-date documentation of the details of the installment process. Anja Göritz has provided tutorials on installation of the free Apache server along with PHP and MySQL, see

http://www.goeritz.net/ati/

Running your own Web server (rather than depending on your institution) confers several advantages (Schmidt, Hoffman, and MacDonald 1997). First of all, you can have physical access to the server, allowing you to directly observe and control it. You can, for example, easily add and delete files to your Web site by moving files from folder to folder; you can disconnect the server, modify it, and restart it.

Second, Web servers of institutions are restricted because they have to fulfill many tasks for different purposes. Consequently, many settings are designed to satisfy as many requirements as possible (one of which is reducing the likelihood of getting the network supervisors into trouble). On your own server, you can install and configure software according to the requirements of your own research (for example, you should change the server's log file format to include the information mentioned in Section 27.7).

Third, you can have greater harm on your access to the server if you operate it yourself. The institution's administrators do in terms of hindering research may outweigh any help they might provide with technical issues. You can try to explain your research goals and provide assurances that you would not do anything to weaken the security of the system or to compromise confidential files. Still, some administrators will resist efforts by researchers to add CGI files that save data to the server, for example, fearing that by error or intent, you might compromise the security of the system. Some Institutional Review Boards (IRBs) insist that administrators deny researchers complete access to the server for fear of compromising privacy and security of data collected by other researchers who use the same server.

Even if you run your own server within an institution you may run into trouble regarding the domain names and IP addresses by which your server and the services on it are accessed. The institution might decide to change or reassign names or delete them. To avoid such problems it is best to register your own domain name (or names) with an independent provider and announce that name to your participants and the research community. In case something goes wrong at your institution you will be able to quickly point the domain name at a different IP address.

In the example below, we will describe how to create and publish Web pages for Internet-based studies using Apple's Macintosh OS X operating system (see Wikibooks n.d.). Things work similarly on PC, but may vary depending on the exact version of the Windows or Unix/Linux operating system and its particular security settings (see Göritz 2006).

27.6.1 PLACE YOUR MATERIALS IN THE DESIGNATED FOLDER

In your private area ("Home") under Mac OS X there is a folder called "Sites" (or "Websites"). Put the folder with your materials (in this example the folder is named, "my_experiment") in the "Sites" folder. Figure 27.7 shows the respective Finder window. No other files need to be installed, if you created your experiment with SurveyWiz, FactorWiz, or WEXTOR. In case you are an advanced user of Internet technology and you would like to use Perl or PHP scripts for database-driven Web studies, you need to configure the system accordingly.

FIGURE 27.7 Web pages go in the Sites folder (or "Websites," in some versions of the operating system), in this case a folder with experimental materials created in WEXTOR.

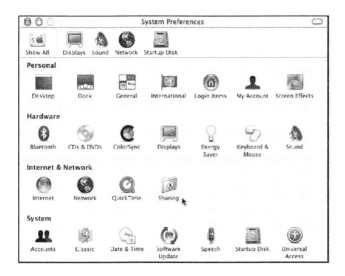

FIGURE 27.8 **(See color insert.)** The second of three clicks to turn on the Mac Server is to click on "Sharing" in the System preferences.

On the Macintosh, Perl, PHP, and mySQL are built-in, but they need to be configured using procedures described in more detail in Section 27.6.4, and in the Web site that accompanies this chapter. For Windows PCs, see the tutorial by Göritz (2006).

27.6.2 TURNING ON THE WEB SERVER IN MAC OS X

The Apache Web Server comes already installed on new Macintosh computers. Turning on the Web server under Mac OS X takes three mouse clicks: First you need to open the System Preferences (Click I), then click on "Sharing" (Click II, see Figure 27.8), and then click on "Personal Web Sharing" (Click III, the "Services" tab will be preselected), as shown in Figure 27.9.

Before you can actually make anything available on the Web using the built-in Apache Web server you need to make sure that your computer is connected to the Internet. However, you can always test your site by "serving" the pages to yourself locally, i.e., view them as if you were surfing in from the Web. Here is how you do this:

- Open a Web browser
- Type "http://localhost/~USERNAME/my_experiment/" into the browser's location window (where USERNAME is your login name).

The exact address will actually be shown at the bottom of the system preferences window displayed in Figure 27.9 (not shown here to preserve privacy).

27.6.3 WHERE TO FIND THE LOG FILES

The default storage for the log files created by the Apache server that comes with Mac OS X 10.5 or newer is /var/log/apache2/access_log (or, in earlier versions of the operating

system: /var/log/httpd/access_log). It is a text file readable in any text editor. A nice freeware application to read log files is LogMaster (check on http://www.versiontracker.com for download). You can directly copy the log file and upload it to Scientific LogAnalyzer at http://sclog.eu (Reips and Stieger 2004) for further analysis (see next section).

To view the log file in Mac OS X, you would open the *Applications* folder, open the *Utilities* folder and double click the *Terminal* application. A terminal window will open up. This window accepts old-fashioned line commands, and like old-fashioned programming, this Unix-like terminal is not forgiving of small details, like spaces, capitalization, and spelling.

Carefully type the following command:

open /var/log/apache2/access_log
(or, on earlier versions of the operating system: open /var/log/httpd/access_log)

Before you hit the return key, look at what you have typed and make sure that everything is exactly correct, including capitalization (here nothing is capitalized) and spacing (here there is a space after the word "open"). If you have made no typos, when you press the return key, a window will open showing the log file for the server.

The default logging format used by Apache is somewhat abbreviated. There is a lot of useful information available in the HTTP protocol that is important for behavioral researchers (see Section 27.7) that can be accessed by changing the log format, for example to include information about the referring Web page and the user's type of operating system and Web browser. Methods for making these changes to the configuration of Apache Server are given in the Web site that accompanies this chapter and in the FAQ at http://wextor.org.

If you created your study with WEXTOR, you are ready to collect data.

If you used SurveyWiz or FactorWiz, and if you want to save data to your own server, rather than download it from the *psych.fullerton.edu* server, you will need to make the adjustments in your HTML page(s) and server. There are two approaches. One is to use a CGI script to save the data, installing a CGI on your own server to replace the generic PolyForm script provided by Birnbaum (2000b); this technique will be described in Section 27.6.4. The other approach

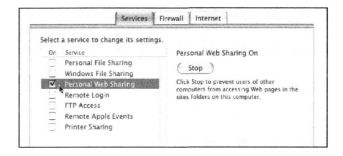

FIGURE 27.9 The third click turns on Web sharing in the Web server in Mac OS X.

is to send the data by "METHOD=GET" to the server's log file.

To use the "GET" method to send data to the log file, find the <FORM> tag of your survey:

```
<FORM METHOD='post' ACTION='http://
psych.fullerton.edu/cgi-win/polyform
.exe/generic'>
```

and change it to

```
<FORM METHOD='GET' ACTION='[address of
next Web page here]'>.
```

Where it says "[address of next Web page here]" you need to specify the name of a Web page on your server, for example, ACTION=ThankYou.htm (but it could even be to the same page). It is conventional to use a page where you thank your participants and provide them with information such as how to contact you for any comments or questions about your study. In Section 27.7 you will be shown how to record and process the form information that is written to the log file each time someone participates in your study.

27.6.4   Installing a Perl Script to Save Data via CGI

The procedures for installing a Perl script with the Apache server for Windows PC are described by Schmidt (2003). This section explains how to use Perl on the Mac OS X.

First, create a folder to hold the data files. Open the icon for the Macintosh hard drive. From the File menu, select *New Folder*. A new folder will appear, which you should name *DataFiles*. Now click the folder once, and press the Apple Key and the letter "i" at the same time, which opens the "Get Info" display. Then click on the pop-up list and choose *Privileges*. Set the first two privileges to *Read and Write* and the third to *Write Only (Dropbox)*.

Next, examine the Perl script in Table 27.4. This is a CGI script, written by Schmidt to emulate the generic PolyForm script used by Birnbaum (2000b, 2001a). This script will save data from a Web form to a new data file in the above folder. You can edit the first three lines to suit your own configuration and needs. For example, you can change the URL in the second line to the address of a "thank you" page on your own server, or you can change the location on your computer where you wish to save the data by changing the third line.

The Perl script in Table 27.4 is also available from the Web site for this chapter, which will save you from typing. You should save this as a Unix text file, with an extension of ".pl" (where "l" is a letter, not a number). For example, you can save it as *save_data.pl*. This file should be saved in the following folder. From your Mac hard drive, open the *Library* folder, then open the *WebServer* folder, and then open *CGI-Executables*. Place *save_data.pl* in this folder.

Now you need to open the Terminal. Open the *Applications* folder and within it open the *Utilities* folder. Double click

TABLE 27.4

CGI Script in Perl that Works with SurveyWiz and FactorWiz.#

```
#!/usr/bin/perl
$redirect_to = "http://psych.fullerton.edu/mbirnbaum/decisions/thanks
 .htm";
$path_to_datafile = "/DataFiles/";
use CGI;

$query = new CGI;
#timestamp the submission
($sec,$min,$hour,$mday,$mon,$year,$wday,$yday,$isdst) =
 localtime(time);
$mon++;

#determine the data filename, open and dump data
$filename = $query->param('00exp');
open(INFO, ">>$path_to_datafile/$filename.data");

foreach $key (sort($query->param))
  {
  $value = $query->param($key);

  #filter out "'s and ,'s
  $value =~ s/\'"/\'/g;
  $value =~ s/,/ /g;
  if ($value !~ /^pf/)
  {
     print INFO "\"$value\", ";
  }
  else
  {
     # filter out items that need to be expanded at submission time pf*
     if ($value =~ /^pfDate/)
     {
        print INFO "\"$mon/$mday/$year\", ";
     }
     if ($value =~ /^pfTime/)
     {
        print INFO "\"$hour:$min:$sec\", ";
     }
     if ($value =~ /^pfRemote/)
     {
        print INFO "\"",$query->remote_addr(),"\", ";
     }
     if ($value =~ /^pfReferer/)
     {
        print INFO "\"",$query->referer(),"\", ";
     }
  }

  #print "$key:$value";
  }
print INFO "\"complete\"\n";
close (INFO);

print $query->redirect($redirect_to);
exit();
```

the *Terminal* program and it will open. Type the following command:

chmod ugo+rwx /Library/WebServer/CGI-Executables/ save_data.pl

Before you hit Return, study what you have typed. Be sure that the capitalization, spacing, and spelling are exactly correct. There should be a space after "chmod" and before "/Library". Hit Return, and a new prompt (new line) appears.

Now, take the survey made by SurveyWiz or FactorWiz (the Driving Survey example in Table 27.2 will do), and find the <FORM> tag. Change this tag to read as follows:

```
<FORM METHOD='post' ACTION='http://
localhost/cgi-bin/save_data.pl>
```

Save your survey in the folder "Sites" as described in Section 27.6.1. You are now ready to collect data. Use your browser to view the survey. Fill it out and push the "submit" button. You should be redirected to the "thank you" page designated on the second line of the Perl Script, and your data should have been saved in the folder, *DataFiles*.

With this one CGI script, you can do all of the experiments and surveys described by Birnbaum (2001a), without needing to install another script. The data will arrive in the designated folder on your server as comma separated values (CSV) files that can be easily imported to many statistical and spreadsheet applications. Methods for working with and analyzing data collected in this format are described in detail by Birnbaum (2001a).

27.7 HOW TO ANALYZE LOG FILES

Scientific investigation relies on the principle of raw data preservation. Raw data need to be saved for scrutiny by other researchers from the community (American Psychological Association 2010), for example to aid in reanalysis or meta analysis. This principle applies to Internet-based research as well, where it can be argued that server log files, properly configured, are the raw data (Reips 2001, 2002b, 2007; Reips and Stieger 2004). In addition to all data sets from full participation, complete log files from a Web server used in an Internet-based investigation contain the following useful information:

1. Data about potential participants who decide not to participate (e.g., the number of people who see the link to a study, but don't click on it).
2. Data about technical conditions during the investigation (i.e., the general Web traffic conditions at the server and the particular conditions of each request by a participant).
3. Data sets for those who provided incomplete information; partial nonresponse may reveal information about potential problems (Reips 2002b).

Not reporting the information listed above, for example because one did not save it, carries the great danger of misleading

scientists with respect to the main effects of variables (Birnbaum and Mellers 1989; Reips 2000, 2002c; Reips and Stieger 2004). Even when drop-out rates are equal in two groups, the observed trend may easily show the opposite of the true effect that one would have found had all data been complete. Because participants find it quite easy to quit Web studies, there can be sizeable attrition in such studies, and this attrition needs to be saved and reported (Frick, Bächtiger, and Reips 2001; Knapp and Heidingsfelder 2001; Reips 2002b, 2007).

Log files need to be configured so that the information needed will be saved. Often the log file format follows a common predefined format, for example the Webstar log file format. This log file format is used by the Web Experimental Psychology Lab and contains information in the following order: CONNECTION_ID DATE TIME RESULT HOSTNAME URL BYTES_SENT AGENT REFERER TRANSFER_TIME SEARCH_ARGS.

To change the log file format on your Web server do the following. On Mac OS X open the Terminal application as described above. Then evoke the pico text editor to edit a file named "httpd.conf." Type

sudo nano /private/etc/apache2/httpd.conf

(If this doesn't work, i.e., you can't find the location of httpd.conf, then type httpd –V and get the location from the list that will appear.)

Scroll down or search for lines beginning with "LogFormat". Below the block of lines you will create a new line by hitting "Return". Then type (or copy from the accompanying Web site or http://wextor.org/wextor/en/faq.php):

```
LogFormat "%P\t%{%m/%d/%y}t\t%{%T}t\t%X\t%h\
    t%v%U\t%B\t%{User-Agent}i\t%{Referer}i\t%T\
    t%q\t%D" my_logfile
```

This directive lists a sequence of field identifiers, each of which tells the server to log a particular piece of information. Fields are separated by tab characters, indicated by "\t".

Next, do a search for "CustomLog." It specifies the location to which the log file will be written. You should see a line similar to:

```
#CustomLog "var/log/apache2/access_log" common
Add the following line below it:
CustomLog "var/log/apache2/access_log" my_logfile
```

Once you have made the changes, save the document (if in the pico text editor) by pressing the control and "O" keys. You will be prompted to confirm the save. Just press return to do so. Finally, quit pico by pressing the control and "X" keys and restart Apache:

apachectl -k graceful

In the format you configured with the LogFormat command each string means the following:

\ : Wildcard used for separating the strings in cases where spaces, '%', or '}' are not used.

t : Tabulator,

%P : The ID of the process that serviced the request.

%{%m/%d/%y} : The date, here in the format Month/Day/Year,

%{%T}: The time, here in the format HH:mm:ss,

%X : Connection status when response is completed (X = connection aborted before the response completed, + = connection may be kept alive after the response is sent, – = connection will be closed after the response is sent),

%h : Remote host (client's computer IP)

%v : The canonical ServerName of the server serving the request.

%U : The URL path requested, not including any query string.

%B : Size of response in bytes, excluding HTTP headers.

%{User-Agent}i : User-Agent information from the header (web browser, operating system).

%{Referer}i : Referer information from the header.

%T : The time taken to serve the request, in seconds.

%q : The query string (prepended with a ? if a query string exists, otherwise an empty string)

%D : The time taken to serve the request, in microseconds.

The following shows an example of a line in the log file, in this case of accessing the Web Experimental Psychology Lab at http://www.wexlab.eu/.

3169 02/25/10 13:43:37 + 130.206.135.23 www
.wexlab.eu/ 253 Mozilla/5.0 (Macintosh; U; Intel
Mac OS X 10.6; es-ES; rv:1.9.2) Gecko/20100115
Firefox/3.6 - 0   775

A platform-independent interactive Web site that helps Internet researchers in analyzing log files is Scientific LogAnalyzer (Reips and Stieger 2004, http://sclog.eu). It was created to meet the needs of those who collect data on the Internet. Scientific LogAnalyzer provides an option for selecting the predefined format mentioned above, and it also contains procedures to identify and process any type of log file. To match a predefined format, the user may also rearrange columns in the log file before uploading it to Scientific LogAnalyzer, which can be done easily in text editors and spreadsheet programs.

Scientific LogAnalyzer has features important to behavioral and social scientists, such as handling of factorial designs, response time analysis, and dropout analysis. Scientific LogAnalyzer was developed to include calculation of response times, flagging of potential multiple submissions, selecting either first or last response from same IP, marking of predefined IP addresses and/or domain names, and free definition of session timeout). The program is highly flexible on the input side (unlimited types of log file formats), while strictly keeping the traditional one-case-per-row output format. Other features include (1) free definition of log file format, (2) searching and identifying any combination of strings (necessary for

organizing conditions in experiment data), (3) computation of approximate response times, (4) a module for analyzing and visualizing dropout, (5) detection of multiple submissions, (6) output in HTML and/or tab-delimited files, suited for import into statistics software, (7) speedy analysis of large log files, and (8) extensive help from an online manual.

27.8 PRETESTING OF INTERNET-BASED RESEARCH

Before an experiment is placed on the Web, it is necessary to perform a number of checks to make sure that the study will yield useful data. First, one should be clear on how the data will be analyzed and that the study will answer the question it is supposed to answer. This check is basic and applies to laboratory research as well as Web research. Those who are unclear on how the data will be analyzed almost never devise a study that can be analyzed.

Second, one should conduct checks of the HTML and CGI script to code and save data to ensure that every possible response is properly coded and recorded in the proper place in the data file. One should check that every radio button (in a given item) functions properly and that answers to one question do not overwrite responses to another item. This is one of the advantages of using a Web service like FactorWiz (Birnbaum 2000b) or WEXTOR (Reips and Neuhaus 2002); these programs save time by automatically creating safe code.

Third, one should test some participants in the laboratory. Observe them as they read and respond to the materials. Ask them to identify aspects of instructions that are unclear. With a few exceptions, Internet-based research does not have a laboratory assistant who can answer questions, so every question must be addressed in advance. Check if people are responding before they have scrolled to see the information that they are supposed to review before responding. One of the second author's students had placed a response device before the material the participant needed to read. During pilot testing, the second author observed a number of people who responded before they had scrolled to make visible what they were supposed to be judging. Analyze the data from the pilot study to see that the coding and analysis will function properly. It is often when analyzing data that students discover problems with their studies. That is why some pilot data should be analyzed before the main study is run.

One can discover a lot by observing participants in pilot research. For example, in a study with the random response method (e.g., Musch, Bröder, and Klauer 2001), participants were supposed to toss two coins and then respond "yes" if both coins were heads, "no" if both coins were tails, and to tell the truth otherwise. The purpose of the technique is to allow an experimenter to assess a population statistic without knowing any person's true answer. For example, most people would be embarrassed to admit that they cheated on their income taxes, but with the random response method there is no way to know if "yes" meant that the person did cheat or that the coins were both heads. If people follow instructions, this method allows the experimenter to subtract 25% "yes" answers (that occurred because of two heads) and 25% "no" (resulting from two tails), and double the remainder to find the correct proportions. For example, if 30% of the group indicated that they cheated on their taxes, it means that 10% of the population cheated. In our pilot test, however, only one of fifteen participants took out any coins, and she asked first if she should actually follow the instructions. The second author urged his student to add stronger instructions and an extra pair of items at the end of the survey asking if the participant had actually used the coins, and if not, why not. About half said they had not followed the instructions, giving excuses such as "lazy" and "I had nothing to hide." This example should make one very concerned about what happens in studies that are not observed, and still more concerned about studies that are launched without pilot testing.

It is also important to pretest the materials with different browsers and systems and with a small monitor, to make sure that everyone will see what they are supposed to see. Otherwise, you would probably run into what the first author coined "configuration error IV" in Internet-based research—the all too common underestimation of the technical variance inherent in the Internet (Reips 2002b; also see Reips 2007; Schmidt 2007). Consider adding items to your research instrument to ask about monitor sizes, volume settings, and other settings that you think might make a difference to the results. Such information can be used to partition the sample for separate analyses. Considerations of delivery of psychophysical stimuli and a discussion of when such settings may or may not matter are reviewed by Krantz (2001) and Krantz and Williams (2010).

27.9 RECRUITING PARTICIPANTS FOR INTERNET-BASED RESEARCH

Participants can be recruited by traditional means, such as course assignments, student "participant" pools, face-to-face requests, word of mouth, posters, flyers, newspaper advertisements, etc. We will focus here on Internet methods such as recruitment via Web site, mailing list, online panel, newsgroup, e-mail, listings, blog, and banner ads. Recruitment for Internet-based studies can be made much more effective by using one or several of the techniques described by Birnbaum (2001a) and by Reips (2000, 2002b, 2002d, 2007).

27.9.1 RECRUITMENT VIA WEB SITES AND SEARCH ENGINES

One natural way to recruit participants is via one's own home page. However, many personal home pages are rarely visited. Besides, visitors of your home page may know too much about your research (or read about it on the page) to be suitable (i.e., naïve) participants for certain studies. (A useful strategy to test whether such self-selection may be biasing your results is the *multiple site entry technique* (Reips 2000, in press), which will be described in Section 27.9.2.4.)

An institution's Home page, for example a university's, may be a better choice for recruiting large numbers of participants.

In addition, an institution's Home page will often convey a more legitimate impression than a personal home page. However, it will not be easy to get agreement to announce your study on an institution's Home page unless a number of administrators have agreed that it is acceptable to do so.

Some of the best places for recruitment are institutionalized Web sites for Internet-based experimenting, such as the Web experiment list (http://wexlist.net), the Web Experimental Psychology Lab (http://wexlab.eu), and the Psychological Research on the Net list (http://psych.hanover.edu/research/exponnet.html) by John Krantz, who published some of the first Web experiments (Krantz, Ballard, and Scher 1997; Welch and Krantz 1996). Some of these Web sites are visited by thousands of potential participants every month (Reips 2001a, 2001b, 2007; Reips and Lengler 2005), and some managers even provide you with a free check of your experiment, before linking it. A link on a Web experiment site may also serve an archiving function, as an example for future studies, and as a reference in publications.

People coming to your study via one of these Web research sites are true volunteers who have already decided that they want to take part in one or more psychology studies, and who chose your study from a list of ways to participate. So, the concerns one might have with respect to students who are participating only to fulfill an assignment are relieved with this source of participants.

To recruit participants by means of search engines, you can enhance your pages to help people find your study. Suppose you wanted to recruit people interested in psychology. To help those who are looking on the Web for "psychology," you can put that word in the title of your site, add it to the early text in your site, and add meta tags including "psychology" as a key word.

Suppose you wanted to recruit people with rare characteristics, such as transvestites or people who suffer from sexsomnia (for a study, see Mangan and Reips 2007). You could include meta tags in your entry page in order to help these people find your study. Table 27.5 shows an example of a Web page that might be used to recruit participants to a survey of memories of early experiences by transvestites. You should

TABLE 27.5
Use of Meta Tags to Recruit via Search Engine

```
<HTML>
<HEAD>
<META NAME="keywords" CONTENT="transvestites,cross-
 dressing,survey,early experiences,psychology,research">
<META NAME="description" CONTENT="research psychologists
 invite transvestites to complete a survey of early life experiences that we
 hope will contribute to understanding this condition">
<TITLE>Survey of Early Experiences of Transvestites</TITLE>
</HEAD>
<BODY>
(Further information on the study and a link to the study would be placed
 here)
</BODY>
</HTML>
```

also consider using an informative title for the first page of your study. Reips (2002b) advises using uninformative page titles and page names for consecutive pages to avoid the possibility that people will find these subsequent pages via search engines and enter the study somewhere in the middle. However, the first page that recruits participants can have an informative title without distracting the participant.

Meta tags can (and should) also be used to keep search engines *away* from all pages except the first page (you may even desire to keep search engines away from that page, if you would like to recruit via other means exclusively). The "robots" tag in Table 27.6 needs to be set to "none," because the routines used by search engines to search the Web for new Web pages are called "robots" (and sometimes "spiders" and "crawlers") and this technique informs them there is nothing for them.

A third important task that can be handled by meta tags is to prevent caches in search engines, mediating servers, and proxy servers from serving old versions of research materials after they have been updated. Caches contain stored files downloaded from the Web that are stored for reuse later. For example, your Web browser may be configured to store HTML code, images, and other media from pages you visited in its own cache. Later, when you request that page again, the Web browser quickly checks in the cache if it holds any of the text and media (e.g., images) you are requesting and displays them to you. This way, the page can be displayed more quickly (than if you had to wait for all the same files to download again), and much unnecessary traffic is avoided.

However, the material loaded from the cache may be outdated: If an experimenter finds an error in her Internet-based study and replaced the Web page on her server, users may continue to see and even be served the old version. Why? Because of caches holding the old version in proxy servers. Proxy servers hold huge caches, but not only for one computer—they hold all of the Web traffic going into and coming out of entire institutions. As you can imagine, there are some interesting analyses that can be performed with data collected on proxy servers (Berker 1999, 2002). Table 27.6 shows two meta tags below the robots tag that will keep proxy servers from caching your Web pages. Web experiments created with Version 2.1 or newer of WEXTOR automatically contain the meta tags described in this section.

TABLE 27.6

Use of Meta Tags in Pages after the Entry Page

```
<HTML>
<HEAD>
<meta name="AUTHOR" content="WEXTOR">
<meta name="ROBOTS" content="NONE">
<meta http-equiv="pragma" content="no-cache">
<meta http-equiv="expires" content="Thursday, 1-Jan-1991 01:01:01 GMT">

<TITLE> </TITLE>
</HEAD>
(the body of the Web page goes here)
```

27.9.2 Recruitment via Other Means

In a survey of the "pioneers" of Web experimenting, Musch and Reips (2000) asked the question "In which media did you announce your experiment?" A surprisingly large number of Web experiments were not announced solely on the Web. Researchers also used newsgroups (18 of 35 experiments), e-mails (15), and search engines (14) to advertise their Web experiments. Only very few researchers used radio (1) or print media (2), although we now know that these offline media can be extremely effective. More recently, it has become possible to quickly recruit many participants via online media. The first author and his students have repeatedly announced Web experiments in http://20min.ch, the online version of "20 minutes," Switzerland's newspaper with the largest circulation nationally. Within a few days, each of the experiments was visited by several thousand participants.

Birnbaum (2001a) gives a number of suggestions regarding recruitment. He recommends against sending e-mail messages to people who do not want to get e-mail from you for reasons we will describe in Section 27.9.3 on ethics below. A better method for recruiting from a listserv is to contact the organization that runs the listserv and ask this organization for help. If you can convince the organization that your research is serious and will be of value to the community that the organization serves, you can often get good help from them. They can post an announcement in their Web site, send invitations in their newsletter, and even post messages for you to their members via e-mail. For example, to recruit transvestites, one could contact organizations to which these people belong, show that your research would be of interest or benefit to the members, and ask the organization to recruit for you.

This suggestion was adopted by Drake (2001), a graduate student who was working with Birnbaum at the time. She wanted to recruit elderly people with interests in genealogy. She realized that members of her target group belong to organizations that have an Internet presence. Drake contacted an organization, which saw the potential interest and value of her research to their members and agreed to send an e-mail to the members on her behalf, vouching for her research and inviting participation. Within one week, she had more than 4000 completed data records and many messages of encouragement and support (Drake 2001).

27.9.2.1 Newsgroups

In the early days of the Internet, communications consisted largely of text-based materials. Many people had very slow connections and logged onto virtual *bulletin boards* via modem to upload and download messages. One of the largest bulletin board networks was USEnet. Most of the USEnet newsgroups were saved and made accessible on the Web by deja.com (playing on the term *déjà vu*), a service that was later acquired by Google. Google now lists all available newsgroups as *Google groups* (http://groups.google.com/) and makes them available for searches. For example, if you surf to Google groups and type in the words "Web experiment survey" the two first postings listed are those originally posted to the newsgroup "sci.psychology.research" by Musch and Reips, inviting Web experimenters to their survey mentioned above.

If you use newsgroups for recruitment of participants, then be aware of the following points:

- Your posting will remain indefinitely, until the newsgroup might be erased from the Internet (if the study has an ending date, you should include that in your post).
- It will take several days until your posting is widely available.
- Make sure your posting is considered appropriate in the newsgroups to which you post—ask the moderator first, if the newsgroup is moderated.

27.9.2.2 Participant Pools and Panels

Many colleges have established pools of students who volunteer to participate in studies. In some cases, the researcher can request a "stratified" sample of participants, stratified with respect to gender, for example. Such a traditional participant pool may be an effective way for recruitment in Web studies, especially at large colleges.

However, student samples are not stratified with respect to age or education, nor are students heterogeneous with respect to many other variables (Birnbaum 2001a; Reips 2000). Most are between 18 and 22 years of age; all are graduates of high school, and none are graduates of college. Psychology pools now contain about two-thirds females, and at many schools, students who take psychology are more likely than the typical student to be undecided majors. Birnbaum (1999b) wanted to study correlates of education (especially education in decision making) and therefore needed to recruit off campus to obtain a sample that would show wide variation in education.

Using the Internet, a number of academic researchers and market researchers have created participant pools of people with more heterogeneous characteristics than college students by means of the Internet (Baron and Siepmann 2000; Smith and Leigh 1997; Göritz 2007, 2009). If nurtured properly, these "online panels" or "online research panels" of volunteers for Web-research are a means of widely distributing easy access to a wide range of participants.

27.9.2.3 E-Mail

One very effective way of recruiting participants involves sending e-mails to mailing lists of people who want to receive your mail. At a conference (SPUDM, Zürich 2003), the first author heard an interesting paper on the first day of the conference and decided to replicate that study overnight. He was able to include the results in his talk on the second day of the conference, in order to demonstrate how efficient Web-based research can be in comparison with the methods used by the original authors (see link to Reips' SPUDM presentation on companion Web site at http://wexlab.eu/hcihandb). Within 8 hours, complete data sets from 162 participants (compared

to 64 in the original study) were recorded in the Web experiment, most of which were recruited via three mailing lists.

27.9.2.4 Banner Ads

People who visit tattoo and piercing sites may have different personalities than those who visit traditional family values sites. Buchanan (2000, 2001) has exploited this insight to form criterion groups from people who are recruited by different methods, in order to validate personality tests. He has had some success comparing people recruited from different user groups. This technique is sometimes called the *multiple site entry technique* (Reips 2000, 2002b), which involves recruiting people via several different methods or sources and comparing data between these groups. Buchanan (pers. comm., Oct. 10, 2002) reported, however, that response to banner ads has been very meager and probably not worth the money (for similar results, see Tuten, Bosnjak, and Bandilla 2000). Because banners usually represent commercial advertising, a banner ad for a scientific study is hard to distinguish from a deceptive come-on for a commercial message.

27.9.2.5 Password Techniques

Passwords can be used to control entry to a study or survey, or they can be used to validate data from members of a voting body. Passwords are one way to determine if one person is trying to "stuff the ballot box" with multiple submissions of the same vote.

Password techniques (Reips 2000; Schmidt 2000) can be used to guarantee authenticity and originality of a participant's identity. In an experimental study on Internet-based versus paper-and-pencil-based surveying of employees, Reips and Franek (2004) printed anonymous individual codes for use in the (mailed) invitation to 655 employees. From the log analysis of the Internet-based data, the following could be determined:

- Whether any codes were used repeatedly (showing multiple submissions, and permitting the investigators to include the first set only).
- Whether people without code tried to access the survey (using wrong codes or no codes).

There are many other methods for authenticating data (see Schultz, this volume) and detecting or preventing multiple submissions of data. These are summarized in a number of papers, including Birnbaum (2004a, Table 1).

27.9.3 Ethics and Etiquette in Internet Science

It is difficult to injure someone in research via the Web, except by dishonesty, so the fundamental ethical principle for Web-based research is honesty. If you promise some benefit for participation, then you must follow through and provide that benefit. For example, if you promise to pay participants, then you must pay them. Failure to do so is fraud. If you promise to give participants a report of the results by a certain date, then you must provide them the report. A good way to do this, incidentally, is to post the results to the Web and send the URL to those who requested the results (but do not send attachments, see below). If you promise to provide a chance in a lottery to win a prize, then you must follow through and run the lottery and distribute the prizes that were promised.

27.9.3.1   Taking Precautions

If participation is not anonymous, and if you promise confidentiality, then you must do everything you can to secure the data and prevent them from becoming public. Birnbaum (2001a) lists a number of steps to provide security of the data on the server. Keep the door to the server locked, and do not leave keys or passwords around. Record only what is necessary, keep personal information separate from other data, if possible, and remove identifying information from stored data as soon as possible. The server should be kept up to date with respect to security from hackers.

Although Musch and Reips (2000) reported that early Web experimenters had not found hackers to be a problem, the potential is certainly there for an unfortunate incident. Reips (2002b) observed a high rate of insecure servers in Web studies ("configuration error I"). During routine checks of Web experiments by researchers who applied for inclusion in the Web Experimental Psychology Lab, Reips was able to download openly accessible data files in several cases. Apparently, many researchers were not aware of the vulnerabilities of their Web servers and operating systems, and were surprised when he sent portions of their confidential data files to them. On the basis of statistics permanently collected by http://www.securityfocus.com ("bugtraq"), Reips (2002b) also noted the better track record of the Mac OS against vulnerabilities to hackers and viruses, compared to Windows and other operating systems. Depending on the operating system used, an Internet scientist will have to invest less or more time in keeping up with newly discovered vulnerabilities and respective security developments. Open source software is often more secure than professional software because vulnerabilities have been discovered and corrected by a large community. Professional developers have only a small number of people working on problems and have mixed motives with respect to informing the public about security concerns.

It is good practice to avoid deception in any Web-based research, at all costs. One of the ethical reasons to do this is to keep a good reputation for Internet-based research. An ethics review board is concerned for the ethical treatment of participants, and certain deceptions would not be considered harmful to participants. However, studies involving deception might be harmful to other researchers who do not practice deception if Internet research should acquire a bad reputation. Therefore, one should consider the impact of one's research on other scientists (and indirectly on society) before one does anything to give Internet-based research a bad name. For more information on issues of deception and ethics of Web research, see Birnbaum (2004b) and Barchard and Williams (2008).

In addition to ethical concerns in recruiting, there are issues of good manners and good taste. It is considered impolite to send people attachments they did not request. People now fear to open them because they may carry a commercial message, a worm, or a computer virus. And large attachments can fill up mailboxes, can be slow to download, and are generally seen as a kind of injury (or threat of injury) that you have inflicted on your recipients. Keep in mind that computers break down all the time. Suppose you opened an attachment from someone, and then your computer broke down. It would be natural to assign blame and seek justice, even though the event was mere coincidence. By means of the Internet, it is easy to reach a large number of people for good or ill, and therefore it is possible to annoy if not anger large numbers of people.

Instead of sending attachments, send a link to the file that you have uploaded to the Web.

In your messages recruiting participants or reporting availability of results, you should include your correct name, e-mail and postal address, telephone number, and affiliation. Behaving politely on the Internet is not only a nice thing to do, it is also a necessity for the Internet researchers to avoid being confused with purveyors of SPAM, spoofers (senders of e-mail from fake addresses), or even worse. The impolite and ignorant behaviors of a few spammers and malicious hackers have already destroyed much of the good spirit and free flow of communication on the Internet. E-mail is now filtered and even innocent messages and innocent attachments are disappearing into e-mail void, as institutions try to protect themselves from the massive number of unwanted or unsolicited e-mails.

27.9.3.2   Polite Recruitment

Similarly, it is important to be very careful when recruiting participants for your Internet-based study. Let others check the wording of your invitation, before you send it off or post it. If you recruit participants via a mailing list or newsgroup, then first have the moderator of the list approve your posting. An even better solution is to ask someone else, such as the manager of a group, to send the announcement for you. That way, people will know that the announcement is legitimate, and if they consider your invitation to be spam, they will blame the moderator's judgment rather than yours.

Do not do anything that resembles a chain letter, commercial advertising, or spamming. Internet researchers have been "blacklisted" and "e-mail bombed" for recruiting too aggressively. Such cases are not only embarrassing and time-consuming (responding to "flames" or angry messages attacking one's manners), they also pose a threat to the reputation of the entire community. If you are in doubt that people want to receive your message, it is probably best not to send it. Consider recruiting participants in the least intrusive manner, which is probably done by posting the announcement at designated Web sites mentioned earlier in this chapter. These sites are explicitly designated to potential participants to find studies and to researchers to find participants.

A set of considerations ("rules") for proper behavior on the Internet can be found at http://www.albion.com/netiquette/ and Internet research specific advice is available from Michalak and Szabo (1998).

27.10 WHERE TO LEARN MORE

In addition to journals and professional books on the topic, there are two organizations holding conferences that provide good opportunities to meet and interact with people involved in psychological Internet-based research. These are the General Online Research Conference (GOR), established in 1997 and hosted jointly by the German Society for Online Research (DGOF) and local organization teams that differ from year to year. More than 350 participants attended each GOR conference in the past few years.

The central purpose of GOR conferences is to provide a venue for the presentation of empirical studies that examine the Internet as both an instrument for, and an object of, scientific investigation. Topics of investigation presented include (but are not limited to): quality of Internet-based research, Internet research methods in market research and polling, computer-mediated communication, Web experiments, and marketing on the Internet.

Society for Computers in Psychology (SCiP) meetings take place on an annual basis, in mid-November and are scheduled to occur in the same location as the Psychonomic Society and Society for Judgment and Decision Making (JDM) meetings, on the day preceding the annual meeting of the Psychonomic Society. This organization is somewhat older than GOR, and is devoted not only to Internet-based research but also to other applications of computers in psychological research and teaching. The primary journal of this organization is *Behavior Research Methods*. Another journal we recommend is the *International Journal of Internet Science*, a free open access journal available from http://www.ijis.net.

REFERENCES

American Psychological Association. 2010. *Publication Manual of the American Psychological Association*, 6th ed. Washington, DC: Author.

Barchard, K. A., and J. E. Williams. 2008. Practical advice for conducting ethical online experiments and questionnaires for United States psychologists. *Behavior Research Methods* 40: 1111–1128.

Baron, J., and M. Siepmann. 2000. Techniques for creating and using web questionnaires in research and teaching. In *Psychological experiments on the Internet*, ed. M. H. Birnbaum, 235–265. San Diego, CA: Academic Press.

Berker, T. 1999. Online-offline: Methodologische Erträge einer Internetnutzungsstudie zur Klärung einer Zentralkategorie des Online-Research [On-line versus off-line: Methodological earnings of a study on Internet use devised to clarify a central category of on-line research]. In *Current Internet Science—Trends, Techniques, Results. Aktuelle Online Forschung—Trends, Techniken, Ergebnisse*, eds. U.-D. Reips et al. Zürich: Online Press, available at http://dgof.de/tband99/.

Berker, T. 2002. World Wide Web use at a German university—computers, sex, and imported names: results of a log file analysis. In *Online Social Sciences*, eds. B. Batinic, U.-D. Reips, and M. Bosnjak, 365–382. Göttingen, Germany: Hogrefe.

Birnbaum, M. H. 1999a. How to show that 9 > 221: Collect judgments in a between-subjects design. *Psychological Methods* 4(3): 243–249.

Birnbaum, M. H. 1999b. Testing critical properties of decision making on the Internet. *Psychological Science* 10: 399–407.

Birnbaum, M. H. ed. 2000a. *Psychological Experiments on the Internet*. San Diego: Academic Press.

Birnbaum, M. H. 2000b. SurveyWiz and FactorWiz: JavaScript Web pages that make HTML forms for research on the Internet. *Behavior Research Methods, Instruments, & Computers* 32(2): 339–346.

Birnbaum, M. H. 2001a. *Introduction to Behavioral Research on the Internet*. Upper Saddle River, NJ: Prentice-Hall.

Birnbaum, M. H. 2001b. A Web-based program of research on decision making. In *Dimensions of Internet Science*, eds. U.-D. Reips and M. Bosnjak, 23–55. Lengerich, Germany: Pabst Science.

Birnbaum, M. H. 2004a. Human research and data collection via the Internet. *Annual Review of Psychology* 55: 803–832.

Birnbaum, M. H. 2004b. Methodological and ethical issues in conducting social psychology research via the Internet. In *Handbook of Methods in Social Psychology*, eds. C. Sansone, C. C. Morf, and A. T. Panter, 359–382. Thousand Oaks, CA: Sage.

Birnbaum, M. H. 2007. Designing online experiments. In *The Oxford Handbook of Internet Psychology*, eds. A. Joinson et al., 391–403. Oxford, UK: Oxford University Press.

Birnbaum, M. H., and B. A. Mellers. 1989. Mediated models for the analysis of confounded variables and self-selected samples. *Journal of Educational Statistics* 14: 146–158.

Birnbaum, M. H., and S. V. Wakcher. 2002. Web-based experiments controlled by JavaScript: an example from probability learning. *Behavior Research Methods, Instruments, & Computers* 34: 189–199.

Buchanan, T. 2000. Potential of the Internet for personality research. In *Psychological Experiments on the Internet*, ed. M. H. Birnbaum, 121–140. San Diego, CA: Academic Press.

Buchanan, T. 2001. Online personality assessment. In *Dimensions of Internet Science*, eds. U.-D. Reips and M. Bosnjak, 57–74. Lengerich, Germany: Pabst Science.

Dillman, D. A., and D. K. Bowker. 2001. The Web questionnaire challenge to survey methodologists. In *Dimensions of Internet Science*, eds. U.-D. Reips and M. Bosnjak, 159–178. Lengerich, Germany: Pabst Science.

Dillman, D. A., J. D. Smyth, and L. M. Christian. 2008. *Internet, Mail, and Mixed-mode Surveys: The Tailored Design Method*. New York: Wiley.

Drake, P. J. W. 2001. Successful aging: investment in genealogy as a function of generativity, mobility and sense of place. Master of Arts Thesis. California State University, Fullerton.

Eichstaedt, J. 2001. An inaccurate-timing filter for reaction-time measurement by JAVA-applets implementing Internet-based experiments. *Behavior Research Methods, Instruments, and Computers* 33: 179–186.

Frick, A., M. T. Bächtiger, and U.-D. Reips. 2001. Financial incentives, personal information, and drop-out in online studies. In *Dimensions of Internet Science*, eds. U.-D. Reips and M. Bosnjak, 209–219. Lengerich: Pabst Science.

Fraley, R. C. 2004. *How to Conduct Behavioral Research over the Internet: A Beginner's Guide to HTML and CGI/Perl*. New York: The Guilford Press.

Francis, G., I. Neath, and A. M. Surprenant. 2000. The cognitive psychology online laboratory. In *Psychological Experiments on the Internet*, ed. M. H. Birnbaum, 267–283. San Diego, CA: Academic Press.

Funke, F., and U.-D. Reips. 2007. Messinstrumente und Skalen [Measuring devices and scales]. In *Online-Forschung 2007: Grundlagen und Fallstudien*, eds. M. Welker and O. Wenzel, 52–76. Köln: Herbert von Halem.

Garrett, J. J. 2005. Ajax: A new approach to Web applications. http://www.adaptivepath.com/ideas/essays/archives/000385 .php (accessed Feb. 8, 2010).

Göritz, A. S. 2006. Web-based data collection. Retrieved April 10, 2010, from http://www.goeritz.net/ati/material.pdf.

Göritz, A. S. 2007. Using online panels in psychological research. In *The Oxford Handbook of Internet Psychology*, eds. A. N. Joinson et al., 473–485. Oxford, UK: Oxford University Press.

Göritz, A. S. 2009. Building and managing an online panel with PHP PanelAdmin. *Behavior Research Methods* 41(4): 1177–1182.

Göritz, A. S., and M. H. Birnbaum. 2005. Generic HTML form processor: a versatile PHP Script to save Web-collected data into a MySQL database. *Behavior Research Methods* 37: 703–710.

Hardin, C., and M. H. Birnbaum. 1990. Malleability of "ratio" judgments of occupational prestige. *American Journal of Psychology* 103: 1–20.

Knapp, F., and M. Heidingsfelder. 2001. Drop-out analysis: effects of the survey design. In *Dimensions of Internet Science*, eds. U.-D. Reips and M. Bosnjak, 221–230. Lengerich, Germany: Pabst Science.

Krantz, J. H. 2001. Stimulus delivery on the Web: What can be presented when calibration isn't possible? In *Dimensions of Internet Science*, eds. U.-D. Reips and M. Bosnjak, 113–130. Lengerich, Germany: Pabst Science.

Krantz, J. H., J. Ballard, and J. Scher. 1997. Comparing the results of laboratory and World-Wide Web samples on the determinants of female attractiveness. *Behavior Research Methods, Instruments, & Computers*, 29: 264–269.

Krantz, J. H., and R. Dalal. 2000. Validity of Web-based psychological research. In *Psychological Experiments on the Internet*, eds. M. H. Birnbaum, 35–60. San Diego, CA: Academic Press.

Krantz, J. H., and J. E. Williams. 2010. Using graphics, photographs, and dynamic media. In *Advanced Methods for Behavioral Research on the Internet*, eds. S. Gosling and J. A. Johnson, 45–61. Washington, DC: American Psychological Association.

Mangan, M., and Reips, U.-D. 2007. Sleep, sex, and the Web: Surveying the difficult-to-reach clinical population suffering from sexsomnia. *Behavior Research Methods, 39*, 233–236.

McGraw, K. O., M. D. Tew, and J. E. Williams. 2000. PsychExps: An on-line psychology laboratory. In *Psychological experiments on the Internet*, ed. M. H. Birnbaum, 219–233. San Diego, CA: Academic Press.

Michalak, E. E., and A. Szabo. 1998. Guidelines for Internet research: an update. *European Psychologist* 3(1): 70–75.

Musch, J., A. Bröder, and K. C. Klauer. 2001. Improving survey research on the World-Wide Web using the randomized response technique. In *Dimensions of Internet Science*, eds. U.-D. Reips and M. Bosnjak, 179–192. Lengerich: Pabst Science Publishers.

Musch, J., and U.-D. Reips. 2000. A brief history of Web experimenting. In *Psychological Experiments on the Internet*, ed. M. H. Birnbaum, 61–87. San Diego, CA: Academic Press.

Parducci, A. 1995. *Happiness, Pleasure, and Judgment*. Mahwah, NJ: Lawrence Erlbaum.

Reips, U.-D. 1995. The Web experiment method. http://www.uni-tue bingen.de/uni/sii/Ulf/Lab/wwwExpMethod.html, now at http:// wexlab.eu/wwwExpMethod.html (accessed Sept. 1, 1995).

Reips, U.-D. 1997. Das psychologische Experimentieren im Internet. In *Internet für Psychologen*, ed. B. Batinic, 245–265. Göttingen, Germany: Hogrefe.

Reips, U.-D. 1999. Online Research with children. In *Current Internet Science—Trends, Techniques, Results*, eds. U.-D. Reips et al., http://gor.de/gor99/tband99/pdfs/q_z/reips.pdf. Zürich: Online Press.

Reips, U.-D. 2000. The Web experiment method: Advantages, disadvantages, and solutions. In *Psychological experiments on the Internet*, ed. M. H. Birnbaum, 89–117. San Diego, CA: Academic Press.

Reips, U.-D. 2001a. Merging field and institution: running a Web laboratory. In *Dimensions of Internet Science*, eds. U.-D. Reips and M. Bosnjak, 1–22. Lengerich, Germany: Pabst Science Publishers.

Reips, U.-D. 2001b. The Web Experimental Psychology Lab: Five years of data collection on the Internet. *Behavior Research Methods, Instruments, & Computers* 33: 201–211.

Reips, U.-D. 2002a. Context effects in Web surveys. In *Online Social Sciences*, eds. B. Batinic, U.-D. Reips, and M. Bosnjak, 69–79. Göttingen, Germany: Hogrefe & Huber.

Reips, U.-D. 2002b. Internet-based psychological experimenting: Five do's and five don'ts. *Social Science Computer Review* 20(3): 241–249.

Reips, U.-D. 2002c. Standards for Internet-based experimenting. *Experimental Psychology* 49(4): 243–256.

Reips, U.-D. 2002d. Theory and techniques of conducting Web experiments. In *Online Social Sciences*, eds. B. Batinic, U.-D. Reips, and M. Bosnjak, 219–249. Seattle: Hogrefe & Huber.

Reips, U.-D. 2006. Web-based methods. In *Handbook of Multimethod Measurement in Psychology*, eds. M. Eid and E. Diener, 73–85. Washington, DC: American Psychological Association.

Reips, U.-D. 2007. The methodology of Internet-based experiments. In *The Oxford Handbook of Internet Psychology*, eds. A. Joinson et al., 373–390. Oxford: Oxford University Press.

Reips, U.-D. 2010. Design and formatting in Internet-based research. In *Advanced Internet Methods in the Behavioral Sciences*, eds. S. Gosling and J. A. Johnson, 29–43. Washington, DC: American Psychological Association.

Reips, U.-D. in press. Using the internet to collect data. In *APA Handbook of Research Methods in Psychology*, eds H. Cooper, P. Camic, R. Gonzalez, D. Long, and A. Panter. Washington, DC: American Psychological Association.

Reips, U.-D., and M. Bosnjak, eds. 2001. *Dimensions of Internet Science*. Lengerich, Germany: Pabst Science.

Reips, U.-D., T. Buchanan, A. Joinson, and C. Paine. 2010. Internet questionnaires in e-health contexts: Non-response to sensitive items. Manuscript submitted for publication.

Reips, U.-D., and L. Franek. 2004. Mitarbeiterbefragungen per Internet oder Papier? Der Einfluss von Anonymität, Freiwilligkeit und Alter auf das Antwortverhalten [Employee surveys via Internet or paper? The influence of anonymity, voluntariness and age on answering behavior]. *Wirtschaftspsychologie* 6(1): 67–83.

Reips, U.-D., and F. Funke. 2008. Interval level measurement with visual analogue scales in Internet-based research: VAS Generator. *Behavior Research Methods* 40: 699–704.

Reips, U.-D., and J. Krantz. 2010. Conducting true experiments on the Web. In *Advanced Internet Methods in the Behavioral Sciences*, eds. S. Gosling and J. Johnson, (193–216). Washington, DC: American Psychological Association.

Reips, U.-D., and R. Lengler. 2005. The Web Experiment List: A Web service for the recruitment of participants and archiving of Internet-based experiments. *Behavior Research Methods* 37: 287–292.

Reips, U.-D., and C. Neuhaus. 2002. WEXTOR: A Web-based tool for generating and visualizing experimental designs and procedures. *Behavior Research Methods, Instruments, & Computers* 34: 234–240.

Reips, U.-D., and S. Stieger. 2004. Scientific LogAnalyzer: A Web-based tool for analyses of server log files in psychological research. *Behavior Research Methods, Instruments, & Computers* 36: 304–311.

Rose, B. J., and M. H. Birnbaum. 1975. Judgments of differences and ratios of numerals. *Perception & Psychophysics* 18: 194–200.

Schmidt, W. C. 1997a. World-Wide Web survey research: Benefits, potential problems, and solutions. *Behavioral Research Methods, Instruments, & Computers* 29: 274–279.

Schmidt, W. C. 1997b. World-Wide Web survey research made easy with WWW Survey Assistant. *Behavior Research Methods, Instruments, & Computers* 29: 303–304.

Schmidt, W. C. 2000. The server-side of psychology Web experiments. In *Psychological Experiments on the Internet*, ed. M. H. Birnbaum, 285–310. San Diego, CA: Academic Press.

Schmidt, W. C. 2007. Technical considerations when implementing online research. In *The Oxford Handbook of Internet Psychology*, eds. A. Joinson et al., 461–472. Oxford: Oxford University Press.

Schmidt, W. C., R. Hoffman, and J. MacDonald. 1997. Operate your own World-Wide Web server. *Behavior Research Methods, Instruments, & Computers* 29: 189–193.

Schultz, E. E. this volume. Web security, privacy, and usability. In *Handbook of Human Factors in Web Design*, 2nd ed., eds. K.-P. L. Vu and R. W. Proctor, 663–676. Boca Raton, FL: CRC Press.

Schwarz, S., and U.-D. Reips. 2001. CGI versus JavaScript: A Web experiment on the reversed hindsight bias. In *Dimensions of Internet Science*, eds. U.-D. Reips and M. Bosnjak, 75–90. Lengerich: Pabst.

Smith, M. A., and B. Leigh. 1997. Virtual subjects: Using the Internet as an alternative source of subjects and research environment. *Behavior Research Methods, Instruments, & Computers* 29: 496–505.

Smyth, J. D., D. A. Dillman, L. M. Christian, and M. J. Stern. 2006. Effects of using visual design principles to group response options in Web surveys. *International Journal of Internet Science* 1: 6–16.

Stevenson, M. K. 1993. Decision making with long-term consequences: Temporal discounting for single and multiple outcomes in the future. *Journal of Experimental Psychology: General* 122: 3–22.

Stieger, S., and U.-D. Reips. 2010. What are participants doing while filling in an online questionnaire: A paradata collection tool and an empirical study. *Computers in Human Behavior* 26: 1488–1495.

Tuten, T. L., M. Bosnjak, and W. Bandilla. 2000. Banner-advertised Web surveys. *Marketing Research* 11(4): 17–21.

Tuten, T. L., D. J. Urban, and M. Bosnjak. 2002. Internet surveys and data quality: A review. In *Online Social Sciences*, eds. B. Batinic, U.-D. Reips, and M. Bosnjak, 7–26. Seattle, WA: Hofrefe & Huber.

Welch, N., and J. H. Krantz. 1996. The world-wide Web as a medium for psychoacoustical demonstrations and experiments: Experience and results. *Behavior Research Methods, Instruments, & Computers* 28: 192–196.

Wikibooks, n.d. Mac OS X Tiger/Using your Mac as a Web Server. http://en.wikibooks.org/wiki/Mac_OS_X_Tiger/Using_your_Mac_as_a_Web_Server (accessed March 24, 2010).

Wikipedia. 2010. Ajax (programming). http://en.wikipedia.org/w/index.php?title=Ajax_(programming)&oldid=352212625 (accessed Jan. 27, 2010).

Williams, J. E., K. O. McGraw, and M. D. Tew. 1999. Undergraduate labs and computers: The case for PsychExps. *Behavioral Research Methods, Instruments, and Computers* 31(2): 287–291.

FIGURE 3.3 (left) Project Cenote and (right) Vodafone 360.

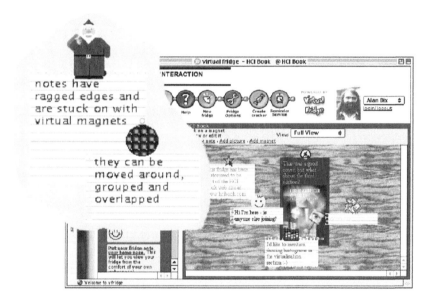

FIGURE 3.13 The vfridge: a virtual fridge door on the Web.

FIGURE 3.14 Safari top sites.

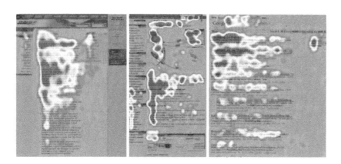

FIGURE 8.3 Examples of heat maps showing the F-pattern; red areas received the most fixation. (From Nielsen, J. 2006. F-shaped pattern for reading Web content. Alertbox. http://www.useit.com/alertbox/reading_pattern.html. With permission.)

FIGURE 8.4 Example of fixed-width design.

FIGURE 8.5 Example of variable-width, or fluid, design.

FIGURE 8.8 Illustration of anti-aliasing applied to the letter "e." On the left is the original version, greatly enlarged. On the right is the same letter with anti-aliasing applied.

A – Plain

Number	Name	Diameter	Dimensions	Dist from sun	Date discovered	Discoverer
1	Ceres	952	9.75×909	2.766	Janurary 1, 1801	Piazzi G.
2	Pallas	532	570×525×500	2.773	March 28, 1802	Olbers, H. W.
4	Vesta	530	578×560×458	2.361	March 29, 1807	Olbers, H. W.
10	Hygiea	407	500×385×350	3.137	April 12, 1849	de Gasparis, A.
704	Interamnia	326	350.4×303.7	3.067	October 10, 1910	Cerulli, V.
52	Europa	302	360×315×240	3.101	February 4, 1858	Goldschmict, H.
511	Davida	289	357×294×231	3.17	May 30, 1903	Dugan, R. S.

D – Triple striped

Number	Name	Diameter	Dimensions	Dist from sun	Date discovered	Discoverer
1	Ceres	952	9.75×909	2.766	Janurary 1, 1801	Piazzi G.
2	Pallas	532	570×525×500	2.773	March 28, 1802	Olbers, H. W.
4	Vesta	530	578×560×458	2.361	March 29, 1807	Olbers, H. W.
10	Hygiea	407	500×385×350	3.137	April 12, 1849	de Gasparis, A.
704	Interamnia	326	350.4×303.7	3.067	October 10, 1910	Cerulli, V.
52	Europa	302	360×315×240	3.101	February 4, 1858	Goldschmict, H.
511	Davida	289	357×294×231	3.17	May 30, 1903	Dugan, R. S.

B – Double striped

Number	Name	Diameter	Dimensions	Dist from sun	Date discovered	Discoverer
1	Ceres	952	9.75×909	2.766	Janurary 1, 1801	Piazzi G.
2	Pallas	532	570×525×500	2.773	March 28, 1802	Olbers, H. W.
4	Vesta	530	578×560×458	2.361	March 29, 1807	Olbers, H. W.
10	Hygiea	407	500×385×350	3.137	April 12, 1849	de Gasparis, A.
704	Interamnia	326	350.4×303.7	3.067	October 10, 1910	Cerulli, V.
52	Europa	302	360×315×240	3.101	February 4, 1858	Goldschmict, H.
511	Davida	289	357×294×231	3.17	May 30, 1903	Dugan, R. S.

E – Single striped

Number	Name	Diameter	Dimensions	Dist from sun	Date discovered	Discoverer
1	Ceres	952	9.75×909	2.766	Janurary 1, 1801	Piazzi G.
2	Pallas	532	570×525×500	2.773	March 28, 1802	Olbers, H. W.
4	Vesta	530	578×560×458	2.361	March 29, 1807	Olbers, H. W.
10	Hygiea	407	500×385×350	3.137	April 12, 1849	de Gasparis, A.
704	Interamnia	326	350.4×303.7	3.067	October 10, 1910	Cerulli, V.
52	Europa	302	360×315×240	3.101	February 4, 1858	Goldschmict, H.
511	Davida	289	357×294×231	3.17	May 30, 1903	Dugan, R. S.

C – Lined

Number	Name	Diameter	Dimensions	Dist from sun	Date discovered	Discoverer
1	Ceres	952	9.75×909	2.766	Janurary 1, 1801	Piazzi G.
2	Pallas	532	570×525×500	2.773	March 28, 1802	Olbers, H. W.
4	Vesta	530	578×560×458	2.361	March 29, 1807	Olbers, H. W.
10	Hygiea	407	500×385×350	3.137	April 12, 1849	de Gasparis, A.
704	Interamnia	326	350.4×303.7	3.067	October 10, 1910	Cerulli, V.
52	Europa	302	360×315×240	3.101	February 4, 1858	Goldschmict, H.
511	Davida	289	357×294×231	3.17	May 30, 1903	Dugan, R. S.

F – Two colour striped

Number	Name	Diameter	Dimensions	Dist from sun	Date discovered	Discoverer
1	Ceres	952	9.75×909	2.766	Janurary 1, 1801	Piazzi G.
2	Pallas	532	570×525×500	2.773	March 28, 1802	Olbers, H. W.
4	Vesta	530	578×560×458	2.361	March 29, 1807	Olbers, H. W.
10	Hygiea	407	500×385×350	3.137	April 12, 1849	de Gasparis, A.
704	Interamnia	326	350.4×303.7	3.067	October 10, 1910	Cerulli, V.
52	Europa	302	360×315×240	3.101	February 4, 1858	Goldschmict, H.
511	Davida	289	357×294×231	3.17	May 30, 1903	Dugan, R. S.

FIGURE 8.12 Table designs examined by Enders (2008). Single striped was rated the most useful. (From Enders, J. 2008. Zebra striping: more data for the case. *A List Apart*. http://www.alistapart.com/articles/zebrastripingmoredataforthecase (accessed November 6, 2010). With permission.)

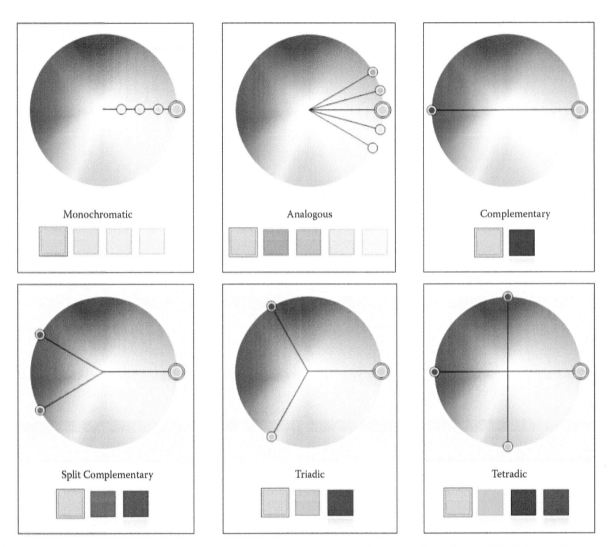

FIGURE 8.13 Examples of six color schemes defined by relationships on the color wheel. Created using ColorSchemer Studio, Version 2.0 (http://www.ColorSchemer.com).

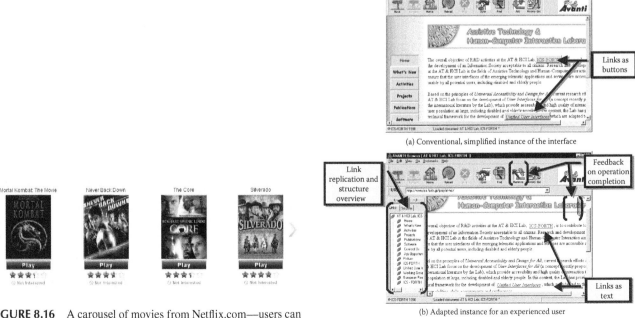

FIGURE 8.16 A carousel of movies from Netflix.com—users can scroll left and right to view different movies. (From Netflix.com. With permission.)

(a) Conventional, simplified instance of the interface

(b) Adapted instance for an experienced user

FIGURE 9.4 Adapting to the user and the context of use.

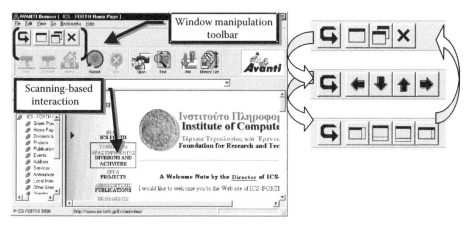

(a) Scanning for switch-based interaction   (b) Window manipulation toolbar

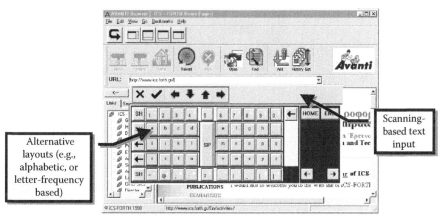

(c) On-screen, "virtual" keyboard.

FIGURE 9.5   Instances for motor-impaired users.

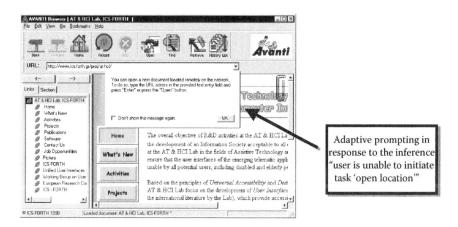

FIGURE 9.6   Awareness prompting.

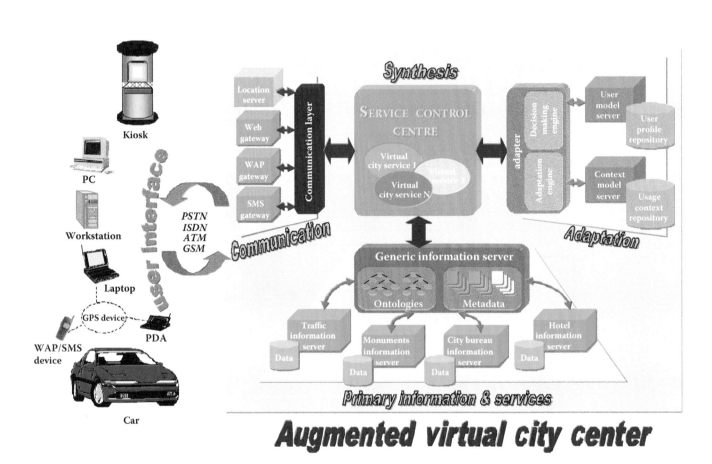

FIGURE 9.7 Overall architecture of the PALIO system.

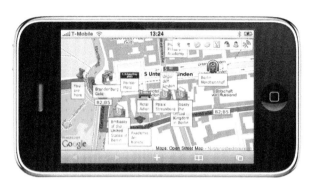

FIGURE 11.2 Results of Powerset search "What did Van Gogh paint?"

FIGURE 11.3 DBPedia Mobile, displaying information from multiple linked data sources on the Brandenburg Gate, located in Berlin. (From Becker, C. and Bizer C. Exploring the Geospatial Semantic Web with DBPedia Mobile. *Journal of Web Semantic*, Vol 7(4) 278–286. Elsevier 2009. With permission.)

FIGURE 11.6 TopBraid Composer visualization of the food ontology (http://www.w3.org/TR/2003/PR-owl-guide-20031209/food) class hierarchy.

FIGURE 11.9 Data-gov Wiki application: Worldwide earthquake viewer.

FIGURE 11.10   Data-gov Wiki application: Comparing U.S. (USAID) and U.K. (DFID) global foreign aid.

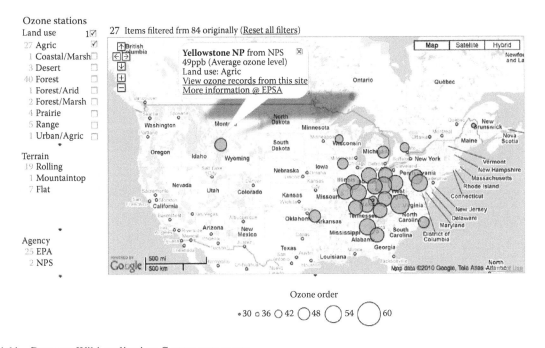

FIGURE 11.11   Data-gov Wiki application: Castnet ozone map.

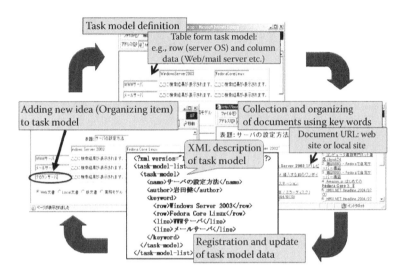

FIGURE 12.10   Task model sharing using an example of a table form.

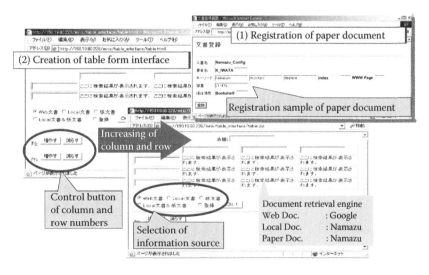

FIGURE 12.11   (1) Registration of paper documents and (2) interface based on table form task model.

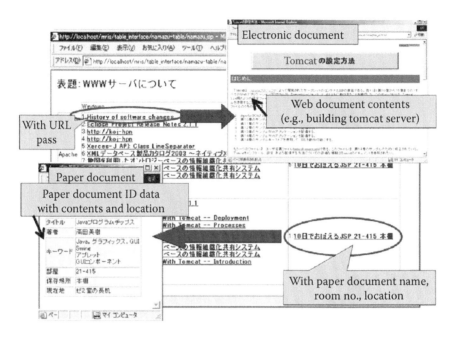

FIGURE 12.12   Example of document retrieval result.

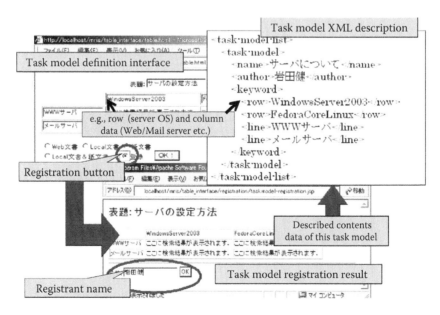

FIGURE 12.13   Registration of task model.

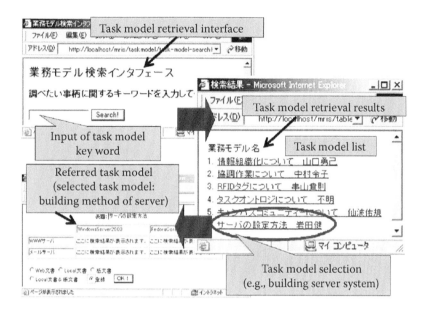

FIGURE 12.14 Retrieval and referencing of task model.

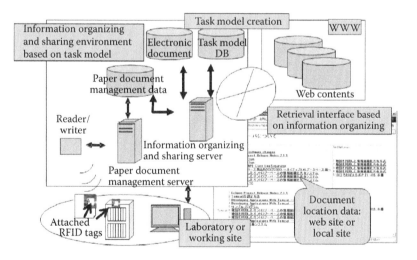

FIGURE 12.15 Seamless document management systems overview.

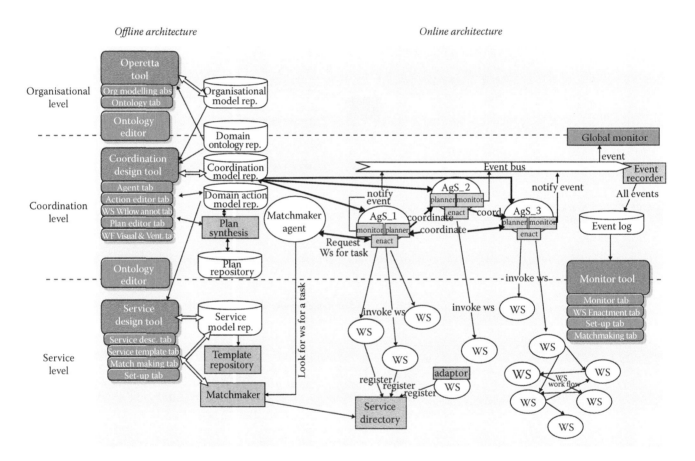

FIGURE 13.8 The ALIVE toolset.

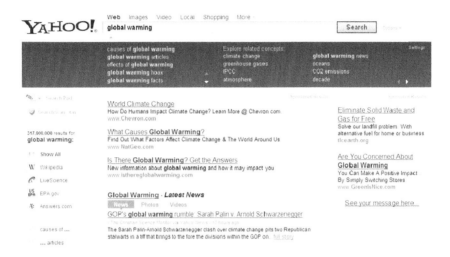

FIGURE 14.3 Screenshot of Yahoo!'s Search Assist interface with the "sliding tray" including dynamic term suggestions (in the left column) and related concepts (in the middle and right columns) for the query "global warming."

FIGURE 14.5 Screenshot of the MrTaggy user interface with (left) related tags section, (right) search results list displaying tags related to the search results, and upward and downward thumbs to provide relevance feedback.

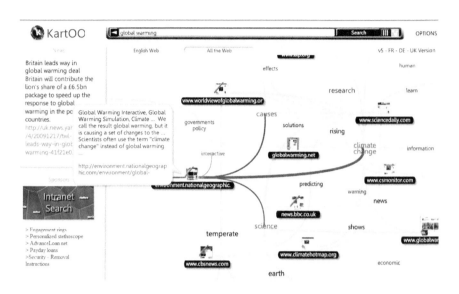

FIGURE 14.9 Screenshot from the graphical-overview interface of the metasearch engine Kartoo for the query "global warming."

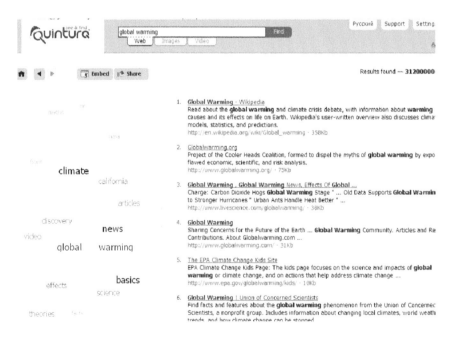

FIGURE 14.10 Screenshot from the tag cloud overview interface of the search engine Quintura for the query "global warming."

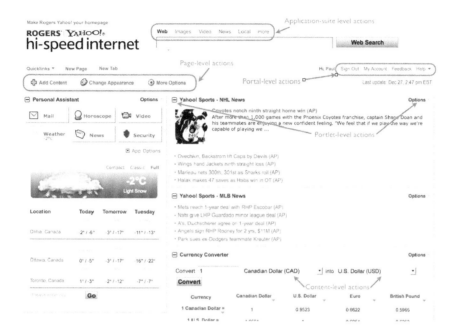

FIGURE 15.1 Portal page showing four levels of user interaction.

FIGURE 15.3 *My msn* personal portal Home page.

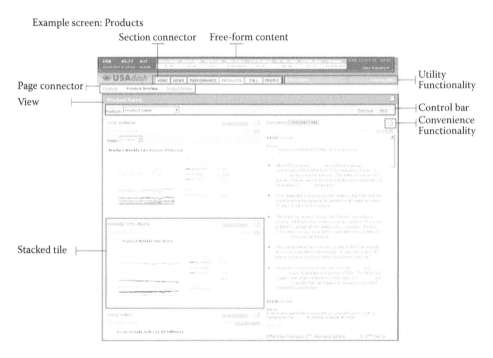

Example screen: Products

Section connector   Free-form content

Page connector

View

Utility
Functionality

Control bar
Convenience
Functionality

Stacked tile

FIGURE 15.10 Sample portal page with blocks and connectors. (From Lamantia, J. 2006. The challenge of dashboards and portals. Boxes and Arrows. http://www.boxesandarrows.com/view/the-challenge-of. With permission.)

FIGURE 15.13 Maximized portlet, feigning appearance of super-imposed window.

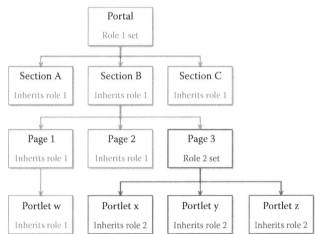

FIGURE 15.15 Inheritance of role assignments for securing portal blocks.

FIGURE 19.4 Screen capture showing insufficient text contrast.

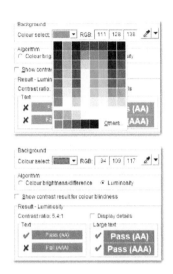

FIGURE 19.5 Colour Contrast Analyser with color picker.

FIGURE 19.6 RNIB screen capture in full color.

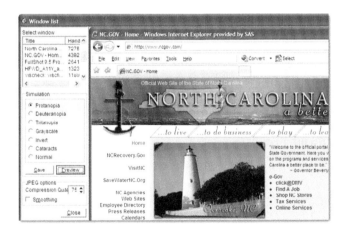

FIGURE 19.7 RNIB with red–green simulation.

FIGURE 19.8 Colour Contrast Analyser—color blind simulation.

(a)

(b)

FIGURE 21.3 Examples of (a) standard usability laboratory test setup and (b) remote usability laboratory test viewed on projection screen.

FIGURE 21.4 Computer monitor with WebCam for user testing.

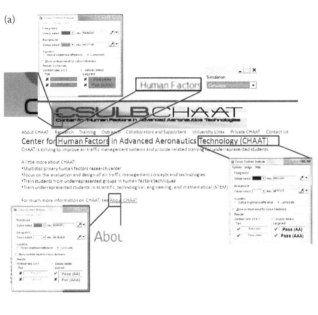

FIGURE 21.6 (a) Use of the Web accessibility toolbar to test a Web site for accessibility concerns. Specific windows were used to test for contrast and to simulate cataracts. (b) Use of the Web accessibility toolbar to test for color deficiencies and color schemes. Courtesy of Ariana Kiken.

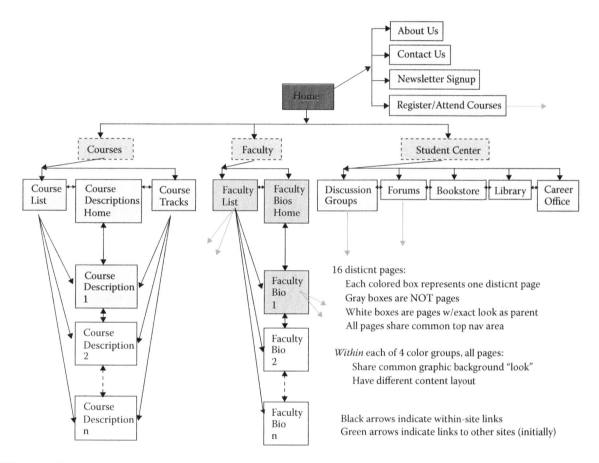

FIGURE 22.9 Case study Web site: initial information architecture specification—page flowchart.

FIGURE 22.10   Case study Web site: information architecture wireframe of the Home page.

FIGURE 22.11   Case study Web site: information architecture wireframe of an internal page.

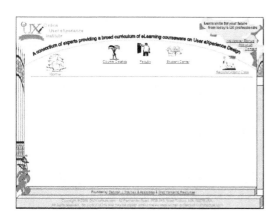

FIGURE 22.14   Case study Web site: early stakeholder mock-up of the vision for graphic design.

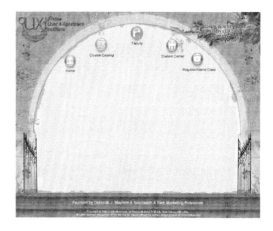

FIGURE 22.15   Case study Web site: early iteration of Home page by graphic designer.

FIGURE 22.16   Case study Web site: mock-up providing feedback from stakeholders on early Home page graphic design.

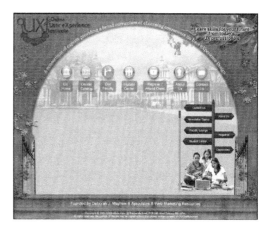

FIGURE 22.17   Case study Web site: iteration of the Home page by the graphic designer.

FIGURE 22.18 Case study Web site: later Home page iteration by the graphic designer.

FIGURE 22.19 Case study Web site: iteration of the Home page callout by the graphic designer.

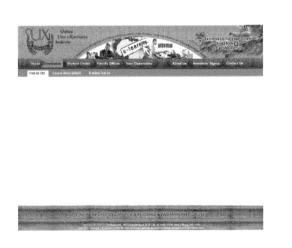

FIGURE 22.20 Case study Web site: later internal page by the graphic designer.

FIGURE 22.23 Case study Web site: page design standards wireframe—Home page above the fold.

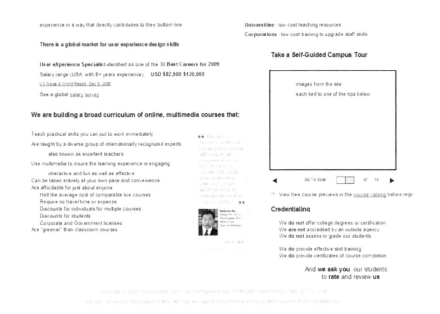

FIGURE 22.24 Case study Web site: page design standards wireframe—Home page below the fold.

FIGURE 22.25 Case study Web site: page design standards wireframe—internal page above the fold.

FIGURE 22.26 Case study Web site: page design standards wireframe—internal page below the fold.

FIGURE 22.27 Case study Web site: late page design standards graphic design—Home page.

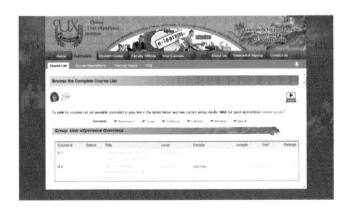

FIGURE 22.28 Case study Web site: late page design standards graphic design—Home page content callout.

FIGURE 22.29 Case study Web site: late page design standards graphic design—internal page template.

Page standards style guide			
Element	Style	Look	Behavior
Internal page body background	White		
Top level content header	Gray background fading into brick	Browse the Complete Course List	
	Bold black text, Arial 14		
Second level content header	Dark green background fading into leaves	Group: Introductory courses	
	Bold black text, Arial 12		
Text links to internal pages	Traditional blue text, no underline	Link	Links to internal page
Text links to external sites	Superscript right pointing caret at end of link text, also blue	Link^	External sites should open in separate tabs/windows rather than overlay OUXI.com pages
Tables	Light gray background with bold teal Arial 12 text for column caption row	Courses Course Title	Sortable by clicking on captions
	White background for other rows with normal black text	R 2 Learning Styles	
	Gray internal cell border lines		
	12 point Arial black normal text in cells		
Download links	Button, dark teal background, bold white text Arial 12	DOWNLOAD pdf of complete course catalog	Takes on text underline when pointing to it, initiates download
Text representing instructions	Preceded by a superscript sized label "Tip" in bold dark teal, otherwise normal text	^Instructions	
Callouts - header/title	Gray background fading into brick	Why should you consider User Design training?	
	14 point Arial bold black text		
Callouts - sub headers	A leaf as bullet point, text in 12 point dark green bold Arial font	New World tips the balance and	
Callouts - body text	12 point normal Arial font	"To gauge the quality of a experiences, Forrester F reviewed 1,200 web adve	

FIGURE 22.30 Case study Web site: page design standards style guide.

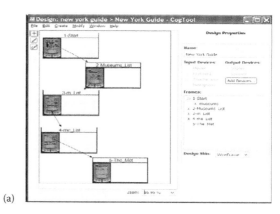

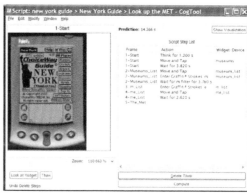

FIGURE 22.31 Case study Web site: a page from the detailed UX design.

FIGURE 23.6 Screenshots of CogTool for keystroke-level models of a task: (a) storyboard and (b) task script and computation of task performance time.

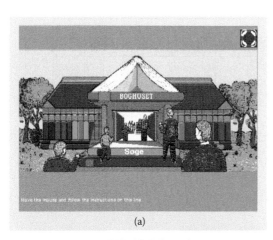

(a)

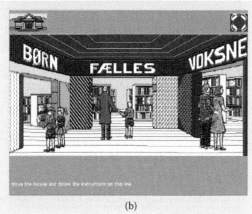

(b)

FIGURE 24.7 The hierarchically nested spatial structure provided in the BookHouse interface. (a) The global metaphor: a virtual library. (b) One of several intermediate metaphors corresponding to locations in the virtual library: the hallway used to select a portion of the database.

(a)

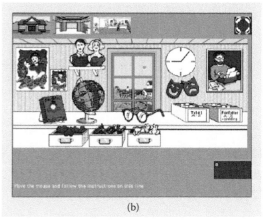

(b)

FIGURE 24.8 Intermediate and local metaphors in the BookHouse interface. (a) The room used to specify a database search strategy: each person/activity combination represents a different strategy. (b) Another "scene" from the Bookhouse. The local metaphors (i.e., the objects in the scene) represent tools for specifying dimensions in the AMP scheme form categorizing fiction. These tools can be used to "analytically" narrow the search to identify potentially interesting books.

FIGURE 27.1   The Web experiment and Web survey lists at http://wexlist.net.

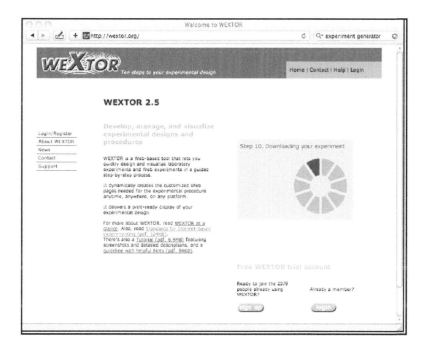

FIGURE 27.4   The entry page to WEXTOR at http://wextor.org.

FIGURE 27.8 The second of three clicks to turn on the Mac Server is to click on "Sharing" in the System preferences.

FIGURE 28.1 BareNecessities.com product page. (2010 © BareWeb, Inc., Bare Necessities®. Used with permission.)

FIGURE 28.2 MyShape "Personal Shop™" personalized shopping. (2010 © MyShape, Inc. All rights reserved. Used with permission.)

FIGURE 34.3 Cover flow interaction on Apple iPhone.

FIGURE 34.4 Map application on Blackberry Storm.

FIGURE 34.5 Zoomed view of a Web page on a 480 × 360-pixel touch screen.

FIGURE 35.1 THQ*ICE, Dragonica Online and their respective logos are trademarks and/or registered trademarks of THQ*ICE. All rights reserved. All other trademarks, logos, and copyrights are property of their respective owners. © 2009 THQ*ICE.

FIGURE 35.2 Warhammer 40,000: Dawn of War II © Games Workshop Limited 2009. Dawn of War, Games Workshop, Warhammer, 40K, the foregoing marks' respective logos and all associated marks, are either ®, TM and/or © Games Workshop Ltd 2000–2009. Used under license. All rights reserved.

FIGURE 35.3 Developed by Volition, Inc. THQ, Volition, Inc., Red Faction: Guerrilla and their respective logos are trademarks and/or registered trademarks of THQ Inc. All rights reserved. All other trademarks, logos, and copyrights are property of their respective owners. © 2009 THQ Inc.

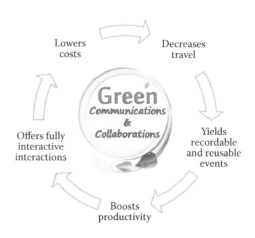

FIGURE 36.1 Organizations see the benefits of green communications and collaborations.

28 Designing E-Commerce User Interfaces

Lawrence J. Najjar

CONTENTS

28.1 INTRODUCTION

E-commerce is growing rapidly. In the United States from 2002 to 2007, online sales increased at a rate of over 23% each year (U.S. Census Bureau 2009) and annual e-commerce sales are now about $204 billion (Nielsen 2008b). The percentage of U.S. retail sales that are made online also increased from 2.8% to 3.2% (U.S. Census Bureau 2009). By 2012, U.S. e-commerce annual sales are expected to be $229 billion (Poggi 2009).

Worldwide, 85% of people with Internet connections have made an online purchase, representing a 40% increase in the past 2 years (Nielsen 2008a). More than half of Internet users now purchase from an e-commerce site at least once a month (Nielsen 2008a). In 2012, almost $1 trillion may be spent annually worldwide on e-commerce purchases (BuddeComm 2008).

By providing quick, convenient access to products, product information, and prices, the Web is turning products into commodities. The ease of use of e-commerce sites is a way to increase customer satisfaction, differentiate a site, increase market share, and enhance a brand (Manning, McCarthy, and Souza 1998; Tsai 2009a). Seventy-nine percent of users named easy navigation as the most important characteristic of an e-commerce site (Lake 2000). Poor navigation and a long and confusing checkout process were some of the reasons people gave for abandoning their shopping carts (Global Millennia Marketing 2002).

When an e-commerce site is easy to use, sales can increase. For example, after improving usability, a major retailer got a US$3,000,000 improvement in annual revenue (Spool 2009), IBM got a 400% increase in sales on IBM.com (Battey 1999; Tedeschi 1999), and Digital Equipment Corporation reported an 80% increase in revenue (Wixon and Jones 1992). By making the products easier to access, removing unnecessary graphics, and making product information easier to scan, Liz Claiborne's Elisabeth.com tripled the rate at which lookers became buyers (Tedeschi 2002a).

Competing sites are only a click away. An e-commerce site that is easy to use can build consumer loyalty (Najjar 1999). Loyalty is essential. Two of every three online sales are made by consumers who knew where they wanted to make their purchases (ActivMedia Research 2000). Sixty percent of users mostly buy from the same e-commerce site (Nielsen 2008a). For example, on Zappos.com, 75% of purchases are from returning customers (Wroblewski 2009). Once users start buying online, they increase the amount of their purchases each year (ePaynews 2003a, 2003b; Hansell 2002).

28.2 PROCESS

A good design process can be more important than a good designer. A good design process is driven by the needs and preferences of users, identifies and controls requirements so design objectives stay consistent, involves users and clients early and throughout the process, and uses repeating design—user feedback—improve design cycles to maximize usability (see Mayhew, this volume). To design efficiently, start with high-level design and work to low-level design, use tools that allow you to make design changes quickly and easily, and communicate your designs using media that allow users and clients to immediately understand the user interface (Najjar 2002).

The following efficient steps for user interface design evolved from many years of doing commercial user interface design work under significant time pressures. These steps emphasize getting high-quality work done in a short period of time. Interaction designers, rather than visual designers, typically perform this work.

1. Define the purpose: Identify the purpose for the site (e.g., generate revenue, change the brand's image). Determine how the e-commerce site will be different from and better than competing sites. Because the purpose drives the design, document the purpose in a sentence or two and get it approved by the clients.

2. Define the users: Identify the users, their objectives, priorities, terminology, contexts (e.g., home, work), computers, display sizes, display resolutions, browsers, and connection speeds. Design to meet the needs of the users.

3. Define the functional requirements: Identify and prioritize the functions users want on the site. Use tools such as focus groups, interviews (including interviews with your clients), competitive assessments, and contextual inquiries to gather functional requirements. For example, "Registration" may be a functional requirement. Prioritize the functions using criteria that include value to the user, value to the business, differentiation from competitors' sites, and ease of implementation. Include some functions that may get low priority, but are essential (e.g., "Contact Us"). If there are too many functions for the planned project schedule, move some of the functions into later projects.

4. Write use cases: Break down the functions into more detailed user tasks. For example, break down the "Registration" function into use cases for "Show confirmation," "Edit registration," "Remove registration," and "Show error message." Prioritize the use cases. Working with software engineers, identify systems (such as databases) that are involved with each task. Use this information to scope the project and to move some use cases into future projects.

5. Develop site structure diagrams: Using products such as Microsoft Visio or the Omni Group's OmniGraffle, draw diagrams with boxes and arrows that show how the sections and subsections of the application are organized and named. If there are more than about 25 sections, use an indented list instead of boxes and arrows. Review and make changes based on feedback from the clients.

6. Create static page drawings: Because they are quick, easy, and cheap to change, use a tool like Visio and its Software Common Controls shapes to draw the interaction design of most screens. Do not show the visual design. Your goal is to figure out how the e-commerce site works, not how it looks. Form follows function (American Heritage Editors 2000). Define the functional user interface first, and then design the visual interface. Document most of the use cases with a series of page drawings. Show which user interface controls and what information is on each page. Put interaction notes in the margins of the drawings via callouts. Improve the drawings with frequent design reviews with the client and domain experts.

7. Get user feedback: To get early feedback on the interaction design, set up reviews with representative users. Print out the page drawings without the callouts. Ask users to perform about five typical use case tasks and to tell you what they would do on the paper pages to perform each task. Request that users think out loud so you can get more ideas to improve the interaction design. Ask what they like, don't like, and would change. Use the early user feedback to improve the drawings.

8. Build interactive mockups: Interactive user interface mockups are a powerful way to show the proposed design to clients. If the schedule allows, develop an interactive, hypertext markup language (HTML) mockup that shows the organization of the application, the functions on each page, and how each function works (Najjar 2000). Show functionality. Show entry fields, buttons, drop-down menus, hyperlinks, confirmation windows, and error messages. Do not show graphics (Fuccella and Pizzolato 1999) because they distract reviewers from the functional design and take too long to create and change.

Do not take the time to create robust, complete, professional HTML. Instead, use an HTML authoring shell such as Macromedia Dreamweaver. Show fake data. Do not connect the HTML to actual databases. Like the page drawings, the purpose of the interactive mockup is to show clients how the site may work, not how it may look.

Conduct iterative design reviews of every representative page of the mockup with the clients. Put the interactive mockup behind a firewall so that remote clients can access it during teleconference design reviews, examine it in detail, and show it to their colleagues. Once the clients approve the first two or three pages of the mockup, the visual designers can create their visual compositions. The "comps" are based on the approved interaction design but use the company's approved color palette, logos, and images to add emotion and branding.

9. Test the usability of the interactive mockups: Add graphics to several typical, important, or hard-to-use pathways in the mockup. Get representative users and test the usability of the mockup. Take this user feedback and improve the interactive mockup. Ideally, perform this test-redesign cycle several times before programmers write a single line of code. That way, you know the user interface is good and the user interface design drives the programming.

10. Write functional design specifications: To allow the front-end programmers to do their work, capture an image of each page in the interactive mockup and place it into a Microsoft Word document. List each of the controls on the page. Then describe how each control works (e.g., default state, available choices, result of control activation, changes that occur if users are registered, changes that occur when users perform each action, error messages). Work with your programmers to make sure you give them the information they need to bring the design to life. Because you quickly and cheaply iterated the design based on feedback from users and clients, the programmers should be able to do their work right the first time.

11. Perform user acceptance tests: Perform a usability test of the final version of the site. Connect the site to databases. Add the graphics. Look carefully at download times and the alignment of images and text in different browsers. Make sure there are no user interface dead-ends. Because you already iterated the design with representative users and the clients, this test should go well. Make any needed changes to the final code, then "go live."

Other steps you might want to add to the process include reviews of competitor's sites and a usability expert review of the existing site. After the site is live, regularly review site metrics (such as where users exit the site) to identify opportunities for improvement.

28.3 DESIGN

Designing the user interface for an e-commerce site is very challenging. E-commerce sites must accommodate nearly all users, include a significant amount of user interactivity, and still be easy to use. Major sections to design include the overall page format (see Tullis, Tranquada, and Siegel, this volume), navigation, catalog, registration, personalization, checkout, and customer service. Also, use social media, social networking, search engine optimization, and site analytics

to promote the site and make it more attractive to potential shoppers. Design for users with disabilities.

28.3.1 PAGE FORMAT

Do not use plug-ins like Flash (Ragus 2000b, 2000c; Nielsen 2000) that users have to download. Do not use Java because it requires users to download a special runtime environment and to enable Java, plus there are Java compatibility issues with different Web browsers.

Horizontal scrolling is annoying and makes users work too hard. Except for a product comparison tool, never require users to scroll horizontally. Avoid forcing users to scroll vertically on the Home page. However, it is acceptable to put closely related information (e.g., product details, checkout fields) on a vertically scrollable page (Sacharow and Mooradian 1999). Put important information "above the fold" (above the vertical scroll line) so users can see it immediately.

To make it easy for users to interact with the Web site, format the pages so that user interface elements are in familiar locations (Bernard 2001). Put the return-to-Home hyperlink in the top, left corner. To make it serve double duty as a branding element, use the company's logo as the hyperlink. Put global navigation controls for the major sections of the site across the top of the page (Pastore 1999). Place local navigation controls that work inside each major section along the left side of the page. Put cross-sell and up-sell promotions on the right side or across the bottom of each product page. Locate the search entry field and a toll-free customer support telephone number near the top of every page.

Users need to know what is in their shopping carts. Except for the shopping cart page and checkout pages, put a shopping cart summary on each page. In the shopping cart summary, show a short hyperlinked product name for each item, the quantity of each item, the price, and a cost subtotal (Chaparro 2001; Pastore 1999; Ragus 2000a). Also include a link to the complete shopping cart, the wish list, and the checkout.

To easily accommodate registered members, include sign-in entry fields on the Home page and a sign-in hyperlink on every page. Display the user's name (e.g., "John Doe") or sign-in name (e.g., "john.doe") inconspicuously near the top of the Home page so that registered users know they are recognized and are receiving member benefits such as express checkout. Be sure to put a "Contact Us" hyperlink on every page, perhaps at the bottom. A good way to attract new users is to put on each product page a link to a simple referral form in which users need to enter only sender and receiver names and e-mail addresses (e.g., Kohls.com's "E-mail to a friend") (Reichheld and Schefter 2000). E-commerce users are very concerned about providing their personal information online. To reduce this concern, provide links to your privacy and security policies at the bottom of every page, but also show the links more obviously where users are entering private information in the content area of the registration and checkout pages (Stanley, McCarthy, and Sharrard 2000).

Design the user interface to encourage purchases. On the Home page and the first page of each major section, include promotions for products such as products that were highly rated by your customers and will appeal to most of your visitors. Show the promotional product names, images, very short descriptions, and prices. Design the product names and images to link to the complete product description pages.

28.3.2 NAVIGATION

To make it easy for users to move through your site, design navigation that is simple, intuitive, and obvious. Put the navigation controls in the same locations on each page. Use navigation to tell users where they are, how they got there, and where else they can go (Fleming 1998). This is especially helpful to users who arrive at the page not from the Home page but via a search or hyperlink. Provide "breadcrumb" navigation on the site (Rogers and Chaparro 2003). Breadcrumbs are small, hyperlinked page titles at the top of each page, usually above the title of the current page. These hyperlinks show the location of the page in the site organization and often match page titles that users came through to get to the current page (e.g., "Home > Men's Apparel > Shirts >"). Like the breadcrumbs dropped by Hansel and Gretel (Grimm and Grimm 1999), the breadcrumb navigation controls allow users to easily retrace their steps.

To make the site more inviting, provide up to about seven intuitive names for the major sections of the site. Familiar names make it easier for users to quickly browse through the site to a desired product. These names serve as the global navigation controls. Design the navigation so that users can browse to any product in five clicks or fewer (Tracy 2000). It may be better to provide more category names at each level (a broad design) than to provide more levels to click through (a deep design) (Selingo 2000). On a product category landing page (such as "Men's Shirts"), allow users to filter products in a category by useful measures such as specific features, price ranges, brands, discounted products, and customer ratings. For example, DSW.com allows users to filter products by gender, clearance, luxury, size, width, color family, brand, price range, material, and heel height. Also, provide specialized browse functions to meet user needs. For example, RedEnvelope.com and 1-800-FLOWERS.com list products by "Occasion."

Search is an extremely important navigation technique. Forty-three percent of users (eMarketer 2001) said that search was the most important online shopping feature. Unfortunately, many e-commerce sites do not design, maintain, or even evaluate their search functions (Hagen, Manning, and Paul 2000). One study (Gunderson 2000) found that 42% of e-commerce sites had inadequate search functions. Another study (Nielsen and Tahir 2001) found that 36% of the time users could not find what they were looking for when they used the search function on 25 e-commerce sites.

To improve search, use metatag tools, thesauruses, alternate spellings, and database search engines. For example, the advanced search function on TowerRecords.com allows users to enter a variety of information (e.g., title, cast/crew). The Tower Records search engine doubled the rate at which users

made purchases (Guernsey 2001). Allow users to search by product name, product category, brand, model/item number, and price (Consumer Reports 2003). To make it easy for shoppers to focus their search on one area of your site, to the left or right of the search field display a drop-down menu that lists the major sections of the site like Walmart.com does. If your site includes products that are very similar, such as electronics or shoes, provide a filter to focus search. For example, Zappos.com has a helpful page that allows users to search for shoes by specific style, size, width, color, and price.

To make the site convenient for users, always try to get the most relevant hits in the first page of search results. On the search results page, display the searched-for keywords and allow users to perform another search, refine the search results, and sort the search results using helpful product attributes such as price range (e.g., NetGrocer.com), style, size, and customer review rating. Let users filter the results using familiar factors such as best selling and department name. For example, Sidestep.com allows users to click on checkboxes to filter flight search results by number of stops, flight periods, airlines, and airports. To make it easier to browse the search results, let users click on a drop-down menu to increase the number of items displayed on each search results page (e.g., 25, 50, 100, All).

28.3.3 CATALOG

Make it easy for users to see products. Never require users to register to see the product catalog. Avoid requiring users to select a city or enter a zip code to see a product catalog or the availability of a product (e.g., Lowes.com requires users to enter a zip code to see product prices and availability). Instead, let users get directly into the product catalog without performing extra steps. Allow users to immediately see product prices and availability (e.g., HomeDepot.com). To make it easy for users to browse for products, organize the product catalog the way users expect the products to be organized (e.g., organize clothing by gender). Because users only look at the first two or three pages of a list of products, allow users to narrow down a long list of products by using helpful filtering tools (Nielsen and Tahir 2001). For example, allow users to reduce a list of shoes by showing only a particular shoe size or style. Provide a way to sort the products by highest customer product rating.

Users spend 6 minutes or more on helpful product pages (Freedman 2008a). On each product page, show a small image of the product, provide links to a larger image and other images, and allow users to zoom in and move the image to see product details (Freedman 2008a). Possibly using tabs below the product image, display a textual product summary, specifications, and customer reviews and ratings. Provide engaging how-to videos, ideally associated with specific products as HomeDepot.com does.

To avoid frustrating users, show only products that are in stock. If you cannot do this, clearly identify the color or product versions that are out of stock. Offer to send an e-mail note to users when the product is available. To prevent user confusion, never allow users to put out-of-stock products into the shopping cart.

To make them easy to find on product pages, display graphical "Add to shopping cart" and "Check out" buttons with larger fonts and brighter colors. When users put a product into the shopping cart, update the cart summary at the top of the page and provide a confirmation window or expand-down window (such as on Gap.com) with a message like "You successfully put the product into your shopping cart." In case members prefer to make a purchase later, display a link on the product page below the "Add to shopping cart" button to move the product into a member wish list. Also, to improve sales, provide a link that allows users to e-mail the page to someone else who may purchase the product. To encourage comparison shoppers to complete the purchase on your site, provide regular gift-with-purchase promotions as BarnesandNoble.com does. Using personalization information, show tailored product suggestions or one related cross-sell, one more expensive up-sell product, or related products on each product page. Shoppers love a bargain. If the product is on sale, show the original price, the current price, and the cost savings. To make it easy for users to make a purchase decision, display a link to the site's guarantee (e.g., 30 day money-back guarantee), describe any promotions (such as free shipping for purchases over $50 or a limited-time offer), and show an estimated shipping cost for the product. Because the inability to talk to someone to get questions answered was the biggest online shopping frustration (Opinion Research Corporation 2009), include a toll-free customer support telephone number so users can get answers to any questions that may delay their purchases.

BareNecessities.com has a good product page (Figure 28.1). The page shows an attractive image of the product, several alternative images, image zoom, available colors, product details, the price, cost savings, a size chart, other matching products, and cross-sell products. There are links to send the page to a friend, the return policy, shipping information, and privacy and security measures. The page also shows a toll-free customer support number, the number of items in the shopping cart, and a link to the full shopping cart.

The Crutchfield.com electronics product page shows the number of customer service representatives currently available to take a toll-free call, plus add to cart, add to wish list, free scheduled shipping, payment options, and a tremendous amount of helpful product information including customer reviews and "What's in the box."

Make the member wish list work like the shopping cart. In the wish list, show product names and thumbnail images that link back to the product description pages. Allow users to change product quantities, remove products, update the wish list, and move products from the wish list to the shopping cart.

Since 34% of users say a product comparison tool increases the chances that they will buy from a site (eMarketer 2001), allow users to select and compare products side-by-side on important, differentiating features (e.g., eBags.com's compare selected items tool). Do not limit the number of products

FIGURE 28.1 (**See color insert.**) BareNecessities.com product page. (2010 © BareWeb, Inc., Bare Necessities®. Used with permission.)

that can be compared. Provide links from the comparison back to each product's detail page. Make it easy for users to remove products from the comparison.

To make the site more helpful, interesting, and "sticky," provide customer reviews via tools like Bazaarvoice and PowerReviews. Nearly half of users look at customer product reviews when making a purchase (Palmer 2009). Allow users to enter simple product ratings (such as 1 to 5 stars), comments on products—both positive and negative, and to rate the value of other comments. Remove only offensive entries. Show the number of reviews for a product and the average rating. Let users sort ratings various ways (such as newest, oldest, highest, lowest, highest value rating), find other product reviews on the site by the same author, and search the reviews (such as for mention of a particular product feature). Allow users to flag inappropriate content. Promote the highest rated products via special sales or a product category (e.g., "Top-rated products" on REI.com).

Many shoppers research products online before making a purchase. One study found that 79% of online shoppers rarely or never make a purchase without getting product information first (Schiff, n.d.). If the product information is poor, 72% of online shoppers will go to a competing site to get the product details they want. So, provide detailed product specifications on each catalog page, but reduce visual clutter by using a separate tab to display them.

Because gift certificates are very popular, include them in the product catalog. Allow customers to e-mail gift certificates and to mail plastic stored value gift cards. Gift certificates can be very profitable because about 15% of gift certificates do not get redeemed (Tedeschi 2002c). One study (Nielsen 2007) found that 30% of e-mailed gift certificates were deleted as spam.

28.3.4 REGISTRATION

The more streamlined the registration process, the more likely users will register and buy (Agrawal, Arjona, and Lemmens 2001). Even when they are on it, most users will not read or complete the registration page, and half the users leave on each succeeding registration page (Sacharow and Mooradian 1999). To reduce the number of registration user entries and to make it easier for users to remember their sign-in names, require users to enter only an e-mail address and a password. Ask users to make explicit clicks on checkboxes to give permission to send e-mail notifications (e.g., sales, new products) (Charron et al. 1998) and to leave a cookie (e.g., "Remember me when I return"). Provide a link to the privacy and security policies.

Gather other user information during checkout. If you require users to register before checking out, 34% of them may abandon their shopping carts (Goldwyn, n.d.). Instead, let unregistered users enter shipping and billing information during checkout. At the end of checkout, tell users the benefits of registering (e.g., quicker future checkouts, personalization, order status, wish list, sale notifications), ask unregistered users to register by utilizing the information they just entered during checkout, and ask unregistered users to provide only a password (Nielsen 1999).

Give users control of their personal information. Allow users to edit the registration information and to unregister. For security reasons, never show the entire credit card number; show only *s and the last four digits.

28.3.5 PERSONALIZATION

Passive personalization is based on the user's registration information, purchase history, and browsing history. Active personalization asks users to answer questions like clothing size and style preferences. With this information, you can tailor the e-commerce site for each user. Personalization is very powerful. It can provide a more compelling user experience, reduce an overwhelming number of choices (e.g., reduce the 100,000 Netflix.com movies to a subset that is likely to interest the user), increase sales (Grau 2009), and lead to higher conversion rates, more repeat visits, greater loyalty, and stronger brands (Agrawal, Arjona, and Lemmens

2001; Cooperstein et al. 1999; Freedman 2008b; Reichheld and Schefter 2000).

Display the user's name at the top of the page. Personalize the suggested products on the Home page, category pages, and product pages. List the products the user viewed most recently. One site (BirkenstockExpress.com) lets users see the shoes that are available in the users' style, gender, arch, material, color, and size. As shown in Figure 28.2, the MyShape.com fashion site allows users to record their style preferences, body measurements, and fit preferences then suggests tailored wardrobe tips, provides a "Personal Shop," and displays search results that match the user's profile.

To make them more effective and less annoying, personalize the e-mail notifications you send to registered users. Send notifications only for products and services that interest each registered user. Allow users to forward the promotional e-mail to a friend or their social network (eMarketer 2009b). Also, reward members by sending them unique offers. For example, the eBags reward club sends five offers per year to members. To comply with the U.S. Controlling the Assault of Non-Solicited Pornography and Marketing Act of 2003 (CAN-SPAM), allow recipients to opt out of receiving your promotional e-mails and remove them from your distribution list within 10 days (Federal Trade Commission 2008). In Europe, recipients must opt in to receive promotional e-mails (L-Soft, n.d.).

Consider providing a way for users to be part of a community (e.g., Wine.com's free "Community" and fee-based "Wine Clubs"). This feature may improve user loyalty, user time on

FIGURE 28.2 (**See color insert.**) MyShape "Personal Shop™" personalized shopping. (2010 © MyShape, Inc. All rights reserved. Used with permission.)

the site, and site revenue. After users join an online community, offer tailored information, tools, and message boards.

28.3.6 Checkout

On the shopping cart page, provide users all the information they need to complete the purchase. Show hyperlinked product names and thumbnail images so users can link back to see product details. Show entry fields with quantities that users can change. Provide links for removing products (use "Remove" rather than forcing users to enter a quantity of "0" for a product) and moving products into the user's wish list. Place convenient buttons above and below the shopping cart list to refresh the cart page. Display the price of each product and a total. Users are more likely to complete checkout if they know the shipping costs before checking out (Campanelli 2009; Hill 2001). So, allow users to enter a zip code and provide a drop-down menu of shipping choices (such as FedEx, UPS, and USPS), delivery speeds (such as next day, 2 business days, 3 business days, and regular 7 to 10 business days), and costs. A sample shipping menu choice might be "FedEx Ground, 5 business days, US$5.00, Continental US only." To encourage users to increase their order sizes and to complete their purchases, offer free shipping for purchases over a specific amount, such as $50. If you include temporary discount coupon codes in your marketing materials, display a field for users to enter discount coupon codes. Allow users to recompute and display the order total (including accurate shipping costs, discount codes, and taxes). Do not show cross-sells or up-sells on shopping cart and checkout pages because they clutter the page, take users away from the checkout process, and may distract users from completing their purchases.

Include a "Continue shopping" button that returns users to the page from which they came to the shopping cart. Display a graphical "Check out" button with large font and bright colors. Also provide reassuring links to pages that describe your guarantees, return policy, and privacy and security measures.

If registered members leave the site without checking out, automatically save their shopping cart contents for up to 90 days (e.g., Amazon.com). To encourage users to purchase the items they left in their shopping carts, send an e-mail alert when an item is close to being sold out, as Zappos.com and Overtock.com do.

In one study (Rehman 2000), checkout was the reason 40% of users failed to complete an online order. Twenty-seven percent of users abandoned an order because the site required them to complete cumbersome forms (Sacharow and Mooradian 1999). So, as described in the "Registration" section, do not require users to register before checking out (Rehman 2000).

On the checkout page, provide entry fields so registered members can conveniently sign in. If the users are signed-in members, automatically fill in the checkout fields using the members' registration information. If the billing address is the same as the shipping address, do not require new users to enter the same information again in the billing address fields.

Instead, allow users to click a checkbox near the billing address for "Use shipping address for billing address." Then, to keep the user interface very simple and obvious, refresh the page and show the shipping information in the billing address fields.

Try to put all the checkout fields on a single, vertically scrollable page. This reduces user confusion and makes it easier to change the order. To allow users to review and change their selected products, provide a link from the checkout page back to the shopping cart page. If your e-commerce company also has a brick-and-mortar store, provide a checkbox so customers on the e-commerce site can elect to pick up online orders at the nearest store (e.g., REI.com, Sears .com) and avoid shipping charges (Tedeschi 2002b). Display the shipping drop-down menu and the choice the user made on the shopping cart page so users can change their shipping choices. Be sure to ask for an e-mail address so you can send an order confirmation.

Identify required fields with a red asterisk and a footnote explaining that the fields with red asterisks are required. When users make an error or leave empty a required field, show the checkout fields page, put an obvious error message at the top (e.g., "We had a problem processing this page. Will you please fix the fields that are marked with a red 'Problem?' "), then above each field that had an error show red "Problem" text and briefly explain what the trouble was and how to fix it.

Make it easy to take users' money. Provide easy, safe, and reliable ways to pay. One study found that sites offering four or more ways to pay in addition to credit cards increased their conversion rates (Ward, n.d.). Design checkout to accept several payment methods (e.g., Bill Me Later, credit cards, debit cards, eBillme walk-in, eCheck, e-mailed gift certificates, Google checkout, PayPal, and stored value gift cards). Allow users to place and pay for an order via a toll-free telephone number. For credit cards that require a security code, display a link to a popup window that explains via text and images where to find the security code on different cards.

Twenty percent of users said they stopped an online purchase because they felt the site was not secure (Hill 2001). Get the site's privacy and security certified by consumer groups, such as TRUSTe or BBBOnLine, and show their logos on the checkout page. Use Secure Socket Layer (SSL) to encrypt sensitive personal information. Also, provide links that promote consumer protection features such as privacy policy, security protection, a no-questions-asked return policy, delivery guarantees, and customer service e-mail response time guarantees (Agrawal, Arjona, and Lemmens 2001; Rhodes 1998).

After users enter their checkout information, provide a complete, read-only purchase summary. If users want to make a change, provide a "Change" button that takes users back to the single, vertically scrollable, editable, checkout information page. To avoid forcing users to retype information, retain the information users entered earlier on the checkout information page. After the system accepts the order, provide an order confirmation that includes the products, product quantity, cost breakdown, cost total, order number, payment method, last four digits of payment card (if applicable), instructions for canceling the order, directions for tracking the order and shipment, expected delivery date, customer service information (e.g., e-mail, telephone number), and a way to print the page (Ragus 2000a). Ask unregistered users if they would like to register. Send a confirmation e-mail that includes all the order confirmation information. After the product ships, send another notification with the planned delivery date, return/exchange information, and a link to the shipper's delivery tracking status for the product.

28.3.7 Customer Service

Design the site to provide outstanding customer service. The best support is available 24 hours a day, 7 days a week, in real-time (Agrawal, Arjona, and Lemmens 2001; InternetWeek 2002). Display a toll-free customer service telephone number at the top of every page. Provide FAQs (questions and answers to frequently asked questions) on topics such as checkout, privacy, security, and returns.

On the "Contact Us" page, provide the toll-free telephone number, e-mail address, live chat (e.g., Godiva.com Live Assistance), "call me now" function that allows a user to request a call from customer service (e.g., Amazon.com Help > Contact Us > Customer Service > Phone), facsimile number, and mailing address (not a post office box). Users prefer live chat to customer service for simpler questions on topics such as order status or shipping choices (eMarketer 2009a). For complicated or sensitive information, technical support, and questions when large amounts of money are involved, users prefer to talk to someone on the phone. On the "Contact Us" page, make your site more personal by showing a photo of the support staff and the building.

Make returns as easy as possible. If the company has an online site and brick-and-mortar stores, allow customers to return online purchases to the stores. With each shipped order, provide a preprinted shipping label (and possibly a mailing envelope) so it is easy for users to mail a return (e.g., Nordstrom.com). Zappos.com is the world leader on returns. They provide free shipping for returns 365 days from the purchase, free shipping on the next order, choice of shipper, and the ability to print out a return shipping label to use on the original shipping box.

28.3.8 Access for Disabled Users

About 10% of the world's population (or about 650 million people) are disabled (United Nations, n.d.). In the United States, 20% of the population experience some form of disability (McNeil 1997; U.S. Census Bureau 2000). Almost everyone will experience a temporary disability due to illness, accident (such as a broken arm), or circumstance (such as a loud or dim environment) (Martin 2002; McNeil 1997).

The Americans with Disabilities Act (U.S. Department of Justice 1990) (28 C.F.R. Sec. 36.303; 28 C.F.R. Sec. 35.160) and section 508 of the Rehabilitation Act Amendments of 1998 (General Services Administration 2003) require that electronic and information technology (such as Web pages and software applications) for U.S. federal sites be accessible to people with disabilities, unless doing so would impose an undue burden on the department or agency. Many state (Georgia Tech Research Institute 2006) and city governments have similar laws (Williams 2001). Other countries and international organizations also encourage or require accessibility of Web sites (International Organization for Standardization ISO/TS 16071: 2003; ISO 9241-171: 2008; ISO/IEC 24786: 2009; W3C, n.d.(a)).

However, in the United States, accessibility is not required for private, e-commerce Web sites (Schwartz 2002). Lawsuits were filed against sites such as Priceline.com, Ramada.com (International Center for Disability Resources on the Internet 2004), and Southwest.com (McCullagh 2002), but the rulings did not require other U.S. e-commerce sites to be accessible.

Accessibility is good business. By designing your site to be accessible (see Stephanidis and Akoumianakis, this volume; Caldwell and Vanderheiden, this volume) you increase the number of people who can purchase from the site. Plus, the simple designs required for accessibility often improve ease of use and work well on other devices such as Web-enabled cellular telephones and personal digital assistants. To improve accessibility, design user interfaces that conform to the World Wide Web Consortium's Web Content Accessibility Guidelines (W3C, 2008; W3C, n.d.(b)) and perform accessibility evaluations with representative users with disabilities.

28.3.9 Social Media

Social media includes customer reviews, product ratings, blogs, and discussion boards. Use social media to improve your relationship with customers, identify opportunities to improve products, and enhance your brand.

People trust their peers. Reviews, ratings, and discussions by other customers can be perceived as more trustworthy than information from the retailer.

Create and frequently update a blog with interesting information and limited offers. Instead of asking the CEO to write the blog entries, ask the staff to share their tips (e.g., Walmart's checkoutblog).

To maintain that trust, do not compensate bloggers from independent sites for their opinions. If you provide free product samples to bloggers, insist that the bloggers include that information in their postings. Do not censor customer reviews or discussion boards, except to remove offensive content.

Respond to positive and negative comments in selected product reviews and discussions. These responses show that you are listening, build a sense of community with your customers, encourage others to contribute, and enhance your brand (Alvarez 2008). Customer inputs can also be a great source for product improvements or new products.

Allow users to submit photos of themselves with your products (e.g., Patagonia.com) and video demonstrations of the products (e.g., Newegg.com), but screen the submissions before posting them. Display this user-submitted content in a separate product page tab (e.g., Description, Specifications, Reviews, Customer Photos). If you have a well-known brand, create a YouTube channel and populate it with professional-quality videos. Or, create a contest in which customers upload product videos onto a specific YouTube channel and viewers vote to determine the winner.

28.3.10 SOCIAL NETWORKING

Social networking is exploding in popularity. Sixty-five percent of worldwide Internet users aged 15 and older visited at least one social networking site during the month of a survey (ComScore 2009b). That was over 734 million people. Almost 20% of U.S. Internet users interact with Twitter or another service to send and receive updates (Fox, Zickuhr, and Smith 2009). Use Twitter to send compelling promotional messages, such as Tiny URLs (Tiny.cc) and coupon codes for limited time or quantity discount products and to drive people to the e-commerce site. For example, Dell posted links to discounted products on a Twitter account and measured over $3 million in sales (Gonsalves 2009).

Facebook.com is the most popular social networking site and the seventh most visited Web site in the world (ComScore 2009a). Encourage people and their friends to build a relationship with your brand by building a Facebook page so you can share helpful information (such as special offers, photographs, and videos) and users can post comments, share their photographs with your products, participate in polls, and get their questions answered, as Patagonia and Victoria's Secret PINK do. Use Flash (e.g., Lacoste Parfums) or videos (WARN Video) to showcase your products. Include product descriptions, frequent links to your online store, a fan offer that displays an e-commerce discount code, and a tab that shows a small Facebook store (e.g., 1-800-FLOWERS Shop). Add links from your online store to your Facebook page and Twitter account.

28.3.11 SEARCH ENGINE OPTIMIZATION

People need to find you to buy from you. Design your e-commerce site to maximize matches with search engines. Your goal should be to get as high as you can on the first page of Google search results. For page titles, meta keywords, meta descriptions, section titles, product names, and product descriptions, use simple, familiar, meaningful phrases that prospective customers are likely to use in a search. For example, Diapers.com changed a section title from "Playtime" to "Toys and Books" (Tsai 2009c). Do not use the site name as the title for every page. Use full product names and model numbers. Encourage others to link to the site by providing useful content in employee expert blogs, product information, and customer reviews. Place customer reviews and their valuable search key words on the product pages rather than on a separate mini-site (Tsai 2009b). Write helpful articles and submit them to Ezinearticles.com and other online article directories (Practical eCommerce 2007).

28.3.12 ANALYTICS

Use analytics to improve site usability. About 73% of e-commerce businesses use Web analytics tools such as Google Analytics, Omniture, or WebTrends to understand user behavior on their sites (Tsai 2009d). Key performance indicators include abandonment (where users leave your site), page stickiness (the rate at which people stay on your site after arriving at a specific page), average page views, percent returning visitors, conversion rate, average order size, and number of new customer product reviews (Nimetz 2006, 2007). This information can help you identify opportunities to improve the user interface (such as on which page most visitors leave the site), to retain visitors, and to encourage visitors to make purchases.

28.4 CONCLUSIONS

To be successful, e-commerce user interfaces must be very easy to use. To improve usability, follow an efficient, iterative, user-centered design process. Use proven user interface designs for browsing and searching the product catalog, registering, personalizing, and checking out of an e-commerce site. Design for accessibility. Use social media and social networking to increase traffic to your e-commerce site. Optimize your site for high placement in search engines. Use site analytics to identify ways to improve your online store.

ACKNOWLEDGMENTS

Lawrence Najjar is a user interface designer with over 25 years of experience. He worked on the Campbell Soup Company intranet, Home Depot's online store, NASCAR .com, and the U.S. air traffic control system. Lawrence has a PhD in engineering psychology from the Georgia Institute of Technology. Portions of this paper appeared in the following works:

Najjar, L. J. 2005. Designing e-commerce user interfaces. In *Handbook of Human Factors in Web Design*, eds. R. W. Proctor and K-P. L. Vu, 514–527. Mahwah, NJ: Lawrence Erlbaum.
Najjar, L. J. 2001. E-commerce user interface design for the Web. In *Usability Evaluation and Interface Design*, eds. M. J. Smith et al., vol. 1, 843–847. Mahwah, NJ: Lawrence Erlbaum.

REFERENCES

ActivMedia Research. 2000. Consumer loyalty online research series. http://www.activmediaresearch.com/consumer_loyalty_online .html.
Agrawal, V., L. D. Arjona, and R. Lemmens. 2001. E-performance: the path to rational exuberance. *The McKinsey Quarterly* 1.
Alvarez, G. 2008. E-commerce and the Web customer experience. Paper presented at Gartner Web Innovation Summit, Los Angeles, CA, September 15–17, 2008.

American Heritage Editors. 2000. *American Heritage Dictionary of the English Language*, 4th ed. [Entry for Henry Louis Sullivan.] Boston: Houghton Mifflin. http://www.bartleby.com/ 61/74/S0877400.html (accessed November 7, 2010).

Battey, J. 1999. IBM's redesign results in a kinder, simpler Web site. *Infoworld*. http://interface.free.fr/Archives/IBM_redesign_results.pdf

Bernard, M. 2001. Developing schemas for the location of common Web objects. *Usability News* 3.1 (Winter). Software Usability Research Laboratory, Wichita State University. http://www.surl.org/usabilitynews/31/web_object.asp.

BuddeComm. 2008. 2008 global digital economy—M-commerce, e-commerce, & e-payments. http://www.budde.com.au/Research/2008-Global-Digital-Economy-M-Commerce-E-Commerce-E-Payments.html (accessed Sept. 2008).

Caldwell, B. B., and G. C. Vanderheiden. this volume. Access to Web content by those with disabilities and others operating under constrained conditions. In *Handbook of Human Factors in Web Design*, 2nd ed., eds. K.-P. L. Vu and R. W. Proctor, 371–402. Boca Raton, FL: CRC Press.

Campanelli, M. 2009. Abandonment issues. http://www.emarketingandcommerce.com/blog/abandonment-issues (accessed June 24, 2009).

Chaparro, B. S. 2001. Top ten mistakes of shopping cart design. Internetworking, 4.1. http://www.internettg.org/newsletter/dec01/article_chaparro.html (accessed Dec. 2001).

Charron, C., B. Bass, C. O'Connor, and J. Aldort. 1998. The Forrester report: making users pay. Forrester Research. http://www.forrester.com/ (accessed July 1998).

ComScore. 2009a. Global Internet audience surpasses 1 billion visitors, according to ComScore. http://www.comscore.com/Press_Events/Press_Releases/2009/1/Global_Internet_Audience_1_Billion (accessed Jan. 23, 2009).

ComScore. 2009b. Russia has world's most engaged social networking audience. http://www.comscore.com/Press_Events/Press_Releases/2009/7/Russia_has_World_s_Most_Engaged_Social_Networking_Audience (accessed July 2, 2009).

Consumer Reports. 2003. E-ratings: a guide to online shopping, services, and information: What we look for. http://www.consumerreports.org/main/detailv2.jsp?CONTENT%3C%3Ecnt_id=871&FOLDER%3C%3Efolder_id=735&bmUID=1043273020029.

Cooperstein, D. M., K. Delhagen, A. Aber, and K. Levin. 1999. Making Net shoppers loyal. Forrester Report: Interviews. http://www.forrester.com/ (accessed June 23, 1999).

eMarketer. 2001. Turning shoppers on(line). http://www.emarketer.com/estatnews/estats/ecommerce_b2c/20010312_pwc_search_shop.html (accessed March 12, 2001).

eMarketer. 2009a. How helpful is live chat? http://www.emarketer.com/Article.aspx?R=1007235 (accessed Aug. 19, 2009).

eMarketer. 2009b. Making your e-mails go viral. http://www.emarketer.com/Article.aspx?R=1007234 (accessed Aug. 19, 2009).

ePaynews. 2003a. Statistics for electronic transactions. Average transaction size by year started using Internet (ActivMedia Research). Total B2C revenues for US, Europe & Asia, 1999–2003 (Fortune Magazine). Comparative estimates of Latin eCommerce, 2000–2003. Pure online transaction sales, 2000–2002. http://www.epaynews.com/statistics/transactions.html.

eMarketer. 2003b. Statistics for online purchases. Predicted online retail sales worldwide, Q4 2002 (GartnerG2). US online retail revenues, 2001–2005 (eMarketer). B2C eCommerce forecast for Europe, 2001–2005 (AMR Research). Percentage of abandoned shopping carts, 2001 (eMarketer). Proportion of abandoned transactions at Web storefronts (Datamonitor).

Number of online US and worldwide purchasers, 1998–2003 (Donaldson, Lufkin and Jenrette). European online market in 2001 to equal USD 3.23 billion (Datamonitor). http://www.epaynews.com/statistics/purchases.html.

Federal Trade Commission. 2008. FTC approves new rule provision under the CAN-SPAM Act. http://www.ftc.gov/opa/2008/05/canspam.shtm.

Fleming, J. 1998. *Web Navigation: Designing the User Experience*. Sebastopol, CA: O'Reilly.

Fox, S., K. Zickuhr, and A. Smith. 2009. Twitter and status updating, Fall 2009. Pew Internet and American Life Project. http://www.pewinternet.org/Reports/2009/17-Twitter-and-Status-Updating-Fall-2009.aspx (accessed Oct. 21, 2009).

Freedman, L. 2008a. Building ecommerce content you can bank on. http://www.e-tailing.com/content/wp-content/uploads/2008/07/demandware_072908_whitepaper.pdf (accessed Nov. 7, 2010).

Freedman, L. 2008b. The executive guide to capturing customers: Personalized product recommendations through the merchant and consumer lens. http://www.e-tailing.com/content/wp-content/uploads/2009/12/mybuys_0609_whitepaper.pdf (accessed Nov. 7, 2010).

Fuccella, J., and J. Pizzolato. 1999. Separating content from visuals in Web site design. Internetworking. http://www.internettg.org/newsletter/mar99/wireframe.html (accessed March 23, 1999).

General Services Administration. 2003. 1998 amendment to section 508 of the Rehabilitation Act. http://www.section508.gov/index.cfm?FuseAction=Content&ID=14.

Georgia Tech Research Institute. 2006. State IT database. http://accessibility.gtri.gatech.edu/sitid/stateLawAtGlance.php (accessed Feb. 23, 2006).

Global Millennia Marketing. 2002. Shopping cart abandonment. http://globalmillenniamarketing.com/press_release_mar_12_02.htm (accessed March 12, 2002).

Goldwyn, C., n.d. The art of the cart: why people abandon shopping carts. http://visibility.tv/tips/shopping_cart_abandonment.html (accessed Nov. 7, 2010).

Gonsalves, A. 2009. Dell makes $3 million from Twitter sales. http://ww7w.informationweek.com/news/hardware/desktop/showArticle.jhtml?articleID=217801030 (accessed June 12, 2009).

Grau, J. 2009. Personalized product recommendations: predicting shoppers' needs. http://www.emarketer.com/Reports/All/Emarketer_2000563.aspx (accessed March 23, 2009).

Grimm, J., and W. Grimm. 1999. *Complete Fairy Tales of the Brothers Grimm*. Minneapolis: Econo-Clad.

Gunderson, A. 2000. Toil and trouble: online shopping is still a muddle. *Fortune*, http://www.fortune.com/fortune/personalfortune/articles/0,15114,37024,00.html (accessed Sept. 4, 2000).

Guernsey, L. 2001. Reviving up the search engines to keep the e-aisles clear. *New York Times*, Feb. 28, 2010.

Hagen, P. R., H. Manning, and Y. Paul. 2000. Must search stink? Forrester Report. http://www.forrester.com/ER/Research/Report/0,1338,9412,00.html (accessed June 2000).

Hansell, S. 2002. Net shopping begins surge for holidays. *New York Times*, Dec. 3, C1.

Hill, A. 2001. Top 5 reasons your customers abandon their shopping carts (and what you can do about it). ZDNet. http://www.zdnet.com/ecommerce/stories/main/0,10475,2677306,00.html (accessed Feb. 12, 2001).

International Center for Disability Resources on the Internet. 2004. Spitzer agreement to make Web sites accessible to the blind and visually impaired. http://www.icdri.org/News/NYSAccessWeb.htm.

International Organization for Standardization. 2003. Ergonomics of human-system interaction—guidance on accessibility for human–computer interfaces (ISO/TS 16071:2003). http://www.iso.ch/iso/en/CatalogueDetailPage.CatalogueDetail?CSNUMBER=30858&ICS1=13&ICS2=180&ICS3=.

International Organization for Standardization. 2008. Ergonomics of human-system interaction—Part 171: Guidance on software accessibility (ISO 9241-171: 2008). http://www.iso.org/iso/iso_catalogue/catalogue_tc/catalogue_detail.htm?csnumber=39080.

International Organization for Standardization. 2009. Information technology—User interfaces—Accessible user interface for accessibility settings (ISO/IEC 24786: 2009). http://www.iso.org/iso/iso_catalogue/catalogue_tc/catalogue_detail.htm?csnumber=41556.

InternetWeek. 2002. Many Fortune 100s ignore Web inquiries. *InternetWeek*, Oct. 29, TechWeb News, http://www.internetwk.com/webDev/INW20021029S0011.

L-Soft, n.d. Opt-in laws in the USA and EU. http://www.lsoft.com/resources/optinlaws.asp (accessed Nov. 7, 2010).

Lake, D. 2000. Navigation: An e-commerce must. *The Standard*, April 17, http://www.thestandard.com/research/metrics/display/0,2799,14110,00.html.

Manning, H., J. C. McCarthy, and R. K. Souza. 1998. Interactive technology strategies: Why most Web sites fail. Forrester Research 3(7), September, http://www.forrester.com/.

Martin, M. 2002. Accessibility breakthroughs broaden Web horizons. Sci.NewsFactor.com. Sept. 17, http://sci.newsfactor.com/perl/story/19417.html.

Mayhew, D. J. this volume. The Web UX design process—a case study. In *Handbook of Human Factors in Web Design*, 2nd ed., eds. K.-P. L. Vu and R. W. Proctor, 461–480. Boca Raton, FL: CRC Press.

McCullagh, M. 2002. Judge: Disabilities act doesn't cover Web. CNET, Oct. 21, http://news.cnet.com/2100-1023-962761.html.

McNeil, J. M. 1997. Americans with disabilities: 1994-95. U.S. Department of Commerce, Economics, and Statistics Administration, http://www.census.gov/prod/3/97pubs/p70-61.pdf.

Najjar, L. J. 1999. Beyond Web usability. Internetworking, 2.2, June. http://www.internettg.org/newsletter/jun99/beyond_web_usability.html.

Najjar, L. J. 2000. Conceptual user interface: A new tool for designing e-commerce user interfaces. Internetworking, 3.3, December, http://www.internettg.org/newsletter/dec00/article_cui.html.

Najjar, L. J. 2002. An efficient Web user interface design process for commercial clients. http://www.lawrence-najjar.com/papers/An_efficient_Web_user_interface_design_process.html (accessed Nov. 7, 2010).

Nielsen. 2008a. Over 875 million consumers have shopped online—The number of Internet shoppers up 40% in two years. http://en-us.nielsen.com/main/news/news_releases/2008/jan/over_875_million_consumers.

Nielsen. 2008b. Quick stats & facts. http://en-us.nielsen.com/main/insights/consumer_insight/issue_13/quick_stats___facts.

Nielsen, J. 1999. Web research: Believe the data. http://www.useit.com/alertbox/990711.html (accessed Nov. 7, 2010).

Nielsen, J. 2000. Flash: 99% bad. http://www.useit.com/alertbox/20001029.html (accessed Nov. 7, 2010).

Nielsen, J. 2007. Wishlists, gift certificates, and gift-giving in e-commerce. Jan. 29, http://www.useit.com/alertbox/wishlist-giftcards.html.

Nielsen, J., and M. Tahir. 2001. Building sites with depth. *New Architect Magazine*, February, http://www.newarchitectmag.com/documents/s=4570/new1013637059/index.html.

Nimetz, J. 2006. Key performance indicators for e-commerce based sites. Sept. 12. http://www.searchengineguide.com/jody-nimetz/key-performance-2.php.

Nimetz, J. 2007. Online key performance indicators in 2007. Feb. 7. http://www.searchengineguide.com/jody-nimetz/online-key-perf.php (accessed Nov. 7, 2010).

Opinion Research Corporation. 2009. Internet shopping increases as consumers hunt for bargains. April 9, http://www.opinionresearch.com/fileSave/Ouch_Point_InternetResearchFinal_Apr0609.pdf.

Palmer, A. 2009. Web shoppers trust customer reviews more than friends. September. http://www.managesmarter.com/msg/search/article_display.jsp?vnu_content_id=1004011176.

Pastore, M. 1999. Online stores lacking. E-tailers should follow lead of offline shops. Feb. 25. http://cyberatlas.internet.com/markets/retailing/article/0,,6061_154021,00.html.

Poggi, J. 2009. Top 50 online retailers revealed. Sept. 3. http://www.thestreet.com/story/10592741/1/top-50-online-retailers-revealed.html?puc=_tscrss.

Practical eCommerce. 2007. Great ecommerce ideas. Dec. 17. http://www.practicalecommerce.com/articles/632-Great-Ecommerce-Ideas.

Ragus, D. 2000a. Best practices for designing shopping cart and check-out interfaces. http://web.archive.org/web/20001211161300/http://dack.com/web/shopping_cart.html (accessed Nov. 5, 2010).

Ragus, D. 2000b. Flash is evil. http://www.dack.com/web/flash_evil.html (accessed Nov. 7, 2010).

Ragus, D. 2000c. Flash vs. HTML: a usability test. http://www.dack.com/web/flashVhtml/.

Rehman, A. 2000. Effective e-checkout design. ZDNet/Creative Good, Nov. 22. http://techupdate.zdnet.com/techupdate/stories/main/0,14179,2638874-1,00.html.

Reichheld, F. F., and P. Schefter. 2000. E-loyalty: your secret weapon on the Web. *Harvard Business Review* 105–113, July–August.

Rhodes, J. S. 1998. 8 quick tips for a more usable e-commerce Web site. *Webword*, Dec. 29. http://webword.com/moving/8quick.html.

Rogers, B. L., and B. Chaparro. 2003. Breadcrumb navigation: Further investigation of usage. *Usability News* 5(2).

Sacharow, A., and M. Mooradian. 1999. Navigation: Toward intuitive movement and improved usability. *Jupiter Communications*, March.

Schiff, J. L., n.d. Ecommerce guide essentials: Profitable product pages. http://www.ecommerce-guide.com/essentials/ebiz/article.php/3710491 (accessed Nov. 7, 2010).

Schwartz, J. 2002. Judge rules on Web sites for the disabled. *New York Times*, Oct. 22, Technology Briefing, C10.

Selingo, J. 2000. A message to Web designers: If it ain't broke, don't fix it. *New York Times*, Aug. 3, D11.

Spool, J. M. 2009. The $300 million button. Jan. 14. http://www.uie.com/articles/three_hund_million_button/.

Stanley, J., J. C. McCarthy, and J. Sharrard. 2000. The Internet's privacy migraine. May. http://www.forrester.com/.

Stephanidis, C., and D. Akoumianakis. this volume. A design code of practice for universal access: Methods and techniques. In *Handbook of Human Factors in Web Design*, 2nd ed., eds. K.-P. L. Vu and R. W. Proctor, 359–370. Boca Raton, FL: CRC Press.

Tedeschi, B. 1999. Good Web site design can lead to healthy sales. *New York Times*, Aug. 30, e-commerce report. http://www.nytimes.com/library/tech/99/08/cyber/commerce/30commerce.html.

Tedeschi, B. 2002a. E-tailers are putting the multimedia tinsel back on their shopping sites this holiday season. *New York Times*, Nov. 18, e-commerce report, C6.

Tedeschi, B. 2002b. Online sales are up for holiday, but the question is just how much. *New York Times*, Dec. 23, C1, C7.

Tedeschi, B. 2002c. There is more cheer in the virtual realm as online gift certificates soar. *New York Times*, Dec. 30, e-commerce report, C5.

Tracy, B. 2000. Easy net navigation is mandatory—viewpoint: Online users happy to skip frills for meat and potatoes. *Advertising Age* (Aug. 16), 38.

Tsai, J. 2009a. Customer satisfaction drops for e-commerce top 100. May 11. http://www.destinationcrm.com/Articles/CRM-News/ Daily-News/Customer-Satisfaction-Drops-for-E-Commerce-Top-100-53817.aspx.

Tsai, J. 2009b. Diaper duty. July 1. http://www.destinationcrm.com/Articles/ Columns-Departments/REAL-ROI/Diaper-Duty-55264.aspx.

Tsai, J. 2009c. Search engineering. July 1. http://www.destinationcrm .com/Articles/Editorial/Magazine-Features/Search-Engineering-55059.aspx.

Tsai, J. 2009d. Omniture tops Forrester wave in Web analytics. Aug. 14. http://www.destinationcrm.com/Articles/CRM-News/ Daily-News/Omniture-Tops-Forrester-Wave-in-Web-Analytics-55499.aspx.

Tullis, T. S., F. J. Tranquada, and M. J. Siegel. this volume. Presentation of information. In *Handbook of Human Factors in Web Design*, 2nd ed., eds. K.-P. L. Vu and R. W. Proctor, 153–190. Boca Raton, FL: CRC Press.

US Census Bureau. 2000. Disability status of persons (SIPP). http:// www.census.gov/hhes/www/disability/sipp/disstat.html.

US Census Bureau. 2009. E-stats e-commerce 2007. http://www .census.gov/econ/estats/2007/2007reportfinal.pdf.

US Department of Justice. 1990. Americans with Disabilities Act of 1990. http://www.ada.gov/pubs/ada.txt.

United Nations, n.d. Factsheet on persons with disabilities. http:// www.un.org/disabilities/default.asp?id=18.

Williams, J. 2001. Making Uncle Sam accessible—and accountable. *Business Week Online*, Sept. 7. http://www.businessweek .com/smallbiz/content/sep2001/sb2001097_766.htm.

W3C. 2008. Roadmap for accessible Rich Internet Applications. Feb. 4. http://www.w3.org/TR/wai-aria-roadmap/.

W3C. n.d. (a). Policies relating to Web accessibility. http://www.w3 .org/WAI/Policy/#US.

W3C. n.d. (b). Web content accessibility guidelines 2.0. http://www .w3.org/WAI/Policy/#US.

Ward, S. n.d. Online payment options for your online business. http://sbinfocanada.about.com/od/onlinebusiness/a/online payment.htm (accessed Nov. 7, 2010).

Wixon, D., and S. Jones. 1992. Usability for fun and profit: A case study of the design of DEC RALLY version 2. Internal report, Digital Equipment Corporation. Cited in Karat, C. 1994. A business case approach to usability cost justification. In *Cost-Justifying Usability*, eds. R. G. Bias and D. J. Mayhew. San Diego, CA: Academic Press.

Wroblewski, L. 2009. Web app summit: delivering happiness. April 27. http://www.lukew.com/ff/entry.asp?807.

29 E-Health in Health Care

François Sainfort, Julie A. Jacko, Molly McClellan,
Kevin P. Moloney, and V. Kathlene Leonard

CONTENTS

29.1   INTRODUCTION

There is little doubt that the U.S. health-care system has its share of problems (Arrow et al. 2009; Sainfort et al. 2008). Although the United States uses some of the most advanced medical technologies, has the largest medical workforce, and spends the largest proportion of its gross domestic product, the World Health Organization (2000, 2009) rated the U.S. health-care system worse than most of the Western world, with respect to quality and performance. Additionally, there has been extensive documentation of widespread errors that have resulted in avoidable injuries to patients and surging health-care costs (Institute of Medicine 2001; Jha et al. 2009). In an effort to rectify several of these problems, there has been a push to develop better health-care information systems. Properly designed and implemented technologies have the ability to improve quality of health care (Classen et al. 2007; Raymond and Dold 2002), decrease health-care costs (Jha et al. 2009; Meyer, Kobb, and Ryan 2002; Vaccaro et al. 2001), prevent medical errors (Agrawal 2009; Institute of Medicine 2001), and support the ever-growing demands placed on the health-care industry by governmental regulation and health-care consumers.

Although there has been support for new technologies in health care, these projections need to be tempered by a recognized need for development and evaluation fed by scientific study. E-health, the conversion of health-care services, products, and information to Web-based technologies has been one proposed solution to many of these problems. However, e-health technologies (thus far) have failed to deliver on these promises. Although many of the barriers that have contributed to the failure of e-health will require larger changes in governmental regulations and policy changes by third-party payers and insurance companies, we believe that human factor and human–computer interaction (HCI) research can improve the success of and support for these technologies. HCI, a branch of human factors, can improve e-health technologies by providing valuable insights into the processes and needs of individuals who use these systems. Characteristics of the users, including their abilities, goals, and perceptions, impact the quality and success of their interaction with technologies. HCI focuses on these issues in order to understand how users interact with technologies and how to increase the usefulness and effectiveness of the system (see Dix and Shabir, this volume).

In this chapter, we provide a review of e-health, discuss how e-health can contribute to the improvement of the U.S. health-care system, and examine barriers to the advancement of e-health. We focus only on the health-care industry of the United States for the sake of brevity, although most of the concepts (particularly with regard to HCI) discussed throughout this chapter can also be applied to health care in other countries. Additionally, we discuss the importance of human factors and HCI in facilitating the development and implementation of e-health technologies in health care.

29.2   THE EVOLVING NEEDS OF INFORMATION SYSTEMS IN HEALTH CARE

Before exploring e-health, and the human factors issues that will drive e-health, it is worthwhile to take a look back at how the health-care industry has changed and how these changes have affected the requirements of health-care information systems. There have been a number of social, technological, economic, and regulatory factors that have contributed to the current state of health care in the United States and the inadequacy of the information technologies in this industry. The Institute of Medicine (2001) outlined four underlying causes of the inadequate quality of health care in the United States, including a relatively poorly designed delivery system, the increase in individuals with chronic conditions, the growing complexity of science and technology, and constraints on utilizing the products of the revolution in information technology. Additionally, increased consumerist behavior is providing more challenges in health care (Sainfort et al. 2008; Tritter 2009). Table 29.1 provides an overview of several factors that have contributed to significant changes in the health-care industry.

These factors have all contributed to the transformation of the U.S. health-care system, resulting in increased demands placed on health-care information systems and related technologies. The majority of health-care information systems currently in use have become insufficient, which has generated interest in creating new applications and systems that take advantage of current technologies. Several of the factors that have created the need for and interest in new technologies will now be discussed.

29.2.1   Regulatory and Economic Changes Affecting Health Care

There have been a number of regulatory and economic factors that have changed health care, ranging from the transformation

---

TABLE 29.1

Factors that Have Transformed the Health-Care Industry

Factor Category	Examples
Regulatory and economic	• Effects of the Medicare, Medicaid, and the HMO act • Rising health-care expenditures • Managed care business model • HIPAA reform • Paradoxical effect of regulations on technology use
Social	• Increasing number of patients with chronic illness • General aging of U.S. population • Push toward preventative medicine • Social demand for control of and access to information

of organizational and delivery models of health care to the changing responsibilities and protocols of health-care providers with respect to the handling of electronic medical data. As will be seen, these changes have had an impact on the delivery of health care, which has ultimately affected the technological needs of health-care organizations.

The nature of the today's health-care delivery system is much different from that of several decades ago. In the 1960s and 1970s, medical treatment was focused on acute care, the payment structure was fee for service, health insurance was scarcely available for those below the middle class, and hospitals and medical centers (largely associated with universities) were the primary settings for health-care delivery (Safran and Perreault 2001; Shi and Singh 2001). To solve many of these economic and accessibility of care issues, the federal government expanded the Social Security Act, creating Medicare and Medicaid (Potter and Longest 1994). However, this dramatically increased governmental spending on health care, which resulted in the creation of the Health Maintenance Organization (HMO) Act of 1973. This encouraged a more competitive, corporate market for health-care delivery, which had the overall effect of spawning a new organizational and business model in the health-care industry.

The managed care model was developed to help manage resource utilization and control costs through the integration of health-care functions (e.g., delivery and payment) within a single organizational setting. Managed care organizations (MCOs) accomplish this by organizing providers into coherent networks, sharing risks, leveraging the services of different facilities, extracting discounts based on large patient volumes, and eliminating payer and insurance intermediaries (Shi and Singh 2001). MCOs naturally transformed into massive integrated delivery networks (IDNs) (see Safran and Perreault 2001, for a review), creating geographically dispersed care networks comprised of alliances between care facilities such as ancillary and ambulatory care facilities, health maintenance and insurance companies, larger health-care facilities (e.g., hospitals and medical centers), and extended care facilities (see Figure 29.1).

Although the transition to the managed care model was largely based on economic reasons, it also created changes in the practices of health-care professionals, affecting the

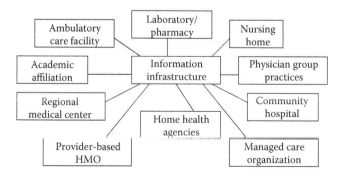

FIGURE 29.1 Organizational structure of the integrated delivery network (IDN).

workflow and practices of physicians and health-care professionals. Primarily, the onset of managed care pushed primary care out of the hospitals and into outpatient clinics and physician practices. As a result, the major burden of health care, especially with the focus on preventative medicine, has fallen onto the primary care physician. So, primary care physicians now tend to have a higher patient volume, resulting in less time spent with patients and more severe knowledge requirements. Additionally, the new IDN organizational model of distributed care facilities requires that health-care organizations meet new levels of communication and data sharing in order to support collaboration. As can be seen, these economic and regulatory factors have had long-range impacts on the way that medicine is practiced and how health care is delivered. As a result, these changes to medical practices have also changed the technological needs of health-care organizations.

Another more recent regulatory factor that has affected health-care organizations is the Department of Health and Human Services' (DHHS) Health Insurance Portability and Accountability Act (HIPAA) of 1996. The HIPAA legislation has dramatically altered provider and health-care organization liability and responsibility with respect to the integrity, confidentiality, and availability of electronic health information (U.S. Department of Health & Human Services 2003a). This legislation impacts the protocols used by health-care organizations for transaction coding, data storage and transmission, documentation standards, and insurance claims and payments. These regulatory factors have also changed the technological needs of health-care organizations, as most current systems do not yet meet the standards set out by these mandates.

Health-care reform and legislation continues to be debated within the U.S. government and among the private sector. Despite the fact that reform is proceeding at a snail's pace, there does seem to be one issue of agreement: Both political parties support health-care information technology (IT). One of the small steps accomplished in 2009 was the authorization of $1.1 billion to conduct research comparing "clinical outcomes, effectiveness, and appropriateness of items, services, and procedures that are used to prevent, diagnose, or treat diseases, disorders, and other health conditions" (Title VIII of the American Recovery and Reinvestment Act of 2009 [Recovery.gov 2009]). Although the passing of this act does not reform the health-care system, it does provide much needed funds for research to promote quality and safety.

29.2.2 Social Changes Affecting Health Care

In addition to the regulatory and economic factors that have changed health-care practices and technological needs of health-care organizations, there are also social changes that have a large impact on health care. Two main categories of these social changes are (1) epidemiological and demographic and (2) attitudinal and consumerist. These epidemiological and demographic factors have restructured the medical needs of the general population, reinforcing the need for the prevention and maintenance paradigms of care. The attitudinal and consumerist factors have reinforced the needs (and demands) for new medical technologies and health-care services to provide more convenience and increased quality of care. These two categories of social factors will be discussed in the following sections.

29.2.2.1 Demographic and Epidemiological Factors

Two major demographic and epidemiological changes in the U.S. population that have affected health care are the increasing number of aged individuals (e.g., 65 years and older) and the increasing prevalence of individuals with one or more chronic ailments. Aging of the population and poor health habits have resulted in increases in the number of individuals with chronic illnesses. As will be seen, these factors are interrelated and have had severe economic consequences for the health-care industry.

The "baby boomer" contingent, numbering 82 million, has been dramatically increasing the proportion of the U.S. population aged 65 years and older since the last millennium and will continue to do so over the next several years (Meyer 2001). Despite the projected doubling of this population by 2030, the nation is unprepared for their health-care needs. Challenges include the increased prevalence of chronic diseases and the depletion of Medicare and Medicaid funds by 2019 (Institute of Medicine 2008). Much of this aging (or aged) population, as well as millions of younger individuals, have one or more chronic ailments, including arthritis, congestive heart disease, cancer, and Parkinson's disease (just to name a few). The prevalence of chronic conditions increases with age, primarily occurring in individuals aged 45 and older, with increasing rates at ages 65 and older (Partnership for Solutions 2002). It is estimated that, in 2000, 125 million individuals in the United States had at least one chronic condition. This number is expected to increase 25% by the year 2020 (Mollica and Gillespie 2003). The cost of health care for individuals with chronic conditions comprises approximately 78–85% of all health-care spending (Mollica and Gillespie 2003; Strohecker 2000). To better explain the severity of this situation for the health-care industry, these 125 million (or so) individuals account for about 80% of $1.5 trillion in health-care expenditures (2002 estimate) in the United States.

Clearly, providing for individuals with chronic conditions will have a strong impact on the economic future of health care. This economic impact has implications on the drive toward technological innovations in health care, as health-care organizations will seek solutions that can provide cost-effective care for these individuals. The increased probability of these individuals to experience an acute event requiring expensive treatment dictates that alternative forms of care be provided to help prevent excessive hospitalization, encourage behavioral modification to promote compliance with treatment orders, and provide more frequent support. The paradigm shift toward preventative medicine dictates that these individuals should be provided with means to receive extended support and care to improve or maintain noncritical

health states. The increasing care needs of these individuals created new technological needs to support this paradigm of continual, active care delivery.

29.2.2.2 Attitudinal and Consumerist Factors

In addition to epidemiological changes, there have been other social changes that have created new technological demands. The attitudinal and consumerist changes include the general trend toward increasing consumer demand of health information and control over one's own health care. Namely, the World Wide Web and communication technologies have reformed the way in which individuals conduct business, communicate, manage their assets and information, and further their education (Jadad 1999; Mittman and Cain 1999). This, in turn, has changed health-care consumer needs, expectations, and behaviors, which has been shown to modify their interaction with physicians and health-care professionals (Giorgianni, Grana, and Sewell 2000) and their knowledge of and demand for new health-care services, treatment options, and insurance plans (Mittman and Cain 1999). Increased consumerism, in the form of increased demands for information, services, and convenience, has raised a number of technological concerns.

Reliance on technologies and the Web for everyday tasks has become commonplace for many individuals. As a result, many individuals seek to use the same technologies that they use for online shopping and investment management for the management of medical records and health insurance and shopping for health-care products. Consumers can browse a multitude of Web sites for information on a specific disease or may even enter their current symptoms to obtain information on possible causes. This type of on-demand health information is improving the medical literacy of the consumer base (Mullner 2002). Individuals are now bringing Internet printouts to their appointments. More research is needed to determine whether or not this change in patient behavior is helpful and how it is perceived by physicians.

Studies have shown the overwhelming extent to which people access health-related information on the Web (Reuters Health 1999). Today, because of increased access to information, people generally have more control over the health insurance plan that they choose, the physician that they see, and even the treatment options that they receive. Information technologies have fueled a change in consumer behavior and expectations—individuals want, demand, and seek out more (and better) choices (Sainfort et al. 2008). These technologies empower consumers and will continue to influence their involvement with and decisions about their health care.

The personal health record (PHR) is another example of increased consumer empowerment in health care. A PHR allows individuals to populate an electronic record with their continuity of care documents, prescription medication lists, health histories, hospital discharge summaries, and laboratory test results. The consumer can often choose to share their information with third parties in order to facilitate wellness or fitness services. Currently, the majority of PHRs available to consumers are unsophisticated. There is very little decision support available such as medication contraindications. Instead of being integrated tools for education and health management, the PHRs are used as a place to store medical data instead of utilizing it (Masterson 2008). Owing to a lack of response from consumers, PHR vendors are starting to shift their focus to a more interactive product. One example of this interactive approach is the use of text message reminders to take a medication at a specified time. Vendors are hoping that by creating more useful and interactive PHRs, patients will be more likely to subscribe.

While PHRs remain in their infancy, other systems are being developed for use in patients' homes on their PCs. These systems are designed to assist self-management of chronic diseases. One example is CHESS (Comprehensive Health Enhancement Support System) developed at the University of Wisconsin-Madison for use with patients who suffer from diseases such as breast cancer, HIV infection, heart disease, Alzheimer's disease, or alcoholism. The HIV/ AIDS CHESS module system includes a library of full-text articles, frequently asked questions and answers, consumer tips, referral system, tools to assess lifestyle choices and risks, online discussion groups, and private messaging with experts. According to an AHRQ study of CHESS patients who had HIV/AIDS, they were better prepared with questions for their providers resulting in 15% shorter visits. Patients who received only standard care had 50% more hospitalizations than the CHESS patients, and their length of stay was 39% longer. Hospitalization costs were reduced by $728 per month during implementation and $222 after implementation of CHESS (Gustafson et al. 1999).

29.2.3 THE SHIFTING TECHNOLOGICAL NEEDS OF HEALTH CARE

As discussed, there have been a number of factors that have contributed to major changes in the technological needs of health-care organizations, ultimately resulting in information systems that are ill equipped to meet these needs. The new technological needs are a result of fundamental changes in the care delivery requirements, financial structure, consumer demands, regulatory forces, or organizational model of health care that dictated changes in the practical needs (e.g., cost savings, extra care for the chronically ill, modified data handling protocols) of the health-care industry.

The geographically dispersed nature of the IDN organizational model necessitates that health-care services entities (e.g., physician practices and insurance providers) become much more collaborative. However, most health-care applications, built upon proprietary platforms and legacy mainframe systems, primarily handle only the needs of the individual provider or department within the health-care organization (Shi and Singh 2001). Information systems, even within a single health-care facility (e.g., a hospital), still rely on complicated and costly interface engines that translate the different data content and formats from the disparate sending and receiving systems (Tang and McDonald 2001). Thus, while the coordination, sharing, and management of information,

resources, and business operations can achieve both increasing quality of patient care and decreasing costs, the information systems currently in place do not support these needs. In fact, even today, it would be more accurate to say that the majority of the IDNs in the United States are still integrating rather than being already integrated (Safran and Perreault 2001). According to the HIMSS-Dorenfest database, only 21% of hospitals had installed a partially integrated inpatient EMR in 2005 (Fonkych and Taylor 2005). Moreover, many health-care organizations still perform a number of data transactions by phone or fax and still employ the use of extra personnel for medical coding and transcription.

Recent regulatory constraints (namely HIPAA) provide an extreme example of technological inadequacy, as health-care information systems are actually noncompliant with governmental mandates. This piece of regulation has required health-care organizations to become compliant with definitive rules for the privacy, confidentiality, standards, and security of health-care data and the transactions performed with this data. To meet these standards, most health-care organizations have had to implement entirely new information systems, often with extraordinary costs.

The social changes, with respect to increased care needs for the chronically ill and the increased demands for consumer control of and access to health-care information, have also created new technological needs. In order to achieve the reduction in the cost of care, increases in the quality of care, and extra education and social support for these individuals, health-care organizations and providers need technological systems that can provide continuous, remote, targeted care to monitor health status, enforce treatment protocols, promote healthy behaviors, and provide support. Increased consumer demands for more convenient health services and more control of their own health-care dictate that consumers are granted easier and more extensive access to health-care information.

In addition, as suggested by the Institute of Medicine (2001), the ever-growing complexity of science and medicine also dictates the need for new health-care information systems. As Sainfort et al. (2008, 662) state, "The sheer volume of new health care science and technologies . . . is large today and has advanced much more rapidly than our ability to use and deliver them in a safe, effective, and efficient way." Advances in the knowledge and research in the genetic, biomedical, biochemical, and other fields have created a need for health-care and medicine to come up with ways to store, share, visualize, communicate, and implement this information. Although these advances bring a great many benefits to improving the knowledge, treatments, drugs, and devices available to health care, they also bring enormous costs in terms of developing new technologies that can support all of these new possibilities.

29.2.4 SUMMARY

The underlying current that has resulted from all of these changes is an increased need for new technologies. Health-care organizations are only beginning to apply information technologies and still do not make much use of the Web and other information technologies (Institute of Medicine 2001; Sainfort et al. 2008). Although health care once had a fee-for-service, point-of-care, acute care delivery model, it now has a financially complex, heavily managed, preventative care model. Consumers once had a passive role in the management of their health care, but they are becoming empowered, demanding decision makers. To meet these increasing technological needs, health-care information systems developers have turned to the Web and other networked technologies. These new technological needs spawned the movement toward e-health.

29.3 THE E-HEALTH MOVEMENT

E-health has become a highly touted, yet poorly supported, solution to many of these deficiencies with the current information systems in health care. E-health has yet to be confined to one clear, concise definition (Eysenbach 2001). Interestingly, some industry professionals feel that "e-health" was a term created to boost corporate and investor interest in these technologies and now serves as a blanket term covering more traditional fields of telehealth, telemedicine, computer-based learning, remote consultation, and remote disease management, among others (Della Mea 2002; De Rouck et al. 2008; Mitchell 1999). For our purposes, e-health refers to the transition of health-care processes, information, transactions, and the like to Web-based (or Web-like) form factors. Within this working definition of e-health are all networked health-care information systems, information repositories, and medical applications whether they are implemented in Intranets, Extranets, or the public domain Internet.

Common e-health technologies include electronic medical records (EMRs), Web sites providing health-care information, order entry and billings systems, among a wide range of other applications and technologies. The reason that interest in e-health, and new health-care technologies, has so greatly increased is twofold: (1) social, regulatory, organizational, and economic changes have dictated new technological needs in the health-care industry (Singer, Enthoven, and Garber 2001); and (2) these new technologies have shown promise in improving health-care services, improving quality of care, decreasing costs, and changing traditional models of health care (Sainfort et al. 2008). Although the former point should follow naturally from the previous section, the potential benefits of these technologies will now be discussed.

29.3.1 WHY THE WEB?

Although information technologies have been in use in health care since the 1960s, the industry only began to recognize the potential benefits of the World Wide Web and other new information technologies in the first few years of the past decade. One major question is why the Web is being considered the platform on which to build the new technological

infrastructure of health care. It has been suggested that one of the ways in which Web applications are different from the failed health-care IT ventures is that Web-based technology solutions engage the most powerful stakeholders in the health-care industry—the millions of health-care consumers (Kleinke 2000). There are characteristics of the Web that make it an attractive infrastructure for e-health technologies. Mittman and Cain (1999) identified some beneficial characteristics, describing the Web as (1) inexpensive, (2) easy to use, (3) more democratic (e.g., provides an effective way to share information and viewpoints by many), (4) boundless (e.g., information can come from and be accessed anywhere), and (5) continually increasing useful functionality. Some additional characteristics and their associated benefits are outlined in Table 29.2.

The overarching domain of e-health incorporates a lot of non-Web-based applications and technologies, which do have their place and use in health care. For example, health-care professionals do use self-contained decision aids that are not networked to a larger database and stand-alone electronic medical records that do not incorporate interfacing with order entry and billing systems. However, making the application and systems Web-based increases the viability of these tools. For example, providing professionals with better (or more) information to aid in decision making can reduce logical errors and increase quality of care by providing care givers with larger, comprehensive databases that are continually improved through evidence-based medicine and then collectively updated and distributed. Additionally, administrative costs can be reduced by interfacing the functionality of patient interview, billing, and order entry systems in order to decrease the amount of time and resources needed to transcribe physician notes, prepare and send paperwork, or look up and enter diagnostic and procedure coding. From these common examples, one can see how the interconnectivity

---

TABLE 29.2

Overview of Potential Benefits of Web-Based Technologies

Characteristic	Example Benefits
Multimedia support/ information visualization	• Better visualization of complex information • Support for multiple media form factors
Universalizability/ cross-compatibility	• Platform/form factor cross compatibility • Immediate/consistent information updating
Flexibility/ customization	• Scalable systems to fit changing needs • Customizable interfaces to suit user needs
Connectivity	• Interfacility collaboration/communication • Variable access controls
Efficiency/inexpensive	• Decreased data reentry • Cost-effective data storage
Ubiquity/familiarity	• Decreased training/familiarity requirements • Decreased implementation/maintenance costs

and flexibility allowed by Web-based technologies can aid in the development of more powerful and useful tools.

29.3.2 E-Health's Brief History

Web-based health-care applications or information systems, or e-health, were the natural result of a combination of factors. The relative success of e-commerce and Web-based transaction paradigms in decreasing business costs, the failure in properly coordinating care and resources between health-care facilities, and the increasing consumer demands for more access and convenience with regards to their health care likely all contributed to players in the health-care industry looking into Web-based technologies for solutions to these problems. Although it is difficult to even approximate a date of e-health's birth, we can use the history of notable IT ventures in the health-care domain as a starting point of Web-based applications in health care.

First, it should be noted that e-health, in the broader sense of technology-enabled health-care tools, largely stemmed from the interest in applications of telehealth or telemedicine (which actually started in the early 1900s; Rosen 1997). The health-care industry became interested in remotely delivered health-care services (e.g., phone consultation, transfer of radiographic imagery) in order to better leverage resources and provide service to rural areas and remotely located consumers, including astronauts (Bashshur and Lovett 1977). Interest in these methods of providing care to patients remotely increased with the advent of the Internet, Web-based, and networking technologies in the 1980s and 1990s, spawning new possibilities for remote consultation and care delivery including telesurgery, videoconferencing, and remote monitoring through sensor technologies.

E-health naturally resulted from these conditions, as new technological innovations were being applied to health care and medicine. Starting in the mid to late 1990s and up to the present, the Web technologies have been used as a springboard for widespread access to "health e-commerce enterprises" (Parente 2000). Currently, use of the Web-based technologies in health-care focuses on: Web sites that provide online services for providers and consumers (Parente 2000; Payton 2003), application of Web-based technologies to provide online health management functionality (Kidd 2008; LeGrow and Metzger 2001), and use of wireless-enabled mobile devices for clinical and administrative functions (Turisco and Case 2001; Wu, Wang, and Lin 2007).

Digital gaming is the newest tool in health care. Virtual realities, simulations, and online games are interactive ways to help patients learn self-management strategies (Hawn 2009). Many interactive Web sites are available for children and adults to learn about their disease. Providers often refer patients to these online games to encourage patient participation in their treatment plan. Table 29.3 provides a taxonomy of common health-related online ventures.

There are some current examples of the adoption and use of new technologies, such as the use of handheld and portable technologies for reference tools, prescriptions, charge capture and billing, documentation and dictation, and functions of the electronic medical record (Roa et al. 2002). The use of personal digital assistants (PDAs) by health-care providers has been on the rise since 1999. Research has demonstrated that the overall adoption rate for physicians ranges between 45 and 85% (Garritty and El Emam 2006). In addition, clinical decision support tools (e.g., protocol and treatment databases, medical research repositories, and analytical decision aid tools) are becoming popular (Metzger and MacDonald 2002). Interactive clinical decision support databases that "close the loop" between clinical practice guidelines (CPG) and end-user physicians are on the horizon. Physicians will be provided with CPGs at the point of care, and will either follow the guideline or disagree and post their reasoning. This information will be collected in massive databases and compiled in order to review and confirm the CPG. It is believed that this technological breakthrough could improve consistency, quality, and safety of patient care while disseminating good practices worldwide (Fox et al. 2009). Although

---

TABLE 29.3

Taxonomy of Current Health-Care Web Sites and Web-Based Ventures

Company Type	Common Services	Examples
Portal	• Provide information and links to health topics • Common ground for seeking and providing information for both consumers and providers	• Medscape • drkoop.com • healthgrades.com
Connectivity	• Provides quality assessments of providers • Web-accessible EMRs • Provides services for data and claims transactions	• Healtheon/WebMD • XCare.net • MedicaLogic
Business-to-business	• Provides marketplace for buying and selling medical equipment • Facilitate procurement/selection of group health insurance plans	• Allscripts • Medical Manager
Business-to-consumer	• Customizable insurance plans • Online ordering of medications and health products	• drugstore.com • PlanetRx

Source: Adapted from Parente, S. T. 2000. Beyond the hype: A taxonomy of e-health business models. *Health Affairs* 19(6): 89–102.

application of Web-based technologies in health care is still not widespread and does not take full advantage of the available technology, providers and professionals are increasingly seeing the utility of these technologies.

29.3.3 Benefits of E-Health

Although there is some contention about how or when (or even if) e-health can fix the ailing U.S. health-care industry, there is little argument that the industry is ailing: health-care costs are outrageous, medical errors are abundant, resources are used inefficiently, and the managed care system (thus far) has generally failed (see Sainfort et al. 2008, for a review). Even though predictions need to be guarded, e-health has the potential to improve many of health care's problems. Researchers and experts have suggested that application of information technologies can benefit health care in several domains such as (1) consumer health, (2) professional education, (3) clinical care, (4) public health, (5) administrative and financial transactions, and (6) research (National Research Council 2000). We briefly discuss each of these application domains. The following sections outline a number of areas in which e-health is well positioned to benefit health care in the near future.

29.3.3.1 Consumer Health

We discuss the consumer health domain first because it is the most likely domain in which the application of Web-based technologies can immediately begin to benefit and reshape health care. In this context, consumer refers to individuals who seek, use, or need health-care information or services (e.g., a patient). Consumer health refers to efforts in giving consumers a more active, direct role in the management of their own health and health care (National Research Council 2000). E-health technologies will contribute to consumer health largely through increasing access to health-care and medical information. Harris (1995, 23) defines consumer health information as "any information that enables individuals to understand their health and make health-related decisions for themselves or their families." While this definition clearly encompasses a lot of non-Web-based information dissemination methods, it should also be noted that these electronic forms of information have a larger potential to be more timely, comprehensive, and accessible to an increasingly larger base of health-care consumers.

In support of this idea of the relative effectiveness of Web-based technologies as information dissemination and access tools, it has been estimated that 70–160 million Americans, or half of the 7 out of 10 Americans who are online, look on the Web for health-related information (Cain et al. 2000; Giorgianni, Grana, and Sewell 2000; Harris Interactive 2007; Taylor 2002; Ybarra and Suman 2005). Trends compiled by Harris Interactive since 2001 show an increase from 29% of the general public to 52% in 2007 who say they frequently or sometimes go online to search for health information. With an estimated 100,000 Web sites devoted to health (Ferguson 2000), health-care consumers have access

to seemingly unlimited health-related information. Previous studies performed by the American Medical Association and the Pew Internet Project, stated that about six million Americans seek health information online on an average day, which exceeds the number of individuals that actually visit health-care professionals (Fox and Rainie 2002). Despite the fact that physicians are the preferred first source of health information for half of all Americans, only 11% use their physician as the first choice of health questions, as compared with 49% who report that they use the Internet first (Hesse et al. 2005) Clearly, consumers are taking advantage of the information provided on the Web. In evidence of this behavioral change is the frequency with which patients now present their physicians with printouts of information from the Web (Giorgianni et al. 2000; Kassirer 2000). An additional way patients are changing their behavior is by using interactive Web sites. Consumers answer questions about their condition and receive customized questions for patients to ask their physician. Research has shown that these Web sites result in favorable responses from both patients and physicians (Hartmann et al., n.d.).

Consumers are using the Web and other technologies to research diseases, signs and symptoms, treatment options, new pharmaceuticals, or even seek advice from other health-care specialists, peers, and professionals online. Additionally, the Web also allows for the education of the online public about health conscious behaviors and provides social support (via virtual communities and support groups) for individuals with chronic illnesses. Consumers are realizing, likely through their experience with the advent of e-commerce and online shopping, that the Web provides them with a means of getting more information, better information, comparative data, knowledge of alternatives, and even sharing information to leverage collective knowledge.

29.3.3.2 Professional Education

As suggested by the Institute of Medicine (2001), one key challenge in medicine is the ever-growing volume and complexity of knowledge, tools, and information in medicine and science. It often proves difficult for clinicians and health-care professionals to stay current on scientific research findings, new pharmaceuticals, effectiveness of treatment protocols, and new diagnostic and treatment tools, let alone all of the new bureaucratic and legislative regulations that are to be enforced. Although in some ways, e-health technologies may contribute to this growing complexity in health care, it can also do much to facilitate the training and education of health-care professionals and consumers.

The education of health-care professionals, ranging from physicians to laboratory technicians, can be greatly enhanced through the use of Web-based technologies. As with consumer education, professional education will be enhanced primarily through the mere availability of more (and better) information through the information superhighway. For example, the Web site of the National Library of Medicine provides a searchable database of more than 10 million references from more than 5000 medical and biomedical journals (U.S. National Library of Medicine 2006). Yet, this is just one Web site out of more than 100,000 health-related Web sites (Benton Foundation 1999; Ferguson 2000). Information availability has led increasingly large numbers of health-care professionals to seek out health-related information online.

The boundless nature of the Web not only provides easy access to more textual information, scientific research results, and clinical outcomes, but it also provides health-care professionals with an opportunity to view videoconference sessions of surgical procedures, examine simulations of medical procedures, and more easily interact with their peers. Additionally, the Web provides unparalleled abilities of information visualization, which has been shown to facilitate learning and knowledge crystallization (see Card 2008, for a review). Although it is true that all of these experiences can be achieved with computers and software, without connection to any network, the Web allows for increased information sharing and collaboration. Web-based technologies allow for health-care professionals, students, and researchers to better leverage their collective knowledge to improve their education.

29.3.3.3 Clinical Care

One of the most tangible and investigated potential benefits of e-health is the use of Web-based technologies (e.g., the Web, e-mail) to enhance the provider-patient interaction. This interaction between patient and physician is the key interaction in health care, as it involves the two major decision makers in the maintenance and management of an individual's health. As previously discussed, use of the Web as a repository of health-related information has helped consumers achieve a more active role in their health-care and health-related decisions. Providing consumers with better, more accessible sources of information has increased the quality of the dialog during encounters and helps patients to better utilize the limited time they have with their physicians (Giorgianni, Grana, and Sewell 2000; MacDonald, Case, and Metzger 2001).

E-health technologies have also provided health-care professionals with clinical support and communication tools, such as electronic medical records, computerized order entry systems, information databases and query tools, electronic decision support tools, warning/alert systems, and sensor and data analysis tools (Sainfort et al. 2008). These applications of technology decrease medical errors and increase the quality of care by assisting the physician in applying the most appropriate treatment protocol, avoiding drug interactions and allergies, keeping up to date on the most recent information on a patient's condition, and supporting the physician in decision making.

Web-based technologies, such as e-mail, have provided health-care professionals with new tools to help facilitate effective communication and increase patient compliance to treatment protocols. For example, the use of e-mail between patients and their physicians helps physicians to remain abreast of a patient's current health status, reinforces

treatment orders and protocols, and allows for a more efficient, flexible outlet for communication with patients who do not need synchronous, face-to-face consultation (MacDonald, Case, and Metzger 2001). In an age where physicians spend less and less time in direct physical contact with their patients, these technologies help provide new tools that transform care delivery from a paradigm based on discrete interaction dependent on physical proximity to one that is based on continuous, remote care delivery. This more continual contact between patient (or consumer) and practitioner has tremendous potential to increase quality of care by allowing for more frequent reinforcement of treatment protocol and health-promoting behaviors, monitoring of health status, and early notification and resolution of potential health issues.

29.3.3.4 Public Health

E-health can improve public health, primarily through using Web-based technologies to increase the collection and distribution of information between public health officials, health-care providers, and public health-related organizations (e.g., the Centers for Disease Control [CDC]). Public health refers to the epidemiological trends, such as spread of infectious disease, outbreaks of food poisoning, or dense areas of the chronically ill, on the community (e.g., city, county, state, and federal levels). These technologies will contribute to public health primarily through the rapid, easy, continual transmission of relevant public health data from health-care practitioners to public health organizations and officials (and vice versa). This transmission of data can help with preventative planning (such as targeting education at affected populations), determining resource allocation and planning (such as redirecting more resources or supplies to needier areas), and general public health surveillance (such as monitoring public health trends for pathogens, communicable disease, and chronic illness).

There have been long-standing issues with the communication within the community, state, and federal levels of public health agencies and between these officials and health-care providers or clinicians (National Research Council 2000). The levels of agencies are organized inefficiently (vertically and disease specific), limiting effective, complete, and timely communication. In fact, much of the reporting from clinicians and medical laboratories is still paper based. In support of the need for Web-based technologies to improve this communication, the use of these technologies has, in some cases, drastically limited the delay of information transmission between health-care providers or medical laboratories and state- and federal-level public health officials. Additionally, this clearly has cost-savings implications to reduce costs by limiting paper waste, streamlining administrative costs, and improving resource allocation. The use of Web based technologies and electronic forms will decrease the rates of reports being lost, misdirected, or misunderstood (because of poor handwriting), helping to feed those agencies responsible for promoting public health with higher-quality information in a more timely fashion.

29.3.3.5 Administrative and Financial Transactions

Streamlining and reducing the costs associated with administrative process and financial transactions is one application domain in which Web-based technologies could easily produce immediate, tangible improvements. For example, researchers have estimated that the use of electronic claims have reduced the cost of these transactions from several dollars to a few cents (McCormack 2000). While the managed care model of health-care delivery was originally designed to control costs, the system has spiraled out of control as the majority of health care in the United States is financed by a huge network of third-party payers that handle the insurance, payment, and reimbursement of health services (National Research Council 2000). The costs of these interactions between consumers with their providers (e.g., health insurance companies, health management organizations) and between these providers with the consumers' care deliverers (e.g., clinicians, hospitals) account for a larger proportion of U.S. health-care expenditures. Streamlining these processes through electronic claim filing and processing can reduce extra costs associated with small and private practices having to afford expensive electronic data interchange (EDI) systems, medical practices and payers having to hire extra personnel to simply transfer information by fax or phone, and the extra processes associated with data reentry to ensure that claim formats are compatible between medical practices and associated payers.

The Web has greatly benefited many nonhealth sectors of the economy by using these technologies to streamline administrative processes and decrease operating costs. It is hoped that many of these advantages will carry over into the health-care domain. While there are a number of forces aligned against this progress (e.g., traditional players of the "old" health care, federal and state legislation, and an ailing economy), there is still a clear interest in developing and trying to market and implement Web-based technologies in health care (Kleinke 2000; Parente 2000). Only time will tell if e-health technologies will succeed where governmental regulation, managed care, and other IT promises have all failed in taming the uncontrollable costs of health care.

29.3.3.6 Research

Computing technologies have provided the medical and scientific research communities with new mass storage capabilities, methods of visualization and management of large amounts of data, and means of collection and analyzing data. Web-based technologies can continue to build upon these strengths by allowing for effective methods of information dissemination and more widespread access to current scientific knowledge. Science and research are primarily performed by individuals or small groups, which can be located all over the world. The way that science advances is for the scientific findings of researchers to be shared with other researchers in an infinite self-feeding process. Web-based technologies are well equipped to facilitate this sharing of information, ensuring that new findings and data can be shared in a much more timely and continuous fashion as opposed to the discrete and lengthy processes of disseminating new results through publication in paper journals and presentations at meetings or conferences. Many domains of scientific research have already realized the benefits of this more rapid and accessible medium for information dissemination, as the number of online publications for research has continually increased.

As discussed with how Web-based technologies can improve professional education, these technologies can also improve medical research by allowing for effective ways to manage and visualize complex information. The flexible, dynamic nature of the Web allows for multimedia presentation of information (such as manipulability of three-dimensional models of proteins or anatomical structures) and the building of large, distributed databases (such as the construction and development of "gene banks"). The increasing volume and complexity of data coming from biomedical and scientific research requires that new tools are developed for researchers to share this information. E-health can help research to advance through the support of these needs.

29.3.4 Web 2.0

The concept of Web 2.0 was introduced in 2004 to reference the second generation of the World Wide Web. Instead of focusing on technical changes to the Web, the term refers to applications that facilitate interactive information sharing, interoperability, user-centered design, and collaboration. The main difference between the World Wide Web and Web 2.0 is the increased interactivity by the consumer. Users participate in developing, managing, and changing content of applications. Some examples of Web 2.0 are Web-based communities, hosted services, Web applications, social networking sites, video sharing sites, wikis, blogs, mashups, and folksonomies.

The history of Web 2.0 began following the crash of the dot com markets. The Web 2.0 Conference in 2004 was based on the premise that the Web is a platform, and software applications are built upon the Web as opposed to upon the desktop (O'Reilly 2005). Some people consider Web 2.0 to be a buzzword; yet, others consider it to be an indicator of the future of the Web. Understanding the concept of Web 2.0 is simple when comparing current Internet sites. For example, the consumer can pay a subscription for content from an encyclopedia Web site in Web 1.0, whereas the consumer can obtain information for free from a Wiki in Web 2.0.

The health-care community needs to be aware of ever changing technologies and their ability to provide on-demand medical information. Key aspects of Web 2.0 in health care include Really Simple Syndication (RSS) to rapidly disseminate new information, blogs to describe new trends, wikis to share knowledge, and podcasts to make information available anytime (McLean, Richards, and Wardman 2007).

29.3.4.1 Medical Blogs and Wikis

The term "blog" is the result of condensing the words "Web logs." Blogs are an important component of Web 2.0

because they allow an increased number of contributors to publish information on the Web. People without knowledge of HTML (hypertext markup language) utilize blogs to self-publish everything from opinion to experiences. Consumers can access these blogs to learn about new medical devices, experiences in clinical trials, and even points of view from practicing physicians.

Medical blogs have become increasingly popular methods for consumers to obtain health information. There are currently many types of medical blogs, ranging from anonymous physicians and their daily activities to consumers testing medical devices. When a consumer finds a blog relevant to their health-care needs, they can track it using RSS, a Web 2.0 language for syndicating content. RSS sends "feeds" to a Web site aggregator, thus streamlining the process for alerting users to new content. The convenient alerting from blogs and the variety of content makes them a concept to watch in the future of Web 2.0.

Wikis are another way that Web 2.0 consumers can get information. The word Wiki is taken from the Hawaiian word for "hurry." A Wiki is a freely accessible and expandable collection of interlinked Web pages. Wikis allow any user to add, remove, or edit content. As with blogs, Wikis are a method of dispersing information rapidly to consumers. A variety of medical Wikis are currently available to consumers. These include topics such as preparing for an avian flu pandemic and new approaches to the classification of diseases.

Despite the advantages of blogs and Wikis, there are some drawbacks. A main controversy is known as the "window of harm." This term refers to the amount of time when incorrect information is listed on a wiki before it is corrected (McLean, Richards, and Wardman 2007). To combat the potential for misinformation, many wiki sites now limit the ability to author to certain groups of people. For example, some heath care wikis only allow contributors who can provide valid medical credentials. As with most Web 2.0 applications, it is up to the consumer to determine whether or not the Web site and its content are reliable.

29.3.4.2 Other Web 2.0 Applications

In addition to blogs and wikis, other Web 2.0 applications are becoming common in health care. Medical schools are now using Podcasts to provide lectures and book chapters or journal readings to students. Podcasts are a mechanism for downloading audio and video content to MP3 and MP4 players. The portability of the podcasts allows students to access their lectures at any time and as many times as they desire. Students save time and money by downloading articles to their MP3 players instead of searching through library stacks or purchasing expensive textbooks. Because of the portability and convenience of Podcasts they are becoming increasingly popular with both students and professors.

Other Web 2.0 applications are catching on in health care. Both social networking sites and health mashups are starting to find applications in medicine. Social networking sites provide a forum for individuals with a common purpose or health issue to connect with each other. Often times these groups will work together to raise money for causes. Other groups provide support through their online community. Popular dieting Web sites are even taking advantage of the popularity of social networking activities by offering similar services to engage the consumers. Another Web 2.0 application, called a mashup, is becoming popular on the Internet. A mashup is a combination of information from a variety of Web sources. One example of a health mashup is a global disease site that uses mapping services overlaid with RSS news feeds. Mashups could have global implications for public health. Despite the popularity of these sites, there is a paucity of research validating their usefulness in health care.

Despite the fact that research into the use and evaluation of Web 2.0 in medicine is still in its infancy, individuals involved in the e-health initiative should be aware of its potential. Although some Web 2.0 applications, such as Podcasts, are currently being used, the current pedagogic evidence is sparse (Boulos, Maramba, and Wheeler 2006). Over the next several years it is expected that research into this area will be expanded, possibly demonstrating that Web 2.0 is in the technological future for virtual collaborative clinical practice.

29.3.5 SUMMARY

Although information systems and closed networks have long been used in the health-care industry, the Web is beginning to make its mark on the reformation of the industry. E-health is beginning to change the current state of the U.S. health-care industry through cost reduction, resource management, and improvements to quality of care, reduction of medical errors, augmentation of provider/consumer education, and facilitation of the preventative model of care. E-health technologies leverage the power, ubiquity, ease, and familiarity of the Web and other technologies to enhance health care. Although these technologies are well positioned to deliver these benefits, there are a number of barriers to their proliferation and success.

29.4 BARRIERS TO THE PROLIFERATION OF E-HEALTH

As demonstrated in previous sections, the Web-based technologies are being targeted at, and even already being used for, a variety of applications in health care. However, use of Web-based technologies in health care is still limited when compared to other sectors of the economy that utilize these technologies, as well as when compared to the broad potential for application in the health-care domain. There are a host of issues that contribute to this industry's sluggish pace in adopting, implementing, and utilizing Web-based technologies. These issues include problems stemming from constrictive regulations and policies, economic issues with the costs of development, maintenance, and implementation, organizational barriers, reservations with the usefulness of technology in the medical community, remaining technological limitations, and societal issues concerning consumer demands and equitable access. Table 29.4 provides an overview of these issues.

It should be noted that all of these issues must be addressed if e-health is to reach its potential in aiding health care in those application domains outlined in the previous section (Section 29.3). We will now provide a brief discussion of these different classes of issues affecting the development and implementation of e-health technologies in health care.

29.4.1 Regulatory Issues

Although several other domains have been able to effectively implement Web-based technologies, health care has been largely unsuccessful. However, health care may very well be the most regulated industry (Fried et al. 2000), which has likely contributed to the health-care industry's constant lag in adopting technology. Conversely, it has also been suggested that federal and state regulations, and not the recognized needs and opportunities for growth, have driven the recent surge of interest and adoption of new technologies by health-care providers (Sainfort et al. 2008). The effects of the regulation and policies governing health-care delivery and business practices are far too complicated to sufficiently cover in this chapter. However, we provide an overview of

some of these issues that impact the ability of the health-care industry to use Web-based technologies.

The effects of the health-care regulatory environment go beyond the much discussed Health Insurance Portability and Accountability Act (HIPAA) put forth by the Department of Health and Human Services (DHHS) in 1996. Although the HIPAA act was well intentioned, it did not account for the advancements that would be made on the World Wide Web and therefore was outdated as it was written. The rapid changes in technology require that HIPAA is constantly updated. Because current information systems are extremely complex and often linked to aging legacy systems it makes compliance nearly impossible to keep up with (Califf and Muhlbaier 2003). Regulatory directives have put the health-care industry in a "Catch-22." Pieces of regulations, such as HIPAA, require that health-care organizations develop and implement new information systems in order to become compliant with definitive rules for the privacy, confidentiality, standards, and security of health-care data and the transactions performed with this data. There are severe monetary and punitive (including years in prison) consequences for noncompliance, negligence, or misuse of patient information (Phoenix Health Systems 2003). This has actually encouraged (or forced) health-care organizations to develop new information systems, including utilization of Web-based technologies.

In addition to federal regulations, providers are also under the jurisdiction of state-level control. States have some control over the coding standards and implementation specifications used for health-care information transactions, privacy of patient information, and regulation of health insurance plans (U.S. Department of Health & Human Services 2003d). Differences between state-specific policies, such as professional licensure laws, limit the ability of health-care organizations to deliver health-care services and transfer information across state lines.

Additionally, on top of this complexity, the policies of the third-party payers and insurance companies may not allow for payment or reimbursement for Web-based services. This leaves little incentive for care providers and health-care organizations to invest in these technologies, if they will not be allowed to fully utilize them and/or not get paid for the services rendered through these systems. Moreover, there is existing legislation originally proposed to stave off many of the potential harms inherent in a competitive health-care market, which also inhibits the development and success of e-health. For example, the antikickback, self-referral (aka the Stark law), and beneficiary inducement laws hinder the ability of e-health in areas such as supporting the integration of health-care organizations within an IDN and fostering the relationships between technology developers and the health-care industry (Fried et al. 2000; Kleinke 2000). These vague, outdated laws are problematic for the open access, flexible, integrated nature of online health-care information systems. Clearly, these regulatory issues affect the ability of health-care organizations to take advantage of the power of Web-based technologies.

TABLE 29.4
Overview of Issues that Affect Health Care and the Progress of E-Health

Category	Examples
Regulatory	• Health Insurance Portability and Accountability Act (HIPAA)
	• Antikickback, self-referral, and beneficiary inducement law
	• Difference in state-level regulations
Economic	• Previous investment in legacy systems
	• Large investments for development and implementation
	• Financial disincentives for care providers
Cultural	• Longstanding history of empty IT promises
	• New tools are not more efficient or easier to use
Organizational	• Uncertainty associated with changing roles/relationships
	• Difficulty determining resulting costs/organizational needs
Societal	• Lack of ethical/legal assurance of online information
	• Limited access to technology (the "digital divide")
	• Contrary demands of increased access and increased security
Technological	• Limitations with bandwidth, latency, and network stability
	• Insufficient data security, privacy, and access controls
	• Lack of standardization of medical data coding schemes

29.4.2 Economic Issues

There are economic issues affecting the use and acceptance of e-health technologies. A number of groups have conducted studies and analyses that argue both for (e.g., Centers for Disease Control and Prevention 1999; Hillestad et al. 2005; Littell and Strongin 1996) and against (e.g., Cassell 1993; Drake, Jaffe, and Fitzgerald 1993; Newhouse 1992; Nitzkin 1996; Sidirov 2006) the financial impact of implementing and using e-health technologies. Economic obstacles to e-health include massive implementation, maintenance, and training costs associated with revamping legacy information systems; complex payment mechanisms and economic relationships; and the lack of payment/reimbursement rules for the delivery of services now available through these new technologies.

As previously alluded to when discussing regulatory issues, health-care organizations have a need and/or desire to implement new information systems. However, most e-health technologies are not "ready to use" and require a substantial, initial investment in terms of development, implementation, and training. Additionally, these organizations will need to establish the in-house facilities for proper service, maintenance, and further development of these systems. This investment, especially in light of the disincentives created by regulation and policy, is objectionable for health-care organizations that want to leverage their previous investments in technology. Additionally, there is an inherent paradox with using new technologies as health-care organizations implement new technologies in an effort to streamline costs and improve resource utilization, while these same technologies create the possibility of new products and services, which tend to increase expenditures on supporting new services, training personnel, and maintenance of systems (Starr 2000).

The general payment structure in health care creates a host of issues that need to be addressed. For example, physician use of the Internet, e-mail, and similar applications has suffered because there are no provisions for these new forms of care in either the governmental programs (i.e., Medicare and Medicaid) or the private sector (Kleinke 2000; Landa 2002). Additionally, because care providers are typically rewarded for treatment rather than prevention and based on actual office visits, it leaves little incentive to invest in technologies that promote overall illness prevention and facilitate remote consultation (National Research Council 2000). As with some of the legislation discussed, service payment and reimbursement models were not created with these new care paradigms in mind. All of these issues create economic barriers to the use and adoption of e-health technologies.

29.4.3 Cultural Issues

There are also cultural issues in the medical industry that have limited the adoption and use of e-health technologies. In the medical community, there has been a long-standing lack of faith that technology will provide viable solutions to solve problems within the health-care industry. This is true for two reasons: (1) there has been a history of highly touted IT solutions that have been oversold and underdeveloped and (2) the new IT tools that have been developed often have not meet the needs of practicing clinicians and hospital administrators. As a result, health-care providers and professionals have often reverted back to paper records and legacy systems that have served as trusted solutions (Kleinke 2000).

The health-care industry commonly views new technology as too expensive, not supportive of current work practices, unusable (or not "user-friendly"), difficult to implement, and maintenance intensive (Schoen et al. 2000; Sainfort et al. 2008). Although physician interest and acceptance of technology is changing, particularly with respect to physician adoption and use of the Web and handheld and wireless technologies, there is still a widely held view that an information system that is not 100% reliable is 0% useful (Kleinke 2000). If physicians require too much time and effort or have to endure too much frustration when trying to use new technologies to do the same things that they have been doing relatively efficiently and effectively for years, they tend to discard these technologies as useless or overly complicated. Instead of advanced IT tools, the pen, paper, telephone, fax, and Post-It note are still tools of choice for providers (Kleinke 2005). To increase the use and adoption rates of these new technologies, these tools need to be designed to support the natural workflow of the user.

29.4.4 Organizational Issues

There are also a number of organizational issues that have prevented use of e-health technologies from becoming more widespread. The major organizational issue affecting e-health is how these new technologies will alter the business relationships and the power structure between consumers, practitioners, and payers. The National Research Council (2000) discussed several sources of uncertainty that health-care organizations might face when considering how to implement a Web-based business strategy including organizational and industry structures for Web-based care and service, the technological capabilities of Web-based systems, the internal policies and procedures that will guide use of these technologies, and the human resources needs dictated by these new systems. All of these sources of uncertainty affect if and how health-care organizations will use e-health technologies.

The complex structure of relationships in the health-care industry has created problems for the implementation of e-health technologies. The use of Web-based technologies will likely cause highly unpredictable changes in the business relationships between and practices of payers, insurance companies, health maintenance organizations, and health-care organizations. For example, the increased connectivity between care providers (e.g., physicians) and benefits managers may influence the treatment protocols chosen. Additionally, the empowerment of consumers by providing them with access to more information through the Internet and Web sites may tip the balance of power and grant consumers with the power to determine medications and procedures covered by the managed care organization or the power to directly access specialists without first going through a generalist. The reality is that health-care organizations and insurers/payers cannot forecast all of the changes that may result from Web-based systems, which creates something highly undesirable for businesses: uncertainty.

29.4.5 SOCIETAL ISSUES

There have also been a number of societal issues that have affected the development of e-health technologies. These issues include the lack of established ethical standards to govern online health-related information, uneven access to information technologies and the Web, and assurance of security and confidentiality of patient information. All of these societal issues will have an enormous impact on the extent to which e-health technologies will be used (especially in the public domain) and the future of governmental regulations governing these development, use, and implementation of these technologies.

There are growing concerns about the lack of governing bodies controlling the quality of health-related content on the Web. The health-care information available to consumers can affect their decisions and behaviors, creating a real need for ensuring the accuracy and quality of this information. Current research has shown that patients use several factors to determine Web site accuracy: the endorsement of the site by a government agency or professional organization, their own perception of reliability, and the understandability of the information (Schwartz et al. 2006). Many assessment tools for quality can be found on the Internet published by a variety of organizations. Often, these tools do not provide their criteria to consumers, making it difficult to tell what the rating is based on (Bernstama et al. 2005). This makes it impossible for consumers to determine the validity of the assessment tool, making them rely on their own perceptions of the organization.

Several e-health ethics initiatives have already begun to explore and propose standards for control of information quality, accuracy, and privacy. For example, eHealth Ethics Initiative (2000) established the eHealth Code of Ethics including general guidelines, such as candor, honesty, quality, informed consent, privacy, professionalism in online health care, responsible partnering, and accountability. Although these guidelines for online health information have received some support, enforcement is likely to be achieved only through governmental regulation (as HIPAA legislation was needed to enforce the handling of patient information).

Limited access to the Web and other technologies for many health-care consumers, especially the numerically large Medicaid and Medicare recipient population, is another societal concern. The availability of access to Web-based and computing technologies is a major societal barrier. The well-documented decrease in access to these technologies for various demographic groups, including certain minority groups and those with low income, has been termed the "digital divide." This has been a topic of general debate with respect to equitable access for education and related domains. The widespread use of e-health technologies may well increase the impact of the digital divide by putting these individuals with limited access to technologies at risk of having limited access to health-care information and actual medical care. In addition to the digital divide, there is also an educational divide. Research has shown that the majority of health information available to consumers on the Web is above the reading level of a significant portion of the U.S. population (Graber, D'Allessandro, and Johnson-West 2002). This also raises issues of equitable access for other individuals, such as those with disabilities who may have limited access due to problems with being able to use these technologies.

A large problem for e-health technologies that has been created by contrary consumer demands for increased access to information while still demanding for the highest levels of privacy and data security. The capabilities of Web-based technologies with respect to privacy and security continue to increase as data encryption, access controls, and authentication security functions continually improve and are tested in more commercial Web-based domains. However, the intrinsic personal and stigmatic nature of health-care or medical information brings the seriousness of confidentiality, data integrity, and security to another level. While these societal issues truly impact only health-care consumers currently, these issues may eventually affect health-care providers and insurance companies as governmental regulation spills into this domain of health care.

29.4.6 TECHNOLOGICAL ISSUES

There are a host of technological issues that have raised serious questions about the capability of Web-based technologies to meet the constraints of regulatory mandates, the needs of health-care providers, the demands of consumers, and the policies of payers and insurance companies. Many of these technological issues have been the source of other issues, such as the lack of sufficient data and transaction security causing societal issues or the lack of system stability and the organizational concern for limited downtime and data loss. While we will not discuss many of these technological issues in depth (e.g., how XML works and why it is being proposed as a good alternative to various coding standards), we will discuss the overall impact of how these technological limitations continue to impact the fate of e-health. The National Research Council (2000) suggested five aspects of Web-based technologies that they believe will impact the ability of these technologies to support application domains of health care (e.g., clinical care or public health). These factors include bandwidth, latency, security, availability, and ubiquity. Although this list addresses many of the technological concerns that will impact many of the issues described previously, we would also like to add coding standardization and implementation with legacy systems.

We believe that issues of bandwidth, latency, and availability will become less important as technology continues to advance and become cheaper. Ubiquity, or access to Web-based technologies, will likely continue to be a problem until highly affordable technological solutions become available or governmental regulations require that the digital divide be narrowed. Security for Web-based transactions has already greatly improved, as previously discussed, largely through the efforts for transaction and information security and privacy in the e-commerce domain. Despite improvements in these issues, e-health must still face issues of implementation of these new systems with legacy systems (also an economic issue) and the continual problem with data coding and structure (also a regulatory issue).

The lack of data standardization (or the overabundance of standardization schemes) in health care is an enormous problem for e-health developers. There is a long history of efforts to develop data and coding standards in health care (see Hammond and Cimmino 2006, for a review). In this case, coding standards refer to the syntactical structure of procedural, diagnosis, and billing events. HIPAA legislation sets some new coding standards for data interchange and transactions were determined, largely on the basis of the established EDI standards. Technologies, like XML, have greatly improved the ability to both meet HIPAA compliance and support the integration of transactions with EDI (the HIPAA standard) and non-EDI systems (Yang and Chang 2002).

Standards intended to structure the clinical content of electronic records for exchange between organizations include the Health Level 7 (HL7) Clinical Document Architecture (CDA) and open EHR. Despite the initiatives to produce data standardization in health care, there are many barriers. One of the most complex challenges is semantic interoperability. SNOMED CT, a concept based ontology, contains over 366,000 health-care concepts organized into hierarchies, with approximately 1.46 million semantic relationships between them, and more than 993,420 terms (Dogac et al. 2006). It is currently unclear if these data structuring techniques can truly integrate all of the different coding standards in health care, which number in the dozens.

Developing systems that are truly integrated, which can transfer data seamlessly between disparate entities and facilities, requires that these systems be able to handle all of the various coding standards for health-care data. This comes with heavy development, implementation, and maintenance costs, which is cause for health-care organizations to hesitate making a complete transfer to a Web-based system. Related to this idea is the desire of many health-care organizations to salvage their initial investments in information systems. As a result, e-health developers will have to develop front-end systems and integration engines that will both consolidate the information from these disparate systems, organize it for use, keep the data structure consistent for transfer back to the original system, and translate the data for transactions with systems with inconsistent coding schemas.

Despite these technological woes, there is hope. The travel industry has succeeded in integrated multiple databases, services, and payment systems fairly seamlessly (Parente 2000), the big three players in the automobile manufacturing industry have established standards for interchange and ordering of parts (Lumpkin 2000), and the e-commerce infrastructure

is a successful means of making online transactions using sensitive information. These examples of success in applying Web-based technologies toward issues representative of those to be faced in health care do hold some promise for the future of e-health. However, the costs (economic and otherwise) of system downtime and data loss or piracy in the health-care domain are unparalleled.

Many of these technological issues will recede as technology continues to advance and become cheaper. However, the extent to which these technological issues hinder the advancement of e-health will be largely determined by the resolution of issues in the regulatory, economic, cultural, organizational, and societal domains.

29.4.7 SUMMARY

These issues are all barriers to rapid, widespread acceptance and implementation of e-health technologies. For e-health technologies to help the U.S. health-care system, these barriers will need to be broken down. Many of these issues will take time to resolve, requiring new legislation, new business and delivery of care models, new payment and reimbursement policies, and new research and development groups to guide standardization rules. Fortunately, research focusing on how consumers, health-care professionals, and other users interact with these new systems can help with some of these roadblocks. Human factors and HCI research can help improve the viability of e-health technologies by helping to ensure that the applications and systems developed support the needs of users, effectively help users to complete their work, and provide insight into the faults of unsuccessful designs.

29.5 HUMAN FACTORS AND HUMAN–COMPUTER INTERACTION ISSUES

As can be seen in the previous section, there are still many barriers in the path of e-health. However, we believe that research in human factors and HCI can help improve the chances for e-health to become more widely implemented by health-care and health-related organizations. Realistically, human factors and HCI will not be able to help resolve (at least directly) all of the issues previously discussed. For instance, the changes to the payment policies of insurance companies and third-party payers, as well as governmental regulations, will likely be required to allow for remote consultation and Web-based care to become viable care practices. However, human factors and HCI research can help to feed these processes of change by illustrating how the tools can be used and how the use of these technologies can improve health care, in terms of increased patient satisfaction, and decreased medical errors. Additionally, HCI research should and can involve cost-effectiveness analysis in order to contribute directly to the improvement of organizational and investor interest in implementing these technologies. HCI research can also reveal changes in procedures, increased productivity, and/or decreased errors that can directly translate into cost savings.

The real value of e-health technologies, as well as the ways to design these systems and applications, can be revealed through human factors and HCI research efforts directed at this domain.

As with all domains in which technologies are used, the human element is vital to the success of e-health systems. Developers must address human factors and HCI issues, such as user capabilities and needs, natural user workflow, and contextual design, when designing these new systems. However, as e-health is still in the developmental stage, much of the focus is on establishing the e-health technological infrastructure and battling many of the regulatory and technological issues that prevent the widespread use of these technologies. As a result, there has been limited research focused on human factors and HCI aspects of these technologies. Jacko, Sears, and Sorensen (2001) reported an early investigation into the perceptions of health-care professionals and students regarding the effectiveness of the Web as an information retrieval tool. The study examined one set of potential e-health users (practitioners) and one type of behavior (information retrieval). Clearly, more research is needed to build on this basis so that broader segments of the system are studied, including other stakeholders and additional interaction behaviors. An example of such research involves the use of Web 2.0 by junior physicians. Hughes et al. (2009) found that despite information credibility awareness, over half of the physicians they followed utilized Web 2.0 applications. Fortunately, researchers can draw upon much of the research that has already been conducted in other work domains, such as computer science, industrial engineering, psychology, education, and cognitive science, to name a few.

When discussing the importance of applying human factors and HCI principles to e-health, the following research areas are particularly relevant: (1) human information processing and cognitive modeling, (2) models of work and system use, (3) universal accessibility, (4) computer-mediated collaboration, and (5) testing and evaluation. The next section focuses on these topics, while emphasizing their implications on the development and proliferation of e-health.

29.5.1 Human Information Processing and Cognitive Modeling

Effective human–computer interaction occurs when this interaction coincides with inherent human processing capabilities (Proctor and Vu 2008). The human information processing approach to understanding human–computer interaction has traditionally focused on developing models of human information processing, based on empirical data collected from users performing tasks, which can be used to characterize and predict human behavior or performance. There has been a long history of efforts in human factors and HCI to develop representative models and cognitive architectures that characterize human cognitive activities and behavior during interaction with computer systems (see Byrne 2008; Yoshikawa 2003; van Rijn et al., this volume, for reviews). Some of these

models include the Model Human Processor (Card, Moran, and Newell 1983), the Executive-Process Interactive Control (EPIC) architecture (Meyer and Kieras 1997), the Adaptive Control of Thought (ACT) model (Anderson, Matessa, and Lebiere 1997), and the GOMS family of models (John and Kieras 1996).

Historically, a variety of problems with the design and implementation of information systems and technologies have stemmed from a failure to properly consider the cognitive needs of users (Tang and Patel 1994). These problems include decreased productivity, increased user frustration, increased error rates, increase user stress, and poor decision making, among a host of problems. Additionally, designers must recognize that human information processing behaviors and capabilities fluctuate with internal and environmental changes that affect the cognitive or sensory behavior of the user. Although there is a large base of HCI research examining cognition and information processing, further research is needed in the health-care domain to understand how these theories and models apply to use of e-health technologies.

29.5.2 Modeling Work and System Use

The inherent complexity and required flexibility of health-care information has contributed to the failure of past IT solutions and partially accounts for why physicians have often reverted back to paper-based charts (Kleinke 2000). Part of the reason that many e-health ventures have failed (especially EMRs) is the fact that these systems have not properly supported the work of health-care professionals. The concept of context and workflow support is the idea that systems and technologies should be designed with consideration for the actual parameters of a user's work, in the context that this work actually occurs. Unfortunately for e-health systems designers, they need a solid understanding of the actual working conditions of health-care professionals and consumers in addition to an understanding of their information processing and cognitive capabilities. Fortunately, however, there are already a number of methods for understanding actual user work.

HCI research has developed a number of methods involving user-centered design or participatory design (see Muller 2008; Norman and Draper 1986, for reviews), which focus on getting representative users involved in the design and development process. Additionally, methods of contextual design or ethnography (see Beyer and Holtzblatt 1998; Blomberg and Burrell 2008; Holtzblatt 2008, for reviews) focus on observing how users behave in their actual work environment. These overall "good practices" during the design life cycle of systems development help to gather valuable insights into user characteristics and behaviors, in the actual work environment, to help guide system requirements needed for design (Mayhew 2008, this volume). HCI research has produced several modeling techniques and methodologies, which model the constraints of the user, work, and environment, that have been shown to have the potential to guide system design, lessen user workload, minimize errors, and improve performance in various domain such as process control, aviation, and manufacturing. These techniques include the hierarchical task analysis (Shepherd 1989) and the abstraction-decomposition space (Rasmussen 1985), among others.

These various methods for investigating the work practices of e-health systems users in a real work environment will help designers and developers to better understand the needs of these users, the constraints of the work environment, and the work practices that define the actual use of these technologies. Examining the wide variety of users, usage scenarios, workflow practices, and contexts of use will help e-health technologies designers and developers to produce applications and systems that can be successfully implemented, achieve high user adoption rates, and benefit the health-care industry.

29.5.3 Universal Accessibility

Given the wide array of users, and their respective needs, knowledge, and abilities, effective e-health system must adhere to the principles of universal design and access. E-health technologies, especially those used by consumers, need to be designed to support varying user functional capacities. The concept of universal accessibility goes beyond the issues of physical accessibility or opportunity to use technologies (as with the digital divide). Universal accessibility refers to the design of information technologies that can be used by all individuals, including individuals with sensory, motor, and cognitive impairments, older and aging individuals, and individuals from different cultures (see Stephanidis 2001, for a review; see also Stephanidis and Akoumianakis, this volume and Rau et al., this volume). This issue has been the focus of countless studies in HCI. An in-depth discussion of the relevance of universal accessibility and "design for all" in health care is outside the scope of this chapter. However, suffice it to say that health care has a seemingly infinite variety of users, with unique needs and abilities. To ensure that these individuals can use e-health technologies and take advantage of all the benefits (e.g., increased care, social support, education) that these technologies can provide, designers and developers must ensure that these systems are accessible to everyone.

To understand the importance of this issue to the prescribed problems of health care, one can consider the need to make e-health tools accessible to individuals with chronic illness and impairment. As previously discussed, the majority of health-care expenditures in the United States can be attributed to individuals with chronic illness. Some of these individuals suffer from various cognitive or sensory disorders, which impact their abilities to interact with technology. For example, people with visual impairments or blindness (one class of chronic illness) number in the millions. If e-health tools are not accessible to, and/or usable by, these individuals, the potential economic and quality of care benefits brought about by the use of these technologies is effectively diminished. On a related note, it has been proposed that many situationally induced impairments (SIIs) result in similar performance decrements as disability-induced impairments (DIIs) (Sears et al. 2003). This further supports recognizing

the unique needs of different classes of users and use scenarios, as even users with normal abilities are sometimes compromised. For example, a normally abled physician working in a highly distracting environment, such as an ER, may not be able to effectively use an e-health tool as would have been possible in a less distracting environment. Research in HCI will help address many of these issues, ensuring that all users have consistent, equal, and reliable access to vital health-care information.

29.5.4 Computer-Mediated Collaboration

Study of collaborative work has been conducted for a few decades, as the advent of networking for information technologies has enabled new methods of interactive working (Grudin 1991). Computer-mediated collaboration, also known as computer-supported cooperative work (CSCW), is the study of the development and use of software that allows for geographic and/or temporal flexibility in the collaboration of individuals to accomplish work (Olson and Olson 2008). The health-care domain is an extremely complicated amalgamation of participants, agents, and stakeholders, each with their own goals, knowledge, strategies, and needs. This makes the much-needed collaboration between players in health care, including professionals, consumers, and organizations, extremely important, yet very difficult. While some research has found that the use of Web-based technologies (e.g., e-mail) in health care can enhance interactions and collaboration, particularly through patient involvement in the health-care process (MacDonald, Case, and Metzger 2001; Safran et al. 1998), other research has noted that much more work needs to be done in the understanding of how technology can be used to support this collaboration (Patel et al. 1999; Patel and Kushniruk 1998).

The support of this distanced (either by time or space) collaboration, in terms of business transactions, care delivery from clinician to patient, or problem solving by a group of health-care professionals, is a fundamental aspect of e-health technologies. The onset of the managed care model of health care, as well as the push toward preventative medicine, has made the communication among health-care entities and between practitioners and consumers a crucial issue. Currently, patients are typically required to see a primary care physician, who will then refer this patient to a specialist as need be. It should also be noted that during this process, insurance companies and third-party payers are at work behind the scenes to handle payment, transaction, and benefits issues. This complex system of information sharing for physician referrals is just one example of the need for researchers to examine how e-health technologies will change the social dynamics and communication of work in health care.

29.5.5 Testing and Evaluation

Finally, the penchant for testing and evaluation is another benefit that HCI research can offer to the development of e-health systems. As long as health-care information systems are being built, there will always be a need (and a desire) to evaluate and improve user interaction with these systems. E-health technologies have suffered from lack of user acceptance because these new tools have largely not supported users or provided users with usable tools. Research in human factors and HCI, directed at e-health technologies, has the potential to remedy many of these issues.

Testing and evaluation are particularly important for e-health applications because of both the potential consequences of poor design and the history of undertesting these technologies before implementation. The so-called "discount usability methods" such as heuristic evaluation (Nielsen 1992), cognitive walkthrough (Wharton et al. 1992), and heuristic walkthrough (Sears 1997) have also been shown to have value in diagnosing potential usability problems with systems and design prototypes (see Cockton, Lavery, and Woolrych 2008; Vu et al., this volume, for reviews). These techniques can be used by e-health developers, in tandem with representative users and subject matter experts (SMEs), to locate issues with the design or functionality of the systems or application that may decrease its usability. These formative evaluations can help to improve e-health products and reverse the longstanding view by health-care professionals that these new technologies are unreliable or not useful.

Owing to the dynamic environment of health-care settings, there is increased demand for innovative solutions for usability testing. Recently, mobile solutions have been utilized for automatic and synchronized recording of user interactions with medical systems and applications (Jaspers 2008). These mobile automatic tools make usability testing *in vivo* a possibility. They also dramatically reduce the amount of time and effort required to analyze data from end users at the conclusion of the study.

Human error identification is a huge realm where testing and evaluation during the design life cycle (e.g., prior to implementation) can help to decrease the well-documented problem of medical errors. Techniques such as Systematic Human Error Reduction and Prediction Approach (SHERPA; Embrey 1986) and Task Analysis for Error Identification (TAFEI; Baber and Stanton 1994) are methods for analyzing work activity and human–machine interaction in order to analyze work tools and processes that lead to errors. Incidents caused by errors or malfunctions are considered to be the cause of many failures that cause patients undue harm. By applying safety management systems (SMS) tools used in areas such as aviation and nuclear energy production to medicine, these errors may be able to be analyzed prior to major accidents occurring (Cacciabuea and Vella 2010). These and other models of human error identification have been shown to have predictive power in assessing the likelihood for errors during interaction (Baber and Stanton 1996; Hollnagel, Kaarstad, and Lee 1998). These techniques can be used to help decrease the number of medical errors that occur during interaction with new or poorly designed technologies.

Although a detailed view of the methods and metrics that are most appropriate for the evaluation of e-health technologies is outside our current scope, Sainfort et al. (2008) provide

a more extensive discussion. As can be seen, the testing and evaluation methods used in human factors and HCI research will help to reveal problems with the design of these technologies. This information can ultimately be used to improve the viability of e-health technologies by ensuring that these new tools actually do what they have been designed to do—decrease medical errors, improve work proficiency, support user needs, improve care, and decrease costs.

29.5.6 Summary

Work in human factors and HCI is important to ensure that e-health systems and applications are usable, accessible, efficient, and supportive as these technologies advance. As with many other domains, such as with automation in aviation (Parasuraman and Riley 1997; Wiener and Curry 1980), the technical capabilities and economic benefits of technology are often the determinants of employing new technologies, while the effects on the human users are often ignored. Human factors and HCI research of e-health technologies will bring the focus back to the users of these new systems and how these technologies can be designed and implemented in order to best support the needs of these users. We have suggested that the focus of this research should be on supporting human information processing needs, modeling and supporting the natural workflow practices, ensuring universal accessibility, examining the inherent collaborations in the health-care process, and developing solid testing and evaluation method. This work will be important in increasing the acceptance rates of technology and ensuring users leverage the power and efficiency of these tools. In early stages of design, and especially as the e-health infrastructure develops further, attention to human factors and HCI issues will become paramount in keeping e-health from simply becoming another health-care IT pipedream.

29.6 PREDICATIONS FOR E-HEALTH TECHNOLOGY IN HEALTH CARE

E-health technologies are becoming more widely used despite the many barriers to the development and implementation of these systems. For example, many health-care providers are already allowing some consumer control and management of health insurance claims over the Web (Faulkner and Gray 2000). We predict that the use of Web-based health-care information systems and applications will continue to increase. More and more health-care organizations, health-related commercial enterprises, individual practitioner, and insurance companies will begin to implement these technologies as they will no longer be able to ignore the inadequacies of their current information systems and the potential benefits of these Web-based technologies. As these health-care entities recognize the potential for these technologies to improve quality of care, decrease medical errors, provide administrative and process streamlining, simplify and speed transactions, and provide health-care professionals with the technological support they need to function in spite of an increasingly complex, information- and process-laden field. The increasing development and ubiquity of Web-based technologies, combined with the ever-increasing numbers of individuals seeking health information and services online, will feed this movement toward a Web-based health-care technological infrastructure. E-health developers and investors will naturally fall in line, feeding on the potential for a financial windfall.

In the previous edition we predicted that the following applications of Web-based technologies would have an immediate impact on the past few years: online consumer information services, online support groups and virtual communities, remote consultation and communication (primarily via e-mail), the online transaction infrastructure between health-care provider groups, third party payers, and insurance/ benefits companies, electronic medical records (EMRs) and other Web-based charts, and Web-based remote monitoring via sensor technologies. Since the last edition, research has demonstrated that patients of all ages are interested in patient portals or online services such as viewing parts of their medical record, messaging with their physician, online medication refills, appointment requests, and billing inquiries (Adler 2005; Liederman et al. 2005; Kummervold et al. 2008). Studies have even demonstrated that patients would pay a small annual fee for the convenience of online services with their primary care physician's office, thus supporting the economic feasibility of patient portals (Adler 2006).

Despite the excitement about cutting-edge technologies, such as biomedical nanotechnologies, we were correct that widespread application of these technologies did not broadly impact medicine. Instead, consumers drove the market for change by increasing their use of online e-commerce Web sites, online support groups, and by demanding Web communication and scheduling. The U.S. health-care system still remains on the verge of major overhaul of information systems as paper charts and legacy systems still permeate the market. Web-based technologies, while still limited by various barriers, continue to be well positioned to make an enormous impact on several of the problems outlined by the Institute of Medicine (2001) and other researchers. On the basis of the progress of past Web technologies in health care, we predict that e-prescribing, electronic information exchange (regional and possibly national), personal health records (PHRs), and Web 2.0 applications will be the largest areas of growth in new technologies in the next several years.

The interconnectivity generated by a Web-based infrastructure has helped the health-care industry support the dispersed nature of the IDN organizational model, support the collaboration required by the managed care model, and enabled practitioners, providers, and public health organizations to provide preventative care and educational materials. A key piece necessary to support the IDN organizational model is interoperability. Organizations such as the International Standards Organization (ISO) Technical Committee 215; Health Level Seven (HL7); Comité Européen de Normalisation (CEN); Digital Imaging and Communications in Medicine (DICOM); GS1; Clinical Data Interchange Standards Con-

sortium (CDISC); and International Health Terminology Standards Developing Organization (IHTSDO) have created standards for planning, data and information models, ontology and terminologies, data definitions and attributes, data exchange, clinical decision support, and query standards. Despite these efforts, true interoperability has not yet been achieved. Standards are not consistently applied and are often overlapping or competing. We predict that over the next several years more focus will be placed on standardization in order to connect information to improve patient health.

The ability for consumers to easily seek continuous, remote care from providers, learn more about their condition or illness, and find social support from virtual communities all on the Web has helped to reinforce the preventative care and health maintenance necessary to better manage the costly health care of the chronically ill masses. The Veteran's Health Administration (VHA) successfully completed a national home telehealth program by the end of 2007. For an annual cost lower than nursing home care the VHA found that the Care Coordination/Home Telehealth (CCHT) patients experienced a 25% reduction in bed days, 19% reduction in hospital admissions, and mean satisfaction score rating of 86% after enrollment at an annual cost per patient (Darkins et al. 2008). The open, flexible nature of the Web will allow the ever-increasing body of health-care and medical knowledge and research to be easily manageable and accessible by health-care professionals, providers, and consumers. This will help to increase consumer education, improve professional education and training, and keep providers and payers informed of the latest research and clinical trials results. The open, public nature of most Web-based technologies will also begin to quell the increasing consumer demand discussed by Sainfort et al. (2008). Web-based technologies can finally provide consumer with the increased convenience to health-care information and services. Finally, as Web-based technologies become more ubiquitous, less cost prohibitive, and more secure, the constraints on using these technologies in health care will start to dissolve.

29.7 CONCLUSION

This chapter provided an overview of e-health, from the historical factors that transformed the nature of health care in the United States to the applications of e-health that have come to be used and for which use will be even more widespread in the near future. We discussed the factors that have led to changes in the technological needs of U.S. health-care industry and provided an overview of e-health, including why the Web has been the proposed platform, and several application domains that have been targets for implementation of Web-based technologies. Then, an overview was provided of the various barriers, including regulatory, economic, organizational, cultural, societal and technological issues, which stand in the path of e-health developers and proponents. Finally, we discussed the role that research and theory in human factors and HCI is playing in improving the viability of e-health technologies.

Despite the controversy surrounding the ability of these Web-based systems and applications to break through the barriers and provide viable solutions to the industry's problems, e-health is rapidly becoming the new technological infrastructure of health care in the United States. There are many hospitals, physician offices, e-Commerce health-care ventures, and insurance companies who are testing and using Web-based information systems and applications. Evidence of e-health's benefits, including cost savings, increased patient satisfaction, and fewer medical errors, indicates that these technologies will receive even more interest from investors and lead decision makers in the industry in the future.

Although new technologies have been created or adopted in order to meet increasing demands of consumers for new and better services and information, many problems still remain. The increasingly stringent regulations by the government about the accessibility, security, format, and storage of health-care and medical data, and the continuing economic drive to decrease health-care costs and support managed care, public health, and the chronically ill still remain as challenges. The passing of the American Recovery and Relief Act of 2009 has released much needed funding in order to improve Health IT. However, the mere availability of funding and new technologies will not ensure e-health's success. Rather, the ability of e-health technologies to deliver on their promises will be a result of attention from researchers in the fields of engineering, computer science, cognitive science, psychology, medicine, and biological sciences to make sure that these systems are designed to support the natural work practices of users, ensure universal accessibility for all user groups, support the cognitive and information processing needs of users, enhance the collaboration between users, and ensure all of these benefits through proper testing and evaluation during the engineering design life cycle. This will be a job for you, the readers and the researchers and practitioners in human factors and HCI.

ACKNOWLEDGMENT

The preparation of the second edition of this chapter was supported by a fellowship provided to the third author by the University of Minnesota Institute for Health Informatics.

REFERENCES

Adler, K. G. 2006. Web portals in primary care: an evaluation of patient readiness and willingness to pay for online services. *Journal of Medical Internet Research* 8(4): e26, doi:10.2196/jmir.8.4.e26.

Recovery.gov. 2009. American Recovery and Reinvestment Act, Jan. 6, 2009. http://frwebgate.access.gpo.gov/cgi-bin/getdoc .cgi?dbname=111_cong_bills&docid=f:h1enr.pdf (accessed Feb. 22, 2009).

Anderson, J. R., M. Matessa, and C. Lebiere. 1997. ACT-R: A theory of higher level cognition and its relation to visual attention. *Human–Computer Interaction* 12: 439–462.

Agrawal, A. 2009. Medication errors: prevention using information technology systems. *British Journal of Clinical Pharmacology* 67(6): 681–686, doi:10.1111/j.1365-2125.2009.03427.x.

Arrow, K., A. Auerbach, J. Bertko, S. Brownlee, L. P. Casalino, J. Cooper, et al. 2009. Toward a 21st-century health care system: Recommendations for health care reform. *Annals of Internal Medicine* 150(7): 493–495.

Baber, C., and N. A. Stanton. 1994. Task analysis for error identification: A methodology for designing error tolerant consumer products. *Ergonomics* 37(11): 1923–1941.

Baber, C., and N. A. Stanton. 1996. Human error identification techniques applied to public technology: Predictions compared with observed use. *Applied Ergonomics* 27(2): 119–131.

Bashshur, R., and J. Lovett. 1977. Assessment of telemedicine: Results of the initial experience. *Aviation Space and Environmental Medicine* 48(1): 65–70.

Benton Foundation. 1999. Networking for better care: Health care in the information age. http://www.benton.org/Library/health/ (accessed Dec. 5, 2001).

Beyer, H., and K. Holtzblatt. 1998. *Contextual Design: Defining Customer Centered Systems*. San Francisco, CA: Morgan Kaufmann.

Bernstama, E. V., D. M. Sheltona, M. Walija, and F. Meric-Bernstamb. 2005. Instruments to assess the quality of health information on the World Wide Web: What can our patients actually use? *International Journal of Medical Informatics* 74: 13–19.

Blomberg, J., and M. Burrell. 2008. The ethnographic approach to design. In *Human–Computer Interaction Handbook: Fundamentals, Evolving Technologies and Emerging Applications*, 2nd ed., eds. J. A. Jacko and A. Sears, 965–990. Mahwah, NJ: Lawrence Erlbaum.

Boulos, M., I. Maramba, and S. Wheeler. 2006. BioMed Central | Full text | Wikis, blogs and podcasts: A new generation of Web-based tools for virtual collaborative clinical practice and education. *BioMed Central*, Aug. 15, http://www.biomedcentral.com/1472-6920/6/41.

Byrne, M. D. 2008. Cognitive architecture. In *Human–Computer Interaction Handbook: Fundamentals, Evolving Technologies and Emerging Applications*, 2nd ed., eds. J. A. Jacko and A. Sears, 93–114. Mahwah, NJ: Lawrence Erlbaum.

Cacciabuea, P., and G. Vella. 2010. Human factors engineering in healthcare systems: The problem of human error and accident management. *International Journal of Medical Informatics* 79(4), 1-17. doi: 10.1016/j.ijmedinf.2008.10.005

Cain, M. M., R. Mittman, J. Sarasohn-Kahn, and J. C. Wayne. 2000. *Health e-People: The Online Computer Experience*. Oakland, CA: Institute for the Future, California Health Care Foundation.

Califf, R., and L. Muhlbaier. 2003. Health insurance portability and accountability act (HIPAA): Must there be a trade-off between privacy and quality of healthcare, or can we advance both. *Journal of the American Heart Association* 108(8): 915–918.

Card, S. K., T. P. Moran, and A. Newell. 1983. *The Psychology of Human–Computer Interaction*. Hillsdale, NJ: Lawrence Erlbaum.

Card, S. 2008. Information visualization. In *Human–Computer Interaction Handbook: Fundamentals, Evolving Technologies and Emerging Applications*, 2nd ed., eds. J. A. Jacko and A. Sears, 509–505. Mahwah, NJ: Lawrence Erlbaum.

Cassell, E. J. 1993. The sorcerer's broom: medicine's rampant technology. *Hastings Center Report* 23(6): 32–39.

Centers for Disease Control and Prevention. 1999. New data show AIDS patients less likely to be hospitalized. June. http://www .cdc.gov/od/oc/media/pressrel/r990608.htm (accessed Nov. 16, 2002).

Cherry, J. C., S. J. Colliflower, and A. Tsiperfal. 2000. Meeting the challenges of case management with remote patient monitoring technology. *Lippincott's Case Management* 5(5): 191–198.

Classen, D. C., A. J. Avery, and D. W. Bates. 2007. Evaluation and certification of computerized provider order entry systems. *Journal of the American Medical Informatics Association* 14(1): 48–55.

Cockton, G., D. Lavery, and A. Woolrych. 2008. Inspection-based evaluations. In *Human–Computer Interaction Handbook: Fundamentals, Evolving Technologies and Emerging Applications*, 2nd ed., eds. J. A. Jacko and A. Sears, 1171–1190. Mahwah, NJ: Lawrence Erlbaum.

Darkins, A., P. Ryan, R. Kobb, L. Foster, E. Edmonson, B. Wakefield, and A. E. Lancaster. 2008. Telemedicine and e-health. 14(10): 1118–1126, doi:10.1089/tmj.2008.0021.

Della Mea, V. 2002. What is e-health (2): The Death of Telemedicine? [editorial]. *Journal of Medical Internet Research* 3(2): e22.

De Rouck, S., A. Jacobs, and M. Leys. 2008. A methodology for shifting the focus of e-health support design onto user needs: A case in the homecare field. *International Journal of Medical Informatics* 77(9): 589–601.

Dix, A. J., and N. Shabir. this volume. Human–computer interaction. In *Handbook of Human Factors in Web Design*, 2nd ed., eds. K.-P. L. Vu and R. W. Proctor, 35–62. Boca Raton, FL: CRC Press.

Dogac, A., T. Namli, A. Okcan, G. Laleci, Y. Kabak, and M. Eichelberg. 2006. Key issues of technical interoperability solutions in eHealth. eHealth Conference 2006. http://www.srdc.com.tr/~asuman/GenevaSemInteroperabilityTalk.ppt (accessed November 9, 2010).

Drake, D., M. Jaffe, and S. Fitzgerald. 1993. *Hard Choices: Health Care at What Cost?* Kansas City, MO: Andrews and McMeel.

eHealth Ethics Initiative. 2000. eHealth code of ethics. Internet Healthcare Coalition. http://www.ihealthcoalition.org/ethics/code0524.pdf (accessed Sept. 15, 2002).

Embrey, D. E. 1986. SHERPA: A systematic human error reduction and prediction approach. Paper presented at the International Meeting on Advances in Nuclear Power Systems (April 1986). Knoxville, TN.

Eysenbach, G. 2001. What is e-health? [editorial]. *Journal of Medical Internet Research* 3(2): e20.

Faulkner and Gray. 2000. *Faulkner & Gray's Health Data Directory, 2000 Edition*. New York: Faulkner & Gray.

Ferguson, T. 2000. From doc-providers to coach-consultants: Type 1 vs. type 2 provider-patient relationships. *The Ferguson Report*. http://www.fergusonreport.com/articles/tfr07-01.htm (accessed Oct. 10, 2002).

Fonkych, K., and R. Taylor. 2005. The state and pattern of HEALTH information technology adoption. Rand Publications. http://www.rand.org/pubs/monographs/2005/RAND_MG409.pdf (accessed Jan. 18, 2010). Fox, J., V. Patkar, I. Chronakis, and R. Begent. 2009. From practice guidelines to clinical decision support: closing the loop. *Journal of the Royal Society of Medicine*. 102(11): 464–473.

Fox, S., and L. Rainie. 2002. Vital decisions: How Internet users decide what information to trust when they or their loved ones are sick. Pew Internet and American Life Project. (accessed Feb. 3, 2003).

Fried, B. M., G. Weinreich, G. M. Cavalier, and K. J. Lester. 2000. E-health: Technological revolution meets regulatory constraint. *Health Affairs* 19(6): 124–131.

Garritty, C., and K. El Emam (2006, May 12). Who's using PDAs? Estimates of PDA use by health care providers: A systematic review of surveys. *National Center for Biotechnology Information*, May 12. http://www.ncbi.nlm.nih.gov/pmc/articles/PMC1550702/ (accessed Jan. 19, 2010).

Giorgianni, S. J., J. Grana, and S. Sewell. 2000. E-health care: The Internet information revolution. In *The Pfizer Journal*, vol. 4, no. 2. New York: Impact Communications.

Goldsmith, J. 2000. How will the Internet change our health system? *Health Affairs* 19(1): 148–156.

Graber, M. A., D. M. D'Allessandro, and J. Johnson-West. 2002. Reading level of privacy policies on Internet health Web sites. *Journal of Family Practice* 51(7): 642–645.

Grudin, J. 1991. CSCW: the development of two development contexts. *Proceedings of the ACM Conference on Human Factors in Computing Systems (CHI'91)* 91–97.

Gustafson, D. H., R. Hawkins, E. Boberg, S. Pingree, R. E. Serlin, F. Graziano, et al. 1999. Impact of a patient-centered, computer-based health information/support system. *American Journal of Preventive Medicine* 16(1): 1–9.

Hammond, W. E., and J. L. Cimmino. 2006. Standards in biomedical informatics. In *Biomedical Informatics: Computer Applications in Health Care and Biomedicine*, 3rd ed., eds. E. H. Shortliffe and J. L. Cimmino, 265–311. New York: Springer.

Harris Interactive. 2007. Harris poll shows number of "cyberchondriacs"—adults 53. Who have ever gone online for health information—increases to an estimated 160 million nationwide. *The Harris Poll* July 31.

Harris, J. 1995. Consumer health information demand and delivery: A preliminary assessment. In *Partnerships for Networked Health Information for the Public* (Summary Conference Report). Washington, DC: Office of Disease Prevention and Health Promotion, U.S. Department of Health and Human Services.

Hartmann, C. W., C. N. Sciamanna, D. C. Blanch, S. Mui, H. Lawless, M. Manocchia, et al., n.d. A Website to improve asthma care by suggesting patient questions for physicians: qualitative analysis of user experiences. *National Center for Biotechnology Information*. http://www.ncbi.nlm.nih.gov/pmc/articles/PMC1794671/ (accessed Jan. 19, 2010).

Hawn, C. 2009. Games for health: the latest tool in the medical care arsenal. *Health Affairs* 28(5): w842.

Hesse B. W., D. E. Nelson, G. L. Kreps, R. T. Croyle, N. K. Arora, B. K. Rimer, and K. Viswanath. 2005. Trust and sources of health information: The impact of the Internet and its implications for health care providers: Findings from the first Health Information National Trends Survey. *Archives of Internal Medicine* 165: 2618–2624.

Hillestad, R., J. Bigelow, A. Bower, F. Girosi, R. Meili, R. Scoville, et al. 2005. Can electronic medical record systems transform health care? Potential health benefits, savings, and costs. *Health Affairs* 24(5): 1103.

Hollnagel, E., M. Kaarstad, and H.-C. Lee. 1998. Error mode prediction. *Ergonomics* 42: 1457–1471.

Holtzblatt, K. 2003. Contextual design. In *Human–Computer Interaction Handbook: Fundamentals, Evolving Technologies and Emerging Applications*, eds. J. A. Jacko and A. Sears, 941–963. Mahwah, NJ: Lawrence Erlbaum.

Hughes, B., I. Joshi, H. Lemonde, and J. Wareham. 2009. Junior physician's use of Web 2.0 for information seeking and medical education: A qualitative study. *International Journal of Medical Informatics* 78: 645–655.

Institute of Medicine. 2001. *Crossing the Quality Chasm: A New Health System for the 21st Century*. Washington, DC: National Academy Press.

Institute of Medicine. 2008. *Retooling for an Aging America: Building the Health Care Workforce*. Washington, DC: National Academy Press.

Jacko, J. A., A. Sears, and S. J. Sorenson. 2001. Framework for usability: Healthcare professionals and the Internet. *Ergonomics* 44(11): 989–1007.

Jadad, A. R. 1999. Promoting partnerships: Challenges for the Internet age. *BMJ* 319: 761–764.

Jaspers, M. W. 2008. A comparison of usability methods for testing interactive health technologies: Methodological aspects and empirical evidence. *International Journal of Medical Informatics* 78(5): 340–353.

Jha, A. K., D. C. Chan, A. B. Ridgway, C. Franz, and D. W. Bates. 2009. Improving safety and eliminating redundant tests: Cutting costs in U.S. hospitals. *Health Affairs* 28(5): 1475–1484.

John, B. E., and D. E. Kieras. 1996. The GOMS family of user interface analysis techniques: Comparison and contrast. *ACM Transactions on Computer–Human Interaction* 3(4): 320–351.

Kassirer, J. P. 2000. Patients, physicians, and the Internet. *Health Affairs* 19(6): 115–123.

Kidd, M. R. 2008. Personal electronic health records: MySpace or HealthSpace? *BMJ* 336: 1029–1030, doi:10.1136/bmj.39567.550301.80.

Kleinke, J. D. 2000. Vaporware.com: The failed promise of the health care Internet. *Health Affairs* 19(6): 57–71.

Kleinke, J. 2005. Dot-Gov: Market failure and the creation of a national health information technology system. *Health Affairs* 24(5): 1246.

Kummervold, P. E., C. E. Chronaki, B. Lausen, H. Prokosch, J. Rasmussen, S. Santana, et al. 2008. eHealth trends in Europe 2005–2007: A population-based survey. *Journal of Medical Internet Research* 10(4): e42, doi:10.2196/jmir.1023.

Landa, A. S. 2002. Telemedicine payment expansion sought. Amednews.com. http://www.ama-assn.org/amednews/2002/07/08/gvsc0708.htm (accessed March 12, 2003).

LeGrow, G., and J. Metzger. 2001. *E-Disease Management.* Oakland: California Healthcare Foundation.

Liederman, E. M., J. C. Lee, V. H. Baquero, and P. G. Seites. 2005. Patient–physician web messaging: The impact on message volume and satisfaction. *Journal of General Internal Medicine* 20: 52–57.

Littell, C. L., and R. J. Strongin. 1996. The truth about technology and health care costs. *IEEE Technology and Society Magazine* 15(3): 10–14.

Lumpkin, J. R. 2000. E-health, HIPAA, and beyond. *Health Affairs* 19(6): 149–151.

MacDonald, K., J. Case, and J. Metzger. 2001. *E-Encounters.* Oakland: California Healthcare Foundation.

Masterson, L. 2008. Two phrases from AHIP: Consumerism and interactivity. HealthLeaders Media for Healthcare Executives—HealthLeaders Media. June 25. http://healthplans.hcpro.com/print/IIEP-214109/Two-Phrases-from-AHIP-Consumerism-and-Interactivity (accessed Jan. 18, 2010).

Mayhew, D. J. 2008. Requirements specifications within the usability engineering life cycle. In *Human–Computer Interaction Handbook: Fundamentals, Evolving Technologies and Emerging Applications,* eds. J. A. Jacko and A. Sears, 917–926. Mahwah, NJ: Lawrence Erlbaum.

Mayhew, D. J. this volume. The web UX design process—a case study. In *Handbook of Human Factors in Web Design,* 2nd ed., eds. K.-P. L. Vu and R. W. Proctor, 461–480. Boca Raton, FL: CRC Press.

McCormack, J. 2000. Group practices find their way to the Internet. *Health Data Management* 8(1): 46–53.

McLean, R., B. H. Richards, and J. Wardman. 2007. The effect of Web 2.0 on the future of medical practice and education: Darwikinian evolution or folksonomic revolution? Health IT

and Viewpoint. August 6. https://www.mja.com.au/public/issues/187_03_060807/mcl10181_fm.pdf (accessed Jan. 18, 2010).

Metzger, J., and K. MacDonald. 2002. Clinical decision support for the independent practice physician. Oakland: California Healthcare Foundation.

Meyer, D. E., and D. E. Kieras. 1997. A computational theory of executive cognitive processes and multiple-task performance: Part 2. Accounts of psychological refractory-period phenomena. *Psychological Review* 104: 749–791.

Meyer, J. 2001. Age: 2000. Census 2000 Brief, October 2001 (C2KBR/01-12). U.S. Census Bureau. http://www.census.gov/prod/2001pubs/c2kbr01-12.pdf (accessed Oct. 4, 2002).

Meyer, M., R. Kobb, and P. Ryan. 2002. Virtually healthy: Chronic disease management in the home. *Disease Management* 5: 87–94.

Mitchell, J. 1999. *From Telehealth to E-health: The Unstoppable Rise of e-health.* Canberra, Australia: National Office for the Information Technology. http://www.noie.gov.au/projects/ecommerce/ehealth/rise_of_ehealth/unstoppable_rise.htm.

Mittman, R., and M. Cain. 1999. *The Future of the Internet in Health Care: Five Year Forecast.* Oakland: California Healthcare Foundation.

Mollica, R. L., and J. Gillespie. 2003. *Care Coordination for People with Chronic Conditions.* Portland, ME: National Academy for State Health Policy.

Muller, M. J. 2008. Participatory design: the third space in HCI. In *Human–Computer Interaction Handbook: Fundamentals, Evolving Technologies and Emerging Applications,* eds. J. A. Jacko and A. Sears, 1061–1082. Mahwah, NJ: Lawrence Erlbaum.

Mullner, R. 2002. The Internet and healthcare: Opportunities and challenges. *Journal of Medical Systems* 26(6): 491–493.

National Research Council. 2000. *Networking Health: Prescriptions for the Internet.* Washington, DC: National Academies Press.

Newhouse, J. P. 1992. Medical care costs: How much welfare loss? *Journal of Economic Perspectives* 6(3): 3–21.

Nielsen, J. 1992. Finding usability problems through heuristic evaluation. In *Proceedings of the ACM Conference on Human Factors in Computing Systems (CHI'92),* eds. P. Bauersfield, J. Bennett, and G. Lynch, 373–380. New York: ACM Press.

Nitzkin, J. L. 1996. Technology and health care—driving costs up, not down. *IEEE Technology and Society Magazine* 15(3): 40–45.

Norman, D. A., and S. Draper. 1986. *User Centered System Design.* Mahwah, NJ: Lawrence Erlbaum.

Olson, G. M., and J. S. Olson. 2008. Groupware and computer-supported cooperative work. In *Human–Computer Interaction Handbook: Fundamentals, Evolving Technologies and Emerging Applications,* eds. J. A. Jacko and A. Sears, 545–558. Mahwah, NJ: Lawrence Erlbaum.

O'Reilly, T. 2005. What is Web 2.0? In *Technology Books, Tech Conferences, IT Courses, News.* Sept. 30. O'Reilly Media.

Parasuraman, R., and V. Riley. 1997. Humans and automation: Use, misuse, disuse, abuse. *Human Factors* 39(2): 230–253.

Parente, S. T. 2000. Beyond the hype: A taxonomy of e-health business models. *Health Affairs* 19(6): 89–102.

Partnership for Solutions. 2002. *Chronic Conditions: Making the Case for Ongoing Care.* Baltimore, MD: Johns Hopkins University.

Patel, V. L., D. R. Kaufman, V. G. Allen, E. H. Shortliffe, J. J. Cimino, and R. A. Greenes. 1999. Toward a framework for computer-mediated collaborative design in medical informatics. *Methods of Information in Medicine* 38(3): 158–176.

Patel, V. L., and A. W. Kushniruk. 1998. Interface design for health care environments: The role of cognitive science. *Proceedings of the 1998 AMIA Annual Symposium*, 1998: 29–37.

Payton, F. C. 2003. e-Health models leading to business-to-employee commerce in the human resources function. *Journal of Organizational Computing and Electronic Commerce* 13(2): 147–161.

Phoenix Health Systems. 2003. HIPAA Primer. http://www.hipaadvisory.com/regs/HIPAAprimer1.htm (accessed March 11, 2003).

Potter, M. A., and B. B. Longest. 1994. The divergence of federal and state policies on the charitable tax exemption of nonprofit hospitals. *Journal of Health Politics, Policy, and Law* 19(2): 393–419.

Proctor, R. W., and K.-P. L. Vu. 2008. Human information processing: An overview for human–computer interaction. In *Human–Computer Interaction Handbook: Fundamentals, Evolving Technologies and Emerging Applications*, 2nd ed., eds. J. A. Jacko and A. Sears, 43–62. Mahwah, NJ: Lawrence Erlbaum.

Rasmussen, J. 1985. The role of hierarchical knowledge representation in decision making and system management. *IEEE Transactions on Systems, Man and Cybernetics* 15: 234–243.

Rau, P.-L. P., T. Plocher, and Y.-Y. Choong. this volume. Cross-cultural web design. In *Handbook of Human Factors in Web Design*, 2nd ed., eds. K.-P. L. Vu and R. W. Proctor, 677–698. Boca Raton, FL: CRC Press.

Raymond, B., and C. Dold. 2002. *Clinical Information Systems: Achieving the Vision*. Oakland, CA: Kaiser Permanente Institute for Health Policy.

Reuters Health. 1999. More Americans seek health information online. Reuters.

Roa, R., D. H. Hoglund, J. Martucci, and S. C. Wilson. 2002. The future of wireless personal digital assistants (PDAs) in healthcare. Paper presented at Healthcare Information and Management Systems Society. http://www.himss.org/content/files/proceedings/2002/sessions/sesslides/sessl008.pdf (accessed Jan. 7, 2003).

Rosen, E. 1997. The history of desktop telemedicine. *Telemedicine Today* 5(2): 16–17, 28.

Safran, C., P. C. Jones, D. Rind, B. Bush, K. N. Cytryn, and V. Patel. 1998. Electronic communication and collaboration in a health care practice. *Artificial Intelligence in Medicine* 12(2): 139–153.

Safran, C., and L. E. Perreault. 2001. Management of information in integrated delivery networks. In *Medical Informatics: Computer Applications in Health Care and Biomedicine*, eds. E. H. Shortliffe and L. E. Perreault, 359–396. New York: Springer-Verlag.

Sainfort, F., J. A. Jacko, P. J. Edwards, and B. C. Booske. 2008. Human computer interaction in health care. In *Human–Computer Interaction Handbook: Fundamentals, Evolving Technologies and Emerging Applications*, 2nd ed., eds. J. A. Jacko and A. Sears, 661–678. Mahwah, NJ: Lawrence Erlbaum.

Sanders, J. H., and P. H. Salter, and M. E. Sachura. 1996. The unique application of telemedicine to the managed healthcare system. *American Journal of Managed Care* 2(5): 551–554.

Schoen, C., K. Davis, R. Osborn, and R. Blendon. 2000. *Commonwealth Fund 2000 International Health Policy Survey of Physicians' Perspectives on Quality*. New York: Commonwealth Fund.

Schwartz, K., T. Roe, J. Northrup, J. Meza, R. Seifeldin, and A. V. Neale. 2006. Family medicine patients' use of the Internet for health information: A MetroNet study. *The Journal of the American Board of Family Medicine* 19(1): 39.

Sears, A. 1997. Heuristic walkthroughs: Findings the problems without the noise. *International Journal of Human–Computer Interaction* 9: 213–234.

Sears, A., M. Lin, J. Jacko, and Y. Xiao. 2003. *When Computers Fade . . . Pervasive Computing and Situationally-Induced Impairments and Disabilities*, 1298–1302. Mahwah, NJ: Lawrence Erlbaum.

Shepherd, A. 1989. Analysis and training in information technology tasks. In *Task Analysis for Human–Computer Interaction*, ed. D. Diaper 15–55. Chichester, England: Ellis Horwood.

Shi, L., and D. A. Singh. 2001. *Delivering Health Care in America: A Systems Approach*. Gaithersburg, MD: Aspen.

Sidorov, J. 2006. It ain't necessarily so: The electronic health record and the unlikely prospect of reducing health care costs. *Health Affairs* 25: 1079–1085.

Singer, S. J., A. C. Enthoven, and A. M. Garber. 2001. Health care and information technology: Growing up together. In *Medical Informatics: Computer Applications in Health Care and Biomedicine*, eds. E. H. Shortliffe and L. E. Perreault, 663–696. New York: Springer-Verlag.

Starr, P. 2000. Health care reform and the new economy. *Health Affairs* 19(6): 23–32.

Stephanidis, C. 2001. *User Interfaces for All: Concepts, Methods, and Tools*. Mahwah, NJ: Lawrence Erlbaum.

Stephanidis, C., and D. Akoumianakis. this volume. A design code of practice for universal access: Methods and techniques. In *Handbook of Human Factors in Web Design*, 2nd ed., eds. K.-P. L. Vu and R. W. Proctor, 359–370. Boca Raton, FL: CRC Press.

Strohecker, J. 2000. HealthWorld online: Promoting healthy living via the Internet. *HealthWorld News*. http://www.healthy.net/media/hwnews/faulknergray.htm (accessed Feb. 3, 2003).

Tang, P. C., and C. J. McDonald. 2001. Computer-based patient-record systems. In *Medical Informatics: Computer Applications in Health Care and Biomedicine*, eds. E. H. Shortliffe and L. E. Perreault, 327–358. New York: Springer-Verlag.

Tang, P. C., and V. L. Patel,. 1994. Major issues in user interface design for health professional workstations: Summary and recommendations. *International Journal of Bio-Medical Computing* 34: 139–148.

Taylor, H. 2002. The Harris Poll #21, May 1, 2002. Cyberchondriacs Update. HarrisInteractive. http://www.harrisinteractive.com/harris_poll/index.asp?PID=299 (accessed Jan. 27, 2003).

Tritter, J. Q. 2009. Revolution or evolution: The challenges of conceptualizing patient and public involvement in a consumerist world. *Health Expectations* 12(3): 275–287.

Turisco, F., and J. Case. 2001. *Wireless and Mobile Computing*. Oakland: California Healthcare Foundation.

U.S. Department of Health and Human Services. 2003a. Health insurance reform: Modifications to electronic data transaction standards and code sets. *Federal Register* 68(34): 8381–8399.

U.S. Department of Health and Human Services. 2003b. Notice of address for submission of requests for preemption exceptions determinations. *Federal Register* 68(47): 11554–11555.

U.S. Department of Health and Human Services. 2003c. Notice of addresses for submission of HIPAA health information privacy complaints. *Federal Register* 68(54): 13711–13712.

U.S. Department of Health and Human Services. 2003d. Health insurance reform: Standards for electronic transactions. *Federal Register* 65(160): 50312–50372.

U.S. National Library of Medicine. 2006. Pub Med: Medline® retrieval on the World Wide Web [fact sheet]. http://www.nlm.nih.gov/pubs/factsheets/pubmed.html (accessed Jan. 18, 2010).

Vaccaro, J., J. Cherry, A. Harper, and M. O'Connell. 2001. Utilization reduction cost savings, and return on investment for the PacifiCare Chronic Heart Failure Program, "Taking Charge of Your Heart Health." *Disease Management* 4(3): 1–10.

van Rijn, H., A. Johnson, and N. Taatgen. this volume. Cognitive user modeling. In *Handbook of Human Factors in Web Design*, 2nd ed., eds. K.-P. L. Vu and R. W. Proctor, 527–542. Boca Raton, FL: CRC Press.

Vu, K.-P. L., W. Zhu, and R. W. Proctor. this volume. Evaluating web usability. In *Handbook of Human Factors in Web Design*, 2nd ed., eds. K.-P. L. Vu and R. W. Proctor, 439–460. Boca Raton, FL: CRC Press.

Weiss, G. 2002. Welcome to the (almost) digital hospital. *IEEE Spectrum* March: 44–49.

Wharton, C., J. Bradford, R. Jeffries, and M. Franzke. 1992. Applying cognitive walkthroughs to more complex user interfaces: Experiences, issues, and recommendations. In *Proceedings of the ACM Conference on Human Factors in Computing Systems (CHI'92)*, eds. P. Bauersfield, J. Bennett, and G. Lynch, 381–388. New York: ACM Press.

Wiener, E. L., and R. E. Curry. 1980. Flight-deck automation: Promises and problems. *Ergonomics* 23(10): 995–1011.

World Health Organization. 2000. The world health report 2000: Health systems: Improving performance. http://www.who.int/whr2001/2001/archives/2000/en/ (accessed April 11, 2003).

World Health Organization. 2009. World health statistics 2009: health expenditure. http://www.who.int/whosis/whostat/EN_WHS09_Table7.pdf (accessed Feb. 23, 2010).

Wu, J., S. Wang, and L. Lin. 2007. Mobile computing acceptance factors in the healthcare industry: A structural equation model. *International Journal of Medical Informatics* 76(1): 66–77.

Yang, D., and C.-H. Chang. 2002. Web services & security: A better approach to HIPAA compliance. http://www.cysive.com/news/Web_Services_And_Security.pdf (accessed March 3, 2003).

Ybarra, M., and M. Suman. 2005. Help seeking behavior and the Internet: A national survey. *International Journal of Medical Informatics* 75: 29–41.

Yoshikawa, H. 2003. Modeling humans in human–computer interaction. In *Human–Computer Interaction Handbook: Fundamentals, Evolving Technologies and Emerging Applications*, eds. J. A. Jacko and A. Sears, 118–146. Mahwah, NJ: Lawrence Erlbaum.

Section X

User Behavior and Cultural Influences

30 Human Factors in Online Consumer Behavior

Frederick A. Volk and Frederic B. Kraft

CONTENTS

Consumer behavior is a complex set of related mental and physical activities that people pursue to conduct the exchange and consumption aspects of their lives (Markin 1974; Peter and Olson 2008). E-commerce solution providers must understand how consumers employ the Internet in their conduct of these exchange and consumption activities. These activities can be conceptualized as responses to various internal (physical and mental) and external (social and physical—including marketing) stimuli. When consumers recognize unmet needs as a result of processing these stimuli, they engage in a sequential (and sometimes reiterative) series of goal-directed mental and physical activities. These include the perception of a deficit or problem, a determination of available solutions for the problem, an evaluation of the alternative solutions, a choice of solutions, and post choice and post consumption processes.

As a whole, these activities have become known as the consumer decision process, an organizing model for many textbooks in consumer behavior. The nature and extensiveness of each stage of this process depends on the skills, experience, attitudes, knowledge, confidence, and behavioral tendencies of the consumer; the goal object (target of acquisition); context of the customer experience; user motivations; and the interaction between two or more of these. We must consider these behaviors within the context of the target product of the behavior and how that target limits or enhances the users' application of the Internet to support their consumer experiences. As providers of e-commerce customer experiences, it is generally our goal to guide potential customers to a positive post consumption experience in a timely manner (both from the customers' and the providers' perspectives).

Some users are already engaged in the consumer decision process when they begin their Internet session, while others have hedonic goals that do not immediately involve the consumer decision process. Each user group and subgroup presents designers of e-commerce solutions with an opportunity to design advertising, product information, and Web-based product representations in a manner that assists users in clarifying their needs, reducing their perceived risk of meeting those needs, and obtaining the products and services that will enable post consumption satisfaction.

30.1 USER CHARACTERISTICS

The chapter begins by examining some of the ways that consumer researchers have categorized e-commerce users in segments for analysis purposes. These research efforts assist us in identifying the variables that influence the different e-commerce usage behaviors and must be understood for successful e-commerce site designs. These variables include demographics, behavioral patterns, and psychological factors. It is important to note that these market segments may change over time as must the design recommendations regarding these segments.

30.1.1 DEMOGRAPHICS

A traditional method of consumer analysis has been the description of the demographic and behavioral characteristics of relevant consumer market segments. Numerous researchers of Internet shoppers have taken this approach with varying results. Variables such as gender, age, income, and education have been shown to be related to Internet consumer behavior. For example, gender is a powerful predictor of purchase behavior across product categories: men are more likely to purchase electronics and computer products, and women are more likely to purchase food, beverages, and clothing (Bhatnagar, Misra, and Rao 2000; Kwak, Fox, and Zinkhan 2002; Rosa, Garbarino, and Malter 2006). There is also evidence that women differentially access content, features, and advertising. In their analysis of three athletic shoe Web sites, McMahan, Hovland, and McMillan (2009) found that men and women responded differently to Web site features, with men spending significantly more time interacting with advertising on the Nike Web site, while women engaged advertising equally on each of three Web sites (i.e., Nike, Reebok, and New Balance). It is important to note however, that even when "interactivity" was similar between men and women, the specific advertising content accessed was different.

There are also differences in e-commerce trust between men and women. In a sample that approximated a Web population, Janda (2008) found that women's purchase intentions were more likely than those of men to be negatively influenced by concerns about sharing personal information. At the same time, there is evidence to suggest that men have a higher degree of self-efficacy, trust, and positive attitude related to e-commerce (Cho and Kho Jialin 2008; Hui and Wan 2007).

This gender effect and the impact of a firms' failure to consider it in site design is well demonstrated in a study by Moss, Gunn, and Kubacki (2007) on the influence of Web site designer gender on men's and women's Web site satisfaction. They found that women were more satisfied with sites designed by women and men were more satisfied with sites designed by men. This does not mean that only men can design optimally for men and women for women; rather, it likely means that designer gender is merely a proxy for user research and understanding the needs of the Web site users. In addition, there is convincing evidence that adults and teenagers of different genders engage the Internet in different ways (Tsao and Steffes-Hansen 2008). Interestingly, in their interview of Scottish families with teenagers between

13 and 15 years old Thomson and Laing (2003) found that although children historically are thought of as primarily experiential users they were still actively involved in purchasing products on the Internet by themselves or through their parents, of course, with an additional persuasion step prior to the purchase. While this highlights the difference in process between teens and adults, both groups are actively involved in the purchase process.

30.1.2 BEHAVIORAL PATTERNS

In designing user interfaces it is useful to develop accurate target user profiles. This provides development teams a framework by which requirements can be generated, discussed, and defined. The current literature focuses on individual characteristics and how they affect attitude development and subsequent behavior. Unfortunately, this type of research is difficult to apply in forming reasonable design requirements. For example, neuroticism is associated with selling and buying online (McElroy et al. 2007) and openness to experience, agreeableness, and conscientiousness are associated to the development of a passion for online buying (Wang and Yang 2008). These findings are interesting but are of little use to designers of e-commerce Web sites. Some useful conceptualizations do exist, however, in terms of user purpose (i.e., experiential versus utilitarian; Hirschman and Holbrook 1982; Holbrook and Hirschman 1982; Jones and Fox 2009) and Web site interaction (Moe 2003). Each categorization implies an approach for improving or optimizing the target groups' experience with the Internet as a commerce medium and can yield financial benefits for the firm that targets their advertising designed differently for these categories (Manchanda et al. 2006).

In an analysis of consumer site traffic patterns, Moe (2003) identified five clusters of behavior that were defined by three categories of dependent measures: session (duration at site, number of pages viewed, and average time spent per page), category (percentage of search result, information related, and Home pages and number of pages across categories, products, brands), and variety (percentage of pages across categories, products, and brands). The first cluster of users, "knowledge builders," spent extensive time on each page and had a limited number of views of product pages. Moe (2003) interpreted these as Internet site users who desired to acquire knowledge rather than purchase a product. This segment is important because it represents users who do not intend to purchase in that given session, yet their behavior indicates the likelihood of future purchase. Because most Internet consumer research has focused on actual buyers, little is known about these nonbuyer types of users and their potential for generating revenue. Some writers have proposed that brick-and-mortar businesses take advantage of the Internet as a medium to support an overall strategy for revenue growth both online and in the physical store. As an example of managing postpurchase satisfaction, a customer who buys a book from the Barnes and Noble Web site can return the book for an in-store credit at any local Barnes and Noble location.

The second cluster of users, "hedonic browsers," spent most of their sessions perusing category and product pages with a high percentage of unique page views (Moe 2003). These first two clusters, hedonic browsers and knowledge builders, represent a unique opportunity for Web designers. Providing knowledge builders with the types of tools that facilitate information acquisition, an information architecture that enables their desire to gain knowledge, and the ability to contrast a firm's offerings objectively with competitors supports them in a way that is consistent with an overall customer centric strategy. Also, the hedonic cluster represents an opportunity for providing these users an online experience that is attractive to the surfer of e-commerce sites. Realtor .com provides an excellent example by offering users an opportunity to sort the home listings associated with a query with the Virtual Tour listings displayed first. This certainly supports these surfers, while not slowing down the knowledge builders. Although these types of implementations for these first two clusters may not translate into sales for that particular Internet session, they are likely to create a positive brand image for the site and its partners.

The third cluster of users, "directed buyers," spent most of their sessions at product-level pages with a high number of repeat views at the product page level. The fourth cluster of users, "deliberate searchers," viewed the most pages and spent most of their sessions in relatively few product categories, with few repeat views at the product level (Moe 2003). A fifth cluster of users, "shallow" users, accessed only two shopping pages and then left the site. The sales conversion rates were 0.01% for the "shallow" and "knowledge building" users, 1.4% for "hedonic browsers," 20.0% for "directed buyers," and 6.4% for "deliberate searchers." The behavior of each of these clusters suggests that consumers have a specific cognitive orientation or mind-set with which they approach a specific Internet session (Dholakia and Bagozzi 2001).

Understanding the demographics and behavioral characteristics of users is one piece of the puzzle and plays a considerable role in the success of an e-commerce effort. Although this section is not meant to be a complete discussion of all the user characteristics that could affect the success of e-commerce solutions, they are the basis for a reasonable set of user experience and business guidelines for facilitating a positive user experience across a number of solution scenarios (Table 30.1). It is important to note that the previous paragraphs related demographic and behavioral groupings to cognitive patterns among the members of these behavioral groups. Although the value of demographic and behavioral analysis has been noted, the psychological characteristics of e-commerce users must also be considered.

30.1.3 Psychological Variables

Numerous psychological variables are associated with goal oriented and problem-solving behaviors by consumers. Some of the most important variables have received considerable attention in the consumer behavior literature. Particularly

---

TABLE 30.1

User Characteristics Design Recommendation

1. E-commerce sites that are primarily transactional in nature should target households with a high income and design for efficiency of purchase process (Kim, Cho, and Rao 2000; Jones and Fox 2009; Moe 2003).
2. A significant portion of online sales are generated by high-income households that have children living in the home (Reibstein 2001; Jones and Fox 2009).
3. Lower-income households use the Internet more experientially than higher-income households. Provide an experiential online environment for those users who have a lower household income (Jones and Fox 2009).
4. Adult men and women and teenage boys and girls use the Internet differently. Design a user experience that is consistent with those differences (Dholakia and Chiang 2003).
5. Provide products and services that are consistent with early adopters/ opinion leaders needs (Kwak, Fox, and Zinkhan 2002).
6. When targeting older populations, spend resources on increasing the accuracy of product representation as opposed to product selection (Jones and Fox 2009).
7. Design for efficient knowledge acquisition to support knowledge builders. This is especially important for those sites that are a part of a multiple-channel business strategy (Moe 2003; Steinfeld and Whitten 1999).
8. Provide hedonic browsers an opportunity for pleasurable e-commerce activities. For example, providing a virtual tour on a real estate Web site may not translate into a referral in that session but will help develop a positive experience association with the site brand for future business (Moe 2003).

---

relevant to our discussion are motivation, personal involvement, perceived risk, and trust.

30.1.3.1 Motivation

Human behavior is goal oriented. The difficulty is in understanding those goals and designing an Internet site that provides user experiences that facilitate consumers' goal accomplishment in a way that is mutually beneficial for consumers and the provider. The specificity of consumer goals varies from instance to instance. For example, a consumer may use the Internet on one occasion to obtain a very specific product (target), such as the latest book on Web usability. On a second occasion, that consumer will be searching for a gift but has no idea what that gift might be, and on a third occasion the same consumer will have a goal of surfing the net and checking his or her favorite online haunts. The motivations represented by each of these occasions represent an opportunity for providers to advertise, sell, and provide support in different ways.

Each person engages the Internet with motivations that are either enhanced or inhibited by the e-commerce sites. Enhancers are attributes of the user experience that encourage users' e-commerce behavior, while inhibitors discourage users from participating in e-commerce activities. Characteristics that enhance user motivation include the availability of good options for products or services, convenience, price, and easy comparison shopping, whereas inhibitors include poor usability, unprofessional presentation, technical difficulties, and perceived risks.

In their cross-cultural sample, SanMartin et al. (2009) categorized these enhancers into five categories that encouraged e-commerce behavior. The first of these categories includes the perceived attributes of the e-commerce provider including payment security, high potential for positive post purchase experience, overall reputation, product offering, well-designed product information (Cyr 2008; Warden, Wu, and Tsai 2006), and "quick" Web site. Other enhancers include a professionally designed and enjoyable Web site (Cyr 2008), convenience (i.e., speed of purchase, easy price comparison, accessibility of site), and lower prices (Cho, Kang, and Cheon 2006; Yang and Lester 2004). Inhibitors include technical difficulties, transaction cost (i.e., delivery cost, poor clams and returns process, and lack of payment security), and the impersonal nature of the Web experience (Cho, Kang, and Cheon 2006; Warden, Wu, and Tsai 2006; Yang and Lester 2004).

The Web Motivation Inventory (WMI) is a 12-item scale that assesses Internet users' behavioral orientation on four dimensions (Rodgers and Sheldon 2002). Shopping motivation is measured by three purchase items; research is measured by a "do research" item, a "get information I need" item, and a "find out things I need to know" item; Web surfing items include three experientially oriented items; and communication motivation is measured by "e-mail other people," "connect with my friends," and "communicate with others" items (Rodgers and Sheldon 2002, 88). Each of these scale dimensions had an acceptable alpha coefficient and reasonable test-retest reliability (Rodgers and Sheldon 2002). Rodgers et al. (2007) confirmed the factor structure of the original WMI in a multicultural sample with some slight modifications in loadings (i.e., e-mail other people and explore new sites also loading on the research motivation). In confirmatory studies of the WMI, Rodgers and Sheldon (2002) found that each of the dimensions predicted evaluations, intentions to click, or feelings of persuasion from corresponding banner advertisements in student and adult populations (Rodgers and Sheldon 2002; Rodgers et al. 2007). In other words, banner advertisements promoting communicating behavior such as e-mail were associated with the communication dimension of the WMI.

Although the WMI is useful in categorizing the motivation for engaging in a given session, it effectively illustrates the limitations presented by conceptualizing consumer behavior as limited to actual purchase behavior. This limitation was addressed by Rodgers et al. (2007) who expanded the set of motives to 12 subcategories including desires for (1) community, (2) entertainment, (3) product trial, (4) information, (5) transaction, (6) game, (7) survey, (8) downloads, (9) interaction, (10) search, (11) exploration, and (12) news. This expanded characterization is more consistent with a broader range of people's current activities on the Internet.

30.1.3.2 Utilitarian versus Experiential Motives

From an e-commerce perspective, consumers either view the Internet as an experiential opportunity or as a utilitarian tool. The proportion of these two types of users is unknown, but there is evidence the online population spends most of their time in experiential activities. For example, young people are more likely to use the hedonic endeavors such as music, games, and videos, while older Internet users are more interested in information and shopping (Jones and Fox 2009). Informed businesses recognize that the Web presents a unique opportunity and attempt to engage users in experiential ways as well as utilitarian. Understanding target users' motivations, whether primarily experiential or utilitarian, can influence how particular users respond to the design (San Jose-Cabezudo, Gutierrez-Arranz, and Gutierrez-Cillan 2009). Experience-oriented consumers are interested in interacting with products because it is fun (Table 30.2). They prefer online experiences characterized by multisensory input and feelings of fantasy and fun (Holbrook and Hirschman 1982). If user registration is associated with highly interactive (i.e., multisensory) experiential features as opposed to merely utilitarian benefits, experiential users will be more likely to view those features favorably and register on a given Web site (Kamali and Loker 2002). It could be argued that the types of consumer experiences that appeal to the experiential consumer are different than those that appeal to the utilitarian consumer. Lepkowska-White and Eifler (2008) found that ease of ordering and experiential characteristics were differentiators for "feeling" product purchases. For experiential shoppers, ease for ordering is important but for potentially different reasons than those users who view e-commerce in a more utilitarian manner. Experiential users need an efficient purchase process to, presumably, maintain their desired psychological state.

The experiential nature of the Internet increasingly permits the creation of experiences that enable telepresence and bricolage (Rosa and Malter 2003; Song, Fiore, and Park 2006). Telepresence refers to the degree that users perceive their presence in a virtual environment. Bricolage is the process of creating something out of what is available. Providing these features has benefits for both consumers and businesses. For consumers, features that improve telepresence may facilitate elaboration and recall, information acquisition, and provide an opportunity to evaluate a wider range of products, more thoroughly increasing the satisfaction for the experientially oriented user (Daugherty, Li, and Biocca 2008; Rosa and Malter 2003; Song, Fiore, and Park 2006; Wood, Solomon, and Englis 2006). For businesses, experiential features enable competitive differentiation, greater user customization, influence brand recognition, and create more satisfaction for consumers who prefer experiential online sessions (Daugherty, Li, and Biocca 2008; Song et al. 2006). All of these factors combine to help develop a positive brand image for the types of online businesses patronized by a broad range users.

Designing for the utilitarian consumer has been the focus of Web site designers for some time. In data collected in 1996, Lohse and Spiller (1999) used regression techniques to determine that 61% of the variance associated with monthly sales was accounted for by navigation mechanisms that reduced the time required for purchasing. Although dated,

TABLE 30.2
Motivation Design Recommendations

1. Associate registration benefits with increased efficiency in information acquisition, purchase, fulfillment, and customer support for utilitarian users (Rodgers and Sheldon 2002).
2. Associate registration benefits with access to features that enrich the communicative and multisensory interactive nature of the consumer experience for experiential users (Holbrook and Hirschman 1982; Rodgers and Sheldon 2002; Song, Fiore, and Park 2006).
3. Provide experiential users with the product representations that increase telepresence (Rosa and Malter 2003; Song, Fiore, and Park 2006).
4. Provide utilitarian users with clean efficient user interfaces, minimizing the time and effort it takes them to attain their target (Lohse and Spiller 1999; Moe 2003; Rodgers and Sheldon 2002).
5. Provide users an opportunity to choose the type of experience that they prefer relative to next-generation interactive experiences (Moe 2003).
6. Do not assume that the experiential user will prefer an experiential purchase process. Regardless of purchase type, allow users to move from the experiential to the utilitarian according to the task. For all users, facilitate efficiency from the decision point of purchase to actual purchase and fulfillment.
7. Provide opportunities for experiential users to make spontaneous purchases outside of the topical scope of their current activity but consistent with the apparent desired multisensory experience of current browsing activity (Baumgartner 2002). For example, a user that is participating in an interactive product customization activity may be predisposed to enjoy other activities that require similar interactivity.
8. For deliberate purchases, offer consumers the types of products and services that are consistent with that product, for example, offering accessories to a cell phone or offering a second book title in the same area as the target book.
9. Make ease of ordering a priority for all users (Lepkowska-White and Eifler 2008).

these results suggest the importance of efficiency to utilitarian consumers. Thus, it is reasonable to use efficiency as a site design principle when considering other aspects of consumer behavior such as information search and alternative evaluation. The fact that customers prefer efficiency during the purchase process is obvious when we consider that even the most experientially oriented consumer is engaged in the utilitarian task of purchase and monetary exchange. However, the focus of online purchasing as the ultimate measure of Web site success has resulted in the understandable emphasis on streamlining the purchase process and a lack of exploration of the importance of experiential motives of consumers who use the Internet. Web site designers should recognize that experiential interactions are a major goal for many consumers in using the Internet.

30.1.3.3 Personal Involvement

Personal involvement is one of the most important concepts in the analysis of the consumer decision process and the various factors that determine its nature. Involvement, as originally conceptualized by Krugman (1965), refers to a person's perception of connections between his or her own life and a product, service, idea, or activity. This perception of the personal relevance of the use, acquisition, or disposal of a product in a particular situation causes people to experience what Celsi and Olson (1988) referred to as "felt involvement." In one study, product involvement as measured by the degree of information required for purchase was reported as the most powerful predictor of purchasing products online (Kwak, Fox, and Zinkhan 2002; Table 30.3).

Consumer involvement is important because it is a motivational state that influences the level of cognitive and behavioral effort consumers expend, as well as the level of affect they experience during the purchasing process (Antil 1984). The level of consumers' involvement influences the amount of attention they pay to communications and how much they think about the product information communicated (Celsi and Olson 1988). Involvement also determines the extent and type of information processing in which consumers will engage (Petty, Cacioppo, and Schumann, 1983), the amount of belief and attitude change produced by communications

TABLE 30.3
Personal Involvement Design Recommendations

1. For high involvement products, take care to provide users with enough information to make an informed decision (Kwak et al. 2002).
2. Hedonic and high involvement products are more likely to have strong affect associated with them, so it may be useful to provide a user access to experiential descriptions and interactive interfaces that reinforce positive imagery (Mowen and Minor 1998; Sherif and Sherif 1967).
3. Frequently repeated purchases should be facilitated by primarily utilitarian or efficiency-oriented interfaces or low involvement user experiences (Lastovicka 1979).
4. For high involvement products, it is important for the site to support an extensive search for information. Users may be building their knowledge in a given product area for future purchase, and site designers should optimize their experience for attaining product knowledge that is consistent with the strengths of their products (Bloch, Sherrell, and Ridgeway 1986; Moe 2003).
5. For high involvement hedonic products, it is important to provide a well-designed experiential customer experience that is consistent with the target products (Hirschman 1980; Peter and Olson 2008; Tauber 1972; Wood, Solomon, and Englis 2006).
6. Saving time is a principle motivation for researching and buying products online. High involvement products typically require an extensive information search. Providers that are selling high involvement products need to allocate more resources to the design of their site to ensure users can acquire information efficiently (Beatty and Smith 1987; Stigler 1961).
7. For low involvement products where users are not motivated to attend to product information, designers should employ hedonic messages or content to encourage the elaboration of the low involvement item (Petty, Cacioppo, and Schumann 1983; Mowen and Minor 1998).
8. The use of graphics for high involvement must be informative in either a multisensory way for hedonic purchases and/or by providing more information to the user than could be efficiently or accurately delivered via text (Netguide 1996).

(Sherif and Sherif 1967), the relationship of attitude development and change to purchase behavior (Ray 1973), and the likelihood of cognitive dissonance following a purchase (Insko and Schopler 1972). Also, Sherif and Sherif (1967) found that affect toward objects of high involvement may be experienced more strongly than affect for those of low involvement, although the reverse is also true in that objects that create high emotions, such as hedonic purchases, may also be more highly involving (Mowen and Minor 1998).

Consumers experience involvement with products and services for two broad reasons. First, their knowledge of a means-end linkage between a product and their own personal goals produces an intrinsic or enduring involvement (Celsi and Olson 1988). The linkage exists because consumers view the consequences of acquiring or using the product or service as a means of reaching these goals. Although this linkage and the resulting involvement level reside in the minds of consumers, some types of products (e.g., frequently purchased commodity-like goods) are more likely to be ones with which people have low involvement, whereas other goods (e.g., high-priced, infrequently purchased and brand differentiated items) are ones with which people are likely to be highly involved (Lastovicka 1979).

A second reason why consumers may feel involvement is the purchasing situation. Purchase situations are temporarily perceived linkages between the purchase and important outcomes (Peter and Olson 2008). The immediate physical or social environment of a purchase may activate a temporary awareness of important goals that a purchase may achieve. These may include purchases that involve temporary emergencies, or purchases that confirm one's shopping acumen, that impress friends, or that take advantage of a sale (Bloch 1982). Once the purchase is made and the temporary means-end linkage has decreased in salience, the level of involvement decreases. An important way that consumers' involvement level is of concern to commercial Web site designers is its influence on the extent that consumers will spend effort to search among sites and to probe succeeding layers of information within a particular site. In addition, involvement level determines the most appropriate design of information presented in a Web site.

Two types of consumer search processes may occur, either "ongoing" search or "prepurchase" search (Bloch, Sherrell, and Ridgeway 1986). Ongoing search, sometimes referred to as browsing or window shopping, may be motivated for two reasons. First, consumers who have high intrinsic involvement with a product category search because of their continuing interest in and desire to build knowledge that may be accessed during future purchase situations. For example, an automobile enthusiast may examine photos of Corvette sports cars on an auto dealer Web site even though no purchase is contemplated in the foreseeable future. Such search is a product-specific phenomenon for members of particular market segments (Bloch, Sherrell, and Ridgeway 1986). This can especially be true for products that are traditionally purchased in face-to-face contexts and for users who have less Internet experience. In Frambach, Roest, and Kirshnan's

(2007) exploration of consumers' channel (online Vs offline) preferences for prepurchase, purchase, and postpurchase activities for mortgages, Internet experience predicted online channel preference across all three phases of the consumer behavior. However, channel accessibility (online or brick and mortar) was only predictive in prepurchase, whereas perceived usefulness was predictive during the purchase stage. All other significant predictors across all stages were related to negative aspects of the offline channel. These consumers require that the information from sources be adequate in breadth and depth.

A second cause of ongoing search is an experiential or hedonic motivation. This form of search is motivated by the enjoyment and stimulation of the activities of the search process (Tauber 1972) or by a desire to seek novelty (Hirschman 1980). This hedonic form of search requires that the "atmospherics" of the search source provide rewards, whether they are from attractive physical store designs that cause consumers to linger or from "sticky" Web sites that provide enjoyable interaction experiences (Peter and Olson 2008). The second type of search is the "prepurchase" search, which is motivated by the consumer's need to acquire information to make a purchase decision. The higher a consumer's involvement with a product or product category, the greater the motivation to acquire information to reduce the risk of a poor purchase (Beatty and Smith 1987). The value of searching, however, is also related to the amount and quality of knowledge consumers already have. The economist Stigler (1961) pointed out that from an economic perspective, consumers should search only while they perceive that the value of search outweighs costs. An important appeal of shopping via the Internet is that it decreases the costs of shopping in terms of time and travel. Therefore, Web page designers must make shopping as easy and straightforward as possible and work toward decreasing the perceived risks in purchasing via this medium.

Another way that the involvement construct is important in consumer analysis is its influence on perceptual processes. Celsi and Olson (1988) found that involvement strongly influences people's motivation to attend to and comprehend information. Not only are highly involved individuals more motivated to process information, but they also have a greater ability to comprehend information because they more easily activate knowledge from long-term memory.

Involvement also determines the role of communication in influencing consumer attitudes and decision making. Krugman (1965) was one of the earliest to note that when consumer involvement is low, attitudes do not change prior to purchase, as in the traditional Hierarchy of Effects (cf. Lavidge and Steiner 1961). Rather, attitudes toward low involvement products change following simple belief change and purchase.

Research on the low involvement hierarchy led to the development of the "elaboration likelihood model" by Petty, Cacioppo, and Schumann (1983), according to which there are two routes to persuasion. The first is the "central route," which corresponds to the traditional hierarchy of effects model of attitude change. This central route occurs when consumers are high in involvement and when they are motivated to process and elaborate on the product information they receive. The provision of relevant product information may lead to attitude change followed by a change in purchase behavior.

The second route to persuasion is the "peripheral route" that occurs when consumers are low in involvement and not motivated to process information. In this case, consumers may be influenced in spite of their low product involvement because they still attend to peripheral, nonproduct information for its entertainment or hedonic value. Such messages encourage higher involvement with the advertising itself, and effective messages might include humor, pleasant music, attractive models, or even soap opera-like stories. Positive feelings generated toward the advertisement itself may later become associated with the brand name sponsored in the ads (Mowen and Minor 1998).

A final point is that the research on low versus high involvement persuasion is very relevant to the type of information and graphics presented by a Web site. For high involvement purchasing, informative data and perhaps pictorial information is mandatory, whereas splashy graphics may be detrimental (Netguide 1996). However, for low involvement items that are not actively sought, Web sites must capture attention and prolong interaction with attention-getting stimuli in order to influence buyers through the low involvement route.

Involvement is fundamental to understanding (1) the motivation of consumers to search for information, (2) consumers' willingness to pay attention to information, (3) consumers' motivation and ability to comprehend information, and (4) the type of information most likely to influence them. Web site developers should be guided by the involvement construct as they determine the stimulus and information characteristics of the Web pages with which consumers will interact (Table 30.3).

30.1.3.4 Purchase Type as Defined by Experiential and Cognitive Dimensions

Purchases can be categorized by the extensiveness of decision making, the level of consumer involvement, and the extent of consumers' emotional processes in the purchase. When the extent of these is known, several implications for Web site design may be derived. Using multidimensional scaling and cluster analysis techniques, Baumgartner (2002) identified eight categories of purchases that were defined by two levels of each of the three dimensions. He classified each purchase as either thinking or feeling, spontaneous or deliberate, and having high or low involvement.

In this context, it is useful to define deliberate purchases as ones made when a consumer approaches the consumer decision process and its associated shopping tasks with a defined purpose or goal (e.g., purchase office supplies). Purchases that require extended decision-making processes are also described as "goal-oriented" consumer purchases. Symbolic purchases include those that enhance a consumer's self-image or social status such as designer clothes or a particular make and model of automobile (Belk, Wallendorf, and Sherry 1989). A purchase that has "multi-sensory, fantasy and emotive ... " evaluative components can be described as hedonic (Hirschman and Holbrook 1982, 92; Holbrook and Hirschman 1982). This classic definition is consistent with Baumgartner's (2002) notion of symbolic purchases.

Spontaneous purchases are unplanned purchases and include promotional, casual, exploratory, and impulsive purchases. These purchases occur in the context of a user's engagement in online activities and are often made impulsively with little consideration. Promotional purchases refer to those unplanned purchases that occur as a result of a vendor's special offer, casual purchases are those that consumers perform without much deliberation, exploratory purchases are made when a user is curious or is seeking some variety, and impulse purchases are those that are executed without any thought beyond the immediate stimulus situation. People who have had the experience of purchasing books on Amazon.com have no doubt received a special offer that consists of the target book with another book on a similar topic or by the same author. Such special offers usually promote a bundled price less than that of purchasing the two books separately. It is also useful to describe spontaneous purchases, like deliberate purchases, in terms of these three bipolar dimensions.

An examination of the several possible purchase categories and their implications for behavior offers useful site design ideas, but because these categories are generalizations of behavior, we must take great care in how we apply them to design recommendations. It is probably most useful to think of each of these purchase types as having more or less of each of the characteristics defining each of the possible categories. For example, a consumer may employ an extended decision-making process for hedonic, symbolic, or even repetitive purchases when participating in the consumer decision process (Belk et al. 1989). Thus, the purchase of a Ford F150 could be rated highly on four purchase-type dimensions. It might require extended decision making, be symbolic relative to image, repetitive because we always buy Fords, and hedonic in terms of pleasure seeking.

That said, this research is extraordinarily useful in conceptualizing products in terms of the interaction between users and their likelihood of feeling certain ways about particular products. By rating our products on these dimensions with our various target user populations, we are able to consider the design of advertising content and nature of desired user experience with our product pages.

30.1.3.5 Perceived Risk and Trust

Perceived risk refers to the degree that a consumer estimates the potential negative outcomes of a given behavior. Is the risk of purchasing a CD from Amazon.com greater than the risk associated with purchasing the same CD on eBay? Reputation, integrity, competence, security control, third party recognition, and ease-of-use all influence users' trust of online providers (Aiken et al. 2007; Awad and Ragowsky 2008; Cheung and Lee 2006; Metzger 2006; Schaninger 1976; Zhang et al. 2009). Gurung, Luo, and Raja (2008) found that perceived risk was not directly related to purchase intentions but was moderated by trust. This does not necessarily mean that risk assessment is not important but that the risk-trust relationship may have different effects on different behaviors. These risks can be described in terms of finances, performance, time, image, and nuisance (Ha 2002):

- Financial or monetary risk is the monetary cost associated with doing business with an online vendor and is comprised of the security associated with the transfer of electronic financial information and implications of dealing with a potentially fraudulent vendor (Teltzrow, Meyer, and Lenz 2007; Wang, Lee, and Wang 1998).
- Performance risk is the degree to which product performance meets the consumer's expectations and is especially of concern for consumers of single purchase, durable, high-cost items (Bhatnagar, Misra, and Rao 2000; Eggert 2006).
- Time risk is the extent to which a particular behavior will increase or decrease goal accomplishment in a given time period and plays a role with regard to the user's external search for information. If a site is poorly designed from either an information architecture or task performance perspective, users may be dissuaded from using the site to search for product information due to perceived time risk (Awad and Ragowsky 2008; Pearson and Pearson 2008; Zhang et al. 2009). It also refers to the time that a product is expected to satiate the identified need.
- Image risk is the degree to which a particular consumer behavior or relationship is congruent with the desired image of the participant.
- Nuisance risk is the degree that participating in a given behavior or exchange of personal information will result in irrelevant vendor communication exchanges (Nam et al. 2006; Wang, Lee, and Wang 1998). Users are more likely to be comfortable sharing their personal information when on sites that are well designed, void of errors, and have more user centered interfaces (Nam et al. 2006). There is some indication that this is of more concern for women than men (Janda 2008).

As e-commerce has matured, there is evidence that even though financial risks are still of concern to online consumers, they are not necessarily predictive of future online purchases (Gurung, Luo, and Raja 2008; Miyazaki and Fernandez 2001). The degree to which marketers and designers take each of these types of risk into account is contingent on the degree that their target customers find these important (Table 30.4). Certainly, there are those marketers who presume an unlimited number of potential customers and market all of their products directly, but most marketers cannot ignore the contribution of many irrelevant and unproductive contacts to increasing consumers' perceived risk. Inasmuch as perceived risk is important, its effect on purchase and purchase intention

is primarily moderated by trust (Gurung, Luo, and Raja 2008). As the Internet matures, it appears that consumers' concerns about security risks relative to online purchasing (i.e., financial risk) are decreasing (Girard, Silverblatt, and Korgaonkar 2002). In contrast, findings also suggest that, as the cost of a product increases, risk plays a larger role in consumers' decision process (Lowengart and Tractinsky 2001). Although these findings appear to conflict, these two concerns relate to two different relationships in our view of perceived risk. Lowengart and Tractinsky (2001) experimentally addressed the relationship between product or service cost and financial concerns over the security of the financial transaction (Girard, Silverblatt, and Korgaonkar 2002).

A consumer will participate in a behavior when the perceived risk of engaging in that behavior is meaningfully less than the perceived benefit or when the expected outcome associated with a behavior is the most positive of a set of negative alternatives. That is, people will participate in behaviors that they believe will represent their best interest.

TABLE 30.4
Perceived Risk Design Recommendations

1. Designing for efficiency of use decreases the perceived time risk associated with navigating to the desired target.
2. Have professionals design the user experience. Customers recognize an amateurish effort when they see it and are likely to infer both security and information risks associated with establishing a relationship associated with the e-tailer.
3. Organize the Web site in a manner that is consistent with real world experience. It permits the user to find the information they desire and leverages current user models.
4. Do not mislead the user by integrating content and advertisements in a manner that does not facilitate differentiation and adds to overall clutter (Ha and McCann 2008).
5. Facilitate user control by displaying ad content in noncompulsive formats (Ha and McCann 2008).
6. Demonstrate benevolence by recommending competitors when it is advantageous to consumers (Olson and Olson 2000).
7. For technically complex products, provide easy-to-understand information to increase the customers' confidence with the performance of the product. The more technically complex, the higher the perceived risk of purchasing the product (Bhatnagar, Misra, and Rao 2000).
8. For ego related products such as sunglasses or perfume, provide a money back guarantee to decrease the image risk for the consumer (Bhatnagar, Misra, and Rao 2000).
9. The design should place emphasis on the reputation of the retailer (Jarvenpaa and Tractinsky 1999).
10. Provide users an opportunity to have a positive personal interaction experience prior to purchase (Miyazaki and Fernandez 2001; Olson and Olson 2000).
11. Share information about shared values that helps the customer identify with the provider (Dowling and Staelin 1994; Olson and Olson 2000).
12. Provide risk handling tools on your site, including independent reviews by experts, richer examination of products when possible, or personal interactions (chat or e-mail) with sales people (Dowling and Staelin 1994; Olson and Olson 2000).

A key issue in today's online environment is the level of perceived risk that is associated with participating in the range of consumer behaviors (Table 30.3). Trust and perceived credibility are two factors that play a considerable role in the perception of risk and how that perception affects consumer behavior.

Trust is the degree to which we permit another an opportunity to harm us by making ourselves vulnerable (Freidman, Kahn, and Howe 2000). The relationship between trust and the perceived trustworthiness of an online vendor is directly related to customer loyalty and purchase intention (Gefen 2002; Lee, Kim, and Moon 2000). Trust is founded on consumers' perception of the online provider's (1) skills and competencies, (2) commitment to serve the customer well, and (3) adherence to acceptable behaviors with regard to the proposed relationship (Gefen 2002). The willingness of consumers to expose themselves to harm by participating in e-commerce behaviors is directly related to their trust in the provider. The amount of trust required is dependent on the behavior of the consumer. In the case of surfing, very little trust is required, but in the case of information search and alternative comparison a consumer must trust the accuracy of the information representation. In the instance of high-trust, users may use a single site as their source for product information and in a low-trust scenario, they will use the information as a single input to a broader information search. High trust is required when users share their personal information during the registration process and provide their credit card number for purchase. The provision of comprehensive information, shared values, and ease of communication are antecedents to developing trusting relationships between consumers and online retailers (Gefen 2002; Grewal, Gottlieb, and Marmostein 1994; Lee, Kim, and Moon 2000).

In a cross-cultural study of consumer trust for an Internet store, Jarvenpaa and Tractinsky (1999) reported that store reputation and perceived retailer size were reliable antecedents of consumer trust. These findings affect financial and nuisance risk through privacy and security antecedents as opposed to evaluation and information antecedents. An increase in trust has been consistently associated with embeddedness (connectedness to the community), reputation, and positive interpersonal relationships between buyer and supplier and ultimately a decrease in perceived risk (Jarvenpaa, Tractinsky, and Vitale 2000; Steinfeld and Whitten 1999). In their study comparing the purchase of computers (high risk) and books (low risk), they reported that the dimensions of the Internet store (e.g., shopping process, reputation, information quality) are more important for the high-risk condition (Lowengart and Tractinsky 2001).

Consumer trust and Web site credibility are negatively related to perceived risk that consumers reduce through a series of risk handling activities (Dowling and Staelin 1994). Primarily, these activities delay purchase by extending the information search process, including looking for a better price or better return policy at another store, consulting other evaluative sources such as friends and family or other advertising media, or merely reconsidering based on internally

developed standards (Dowling and Staelin 1994). Although these strategies are effective in reducing consumer risk, it would obviously benefit e-commerce providers to leverage those relationships that consumers already find comfortable. These include consumers' current relationships with the products, brands, and retailers that they perceive as low risk. Alternatively, facilitating risk handling tools, such as online live help, customer reviews, or seller rating, and employing risk reduction policies with regard to returns and fulfillment tracking can discourage customers from comparison shopping at other stores.

Framing a product in a manner that will reduce the perceived risk of acquisition can make the difference between success and failure. Consumers have a tendency to behave differently under high-risk versus low-risk conditions. Research findings support the notion that under low-risk conditions consumers tend to choose moderately incongruent options (an option that has a higher potential to be inconsistent with the consumers' preferences in terms of performance), but under high-risk conditions they tend to choose the more familiar options (Campbell and Goodstein 2001). For example, if a user is shopping for a gift for her manager, she is much more likely to pick the gift option that has a performance value that she is confident about (familiar) than the option that she thinks might be better performing, but in which she is less sure (unfamiliar). This contrasts with the situation where a consumer is shopping for something for her own use and might be willing to choose the less familiar option because her perception of the risk with regard to the negative outcomes is lower. If the online retailer is overstocked with a moderately incongruous option, it is desirable from the retailer's perspective to present that option in a low-risk format or reduce the perceived risk by employing methods such as taste tests, rebates, and free trials. Comparison between in-store experience and remote purchase environments suggests that a lenient return policy is more important in reducing consumers' perceived risk in remote purchase environments (Wood 2001). Having a positive experience with the Internet is consistently associated with a reduction of perceived risks of purchasing products online (Miyazaki and Fernandez 2001). This certainly suggests the need to attend to design issues that affect consumers' overall assessment of the Internet session in terms of a general favorable or unfavorable view of the provider. In fact, it is reasonable to assert that those sites that have the lowest perceived risk associated with their customer-provider relationships will flourish in the digital environment.

30.2 DECISION PROCESS

30.2.1 Problem or Need Recognition

The psychological variables discussed in the previous sections of this discussion play a major role in the analysis of the consumer decision process (CDP). The CDP is initiated with the consumer's perception of a significant deficit, that is, the realization of a noticeable difference between the person's actual state of affairs and desired state regarding some need. This is required for the initiation of purposeful problem-solving activity by the consumer. The perception of problems may be caused by internal physical or mental state changes or by stimuli in the external social or physical environment. Marketers have paid special attention to the manipulation of marketing stimuli in the consumers' environment and their role in stimulating problem recognition responses (Engel and Blackwell 1982).

The consumers' use of the Internet during the CDP may be triggered as a purposeful response to problem recognition, although Internet usage for reasons unrelated to the CDP may provide stimuli that produce problem recognition. Stimuli at this stage of the CDP include banner and pop-up advertisements and e-mail lists and may be facilitated by software agents that notify a consumer when a potentially desired product meets a pre-identified threshold in terms of price, features, or availability (Scheepers 2001). For the most part, all these methods are intended to attract the attention of users, emphasize an apparent need that the target (product or brand) will satisfy, and increase the probability that users will select the advertisers' product for meeting that need.

Automatic and controlled information processing play a significant role in the design of online advertising (Grunert 1996). Automatic processes fall into two distinct categories, preattentive and automatic (Logan 1992). Although both preattentive and automatic processes are accomplished quickly, effortlessly, and without conscious attention, preattentive processes are those processes that were never in the conscious attention and operate independently of attention (Logan 1992). They are certainly automatic but appear to be more closely linked with the physiological potential of the organism (e.g., pattern recognition). Automatic behaviors are those that, upon first effort, require attentional resources but after practice slowly migrate to requiring less and less attention (navigating a Web page). A behavior would be considered automatic if it is characterized by single-step retrieval from memory (Logan 1988). Controlled processes are effortful, take a sequence of steps, and are sensitive to attentional manipulations. Such skills are concerned with recognizing and categorizing stimuli, and finding appropriate reactions to them. They are "processes concerned with the acquisition, storage, and use of information to direct behavior" (Grunert 1996, 89). Grunert (1996) proposed a framework for advertising strategy and suggested a series of strategies to take advantage of experimental findings in attention, memory, and skill acquisition literature. In addition, Ha and McCann (2008) provide a framework for understanding how the design of online advertisements can contribute to ad clutter, negatively affecting consumers' ability to process information. These information processing concepts provide the basis for a series of design recommendations (Table 30.5). Need recognition may occur in the present, within the context of a users' current online session, or prior to a consumer engaging the digital environment (Moe 2003). Both of these scenarios call for a different approach to Web design. If the user has already entered the information search process, the design

and purpose of the advertising may be very different than when a user is merely surfing the net. For example, it makes more sense for the provider to display advertisements for DVDs and DVD player accessories for a user who is searching for information about DVD players, than for a user who is searching across several electronics categories.

Another consideration when designing online advertisements is that design elements of the ad should permit the user to interact with it in his or her preferred fashion. Yoo and Kim (2005) found that banner ads that had high levels of animation were viewed less favorably than ads with moderate or no

---

TABLE 30.5
Consumer Decision Process Design Recommendations

1. Advertisements that are personally relevant to users will be automatically processed. Present users with advertisements that are relevant to their task, characteristics, and targets (Grunert 1996; Robinson, Wysocka, and Hand 2007).

2. Prior experience with a stimulus will result in automatic activation of that experience into conscious attention. For familiar products, designers should activate those experiences that are likely to reinforce the positive aspects of the product experience (Robinson, Wysocka, and Hand 2007).

3. Novelty and ambiguity decrease the likelihood of initial attention, but once attention is captured they increase elaboration. Create a perceived knowledge gap in your target users. Although this is easier with new products, it is not uncommon for established products or brands to leverage the "new and improved" label to generate interest (Grunert 1996; Robinson, Wysocka, and Hand 2007).

4. Memory is a function of the number of instances that a stimulus is represented and the user can attend (Grunert 1996; Logan 1988).

5. If the brand is unknown, positive attributions can be developed by exposure without engaging controlled cognitive processes (Grunert 1996).

6. Provide a cue that is consistent with the goals of the business, and the strengths of the product or brand to guide the user to curiosity resolution for experientially motivated users (Ha and McCann 2008; Robinson, Wysocka, and Hand 2007).

7. Do not use curiosity-based advertising when consumers perceive the risks of investigating and acquiring that product as high. Under high-risk conditions, a product perceived high in novelty will have a higher perceived risk associated with it (Ha and McCann 2008).

8. In a high user-control environment avoid low user-control methods (e.g., in the midst of editorial content) of displaying your ads (Ha and McCann 2008). This is especially true for sites designed primarily for minor's content (Nairn and Dew 2007).

9. In sites designed primarily for minors, advertising has to be clearly delineated from content (Nairn and Dew 2007).

10. Advertising clutter decreases the performance of all advertising on a Web site. Determine what is clutter in your page designs and design accordingly (Ha and McCann 2008).

11. Large advertisements increase users' perception of clutter. Use smaller advertisements, particularly in content spaces (Ha and McCann 2008).

12. For advertisements that obscure content areas, user control mechanisms should be provided (McCoy et al. 2008).

13. Do not provide user controls for advertisements that do not obscure the content area (McCoy et al. 2008).

animation. There is also some indication that users are more likely to view ads that do not have the appropriate user controls (no controls for ads placed outside of the focus area and more user controls for those ads that obscure the focus area) given their placement on the page (McCoy et al. 2008). By looking at McCoy et al.'s study on ad placement and control and Yoo and Kim's (2005) study of banner ad animation, it can be seen that advertisements that demand attention owing to their design elements when attentional resources should not be required can have a negative impact on users' attitudes toward the advertisements and by extension their site.

In terms of human perception and action, it has been found that click-through rate on banner ads can be increased by simply including the imperative "click here" within the ad copy (Hofacker and Murphy 1998). Although increasing click-through rate is good, the generic click-through does not necessarily translate into more sales. It is important that the right type of customers, those that are likely to buy during their current session or at some future time, click through. For example, if the principal competitive advantage of an offering is price, then the site designer should develop advertising stimuli that will prime users to recognize needs based on criteria on which the firm has a competitive advantage. Most findings regarding advertising do not suggest that a consumer must take immediate observable action in order for that advertising to be effective. Experimental findings indicate that the priming of attributes will result in novice consumers spending more time in their external search on the primed attributes and ultimately influence choice (Mandel and Johnson 2002). Users who are surfing the net may be gathering information for the development of standards for future evaluation through the unintentional processing of stimuli (Adval and Monroe 2002), and the experiential nature of the Web provides opportunity to take advantage of surfer curiosity to improve brand or product position in the information search process (Menon and Soman 2002).

In their study of the role of curiosity in the information search process, Menon and Soman (2002) found that an intriguing stimulus, a teaser for a feature of which the consumer was unaware, resulted in more extensive elaboration and better knowledge acquisition on the trigger topic. Their research also suggests that satisfying curiosity too quickly inhibits elaboration and, as a consequence, learning (Menon and Soman 2002). Further, learning is optimized when a moderate period of time is provided between the curiosity-generating stimulus and curiosity resolving stimulus. This allows the time for the user to generate hypotheses regarding the novel product and enhances elaboration (Menon and Soman 2002). Last, it was reported that curiosity advertising positively affects product evaluation and increases perceived novelty. Using the curiosity trigger can apply across many situations and users but will likely get the best results for the experiential Internet session. At the same time, for some products it is reasonable to suggest that designers can put together an online advertising strategy that may also trigger the curiosity of goal-directed users. This is consistent with the model proposed by Venkatesh and Agarwal (2006) that

suggests that the longer a customer lingers on a site the more likely that customer is to continue to use and purchase on that site.

None of these findings suggest the effectiveness of an advertisement should be measured by click-through rate, but rather the advertisement should be designed in such a way that is consistent with product positioning and grounded in empirical theoretical findings. It is not enough to design an advertisement to attract attention and encourage click-through. It must be designed to encourage the types of behaviors and cognitions that will result in the positive evaluation of the brand, site, and product. This positive evaluation, of course, depends on the degree that the advertisement enhances users' experiences.

The decision on how to enhance the user experience with online advertisements is contingent on the goal of the users' online session, the strategy of the business or agency, the characteristics of the user in terms of their skill, perceptual capabilities, and attitudes (product, risk, and brand), and the position of the online retailer in the overall marketplace. Given that the goal of an ethical business is to provide a beneficial experience for the customer, each of the previous factors must be considered to ensure the ongoing positive consumption experience of the customer and the financial future of the business.

30.2.2 Information Search and Alternative Evaluation

After recognizing a problem, consumers require information for determining a solution. They must either retrieve information stored in their memories (internal search) or acquire new information from their environment (external search). The success of internal search depends on the ability to recall relevant solutions, and the related choice criteria and heuristics for choosing among them. As consumers begin their search for information, there is a tendency for them to be overconfident about their own knowledge. They underestimate their knowledge in low-confidence conditions and overestimate their knowledge in high-confidence conditions (Alba and Hutchinson 2000). Poor knowledge calibration develops for a number of reasons: the tendency to remain confident as the facts of previous experience fade and overemphasis on autobiographical experience (Hoch 2002), poor recall due to misinformation or irrelevant information at encoding and/or recall, the reinforcement and polarization effect of elaboration on currently held beliefs irrespective of their accuracy, and consumers' tendency to overestimate their own performance, to name a few (Alba and Hutchinson 2000). The role of confidence in the information search process is obvious: the more confident the consumer, the less likely the consumer will engage in an extensive information search for alternatives. The less confident he or she is, the more likely the consumer will continue with his or her information search and delay purchase decision. From the solution provider's perspective, the preference is that target customers are confident and well calibrated with regard to their positive evaluation of the information, products, and services offered by the site.

As the user gathers information about a set of alternatives for problem solution, they assess each of the alternatives relative to perceived costs and benefits of attaining one of those options. A user stops the information search process when the perceived cost of gathering further information about the current set of alternatives (depth) or about other potential alternatives (breadth) is greater than the perceived risk associated with choosing one of the available alternatives and the consumers' confidence that an alternative will meet the identified need. Theoretically, the information search and alternative evaluation processes are distinct and sequential. From a behavioral perspective, however, information search and alternative evaluation occur in an iterative loop with each new relevant stimulus.

30.2.2.1 Internal Search

The effectiveness of the internal search for information depends on the consumers' abilities to store information about the set of acceptable options for satisfying the identified stimulus deficit and recalling that information accurately. The source of these memories is the sum of information and advertising stimuli and product use experiences of the consumer.

Drawing on previous experiences is much cheaper for a user compared to the costs of conducting an external search (Hoch 2002). Unfortunately, product experience can, in fact, result in a less than accurate assessment of a product's potential for meeting consumers' needs. That is, consumers have a tendency to recall the set of their experiences with a product rather than the initial experience. For example, a user may have a particularly positive feeling with regard to her product experience because she has developed a set of skills in using that product that effectively meet her higher-order needs. This can occur with products that have a steep learning curve. Experts rarely evaluate a product's usability costs based on the depth of difficulty of their first learning experiences but rather on a more global assessment of their most recent experience relative to the ease of use and functional benefit. The nature of personal experience is nearly always exaggerated in terms of accuracy, lack of bias, and recall (Table 30.6; Hoch 2002). When a consumer recalls product usage, rationalization of that product experience as positive sometimes occurs irrespective of actual experience. Further, users rarely consider their own potentially inaccurate recall of product experience that may be attributable to other experiential factors. Needless to say, the efficacy of our internal search is highly dependent on one's ability to recall in an objective manner. Objectivity based on the recollection of experiential and factual information is not always easy to achieve.

The decision to engage in an external information search is quite different for the user if the need identified is mostly experiential in nature rather than merely utilitarian. That is, in the case of utilitarian products, where perceived risks are low, users are likely to spend little time in the information search process, but with experiential products, where risks are high, consumers are expected to spend a great deal of time in the information search process.

TABLE 30.6

Information Search and Alternative Evaluation Design Recommendations

1. Display utilitarian product information in a manner that maximizes efficiency of target acquisition (Lohse and Spiller 1999).
2. Provide shortcuts for users to minimize the information search process with products that are repeat purchases or consumable items that they have purchased on another occasion (Hoch 2002).
3. Provide consumer tools to evaluate alternatives at any time during the information search process.
4. Create online advertisements that prime information search and alternative evaluation activities. Emphasize those aspects of your product that position the offering favorably relative to your competition for a well-calibrated consumer.
5. Products that have a competitive advantage on the basis of price are more sensitive to negative word-of-mouth information (Ha 2002). Do not provide access to user reviews or ratings if you are competing primarily on price.
6. Users will infer a higher-quality product based solely on a lenient return policy (Wood 2001). Provide a lenient return policy if product quality is one of your competitive advantages.

30.2.2.2 External Search

If no satisfactory solutions can be retrieved, the consumer must acquire information externally. It is at this point that consumer behavior becomes observable and new evaluative activity occurs. Because the iteration from search to evaluation and then back to search can transpire so quickly, they are often temporally indiscernible from a behavioral perspective. As previously discussed, marketers have devised several strategies to improve the likelihood that their brands will be activated in the consumer's memory with each new stimulus instance. The extensiveness of external search depends on the amount of retrievable knowledge, the degree of risk experienced by the consumer, and the perceived personal costs of acquiring the information. Information search plays a major role in reducing the perceived risk of the online consumer, and the perceived cost of acquiring product information via the Internet is critical to the use or disuse of the Internet as an information source (Ha 2002). The degree that information is accessible in terms of design and relevance is an important component of customer satisfaction (Lin 2007).

External search for information can be very brief in cases of spontaneous and repetitive purchases. External resources include personal contacts, online sources, and print or other multimedia sources. The information on which users base their purchases may be integrated with the stimulus that triggered need recognition. For example, an advertisement for a vacation to Jamaica may include the most relevant criteria for the consumer to make the decision (i.e., price, duration). At the same time, deliberate searches for product information may take place across a number of sessions. Ha (2002) found that users rely on brand and word-of-mouth information and customization for registered customers to reduce the perceived costs associated with searching for information and purchasing products online.

One way that consumers decrease the cost of searching for product information is through the use of recommendation systems. These systems range from the very simple price comparison search engines to very complex intelligent agents that consider a user's preferences and history with other agents and operate semi-autonomously as a proxy for the consumer (McDermott 2000). Simple comparison bots are based on search technologies and should be implemented in terms of query and results that are consistent with good user interface design. These bots can provide the user with control entry of relevant search fields and shortcuts for comparison characteristics and sort order. For example, Best Book Buys, Amazon, and Barnes and Noble all provide the consumer with the opportunity identify the lookup field (Title, Keyword, ISBN, or Author) to limit the potential for users to return an unmanageable number of search results (Rowley 2000). Further, Best Book Buys provides users a quick process for comparing books on price, delivery time, and condition (new or used) from a set of online book retailers. It is also important to provide consumers with price, shipping costs, total charges, a rating of the vendor, condition of the item if it is used, shipping method, expected arrival time, and payment options (Rowley 2000).

Target-focused agents make recommendations to consumers based on the target of their current search or purchase. For example, when a consumer accesses Amazon .com, a list of books that are consistent with that customer's previous purchasing experiences are displayed. In a study of context, item-specific information, and familiarity of intelligent agent recommendations for music CDs, participants were willing to pay nearly twice as much for an unfamiliar CD when the agent provided a recommendation in the absence of item-specific information about the unfamiliar option than when the unfamiliar option was presented in the context of a set of familiar options with distinguishing characteristics (Cooke et al. 2002). In the latter condition, participants evaluated these unfamiliar options negatively on the distinguishing characteristics. In Wood, Solomon, and Englis's (2006) study of avatars' influence on the buying behavior of women, they found that those shoppers assigned an avatar were more likely to make a purchase of certain types of products (e.g., lingerie and dress versus raincoat and bathrobe), more satisfied with the Web site, more likely to revisit the Web site, and had higher levels of satisfaction with their purchase.

There is some indication that users do not give equal weight to agent recommendations, and they are more likely to assign a higher weight to extreme positive recommendations as opposed to extreme negative recommendations (Gershoff, Mukherjee, and Mukhopadhyay 2003). On the basis of these findings, researchers (Cooke et al. 2002) tentatively suggest alternative design recommendations for these systems. They should make recommendations in context when (1) the recommendation is known to be preferred by the consumer through extensive purchase history, (2) the recommendation can be presented on meaningful item-specific characteristics that are similar to familiar options, and (3) absent of distinguishing characteristics from familiar options (Cooke et al. 2002). Present recommendations singularly when (1) little is known about the consumer, (2) distinguishing characteristics are available for the consumer to form evaluative contrasts, and (3) the agent can evaluate the item extremely positively (Cooke et al. 2002; Gershoff, Mukherjee, and Mukhopadhyay 2003). Some of these systems may require some effort at the outset to "serve" the customer. Gretzel and Fessenmaier (2006) found that value of this process and ultimately perceived recommendation fit was contingent on the consumers' perception of the relevance, transparency, and effort of the initiation process of the recommendation system.

The role of word-of-mouth information in the information search process is filtered by the users' prior knowledge. That is, if online consumers base their choice of a vendor on familiarity (as opposed to price), they are less likely to engage in an extensive search for reviews of vendor performance and less susceptible to negative word-of-mouth information obtained on the Internet (Chatterjee 2001). In addition, familiarity-based shoppers are also more likely to attribute negative information to unstable factors (Chatterjee 2001). Familiarity with the vendor brand reduces consumers' security risk of doing business with the online vendor. Also, Chatterjee (2001) found that consumers who based their choice of vendor on price were more likely to search for word-of-mouth evaluations of the vendor performance and heed that negative information. As a rule, customers prefer to have all the information associated with any online purchase, including shipping, transaction costs, and taxes.

In a study of consumers' expectations of online information for financial products, providing the monetary details associated with transactions (costs associated with an ongoing relationship) and marketing offers, and easy access to that information were found to be the most important factors in provider selection (Waite and Harrison 2002). These findings suggest that reducing the ambiguity of the financial impact of the customer vendor relationship is important to consumers' satisfaction with online information. Of course, reducing ambiguity in regard to costs of a relationship may not result in increased sales; it merely reduces the uncertainty with regard to the actual costs of the options.

Requiring users' registration during the information search stage has been clearly demonstrated to contribute to consumers' dissatisfaction who rated it as one of the least important attributes of information resources (Chen and Wells 2001; Waite and Harrison 2002). At the same time, if a site has an established relationship with a customer, Ha (2002) found that a customized informational source reduced the antecedents of risk, including privacy, evaluation, and usability (nuisance, financial, performance, and time). This apparently contradictory evidence illustrates the notion that the types of features offered to users depend on the solution, user task, and the customers' desired relationship with the vendor (Chen and Wells 2001; Ha 2002; Waite and Harrison 2002). That is, requiring a user to register early in the relationship for unclear and poorly articulated benefits is presumptuous, but after a trusting relationship is established, registration and customization can create a desirable user experience in terms of efficiency (e.g., one-click ordering) and personalization. This is consistent with Gretzel and Fessenmaier's (2006) study on a travel recommendation system. Requiring users to register during the information search process will result in the majority of the information search traffic seeking information from another source unless a prior trusted relationship has been established. Of course, this could have some advantages from the provider's perspective in terms of managing bandwidth or controlling access to only the most motivated users. In the case of ongoing relationships, users are willing to register to gain access to a customized user experience.

When given the choice between text only, graphics only, or a combination of both, users prefer a combination of text and graphics for the display of product information on the Web (Lightner and Eastman 2002). This may be rooted in the relative importance of accurate product representation to online consumers (Reibstein 2001). More generally, the design of the site must promote customer centric values such as simplicity, value, trust, fulfillment, and support.

30.2.2.3 Search and Product Class

Girard, Silverblatt, and Korgaonkar (2002) reported that the type of product class contributes to online consumer purchase preferences. Products were categorized into search, experience, or credence goods. A search product is an item for which the relevant product attributes can be reliably assessed prior to purchase. Experience products may be either "experience durable" (low purchase frequency goods) and experience nondurable (high purchase frequency goods) (Girard, Silverblatt, and Korgaonkar 2002, 3). Intuitively, the product attributes of experience goods cannot be reliably assessed prior to purchase or the costs associated with information search regarding the experience are higher than actual product experience. A credence product is even more ambiguous to the user, "such that the average consumer can never verify the level of quality of an attribute possessed by a brand or even their level of need for the quality supplied by the brand. That is, consumers will have great difficulty in evaluating the quality level of a product such as vitamins with confidence, or similarly a service such as termite fumigation or surgery" (Girard, Silverblatt, and Korgaonkar 2002, 3). Girard and colleagues found that consumers were more willing to purchase search products than low-frequency durable products. The experience nondurables were least likely to be purchased online.

In a second study of product class, Lowengart and Tractinsky (2001) examined the underlying dimensions related to purchasing books and computers. Interestingly, Lowengart and Tractinsky (2001) characterized a computer as experiential product in contrast to Girard and her colleagues' (2002) characterization of a computer as a search product. Although the methodology for inclusion of the computer in the search category in the first study is documented and robust, Lowengart and Tractinsky (2001) found that the underlying dimensions for the purchase of these two items were different in terms of

generalized risk and financial risk. Although these findings are preliminary and the products chosen for the categories were limited, this conceptual framework may begin to provide us with a useful way for characterizing and prioritizing design requirements. Considering these two studies, it is apparent that product class is a relevant issue when devising an online strategy for a particular product. At the same time, these studies were focused singularly on online purchase, and it is strategy to expect that consumers may use the Internet to engage in prepurchase information search and alternative evaluation or postpurchase customer support or evaluation activities in the future. The Internet as a medium for consumer activity is by no means static, and we can expect providers and consumers to use the technology in unforeseen ways.

30.2.2.4 Alternative Evaluation

As search proceeds, beliefs about alternative solutions (products or services) are developed and evaluative criteria identified. Researchers believe that consumers are typically unable to collect and evaluate all relevant information and do not have the mental processing capacity to handle all that is available. Further, they may be attempting to meet multiple goals with a particular purchase. As a result, consumers may construct new ad hoc evaluation approaches appropriate in the context of the decision (Arnould, Price, and Zinkhan 2002). If experienced enough, they may recall and apply previously successful evaluation methods as they go through the decision process. The evaluation methods used are usually classified as compensatory (usually for high involvement products), noncompensatory (usually for low involvement goods), or simple choice heuristics. Two-thirds of online consumers use the Web as a method to comparison shop (McDermott 2000). Online consumers can access discussion groups, user evaluations of products and vendors, and/or any number of online comparison sites (Scheepers 2001).

It is important to note that the format in which information is acquired may affect the resulting evaluation (Blackwell, Miniard, and Engel 2001). This should be of special concern to Web page designers who may alter the outcomes of evaluation through the method of information presentation. When users have relatively little experience with a target and the purchase task calls for a quality by price trade-off, a user is more likely to select the middle option relative to price and quality (Prelec, Wernerfelt, and Zettelmeyer 1997). The implications are certainly evident from both design and business perspectives. For example, in the case where the e-commerce site provides users with recommendations for purchase, the research suggests that a business should display purchase options in a manner that would highlight those items that are in stock or provide the business with the best margins. From a marketing perspective, this also provides some guidance with regard to desirable product presentation when negotiating with third-party vendors (Table 30.6).

When consumers are aware of a lenient return policy, they infer higher product quality (Wood 2001). Inferences of quality by brand name play a larger role in some product categories than others (Kim and Pysarchik 2000). This suggests that highlighting brand strength in the presentation and evaluation of technology products will result in the positive evaluation of the better brand. There is also evidence that these positive brand evaluations are related to consumers' intention to purchase these brands, and that finding replicated across cultures (Kim and Pysarchik 2000). Therefore, multinational retailers can expect positive brand image to be more important than the product's country of origin in consumers' brand quality inferences.

The Internet provides some unique opportunities with regard to dynamic pricing. Reverse pricing elicitation (name your own price), which gives buyers an active role in setting prices, is not as desirable as approaches in which the vendor provides the user with some referent prices or price ranges generated by users (Chernev 2003). Given that the vendor needs to account for the users' desire to have an available referent price range, it is suggested that designers of these interactions provide users with three referents: a referent range of successful similar bids (when the data exist), a user-generated referent range, or optional prices for view by the user.

There are many other issues in designing the online user experience that are directly related to the information search process and typically discussed in the context of user interface design in general. Readers are able to identify several chapters in this text that support the notion of providing the users with Web-based interfaces that are easy to use, have the necessary attributes that present a professional image, and offer users an opportunity to participate in their desired modes of interactive experience. Unless a product or service is so essential that consumers will overcome any difficulty to obtain it, a poorly designed Web site poses unnecessary costs on consumers that must be eliminated by competent Web site design.

30.2.3 CHOICE AND PURCHASE

Although consumers do not need to identify themselves during initial interactions with a Web site, they are forced to do so at the choice and purchase stages. At these stages, they must have developed a degree of commitment and trust (at least to the level of the price and importance of the product) associated with as many as five different business entities, including the credit card company, the product maker, the product seller, the delivery company, and the technical support company (Novak and Hoffman 1998). With each additional entity in the marketing channel, the possibility of a breakdown in one of the systems prior to product delivery increases cumulatively.

Following the evaluation, when a satisfactory solution to the problem has been identified, the consumer develops an intention to make the purchase. Solution providers must take care to prevent inhibiting factors such as cumbersome ordering procedures or fear of deception from deterring the purchase. Shopping cart design, shipping choices, guarantees, and so on are critical for encouraging the efficient and satisfying consummation of the transaction. Separating the costs of the product and fulfillment, such as shipping, handling,

and taxes on different Web pages, increases concern that the vendor is taking advantage of the consumer with potentially hidden charges. Design considerations include enhancing efficiency of the ordering and registration tasks, while reducing the perceived risk of transaction with regard to security of personal and financial information and establishing a fulfillment relationship with customers (Scheepers 2001).

30.3 POSTPURCHASE PHENOMENA

Purchase and consumption are not the conclusions of the CDP. Several other important phenomena occur following purchasing decisions that influence the long-term effectiveness of marketing efforts. Consumers learn from their purchase experiences and habitualize their decision making. Consumers also evaluate the results of their decision relative to their prepurchase expectations. A negative disconfirmation of their expectations results in dissatisfaction, negative attitude change, and a possible reassessment of their search and evaluation processes.

In some cases, complaining or negative word-of-mouth activities may occur. Howard (1977) described the development of efficient consumer purchasing processes. When unfamiliar with a product category, consumers engage in extensive information search, concept formation, evaluation, and deliberation. As they gain knowledge, they move from extensive problem solving toward much simpler decision processes that draw on accumulated knowledge of choice criteria, product attributes, brands, and brand availability. When buying behavior has become thoroughly routine or habitual, the consumer does little more than a scan of memory to retrieve appropriate purchasing scripts and form buying intentions. When decision making for high involvement purchases has become habitualized to the extent that external search is unnecessary, no new evaluation of alternatives is required, scripted buying intentions exist, and brand loyalty is said to have been established (Engel and Blackwell 1982).

The choice of a particular Web site such as Amazon.com for a consumer's book buying is itself a consumer decision. When consumers have found particular Web sites to be productive sources of information and useful as a shopping tool, they may become brand loyal to the site and "bookmark" them for future use, thus embedding them in their behavioral scripts. This makes the selection of a Web site part of a consumer's routine decision making, and continual reinforcement of this behavior by maintaining site quality is required to ensure the consumer's satisfaction with the site. In general, consumer research findings point to the conclusion that consistent quality and brand advertising are needed to keep consumers loyal to a brand or service (Mowen and Minor 1998).

The fulfillment of consumer expectations has been widely recognized as the basis for consumer satisfaction or dissatisfaction (Woodruff, Cadotte, and Jenkins 1983). If businesses fail to deliver on promises, expectations are not met and dissatisfaction results (Olshavsky and Miller 1972). For example, in December 1999, eToys.com's site was inaccessible to many would-be buyers because of high traffic. For those consumers who were able to place orders, 10% of the orders failed to arrive in time for Christmas (Stankevich 2000). Undoubtedly, negative word of mouth and customer refusal to give the firm a second chance were partially responsible for the subsequent demise of the business.

In the majority of instances of dissatisfaction, consumers do not complain to the business involved but rather discontinue patronage and/or spread negative word-of-mouth information about the business (Oliver 1997). The well-known phenomenon of "buyer's remorse" or cognitive dissonance is another important postpurchase phenomenon. Dissonance is an unpleasant emotional state caused by conflicting perceptions such as a consumer's realization that the best purchase was not made and that an attractive alternative choice was forgone (Mowen and Minor 1998). Dissonance may be relieved by consumers changing their attitude toward a chosen alternative, by undoing the purchase and returning the product, or by deciding never to make such a purchase again.

Web sites that provide supportive information to consumers following purchase can help buyers to reduce their dissonance and can reduce the possibility of negative behavior or attitude change on the part of the buyer. Some marketers are not aware of the benefits of analyzing postpurchase consumer behavior, although many have become concerned with measuring consumer satisfaction (Mowen and Minor 1998). The consumer behavior literature suggests that postpurchase phenomena are of great importance and that postpurchase contact by Web-based sellers, as well as brick-and-mortar retailers, is important in reducing the effects of dissonance, as well as assessing and influencing other postpurchase outcomes, such as satisfaction and repatronage behavior.

The same things that have been found to drive users to an initial purchase (perceived quality, i.e., system, information, service), trust, ease-of-use, the design of advertisements, and professional design are the primary drivers for customers to patronize a site and purchase products in the future (Kuan, Bock, and Vathanophas 2008). Consistent quality and consumer satisfaction are the foundation for a continued relationship with an e-commerce Web site.

30.4 CONCLUSION

This chapter has identified demographic and psychological variables which can be used to group e-commerce site users in market segments. Developers of e-commerce sites must realize first and foremost that any given site cannot be all things to all users, and the only strategically sensible way to target users is to define them as target segments. Thus, e-commerce site developers should begin their task with a clear idea of the psychological and behavioral predilections and distinctive psychological features of their strategic consumer targets. In particular, the motivation for site usage, whether utilitarian or experiential, provides an initial guide for site development. However, two particularly important psychological characteristics, the level of personal involvement with a consumer goal, reflect the amount and type of information that should be provided by a site. An issue of

particular additional importance is the need for e-commerce participants to mitigate a variety of risks experienced when participating in an e-commerce experience. When consumers' reliably familiar methods of reducing purchasing risk are no longer applicable in an on-line marketplace, special care must be taken to design sites in a way that increases the trust and confidence of consumers.

The chapter has highlighted the consumer decision process for the purpose of sensitizing the site designer to the easily forgotten fact that consumer decisions involve more than a one-shot click on the button to enter a credit card number and leave the site. Rather, the consumer decision process involves numerous psychological and behavioral phenomena that must be properly engaged and guided for a successful transaction to occur, and all of these have implications (discussed above) for site design.

This brief summary of consumer behavior research is by no means a complete work with regard to the range of theoretical works that could be used in the development of online consumer experiences. Instead, it is a sampling of applied and theoretical works that can be used to establish either general, or in some cases very specific, design guidelines. Regardless of the foundation of any single consumer e-commerce solution, those who create these online experiences must consider the unique characteristics of the target users, the influence of product class, and how a particular solution integrates with the broad range of consumer behavior both online and offline.

REFERENCES

Aiken, D. K., R. Mackoy, B. Shaw-Ching Liu, R. Fetter, and G. Osland. 2007. Dimensions of Internet commerce trust. *Journal of Internet Commerce* 6(4): 1–25.

Adval, R., and K. B. Monroe. 2002. Automatic construction and use of contextual information for product and price evaluations. *Journal of Consumer Research* 28: 572–592.

Alba, J., and W. Hutchinson. 2000. Knowledge calibration: What consumers know and what they think they know. *Journal of Consumer Research* 27: 324–344.

Antil, J. H. 1984. Conceptualization and operationalization of involvement. In *Advances in Consumer Research*, vol. 11, ed. T. C. Kinnear, 203–209. Provo, UT: Association for Consumer Research.

Arnould, E., L. Price, and G. M. Zinkhan. 2002. *Consumers*. Chicago, IL: Irwin/McGraw-Hill.

Awad, N. F., and A. Ragowsky. 2008. Establishing trust in electronic commerce through online word of mouth: An examination across genders. *Journal of Management Information Systems* 24(4): 101–121.

Baumgartner, H. 2002. Toward a personology of the consumer. *Journal of Consumer Research* 29: 286–292.

Beatty, S., and S. Smith. 1987. External search effort: An investigation across several product categories. *Journal of Consumer Research* 14: 84–92.

Belk, R. W., M. Wallendorf, and J. F. Sherry Jr. 1989. The sacred and the profane in consumer behavior: Theodicy on the odyssey. *Journal of Consumer Research* 16: 1–38.

Bhatnagar, A., S. Misra, and H. R. Rao. 2000. On risk, convenience, and Internet shopping behavior. *Communications of the ACM* 43(11): 98–105.

Blackwell, R. D., P. W. Miniard, and J. F. Engel. 2001. *Consumer Behavior*, 9th ed. Fort Worth, TX: Harcourt Brace.

Bloch, P. H. 1982. Involvement beyond the purchase process. In *Advances in Consumer Research*, vol. 9, ed. A. Mitchell, 413–417. Ann Arbor, MI: Association for Consumer Research.

Bloch, P. H., D. L. Sherrell, and N. M. Ridgeway. 1986. Consumer search. *Journal of Consumer Research* 13: 119–126.

Campbell, M. C., and R. C. Goodstein. 2001. The moderating effect of perceived risk on consumers' evaluations of product incongruity: Preference for the norm. *Journal of Consumer Research* 28: 439–449.

Celsi, R. L., and J. C. Olson. 1988. The role of involvement in attention and comprehension processes. *Journal of Consumer Research* 15: 210–224.

Chatterjee, P. 2001. Online reviews: Do consumers use them? *Advance in Consumer Research* 28:129–133.

Chen, Q., and W. D. Wells. 2001. .com Satisfaction and .com dissatisfaction: one or two constructs? *Advances in Consumer Research* 28: 34–39.

Chernev, A. 2003. Reverse pricing and online price elicitation strategies in consumer choice. *Journal of Consumer Psychology* 13: 51–62.

Cheung, C. M. K., and M. K. O. Lee. 2006. Understanding consumer trust in Internet shopping: A multidisciplinary approach. *Journal of the American Society for Information Science and Technology* 57(4): 479–492.

Cho, C. H., J. Kang, and H. J. Cheon. 2006. Online shopping hesitation. *CyberPsychology and Behavior* 9(3): 261–274.

Cho, H., and S. Koh Jialin. 2008. Influence of gender on Internet commerce: An explorative study in Singapore. *Journal of Internet Commerce* 7(1): 95–119.

Cooke, A. D., H. Sujan, M. Sujan, and B. A. Weitz. 2002. Marketing the unfamiliar: The role of context and item-specific information in electronic agent recommendations. *Journal of Marketing Research* 39: 488–497.

Cyr, D. 2008. Modeling Web site design across cultures: Relationships to trust, satisfaction, and e-loyalty. *Journal of Management Information Systems* 24(4): 47–72.

Daugherty, T., H. Li, and F. Biocca. 2008. Consumer learning and the effects of virtual experience relative to indirect, and direct product experience. *Psychology and Marketing* 25(7): 568–586.

Dholakia, U., and R. Bagozzi. 2001. Consumer behavior in digital environments. In *Digital Marketing*, eds. J. Wind and V. Mahajan, 163–200. New York: Wiley.

Dholakia, R. R., and K. Chiang. 2003. Shoppers in cyberspace: Are they from Venus or Mars and does it matter? *Journal of Consumer Psychology* 13: 171–176.

Dowling, G. R., and R. Staelin. 1994. A model of perceived risk and intended risk-handling activity. *Journal of Consumer Research* 21: 119–134.

Eggert, A. 2006. Intangibility and perceived risk in online environments. *Journal of Marketing Management* 22: 553–572.

Engel, J. F., and R. D. Blackwell. 1982. *Consumer Behavior*. Chicago, IL: Dryden Press.

Frambach, R. T., H. C. A. Roest, and T.V. Krishnan. 2007. The impact of consumer Internet experience on channel preference and usage intentions across the different stages of the buying process. *Journal of Interactive Marketing* 21(2): 26–41.

Friedman, B., P. H. Kahn, and D. C. Howe. 2000. Trust online. *Communications of the ACM* 43(12): 34–40.

Gefen, D. 2002. Reflections on the dimensions of trust and trustworthiness among online consumers. *The DATA BASE for Advances in Information Systems* 33: 38–53.

Gershoff, A. D., A. Mukherjee, and A. Mukhopadhyay. 2003. Consumer acceptance of online agent advice: Extremity and positivity effects. *Journal of Consumer Psychology* 13(1/2): 161–170.

Girard, T., R. Silverblatt, and P. Korgaonkar. 2002. Influence of product class on preference for shopping on the Internet. *Journal of Computer-Mediated Communication* 8(1), http://www.ascusc.org/jcmc/voI8/issue l/girard.html.

Gretzel, U., and D. R. Fesenmaier. 2006. Persuasion in recommender systems. *International Journal of Electronic Commerce* 11(2): 81–100.

Grewal, D., J. Gottlieb, and H. Marmostein. 1994. The moderating effects of framing and source credibility on the price-perceived risk relationship. *Journal of Consumer Research* 21: 145–153.

Grunert, K. 1996. Automatic and strategic processes in advertising effects. *Journal of Marketing* 60: 88–101.

Gurung, A., X. Luo, and M. K. Raja. 2008. An empirical investigation on customer's privacy perceptions, trust and security awareness in e-commerce environment. *Journal of Information Privacy and Security* 4(1): 42–63.

Ha, H. Y. 2002. The effects of consumer risk perception on pre-purchase information in online auctions: Brand, word-of-mouth, and customized information. *Journal of Computer-Mediated Communication* 38(1), http://www.ascusc.org/jcmc/voI8/issue l/ha.html.

Ha, L., and K. McCann. 2008. An integrated model of advertising clutter in offline and online media. *International Journal of Advertising* 27(4): 569–592.

Hirschman, E. C. 1980. Innovativeness, novelty seeking, and consumer creativity. *Journal of Consumer Research* 7: 283–295.

Hirschman, E. C., and H. B. Holbrook. 1982. Hedonic consumption: emerging concepts, methods and propositions. *Journal of Marketing* 46: 92–101.

Hoch, S. J. 2002. Product experience is seductive. *Journal of Consumer Research* 29: 448–454.

Hofacker, C. F., and J. Murphy. 1998. World Wide Web banner advertising copy testing. *European Journal of Marketing* 32: 703–712.

Holbrook, M., and E. Hirschman. 1982. The experiential aspect of consumption: consumer fantasies, feelings, and fun. *Journal of Consumer Research* 9: 132–140.

Howard, J. A. 1977. *Consumer Behavior: Application of Theory*. New York: McGraw-Hill.

Hui, T.-K., and D. Wan. 2007. Factors affecting Internet shopping behavior in Singapore: Gender and educational issues. *International Journal of Consumer Studies* 31: 310–316.

Insko, C. A., and J. Schopler. 1972. *Experimental Social Psychology*. New York: Academic Press.

Janda, S. 2008. Does gender moderate the effect of online concerns on purchase likelihood? *Journal of Internet Commerce* 7(3): 339–358.

Jarvenpaa, S. L., and N. Tractinsky. 1999. Consumer trust in an Internet store: A cross-cultural validation. *Journal of Computer-Mediated Communication* 5(2), http://www.ascusc.org/jcmc/voI5/issue2/jarvenpaa.html.

Jarvenpaa, S. L., N. Tractinsky, and M. Vitale. 2000. Consumer trust in an Internet store. *Information Technology and Management* 1(1–2): 45–71.

Jones, S., and S. Fox. 2009. Pew Internet project data memo. The Pew Internet and American Life Project. http://pewInternet.org/trends.asp (accessed April 27, 2010).

Kamali, N., and S. Loker. 2002. Mass customization: On-line consumer involvement in product design. *Journal of Computer-Mediated Communication* 7(4), http://www.ascusc.org/jcmc/voI7/issue4/loker.html.

Kim, D. J., B. Cho, and H. R. Rao. 2000. Who buys on the Internet? An investigation of consumer characteristics for Internet purchasing behavior. Paper presented at INFORMS Conference on Information Systems and Technology, San Antonio, TX, Nov. 5–8.

Kim, S., and D. T. Pysarchik. 2000. Predicting purchase intentions for uninational and binational products. *International Journal of Retail and Distribution Management* 28: 280–291.

Krugman, H. E. 1965. The impact of television advertising: learning without involvement. *Public Opinion Quarterly* 29: 349–356.

Kuan, H.-H., G. W. Bock, and V. Vathanophas. 2008. Comparing the effects of Website quality on costumer initial purchase and continued purchase at e-commerce Websites. *Behaviour and Information Technology* 27(1): 3–16.

Kwak, H., R. J. Fox, and G. M. Zinkhan. 2002. What products can be successfully promoted and sold via the Internet? *Journal of Advertising Research* 42: 23–38.

Lastovicka, J. L. 1979. Questioning the concept of involvement defined product classes. In *Advances in Consumer Research*, vol. 6, ed. W. L. Wilkie, 174–179. Ann Arbor, MI: Association for Consumer Research.

Lavidge, R. J., and G. A. Steiner. 1961. A model for predictive measurements of advertising effectiveness. *Journal of Marketing* 25: 59–62.

Lee, J., J. Kim, and J. Moon. 2000. What makes Internet users visit cyber stores again? Key design factors for customer loyalty. *Proceedings of the SIGCHI Conference on Human Factors in Computing Systems* 305–312.

Lepkowska-White, E., and A. Eifler. 2008. Spinning the Web: The interplay of Web design features and product types. *Journal of Website Promotion* 3(3/4): 196–212.

Lightner, N. J., and C. Eastman. 2002. User preference for product information in remote purchase environments. *Journal of Electronic Commerce Research* 3: 174–186.

Lin, H.-F. 2007. The impact of Website quality dimensions on customer satisfaction in the B2C e-commerce context. *Total Quality Management* 18(4): 363–378.

Logan, G. D. 1988. Towards an instance theory of automatization. *Psychological Review* 22: 1–35.

Logan, G. D. 1992. Attention and preattention in theories of automaticity. *American Journal of Psychology* 105: 317–339.

Lohse, G. L., and P. Spiller. 1999. Internet retail store design: How the user interface influences traffic and sales. *Journal of Computer Mediated Communication* 5(2), http://www. ascusc.org/jcmc/voI5/issue2/lohse.htm.

Lowengart, O., and N. Tractinsky. 2001. Differential effects of product category on shoppers' selection of Web-based stores: A probabilistic modeling approach. *Journal of Electronic Commerce Research* 2(4): 142–156.

Manchanda, P., J.-P. Dube, K.Y. Goh, and P. K. Chintagunta. 2006. The effect of banner advertising on Internet purchasing. *Journal of Marketing Research* 43: 98–108.

Mandel, N., and E. J. Johnson. 2002. When Web pages influence choice: effects of visual primes on experts and novices. *Journal of Consumer Research* 29: 235–245.

Markin, R. J., Jr. 1974. *Consumer Behavior: A Cognitive Orientation.* New York: Macmillan.

McCoy, S., A. Everard, P. Polak, and D. F. Galletta. 2008. An experimental study of antecedents and consequences of online ad intrusiveness. *International Journal of Human–Computer Interaction* 24(7): 672–699.

McDermott, I. 2000. Shopping bots: Santa's electronic elves. *Searcher: The Magazine for Database Professionals* 10–16.

McElroy, J. C., A. R. Hendrickson, A. M. Townsend, and S. M. DeMarie. 2007. Dispositional factors in Internet use: Personality versus cognitive style. *MIS Quarterly* 31(4): 809–820.

McMahan, C., R. Hovland, and S. McMillan. 2008. Gender and Internet advertising: Differences in the ways males and females engage with and perceive Internet advertising. *Proceedings of the American Academy of Advertising Conference*, 52–55. San Mateo, CA.

Menon, S., and D. Soman. 2002. Managing the power and curiosity of effective Web advertising strategies. *Journal of Advertising Research* 31: 1–14.

Metzger, M. J. 2006. Effects of site, vendor, and consumer characteristics on Web site trust and disclosure. *Communication Research* 33(3): 155–179.

Miyazaki, A. D., and A. Fernandez. 2001. Consumer perceptions of privacy and security risks for online shopping. *Journal of Consumer Affairs* 35: 27–44.

Moe, W. W. 2003. Buying, searching, or browsing: Differentiating between online shoppers using in-store navigational clickstream. *Journal of Consumer Psychology* 13: 29–39.

Moss, G. A., R. Gunn, and K. Kubacki. 2007. Successes and failures of the mirroring principle: The case of angling and beauty Websites. *International Journal of Consumer Studies* 31: 248–267.

Mowen, J. C., and M. Minor. 1998. *Consumer Behavior*, 3rd ed. Upper Saddle River, NJ: Prentice Hall.

Nairn, A., and A. Dew. 2007. Pop-ups, pop-unders, banners and buttons: The ethics of online advertising to primary school children. *Journal of Direct, Data and Digital Marketing Practice* 9(1): 30–46.

Nam, C., C. Song, E. Lee, and C. I. Park. 2006. Consumers' privacy concerns and willingness to provide marketing-related personal information online. *Advances in Consumer Research* 33: 212–217.

Netguide. 1996. Just say no to graphics. *Netguide*, July, 32.

Novak, T. P., and D. L. Hoffman. 1998. Building consumer trust in online environments: The case for information privacy. *Communications of the ACM* 41(12): 41–49.

Oliver, R. L. 1997. *Satisfaction: A Behavioral Perspective on the Consumer.* New York: McGraw-Hill.

Olshavsky, R. W., and J. A. Miller. 1972. Consumer expectations, product performance, and perceived product quality. *Journal of Marketing Research* 9: 19–21.

Olson, J., and G. M. Olson. 2000. i2i Trust in e-commerce. *Communications of the ACM* 43(12): 41–44.

Pearson, J. M., and A. M. Pearson. 2008. An exploratory study into determining the relative importance of key criteria in Web usability: A multi criteria approach. *Journal of Computer Information Systems* 48(4): 115–127.

Peter, J. P., and J. C. Olson. 2008. *Consumer Behavior and Marketing Strategy*, 8th ed. New York: McGraw-Hill/Irwin.

Petty, R., J. Cacioppo, and D. Schumann. 1983. Central and peripheral routes to advertising effectiveness. *Journal of Consumer Research* 10: 155–146.

Prelec, D., B. Wernerfelt, and E. Zettelmeyer. 1997. The role of inference in context effects: Inferring what you want from what is available. *Journal of Consumer Research* 24: 118–125.

Ray, M. 1973. Marketing communications and the hierarchy of effects. In *New Models for Mass Communications*, ed. P. Clarke, 147–176. Beverly Hills, CA: Sage.

Reibstein, D. J. 2001. The Internet buyer. In *Digital Marketing: Global Strategies from the World's Leading Experts*, eds. J. Wind and V. Mahajan, 201–225. New York: Wiley.

Robinson, H., A. Wysocka, and C. Hand. 2007. Internet advertising effectiveness: The effect of design on click-through rates for banner ads. *International Journal of Advertising* 26(4): 527–541.

Rodgers, S., and K. M. Sheldon. 2002. An improved way to characterize Internet users. *Journal of Advertising Research* 1: 85–94.

Rodgers, S., Y. Wang, R. Rettie, and F. Alpert. 2007. The Web motivation inventory: Replication, extension, and application to Internet advertising. *International Journal of Advertising* 26(4): 447–476.

Rosa, J. A., E. C. Garbarino, and A. J. Malter. 2006. Keeping the body in mind: The influence of body esteem and body boundary aberration on consumer beliefs and purchase intentions. *Journal of Consumer Psychology* 16(1): 79–91.

Rosa, J. A., and A. J. Malter. 2003. E(embodied)-knowledge and e-commerce: How physiological factors affect on-line sales of experiential products. *Journal of Consumer Psychology* 13: 63–73.

Rowley, J. 2000. Product searching with shopping bots. *Internet Research* 10: 203–214.

SanJose-Cabezudo, R., A. M. Gutierrez-Arranz, and J. Gutierrez-Cillan. 2009. The combined influence of central and peripheral routes in the online persuasion process. *CyberPsychology and Behavior* 12(3): 299–308.

SanMartín, S., C. Camarero, C. Hernández, and L. Valls. 2009. Risk, drivers, and impediments to online shopping in Spain and Japan. *Journal of Euromarketing* 18: 47–64.

Schaninger, C. M. 1976. Perceived risk and personality. *Journal of Consumer Research* 3: 95–100.

Scheepers, R. 2001. Supporting the online consumer decision process: Electronic commerce in a small Australian retailer. *Proceedings of the Twelfth Australasian Conference on Information Systems.* Coffs Harbour, NSW, Australia.

Sherif, C., and M. Sherif. 1967. Attitudes as the individuals' own categories: The social judgment-involvement approach to attitude and attitude change. In *Attitude, Ego-involvement and Change*, eds. C. W. Sherif and M. Sherif, 105–139. New York: Wiley.

Song, K., A. M. Fiore, and J. Park. 2006. Telepresence and fantasy in online apparel shopping experience. *Journal of Fashion Marketing and Management* 11(4): 553–570.

Stankevich, D. G. 2000. Was the grinch really online? *Discount Merchandiser* 3: 42–44.

Steinfield. C., and P. Whitten. 1999. Community level socio-economic impacts of electronic commerce. *Journal of Computer-Mediated Communication* 5(2), http://www.ascusc.org/jcmc/vol5/issue2/steinfield.html.

Stigler, G. 1961. The economics of information. *Journal of Political Economy* 69: 213–225.

Tauber, E. M. 1972. Why do people shop? *Journal of Marketing* 30: 50–52.

Teltzrow, M., B. Meyer, and H.-J. Lenz. 2007. Multi-channel consumer perceptions. *Journal of Electronic Commerce Research* 8(1): 18–31.

Tsao, J. C., and S. Steffes-Hansen. 2008. Predictors for Internet usage of teenagers in the United States: A multivariate analysis. *Journal of Marketing Communications* 14(3): 171–192.

Thomson, E. S., and A. W. Laing. 2003. The net generation: Children and young people, the Internet and online shopping. *Journal of Marketing Management* 19: 491–512.

Venkatesh, V., and R. Agarwal. 2006. Turning visitors into customers: A usability-centric perspective on purchase behavior in electronic channels. *Management Science* 52(3): 367–382.

Waite, K., and T. Harrison. 2002. Consumer expectations of online information provided by bank Websites. *Journal of Financial Services Marketing* 6(4): 309–322.

Wang, C.-C., and H.-W. Yang. 2008. Passion for online shopping: The influence of personality and compulsive buying. *Social Behavior and Personality* 36(5): 693–706.

Wang, H., M. K. O. Lee, and C. Wang. 1998. Consumer privacy concerns about Internet marketing. *Communications of the ACM* 41(3): 63–70.

Warden, C. A., W.-Y. Wu, and D. Tsai. 2006. Online shopping interface components: Relative importance as peripheral and central cues. *CyberPsychology and Behavior* 9: 286–296.

Wood, S. L. 2001. Remote purchase environments: The influence of return policy leniency on two-stage decision processes. *Journal of Marketing Research* 38(2): 157–169.

Wood, N.T., M. R. Solomon, and B. G. Englis. 2006. Personalization of the Web interface: The impact of Web avatars on users' responses to e-commerce sites. *Journal of Website Promotion* 2(1/2): 53–69.

Woodruff, R. B., E. R. Cadotte, and R. L. Jenkins. 1983. Modeling consumer satisfaction processes using experience-based norms. *Journal of Marketing Research* 20(3): 296–304.

Yang, B., and D. Lester. 2004. Attitudes toward buying online. *CyberPsychology and Behavior* 7(1): 85–91.

Yoo, C. Y., and K. Kim. 2005. Process of animation in online banner advertising: The roles of cognitive and emotional responses. *Journal of Interactive Marketing* 19(4): 18–34.

Zhang, X., V. R. Prybutok, S. Ryan, and R. Pavur. 2009. A model of the relationship among consumer trust, Web design and user attributes. *Journal of Organizational and End User Computing* 21(2): 44–66.

31 Analyzing and Modeling User Activity for Web Interactions

Jianping Zeng and Jiangjiao Duan

CONTENTS

31.1 INTRODUCTION

Great progress has been made in the fields of Web technology development and applications, and it has speeded up integration of the Web with personal and societal uses. Especially, Web 2.0 technology leads to the formation of Web ecology, in which humans, digital environments, and digital resources coexist and will evolve for a long time. Just like in the real-world ecology, human factors play an important role in the Web ecology. People are active in all kinds of activities, such as creating new types of digital contents, interacting with each other, pushing the formation of opinion, etc. By analyzing and modeling user activity for the Web, the how and why of complex phenomena in Web ecology are expected to be revealed gradually.

Online business activity becomes more and more popular as Web technology provides an increasingly comfortable environment for customers and service providers. Different from real-world business, online business can utilize the Web technology to improve the service so that the activity can achieve successful results in a shorter time. For example, online service providers can find potential customers by mining the activity records that are produced by Web users during their visits to online applications. On the other hand, Web users can also benefit from the user activity analysis. For example, recommendation and information filtering can be performed accurately if the interests of Web users are captured on the basis of the user models.

There are many kinds of Web users activities, such as browsing, propagating, clicking, searching, and so on. The characteristics of uncertainty, randomness, and fuzziness in Web user activity make understanding human behavior in Web environments one of the great challenges for computer science of the twenty-first century. However, data-driven exploration of digital traces, such as Web logs and online forums, has provided an important starting point for furthering our understanding of "digital behavior" (Cadez et al. 2003). Hence, on the basis of this consideration, the following section provides several methods to analyze and model Web user activity, such as clicking activity, selective activity, and propagation activity.

31.2 ANALYZING AND MODELING CLICKING ACTIVITY

Clicking activity happens whenever a user clicks a link on a Web page. Some kinds of Web sites, such as news providers, product sellers, etc., usually receive large numbers of clicks. These clicks are in the form of a sequence that usually reflects what the Web user is thinking or as an indicator of what information the user likes to retrieve. The browser tells us nothing about this type of behavior, but this kind of information can be very important to sellers or providers because it allows them to provide more suitable service for the users based on their behavior. For example, links with strong correlation can be placed together for convenient access, which helps Web users find the information they need more quickly. Hence, modeling and analysis of the clicking activity on the Web is useful and valuable (Bucklin and Sismeiro 2003; Montgomery, Srinivasan, and Liechty 2004). It can be used to improve the Web cache performance (Padmanabham and Mogul 1996; Schechter, Krishnan, and Smith 1998), recommend related pages (Dean and Henzinger 1999; Gündüz and Özsu 2003a), improve search engines (Brin and Page 1998; Guo et al. 2009; Piwowarski, Dupret, and Jones 2009), and personalize the browsing experience (Anand and Mobasher 2005).

In the following, we describe how to collect and process the clicking data, how to model the user clicking activity, and how to utilize the models to analyze user clicking.

31.2.1 CLICKING SEQUENCE: DATA FORMAT AND PROCESS

Many kinds of Web servers record the users' clicking activity and save it to a disk file, which is known as "Web log." However, the format and saving mechanism are slightly different for several Web servers that are widely used.

31.2.1.1 Internet Information Services (IIS)

By default, IIS log files are stored in the %windir%\system32\LogFiles\W3SVC# folder on the computer that is running IIS, where %windir% is the drive partition where the Windows OS is installed and # is the number of the site (http://msdn.microsoft.com/en-us/library/ms525410.aspx). For example, the default location of the log file folder for the Web site where Windows is installed on driver C would be as follows: C:\WINNT\system32\LogFiles\W3SVC1.

IIS uses W3C extended log file format to store log by default. However, other formats, such as IIS log file format and NCSA Common log file format are also allowed. The

TABLE 31.1
Useful Fields for Click Analyses

Field Name	Description
Date	The date on which the click activity occurred.
Time	The time that the click activity occurred.
Client IP	That is the IP that the client visiting the server. Note that if the client uses a proxy to visit, then the proxy IP is considered as the client IP.
User name	If anonymous is configured, then "-" will be recorded.
Methods	That is the operations of client execution. "GET" and "POST" are expected.
URI	The resource that the client is visiting. The resource can be Web page, image, script, etc.
User agent	The name of client browser.
Server IP	It is useful when multiple IP are configured in the same server.
Server port	It is usually the service type, which is provided for the clients.

user click activity is recorded as several fields, and some of them are shown in Table 31.1.

31.2.1.2 Tomcat

Different from IIS, tomcat does not attempt to record the visiting information by default. However, it can be configured for this purpose by changing the settings in file ${catalina}/conf/server.xml, where ${catalina} is the directory for installing tomcat.

When a Tomcat Web server is configured in this way, it can record many available details in user activity. This is achieved by setting the log patterns, some of which are listed in Table 31.2.

For example, we can set the log pattern as "%a %t %m %U," which means that we can get the client IP address, visiting date, request method and URL, which is related to a user click.

31.2.1.3 Apache

Similarly, in Apache, Web logs can be recorded. An example is shown as follows.

127.0.0.1 – skyt [10/Oct/2000:13:55:36-0700] "GET / apache_pb.gif HTTP/1.0" 200 2326

TABLE 31.2
Useful Fields for Click Analyses

Pattern of Field	Description
%a	Remote IP address
%A	Local IP address
%m	Request method (GET, POST, etc.)
%p	Local port on which this request was received
%t	Date and time
%U	Requested URL path
%H	Request protocol

It indicates that client user skyt with IP address of 127.0.0.1 issued a request to get file /apache_pb.gif via HTTP/1.0 protocol, that the server finished the process of the request at 10/Oct/2000:13:55:36 and successfully sent the result of 2326 bytes to the client.

31.2.1.4 Processing of Web Log

Before analyzing and modeling user clicks based on Web logs, a log file should be preprocessed. Some of the main concerns are as follows.

Parsing the File

The following is a typical example of a log file. The lines that begin with # are the description of the log file. The line of "#Fields" indicates the fields that are recorded in the file. Starting from the fourth line, a user click is recorded. The line example shows that an anonymous user with the IP address of 172.16.255.255 issued an HTTP GET command for the html file /default.htm at 17:42. Note that the items use spaces as separators. Hence, the record extraction program should utilize this knowledge to achieve correct results.

#Software: Internet Information Services 6.0
#Version : 1.0
#Date: 2001-05-02 17:42:15
#Fields: time c-ip cs-method cs-uri-stem sc-status cs-version
17:42:15 172.16.255.255 GET/default.htm 200 HTTP/1.0

Note that the URI items that Web servers record when a visitor clicks include several resources: HTML pages, ASP and JSP pages, images, Java class files, PostScript documents, and so on. Hence, requesting a page that contains images will generate more than one record in the Web log file. However, only the record whose URI is page type is of benefit to user clicking analyses. Hence, only these types of records should be kept in the final user session data sets.

It should also be pointed out that not all the page type records are generated by Web user clicking. Nowadays, there are many kinds of search engines, such as Google, Yahoo!, Baidu, etc. A crawler, which is an important element of search engines, will automatically launch requests of pages at certain times. This action also generates Web log records, which are not useful in Web user clicking activity analyses. Hence, these automatically generated records should be removed. Often, crawlers request pages in a manner that is different from a Web

user. The characteristics are that the number of pages requested is unbelievably large in a small time interval or the interval between two successive requests is always the same. These features provide a good rule for removing the records.

Another concern is about the recording time stamp. Different log formats use different time zones for the basis of times listed in the logs. W3C Extended format uses Coordinated Universal Time (UTC), which is basically the same as Greenwich Mean Time (GMT). However, other types of log formats use local time.

After parsing the log file, a sequence of user clicking activity can be obtained. Each item in the sequence is usually a vector, including time stamp, method, URI, protocol, etc. It can be denoted as follows.

$$
\begin{aligned}
click &= \{c_1, c_2, \cdots, c_n\} \\
&= \left\{ \begin{bmatrix} t_1 \\ m_1 \\ u_1 \\ p_1 \end{bmatrix}, \begin{bmatrix} t_2 \\ m_2 \\ u_2 \\ p_2 \end{bmatrix}, \cdots, \begin{bmatrix} t_n \\ m_n \\ u_n \\ p_n \end{bmatrix} \right\}
\end{aligned} \quad (31.1)
$$

where n is the length of the sequence. C_i is the ith click event, which is described by clicking time t, used method m, URI u, and protocol p.

User Recognition

There is no explicit user indicator in Web log; hence it is necessary to identify the user before performing Web user modeling and analyzing. In the Web log, the client IP address and user agent can be used to identify the user. Different users can be grouped by IP address and user agent. However, in the case of a large number of users with the same proxy, the identification becomes more complex. A more accurate technique for identifying separate Web users requires either setting cookies or forcing users to register when they arrive at the site. However, this process is not always accepted by users.

Session Recognition

In order to find useful patterns and then provide a sound foundation for user click activity analyzing and modeling, user click events should be grouped into sessions that describe the usage for one time visiting. As far as session recognition is concerned, we need to separate the clicking sequence into several segments, $C_1 = \{c_1, c_2, \ldots, c_{k1}\}$, $C_2 = \{c_{k1}, c_{k1+1}, \ldots, c_{k1+k2}\}, \ldots$, so that each click event in C_i belongs to the same sequence of user visiting.

In an online system, such as a personnel bank service, with user login and logout facilities, a session can be clearly defined as starting from the time of user login and ending at the time the user logs out. However, for many Web-based applications, such as news publishing, Web searching, and so on, when a session starts and ends is unclear because there is no explicit user login and logout event.

Web users may visit a site more than once; hence, by supposing that the time interval between two adjacent visitings should be below a threshold value, a timeout-based method is introduced to divide the page accesses of each user into individual sessions (Cooley, Mobasher, and Srivastava 1999). On the basis of the clicking sequence, the time stamp of consecutive click events which are in a session should fulfill with

$$T(c_{i+1}) - T(c_i) < \xi \qquad (31.2)$$

where ξ is a time interval threshold and TI is the visiting time of click event c.

Despite its simplicity, the method suffers from the difficulty in setting the time threshold. Different users may have different clicking behaviors, so the threshold varies. Even for the same user, the threshold may vary. Catledge and Pitkow (1995) found a 25.5-minute timeout based on their user experiments in 1994, based on the assumption that the most statistically significant events occur within 1.5 standard deviations from the mean. He and Göker reported the results of experiments that used the timeout method on two sets of Web logs (He and Göker 2000). In their experiments, the threshold was set large initially, and then gradually decreased. The authors concluded that a time range of 10 to 15 minutes was an optimal session interval threshold (He and Göker 2000). Many applications and most commercial products use 30 minutes as a default timeout (Cooley, Mobasher, and Srivastava 1999; Gündüz and Özsu 2003a; Sen and Hansen 2003; Xing and Shen 2002).

But the reasons behind choosing that particular amount of timeout are not clear. The timeout-based method considers the notion of a session on the Web just from the time interval, which leads to difficulty in selection of threshold. It lacks a more user-oriented view. Hence, in addition to time interval, some kinds of context information should be incorporated to dynamically identify a session.

A transaction identification method, also known as reference length, is proposed (Cooley, Mobasher, and Srivastava 1999). In this method, the request page issued by user clicking is classified into an "auxiliary" or "content" page. Cooley et al. observed that the time spent on auxiliary pages is usually shorter than that spent on a content page and also that the variance of the times spent on auxiliary pages is smaller than that of times spent on content pages. By making an assumption about the percentage of auxiliary references in a log, then a reference length can be calculated that estimates the optimal cutoff between auxiliary and content references based on a histogram.

Language modeling provides a general strategy for estimating the probability of any sequence—regardless of whether the basic units consist of words, characters, or any other arbitrary alphabet. By introducing smooth methods, such as absolute smoothing (ABS), Good-Turing smoothing (GT), linear smoothing (LIN), and Witten-Bell smoothing (WB), nonzero probability of any sequence, especially those which are not provided in train set, can be assigned. By considering each visited object as a basic unit, like a word or character in natural language, the probability of object sequences can then be estimated using the same language modeling tools (Huang et al. 2004). Then session identification is performed whenever perplexity of the clicking sequence is above a threshold ω.

$$\text{Perplexity}(s) > \omega \qquad (31.3)$$

where

$$\text{Perplexity}(s) = P(s)^{-1/N}$$

and

$$P(s) = \prod_{i=1}^{N} p(u_i \mid u_{i-n+1} \cdots u_{i-1}) \qquad (31.4)$$

u_i is the request objected in clicking sequence, and $p(u_i \mid u_{i-n+1} \cdots u_{i-1})$ $i = 1,2 \ldots, n$ can be estimated in the way as n-gram language modeling.

For a clicking sequence in a session, each object is frequently visited one after another; hence, the perplexity of the sequence is low. However, when a new object that is not considered in the session appears, the perplexity will increase. In this way, a dynamic recognition method based on statistical language models can produce expected results. The method does not depend on time intervals when identifying session boundaries. However, it introduces a new perplexity threshold in identifying session boundaries dynamically. The experiments show that the detection performance is sensitive to the threshold value.

Note that the size of the Web log depends on the Web sites. So the number of user sessions varies. For example, the Web servers from msnbc.com for a 24-hour period typically produce about one million user sessions. The data set is available online at http://kdd.ics.uci.edu/databases/msnbc/msnbc.html. Hence, a large number of sessions should be a serious consideration in computer programming on Web session automatic processing.

31.2.2 Activity Models: Specifications and Effectiveness

Usually, in the browsing process, the server allows anonymous visiting; that is, user name is not included in the Web log. Hence, user identification and user session recognition are necessary to get a user clicking activity set. Then, all kinds of activity modeling, such as Markov model and variations can be used to fit the clicking activity data set.

31.2.2.1 Markov Model for User Clicking

Markov model is a commonly used model for time series. Hence, it can be used to model individual clicking behavior, which is usually in the form of clicking sequence. It is sometimes also known as Markov process, Markov chain, and so on. A Markov chain is simply a sequence of state distribution vectors at successive time intervals, that is, $(S_1, S_2, \ldots S_n)$, where S_i, $i = 1, 2, \ldots, n$, is the state of Markov model.

Markov model describes the transition probabilities between these states, and they are represented by a transition matrix, $P = \{p_{ij}\}$, where $p_{ij} = pr\{St = j | S_{t-1} = i\}$, which means the state probability at t given time $t - 1$.

Hence, the model describes a dynamic process, and the starting of process is described by initial probability distribution on each state, as follows:

$$\pi = (\pi_1, \pi_2, \cdots, \pi_n) \tag{31.5}$$

Modeling a random variable X as Markov model, there is an assumption, which is known as the Markov assumption, to be followed. The assumption is that the state at time t only depends on the state at $t - 1$ and not the state at $t - 1$. That means

$$p(X_t = j | X_{t-1} = i) = p(X_t | X_{t-1}, X_{t-2}, \ldots)$$

Hence, before applying a Markov model to user clicking activity, it is desirable to test whether the assumption is met in that particular case. Considering Web page as state, this assumption means that Web page accesses are "memoryless"; that is, access statistics are independent of events more than one interval ago.

By computing the frequency of clicking page in the Web log, Li and Tian (2003) found that the Markov model, which use clicked pages as states, can provide enough accuracy. The Markov assumption for Web clicking was evaluated, and the Markov representation was found to be feasible (Jespersen, Pedersen, and Thorhauge 2003).

However, the length of clicking sequence in the Web log is often much smaller than that required to estimate accurately the various conditional probabilities for modeling user activity. Hence, these conditional probabilities are commonly estimated by assuming that the sequence of Web pages visited by the user via clicking follows a Markov process. That is, the probability of clicking a URI u_i does not depend on all the URIs in the Web session but only on a small set of k preceding URIs, where $k \ll l$, where l is the length of clicking sequence in user session. As a result, many researches directly model user clicking activity based on Markov model. Several Markov model and variations with different order

can be used as descriptions of user clicking activity on the Web (Sarukkai 2000).

31.2.2.2 One-Step Markov Model

For a given Web site, we suppose that there are K pages. Then, the number of model states is set to $K + 1$. The additional state is the end of each visiting. The other two types of parameters in the user click activity model are described as follows (Dhyani, Bhowmick, and Ng 2003; Sarukkai 2000).

1. *Initial state probabilities.* After preprocessing of the Web log, we get a data set that describes user clicking activity in some sessions for time periods of a given duration. Suppose that the data set contains m entries and is denoted as

$$D = \{D_1, D_2, \ldots D_m\}$$

where D_i is a subset that contains the clicking activity in a session. Each session subset is denoted as

$$D_j = \{d_i\}, i = 1 \ldots, n_j$$

where $d_i = u_i$, which is a URI clicked by user. Each entry in D is sorted by the click time.

Hence, the initial state distribution of the clicking model can be estimated as

$$\pi_0^i = \frac{T(u_i)}{m}, i = 1, 2, \ldots K + 1 \tag{31.6}$$

where $T(u_i)$ is the total times that u_i appears as the first item of the entries in each sub-data set D_i, $(i = 1, 2, \ldots, m)$.

The computation method supposes that the first item URI in each entry is the entrance when user begins to visit the site. So the session recognition method is important to the estimation.

2. *State transition matrix.* State transition matrix is another important parameter of a Markov model for clicking activities. On the basis of the previous acquired sub-data sets, the transition probability is calculated as follows using the maximum likelihood principle,

$$p_{ij} = \frac{\sum_{k=1}^{m} T_k(i, j)}{\sum_{k=1}^{m} T_k(i)} \tag{31.7}$$

where $i = 1, 2, \ldots K + 1$; $j = 1, 2, \ldots K + 1$, $T_k(i, j)$ is the occurrence of state pairs (I, j) and $T_k(i)$ is the occurrence of state I in session k.

In the calculation process, the frequency is considered as the probability of transition. To achieve this goal, usually, large data set and therefore the large sub-data set are expected.

Example

Suppose that a Web site contains 10 pages, denoted as u1, u2, . . . , u10. On the basis of the Web log file, five user sessions are found, and the user clicking sequences are extracted as follows:

D1: u1 u2 u4 u8 u7 u9 u2
D2: u2 u3 u5 u8 u10 u1 u3
D3: u8 u10 u1 u3 u10 u1 u4
D4: u3 u6 u5 u10 u1 u6
D5: u3 u7 u9 u10 u1 u7 u9

Then, $K = 11$, $m = 5$;
$T(u1) = 1$, $T(u2) = 1$, $T(u3) = 2$, $T(u4) = 0$, $T(u5) = 0$, $T(u6) = 0$, $T(u7) = 0$, $T(u8) = 1$, $T(u9) = 0$, $T(u10) = 0$, $T(u11) = 0$. U11 means the end state of each visting. Hence

$$\pi_0^1 = \pi_0^2 = \pi_0^8 = 0.2, \pi_0^3 = 0.4,$$

$$\pi_0^4 = \pi_0^5 = \pi_0^6 = \pi_0^7 = \pi_0^9 = \pi_0^{10} = \pi_0^{11} = 0$$

$$p = \begin{bmatrix} 0 & \frac{1}{6} & \frac{2}{6} & \frac{1}{6} & 0 & \frac{1}{6} & \frac{1}{6} & 0 & 0 & 0 & 0 \\ 0 & 0 & \frac{1}{3} & \frac{1}{3} & 0 & 0 & 0 & 0 & 0 & 0 & \frac{1}{3} \\ 0 & 0 & 0 & 0 & \frac{1}{5} & \frac{1}{5} & \frac{1}{5} & 0 & 0 & \frac{1}{5} & \frac{1}{5} \\ 0 & 0 & 0 & 0 & 0 & 0 & 0 & \frac{1}{2} & 0 & 0 & \frac{1}{2} \\ 0 & 0 & 0 & 0 & 0 & 0 & 0 & \frac{1}{2} & 0 & \frac{1}{2} & 0 \\ 0 & 0 & 0 & 0 & \frac{1}{2} & 0 & 0 & 0 & 0 & 0 & \frac{1}{2} \\ 0 & 0 & 0 & 0 & 0 & 0 & 0 & 0 & 1 & 0 & 0 \\ 0 & 0 & 0 & 0 & 0 & 0 & \frac{1}{3} & 0 & 0 & \frac{2}{3} & 0 \\ 0 & \frac{1}{3} & 0 & 0 & 0 & 0 & 0 & 0 & 0 & \frac{1}{3} & \frac{1}{3} \\ 1 & 0 & 0 & 0 & 0 & 0 & 0 & 0 & 0 & 0 & 0 \\ 0 & 0 & 0 & 0 & 0 & 0 & 0 & 0 & 0 & 0 & 0 \end{bmatrix}$$

Actually, in the preprocessing of the user session data set, the clicking time is ignored. However, the time duration that a user spends on a page is a critical indicator of user actions. If the user is interested in the page, he probably will spend more time on the page, which results in a larger time duration. Time duration is incorporated into the Markov model of user clicking sequence by the self-transition probability (Graja and Boucher 2003). For a given state *I* with self-transition coefficient p_{ii}, it can be inferred that the probability density of the state time duration *d* is described by geometric distribution (Graja and Boucher 2003), given by

$$p_i(d) = (p_{ii})^{d-1}(1 - p_{ii}) \tag{31.8}$$

That means a Markov model can only generate the state time duration that is subjected to geometric distribution. However, is it the case in Web user clicking activity?

We suggest that a heavy-tailed distribution can be employed to describe the duration distribution more effectively. It is reasonable to suppose that in the duration, a Web user performs reading the page and thinking about its content. In the study of self-similarity of Web traffic via ON/OFF modeling, it is found that the ON time corresponds to the transmission duration of individual Web files and OFF time corresponds to periods when a workstation is not receiving Web data (Crovella and Bestavros 1997), which can be due to the several reasons. First, the workstation may have just finished receiving one Web page and is busy interpreting and formatting it before requesting the next component. On the other hand, the workstation may not receive data because the user is inspecting the results of the last transfer. These considerations indicate that the heavy-tailed nature of OFF times is primarily due to inactive time that results from user-induced delays and is usually due to user reading and thinking time (Crovella and Bestavros 1997). The heavy-tailed distribution of user think times also seems to be a feature of human information processing (Pitkow and Recker 1994). It is also found that when individuals execute tasks based on some perceived priority, the timing of the tasks will be heavy tailed, most tasks being rapidly executed, while a few experiencing very long waiting times (Vázquez et al. 2006).

Hence, the time duration can be better described by a heavy-tailed distribution:

$$P[X > x] \sim x^{-a}, 0 < a < 2$$

The simplest heavy-tailed distribution is the Pareto distribution, with probability mass functions

$$p(x) = ak^a x^{-a-1}, a, k > 0, x \geq k \tag{31.9}$$

But how can a Markov model be designed that can generate state time duration with heavy-tailed probability distribution?

A more principled and effective approach is to model the probability density of the individual state durations explicitly, using a semi-Markov model (Chen and Cooper 2002). In this kind of model, the self transition probability can be set to zero (Levinson 1986) or nonzero (Chen and Cooper 2002), and an explicit probability density with $p(x)$ is specified for the duration of each state. Then, the semi-Markov model runs in a manner as follows (Graja and Boucher 2003); for a given state, the amount of time the model spends in it is determined by the state duration densities. The transition probability matrix only determines the probability of the next state once this time has elapsed.

Once the Markov model for user activity is set up, the state probability at any time *k* can be calculated because it can be used to predict user clicking activity (Dhyani, Bhowmick and Ng 2003).

$$\pi_j^k = \sum_i p(s_n = j \mid s_0 = i) p(s_0 = i) \quad (31.10)$$

That is,

$$\pi_j^k = \sum_i p_{ij}^{(k)} \pi_i^0 \quad (31.11)$$

where $p_{ij}^{(k)}$ is also known as the matrix of k-step transition probabilities, which is the kth power of the one-step transition matrix P. $p_{ij}^{(k)}$ can be considered as the probability that a user navigates from page I to j after k times clicks. Especially, as k becomes infinitely large, the k-step transition probabilities are less relevant than the initial state distribution. Thus, the rows of P^k become identical to each other and the state distribution becomes steady. In this case, this distribution can be considered as the steady user clicking activity.

31.2.2.3 Markov Variation-Based Model for Clicking

Although a one-step Markov model is simple in describing Web user clicking activities, the main drawback comes from the order of the Markov model. A lower-order Markov model as clicking sequence is not sufficient and cannot successfully predict the future state distribution because the model depends only on the previous click. It is desired to extend to a high-order Markov model and make the current click depend on more past clicks. Second-order and higher-order Markov models can work better than the one-step model and provide better predictive accuracy if the given sequence has been trained into the model (Deshpande and Kaprypis 2004; Sen and Hansen 2003). However, a high-order Markov model may introduce some complexity because of the transition matrix. In a higher-order model, states are different combinations of the clicks observed in the Web log, and thus the number of states tends to rise exponentially as the model order increases (Deshpande and Kaprypis 2004). For example, if there are 3 pages in a Web site, p1, p2, and p3, then for a two-order Markov model, the set of states contains $3 \times 3 = 9$ items, as follows:

(p1 p1) (p2 p2) (p3 p3) (p1 p2) (p1 p3)
(p2 p3) (p2 p1) (p3 p1) (p3 p2).

For a third-order Markov model, the state number is $3 \times 3 \times 3 = 27$. While for k-order model, the state number is 3^k.

Another drawback is that the state transition matrix of a high-order Markov model is very sparse, and thus for a given clicking sequence, the next states cannot be found. This is also known as the low coverage problem (Xing and Shen 2002). Furthermore, as the order increases, the coverage becomes lower. As a result, the click action prediction is not suitable for all sequences.

A tree-like data structure that stores the sequence of pages accessed by the users is built (Schechter, Krishnan, and Smith 1998). A tree-like structure Markov model (TMM) is

proposed to reorganize the high-order states (Xing and Shen 2002). Each node in the tree is a visited page, and each branch is the path to the node. Then, for a given clicking sequence, the prediction method based on TMM just extracts the subsequence whose length equals to the depth of the tree, and then the child node can be chosen as the prediction results. Then hybrid-order tree-like Markov model (HTMM) is proposed to overcome the low coverage of TMM in matching all access sequences. The similar method to overcome the problem of low coverage on the test set is also performed based on train varying order Markov models and then combining them for prediction using the highest-order Markov model as preference (Pitkow and Pirolli 1999). Experiments on 101,020 records on a 378-page education Web site using 1–7 order HTMMs showed that HTMM is more scalable than traditional Markov models, and it provides efficient modeling and prediction of Web user clicking and access behavior (Xing and Shen 2002).

As an intermediate approach, several one step Markov models are integrated to achieve the same performance as a second-order Markov model, while keeping lower computational resources (Cadez et al. 2003; Sen and Hansen 2003).

By considering to pages as grouped into classes, each class is represented by a one-step Markov model. Let C denote the set of classes and π_{il} be the probability that page I belongs to class $I \in C$. Then, the probability of transition probability from I, j to k is represented by a mixture Markov model, as follows (Sen and Hansen 2003):

$$p(S_{t+1} = k \mid S_t = j, S_{t-1} = i) = \sum_{I \in C} \pi_{il} p_I(k \mid j) \quad (31.12)$$

In the model, the parameters that should be assigned include the number of classes, the probability that each page belongs to each class, and the one step transition probability in each class Markov defined by p_I. The Expectation Maximization (EM) algorithm is employed to get the parameters for cases in which the number of classes is known. EM is carried out by initially allocating each page randomly to a class, and it is created to maximize the likelihood of the model.

The same as in hybrid-order tree-like Markov model, selective Markov model is proposed to select parts of different order Markov models intelligently so that the resulting model has a reduced state complexity, while retaining higher coverage and maintaining a high predictive accuracy (Deshpande and Kaprypis 2004). Selective model is achieved by pruning the states of the All-Kth-Order Markov model, which is a model that contains different order Markovs from 1 to K. Three schemes for pruning are (1) frequency pruning, (2) confidence pruning, and (3) error pruning.

The key idea behind the pruning techniques is that many of the states of the different order Markov models can be eliminated without affecting the performance of the overall scheme. Frequency pruning is based on the assumption that

URI states that occur with low frequency in the training data set will contribute low prediction accuracies, because the conditional transition probabilities of the model are not reliable based on the maximum likelihood estimations.

With confidence pruning, a decision is made whether the transition probabilities from a state S to other states are significantly different. Only those states for which the transition probability differences are significant are retained, because the uncertainty in the probabilities is lower from an information theory point of view. In the context of user clicking, suppose there are two states, S1 and S2, and the transition probability to next state, is shown as in Table 31.3. It can be inferred that the predicted result based on S2 is more reliable than that based on S1 because the transition probabilities are almost the same.

By error pruning, it is meant that only the states of the user model that have the smallest estimated predicted error rate are kept. This is achieved by means of validation test. For a given training set, parts of the records are selected as the validation set. During the training step, the model is inferred from the training set but excluding the validation set. During the validation step, the model is tested using the validation set, and the predicted error rate is estimated for each state. However, n-fold validation with large n should be performed so that the error rate can be correctly estimated. As a result, this may increase the complexity of the prediction algorithm.

With the methods of the frequency and confidence pruning, a corresponding threshold should be assigned to determine which states are to be kept. However, selection of the optimal value is data set dependent. On the other hand, the problem of modeling users on the Web is an active research field that covers a wide range of problems. However, in the above discussion, only the prediction problem is accounted for and the user model is optimized for this consideration. Next, we discuss Web user cluster based on user clicking sequence. The goal is to group users with similar clicking habit into the same cluster.

The challenge lies in the user dynamic and heterogeneous behaviors. Suppose that different users lie in different clusters, where each cluster has a different Markov model. Model-based clustering is employed to achieve the user clusters despite the different lengths of user clicking sequences (Cadez et al. 2003).

In the model-based clustering approach, each clicking sequence in a user session is assumed to be generated in the following ways (Cadez et al. 2003):

1. A Web user arrives at the Web site and is assigned to one of K clusters with some probability.
2. Given that a user is in a cluster, his or her behavior in the session is generated from some statistical model specific to that cluster.

On the basis of the above discussion, the clicking sequence is important, and basically a one-order Markov model is employed to describe the patterns of a cluster. However, there are many pages in a Web site, which can lead to a huge state space; hence, categories of pages are introduced to the model representation and considered as a state of the Markov model. The categories should be developed for site analysis prior to Markov modeling. For example, the categories can be based on page content, such as news, technology, local, opinion, on-air, weather, health, business, sports, and travel.

According to the generation process, for a given clicking sequence $\mathbf{x}$ that a user generates in a session, the probability that x is generated by the process is expressed as

$$p(\mathbf{x} \mid \theta) = \sum_{i=1}^{K} p(c_k \mid \theta) p_k(\mathbf{x} \mid c_k, \theta) \qquad (31.13)$$

where θ is the parameters of the generation process, c_k is the cluster assignment of the user, $p(c_k \mid \theta)$ is the probability that the user is assigned to cluster c_k, and $p_k(\mathbf{x} \mid c_k, \theta)$ is the model describing the distribution for the clicking activity for users in the kth cluster.

As discussion above, the component model $p_k(\mathbf{x} \mid c_k, \theta)$ is described with a one-order Markov model whose state is the category of corresponding clicked pages. Hence, the generation process $p(\mathbf{x} \mid \theta)$ is a mixture of one-order Markov model. The parameter $\theta = \{\varphi, \theta^I, \theta^T\}$, where, φ is a vector of K mixture weights $p(c_k \mid \theta)$, θ^I is a set of K initial probability vectors with length as the number of categories, and θ^T is a set of K transition probability matrices. The parameters $\theta = \{\varphi, \theta^I, \theta^T\}$ is learned from a given data set $D = \{\mathbf{x}_1, \mathbf{x}_2, \ldots, \mathbf{x}_n\}$ that contains all sessions of all users by using EM algorithm with Dirichlet distribution as the prior distribution of the parameters.

A similar method is also employed by Pallis, Angelis, and Vakali (2005) to cluster users, while different methods are used in determining an optimum K. In mixture models, Markov transition is only represented for pages within a given category, while a Web user may transit among different component categories. A hidden-Markov model (Rabiner 1989) can be used to learn categories and category transitions simultaneously.

31.3 ANALYZING AND MODELING THE SELECTIVE ACTIVITY

Online community forums are one of the most important Web 2.0 applications, and they provide a network social environment that enables Web users to get together, read

---

TABLE 31.3
Transition Probability

	S3	S4	S5	S6
S1	0.2	0.2	0.3	0.3
S2	0.1	0.2	0.3	0.4

articles, discuss topics, and share opinions freely. Typically, Web forums have tens or hundreds of boards, which group many related topics together. Web users are attracted to continuously publish new articles and to join the discussion by replying to articles. As a result, Web forums can exhibit some remarkable characteristics (Zeng and Zhang 2009a), such as self similarity, burst, etc. These features are due to kinds of complex Web user actions, such as visiting, browsing, replying, following and so on in the forum community. The user activity in Web forums has attracted much attention recently (Danescu-Niculescu-Mizil, Kossinets, and Kleinberg 2009; Goyal, Bonchi, and Lakshmanan 2008; Shi et al. 2009; Wu and Li 2007).

In this section, we focus on the selective activity that makes a Web user select an article or topic and decide whether to join the discussion by replying to articles. On the basis of the selective activity in Web forums, much useful information can be mined from the publishing and replying actions. The information can be of benefit to business activity. For example, advertisements can be delivered more precisely to the Web users who are interested in the advertised products by investigating the topics on which they have published or replied. Product or service experts in different aspects can be discovered by examining the selective activity. On the other hand, knowing why and how people publish and reply on topics can help researchers from different domains understand many sociological aspects of human behavior. Hence, it is necessary to model and analyze the selective activity in Web forums.

31.3.1 Framework for Analyzing and Modeling User Selective Activity on Web Forums

It is difficult to examine the selective activity of Web users by investigating their clicking habits or exploring the Web logs as presented in previous sections, because these methods cannot infer what kinds of topics the user will select. User interests toward information are considered as the most important factor that influences user activity, such as selection of information and participation behavior (Shi et al. 2009). User interests are defined as a set of terms and weights, and the terms can be unigram or bigram (Singh, Murthy, and Gonsalves 2008). Because in the space of terms, topics are considered as an organization structure of these terms, which can achieve more meaningful results, user interests are then described by different topics which consist of these terms (Kim and Chan 2008). Hence, automatic topic detection and analysis are the essential task to model the selective activity of Web users.

In the process of modeling user selective activity based on user interest, there are several concerns, some of which are listed as follows:

1. How to capture the articles that a Web user has published or replied to on a given forum? Usually, the information along with the Web user activity and personal information are stored in databases or special files. Public access to these files is not allowed because of personnel privacy. New articles and messages are continuously produced on Web forums. Hence, automatically capturing these articles is the only choice.

2. How to extract the useful information from the Web page that includes the user article? User articles are embedded into a Web page that also contains other messages, such as recommendation links, advertisements, and other user articles. The page format varies with different Web forums. It is desired to design an extraction method that can be suitable to most types of Web forums.

3. How to recognize the entities and related words in each article? Selective activity of Web users is usually triggered by the entities and related words in the articles. For example, an article about a football game may contain the names of athletes. Interest in the athletes may be the main reason why a Web user replies to the article. However, it is difficult to extract the named entity, especially for Chinese articles.

4. How to set up a model that can represent the Web user's selective action? There might be a large number of words including named entities and related, and the relationship among these words is difficult to describe in a compact model.

5. How to keep the user activity model active with changes in Web user actions? It is reasonable to assume that the selective activity of a Web user evolves with time dynamically. A model that represents the activity should keep up with the change so that it can describe the latest activity features.

These technique issues have been explored from different points of view, such as Web information extraction (Crescenzi, Mecca, and Merialdo 2002), named entity recognition (Bender, Och, and Ney 2003), etc. Here, a framework for analyzing and modeling selective activity for Web forum users is described. The framework integrates several techniques for dealing with the above problems and provides a platform for investigation of Web forum user selective activity. The framework is shown as in Figure 31.1.

Suppose that a Web user in an online forum is to be analyzed and modeled. The task can then be planned in the

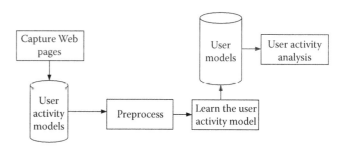

FIGURE 31.1 Framework for selective activity analysis.

framework, and several steps, which are explained below, should be performed to achieve the goal.

1. Automatically capturing the Web pages from the online forum is achieved by a crawler (Menczer, Pant, and Srinivasan 2004), a kind of program that can issue HTTP commands to the forum and capture the response data stream to save as a local file. The crawler acts as a forum user by browsing the articles, selecting the articles, and sliding to next page and the previous page. The local file is usually in HTML format. However, to investigate the user's activity more accurately, a series of pages that fall in long continuous time range should be captured to local.

2. Preprocess the local files. The goal is to extract the structure information, which contains fields of Web user identification, time stamp of user article, the title of user article, text of user article, and action type. An HTML page is parsed to produce these fields by Web information extraction methods (Kayed and Shaalan 2006), such as Tag tree based method, visualization based method, etc. However, this process is usually forum dependent because there are no standards to restrict the page organization of different Web forums to follow. Web users can publish an article or reply to other users' articles. These two actions are important in distinguishing the selective activity. Hence the action type should be labeled in case of missing such flag in some forums.

3. Model the selective activity. After preprocessing the local files, a data set that is described by five fields is generated. The data set is used to train a selective activity model for the user. Different kinds of modeling methods require further processing in the field of text of user articles. For example, as is mentioned above, the words of named entities in text of user articles are considered as useful features in describing the model; hence, they should be extracted from the text. Other considerations do not distinguish the word type. The model not only describes the probability distribution of these words but also relationship among these words. The model training method generates the distribution and estimates the relationship from the data set.

4. Analyze the user activity based on the model. On the basis of the user model, multidimensional data analysis methods can be introduced to check the degree to which the user is willing to select a given article that contains words the same as a subdimension space. The degree is described by a probability value. The higher the value, it is considered that the user is interested in the article text and is more likely to reply to such an article. Hence, the selective activity model also describes the user interest in Web forum articles because the user usually publishes or replies according to his or her interest.

31.3.2 Capture and Preprocess the Pages in the Web Forums

A crawler retrieves a page by launching an HTTP command to Web forums. Some program languages encapsulate this process into a class, which greatly simplifies the developing process. For example, in Java, class of httpURLConnection can be used to make a single request with no care of the underlying network connection to the Web forum server. The methods used to access the contents after the connection is made to the server are getContent, getInputStream, and getHeaderField. By using these methods, a crawler can get the remote pages in the same way as reading local files. By finding and extracting the links in a page, other related pages can be iteratively retrieved and stored in the same way. However, large numbers of new articles appear every day. A crawler should utilize technology to improve the performance in pages access. For example, a parallel process (Cho and Garcia-Molina 2002) is employed to get many pages simultaneously.

In the preprocessing, fields' value extraction is one of the most important tasks. These fields locate at different places with HTML tag. Web extraction information methods are employed to achieve this goal. ROADRUNNER (Crescenzi, Mecca, and Merialdo 2002) and IEPAD (Chang and Lui 2001) are the tools that use tags on the generation of HTML templates. IEPAD uses repeating patterns of closely occurring HTML tags to identify and extract records of Web document.

31.3.3 User Selective Activity Model

The model describes the probability distribution of feature words that characterize the user selective activity. Hence, in modeling the activity, selection of feature words and determination of the probability distribution of words are the two important aspects, which is consistent with the goal of topic detection and tracking (TDT; Allan et al. 1998). Thus, the techniques in TDT can be employed to infer user selective activity model from the data set which is extracted from online forum.

In TDT, an article is supposed to contain several topics with each one describing different aspects of the event in the article. Hence, the article is expressed as a mixture probability distribution P_1 over topic space, under the assumption (**ASSUMPTION 1**) that topics relationship can be ignored. However, topic is a hidden concept, so modeling just on the topic level cannot reveal the articles content. Hence, topic is defined as a probability distribution P_2 in word space, similarly with the assumption (**ASSUMPTION 2**) that the relationship among words can be ignored, also known as "bag of words" assumption (Blei, Ng, and Jordan 2003).

It should be noted that before TDT research framework is proposed, an article or document is directly described by the probability distribution over word space. Under **ASSUMPTION 2**, a unigram model describes the distribution of the words W in a document with a multinomial distribution, as follows:

$$p(W) = \prod_{i=1}^{N} p(w_i) \qquad (31.14)$$

where w_i is the word in the document.

"Concept," which is similar with "topic," is introduced into the text model, latent semantic indexing (LSI) (Deerwester et al. 1990), and it is found that it is useful. Thus document, topic, and word become the three important units in the fields of topic modeling. Much research is performed to seek the distributions P_1 and P_2.

Some of the well-known topic models, such as mixture unigram model (Nigam et al. 2000), pLSI (Hofmann 1999), LDA (Blei, Ng, and Jordan 2003), finite mixture model (Morinaga and Yamanishi 2004), etc., are based on the assumption of **ASSUMPTION 1** and **ASSUMPTION 2**.

For a given data set that is extracted from online forum, it usually contains many pieces of document. By using mixture unigram model, the assumption (**ASSUMPTION 3**) is that there is only one topic z in each document. Then, the probability of a document with words W, with respect to the model, is represented as

$$p(W) = \sum_z p(z) \prod_{i=1}^{N} p(w_i \mid z) \qquad (31.15)$$

where $p(z)$ is the topic distribution over the data set.

Probabilistic latent semantic indexing (pLSI) model attempts to relax the simplifying assumption **Assumption 3**, which makes it more reasonable, especially for modeling a large collection of documents. The model describes the joint probability distribution of document d and word w as follows:

$$p(d, w) = p(d) \sum_z p(w \mid z) p(z \mid d) \qquad (31.16)$$

However, it suffers from linearly growing of the number of parameters with the number of training documents. LDA overcomes the problem by treating the topic mixture weights $p(z \mid d)$ as a k-parameter hidden random variable rather than a large set of individual parameters that are explicitly linked to the training set. The main idea is to employ a Dirichlet distribution to assign the weights (Blei, Ng, and Jordan 2003).

From a point of view of computational efficiency in dealing with enormous words, **Assumption 2** provides a good foundation. However, it is unrealistic. Word order is useful in helping topic inference, producing better models, and improving the ability of topic model to predict previously unobserved words in trained documents. To get rid of the assumption by introducing the relationship among words, K-gram text models are proposed under the assumption (**Assumption 4**) that words are subjected to the relationship defined by a K-order Markov.

HMMLDA model (Griffiths, Steyvers, and Blei 2005) is a pioneer in attempting to rule out **Assumption 2**. The new model takes consideration of both short-range syntactic dependencies and long-range semantic dependencies between words. Its syntactic part is a Hidden Markov Model, and the semantic component is a topic model (LDA). Topic model that integrates bigram-based and topic-based approaches LDA is proposed in Wallach (2006). The probability distribution of words given a topic p(w|z) is replaced by p(w_t|w_{t-1}, z), which is a conditional bigram model. Experiments show that the predictive accuracy of the new model, especially when using the prior **Assumption 2**, is significantly better than that of LDA.

Hidden Topic Markov Model (HTMM) (Gruber, Rosen-Zvi, and Weiss 2007) is proposed to relax **Assumption 1** by assuming that the topics of words in a document form a Markov chain and that topic transitions can only occur between sentences. It shows that incorporating this dependency is expected to learn better topics and to disambiguate words that can belong to different topics. HTMM binds together parameters of different documents via a hierarchical generative model, which is similar to the LDA model. A transition probability between hidden topics is assumed to depend on multinomial random variable and a topic transition variable. A hidden Markov model based topic model is proposed based on the word sequence and incorporating with a variable space of words. Thus, transition probability between topics can be described, and also the model can handle large numbers of words (Zeng and Zhang 2007a, 2009b).

Basically, topics are hierarchical. PAM, hLDA, and h-PAM are the typical examples of hierarchical topic models. In hLDA model (Blei et al. 2004), each document is assigned to a path through the topic tree, and each word in a given document is assigned to a topic at one level of that path. hLDA is a tree structure, so only one topic at each level is connected to a given topic at the next lower level. Hierarchical PAM (hPAM) is an extension of the basic PAM structure (Li and McCallum 2006) to represent hierarchical topics. Every node in hPAM is associated with a distribution over the vocabulary not just the nodes at the lowest level. Two variations of hPAM, PAM1, and PAM2 are also presented by Mimno, Li, and McCallum (2007). A new hierarchical topic extraction algorithm based on grain computation is proposed (Zeng, Wu, and Wang 2010). The granularity of a topic is clearly defined, and a method to calculate the granular value based on the word frequency distribution is also provided. An algorithm to generate a hierarchically meaningful topic description is proposed, with focus on the compact topic structure representation and the automatic granular number decision. As an example, the topics that have been found from an online news report are shown in Table 31.4. The topics are organized as two-hierarchical structures, each of them have three and two subtopics, respectively.

The performance of a user selective activity model can be evaluated by the measures as follows.

31.3.3.1 Granularity

A user might possess interests at different abstraction levels—the higher-level interests are more general, while

TABLE 31.4
Topic Description

Feature Words	Grain 1		
	Topic 0	Topic 1	Topic 2
Sport	0.014	0.011	0.011
Ball fans	0.013	0.012	0.006
Supply	0.013	0.010	0.002
Indicate	0.014	0.012	0.001
Recommendation	0.029	0.024	0.001
Game	0.014	0.082	0.195
Reporter	0.079	0.051	0.004

	Grain 2	
	Topic 3	Topic 4
National	0.188	0.012
Possibility	0.060	0.034
Game season	0.019	0.696
Become	0.082	0.070
Football	0.170	0.001
League matches	0.123	0.065
Football Association	0.193	0.0

the lower-level ones are more specific. During a Web browsing session, general interests are in the back of one's mind, while specific interests are the current foci (Kim and Chan 2008). Selective activity depends on the user interests, which also may lead to the granularity in selection of different topics.

31.3.3.2 Ability to Change with Time

A user can exhibit different kinds of interests at different times, which provides different contexts underlying a user's information selection behavior (Kim and Chan 2008). For example, a student in computer science might be interested in the online discussion on Java program when she begins to study computer science. Her interest might turn to software architecture when she becomes a senior student. Hence, the selective activity based on user interests should change dynamically with time.

31.4 ANALYZING AND MODELING PROPAGATION ACTIVITY

A news article that appears on one Web site can soon spread to other Web sites, and as time goes on, it can also be seen in some blogs, online forums, etc. Different kinds of messages seem to spread in very different ways and exhibit different results, especially in the influence scope. In the process of information propagation, Web user activity plays an important role in contribution to the final influence effect. It is useful to model and analyze Web user propagation activities because we can estimate the influence scope before the message begins to spread and we can estimate the propagation path, so that we can improve the Web design by adopting business activity on the related pages to achieve maximum influence scope. Many other applications can benefit from

the information propagation activity. In this section, we describe the information propagation model for Web users and the specific modeling methods for information spreading among blogs and community forums.

31.4.1 INTRODUCTION

As the technology of Web 2.0 appears, it provides more flexible methods for Web users to read and write on the Internet. It can often be observed that once some piece of information appears on a Web site, it can be found in other sites after some time, and finally, the number of Web sites that publish the information seems to be large. This kind of phenomenon is mainly due to the Web user propagation activity.

As an example, we selected seven topics about several aspects of national news in China that were first published on some main Web sites on November 1, 2007. These topics are described in Table 31.5.

Then, starting from November 1, 2007, we continuously observed the URL of Web sites that have published the topics in China by searching from http://www.google.com every day, and then we counted the number of distinctive URLs, which can be considered as the whole propagation scope of the topics during the period from November 1 to subsequent days. Thus, we get the whole propagation scope from November 1 to 20 (20 days) for all of these topics, and the result is shown in Figure 31.2 (Zeng et al. 2009).

The observation value on each day shows that the topic propagation scope is increasing in the initial stages. However, these topics show different propagation dynamics. For example, the spreading scope of the seventh topic increases very quickly, with more than 100 Web sites publishing it. However, several topics, such as the fourth and third, cause less attention from Web users and get much lower propagation scope and propagation speed. No matter what the propagation activity in the initial stages, the final propagation scope of these topics tends to reach a stable status, with different numbers of Web sites affected. However, it is interesting to see that although there are many thousands of Chinese Web sites, the number of infected Web sites is small.

TABLE 31.5
Description of Selected Topics

Topic	Topic Description
1	Oil price rises; compensation policy is expected to appear next week in Beijing.
2	Huawei company responds to the controversy human resource.
3	Scholars gather at Xiamen to discuss utilization of sea resource.
4	New leaders of culture and economic center of Fujian province are elected.
5	The first forum on stock futures index will be held in Beijing on November 11.
6	Experts predict that future economics of China will increase at high rate.
7	China decides the location of 13 new nuclear power plants.

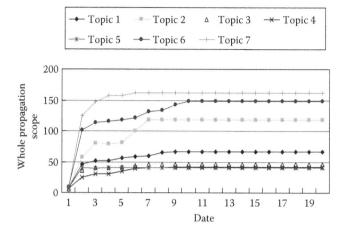

FIGURE 31.2 Whole propagation scope from the Internet.

It is useful to analyze the propagation process, with emphasis on the estimation of final propagation scope of different topics. For example, if we estimate that the final propagation scope of topic 7 will be large once it appears on the Internet, then we can set businesses advertisement on the related pages to get attention from Web users during the propagation process. Thus, the advertisement can be more effective.

In the process of information propagation, Web users play an important role. They select the Web sites and browse the messages in which they are interested. Furthermore, Web users may publish the message in other Web sites if they feel that many people will be interested with the message. As a result, the propagation goes on.

To achieve this goal of analyzing propagation activities, the individual activity of information propagation should be first set up. However, because the propagation activity of Web users varies, a Web user propagation model is usually based on several assumptions, such as user interest is independent of the other users, and so on.

As far as the example is concerned, it is clear that interest in the topic's content of each Web user determines the information propagation process and results. Topics 6 and 7 are about China economic development, and they get much focus from public, so the spreading is faster than others. In contrast, topics 3 and 4 are news about local provinces of China, and they received little attention from the public; as a result, the spreading is much slower and the final scope is smallest. On the other hand, no more new Web sites would like to publish the news about the topics after some times, so we can infer from this example that Web user interest in spreading news information tends to become smaller as time goes on because of the news's effectiveness for a given period of time.

31.4.2 Epidemic Model

Information propagation is usually considered as the same as the epidemic infection process. In traditional epidemic infection process SIS, each individual is supposed to have two states, i.e., "susceptible" and "infected." An individual

who is healthy is in a susceptible state, and the state can be changed into "infected" with a probability p1. On the other hand, an infected individual can also be transformed into a susceptible state with probability p2.

Another classical disease-propagation model in epidemiology is SIR/SIRS. An individual is first susceptible (S) to the disease. If then exposed to the disease by an infectious contact, the person becomes infected (I) with some probability. The disease can then be removed, and the person is subsequently recovered (R). A recovered individual is immune to the disease for some period of time. SIR models diseases in which recovered hosts are never again susceptible to the disease—as with a disease conferring lifetime immunity, like chicken pox, or a highly virulent disease from which the host does not recover—while SIRS models the situation in which a recovered host eventually becomes susceptible again, as with influenza (Adar and Adamic 2005).

Although the framework of SIS, SIR/SIRS is proposed based on disease infection, it is also considered as an extension to other types of information. Hence, as far as Web information propagation is concerned, the SIS model can be applied to model the propagation process by considering a Web user as an individual. For example, in blogspace, a blog owner who has not yet written about a topic is considered as in susceptible state. However, after she read the topic on the blog of a friend, she may decide to write and publish articles about the topic on her blog. As result, she is changed into the state of "infected." However, after that, if she does not publish new articles about the topic on his blog, she is changed into "susceptible" state. On the other hand, from the SIR/SIRS point of view, the owner may read new articles about the topic later, and write something about it from a new perspective, so the owner becomes susceptible again.

Obviously, an epidemic process happens only on a particular network that is constituted by individuals. Different types of network may have an effect on the propagation dynamic. Different connections between these individuals may lead to networks with different degree distribution, cluster coefficient, length of shortest path, and so on. Research has been done on "fully mixed" or "homogeneous" networks in which a node's contacts are chosen randomly from the entire network. In case of Web propagation, the network can be considered as the friendship in blog, and it is more likely a kind of small-world network and free-scaled network (Barabási and Albert 1999).

Despite the simplicity of the traditional epidemic model in providing an assumption framework for investigating epidemic spreading, the main drawback lies in, only several few transition probabilities are very limited in describing many complicated spreading scenarios. In blogspace, many topics propagate without becoming epidemics, so such models would be always successful.

31.4.3 Information Propagation on Blogs

The owners of blogs are connected and formed a network, and, usually, this is a kind of friend network. A propagation model for blogs based on user actions of reading and writing is proposed for modeling topic propagation on blogs (Gruhl et al. 2004). Information propagation on the network is based on a directed graph, as shown in Figure 31.3.

Each edge (v,u) is labeled with a copy probability $k_{v,u}$. When author v writes an article at time t, each node u that has an arc from v to u writes an article about the topic at time $t + 1$ with probability $k_{u,v}$. However, for example, if an article about the topic is found on u1, we cannot confirm that it is copied from v if other articles about the topic have appeared on u4 and u2. To deal with this ambiguous situation, an assumption is made that the copy influence is independent of the history of whether any other neighbors of u1 have written on the topic. In this case the influence is defined by an Independent Cascade Model (ICM; Goldenberg, Libai, and Muller 2001). If the assumption is omitted, the influence is known as the General Cascade Model (GCM).

As is known, a Web user may visit certain blogs frequently, while other blogs infrequently. To feature this kind of reading action on neighbor blogs, an additional edge parameter $r_{u,v}$, denoting the probability that u reads v's blog on any given day is introduced.

Hence, given a set of N nodes, corresponding to the authors, the propagation process begins when node v has written about the topic, and then at each successive step, other authors decide whether to write about the topic or not under SIR framework, in which authors do not write multiple postings on the topic. The propagation process is described as follows (Gruhl et al. 2004):

1. Compute the probability that the topic will propagate from v to a neighbor u. The probability is equal to reading probability $r_{u,v}$ that u reads the topic from node v on any given day.
2. With probability $k_{u,v}$, Web user u will choose to write about it.
3. If u reads the topic and chooses not to copy it, then u will never copy that topic from v; there is only a single opportunity for a topic to propagate along any given edge.

For such a propagation model, there are two types of parameters, $k_{u,v}$ and $r_{u,v}$. Given a network with N blog owners, the number of edges is $N(N-1)/2$; hence the number of

parameters is $2N(N-1)$. However, there are many zero value parameters because blogs are not fully connected. Hence, the goal is to infer relevant edges for which the values are not zero among a candidate set of $O(n^2)$ edges. EM algorithm is employed to estimate the parameters based on a traversal sequence of blog topic time stamp, i.e., $[(u_1, t_1), (u_2, t_2), \ldots, (u_k, t_k)]$, where u_i is the URL identifier for blog i and t_i is the first time at which blog u_i contained a reference to the topic.

By using this propagation activity model of the Web log owner, several task analyses can then be performed. The first one is to find a set of k owners as the initial activation of topic so that the topic can spread to a maximum number of blogs in the network after the propagation process. The question is also known as the influence maximization problem. For a propagation activity in which influence is based on ICM, the expected number of nodes f(A) influenced by a given initial set A of nodes should be computed. However, it is difficult to compute f(A) with an efficient method, and it is usually obtained by simulating the random process many times (Kempe, Kleinberg, and Tardos 2003). Two natural special cases of the ICM are proposed to get an approximate computation of influence f(A) when the number of network nodes is extremely large (Kimura1 and Saito 2006). The two cases are Shortest-Path Model (SPM) and the generalization version of SPM. The SPM is a special case of the ICM such that each node is activated only through the shortest paths from the initial active set. Namely, the SPM is a special type of the ICM where only the most efficient information spread can occur.

Another case is to discover information propagation paths, indicating that a sequence of blogs frequently propagate information sequentially. This knowledge can be useful in various applications of online campaign. By using the above read/write propagation model, a transmission graph is created, in which each directed edge in the graph has a probability value $r_{u,v}k_{v,u}$, indicating the information propagation from a node to another. The method to produce a transmission graph is just based on the Web blog owner activity of read and write. Other methods to incorporate the hyperlinks, post content, etc., are also proposed to infer the propagation path among blog owners (Adar and Adamic 2005).

One obvious method is based on the assumption that the hyperlink relationship between two blogs indicates an information propagation path from one blog to another one. Hence, the hyperlinks in a post are used to find the path. However, many blogs do not present the source of their information.

Thus, link inference techniques to find graph links that are not explicit are proposed to take advantage of data describing historical, repeating patterns of "infection" (Adar and Adamic 2005). To accurately predict links on which an infection may travel, different kinds of available features are used (Adar and Adamic 2005):

- The number of common blogs explicitly linked to by both blogs (indicating whether two blogs are in the same community)
- The number of nonblog links (i.e., URLs) shared by the two

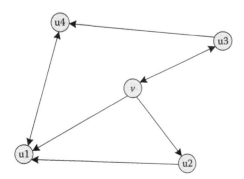

FIGURE 31.3   Blog propagation network.

- Text similarity
- Order and frequency of repeated infections. Specifically, the number of times one blog mentions a URL before the other and the number of times they both mention the URL on the same day
- In-link and out-link counts for the two blogs

An approach based on analyzing the content of blogs is proposed (Stewart et al. 2007). After detecting trackable topics of blogs, a blog community is modeled as a blog sequence database. The discovery of information propagation paths is formalized as a problem of frequent pattern mining based on the closed world assumption that, in a given blog community, all posts about a topic except the first one are a result of propagation within the community. The processing steps are as follows:

1. Blog posts are collected in a given time period by means of a crawler, and all the posts are preprocessed into a text set.
2. A set of topics that contains k topics is detected from the text set by using LDA topic model.
3. For each post in the text set, k numbers of probabilities that the post belongs to the topics are calculated. Then each post is assigned a cluster label corresponding to the topic that has the highest probability of the post.
4. Given a cluster, the blog identities and time points of each post are selected to form a topic blog sequence. Hence, for all clusters, a blog sequence database can be created in this way.
5. Information propagation paths are mined from the database with respect to given threshold values of support and strength by using IDP-Miner, which is based on frequent item set mining and frequent subsequence mining algorithm (El-Sayed, Ruiz, and Rundensteiner 2004; Han, Pei, and Yin 2000).

Considering the relation between the propagation model based on read-write activity, and the content based method, the content about a topic on a blog can be considered as the result of the owner reading from other blogs, then organizing the content and publish it on the blog. Hence, the read-write activity model is more essential for capturing the Web user (blog owner) propagation activity. The model accuracy will be improved when incorporating content into the read-write model. Next, a propagation model with much insight into Web user activity toward community forums is described.

31.4.4 Information Propagation on Community Forums

Online Web community forums have accumulated large numbers of Web users. Many articles that can be grouped into topics are created continuously. In forums, Web users can read, write, and reply to the articles and then become the trigger of information propagation by copying articles from and to.

The main process in the Internet topic propagation is as follows. First, an article about a topic appears on one or more Web sites. Then more and more people visit the Web sites and copy or reference the article to other Web sites; in this way, topic spreading goes on. Topic spreading from Web site A to Web site B is performed by the users in B who has the right to post the related articles about the topics. On the other hand, topic propagation over the Internet may be influenced by the attention of visiting people. In general, a topic that gets less attention from people will spread to other Web sites with a much lower probability. However, as more and more people pay attention to it, the topic may spread to other Web sites with a higher probability. During this process, Web user activity of reading, writing, accumulating, and topic creating is concerned.

To account for the attention of Web users visiting a Web site, a variable ns is assigned to each of the nodes in the network to describe the state of a node, which reflects how much people pay attention to the topic on the Web site. Suppose that there are four stages, i.e., "no attention," "present," "growing," and "descending" (Zeng and Zhang 2007a). Thus, ns can take the four values. At different stages of the topic, the number of postings about a topic varies, and it attracts different amounts of attention from visiting people.

It can be observed that some topics that appeared on a Web site may soon turn to the descending stage and then no attention, whereas other topics will be in the growth stage for a long time before it goes to no attention. Hence, the topic states can transit to others with a probability. Suppose that the activity of the stochastic variable ns is modeled by a Markov process.

Then, the topic propagation model includes three basic processes, which are shown as follows. Process 2 and Process 3 iteratively occur during the propagation process (Zeng et al. 2009).

Process 1: Topic appears. Topic appears on any nodes on the network.

Process 2: Topic spreads. Whether node m is infected with a topic or not depends on the total infection strength of nodes around the node, that is,

$$\sum_i wr_{im} \times x_j(i) > y$$

where wr_{im} is the weight of link from node i to node m, $x_j(i)$ is the infection coefficient of the node i at state j, and y is the expected threshold.

Here, $j \in \{ns_0, ns_1, ns_2, ns_3\}$ means the current topic state. For a node m which has k neighbor nodes, the weight from i to m is computed as follows,

$$wr_{im} = \frac{w_{im}}{\displaystyle\sum_{j=1}^{k} w_{jm}}$$

where w_{im} is the number of articles which are copied from node i to m.

Process 3: Node state changes. The state of each node changes according to the transition model, that is,

$$V(t + 1) = A \times V(t)$$

where $V(t)$ is the state vector of the topic at time t on Web sites. It means the probability of the node at each state, and it can be denoted as,

$$V(t) = [v_{ns_0}(t), v_{ns_1}(t), v_{ns_2}(t), v_{ns_3}(t)]$$

The following measurements are defined to describe the dynamic properties of topic propagation.

1. Whole propagation scope is the number of nodes which have ever been infected with the topic. So long as a node was infected, it is included in the whole propagation scope in spite of the final state it is at.
2. Active propagation scope is the number of nodes which is currently not at state ns0.

Simulation experiments using the propagation model show that *AIA* impacts the whole propagation scope. As *AIA* increases, the healthy nodes are more difficult to turn to infection state; as a result, the number of infected nodes decreases, and the whole propagation scope decreases, too. The maximum of active propagation scope is dependent on the initial infected node number. As the number of initial infection nodes increases, the number of infected nodes in each step will also increase. On the other hand, although different numbers of initial infected nodes are set, the active propagation scope finally turns to zero. This is in accord with observations from the real world; that is, most topics, especially news topics, on the Internet will not be mentioned at last.

31.5 SUMMARY

The major methods that can be used to analyze and model user activity for Web interactions are summarized in this chapter. Three kinds of user activity, i.e., clicking activity, selective activity, and propagation activity, are discussed in detail. For each activity, the steps for modeling user activity are introduced. These steps usually include acquiring data that are useful in distinguishing user activities, converting data into a suitable format that can be used to train a user model, and inferring a user model by computer algorithms. Specially, each of the methods is summarized as follows.

1. User clicking activity is recorded in Web log that is gradually updated by Web servers. By parsing the log records, identifying users, and recognizing sessions, clicking sequence is separated into segments.

Then, all kinds of activity modeling, such as Markov model and variations, are employed to generate compact description for user clicking activity.

2. Selective activity in Web forums can be observed from the postings that are published by users. By collecting the Web pages that include the postings via Crawler programs, parsing HTML pages, and extracting information about user post text, a data set that is related with the user selective activity is produced. Then, topic modeling techniques under different assumptions are employed to infer user selective activity from the data set.

3. Propagation activity is more difficult to be analyzed. From an epidemic point of view, the states of "susceptible," "infected," and "recovery" are introduced into the topic propagation process. By modeling individuals with these states, a simulation model that consists of larger numbers of such individuals are set up to examine the propagation activity.

REFERENCES

Adar E., and L. A. Adamic. 2005. Tracking information epidemics in blogspace. In *Proceedings of Proceedings of the 2005 IEEE/ WIC/ACM International Conference on Web Intelligence.* (Compiegne, France, Sept. 19–22), 207–214.

Allan, J., J. Carbonell, G. Doddington, J. Yamron, and Y. Yang. 1998. Topic Detection and Tracking Pilot Study: Final report. In *Proceedings of the DARPA Broadcast News Transcription and Understanding Workshop.* (Lansdowne, VA, Feb. 8–11), 194–218.

Anand, S. S., and B. Mobasher. 2005. Intelligent techniques for Web personalization. *In Proceedings of IJCAI Workshop on Intelligent Techniques for Web Personalization* (Acapulco, Mexico, Aug. 11), 1–36.

Barabási, A.-L., and R. Albert. 1999. Emergence of scaling in random networks. *Science* 286: 509–512.

Bender, O., F. J. Och, and H. Ney. 2003. Maximum entropy models for named entity recognition. In *Proceedings of Conference on Computational Natural Language Learning* (Edmonton, Canada, May 31 to June 1), 148–151.

Blei, D. M., T. L. Griths, M. I. Jordan, and J. B. Tenenbaum. 2004. Hierarchical topic models and the nested Chinese restaurant process. In *Advances in Neural Information Processing Systems*, eds. S. Thrun, L. Saul, and B. Scholkopf, 17–24. Cambridge, MA: MIT Press.

Blei, D. M., A. N. Ng, and M. I. Jordan. 2003. Latent Dirichlet allocation. *Journal of Machine Learning* 3: 993–1022.

Brin, S., and L. Page. 1998. The anatomy of large-scale hypertextual Web search engine. In *Proceedings of the 7th International World Wide Web Conference* (Brisbane, Australia, April 14–18), 107–117.

Bucklin, R. E., and C. Sismeiro. 2003. A model of Web site browsing behavior estimated on clickstream data. *Journal of Marketing Research* 3: 249–267.

Cadez, I. V., D. Heckerman, C. Meek, P. Smyth, and S. White. 2003. Model-based clustering and visualization of navigation patterns on a Web site. *Data Mining and Knowledge Discovery* 4: 399–424.

Catledge, L. D., and J. E. Pitkow. 1995. Characterizing browsing strategies in the World Wide Web. In *Proceedings of the 3rd International World Wide Web Conference* (Darmstadt, Germany, April 10–14), 1065–1073.

Chang, C. H., and S. C. Lui. 2001. IEPAD: Inform extract based on pattern discovery. In *Proceedings of the 10th International World Wide Web Conference* (Hong Kong, China, May 1–5).

Chen, H., and M. D. Cooper. 2002. Stochastic modeling of usage patterns in a Web-based information system. *Journal of the American Society for Information Science and Technology* 7: 536–548.

Cho, J., and H. Garcia-Molina. 2002. Parallel crawlers. In *Proceedings of the 11th International World Wide Web Conference* (Honolulu, Hawaii, May 7–11), 124–135.

Cooley, R., B. Mobasher, and J. Srivastava. 1999. Data preparation for mining World Wide Web browsing patterns. *Knowledge and Information Systems* 1: 5–32.

Crescenzi, V., G. Mecca, and P. Merialdo. 2002. RoadRunner: Automatic data extraction from data-intensive Web sites. In *Proceedings of SIG on Management of Data* (Madison, WI, June 3–6), 624.

Crovella, M. E., and A. Bestavros. 1997. Self-similarity in World Wide Web traffic: Evidence and possible causes. *IEEE/ACM Transactions of Networking* 6: 835–846.

Danescu-Niculescu-Mizil, C., G. Kossinets, and J. Kleinberg. 2009. How opinions are received by online communities: A case study on Amazon.com helpfulness votes. In *Proceedings of the 3rd International World Wide Web Conference* (Madrid, Spain, April 20–24), 141–150.

Dean, J., and M. R. Henzinger. 1999. Finding Related Pages in World Wide Web. In *Proceedings of the 8th International World Wide Web Conference* (Toronto, Canada, May 11–14), 1467–1479.

Deerwester, S., S. T. Dumais, G. W. Furnas, T. K. Landauer, and R. Harshman. 1990. Indexing by latent semantic analysis. *Journal of the American Society for Information Science* 41: 391–407.

Deshpande, M., and G. Kaprypis. 2004. Selective Markov models for predicting Web page accesses. *ACM Transaction on Internet Technology* 2: 163–184.

Dhyani, D., S. S. Bhowmick, and W. K. Ng. 2003. Modelling and predicting Web page accesses using Markov processes. In *Proceedings of 14th International Workshop on Database and Expert Systems Applications* (Prague, Czech, Sept. 1–5), 332.

El-Sayed, M., C. Ruiz, and E. A. Rundensteiner. 2004. Fs-miner: Efficient and incremental mining of frequent sequence patterns in Web logs. In *Proceedings of the 6th ACM International Workshop on Web Information and Data Management* (Washington, DC, Nov. 12–13), 128–135.

Goldenberg, K. J., B. Libai, and E. Muller. 2001. Talk of the network: A complex systems look at the underlying process of word-of-mouth. *Marketing Letters* 12: 211–223.

Goyal, A., F. Bonchi, L. V. S. Lakshmanan. 2008. Discovering leaders from community actions. In *Proceedings of the 17th ACM Conference on Information and Knowledge Management* (Napa, California, Oct. 26–30), 499–508.

Graja, S., and J.-M. Boucher. 2003. Markov models for automated ECG interval analysis. In *Proceedings of IEEE International Symposium on Intelligent Signal Processing* (Budapest, Hungary, Sept. 4–6), 105–109.

Griffiths, T. L., M. Steyvers, and D. M. Blei. 2005. Integrating topics and syntax. In *Advances in Neural Information Processing Systems*, eds. K. Saul, Y. Weiss, and L. Bottou, 537–544. Cambridge, MA: MIT Press.

Gruber, A., M. Rosen-Zvi, and Y. Weiss. 2007. Hidden topic Markov models. In *Proceedings of AISTATS'07 Society for Artificial Intelligence and Statistics* (San Juan, PR).

Gruhl, D., R. Guha, D. Liben-Nowell, and A. Tomkins. 2004. Information discussion through blogspace. In *Proceedings of 13th International Conference on World Wide Web* (New York, May 19–21), 491–501.

Gündüz, Ş., and M. T. Özsu. 2003a. A user interest model for Web page navigation. In *Proceedings of International Workshop on Data Mining for Actionable Knowledge* (Seoul, Korea, April), 46–57.

Gündüz, Ş., and M. T. Özsu. 2003b. A Web page prediction model based on click-stream tree representation of user behavior. In *Proceedings of the Ninth ACM SIGKDD International Conference on Knowledge Discovery and Data Mining* (Washington, DC, Aug. 24–27), 535–540.

Guo, F., C. Liu, A. Kannan, T. Minka, M. Taylor, and Y. M. Wang. 2009. Click chain model in Web search. In *Proceedings of 13th International Conference on World Wide Web* (Madrid, Spain, April 20–24), 11–20.

Han, J., J. Pei, and Y. Yin. 2000. Mining frequent patterns without candidate generation. In *Proceedings of the 2000 ACM SIGMOD International Conference on Management of Data* (Dallas, Texas, May 15–18), 1–12.

He, D. Q., and Göker, A. (2000). Detecting Session Boundaries from Web User Logs. In *Proceedings of the BCS/IRSG 22nd Annual Colloquium on Information Retrieval Research.* (Cambridge, UK, 5–7, Apr.), 57–66.

Hofmann, T. 1999. Probabilistic latent semantic indexing. In *Proceedings of the Twenty-Second Annual International SIGIR Conference* (Berkeley, California, Aug. 15–19), 35–44.

Huang, X. J., F. C. Peng, A. J. An, and D. Schuurmans. 2004. Dynamic Web log session identification with statistical language models. *Journal of the American Society for Information Science and Technology* 14: 1290–1303.

Jespersen, S., T. B. Pedersen, and H. Thorhauge. 2003. Evaluating the Markov assumption for Web usage mining. In *Proceedings of the 5th ACM International Workshop on Web Information and Data Management* (New Orleans, LA, Nov. 7–8), 82–89.

Kayed, M., and K. F. Shaalan. 2006. A survey of Web information extraction systems. *IEEE Transactions on Knowledge and Data Engineering* 10: 1411–1428.

Kempe, D., J. Kleinberg, and E. Tardos. 2003. Maximizing the spread of influence through a social network. In *Proceedings of the Ninth ACM SIGKDD International Conference on Knowledge Discovery and Data Mining* (Washington, DC, Aug. 24–27), 137–146.

Kim, H. R., and P. K. Chan. 2008. Learning implicit user interest hierarchy for context in personalization. *Applied Intelligence* 28: 153–166.

Kimura1, M., and K. Saito. 2006. Tractable models for information diffusion in social networks. In *Proceedings of 10th European Conference on Principles and Practice of Knowledge Discovery in Databases* (Berlin, Germany, Sept. 18–22), 259–271.

Li, W., and A. McCallum. 2006. Pachinko allocation: DAG-structured mixture models of topic correlations. In *Proceedings of the 23rd International Conference on Machine Learning* (Pittsburgh, Pennsylvania, June 25–29), 577–584.

Li, Z., and J. Tian. 2003. Testing the suitability of Markov chains as Web usage models. In *Proceedings of 27th Annual International Computer Software and Applications Conference* (Dallas, Texas, Nov. 3–6), 356–361.

Levinson, S. E. 1986. Continuously variable duration hidden Markov models for automatic speech recognition. *Computer Speech and Language* 1: 29–45.

Menczer, F., G. Pant, and P. Srinivasan. 2004. Topical Web crawlers: Evaluating adaptive algorithms. *ACM Transactions on Internet Technology* 4: 378–419.

Mimno, D., W. Li, and A. McCallum. 2007. Mixtures of hierarchical topics with Pachinko allocation. In *Proceedings of the 24th International Conference on Machine Learning* (Corvallis, Oregon, June 20–24), 633–640.

Montgomery, A. L., S. B. Li, K. Srinivasan, and J. C. Liechty. 2004. Modeling online browsing and path analysis using clickstream data. *Marketing Science* 4: 579–595.

Morinaga, S., and K. Yamanishi. 2004. Tracking dynamics of topic trends using a finite mixture model. In *Proceedings of the Tenth ACM SIGKDD International Conference on Knowledge Discovery and Data Mining* (Seattle, WA, Aug. 22–25), 811–816.

Nigam, K., A. McCallum, S. Thrun, and T. Mitchell. 2000. Text classification from labeled and unlabeled documents using EM. *Machine Learning* 2/3: 103–134.

Padmanabham, V., and J. Mogul. 1996. Using predictive prefetching to improve World Wide Web latency. *Computer Communications Review* 3: 22–36.

Pallis, G., L. Angelis, and A. Vakali. 2005. Model-based cluster analysis for Web users sessions. In ISMIS (LNAI 3488), eds. M.-S. Hacid et al., 219–227. New York: Springer.

Pitkow, J., and P. Pirolli. 1999. Mining longest repeating subsequence to predict World Wide Web surfing. In *Proceedings of 2nd USENIX Symposium on Internet Technologies and Systems* (Boulder, Colorado, Oct. 11–14), 139–150.

Pitkow, J. E., and M. M. Recker. 1994. A simple yet robust caching algorithm based on dynamic access patterns. In *Proceedings of the 2nd WWW Conference*, Chicago.

Piwowarski, B., G. Dupret, and R. Jones. 2009. Mining user web search activity with layered Bayesian networks or how to capture a click in its context. In *Proceedings of the Second ACM International Conference on Web Search and Data Mining* (Barcelona, Spain, Feb. 9–12), 162–171.

Rabiner, L. R. 1989. A tutorial on hidden Markov models and selected applications in speech recognition. *Proceedings of the IEEE* 77: 257–286.

Sarukkai, R. R. 2000. Link prediction and path analysis using Markov chains. *Computer Networks* 1/6: 377–386.

Schechter, S., M. Krishnan, and M. D. Smith. 1998. Using path profiles to predict http requests. In *Proceedings of 7th International World Wide Web Conference* (Brisbane, Australia, April 14–18), 457–467.

Sen, R., and M. Hansen. 2003. Predicting a Web user's next access based on log data. *Journal of Computational Graphics and Statistics* 1: 143–155.

Shi, X. L., J. Zhu, R. Cai, and L. Zhang. 2009. User grouping behavior in online forums. In *Proceedings of the 15th ACM SIGKDD International Conference on Knowledge Discovery and Data Mining* (Paris, France, June 28 and July 1), 777–786.

Singh, S. R., H. A. Murthy, and T. A. Gonsalves. 2008. Determining user's interest in real time. In *Proceeding of the 17th International Conference on World Wide Web* (Beijing, China, April 21–25), 1115–1116.

Stewart, A., L. Chen, R. Paiu, and W. Nejdl. 2007. Discovering information diffusion paths from blogosphere for online advertising. In *Proceedings of the 1st International Workshop on Data Mining and Audience Intelligence for Advertising* (San Jose, California, Aug. 12), 46–54.

Wallach, H. 2006. Topic modeling: beyond bag of words. In *Proceedings of the 23rd International Conference on Machine Learning* (Pittsburgh, Pennsylvania, June 25–29), 977–984.

Wu, Z., and C. Li. 2007. Topic detection in online discussion using non-negative matrix factorization. In *Proceedings of the 2007 IEEE/WIC/ACM International Conferences on Web Intelligence and Intelligent Agent Technology* (Silicon Valley, CA, Nov. 2–5), 272–275.

Xing, D. S., and J. Y. Shen. 2002. A mew Markov model for Web access prediction. *Computing in Science and Engineering* 6: 34–39.

Zeng, J. P., and S. Y. Zhang. 2007a. Predictive model for Internet public opinion. In *Proceedings of International Conference on Fuzzy Systems and Knowledge Discovery* (Haikou, China, Aug. 24–27), 7–11.

Zeng, J. P., and S. Y. Zhang. 2007b. Variable space hidden Markov model for topic detection and analysis. *Knowledge-based Systems* 7: 607–613.

Zeng, J. P., and S. Y. Zhang. 2009a. Modeling the self-similarity of Web-based forum (in Chinese). *Computer Engineering* 6: 63–66.

Zeng, J. P., and S. Y. Zhang. 2009b. Incorporating topic transition in topic detection and tracking algorithms. *Expert Systems with Applications* 1: 227–232.

Zeng, J. P., C. R. Wu, and W. Wang. 2010. Multi-grain hierarchical topic extraction algorithm for text mining. *Expert Systems with Applications* 4: 3202–3208.

Zeng, J. P., S. Y. Zhang, C. R. Wu, and X. W. Ji. 2009. Modelling the topic propagation over the Internet. *Mathematical and Computer Modelling of Dynamical Systems* 1: 83–93.

Vázquez, A., J. G. Oliveira, Z. Dezsö, K. I. Goh, I. Kondor, A.-L. Barabási 2006. Modeling bursts and heavy tails in human dynamics. *Physical Review E* 73: 036127.

32 Web Security, Privacy, and Usability

E. Eugene Schultz

CONTENTS

32.1 INTRODUCTION

Information security (often also somewhat less appropriately called "computer security") involves protecting the confidentiality of data stored in computers and transmitted over networks, integrity of data, applications, systems, and network devices, and accessibility of data, applications, databases, systems, network services, and so forth (Bernstein et al. 1995). Information security professionals have also become increasingly interested in the goal of nonrepudiation or nondeniability, which means preventing individuals who have initiated electronic transactions from denying that they have done so, and auditability, which means ensuring that each user's actions are recorded so that all users can be held accountable for their actions. "Privacy" refers to individuals being able to control what kinds of information is being collected about them, find out if this information is being suitably protected against unauthorized access, and being able to "opt out" if they do not want others to collect, process, and store this information.

Over the past decades, information security has grown from a small area to a major one within computer science and information technology. Computer and networking technology has expanded to the point that organizations use and depend on computers and networks that what is stored on and processed by computers and then transmitted over networks can literally be worth millions upon millions of dollars, as evidenced by empirical surveys. For example, a recent Computer Security Institute survey of 522 information practitioners found that reported security-related losses by the respondents' organizations during 2008 totaled over $150 million (Richardson 2009). According to the Internet Crime Complaint Center (IC3), the cost of reported Internet security incidents has increased from approximately $200 million in 2006 to approximately $290 million in 2008 (IC3 2009). This chapter explains the basics of information security—particularly how it is applied to Web security and privacy, describes how Web security and privacy are most often breached, presents an analysis of the relationship between human factors issues and Web security and privacy, and offers recommendations for improving usability in Web security and privacy.

32.2 TYPES OF SECURITY-RELATED INCIDENTS

A security incident is one in which an actual or possible adverse outcome due to a breach or bypass in a security mechanism has occurred (Schultz and Shumway 2001). One of the best-known types of security incidents is unauthorized access to systems (commonly known as a "hacker attack") in which an attacker guesses or cracks the password for one or more accounts on a system or exploits a vulnerability* in a program to gain access. Another common type of incident is a denial of service (DoS) attack in which the perpetrator causes a computer or application to crash or causes a computer, application, or network to slow down to the point of being disruptive. In still another, the integrity of a system or data stored on the system or an application is changed without authorization. Web defacements, unauthorized alteration of the content of one or more pages on a Web server, are a very common kind of integrity violation attack. Hoaxes, false information about incidents, vulnerabilities in systems, malicious code such as viruses (self-reproducing programs that spread because of actions of users) and worms (self-reproducing programs that spread independently of users), and scams (attempts to financially profit by using e-mail or Web sites to convey bogus information, often in the form of some kind of investment opportunity), extortion plots and electronic harassment (i.e., sexual harassment through sending e-mail messages) activity are other kinds of incidents.

Over the past few years the most serious type of Web-related security incident is the data security breach. In this type of incident a perpetrator gains unauthorized access to personally identifiable information (PII), which includes personal and financial information, stored on the Web server or on a database that the Web server access in a "two-tier architecture." This kind of incident is explained in more detail shortly.

Today's computing systems, networks, and applications (the majority of which are now Web based) have become extremely sophisticated and complex. One unfortunate result is an escalation of the number of vulnerabilities in all of them that can be exploited. This plethora of vulnerabilities in combination with the range and existence of easy-to-use hacker tools freely available on the Internet and the vast numbers of people (a non-trivial portion of whom are, unfortunately, unscrupulous) who use computers that connect to the Internet have resulted in an almost unlimited potential for costly and disruptive security breaches.

32.3 WHY PROTECTING WEB SERVERS IS SO DIFFICULT

To say that protecting computer systems, networks, data, and applications from attacks is a challenging undertaking that

requires considerable planning and resources is an understatement. But protecting Web servers is in many respects the most difficult of all. Among the most fundamental challenges in keeping Web servers secure is the fact that both the existence and location of most Web servers are deliberately advertised to the entire Internet user community. Furthermore, most Web servers allow anonymous access, that is, access without any requirement for individuals to establish their identities. Only a few require subscription accounts, accounts that require each user to authenticate Web access by entering a password that is unique to that user's account. Authentication means providing the required level of assurance that a user is who that person claims to be for the purpose of accessing a system, network, and/or application. Although the identities of subscription Web servers can be guarded closely if desired, the fact that anonymous Web servers must be widely known if they are to fulfill their purpose has the unfortunate side effect of alerting a wide range of potential attackers around the world to another potential target. Additionally, many automated ways of attacking frequently used Web services and protocols now exist. Many of these methods are embodied in attack tools that often require little more than entering the IP address of the to-be-attacked Web server and choosing the type or scope of attack to launch. Furthermore, new Web attacks are constantly surfacing. The many vulnerabilities in Web servers and applications provide fertile ground for those determined to discover and exploit them.

The complexity of Web servers provides yet another challenge—it makes defending them more difficult. In addition, organizations frequently quickly roll out Web servers to "beat the competition" without considering security needs. The result is Web servers and applications that are too often wide open to attack. Finally, attacking Web servers is very popular within the "black hat" community right now; the major motive is profit (Schultz 2006; Pistole 2009). Web sites such as http://www.attribution.org post reports of successful Web attacks, thus giving recognition to Web hackers and motivating others to match their feats, although staying as clandestine as possible to evade detection and ultimately to avoid law enforcement investigation and prosecution is becoming an increasingly used strategy (Schultz 2006).

32.4 THE BASICS OF WEB FUNCTIONALITY

Understanding the basics of Web functionality is essential to understanding Web service and privacy concerns. The fundamental protocol to Web services is the Hypertext Transfer Protocol (HTTP), a type of "command and control protocol" that manages all interaction between browsers (clients) and servers. Interaction involves two fundamental types of transactions, GET (when a browser asks the server to retrieve some information and send it back) and POST, which enables the browser to send some information to the server. Browsers in reality have no control over when there will be a GET and when there will be a POST (Cox and Sheldon 2000). Furthermore, Web communication is "stateless"; each interaction between

---

* A vulnerability is a flaw in a program that allows someone to exploit it by bypassing one or more security mechanisms.

browsers (clients) and servers is an independent transaction, meaning that if the browser sends a GET request to a server, the server itself retains no information about any previous GET's or POST's that may have occurred.

A Web server is an implementation of HTTP that allows Web applications to run. The stateless nature of Web interaction presents a problem for Web servers, which often need information from previous interactions to deal with current interactions. Web session management mechanisms have been developed as a solution. The server can create a session identifier early in the user interaction sequence and then transmits the ID to the browser, ensuring that the browser sends the identical ID with every future request, thereby linking each transaction with unique session data. Of the ways of maintaining sessions in this manner, "cookies" have proven to be the most popular. A cookie is an object that contains information that a Web server sends to a client for future use.* Cookies are most often used to hold information about user preferences, but they also can contain PII such as credit card numbers, dates of birth, social security numbers, and so forth.

A "Web client" (commonly known as a "Web browser" or simply as a "browser") is software used to access Web servers. Web clients are by design less sophisticated in functionality than are Web servers, although Web clients typically contain a great deal of the software involved in human–computer interaction. Web clients are a cause of considerable security-related concern because Web servers are in control of client-server interactions, frequently independently of user awareness and consent. For example, "executable content languages" (XCLs), also called "mobile code," are a class of executables that Web servers frequently download into browsers at certain points in client-server interactions. ActiveX, Java applets, JavaScript, and Visual Basic Script are four of the most common types of XCLs. Web browsers generally are passive while Web servers send XCLs to them, which afterward are executed within the browsers. Although some types of XCLs, particularly Java applets, have built-in security constraints to prevent them from doing something malicious (such as launching attacks against other systems on the network if applets have been remotely downloaded), most have few if any constraints. ActiveX, for example, does not limit how its code executes, potentially allowing an ActiveX control to access and start applications on the system that houses a browser, initiate network connections, and so forth.

In the World Wide Web today, Web pages are usually built using the Hyptertext Markup Language (HTML). HTML consists of tags (special notations that determine how browsers should display Web pages, including text centering, font selection, locations of images, and so forth), code, fields, and other elements needed to organize the data within each Web page. Hidden fields within HTML pages (fields that are hidden from view) are designed primarily to obviate the need for users to reenter data on each form that is presented but also to keep sensitive information such as users' social security numbers from users. Hiding content in this manner is at best a superficial control measure; most browsers offer options that allow any user to view hidden fields. The Extensible Markup Language (XML) is another type of metalanguage in which Web developers or Webmasters can build specialized messages that express sophisticated interactions between clients and services or between components of a service. In short, it enables people to construct their own language tailored to their individual requirements. Although HTML is currently used more than any other metalanguage, XML is the heir apparent because of its very sophisticated capabilities. Unfortunately, XML is also beset with security-related vulnerabilities that allow attackers the ability to engage in unauthorized and malicious actions such as creating bogus XML pages and pointing to them from a legitimate page (Ilioudis, Pangalos, and Vakali 2001) and injecting data of a perpetrator's choice into XML streams that are transmitted across networks (Bhalla and Kazerooni 2007). HTML is not so perfect either; however, many of today's attacks against Web servers and applications involving inserting malicious code in hidden fields and elsewhere within HTML forms.

32.5 COMMON TYPES OF WEB-RELATED INCIDENTS

A wide range of choices regarding the particular Web server to be used—Apache, Internet Information Server (IIS), the Netscape Web Server, Wusage, Domino, Websphere, and many others—is available. No matter which is used, that Web server will by default at least to some degree be vulnerable to a variety of remote attacks, including Web page defacements, denial of service attacks, a variety of privacy infringement attacks, session hijacking attacks, password attacks, buffer overflow attacks, and mobile code attacks, all of which will now be covered in this section.

32.5.1 WEB PAGE DEFACEMENTS

Web page defacements, unauthorized alteration of the content of one or more pages on a Web server, are a very common kind of Web attack. As mentioned earlier, the fact that Web defacements are frequently reported to sites such as http://www.attrition.org not only makes these attacks quite visible but also motivates attackers to perpetrate these types of attacks. The most common target is home pages, pages that users first see when they reach a Web site. Defacement styles vary considerably—some attackers completely overwrite the content of Web pages (sometimes with a message that attacks the organization that hosts the Web site or one of its employees as well as a range of new graphic images), whereas others simply add a few lines of text to prove that a certain attacker or hacking group did the defacement. One of the most serious downsides to Web defacements is that they can lead to

---

* There are two basic types of cookies, "persistent" and "nonpersistent" cookies. Persistent cookies are usually written to the hard drive of the machine that runs a browser; as such, they are available the next time that machine boots. Nonpersistent cookies are written into the memory of the machine that runs a browser and are generally purged when that machine boots.

unauthorized alteration of critical information such as pricing information, leading to all kinds of complications (such as lawsuits over prices of commodities advertised via the Web and embarrassment). Web defacements also attract the attention of the press, causing the organization that has a defaced Web server embarrassment and/or public relations setbacks.

32.5.2 Denial of Service Attacks

Another type of frequently occurring security-related Web incident is a DoS attack in which an attacker or malicious code crashes a Web server or a Web application or causes the Web server to slow down so much that it seriously disrupts the functionality of the Web site. Although intensive intervention efforts can allow the victim Web site to be running once again in a short period of time, even a small amount of Web site down time often translates to major business losses, operational disruptions and embarrassment. Many organizations rely on Web servers as a means of allowing customers to order products and services, make financial transactions, and so forth. Additionally, if Web sites are down for even a few minutes, self-appointed vigilantes may notice the problem and make negative Web postings, inform the press, and so forth. Patching vulnerabilities that are discovered in Web servers and applications and screening out abnormal input to them that can result in exploitation of vulnerabilities are the best preventative measures for DoS threats.

32.5.3 Compromise of Personally Identifiable Information (PII)

Still another type of Web security compromise involves infringement of privacy. Attackers may glean a wide range of information about individuals from poorly secured Web servers and applications. Types of PII likely to be exposed include credit card numbers, addresses of residences, phone numbers, e-mail addresses, social security numbers, user preferences, and much more.

Privacy infringement can be accomplished in many ways. Virtually anyone may, for instance, read the information contained in other users' cookies, provided of course that cookies are in cleartext, i.e., they are not encrypted. Attackers must simply invent a method to access and then copy the information in cookies, something that is greatly facilitated by the existence of widely available hacker tools on the Internet. Many of these tools require virtually no skill or knowledge on the part of the attackers, who in many cases need only to run the tool, entering the IP address of the system to be attacked and then clicking on Start. Information in HTML or XML pages is likewise vulnerable to compromise. For example, virtually anyone who knows how to read data within hidden HTML fields can access and read these data unless they are encrypted. Weak passwords for Web access accounts can also lead to privacy compromises. Attackers are more likely to guess or crack weak passwords, enabling them to access Web accounts and to glean information accessible

via these accounts. Furthermore, scripting languages such as Visual Basic may allow users to discover code paths associated with the selection of user options. Attackers and can also construct XCLs that once downloaded into browsers and can find files that hold personal information, read this information, and send it to an address of the attackers' choice. Finally, well-known vulnerabilities in Microsoft's Internet Explorer browser in Windows systems can allow an attacker to not only achieve unauthorized access to a user's cookies, but also to change values in these cookies without consent or knowledge on the user's part. The attacker needs only to construct a specially formed URL (Universal Resource Locator) to access other users' cookies in this manner.

The most massive data security breach to date involved Heartland Payment Systems. Nearly 100 million credit cards from at least 650 financial services companies were compromised (Acohido 2009; Claburn 2009). Digital information within the magnetic stripe on the back of credit and debit cards was copied by a malicious tool that captured all keystrokes and other input and output that went in and out of Heartland systems that stored, processed, and/or transmitted cardholder data. Perpetrators harvested this information and used it to create counterfeit credit cards. Although accurate loss figures are not available to the public, losses experienced to date are believed to be in the hundreds of millions, and scores of lawsuits are pending at the time of this writing.

Defending against privacy compromises is not easy because such a wide range of vulnerabilities that result in privacy compromises exists. Solutions such as encrypting all data transmitted between the Web server and client, encrypting all cookies, avoiding using hidden HTML fields to store sensitive, personal information, using scripting languages that do not allow discovery of code paths, and limiting or stopping altogether the execution of XCLs are widely accepted (but not widely used) measures used to defend against privacy compromises.

Exploitation of vulnerabilities is not the only way that privacy can become compromised, however. Every cookie that a Web server creates is under the control of the server. As mentioned earlier, cookies contain a variety of information about Web users. Many organizations that host Web sites (and thus have control over Web servers) routinely gather the information contained in cookies for marketing and other purposes. The amount of information concerning individuals that some organizations have collected and are still collecting is alarming; the fact that in many cases the systems and databases that store this information are often vulnerable to unauthorized access in many ways only exacerbates the problem. Privacy legislation in many European countries such as the Datenschutz Law in Germany as well as the European Union Privacy Act help protect against practices such as harvesting information contained in cookies. But most countries around the world have no such legislation, and if they have legislation, it is too often inadequate in dealing with the problem (as in the case of the United States, which has only a few and very weak privacy protection laws, that do not cover important issues such as harvesting PII from Web sites).

An important caveat about the privacy-related concerns regarding cookies is that cookies are not by any means any kind of all-powerful mechanisms—they are not in fact even programs. They do not gather all the information about individuals that they can. Cookies are really nothing more than repositories of information that users directly provide when they engage in tasks such as completing fields in Web forms. If users do not enter data such as their social security numbers at some point in interacting with Web servers, cookies will not hold this information. This, of course, suggests that users should be provided with prompts that warn them every time that they enter private information in each field in a Web form, although the expediency of overwhelming users with such prompts while they are engaged in Web interaction tasks that require input of PII is at best questionable. Additionally, cookies can be transmitted only to and from the Web server or domain (such as xyz.net) that actually created the cookie in the first place. This means that another Web server cannot simply request cookies. Thus, although cookies represent a genuine threat to privacy, some inherent mechanisms help contain this threat.

32.5.4 Exploitation of Cross-Scripting Vulnerabilities

A certain class of vulnerabilities called "cross-scripting vulnerabilities" enables attackers to obtain an unauthorized connection to a user's browser via the Web server to which the user is connected. Once connected to the browser, attackers can potentially access a range of data and applications on the system that runs the browser. This class of vulnerabilities poses an unusually high level of security risk in that it allows unauthorized access to a such a wide range of user resources—cookies, files, applications, and so forth—thereby potentially exposing a considerable amount of PII. Installing a patch for each cross-scripting vulnerability that is discovered is the proper antidote.

32.5.5 Session Hijacking

In another kind of attack, a "session hijacking" attack, attackers monitor network traffic to steal session ID data. In Web interactions, every visit to a Web site results in the creation of a "session" that allows continuous exchange of data between the client and server. Session ID data (as the name implies) include information that Web servers need to uniquely identify and track each session in an attempt to deal with each client separately. If attackers obtain session ID data, however, they can create another Web server connection that goes to the same Web pages with the same access as a legitimate user, allowing them to take over a user's session to the Web server. This is also a very serious problem in that once the user's session has been stolen, an attacker can now do whatever the legitimate user can do with respect to the session.

Avoiding including session ID data in URLs is an effective way to avoid session hijacking.

32.5.6 SQL Injection Attacks

Relational Database Management Systems (RDBMSs) normally have built-in controls to prevent unauthorized access to the data they hold. In poorly coded database applications, however, attackers may be able to send Structured Query Language (SQL) statements to a database, resulting in execution of malicious commands that reveal database content (e.g., PII) to the attackers and allowing them to edit/modify the protected data. This kind of attack is known as an "SQL injection attack." The worst case scenario is when attackers are able to exploit SQL injection vulnerabilities to compromise the security of the database host machine, i.e., the one on which the Web server runs and then uses as a "springboard" to penetrate other systems within the same network. Having a Web application thoroughly validate all input it receives from clients is the best way to thwart SQL injection attacks.

32.5.7 Password Attacks

Still another similar type of attack involves guessing passwords to accounts on subscription Web servers to gain the same access to the Web server that a legitimate user would obtain. Alternatively, an attacker can use a hacking tool, a "password cracker," to generate possible passwords and then compare them to entries in the password file of a system. As in the case of session hijacking, the attacker who gains access to an account through password guessing or a password cracking tool can do whatever the legitimate user of the account can. Creating an information security policy (a statement of an organization's information security requirements, particularly with respect to what does and does not constitute acceptable use) that requires users to create and use difficult-to-guess passwords is a good countermeasure for password attacks. Tools (called "password filters") that prevent users from entering easy-to-guess passwords are even more effective. Using password cracking tools to find weak passwords and then requiring users to change these passwords to stronger ones is another effective countermeasure. Still another is requiring users to select passwords based on passphrases, strings of alphanumeric characters and special characters such as "&" taken from highly memorable phrases. So, for example, the phrase "take me out to the ball game, take me out to the crowd" will produce the following very difficult-to-crack password: tmottbg,tmottc. The best solution, however, is to implement "one-time" passwords in which a unique password is created for every logon attempt by every user. The password is displayed to the user during each login attempt; the user must simply enter the displayed password using a keyboard or another input device. This login method makes password stealing and password cracking attacks futile, because any stolen or cracked password will not be the password to be entered during any login attempt. Given the abundance of illegally deployed keyboard and network sniffers and the widespread availability of powerful password cracking tools, using one-time passwords is becoming a necessity rather than simply a good option.

32.5.8 BUFFER OVERFLOW ATTACKS

Another kind of Web attack is a "buffer overflow" attack in which an attacker sends an excessive amount of input to a Web server. If the Web server does not have sufficient memory to hold the input, the input can overflow the buffer, potentially causing commands that the attacker has inserted in the overflow portion of the input to be executed without authorization. Results can be catastrophic—in some cases an attacker can capitalize upon a buffer overflow condition to gain complete control over a Web server. Another possible outcome is DoS in the form of the application or Web server crashing. The best solutions for buffer overflow attacks are to have application developers allocate (reserve) considerably more memory than appears necessary by using memory allocation commands that do this and also to have the Web server reject unusual input, such as an excessive number of characters.

32.5.9 MALICIOUS MOBILE CODE ATTACKS

The final kind of Web-related attack considered here is an attack in which malicious mobile code (or an XCL) is downloaded into a Web browser. Once the downloaded mobile code executes, a variety of undesirable outcomes is possible. Some types of malicious mobile code, for instance, glean information about users from "Java wallets," objects used to hold information such as credit card and social security numbers to make electronic business transactions easier for users. Although obviating the need for users to enter this information during transactions is advantageous from a human factors perspective, the fact that carefully constructed mobile code can glean this information is highly undesirable from a security perspective. Other types of malicious mobile code open one window after another until the system that houses a Web browser crashes, or produce annoying sound effects, or initiate long-distance phone calls using a model within the system that houses a Web browser, and so forth. As mentioned earlier, the fact that in most cases the Web browser cannot control whether or not XCLs will be downloaded makes this problem potentially very serious.

32.6 HUMAN FACTORS ISSUES IN WEB SECURITY AND PRIVACY

The almost universal phenomenon of user resistance to security-related tasks such as authentication (proving one's identity such as through entering a password for the purpose of access), setting file permissions, and inspecting system logs for signs of misuse suggests that human factors issues play a large role in the ability to perform tasks that improve security and privacy. Systems with inadequate usability design are apt to cause more user resistance than systems with adequate usability design (Al-Ghatani and King 1999). Security control measures may introduce additional usability barriers for system administrators and users, barriers that result in systems that need to be patched and configured for better security being left in an insecure, vulnerable condition—an easy target for attackers. The study of the relationship between human factors and information security itself is, however, still in its relative infancy. Whitten and Tygar (1999) identified usability hurdles in using a popular encryption program, PGP (Pretty Good Privacy). Proctor et al. (2000) showed how task analysis could be applied to different user authentication tasks to obtain an estimate of the difficulty of performing each task. Schultz et al. (2001) developed a taxonomy of security-related tasks that included a delineation of the major usability hurdles for each type of task. Wool (2004) found that numerous usability problems exist in frequently used firewall products. For example, firewalls can be configured to selectively filter traffic on the basis of packet header information, but the user interfaces on several firewalls are counterintuitive with regard to the particular direction (incoming versus outgoing) of network traffic to which filtering rules apply, resulting in errors in rule configurations. Vu et al. (2007) showed that creating passphrases yielded more crack-resistant passwords only when users were told to embed a digit and special character into the passphrase. Embedding a digit and special character also resulted in *less* ability to remember passwords. Werlinger et al. (2008) identified numerous usability problems in using intrusion detection systems (IDSs). Unfortunately, the relatively few studies and analysis papers that have been published to date do not go very far in addressing the many unsolved issues concerning the relationship between human factors and effectiveness of information security measures.

The relationship between human factors and Web security is even less understood. Clark, van Oorschot, and Adams (2007) found a substantial number of usability hurdles in connection with using tools for anonymous Web surfing, but did not investigate the relationship between security/privacy and usability per se. Proctor et al. (2002b) have identified factors that affect the effectiveness of Web content, such as the information that needs to be extracted, how that information needs to be stored and retrieved, and how information should be presented to users. Accessing and using Web sites and securing them are, however, for the most part two completely different types of tasks. Securing Web sites requires rather complex knowledge and skills not required of Web users, as discussed shortly.

Web security requires securing Web servers, Web applications, ensuring privacy, securing data sent over networks, securing the operating systems that host Web servers, and ensuring that meaningful options are available to users via interaction with browsers. Human factors issues and challenges in each of these areas will now be discussed.

32.7 HUMAN FACTORS CHALLENGES IN SECURING WEB SERVERS

Individuals who set up a Web site seldom engage in all the effort necessary to build a custom Web server "from scratch." Developing a Web server in this manner requires considerable knowledge and effort (and thus ultimately entails considerable delay and expense for the organization that owns the Web site for which the Web server will be deployed). Instead, individuals needing to implement a Web site typically choose from preexisting Web servers such as the ones mentioned earlier, customizing them to meet their particular needs.

Of all the Web servers today, the one that requires by far the least effort to create is the IIS Web server. This Web server is bundled (included) with server versions of Windows operating systems (e.g., Windows Server 2003 and 2008) and can also be run on workstation versions of these operating systems (e.g., Windows Vista and Windows 7). From a human factors perspective, IIS might superficially seem almost ideal. Little effort is required to create a Web site using this server—the directories, executables, accounts, and so forth necessary for creating a Web site are built-in, requiring almost no human intervention. Default parameters are provided en masse, relieving Webmasters of the need to consult manuals and help pages to determine which parameters and values are appropriate. To build a minimal Web site, all one must do is create HTML or other pages, select one as the home (default) page, and link the other pages in the desired order. Furthermore, additional related utilities such as Active Server and Front Page eliminate large parts of many task sequences in implementing Web pages as well as modifying them and maintaining a Web site.

The default level of IIS security in versions prior to IIS 6.0 was not very adequate. The opposite is true of IIS 6.0 and 7.0; default parameters and settings are now generally sufficient for most organizations. Raising security to an even higher level (e.g., the level required by most financial institutions), however, requires that Webmasters change a large number of Web- and system-specific parameters, install a large number of patches (an ever-ongoing task), and so forth (Schultz 2001). Although the graphical user interface (GUI) that Microsoft has designed for this Web server is for the most part effective from a human–computer interaction perspective, the sheer number of task steps and knowledge required to harden IIS Web servers is challenging.

From a human factors perspective the Apache Web server stands in striking contrast to the IIS Web server. Apache is also bundled with a number of operating systems (e.g., RedHat Linux, Solaris, and many others). In comparison to IIS, it is more difficult for an amateur Webmaster to create a Web site using Apache. The reason is that this Web server requires selection of desired modules (some of which* are far more

important to security than others) and then compilation of the source code and then configuration of a number of parameters distributed in various files throughout the system that houses the Web server (Schultz 2002). Interestingly, however, the default parameters in Apache have for many years been more conducive to security than are the default parameters in pre-5.0 versions of IIS Web servers.† Additionally, the Apache Web server has special features such as constraints called "directives" that prevent Web users from being able to access certain directories and files and from running certain dangerous programs and services. The syntax of "directives" is quite straightforward, although the fact that text lines must be formatted in a particular manner is not terribly conducive to usability. Shown immediately below, for example, is the format of a directive that prevents Apache Web users from being able to list the contents of a particular directory:

```
<Directory_path>
Options -Indexes
</Directory>
```

All things considered, therefore, a complex relationship between usability and security in Web servers exists. Usability is high for a default deployment of the IIS Web server, and the default level of security is good in recent versions of IIS, but high levels of security are possible only if the Webmaster engages in numerous tasks that require specialized knowledge. Proper Apache Web server installation is more difficult from a usability perspective, and hardening this type of Web server also requires specialized knowledge, but configuration files in which one erroneous keystroke makes the difference between a security feature working properly or not must be correctly edited if the server is to be hardened.

32.8 HUMAN FACTORS CHALLENGES IN WRITING SECURE WEB APPLICATIONS

Securing Web servers is important, but doing so is only part of achieving total Web security. Another important consideration is securing Web applications that run on Web servers. Web servers can be secure, but if the applications are not secure, malicious users can exploit vulnerabilities in applications to accomplish a range of dire outcomes, including financial fraud, unauthorized gleaning personal information about other users, application crashes, loss of control of Web servers to attackers, and so forth. The state of the art of securing Web applications has generally improved over the years as the result of organizations having suffered the consequences of running insecure Web applications as well as the increased availability of information concerning how to secure these applications.

---

* The mod_ssl module, for example, supports Secure Sockets Layer (SSL) encryption, a commonly used type of encryption that helps protect information sent between Web servers and browsers from being read by unauthorized individuals.

† This statement is true for all versions of IIS between 1.0 and 5.1. Microsoft, however, completely re-implemented its IIS Web server in IIS version 6.0 and carried the security improvements into IIS 7.0, such that IIS is now considerably more secure by default. IIS 6.0 and Apache's newest version, 3.1.27, appear in fact to be very comparable in default security levels.

Unfortunately, the security of Web applications and usability considerations too often do not go together very well. The main reason is that the majority of Web applications are implemented in PHP (Hypertext Preprocessor) or Perl. The fact that PHP is difficult to learn and use is reflected by the widely used nickname for this acronym— "people hate programming." PHP is structured in terms of "scripting blocks" that can be inserted anywhere within a file that constitutes a Web page. Each block begins with <?php and ends with ?>. A variety of PHP commands and HTML code can then be inserted between the start and end of the block. For example, the following PHP script must be written to simply send the rudimentary message "Welcome to my site" to a browser:

```
<html>
<body>

<?php

Echo "Welcome to my site";

?>

<html>
<body>
```

Perl is also difficult to learn and use—in many ways because it has an even more perplexing syntax accompanied by a wide range of precise conventions. Consider, for example, the following Perl expression:

$number=~/^[\d-]+{1,12}$/ || die "Non-allowed characters in input [0]";

The start of this string *($number=~)* in essence means that phone numbers that are entered must adhere to the rules that follow. \d means that numerals are acceptable input— means that hyphens in the phone number that users enter are also acceptable. ^[\d-]+ signifies that any permitted characters (in this particular instance, numerals) are allowed, beginning at the start of every line. *{1,12}* means the amount of input must be between 1 and 12 characters—otherwise, it will be rejected. This length constraint is especially critical because it guards against the potential for buffer overflows due to excessively long input strings. *$* means that when the string comparison is complete, the end of the line has been reached. This final constraint stops an attacker from appending commands or other types of dangerous input after the last character of input. If an input string does not adhere to all these rules, the program quits and displays "Non-allowed characters in input."

The above examples poignantly illustrate the usability problems involved in writing secure Web applications. Given the difficulty in writing applications in PHP and Perl (and other languages, too), developing Web applications is usually in and of itself a rather arduous task. Yet truly secure

Web applications need not only to perform input checking, but they must also (among other things):

- Be modular with well-defined modules that rescind any elevated privilege levels when transitioning from one routine to another,
- Run with the minimum privilege levels that are needed (so that if an attacker exploits a vulnerability somewhere in the application, the likelihood that the attacker will gain superuser privileges is lessened),
- To a maximum extent avoid making dangerous system calls that can allow access to system files or memory or that could allow someone who accesses a Web server the ability to execute commands on the system that supports the Web server,
- Encrypt sensitive information,
- Filter out state and environment variables as well as excessively long input strings (as explained earlier).

It is little wonder, then, that writing special statements and routines for the sake of security is often pushed aside for the sake of completing applications within an allocated period of time.

Not all Web applications are based on PHP and Perl scripts, however. Functions such as Active Server in IIS and Web development tools such as FrontPage and ColdFusion can help a Web developer much more quickly and easily create Web pages and Web site functionality. Interestingly, a large number of security-related vulnerabilities have been found in both Active Server and the aforementioned development tools. Additionally, achieving the degree of precision in controlling security as in the Perl script shown above is generally not possible, something that shows that good usability and Web security appear to at least some degree be orthogonal to each other.

32.9 HUMAN FACTORS CHALLENGES IN ENSURING PRIVACY

Ensuring privacy requires a large number of measures and is thus not trivial to achieve. Cookies, for example, pose a particularly difficult problem for privacy, but it is possible to encrypt cookies, ensuring that people other than a particular user can read the information in any particular cookie. Encrypting cookies requires a Web application developer to use one or more command(s) to do so, something that is well within the knowledge domain of most Web developers. Other privacy-related challenges, such as protecting information in HTML hidden fields, are addressed by finding other ways to store and retrieve user-related variables, something that generally requires a higher than average amount of programming skill. Privacy enhancement modules for Web transactions with many default values that would protect user/customer data and that also would allow inspection of these values in an optimal manner (e.g., via a spoke display depicting how far from desired values each of a number of critical security parameters is) would be a welcome addition from a human factors standpoint. Unfortunately, no such modules currently exist.

32.10 HUMAN FACTORS CHALLENGES IN SECURING DATA SENT OVER NETWORKS

Cleartext data going over a network comprise a major security threat. Hardware devices and programs can easily capture the data, making them available to perpetrators. In subscription Web sites, users must enter their passwords to gain access to Web servers. If the passwords traverse the network in cleartext, they are subject to being captured and then used to gain unauthorized access to user accounts. PII such as credit card and social security numbers can be captured in the same manner.

The most frequently used solution for protecting data sent to and from Web servers is encryption. Encryption means scrambling characters using a special algorithm so that only individuals who possess a particular object called a "key" can unscramble them (Schneier 1998). The most often used type of encryption for Web traffic is SSL (Secure Sockets Layer) encryption. Recent, versions of Web servers have built-in SSL and newer operating systems have built-in SSL, but users and/or system administrators generally must perform tasks such as generating encryption keys. System administrators and/or Webmasters must engage in a number of actions (e.g., by entering commands or interacting via a graphical interface) to configure SSL properly. Just as we have seen previously, Web security, this time in the form of encrypted network transmissions, is difficult to achieve from a human factors point of view in that specialized knowledge and entry of precise commands or knowing what components of a graphical interface are both required. And whenever encryption is used, key management (ensuring that keys are stored properly and that additional copies are available if needed) becomes a necessity; if a key is destroyed or damaged, encryption and decryption fail, often with catastrophic consequences (e.g., not being able to read business-critical files that have been encrypted). Key management also involves a number of human-in-the-loop tasks, adding substantially to the complexity of using encryption. Someone must at a minimum, for instance, perform a number of tasks such as searching for each user's encryption key, verifying that it is indeed the correct key for that user, copying that key to some medium (e.g., a CD or USB storage device), labeling the medium correctly, and so on.

Although SSL has proven effective in protecting data sent over the network, critics have pointed out that there are a number of inherent limitations in SSL that render this encryption protocol less than adequate when activities such as business-to-business (B2B) transactions, monetary transfers, and credit card debits are involved. Several credit card companies, Microsoft, and others created an alternative to SSL, the Secure Electronic Transaction (SET) protocol, to provide an integrated way of handling security for such transactions. Among other things, set provides strong user authentication, validation of credit card numbers,

and authorization of specific transactions. SET not only encrypts all information sent over the network, but it also hides information about individuals from merchants and the nature of any purchases from the banks or credit card companies that process transactions. Although SET in principle provides a much more secure, private, and comprehensive method of handling transactions, SET's popularity has waned dramatically in recent years. The principle reason is that SET has inherent usability liabilities to the point that many users never learned to use this protocol in the first place or learned how to use it but quit using it because of the difficulty of usage. To simply initiate a SET transaction, for example, the user must request and then complete a certificate (an electronic data structure used to identify individuals and systems and to transmit encryption keys), something that requires literally dozens of discrete interaction steps. Listing all the steps in a complete SET transaction would require more space in this chapter than is available. SET is another case in point that security and usability often do not coexist very well.

As discussed previously in this chapter, one-time passwords provide a strong solution for protecting against the unauthorized capture of passwords sent by users attempting to remotely logon to Web servers. Many different versions of one-time password solutions are available; regardless of the particular solution, human factors problems generally abound in this method of authentication. One of the most frequently used types of one-time password tools is one in which a program generates a list of one-time passwords for a series of logons on a per user basis. This list contains columns of number-password pairs, each of which is good for one and only one logon. Figure 32.1 contains a hypothetical one-time password list for 10 successive logons. On each logon attempt the system to which the user is allowed access displays a prompt consisting of the logon number (corresponding to one of the numbers in the left column in Figure 32.1). The user must enter the password with which this number is paired. So, for example, referring to Figure 32.1 it is easy to see that the user needs to enter "SD1e$76yF" on the first logon attempt, "yL%5U1VCx" on the second, "3*sAl@z4" on the third, and so on.

Remembering passwords that are difficult-to-crack is typically more difficult for users than remembering trivial (and

1	SD1e$76y
2	y%5U1VCx
3	3*sAl@z4
4	>E0t6r*9
5	8$w3Ce7%
6	ws1&MeA
7	Lur5edF<
8	83@aMED%
9	#trie8v*8
10	?sHI$&ne

FIGURE 32.1 A hypothetical one-time password list.

TABLE 32.1

A Taxonomy of Security-Related Tasks and Associated Usability Issues

Type of Task	Type of Threat Countered	Usability Issues
Identification and authentication	Masquerading as another user; repudiation	Willingness of users to adopt; ease of using method/device
Data integrity	Unauthorized deletion and/or changes	Install and maintain appropriate software; control access rights and privileges
Data confidentiality	Unauthorized disclosure and/or possession	
Data availability	Unauthorized deletion of data and/or the databases/ programs used to store and retrieve them; denial of service attacks	Control access rights and privileges
		Protection provided by system manage backup media; ease of implementing
System integrity	Unauthorized deletion and/or changes to system data/configuration files; theft; denial of service attacks	Inspection by administrators; detection by software
Intrusion detection	Unauthorized access to systems; denial of service attacks	Inspection by administrators; detection by software; ease of implementation

thus easy to crack) passwords (Proctor et al. 2002a). A password such as "safeplace" would thus be considerably easier to remember, for example, than would "4hFd*andbX," although the latter would be considerably more difficult to crack. One-time passwords, however, for the most part obviate the need to choose good passwords in that by the time an attacker or password cracking tool can determine what a password is, that password is likely to have been already used. One-time passwords thus potentially solve a range of human factors problems related to user memory. At the same time, however, one-time passwords create new, nontrivial human factors problems. One-time password lists are, for example, simple columnar displays plagued with well-documented usability problems such as proneness to visual vertical transposition errors in which users enter the password for a preceding or succeeding logon on a particular logon attempt. Additionally, passwords for each logon attempt almost invariably consist of alphabetical and numeric sequences that do not resemble dictionary words. As such, users tend to enter them more slowly and with more errors than simpler, more meaningful passwords. Still once again, human factors and security are in the opposite direction from each other.

32.11 HUMAN FACTORS CHALLENGES IN SECURING SYSTEMS THAT HOST WEB SERVERS

Web servers run on a wide variety of operating systems—Windows XP, Windows Vista, Windows 7, Windows Server 2003/2008, Linux, Unix, Macintosh, z/OS, AS/400, and many others. It is also critical to ensure that the operating system on which any Web server runs is secure. Failing to secure the operating system, but making the Web server as secure as possible does not work; attackers will be able to exploit operating system vulnerabilities to reach the Web server without authorization. Security guidelines for the major operating systems that are currently used are posted at http://www.cisecurity.org/.

The particular measures needed to secure an operating system to a large degree depend on each operating system

in question. Some overlap nevertheless exists. The taxonomy for human factors in information security previously developed by Schultz et al. (2001) applies especially well to operating systems. This taxonomy, shown in Table 32.1, includes six major types of security-related tasks, identification and authentication, ensuring data integrity, ensuring data confidentiality, ensuring data availability, ensuring system integrity, and detecting intrusions and misuse. Each of these tasks has associated usability issues; some tasks such as identification and authentication present special challenges in that they must be performed by users, who although often insufficiently trained must complete rather complex behavior sequences without any errors to achieve success in a given task.

Additionally, ensuring system integrity too often involves tasks that require perfectly precise (and thus error intolerant) entries in configuration files. Certain entries in Unix and Linux systems enable system administrators to learn whether anyone has engaged in actions that have threatened system integrity. Consider, for example, the entries for the critical /etc/syslog.conf file (which controls the type, amount and destination of system logging in Unix and Linux systems) shown in Figure 32.2 below.

The first line (which begins with "#") is a comment line that does not affect the level of logging that is captured, but appears purely for the purpose of providing context to whomever reads the entries in this file. The second line in essence means that all events that happen in a system that are at the priority of "errors" or higher in an eight-level hierarchy of auditing priorities,* all events that happen in the operating system's kernel (the most basic, innermost part of an operating system) with the priority of "debugging" or higher, and all authentication-related events that have "notice" priority or higher will be sent to the console (the terminal itself). A slight error, such as inserting a colon where a semicolon goes or inserting an extra space in the entries in either the left- or right-hand column, renders an entry completely meaningless. Needless to say, the need for a more usable

---

* Priorities of logging are (from highest to lowest) emerg, alert, crit, err, warning, notice, info, and debug.

# /etc/syslog.conf	
*.err;kern.debug;auth.notice;	/dev/console
*.info	/var/adm/messages
*.info	@loghost.domain.com
*.alert;kern.err;daemon.err;user.none	operator
*.alert;user.none	root

FIGURE 32.2 Entries in the /etc/syslog file used to control levels of system logging.

syntax for system configuration entries or possibly a graphical user interface with pictorial representations of the results of choosing any particular level of logging remains a very high human factors-related priority in most operating systems today.

The main point in this case is that when operating systems are considered, human factors becomes even more important to Web security because of the need for humans (in this case, system administrators) to intervene in systems to provide reasonable assurance that vulnerabilities and/or faulty configurations in systems will not be the vector through which Web servers that run on them are compromised. Unfortunately, the computing world as a whole relies almost completely on well-experienced system and network administrators who over time have mastered idiosyncrasies in syntax and other usability hurdles to be able to configure the systems that house Web servers properly for the sake of security. The issue of having beginning or junior system administrators is, however, a completely different one—one in which effective usability design could potentially make a huge difference.

32.12 HUMAN FACTORS CHALLENGES IN USER INTERACTION WITH BROWSERS

Using browsers also presents a number of human factors challenges. Browsers typically notify users of conditions in which security could potentially be threatened or when changes that might affect the user could occur. The Internet Explorer browser is one of the best examples. This browser frequently displays warnings (see Figure 32.3 below) whenever a Web server attempts to load content such as XCLs, certain security features are about to be bypassed, or other potentially negative events are about to occur if a user continues with certain actions or choices, although the content of such dialog boxes may be superficially easy for users to understand. Additionally, the pure frequency with which they appear and the nature of their content (which may warn users that yet another cookie is about to be downloaded) tends to wear users out.

Furthermore, users generally have little knowledge concerning what "good" and "bad" Web content is, what a malicious cookie is, or what aspects of a particular XML constitute "dangerous" or "safe." Consequently, users frequently resort to simply ignoring or possibly turning off all such warnings. Turning off these warnings has ironically

for most versions of browsers become simpler from a user point of view! Similarly, browsers may offer users choices of types and levels of encryption. The average user has no idea whatsoever of the meaning of these choices. The same principle also applies to types of XCLs that may or may not be downloaded and executed in a user's browser. The typical user has no idea whatsoever what an ActiveX control or a Java applet is; the fact that dialogue boxes that warn users of the imminent downloading of an XCL offer virtually no explanation of the relative dangers of each type effectively take controlling the downloading of these executables out of the hands of users.

Finally, critical vulnerabilities in SSL exist. Some versions, especially SSL 2.0, have worse vulnerabilities than others, but a recently discovered vulnerability in all versions in SSL can allow attackers to take control of SSL-encrypted sessions and to insert malicious content into them. Although organizations generally have system administrators that are capable of installing patches for these vulnerabilities on the behalf of users, the opposite is true for small organizations and home users. Patching vulnerabilities in browsers generally requires a multiple number of nonintuitive steps; very few users and novice system administrators are capable of performing these nonintuitive steps without training.

Better usability engineering would result in considerably more effective user interaction with browsers in matters that affect security. Warnings concerning cookies that are to potentially be downloaded into browsers should, for example, be made more simple and meaningful to users. Giving users a simple desktop option to designate certain "trusted" Web sites from which cookies will always be downloaded without interruption to the user is a step in the right direction. Warnings could then be presented whenever cookies from other sites are about to be downloaded. Microsoft's Internet Explorer browser comes closest to this prescription of any well-used Web browser. If a user brings up this browser (by double-clicking on the icon for Internet Explorer on the desktop) and pulls down the Tools menu in the menu bar at the top to Internet Options and then clicks on the Privacy tab at the top of the dialog box that appears, the form shown in Figure 32.4 below appears:

FIGURE 32.3 An Internet Explorer warning of potentially malicious Web content.

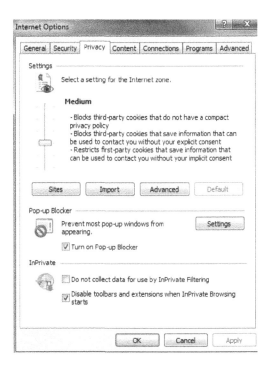

FIGURE 32.4 The privacy control dialog box in the Internet Explorer browser.

In the dialog box in Figure 32.4 users can manipulate a slide bar to select privacy levels anywhere between high and low. Note that a medium level of privacy has been selected. This level not only blocks third-party cookies from sites that do not have a concisely stated privacy policy (something that, unfortunately, most users are not likely to genuinely understand) but also prevents cookies from third-party sites that glean personal information without the user's consent from being downloaded and restricts first-party cookies that utilize such information without the user's consent. The fact that the menu depth is shallow, not deep, makes these options easier for users to find (Schultz and Curran 1986). The slide bar method of selecting privacy levels not only provides an intuitive method of setting the privacy level, but it also allows users to explore the outcome of choosing different privacy levels before they make a selection.

A major usability limitation is the understandability of the options—many users do not really know what a "cookie" is, how "third-party" differs from "first-party," and so forth. Another potentially more serious limitation is that over time users tend to develop a mental framework often called "implicit trust" in interacting with computing systems—they assume that the outcome of interacting with systems will be favorable to them. So when a dialog box that warns them of some potentially negative consequence is presented, they presume that they can continue with the interaction sequence without something bad happening. They thus almost always click on "yes" when presented with the options of continuing or discontinuing the interaction sequence. Finally, users tend to become weary after seeing dozens of warnings and dialog boxes related to privacy and security to the point that they

functionally ignore them, clicking "yes" to just get them out of their way. Nevertheless, despite these usability problems, in general the interaction methods available to users who use the Internet Explorer are for the most part compatible with principles of effective human–computer interaction.*

Color-coding dialog box messages according to the potential for security and/or privacy compromise would be an effective solution for some of the usability problems with commonly used browsers. Alternatively, a user could select once and for all the types of information (such as social security numbers) that can and cannot be contained in cookies, thereby eliminating the need to constantly allow or disallow a new cookie from being downloaded. Additionally, instead of having to traverse through several layers of menus to disable XCLs from being downloaded, the desktop should make this option available. Each type of XCL should be color-coded to clearly indicate its relative danger, with ActiveX (the most dangerous) in bright yellow and Java (one of the safer types of XCLs) in a green-yellow color. Better defaults would also obviate or in some cases substantially reduce the need for user interaction altogether. Browsers should not, for example, even offer encryption methods that provide weak security, thereby sparing users from having to make choices that are for all practical purposes non-viable.

32.13 CONCLUSION

This chapter has explored how usability and Web security and privacy are interrelated. Failure to consider usability issues can result in tasks that need to be performed for the sake of Web security and privacy not getting done or being done improperly. One of the unfortunate outcomes is the presence of vulnerabilities that can be exploited by perpetrators of computer crime. Unfortunately, we have seen that security and usability needs can be and are often conflicting—that higher security levels are too often associated with the performance of long, complex, and unintuitive task sequences.

One straightforward solution is to simply elevate the default level of security in Web servers. Web server developers are reluctant to adopt this solution, however, because raising the default level of security often results in malfunction in Web servers and possibly often also in the systems on which the Web servers run. Software purchasers tend to avoid products that do not work as desired out-of-the-box, yet raising the security level of default settings would obviate the need for Webmasters to change so many settings and take other measures that improve security.

Additionally, Web server vendors could offer simple, user-intuitive settings that result in groups of related security

---

* Note also that if a user selects Internet Options and then clicks on Security, the user can also designate "Trusted Sites," Web sites that users feel will not damage their systems, as well as "Restricted Sites," sites that for various reasons (for example, that site has been known to download malicious code into browsers) cannot be trusted to download anything into the users' browsers. Once again, the human–computer interaction methods necessary to designate Trusted and Restricted Sites are for the most part very simple and intuitive.

parameters settings. So, for example, a Web server could have a setting for overall security level—high, medium, and low. In this scheme the high setting would result in scores of settings that would tighten file permissions on Web-related files to allow nothing more than Read access to anyone but Webmasters and system administrators, lower the privilege level with which the Web server and its applications run, increase the amount of logging of Web transactions to the maximum, keep users from being able to submit any kind of input, and so on. In contrast, choosing the low setting would result in a group of settings that would correspond to out-of-the box settings. Ideally, Webmasters would have a graphical interface through which they could use a pointing device to select the desired level. This type of solution has been implemented in recent versions of the Linux operating with enormous success. System administrators can use a graphical interface to set a firewall (traffic blocking) function to high, medium, or low security, thereby greatly simplifying network security. To do this, system administrators must simply enter *setup* and then choose the menu option named "Firewall Configuration," and finally choose either "High," "Medium," or "Low."

User interaction with browsers could be substantially improved if human–computer interaction methods such as color coding of the relative danger of various conditions such as downloading of cookies and elimination of excessive menu depth were used. In particular, privacy should be put in the hands of each user through improved interaction methods. Allowing simple, up-front specification of what the user will and will not tolerate in terms of potential privacy compromise would be a significant step forward although, as mentioned previously, Microsoft's Internet Explorer browser comes closer to fulfilling these requirements than does any other widely used browser. At the same time, however, even with the use of the best human factors methods, Web interaction security is likely to be hindered by the fact that users tend to implicitly trust their interactions with Web servers and applications and thus tend to thoughtlessly approve each interaction step without considering negative security implications. Security awareness training for users rather than better human factors design is most likely to successfully address this problem.

Authentication is critical to security in Web servers and applications, yet authentication methods are often too difficult for most users. Password entry–dependent methods often require special combinations of characters and symbols that are extremely difficult for users to remember. In Web-based financial transactions an increasing number of businesses are requiring customers to authenticate to Web applications by inserting a special card such as a credit card with embedded computer chips in a special reader device provided to each customer especially for electronic transactions. Although this type of authentication method is much stronger than password-based authentication, task analyses conducted by researchers have shown that this kind of method too often involves an excessive number of interaction steps. Improving the strength of authentication while at the same time making

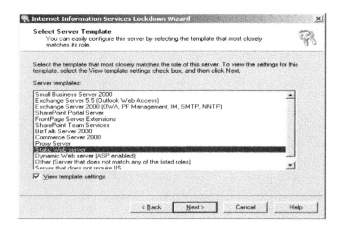

FIGURE 32.5 A dialog box from Microsoft's IIS Lockdown Tool.

authentication tasks as simple and intuitive as possible for users remains a huge challenge not only in interaction with Web servers, but also in human–computer interaction in general.

Finally, more security-enhancing routines that can be easily plugged into current systems and applications and tools that run on Web servers and in connection with applications that run on the Web servers could also help usability considerably provided that they minimized the number and complexity of steps needed to elevate security. So, for example, instead of having to write security routines for Web applications in the difficult PHP and Perl languages, a programmer would simply have to locate the appropriate input-filtering routine and integrate it into one or more of the applications that run on any Web server. Microsoft has developed one such tool, the IIS Lockdown Tool, for its IIS Web server. This tool is an easy to install tool that weeds out malformed URL requests, preventing a wide range of attacks and precluding the need to write custom filters (see Figure 32.5). Better yet, having something such as an integrated toolkit from which any particular component tool could be selected from a simple graphical display would be a gigantic step in the right direction. These and other solutions are likely to emerge as individuals and organizations eventually realize just how great the scope of the usability problem is and how much time and resources they could save in creating and maintaining Web servers if security-related task sequences were made more intuitive. So the real message of this chapter is by no means of "gloom and doom," but rather of opportunity. A number of effective security-related human–computer interaction techniques already exist, as pointed out in various parts of this chapter. Security and usability are for the most part not in reality inherently incompatible; the problem is that the preponderance of current security-related human–computer interaction tasks have simply not been designed well from a human factors perspective. Devoting the time and effort to reengineer these tasks would result in great benefits, both to organizations and individuals, especially considering the ever-growing worldwide threats to computing systems, data, applications, and networks.

REFERENCES

Acohido, B. 2009. Hackers breach Heartland Payment credit card system. *USA Today*, Jan. 23, 2009.

Al-Ghatani, S. S., and M. King. 1999. Attitudes, satisfaction and usage: Factors contributing to each in the acceptance of information technology. *Behaviour and Information Technology* 18: 277–297.

Bhalla, N., and S. Kazerooni. 2007. Web services vulnerabilities. In *Proceedings of 2007 Black Hat Conference* (Amsterdam, Netherlands). http://www.blackhat.com/presentations/bh-europe-07/Bhalla-Kazerooni/Whitepaper/bh-eu-07-bhalla-WP.pdf.

Claburn, T. 2009. Heartland Payment Systems hit by data security breach. *Information Week*, Jan. 20, 2009. http://www.informationweek.com/news/security/attacks/showArticle.jhtml?articleID=212901505.

Clark, J., P. C. van Oorschot, and C. Adams. 2007. Usability of anonymous web browsing. An examination of Tor interfaces and deployability. In *Proceedings of the 3rd Symposium on Usable Privacy and Security*, 41–51.

Cox, P., and T. Sheldon. 2000. *The Windows 2000 Security Handbook*. Berkeley, CA: Osborne.

Internet Crime Complaint Center. 2009. 2008 Internet Crime Report. http://www.ic3.gov/media/annualreport/2008_IC3Report.pdf.

Ilioudis, C., G. Pangalos, and A. Vakali. 2001. Security model for XML data. In *Proceedings of the 2nd International Conference on Internet Computing* 1: 400–406.

Jendricke, U., and D. T. Markotten. 2000. Usability meets security—the identity-manager as your personal security assistant for the Internet. In *Proceedings of the 16th Annual Computer Security Applications Conference*.

Pistole, J. S. 2009. Statement before the Senate Judiciary Committee, February 11, 2009. http://www.fbi.gov/congress/congress09/pistole021109.htm.

Proctor, R. W., M. C. Lien, G. Salvendy, and E. E. Schultz. 2000. A task analysis of usability in third-party authentication. *Information Security Bulletin* 5: 49–56.

Proctor, R. W., M. C. Lien, K.-P. L. Vu, E. E. Schultz, and G. Salvendy. 2002a. Improving computer security for authentication of users: Influence of proactive password restrictions. *Behavior Research Methods, Instruments and Computers* 34: 163–169.

Proctor, R. W., K.-P. L. Vu, G. Salvendy, et al. 2002b. Content preparation and management for Web design: Eliciting, structuring, searching, and displaying information. *International Journal of Human–Computer Interaction* 14: 25–92.

Richardson, R. 2009. 2008 CSI Computer Crime and Security Survey. http://i.cmpnet.com/v2.gocsi.com/pdf/CSIsurvey2008.pdf.

Schneier, B. 1998. *Applied Cryptography*, 2nd ed. New York: John Wiley.

Schultz, E. E. 2001. IIS web servers: It's time to just be careful. *Information Security Bulletin* 6: 17–22.

Schultz, E. E. 2002. Guidelines for securing Apache Web servers. *Network Security* 8: 8–14.

Schultz, E. E. 2006. Where have viruses and worms gone? New trends in malware. *Computer Fraud and Security*, July 2006, 2–7.

Schultz, E. E., and P. S. Curran. 1986. Menu structure and ordering of menu selections: Independent or interactive effects? *SIGCHI Bulletin* 18: 69–71.

Schultz, E. E., R. W. Proctor, M. C. Lien, and G. Salvendy. 2001. Usability and security: An appraisal of usability issues in information security methods. *Computers and Security* 20: 620–634.

Schultz, E. E., and R. Shumway. 2001. *Incident Response: A Strategic Guide for Handling Security Incidents in Systems and Networks*. Indianapolis: New Riders.

Vu, K.-P., R. W. Proctor, A. Bhargav-Spantzel, B.-K. Tai, J. Cook, and E. E. Schultz. 2007. Improving password security and memorability to protect personal and organizational information. *International Journal of Human–Computer Studies* 65: 744–757.

Werlinger, R., K. Hawkey, K. Muldner, and P. Jaferian. 2008. The challenges of using an intrusion detection system: Is it worth the effort? In *Proceedings of the 4th Symposium on Usable Privacy and Security*, 107–118.

Whitten, A., and J. D. Tygar. 1999. Why Johnny can't encrypt: A usability evaluation of PGP 5.0. In *Proceedings of 8th USENIX Security Symposium*, Usenix Association.

Wool, A. 2004. The use and usability of direction-based filtering in firewalls. *Computers and Security* 23: 459–468.

33 Cross-Cultural Web Design

Pei-Luen Patrick Rau, Tom Plocher, and Yee-Yin Choong

CONTENTS

33.1　INTRODUCTION

Around the late 1980s, the Internet was ready to become the medium within which Tim Berners-Lee created the World Wide Web (hereinafter referred to as Web) that has become the most popular medium of communication around the world. With this new medium, users can now work in an even more mixed computing environment with easy access to Web sites or Web applications remotely, as opposed to the traditional environment where people mainly interact with computers in their office. Cloud computing services provide almost unlimited computational resources to support users connecting virtually to software applications and services.

Many facets of human performance with computers are common across cultures. We should not ignore these behavioral commonalities or the standard and more general design guidelines that support them. For example, the ability of the eye to track movement on the screen is common for all people. However, with the emerging international market, various cultural characteristics of Web users have become significant issues to those international companies who are trying to design user interfaces on the Web. Therefore, this chapter emphasizes those things that distinguish different cultures and affect the behaviors and preferences of their computer and Internet users. People with different cultural backgrounds think and behave differently. Cultural differences such as language, thinking style, communication style, and social relation may affect Web use. Making the Web into a usable environment supporting users with different cultural backgrounds remains as a challenge for many Web user interface designers. Until recent years, little attention has been given to the effects of cultural differences on Web usability. Research on cultural differences from various perspectives such as linguistic, cognitive style, cultural patterns, and models of cultures has started to shed some light in the area of human–computer interaction (HCI) for cross-cultural design.

This chapter provides HCI researchers and Web design practitioners a comprehensive perspective on cultural effects for Web design. It reviews past literature for Web usability as well as cross-cultural studies and categorizes them into four major dimensions: cognitive, affective, perceptual, and functional (Figure 33.1). Design guidelines are developed and organized as languages and format, presentation, graphics, cultural preferences, information architecture, searching, and interaction. Recent trends on older adults, Web 2.0, mobile Internet, and usability testing methodologies are also introduced and discussed. Suggestions for conducting international usability testing are provided.

33.2　CROSS-CULTURAL INFLUENCES ON DESIGN

33.2.1　Cognitive Dimension

From the research and theorizing of Liu (1986), Hall (1984, 1989, 1990), Hall and Hall (1990) and, more recently, Nisbett

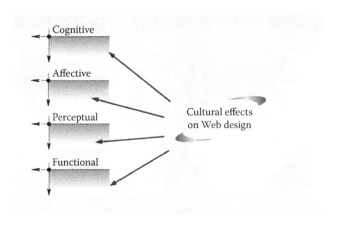

FIGURE 33.1　Cultural dimensions affecting Web design.

et al. (2001) and Nisbett (2003), we now understand many of the fundamental differences in cognitive behavior between people of different national cultures. These differences in cognitive style seem particularly magnified when Asian and Western cultures are compared (Nisbett 2003; Nisbett et al. 2001). The classic paper by Liu (1986) was the first attempt at describing a Chinese cognitive style and the experiential factors that shape Chinese cognitive style during development: the family order, the Chinese educational system, and the nature of the Chinese language. More recently, Nisbett (2003) and his colleagues in Asia (Nisbett et al. 2001) have developed a significant body of experimental evidence characterizing fundamental differences between Easterners and Westerners in reasoning style. Hall (1984, 1989, 1990) and Hall and Hall (1990) wrote extensively about the dimension of time cognition, how it was expressed in many different behaviors of daily life, and how it varied across national cultures. He also observed and wrote extensively about communication styles and the tendency in different cultures to provide more or less context when they shared information and to require more or less context when they made decisions.

So, what do these theoretical and basic research findings on cross-cultural differences in cognition mean for user interface and Web site designers? Unfortunately, not much yet, for we still have to translate the theory and research into sound principles for cognitive design of cross-cultural user interfaces. However, there are several new research initiatives that promise to provide us with design guidelines in the near future.

To date, most research and practice in cross-cultural user interface design has focused on the internationalization and localization of display codes, with such features as formats, colors, icons, and graphics (Liang and Plocher 2003). The cognitive elements of the user interface, such as information architecture and user interaction, mostly have been neglected. Yet, there is great potential to enhance ease of use and pleasure of Web interactions if we better understand how to design Web sites to be more compatible with the cognitive styles and patterns of culturally diverse user populations.

33.2.1.1 Organizing and Searching Information

The information architecture of a user interface, the "deep structure" that organizes information for navigation and searching, is fundamental to ease of use and pleasure. It also appears to be a locus of cultural effects.

Choong (1996) showed that different cultures often focus on different attributes of the same items or objects. For example, Americans tend to focus on functional attributes, whereas Chinese tend to focus on thematic attributes. As a result, Chinese and Americans tend to group items in fundamentally different ways. Choong conducted experiments that showed Chinese and American users of a Web department store performed better if the contents of the store were organized in a manner that was consistent with their natural way of organizing objects, functional for Americans and thematic for Chinese. So, for example, Americans would prefer to see products in a department store organized by function:

cleaning supplies, linens, and furniture. Chinese prefer to organize products by themes, in the case of a department store, the different rooms of a house: kitchen, bathroom, and bedroom. Products with the same general function are located in multiple places in the store depending on where they are used.

People categorize items of information, objects, and functions according to perceived similarities and differences. If the items have two or more attributes, then there is a basis for variation in how they are grouped together. Some people will emphasize a certain attribute and sort items into categories accordingly. Others will focus on another attribute and sort accordingly. The more attributes of the items or objects, the more possibilities for grouping. Choong's (1996) research highlights that different cultures often focus on different attributes of the same items or objects. The result is that they often group the items differently. Now assume that the items are pieces of information to be displayed on a Web site or functions to be exercised from a Windows application, then how the items are grouped together as pages and links or buttons and menus becomes very important. Culture will influence what groups of items or functions should be placed together in the user interface and how they should be labeled. Significant benefits in usability and satisfaction will result if the information architecture is deliberately designed to be easily perceived and understood by a broad range of cultures. If this is not possible, then provide the users some options from which they can select an information structure that is more consistent with their cultural preferences. Engineers and designers should not assume that their concept for grouping information and functions would necessarily make sense to people in other cultures.

Preferred categories and information structure can be determined by card sorting. Card sorting can be done by hand with paper note cards or with a computer program (see Vu, Zhu, and Proctor, this volume). Anderson, Anderson, and Deibel (2004) describe a method for computing "edit distance" from card sort data as a method for revealing the underlying categories and concepts. Their software tools for conducting edit distance analysis are available on the Web at http://www.cs.washington.edu/research/edtech/CardSorts/.

Alternative ways of structuring information for search were also studied by Zhao (2002). She first classified users as polychronic or monochronic (see next subsection) by means of a standard survey instrument. Then, she measured their speed of performance on an information search and retrieval tasks using two different information structures, hierarchical and network. The hierarchical information structure was basically a tree. It placed information in categories and subcategories, with the vertical structure going up to six levels deep. Crossing over from one vertical branch of the tree to another was possible only at the topmost level. The network information structure allowed more flexible searching based on relationships between items of information, regardless of their place in a hierarchy of categories. Performance of polychronic users was significantly faster using a network information structure. In contrast, monochronic users performed significantly better using a hierarchical information structure.

Rau and Liang (2003a) used a survey designed by Plocher et al. (2001) to classify Web users as either high or low context on Hall's communication style dimension. They postulated that communication style would affect how people interact with information systems, particularly nonlinear, hypertext systems such as the Web. In their experiments, they found that high context people browsed information faster and required fewer links to find information than did low context users. However, high context users also had a greater tendency to become disoriented and lose their sense of location and direction in hypertext. Low context users were slower to browse information and linked more pages but were less inclined to get lost. They pointed to well-designed navigational supports to combat the tendency toward disorientation of users in high context cultures.

33.2.1.2 Time Cognition

Hall's (1984) classic ethnographic observations showed that different cultures have different attitudes toward time. This is reflected in many aspects of peoples' lives, including how they adhere to schedules, approach the tasks of their job, and cope with competing task demands (Bluedorn et al. 1999; Haase, Lee, and Banks 1979; Kaufman-Scarborough and Lindquist 1999; Lindquist, Knieling, and Kaufman-Scarborough 2001). The user's time orientation also affects how they perform tasks supported by computers and information systems. There are two ways in which people understand time: monochronic and polychronic.

Monochronic time is dominant in Germany, the United Kingdom, the Netherlands, Finland, the United States, and Australia (Hall 1984; Hall et al. 1990). Cultures with a monochronic time orientation treat time in a linear manner. Time is divided into segments that can be easily scheduled and "spent." Monochronic people prefer to follow clear rules and procedures. They prefer to work on one task at a time and are frustrated when other competing tasks disrupt that focus.

Monochronic users search for information in hypertext in a deliberate and linear manner, making more links than polychronic users to find the same information (Rau 2001). Hence, they are slower at searching hyperspaces than polychronic users. Monochronic users also tend to focus on one task or one application even though they may have other tasks running and waiting for attention. They have a narrower view of the overall situation or activity and may miss significant events related to the waiting tasks (for example, alarms). Clear procedures are important to monochronic users. They are less inclined to invent procedures in new situations or where standard procedures are not available.

Polychronic time is dominant in Italy, France, Spain, Brazil, and India (Hall 1984; Hall et al. 1990). In contrast to monochronic people, polychronic people perceive time in a less rigid, more flexible way. Adhering to rules, procedures, and schedules is not as important to them. They prefer a work environment in which many tasks are going on at one time and they must switch between tasks as they deal with competing demands.

Polychronic users are more inclined to switch back and forth between tasks and applications. They have a broader view of the overall situation or ongoing activity. But they are prone to task switching errors. When they try to resume a task, they may "forget where they left off," resuming at the wrong place in the procedure or process. Standard procedures are less important to polychronic users, and they are more inclined to invent procedures to deal with new situations.

Time orientation, monochronic or polychronic, is deeply rooted in culture. Within any one culture, there will be significant variation in time orientation and related behaviors, but one style will tend to be dominant. That being said, it appears that time cognition can be influenced by factors other than national culture. Zhao et al. (2002) found that the natural or preferred time orientation of Chinese industrial workers (monochronic) was quite different from what they displayed on the job (polychronic). During a debriefing of the study, participants revealed that there were many factors at work in the Chinese industrial workplace, and in society that simply made polychronic behavior more adaptive.

33.2.2 Affective Dimension

33.2.2.1 Colors

Colors and combination of colors have different meanings in different cultures. The aspects of culture consist of physical and mental aspects (Kuruso 2001). Use of color is one kind of surface characteristic that can be observed physically. Color could be used as a redundant cue to reduce ambiguity and to maximize information in Web design (Marcus and Gould 2000). Prabhu and Harel (1999) studied users' needs and preferences for digital imaging products in Japan and China. Japanese men preferred single color fonts, simple fonts without emphasis on all the three lines of help, whereas Japanese women, Chinese men, and Chinese women preferred multiple colors, highlighted, or emphasized fonts. Also, Japanese preferred pastel colors for both the welcome screens and the interaction screens. Though Chinese men preferred Chinese colors, preference for women was mixed between Chinese and Japanese pastel colors.

Minocha, French, and Smith (2002) offered informal observations and analysis for some e-Finance sites in India and Taiwan. Choice of colors was one of the major attracting factors for Indian e-Finance Web sites. Three e-Finance Web sites in India were studied for Indian cultures, values, and customs: (1) ICICI Bank, the second largest commercial bank in India and pioneered Internet banking so far, (2) Allahabad Bank, an interesting and prominent example of a dual language (Hindi and English) site, and (3) the State Bank of India, India's largest commercial bank. The authors analyzed the Web site of ICICI Bank as the representative Indian choice of colors. For Indian users, use of red is associated with vitality, energy, prosperity, and health. Red is considered stimulating and shows ambition and initiative. In religious ceremonies and marriages, the guests dress in red colored clothes. Besides, use of saffron is considered auspicious amongst the Hindus, Sikhs, Jains, and Buddhists. The combination of red and saffron can be considered to signify prosperity and growth for current and prospective customers.

33.2.2.2 Graphics

Graphical design is another surface characteristic in affective dimension. Minocha, French, and Smith (2002) identified language and localized graphical design as cultural issues for Taiwanese e-Finance Web sites. For SinoPac bank, the "double-fish" sign can be seen to have strong bounds. It serves as a visual reminder to local residents who are familiar with the physical decoration of the building and conveys the message of "prosperity" (the fish)—something will always be left over each year after a Chinese New Year. For Grand bank, the graphical design shows strong Japanese influences and can be related to the success of a particular Taiwanese "Hello Kitty" credit card—directed mainly toward young women. The use of cartoon iconography forms part of a wider Asia-Pacific shopping cultural phenomenon.

Prabhu and Harel (1999) attempted to study and understand users' needs and preferences for digital imaging products in Japan and China. They found that Japanese preferred harmonious, balanced, and sensuous design, whereas Chinese had mixed preferences for Chinese and Japanese screen layouts. Also, Japanese men preferred either the standard Macintosh or standard Microsoft Windows trashcans, whereas women preferred the Japanese trashcan. For all other icons and metaphors tested, both men and women preferred Japanese designs. Both Chinese men and women preferred Chinese icons for all the icons and metaphors.

Fukuoka, Kojima, and Spyridakis (1999) investigated Japanese and American users' preference on first impression and perceived effectiveness for instructional formats that varied in the design of illustrations. All participants, including 13 Americans and 16 Japanese, preferred to use a format with illustrations (the full, half, or overview format) and no one preferred to use the text-only format. These results imply that the assumption that Japanese readers like illustrations more than American readers is incorrect when it comes to the design of user manuals. Both American and Japanese subjects believed that the formats with both text and illustrations would help them follow procedures more easily. It was also found that both American and Japanese participants had similar attitudes toward the cartoon graphics. Use of culturally specific symbols and iconography is significant in cross-cultural Web design.

Rau, Gao, and Liu (2007) studied the effect of rich Web pages with many banners and floating animations on visual search between Chinese users and German users. Chinese Web portals typically offer all the available information on the Home page and very long Web pages with strong visual stimuli such as animation and colorful texts. This design style can be considered a big usability problem by many Web design experts and textbooks. Two experiments were conducted to test the effect of Web portal design (rich and simple) on visual search performance for Chinese and German participants and the effects of static animations (leader boards, couplets, and large squares) and floating animations (moving down, moving up/down, and random movement) on visual search performance. Generally, both Chinese and German users' satisfaction for floating animations was significantly lower than for pages with no animations. So even though Chinese users are used to rich Web design, they still prefer simple Web design. Their experience with rich Web portals does not help Chinese users perform better than German users, but such experience makes Chinese users less sensitive to the difference between rich and simple Web designs.

Gould (2001) analyzed the use of text and graphics for cross-cultural Web design based on few cross-cultural theories such as Hall's notion of high/low context communication, and monochronic/polychronic, Hofstede's work dimensions, and Kress and van Leeuwen's (1996) grammar of visual design. Gould (2001) proposed a grammar model that consists of five properties of graphics: representation and interaction, size of frame and social distance, perspective and naturalistic/subjective point of view, horizontal angle and personal involvement, and vertical angle and power.

33.2.3 Perceptual Dimension Metaphors

One of the most recommended techniques for user interface design is the use of metaphor. Metaphor enables users to associate their real-world knowledge and experience with the target computer system to ensure a smooth knowledge transfer. Most metaphors use visual and conceptual representations of major user objects and their associate actions. A good metaphor can help users connect what they don't know with what they do know. The connections made by users rely greatly on users' capability of recognizing the patterns of the objects presented. However, as pointed out by Fernandes (1995), the look and feel of the real world vary from place to place, and people have internalized perceptions of what looks local and what looks foreign. Even if a metaphor was found to be applicable, there is always the problem of objects used within the metaphor. Common everyday objects are not the same everywhere in the world. For example, while postal service may be a common concept around the world, the physical mailboxes are very different (e.g., color and shape) all over the world. When presenting a U.S. rural mailbox, it cannot be assumed that people from various cultures will perceive and recognize it as a mailbox. Initial users of the Mac in the United Kingdom perceived the trashcan as a mailbox (Fernandes 1995). A very common concept being used in Web design is the concept of a "Home" page that denotes the starting point of a Web site or a Web application. Even as "home" is such a common concept known to most people, the graphical representation could vary so vastly across different cultures around the world as shown in Figure 33.2.

Everyday objects provide design inspiration and are commonly used in user interfaces, but they can look differently or be perceived differently in other parts of the world. Examples of those objects are mailboxes, paper sizes, telephones, office supplies, and signs.

As mentioned earlier, culture can be broken into four main components: values, rituals, heroes, and symbols (Hofstede 1997). Fernandes (1995) also pointed out that culture is significant to user interface design because quite often a design will affect several of these areas consciously or unconsciously. If the interface does not take into account cultural symbols, heroes, and rituals, misunderstanding can result. If an interface insults people's values, the impact can be far worse. One example of the potential problems with symbols is hand gestures. The same hand gesture can mean differently, sometimes the opposite, for different cultures. For example, "thumbs up" means "good" in the United States, but it is insulting in Australia. Because hands can be so expressive, they can easily make their way into user interfaces. If using hand gestures cannot be avoided, it is important to make sure that the gestures will not be insulting to the target locales.

33.2.4 FUNCTIONAL DIMENSION

Although Web applications have become more and more popular, relatively little research has studied the functionality for cross-cultural Web design. One cultural aspect that could affect cross-cultural Web design is the uncertainty avoidance. Cultures low in uncertainty avoidance would emphasize complexity with maximal choices, and acceptance of wandering and risk, providing mental models and help systems focusing on understanding underlying concepts rather than narrow tasks (Marcus et al. 2001). Lee (2001) and K. P. Lee and A. Harada (unpublished paper, 1999) studied what role culture plays in people's way of interacting with products. They found that American respondents want to get control of user interface more compared with Korea and Japan, whereas Korean and Japanese respondents tend not to

accept uncertainty. This research utilized the Web as a tool for cultural study.

The need for different types of functionality may vary from culture to culture. Plocher and Zhao (2002) investigated elderly Chinese for the differences in the living environment, their values and their concerns about living independently in urban China by the photo interview method. From the viewpoint of the American investigator, some surprising findings emerged. These included the prevalence of computer technology in the home and the omnipresence of various electronic devices and appliances. Emphasis was placed on physical activity for good health, rather than medication, and concerns about physical security were common. Comfort and convenience ranked rather low on the list of concerns compared to health and safety. All of these findings should inform any attempt to conceive of new electronic products for this particular population of Chinese.

Wang (2000) compared and analyzed visualized information to explore differences in the use of graphics to convey scientific and technical information in China, a high-context culture, and in the United States, a low-context culture. The author argued that when presenting a new idea to general readers, the Chinese tend to provide more contextual information, whereas the Americans tend to be direct. This difference results in different emphases on manual design. The American manuals emphasize task performance. Illustrations are more detailed, larger in size, and prominently marked. The Chinese manuals, with tables and a wiring layout, give greater weight to technical information.

The physical ability, cognitive ability, and mental load capacity of the older adults decrease with age and thus can cause them difficulties in browsing Web sites. Though aging is universal, there may be cultural differences on Web site usage among senior users. In a study of a small group of older users in Malaysia (Hisham and Edwards 2007), older users complained that the icons used on standard Web browsers were not intuitive to them. In Sangangam and Kurniawan's (2007) survey, in response to the question about which Web site they were browsing, 32% of Thai older adults and 47% of UK older adults chose the name shown on the title bar, while 36% of Thai older adults and 15% of UK older adults chose a URL shown on the address bar. When asked which object indicates a link, 81% of the Thai subjects said that the browser's animated logo provided useful information, but only 49% of UK subjects agreed with this. Special considerations for older adults include bigger fonts and line spaces, simple information architecture and page layout, and use of graphics and animations. However, there are concerns of using animations. A study of effects of moving and zooming in flash animations on performance (time and error) and subjective perception (satisfaction, vision fatigue, and workload) of 18 older adults in China found that performance time without animations was 66.6% faster than with animations and the vision fatigue with animations were 50.3% higher than no animation mode (Wang et al. 2007).

FIGURE 33.2 "Home" is different around the world.

Recently, researchers have started to investigate the cross-cultural differences in functionality of Web 2.0 applications such as Social Networking Services (SNSs), Wikipedia, and Blogs. One approach is to compare the design of localized Web sites. Fogg and Lizawa (2008) compared the persuasive design of two SNS Web sites, Facebook from the United States and Mixi from Japan. Their observations showed that Facebook's persuasive design is more assertive and mechanistic, while Mixi's approach is subtle and indirect. They argued that these persuasion styles seem to map generally to cultural differences between the United States and Japan. Marcus and Krishnamurthi (2009) examined user interface features of social networking Web sites in Japan, Korea, and the United States. They found that Web sites such as Facebook designed for users in low uncertainty avoidance cultures such as the United States included ambiguous features like "People You May Know." In contrast, Web sites in high uncertainty avoidance cultures like Japan and Korea did not implement similar features.

The second approach is to compare the behavior, perception, and attitudes of Web 2.0 users from different cultures. Lewis and George (2008) surveyed users' deception behavior in social networking Web sites. They found that although Koreans were more deceptive using face-to-face communication than Americans, Koreans did not exhibit more deceptive behavior than Americans using computer-mediated communication. They also found differences in the preferred topics of each group's deception. The Koreans tended to provide incorrect information about their job, their salary, and their physical appearance, whereas Americans tended to not be truthful about where they lived, their age, and their interests. The study also investigated the relation between Hofstede's five cultural dimensions and deceptive behavior in social networking Web sites. It found that more masculine individuals tended to be more deceptive than individuals scoring higher on feminine cultural values.

Pfeil, Zaphiris, and Ang (2006) explored cultural differences in collaborative authoring behavior of Wikipedia, also within the frame of Hofstede's five dimensions. In their results, the higher the individualism of a country, the less likely its users are to add or clarify information. The lower the individualism of a country or the higher the masculinity, the more its users are inclined to make contributions such as adding and clarifying information. Shang, Chen, and Hung (2008) explored the impacts of individualism-collectivism orientation on perceived self-efficacy in blogging and attitudes toward blogging. The results of an online survey indicated that users' cultural orientation of individualism-collectivism influences blogging through the mediation of perceived self-efficacy. Users of individualistic cultures feel most efficacious and perform best under an individual oriented system while users of collectivistic cultures feel themselves most efficacious under a group-oriented system. In addition, self-reliance and the supremacy of individual goals increase blogging but competitiveness, solitary work preference, and the supremacy of individual interests decrease blogging.

33.3 CROSS-CULTURAL WEB DESIGN GUIDELINES

Although designing for the Web interface is relatively new compared to conventional GUI (graphical user interface) design, the key components that contribute to a good user interface can be applied to Web design. The following sections list design guidelines that should be considered when designing the Web cross culturally. A summary table of the guidelines is provided in the Appendix.

G1 LANGUAGES AND FORMAT

In 1996, as much as 80% of Web users spoke English as their first language (Luna, Peracchio, and de Juan 2002). It is estimated that only 30% of Web users have English as their first language (Crockett 2000). Designing Web sites in languages other than English will become increasingly common and the use of language an increasingly critical usability issue.

Luna, Peracchio, and de Juan (2002) argued for multiple language support in international Web sites. They point out that Web sites presented in the user's second language present an additional cognitive processing load. This creates a mismatch between the cognitive effort required to process the site and the amount of cognitive resources or effort the consumer is willing to expend (Peracchio and Meyers-Levy 1997). They believe that the result frequently is a lost consumer of that Web site.

DeGroot (1991) provides a theoretical rationale for Luna, Peracchio, and de Juan's (2002) position. Their theory of language and culture is based on the semantic link between language and culturally specific concepts and values. Words map to concepts with culturally specific attributes or features. But the features associated with a word may not be the same in different languages. For example, the word "dinner" in American English means an evening meal that is quick. In French, the word for dinner "comida" has quite different associations of a long evening meal with family and conversation. This has led Luna et al. to propose a Web design guideline that relates language to Web site content. They suggest that allowing the user to select a local language is not sufficient. Rather, the content that follows the selection of a local language must be semantically consistent with the concepts and values of the culture. So, a site in which all text has been translated into the user's first language may fall short of the mark if the content (images, etc.) linked to the text is not also modified to reflect the user's culture. For example, the user has to work too hard to connect the text with the less culturally familiar concepts presented in the image of the Web site. Conversely, a Web site with text in the user's second language may be rendered more effective if the text is linked to content that reflects the user's own culture, rather than the culture associated with that second language.

Aside from the problems of language cognition and how to enhance it, there are a myriad of design issues associated with simply presenting different languages on Web pages. Some guidelines about the use of language in Web site design follow below.

G1.1 Use Unambiguous Language

Ambiguous or contextually inappropriate translations confuse users, causing frustration, errors, inefficiency and, in some cases, even hazards.

Select technology jargon words with care.

A study by Zhao et al. (2003) found that more than a quarter of the terms contained in a standard English glossary of petrochemical technology had two or more possible translations into Chinese, depending on the region and the industrial facility. Röse, Liu, and Zühlke (2001, 2002) found that, on average, only 47% of Chinese industrial machine operators correctly understood the English computer jargon words presented to them.

Make sure words can be translated to appropriate context.

In some languages, such as Chinese, the same word can be used as a verb and a noun. For example, the translation of "communications" and "communicate" is exactly the same. As a menu item or label in the user interface, the term "communications" refers to a thing or class of things (nouns), whereas "communicate" implies action initiation (verb). These are very different functions to the user. To avoid user confusion in these cases, choose clearly unambiguous words.

Avoid abbreviations.

Few abbreviations are truly international in their meaning. Avoid abbreviations in screen text. Röse, Liu, and Zühlke (2001, 2002) found that on average, only 27.5% of Chinese industrial machinery operators recognized abbreviated English computer jargon words. Words such as ESC, DEL, and AC are examples of obscure abbreviations.

Use simplified English.

For English language screens intended for international use, simplify your use of English language by applying a rigorous editing system such as Association Europeenee des Constructeurs de Material Aerospatial (AECMA) Simplified English. According to Mills and Caldwell (1997), Simplified English will

1. Reduce the amount of screen text
2. Reduce the level of English proficiency needed to read the screens
3. Reduce future translation costs
4. Reduce the number of reading errors made by non-native English speakers

Applying this system produces screen text with
1. Limited vocabulary (900 words)
2. Consistency (one and only one word for every idea)
3. Simple active verb tenses
4. Short sentences
5. Clear procedural guidance, with only one instruction per sentence

Evaluate the reading level of each screen using a software tool such as Correct Grammar™ from Lifetree Software. Simplify verb tenses and shorten sentences until the reading level approaches grade level 6.

G1.2 Allow Extra Space for Text

In European languages, such as German, words tend to be long strings of characters. They require more space in a box or on a button. In contrast to European languages, Chinese words are short. But the characters are very complex, with up to 20 line strokes in a single character. They require *more pixels or points* than European languages to render clearly on a display screen. Therefore, rendering Chinese characters on a display screen will usually require a larger font size than rendering characters of English or a European language.

For European languages:

Allow an extra 40–50% of line space in text fields, menus, and on buttons to accommodate future translations into European languages.
Avoid narrow columns of text to reduce the amount of word-wrapping in languages characterized by frequent long words.

For Chinese language:

Use a minimum font size large enough to occupy a 12 × 12-pixel matrix on the display (Tian 1987; Chen 1989).

Example: Table 33.1 illustrates the differences in text box sizing requirements for a sample of words in several different languages:

G1.3 Accommodate Text Reproduction Methods

If you anticipate that the users will have a need to print information from the screen and then transmit it by facsimile, an even larger font size may be called for on the screen to ensure the characters remain clear after printing and faxing. When designing a screen layout that includes text, anticipate this requirement by allowing sufficient height in text boxes, menu lists, and on buttons to accommodate complex characters of certain languages like Chinese.

G1.4 Do Not Embed Text in Icons

If text is used in icons, the icon will have to be reprogrammed as part of the language translation effort. Recoding icons adds cost and skill requirements to any translation effort. Design icons that clearly convey their intended meaning without any text.

TABLE 33.1
Examples of Text Box Sizing Requirements by Different Languages

English	File	Edit	View	Print	Help
German	Datei	Editieren	Anzeige	Drucken	Hilfe
French	Fichier	Edition	Visualisation	Impression	Aide
Italian	File	Editare	Visualizzare	Stampare	Aiuto
Spanish	Archivo	Editar	Ver	Imprimir	Ayuda
Chinese (Simplified)	文件	编辑	视图	打印	帮助

G1.5 Use an Appropriate Method of Sequence and Order in Lists

For alphabetic languages, such as English, the choices in a list are presented in alphabetical order. For languages based on the Latin alphabet, there may be specific collating requirements that are unfamiliar to English-speaking people.

> For Spanish, **ñ** comes between **n** and **o**, and **ch** needs to be treated as one letter.
> For French, character variants are treated as equivalent, such as (**c**, **ç**) and (**a**, **à**, **á**, **â**).
> Different countries may treat the same character differently. For example, **Ä** is sorted as equivalent to **A** in Germany and France; but in Sweden and Finland, **Ä** is treated as a distinct character and is sorted after **Z**.

Collating ideographic characters (such as Chinese Hanzi and Japanese Kanji) is more complex than sorting Latin characters. There are four different methods of collating.

Radicals: radicals are the root forms of a character that give the character its basic meaning. The radical collating sequence sorts according to the radicals that make up the character. If there is more than one character with the same radical, then these similar characters are further sorted by number of strokes that make up the character.
Number of strokes: characters are sorted by number of strokes that make up the character. If more than one character has the same number of strokes, these characters are further sorted by radicals.
Phonetic sequence: characters are sorted according to the sequence in which they appear in a phonetic alphabet.
Frequency of use: place the most frequently used or most important items first followed by decreasingly important items.

For Arabic and Hebrew characters, both are single-case languages, so there is no collation between uppercase and lowercase characters.

G1.6 Linguistic Differences

Avoid combining UI objects/messages into a flow phrase or sentence. Use a "subject:predicate" arrangement for composite messages. For example, the Find File function in Figure 33.3a below combines UI objects in a sentence format, which will make the translation/localization effort difficult.

Figure 33.3b shows the preferred redesign of the Find File function into a "subject:predicate" format.

G1.7 Case

Some scripts do not have the concept of upper and lower case characters, such as Chinese, Japanese, Arabic, and Hebrew. Be aware that some programming functions for case conversions will not operate properly on certain scripts. There are

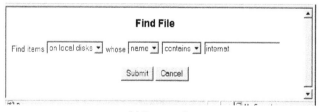

(a) Sentence format

(b) "Subject: predicate" format

FIGURE 33.3 Composite messages in (a) sentence format and (b) "subject:predicate" format.

also problems when converting cases for some scripts, for example,

> Straße (German for "street") becomes STRASSE when capitalized
> être (French) becomes ETRE when capitalized; however, it is not an issue for Canadian French.

Avoid using upper/lowercase characters as distinguishing factors among textual components on the user interface because they cannot be translated for single-case scripts.

G1.8 TEXT DIRECTIONALITY

Written languages can be presented in different directions: horizontally or vertically, and either one can be left to right or right to left (unidirectional) or bidirectional. Some examples for text directionality are listed below.

Horizontal, unidirectional (left to right): English, most European languages.
Mixing horizontal and vertical directions: Chinese, Japanese, and Korean support both horizontal and vertical text orientations. For horizontal presentation, it is most common to read/write from left to right, although right-to-left orientation is acceptable. For vertical orientation, it always reads/writes top to bottom, then flows from right to left. It is very common to find a mixture of vertical and horizontal text in newspapers.
Bidirectional: Languages based on the Arabic and Hebrew scripts are bidirectional. Those languages are read/written from right to left, with embedded Latin text and all numbers running from left to right.
Vertical only: Mongolian is an unusual script because it can only be written in a vertical orientation and it reads/writes top to bottom, then flows left to right.

Text directionality can affect many layout considerations including menu orientation, labeling, scrolling, and the arrangement of items in arrays such as "thumbnails." These are discussed below in Section G2.

G1.9 LINGUISTIC BOUNDARIES, TEXT WRAPPINGS, AND JUSTIFICATIONS

Linguistic boundaries: Not all languages insert spaces between words; for example, Chinese, Japanese, and Thai do not use any spaces to separate characters. To end a sentence, not all languages use the same symbol (e.g., period "." or question mark "?" as in English).
Text wrappings and justification: Text wrappings and justification can follow very different rules for various writing systems. Typical variations in line wrapping include wrapping whole words or breaking the word where it meets the margin (e.g., hyphenate words at the syllabic boundaries). Typical variations in justification include stretching the interword spacing, stretching

TABLE 33.2
Punctuation Symbols Used by Different Languages

	Punctuation		
	Question Marks	Periods	Quotes
Language/	French?	English.	"English"
Symbol	¿Spanish?	Japanese.	«French»
	Greek;	Chinese.	。Japanese。
		Hindi ǀ	

the intercharacter spacing across the whole line, and stretching the baseline of joined characters.

However, the wrapping of whole words or breaking at syllabic boundaries does not always work the same way for all languages using Latin script. For example, in German, hyphenation can actually change the spelling of words, such as "ck" becomes "kk" when split ("Zucker" turns to "Zuk-ker").

For Chinese and Japanese, text can be wrapped anywhere because there are no spaces between words or sentences. However, certain rules that specify some characters cannot begin or end a line need to be followed (called "Kinsoku rules" in Japanese). Justification is usually unnecessary because the characters are monospaced. Where justification is needed, the intercharacter spacing is stretched.

Punctuation: Be aware that different languages use different punctuation symbols. For example, Table 33.2 shows several punctuation symbols used by different languages:
Symbols: Symbols are characters like the / (slash), the # (pound sign or number sign), the @ (at sign), the ' (single quotation marks), the "" (double quotation marks), and the & (and sign). In English, these characters are used to mean a variety of things. Whenever you use these symbols, make sure they will be understood by the target users or they can be translated properly.

G1.10 CONSIDER PRESENTATION (FONT SIZE, FONT STYLE, LINE SPACING, AND WORD SPACING) OF TEXT FOR THE CHINESE LANGUAGE

Jin, Zhu, and Shen (1988) compared the visual recognition performance of Chinese users for four Chinese font styles: Song, Hei (Black), Zheng Feng Song, and Chang Fang Song. The results indicated that the successful recognition rates for Song and Hei fonts ranged from about 95 to 98%. Cai, Chi, and You (2001) studied the legibility threshold of Chinese characters in three styles: Ming, Kai, and Li font styles. The results showed that character style and number of strokes both have a significant impact on the legibility threshold. Ming is the most legible among the three styles, and Kai is also significantly more legible than the Li style.

Shen et al. (1988) studied the effect of Chinese character stroke width on reading efficiency. The results indicated that the stroke widths of 0.3–0.5 mm for 5 mm height Chinese characters are associated with the highest successful rate of visual recognition. Shieh, Chen, and Chuang (1997) found that characters of higher frequency and fewer strokes were identified more accurately on a visual display terminal (VDT). Also, red on green was ranked inferior to color combinations generally used in computer software.

G1.11 Be Aware that Variations Exist within the Same Language

Goonetilleke, Lau, and Shih (2002) investigated visual search strategies and eye movements when searching Chinese character screens. The results showed that Hong Kong Chinese use predominantly horizontal search patterns, while the Mainland Chinese change their search pattern depending on the layout presented.

G2 PRESENTATION AND LAYOUT

The orientation of information presented on the Web has predominantly been of Western styles in which their users hold a reading style as left to right, horizontally. As mentioned previously, there are some languages with different orientations that could impose different reading styles on users, thus affect their expectation of the information presented on a user interface.

G2.1 Provide Natural Layout Orientation for Information to Be Scanned

The arrangement of information on the screen affects how efficiently and comfortably people can scan, search for, and find the information. How efficiently people scan the screen and search for specific items of information is most related to the direction of their language.

In Traditional Chinese and Japanese languages, text is printed in columns, with breaks between the columns. It is read from top to bottom and horizontally (from right to left), following a typical "N" pattern. In cultures like Hong Kong and Taiwan, where the Traditional Chinese form is still used, people have a strong tendency to scan the screen "across the columns" or horizontally when searching for a specific piece of information. In contrast, Simplified Chinese can be printed either in rows or in columns, although it is mostly printed in rows and read from left to right and top to bottom (a "Z" pattern). As a result, people in Mainland China tend to scan the screen in a more vertical pattern. However, being accustomed to seeing both patterns of printed language, Mainland Chinese are able to adapt their search patterns based on the layout of the information (Lau, Goonetilleke, and Shih 2001; Goonetilleke, Lau, and Shih 2002). In a horizontal layout they search vertically or "row by row." In a vertical layout they search horizontally or "column by column."

Arrange the information on the screen in a way that is compatible (rows or columns) with the user's language so it can be scanned in a comfortable and efficient manner. This guideline applies to screens that display items of information with similar properties. For example, it applies to two-dimensional menus such as the item menus found on many online stores and thumbnail photo menus. It also applies any time you display a large field of data on the screen and the user's task is to search for a specific item. The guideline may also apply to searching Web pages, but there currently is no validation for that.

G2.2 For Menu Design, Provide Orientation Compatible with the Language Being Presented

It is common to have a menu bar placed horizontally in a user interface. However, as reported by Dong and Salvendy (1999), for Chinese users in Mainland China, users responded faster when presented with a vertical Chinese menu bar, whereas the same group of users responded faster when presented with a horizontal English menu bar.

The conventional guideline for displaying options in a pull-down menu is vertical orientation. Each item in the pull down list of options is separated by a horizontal "line break." This is a natural way to present and read lists in languages such as English. However, some languages such as the Traditional Chinese used in Hong Kong and Taiwan are mostly written starting at the top right corner of the page. Text is written in columns from top to bottom. Natural breaks in the text are the spaces or "column breaks" between the columns. The most natural menu orientations for users of these vertically oriented languages are *horizontal* (Shih and Goonetilleke 1997, 1998).

G2.3 Text Direction, Labeling, and Scrolling

Conventional guidelines use a *left-to-right* orientation or left justification for labeling text boxes, presenting text in text boxes, scrolling text within a text box, and presenting a series of control buttons. The guideline assumes that people read left to right. However, some languages such as Arabic are read from *right to left*. If you are designing a user interface that will have to accommodate a "right-to-left" language, you should adapt the display features that assume a certain text direction. For languages with a right-to-left orientation,

1. Place text box label to the right of the box
2. Right justify the text within the box
3. Scroll text in the box so it can be read from right to left
4. Move the insertion point from right to left in front of the leading character
5. If two or more buttons are used in the controls, and their order or frequency of use is important, then place them in a right-to-left order, with the most important one placed on the right.

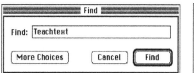   

(a) English (left to right)   (b) Arabic (right to left)

FIGURE 33.4   The Find function in (a) English and (b) Arabic.

6. Use right-to-left orientation to imply the order or sequence of items.
7. If an icon is associated with a line of text, position the icon consistent with reading direction (to the right of text item)

Figures 33.4a and 33.4b illustrate the same Find function displayed with different orientations.

G3   GRAPHIC DESIGN

Graphic design is a form of visual communication that uses visual elements such as image, color, form, shape, and typography as a unique type of language—visual language. While Horton (1994) stated that one of the benefits of using icons is to reduce the needs of text translation when marketing products globally; other researchers also pointed out that visual language can be problematic when it's used to communicate across cultures. The meaning of the image is highly negotiable in a cultural system (Fu 2001), and icons can be effective in one culture but offensive when used in another culture (Shirk and Smith 1994). As cultures can play a key role in the effectiveness of visual communication, the International Telecommunication Union proposed a four-step methodology for the design, evaluation, and standardization of international symbols, pictograms, and icons in the telecommunication context (ITU-T Telecommunications Standardization Sector of ITU 1994). Graphical representation has been mainly designed and tested in the West, but often times it is targeted for international use. While there have been no conclusive design guidelines of icons for cross-cultural uses, a few noteworthy studies, described below, have been published.

G3.1   ICONS DESIGNED AND TESTED WELL IN ONE REGION MAY NOT BE ACCEPTED BY PEOPLE IN OTHER REGIONS

A series of international studies on telecommunications icons designed and tested by ETSI (European Telecommunication Standards Institute) indicated that people from five Southeast Asian countries encountered difficulties of recognizing and using the symbols (Piamonte, Ohlsson, and Abeysekera 1997; Piamonte, Abeysekera, and Ohlsson 1999). Another study investigated the efforts by two design teams (one in United States and one in Singapore) working independently on the GUIs of two pager/PDA products targeted for the Chinese

market (Tham and Tan 1999). The icons designed by the Asian team used caricature or "cartoon" style in their icons. The American icons were more conventional computer icons. Each team relied on images and symbology from their respective cultures. Results showed that users in the target market (China and Singapore) preferred the design by the Singapore team. Other researchers (Plocher, Garg, and Wang 1999) also reported findings on difficulties of recognizing application tool bar icons by the Chinese users for the process control workstations developed mainly in the United States and in Australia.

G3.2   WHEN DESIGNING ICONS, PROVIDE A COMBINATION OF TEXT AND PICTURE

If alternative methods (e.g., text only or picture only) have to be adopted due to design constraints, impacts to the usability of the alternatives due to cultural differences will need to be considered. Choong and Salvendy (1998) examined the performance differences between the American and the Chinese users in recognizing icons presented in different modes (combined, text only, and picture only). The results indicated that it is favorable to provide a combined presentation mode since the performance with a combined mode can be at least as good as or better than the performance with either an alphanumeric or a pictorial mode. However, when testing with the alternatives, the study found that American users performed better with text only presentation and Chinese users performed better with picture only presentation. Similar findings have been reported that bimodal (text and picture) Chinese icons provide the best appropriateness and meaning for Chinese users (Kurniawan, Goonetilleke, and Shih 2001).

G3.3   CAREFULLY EXAMINE THE TEXTUAL COMPONENT IN THE GRAPHICS ON THE WEB INTENDED FOR A GLOBAL AUDIENCE

The use of words and abbreviations should be minimized since those graphics with embedded text will require translation. There are some cases where individual alphabets appear in icons as an intrinsic part of the design (e.g., the Font icons in Microsoft Word uses "A," "B," "I," and "U"). These types of icons will need to be replaced with appropriate words in the language of the target users. However, sometimes it is not possible to simply replace the letter because the target language does not have alphabetic characters, for example, the Chinese writing script is ideographic rather than alphabetical.

G3.4 Pay Attention to Graphics with Culture-Specific Meanings or Associations Such as Visual Puns, Verbal Analogies, Gestures and Body Parts, Religious Symbols, Animals, and Colors

The graphics on any Web site should support the verbal site content and can be used to decode it, especially for a site with potential international visitors. For example, some software uses a mouse in the icon representing the point-and-click device. In some languages, the name for the pointing device is not the same as that of a small rodent. This icon will then need to be completely redesigned for users with other languages who do not refer to the device as a "mouse." Luna, Peracchio, and deJuan (2002) found that the level of graphic congruity (i.e., "making sense") could moderate language effects on attitudes toward cross-cultural Web sites. In a high-congruity Web site, translation to the site visitors' first language may not be essential, if the visitors speak English with a sufficient level of fluency.

G3.5 Consider Reading and Scanning Direction of the Target Users

The accustomed scanning direction of the user influences his or her interpretation of the sequence of events in a graphic, the relative importance and virtue of objects in a graphic, and the recognition of objects representing text. As suggested by Horton (1994), certain steps can be taken to avoid misperception or misinterpretation of graphics caused by differences in users' reading directionality. Those steps include provide indication (e.g., arrow) for the sequence of objects, use vertical sequence, flip the graphics if appropriate, and draw images symmetrically.

G4 CULTURAL PREFERENCES FOR COLORS

Colors and combination of colors are associated with different meanings in different cultures. Web designers have used colors to attract users and to represent certain culture background. The color preferences of adults are usually influenced by learned responses including different color meaning associations in other cultures.

G4.1 Use Consistent Color Associations

The following color associations are similar in many cultures (Courtney 1986; Kyrnin 2002; Liang et al. 2000; Luximon, Lau, and Goonetilleke 1998; Röse, Liu, and Zühlke 2001, 2002). Therefore, it is suggested that Web designers conform use of the colors to their associated meanings:

Danger—Red
Go—Green

Hot—Red
Stop—Red
Safe—Green
Caution—Yellow

G4.2 Notice Ambiguous Color Associations

The following color associations are similar in some cultures (Courtney 1986; Kyrnin 2002; Liang et al. 2000) but different in others. Most colors have some ambiguity associated with their meaning and should be avoided for signaling important concept unless they are combined with other coding such as text or icon or both. Designers should be aware of ambiguities and use color coding with caution.

Cold White: Chinese, Japanese
      Blue: American

On Green: Chinese
    Red: American

G4.3 Notice Special Color Associations

The following color associations are very special in some cultures (Courtney 1986; Kyrnin 2002):

Sacred—Saffron: Hindus, Sikhs, Jains, and Buddhists
Mourning—Brown: Indian
Cheating—Green: Chinese
Holiness—Blue: Jews

G5 INFORMATION ARCHITECTURE, SEARCHING, AND INTERACTION

G5.1 Information Should Be Organized in Association with Target Users' Cultural Traits

As noted earlier, people around the world hold different thinking styles and the differences in thinking could affect their performance interacting with computers. Choong (1996) and Choong and Salvendy (1999) research highlights that different cultures often focus on different attributes of the same items or objects. A Web site should be structured to reflect the target users' tasks and their views of the information space to facilitated human information processing. Culture will influence what groups of items or functions should be placed together as pages and how links on a Web site or buttons and menus on the user interface should be labeled.

Luna, Peracchio, and deJuan (2002) suggest that Web sites should be structured to conform to the target cultures, e.g., the site could have a hierarchical or a search-based structure, depending on whether the target visitors belong to a high-context culture (e.g., Japan) in which hierarchical structures might be preferred or a low-context culture (e.g., Germany) in which search-based structures might be preferred.

G5.2 THE NAVIGATION OF A WEB SITE SHOULD BE DESIGNED TO MEET USERS' EXPECTATIONS BY CLEARLY INDICATING WHERE THEY ARE, WHERE THEY HAVE BEEN, WHAT THEY CAN ACCESS, AND HOW THEY CAN PROCEED

It is important to provide users with a culturally congruent site by offering links to pages that address the respective values, symbols, heroes, and rituals of a particular culture (Luna, Peracchio, and deJuan 2002). The culturally appropriate navigation patterns will lead to less confusing and more satisfactory user experience.

G5.3 PROVIDE SEARCHING MECHANISMS

Researchers have indicated the significance of searching mechanisms for Web design. Morkes and Nielsen (1997) suggested that designers provide search mechanisms and structure information to facilitate focused navigation on all Web sites. They found that 79% participants scanned text and only 16% read word for word. Nielsen (1997) found that only 10% of Web users would scroll a navigation page to see any links that were not visible in the initial display. Zhao (2002) studied the effect of time orientation on browsing performance and found that polychronic participants performed browsing tasks faster than neutral participants when participants are not familiar with the information architecture of the browsing materials.

G5.4 PROVIDE BOTH SEARCH ENGINE AND WEB DIRECTORY TO SUPPORT DIFFERENT NEEDS OF USERS

Users with different cultural background may have different needs for searching mechanisms. Most Web sites have two types of search mechanisms built in: Web directories and search engines. Fang and Rau (2003) examined the effects of cultural differences between the Chinese and the Americans on the perceived usability and search performance of Web portal sites. Chinese participants tended to use keyword search to start a task. If that failed after one or more trials, they would then try to browse the categories to complete the task. American participants tended to browse categories in the beginning of a task. They might use keyword search to supplement category search. Some Chinese participants reported that they tended to use keyword search first when the search task was difficult. Choosing keywords was relatively easier than going through several levels of categories for difficult search tasks for some Chinese participants. American participants were likely to use keyword search when they believed that they knew exactly what they were looking for.

G5.5 CULTURE INFLUENCES THE SATISFACTION OF USERS FOR SEARCHING TASKS

Fang and Rau (2003) found that Chinese participants were less satisfied with their searching performance than their American counterparts, even though no significant difference was found on their browsing performance for most of the searching tasks. The differences in attribution for American and Chinese users may explain the differences in satisfaction. The Chinese tend to attribute consequence of events more internally than the Americans. The Chinese participants might think that if they had tried harder or had paid more attention than they did in the test, they would have done better.

G5.6 PROVIDE POSSIBLE OUTCOMES AND RESULTS OF OPERATIONS AS MUCH AS POSSIBLE FOR ASIAN USERS OR USERS IN HIGH UNCERTAINTY AVOIDANCE CULTURES

High-uncertainty avoidance cultures would emphasize simplicity in information, clear mental models, preventing users from disorientation, and redundant cues (Marcus and Gould 2000). Asian countries and some countries in the Middle East follow high power distance in daily life so that probabilities and "what-ifs" associated with different procedural options are recommended (Plocher, Garg, and Krishnan 1999).

G5.7 PROVIDE EXTRA NAVIGATIONAL AIDS FOR JAPANESE, ARABIC, AND MEDITERRANEAN USERS OR USERS IN HIGH-CONTEXT COMMUNICATION STYLES

Rau and Liang (2003b) investigated the effects of communication style on user performance in browsing a Web-based service. Context refers to the amount of information packed into a specific instance of communication. Low-context communication contains a lot of information on the subject of the communication (Hall 1984; Hofstede 1980; Plocher, Garg, and Krishnan 1999). Prior knowledge of the subject matter is not assumed. In contrast, a high-context communication is terse and the recipient's prior familiarity with the subject matter is assumed. Culturally, the Japanese, the Arabs, and the people from the Mediterranean region are high context, whereas Americans, Germans, Swiss, Scandinavians, and other northern Europeans are low-context (Hall 1984; Hofstede 1980; Plocher, Garg, and Krishnan 1999). Participants with high-context communication style were more disorientated during browsing than were those with low-context communication style (Rau and Liang 2003b).

G6 MOBILE WEB

Mobile use of the Internet and Web is becoming common across the globe. However, there are very few studies about cross-cultural Web design for mobile devices.

G6.1 DESIGN THE WEB TO SUIT TAILORED OPERATOR PORTALS OR FULL WEB BROWSERS, DEPENDING ON LOCAL USERS' PREFERENCES AND INFRASTRUCTURES

Kaikkonen (2008) studied users' usage patterns of full Web sites and sites tailored for mobile browsers. A global online survey and a set of interviews in Hong Kong, London, and New York found that Asian users often use mobile tailored operator portals, whereas European and American users are more familiar with full Web browsers and the newest mobile devices. The results also indicated that the mobile Internet has become popular in Hong Kong, where the majority of users are females with nontechnical backgrounds. In contrast, in Europe and North America mobile Internet users are tech-savvy, early adopters.

33.4 INTERNATIONAL USABILITY EVALUATION

33.4.1 The Moderator and the Test User: A Social Interaction

Usability testing inherently involves at least some human social interaction between test moderator and test user. Social and cultural norms should affect this interaction in a similar manner to the way that they affect other interpersonal interactions. There is a growing literature on how Easterners and Westerners react, usually predictably, in usability tests. A number of best practices now can be described that help to mitigate cultural bias in usability tests resulting from social interaction effects.

First, it is a good practice to use evaluators, moderators, or interviewers from the same culture as the test users. Vatrupu and Pérez-Quiñones (2006) studied how test participants from different cultures behaved in a structured interview setting in which the participant's task was to comment on a Web site. Indian participants found more usability problems and made more suggestions with an Indian interviewer than with an Anglo-American interviewer, but the comments they made to the Anglo-American interviewer tended to be more positive than negative. With an Anglo-American conducting the interview, Indians also were reluctant to discuss culture-related problems with the Web site and kept their comments quite general. The participants were more detailed and candid with the Indian interviewer. Furthermore, Yammiyavar, Clemmensen, and Kumar (2008) found that when subjects were paired with evaluators from the same culture, they used more head and hand gestures to communicate than if the evaluator was from a different culture, providing a richer source of non-verbal data to analyze. Sun and Shi (2007) studied how using one's primary versus secondary language (English versus Chinese in this case) in a think aloud test affected the process of the test. Chinese evaluators speaking Chinese to Chinese test users gave more help to users and a more complete introduction to the product being tested, and encouraged users more frequently than did Chinese evaluators speaking English during the test.

A second good practice is to avoid pairing moderators and test users who differ in their perceived status or authority. Particularly in cultures with high power distance, such as China and Malaysia, the behavior of both the test user and the test evaluator are affected by perceived differences in status or authority. Participants in high power distance cultures do not challenge or question the evaluator or experimenter because of the perception of the facilitator as a person of authority (Burmeister 2000). Yeo (1998) illustrated this with an example of a usability test

conducted in Singapore, in which a participant broke down and cried from frustration. A post-test interview revealed that the participant's behavior was in part due to the Eastern culture, in which it is not acceptable to criticize the designer openly, because it may cause the designer to lose face. Evers (2002) evaluated cultural differences in understanding of a virtual campus Web site across four culturally different user groups (England, North America, Netherlands, and Japan) by using the same methods for each group. The results indicated that Japanese participants who were secondary school students felt uncomfortable speaking out loud about their thoughts and seemed to feel insecure because they could not confer with others to reach a common opinion.

But the effect of culture can go both ways. In a study of think aloud tests, Sun and Shi (2007) found that evaluator's behavior is also affected by differences in level of perceived authority. When the evaluator's academic title or rank was higher than that of the users, the evaluator tended to more frequently ask the user what he or she was thinking during the test. The evaluator also tended not to provide the user with more detailed instructions during the test.

The third guideline is to train evaluators to combat the "conversational indirectness" of Asian users. Easterners will tend to seek a compromise and be indirect when evaluating user interfaces. For example, Herman (1996) studied cultural effects on the reliability of objective and subjective usability evaluation. The results of objective and subjective evaluation correlated poorly in Herman's study. The Asian participants were less vocal, very polite, and disinclined to express negative comments in front of observers, so that the results of subjective evaluation tended toward the positive despite clear indication of poor user performance. Herman's solution was to invite test participants to work in pairs to evaluate the interface and make the usability test more of a peer discussion session.

Shi (2008) conducted observations of usability tests in China, India, and Denmark and like Herman, also noted that Chinese users often kept silent and did not speak out actively, particularly in formative evaluations. Shi (2008) and Clemmensen et al. (2009) explained this observation in terms of Nisbett's (2003) cultural theory of Eastern and Western cognition. Accordingly, Chinese people tend to have a holistic process for thinking as opposed to the more analytic style of Westerners. Holistic thought is not as readily verbalized as analytic thought. So, the conclusion is that Chinese users in a think aloud test situation are thinking about the user interface in holistic terms and simply have a more difficult time putting those holistic thoughts into spoken words.

Shi recommended that evaluators receive special training to conduct testing with Chinese users that is based on think aloud methods. Evaluators should be trained to use reminders and questions, "digging deeper probes," to get the users to talk. For example, evaluators in Shi's study reported that if they knew that users were looking for some object or feature on the screen, they would ask, "what are you looking for?" and then the user would tell them immediately about what they were looking for. This method of asking related questions to encourage speaking aloud was found to be more natural than just asking people to "keep talking." Evaluators also should be trained to be patient with Eastern participants as they pause and think in between verbalizations (Shi 2009). But perhaps the real implication of this reticence of Eastern users to put thoughts to words in the test situation is that an alternative to the think aloud method should be considered.

Shi (2009) found no significant differences in the set of usability problems found by Chinese and Danish evaluators. However, their ratings of the severity of the usability problems did differ significantly. Chinese evaluators rated problems less severely than Danish evaluators and often rated problems in the middle of the 5-point severity scale. Shi (2009) suggests that if problem severity rating is part of the test that a 4-point, rather than 5-point, scale be used to prevent middle-of-the-scale ratings.

33.4.2 Participant and Evaluator Recruiting

Recruiting participants with similar background in different places at the same time has made international usability evaluation very difficult. The experimenters have options to carry out international usability evaluation such as going to the foreign country, running the test remotely, hiring a local usability consultant, or asking staff in a local branch office (Nielsen 1990, 2003). Many researchers (Choong and Salvendy 1998, 1999; Dong and Salvendy 1999; Fang and Rau 2003; Fukuoka, Kijima, and Spyridakis 1999; Prabhu and Harel 1999; Evers 2002) chose to recruit participants in two or more countries by going to the countries. The Web has made conducting international usability evaluation a new option. Lee and Harada (unpublished paper, 1999) conducted an evaluation by recruiting participants on the Web, but they found it difficult to recruit participants from countries with no experimenters present.

Clemmensen et al. (2007) address the problem of "hidden user groups," groups of people who represent significantly different target user segments within the same culture. They suggest that test planners attempt to balance out potential hidden user groups within user segments. For example, users who are accustomed to foreigners and adapt quickly to international test conditions should be balanced off by users who are not accustomed to foreigners. Traditional and culturally sensitive users such as those one might find in rural areas need to be balanced in the pool of test participants by more modern, urbanized people who are less influenced by a country's local culture. To avoid missing critical usability problems during the test, they recommend that evaluators be chosen carefully from evaluator groups suitable to the members of the identified hidden user groups.

33.4.3 Language

33.4.3.1 Verbal Language-Using Interpreters and Translators

Language is a significant factor in international usability evaluation. Without translated content, the target audience may be limited to Web users with a certain education and social background. Nielsen (2003) indicated that displaying in the user's native language, character set, and notations, and translating user interface and documentation in the user's native language, are the first two concerns for international usability testing. Rau and Liang (2003b) conducted a card-sorting test with Chinese users in Taiwan for an international Web site. Even though all the information items were translated into Chinese, a dictionary and instructions were available if participants had any question about the meaning of these items. Also, anything offensive in the user's cultural background should be avoided. The testing materials and procedure should accommodate the way business is conducted and the way people communicate in various countries (Nielsen 2003). Interpreters are necessary if the evaluators are not able to speak the user's native language well. The evaluators need to avoid any important information lost in translation, so that videotaping or audio recording are useful for further analysis. Nielsen (2003) suggests meeting interpreters beforehand and reminding them that they should not help users during the test.

33.4.3.2 Nonverbal Language

Nonverbal language displayed in the form of hand and head gestures is commonly used by test participants as an occasional substitute for verbal language or to elaborate on or supplement it. Hand and head gestures also often communicate the comfort level of the participant and his or her readiness to communicate with the evaluator. Observing and analyzing gestures during a test can provide a rich source of data that adds context to the verbal data being recorded.

Yammiyavar, Clemmensen, and Kumar (2008) questioned whether users from different cultures exhibited similar or different patterns of nonverbal communication including the type, frequency, and usage of gestures. They analyzed the occurrence of gestures in video recordings of think aloud tests that used subjects from Denmark, India, and China. Gestures were grouped into four types (Ekman and Friesen 1969): emblems, illustrators, adapters, and regulators. Emblems replace words with gesture-based signs like nods of head for "yes" or a V sign for victory. They tend to be culture specific. Illustrators include such actions as banging on the table or sketching shapes in the air. They help the subject to verbalize their thoughts. Yammiyavar et al. discovered that these can be quite important markers for usability problems because illustrator gestures frequently precede the verbalization of a usability problem to the evaluator. Adapter gestures are actions of the body that convey feelings of pressure or discomfort, for example, cracking one's knuckles, tapping one's feet, stroking hair or chin while in deep contemplation, and "squirming" in one's seat. They indicate the subject's comfort level with the test situation. Finally, subjects control the flow of conversation with the evaluator by using regulator behaviors, such as nodding the head up and down to indicate agreement.

Yammiyavar, Clemmensen, and Kumar (2008) found that the frequency of using gestures during the test was not significantly different across the three cultural groups. This contradicts the popular belief that Indians, for instance, use more gestures to communicate than other cultural groups. Also, regulator gestures appeared to be used similarly across the cultures studied but with some tendency for Chinese to use them the least. In contrast, there were significant differences across the cultural groups in the specific emblem gestures used to replace words and in adapter behaviors. Certain illustrators appeared to be culture specific, as well. The researchers concluded that there is a need to benchmark gestures used in these different cultures and their meanings and then provide those to usability test evaluators to guide observations and understand what they observe.

The facial expression associated with surprise often is used as a marker by evaluators to indicate that a usability problem has been detected by the user. Clemmensen et al. (2009) have questioned the validity of this practice in cross-cultural usability tests. From Nisbett's (2003) theory of cultural cognition, they hypothesize that Easterners will experience less surprise than Westerners when presented with inconsistencies in user interfaces. With their logical, analytic orientation, Westerners tend to focus on fewer causes of observed events, while Easterners, with their holistic orientation, tend to consider more causal factors as well as the context of the event. As Clemmensen et al. (2009) point out, this makes it easier for them to identify a rationale for why the event occurred in the way it did, resulting in less surprise.

33.4.4 Instructions, Tasks, and Scenarios

Instructions to the test participant can vary significantly in how much contextual information they provide. At the one extreme, instructions are strictly focused on the task to be performed with the application being tested. No explanation is given about the purpose of the application, when you might use it or why. At the other extreme, the explanation of the task is embedded in a rich context provided by a real life scenario. Clemmensen et al. (2009) suggest that Westerners, with their tendency to focus on the central elements presented such as the details of the task, will be able to obtain that in either presentation of instructions. Easterners, however, will find the stark instructions lacking. With their holistic style of thinking, they will prefer to have the task explanation embedded in the context of a real-life scenario. The recommendation from Clemmensen et al. (2009) is that if cross-cultural testing is to be conducted, then the test planner should consider adapting the instructions to the cultures of the participants. Planners might want to have different versions of the test protocol prepared that include different types and amounts of background information. A classic example of adapting test scenarios to the target culture is from Chavan (2005), who engaged Indian test participants by embedding the test tasks in "Bollywood" scenarios. The method capitalizes on the popularity in India of watching Bollywood movies and the fun of openly critiquing them. The test participant is asked to imagine a dramatic scenario similar to a movie script, perform the desired tasks using the application in that context, and critique it.

33.4.5 Questioning Universal Constructs of Usability

Underlying much of what is done in user-centered design is the notion that people worldwide understand the fundamental attributes of "good usability" in the same way—effectiveness, ease of use, visual appearance, efficiency, satisfaction, fun, and nonfrustration (Frandsen-Thorlacius et al. 2009). As the preceding sections of this chapter have shown, we may culturally adapt specific local instantiations of a user interface to the preferences of local cultures. But it is assumed that the local instantiations are all done with the goal of enhancing the universal construct of usability as defined by these basic attributes.

Recent research questions this assumption of universal constructs of usability (Frandsen-Thorlacius et al. 2009; Hertzum et al. 2007). Frandsen-Thorlacius et al. sampled 412 users from China and Denmark to determine how they understood and prioritized attributes of usability. Chinese users placed greater value on visual appearance, satisfaction, and fun than Danish users. Danish users valued effectiveness, efficiency, and lack of frustration more highly than Chinese users. Clearly, dimensions of usability were weighted differently for these two cultural groups in the study. Hertzum et al. (2007) used repertory grid interviews, a method for identifying how users give meaning to their experience, to explore how personal constructs of usability differed between people from three different cultures, Denmark, China, and India. They found that some of the constructs verbalized by study participants were consistent with common notions of usability such as ease of use and were important at least to some degree to participants from all three cultures. But other constructs differed from commonly used attributes of usability, for example, the attribute of "security." The most important usability attributes for Chinese subjects involved issues of security, task types, training, and system issues. In contrast, Danish and Indian participants focused on more traditional aspects of usability such as "easy-to-use," "intuitive," and "liked." None of the Chinese subjects verbalized these as primary constructs of usability.

These two studies are just a start in understanding what aspects of usability people value in different cultures. However, they raise questions about how user interfaces are currently designed and how they are tested across cultures. If the testing assumes that everyone values the same attributes of usability, then usability test participants from all cultures would be expected to find the same number of usability problems related to the same usability attributes. The two studies reviewed here raise the possibility that test participants in one culture may not identify problems related to a particular attribute of usability simply because that attribute is not as important to them as it is to a participant from another culture. The absence of verbalized problems in a particular area of the user interface design could be misinterpreted as an indication that no problems exists. Also, from a business perspective, one would have to question why a company would invest scarce development dollars to perfect aspects of a user interface that are not particularly important to a targeted user group. These findings should make companies aware that basing usability requirements for a global product on just one or two cultural groups runs the risk of minimizing attributes that turn out to be important to a second, third, fourth, *n*th, cultural group of users somewhere in their market space, perhaps even to the majority of potential product users. Perhaps a means to identify and weight what attributes of usability are important to test subjects must become a standard part of global usability testing methodology. And the global locations for usability testing should reflect the segments of the intended global market.

33.5 CONCLUSION

This chapter had two goals. One goal was to summarize most of the studies done in the cross-cultural design of HCI and to provide a common repository of information and references for researchers. The other goal was to provide Web design practitioners with a set of design guidelines for cross-cultural Web designs.

Four major dimensions affecting cross-cultural user interface design were discussed: cognitive, affective, perceptual, and functional. Cross-cultural Web design guidelines were extracted from books and research papers and discussed in detail for five areas: languages and format; presentation and layout; graphic design; cultural preferences; information architecture, searching, and interaction; and mobile phone Web design. Some of the guidelines are well supported by research and best industry practices; others are still at an early stage of research and should be applied more carefully.

Recent research on cross-cultural usability testing also was reviewed. This research indicates that in cross-cultural usability testing great care must be taken in pairing evaluators and test subjects. Evaluators and subjects should be at the same perceived level of authority and speak the same language. Evaluators working in Asian cultures where test subjects might be reluctant to openly express their positive and negative opinions should be trained in special techniques to elicit feedback. Evaluators also must be trained to recognize culturally specific gestures and body language. Finally, emerging research suggests that usability may mean different things in different cultures; e.g., there may be no universal construct of usability. Thus, the test designer must be sensitive to the specific aspects of usability that are valued by the local culture. Finally, there are areas that are still in great need of cross-cultural investigation specific to Web design. These areas include information architecture, interaction and navigation, searching behaviors, graphical images and affect, and the impact of newly developed technologies (e.g., voice recognition/searching for accessing the Web, personal communication devices design challenges). New technologies such as multitouch pose new design challenges in facilitating cooperative interaction among two or more users. Research is needed to understand how users in different cultures will utilize these new technologies and work and play cooperatively. A thorough understanding of these issues and applications of new guidelines in these areas will enhance cross-cultural user interface design and bring us closer to realizing the full potential of the Internet.

REFERENCES

Anderson, R., R. Anderson, and K. Deibel. 2004. Analyzing concept groupings of introductory computer programming students. Technical Report, University of Washington, November.

Bluedorn, A. C., T. J. Kalliath, M. J. Strube, and G. D. Martin. 1999. Polychronicity and the inventory of polychronic values (IPV): The development of an instrument to measure a fundamental dimension of organizational culture. *Journal of Managerial Psychology* 14(3–4): 205–230.

Burmeister, O. K. 2000. Usability testing: revisiting informed consent procedures for testing Internet sites. Paper presented at 2nd Australian Institute of Computer Ethics Conference, Canberra, Australia, December.

Cai, D. C., C. F. Chi, and M. L. You. 2001. The legibility threshold of Chinese characters in three-type styles. *International Journal of Industrial Ergonomics* 27(1): 9–17.

Chavan, A. L. 2005. Another culture, another method. In *Proceedings of the 11th International Conference on Human–Computer Interaction*, CD-ROM. Mahwah, NJ: Lawrence Erlbaum.

Chen, J. 1989. Experiments on effects of dot-matrix size and its format of Chinese characters upon VDT visual performance. MA thesis, Hangzhou University.

Choong, Y. Y. 1996. Design of computer interfaces for the Chinese population. Doctoral Dissertation, Purdue University, August.

Choong, Y. Y., and G. Salvendy. 1998. Design of icons for use by Chinese in Mainland China. *Interacting with Computers* 9: 417–430.

Choong, Y. Y., and G. Salvendy. 1999. Implications for design of computer interfaces for Chinese users in Mainland China. *International Journal of Human Computer Interaction* 11(1): 29–46.

Clemmensen, T., Q. Shi, J. Kumar, H. Li, X. H. Sun, and P. Yammiyavar. 2007. Cultural usability tests—how usability tests are not the same all over the world. In Usability and Internationalization, Part I, HCII 2007, vol. 4559, *Lecture Notes in Computer Science*, ed. N. Aykin, 281–290. Berlin: Springer-Verlag.

Clemmensen, T., M. Hertzum, K. Hornbæk, Q. Shi, and P. Yammiyavar. 2009. Cultural cognition in usability evaluation. *Interacting with Computers* 21(3): 212–220.

Courtney, A. J. 1986. Chinese population stereotypes: color associations. *Human Factors* 28(1): 97–99.

Crockett, R. O. 2000. Surfing in tongues. *Business Week*. Dec. 11, 18.

DeGroot, A. 1991. Bilingual lexical representations: A closer look at conceptual representations. In *Orthography, Phonology, Morphology, and Meaning*, eds. R. Frost and L. Katz, 389–412. Elsevier: Amsterdam.

Dong, J., and G. Salvendy. 1999. Designing menus for the Chinese population horizontal or vertical? *Behaviour & Information Technology* 18(6): 467–471.

Ekman, P., and W. V. Friesen. 1969. The repertoire of nonverbal behavior: Categories, origins, usage, and coding. *Semiotica* 1: 49–98.

Evers, V. 2002. Cross-cultural applicability of user evaluation methods: A case study amongst Japanese, North-American, English and Dutch users. Paper presented at CHI 2002, Minneapolis, MN, April 20–25.

Fang, X., and P. L. P. Rau. 2003. Culture differences in design of portal sites. *Ergonomics* 46(1–3): 242–254.

Fernandes, T. 1995. *Global Interface Design: A Guide to Designing International User Interfaces*. Chestnut Hill, MA: AP Professional.

Fogg, B. J., and D. Lizawa. 2008. Online persuasion in Facebook and Mixi: A cross-cultural comparison. Paper presented at the 3rd International Conference on Persuasive Technology, Oulu, Finland, June 4–6.

Frandsen-Thorlacius, O., K. Hornbæk, M. Hertzum, and T. Clemmensen. 2009. Non-universal usability?: A survey of how usability is understood by Chinese and Danish users. In *CHI2009*, 41–50. New York: ACM Press.

Fu, L. 2001. When West meets East: the intercultural communication challenge for graphic design in a global context. *Bulletin of the 5th Asian Design Conference* (Seoul, Korea, Oct. 13–15).

Fukuoka, W., Y. Kojima, and J. H. Spyridakis. 1999. Illustrations in user manuals: preference and effectiveness with Japanese and American readers. *Technical Communication*, 2nd Quarter, 1999.

Goonetilleke, R. S., W. C. Lau, and H. M. Shih. 2002. Visual search strategies and eye movements when searching Chinese character screens. *International Journal of Human–Computer Studies* 57: 447–468.

Gould, E. W. 2001. More than content: Web graphics, cross-cultural requirements, and a visual grammar. Paper presented at HCI International 2001 Conference, New Orleans, LA, August 5–10.

Haase, R. F., D. Y. Lee, and D. L. Banks. 1979. Cognitive correlates of polychronicity. *Perceptual and Motor Skills* 49: 271–282.

Hall, E. T. 1984. *Dance of Life: The Other Dimension of Time*. Yarmouth, ME: Intercultural Press.

Hall, E. T. 1989. *Beyond Culture*. New York: Anchor Press.

Hall, E. T. 1990. *The Hidden Dimension*. New York: Anchor Press.

Hall, E. T., and M. Hall. 1990. *Understanding Cultural Differences: Germans, French and Americans*. Yarmouth, Maine: Intercultural Press, Inc.

Herman, L. 1996. Toward effective usability evaluation in Asia: Cross cultural differences. *Interact '96/OZCHI* 1–7, 135.

Hertzum, M., T. Clemmensen, K. Hornbæk, J. Kumar, Q. Shi, and P. Yammiyavar. 2007. *Usability Constructs: A Cross-Cultural Study of How Users and Developers Experience Their Use of Information Systems*, vol. 4559, *Lecture Notes in Computer Science*, 317–327. Springer: New York.

Hisham, S., and A. D. N. Edwards. 2007. Incorporating culture in user-interface: A case study of older adults in Malaysia. Paper presented at the 18th ACM Conference on Hypertext and Hypermedia, Manchester, UK, Sept. 10–12.

Hofstede, G. 1980. *Culture's Consequences: International Differences in Work-related Values*. London: Sage Publications.

Hofstede, G. 1997. *Cultures and Organizations: Software of the Mind*. New York: McGraw-Hill.

Horton, W. 1994. *The Icon Book: Visual Symbols for Computer Systems and Documentation*. New York: John Wiley.

ITU-T Telecommunication Standardization Sector of ITU. 1994. Procedures for designing, evaluating and selecting symbols, pictograms and icons. ITU-T Recommendation F.910. Geneva: International Telecommunication Union.

Jin, W., Z. Zhu, and M. Shen. 1988. The effect of different typefaces of Chinese characters on recognition. Engineering Psychology Reports, 2, Hangzhou University, 96–100.

Kaikkonen, A. 2008. Full or tailored mobile Web—where and how do people browse on their mobiles? Paper presented at the International Conference on Mobile Technology, Applications & Systems 2008 (Mobility Conference), Ilan, Taiwan, Sept. 10–12.

Kyrnin, J. 2002. Web design and color symbolism. http://webdesign.about.com/od/color/a/bl_colorculture.htm (accessed November 15, 2010).

Kaufman-Scarborough, C., and J. D. Lindquist. 1999. Time management and polychronicity: Comparisons, contrasts, and insights for the workplace. *Journal of Managerial Psychology* 14(3-4): 288–312.

Kress, G., and T. van Leeuwen. 1996. *Reading Images: The Grammar of Visual Design*. London: Routledge.

Kurniawan, S. H., R. S. Goonetilleke, and H. M. Shih. 2001. Involving Chinese users in analyzing the effects of languages and modalities on computer icons. In *Universal Access in HCI: Towards an Information Society for All*, ed. C. Stephanidis, 491–495. Mahwah, NJ: Lawrence Erlbaum.

Kurosu, M. 2001. Cross-cultural interface design: What, so what, now what? HCI International 2001 Conference, New Orleans, LA, August 5–10.

Lau, P. W.-C., R. S. Goonetilleke, and H. M. Shih. 2001. Eye-scan patterns of Chinese when searching full screen menus. In *Universal Access in HCI: Towards an Information Society for All*, ed. C. Stephanidis, 367–371. Mahwah, NJ: Lawrence Erlbaum.

Lee, K. P., and A. Harada. 1999. A study of the cultural effects on user-interface design: with the emphasis on the cross-cultural usability testing through World Wide Web. Unpublished paper. September.

Lee, K. P. 2001. Culture and its effects on human interaction with design: With an emphasis on cross-cultural perspectives between Korea and Japan. PhD Diss. Institute of Art and Design, University of Tsukuba, Japan.

Lewis, C. C., and J. F. George. 2008. Cross-cultural deception in social networking sites and face-to-face communication. *Computers in Human Behavior* 24(6): 2945–2964.

Liang, S.-F. M., and T. Plocher. 2003. Towards a cross-cultural human–computer interaction. Paper presented at the International Ergonomics Association XVth Triennial Congress, Seoul, South Korea, August.

Liang, S.-F. M., T. A. Plocher, P. W. C. Lau, Y. T. B. Chia, N. Rafi, and T. H. R. Tan. 2000. Cross-cultural stereotypes of colors and graphics in Honeywell process control workstations. Paper presented at 4th Biannual Asia-Pacific Conference on Computer–Human Interaction, Singapore, November.

Lindquist, J. D., J. Knieling, and C. Kaufman-Scarborough. 2001. Polychronicity and consumer behavior outcomes among Japanese and U.S. students: A study of response to culture in a U.S. university setting. Proceedings of the Tenth Biennial World Marketing Congress, City Hall Cardiff, UK.

Liu, I. M. 1986. Chinese cognition. In *The Psychology of the Chinese People*, ed. M. H. Bonds, 73–106. New York: Oxford University Press.

Luna, D., L. A. Peracchio, and M. D. deJuan. 2002. Cross-cultural and cognitive aspects of web site navigation. *Journal of the Academy of Marketing Science* 30(4): 397–410.

Luximon, A., W. C. Lau, and R. S. Goonetilleke. 1998. Safety signal words and color codes: The perception of implied hazard by Chinese people. In *Proceedings of the 6th Pan-Pacific Conference on Occupational Ergonomics*, Japan, July 21–23.

Marcus, A. 2001. Cross-cultural user-interface design. In *Proceeding of the Human Computer Interface Internat (HCII) Conference*, vol. 2, eds. M. J Smith and G. Salvendy, 502–505. Mahwah, NJ: Lawrence Erlbaum.

Marcus, A., and E. M. Gould. 2000. Crosscurrents: cultural dimensions and global Web user interface design. *Interactions* 7(1): 32–46.

Marcus, A., and N. Krishnamurthi. 2009. Cross-cultural analysis of social network services in Japan, Korea, and the USA. In *Internationalization, Design and Global Development*, ed. N. Aykin, 59–68. Berlin: Springer.

Mills, J. A., and B. S. Caldwell. 1997. Simplified English for computer displays. In *Advances in Human Factors and Ergonomics. Design of Computing Systems: Cognitive Considerations*, eds. G. Salvendy, M. J. Smith, and R. J. Koubek, 133–136. Amsterdam: Elsevier.

Minocha, S., T. French, and A. Smith. 2002. eFinance localisation: an informal analysis of specific eCulture attractors in selected Indian and Taiwanese sites. In *Proceedings of the 4th International Workshop on Internationalization of Products and Systems*, ed. D. Day.

Morkes, J., and J. Nielsen. 1997. Concise, scanable and objective: How to write for the Web. http://www.useit.com/papers/webwriting/writing.html (accessed November 15, 2010).

Nielsen, J. 1990. Usability testing of international interfaces. In *Advances in Human Factors/Ergonomics*, vol. 13, *Designing User Interfaces for International Use*, ed. J. Nielson, 39–44. New York: Elsevier Science.

Nielsen, J. 1997. Changes in Web usability engineering since 1994. http://www.useit.com/alertbox/9712a.html (accessed November 15, 2010).

Nielsen, J. 2003. International Usability Testing. http://www.useit.com/papers/international_usetest.html (accessed November 15, 2010).

Nisbett, R. E. 2003. *The Geography of Thought: Why We Think the Way We Do*. New York: Free Press.

Nisbett, R. E., K. Peng, I. Choi, and A. Norenzayan. 2001. Culture and systems of thought: Holistic vs. analytic cognition. *Psychological Review* 108: 291–310.

Peracchio, L. A., and J. Meyers-Levy. 1997. Evaluating persuasion-enhancing techniques from a resource matching perspective. *Journal of Consumer Research* 24: 178–191.

Pfeil, U., P. Zaphiris, and C. S. Ang. 2006. Cultural differences in collaborative authoring of Wikipedia. *Journal of Computer-Mediated Communication* 12(2006): 88–113.

Piamonte, D. P., K. Ohlsson, and J. D. A. Abeysekera. 1997. Evaluating telecom icons among Asian countries. In *Advances in Human Factors/Ergonomics*, vol. 21A, *Design of Computing Systems: Cognitive Considerations*, eds. G. Salvendy, M. J. Smith, and R. J. Koubek, 169–172. New York: Elsevier.

Piamonte, D. P. T., J. D. A. Abeysekera, and K. Ohlsson. 1999. Testing videophone graphical symbols in Southeast Asia. In *Proceedings of the 8th International Conference on Human–Computer Interaction (HCI99), Munich*, 793–797. London: Lawrence Ehrlbaum Associates.

Plocher, T. A., C. Garg, and K. Krishnan. 1999. Cross-cultural issues in the business aviation cockpit. Technical report prepared for Honeywell Business and Commuter Aviation Systems, June.

Plocher, T. A., C. Garg, and C. S. Wang. 1999. Use of colors, symbols, and icons in Honeywell process control workstations in China. Joint technical report prepared with Tianjin University, December.

Plocher, T. A., and C. Zhao. 2002. Photo interview approach to understanding independent living needs of elderly Chinese: A case study. Paper presented at the 5th Asia-Pacific Conference on Computer–Human Interface, Beijing, November 1–4.

Plocher, T. A., C. Zhao, S. M. Liang, X. Sun, and K. Zhang. 2001. Understanding the Chinese user: Attitudes toward automation, work, and life. In *Proceedings of the 9th International Conference on Human–Computer Interaction* (New Orleans, LA, August 5–10), 524–528.

Prabhu, G., and D. Harel. 1999. GUI design preference validation for Japan and China—a case for KANSEI engineering. In *Proceedings of the 8th International Conference on Human–Computer Interaction (HCI99)* (Munich), 521–525.

Rau, P.-L. P. 2001. Cross-cultural user interface research and design with emphasis on Asia: Chinese users in Taiwan. Final Report to Honeywell Singapore Laboratory, October.

Rau, P.-L. P., and S.-F. M. Liang. 2003a. A study of cultural effects on designing user interface for a Web-based service. *International Journal of Services Technology and Management* 4(4–6): 480–493.

Rau, P. L. P., Q. Gao, and J. Liu. 2003b. Internationalization and localization: evaluating and testing a Web site for Asian users. *Ergonomics* 46(1–3): 255–270.

Rau, P. L. P., Q. Gao, and J. Liu. 2007. The effect of the rich Web portal design and floating animations on visual search, *International Journal of Human-Computer Interaction* 22(3): 195–216.

Röse, K., L. Liu, and D. Zühlke. 2001. Design issues in Mainland China and western Europe: Similarities and differences in the area of human–machine-interaction design. In *Systems, Social and Internationalization Design Aspects of Human–Computer Interaction* (New Orleans, LA, August 5–10), M. J. Smith and G. Salvendy, 532–536. London: Lawrence Ehrlbaum Associates.

Röse, K., L. Liu, and D. Zühlke. 2002. Analysis and Structuring of the Interaction areas for a Chinese User Interface. In *WWDU 2002, Work with Display Units, World Wide Work*, eds. H. Luczak, and A. Cakir, 58–60.

Sangangam, P., and S. Kurniawan. 2007. A comparative study of Thai and UK older Web users. In *Usability and Internationalization, Part I, HCII 2007*, ed. N. Aykin, 596–605. Berlin: Springer.

Shang, R.-A., Y.-C. Chen, and C.-C. Hung. 2008. Toward a cultural phenomenon of blogging—the impacts of individualism—collectivism and self-efficacy. Paper presented at the 12th Pacific Asia Conference on Information Systems, Suzhou, Peoples Republic of China, July 3–7.

Shen, M., Z. Zhu, W. Jin, P. Dai, and X. Han. 1988. The effect of Chinese character stroke width on reading efficiency. *Engineering Psychology Research Reports, Hangzhou University* 2: 110–114.

Shi, Q. 2008. A field study of the relationship and communication between Chinese evaluators and users in thinking aloud usability tests. In *NordiCHI 2008* (Lund, Sweden, Oct. 18–22), 344–352.

Shi, Q. 2009. An empirical study of thinking aloud usability testing from a cultural perspective. Ph.D. Thesis. LIMAC PhD School, Programme in Informatics, Copenhagen Business School (CBS), Copenhagen, Denmark. November.

Shi, Q., and T. Clemmensen. 2008. Communication patterns and usability problem finding in cross-cultural thinking aloud usability testing. In *CHI '08 Extended Abstracts on Human Factors in Computing Systems* (Florence, Italy), 2811–2816. New York: ACM Press.

Shieh, K. K., M. T. Chen, and J. H. Chuang. 1997. Effects of color combination and typography on identification of characters briefly presented on VDTs, *International Journal of Human–Computer Interaction* 9(2): 169–181.

Shih, H. M., and R. S. Goonetilleke. 1997. Do existing menu design guidelines work in Chinese? In *Advances in Human Factors/Ergonomics*, vol. 21A, *Design of Computing Systems: Cognitive Considerations*, eds. G. Salvendy, M. J. Smith, and R. J. Koubek, 161–164. New York: Elsevier.

Shih, H. M., and R. S. Goonetilleke. 1998. Effectiveness of menu orientation in Chinese. *Human Factors* 40(4): 569–576.

Shirk, H. N., and H. T. Smith. 1994. Some issues influencing computer icon design. *Technical Communication*, Fourth Quarter, 680–689.

Sun, X., and Q. Shi. 2007. Language issues in cross cultural usability testing: A pilot study in China. *Lecture Notes in Computer Science* 4560: 274–284.

Tham, M. P., and K. C. Tan. 1999. Challenges in designing user interfaces for handheld communication devices: A case study. In *Proceedings of the 8th International Conference on Human–Computer Interaction (HCI99)* (Munich), 808–812.

Tian, Q. H. 1987. An experimental study on legibility of the dot-matrix sizes of Chinese characters. B.A. thesis, Hangzhou University.

Vatrapu, R., and M. Pérez-Quiñones. 2006. Culture and usability evaluation: The effects of culture in structured interviews. *Journal of Usability Studies* 1(4): 156–170.

Vu, K.-P. L., W. Zhu, and R. W. Proctor. this volume. Evaluating web usability. In *Handbook of Human Factors in Web Design*, 2nd ed., eds. K.-P. L. Vu and R. W. Proctor, 439–460. Boca Raton, FL: CRC Press.

Wang, Q. 2000. A cross-cultural comparison of the use of graphics in scientific and technical communication. *Technical Communication*, 4th Quarter.

Wang, L., H. Sato, L. Jin, P.-L.P. Rau, and Y. Asano. 2007. Perception of movements and transformations in flash animations of older adults. In *Human Computer Interaction, Part I, HCII 2007*, ed. J. Jacko, 966–975. Berlin: Springer.

Yammiyavar, P., T. Clemmensen, and J. Kumar. 2008. Influence of cultural background on non-verbal communication in a usability testing situation. *International Journal of Design* 2(2), special issue, 31–40.

Yeo, A. 1998. Cultural effects in usability assessment, doctoral consortium, *Proceedings of the Conference on CHI 98 Summary: Human Factors in Computing Systems*, 71–75. New York: ACM Press.

Zhao, C. 2002. Effect of information structure on performance of information acquiring: A study exploring different time behavior: monochronicity/polychronicity. PhD diss., Institute of Psychology, Chinese Academy of Sciences, May.

Zhao, C., T. Plocher, Y. Xu, R. Zhou, X. Liu, S.-F. M. Liang, and K. Zhang. 2002. Understanding the polychronicity of Chinese. In *Proceedings of the APCHI2002, 5th Asia-Pacific Conference on Computer Human Interaction*, vol. 1, 189–195. Beijing, China: ACM Press.

Zhao, C., V. Riley, T. Plocher, and K. Zhang. 2003. Cross-cultural interface design for Chinese users. Paper presented at International Ergonomics Association XVth Triennial Congress, Seoul, South Korea, August.

APPENDIX A
Cross-Cultural Web Design Guidelines Summary

Category	Guidelines	Dimension	Supporting Research/ Best Practice
G1 Language and format	G1.1 Use unambiguous language	Cognitive	Zhao et al. (2003), Röse, Liu, and Zühlke (2001, 2002), Mills and Caldwell (1997)
	G1.2 Allow extra space for text	Perceptual	Tian (1987), Chen (1989)
	G1.3 Accommodate text reproduction methods	Perceptual	Best practice
	G1.4 Do not embed text in icons	Cognitive and perceptual	Best practice
	G1.5 Use an appropriate method of sequence and order in lists	Cognitive	Best practice
	G1.6 Linguistic differences	Cognitive	Best practice
	G1.7 Case	Cognitive	Best practice
	G1.8 Text directionality	Cognitive and perceptual	Best practice
	G1.9 Linguistic boundaries, text wrappings, and justifications	Cognitive	Best practice
	G1.10 Consider presentation (font size, font style, line spacing, word spacing) of text for the Chinese language	Perceptual	Jin, Zhu, and Shen (1988), Cai, Chi, and You (2001), Shen et al. (1988), Shieh et al. (1997)
	G1.11 Be aware that variations exist within the same language	Cognitive	Goonetilleke, Rau, and Shih (2002)
G2 Presentation and layout	G2.1 Provide natural layout orientation for information to be scanned	Cognitive and perceptual	Lau, Goonetilleke, and Shih (2001), Goonetilleke, Rau, and Shih (2002)
	G2.2 For menu design, provide orientation compatible with the language being presented	Cognitive	Dong and Salvendy (1999), Shih and Goonetilleke (1997, 1998)
	G2.3 Text direction, labeling, and scrolling	Cognitive	Best practice
G3 Graphic design	G3.1 Icons designed and tested well in one region may not be accepted by people in other regions	Perceptual and affective	Piamonte, Ohlsson, and Abeysekera (1997), Piamonte, Abeysekera, and Ohlsson (1999), Tham and Tan (1999), Plocher, Garg, and Wang (1999)

(continued)

APPENDIX A (Continued)
Cross-Cultural Web Design Guidelines Summary

Category	Guidelines	Dimension	Supporting Research/ Best Practice
	G3.2 When designing icons, provide a combination of text and picture	Cognitive and perceptual	Choong and Salvendy (1998), Kurniawan, Goonetilleke, and Shih (2001)
	G3.3 Carefully examine the textual component in the graphics on the Web intended for global audience	Cognitive	Best practice
	G3.4 Pay attentions to graphics with culture-specific meanings or associations such as visual puns, verbal analogies, gestures and body parts, religious symbols, animals, and colors	Affective and cognitive	Luna, Peracchio, and deJuan (2002)
	G3.5 Considering reading and scanning direction of the target users	Cognitive	Horton (1994)
G4 Cultural preference for colors	G4.1 Use consistent color associations	Cognitive and perceptual	Courtney (1986), Kyrnin (2002), Liang et al. (2000), Luximon, Lau, and Goonetilleke (1998), Röse, Liu, and Zühlke (2001, 2002)
	G4.2 Notice not clear-cut color associations	Cognitive and perceptual	Courtney (1986), Kyrnin (2002), Liang et al. (2000)
	G4.3 Notice special color associations	Affective	Courtney (1986), Kyrnin (2002)
G5 Information Architecture, Searching, and Interaction	G5.1 Information should be organized in association with target user's cultural traits	Cognitive	Choong (1996), Choong and Salvendy (1999), Luna, Peracchio, and deJuan (2002)
	G5.2 The navigation of a Web site should be designed to meet users' expectation by clearly indicating where they are, where they have been, and what they can access to and how they can proceed	Cognitive and functional	Luna, Peracchio, and deJuan (2002)
	G5.3 Provide searching mechanisms	Functional	Morkes and Nielsen (1997), Nielsen (1997), Zhao (2002)
	G5.4 Provide both search engine and Web directory to support different needs of users	Functional	Fang and Rau (2003)
	G5.5 Culture influences the satisfaction of users for searching tasks	Affective	Fang and Rau (2003)
	G5.6 Provide possible outcomes and results of operations as much as possible for Asian users or users in high uncertainty avoidance cultures	Cognitive and affective	Marcus and Gould (2000), Plocher, Garg, and Krishnan (1999)
	G5.7 Provide extra navigational aids for Japanese, Arabic, and Mediterranean users or users in high-context communication style	Functional and cognitive	Rau et al. (2003b), Hall (1984), Hofstede (1980), Plocher, Garg, and Krishnan (1999)
G6 Mobile Web	G6.1 Design the Web to suit tailored operator portals or design the Web to suit full Web browsers depending on local users' preference and infrastructure	Functional	Kaikkonen (2008)

Section XI

Emerging Technologies

34 Mobile Interface Design for M-Commerce

Shuang Xu and Xiaowen Fang

CONTENTS

34.1 INTRODUCTION

Mobile devices, also known as handheld devices or personal digital assistants (PDAs), refer to portable computing devices that support certain information input and output methods. The first mobile device was introduced by Bell Labs in 1946 to enable wireless communications. The era of mass use of wireless communications, however, started 30 years later after Martin Cooper of Motorola had his patent for cellular handset technology granted in 1975 (see a timeline overview of Motorola [n.d.] history). From the mid 1980s to the mid 1990s, mobile devices were categorized as either communication oriented (e.g., cell phones) or information centric (e.g., PDAs, PalmPilot, and Pocket PC). Early adopters of these advanced gadgets were mostly people who had to spend their business hours away from a conventional office environment (Jones and Marsden 2006).

Over the past two decades, the difference between the two categories of mobile devices has been blurred. With Wireless Application Protocol, cell phones are able to provide mobile access to a variety of online and offline information. The development of mobile processors, equipped with rapidly increasing storage capacities, has allowed cell phone users to be entertained by a growing list of features in addition to basic communication means. On the other hand, handheld information devices have evolved into smart phones that support features such as wireless communications, voice recognition, and multimedia messaging. In the near future, many sophisticated functionalities will become built-in standards on mobile devices to keep users informed, connected, and entertained any time, any where.

Regardless of the astonishing development of wireless technology, the inherited usability issues on mobile devices have been confronting user experience researchers with interaction design challenges. Small screen size, limited computing speed, battery capacities, etc., have hindered, and will continue to hinder, the improvement of mobile user experience. This chapter is organized as follows. We begin with a review of the various interaction techniques proposed to address the usability constraints on mobile devices. We then discuss the impact of Internet and Web applications on users' mobile experiences, followed by a summary of design strategies and guidelines identified in previous research studies. In this chapter, we endeavor to outline the current state of the development of mobile interaction design with a focus on users' Internet experiences for design researchers in academia and designers in industry.

34.2 MOBILE INTERACTION

Mobile devices such as cell phones and PDAs offer convenience to access information and entertainment on the go but also bring design challenges to information input and output techniques. The common form factors of mobile devices constrain screen space to a small fraction of what is available on a desktop. Visual displays on mobile devices can easily become cluttered with information and widgets. The traditional small keypad has also significantly limited the efficiency and effectiveness of information input on mobile devices. Tasks such as text entry, navigation, information selection, and retrieval become inevitably difficult for mobile users.

Over the past decades, various techniques have been proposed to improve users' interaction experiences on mobile devices. This section provides an overview of traditional and advanced information input and presentation techniques available on mobile devices.

34.2.1 INPUT TECHNIQUES

Mobile devices play an important role in supporting people's everyday activities. Since the first text message was sent to a mobile phone in 1992 (GSM Association 2000), mobile text message transmission rates have exploded. According to the Semi-Annual Wireless Industry Survey conducted by the Cellular Telecommunications Industry Association (Semi-Annual Wireless Industry Survey, n.d.), about 3.5 billion text messages were sent every day in 2008 by over 270 million wireless subscribers in the United States alone. The international standard keypad on mobile phones places three or four English letters on each of the eight keys (Grover, King, and Kuschler 1998). For text entry, a mobile keypad is clearly not as efficient as a traditional QWERTY keyboard. The uptake of mobile messaging indicates that humans will continuously try to adapt tools to their needs by overcoming usability difficulties on these tools.

34.2.1.1 Multitap and T9

Introduced as a wireless communication tool, mobile devices inherit the interface design of landline phones to support basic dialing functions. A traditional mobile keypad has only 12 keys (0–9, *, and #) that can be used for text entry, with additional navigation keys (four arrow keys and one Select key) and functional keys (power on/off, dial, etc.), as shown in Figure 34.1. Apparently, this form factor is not designed as an optimal interface for typing.

The most commonly used methods for text entry on mobile devices are Multitap and the predictive text input system T9. A Multitap system divides the alphabet onto eight separate keys from numbers 2 to 9. The system records the number of times an individual key is pressed and matches it to one of the letters assigned to that key. For example, letters T, U, and V are assigned to the number key 8. If the 8-key is pressed three times in quick succession, the system recognizes the user input as the letter "V."

Although Multitap is an intuitive and straightforward typing method commonly used by mobile users, it has inherent usability issues. As more than three or four letters are assigned to each key, the letter segmentation becomes difficult especially for words that contain successive letter. For example, a key sequence of 222222 can be interpreted as "BAC" (22-2-222) or "CAB" (222-22-2). A time-out process is introduced to resolve this problem. To differentiate two letters that share the same key, a user is required to wait a short period of time before pressing the same key for the next letter. If a user presses another key, the system will immediately cancel the time-out timer and accept the currently displayed letter. Multitap has been criticized for being slow and cumbersome in general. Meanwhile, it is hard to predict the workload based on the number of letters, because the typing performance of Multitap varies greatly depending on the placement of letters within their individual key mapping sequences. It may involve drastically different amounts of

FIGURE 34.1   Keypad of Motorola RAZR v3.

key presses to enter different words that have the same number of letters. For examples, using Multitap, it takes five key presses to type the word "game," but it takes 14 key presses to type "loss."

To enable faster and easier text entry, predictive text input systems were proposed to allow the user to press one key for each letter. The most widely adopted predictive text input system was developed by Tegic Communications and licensed to mobile phone manufactures under the name T9 Text Input, and thus often referred as T9 (Grover, King, and Kuschler 1998). In order to determine the intended word entry, T9 compares the sequence of ambiguous keystrokes to words in a large database. This database usually contains word usage frequency information. Therefore, while there might be a list of words that match the given key sequence, the system will first present the most frequently used word among these candidates. A typical interface design for T9 also displays the remaining word candidates and allows the user to press an arrow key to select the intended word. For example, the key presses of 4663 may display both word candidates "home" and "good" for the user to choose from. The user can use Multitap to add a word that is currently not in the database. This added word will be available for selection if the user tries to type it using T9 the next time.

In order to provide a superior performance of word prediction on predictive text input systems, it is critical to define the right database size. The size of the word database affects both the chances that the desired word is not on the list and the number of presses a user needs to navigate through the word candidate list. Using a smaller database, it reduces the chances of potential conflicts between words that share the same keystroke sequence but increases the probability of missing the intended word in the database. On the other hand, a larger database may include all potential word candidates but may require the user to select among a longer list of candidate words. In contrast to their intuitive but cumbersome experience with Multitap, T9 users will typically experience a learning curve before getting used to this typing method. A commonly encountered usability problem with T9 is the word stability and visual feedback. As the system cannot predict whether the user has finished entering a word or if she is still in the process of constructing a word, T9 always matches the current key sequences to a completed word in its database. This mechanism results in the fact that the displayed word on the screen changes with each additional letter that is entered. James and Reischel (2001) indicate that if explained correctly, a user will get used to the visual feedback because most words stabilize very quickly. Also, T9 users rarely look at the candidate words displayed on the screen until they have finished typing the entire word. Instead, most users concentrate on identifying the next key to press.

Comparing the typing performances using Multitap and T9, there have been various estimations. On average, 14.9–20.8 words per minute (wpm) has been expected for Multitap and 17.6–40.6 wpm has been expected for predictive text system (Dunlop and Crossan 2000; Silfverberg, MacKenzie, and Korhonen 2000). These speed rates pale when compared to the average typing speed of 80 wpm on a standard QWERTY keyboard (Card, Moran, and Newell 1983). One of the recent interface design trends is to apply the ubiquitous QWERTY keyboard to the mobile paradigm. Despite the advantages such as familiarity and faster typing speed, there are also usability issues with this solution. First, a hard QWERTY keyboard is bulky and will increase the dimensions of the mobile device. Second, touch typing on a soft QWERTY keyboard requires more physical effort and often results in lower efficiency and accuracy. Last, but definitely not least, many mobile devices are designed for single-handed operation, while the key arrangement on a QWERTY keyboard is designed for typing with both hands (MacKenzie and Soukoreff 2002). Therefore, researchers have been actively seeking for alternative input techniques to improve the effectiveness of mobile users' interaction experiences.

34.2.1.2 Touch-Based Interaction

Touch-based interaction enables mobile users to select and enter information directly and intuitively. To support this interaction paradigm, manufacturers have equipped some mobile devices with pen-based touch screens. These devices allow mobile users to select and enter information with a stylus. Some of the research has investigated different selection techniques on pen-based mobile interfaces (Ren and Moriya 1997, 1998, 1999). While using a stylus improves the accuracy of mobile interaction, there are usability problems associated with this interaction style (Karlson, Bederson, and Contreras-Vidal 2007; Pascoe, Ryan, and Morse 2000). First, there is no tactile feedback when selecting targets on a mobile screen with a stylus. Second, retrieving the stylus of a pen-based device takes time. It also requires the user to hold the device with one hand and manipulate the stylus with the other hand. Additionally, using a stylus to interact with a mobile device becomes cognitively demanding and physically difficult when a user is moving or performing multiple tasks simultaneously (MacKay et al. 2005). Many pen-based devices utilize sensing technologies that also track touch input. This makes one-handed operation possible by supporting direct touch input with a thumb. Unfortunately, pen-based mobile interfaces usually contain small graphical targets, which make finger interaction slow and error prone (Vogel and Baudisch 2007).

Ideally, mobile interaction should support single-handed operations, with the thumb being used for interacting with targets on the screen. Direct thumb touch input on mobile devices is intuitive and fast (Forlines et al. 2007). With the introduction of the Apple iPhone, more and more mobile devices are now fitted with touch screens that are designed for direct finger touch input. While touch is a compelling input modality for mobile interaction, there are three fundamental problems with direct finger touch input. The *occlusion* happens when the selecting target is smaller than the size of the finger contact area, which further prevents users from receiving visual feedback or confirmation. This problem is more pronounced for one-handed operation because the thumb pivots around the joint and can hide half of a mobile screen. *Accuracy* is also a problem commonly encountered during touch screen interactions. Early investigation on touch screens showed that to reach 99% selection accuracy, it required the target size to be 26 mm^2 for seated users and 30 mm^2 for standing users (Hall et al. 1988). Parhi, Karlson, and Bederson (2006) reported that 9.2 mm is the minimum size for targets to be accurately accessible with the thumb. Some mobile devices, such as the iPhone, rely on a limited set of large buttons at the price of reduced number of interactive targets. This approach, therefore, is not appropriate for the increasing number of features and functions available on mobile devices. Last, target *accessibility* becomes problematic as the borders of the touch screen are difficult to reach with the thumb because the morphology of the hand constrains thumb movements (Roudaut, Huot, and Lecolinet 2008).

Researchers have proposed various solutions to address the above usability issues. Moving the input surface to the back of the device has been considered as an effective solution to the occlusion problem, as implemented in *BehindTouch* (Hiraoka, Miyamoto, and Tomimatsu 2003), *HybridTouch* (Sugimoto and Hiroki 2006), and *Under the Table* (Wigdor et al. 2006). However, it introduces a new occlusion since the device prevents the user from seeing her hands. *LucidTouch* (Wigdor et al. 2007) addresses the new occlusion problem by making the user's hands and fingers visible while she is touching the back of the device. But it is not currently available on any mobile devices. Inspired by the *take-off* paradigm (Potter, Weldon, and Shneiderman, 1988), *OffsetCursor* (Sears and Shneiderman 1991), *MagStick* (Roudaut, Huot, and Lecolinet 2008), and *Shift* (Vogel and Baudisch 2007) were proposed to alleviate the accuracy problem by displaying the cursor or target area at a distance from the contact point. But these approaches further limit the accessible area on the display. *Thumbspace* (Karlson and Bederson 2007) was designed to address the accessibility problem in the border areas, but it does not prevent occlusion in the center of the screen. While many different solutions have been proposed to improve the usability of mobile touch interaction, individually none of these existing techniques has successfully addressed all of the problems.

Recently, capacitive sensing technology has been introduced to overcome the limitations of mobile touch screens. By adding a touch-sensitive layer on top of the buttons, it is possible to capture users' finger touch input on the keypad. Because the interaction takes place on the mobile keypad, it does not occlude the visual presentations on the screen and neither is it necessary to increase the graphic target size for accurate selections. Several touch-sensing user interfaces (e.g., the *TouchTrackball*, the *Scrolling TouchMouse*, and the *On-demand Interface*) have been developed and demonstrated by Hinckley, Hinckley, and Sinclair (1999) and Hinckley et al. (2000). These input devices use unobtrusive capacitive sensors to detect users' touch input without requiring pressure or mechanical actuation of a switch. Clarkson et al. (2006) add a layer of piezo-resistive force sensors to support continuous measurement of the force exerting on each key. It enables applications such as smooth zooming into image in proportion to the force on a button. But the sensors do not distinguish between pressure received in a pocket from the touch of fingers. Rekimoto et al. present a series of capacitive sensory-enhanced input devices such as SmartSkin (Rekimoto 2002), ThumbSense (Rekimoto 2003), SmartPad (Rekimoto, Ishizawa, and Oba 2003), and PreSense (Rekimoto et al. 2003). In these papers, the authors discussed various ideas for potential mobile applications to provide preview information before execution, to enable text input with touch-sensitive keypad, and to recognize finger motions or gestures as different input commands in addition to conventional keypad inputs. Using capacitive sensing technology to enhance traditional input techniques can potentially alleviate mobile interaction constrains. However, the touch interaction gestures must be designed with caution. It is suggested that touch-based mobile interaction design focus on (1) conforming to the conventional interaction and navigation paradigms users are familiar with and (2) providing means for users to discover and adopt innovative interaction techniques (Xu 2010).

As discussed above, a large body of research work has focused on introducing touch-based interaction to mobile devices. The current tactile interaction technology provides a direct and intuitive means for users to interact with their mobile devices more effectively. However, further research is expected to address fundamental usability issues, such as visual occlusion, selection accuracy, and target accessibility, associated with touch-based mobile interaction.

34.2.1.3 Speech Recognition Technology

Most interaction designs of mobile devices are based on the graphic user interface (GUI) design of desktop computing systems (Lumsden and Brewster 2003). In contrast to desktop users who usually sit in front of a computer and focus their full visual attention on the screen, mobile users are often in motion. When a user is interacting with her mobile device while walking, running, or driving, it is difficult to allocate her attentional resources to the visual interface (Brewster 2002). As mobile devices grow smaller and as in-car computing platforms become more common, traditional interaction methods seem impractical and unsafe in a mobile environment such as driving (Alewine, Ruback, and Deligne 2004). Many mobile device manufacturers are seeking hands-free and eyes-free interaction solutions that can overcome the mobile usability constraints.

Speech-based user interfaces provides a natural and efficient interaction paradigm that is similar to human-to-human communication, with a significantly improved input speed close to 125–150 wpm (Feng, Karat, and Sears 2005). To deliver a new level of convenience and accessibility in human–computer interaction, automatic speech recognition (ASR) has been utilized in a wide range of applications (Rudnicky, Lee, and Hauptmann 1994). These applications vary from dictation systems on personal computers, customer service call routing, to voice-based information inquiry systems. Currently, speech recognition technology embedded in mobile devices supports simple voice commands such as dialing a number, searching and playing songs, or composing messages. With the advancement of user-independent speech recognition technology, voice-based mobile interface is expected to enable information entry via speech dictation (Basapur et al. 2007).

Owing to the limited memory and processing power, it has been a challenge to improve the performance of speech recognition engines on mobile devices. Automatic speech recognition typically involves extensive computation. Compared to desktop computers, mobile phones have modest computing resources and battery power. Network-based speech recognition has been proposed as a potential resolution, where the recognition engine resides on the server. However, speech signals transferred over a wireless network tend to be noisy with occasional interruptions. Additionally, network-based solutions may not be well suited for applications that need to retrieve information stored on the mobile device (Marcussen 2003). An alternate solution to improving the accuracy of speech recognition on mobile devices is to utilize the information of a user's everyday activities. For example, using contacts information retrieved from the address book or the recent call history can help the recognition engine identify intended speech input more quickly and accurately. Context-aware systems often use probabilistic detection algorithms that require a large amount of training data (Intille et al. 2004), which makes them less applicable for mobile devices. Improving users' voice interaction experience on mobile devices is also challenged by the mobile environment, where there is usually higher background noise and the user may not be able to speak loud and clear in public.

Considering the growing needs for voice-based mobile interfaces and the status of current mobile speech recognition technology, researchers have been actively exploring different error handling mechanisms. In general, user-initiated error correction methods can be categorized into four types: (1) respeaking the misrecognized word, (2) replacing the wrong word by typing, (3) choosing the correct word from a list of alternatives, and (4) using multimodal interaction that supports various combinations of the above methods. As one of commonly used repair strategies in human-to-human communication, respeaking is believed to be the most intuitive correction method (Larson and Mowatt 2003; Robbe, Carbonell, and Valot 1994). However, given the same speaker, device, and environmental conditions, respeaking does not necessarily improve the accuracy of the mobile recognition. Some researchers indicate that eliminating alternatives that are known to be incorrect can increase the recognition accuracy of respeaking (Ainsworth and Pratt 1992; Murray, Frankish, and Jones 1993). A multimodal error correction strategy has been proposed by Sturm and Boves (2005), with a speech overlay that recognizes pen and speech input. This interface is perceived to be more effective and less frustrating as the participants feel more in control. Other research (Oviatt 1999) also shows that redundant multimodal (speech and manual) input can increase interpretation accuracy on a map interaction task. Although the multimodal error correction seems to be most effective among other techniques, it is

not an optimal solution for the mobile devices. Users usually have limited attentional resources in mobile contexts (such as driving) where speech interaction is mostly appreciated. Thus, it is difficult for a user to type or make selections to correct speech recognition errors on mobile devices.

34.2.1.4 Gesture Input

A number of gesture-based interaction techniques have been introduced to mobile devices recently. Gestural interfaces make an important departure from the traditional interaction paradigm available on mobile devices. Using gestures as a new modality of information input allows mobile users to effectively and intuitively express themselves with an improve perception of playfulness. In general, the research on gesture interaction design can be summarized in two categories: designed gestures and natural gestures (Cassell 1998; Hummels and Stappers 1998).

With designed gestures, a user is expected to follow a set of predefined gesture modes for information entry. Depending on the intuitiveness and complexity of the designed gestures, it may require some effort to learn and memorize. Some successful examples include a series of handwriting recognition systems such as Graffiti on Palm OS (Butter and Pogue 2002) and Motorola's WisdomPen. Traditional gesture-based handwriting recognition systems allow users to write from left to right. This becomes difficult on mobile devices because users will run out of space after a few letters. Graffiti introduces a different approach by enabling users to write letters on top of each other, lifting the pen between letters. This requires that each letter must be completed in one continuous stroke. As a result, users need to learn new ways to write letters such as "f," "i," "t," "x," etc. Figure 34.2 shows the alphanumeric gesture strokes used in the original graffiti.

The new way of writing was quickly adopted by Palm users because it was easy to learn and remember. It also significantly improved recognition accuracy because of the simplified gesture strokes. The recent advancement of gesture recognition technology has unlocked the power of mobile

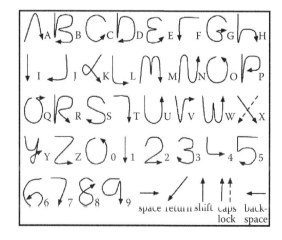

FIGURE 34.2 Alphanumeric gestures defined in graffiti. This figure is available from the Wikimedia Commons, http://en.wikipedia.org/wiki/File:Palm_Graffiti_gestures.png.

input techniques to support handwriting in character-based languages such as Chinese, Japanese, and Korean.

Natural gestures, on the other hand, focus on understanding how users visualize digital data and provide interaction paradigms that are similar to how users interact with physical objects. A large body of research work has explored the intuitiveness and effectiveness of gesture interaction on mobile devices. Table 34.1 summarizes a list of commonly used gestures and their recommended applications on mobile devices (Bhandari and Lim 2008; Fagerberg, Stahl, and Hook 2003; Huot and Lecolinet 2006; Linjama and Kaaresoja 2004).

Gesture-based interaction improves mobile experiences by empowering users to explore and accomplish tasks in ways the conventional mobile modalities do not allow. It is suggested that in the near future, users should be able to customize gesture inputs on their mobile devices to make interaction more natural, intuitive, and enjoyable (Bhandari and Lim 2008). However, there are potential limitations of gesture interaction. First, users must memorize each gesture and reproduce them accurately for the device to recognize the intended inputs. Second, the growing number of functionalities available on mobile devices makes it difficult, if not impossible, to map individual control to a unique gesture entry.

Researchers in academia and industry are continuously investigating novel methods of interaction to counter the limitations posed by small screens and keypads of mobile devices. In the next section, we will discuss various information presentation techniques proposed to alleviate the usability issues associated with mobile interaction.

34.2.2 Information Presentations

As we mentioned previously, one major usability constraint with mobile devices is that the screen real estate is significantly reduced as compared to desktop computing systems. Mobile screen cannot be enlarged as the devices must be made physically portable to fit into the hand or pocket easily. Being light and compact, mobile devices present a challenge for interface designers, particularly on the design of information presentation.

Mobile devices have limited display size, resolution, memory, and processing power. Applying conventional graphical techniques of interface design on desktop computers to handheld devices can often result in legibility issues such as small text that is difficult to read, cluttered graphical elements, and minimum contextual information or status feedback. To tackle the information presentation challenge on mobile devices, researchers have investigated various interaction techniques. In this section, we discuss these research efforts in three areas: (1) innovative visual presentations, (2) context awareness, and (3) multimodal information presentations.

34.2.2.1 Innovative Visual Presentations

The past decade has witnessed a drastic improvement of the hardware performance of mobile devices, from display resolution, storage capacities, computing speed, and battery life to network connection bandwidth. It is probably safe to predict that evolutionary breakthroughs of mobile hardware development in the next decade will allow users to access more advanced features and functions on the go. These highly anticipated improvements, however, are limited by the physical dimensions and weight of a portable device. Portable devices are expected to easily fit into pockets or held in one hand. In the near future, mobile devices may become even smaller and lighter to further facilitate portability so that a user can wear it on her wrist or carry it on a keychain.

On one hand, the physical size of a mobile device limits the maximum size of its screen display, as long as the display is embedded as part of the device. On the other hand, reductions of the size of displayed information are limited

---

TABLE 34.1
Commonly Used Gesture Modes and Their Mobile Applications

Gesture Mode	Characteristics	Mobile Applications
Tapping on the screen	The simplest and easiest gesture, similar to clicking on a button	• Media control (play/stop, pause/resume) • Take pictures on camera phone • Target selection/activation in general
Brushing off the screen	Mimicking physical maneuver as if the user is actually touching the digital data	• Data browsing (next/previous image) • List scrolling • Move a pan window on a target that is partly displayed on the current screen
Tilting the device	To control the position of the target, as if it is an object trapped in the device	• Move highlight/cursor in four directions • Continuous scrolling (up and down) • Games (control movement of an object)
Shaking the device	Randomization	• Shuffle a play list • Advanced image editing effect (e.g., blur)
Touch and drag diagonally	Similar to resizing a Windows object	• Zoom in/out of an image, photo, or map • Resize the displayed dimension • Connect/separate two digital objects
Moving the device toward/away from the user	Simulating the visual or auditory experience with a physical object in real world	• Zoom in/out of an image, photo, or map • Turn audio volume up/down

by humans' capability of discerning small target sizes on the screen. Kamba et al. (1996) have suggested that a practical threshold of legibility for most users lies between 3.25 mm (9 points) to 4.17 mm (12 points). As the screen shrinks in size, inevitably less information may be displayed to keep the target size above the legibility threshold. An effective mobile interface design must balance between the two opposing needs: the need to further reduce screen size for better portability and the need to keep the screen large enough so that a user can possibly recognize and interact with the presented information. A wide range of information visualization theories have been explored by researchers and designers in order to maximize the available screen space for content display on mobile devices. In the remainder of this section, we will categorize and review different visual presentation techniques that have been proposed to improve the effectiveness of mobile information display.

34.2.2.1.1 Perspective Presentations

One approach to present a large amount of information in an easy-to-understand and easy-to-interact fashion is to simultaneously reveal the detailed information and its context. Cone Tree is an example that uses rotation animation to illustrate information structure. It presents data hierarchies in three-dimensional (3D) cone-shaped graphics. User selects areas of interest by rotating a cone to bring the data into the brightest light and closest angle. It is found that this interaction paradigm significantly reduces cognitive load and expedites users' performance from several seconds to less than a second (Robertson, Mackinlay, and Card 1991). Similarly, Hyperbolic Geometry (Lamping, Rao, and Piroli 1995) displays the focal point within a data hierarchy in a large bounded space and its relevant context in a smaller bounded space. Perspective Wall (Mackinlay, Robertson, and Card 1991) indicates relations between different information sections within the same document on two adjacent walls. The semantic relations are preserved by the relative positions of these walls. On the basis of the perspective presentation theories, Wang and Sajeev (2007) proposed a Roller interface to make Web browsing more effective on mobile devices. In recent years, the above information visualization techniques have been seamlessly fused with advanced interaction

technologies and delivered a new level of mobile interaction experience. Figure 34.3 shows the Cover-Flow presentation on the iPhone, where a user can touch and browse through the digital music albums.

34.2.2.1.2 Zooming Presentations

Zooming is another effective strategy that has been used for displaying a large volume of data on a limited screen area. This technique uses human capacity for selective omission of information and aggregation of experienced information into abstract forms. Fisheye distortion techniques were initially introduced by Spence and Apperley (1982) to present information in the way that the focus items were enlarged and peripheral items were shrunk. On the basis of this theory, Furnas (1986) further proposed the "degree of interest" by calculating the relevance value of each information item in the space. The relevance value is used to decide the size and visibility of that item. Fisheye visualizations have been successfully applied to a wide range of computing interfaces (Benderson 2000; Benderson, Meyer, and Good 2000; Rao and Card 1994; Sarkar and Brown 1992). Flip Zooming was proposed by Björk et al. (1999) to display Web pages and personal information on mobile devices. This presentation model consists of one medium-sized focus page and several small pages in the periphery that can be used to facilitate navigation. DateLens (Benderson et al. 2004) is a mobile calendar interface that uses Fisheye representation to display tabular data with customizable views and integrated search capability.

Zooming interaction technique allows the user to focus on and zoom into any part of the content. The smooth animation during the zoom maintains the semantic connection between the focused information and its context. Galaxy of News (Rennison 1994) combines the zooming techniques with dynamic restructuring of data hierarchies to improve information browsing performance. Zooming interaction has been widely applied to map applications on mobile devices. Figure 34.4 illustrates the map views at different zooming levels on the Blackberry Storm.

Another successful application of the zooming technique on mobile devices is ZenZui, patented and launched by Microsoft (Microsoft 2007). This zoomable interface organizes Web and personal content in 36 tiles arranged in a grid layout. These tiles can be panned and zoomed in or out using the device's touch screen or keypad. When activating one of the tiles, it shows a small interface that follows the same directional pad control for user interaction.

34.2.2.2 Context Awareness

Context awareness has recently become a popular research field in mobile usability. The idea behind context awareness computing is to take advantage of environmental information such as current location, activities, time, temperature, and so on. On the basis of this additional knowledge, computing devices are able to present information dynamically and intelligently to meet users' needs. For examples, using information retrieved from the calendar application, a mobile device

FIGURE 34.3 (See color insert.) Cover flow interaction on Apple iPhone.

FIGURE 34.4   **(See color insert.)** Map application on Blackberry Storm.

can automatically switch to vibration mode when the user is in a meeting. Location-aware applications enable the mobile device to recommend restaurants based on where the user currently is. Context awareness technology allows the computing systems to present relevant information to a user when it is needed. It is thus considered as a promising approach to resolve the interaction design challenges on mobile devices.

Context generally refers to the information that can be used to characterize the situation of an entity (Dey 2001). A piece of information becomes context if it is used to interpret the current usage situation. Therefore, a system is context aware if it collects and interprets contextual information to provide relevant functionality or services to the user. The amount of the information mobile users are able to save on their devices is growing exponentially in recent years. Mobile devices have become capable of processing a wide range of different types of information, from contacts, notes, ring tones, images, video clips, and messages, to Web site bookmarks. Additionally, the rapid increase of storage capacity allows users to store large amount of information on their mobile devices. This presents a great opportunity for researchers and designers to create intelligent mobile user experience based on the abundant contextual resources. Meanwhile, it brings challenge to define and retrieve relevant content when and where this information is needed by the user.

Research in mobile context-awareness has been heavily focusing on system architectures, sensing technologies, and context recognition (Mäntyjärvi 2003). Among the fewer researchers that have concentrated on the usability issues associated with context-based interaction design, Sorvari et al. (2004) have summarized mobile contexts into the following categories: (1) Physical context such as time, location, temperature, etc.; (2) Social context such as social groups, friends, social activities or events, etc.; and (3) Mental context such as emotional feelings, mood, current activities, etc. Dey (2001) suggests that a context aware application support three types of functionalities on mobile devices: (a) Presentation of information and services to a user, (b) automatic execution of a service for a user, and (c) tagging metadata to support future retrieval. Häkkilä and Mäntyjärvi (2006)

have proposed design guidelines to improve the usability of context-aware mobile applications, as categorized below:

- *Uncertainties* are inevitable in context-based applications. Designers shall deal with caution whether to ask for user's confirmation before automatic execution.
- *Personalization* is the ultimate goal of introducing context aware applications. Designers shall take into account user's individual preferences, as well as how mobility will affect these preferences.
- *Privacy* becomes a concern when user's personal information might be automatically collected and shared. Unwanted information being pushed to mobile device can be perceived as a violation to user's personal space.
- *User control* needs to be protected and reserved in context-based applications. Regardless of the advance of intelligent user interfaces, users should have control of what they need over the device. Designers shall provide different levels of automation to suit individual user's needs.
- *System status* shall always be visible so that a user is aware of what information has been collected and/or shared, how decision was made, recognition accuracy of the contextual metadata, etc.

Introducing context awareness technology to support information presentation improves the effectiveness and efficiency of mobile interaction. However, there is apparently a trade-off between privacy intrusion and user benefit. For example, Global Positioning System (GPS) embedded in mobile phones identifies the exact location of the device carrier and has been previously used for surveillance against criminals. Combined with 3G wireless networks, GPS has become one of the most desired features on mobile devices, especially for travelers and people in emergency. Therefore, context aware interfaces will be likely embraced by mobile users if the perceived benefit outweighs the potential loss of personal information.

34.2.2.3 Multimodal Information Presentations

Nonvisual information presentation has become an interesting research area for mobile interaction designers. As we discussed in Section 34.2.1.3, it is difficult for mobile users to allocate undivided visual attention to the screen display when they are on the go. In many mobile situations, visual presentation alone may not be appropriate or adequate. On the other hand, humans are capable of processing information perceived from different sensory channels simultaneously. While looking at the primary task, a user is also able to attend to information presented in other modalities using her remaining sensory registers. This justifies the benefits of developing multi-modal information presentations on mobile devices. For example, a user can "read" a received text message by listening to the text-to-speech output of that message while keeping her mobile phone in the pocket. A tactile vibration reminds her of an incoming phone call without interrupting her auditory perception of the text message.

Designing effective multimodal information presentations can be challenging because of the interference and distraction during the processing of concurrently received information. Previous research in cognitive psychology has shown that visual and auditory perceptual processing are closely linked (Eimer 1999). Mental integration of disparate information from different modalities can cause heavy cognitive workload. Switching attention among sensory channels may also result in degradation of task performance (Cook et al. 1997). Wickens, Gordon, and Liu (1998) have concluded that the amount of shared attentional resources is one of the main factors that determine how well people can divide their attention between tasks. Baddeley (1986) suggests that imagery spatial information and verbal information can be concurrently held in different working memory systems. On the basis of the Baddeley's model, Xu et al. (2008) have proposed a dual-modal presentation to convert texts into distinct graphics and speech output on small screen devices. Their evaluation results indicate that participants' task performance and satisfaction have been improved on the proposed interface.

A large body of multimodal interaction research has concentrated on providing effective cross-modality feedbacks. Mobile devices are now equipped with audio and vibrotactile capabilities that enable a user to receive information using senses of hearing and touch. Tactile or auditory feedback alone can often go unnoticed, considering the mobile context where the device is usually kept in the pocket or purse and the background is noisy. Cross-modal presentation uses different sensory channels to deliver the same information. The concept of sensory substitution refers to using one modality to deliver information that is usually presented in a different modality. On the basis of this concept, feedbacks or alerts can be encoded in both auditory and tactile presentations, which enable a mobile device to deliver the information in the modality channel that is most appropriate for the context (Hoggan and Brewster 2007). New terms such as earcons (Brewster, Wright, and Edwards 1993) and tactons (Brown, Brewster, and Purchase 2005) are created to refer to the counterparts of icons presented in auditory and tactile modes, respectively. Hoggan and Brewster have investigated users' recognition performance of cross-modal icons presented with variable parameters such as the rhythm, the roughness, and the spatial location. Their results suggest that it is possible to present cross-modal feedbacks in alternative modalities that allow users to recognize these information cues effectively with minimum training.

In summary, multimodal interfaces have the potential to alleviate the usability constraints on mobile devices by incorporating a number of interaction technologies (e.g., speech recognition, eye or head tracking, 3D gestures). The integration of multiple input and output methods transforms the mobile interaction into a natural and optimal paradigm that is similar to human communication. Implementation of multimodal interfaces on mobile devices, however, is more difficult than on desktop computing systems. Because hardware resources available on mobile devices are limited in various respects, as we discussed earlier. In the next section, we will review the current status of Web-based applications on mobile devices, as well as how advanced interaction technology and wireless network have jointly contributed to the improvement of mobile experience.

34.3 WEB APPLICATIONS ON MOBILE DEVICES

The advance of wireless technology in the past two decades has enabled mobile users to access the World Wide Web any time, any where with their handheld devices. Mobile access to the Internet is an exciting addition to the current use of the Web because it satisfies users' information needs when they are on the go. Wireless broadband networks have recently experienced drastic development. With a 3G networks coverage of 97% in the United States (CostQuest Associates 2008), network carriers have been moving forward with the deployment of 4G networks such as LTE and WiMax that allow bandwidth of over 100 Mbit/s. Equipped with broadband connections, higher resolution, larger screen display, and faster computing speed, mobile devices have evolved into a powerful platform for information retrieval and sharing. In this section, we will discuss how mobile users' Internet experiences have been transformed by the recently development of wireless technology. Section 34.3.1 reviews user experience research of social network and communication on mobile devices. Section 34.3.2 provides an in-depth discussion of the current usability research on mobile commerce.

34.3.1 SOCIAL NETWORKS AND COMMUNICATION

In the past few years, online social networks (OSN) have become a phenomenal success on the Internet. Different from traditional Internet applications such as e-mail and instant messaging, OSNs provide additional value beyond facilitating one-on-one communications. An online social network creates a virtual community that is made of individuals or groups connected via one or multiple types of interdependency, such as friendship, common interests, and other relationships. Members of a social network community can post and share online profiles that often map to their real life interests, personalities, or activities. By sharing members' social information, OSNs enable users to exchange information, view friends' social networks, and look for friends who may share the same interests or background. In the remaining of this section, we discuss the characteristics of Web-based online social networks in Section 34.3.2.1, followed by a review of mobile social network applications in Section 34.3.2.2.

34.3.1.1 Online Social Networks

Web-based online social networks, such as MySpace, Facebook, LinkedIn, and Meetup, have attracted a massive audience over the past few years. Golder, Wilkinson, and Huberman (2007) reported that about 90% of the college students in the United States use one or more online social network on a regular basis. It is not surprising that Facebook and MySpace are listed among the top ten most frequently visited Web sites on the Internet (Dwyer, Hiltz, and Passerini 2007).

Similar to real-world social networks, the quick spread of information enables the user base of an online social network to undergo exponential growth. In addition, large OSNs have recently opened platforms for developers globally. For example, since May 2007, developers have been able to create applications that either augment Facebook's functionality or appear as front-end to other Web-based services. Facebook informs a user what applications her friends have installed and used. In return, these third-party applications have brought significant traffic to Facebook. Within a year, there have been about 866M installations of 16.7M distinct applications developed by approximately 200K people who utilize the platform (Nazir, Raza, and Chuah 2008).

A large body of user experience research has probed the nature of online social networks. Some have examined users' concerns of identity and privacy. Dwyer, Hiltz, and Passerini (2007) indicate that Facebook users have expressed a high level of trust and willingness to share personal information, knowing their profiles are searchable by other Facebook members. Their results conform to previous research findings of collaborative work and virtual community (Wellman 1996; McKenna and Green 2002) from a sociological perspective. In general, online social networking focuses on building connections with people whom you may want to stay in touch with. Most OSN Web sites foster this kind of relationship development by allowing users to track other people of their communities. Such surveillance function can be categorized as "social searching" or "social browsing" (Lampe, Ellison, and Steinfield 2006). Social searchers may use an OSN site to learn more about people they have previously met offline. Social browsers, on the other hand, are more interested in finding people or groups that they would like to establish an offline connection with. In the example of Facebook, members can use the site as a tool for maintaining and intensifying relationships initiated offline. Other OSNs, such as Meetup,

facilitate searching for new friends and building online connections with a strong potential of turning this new relationship offline in real life.

Motivated to understand the reasons behind the popularity of online social networks, Hart et al. (2008) have evaluated the Facebook experience from users' perspectives. The results of their evaluation show that Facebook performs poorly when measured by traditionally defined Web usability heuristics (Nielsen and Molich 1990). Its apparent usability issues in areas such as consistency, following standards, error prevention, and recognition cannot explain its commercial success. In comparison, 85% of the participants rated positively in their user evaluation on the perceived ease of use on Facebook. The authors also indicate that those most frequently used Facebook features (such as browsing/sharing photos, checking/creating profiles, reading/writing on the Wall, finding/adding friends, etc.) are highly associated with positive experiences such as fun/playfulness, enjoyment, excitement, self-expression, and curiosity. The authors attribute the success of online social networks to the following factors:

- The pleasure and connectivity offered by OSNs satisfies human needs for effective social interactions.
- Curiosity and fun have also emerged as an important element in the definition of positive user experience.
- Identification and self-expression allow users to individually represent themselves via online personas that can be searched by and shared with other friends.
- Serendipity refers to the positive surprise of making discoveries. OSNs bring serendipity into user experiences by encouraging people to reconnect to distant friends.
- Privacy remains a concern among the OSN users. However, it is often perceived as the cost of being able to access other members' information and recent activities.

The conflict between traditional usability evaluation results and user experiences summarized in the above study calls for a more holistic method of assessment that may require redefining usability to measure new user experience supported by modern technology.

34.3.1.2 Mobile Social Network Applications

The surge of social network phenomena on the Internet and the increasing number of available social network services have been a driving force of the recent development of mobile social network application. Typically, a mobile social network application builds virtual communities that allow members to share common interests and social activities via mobile devices. A large number of various mobile social network applications have been proposed in the past few years, such as MobShare (Sarvas et al. 2004), Plog (Gossweiler and Tyler 2004), MobiSNA (Gou et al. 2009), JuiceCaster

(JuiceCaster 2009), and EZMoBo (Lai 2007). These mobile applications support social networking by allowing users to exchange information such as text messages, instant chatting, blogging, photos, and video clips.

Using social network applications in a mobile environment is essentially different from accessing traditional OSN Web sites via a mobile device. A mobile environment can be defined by its major components such as mobile devices, networking technologies, protocols, middle ware, wireless services, and users in mobile context. Social networking in a mobile environment allows mobile nodes to communicate with each other without connecting to a centralized OSN server. Mobile users can take advantage of the unique characteristics of mobile social networking to share information with other mobile users in a peer-to-peer manner. Such opportunistic exchange of content is made possible by networking technologies including Bluetooth, Near Field Communication (NFC), Wireless Local Area Networks (WLAN), and General Packet Radio Services (GPRS). These technologies allow mobile devices to detect and communicate with other devices within discoverable distance. MobiClique (Pietiläinen et al. 2009) is proposed as a mobile social application that allows users to build and maintain their networking through opportunistic encounters in real life. Relying exclusively on opportunistic connections between mobile devices, MobiClique enables communications based on proximity and social compatibility in the physical world. For two neighboring devices, if the user profiles share certain amount of commonality in predefined interests or relationship, both users will be alerted and they can further decide whether to connect to each other and exchange information.

Research on the mobile social networking applications has mostly focused on the design of system architecture. Very few studies have examined the current development of mobile social networking from users' perspectives. Conducting user-centered research experiments in this area can be challenging because it requires a set of critical elements such as a proper social environment, a simple yet sufficient social networking application, and strategy to handle the limitations cast by mobile contexts. Pietiläinen et al. (2009) assert that the physical communication based on real-life contacts and relationship will remain as the essential part of humans' social lives. The unique capabilities introduced by virtual social communities will allow online and mobile social networking to become an important entity complementary to the traditionally defined social communications.

34.3.2 Mobile Commerce

In contrast to the astonishing development of social networking, the convergence of mobile Internet and wireless communications has not yet resulted in major growth in mobile commerce (m-commerce). Consumer adoption of m-commerce has been slow, even in countries that have broadly adopted wireless technology (Anckar and D'Incau 2002). The enterprise and business use of wireless technology holds greater promise, but it demands transformation of business process

and infrastructure (Kalakota and Robinson 2001). Poor usability of mobile Internet sites and wireless applications for commerce activities stands out as a major obstacle for the adoption of mobile solutions. Such difficulty discourages users from accessing mobile Internet sites (Chan et al. 2002) or choosing m-commerce as a distribution channel (Shim, Bekkering, and Hall 2002). Even with the latest 3G phones in Japan, consumers still find the small screen display and small buttons on these devices difficult to use (Belson 2002).

Researchers suggest that interface developers need to consider the interaction among user tasks, form factors, and purposes of applications (Chan et al. 2002). Application developers should also consider the interaction between the context of the environment and the device (Johnson 1998). Some researchers question whether the existing user interface design guidelines are still applicable for mobile application development or if new ones should be created (Tarasewich, Nickerson, and Warkentin 2002). A comprehensive methodological and comparative framework for evaluating the usability of m-commerce applications is also necessary (Siau, Lim, and Shen 2001).

The objective of this section is to provide a critical analysis of usability issues confronting the interface design, development, deployment, and adoption of m-commerce applications. We focus primarily on the Internet-based solutions in North America. In addition, we examine the unique characteristics of mobile technology that affect user interface design, discuss the incorporation of user interface in the development of m-commerce applications, and suggest topics for future research and development.

34.3.2.1 Defining Mobile Commerce

Mobile commerce broadly refers to the use of wireless technology, particularly handheld mobile devices, and mobile Internet to facilitate transaction, information search, and user task performance in consumer, business-to-business, and intra-enterprise communications (Chan and Fang 2001, 2003). These activities are typically grouped under the term "m-business." Such a broad definition is consistent with several proposed m-commerce frameworks. One such framework presents 12 classes of m-commerce applications, ranging from retail and online shopping, auction, mobile office, and entertainment to mobile inventory emphasizing the potential of mobile B2B and intra-enterprise applications (Varshney and Vetter 2001). Another framework groups m-commerce into goods, services, content for consumer e-commerce, and activities among trading partners (Kannan, Chang, and Whinston 2001). M-commerce applications can also be grouped as communication and interaction, value added service, information and data access, remote control and decision support, transactions, and entertainment (Lehner and Watson 2001). While successful m-commerce business models in North America are rare, until recently, there has been a belief in the potential for wireless technology in enhancing a broad range of commerce processes and activities, particularly in the B2B and intra-enterprise arena.

There are two visions for the potential and opportunities of m-commerce (Waters 2000). One perspective argues that the mobile, wireless channel should be viewed as an extension of the current e-commerce channel or part of a company's multichannel strategies for reaching customers, employees, and partners. The second and more radical view suggests that mobile commerce can create new markets and new business models. In recent years, the selective adoption of wireless technology has indicated that the first perspective is more realistic.

Several studies provide empirical support for the first perspective. Consumers have shown relatively low willingness to use m-commerce, but adopters of e-commerce are more likely to embrace this new technology (Anckar and D'Incau 2002). Another study shows that major e-commerce sites implement their mobile Internet sites as an extension of wired e-commerce to support existing customers (Chan et al. 2002). The perceived difficulty of use can affect the consumer's choice of m-commerce as a distribution channel (Shim et al. 2002). These findings suggest that in a multichannel environment, m-commerce supplements e-commerce instead of becoming a substitute for e-commerce.

Enterprise and business applications of m-commerce technologies seem to hold greater promise because it is easier for companies to standardize and customize the applications and the devices to enhance current work processes. Except for the retail industry sector, most industries view m-commerce as being vital for growth and efficiency strategies, but not necessarily for generating new revenue (Ernst and Young 2001). Most mobile applications implemented in the United States are for business purposes, extending or enhancing existing work processes and business models geographically (Jarvenpaa 2001). More comprehensive integration of the wireless platform in an enterprise requires significant structural transformation (Kalakota and Robinson 2001). It will be challenging for companies to undertake significant process redesigns and breakthrough strategies necessary for transitioning the organization into a multidevice, multichannel computing environment. Such transformation increases an organization's capability for real-time interaction with its customers, employees, and suppliers.

An essential goal of m-commerce is to search for mobile values for individual users (Keen and Mackintosh 2001). Anckar and D'Incau (2002) presented a framework that differentiates the value offered by the wireless Internet technology (wireless value) from the value arising from the mobile use of the technology (mobile value). Wireless values are best represented by convenience, cost savings, and cell phones. Users can attain the wireless value by using mobile devices to perform tasks available through e-commerce. Services that deliver strong mobile values make m-commerce a dominant channel. These services meet the following five needs:

1. Time critical needs and arrangement
2. Spontaneous needs and decisions, such as auction, e-mail, and news
3. Entertainment needs

4. Efficiency needs and ambitions
5. Mobility related needs

In their survey of consumers in Finland, they found that, at present, consumers are most interested in services with high mobile values that meet time spontaneous and time critical needs. Furthermore, e-commerce users are more likely to adopt m-commerce services. These research findings have implications for cross-channel coordination and highlight the relationship between e-commerce and m-commerce.

34.3.2.2 Usability Research for Mobile Commerce

Usability gauges the quality of a user's experience in interfacing with a product or system, be it a Web site, a software application, mobile technology, or any user-operated device. Generic usability principles (Nielsen 2001) have guided the development of e-commerce applications:

- **Ease of learning**: How fast can a novice user learn the interface sufficiently well to perform basic tasks?
- **Efficiency of use**: Once an experienced user has learned to use the system, how fast can he or she accomplish tasks?
- **Memorability**: Can a past user remember enough to use the system more effectively next time?
- **Error frequency and severity**: How often do users make errors, how serious are these errors, and how easy is it to recover from an error?
- **Subjective satisfaction**: How much does the user *like* using the system?

Industry consultants incorporated these principles into a set of guidelines for e-commerce Web site design. Guidelines developed by the Nielsen/Norman Group (Nielsen et al. 2000a, 2000b, 2000c, 2000d) focus on the design of category pages, the checkout and registration process, product pages, and user's trust. Several unpublished research papers (Hurst and Gellady 2000; Hurst and Terry 2000; Rehman 2000) also examined customer experience in the e-commerce process and suggested a broad set of design guidelines:

- *Home page*. Web pages should be clean and not cluttered with text and graphics. Horizontal scrolling should be avoided.
- *Navigation*. Text on the links or buttons should be descriptive and self-explanatory. Links to another product related Web site should be direct.
- *Categorization*. Products should be categorized meaningfully with no more than three levels in depth.
- *Product information*. Accurate, consistent, and detailed descriptions of products should be provided along with full pictures. Inventory information and related charges should be presented up front. The size of products should be shown in a measurable and comparable way.

- *Shopping cart.* There should be a link directing the customer back to the page he/she left in order to resume shopping.
- *Checkout and registration.* The vendor should only ask for necessary and meaningful information, such as name and address, omitting marketing questions. Customers should be allowed to browse the site without logging in.
- *Customer service.* Customers should be provided with a 1-800 telephone number on every page of the site.

Contrary to these broad guidelines, formal usability studies tend to focus on specific tasks or specific user behaviors in the e-commerce context. One study found that enjoyment in using an electronic system and peer norms significantly contributed to the online shopping experience (Henderson, Rickwood, and Roberts 1998). It was possible to manipulate the visual design factors of the customer interface in order to induce a target emotion, such as trustworthiness (Kim and Moon 2000). The use of a combination of navigation features (neighborhood, top, and index) generated the optimal link structure and could increase the degree of shopping pleasure and convenience (Kim and Yoo 2000). These studies did not validate any design guidelines and focused on very narrow aspects of e-commerce sites, which typically are feature and information rich.

There have been quite a few usability studies on wireless applications, focusing on various aspects such as form factors, use contexts, and sociotechnical issues. Diverse form factors offer different functionalities and have different interface requirements. Chan et al. (2002) systematically reviewed 10 wireless Web sites—ranging from travel, financial services, retail, news, and Internet portals—across three form factors: wireless Palm, WAP phones, and Pocket PCs. They found that user tasks for the wireless sites were designed with steps similar to the wired e-commerce sites and primarily for experienced users. Many usability problems, such as long downloads and broken connections, information overload, and excessive horizontal and vertical scrolling, are common to all three form factors. They pointed out that interface design flaws are platform independent, but the more limitations imposed on the form factors, the more acute the design problems become.

Context factors have special impact on wireless usability. A study conducted by Kim et al. (2002) reveals that three use context factors—hand (one or two hands), leg (walking or stopping), and colocation (alone or with others)—result in different usability problems. Users were more likely to report problems of site structure when they accessed wireless sites with one hand instead of two hands. When accessing wireless sites while moving rather than stopping, users experienced more difficulty with site representation. Content presentation also poses more usability problems for those who were stopping or alone. Therefore, the design of user interface has to consider various use contexts.

On the basis of the results of a study of 19 novice wireless phone users who were closely tracked for the first 6 weeks

after service acquisition, Palen and Salzman (2002) described the wireless telephony system as having four socio-technical components: hardware, software, "netware," and "bizware." They indicated that each of these four components has to be designed as user-friendly. This research suggests a systems-level usability approach. In another study, Perry et al. (2001) presented a study of mobile workers that highlighted different facets of accessing remote people and information anytime, anywhere. They identified four key factors in mobile work: the role of planning, working in "dead time," accessing remote technological and informational resources, and monitoring the activities of remote colleagues. Mobile users access information from different sources and often experience a wide range of network connectivity. Ebling, John, and Satyanarayanan (2002) conducted a study addressing the importance of translucence in mobile computing systems. Their findings suggest that with the presence of the illusion of connectivity provided by translucent cashing even when network performance is poor or nonexistent, novice users performed almost as well as experienced users. Kim and Kim (2003) proposed an adoption model for the mobile Internet consisting of seven critical factors. An online survey was conducted on the basis of this model to compare continuers and discontinuers. The survey results show that discontinuers are more sensitive to usefulness and social influences in using mobile Internet services, while continuers are more sensitive to ubiquitous connectivity. Lee and Benbasat (2003) proposed a reference framework for m-commerce interfaces. This framework consists of seven design elements: context, content, community, customization, communication, connection, and commerce. Venkatech, Ramesh, and Massey (2003) assessed usability guidelines for Web and wireless sites and found that ease of use was significantly more important in wireless contexts.

34.3.2.3 User Tasks

In a stationary environment, users can fully focus on their tasks or spend hours exploring a regular Web site; rich information is far more important than time. M-commerce assumes that users may access the Internet or wireless applications away from a desktop computer either on the move or stationary. Services that emphasize mobile values, time critical and spontaneous needs, such as checking flight schedules, checking stock prices, submitting bids for auction, add more value for m-commerce users. Mobile users can spare only limited time and cognitive resources in performing a task. Three issues emerge in the design of tasks for m-commerce applications.

First, in developing m-commerce applications, one should determine what tasks are essential to mobile users and what tasks are suitable for wireless applications. Because of the user's mobility, the interaction between the user and the mobile device is usually very short. It is important to design tasks in such a way that users can perform them in a timely manner. Currently, most wireless Web sites only allow users to perform simple tasks such as checking flight status and searching for a movie. The latest Gartner research report on consumer mobile applications (Gartner Inc. 2009) suggests adoption of the following mobile applications:

- Is rising: mobile virtual worlds, mobile sports and fitness, mobile portal VoIP (voice over Internet protocol), and rich communication suite.
- Is at their peak: VoIP wireless WAN (wide area network), mobile health monitoring, NFC (near-field communication) payment, and mobile application stores.
- Is sliding through of disillusionment: mobile advertising, mobile widgets, bar code on mobile, voice-to-text conversion services, mobile instant messaging, mobile ticketing, mobile gambling, mobile presence, VoIP over WLAN (wireless local area network), WAP/WEB payment, mobile TV broadcasting, mobile banking, and mobile communication.
- Is climbing up to a plateau of productivity: mobile browsers, mobile e-mail services, mobile music streaming, SMS (short message service) payment, full-track music download, mobile TV streaming, location-based services, and personal navigation.

Second, how to handle differences in interface design for expert and novice users should be considered. Many wireless Web sites currently assume that users have experience with regular Web sites and are familiar with the task flow (Chan et al. 2002). The wireless site is basically a simplified version of its regular counterpart. This strategy makes sense in that it allows existing customers to transfer knowledge about a Web site to the wireless site. However, this strategy fails to accommodate the needs of novice users who have never visited the regular Web site. For example, many wireless sites ask users to log in on the main screen without instructing them how to create a new account.

Third, the best use of location-based technology to support user task performance for m-commerce presents additional challenge. Mobile networks can pinpoint a caller's location and supply that information to providers of geographically targeted services. The question is when to use the location information. On the basis of the location theory, Mennecke and Strader (2001) suggested that location-based technology should be used for applications for geographical differentiation and for low-involvement purchases that do not require extensive information search. Further research is needed in the proper use of location-based technology to enhance mobile tasks.

Tasks that can be performed on handheld devices are different from what can be executed on regular computers. As suggested by Anckar and D'Incau (2002), mobile applications must address the following five most common mobile values: (1) time critical needs and arrangement; (2) spontaneous needs and decisions, such as auction, e-mail, and news; (3) entertainment needs; (4) efficiency needs and ambitions; and (5) mobility related needs. In a separate study, Xu et al. (2003) identified five factors contributing to user's preference of tasks to be performed on handheld devices: perceived usefulness, perceived ease of use, perceived playfulness,

complexity, and perceived security. This study also found that users would like to

- Perform tasks that will meet their mobile needs such as sending/receiving e-mails, and checking flight status
- See an easy and user-friendly interface with considerations of all the constraints imposed by handheld devices and mobility
- Use handheld devices to entertain themselves in their free time

In addition, users are concerned about the security of online transactions.

In a subsequent study, Fang et al. (2003a) conducted a survey to validate the five factors contributing to user's preference of tasks to be performed on handheld devices: perceived usefulness, perceived ease of use, perceived playfulness, complexity, and perceived security. The results from this study showed:

- Perceived usefulness, perceived security, perceived playfulness, and perceived ease of use were positively correlated with user intention to use handheld devices.
- Perceived usefulness, perceived security, and perceived playfulness were the three most important factors affecting user intention to use handheld devices. Together they accounted for 30% of the total variance in user intention.
- Tasks providing communication functions and useful information are more suitable for handheld devices.
- Users are less inclined to use mobile channels to conduct transactions, even for tasks with added mobile values, such as stock trading and online banking.

Benou and Bitos (2008) investigate the environment in which mobile applications operate, identify possible categories of them, and propose the following guidelines for their development process:

- *Management-related guidelines*: (1) Managers should encourage the generation of novel ideas regarding the m-commerce solutions. (2) The undertaking of innovative m-commerce solutions does not only aim at generating revenues but also diffusing the reputation, acquiring knowledge, and gaining competitive advantage. (3) The roles and the interests of the participating entities should be carefully determined, as well as the values that they will share in "a win to win situation." (4) Companies should form inter-organizational ties during the development of m-commerce solutions to complement their cognitions and competitive advantages.
- *Development process-related guidelines*: (1) The alternative technologies, in each mobile case,

should be carefully compared with each other and the suitable tools, platforms, protocols, devices and networks should be selected. (2) In the development of m-commerce information systems, prototyping development approaches should be adopted. (3) For the development of m-commerce information systems, rapid development approaches should be selected. (4) During the m-commerce application's design, techniques of separation of context and adaptivity concerns from the core functionalities should be searched out. (5) The structure of a mobile program should be described in terms of tasks and subtasks and their navigation order. (6) The logic of an application should be materialized through distinct services that can be bound into a presentation unit. (7) For the user requirements gathering, repetitive and scenario-driven approaches should be adopted. (8) Field test and evaluation planning should be carefully designed and executed. (9) The mobile applications should be refined and adapted to the working environment through an iterative process and repeated feedback loops.

- *User-related guidelines*: (1) During the mobile application design it is crucial to identify the potential target user groups, their age, cultural background, cognitive capabilities, as well as their information and communication needs. (2) The widespread adoption of mobile applications is highly dependent on their usability in terms of ease to learn and ease to use. (3) The mobile application should offer clearly value-added services, such as saving time and increasing efficiency. (4) Simple solutions will work better than complicated solutions. (5) Only the information that is absolutely critical should be displayed. (6) During the mobile applications design the minimization of clicks should be sought. (7) Convenient user interfaces, such as voice or sound interfaces, should be investigated. (8) There should be a matching between the mobile applications services and the tasks that the mobile user tries to complete. (9) Personalized services, tailored to users' preferences, could be provided.

- *Context related guidelines*: (1) The context can be captured and exploited in order to provide adaptive mobile applications. These applications recommend content and services to the user or provide suitable information and services in each situation. (2) During the mobile applications' design performance issues, regarding the computing devices and the network should be considered. (3) Mass information dissemination can be achieved through mobile networks to users with same interests or in the same location, for example, stock prices or advertising messages. (4) The mobile applications should provide consistent functionality and user interfaces through the different devices in which they run.

After investigating three companies' adoption and use of mobile data solutions for sales automation, freight tracking, and service support, Doolin and Ali (2008) find that the relative advantage of the technological innovation and the information intensity of the company were the most important factors influencing adoption. Other factors that appeared to influence adoption included the compatibility of the technology with the company's business approach, the presence of top management support, and the degree of organizational readiness. Environmental factors such as competition within the industry or business partner influence seemed less influential for these pioneers of mobile technology use in supply-side activities. On the basis of a content analysis of online user reviews that was followed by structural equation modeling, Gebauer, Tang, and Baimai (2008) found four factors to be significantly related with overall user evaluation, namely functionality, portability, performance, and usability.

Taylor et al. (2008) tracked 11 early U.S. mobile Internet adopters and uncovered through content analysis distinct motivations and behaviors.

- Motivations: awareness, time management, curiosity, diversion, social connection, and social avoidance.
- Behaviors: status checking, browsing, information gathering, fact checking, in-the-moment, planning, transaction, and communication.

The Delphi surveys conducted by Xu and Gutierrez (2006) suggest that the short message service (SMS) and a killer portfolio were the two most likely killer applications of m-commerce. Additionally, four factors—convenience, ease of use, trust, and ubiquity—were identified as the most important to m-commerce success. Yang (2009) found that the primary factors associated with resisting mobile banking technologies include concerns over system configuration, security, and basic connection fees.

34.3.2.4 Future Research Directions

Wireless technology for m-commerce may evolve in two areas. For intra-enterprise and business-to-business uses, wireless technology provides location-aware and mobility-aware solutions for mobile workers. The range of possibilities is broad. Content distribution can be integrated with the enterprise systems. Context-based applications, interfaces, functionality, and even devices can be customized according to the mobile tasks and user groups. This approach makes application development, deployment, and integration easier to manage. In contrast, it is far more challenging to manage the design, development, and deployment of wireless applications for customers. The technology's capability for personalization seems to be the strongest argument for mobile Customer Relationship Management (m-CRM) services to enhance customer retention (Chan and Lam 2004). A careful mapping of tasks, data, form factors, and the CRM process will become essential for user interface design. The consumer e-commerce Web sites will need to focus on the selection of tasks that are most suitable for the wireless channel and demonstrate mobile values, especially for experienced users. Such mapping process requires a good understanding of the CRM strategy, user's preference, and constraints imposed by a mobile environment. For enterprise adoption, consolidating the wireless platforms and form factors will facilitate interface design. In either case, research to improve usability for mobile commerce is essential.

34.4 DESIGN FOR MOBILE INTERNET

As discussed in the previous section, mobile devices and wireless technology have advanced rapidly and provided users with mobile access to a wide range of online information. While mobile devices in general provide a convenient medium for keeping people connected, informed, and entertained with information retrieved from the Internet, designing mobile interaction and applications for Web content remains challenging. In this section, we will summarize key issues in the mobile Web interface design, followed by reviews of current design strategies and guidelines proposed to improve users' Internet experiences on mobile devices.

34.4.1 Usability Issues

Mobile Internet refers to the use of the Internet via handheld devices. Mobile Internet is different from stationary Internet in two main aspects: (1) various contexts and (2) limited system resources. Several usability constraints have made users' mobile access to the Internet difficult. Web content rendered and displayed on small screens can often result in confusing presentation and cumbersome interaction on mobile devices. Time-consuming and error-prone information entry methods used on mobile devices also significantly degrade users' mobile Internet experiences. Compared to wired connections, downloading Web information to mobile devices is still much slower. We have provided an in-depth discussion on general usability issues of mobile interaction techniques in Section 34.2. In the remainder of this subsection, our discussion will focus on (1) Web content and (2) site structure in the interface design for mobile Internet.

34.4.1.1 Web Content

Regardless of the significantly improved quality of display resolution, the small screen size continues to limit mobile devices' ability of presenting information as effectively as desktop computers. Rich interaction experiences of Web browsing have been supported by technologies such as flash animation, Ajax, high-speed broadband connection, and larger flat-screen monitors. In comparison, mobile Internet experience is much less convenient and enjoyable. Most Web sites on the Internet today have not catered to support mobile platforms. Accessing regular Web pages on mobile devices often causes frustrating experiences as the user spends large amount of time and effort on extensive scrolling. It is challenging to prepare content for a wireless application or to convert a regular Web site into an effective presentation on mobile devices. Web design guidelines usually embrace the development of rich information presentation and interaction. For mobile devices, however, designing a wireless Web site should focus on the effectiveness of the user interface by prioritizing information, simplifying site structures, and utilizing mobile interaction techniques. To support substantial amount of content, the mobile Web application should provide excellent navigation design, as well as effective help and search features (Lee, Schneider, and Schell 2004).

34.4.1.2 Site Structure

Comparing to their Web browsing on desktop computers, mobile users spend much less time on wireless applications. Content organization is therefore critical to the improvement of the efficiency. Information in most wireless applications is organized into a hierarchy. Early studies have discussed the hierarchy design issue of computer menu systems (Jacko and Salvendy 1996; Jacko, Salvendy, and Koubek 1995). These studies generally assume the response time of each step is constant and users have some basic training before using the system. However, wireless applications require much more time and effort to connect to the server and to download a page. A flatter hierarchical structure with fewer steps would allow users to review more options in the same step and to locate the desired information in less time. Some recent studies on user preferred methods of navigation on small screens indicate that users favor scrolling over pagination (Rabin and McCathieNevile 2006). It is recommended in the W3C "Mobile Web best practices" that a mobile Web site should not contain too many links on one page, but each page must be easily reachable. Balancing the depth and breadth of the hierarchy design is critical for the usability of a mobile Web site. A theoretical framework of menu design for wireless applications is yet needed in order to develop effective and efficient interface design for a particular Web application on mobile devices.

Overall, designing Web content and applications for mobile devices is challenging. In the following subsections, we will review the current design strategies to improve mobile Web browsing (in Section 34.4.2) and design guidelines identified in previous usability research on mobile Internet (in Section 34.4.3).

34.4.2 Improving Mobile Web Browsing

Mobile browsing has become an important means of Internet access. However, Web pages are designed for desktop computer displays. Mobile information searching and browsing can be frustrating when a user has to scroll both vertically and horizontally on the small screen of a mobile device to find the desired content. Three general solutions, as discussed in details below, are commonly used to cope with this constraint and to improve the usability of Web browsing on small-screen devices: (1) device-specific authoring, (2) sever-based automatic re-authoring, and (3) client-side navigation techniques.

34.4.2.1 Device-Specific Authoring

The first solution is straightforward. It requires all mobile Web pages to be redesigned and reconstructed using standard formats such as Wireless Markup Language (WML) prior to being presented on mobile devices. The major disadvantage of this method, however, is the cost of developing a mobile version of the entire Web site. Moreover, it is very difficult and complex to maintain consistent content for the mobile and regular versions of the same Web site. Therefore, this solution has not been well adopted by the vast majority of Web sites.

34.4.2.2 Server-Based Automatic Re-Authoring

The second solution is more practical. It employs a proxy server to automatically adapt existing Web pages from regular presentation to their mobile versions. This method allows Web content to be converted on-the-fly for mobile devices and therefore does not impose additional cost on the design and development of the original Web site. A wide variety of approaches have been deployed to improve the efficiency and accuracy of such server-based Web content conversion. Early research prototypes of automatic re-authoring techniques focused on page reformatting and page scaling. Examples of page reformatting method include the Power Browser (Buyukkokten et al. 2000), which removes images and unwanted space in the presented textual summary, and the WEST browser (Björk et al. 1999), which uses visualization technique to break Web pages into a stack of tiles that fit the mobile screen. Prototypes of page scaling usually create an image map thumbnail of a Web page based on its semantic content. A user needs to zoom to a block of the thumbnail and read the scaled content, which can be poorly legible on the mobile device. Heidi and Baudisch (2005) have proposed a solution that combines the above page scaling and text summarization techniques. Their prototype allows a user to read the full text when she zooms to a scaled area of the Web page.

The Related Content Detection, an approach based on semantic conversion, identifies closed related information to form a topic and then presents the formed topics in a user-friendly format for mobile devices. This approach can often be categorized as either structure based or layout based. The structure-based approach analyzes the Web content by parsing the HTML elements (Gupta et al. 2003; Baluja 2006), while the layout-based approach focuses on the segmentation of the Web page layout (Chen et al. 2005). However, if the Web pages were unstructured or the layout templates were inconsistent, an inaccurate or false detection could happen and reduce the efficiency of these approaches. Ahmadi and Kong (2008) have proposed a hybrid approach that combines the analysis of both the structural and visual layout information of a Web page. The result of their evaluation of this approach shows that the browsing usability is significantly improved.

Other researchers have integrated the automatic re-authoring with really simple syndication (RSS) feeds (Blekas, Garofalakis, and Stefanis 2006) to improve the reformatted Web content. RSS is a family of XML file formats and is increasingly used in news and other information-centric Web sites for information updates and summary. The proposed technique can scan a Web site that has RSS feed, remove multimedia information (e.g., image, video, and animation), and present the filtered information in a simpler and more comprehensive format on mobile devices.

Overall, the automatic re-authoring techniques do not require the collaboration of Web page authors and therefore are more applicable than the device-specific authoring. Because the layout design is significantly changed after the reformatting, however, these techniques have often been criticized for the difficulty in recognizing the original layout design and the failure in providing consistent cross-platform browsing experiences.

34.4.2.3 Client-Side Navigation Techniques

The third solution, client-side navigation technique, is made possible by the recent development of mobile device technology. With this solution, the entire Web content is delivered to a mobile browser, which further decides how to render and present the page properly for the specific mobile device. The Opera Mobile Browser (http://www.opera.com/mobile/) is an application of small screen rendering (SSR) technology. By presenting the Web content in a single column, this technology can properly adjust the width of the displayed area and save users from horizontal scrolling. Web authors can use CSS to affect the way their Web content is rendered by SSR technology.

Although single-column view supported by mobile browsers facilitates the reading process, it also results in a larger amount of vertical scrolling. In order to reduce users' efforts on vertical scrolling and also to provide them with appropriate visual context, researchers have explored the applications of advanced visualization technologies and mobile interaction techniques. Such mobile browsers typically provide a zoomed-out version of the Home page of a Web site or a summarized overview of the Web content on the Home page. The miniature version of a Web page requires a user to further zoom in for a close-up look at the desired content.

Figure 34.5 shows an example of the zoomed view of a regular Web page displayed on a touch screen with a resolution of 480 × 360 pixels, 184 ppi. Although the touch-based gesture interaction makes zooming and panning much easier on mobile devices, the legibility of the content on the zoomed-out Web page is very low. Screen resolutions of mobile phones vary from 128 × 128 to 480 × 360 pixels. A Web site with rich content usually requires scrolling even on a computer monitor with a resolution of 1024 × 768 pixels. Resizing a regular Web page to fit into a mobile screen often significantly reduces the readability of the page content. When the headlines on a Web page become unreadable, mobile users are forced to use hunt-and-peck strategies that can be time consuming and frustrating. Baudisch et al. (2004) propose a collapse-to-zoom solution to address the legibility issue. In addition to the zooming capability, collapse-to-zoom allows a user to collapse areas that are identified as irrelevant, such as menus, archived information, and online advertisements.

FIGURE 34.5 (**See color insert.**) Zoomed view of a Web page on a 480 × 360-pixel touch screen.

The collapsed Web content allows the remaining information to expand in size with more details displayed. This filtered presentation helps a user identify relevant information more effectively.

Client-side browsing methods require the code of an entire Web page to be downloaded to the mobile device. Therefore, some disadvantages of this strategy include (1) the memory requirement on the device side for storing and executing the code and (2) the potential cost of downloading rich Web content, because mobile users are usually charged based on the amount of data transmission.

34.4.3 GUIDELINES FOR MOBILE INTERNET DESIGN

While resolution and color quality of mobile devices will be continuously improved, the limited screen size will always be a usability constraint for mobile interaction designers. Hardware technology development has increased the number of pixel per inch for the small displays, which does improve the legibility of texts and the sharpness of images. However, limited by humans' visual perception, extremely small presentation of text or graphics on mobile devices will be useless. In order to provide mobile users with truly ubiquitous access to the Internet, design guidelines addressing potential usability problems on mobile devices are essential.

Early usability studies on mobile Internet experiences have focused on resolving the design constraints imposed by bandwidth limitations and small screen size of handheld devices. Along this line, Buchanan et al. (2001) have summarized design guidelines based on their research on mobile Internet usability. To improve users' Internet experiences on small-screen devices, they suggest the following usability guidelines to mobile content authors and interface designers:

- Provide direct and simple access to desired Web content;
- Use simple hierarchies to minimize page-to-page navigation;
- Reduce the amount of vertical scrolling by simplifying the displayed information; and
- Reduce the number of keystrokes a user is expected to perform.

After systematically reviewing 10 wireless Web sites on three different mobile form factors, Chan et al. (2002) recommended eight design guidelines: (1) avoid scrolling, (2) use a flat hierarchy, (3) design a navigation system consistent with a regular Web browser, (4) design a back button, (5) provide a history list, (6) provide indication of signal strength, (7) reduce user's memory load, and (8) limit the search scope to improve search efficiency. On the basis of the findings from another cross form factors study, Fang et al. (2003b) proposed the following user-centered design guidelines, as summarized in Table 34.2.

Users' mobile Web browsing experience was evaluated in comparison to desktop Web browsing in a recent usability study (Shrestha 2007). Results from this study indicate the mobile Web browsing experience is not yet satisfactory as users expect similar experience as on a desktop. On the basis of this study, Shrestha has recommended that the following considerations be applied to limit scrolling and improve readability of the Web pages designed for mobile browsers:

- The font and background colors shall be chosen carefully to avoid hampering the visibility of hyperlinks;
- The input fields and form controls shall be labeled properly for easy identification;
- Links shall be grouped appropriately and labeled with a self-explanatory title to facilitate recognition and navigation;
- Text shall be used for links and buttons where possible;
- Use of images shall be reduced as large images increase the page loading time, and small images (as in dimension or resolution) are hard to interpret; and
- Search function shall always be available as a means to avoid time-consuming scrolling or navigation.

Jones and Marsden (2006) indicate that two aspects are involved in the usability of mobile browsing. A successful mobile Web site should enable (1) efficient navigation and (2) easy comprehension of the presented information. They emphasize some high-level requirements, as listed below, to improve the usability of mobile Internet design:

TABLE 34.2
User-Centered Design Guidelines for Mobile Applications

Category	Design Guidelines
Input	• Input on handheld devices must be minimized.
	• Handwriting recognition is faster but less accurate compared to soft keyboard.
Display	• Short and concise displays are preferred on handheld devices. A large amount of graphic and audio/video information is less effective for handheld devices.
	• Features indicating security protection must be prominent on handheld devices.
	• Icons with descriptive and accurate labels are preferred.
	• Hyperlinks and buttons must be prominent.
	• Scrolling should be avoided if possible. If scrolling cannot be avoided, use "page up" and "page down" for efficient scrolling.
	• Items in a drop-down menu should be sorted.
	• Buttons must be clearly separated from each other.
	• Action buttons in a form should not be hidden by scrolling.
	• Widely accepted short names are appropriate for the items in a drop-down menu.
Navigation	• Wireless Web sites must be consistent with regular Web sites.
	• Exit from an application on handheld devices must be visible.
	• A hierarchical information structure is helpful and preferred.
	• A history list is important in the navigation process.
	• Buttons and menus must be presented on the top of a Web page on handheld devices.
Feedback	• For successful submissions or tasks involving multiple steps (pages), confirmation is necessary.

• *Always present an overview of the Web content.* Mobile Internet design shall aim to allow users to quickly identify whether the content seems promising, without futilely navigating through information that does not meet their needs. Users indicate unique reading behaviors on mobile devices: (1) they prefer concise and accurate writing style, (2) they look for summarized content before committing to the details, and (3) they skim-read information so that they can quickly identify focal points and assess relevance. Therefore, an overview can effectively help users locate useful or interesting information during mobile browsing.

• *Provide focused and direct access.* Mobile users often prefer direct access to navigate to the target content. Several commonly used strategies are (1) provide navigation schemes that allow users to move purposefully toward the information or action they need, (2) present a small set of paths to the content, and avoid overwhelming users by giving access to everything that is available, and (3) allow search functions to enable point-to-point mobile navigation.

• *Reduce scrolling.* Scrolling activities inevitably interrupt users' primary task. Horizontal scrolling shall be eliminated in general. To reduce the amount of vertical scrolling, Jones and Marsden further suggest the following mechanisms: (1) ensure the accessibility of navigation features by placing them in a fixed place near the top of each page, (2) prioritize and locate important information on the top

of the page, and (3) reduce the amount of content on each page and make information task-oriented rather than verbose.

In summary, various design guidelines have been proposed and evaluated by usability researchers. Broadband wireless networks have enabled mobile users to access an increasingly large amount of heterogeneous online information. On the other side, a mobile device has inherent constraints in screen size and system resources to maintain its portability. Standard mobile Internet design guidelines are hard to develop because of the disparity in form factors and wireless communication protocols used by various mobile devices. Thus, improving users' mobile Internet experience will continuously challenge interaction designers and researchers.

34.5 CONCLUSIONS

To address the inherited usability constraints on mobile devices, a range of various information interaction and presentation techniques have been proposed by researchers in the field of mobile user experiences. With the advances in mobile technology, from significantly improved wireless network bandwidth to computing resources, mobile devices have revealed their potential of offering users ubiquitous access to information and entertainment. User interface design for mobile Internet, however, remains a challenge to the usability researchers and content providers. Techniques in automatic Web content conversion such as text summarization, page reformatting, and zoomable presentations have partly alleviated the usability issues with mobile Internet.

Users need to be able to surf the Internet on their mobile devices with an enjoyable experience similar to their regular Web browsing experiences. To improve mobile users' Internet experience, exploring alternative methods for interface design and evaluation will become necessary. We suggest the following four directions for user experience researchers to improve mobile usability in the near future. First, requirement analysis has to focus on the context of mobile users' behaviors and tasks. Contextual inquiry and other methods may be developed to facilitate the understanding of interaction between mobility and usability. Second, usability testing has to be conducted with an understanding of contextual variables besides user behavior. Third, mapping form factors, user tasks, data needs, and content across multiple channels and platforms is important to synchronize content and coordinate functionality in a distributed system. Fourth, user-centered design guidelines for mobile applications will be critical. These research directions require a fresh look at the methodology in use and determine new ways of incorporating user interface design and usability testing for mobile application development.

REFERENCES

Ahmadi, H., and J. Kong. 2008. Efficient Web browsing on small screens. *Proceedings of the AVI'06*, 23–30. New York: ACM Press.

Ainsworth, W. A., and S. R. Pratt. 1992. Feedback strategies for error correction in speech recognition systems. *International Journal of Man–Machine Studies* 36(6): 833–842.

Alewine, N., H. Ruback, and S. Deligne. 2004. Pervasive Speech Recognition. *IEEE Pervasive Computing* 3(4): 78–81.

Anckar, B., and D. D'Incau. 2002. Value creation in mobile commerce: findings from a consumer survey. *Journal of Information Technology Theory and Application* 4(1): 43–64.

Baddeley, A. D. 1986. *Working Memory*. New York: Oxford University Press.

Baluja, S. 2006. Browsing on small screen: Recasting Web-page segmentation into an efficient machine learning framework. In *Proceedings of the International World Wide Web Conference'06*, 33–42. New York: ACM Press.

Basapur, S., S. Xu, M. Ahlenius, and Y. S. Lee. 2007. User expectations from dictation on mobile devices. In *Human–Computer Interaction*, Part II, vol. 4551, *Lecture Notes in Computer Science*, ed. J. Jacko, 217–225. Berlin: Springer.

Baudisch, P., X. Xie, C. Wang, and W.-Y. Ma. 2004. Collapse-to-zoom: viewing Web pages on small screen devices by interactively removing irrelevant content. In *Proceedings of ACM Symposium on User Interface Software and Technology*, 91–94. New York: ACM Press.

Belson, K. 2002. Japan is slow to accept the latest phones. *The New York Times*, April 22, C4.

Benou, P., and V. Bitos. 2008. Developing mobile commerce applications. *Journal of Electronic Commerce in Organizations* 6(1): 6–8.

Benderson, B. B. 2000. Fisheye menus. *Proceedings of ACM Symposium on User Interface Software and Technology, CHI Letters* 2(2): 217–225.

Benderson, B. B., A. Clamage, M. P. Czerwinski, and G. G. Robertson. 2004. DateLens: A Fisheye calendar interface for PDAs. In *Proceedings of the SIGCHI Conference on Human*

Factors in Computing Systems, 90–119. New York: ACM Press.

Benderson, B. B., J. Meyer, and L. Good. 2000. Jazz: an extensible zoomable user interface graphics toolkit in Java. *Proceedings of ACM Symposium on User Interface Software and Technology, CHI Letters* 2(2): 171–180.

Bhandari, S., and Y.-K. Lim. 2008. Exploring gestural mode of interaction with mobile phones. In *Proceedings of the SIGCHI Conference on Human Factors in Computing Systems*, 2979–2984. New York: ACM Press.

Björk, S., L. E. Holmquist, J. Redström, I. Bretan, R. Danielsson, J. Karlgren, and K. Franzénk. 1999. WEST: A Web browser for small terminals. *Proceedings of ACM Symposium on User Interface Software and Technology, CHI Letters* 1(1): 187–196.

Blekas, A., J. Garofalakis, and V. Stefanis. 2006. Use of RSS feeds for content adaptation in mobile Web browsing. In *Proceedings of the International World Wide Web Conference'06*, 79–85. New York: ACM Press.

Brewster, S. A. 2002. Overcoming the lack of screen space on mobile computers. *Personal and Ubiquitous Computing* 6(3): 188–205.

Brewster, S. A., P. C. Wright, and A. D. N. Edwards. 1993. An evaluation of Earcons for use in auditory human–computer interfaces. In *Proceedings of the SIGCHI Conference on Human Factors in Computing Systems*, 222–227. New York: ACM Press.

Brown, L. M., S. A. Brewster, and H. C. Purchase. 2005. A first investigation into the effectiveness of Tactons. In *Proceedings of the Conference on World Haptics*, 167–176. Washington DC: IEEE Computer Society.

Buchanan, G., S. Farrant, M. Jones, H. Thimbleby, G. Marsden, and M. Pazzani. 2001. Improving mobile Internet usability. In *Proceedings of the 10th International World Wide Web Conference on World Wide Web*, 673–680. New York: ACM Press.

Butter, A., and D. Pogue. 2002. *Piloting Palm*, 62–66. New York: John Wiley.

Buyukkokten, O., H. Garcia-Molina, A. Paepcke, and T. Winograd. 2000. Power browser: Efficient Web browsing for PDAs. In *Proceedings of the SIGCHI Conference on Human Factors in Computing Systems*, 430–437. New York: ACM Press.

Card, S. K., T. P. Moran, and A. Newell. 1983. *The Psychology of Human–Computer Interaction*, 259–297. Mahwah, NJ: Lawrence Erlbaum.

Cassell, J. 1998. A framework for gesture generation and interpretation. In *Computer Vision in Human Machine Interaction*, eds. R. Cipolla and A. Pentlan. New York: Cambridge University Press.

Chan, S., and X. Fang. 2001. Usability issues for mobile commerce. In *Proceedings of the 7th Americas Conference on Information Systems*, 439–442.

Chan, S., and X. Fang. 2003. Mobile commerce and usability. In *Advances in Mobile Commerce Technologies*, eds. K. Siau and E. Lim, 235–257. Hershey, PA: Idea Group.

Chan, S., X. Fang, J. Brzezinski, Y. Zhou, S. Xu, and J. Lam. 2002. Usability for mobile commerce across multiple form factors. *Journal of Electronic Commerce Research* 3(3): 187–199.

Chan, S., and J. Lam. 2004. Customer relationship management on Internet and mobile channels: A framework and research direction. In *E-Commerce and M-Commerce Technologies*, ed. C. Deans, 2212–2232. Hershey, PA: Idea Group.

Chen, J., X. Xie, W. Ma, and H. Zhang. 2005. Adapting Web pages for small-screen devices. *IEEE International Computing* 9(1): 50–56.

Clarkson, E., S. Patel, J. Pierce, and G. Abowd. 2006. Exploring continuous pressure input for mobile phones. GVU Technical Report. http://hdl.handle.net/1853/13138 (accessed Oct. 16, 2009).

Cook, M. J., C. Cranmer, R. Finan, A. Sapeluk, and C. Milton. 1997. Memory load and task interference: Hidden usability issues in speech interfaces. *Engineering Psychology and Cognitive Ergonomic* 3: 141–150.

CostQuest Associates. 2008. U.S. 3G Mobile Wireless Broadband Competition Report. http://costquest.com/costquest/docs/CostQuest_3G_Competition_Report.pdf (accessed Nov. 8, 2009).

Dey, A. K. 2001. Understanding and using context. *Personal and Ubiquitous Computing Journal* 5(1): 4–7.

Doolin, B., and E. A. H. Ali. 2008. Adoption of mobile technology in the supply chain: An exploratory cross-case analysis. *International Journal of E-Business Research* 4(4): 1–15.

Dunlop, M. D., and A. Crossan. 2000. Predictive text entry methods for mobile phones. *Personal Technologies London* 4(2): 134–143.

Dwyer, C., S. R. Hiltz, and K. Passerini. 2007. Trust and privacy concern within social networking sites: A comparison of Facebook and MySpace. In *Proceedings of the 13ᵗʰ American Conference on Information Systems*, 339–350. Boston: Curran Associates, Inc.

Ebling, M. R., B. E. John, and M. Satyanarayanan. 2002. The importance of translucence in mobile computing systems. *ACM Transactions on Computer–Human Interaction* 9(1): 42–67.

Eimer, M. 1999. Can attention be directed to opposite locations in different modalities? An ERP study. *Clinical Neurophysiology* 110: 1252–1259.

Ernst and Young. 2001. Global online retailing: An Ernst and Young special report. Unpublished report. Cap Gemini Ernst and Young.

Fang, X., S. Chan, J. Brzezinski, and S. Xu. 2003a. A study of task characteristics and user intention to use handheld devices for mobile commerce. In *Proceedings of the 2ⁿᵈ Annual Workshop on HCI Research in MIS*, 90–94. Seattle, WA, December 12–13, 2003.

Fang, X., S. Chan, J. Brzezinski, S. Xu, and J. Lam. 2003b. User-centered guidelines for design of mobile applications. Unpublished research report.

Fagerberg, P., A. Stahl, and K. Hook. 2003. Designing gestures for affective input: an analysis of shape, effort and valence. In *Proceedings of Mobile Ubiquitous and Multimedia*, 57–65. New York: ACM Press.

Feng, J., Karat, C. M., and Sears, A. 2005. How productivity improves in hands-free continuous dictation tasks: Lessons learned from a longitudinal study. *Interacting with Computers* 17(3): 265–289.

Forlines, C., D. Wigdor, C. Shen, and R. Balakrishnan. 2007. Direct-touch vs. mouse input for tabletop displays. In *Proceedings of the SIGCHI Conference on Human Factors in Computing Systems*, 647–656. New York: ACM Press.

Furnas, G. W. 1986. Generalized Fisheye views. In *Proceedings of the SIGCHI Conference on Human Factors in Computing Systems*, 16–23. New York: ACM Press.

Gartner, Inc. 2009. Hype cycle for consumer mobile applications.

Gebauer, J., Y. Tang, and C. Baimai. 2008. User requirements of mobile technology: Results from a content analysis of user reviews. *Information Systems and E Business Management* 6: 361–384.

Golder, S., D. Wilkinson, and B. Huberman. 2007. Rhythms of social interaction: messaging within a massive online network. In *Proceedings of International Conference on Communities and Technologies*, 41–66. London: Springer.

Gossweiler, R., and J. Tyler. 2004. Plog: easily create digital picture stories through cell phone cameras. In *Proceedings of International Workshop on Ubiquitous Computing*, 94–103. Porto, Portugal: INSTICC Press.

Gou, L., J.-H. Kim, H.-H. Chen, J. Collins, M. Goodman, X. Zhang, and C. L. Giles. 2009. MobiSNA: A mobile video social network application. In *Proceedings of the 8ᵗʰ ACM International Workshop on Data Engineering for Wireless and Mobile Access*, 53–56.

Grover, D. L., M. T. King, and C. A. Kuschler. 1998. Reduced keyboard disambiguating computer. Patent US5818437. Tegic Communications, Inc.

GSM Association. G-Mail growth: global surge continues. Press Release. http://www.gsmworld.com. (accessed Nov. 8, 2009).

Gupta, S., G. Kaiser, D. Neistadt, and P. Grimm. 2003. DOM-based content extraction of HTML documents. In *Proceedings of the International World Wide Web Conference '03*, 207–214. New York: ACM Press.

Hall, A., J. Cunningham, R. Roache, and J. Cox. 1988. Factors affecting performance using touch-entry systems: Tactual recognition fields and system accuracy. *Journal of Applied Psychology* 73: 711–720.

Häkkilä, J., and J. Mäntyjärvi. 2006. Developing design guidelines for context-aware mobile applications. In *Proceedings of the 3ʳᵈ International Conference on Mobile Technology, Applications and Systems*, 1–7. New York: ACM Press.

Hart, J., C. Ridley, F. Taher, C. Sas, and A. Dix. 2008. Exploring the Facebook experience: A new approach to usability. In *Proceedings of the 5ᵗʰ Nordic Conference on Human–Computer Interaction*, 471–474.

Heidi, L., and P. Baudisch. 2005. Summary thumbnails: Readable overviews for small screen Web browsers. In *Proceedings of the SIGCHI Conference on Human Factors in Computing Systems*, 681–690. New York: ACM Press.

Henderson, R., D. Rickwood, and P. Roberts. 1998. Beta test of an electronic supermarket. *Interacting with Computers* 10: 385–399.

Hiraoka, S., I. Miyamoto, and K. Tomimatsu. 2003. BehindTouch, a text input method for mobile phones by the back and tactile sense interface. *Information Processing Society of Japan, Interaction* 2003: 131–138.

Hinckley, K., J. Pierce, M. Sinclair, and E. Horvitz. 2000. Sensing techniques for mobile interaction. In *Proceedings of ACM Symposium on User Interface Software and Technology*, 91–100. New York: ACM Press.

Hinckley, K., and M. Sinclair. 1999. Touch-sensing input devices. In *Proceedings of the SIGCHI Conference on Human Factors in Computing Systems*, 223–230. New York: ACM Press.

Hoggan, E., and S. Brewster. 2007. Designing audio and tactile crossmodal icons for mobile devices. In *Proceedings of the SIGCHI Conference on Human Factors in Computing Systems*, 162–169. New York: ACM Press.

Hummels, C., and P. J. Stappers. 1998. Meaningful gestures for human computer interaction: Beyond hand postures. In *Proceedings of the 3ʳᵈ IEEE International Conference on Automatic Face and Gesture Recognition*, 591–596. Washington: IEEE Computer Society Press.

Huot, S., and E. Lecolinet. 2006. SpiraList: A compact visualization technique for one handed interaction with large lists on mobile devices. In *Proceedings of the 4ᵗʰ Nordic Conference on Human–Computer Interaction: Changing Roles*, 445–448.

Hurst, M., and E. Gellady. 2000. White paper one: Building a customer experience to develop brand, increase loyalty and grow revenues. Unpublished report.

Hurst, M., and P. Terry. 2000. The dotcom survival guide. Unpublished report.

Intille, S. S., L. Bao, E. M. Tapia, and J. Rondoni. 2004. Acquiring in situ training data for context-aware ubiquitous computing applications. In *Proceedings of the SIGCHI Conference on Human Factors in Computing Systems*, 1–8. New York: ACM Press.

Jacko, J. A., G. Salvendy, and J. R. Koubek. 1995. Modelling of menu design in computerized work. *Interacting with Computers* 7(3): 304–330.

Jacko, J. A., and G. Salvendy. 1996. Hierarchical menu design: Breadth, depth and task complexity. *Perceptual and Motor Skills* 82(3): 1187–1201.

James, C. L., and K. M. Reischel. 2001. Text input for mobile devices: Comparing model prediction to actual performance. In *Proceedings of the SIGCHI Conference on Human Factors in Computing Systems*, 365–371.

Jarvenpaa, S. 2001. Developing mobile commerce capabilities. Presentation to the SIM Advanced Practices Council, May 7–8.

Johnson, P. 1998. Usability and mobility: Interactions on the move. In *Proceedings of the 1st Workshop on Human Computer Interaction with Mobile Devices*, ed. C. Johnson. Glasgow, Scotland: GIST Technical Report G98-1.

Jones, M., and G. Marsden. 2006. *Mobile Interaction Design*. New York: John Wiley.

JuiceCaster. 2009. Juice Wireless Inc. http://www.juicecaster.com/ (accessed Nov. 8, 2009).

Kalakota, R., and M. Robinson. 2001. *M-Business: The Race to Mobility*. New York: McGraw-Hill.

Kamba, T., S. A. Elson, T. Harpold, T. Stamper, and P. Sukaviriya. 1996. Using small screen space more efficiently. In *Proceedings of the SIGCHI Conference on Human Factors in Computing Systems*, 383–390. New York: ACM Press.

Kannan, P., A. Chang, and A. Whinston. 2001. Wireless commerce: Marketing issues and possibilities. In *Proceedings of the 34th Hawaii International Conference on System Sciences*, 1–6. Washington DC: IEEE Computer Society Press.

Karlson, A., and B. Bederson. 2007. ThumbSpace: Generalized one-handed input for touchscreen-based mobile devices. In *Proceedings of the SIGCHI Conference on Human Factors in Computing Systems*, 324–338. New York: ACM Press.

Karlson, A., B. Bederson, and J. Contreras-Vidal. 2007. *Understanding on User Interface Design and Evaluation for Mobile Technology*. Hershey, PA: Idea Group.

Keen, P., and R. Mackintosh. 2001. *The Freedom Economy*. Berkeley, CA: Osborne/McGraw-Hill.

Kim, H., and J. Kim. 2003. Post-adoption behavior of mobile Internet users: A model-based comparison between continuers and discontinuers. In *Proceedings of the 2nd Annual Workshop on HCI Research in MIS*, 95–99.

Kim, H., J. Kim, Y. Lee, M. Chae, and Y. Choi. 2002. An empirical study of the use contexts and usability problems in mobile Internet. In *Proceedings of the 35th Annual Hawaii International Conference on System Sciences*, 1–10.

Kim, J., and J. Moon. 2000. Designing towards emotional usability in customer interfaces—trustworthiness of cyber-banking system interfaces. *Interacting with Computers* 10: 1–29.

Kim, J., and B. Yoo. 2000. Toward the optimal link structure of the cyber shopping mall. *International Journal of Human–Computer Studies* 52: 531–551.

Lai, C.-H. 2007. Understanding the design of mobile social networking: the example of EzMoBo in Taiwan. *Journal of Media and Culture* 10(1). http://www.journal.media-culture.org.au/0703/08-lai.php.

Lampe, C., N. Ellison, and C. Steinfield. 2006. A face(book) in the crowd: social searching vs. social browsing. In *Proceedings of the 20th Conference on Computer Supported Cooperative Work*, 167–170. New York: ACM Press.

Lamping, J., R. Rao, and P. Piroli. 1995. A focus+context technique based on hyperbolic geometry for visualizing large hierarchies. In *Proceedings of the SIGCHI Conference on Human Factors in Computing Systems*, 401–408.

Larson, K., and D. Mowatt. 2003. Speech error correction: The story of the alternates list. *International Journal of Speech Technology* 6(2): 183–194.

Lee, Y. E., and I. Benbasat. 2003. Interface design for mobile commerce. *Communications of the ACM* 46(12): 49–52.

Lee, V., H. Schneider, and R. Schell. 2004. *Mobile Applications: Architecture, Design, and Development*. Englewood Cliffs, NJ: Prentice-Hall.

Lehner, F., and R. Watson. 2001. From e-commerce to m-commerce: research directions. Unpublished paper.

Linjama, J., and T. Kaaresoja. 2004. Novel, minimalist haptic gesture interaction for mobile devices. In *Proceedings of the 3rd Nordic Conference on Human–Computer Interaction*, 457–465. New York: ACM Press.

Lumsden, L., and S. Brewster. 2003. A paradigm shift: alternative interaction techniques for use with mobile and wearable devices. In *Proceedings of the Conference of the Centre for Advanced Studies on Collaborative Research*, 197–210. Lebanon: IBM Press.

MacKay, B., D. Dearman, K. Inkpen, and C. Watters. 2005. Walk 'n scroll: A comparison of software-based navigation techniques for different levels of mobility. In *Proceedings of MobileHCI'05*, 183–190. New York: ACM Press.

MacKenzie, I. S., and R. W. Soukoreff. 2002. Text entry for mobile computing: Models and methods, theory and practice. *Human–Computer Interaction* 17: 147–198.

Mackinlay, J., G. Robertson, and S. Card. 1991. The Perspective Wall: Detail and context smoothly integrated. In *Proceedings of the SIGCHI Conference on Human Factors in Computing Systems*, 173–179. New York: ACM Press.

Mäntyjärvi, J. 2003. *Sensor-based Context recognition for Mobile Applications*. VTT Publications: 511, 118 pp. + app. 60 pp.

Marcussen, C. H. 2003. Mobile phones, WAP and the Internet—the European market and usage. http://www.crt.dk/UK/Staff/chm/wap.htm (accessed Oct. 16, 2009).

McKenna, K. Y. A., and A. S. Green. 2002. Virtual group dynamics. *Group Dynamics: Theory, Research, and Practice* 6(1): 116–127.

Mennecke, B., and T. Strader. 2001. Where in the world on the Web does location matter? A framework for location based services in m-commerce. In *Proceedings of the 7th Americas Conference on Information Systems*, 450–455.

Microsoft News Press. 2007. Microsoft backs new technology company: ZenZui aims to change the way people use mobile devices. http://www.microsoft.com/presspass/press/2007/mar07/03-26ZenZuiPR.mspx (accessed Oct. 16, 2009).

Motorola, n.d. A timeline overview of Motorola history. www.motorola.com/mot/doc/6/6800_MotDoc.pdf (accessed Oct. 30, 2009).

Murray, A. C., C. R. Frankish, and D. M. Jones. 1993. Data-entry by voice: facilitating correction of misrecognitions. In *Interactive Speech Technology: Human Factors Issues in the Application of Speech Input/Output to Computers*, eds. C. Baber and J. M. Noyes, 137–144. Bristol, PA: Taylor & Francis.

Nazir, A., S. Raza, and C.-N. Chuah. 2008. Unveiling Facebook: A measurement study of social network based applications.

In *Proceedings of the 8th ACM SIGCOMM Conference on Internet Measurement*, 43–56.

Nielsen, J. 2001. What is usability? http://www.zdnet.com/devhead/stories/articles/0,4413,2137671,00.html. (accessed Nov. 1, 2009).

Nielsen, J., and R. Molich. 1990. Heuristic evaluation of user interfaces. In *Proceedings of the SIGCHI Conference on Human Factors in Computing Systems*, 249–256. New York: ACM Press.

Nielsen, J., S. Farrell, C. Snyder, and R. Molich. 2000a. E-commerce user experience: category pages. Unpublished report. Nielsen Norman Group.

Nielsen, J., S. Farrell, C. Snyder, and R. Molich. 2000b. E-commerce user experience: Checkout and registration. Unpublished report. Nielsen Norman Group.

Nielsen, J., S. Farrell, C. Snyder, and R. Molich. 2000c. E-commerce user experience: product pages. Unpublished report. Nielsen Norman Group.

Nielsen, J., S. Farrell, C. Snyder, and R. Molich. 2000d. E-commerce user experience: trust. Unpublished report. Nielsen Norman Group.

Oviatt, S. 1999. Mutual disambiguation of recognition errors in a multimodal architecture. In *Proceedings of the SIGCHI Conference on Human Factors in Computing Systems* 576–583.

Palen, L., and M. Salzman. 2002. Beyond the handset: designing for wireless communications usability. *ACM Transactions on Computer–Human Interaction* 9(2): 125–151.

Pascoe, J., N. Ryan, and D. Morse. 2000. Using while moving: HCI issues in fieldwork environments. *ACM Transactions on Computer–Human Interaction* 7(3): 417–437.

Parhi, P., A. Karlson, and B. Bederson. 2006. Target size study for one-handed thumb use on small touchscreen devices. In *Proceedings of MobileHCI'06*, 203–210. New York: ACM Press.

Perry, M., K. O'Hare, A. Sellen, B. Brown, and R. Harper. 2001. Dealing with mobility: Understanding access anytime, anywhere. *ACM Transactions on Computer–Human Interaction* 8(4): 323–347.

Pietiläinen, A.-K., E. Oliver, J. LeBrun, G. Varghese, and C. Diot. 2009. MobiClique: Middleware for Mobile Social Networking. In *Proceedings of the 2nd ACM SIGCOMM Workshop on Online Social Networking*, 49–54. New York: ACM Press.

Potter, R., L. Weldon, and B. Shneiderman. 1988. Improving the accuracy of touchscreens: An experimental evaluation of three strategies. In *Proceedings of the SIGCHI Conference on Human Factors in Computing Systems*, 27–32.

Rabin, J., and C. McCathieNevile. 2006. Mobile Web Best Practices 1.0. W3C Working Draft, vol. 13.

Rao, R., and S. K. Card. 1994. The table lens: Merging graphical and symbolic representations in an interactive focus+context visualization for tabular information. In *Proceedings of the SIGCHI Conference on Human Factors in Computing Systems*, 318–322. New York: ACM Press.

Rehman, A. 2000. Holiday 2000 e-commerce. Unpublished report. New York: Creative Good.

Rekimoto, J. 2002. SmartSkin: An infrastructure for freehand manipulation on interactive surfaces. In *Proceedings of the SIGCHI Conference on Human Factors in Computing Systems*, 113–120. New York: ACM Press.

Rekimoto, J. 2003. ThumbSense: Automatic input mode sensing for touchpad-based interactions. In *Proceedings of the SIGCHI Conference on Human Factors in Computing Systems*, 852–853. New York: ACM Press.

Rekimoto, J., T. Ishizawa, and H. Oba. 2003. SmartPad: A finger-sensing keypad for mobile interaction. In *Proceedings of the SIGCHI Conference on Human Factors in Computing Systems*, 850–851.

Rekimoto, J., T. Ishizawa, C. Schwesig, and H. Oba. 2003. PreSense: interaction techniques for finger sensing input devices. In *Proceedings of ACM Symposium on User Interface Software and Technology*, 203–212.

Ren, X., and S. Moriya. 1997. The best among six strategies for selecting a minute target and the determination of the minute maximum size of the targets on a pen-based computer. In *Proceedings of IFIP Interact '97*, 85–92.

Ren, X., and S. Moriya. 1998. The influence of target size, distance and direction on the design of selection strategies. In *Proceedings of BCS HCI '98*, 67–82. New York: ACM Press.

Ren, X., and S. Moriya. 1999. A state transition model representing pen-based selection strategies. In *Proceedings of IFIP Interact '99*, 57–58. Washington DC: IEEE Computer Society Press.

Rennison, E., 1994. Galaxy of news. In *Proceedings of ACM Symposium on User Interface Software and Technology*, 3–12. New York: ACM Press.

Robbe, S., N. Carbonell, and C. Valot. 1994. Towards usable multimodal command languages: Definition and ergonomic assessment of constraints on users' spontaneous speech and gestures. In *Proceedings of the International Conference on Spoken Language Processing*, 1655–1658. New York: ACM Press.

Robertson, G. G., J. D. Mackinlay, and S. K. Card. 1991. Cone trees: Animated 3D visualizations of hierarchical information. In *Proceedings of the SIGCHI Conference on Human Factors in Computing Systems*, 189–194. New York: ACM Press.

Roudaut, A., S. Huot, and E. Lecolinet. 2008. TapTap and MagStick: Improving one-hand target acquisition on small touch-screens. *AVI, Napoli, Italy*, 146–153.

Rudnicky, A. I., K.-F. Lee, and A. G. Hauptmann.1994. Survey of current speech technology. *Communications of the ACM* 37(3): 52–57.

Sarkar, M., and M. H. Brown. 1992. Graphical Fisheye views of graphs. In *Proceedings of the SIGCHI Conference on Human Factors in Computing Systems*, 83–91. New York: ACM Press.

Sarvas, R., M. Viikari, J. Pcsoncn, and H. Ncvanlinna. 2004. MobShare: controlled and immediate sharing of mobile images. In *Proceedings of the 12th Annual ACM International Conference on Multimedia*, 724–731.

Sears, A., and B. Shneiderman. 1991. High precision touchscreens: Design strategies and comparisons with a mouse. *International Journal of Man–Machine Studies* 34(4): 593–613.

Semi-Annual Wireless Industry Survey, n.d. CTIA—the Wireless Association® announces semi-annual wireless industry survey results. http://www.ctia.org/media/press/body.cfm/prid/1811 (accessed Oct. 9, 2009).

Shim, J. P., E. Bekkering, and L. Hall. 2002. Empirical findings on perceived value of mobile commerce as a distributed channel. In *Proceedings of the 8th Americas Conference on Information Systems* 1835–1837. Dallas: AIS Press.

Shrestha, S. 2007. Mobile Web browsing: Usability study. In *Proceedings of the Mobility '07*, 187–194. New York: ACM Press.

Siau, K., E. Lim, and Z. Shen. 2001. Mobile commerce: Promises, challenges, and research agenda. *Journal of Database Management* 12(3): 4–13.

Silfverberg, M., I. S. MacKenzie, and P. Korhonen. 2000. Predicting text entry speed on mobile phones. In *Proceedings of the SIGCHI Conference on Human Factors in Computing Systems*, 9–16. New York: ACM Press.

Sorvari, A., J. Jalkanen, R. Jokela, A. Black, K. Koli, M. Moberg, and T. Keinonen. 2004. Usability issues in utilizing context metadata in content management of mobile devices. In *Proceedings of the SIGCHI Conference on Human Factors in Computing Systems*, 357–363. New York: ACM Press.

Spence, R., and M. Apperley. 1982. Date base navigation: An office environment for the professional. *Behavior and Information Technology* 1(1): 43–54.

Sturm, J., and L. Boves. 2005. Effective error recovery strategies for multimodal form-filling applications. *Speech Communication* 45(3): 289–303.

Sugimoto, M., and K. Hiroki. 2006. HybridTouch: An intuitive manipulation technique for PDAs using their front and rear surfaces. In *Proceedings of MobileHCI '06*, 137–140. New York: ACM Press.

Tarasewich, P., R. Nickerson, and M. Warkentin. 2002. Issues in mobile e-commerce. *Communications of the Association for Information Systems* 8: 41–64.

Taylor, C. A., O. Anicello, S. Somohano, N. Samuels, L. Whitaker, and J. A. Ramey. 2008. A framework for understanding mobile Internet motivations and behaviors. In *Proceedings of the SIGCHI Conference on Human Factors in Computing Systems*, 2679–2684. New York: ACM Press.

Varshney, U., and R. Vetter. 2001. A framework for the emerging mobile commerce applications. In *Proceedings of the 34th Hawaii International Conference on System Sciences*, 1–10. Washington DC: IEEE Computer Society Press.

Venkatech, V., V. Ramesh, and A. P. Massey. 2003. Understanding usability in mobile commerce. *Communications of the ACM* 46(12): 53–56.

Vogel, D., and P. Baudisch. 2007. Shift: a technique for operating pen-based interfaces using touch. In *Proceedings of the SIGCHI Conference on Human Factors in Computing Systems* 657–666. New York: ACM Press.

Wang, L., and A. S. M. Sajeev. 2007. Roller interface for mobile device applications. In *Proceedings of the 8th Australasian User Interface Conference*, 7–13. Darlinghurst, Australia: Australian Computer Society, Inc.

Waters, R. 2000. Rival views emerge of wireless Internet. *Financial Times FT-IT Review*, March 1, 1.

Wellman, B. 1996. For a social network analysis of computer networks: a sociological perspective on collaborative work and virtual community. In *Proceedings of ACM SIGCPR/SIGMIS Conference on Personnel Research*, 1–11. New York: ACM Press.

Wickens, C. D., S. E. Gordon, and Y. Liu. 1998. *An Introduction to Human Factors Engineering*. New York: Addison Wesley Longman.

Wigdor, D., C. Forlines, P. Baudisch, J. Barnwell, and C. Shen. 2007. LucidTouch: A see-through mobile device. In *Proceedings of the ACM Symposium on User Interface Software and Technology*, 269–278. New York: ACM Press.

Wigdor, D., D. Leigh, C. Forlines, S. Shipman, J. Barnwell, R. Balakrishnan, and C. Shen. 2006. Under the table interaction. In *Proceedings of the ACM Symposium on User Interface Software and Technology*, 259–268. New York: ACM Press.

Xu, G., and J. A. Gutierrez. 2006. An exploratory study of killer applications and critical success factors in m-commerce. *Journal of Electronic Commerce in Organizations* 4(3): 63–79.

Xu, S. 2010. Usability issues in introducing capacitive interaction into mobile navigation. Under review of the *14th International Conference of Human Computer Interaction*.

Xu, S., X. Fang, S. Chan, and J. Brzezinski. 2003. What tasks are suitable for handheld devices? In *Proceedings of the 10th International Conference on Human–Computer Interaction*, vol. 2, eds. C. Stephanidis and J. Jacko, 333–337. Mahwah, NJ: Lawrence Erlbaum.

Xu, S., X. Fang, J. Brzezinski, and S. Chan. 2008. Development of a dual-modal presentation of texts for small screens. *International Journal of Human–Computer Interaction* 24(8): 1–18.

Yang, A. S. 2009. Exploring adoption difficulties in mobile banking services. *Canadian Journal of Administrative Sciences* 26(2): 136–149.

35 Human Factors in the Evaluation and Testing of Online Games

Karl Steiner

CONTENTS

35.1 INTRODUCTION

35.1.1 POPULARITY OF VIDEO GAMES AND ONLINE GAMES

The size and scope of the video game industry has grown dramatically over the past several years. In 2007, the market for games (excluding hardware) was over 9.5 billion, a figure rivaling the annual box office take in the film industry (Bangeman 2008). In 2008, the video game market was even larger, with game sales of over $11.7 billion and almost 300 million individual games sold (Entertainment Software Association 2009).

Driving this growth are dramatic increases in the number and diversity of people playing games. Many will not be surprised at reports that video games are extremely popular among older children and teens. In recent surveys, 97% of all American teens reported playing video games, and on any given day, 50% reported having played within the last day (Lenhart et al. 2008). However, the number of adults playing video games is also large and growing. According to one recent report, over 50% of American adults report playing video games (Lenhart, Jones, and Macgill 2008), and the average age of the American video game player is 35 (Entertainment Software Association 2008). Almost one in five adults play video games on a near daily basis, and the adult age group most likely to play video games daily are those aged 65 and over (Lenhart, Jones, and Macgill 2008).

The game market has responded to the growing population of players by becoming increasingly diverse. Players can find games to suit various tastes, such as sports simulations on the Xbox 360, fantasy role-playing on the PS3, complex strategy games on the PC, health and fitness training on the Wii, and numerous others. As another example of this diversity, the NPD Group/Retail Tracking Service breaks down the video game market into the following genres (though fans and game professionals might debate the fine points of their categorization):

- Action
- Adventure
- Arcade
- Children's entertainment
- Family entertainment
- Fighting
- Flight
- Role playing
- Shooter
- Sports games
- Strategy
- Other games/compilations

35.1.2 Human Factors in Video Games

As the video game industry has matured and the market has begun to cater to larger and more diverse groups of players, the already substantial usability and playability challenges in video games have only increased. In many ways, the maturation of video games usability is following a pattern similar to PC software usability. In the earliest days of the PC industry, PC software was used primarily by professionals and hobbyists who possessed significant technical skills and had an interest and willingness to learn arcane commands. However, as the user base grew, the practical and marketing importance of having a usable and intuitive software product also increased. Users became less tolerant of needless complexity, and usability became a competitive differentiator.

Similarly, in the early days of the game industry, players were more tolerant of complex menu structures, arcane keyboard commands or control schemes, or unforgiving level designs that punished players with a permanent death and the dreaded "Game Over." Gaining mastery of these elements

was seen as a badge of honor by some. Perhaps most significantly, if players did not like these elements, there were not many other options from which to choose. However, as the sophistication of gamers increased and the pool of gamers became dramatically larger, the market began to demand better experiences for gamers. The form of these playability improvements covered many areas, including traditional usability, suitability to audience and platform, challenge, and entertainment value, among others.

Usability is as important (if not more important) for video games than it is with PC software and Web sites. As with PC software, many video games incorporate a variety of screens and menus. These options allow players to control and customize various aspects of their game, their characters, and their gameplay. Just as with PC software, players expect that their interactions with these screens and menus will be efficient and effective. In fact, in many cases, game players have even higher expectations regarding efficiency in feature access because many games involve a real-time element where every second interacting with a menu might mean the difference between victory and defeat.

As with any software product, suitability for audience is also an important factor. In 2008, Family Entertainment was the most popular genre of video games, beating genres such as Sports, Strategy and Shooters. Games with the rating of E for Everyone 10+ were more popular than those rated T for Teens or M for Mature Audiences (Entertainment Software Association 2008). Designing products specifically for young children is different than designing for adults, and trying to design a flexible product meeting the needs of a broad audience can introduce additional challenges and complexities. Where many game designers could once assume that their audience was composed of players who approached the game with backgrounds and experiences similar to their own, now designers are often creating games for players who may be younger or older than themselves, less experienced with games, and bringing different assumptions and expectations to their gameplay. This increases the need for usability and playability evaluation to ensure a suitable fit between game and audience.

Technology also plays a driving role in video game usability. Unlike the business PC industry that has largely standardized on the Windows Operating System which dictates many aspects of the user experience, the world of game development continues to support multiple (sometimes dissimilar) standards. PC games utilizing the keyboard and mouse have different conventions from console games using the controller, and there are even differences within the same platform. Three-dimensional (3D) PC games (such as shooters) frequently have different conventions and control schemes from browser-based games. In the current generation of game consoles, the PS3 and Xbox 360 offer similar but not entirely comparable controls, while the Wii provides a unique input device. New games featuring novel input devices also continue to impact the market (e.g., the guitar-shaped input device for Guitar Hero, the Wii Balance Board for Wii Fit, etc.). Again, the constant utilization of novel approaches to input and user interface make game usability a constant necessity and challenge.

Making the notion of a successful player experience more complex are some additional playability concerns, such as challenge, pace, and fun (Pagulayan et al. 2002). Whereas ease of use is generally recognized as desirable in PC software (Nielsen 1994), game players often desire a certain element of challenge in their games. Designers must ensure that the right elements of the game are challenging (e.g., beating a boss should be challenging but opening menus to save your game should not) and must provide a suitable level of challenge that is not too easy but yet not too difficult. Unlike PC software where predictability is an asset, video games must also provide an appropriate pace of new situations and details. Perhaps most importantly, players choose to play a game and expect to be rewarded with an enjoyable experience.

Web-based game delivery and online play provide new features and opportunities for game players, but they also introduce new complexities. The online components add a variety of usability and playability challenges, from the technical (providing feedback on connection status and communications lag), recreational (can the game maintain meaningful and balanced competition or cooperation with other players), to the social (can players communicate and compete in enjoyable and safe ways).

To meet the usability and playability demands of this increasingly challenging environment, many video game publishers and studios now routinely incorporate some form of user experience evaluation for their games. In this chapter, we provide a review of the common and emerging techniques for evaluating game usability and playability, with a special emphasis on Web-based and online games.

35.1.3 Unique Aspects of Online Games

Just as Internet access is becoming more common, online gaming is also increasing in popularity. More people are using the Internet, and Internet users are more likely to play games (both online and off). A survey conducted by the Pew Research Center in 2008, found that of the 75% of the adults interviewed who reported using the Internet, 64% also reported playing games. By comparison, only 20% of the adult non-Internet users reported playing video games. Almost a quarter (23%) of adults surveyed reported playing games online as did 76% of teens (Lenhart, Jones, and Macgill 2008).

While Web-based games and other online games must support the same aspects of playability as other games (ease of use, challenge, pace, fun), the fact that they are online provides new areas of design and play opportunity, such as online competition, real-time communication and collaboration, or content sharing.

Competition is a familiar element of many games. Players may be placed in environments where they can compete in real-time, attempting to complete goals, get higher scores, or otherwise defeat an opponent. Some online games support asynchronous competition, where players compete against the computer, and their score or best time are posted to a public Leaderboard where others can view their success. This recognition can be an important motivator for some players.

Not all online games are competitive. In some games, players work together to achieve goals, often working against computer-managed, artificial-intelligence opponents (or bots). Educational games may also be designed specifically to encourage beneficial types of communication (Steiner and Moher 1994, 2002). Effective coordination of activities in a virtual environment requires a number of common game features, including voice or text chat, awareness of other players' locations, mechanisms for joining or following colleagues, as well as a host of other game-specific features.

Some games have expanded the role of players from consumers of game experiences to creators. A number of games now feature sophisticated tools that allow players to create their own playable experiences that can be shared with other online users. Players may create their own levels for platform games, complete adventures for role-playing games, or generate sophisticated backstories for their favorite sports personalities that unfold over the course of a simulated career.

In addition to the online features that directly support the play experience, a number of online features provide additional touch points for players of both on and offline games. Game-specific e-commerce tools exist to facilitate easy download and update of both PC and console games (and to prevent piracy of games). Leaderboards may be accessible from outside a game, allowing a player to keep tabs on the activities of colleagues and opponents even when not playing. Items, equipment, and even fully developed characters for use in games, are sold for real-world cash in online auctions. A variety of Web-based forums exist for most games (online or off), providing support as well as a sense of community for players of a particular game.

35.2 ONLINE GAME EVALUATION AND TESTING

35.2.1 Evaluation Techniques

To evaluate and improve player experiences, a variety of techniques are commonly employed by video game studios and publishers. Many of these techniques originated in the world of PC software (or commercial product testing) and were then adapted for use with video games. Drawing on personal experiences managing usability testing for THQ Inc. (a leading worldwide developer and publisher of interactive entertainment software), as well as professional and research literature, the following section describes some of the most common game testing techniques, including heuristic evaluation, usability testing, playability testing, and game analytics.

35.2.2 Heuristic Evaluation (Expert Review)

35.2.2.1 Description

Heuristic evaluation is a technique for reviewing user interfaces popularized by Nielsen (1994). Typically, one or more evaluators will independently review a user interface by applying a set of "rules of thumb" or heuristics. Following their evaluation, results from each evaluator are compiled, duplicated issues are merged, and a final list of issues is generated and prioritized. Heuristic evaluations can be conducted relatively quickly and require few resources beyond the time of the evaluators. Nielsen described heuristic evaluation as a "discount usability" technique because of its relatively low cost. In addition to use as a tool for evaluation, heuristics can also be applied by designers, as a way of anticipating potential issues. The heuristics provide a checklist of areas to consider as designers complete their work.

While this technique is relatively quick and requires few resources, it does have limitations. Desurvire (1994) reported that heuristic evaluation identified roughly 42% of the problems found in usability testing, though some practitioners report higher success rates (depending on the application domain, the complexity of issues, and the number of reviewers). There are also differences in the type of issues identified. Heuristic evaluations tend to be good at finding more common user interface issues, where usability testing is better at finding the deeper and more subtle issues related to actual usage and interaction.

Heuristic evaluation has been widely adopted by software usability professionals, has been the subject of much analysis by game researchers, and is becoming more common in the analysis and design of commercial video games. There have been a number of sets of heuristics for evaluation of video games proposed by researchers. Federoff proposed a set of 28 heuristics in 3 categories (game interface, game mechanics, and gameplay) based on a review of game literature as well as work with a game publisher (Federoff 2002). Desurvire, Caplan, and Toth (2004), proposed a set of 43 heuristics across 4 categories: Gameplay, Game Story, Mechanics, and Usability in a set called Heuristic Evaluation for Playability (HEP). This work was updated to address differing genres and game goals and resulted in a new set of heuristics called PLAY (Desurvire and Wiberg 2009). The PLAY list categorizes heuristics into four categories: Gameplay, Game Coolness/ Entertainment/Emotional Connection, Game Usability and Mechanics, Game Beginning and Tutorial Levels, and an optional fifth category Game Story Immersion. Some sample heuristics from the PLAY list include

- Gameplay–A2. The player should not experience or be penalized repetitively for the same failure.
- Gameplay–C2. The game provides clear goals, presents overriding goals early as well as short-term goals throughout gameplay.
- Entertainment–C1. The game offers something different in terms of attracting and retaining the players' interest.
- Entertainment–D1. The game utilizes visceral/ audio/visual content to immerse the player further into the game.
- Game Mechanics–B1. The game controls are consistent within the game and follow standard conventions except where a novel mechanic brings increased fun.

- Game Mechanics–B2. Status score indicators are seamless, obvious, available, and do not interfere with gameplay.
- Tutorial–A1. Seamless game play, learning, and development of tool mastery is seamlessly offered.
- Tutorial–E1. Gameplay is fun while teaching the player how to play—teaches and entertains.

Pinelle, Wong, and Stach (2008) proposed a set of 10 heuristics based on an analysis of common usability issues identified in game reviews. In an analysis of one aspect of gameplay, Sweetser and Wyeth (2005) described a set of 8 elements for evaluating how well a game supports a sense "flow."

There have also been heuristics developed specifically for online and mobile games. Pinelle et al. (2009) created a set of specific guidelines for networked games (Network Game Heuristics or NGH):

1. Provide simple session management.
2. Provide flexible matchmaking.
3. Provide appropriate communication tools.
4. Support coordination.
5. Provide meaningful awareness information.
6. Give players identifiable avatars.
7. Provide protected training for beginners.
8. Support social interaction.
9. Reduce game-based delays.
10. Manage bad behavior.

On the basis of an analysis of mobile device usage patterns, Korhonen and Kovisto (2006) proposed a set of 11 game heuristics specifically for mobile devices:

H1: Don't waste the player's time.
H2: Prepare for interruptions.
H3: Take other persons into account.
H4: Follow standard conventions.
H5: Provide gameplay help.
H6: Differentiation between device UI and the game UI should be evident.
H7: Use terms that are familiar to the player.
H8: Status of the characters and the game should be clearly visible.
H9: The player should have clear goals.
H10: Support a wide range of players and playing styles.
H11: Don't encourage repetitive and boring tasks.

35.2.2.2 Considerations for Online Games

A key for effectively using this technique to evaluate a game is to agree on a set of heuristics. The literature provides a number of credible heuristics for different genres, game modes, and game problems including online and mobile. Practitioners can use these as a starting point and tailor the heuristics to meet the needs of their specific game.

Heuristic evaluations can be an efficient and cost-effective technique for evaluating online games throughout the design and development process. Because heuristic reviews rely on experts' evaluations of an interface, in some cases limited reviews can even be conducted in the design phase of development, provided that sufficient design documentation or design wireframes are available to clearly communicate design intentions. This approach is well suited to Web-based games that employ more traditional browser-based design conventions, because interactions and states of the screen can be clearly represented in design documentation.

While heuristic evaluations can successfully identify many issues in a cost-effective manner, they do not effectively capture all of the nuances regarding player behaviors or responses that can be gathered through other techniques such as usability testing or playability testing (Desurvire, Caplan, and Toth 2004). As such, heuristics can perhaps best be viewed as a complementary technique rather than a replacement for other forms of evaluation.

35.2.3 Heuristic Evaluation Case Study

35.2.3.1 Game

Cars is a racing game for the Wii based on the popular Pixar movie of the same name. The game was developed by Rainbow Studios. The target audience is younger game players. The game features motion input with the Wii remote, the ability to race against other humans or computer opponents, and a variety of tracks on which to race.

35.2.3.1.1 Test Goal/Design Questions

The team was interested in getting usability feedback in the early (pre-Alpha) stage of development. Because the game was designed for a younger audience, it was important to ensure that players were able to understand options and menus, as well as actually control their cars using the Wii remote.

35.2.3.1.2 Test Approach

In order to thoroughly examine deeper gameplay issues as well as user-interface oriented issues, both usability tests and heuristic evaluations were conducted (according to a plan developed and led by Heather Desurvire of Behavioristics). Two game researchers performed the heuristic evaluation using the PLAY heuristics. A different game researcher then conducted a usability test with 10 participants.

35.2.3.1.3 Findings

Both test approaches successfully identified a number of issues. The usability test identified more issues overall, as well as more issues not covered in the heuristic analysis. While most of the issues found by the heuristic evaluation were also duplicated by findings from the usability test, there were some unique findings from the heuristic evaluation, including control issues related to driving backward as well as discovering control combos.

During the usability testing, players performed a number of activities but were not asked to try driving backward. No

players spontaneously performed the move, and because it was not otherwise tested, no related issues were observed during the usability testing. However, the heuristic evaluation revealed that this feature was never introduced or demonstrated to players (Heuristic: Demonstrate actions and reinforcement) and could cause potential difficulties as a result.

Similarly, there were a number of other control options that were unclear or unexplained, including the use of combos. Players had no way to review or look-up specific combos. While this issue was not identified during usability testing, the review of the heuristics (Heuristic: Information on demand and in time) identified this as a potential issue.

35.2.3.1.4 Outcomes

As a result of both the PLAY heuristic evaluation and the usability test, the designers made significant changes based on the recommendations. Changes to tutorials (covering driving backward) and online resources (including control combos) were made based on the heuristic evaluation findings. A number of other gameplay features were modified as a result of the usability testing as well, such as use of the Wii Remote, menu structures, and racing track design. The combination of the two test techniques (heuristic evaluation and usability testing) provided broader coverage than one technique alone.

35.2.3.2 Summary

Heuristic evaluation is a relatively inexpensive and rapid way to evaluate a game, even in early stages of design or development. When successfully administered, the technique provides lists of prioritized issues or opportunities for improvement in the game. The technique can be an effective supplement to other techniques, providing feedback earlier or more frequently than if a team relied solely on more resource-intensive techniques. However, heuristic analyses generally do not find all types of issues equally well. Because the technique does not involve actual game players, it does not provide insights into player attitudes or actual player behaviors. As such, heuristics can perhaps best be viewed as a complementary technique rather than a replacement for other forms of evaluation.

35.2.4 Usability Testing (Observing Players)

35.2.4.1 Description

Usability testing refers to techniques where users of a software product are observed conducting activities with the software (Rubin 1994). Given the focus on observation, usability testing typically supplies rich feedback on what users do, where they encounter issues or difficulties, and the context around those issues or difficulties. Testing in a dedicated laboratory with multiple cameras and one-way glass provides benefits for observers and facilitators but is not necessarily required for running a successful test. Usability testing offers rich observational feedback but can also be time and resource intensive given its focus on close observation and analysis.

This deep analysis and focus on not just where issues occur but why they occur has made usability testing an important technique in the evaluation of video games. Many video game studios and publishers conduct some form of usability testing, and there is a growing body of literature discussing case studies and specific benefits and issues in game usability testing.

Fulton and Medlock (2003) and Pagulayan et al. (2002) describe user testing experiences at the Microsoft Games Studios. Fulton and Medlock explain their approach to testing and explain strengths of usability testing for games, such as the ability to balance subjective player statements against objective observations of their behavior, provide qualitative feedback, and evaluate responses to an existing interface (as opposed to generating ideas for new enhancements or features). In further describing usability testing at the Microsoft Games Studios, Pagulayan et al. mention gameplay issues well suited to the technique (e.g., expectancies, efficiency, and performance interactions). These issues frequently involve certain areas of games, such as screens for navigating options and controlling settings (the game shell) and in the control schemes. Both authors make the point that usability testing (like other techniques) has pros and cons, and no single technique can serve all test needs. Laitinan (2005) provides a description of the usability test approach used on the game Shadowground (similar to the approach described earlier), and describes representative usability findings, as well as the severity of issues (almost 30% being classified as severe or catastrophic). Laitinan (2006) also conducted a study in collaboration with a professional game development studio that concluded that successful game usability tests can provide findings that are both novel (previously unknown) and useful to game development teams. Pinelle, Wong, and Tadeusz (2008) conducted an analysis of usability issues as reported in video game reviews and used this information to compile profiles of issues by genre, including a summary of high- and low-incidence issues. A series of usability tests on massively multiplayer online (MMO) games was conducted in order to identify usability issues common to new players. Significant issues found in this study included difficulties with installation and setup, difficulties finding and using help, misunderstandings regarding the modal combat common in MMOs, and failures to communicate with both non-player characters and human players within the game.

A strength of usability testing is the ability to customize the test format and structure to address specific test goals and answer specific questions. While usability testing can provide rich feedback, the post-test analysis, prioritization and implementation of changes can require time (a scarce commodity in many game development projects). The Rapid Iterative Testing and Evaluation (RITE) technique is one approach to ensuring that issues identified in the test process achieve buy-in and receive quick remedial actions. In this approach, stakeholders commit time and resources in advance so that once issues are identified, changes can be made immediately, and then retested, with iterations occurring daily over the course of several days (Medlock et al. 2002).

Wizard of Oz testing is another modification of usability tests with implications for game testing. In order to test software or other interactive products before they are complete,

a human can simulate the behaviors of the product (often in a manner where the user is unaware that they are interacting with a person instead of a product). This technique is referred to as a Wizard of Oz approach (a reference to L. Frank Baum's book *The Wizard of Oz* in which a wizard secretly manipulated machinery from behind a curtain) and has been used in various studies of PC software. The technique was also used to successfully test the design of a video game that responded to recognition of player body movements before the vision-recognition system was completed (Hosniemi, Hamalainen, and Turkki 2004).

35.2.4.2 Discussion and Examples

As at many game development studios and publishers, usability testing is a commonly used technique at THQ, for both online and offline games. Depending on the test goals and available resources, specific methods may vary with regards to test setting and structure.

In our own studies at THQ, the most common test setting is our dedicated laboratory, which provides an effective and relatively unobtrusive way to observe participants. Participants sit in one room with a game system (PC or console) and several cameras. A facilitator and observers sit in a separate room where the game image can be viewed and recorded, along with images of the players' facial expression (and in some cases, images of the controller or keyboard and mouse). The ability to closely observe and record these different views is a significant benefit of the dedicated laboratory setting.

Session recordings can serve a number of valuable research and communication purposes. The session recordings provide a means for reviewing player behavior and comments after the test to ensure that specific details were fully understood. Recorded sessions can also be made available to team members who could not observe the sessions directly, and highlight reels can be prepared for quick review. Having a recording also allows a deeper level of analysis. The ability to rewind and observe frame-by-frame activities allows specific situations to be better analyzed, as well as for quantitative analysis to be employed using precise data collection protocols. The recordings can also serve as a historical archive. When new team members come on-board or when other team members want to better understand player behaviors, they can refer to the recordings for insights into how players actually experience the product.

Because of these benefits, our preferred test location is in our usability labs. However, successful tests can also be conducted in less formal settings. In some situations, we have found it more effective to bring the test to our participants rather than bringing the participants to our laboratory. We have conducted sessions in *ad hoc* test locations (classrooms, conference rooms, etc.) where a facilitator may sit next to a participant or even observe interactions remotely using screen-sharing software. Because many factors in a "field" test cannot be controlled as well as in a laboratory study, it becomes even more important to have a plan that focuses on the test elements that can be controlled, such as recruiting, data gathering, and analysis methods.

A concern with any test is the determination of what can and should be tested and selecting a test structure and data collection scheme accordingly. It is common to explore the behaviors of new players using a game for the first time in usability tests. This is important feedback to receive and is a natural outcome of the basic test structure where players typically sit down to play a game they have never seen before. However, almost all games feature many hours of gameplay and introduce newer and more complicated features as the game progresses. In order to observe the behaviors of experienced players in later levels, we have employed test structures that have called for players to return over several days, progressing through the game, and acquiring more experience as they proceed.

35.2.4.3 Usability Testing and Focus Discussions Case Study

35.2.4.3.1 Game

Dragonica Online is a free-to-play massively multiplayer online game (MMO). The game features a unique combat system, as well as a whimsical visual style (see Figure 35.1). The game was originally designed for the Korean market, then localized for the U.S. market (and other regions) by THQ*ICE.

35.2.4.3.2 Test Goal/Design Questions

Part of the appeal of MMOs is the opportunities to participate in quests and missions as part of a group. Dragonica Online supports a number of features that enable group interaction, including a system for entering instanced dungeons as a party, commands for finding and following friends, and in-game chat. In order to be successful, especially with new users, these features needed to be intuitive and highly usable. Early feedback suggested that some features were not being used as expected, so a usability test format was chosen to confirm if there were issues and to explore the underlying causes.

35.2.4.3.3 Test Approach

In order to discover potential barriers to the use of the online group tools in Dragonica Online (as well as any other issues), we conducted several usability test sessions. Groups of four players (the largest possible party size) were brought into the laboratory, allowed to complete an online tutorial covering game basics, and then asked to perform various group-oriented activities. Participants were physically separated so that they could not communicate other than by the use of in-game tools. In-game activities were observed as well as recorded. Following the test sessions, players engaged in focus discussions to clarify and expand on the observations of in-game issues.

35.2.4.3.4 Findings

Along with some general usability issues related to solo play, the test identified opportunities for improvement to the controls and UI for the group tools (such as party formation, assignment of party leadership, and awareness of locations of party members). Players generally had clear goals of what they wanted to accomplish but were not initially sure of how to successfully use some of the in-game tools to meet their goals. For example, several players sought a command that would let them form a party. Many overlooked the button

FIGURE 35.1 (See color insert.) THQ*ICE, Dragonica Online, and their respective logos are trademarks and/or registered trademarks of THQ*ICE. All rights reserved. All other trademarks, logos, and copyrights are property of their respective owners. © 2009 THQ*ICE.

marked "Invite," and some (incorrectly) assumed the command they wanted was "Follow." Players were also unclear on where parties could be formed.

35.2.4.3.5 Outcomes

On the basis of the test findings, the developers were able to make modifications to the UI to make the identified aspects of party management more obvious and usable. For example, instructions describing where and how to form parties were made clearer, menu labels were changed to be more obvious ("Invite" became "Party Invite"), and other UI changes were made to reduce confusion.

35.2.4.4 Considerations for Online Games

Depending on the mode of play, conducting usability tests on online games can be very similar to testing other types of games. For instance, many online games incorporate a tutorial that is completed before beginning regular play. While this is an important part of players' introduction to the game, if it is delivered without requiring interaction with other players, it can be tested as an essentially stand-alone element of gameplay. Games with an asynchronous online mode (e.g., where aspects of the game are delivered via browser-based Web pages but where players do not interact in a real-time manner with other players) can also often be tested in the same manner as a fully offline game. While some asynchronous features may involve updates based on activities of other players (such as updating scores on a Leaderboard or receiving gifts in a social game on Facebook or MySpace), these events can often be simulated by the game prototype being tested, or managed using a Wizard of Oz approach. However, online games that include synchronous interaction with other players require some additional test planning considerations.

When testing an online game that involves real-time interactions by multiple players, it can be valuable to create a situation that provides an appropriate number of play partners. These additional players can be other test participants or they could be confederates of the facilitator. For example, in the MMO game Dragonica Online, groups of up to four players may enter dungeons together, coordinating activities in order to defeat opponents. In order to test this game, groups of four players were brought to the laboratory. Players were seated at physically separated workstations and wore headphones to allow them to hear their game audio (while making it difficult to communicate in any way other than by using the in-game chat). Observers in the control room were able to see the screens of all four participants in order to follow their activities and interactions. When testing a prototype of a new first-person shooter, we wanted observations of players in a group "Death Match." Because we wanted to examine responses to certain situations, confederates controlled some of the in-game opponents in order to create the specific situations called for by the test plan.

35.2.4.5 Summary

Usability testing is a good technique for gathering qualitative, behavioral feedback from players. The use of think-aloud protocols and the opportunity to ask participants probing questions allows insights into not just "what" players do but "why" they do it. Results are typically reported as a set of observed behaviors (such as player failures to accomplish goals, areas of difficulty, or unexpected activities), and where possible, explanations of causes of these behaviors. This information is particularly valuable in determining solutions for the issues that are identified. Large numbers of players are generally not needed in order to identify most issues, and given the smaller samples sizes, this technique is less frequently used for quantitative analysis.

35.2.5 PLAYABILITY TESTING (TEST AND SURVEY)

35.2.5.1 Description

Another approach to testing video games involves allowing larger groups of players to experience the game and collecting feedback through standardized, more easily quantifiable forms such as surveys. For the purpose of this paper, we will refer to this technique as playability testing, though there is not a widely accepted terminology for this form of test. Similar techniques have been referred to by some as play tests (Davis, Steury, and Pagulayan 2005), multiple-user simultaneous testing or MUST (Nielsen 2007), and others have made a distinction between early-development play tests (similar to usability tests) and late-development play tests (similar to playability tests) (Luban 2009). While usability tests are frequently used to gather observational feedback on where and why players experience issues, the playability test format is good for gathering quantitative feedback useful in making comparisons, such as comparing a game to a competitive title, comparing groups of players, or comparing one version of a game to another. It is also useful in measuring player satisfaction feedback related to different features, levels, or encounters within a game.

Davis, Steury, and Pagulayan (2005) describe the general playability method used at Microsoft and contrasts this technique with usability testing, focus groups, and Beta testing. Characteristics of the Microsoft playability approach include a focus on initial experience (the first hour of game); use of larger, targeted player samples; use of standardized questionnaires; output focused on identifying issues; and low-cost, rapid results turn-around. Luban (2009) describes testing techniques used at Ubisoft, where testing of games later in the development process begins to emphasize quantitative data from larger numbers of test participants. In addition to the importance of test structure and data collection, Luban also emphasizes the importance of recruiting the right test participants.

As with other techniques, a number of variations exist for playability tests. It is common practice for many online games to make late-development versions of the game available to select volunteers as part of an Alpha or Beta testing program. Often participants are expected to participate in a survey as part of the Beta process. While this type feedback from Alpha and Beta test participants can be extremely valuable, analysts should keep in mind that the self-selected nature of the participants may separate them from average gamers, and the longer timeframes they play may make their feedback different from first-time players. Also, the uncontrolled differences found between players in an Alpha or Beta, their play settings, and their play habits may make comparison between individuals, test groups, or games difficult.

Another variation of the quantitative data gathering playability test approach is to use physiological data collected from players instead of or in addition to player responses to questionnaires. If used successfully, it is hoped that the physiological data can provide insights into the players' physical/emotional state at finer granularity than what is self-reported by a participant. Because this is a relatively novel data collection technique, many are still working to validate whether physiological results match up with players self-reported emotional responses to games. For example, one paper describes guidelines for the use of physiological data gathering (galvanic skin response, electrocardiograph, electromyography of the jaw, and respiration) with games, as well as two studies examining the application of this data (Mandryk, Inkpen, and Calvert 2006). The first of these studies attempted to correlate user satisfaction ratings with physiological data at different difficulty levels; however, the study had inconclusive results due to methodological issues (trying to correlate overall satisfaction point data with time-based physiological data). A second study correlating self-rated satisfaction under two conditions (playing against AI versus players) was more successful. The author cautions that the physiological data can provide additional insight into player emotional responses but does not provide enough information about the player and context when considered on its own. Another study by Nacke and Lindley (2008) successfully correlated physiological responses to different levels designed to emphasize boredom, immersion, and flow.

35.2.5.2 Discussion and Examples

Playability testing is another common test technique at THQ where we leverage the quantitative data to provide another view of the player experience (in addition to the observational, qualitative feedback from usability testing). We often employ this technique to evaluate games that are getting closer to release or when we want to quantitatively compare different games or gameplay elements.

THQ has separate rooms for play testing and usability testing. The regular play test rooms have been configured to seat up to 12 simultaneous users at individual workstations featuring separate screens for console games and for PCs. Note that in addition to supporting testing of PC games, the PCs at each workstation serve dual duty as data collection devices for administering online surveys. We have employed even larger rooms for occasional specific test needs.

The quantitative data gathered from the playability tests allows us to make comparisons between games (e.g., between one game and a competitive title), between versions of the same game over time (e.g., between version 1 and an updated version 2), and between groups of players (e.g., between experienced players and novices). As with any quantitative

analysis, an important consideration in making a fair comparison is to isolate variables. To minimize variance, we administer standardized questionnaires to ensure similar wording, order, and presentation of questions, and we strive to provide a test structure that is identical or as close as possible from one test group to another. For example, when comparing one version of a game to another, we seek to provide similar players with similar amounts of playtime, performing similar activities.

35.2.5.3 Iterative Playability Testing Case Study

35.2.5.3.1 Game

Dawn of War II (DoW II) is a real-time strategy (RTS) game based on the Games Workshop Warhammer 40,000 universe (see Figure 35.2). Both the original Dawn of War and DoW II games were developed by Relic Entertainment. In addition to a single-player story mode, the game supports several online player-versus-player (PvP) modes.

35.2.5.3.2 Test Goal/Design Questions

DoW II introduced or expanded on a number of new play mechanics that were novel or uncommon in the RTS genre, emphasizing attention on a small number of troops rather than management of large armies, as well as acquiring two unique types of in-game resources (Power and Requisition). The developer requested assistance in improving the introduction of these new gameplay concepts to new players, as well as identifying and improving any other PvP usability issues (especially for new players). In addition to identifying issues, we wanted to validate that proposed solutions actually improved the player experience.

35.2.5.3.3 Test Approach

Several rounds of iterative test and update were conducted with groups of players in both the story mode and the PvP mode. Players played the game and completed surveys indicated relative satisfaction levels specific features and the game overall, as well as areas of difficulty. After one round of testing, high-priority issues were identified, and changes were made. Then the process was repeated with the new build and a new set of players, using the same surveys to track changes in player satisfaction and issue identification across builds.

35.2.5.3.4 Findings

On the basis of the survey results in the initial round of testing, issues were identified with some of the new gameplay concepts. For example, some PVP players were not clear on how to acquire and manage the different resource types and differences between controlling resources and controlling victory points. Players also encountered difficulties reviving a hero. Rather than clicking a revival status indicator on the side of the screen, most players clicked instead on the defeated hero. Over the course of testing, specific enhancements were implemented to address the issues (such as new tips explaining control and victory points, visual changes

FIGURE 35.2 **(See color insert.)** Warhammer 40,000: Dawn of War II © Games Workshop Limited 2009. Dawn of War, Games Workshop, Warhammer, 40K, the foregoing marks' respective logos and all associated marks, are either ®, TM, and/or © Games Workshop Ltd 2000–2009. Used under license. All rights reserved.

making type and status of control points clearer, relocating a revival button directly above the defeated hero, etc.), and resulting player ratings were tracked and analyzed.

35.2.5.3.5 Outcomes

The surveys helped to pinpoint areas of difficulties encountered by players. Over time, we were able to track the impact of changes to the game in both the story mode and the online PvP. In the final rounds of testing, players rated their understanding of key issues (such as resource management and victory conditions) more favorably and were consistently experiencing fewer difficulties.

35.2.5.4 Considerations for Online Games

In many ways, playability tests are well suited to testing online games. In particular, the larger sample sizes, quantitative feedback, and test efficiencies make these tests an attractive approach for testing online games with large, competitive populations of players.

Playability tests are typically conducted with larger groups (playing either individually or together). Because many of the online games we test have modes for 4, 6, 8, or even up to 60 simultaneous players in one match or round of gameplay, it is beneficial to have a technique that allows for efficiently managing multiple groups of the necessary number of players.

Quantitative feedback can be particularly valuable for online games. Because many online games involve participation of large numbers of players, being able to quantify and understand the magnitude of an issue (in terms of number of players affected, and the severity of the impact in terms of scores, deaths, or failures) can help the developer prioritize issues. Having quantitative results can also help in the visualization of results in order to pinpoint anomalies, trends, or

patterns. Tracking quantitative feedback also facilitates comparisons that can be useful when assessing the relative performance of a proposed (or implemented) change.

35.2.5.5 Summary

Playability testing, or testing larger groups of players and administering standard questionnaires, is a technique best suited to the collection of quantitative feedback. Because the primary data collection method with this technique is self-reported responses, some actual gameplay details may be missed or misreported. These issues are better captured via usability tests. However, for efficiently gathering feedback to facilitate quantitative comparison of one game to another, between groups of players, or between other controlled groups, the playability test format is ideal. In addition, this technique can be effectively used to evaluate player opinions of fun and level of difficulty with more mature versions of a game.

35.2.6 Video Game Analytics (Test and Track)

35.2.6.1 Description and Literature

Just as popular Web sites have turned to increasingly sophisticated Web usage statistics (or Web analytics) to evaluate the behavior of their users, so too have games turned to logging and analysis to track patterns in the behavior of their players. Tracking video game player statistics is in many ways similar to tracking Web statistics. In order to track statistics, a video game is modified to create a log of player activities and/or outcomes of player activities. For example, the log could include every click of a controller, every movement, every hit, every death, every win or loss. The log is then saved to a local file (if the game is a stand-alone) or saved to a centralized database (if it is an online game). Once the data have been saved, it can be analyzed for patterns and anomalies, such as activities or locations that lead to unexpected player deaths in an FPS, or quests with unusually high success or failure rates in an MMO.

While access to such data for analysis can be valuable, it is important to ensure that the right data are tracked. For example, researchers should consider if the tracked data elements help to answer important questions regarding player behavior. Having too much of the wrong data can confuse the issue and lead to confusion or overanalysis. Starting any video game analytics effort with a clear understanding of the goals and expected outcomes of the initiative is imperative to alleviate some of these issues. For example, Romero describes the following steps to ensuring effective video game analytics projects: instrument (apply automated data collection) to answer questions, define intent (declare and measure), and ensure that research staff works with designers to analyze the data (Romero 2008).

Given the popularity of MMOs, and the availability of rich data sets based on their online usage, it should be no surprise that they are well studied by both their publishers and the academic community. Updates and changes to gameplay are constantly made to MMOs based on analytics. While some publishers keep all or some of this information confidential, some recent academic publications shed some light on play patterns in commercial MMOs. For example, one study tracked player activities in the popular World of Warcraft, specifically examining the size of the player base and how it changes over time, the frequency with which players arrive and depart the game (churn), the location of players in the game world, and the movement of players over time (Pittman and Dickey 2007).

35.2.6.2 Discussion and Examples

At THQ, many of our online games already have the necessary modification to allow them to track player behaviors, so we find that this technique is often used with online games. While the player tracking is often put in place to support analytics during a Beta period or once the game has gone live, it is also extremely useful for testing prior to (or concurrent with) a Beta test. While valuable data are available from Beta test usage, we also conduct laboratory-based tests prior to or sometimes concurrent with Beta testing. Testing in a laboratory allows us better experimental control over certain variables. For example, we can correlate tracked behaviors with player profiles that are provided as part of the recruiting process or gathered through questionnaires completed by test participants. We also have the opportunity to better control other variables, such as the amount of time players spend in certain activities, the order of activities, and so on. While the data elements being tracked in the live game and the test sessions of the game are often similar or identical, the test allows us to control or manipulate other variables in ways that are not possible in a Beta or with the live game.

In practice, we find that tracked data are often a supplementary form of data. When we decide to analyze logs of

player behaviors, it is almost always in addition to another form of feedback, such as survey or in some instances usability feedback. The more typical case is where data logs are correlated with quantitative information gathered in a playability test. For example, we might attempt to correlate player satisfaction levels reported on a survey with frequency of success or failure as reported in the logs. However, the logs can also be useful in gaining a deeper understanding of the causes of observations in a usability test. For example, if a player becomes frustrated during a usability test due to their inability to complete an activity, and the source of their failure is not obvious on the basis of direct observation and review of the session recording, a review of specific player interactions (as available in the log) might clarify why they encountered this particular difficulty.

35.2.6.3 Analytics Case Study

35.2.6.3.1 Game

Red Faction: Guerrilla (RF:G) is an open-world game with several online player-versus-player (PvP) modes developed by Volition, Inc. In additional to many features common to PvP in other first-person shooter (FPS) games, RF:G provides players a fully destructible environment, as well as a variety of items that confer special abilities on players, giving players opportunities for unique gameplay experiences (see Figure 35.3).

35.2.6.3.2 Test Goal/Design Questions

In order to balance and tune the gameplay, the developer wanted to better understand how players responded to various features in a Beta build of the game. In particular, the developer wanted to better understand the frequency with which players found and used certain in-game items, as well as to correlate and compare item usage with initial satisfaction ratings, comparing hardcore FPS players and gamers with less experience. Additional playability issues were also explored as well.

35.2.6.3.3 Test Approach

Groups of players, both hardcore FPS players and less experienced players, were brought in and instructed to participate in multiple PvP matches. Following each match, participants completed a survey identifying issues and gameplay highlights, as well as tracking relative satisfaction across a number of factors. The system was also instrumented to capture logs of player actions (such as movement, kills/deaths, and pickup and use of items). Following the sessions, the survey results and data logs were analyzed for trends and differences.

35.2.6.3.4 Findings

Both groups encountered similar issues with certain controls and gameplay features (such as the scoring mechanic in one of the PvP modes). While both groups viewed the PvP gameplay in the Beta positively, specific differences were found between the groups with regards to their awareness and satisfaction with some of the unique elements of gameplay introduced by the items and destructible environment. Also,

FIGURE 35.3 **(See color insert.)** Developed by Volition, Inc. THQ, Volition, Inc., Red Faction: Guerrilla and their respective logos are trademarks and/or registered trademarks of THQ Inc. All rights reserved. All other trademarks, logos, and copyrights are property of their respective owners. © 2009 THQ Inc.

the data helped to identify imbalances in the usage of certain weapons and items, as well as some unexpected player satisfaction issues. For example, although it did not have the highest kill/death ratio (a measure of weapon effectiveness), players still rated the rocket launcher/jetpack combination as the most satisfying.

35.2.6.3.5 Outcomes

As a result of the findings from this test, the developer was able to modify elements of the UI and gameplay to correct some of the issues and misunderstandings encountered by all players. Also, training and community content was updated to better introduce the unique aspects of the fully destructible environment in RF:G. While some aspects of the items usage patterns were a result of the test format (first hour of gameplay), other factors were due to the location and placement of items within the levels. Level designs changes were made as a result of the play test feedback to make certain items more obvious or attractive to players.

35.2.6.4 Considerations for Online Games

Online games are a natural fit for the logging and analysis of player data. By their nature, most games already have an infrastructure for gathering such data. So the most significant tasks for the player researcher is to work with the design and management teams to identify what measures are likely to be meaningful and useful early in the project to ensure that they can be easily tracked once testing (Playability, Beta, or other forms) begins.

Data availability is another consideration. Researchers and other team members will need some mechanism for viewing and analyzing the data. This can take the form of a dynamic dashboard showing current key metrics, regular reports showing similar data, or access to the raw data through spreadsheets or data mining and visualization tools. Again, early planning with technical team members to ensure that the necessary levels of information are available to appropriate team members is an important activity.

The opportunity to correlate player behaviors with other profiling information may be important. For example, if Beta participants complete questionnaires, are there ways to correlate their survey responses with their tracked in-game activities. Being able to track individual players in some form can be important in such situations, though researchers must also take care to respect any privacy policies or concerns in gathering or disseminating results.

35.2.6.5 Summary

Computer-generated logs of player behaviors are another valuable asset in evaluating video game playability. Conducting analysis of these logs (gathered during Beta or during test sessions) prior to the release of a game gives another quantitative view of player behaviors. Tracked data often include time-stamped logs of player activities, movements through the system, and important outcomes of activities (victories, death, successful completions, failures, etc). Correlating these data with survey data is a particularly effective combination for examining the impact of and player response to various aspects of gameplay.

35.3 CONCLUSION

Video games, especially online video games, present a number of unique usability challenges. These challenges can vary based on the type of game, the target player population, the phase of development, and any of a number of other factors.

To best evaluate and address these challenges, game teams should be able to draw on any of the current techniques in evaluating game usability, such as heuristic analysis, usability testing, playability testing, game analytics, or some combination of these. Correctly applying these techniques is an important element of understanding how gamers actually experience our game, and ultimately, a fundamental tool in the process to make better games.

ACKNOWLEDGMENTS

I would like to thank the many people who assisted in our testing, reviewed versions of this chapter, or contributed to the case studies, including Ben Serviss of THQ*ICE, Brenda Ramos of THQ, Brooke White of Volition, Dina Burton Steiner, Heather Desurvire of Behavioristics, Jeff Lydell of Relic Entertainment, Kenneth Schroeder, Paul Sherman of ShermanUX, Richard Browne of THQ, Tom Smith of Disney Interactive Studios, and Travis Riffle of Rainbow Studios. The views and opinions expressed in this chapter are those of the author and do not reflect an official position of THQ, Inc.

REFERENCES

Bangeman, E. 2008. Growth of gaming in 2007 far outpaces movies, music. Ars Technica. http://arstechnica.com/gaming/news/2008/01/growth-of-gaming-in-2007-far-outpaces-movies-music.ars (accessed Nov. 11, 2009).

Davis, J., K. Steury, and R. Pagulayan. 2005. A survey method for assessing perceptions of a game: The consumer playtest in game design. *Game Studies* 5(1).

Desurvire, H. 1994. Faster, cheaper!! Are usability inspection methods as effective as empirical testing. In *Usability Inspection Methods*, eds. J. Nielsen R. Mack, 173–202. New York: John Wiley.

Desurvire, H., and C. Wiberg. 2009. Game usability heuristics (PLAY) for evaluating and designing better games. Paper presented at HCI 2009. San Diego, CA, July 19–24.

Desurvire, H., M. Caplan, and J. Toth. 2004. Using heuristics to improve the playability of games (HEP). Paper presented at CHI Conference. Vienna, Austria, April 24–29.

Entertainment Software Association. 2009. Computer and video game industry tops $22 billion in 2008. http://www.theesa.com/newsroom/release_detail.asp?releaseID=44 (accessed Jan. 28, 2009).

Entertainment Software Association. 2008. Essential facts about the computer and video game industry. http://www.theesa.com/facts/pdfs/ESA_EF_2008.pdf (accessed Nov. 12, 2009).

Federoff, M. 2002. Heuristics and usability guidelines for the creation and evaluation of FUN in video games. Thesis. Indiana University, Bloomington, IN, December.

Fulton, B., and M. Medlock. 2003. Beyond focus groups: Getting more useful feedback from consumers. In *Proceedings of the Game Developers Conference* (San Jose, CA, February). http://www.mgsuserresearch.com/publications/.

Hosniemi, J., P. Hamalainen, and L. Turkki. 2004. Wizard of Oz prototyping of computer vision based action games for children. In *Interaction Design and Children*, 27–34. New York: ACM Press.

Korhonen, H., and E. Koivisto. 2006. Playability heuristics for mobile games. In *MobileHCI 06*, 9–16. New York: ACM Press.

Laitinen, S. 2005. Better games through usability evaluation and testing. Gamasutra. June 23. http://www.gamasutra.com/features/20050623/laitinen_01.shtml (accessed Aug. 28, 2008).

Laitinen, S. 2006. Do usability expert evaluation and test provide novel and useful data for game development? *Journal of Usability Studies* 1(2): 64–75.

Lenhart, A., S. Jones, and A. Macgill. 2008. Adults and video games. Pew Internet and American Life Project. Dec. 7. http://pewinternet.org/Reports/2008/Adults-and-Video-Games.aspx (accessed Nov. 8, 2009).

Lenhart, A., J. Kahne, E. Middaugh, A. Macgill, C. Evans, and J. Vitak. 2008. Teens, video games and civics. Pew Internet & American Life Project. Sept. 16. http://www.pewinternet.org/Reports/2008/Teens-Video-Games-and-Civics.aspx (accessed Nov. 8, 2009).

Luban, P. 2009. The silent revolution of playtests (parts 1 and 2). Gamasutra. April. http://www.gamasutra.com/view/feature/3985/the_silent_revolution_of_.php (accessed Nov. 8, 2009).

Mandryk, R., K. Inkpen, and T. Calvert. 2006. Using psychophysiological techniques to measure user experience with entertainment technologies. *Behavior and Information Technology* 25(2): 141–158.

Medlock, M., D. Wixon, M. Terrano, R. Romero, and B. Fulton. 2002. Using the RITE Method to improve products: A definition and a case study. Orlando, FL: Usability Professional Association.

Nacke, L., and C. Lindley. 2008. Flow and immersion in first-person shooters: measuring the player's gameplay experience. In *FuturePlay*, 81–88. New York: ACM Press.

Nielsen, J. 1994. *Heuristic Evaluation*. New York: John Wiley.

Nielsen, J. 2007. Multiple-User Simultaneous Testing MUST (Jacob Nielsen's Alertbox). http://www.useit.com/alertbox/multiple-user-testing.html (accessed Feb. 1, 2010).

Pagulayan, R. J., K. Keeker, D. Wixon, R. Romero, and T. Fuller. 2002. User-centered design in games. In *Handbook for Human–Computer Interaction in Interactive Systems*, eds. J. Jacko and A. Sears, 883–906. Mahwah, NJ: Lawrence Erlbaum.

Pinelle, D., N. Wong, and T. Stach. 2008. Heuristic evaluation for games: Usability principles for video game design. In *CHI 2008*, 1453–1462. New York: ACM Press.

Pinelle, D., N. Wong, and S. Tadeusz. 2008. Using genres to customize usability evaluations of video games. *FuturePlay*, 129–136. New York: ACM Press.

Pinelle, D., N. Wong, T. Stach, and C. Gutwin. 2009. Usability heuristics for networked multiplayer games. In *GROUP09*, 169–178. New York: ACM Press.

Pittman, D., and C. Dickey. 2007. A measurement study of virtual populations in massively multiplayer online games. In *NewGames '07*, 25–30. New York: ACM Press.

Romero, R. 2008. Successful instrumentation: Tracking attitudes and behaviors to improve games. Paper presented at Game Developers Conference. San Jose, CA, Feb. 28–Mar. 4.

Rubin, J. 1994. *Handbook of Usability Testing: How to Plan, Design and Conduct Effective Usability Tests*. New York: John Wiley.

Steiner, K., and T. Moher. 1994. A comparison of verbal interaction in literal and virtual shared learning environments. In *Conference on Human Factors in Computing Systems (CHI)*, 97–98. New York: ACM Press.

Steiner, K., and T. Moher. 2002. Encouraging task-related dialog in 2D and 3D shared narrative workspaces. In *Collaborative Virtual Environments (CVE)*, 39–46. New York: ACM Press.

Sweetser, P., and P. Wyeth. 2005. GameFlow: A model for evaluating player enjoyment in games. *ACM Computers in Entertainment*, Article 3A.

36 What Is in an Avatar? Identity, Behavior, and Integrity in Virtual Worlds for Educational and Business Communications

Keysha I. Gamor

CONTENTS

36.1 INTRODUCTION

There is no single, agreed upon definition of "virtual world." However, all definitions acknowledge that a virtual world is an online simulation of either a real or fantasy world environment populated by "avatars," which are pictorial or graphical representations of the human participants. A virtual world can also be described as "a synchronous, persistent network of people, represented as avatars, facilitated by networked computers" (Bell 2008, 2). EDUCAUSE, a nonprofit association concerned with leveraging technology to improve higher education, defines a virtual world simply as an "online environment whose 'residents' are avatars representing individuals participating online" (the EDUCAUSE Learning Initiative 2006, 1). Still other definitions that address the specific affordances of this modality help us understand the potential of the technology as well. Examining popular virtual world applications can help frame an understanding of virtual worlds as "online 3-D virtual worlds ... within which residents are able to establish identities (avatars), explore, create and communicate. [Further, a virtual world may] lend itself well to social networking, collaboration and learning" (IEEE).

Regardless of the number of varying definitions of virtual worlds, it is clear that virtual interaction and communication are here to stay. Virtual collaboration through the use of avatars is a reality for the future of education, business, and social interactions. With advances in technology continually yielding more realistic avatars, an examination of the role of the avatar is an important human factors issue to reinforce the integrity and enhance the experience of future virtual interactions. This chapter explores the avatar as a navigational tool, a representation of self, and an ambassador of one's educational institution, company, or organization to the virtual world.

36.2 VIRTUAL WORLDS AS COLLABORATIVE TOOLS

Virtual worlds are graphically rich tools that support first-person, individual exploration and group collaboration. Although traditional distance learning modalities offer many benefits, such as flexibility and cost savings, common complaints students have about distributed or distance learning is a feeling of being disconnected from the other participants or of lacking peer interaction or of lacking a sense of community (Jung et al. 2002; Shin 2003; Valentine 2002). However, virtual worlds can help to address these challenges through the manner in which they enable synchronous and asynchronous learning. Specifically, virtual worlds facilitate a learner-centered approach wherein participants can determine when and how to navigate through the learning experience—all in a graphically and contextually rich environment with access to many, if not all, the same collaboration tools participants have become accustomed to in the real world. Thus, participants may use virtual worlds to do things with a greater level of depth, involvement, engagement, and interaction than other mediums. Participants and learners can use virtual worlds to familiarize themselves with content; practice processes, procedures, demonstrations, and problem-solving and decision-making activities; conduct self-assessments; and craft self-remediation approaches. Similarly, participants may also use virtual worlds to test their understanding of content by sharing their views with peers, negotiating meaning/understanding together with peer groups, and examining the impact of others' interpretations in context, thereby supporting problem identification (Chin and Williams 2006; Merrill 2007; Jonassen 2000; Palloff and Pratt 2000). Research supports the notion that being able to identify a problem is critical in becoming an expert at solving them (Dede 2005, 2007). Virtual worlds provide a unique environment for problem finding because they provide a multisensory, immersive, graphically rich way to represent an authentic environment, communicate, collaborate, coexist, and co-create with other participants.

With all these benefits, it is no surprise that virtual worlds are being used in multiple settings, including education and business. Virtual worlds offer the added benefit of cost savings. Robert Gehorsam (2009), from Accenture Technology Labs, maintains that for corporate training, virtual worlds decrease training costs, by decreasing associated travel costs and offering flexibility. Supporting Gehorsam's stance, according to ThinkBalm (2009), a leader in immersive Internet research, respondents to their 2008/2009 survey indicated that the main reasons for investing in virtual worlds were to foster collaboration of geographically dispersed staff, save money, and motivate employees. While cost savings alone is not a sufficient justification for investing in virtual worlds, when "green communication" is aligned with the other benefits of virtual worlds, then suddenly a more complete vision of early twenty-first century collaboration potential emerges. Moreover, given the current global economic downturn, in what I call "green communication" (a lower carbon footprint as a result of less travel and minimal consumption of natural resources, easily recorded and reusable interactions, time saving, etc.) has become a necessity, not a nicety (Wilson 2008).

Figure 36.1 illustrates the benefits that organizations may seek through what I call "green communications and collaborations." In the ThinkBalm study previously referenced, nearly 60% of respondents indicated that their immersive technology project was the least expensive of their collaboration

FIGURE 36.1 (**See color insert.**) Organizations see the benefits of green communications and collaborations.

alternatives and nearly 30% reported that they recouped their investment in less than 9 months from the launch date. The same study indicated that 94% of the respondents believed their projects were successful. Increases in productivity are generally calculated based on avoiding travel time and just-in-time refresher learning, so with well-planned meetings and events and easily accessible archives, positive gains are expected.

Although videoconferencing, Web conferencing, teleconferencing, and other online collaboration methods are already incorporated into most education and business collaboration toolkits and have helped many organizations realize cost savings, virtual worlds embody the power of all these, plus more. Virtual worlds provide a level of engagement that the other options alone cannot rival because virtual worlds enable human communication and interactivity similar to that of the real world (Castranova 2007). The engagement centers on the avatar, the mechanism that enables users to enter and operate in-world. Typically, an avatar will interact with the virtual world itself (navigating through and exploring the three-dimensional [3D] graphical environment). An avatar can also collaborate and interact with other avatars—other participants who are also in the 3D graphical environment. Virtual objects exist for the purpose of being acted upon; therefore avatars can become part of, manipulate, modify, and interact with virtual objects. As shown in Figure 36.2, engaging in a virtual world through the avatar enables multi-faceted interactions that potentially involve contextually rich, authentic, and meaningful content and experiences.

Indeed, communication and interaction are just two of the types of collaboration that can occur. Research shows that collaboration is a powerful instructional tactic (planned activities) and learning strategy (participant-initiated activity) in virtual environments (Heiphetz and Woodill 2010; Roberts 2004). These collaborative interactions are but a few of the ways in which an experience in virtual worlds can exploit the strength of the tool. Virtual worlds represent useful environments for collaboration, risk mitigation, or "collective problem resolution via mediated interaction" (Dede 2007) because they enable both realistic and "fictional"

Interactions in
virtual worlds

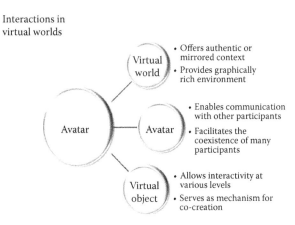

FIGURE 36.2 Interactions available to each avatar in a virtual world.

contexts for learning. In this sense, one can create or duplicate a real-life environment that is only as fictional as to the degree which all risk is removed and principles of physics are defied. The prime learning objectives and context remain, but the potential for loss, harm, and discouragement are diminished or eliminated in order to facilitate more in-depth experiences, understanding, and awareness. The fictional context, however, offers an opportunity to do what cannot otherwise be done. For instance, in no context other than a virtual world can an individual "walk through" or "become part of" a particle, data set, or organism. Certainly, in no other context could such an exploration accommodate a large group of individuals all interacting simultaneously. Further, in no other context could geographically dispersed individuals construct an object together in a tangible, graphically rich environment where the geographical dispersion is replaced by a sense of "thereness" that creates a fluid interaction much like that experienced in real-life interactions. Such life-like interactions leave the participants with a perception of having "been someplace," "experienced something firsthand," and "connected with others." It is these benefits that make virtual worlds attractive collaborative tools for education and business communication because the sense of distance and disconnectedness are diminished.

Virtual worlds also have the benefit of affording participants a place to go in order to experience something firsthand as an individual and collaboratively with other people. Regardless of which platform, there are six affordances most virtual worlds have in common (Virtual World Review 2009; Federation of American Scientists 2009):

- Co-existence—allows many users to participate simultaneously in a shared environment
- Graphical user interface—offers visual depiction of and means of interaction with environment
- Presence—affords real-time interaction; direct and indirect interaction/synchronous and asynchronous interaction
- Co-creation—supports content development or modification

- Collaboration—encourages development of in-world groups and teams, creation of meaning, and connections to expert resources
- Persistence—maintains 24/7 existence, regardless of user login status; the presence and processing of synchronous and asynchronous interactions and contributions of all avatars and objects within the world

However, as shown in Figure 36.3, they all seem to converge in such a way that the most unique affordance, persistence, becomes a reality through the avatars, environment, and objects within the virtual world. Figure 36.3 represents the conceptual framework that depicts how virtual worlds may enable experiences that are meaningful and successful.

Maximizing each of these inherent virtual world affordances may support a better learning experience. Cory Ondrejka (2008), researcher from Linden Lab, credits virtual worlds with helping to create a culture of sharing and pervasive learning in the company by bringing people together in a dynamic workspace. The How People Learn framework (Bransford, Brown, and Cocking 1999), developed as a useful construct to consider when designing Web-based learning opportunities, seems to support Ondrejka's assertion. According to the framework (see col. 1 Table 36.1), there are four optimal learning conditions centered on the participant, knowledge, assessment, and community that must be taken into account in the design of a successful online learning environment. Focus on the participant is known as "learner-centered" instruction. Learner-centered instruction makes learner attributes priority such as culture, interests, language, experiences, etc. Equally important in an optimal learning environment, knowledge-centered instruction builds upon the existing knowledge of learners as well as the ways of knowing in a particular discipline, as springboards to building new knowledge. Assessment-centered learning is pivotal in the development of an optimal learning environment owing to its focus on ensuring that what is assessed aligns with the learning goals. In addition, assessment-centered learning

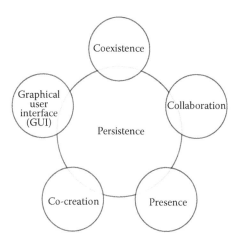

FIGURE 36.3 Affordances of virtual worlds work together to create a persistent learning environment.

TABLE 36.1

How People Learn Framework and the Affordances of Virtual Worlds

How People Learn Framework	Predominant Affordances of Virtual Worlds	Virtual World Design Considerations
Learner centered	Presence	Create individual and group activities
	Coexistence	Plan for synchronous and asynchronous interactions
	Persistence	
Knowledge centered	Graphical user interface persistence	Develop interactive objects beyond basic presentation slides and videos
		Make use of notes, basic building capabilities, and problem identification activities
		Look for platforms that enable remote ways to stay connected to the world via communication/interactions, file portability, asset ownership
Community centered	Co-creation	Include activities and opportunities for multiple perspectives to converge
	Collaboration	
	Persistence	Exploit tried and true instructional strategies that foster collaboration
		Reward collaboration
Assessment centered	Co-creation	Use synchronous/asynchronous learning opportunities
	Collaboration	Capitalize on avatars, objects, and the environment's persistent nature
	Coexistence	
	Persistence	

environments provide opportunities for feedback, reflection, and revision. While the importance of feeling included has been well documented as a critical variable in student achievement (Brown and Campione 1994; Church, Elliot, and Gable 2001; Cobb, Yackel, and Wood 1992; Dalton, Elias, and Wandersman 2001), community-centered learning has earned new recognition with the advent of social media (Alexander 2006; Anderson 2007; Brown and Adler 2008). Peer-to-peer, participant-to-instructor, and participant-to-expert communication are synergistic interactions that enhance participant motivation and achievement (Bransford, Brown, and Cocking 1999). Together, these four conditions help to optimize the learning experience. All four of these conditions are inherent in the nature of collaboration that virtual worlds support.

Subsequently, for the purposes of this chapter, the affordances of virtual worlds are examined through the lens of Bransford's theories on how people learn (Table 36.1). While some of the affordances supporting persistence may not be exclusive to virtual worlds as a technology medium, the concept of persistence, as defined in this chapter is unique to virtual worlds and ushers in new implications for communication in education and business organizations.

Understanding how people learn and how virtual worlds can support learning and collaboration positions educators and businesses to begin virtual world projects with a clear mission and set of goals that may improve chances of experiencing successful implementations. Collaborating with others, rather than participating in learning events on one's own, fosters peer-to-peer coaching, transfer of knowledge, and knowledge sharing, thus building communities and educational/business environments based on cooperation.

Effective online communication through cooperation and collaboration are critical in establishing this type of organizational culture. Virtual worlds enable the level of engagement, knowledge sharing, and depth of experience that many organizations are seeking. However, these experiences can only be realized when one's online identity and behavior invokes trust—real or virtual. As avatars are the mechanism through which collaboration occurs, it is all the more critical that special attention is given to the avatar's role in a virtual world.

The remainder of this chapter focuses on the avatar as the agent of communication in education and business interactions in virtual worlds. This chapter explores human factors issues around the avatar as a navigational tool, a representation of self, and an organization's ambassador to the virtual world. Through its identity, behavior, and its ability to invoke trust, online educational and business experiences will realize new social implications in communication.

36.3 WHAT IS AN AVATAR?

What is an avatar, and why would anyone want one? To participate in virtual worlds, each participant must have an avatar, as it is the mechanism through which navigation and interactions occur. Avatars have the task of being the social communicative agent through which participants transmit all communication cues, both verbal and nonverbal (Bente et al. 2008). The characteristics of avatars can be as varied as humans, but there are some key attributes that are important in order to convey credibility and to gain trust in and access into various communities. Avatar as a concept is not new; neither is virtual world; however, the contexts in which the two concepts are used today are still evolving. Currently, one may observe avatars used in entertainment and serious games, virtual worlds, e-mail, and social networks. Even within the scope of current avatar use, the manner and characteristics differ. This section specifically examines avatars in virtual worlds.

The Association of Virtual Worlds Blue Book helps people get started in virtual worlds by first explaining what an avatar is through its etymology: "Avatar comes from Hindu mythology and means the incarnation of a divine being. But in the virtual world an avatar is an icon or representation of a user" (Association of Virtual Worlds 2008, p. 5). Also in the context of a virtual world, avatar is described as the "fullest possible expression of self in the online environment" (Castronova 2007, 51). Avatars are characters, representations, or expressions of oneself created for the purpose of communication and interaction; they have names and they develop personalities and represent moods within various locations (Meadows 2008; Oravec 1996). Owners of avatars can choose to create avatars that are close approximations of who they are or they can experiment with multiple identities through a collection of different avatars that are designed for various purposes based on mood, location, intent, event, or other factors (Galanxhi and Nah 2007; Holzwarth, Janiszewski, and Neumann 2006; Jin 2009).

Researchers agree that the act of selecting an avatar is as significant as deciding which clothes to wear to an interview or other important meeting (Lee, Kozar, and Larsen 2007; Meadows 2008; Oravec 1996; Taylor 2003). Having access to some level of customization (plasticity) of the avatar is a unique property of virtual world avatars (Balienson et al. 2006) and is a key characteristic that separates an avatar from an "agent." In the early days of video games, a player was represented by some fixed character like Pac man, Mario, or in some cases an unnamed object. Operating a representative token, in this sense, is very similar to having a game piece in a board game, but with even less choice involved. Studies suggest that the act of customization positively correlates to owner attachment and motivation (Blascovich 2002; Meadows 2008; Slater and Steed 2002; Taylor 2002). Although some may find this unbelievable, it is at least understandable. For instance, when Second Life, a popular social entertainment virtual world first launched in 2003, new users were assigned an avatar based on the gender they selected. The avatar was generically clothed, making the new user's first mission to acquire acceptable clothing. Users also had fewer options for customizing their avatars beyond the look of the system default. Today, Second Life avatars are still provided with default settings already in place. Everything from hair style, facial features, and clothing are preselected. During the orientation, new users learn how to modify their appearance. Still, the first mission many users undertake is to change their outfit and overall look to something that appeals to them and distinguishes them from "newbies." The act of customizing an avatar takes time, effort, interest, and involvement, and in Second Life, if the new user would like current fashion, then shopping is

often required. Indeed, as the new user meets new people, new clothing or other objects may be given to them. Users are motivated to tailor their avatars once they see the level of detail and customization that is possible. Being able to make modifications to one's own avatar is added to the top of the list of skills that many users want to learn. Diversity in Second Life is evolving and adding to the seemingly endless variations among avatars. Initially, users could only appear as a Caucasian male or female until additional skin tones were added. Now, more skin tones, hair styles and textures, and facial features are available. Users can also either purchase customized, complete avatars or build one themselves. Today, avatars in Second Life can appear as human, animal, fantasy creature, or anything one can create.

In Teleplace, a virtual world that is designed for business use, avatar selection is somewhat fixed—as is the case in most other virtual world platforms currently available. All avatars in Teleplace are preset: appearance, clothing, and gender are fixed. For example, users can appear as a business woman or man in a suit or military fatigues. Initially, in Teleplace (then Qwaq), human avatar options only included Caucasian males and females with different hair color. However, new users have the option to choose an inanimate object such as a yellow balloon or a box figure that can then be customized with a company logo and/or photo, graphic, or other desired visual representation. Recently, Teleplace has been working on offering a more diverse selection of avatars that represent more ethnic, job role, and physical diversity. The initial selections of avatars among virtual world platform developers were based, at least partly, on the assumption that as long as a user had a representative player, then he/she was involved and included. However, it seems that owning an avatar and connecting with that avatar are two issues that are not quite that simple.

To state that an avatar is the embodiment of expression by the owner would be a true, albeit, overly simplistic definition of what an avatar is, as research has indicated that the concept of avatar has many far reaching implications that need to factor into how it is described. Detailing all of the literature is beyond the scope of this chapter, but a high level overview of a select group of findings is helpful in understanding the power that lies within the avatar.

The avatar is the conduit through which participants experience and contribute to the virtual world, so they tend to want their avatars to be able to have as many expressions as there are in the "real or physical world" (Meadows 2008). There is the inclination toward our human need for visual cues in order to help make meaning and decisions based on communication with other avatars in virtual worlds (Galanxhi and Nah 2007; Nowak 2004; Nowak and Rough 2006; Weyers et al. 2009). To address this need, owners can go in search of gestures to buy and/or apply to their avatars. Visual and/or auditory gestures are available in some virtual worlds, such as Second Life. Gestures add a sense of realism and believability to the avatar. It helps other participants make a connection between the avatar and the person behind it. There is a person behind every avatar.

In a virtual world, the avatar is also both a navigational and experiential tool. Consequently, there are many human factors to consider; however, after a brief overview of avatars as navigational tools, this chapter focuses on the avatar as an experiential tool.

As a navigational tool, avatars move about through several methods: key strokes, teleportation, flying, and direction keys. Currently this rather rudimentary method of navigating, which is also reliant on bandwidth, can result in unrealistic and/or unintended movement and user frustration. Additionally, current avatar navigation can contribute to varying degrees of "simulator sickness," such as light headedness, nausea, dizziness, and/or in severe cases, vomiting. While simulator sickness is believed to occur in video games and fully immersive virtual reality, there have been a few reports of dizziness and lightheadedness in virtual worlds. It is believed that this sort of response is more likely to occur when there is rapid change in "head on" direction, resulting in a mismatch between perception caused by visual motion and physical vestibular cues. Most research conducted on simulator sickness is based on fully immersive virtual reality where participants use head mounted displays and other peripheral devices; however, simulator sickness in other virtual environments has been researched and observed as well (Bonato et al. 2008; Jerome 2007; Kolasinski 1995; Savage 2007). It has been suggested that graphics or hardware quality affect the potential for simulator sickness. More research is needed to determine the factors that impact users' experiences in 3D virtual environments.

Because this chapter focuses on the experiential aspect of the avatar, emphasis is placed on the avatar as a representation of self and a representative of the organization with which the avatar is affiliated. Whether for social or business purposes, people use virtual worlds for personal and group experiences. As representations of self, participants ascribe personal connections that enable them to engage in the virtual space as an extension, alternative, or augmentation of the real world, which seems to increase the sense of presence or telepresence. Thus, we see the adherence to social norms and behaviors, such as observance of personal space, "eye" contact, and attention to appearance, emotions, and gesturing, typically seen in face-to-face interaction. As representatives of their respective organizations, avatars are entrusted as ambassadors of the organization, bringing with them their own individual idiosyncrasies—just like in the real world. Because there is much left unsaid and unseen in the virtual world, avatars carry a sizable burden to accurately portray oneself and organization.

Although there is not much information on the influence of the avatar in communication and interaction, well-established research indicates that the primary goal in online communication and interaction is to determine the level of trustworthiness of other participants by reducing uncertainty. This is known as the reduction theory, which further posits that in addition to trying to interpret current avatars' behaviors, online users also attempt to predict how they will behave in the future (Nowak and Rauh 2006). Other theories should be examined to help construct a more comprehensive

prediction of human behavior and perception processes conducted via avatars in virtual worlds. For instance, cognition-based trust theory submits that rational judgment is used to determine credibility, competence, reliability, and knowledge (Kanawattanachai and Yoo 2002; Meyerson, Weick, and Kramer 1996). Affect-based trust refers to the emotional bond between individuals or confidence in individuals to share and protect one's own interests (McAllister 1995). Although there are conflicting findings in the limited literature that exists today, they all agree that trust is a critical factor in establishing meaningful online relationships. Trust is important in reaching a level of comfort that enables open, free communication and interaction. Given that the avatar is the primary mechanism of communication in virtual worlds, determining its credibility is the first step in moving an online relationship forward.

Behavior in virtual worlds is similar to that of the real world in that humans rely heavily on visual cues. Indeed, the nonsighted community also has well-established ways of creating images using other senses such as touch and sound. In virtual worlds, users also rely on visual cues to shape their perceptions. Much like we develop first impressions in the real world, we do the same in virtual worlds—through direct and indirect interaction. Consistent behaviors help to build more lasting perceptions that have some bearing on relationship development and maintenance. Research further suggests that different types of avatars have different effects on how the person behind the avatar is perceived. In other words, the perceptions the avatar evokes are transferred to the real person (Rauh, Polonsky, and Buck 2004). Indeed, if a determination can be made about specific characteristics and the influence they tend to have, then characteristics of avatars can be modified to elicit the desired effects, impressions, and reactions. More research is needed to determine which characteristics have what impact.

Although research has not been consistent on the effects of specific types of images, studies have consistently shown that avatars have significant effects on the perception process (Bente et al. 2008; Nowak and Rauh 2006). Some believe that the instinctive action of sorting objects into categories, such as inanimate objects, animals, or humans, in order to make meaning is also applicable to virtual interactions (Kunda 1999). Researchers suggest that androgyny and anthropomorphism (concrete factors) and avatar behavior and attractiveness (abstract factors), significantly influence sense of homophily and positive perceptions of avatars (Nowak and Rauh 2006). Nowak and Rauh found that visible characteristics were more important to users, and androgyny was more significant than anthropomorphism. They also found that users tended to prefer human avatars, suggesting that users found them to be more attractive, homophilous, and believable.

Researchers found that gender differences are exaggerated in virtual worlds and that people generally preferred avatars that looked like them (Biocca and Nowak 2002). Research has also suggested that the gender of the person behind the avatar can be determined by the way the avatar is created. Meadows (2008) found that men who chose female avatars overly exaggerate the female's features, whereas females who chose female avatars selected normal features. Although Biocca and Nowak found that the most attractive avatars were considered the most credible and homophilous, given Meadows' findings, it is unclear as to whom the avatar was intended to attract in the first place. Some people anecdotally report that females are treated nicer in virtual worlds, whereas it is more difficult to make friends if you are male.

As is the case in entertainment and social virtual worlds, isomorphism between the avatar and its owner also has social implications for serious collaborations as well. Still, in their study, Nowak and Rauh (2006) concluded that the more anthropomorphic, lowest androgynous avatars were considered the most attractive, credible, and homophilous, although the reasons are still unclear. The fact that users report preferences for avatars that are like them suggests that other characteristics may also be affected by "similarity preferences." More research is needed to explore these issues, as well as other diversity and socio-emotional issues in virtual worlds if they are to assume a more prominent place in the business and education collaboration toolkit.

Because the use of avatars in virtual worlds is the standard method of navigation and interaction, examining the relative impact they have on the virtual world experience may help organizations develop policies that will shorten implementation time and smooth the transition from traditional collaborative tools to 3D, immersive collaborative tools.

36.4 AVATAR AND COMMUNICATIONS

The role of the avatar expands beyond just a navigational tool and representation-based icon—it is an experience-based entity that carries far-reaching implications for the owner of the avatar and the world in which the avatar navigates. As reports about personal responsibility and accountability and social media have become central topics in the news and research, the added dimension that virtual worlds enable through the avatar brings about a whole other level of risk for the individual and the organization. Although self-expression and creativity are foundational to virtual worlds, the use of virtual worlds in most organizations does not require it. Given the risk, perhaps avatar operating guidelines are in order. First, through an examination of the key attributes of the avatar (Figure 36.4), their potential usefulness is revealed.

36.4.1 IDENTITY

The initial perception of an individual online is presented via the avatar. One's appearance and attire connote something about the individual, whether intended or not. While in online social settings this may not be as critical, in more serious settings such as business or formal education, identity could have far reaching implications beyond the graphical representation and the individual it represents. The avatar,

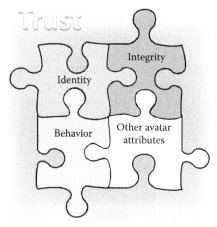

Key avatar attributes for trust

FIGURE 36.4 Key avatar attributes necessary to engender trust.

then, is not just representative of the individual; it is also intertwined with the organization with which the individual is affiliated. This point raises many questions, such as does the organization have the right and responsibility to ensure that the interactions and communications in its world are representative of the organization's brand identity or real world presence? Pillsbury, Winthrop, Shaw, Pittman LLP (2009, 17), a leading firm in virtual law, asserts that "many . . . issues will arise as . . . organization[s] . . . explore virtual worlds, including questions about the ownership of your company's virtual items and property, issues regarding your employees' on and off-clock behavior . . . , and concerns about replicating real-world intellectual property in virtual worlds."

36.4.2 BEHAVIOR

Behavior goes hand-in-hand with identity. Appearance and attire are one thing, but the actions of the avatar are equally, if not more, important. Kolbert (2001) reported avatars stripping clothes off and shouting profanities to express protest against issues and events occurring inside the virtual world. Human interaction elicits the same behavior in the virtual world that is evidenced in the real world. As long as there is human interaction, humans will respond in ways that are, for the most part, predictable. However, the perception of anonymity may reduce inhibitions, as people seek out ways to have their voices heard. There is a human behind every avatar, so organizations need to anticipate basic human responses to in-world events.

Through chat using voice over Internet protocol (VoIP), avatars can enjoy some of the familiar comforts of communication. Mild utterances can have similar effects as they do in the real world, except you do not have the non-verbal cues to support the nature and intent of the utterances. Although avatars have some crude capability of communicating nonverbal cues, by and large, considerable improvement is needed to better simulate real-life human expression. This is yet another reason why efforts are needed to attempt to achieve an accurate portrayal of self and organization from the outset.

36.4.3 Integrity

We have all heard of "truth in lending." In virtual worlds, there is the need for "truth in representation." Representation of user and the organization are embodied in the avatar. Can you imagine collaborating with someone online and then meeting them in person only to find out that they are the very opposite of how they represented themselves online? Although the gender, race, creed, national origin, or personal data about an individual should not matter, the fact is it is human to wonder why a person elected to portray themselves as they do. In life, we do not have any power over what we are born into, but in the virtual world, we have the choice to be reborn into any state we want—as both human and nonhuman objects alike. It is true that, for now, speculation prevails, but research needs to focus on the effects of online, virtual communication and relationships on real-life business decisions (Feldon and Kafai 2007; Heiphetz and Woodill 2010; Nowak 2004; Nowak and Rauh 2006).

Just as organizations bear some responsibility for student and employee behavior in the real world, they too are exposed to similar (if not the same) risks in the virtual world. There is a time and place for identity experimentation, and unlike a social setting, the business context affords little room for this, and by its nature requires a "truth in representation" as part of standard policies and procedures, an Avatar Code of Ethics of sorts.

36.5 ORGANIZATIONAL POLICIES AND CONSIDERATIONS ON AVATARS IN FORMAL COLLABORATIONS

To launch a virtual world project on a foundation designed for success, an examination of the purpose, mission, and overall perception needs to be recognized early on. During the analysis phase, an organization should not only examine the perceived need for a virtual world implementation, but also look at organization-wide implications that such an undertaking may have. Through the use of a process model such as the Analysis, Design, Development, Implementation, and Evaluation (ADDIE) process model that is well known in instructional design communities, requirements can be identified.

The purpose of the analysis phase in the ADDIE process model is to consider requirements or other issues that may provide direction on what, if any, intervention is necessary. Assuming an instructional solution is required, and once a virtual world tool is identified as part of a solution, then it comes down to deciding on which tool to use. (Certainly, due diligence is required in order to arrive at the previously mentioned assumed solutions.)

To get started, however, designers may benefit from asking these questions from the partial checklist shown in Figure 36.5 to reassess their preliminary conclusions (Gamor 2010). These questions represent basic inquiry necessary in any technology-based training design initiative. The "traditional" analysis phase will address only part of the considerations of a virtual world project. Some may find that ADDIE is ineffective as a stand-

Analysis Checklist

√ What learning goals does my current learning design fail to address? Or, where might the unique affordances of virtual worlds enhance my existing curriculum?

√ What specific instructional strategies are optimal at addressing the failed learning goals?

√ Using my current toolset, how can I design an intervention to implement the required instructional strategies?

√ Is there a gap between my current toolset and the instructional strategies I need to implement? If no, then use the appropriate tool from the current toolset. If yes, then consider new tools, such as virtual worlds.

√ With what existing instructional interventions much new tools or platforms interface?

FIGURE 36.5 A thorough analysis checklist may assist in procurement decisions (Gamor 2010).

alone model for designing virtual learning environments and events. On the one hand, if ADDIE is approached as a linear process model (which it is not), then it will most certainly be insufficient. On the other hand, if ADDIE is approached as an iterative process model, then designers will experience more satisfaction with the guidance it may provide. In any case, ADDIE will need the support of other design models to provide an overall approach to creating virtual environment experiences. Contemporary e-learning, digital narrative, film production, and gaming design models could provide useful insight into developing meaningful, effective virtual environment experiences (Gamor 2010).

Because ADDIE is not intended be a detailed, step-by-step process model, the requirements analysis must also recognize the other considerations to take into account. Completing an IT infrastructure analysis, audience analysis, and a job/task/learning experience analysis are a few of the considerations that must be made. Each of these analyses will help to further identify and refine needs that will influence design of the world and guidelines necessary for successful operations. Augmenting the ADDIE process model with Khan's e-learning framework, for example, may help identify most of the critical questions that should be addressed because ADDIE, in general, is neither detailed nor prescriptive (Gamor 2010).

Khan's e-learning framework is applicable to virtual worlds and augments the ADDIE process model by providing more detailed phases and specific life cycle steps than the ADDIE model. In addition to being very detailed, Khan's framework is also suitable for most any rapid prototyping, spiral development process. The eight dimensions of Khan's framework (Figure 36.6) represent areas requiring consideration early in the e-learning development process and revisited throughout the project's life cycle as part of a continuous improvement strategy. These include the following eight dimensions:

- *Institutional* concerns administrative issues, academic considerations, student services, philosophies, goals, mission, and objectives.

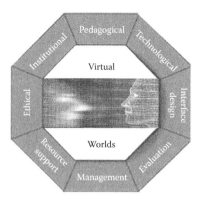

FIGURE 36.6 Khan's e-learning model applied to virtual worlds. (Framework modified with permission of author. Center image used with permission from E-Discovery Bytes.)

- *Pedagogical* refers to instructional approaches, strategies and methodologies associated with teaching and learning.
- *Technological* addresses infrastructure, interoperability standards, security, IT policy, etc.
- *Interface* design examines human factors, instructional relevance, as well as branding consistency and appropriateness.
- *Evaluation* involves assessment, evaluation, performance, behaviors, knowledge transfer, etc.
- *Management* relates to maintaining (servicing, trouble-shooting, help desk support, licensing, updates, etc.) the learning environment.
- *Resource Support* entails securing all the roles necessary to successfully design, develop, implement, evaluate, and maintain the learning environment.
- *Ethics* includes determining codes of conduct, rules, policies, procedures for operating the learning environment (Khan 2005, 2007; Gamor 2010, forthcoming).

Each of the areas delineated in the framework must be examined before selecting a virtual world application (Voorhees and Dawley 2008) and planning implementation. The analysis phase must identify all limitations and restrictions in order to avoid acquiring a tool that cannot be used and a purpose against which virtual worlds may be ineffective.

The ADDIE process model frames the method for identifying the performance problem or organizational challenge; ensuring that the problem is, in fact, education, training, and communication related; and illuminating the various aspects of an organization that need to be factored into the overall implementation strategy. While the ADDIE model alone has been found to be ineffective for designing virtual worlds and virtual world experiences, Khan's framework, augments ADDIE by providing specific guidance on its steps vis-à-vis his eight dimensions (Gamor, forthcoming). These eight dimensions help avoid missing any critical success factors when designing a virtual world project. Solution sets resulting from these analyses can be assessed only after the weaknesses in the education, training, or communications-centered program are thoroughly identified.

Neglecting the analysis phase poses the same risks inherent in any solutions-based project, regardless of the anticipated outcome or desired solution. Requirements must drive solutions definition and tool adoption and implementation and not the other way around. Virtual worlds, like any other tool, should be analyzed as an "enabler" for the specific solution, not as an end unto itself. In addition, affordances and potential uses of virtual worlds and other tools should be factored into the examination of prospective solution sets in order to select the solution with the highest probability of success (Gamor 2010).

The purpose of the analysis phase in the ADDIE process model is to consider requirements or other issues that may provide direction on what, if any, virtual world solution is necessary. Assuming a need for a virtual world exists, parameters of design or concepts of operations guidelines must be developed. (Certainly, due diligence is required in order to arrive at the previously mentioned assumed solutions.)

Given the lessons learned thus far, some recommendations have emerged that will guide implementers concerned with human factors in virtual worlds. Communication and interaction are the focus of the human factors issues discussed in this chapter. As such, the recommendations provide guidance on establishing guidelines that may enhance the human experience, ease the transition into, and prepare the organizational culture for life in the virtual world.

36.5.1 ESTABLISH ORGANIZATIONAL CULTURE

The organization's mission and goals should be evident in the virtual world environment established for its users. Creating a world with the necessary limitations built-in will help to avoid problems in the future. Enable organizational-appropriate ways to express creativity and leadership, entrepreneurship and teamwork, individuality and cooperation. Organizational identity and individual expression should be foundational in the educational or business virtual world; however, if environments are too limiting then the effort may be at risk. Humans will create a way to create, express, or build. Virtual worlds sponsors should make this possible within the limitations of the capabilities built into the educational or business virtual world. Designers should be armed with requirements data (resulting from a requirements analysis report) that delineate the capabilities necessary for the environment. This will ensure that designers provide the right capabilities needed for the goals of the virtual world initiative.

36.5.2 ASSIST EMPLOYEES IN MAKING A REAL CONNECTION TO THEIR AVATAR

Understanding the role of the avatar as a representation of oneself is important in assisting students, staff, and employees in being diligent about ensuring that behavior and actions represent the person that they are in the real world. Enable customization within the realm of the organization's identity but limit the scope of customization to what is needed for the nature of the collaboration intended for the virtual world. Research indicates that the more time a person spends customizing, outfitting, and using their avatar, the more they identify with, feel close to, or perceive the avatar as being a natural extension of self. Organizations would do well to discover platforms that offer a broad range of expressions appropriate for the educational or business environment.

36.5.3 Establish Guidelines for Avatar Development

To avoid distraction, organizations may wish to establish guidelines for developing avatars that will be used for business purposes. They may also find it prudent to stipulate that avatars developed for use at an educational or business organization are only used for such purposes, thus reducing the risk that the avatars having become known as representatives of the institutions where the owners study/work can be used for other social or entertainment purposes.

36.5.4 Establish Clear Rules of Engagement and Map Them to Existing Corporate or Organizational Policy

Because interactions in the world are representative of or in place of traditional face-to-face interactions, corporate expectations on behavior, attire, and general conduct should remain consistent with that which is required in the real world. As virtual communications replace some of the face-to-face interaction, it is critical that the organization's message and the avatar (as the emissary) remain congruent. It is necessary to bear in mind the identified objective when using a virtual world for education and/or business purposes and develop policy around those objectives. Maintaining alignment between standard corporate policies and online corporate policies may help mitigate some concerns associated with various entertainment, gaming, and social perceptions of virtual worlds.

Immersive learning environments are going to be as integral to twenty-first century literacy as the Internet was to late twentieth century literacy. As such, educational and business institutions are experimenting with the tools, endeavoring to discover lessons learned in order to capitalize on the benefits they intuitively perceive are achievable. While more research is necessary, studying the role of the avatar, including its identity, behavior, and integrity, is essential for conducting successful communication and interaction with trustworthy outcomes that transcend the virtual world.

REFERENCES

Anderson, P. 2007. What is Web 2.0? Ideas, technologies and implications for education. JISC Technology and Standards Watch. http://citeseerx.ist.psu.edu/viewdoc/download?doi=10.1.1.108.9995&rep=rep1&type=pdf (accessed Dec. 1, 2009).

Alexander, B. 2006. Web 2.0: A new wave of innovation for teaching and learning. *EDUCAUSE Review* 41(2), 32–44. http://www.educause.edu/apps/er/erm06/erm0621.asp (accessed Dec. 1, 2009).

Association of Virtual Worlds. 2008. *The blue book: A consumer guide to virtual worlds* (4th ed.). Retrieved April 12, 2009 from http://www.associationofvirtualworlds.com/pdf/Blue%20Book%204th%20Edition%20August%202008.pdf.

Bailenson, J. N., J. Blascovich, A. C. Beall, and J. M. Loomis. 2001. Equilibrium revisited: Mutual gaze and personal space in virtual environments. *Presence: Teleoperators and Virtual Environments* 10: 583–598.

Bell, M. 2008. Toward a definition of "virtual worlds." *Journal of Virtual Worlds Research* 1(1): 2–5.

Bente, G., S. Ruggenberg, N. Kramer, and F. Eschenburg. 2008. Avatar-mediated networking: Increasing social presence and interpersonal trust in net-based collaborations. *Human Communication Research* 34(2): 287–318.

Biocca, F., and K. Nowak. 2002. Plugging your body into the telecommunications system: mediated embodiment, media interfaces, and social virtual environments. In *Communication Technology and Society: Audience Adoption and Uses*, eds. D. Atkin and C. Lin, 407–447. Cresskill, NJ: Hampton Press.

Blascovich, J. 2002. Social influence within immersive virtual environments. In *The Social Life of Avatars: Presence and Interaction in Shared Virtual Environments*, ed. R. Schroeder, 127–145. New York: Springer.

Bransford, J., A. Brown, and R. Cocking. 1999. How people learn: Brain, mind experience and school. http://cde.athabascau.ca/online_book/pdf/TPOL_chp02.pdf (accessed Nov. 1, 2008).

Brown, A. L., and J. C. Campione. 1994. Guided discovery in a community of learners. In *Classroom Lessons: Integrating Cognitive Theory and Classroom Practice*, ed. K. McGilly, 229–270. Cambridge, MA: MIT Press.

Brown, J. S., and R. P. Adler. 2008. Minds on fire: Open education, the long tail, and learning 2.0. *EDUCAUSE Review* 43(1): 1–19.

Bonato, F., A. Bubka, S. Palmisano, D. Phillip, and G. Moreno. 2008. Vection change exacerbates simulator sickness in virtual environments. *Presence* 17(3): 283–292.

Castronova, E. 2007. *Exodus to the Virtual World: How Online Fun Is Changing Reality*. New York: Palgrave McMillan.

Chin, S., and J. Williams. 2006. A theoretical framework for effective online course design. *Journal of Online Learning and Teaching* 2(1).

Church, M. A., A. J. Elliot, and S. L. Gable. 2001. Perceptions of classroom environment, achievement goals, and achievement outcomes. *Journal of Educational Psychology* 93: 43–54.

Cobb, P., E. Yackel, and T. Wood. 1992. A constructivist alternative to the representational view of mind in mathematics education. *Journal for Research in Mathematics Education* 19: 99–114.

Dalton, J. H., M. J. Elias, and A. Wandersman. 2001. *Community Psychology: Linking Individuals and Communities*. Belmont. CA: Wadsworth.

Dede, C. 2005. Planning for "neomillennial" learning styles: Implications for investments in technology and faculty. In *Educating the Net Generation*, eds. J. Oblinger and D. Oblinger, 226–247. Boulder, CO: EDUCAUSE Publishers.

Dede, C. 2007. Reinventing the role of information and communications technologies in education. *Yearbook of the National Society for the Study of Education* 106: 11–38.

Federation of American Scientists. FAS virtual worlds whitepaper. http://vworld.fas.org/wiki/FAS_Virtual_Worlds_Whitepaper (accessed March 30, 2009).

Feldon, D., and Y. Kafai. 2007. Mixed methods for mixed reality: Understanding user's avatar activities in virtual worlds. *Educational Technology Research and Development* 56(5/6): 575–593.

Galanxhi, H., and F. F. Nah. 2007. Deception in cyberspace: A comparison of text-only vs. avatar-supported medium. *International Journal of Human–Computer Studies* 65: 770–783.

Gamor, K. 2010. Adopting virtual worlds in ADL: The criticality of analysis. In *Learning on Demand: ADL and the Future of eLearning*, eds. B. Wisher and B. Khan, 177–194. Alexandria, VA: ADL.

Gamor, K. forthcoming. Exploiting the power of persistence in virtual worlds. In *User Interface Design for Virtual Environments: Challenges and Advances*, ed. B. Khan. Washington, DC: McWeadon.

Gehorsam, R. 2009. Corporate training in virtual environments. In *The PSI Handbook of Virtual Environments for Training and Education*, eds. J. Cohn, D. Nicholson, and D. Schmorrow, 413–419. Wesport: Praeger Security International.

Heiphetz, A., and G. Woodill. 2010. *Training and Collaboration with Virtual Worlds: How to Create Cost-Saving, Efficient, Engaging Programs*. New York: McGraw-Hill.

Holzwarth, M., C. Janiszewski, and M. M. Neumann. 2006. The influence of avatars on online consumer shopping behavior. *Journal of Marketing* 70: 19–36.

Institute of Electrical and Electronics Engineers, n.d. IEEE islands in Second Life. http://www.ieee.org/web/volunteers/tab/secondlife/index.html (accessed May 3, 2009).

Jonassen, D. 2000. Toward a meta-theory of problem solving. *Educational Technology: Research & Development* 48(4): 63–85.

Jerome, C. 2007. Effects of spatial and non-spatial multi-modal cues on orienting of visual-spatial attention in an augmented environment. Army Research Institute for the Behavioral and Social Sciences. Alexandria, VA.

Jin, S.-A. A. 2009. The roles of modality richness and involvement in shopping behavior in 3D virtual stores, *Journal of Interactive Marketing* 23(3): 234–246.

Jung, I., S. Choi, C. Lim, and J. Leem. 2002. Effects of different types of interaction on learning achievement, satisfaction and participation in web-based instruction. *Innovations in Education and Teaching International* 39(2): 153–162.

Kanawattanachai, P., and Y. Yoo. 2002. The dynamic nature of trust in virtual teams. *Journal of Strategic Information Systems* 11(3): 187–213.

Khan, B. 2005. *Managing E-learning Strategies: Design, Delivery, Implementation and Evaluation*. London: Information Science Publishing.

Khan, B. H., ed. 2007. Flexible learning in an open and distributed environment. In *Flexible Learning in an Information Society*, 1–17. Hershey, PA: Information Science Publishing.

Kolasinski, E. 1995. Simulator sickness in virtual environments. Army Research Institute for the Behavioral and Social Sciences. Alexandria, VA.

Kolbert, E. 2001. Pimps and dragons: How an online world survived a social breakdown, *The New Yorker*, May 28, http://www.btinternet.com/~braxfield/blogimages/UltimaOnline.htm.

Kunda, Z. 1999. *Social Cognition: Making Sense of People*. Cambridge, MA: MIT Press.

Lee, Y., K. Kozar, and K. Larsen. 2007. Avatar e-mail versus traditional e-mail: Perceptual difference and media selection difference. *Decision Support Systems* 46: 451–467.

McAllister, D. J. 1995. Affect- and cognition-based trust as foundations for interpersonal cooperation in organizations. *Academy of Management Journal* 38: 24–59.

Meadows, M. S. 2008. *I, Avatar: The Culture and Consequences of Having a Second Life*. Berkeley, CA: New Riders.

Merrill, D. 2007. A task-centered instructional strategy. *Journal of Research on Technology in Education* 40(1): 5–22.

Meyerson, D., K. E. Weick, and R. M. Kramer. 1996 Swift trust and temporary groups. In *Trust in Organizations: Frontiers of Theory and Research*, eds. R. M. Kramer and T. R. Tyler, 166–195. Thousand Oaks, CA: Sage Publications.

Nowak, K. 2004. The influence of anthropomorphism and agency on social judgment in virtual environments. *Journal of Computer-Mediated Communication* 9(2).

Nowak, K., and C. Rauh. 2006. The influence of the avatar on online perceptions of anthropomorphism, androgyny, credibility, homophily, and attraction. *Journal of Computer-Mediated Communication* 11(1): 153–178.

Ondrejka, C. 2008. Education unleashed: Participatory culture, education, and innovation in second life. In *The Ecology of Games: Connecting Youth, Games, and Learning*, ed. K. Salen, 229–251. Cambridge, MA: MIT Press.

Oravec, J. 1996. *Virtual individuals, virtual groups: Human dimensions of groupware and computer networking*. Cambridge: Cambridge University Press.

Palloff, R., and K. Pratt. 2000. *Making the transition: Helping teachers to teach online*. Paper presented at EDUCAUSE: Thinking it through. Nashville, Tennessee, October 10–13. http://net.educause.edu/ir/library/pdf/EDU0006.pdf (accessed Oct. 10, 2009).

Pillsbury, Winthrop, Shaw, Pittman LLP. 2009. *10 Frequently Asked Questions About Virtual Worlds*, 10–13. New York: Pillsbury, Winthrop, Shaw, Pittman LLP.

Roberts, T. 2004. *Online Collaborative Learning: Theory and Practice*. Hershey, PA: Information Science Publishing.

Rauh, C., M. Polonsky, and R. Buck. 2004. Cooperation at the first move: Prust, emotional expressiveness and avatars in the prisoner's dilemma game. *Journal of Computer-Mediated Communication* 11(1).

Savage, R. 2007. Training wayfinding: Natural movement in mixed reality. Army Research Institute for the Behavioral and Social Sciences. Alexandria, VA.

Shin, N. 2003. Transactional presence as a critical predictor of success in distance learning. *Distance Education* 24(1): 69–86.

Slater, M., and A. Steed. 2002. Meeting people virtually: Experiments in shared virtual environments. In *The Social Life of Avatars: Presence and Interaction in Shared Virtual Environments*, ed. R. Schroeder, 146–171. New York: Springer.

Taylor, T. 2003. Multiple pleasures: Women and online gaming. *Convergence* 9(1): 21–46.

Taylor, T. L. 2002. Living digitally: Embodiment in virtual worlds. In *The Social Life of Avatars: Presence and Interaction in Shared Virtual Environments*, ed. R. Schroeder, 40–62. New York: Springer.

The EDUCAUSE Learning Initiative. (2006). *7 things you should know about virtual worlds*. Retrieved June 2, 2009 from http://www.educause.edu/ELI/7ThingsYouShouldKnowAboutVirtu/156818.

ThinkBalm. 2009. ThinkBalm immersive Internet business value study, Q2 2009. May 26. http://www.thinkbalm.com/wp-content/uploads/2009/05/thinkbalm-immersive-internet-business-value-study-final-5-26-092.pdf (accessed Oct. 10, 2009).

Valentine, D. 2002. Distance learning: Promises, problems, and possibilities. *Online Journal of Distance Learning Administration* 5(3).

Virtual World Review. 2009. What is a virtual world? http://www.virtualworldsreview.com/info/whatis.shtml (accessed April 6, 2009).

Voorhees, A., and L. Dawley. 2008. Evaluating SL course experience: A learner's evaluation and faculty response. Paper presented at the Annual Conference of Association for Educational Communications and Technology, November, Orlando, FL.

Weyers, P., A. Muhlberger, A. Kund, U. Hess, and P. Pauli. 2009. Modulation of facial reactions to avatar emotional faces by nonconscious competition priming. *Psychophysiology* 46: 328–335.

Wilson, A. 2008. Telepresence: Reducing the impact of business travel. Whitepaper. http://www.awarenessintoaction.com/article.php?url=telepresence-reducing-the-impact-of-business-travel (accessed Oct. 23, 2009).

Section XII

Return on Investment for the Web

37 Determining the Value of Usability in Web Design

Andrea Richeson, Eugenie Bertus, Randolph G. Bias, and Jana Tate

CONTENTS

37.1 INTRODUCTION

In their 1994 book *Cost-justifying Usability*, Bias and Mayhew predicted that in 10 years usability would be generally accepted and return on investment (ROI) would be commonly provided to justify usability work. By the time of the 2005 edition, the Web was more established as a part of everyday life, but usability work was still not routinely integrated into development efforts. Fifteen years after the first Bias and Mayhew book, and six years after the publication of an earlier version of this chapter, usability professionals are still working to integrate usability efforts routinely into the product development life cycle, with less than total success.

ROI concerns are just as relevant today as they have ever been. Today, the overall world economic system is improving but still in crisis. Unemployment in the United States stands at 10.2% and 9.3 million are underemployed, with an average work week of 33 hours (Bureau of Labor Statistics 2009a). At the same time, the nonfarm business sector has seen a jump in the annual productivity rate, which in the third quarter of 2009 reached 9.5% (Bureau of Labor Statistics 2009b). Businesses are producing more, but with fewer employees and in less time. Our organizations' budgets are under increased scrutiny and the need for projects to show a return on investment is obvious.

Usable sites are not yet the standard, and the value of usability work in helping to produce better sites and better return on investment is not yet self-evident to all, or even most who hold the Web development purse strings. The Usability Professionals' Association (UPA) conducted a survey of its members in 2005. The resulting report noted that while user-centered design practices could be shown to be very cost-effective compared to other parts of the development process, the business case for usability had not been adequately made and usability teams were often easy targets for budget cuts during tough economic times (Gunther 2006).

Despite substantial evidence that usability efforts are not yet routinely integrated into the development process, there are some reports of progress. A July 2007 report based on an online survey of more than 700 respondents in the United Kingdom tells us that while only a quarter of organizations are "extremely committed" to providing the best possible user experience, 72% of responding UK organizations were planning to increase their usability budget over the next 12 months. The report also notes that respondents listed improved perceptions of brand, increased conversion rates, and greater customer loyalty and retention as the top three benefits of an investment in usability (Econsultancy 2007). To justify usability efforts, professionals still need to justify the value of usability work.

37.2 THE PROGRESSION OF TECHNOLOGY

In the early 1990's, organizations were only beginning to use the Web for research, marketing, and commercial transactions. As broadband access became more affordable, content and functionality became more sophisticated, and the numbers of Web sites and Web users grew exponentially. All of these factors have implications for usability.

37.2.1 Generations Online

In the United States, Internet use is now commonplace across a wide set of demographics. The numbers related to how different generations use the Internet tell an interesting story. According to a study done as part of the Pew Internet & American Life Project, the biggest increase in Internet use since 2005 is seen in people 70 to 75 years old. While just over one quarter (26%) of 70–75 year olds was online in 2005, 45% of that age group is currently online. Not surprisingly, different generations use the Internet for different purposes. Compared with teens and young adults, older adults use the Internet less for socializing and entertainment and more as a tool to search for information, to read and send e-mail, and to purchase products online (Jones and Fox 2009).

37.2.2 The Global Penetration of Technology and Internet Access

Overall, Internet World Stats reports that there is now about a 73.9% penetration of Internet use in the North American population, with a growth rate of 132.9% from 2000 through 2009. Asia was reported to be at 18.5% penetration with 516.1% growth rate in 2000–2009 (Schonfeld 2009). Internet World Stats reports that from December 1995 to December 1997, the number of users on the Internet jumped from 16 million to 70 million. The numbers increased to 248 million by December 1999, surpassed one billion in December 2008, and are now around 1.5 billion. During this time period, Internet World Stats also reports that the percentage of the world population on the Internet moved from .4% to an impressive 23.5% (Internet World Stats Usage and Population Statistics 2009).

It was often an uphill battle to get organizations to invest in the Web, but those efforts were well rewarded as the Web went from cutting-edge to indispensable. Today, organizations are trying to figure out how and when to use social media, grappling with Web 2.0 concepts to make their sites more interactive, and considering how to adapt their Web content to work on mobile devices. All of these statistics should provide ample incentive for Web designers and developers to ensure that the functionality in their sites can be found, learned, and used by this ever-burgeoning population of potential users.

37.3 MOBILE PROLIFERATION

Mobile phone use has undergone a significant increase in the past decade. In September of 2008, Secretary-General Hamadoun Touré of the International Telecommunication Union (ITU) announced that worldwide mobile cellular subscriptions were likely to hit the 4 billion mark by the end of 2008. The ITU also noted that between 2000 and 2008 growth in the market averaged 24% per year, with an estimated 61% market penetration by the end of 2008. One third of all mobile phone subscriptions included in their estimates were held in China, India, Russia, and Brazil (International Telecommunication Union 2008).

37.3.1 Mobile Internet Use

When considering the crucial role usability work can play in the success of Web sites and Web applications, it may be useful to know what the usage statistics are specifically for Internet access on mobile phones. The Pew Internet & American Life project reported in late 2008 that the mobile device would be the primary connection tool to the Internet for most people in the world in 2020 (Rainie and Anderson 2008). Nielsen Mobile (2008) reported in July of 2008 "in the United States, Mobile Internet has become a mass medium.

As of May 2008, there were 40 million active users of the Mobile Internet in the United States, based on past 30-day usage. And this is just a subset of the 95 million U.S. mobile users who subscribed to the service but do not necessarily use it." The path toward critical mass of mobile phone proliferation and mobile Internet access has obvious implications for bridging the digital divide, providing open government, extending e-learning opportunities, and providing for mobile e-commerce.

37.3.2 WHERE USABILITY PROFESSIONALS FIT IN

With the extension of technology and Internet access into new geographical and generational markets, usability efforts will continue to have a big impact on helping organizations meet their goals. Usability professionals assist organizations in reaching their audiences. They wear a variety of hats and go by several different titles. Common titles for professionals who may have human factors as a primary responsibility include Human Factors Practitioner, User Experience Practitioner, Interface Designer, Usability Practitioner, User-Centered Design Practitioner, Information Architect, Interaction Designer, Usability Manager, and Web Designer (Usability Professionals Association 2009). The rest of this chapter concentrates on how professionals in the usability field can justify investment in usability work, with an emphasis on benefits and how those benefits can make the organization more profitable and effective in reaching its goals.

It is not obvious to everyone that there is a need to focus on the benefits that usability professionals bring to the table. In short, businesses, nonprofit organizations, and governmental organizations are obliged to focus on the bottom line. Even when organizations are flush with resources, it is only smart business to show the value of usability work as it relates to an organization's goals, strategies, and objectives; in tougher economic times it is critical. And indeed, in any economy, it is helpful to employ a cost-benefit analysis approach to identify which methods will likely yield the best ROI at any particular stage in the development of any particular Web site or Web application.

There are several reasons to quantify benefits of usability work. You may need to justify current staffing levels or increase funding for usability positions. An explanation of the benefits of usability work might be used to influence if and when usability work should be integrated into the development cycle or to more generally gain social acceptance for usability work in the organization. Once social acceptance for usability work is achieved, practitioners might use benefits analysis to target usability work of a particular type.

37.3.2.1 The Risks of Amateur Usability Work

> "The only thing more expensive than hiring a professional, is hiring an amateur."—Red Adair

A well-designed site, or an intuitive application, is transparent and allows its users to carry out their tasks without attention to the interface. It looks easy, which can lead people to think anyone can do it. In fact, design is hard. The good user interface, like the good referee or umpire, is the one that is *not* noticed, the one that lets the transaction, or the game, proceed smoothly, without notice; the good user interface (UI) is one whose simplicity is noticed only when the interface itself is intentionally examined.

We have written elsewhere (Bias 2003) on the dangers of amateur usability. Our belief is that if someone does a poor job of usability engineering a Web UI, this is not discovered until the site or product is live, and people leave your site without completing their task, inundate the customer support lines, or otherwise have a generally bad user experience. And doing a good job of human factors engineering or usability engineering a design usually requires training; the best intentioned amateur usability "professional" is still less likely than the trained usability professional to employ the correct methods, and employ them well, in the collection of user requirements or the evaluation of emerging designs. Plus, that amateur is less well equipped to advocate for the data that he or she has gathered and thereby be a valuable change agent in the design and development of the site or application.

The implications are clear: The cost savings of hiring an amateur are easy to measure; the benefits of hiring a professional are more difficult to measure and are often realized only over time. And so there is a danger in the purse-string holders being "penny wise and pound foolish" and simply minimizing costs (hiring the cheapest possible usability help) without regard for the possible impact on subsequent potential benefits.

37.3.2.2 Developing for the Web Is Different than Traditional Software Development Efforts

Organizations and professionals with extensive experience in software development may not see the need for or benefit of usability work on Web sites. To justify usability work, it is helpful to recognize that the risks of losing customers and market-share are different for Web sites than for traditional software and even for some Web applications.

Web sites work under a different profit model than traditional software. Web sites often make money by creating and maintaining site traffic. Advertisers make money only if visitors stay on the site and return to it. As noted by Mayhew and Tremaine (2005) in *Cost-justifying Usability*, there is little investment of time or money for a customer to try a new Web site. Because of the low cost to switch to a competitor's site, poor usability can drive customers to the competition or perhaps back to traditional avenues of support. Either way, the company takes a financial hit, either in reduced ad revenue, in lost sales, or in the lost opportunity to lower in-house costs of providing support.

Web applications may have the same risks as traditional software projects if they must be purchased before use. As with traditional software, when Web applications are purchased, there are inherent costs to switching to another application, including the cost of the application itself, time and personnel costs to make purchases, the costs of implementation and training, and the costs of changing business practices based on the application's functionality.

37.4 USING RETURN ON INVESTMENT TO SHOW THE VALUE OF USABILITY WORK

Return on investment (ROI) arguments can be useful tools when making the case for usability work on the Web; however, there is substantial disagreement about the appropriateness or value of using ROI to justify usability work. Don Norman, as part of a CHI 2002 conference (Shneiderman et al. 2002), expressed his frustration with the HCI community, saying that HCI is considered a secondary profession because there is too much focus on criticism of products and not enough emphasis on the creation of products. Norman suggested that HCI professionals must become designers and learn to speak in "the financial language of ROI and NPV (Return on Investment and Net Present Value—the time value of money)."

On the other hand, Daniel Rosenberg of Oracle Corporation has been an outspoken critic of the ROI approach to cost-justifying user-centered design (UCD), saying in 2005 (Dray et al. 2005), "the traditional ROI approach to defining and measuring the value of usability is not an effective way of demonstrating the added value that UCD provides."

Scott Hirsch (2003) of Adaptive Path addressed this kind of opposition to the use of ROI approach, noting that while poorly substantiated and overgeneralized claims of ROI are the great red herring of ROI research, there is a place for the approach at the project level. Hirsch states that we should "rid ourselves of these top-down, macro-level assertions and get down to the real work of analyzing specific usability interventions at the project level. Only through rigorous and in-depth analysis can larger patterns emerge and applications be developed." As one would expect, different organizations value and have success with different tools and strategies. ROI measurements required for a project at a Fortune 500 company might not be expected or valued at a government agency or small start-up firm. An information-only site will have different metrics of "success" and benefit than will an e-commerce site. Regardless of the sizes or type of the organization, there are striking examples of the successful use of ROI to gain approval for usability work.

Tobias Herrman (2006), the head of Team User Experience within mobilkom austria, Austria's largest telecom and mobile service company, wrote about the value of ROI in proving the business impact of user experience activities to management level: "In retrospect, it was by far the strongest driver for a sustainable integration and long-term partnership with product management—and I'm convinced this also applies to many other companies. Finally, top management always wants to know whether or not their investment paid off, most suitably in financial figures. So, how did we achieve it? As in many other cases, a tailored return on investment."

Jeff Herman (2004) of eBay's User Experience and Design Group said in 2004 that all project proposals at his company must have an ROI analysis to even be considered for approval. He credits the creation and communication of the business case for user experience (UE) projects as a key to his team's success in sponsoring UE projects. Finally, James E. Nieters

of Cisco noted in 2007 that the UXD Group's contribution to revenue rose from $50 million to almost $3 billion after converting the group to an Internal Consultancy/Focus Team model (Nieters 2007).

37.4.1 ROI Work Can Be Done at Different Levels of Complexity/Intensity

When making a case with ROI, it is most meaningful to show the impact that usability work has had on a particular project. Optimally, one would compare a single project before and after usability work. When that is not possible, it can be helpful to compare different projects that have had usability work with those that have not had usability work.

Organizations may have one developer covering design, coding, implementation and project management. Other organizations may have these tasks covered by multiple people. Whatever the case, it is important to assess what level of effort is required to gather the data needed to address the goal of the ROI effort. For example, questions that may be asked include the following: Do you have a lot of time and resources available to perform an ROI analysis, and do you need very accurate measures? If this is the case, you might measure multiple metrics, such as task completion time, user satisfaction, documentation preparation time, training materials preparation time, and a comparison of revenue numbers and development budget. Do you have less time and fewer resources available, and can you use approximate measures? If this is the case, you could measure user satisfaction and compare it to revenue numbers. Indeed, what you measure in the way of "benefits," in the cost-benefit analysis, will depend on many factors such as revenue model, type of site/product, and particular phase in the development cycle. We address this more in a later section.

In the fall of 2009, the authors sent out a survey about cost-justifying usability work to a variety of groups, including members of UPA, the Web Content Managers list, State of Texas IT managers and developers, and members of Knowability's accessibility list. Of the respondents, 67% stated that they found value in the use of ROI for securing usability support in their organizations, but 74% did not use ROI methods to describe the benefits and costs of usability projects. Of those, 42.9% indicated that they did not use ROI methods because they did not know how. Many of the respondents used methods other than formal ROI analysis to justify usability work, including site traffic and customer satisfaction monitoring, cost-benefit analysis, competitive analysis, risk-benefit analysis, and business case development.

37.4.2 Communication Strategy

No matter the method used to show benefit, it is critical to know how to communicate those benefits most effectively to the organization. When planning your sales pitch for usability work, you must clearly identify the target audience for the cost-benefit argument. What are their most important goals? How do they communicate, and what do they value?

Rohn (2005, chap. 7) phrased it well:

"One of the ironies of the role of UE professionals is that the field of usability focuses on understanding users and their requirements, yet very few UE professionals practice this principle on one of the most important influences on their work: their employer. . . . Understanding the culture, value system, and business goals of the company should be the first order of business for a UE department and its members."

This idea that the success of an ROI argument depends on the people involved and the goals and challenges of the particular organization was addressed at the CHI conference in 2005 in a discussion regarding the value of ROI as a persuasive tool to support user-centered design efforts. Panelist Clare-Marie Karat reiterated her views from chapter 4 of *Cost-justifying Usability* when she noted that to be effective, the ROI arguments must be received by a person with, "strategic business vision or at least be a pragmatist who will experiment with new methods." The other factors Karat considered critical to the success of ROI arguments were the ability to tie the ROI arguments to business goals and strategies of the organization and the extent to which ROI arguments assist the organization in addressing global challenges (see Dray et al. 2005).

The 2005 discussion was cited and expanded the following year, at CHI 2006, by a panel discussing the use of rhetoric and argumentation to "help advance the case for UCD on organizational and project levels in various contexts and organizations." The panel noted that objective approaches such as cost-justification might be more effective when paired with arguments that appeal to the audience's emotions and sense of authority. The panel members also noted that subjective arguments should be crafted based on an understanding of the audience's values, communication style, and motives (Jones et al. 2006).

37.4.3 HISTORICAL STUDIES SHOW THE BENEFITS OF USABILITY WORK

When crafting a communications strategy, it is often helpful to cite instances where usability work makes an impact on the bottom line. General industry trends can be helpful in building support for usability work in an organization. Usability work on a site can save time, money, and retain and build trust. The work, of course, is not always cheap, but an investment in usability work, especially early in the development process, can save effort down the road. As Cameron Hayne noted about cost justifying usability testing, "There is no free lunch. But sometimes if you eat a good breakfast, you won't need to spend as much money on lunch."

To begin to develop a cost-benefit argument, draw upon some historical work in the field of usability. There have been many papers and essays written on cost-justifying usability. The book *Cost-Justifying Usability* (Bias and Mayhew 1994) brought together practitioners who shared their experiences and statistics on cost justification of products on which they worked. Although they wrote the book prior to the Web boom, the lessons learned from their experiences are still relevant to the designer of a Web user interface.

One chapter in the book lays out a compelling argument on how usability can impact the development of software, stating that the user interface can be up to 60% of the system code (Karat 1994). Additionally, the interface can require 40% of the development time (Wixon and Jones 1992, as cited in Karat 1994). Budget estimates show that software projects will frequently exceed their budget and time estimates. The most common reasons given for these overruns are related to usability (Lederer and Prassad 1992). Poor communication and understanding of user needs, or overlooked user tasks, are two examples. Given the level of commitment for the development of the user interface, it stands to reason that appropriate effort should be applied to design it well.

A widely cited estimate is that for every dollar invested in usability at the definition phase of a project, it would cost $10 to implement the same feature during development, and $100 to implement after release (Boehm 1981; Pressman 1992). Additionally, Pressman claims that 80% of the cost of software development occurs post-release, in the form of fixes to missed or incorrect requirements. Incorporating usability into the development process early in the development cycle would help alleviate some of these issues, by uncovering user tasks and needs earlier in the development life cycle. Overall, utilizing usability methods can reduce the product development cycle by 33 to 50% (Bosert 1991).

Although most arguments for incorporating usability into the software development life cycle center around introducing the practices early, studies have demonstrated that even methods that occur late in the cycle can add benefit. LaPlante (1992) found that on average a usability test pinpoints many issues and results in 70 to 100 usability recommendations. If only half of these recommendations were addressed, this would improve the usability of the interface by over 50% (Landauer 1995).

This information is useful for any development of user interfaces, including a Web interface. In a paper describing how Web sites can achieve a competitive advantage, Rhodes (2000) claims that the first tenet of survival is to have the better user experience. He states that if all else is equal, then the company with the easier-to-use site will win the market. How much of a market is that? In 1998, it was estimated that more than 44 million people in the United States had already made purchases online, and 37 million more said they would be making purchases soon (Wildstrom 1998). By 2002 the number of online purchases increased from those in 2001 by 37% to 358.6 million totaling $47.98 billion (Fitzgerald 2003). The Pew Internet in American Life Project reported in their 2008 report entitled "Online shopping" that 66% of online Americans have purchased a product online. The study also reports that at over half of American internet users are frustrated with the online shopping experience and that trust issues including online safety, convenience, and efficiency drive down the numbers of shoppers by 2, 3, and 7%, respectively (Horrigan 2008).

In a study of the role of security, privacy, usability, and reputation in the development of online banking, Casaló, Flavián, and Guinalí (2007) found "positive and significant effects of perceived Web site usability and reputation" on consumer trust in the context of online banking. This finding was also confirmed in a small study of three Australian government health Web sites (Fisher et al. 2008). Fisher et al. found that, "ensuring Web sites are easy to use contributes to the level of trust users have in a Web site."

The 2007 e-commerce multi-sector "E-Stats" report was released by the U.S. Census Bureau (2009). The E-Stats report said that in 2007, e-commerce grew faster than total economic activity in three of the four major economic sectors covered by the report. However, change over time in the e-commerce share of each sector's overall shipments, sales, or revenues continues to be gradual. Retailers' e-commerce sales increased by 18.4%. As a share of total retail sales, however, e-commerce sales remained modest—3.2% ($127 billion), up from 2.8% ($107 billion) in 2006. In 2007, as in prior years, business-to-business (B-to-B) activity—by definition here, transactions by Manufacturers and Merchant Wholesalers—accounted for most e-commerce (93%).

In the second quarter of 2009, online sales were estimated at 3.6% of all retail sales, according to the U.S. Census Bureau's "Quarterly Retail E-Commerce Sales 2nd Quarter 2009" (U.S. Department of Commerce 2009). Online sales still constitute a relatively small fraction of the overall amount consumers spend on retail purchases. That said, there is a tremendous potential for online venues to drive sales at both online and in-person storefronts, making it very important to make information about products and services easy to find and secure.

In 2002, Nielsen (2003) reported that a 10% investment of development budget in usability could improve ease-of-use by 135%, increase conversion rates by 100% and increase user productivity by 161% on average. In his January 2008 AlertBox, Nielsen (2008) reported that usability returns for redesign efforts were declining, but still strong, at 83%, compared to the same measurement from 2002. Nielsen contends that the ROI remains high because the costs of doing usability work are still small relative to the gains it produces. While this number cannot be used to justify a specific in-house project, it is a useful signal of the continued value of usability work.

A recent video commissioned by Usability.gov described significant improvements to federal Web sites as a result of usability work (Howcast 2009). For example, the video included a case study showing that implementing user-centered design (UCD) on the Centers for Disease Control Web site led to a 26% increase in user success on top tasks and a 70% increase in user satisfaction. Another case study presented in the video noted that the Federal Aviation Administration saves $2 million per year by making their customers' top tasks easy to complete on their redesigned Web site.

A Federal Emergency Management Agency site redesign that included an extensive user-centered design process resulted in a site that improved user performance by 93%, reduced the time it took users to find information by nearly 50%, and improved user satisfaction from 49% to 71% (Federal Emergency Management Agency 2009). In addition, the US Computer Emergency Readiness Team (US-CERT 2010) completed a baseline and then a comparison test after a variety of changes to the site were made. They found that overall, technical users' success improved 24%, their satisfaction improved 16%, nontechnical (home) users' success improved 20%, and their satisfaction improved 93%.

37.4.4 Context Is King

37.4.4.1 Be Accurate in Your Assumptions

The research generally shows compelling return on investment for usability work but to make a convincing argument, it is advisable to show evidence relevant to a specific project. One must show potential gains for one's organization. To show how usability work might benefit a project and an organization, assumptions must be attainable and measurable. One assumption often made, for example, is that all time savings for an internal organization will be put toward productive enterprise. For example, if the process to complete an annual benefits selection is shortened by two hours, it cannot be assumed that employees will use all of that time for productive work. Even given the current state of the usability world (take, for instance, Nielsen's 2008 estimate of 83% average ROI), such assumptions should be used carefully. Certainly, as projections get more accurate, and as margins get smaller, usability specialists will need to perform more detailed analyses and rely on fewer assumptions.

37.4.4.2 All ROI Is Local

As a 2004 report from Adaptive Path points out (Hirsch, Fraser, and Beckman 2004), the use of general ROI numbers to justify usability work may not be meaningful when presenting a business case to decision makers. Adaptive Path notes, "research showed that valuation methods can help managers justify resource increases (but) it's impossible to measure ROI for user experience with a simple equation that can be applied across a wide swath of companies and projects. Nor is there a specific number that represents the general value of user experience."

Some projects may have significant low-hanging fruit, where usability professionals can achieve quick and cheap big wins. In other projects, the gains might be hard won and have less impact on the bottom line. Each project, each organization, and each usability professional enters the process with different challenges and levels of sophistication. It is best to avoid trying to use a one-size-fits-all approach to establishing ROI for your projects.

Let us offer one hypothetical example to illustrate this point. Imagine two Web development teams developing two Web sites. Team A, which includes one usability professional, is building an e-commerce Web site. This is the first release of this new site, and there are various competitors in place already. There will be no new "functionality" on this site; the one variable that will likely make or break this site's success is usability. And Team A's usability professional has 15 years of experience successfully supporting e-commerce sites. There is no predecessor site from this company.

Compare with that Team B, and their brand new usability person, who has little experience and just transferred over from another department. This team is developing a new version, version 4, of their company's e-commerce site. Version 4 is going to entail just a small number of changes from their successful version 3. The development of earlier versions entailed extensive user requirements gathering. The company has been systematic about conducting usability evaluations of all emerging releases, and there is a closed feedback loop between the development team, and the customer support team that employs a well-founded, granular set of codes for particular "user errors" to help inform each next release.

All things being equal, the usability professional on Team A will be expected to make a much larger contribution to Web Site A than will the usability professional on Team B. The ROI for the usability dollar an hour will likely be much greater for Team A.

37.4.5 The Costs and Benefits that Make Up an ROI Case

37.4.5.1 Measuring Costs

The cost of providing usability assistance to the project may include a variety of variables. For example,

- Loaded salary per hour, typically including salary, benefits, office, supplies, etc. (multiply by estimated hours)
- Processes required, such as
  - Usability requirements gathering and site/product testing
  - Facility
  - Recruitment
  - Participant reimbursement and costs
  - Development team time spent reviewing the test plan, attending test sessions, and responding to requested design fixes

37.4.5.2 Measuring Benefits

Donahue's (2001) four steps to measuring benefits include

1. Select usability criteria for measurement
2. Determine the appropriate unit for measure
3. Make reasonable assumptions of the benefit magnitude
4. Translate the benefit into monetary value

The first step is often the most difficult. Not only are business priorities and goals sometimes unclear but also establishing a direct link between goals and usability metrics can prove challenging. One-to-one causal connections are rare;

correlations are more likely to be found. For example, how might ease-of-use improvements impact sales? How do you know for sure? What other impacts are there on the variability of sales?

37.5 THE PROBLEM OF CONFOUNDS

One problem with measuring the benefits of any usability effort is that as usability improvements are made, there are likely other changes made at the same time. Perhaps there are changes to the marketing message, additional products in the product list, or additional sales force. Any variables that change at the same time as the usability improvements will make impossible the unequivocal attribution of increased revenues or decreased costs to any particular variable. At the very least, there is a confound with time; the old design was available to the site visitors then (possibly before an act of terrorism, the introduction of a new personal music player, or before an economic downturn), and the new design is available to the site visitor only later.

This problem of confounds should not deter pursuit of a cost-benefit analysis approach to usability engineering. The time confound can be mitigated by serving two or more parallel designs and randomly assigning each visitor to one or the other of competing designs. And while other benefits may not often be uniquely associated with a particular change, the Web development professional can make reasonable guesses as to what rough percentage may be attributable to a particular change. Have there been similar, one-at-a-time changes in the past (e.g., just a marketing message change) with a certain ROI, either within this company or elsewhere across the Web?

The fact of the matter is that the state of the art of Web design, while improving, is at such at point that usability improvements usually prove to yield substantial benefits. We were once asked by the vendor responsible for the Web site of a major rental car company to conduct an emergency usability study on their current Web site; they had usability problems and they knew it. In five days we designed and conducted a usability test and presented our results to them, and they immediately implemented some of our findings. The only changes they made, that Monday, were those driven by our usability test. On day one of the new design the site realized a $200,000 per day increase in revenue. And so, the payback period for their low-five-figure usability test was "before lunch."

37.5.1 The Problem of Projections

It is possible to measure benefits for a project after it has been completed. Indeed, this is what we must do, if we are to show that the usability effort we employed for this past project was worth it, and what the ROI was for our usability dollar and hour invested. But, of course, those ROI data are for a project that was already approved. How shall we justify a particular level of usability support for the *next* project? We must count on projected improvements, based on historical data.

Just as the software developers must project, in advance, how many lines of code and/or labor it will take to develop the site or app, the usability professional may be called on to project how much the usability engineering will cost and what benefits it will likely produce. Those earlier retroactive analyses will inform these projections, and, indeed, these projections are the best justification for taking the time to perform those historical cost-benefit analyses. Will these projections be perfect? Likely not. Though the experienced usability professional should be as good at these relative projections as the Web developers are at projecting development costs or as the market research team is at projecting how many more licenses or visits certain additional functionality will fetch. That experienced usability professional will consider such things as how vital the user experience is to the success of a project, how much usability engineering went into a previous version of the site or app, how much change the UI will undergo, what sorts of changes there have been in competitor sites/apps and thus the likely amount of changes in users' mental models, the development schedule, and the availability of possible test participants, and the collection of usability methods that might be employed.

It is sometimes important to "go back to the actuals" and confirm that your projections were close, or conservative. As Mayhew and Tremaine (2005) point out in *Cost-justifying Usability*, this is particularly true for organizations just experimenting with usability engineering techniques. You may find that once you traverse one or two cycles like this (projection of costs and benefits, then measurement of actual ROI via attention to actual costs and benefits) that your management and/or development teams gain faith in your projections and your can skip the retroactive ROI check.

Note here that while the costs are this year's costs, sometimes the benefits take time to be realized. We will address this again in the section below on the time value of money.

37.5.2 The Validity of Measurements

Although it is possible to take measurements such as time on task and show reductions, based on the results of usability work, it is key that all assumptions of benefit are measurable. For example, though potential productivity gains might well be shown for a rework of the organization's Intranet, be careful about assuming that all of the time gained will be used for productive work. Hirsh, Fraser, and Beckman (2004) described these kinds of calculations as "hypothetical," saying that the problem with them is, "in part because there can be no accountability for delivering on the expectations set."

While the focus of much research is on e-commerce sites, many of the benefits e-commerce sites experience are also applicable to .gov, .org, and .edu sites. The potential benefits for ROI are often characterized as "hard ROI" and "soft ROI" (Wright 2002). Hard ROI is easier to measure and is related to dollar values, whereas soft ROI is harder to measure and is related to marketing and branding goals that involve changes in customer perception. Both kinds of

benefits are listed below. When choosing which benefits to target, take the advice of Tom Tullis, Vice President of User Insight at Fidelity Investments, who said in a recent article advising usability professionals what to do in a down economy, that you need to "find out what metrics matter to the project you're involved in or to the senior management of the company" (Tullis 2009).

37.5.3 Different Classes of Potential Benefits for Different Stakeholders

Below are listed different types of stakeholders and the benefits that they may derive from incorporating usability into their Web site development.

- Benefits to the Web design/development team
  - Development efficiencies, quicker to market, slashing redevelopment costs
  - Return business
  - Increased authority in the organization
  - Employee retention
- Benefits to the sponsoring organization
  - Increased number of online sales
  - Increased dollar volume per transaction
  - Increased volume of non-online traffic and sales
  - Increased conversion rates from visitor to customer
  - Increased site traffic (number of visitors)
  - Increased time on site (time per visit)
  - Increased ad revenue
  - Increased market share
  - Halo effect (wash-over positive attitudes of customers)
  - Trade press praise
  - Decreased customer support burden
  - Decreased need for training
  - Decreased need for documentation
  - Lower maintenance costs
  - Increased customer and citizen involvement
  - Increased awareness of services
  - Increased use of services (eGov, .edu, .org)
  - Increased understanding of information
  - Increased trust
  - Improved internal communication
- Benefits to the user/visitor
  - Increased efficiency/throughput
  - Decreased time in training/learning
  - Increased access to information; more than increasing efficiency, with good usability/IA/design, the universe of what's available to the user is expanded; the "Oh, I had no idea that information was available here" moment

37.5.3.1 Specific Measurements

To strengthen your ROI argument, you can conduct A/B studies, comparing the site as it was before usability improvements versus the site after usability improvements have been made. The tests can be carried out prelaunch or postlaunch. For example, say you perform a usability test of a high-fidelity prototype of an emerging site, and you learn that 20% of test participants had to call the simulated help desk to complete Task A. Then, upon working with the design team to redesign the UI and retest, you find that no one in the new test needed help in carrying out Task A. It is a simple and reasonable matter to multiply the number of projected site visitors times .20 and then multiply this number times the average cost of a customer support phone call for your company to project the added benefit of that one iteration of usability testing and the discovery of that one problem.

37.5.3.2 Meeting the Goal of Decreasing Costs

Decreased costs can be measured in a variety of ways. For example, it is possible to project the number of support calls to the help desk avoided by the redesign, per user, multiplied by the number of users. Related to decreased support calls and decreased expenditures on developing and providing training are task completion rates, ease-of-use, error rate, and learning rate. Reduced development costs are related to decreased errors introduced by the design, decreased number of redesigns, decreased cost of fixing problems when they are discovered relatively early in the development cycle, decreased time to market, and decreased number of patches.

37.5.3.3 Meeting the Goal of Increasing Revenue

For e-commerce sites, you can measure conversion rate and rate of completion in a shopping cart. If conversions are improved (more people to the shopping cart) and drop-offs decreased (people not following-through with the purchase), then greater sales can be assumed. Another measure related to increased sales would be increased repeat customer visits. All types of sites need to attract and retain their visitors. Measurements related to visitor attraction and retention include increased site traffic, increased satisfaction, increased trust, decreased frustration, decreased task completion time, and increase in the number of functions used.

37.5.3.4 Metrics

Using conservative, project-specific estimates is critical to the successful use of ROI measurements. Trust is built when reasonable goals are set and those goals are met or exceeded. Increased trust helps build the reputation of the usability team and helps to insure repeat and new business in the organization. It is wise also to remember that not all gains can be attributable to usability improvements and not all productivity gains will be used to improve productivity/performance in other areas.

Credibility is established with your audience by using measurements that are specific and quantifiable and then show the measurements in a way that is understandable and reproducible. The definitions used should be precise and unambiguous about what is being measured and how it is measured. For example, to show an increase in conversion rates after the release of a usability-enhanced product, define

that specifically as the comparison of the number of people who purchased on the site divided by the number of people who entered the site on date "X" before the usability improvements versus the number of people who purchased on the site divided by the number of people who entered the site on date "Y" after the usability improvements. This will allow anyone looking at the data to be able to understand what the measure was exactly, including when and how it was taken. This helps establish credibility with the reader. A recipient of this information can look and see if conversion rates at the slowest sales time and the highest time were compared, or any other possible confounding factors.

Inherent in most of these examples of benefits is the idea that a metric is performing better after the usability enhancement to the Web site. The "better than" assertion can only be made if there is something to compare. The best comparison is to the Web site prior to the usability enhancements being incorporated. Take a baseline of the metrics so a direct comparison can be made of the numbers of the current functionality to the new Web site design. It is possible to compare metrics when there are no before and after measurements; however, the most compelling and clearest comparison is a before and after look. Other possibilities for comparison might be comparing two different products, where one has usability integrated into the process and the other does not.

The findings should be communicated in the language that matches the expectations of the intended audience. In many situations, the information will be provided to people with a business or financial background. A simple measure of benefits minus costs is not a convincing argument for a financial person. Financial people are interested in how long it takes to make an ROI and whether investing in usability work would yield more benefit than investing elsewhere (in other projects, investing money in the bank, etc.). To develop a convincing argument, one must learn to present the usability worth in terms of business values, such as the net present value (NPV) or the internal rate of return (IRR). We will now introduce some financial concepts that will help communicate the worth of usability in terms that a financial person would appreciate.

37.5.3.5 Financial Management and ROI

The primary goal of financial management is to maximize the value of the company. With that, management is continually faced with the question: How should the company, facing uncertainty over future market conditions, invest its capital? These strategic decisions are focused on investments in a variety of real assets, which include tangible assets, such as plant and machinery, and intangible assets, such as management contracts and patents. For the computing or the software industry in particular, these investments include networks, engineers, programmers, and usability experts. Technology companies invest in these assets with the idea that the assets' expected marginal benefits exceed their expected marginal costs. Valuing these benefits and costs can be difficult because they occur today and in the future. In this section of the chapter, we detail the basic steps for understanding how assets are valued and their relationship to the ROI of usability work.

37.5.3.6 The Time Value of Money

Financial managers must make decisions to invest the organization's money on real assets that produce cash flows over a fixed time horizon. It is management's hope that today's value of these future cash flows exceeds the amount invested in the assets. It will be demonstrated that the value of any asset is a function its cash flows and the challenge for managers is that the cash flows occur at different periods of time with varying degrees of uncertainty. To account for this problem, managers need to understand how asset prices are determined. They need a theory of value (Ross, Westerfield, and Jordan 2001).

An asset's value is the sum of cash flow plus risk plus time. The "time value of money" is the economic theory that analytically determines this relationship. The adage "money today is worth more than money tomorrow" defines, in simple terms, the time value of money theory. Thus, increasing revenue as a result of an investment in usability work does not necessarily mean that the investment was profitable. One must determine if the investment in usability work was the optimal investment at the time. In other words, were there other investment options that could have yielded the same or more return on investment?

For example, if the organization invests $100k today at a 10% return, then in one year, that $100k is worth $110k. This is why money today is worth more than money tomorrow.

Payout yield = FV
Original principal = PV
Interest rate = r

Substituting these expressions into the above paragraph, we have

$$FV = PV + PVr.$$

Factoring yields

$$FV = PV(1 + r).$$

The future value of a 1-year investment is equal to the original principal, PV, times an appreciation factor, $(1 + r)$.

The calculations get trickier with multiyear investments. An investment for two years, assuming you reinvest the interest earned in year one:

$$100k + 10k + 10k + 1k = 121k$$

So far, we have described how to determine the value of a lump sum of money at a future point in time. So, you can calculate what a manager might expect to receive if they invest at a given rate of return over time.

37.5.3.7 Opportunity Cost of Capital

The principles of the time value of money show that the market value of any asset is simply the present value of its expected future payoffs. An important element in calculating this value is the rate of return, r, which is referred to as the discount rate or the opportunity cost of capital. This rate of return is referred to as the opportunity cost of capital because it measures in percentage returns the implicit cost of the managers' best alternative or what your manager would forgo to invest the same amount of money in usability work. That is, managers are continually faced with multiple investment opportunities, and once they commit their capital to a particular venture, they forego the earnings they could have earned from their next best alternative. This implied cost is called an *opportunity cost*. The opportunity cost is used to value comparable investments. A comparable investment is an investment with comparable risk and yield.

37.5.3.8 Net Present Value

Economic theory tells us that companies achieve their goals when managers maximize a firm's *economic* profit, which is defined as cash revenues minus cash costs minus the opportunity cost of using firm resources elsewhere.

Profit = Revenue – Cost
Economic profit = Revenue – Cost – Opportunity cost

Note the objective of a manager is not to maximize profit but economic profit. Because opportunity cost measures the value of the financial resources the owners devote to the firm, if the firm maximizes economic profit it earns revenues that cover both actual cash costs and implied costs. In essence, a company that earns a positive economic profit produces goods and services that society values more highly than all the resources consumed in production. Quantitatively, a firm's economic profit is determined by the net present value (NPV) of its investments.

Formally, we define NPV as

$$\text{NPV} = CF_0 + \sum_{i=1}^{n} \frac{CF_i}{(1+r)^i},$$

where CF_0 is the initial cash outlay, CF_i is the after-tax net marginal benefit at the end of period i, and r is the projects' appropriate discount rate. The expression for NPV can be viewed as an expression for economic profit where the CF_i's represent cash revenues minus cash costs and the discount factor represents the opportunity cost.

Net present value is the difference between an investment's market value and its costs in terms of actual dollars. It is a measure of how much value is created in today's dollars. Measuring investments in terms of NPV leads to investing resources to their most productive use.

37.5.3.9 Internal Rate of Return

NPV measures, in absolute dollar terms, the amount of value a project will earn for a firm. Using this criterion guides managers to efficiently allocate a firm's capital; however, at times, managers prefer to talk in terms of returns instead of absolute dollars. Internal rate of return (IRR) is an expression of returns rather than absolute dollars. The IRR is a rate of return for a project's cash flows that makes the NPV equal to zero. Intuitively, it is a rate of return that balances a project's net marginal benefits to its net marginal costs.

Formally, the solution for the IRR is obtained from

$$CF_0 + \sum_{i=1}^{n} \frac{CF_i}{(1+IRR)^i} = 0 \,.$$

If the IRR is greater than a project's opportunity costs, then it is said to earn excess returns, adding value to an organization.

37.5.3.10 The Cost of Ignoring Usability

We have offered some reasons why a team should decide to improve the usability of their Web site or Web-based application. We have also offered some support for why a cost-benefit analysis approach may be a good way to justify that usability work. It is easy to quantify the cost of usability efforts, and somewhat harder, but still possible, to quantify the benefits. Even more challenging is quantifying the costs of a poorly designed UI. Beyond the poor user experience, and its domino effects of wash-over dissatisfaction with a company's entire brand, consider the prospects of poor reviews in the trade press, or perhaps a law suit (successful or not) for lost time or money supposedly caused by a company's poorly designed site.

Like all professionals, usability specialists are proud of what we can contribute to a project, and we likely have a tendency to think that our discipline is one of the most important disciplines represented at the Web development table. But as with any discipline, if we oversell our contribution, if we promise more return than we produce, we will soon lose credibility. And so, we believe there is value in providing conservative, empirically based projections of the ROI a team can expect for their investment in usability support. An additional value in examining "the actuals" is to demonstrate that our projections were good, and were, in fact, conservative.

There are other possibilities. Maybe you will have a vice president who is convinced of the ongoing value of usability and who will provide implicit and explicit support of your usability work without question and at the level you would propose. Maybe you are adept at a "storytelling approach" and the development team you support will be moved by a "thought experiment" as you attempt to secure support for a usability professional or team. Maybe the demands of your project and the attitudes of your purse-string holders will conspire to make futile any sort of attempt to secure usability resources. But if none of these scenarios applies the situation, you should consider a systematic and quantified focus on costs and benefits, and the expression of your argument in a metric that any subdiscipline in your organization can understand: Dollars.

ACKNOWLEDGMENT

This chapter incorporates and builds on the original work by Eugenie Bertus and Mark Bertus, which was included in the first edition of this book. Mark Bertus was a remarkable person and a great teacher. This chapter is dedicated to his memory.

REFERENCES

Bias, R. G. 2003. The dangers of amateur usability engineering. In *Usability in Practice: Avoiding Pitfalls and Seizing Opportunities* (Long Beach, CA, October 2003), ed. S. Hirsch, 423–433. American Society of Information Science and Technology.

Bias, R. G., and D. J. Mayhew, eds. 1994. *Cost-justifying usability*. San Diego, CA: Academic Press.

Bias, R. G., and D. J. Mayhew, eds. 2005. *Cost-justifying Usability: An Update for the Internet Age*, 2nd ed. San Diego, CA: Morgan Kaufmann.

Boehm, B. W. 1981. *Software Engineering Economics*. Englewood Cliffs, NJ: Prentice-Hall.

Bosert, J. L. 1991. *Quality Functional Deployment: A Practitioner's Approach*. New York: ASQC Quality Press.

Bureau of Labor Statistics. 2009a. Economic news release. — Employment situation summary. The employment situation. USDL Publication USDL-09-1583. http://www.bls.gov/news .release/empsit.nr0.htm (accessed December 15, 2009).

Bureau of Labor Statistics. 2009b. Productivity and costs. Third quarter 2009, Revised. USDL Publication USDL-09-1478. http:// www.bls.gov/news.release/pdf/prod2.pdf (accessed December 15, 2009).

Casaló, L.V., C. Flavián, and M. Guinalí. 2007. The role of security, privacy, usability and reputation in the development of online banking. *Online Information Review* 31(5): 583–603.

Donahue, G. M. 2001. Usability and the bottom line. *IEEE Software*, January/February, 31–37.

Dray, S., C. Karat, D. Rosenberg, D. Siegel, and D. Wixon. 2005. Is ROI an effective approach for persuading decision-makers of the value of user-centered design? In *Extended Abstracts on Human Factors in Computing Systems. Conference on Human Factors in Computing Systems*, 1168–1169. New York: ACM Press.

Econsultancy. 2007. Usability and user experience report 2007. http://econsultancy.com/reports/usability-and-user-experience-report-2007 (accessed December 15, 2009).

Federal Emergency Management Agency (FEMA). 2009. The new citizen-centric, user-friendly FEMA website. http://www.fema .gov/media/site_case_study.shtm (accessed December 15, 2009).

Fitzgerald, T. 2003. New Media. Online retail begins to look like offline: With sales jump, price-cutting, and poorer service. *Media Life*. http://www.medialifemagazine.com/news2003/jan03/jan06/1_ mon/news1monday.html (accessed December 15, 2009).

Fisher, J., F. Burstein, K. Lynch, and K. Lazarenko. 2008. "Usability + usefulness = trust": An exploratory study of Australian health Web sites. *Internet Research* 18(5): 477–498.

Gunther, R. 2006. UPA usability in the enterprise project: An analysis of ROI measurement and metrics across usability professionals. http://www.usabilityprofessionals.org/usability_resources/ usability_in_the_real_world/roi-measurement-survey-2005 .doc (accessed December 15, 2009).

Herman, J. 2004. A process for creating the business case for user experience projects. Session: Late breaking result papers table of contents. In *Conference on Human Factors in Computing Systems*, 1413–1416. New York: ACM Press.

Herrman, T. 2006. Corporate UX: Bringing value to the mobile industry. *Interactions* 13(4), July/August.

Hirsch, S. 2003. The red herring of usability ROI, a review of: BayCHI October Program Meeting, Oct. 17, message posted to http://netnow.blogspot.com/2003_10_01_netnow_archive .html#106642724548276267.

Hirsch, S., J. Fraser, and S. Beckman. 2004. Adaptive path reports: Leveraging business value: How ROI changes user experience. http://www.adaptivepath.com/ideas/reports/businessvalue/ apr-005_businessvalue.pdf (accessed December 15, 2009).

Horrigan, J. 2008. Online shopping. Pew Internet & American Life Project. Feb. 13. http://www.pewinternet.org/Reports/2008/ Online-Shopping.aspx (accessed December 15, 2009).

Howcast. 2009. How to increase usability of government Websites and boost your ROI. Dec. 23. http://www.howcast.com/ videos/241198-How-To-Increase-Usability-Of-Government-Web sites-and-Boost-Your-ROI/ (accessed December 15, 2009).

International Telecommunication Union. 2008. Press Release. Worldwide mobile cellular subscribers to reach 4 billion mark late 2008. Sept. 25. http://www.itu.int/newsroom/press_ releases/2008/29.html (accessed December 15, 2009).

Internet World Stats Usage and Population Statistics. 2009. Internet growth statistics. http://www.internetworldstats.com/emar keting.htm (accessed December 15, 2009).

Jones, C. P., S. J. Robinson, N. Sabadosh, D. Bishop, and S. Koyani. 2006. CHI '06 Extended Abstracts on Human Factors in Computing Systems. Session: SIGs. Paper presented at Conference on Human Factors in Computing Systems. ACM, Montréal, Québec, Canada.

Jones, S., and S. Fox. 2009. Generations Online in 2009. Pew Internet & American Life Project. Jan. 28. http://www.pew internet.org/Reports/2009/Generations-Online-in-2009.aspx (accessed December 15, 2009).

Karat, C. 1994. A business cost approach to usability cost justification. In *Cost-justifying Usability*, eds. R. G. Bias and D. J. Mayhew, 45–70. San Diego, CA: Academic Press.

Karat, C. 2005. Why measure the cost benefit of human factors. In *Cost-justifying Usability: An Update for the Internet Age*, 2nd ed., eds. R. G. Bias and D. J. Mayhew, 103–142. San Diego, CA: Morgan Kaufmann.

Landauer, T. K. 1995. *The Trouble with Computers: Usefulness, Usability, and Productivity.* Cambridge, MA: MIT Press.

LaPlante, A. 1992. Put to the test. *Computer World* 27, 75.

Lederer, A. L., and J. Prassad. 1992. Nine management guidelines for better cost estimating. *Communications of the ACM* 35(2): 51–59.

Mayhew, D. J., and M. M. Tremaine. 2005. A basic framework. In *Cost-justifying Usability: An Update for the Internet Age*, 2nd ed., eds. R. G. Bias and D. J. Mayhew, 41–101. San Diego, CA: Morgan Kaufmann.

Nielsen, J. 2003. Jakob Nielsen's alterbox. Return on investments for usability. Jan. 7. http://www.useit.com/alertbox/20030107 .html (accessed December 15, 2009).

Nielsen, J. 2008. Jakob Nielsen's alertbox. Usability ROI declining, but still strong. Jan. 22. http://www.useit.com/alertbox/roi.html (accessed December 15, 2009).

Nielsen, J. 2008. Critical mass: The Worldwide state of the mobile Web. July. http://mmaglobal.com/uploads/NielsenMobile_ Mobile%20Internet_Critical%20Mass_July%202008.pdf (accessed December 15, 2009).

Nieters, J. E. 2007. The internal consultancy model for strategic UXD relevance. Paper presented at Conference on Human Factors in Computing Systems. ACM, San Jose, CA.

Pressman, R. S. 1992. *Software Engineering: A Practitioner's Approach.* New York: McGraw-Hill.

Rainie, L., and J. Anderson. 2008. The future of the Internet III. Pew Internet & American Life Project. Dec. 14. http://www .pewinternet.org/PPF/r/270/report_display.asp (accessed December 15, 2009).

Rhodes, J. S. 2000. Usability can save your company. http://web word.com/moving/savecompany.html (accessed December 15, 2009).

Rohn, J. A. 2005. Cost-justifying usability in vendor companies. In *Cost-justifying Usability: An Update for the Internet Age*, 2nd ed., eds. R. G. Bias and D. J. Mayhew, 185–214. San Diego, CA: Morgan Kaufmann.

Ross, S., R. Westerfield, and B. Jordan. 2001. *Essentials of Corporate Finance,* 4th ed. New York: McGraw-Hill.

Schonfeld, E. 2009. ComScore: Internet population passes one billion; top 15 countries. Jan. 23. http://www.techcrunch .com/2009/01/23/comscore-internet-population-passes-one-billion-top-15-countries/ (accessed December 15, 2009).

Shneiderman, B., S. Card, D. A. Norman, M. Tremaine, and M. M. Waldrop. 2002. Panel Session: CHI@20: Fighting Our Way from Marginality to Power. Paper presented at Conference on Human Factors in Computing Systems, ACM, Minneapolis, MN.

Tullis, T. 2009. Tips for usability professionals in a down economy. *Journal of Usability Studies* 4(2): 60–69.

U.S. Census Bureau. 2009. E-Stats. May 28. http://www.census.gov/ econ/estats/2007/2007reportfinal.pdf (accessed December 15, 2009).

U.S. Computer Emergency Readiness Team (US-CERT). 2010. Usability lessons learned. Jan. 22. http://www.us-cert.gov/ usability/ (accessed December 15, 2009).

U.S. Department of Commerce. 2009. U.S. Census Bureau news. Aug. 17. http://www.census.gov/retail/mrts/www/data/pdf/09Q2 .pdf (accessed December 15, 2009).

Usability Professionals Association. 2009. Resources: About usability. Job titles for usability professionals. http://www.usability professionals.org/usability_resources/about_usability/about_ usability_professionals.html (accessed December 15, 2009).

Wildstrom, S. 1998. A computer user's manifesto. *Business Week*, Sept. 28, http://www.businessweek.com/1998/39/b3597037 .htm (accessed December 15, 2009).

Wixon, D., and S. Jones. 1992. Usability for fun and profit: A case study of the design of DEC RALLY version 2. Internal Report, Digital Equipment Corporation.

Wright, A. 2002. Designing for the bottom line. The selling points of hard and soft ROI. Dr. Dobbs. Jan. 1. http://www.ddj.com/ architect/184413282 (accessed December 15, 2009).

Index